To Cynthia

About the Author

Steve Chapra teaches in the Civil, Environmental, and Architectural Engineering Department at the University of Colorado. His other books include *Numerical Methods for Engineers* and *Introduction to Computing for Engineers.*

Dr. Chapra received engineering degrees from Manhattan College and the University of Michigan. Before joining the faculty at Colorado, he worked for the Environmental Protection Agency and the National Oceanic and Atmospheric Administration and was an Associate Professor at Texas A&M. His general research interests focus on surface water-quality modeling and advanced computer applications in environmental engineering. His research has been used in a number of decision-making contexts, including the 1978 Great Lakes Water Quality Agreement.

He is active in several professional societies and has taught over 40 workshops on water-quality modeling in the United States, Mexico, Europe, and South America. He has received a number of awards for his scholarly contributions, including the 1993 Rudolph Hering Medal (ASCE) and the 1987 Meriam-Wiley Distinguished Author Award (American Society for Engineering Education). He has also been recognized as the outstanding teacher among the engineering faculties at both Texas A&M (1986 Tenneco Award) and the University of Colorado (1992 Hutchinson Award).

Dr. Chapra was originally drawn to environmental engineering because of his love of the outdoors. He is an avid fly fisherman and hiker. His primary professional goal is to apply engineering, mathematics, and computing to maintain a high-quality environment in a wise and cost-efficient fashion.

Surface Water-Quality Modeling

McGraw-Hill Series in Water Resources and Environmental Engineering

Consulting Editor

George Tchobanoglous

Bailey and Ollis: *Biochemical Engineering Fundamentals*
Bouwer: *Groundwater Hydrology*
Canter: *Environmental Impact Assessment*
Chanlett: *Environmental Protection*
Chapra: *Surface Water-Quality Modeling*
Chow, Maidment, and Mays: *Applied Hydrology*
Davis and Cornwell: *Introduction to Environmental Engineering*
Eckenfelder: *Industrial Water Pollution Control*
LaGrega, Buckingham, and Evans: *Hazardous Waste Management*
Linsley, Franzini, Freyberg, and Tchobanoglous: *Water Resources and Engineering*
McGhee: *Water Supply and Sewerage*
Mays and Tung: *Hydrosystems Engineering and Management*
Metcalf & Eddy, Inc.: *Wastewater Engineering: Collection and Pumping of Wastewater*
Metcalf & Eddy, Inc.: *Wastewater Engineering: Treatment, Disposal, Reuse*
Peavy, Rowe, and Tchobanoglous: *Environmental Engineering*
Sawyer and McCarty: *Chemistry for Environmental Engineering*
Tchobanoglous, Theisen, and Eliassen: *Solid Wastes: Engineering Principles and Management Issues*
Tchobanoglous, Theisen, and Vigil: *Integrated Solid Wast Management: Engineering Principles and Management Issues*

SURFACE WATER-QUALITY MODELING

Steven C. Chapra
University of Colorado at Boulder

Boston, Massachusetts Burr Ridge, Illinois Dubuque, Iowa
Madison, Wisconsin New York, New York San Francisco, California St. Louis, Missouri

WCB/McGraw-Hill

A Division of The ***McGraw-Hill*** *Companies*

SURFACE WATER-QUALITY MODELING

This book is printed on acid-free paper.

5 6 7 8 9 BKM BKM 0 9 8 7 6 5 4

ISBN 0-07-011364-5

This book was set in Times Roman by Publication Services, Inc.
The editors were B. J. Clark, David A. Damstra, and James W. Bradley.
The production supervisor was Denise L. Puryear.
The cover was designed by Karen K. Quigley.
Project supervision was done by Publication Services, Inc.

Chapra, Steven C.
Surface water-quality modeling / Steven C. Chapra.
p. cm.
Includes bibliographical references and index.
ISBN 0-07-011364-5—ISBN 0-07-843306-1
1. Water quality—Mathematical models. I. Title.
TD365.C48 1997
628.1'61'015118—dc20 96-15461

http://www.mhcollege.com

CONTENTS

Preface xvii

PART I Completely Mixed Systems 1

LECTURE 1 Introduction 3
1.1 Engineers and Water Quality 4
1.2 Fundamental Quantities 6
1.3 Mathematical Models 10
1.4 Historical Development of Water-Quality Models 14
1.5 Overview of This Book 19
Problems 20

LECTURE 2 Reaction Kinetics 24
2.1 Reaction Fundamentals 24
2.2 Analysis of Rate Data 29
2.3 Stoichiometry 38
2.4 Temperature Effects 40
Problems 42

LECTURE 3 Mass Balance, Steady-State Solution, and Response Time 47
3.1 Mass Balance for a Well-Mixed Lake 47
3.2 Steady-State Solutions 52
3.3 Temporal Aspects of Pollutant Reduction 57
Problems 62

LECTURE 4 Particular Solutions 65
4.1 Impulse Loading (Spill) 66
4.2 Step Loading (New Continuous Source) 68
4.3 Linear ("Ramp") Loading 70
4.4 Exponential Loading 71
4.5 Sinusoidal Loading 73
4.6 The Total Solution: Linearity and Time Shifts 76
4.7 Fourier Series (Advanced Topic) 80
Problems 83

LECTURE 5 Feedforward Systems of Reactors 86
5.1 Mass Balance and Steady-State 86
5.2 Time Variable 91

5.3 Feedforward Reactions 95
Problems 99

LECTURE 6 Feedback Systems of Reactors 101
6.1 Steady-State for Two Reactors 101
6.2 Solving Large Systems of Reactors 103
6.3 Steady-State System Response Matrix 107
6.4 Time-Variable Response for Two Reactors 111
6.5 Reactions with Feedback 113
Problems 117

LECTURE 7 Computer Methods: Well-Mixed Reactors 120
7.1 Euler's Method 121
7.2 Heun's Method 124
7.3 Runge-Kutta Methods 126
7.4 Systems of Equations 128
Problems 131

PART II Incompletely Mixed Systems 135

LECTURE 8 Diffusion 137
8.1 Advection and Diffusion 137
8.2 Experiment 138
8.3 Fick's First Law 141
8.4 Embayment Model 143
8.5 Additional Transport Mechanisms 149
Problems 153

LECTURE 9 Distributed Systems (Steady-State) 156
9.1 Ideal Reactors 156
9.2 Application of the PFR Model to Streams 164
9.3 Application of the MFR Model to Estuaries 168
Problems 171

LECTURE 10 Distributed Systems (Time-Variable) 173
10.1 Plug Flow 173
10.2 Random (or "Drunkard's") Walk 177
10.3 Spill Models 180

10.4 Tracer Studies 186
10.5 Estuary Number 189
Problems 190

LECTURE 11 Control-Volume Approach: Steady-State Solutions 192
11.1 Control-Volume Approach 192
11.2 Boundary Conditions 194
11.3 Steady-State Solution 195
11.4 System Response Matrix 197
11.5 Centered-Difference Approach 198
11.6 Numerical Dispersion, Positivity, and Segment Size 201
11.7 Segmentation Around Point Sources 207
11.8 Two- and Three-Dimensional Systems 208
Problems 209

LECTURE 12 Simple Time-Variable Solutions 212
12.1 An Explicit Algorithm 212
12.2 Stability 214
12.3 The Control-Volume Approach 215
12.4 Numerical Dispersion 216
Problems 221

LECTURE 13 Advanced Time-Variable Solutions 223
13.1 Implicit Approaches 223
13.2 The MacCormack Method 229
13.3 Summary 230
Problems 232

PART III Water-Quality Environments 233

LECTURE 14 Rivers and Streams 235
14.1 River Types 235
14.2 Stream Hydrogeometry 238
14.3 Low-Flow Analysis 243
14.4 Dispersion and Mixing 245
14.5 Flow, Depth, and Velocity 247
14.6 Routing and Water Quality (Advanced Topic) 250
Problems 257

LECTURE 15 **Estuaries** 260

15.1 Estuary Transport 260
15.2 Net Estuarine Flow 262
15.3 Estuary Dispersion Coefficient 263
15.4 Vertical Stratification 270
Problems 272

LECTURE 16 **Lakes and Impoundments** 276

16.1 Standing Waters 276
16.2 Lake Morphometry 278
16.3 Water Balance 282
16.4 Near-Shore Models (Advanced Topic) 287
Problems 293

LECTURE 17 **Sediments** 295

17.1 Sediment Transport Overview 295
17.2 Suspended Solids 297
17.3 The Bottom Sediments 302
17.4 Simple Solids Budgets 304
17.5 Bottom Sediments as a Distributed System 307
17.6 Resuspension (Advanced Topic) 312
Problems 315

LECTURE 18 **The "Modeling" Environment** 317

18.1 The Water-Quality-Modeling Process 317
18.2 Model Sensitivity 327
18.3 Assessing Model Performance 335
18.4 Segmentation and Model Resolution 339
Problems 341

PART IV **Dissolved Oxygen and Pathogens** 345

LECTURE 19 **BOD and Oxygen Saturation** 347

19.1 The Organic Production/Decomposition Cycle 347
19.2 The Dissolved Oxygen Sag 348
19.3 Experiment 351
19.4 Biochemical Oxygen Demand 353
19.5 BOD Model for a Stream 355

	19.6 BOD Loadings, Concentrations, and Rates	357
	19.7 Henry's Law and the Ideal Gas Law	360
	19.8 Dissolved Oxygen Saturation	361
	Problems	365
LECTURE 20	**Gas Transfer and Oxygen Reaeration**	367
	20.1 Gas Transfer Theories	369
	20.2 Oxygen Reaeration	376
	20.3 Reaeration Formulas	377
	20.4 Measurement of Reaeration with Tracers	384
	Problems	386
LECTURE 21	**Streeter-Phelps: Point Sources**	389
	21.1 Experiment	389
	21.2 Point-Source Streeter-Phelps Equation	391
	21.3 Deficit Balance at the Discharge Point	391
	21.4 Multiple Point Sources	393
	21.5 Analysis of the Streeter-Phelps Model	396
	21.6 Calibration	398
	21.7 Anaerobic Condition	399
	21.8 Estuary Streeter-Phelps	401
	Problems	403
LECTURE 22	**Streeter-Phelps: Distributed Sources**	405
	22.1 Parameterization of Distributed Sources	405
	22.2 No-Flow Sources	407
	22.3 Diffuse Sources with Flow	410
	Problems	417
LECTURE 23	**Nitrogen**	419
	23.1 Nitrogen and Water Quality	419
	23.2 Nitrification	421
	23.3 Nitrogenous BOD Model	424
	23.4 Modeling Nitrification	426
	23.5 Nitrification and Organic Decomposition	428
	23.6 Nitrate and Ammonia Toxicity	430
	Problems	432
LECTURE 24	**Photosynthesis/Respiration**	433
	24.1 Fundamentals	433

24.2 Measurement Methods 437
Problems 448

LECTURE 25 Sediment Oxygen Demand 450
25.1 Observations 451
25.2 A "Naive" Streeter-Phelps SOD Model 455
25.3 Aerobic and Anaerobic Sediment Diagenesis 457
25.4 SOD Modeling (Analytical) 459
25.5 Numerical SOD Model 470
25.6 Other SOD Modeling Issues (Advanced Topic) 474
Problems 480

LECTURE 26 Computer Methods 482
26.1 Steady-State System Response Matrix 482
26.2 The QUAL2E Model 486
Problems 500

LECTURE 27 Pathogens 503
27.1 Pathogens 503
27.2 Indicator Organisms 504
27.3 Bacterial Loss Rate 506
27.4 Sediment-Water Interactions 510
27.5 Protozoans: *Giardia* and *Cryptosporidium* 512
Problems 516

PART V Eutrophication and Temperature 519

LECTURE 28 The Eutrophication Problem and Nutrients 521
28.1 The Eutrophication Problem 522
28.2 Nutrients 522
28.3 Plant Stoichiometry 527
28.4 Nitrogen and Phosphorus 530
Problems 533

LECTURE 29 Phosphorus Loading Concept 534
29.1 Vollenweider Loading Plots 534
29.2 Budget Models 536
29.3 Trophic-State Correlations 539

29.4 Sediment-Water Interactions 545
29.5 Simplest Seasonal Approach 551
Problems 558

LECTURE 30 Heat Budgets 560
30.1 Heat and Temperature 561
30.2 Simple Heat Balance 563
30.3 Surface Heat Exchange 565
30.4 Temperature Modeling 571
Problems 575

LECTURE 31 Thermal Stratification 577
31.1 Thermal Regimes in Temperate Lakes 577
31.2 Estimation of Vertical Transport 580
31.3 Multilayer Heat Balances (Advanced Topic) 585
Problems 588

LECTURE 32 Microbe/Substrate Modeling 590
32.1 Bacterial Growth 590
32.2 Substrate Limitation of Growth 592
32.3 Microbial Kinetics in a Batch Reactor 596
32.4 Microbial Kinetics in a CSTR 598
32.5 Algal Growth on a Limiting Nutrient 600
Problems 602

LECTURE 33 Plant Growth and Nonpredatory Losses 603
33.1 Limits to Phytoplankton Growth 603
33.2 Temperature 605
33.3 Nutrients 607
33.4 Light 609
33.5 The Growth-Rate Model 612
33.6 Nonpredatory Losses 613
33.7 Variable Chlorophyll Models (Advanced Topic) 615
Problems 621

LECTURE 34 Predator-Prey and Nutrient/Food-Chain Interactions 622
34.1 Lotka-Volterra Equations 622
34.2 Phytoplankton-Zooplankton Interactions 626
34.3 Zooplankton Parameters 629

34.4 Nutrient/Food-Chain Interactions 629
Problems 631

LECTURE 35 Nutrient/Food-Chain Modeling 633
35.1 Spatial Segmentation and Physics 633
35.2 Kinetic Segmentation 634
35.3 Simulation of the Seasonal Cycle 637
35.4 Future Directions 641
Problems 642

LECTURE 36 Eutrophication in Flowing Waters 644
36.1 Stream Phytoplankton/Nutrient Interactions 644
36.2 Modeling Eutrophication with QUAL2E 649
36.3 Fixed Plants in Streams 658
Problems 663

PART VI Chemistry 665

LECTURE 37 Equilibrium Chemistry 667
37.1 Chemical Units and Conversions 667
37.2 Chemical Equilibria and the Law of Mass Action 669
37.3 Ionic Strength, Conductivity, and Activity 670
37.4 pH and the Ionization of Water 672
37.5 Equilibrium Calculations 673
Problems 676

LECTURE 38 Coupling Equilibrium Chemistry and Mass Balance 677
38.1 Local Equilibrium 677
38.2 Local Equilibria and Chemical Reactions 680
Problems 682

LECTURE 39 pH Modeling 683
39.1 Fast Reactions: Inorganic Carbon Chemistry 683
39.2 Slow Reactions: Gas Transfer and Plants 686
39.3 Modeling pH in Natural Waters 689
Problems 691

PART VII Toxics 693

LECTURE 40 Introduction to Toxic-Substance Modeling 695

40.1 The Toxics Problem 695
40.2 Solid-Liquid Partitioning 697
40.3 Toxics Model for a CSTR 700
40.4 Toxics Model for a CSTR with Sediments 705
40.5 Summary 713
Problems 713

LECTURE 41 Mass-Transfer Mechanisms: Sorption and Volatilization 715

41.1 Sorption 715
41.2 Volatilization 727
41.3 Toxicant-Loading Concept 732
Problems 737

LECTURE 42 Reaction Mechanisms: Photolysis, Hydrolysis, and Biodegradation 739

42.1 Photolysis 739
42.2 Second-Order Relationships 751
42.3 Biotransformation 751
42.4 Hydrolysis 753
42.5 Other Processes 755
Problems 756

LECTURE 43 Radionuclides and Metals 757

43.1 Inorganic Toxicants 757
43.2 Radionuclides 758
43.3 Metals 761
Problems 768

LECTURE 44 Toxicant Modeling in Flowing Waters 769

44.1 Analytical Solutions 769
44.2 Numerical Solutions 778
44.3 Nonpoint Sources 779
Problems 782

LECTURE 45 Toxicant/Food-Chain Interactions 784

45.1 Direct Uptake (Bioconcentration) 785
45.2 Food-Chain Model (Bioaccumulation) 788
45.3 Parameter Estimation 790
45.4 Integration with Mass Balance 794
45.5 Sediments and Food Webs (Advanced Topic) 795
Problems 797

Appendixes 798

A Conversion Factors 798
B Oxygen Solubility 801
C Water Properties 802
D Chemical Elements 803
E Numerical Methods Primer 805
F Bessel Functions 817
G Error Function and Complement 820

References 821

Acknowledgments 834

Index 835

Preface

This is an exciting time to be involved in water-quality modeling. Today increased national and international concern for the environment, coupled with tightly limited national and local budgets, are generating increased interest in rational, economical approaches for water-quality management. At the same time, hardware and software advances are making computer-modeling frameworks more comprehensive and easier to use.

On the positive side these trends could result in models contributing more effectively to improved water-quality management. On the down side widespread and easy use of models could lead to their being applied without insight as "black boxes."

The main thesis of this book is that models must be applied with insight and with regard to their underlying assumptions. A variety of features in the book maximize such a "clear box" approach to water-quality modeling, including:

- *Lecture format.* This design was adopted for two primary reasons. The first, and obvious, reason is to provide a format that facilitates the use of the text for a course in water-quality modeling. The format could be especially useful to someone teaching the course for the first time. The second reason relates to the way in which students assimilate information. In my teaching I've found that students like to receive new material in manageable units. By limiting the exposition to short lecture-chapters, the student is provided a well-prescribed body of material that can be mastered at one or two sittings. In addition the format is useful for those attempting to study the material outside the classroom.
- *Theory and applications.* Most formulations and techniques presented in the book are accompanied by an explanation of their origin and/or theoretical basis. Although the book points toward numerical, computer-oriented applications, wherever possible, strong use is made of analytical solutions to elucidate the underlying behavior of the computer algorithms. In addition extensive worked examples are employed to relate the theory to applications and to illustrate the mechanics and subtleties of the computations.
- *Strong computer orientation.* Details of algorithms and numerical methods underlying the models are provided in the text and in appendixes.

In keeping with the lecture format, I've adopted the active voice and first-person singular in an attempt to speak directly to the students as I do in my lectures. In addition, though I avoid the editorial "we," I employ the interactive "we" extensively to reflect the active participation I try to stimulate in my lectures. Although I might risk alienating more traditionally minded readers, I feel that the risk is worth taking in order to better connect with students.

The book is organized into seven major parts. The first two cover modeling fundamentals, including background material on mathematics, numerical methods, reaction kinetics, diffusion, etc. The review of fundamentals at the beginning of the course serves two purposes. First, it provides a common foundation of mathematical and general modeling knowledge upon which the rest of the course can be built. Thus it allows the students to approach the remainder of the material in a more efficient manner and with a common perspective based on mathematical intuition. Second, it serves to orient the science and engineering students from disciplines outside civil and chemical engineering to new ideas such as mass balance, chemical kinetics, and diffusion.

Part III is designed to introduce the student to some of the environments commonly encountered in water-quality modeling: streams, estuaries, lakes, and sediments. Most of the material deals with physical aspects of these environments. In addition the final lecture involves issues that are commonly encountered in applying models. This description of the "modeling environment" includes information on model calibration, verification, and sensitivity analysis.

The following four parts deal with major water-quality-modeling problems: oxygen/bacteria, eutrophication/temperature, chemistry/pH, and toxics.

It should be noted that although the latter sections of the book are more problem-oriented, they are not devoid of theoretical development. However, such material is introduced in a "just-in-time" fashion when the student would be more motivated by a particular problem context. For example gas-transfer theory is introduced in the lectures on oxygen modeling and predator-prey kinetics in the lecture on nutrient/food-chain modeling.

In summary this book has been written primarily for students. They are the individuals who will determine whether or not water-quality models will be used more effectively in the future. If this book helps educate them properly, the cause of rational water-quality management will be well served.

ACKNOWLEDGMENTS

This book is the result of teaching classes in water-quality modeling over the past 12 years at Texas A&M University and the University of Colorado. So, first and foremost, I would like to thank my students for their challenging questions: in particular, James Martin, Kyung-sub Kim, Rob Runkel, Jean Boyer, Morgan Byars, Chris Church, Marcos Brandao, and Laura Ziemelis. Several students in my modeling class in the spring of 1995 gave me lots of suggestions and detected many errors and typos in the first draft. These are Chrissy Hawley, Jane Delling, Craig Snyder, Dave Wiberg, Arnaud Dumont, Patrick Fitzgerald, Hsiao-Wen Chen, Will Davidson, and Blair Hanna. My Michigan students during the winter of 1996, Ernie Hahn, Murat Ulasir, and Nurul Amin, also provided insight and many suggestions.

A number of colleagues have taught me a great deal about water-quality modeling. These include my good friends and teachers Drs. Ray Canale (Michigan); Ken Reckhow (Duke); and Bob Thomann, Donald O'Connor, Dom Di Toro (Manhattan College); and Rick Winfield (Bell Labs). I would also like to thank my friends in other countries for introducing me to water-quality contexts and modeling problems outside the United States. These include Enrique Cazares (Mexico); Monica and Rubem Porto, Ben Braga, Mario Thadeu, Rosa and Max Hermann (Brazil); Kit Rutherford (New Zealand); Koji Amano (Japan); Hugh Dobson (Canada); Cedo Maximovich (Serbia); Sijin "Tom" Lee and Dong-Il Seo (Korea); Spiro Grazhdani and Sokrat Dhima (Albania); and Pavel Petrovic (Slovakia).

I would like to acknowledge thoughtful reviews of this manuscript and discussion on modeling over the years with my colleagues: in particular, Marty Auer (Michigan Tech) and Joe De Pinto (SUNY Buffalo) made many suggestions and used the draft manuscript for their classes. Other colleagues I would like to acknowledge are Bill Batchelor and Jim Bonner (Texas A&M); Greg Boardman (VPI); Dave Clough, Marc Edwards, Ken Strzepek, and Joe Ryan (Colorado); Tom Cole (WES), John Connolly (Hydroqual); Dave Dilks and Paul Freedman (LimnoTech); Steve Effler (Upstate Freshwater Institute); Linda Abriola, Walt Weber, Nick Katapodes, Jeremy Semrau, and Steve Wright (Michigan); Glenn Miller (Nevada-Reno); Wu-Seng Lung (Virginia); Chris Uchrin (Rutgers); Andy Stoddard, Sayedul Choudhury, and Leslie Shoemaker (TetraTech); Jory Oppenheimer and Dale Anderson (Entranco); Mike McCormick and Dave Schwab (NOAA); Kent Thornton (FTN Associates); Carl Chen (Systech); Bob Broshears, Diane McKnight, and Ken Bencala (USGS); Terry Fulp and Bruce Williams (Bureau of Reclamation); and Hira Biswas and Gerry LaVeck (EPA). Willy Lick and Zenitha Chroneer (UCSB) and Bill Wood (Purdue) provided information and guidance on the topic of sediment resuspension. I would also like to thank B. J. Clark, Jim Bradley, David Damstra, and Meredith Hart of McGraw-Hill for expediting this project. Obviously, while I am grateful to the above and acknowledge their insights and assistance, I alone am responsible for any errors you might discover in this volume.

Last, but not least, I'd like to express my gratitude to my family, Cynthia, Christian, and Jeff, for their support and love. Their encouragement, as well as their tolerance of my long and somewhat eccentric working hours, made this book possible.

Steve Chapra

Cologne

In Köln, a town of monks and bones,
And pavements fanged with murderous stones,
And rags, and hags, and hideous wenches,
I counted two-and-seventy stenches.
All well defined, and separate stinks!
Ye nymphs that reign o'er sewers and sinks,
The river Rhine, it is well known,
Doth wash your city of Cologne;
But tell me, nymphs, What power divine
Shall henceforth wash the river Rhine?

—Samuel Taylor Coleridge (1772-1834)

PART I

Completely Mixed Systems

Part I is devoted to introducing you to modeling fundamentals. To keep the explanation uncomplicated the focus is on continuously stirred tank reactors (or CSTRs). *Lecture 1* provides an overview of water-quality modeling and an introduction to fundamental quantities and units. *Lecture 2* focuses on reaction kinetics. In particular I present methods for estimating reaction rates.

The next two lectures deal with fundamental analytical solution techniques for a CSTR. *Lecture 3* shows how a mass balance can be developed for such a reactor. It also illustrates how a steady-state solution is derived and interpreted. This lecture presents the simplest time-variable solution—the general solution—and illustrates the concepts of an eigenvalue and response time as metrics of a system's temporal response characteristics. *Lecture 4* explores time variable solutions for a CSTR subject to a number of idealized loading functions. These include step, impulse, linear, exponential, and sinusoidal loadings.

In Lecs. 5 and 6 I begin to show how more complicated systems can be simulated using groups of CSTRs. *Lecture 5* deals with reactors connected in series. *Lecture 6* is devoted to reactors with feedback and introduces matrix algebra to expedite the analysis of multireactor systems. In particular I present the steady-state system response matrix as a means to analyze and interpret such systems.

To this point all the material has dealt with closed-form, analytical solutions. *Lecture 7* is devoted to numerical methods for solving modeling problems with the computer. Emphasis in this lecture is placed on three techniques for simulating the dynamics of individual and linked CSTRs: the Euler, Heun, and fourth-order Runge-Kutta methods.

Completely Mixed Systems

LECTURE 1

Introduction

LECTURE OVERVIEW: I provide an introduction to water-quality modeling. After a review of the historical origins of water and wastewater engineering, a brief discussion of major variables and units is presented. Then I formally define what is meant by a water-quality model and discuss some ways in which such models are applied. Finally I review the historical development of water-quality modeling and outline the remainder of this set of lectures.

> One day a king arrived at a town inhabited by the sightless. Along with his troops, he brought a large elephant. Several of the inhabitants visited the king's encampment and tried to gather information by touching some part of the beast. When they returned, their fellow citizens gathered around to learn about its form and shape. The man who felt the trunk said: "An elephant is long, thick, and flexible, like a snake." One man who had touched the tail countered: "No, it is thin like a rope." Another who had felt its leg said: "It is solid and firm, like a tree trunk." Finally, an individual who had climbed on its back remarked: "You're all wrong, an elephant is rough and textured like a brush." Each had been misled by feeling one part out of many. All imagined something incorrect because they did not perceive the whole elephant.
>
> Paraphrased from
> *Tales of the Dervishes*
> Shah (1970)

Now, aside from the fact that elephants tend to generate large quantities of solid waste, this parable might seem to have little relevance to water-quality modeling. In fact it has great pertinence to the task before us. As will be clear from the following lecture, models were originally developed as problem-solving tools. However, beyond their utility in remediating pollution problems, they have a broader function; that is, they provide us with a means to visualize "the big picture." In essence the mathematical model offers a quantitative framework to integrate the diverse physical, chemical, and biological information that constitute complex environmental systems. Beyond solving a particular pollution problem, models provide a vehicle for an enhanced understanding of how the environment works as a unit. Consequently they can be of great value in both research and management contexts. I wanted to start with the elephant parable to underscore this point. You'll get a lot more out of this book if you try to keep "the whole elephant" in mind as you proceed.

1.1 ENGINEERS AND WATER QUALITY

Engineers, particularly civil engineers, have been the principal developers of the field of water-quality modeling. This might seem odd considering that civil engineers are most commonly associated with areas such as structural design and transportation. In fact, rather than being environmentally oriented, civil engineers are sometimes stereotyped as hard-hatted individuals who toss drink containers out the window of their pickup trucks while simultaneously trying to hit any small mammal that crosses their path.

Although such stereotypes are gross oversimplifications (for example many civil engineers do not drive pickups), one would more likely expect such scientists as chemists, biologists, and ecologists to be at the forefront of this area. Further, even within engineering, other disciplines such as chemical engineering might seem more naturally attuned to the environment. The reason for the strong involvement of civil engineers lies within the historical problem contexts that first drew attention to environmental problems.

Originally there were only two types of engineers: military and civilian (or civil). As the label suggests, military engineers were concerned with the technology of war; that is, they built fortifications, naval vessels, and weapons. All other technological concerns, such as roads and housing, generally fell within the civil realm.

With the onset of the industrial revolution, new technologies began to develop and specialized engineering skills became necessary. Hence disciplines such as electrical, mechanical, and chemical engineering matured and broke off from civil engineering. By the early twentieth century, civil engineers were primarily responsible for the building of structures, transportation systems, and major public works projects such as dams and aqueducts.

Stimulated by the awareness that water-borne pathogens were one of the prime causes of disease, civil engineers began to design urban water and wastewater systems (Fig. 1.1) in the late nineteenth century. Consequently civil engineers became involved in the development of water treatment plants, distribution networks, and wastewater collection systems. The design of these projects was fairly straightforward, because the goals were so well defined. The goal was to deliver an adequate quantity of potable water to the urban populace and to safely carry off its wastes.

In contrast, the question of what to do with the waste was a more ambiguous proposition. At first, municipalities discharged their raw sewage directly to receiving waters. It was immediately observed that such action would ultimately transform rivers, lakes, and estuaries into large sewers. Hence wastewater treatment plants

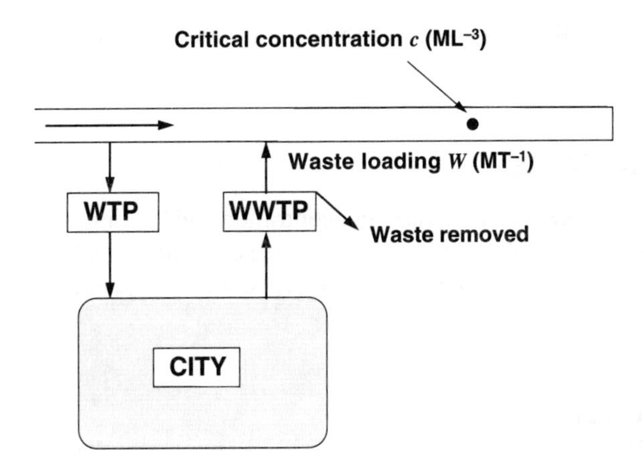

FIGURE 1.1
An urban water-wastewater system. A water treatment plant (WTP) purifi es river water for human consumption. A wastewater treatment plant (WWTP) removes pollutants from sewage to protect the receiving water.

began to be constructed. However, it was soon recognized that treatment could range from simple sedimentation to costly physical/chemical treatment. In the extreme case the latter might actually result in an effluent that was more pristine than the receiving water but at an exorbitant cost. Clearly both extremes were unacceptable. Consequently some design goal had to be established that would protect the environment adequately but economically.

Thus it was decided that waste treatment should be based on producing an effluent that induced an acceptable level of water quality in the receiving water. However, to determine this proper level of treatment, it was necessary to predict water quality as a function of waste loading. As depicted in Fig. 1.1, a linkage had to be established between the waste loading W and the resulting concentration c in the water body. Therefore civil engineers began to develop mathematical models for that purpose.

Today water-quality management has moved well beyond the urban point-source problem to encompass many other types of pollution. In addition to wastewater, we now deal with other point sources such as industrial wastes as well as nonpoint inputs such as agricultural runoff. However, as depicted in Fig. 1.2, a

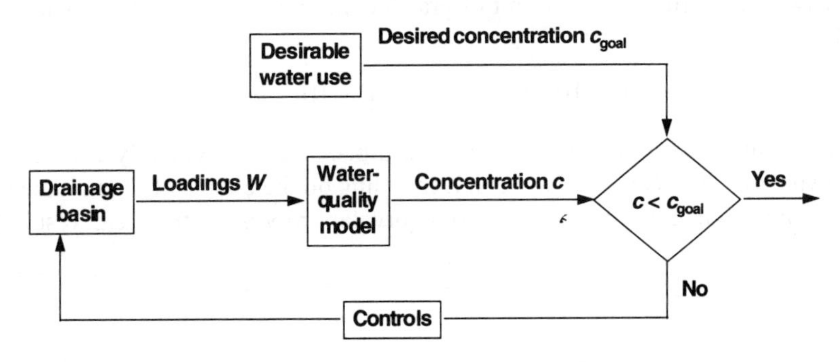

FIGURE 1.2
The water-quality management process.

water-quality model still provides the essential link to predict concentration as a function of loadings.

1.2 FUNDAMENTAL QUANTITIES

In the foregoing discussion I introduced the notions of concentration and loading. Before proceeding further, I would like to take a closer look at these quantities and how they are represented numerically. I also define some other fundamental quantities used in water-quality modeling.

1.2.1 Mass and Concentration

In water-quality modeling, the amount of pollutant in a system is represented by its mass. Such a property is formally referred to as an ***extensive property***. Other examples are heat and volume. All these characteristics are additive.

In contrast, a quantity that is normalized to a measure of system size is referred to as an ***intensive property***. Examples include temperature, density, and pressure. For our problem context, the intensive property is mass concentration defined as

$$c = \frac{m}{V} \tag{1.1}$$

where m = mass (M)[†] and V = volume (L^3). The utility of concentration lies in the fact that, as with all intensive quantities, it represents the "strength" rather than the "quantity" of the pollution. As such it is preferable to use concentration as an indicator of impact on the environment.

An example from everyday life is useful in explaining the difference. One, two, or three lumps of sugar can sweeten your coffee in varying degrees, depending on the size of your mug. The number of lumps is analogous to mass, as the sweetness is to concentration. As an "organism" you are usually more concerned about the sweetness than the number of lumps.

Concentration is conventionally expressed in metric units. The mass in Eq. 1.1 is expressed in the fundamental unit of grams combined with the prefixes in Table 1.1. Thus

$$1 \times 10^3 \text{ mg} = 1 \text{ g} = 1 \times 10^{-3} \text{ kg}$$

The volume unit is not as straightforward because it is typically expressed in one of two ways: liters or cubic meters. Depending on the choice of the volume unit, this can lead to some confusion, because equivalent representations can result. For example

---

[†] Initially I define units in terms of the fundamental dimensions: L = length, M = mass, and T = time. Later I will introduce specific metric units.

TABLE 1.1
SI (International System of Units) prefixes commonly used in water-quality modeling†

Prefix	Symbol	Value
kilo-	k	10^3
hecto-	h	10^2
deci-	d	10^{-1}
centi-	c	10^{-2}
milli-	m	10^{-3}
micro-	μ	10^{-6}
nano-	n	10^{-9}

†A complete list of prefixes is included in App. A.

$$\frac{\text{mg}}{\text{L}}\frac{10^3\ \text{L}}{\text{m}^3}\frac{\text{g}}{10^3\ \text{mg}} = \frac{\text{g}}{\text{m}^3}$$

Thus the unit mg/L is identical to g/m^3.

This situation is further complicated because, for the dilute aqueous solutions common in most surface waters, concentration is sometimes expressed on a mass basis. This conversion is predicated on the fact that the density of water is approximately equal to 1 g/cm^3:†

$$\frac{\text{g}}{\text{m}^3} = \frac{\text{g}}{\text{m}^3 \times (1\ \text{g/cm}^3)}\frac{\text{m}^3}{10^6\ \text{cm}^3} = \frac{\text{g}}{10^6\ \text{g}} = 1\ \text{ppm}$$

where ppm stands for "parts per million." Other identities are summarized in Table 1.2.

TABLE 1.2
Some water-quality variables along with typical units

Variables	Units
Total dissolved solids, salinity	g L^{-1} ⇔ kg m^{-3} ⇔ ppt
Oxygen, BOD, nitrogen	mg L^{-1} ⇔ g m^{-3} ⇔ ppm
Phosphorus, chlorophyll *a*, toxics	μg L^{-1} ⇔ mg m^{-3} ⇔ ppb
Toxics	ng L^{-1} ⇔ μg m^{-3} ⇔ pptr

EXAMPLE 1.1. MASS AND CONCENTRATION. If 2×10^{-6} lb of salt is introduced into 1 m^3 of distilled water, what is the resulting concentration in ppb?

†Following this introductory discussion, units of measure will appear in exponential form throughout the book. Thus mg/L is shown as mg L^{-1}, g/m^3 as g m^{-3}, and so on.

Solution: Applying Eq. 1.1, along with the conversion factor for pound to gram from App. A (1 lb = 453.6 g), yields

$$c = \frac{2 \times 10^{-6}}{1\ \text{m}^3}\left(\frac{453.6\ \text{g}}{\text{lb}}\right) = 9.072 \times 10^{-4}\text{g m}^{-3}$$

Converting to the desired units,

$$c = 9.072 \times 10^{-4}\text{g m}^{-3}\left(\frac{10^3\ \text{mg}}{\text{g}}\ \frac{\text{ppb}}{\text{mg m}^{-3}}\right) = 0.9072\ \text{ppb}$$

As in Example 1.1, it is conventional to use units that express the resulting concentration with a magnitude between 0.1 and 999. Thus water-quality variables are typically expressed in particular units, as outlined in Table 1.2.

1.2.2 Rates

Properties that are normalized to time are commonly referred to as ***rates***. We now discuss several rates that are key to understanding water-quality modeling (Fig. 1.3).

Mass loading rate. As in Fig. 1.1, waste discharges are typically represented by the mass loading rate W. If the mass m of pollutant is determined over a time period t, then the loading rate can be simply computed as

$$W = \frac{m}{t} \tag{1.2}$$

Many of these loadings enter receiving waters as ***point sources***; that is, through conduits such as pipes or channels. For such cases the loading rate is actually

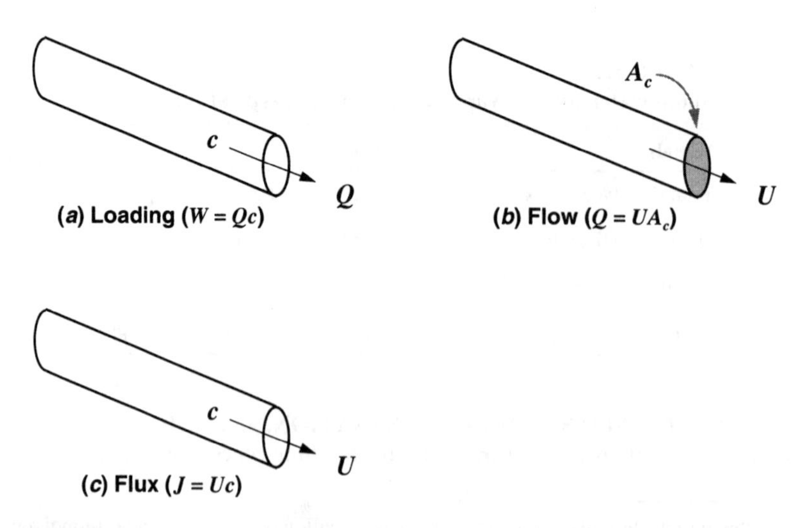

FIGURE 1.3
Three fundamental rates used extensively in water-quality modeling.

determined by measuring the concentration along with the volumetric flow rate of water in the conduit, Q ($\text{L}^3\,\text{T}^{-1}$). The mass loading rate is then computed as (Fig. 1.3*a*)

$$W = Qc \tag{1.3}$$

Volumetric flow rate. For steady flow the flow rate is often calculated with the ***continuity equation*** (Fig. 1.3*b*)

$$Q = UA_c \tag{1.4}$$

where U = velocity of water in the conduit ($\text{L}\,\text{T}^{-1}$) and A_c = cross-sectional area of the conduit (L^2).

Mass flux rate. The term ***flux*** is used to designate the rate of movement of an extensive quantity like mass or heat normalized to area. For example the mass flux rate through a conduit can be calculated as

$$J = \frac{m}{t\,A_c} = \frac{W}{A_c} \tag{1.5}$$

By substituting Eqs. 1.3 and 1.4 into Eq. 1.5, mass flux can also be expressed in terms of velocity and concentration by (Fig. 1.3*c*)

$$J = Uc \tag{1.6}$$

EXAMPLE 1.2. LOADING AND FLUX. A pond having constant volume and no outlet has a surface area A_s of 10^4 m^2 and a mean depth H of 2 m. It initially has a concentration of 0.8 ppm. Two days later a measurement indicates that the concentration has risen to 1.5 ppm. (*a*) What was the mass loading rate during this time? (*b*) If you hypothesize that the only possible source of this pollutant was from the atmosphere, estimate the flux that occurred.

Solution:

(*a*) The volume of the system can be calculated as

$$V = A_s H = 10^4\ \text{m}^2\ (2\text{m}) = 2 \times 10^4\ \text{m}^3$$

The mass of pollutant at the initial time ($t = 0$) can be computed as

$$m = Vc = 2 \times 10^4\ \text{m}^3\ (0.8\ \text{g m}^{-3}) = 1.6 \times 10^4\ \text{g}$$

and at $t = 2$ d is 3.0×10^4 g. Therefore the increase in mass is 1.4×10^4 g and the mass loading rate is

$$W = \frac{1.4 \times 10^4\ \text{g}}{2\ \text{d}} = 0.7 \times 10^4\ \text{g d}^{-1}$$

(*b*) The flux of pollutant can be computed as

$$J = \frac{0.7 \times 10^4\ \text{g d}^{-1}}{1 \times 10^4\ \text{m}^2} = 0.7\ \text{g}(\text{m}^2\ \text{d})^{-1}$$

Sometimes the concept of rate can be confused with quantity (in our case mass or concentration). Box 1.1 provides one example of the problems that can occur when this happens.

BOX 1.1. Rate versus Quantity of Great Lakes Water

In the early 1980s, droughts in the western United States prompted the government to seek out additional sources of water for this region. At one point the notion of building a pipeline from the Laurentian Great Lakes to the west was actually entertained.

At face value the plan seemed to have merit because the Great Lakes contain about 21.71×10^{12} m^3 of water. This volume corresponds to roughly 20% of the world's liquid, fresh, surface water! Consequently it seemed to make "common" sense that the Great Lakes had plenty to spare.

At first, arguments revolved around whether the pumping costs would be prohibitive (it's seriously uphill to get to and across the Rockies). However, there was a more fundamental reason why the scheme was flawed. This involved the distinction between the quantity of water and its flow rate.

Although 20% of the world's water might be held in the system, the volume is not a valid measure of how much water could be spared. The actual generation rate is reflected by the flow rate of water out of the system. Interestingly the lakes that were farthest west (Michigan and Superior) have small outflows (together only about 100×10^9 m^3 yr^{-1}). It is only when you travel east to the most downstream lake (Ontario) that the outflow into the St. Lawrence River reaches a respectable 212×10^9 m^3 yr^{-1}. Further, this flow helps to clean the lakes by flushing pollutants. Thus, seen from this perspective, the Great Lakes did not seem so attractive and the scheme died.

In essence the proponents of the plan confused the meanings of size and rate. Although the Great Lakes have lots of water (as reflected by their size), they cannot spare much of their outflow or "harvest" rate of water. Further, if a pipeline had been built and water pumped at a rate in excess of the average outflow, the Great Lakes would have rapidly become known as the "Great Mud Flats."

So what general lessons does this story offer? First, as Einstein said,

> Common sense is what makes you think the world is flat.

For some individuals a "rate" is not a "common sense" concept. Therefore, as outlined above, irrational judgments can result. Beyond that the distinction between size and rate will be critical to understanding the models described in the remainder of this text. I've included this story to help you appreciate the distinction between these two types of measures.

1.3 MATHEMATICAL MODELS

Now that some fundamental quantities have been defined, we can discuss mathematical models. According to the American Heritage Dictionary (1987), a model is a small object usually built to scale, that represents another often larger object. Thus models typically represent a simplified version of reality that is amenable to testing.

In our case we do not build physical models, but we use mathematics to represent reality. We can therefore define a ***mathematical model*** as an idealized formulation that represents the response of a physical system to external stimuli. Thus, in the context of Fig. 1.1, a mathematical model was needed to compute quality (the response) in the receiving water (the system) as a function of treatment plant effluent (the stimuli). Such a model can be represented generally as

$$c = f(W; \text{physics, chemistry, biology}) \tag{1.7}$$

According to Eq. 1.7 the cause-effect relationship between loading and concentration depends on the physical, chemical, and biological characteristics of the receiving water.

The remainder of these lectures will deal with refinements of Eq. 1.7. One very simple step in this direction is to employ a linear relationship to formulate Eq. 1.7 in mathematical terms as

$$c = \frac{1}{a} W \tag{1.8}$$

where a = an assimilation factor ($L^3 T^{-1}$) that represents the physics, chemistry, and biology of the receiving water. Equation 1.8 is called "linear" because c and W are directly proportional to each other. Consequently if W is doubled, c is doubled. Similarly if W is halved, c is halved.

1.3.1 Model Implementations

Equation 1.8 can be implemented in several ways:

1. *Simulation mode.* As expressed in Eq. 1.8, the model is used to simulate system response (concentration) as a function of a stimulus (loading) and system characteristics (the assimilation factor).
2. *Design mode I (assimilative capacity).* The model can be rearranged to yield

$$W = ac \tag{1.9}$$

   This implementation is referred to as a "design" mode because it provides information that can be directly used for engineering design of the system. It is formally referred to as an "assimilative capacity" computation because it provides an estimate of the loading required to meet a desirable concentration level or standard. Thus it forms the basis for wastewater treatment plant design. It should also clarify why a is called an "assimilation factor."
3. *Design mode II (environmental modification).* A second design implementation is

$$a = \frac{W}{c} \tag{1.10}$$

   In this case the environment itself becomes the focus of the remedial effort. Equation 1.10 is formulated to determine how, for a given loading rate, the environment might be modified to achieve the prescribed standard. This type of application is needed when affordable treatment (that is, reduction in W) is not adequate to meet water-quality standards. Examples of environmental modifications would be dredging of bottom sediments, artificial aeration, and flow augmentation.

EXAMPLE 1.3. ASSIMILATION FACTOR. Lake Ontario in the early 1970s had a total phosphorus loading of approximately 10,500 mta (metric tonnes per annum, where a metric tonne equals 1000 kg) and an in-lake concentration of 21 μg L^{-1} (Chapra and Sonzogni 1979). In 1973 the state of New York and the province of Ontario ordered a reduction of detergent phosphate content. This action reduced loadings to 8000 mta.

(*a*) Compute the assimilation factor for Lake Ontario.

(*b*) What in-lake concentration would result from the detergent phosphate reduction action?

(*c*) If the water-quality objective is to bring in-lake levels down to 10 μg L^{-1}, how much additional load reduction is needed?

Solution:

(*a*) The assimilation factor can be calculated as

$$a = \frac{W}{c} = \frac{10{,}500 \text{ mta}}{21\ \mu\text{g L}^{-1}} = 500 \frac{\text{mta}}{\mu\text{g L}^{-1}}$$

(*b*) Using Eq. 1.8, in-lake levels from the phosphorus reduction can be calculated as

$$c = \frac{W}{a} = \frac{8000 \text{ mta}}{500 \dfrac{\text{mta}}{\mu\text{g L}^{-1}}} = 16\ \mu\text{g L}^{-1}$$

(*c*) Using Eq. 1.9,

$$W = ac = 500 \frac{\text{mta}}{\mu\text{g L}^{-1}} 10 \frac{\mu\text{g}}{\text{L}} = 5000 \text{ mta}$$

Therefore an additional 3000 mta would have to be removed.

Regardless of the mode of implementation, the effectiveness of the model is contingent on an accurate characterization of the assimilation factor. As in Example 1.3, data provides one way to estimate the factor. A primary goal of the following lectures will be to delineate other ways in which this factor can be determined. To place these efforts into proper perspective, we will briefly review the major organizing principle that is used to do this—the conservation of mass.

1.3.2 Conservation of Mass and the Mass Balance

Traditionally, two approaches have been employed to estimate the assimilation factor:

- *Empirical models* are based on an ***inductive*** or data-based approach. This was the approach employed in Example 1.3 for a single lake. More often it involves obtaining values of W and c from large numbers of systems that are similar to the receiving water in question. Regression techniques can then be employed to statistically estimate the assimilation factor (Fig. 1.4).
- *Mechanistic models* are based on a ***deductive*** or theoretical approach. This involves the use of theoretical relationships or organizing principles. For example

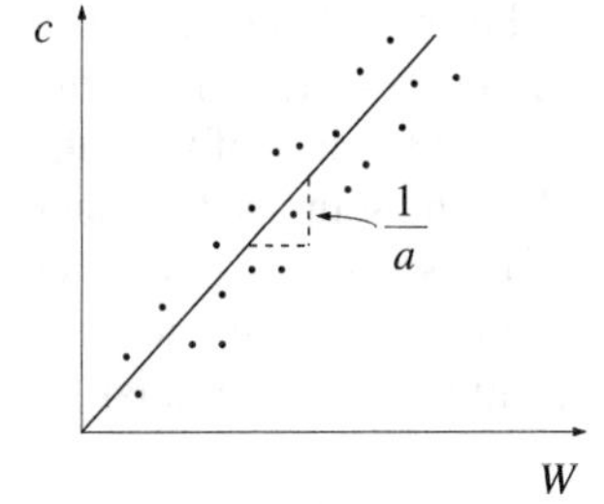

FIGURE 1.4
Empirical water-quality models use data from many water bodies to statistically estimate the cause-effect relationship between loading and concentration.

much of classical engineering is based on Newton's laws, in particular his second law: $F = ma$. In addition the great conservation laws are commonly employed as organizing principles for much engineering work.

Although empirical approaches have proved valuable in certain water-quality contexts such as lake eutrophication (see Reckhow and Chapra 1983 and Lec. 29 for a review), they have some fundamental limitations. Consequently I will adopt primarily a mechanistic approach for most of the remainder of these lectures.

Mechanistic water-quality models are based on the ***conservation of mass***; that is, within a finite volume of water, mass is neither created nor destroyed. In quantitative terms the principle is expressed as a mass-balance equation that accounts for all transfers of matter across the system's boundaries and all transformations occurring within the system. For a finite period of time this can be expressed as

$$\text{Accumulation} = \text{loadings} \pm \text{transport} \pm \text{reactions} \tag{1.11}$$

Figure 1.5 depicts mass conservation for two hypothetical substances that flow through and react within a volume of water. The movement of matter through the volume, along with water flow, is termed ***transport***. In addition to this flow, mass is gained or lost by transformations or ***reactions*** of the substances within the volume. Reactions either add mass by changing another constituent into the substance being modeled or remove mass by transforming the substance into another constituent, as in Fig. 1.5, where X reacts to form Y. Finally the substance can be increased by external ***loadings***.

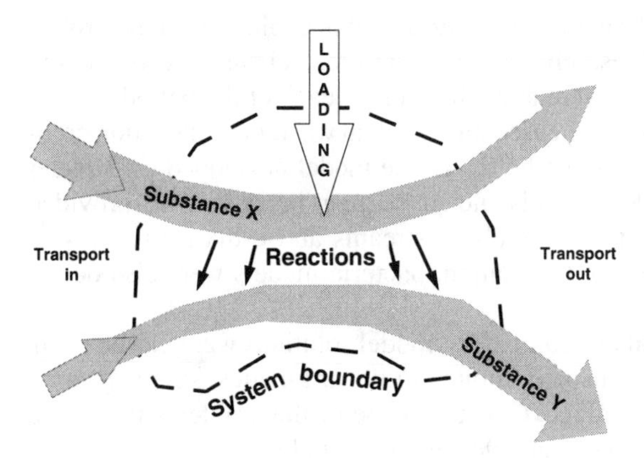

FIGURE 1.5
A schematic representation of the loading, transport, and transformation of two substances moving through and reacting within a volume of water.

By combining all the above factors in equation form, the mass balance represents a bookkeeping exercise for the particular constituent being modeled. If, for the period of the calculation, the sources are greater than the sinks, the mass of the substance within the system increases. If the sinks are greater than the sources, the mass decreases. If sources are in balance with sinks, the mass remains at a constant level and the system is said to be at ***steady-state*** or dynamic equilibrium. The mathematical expression of mass conservation, therefore, provides a framework for calculating the response of a body of water to external influences.

Since the system in Fig. 1.5 includes two substances, separate mass balances should be written for X and Y. Each should include mathematical terms to account for the transport of the substances into and out of the system. In addition the balance for X should include a term to reflect the loss of X to Y by reaction. Likewise, the equation for Y should include the same term but with a positive sign to reflect the gain of mass by Y due to the same process. Finally the balance for X should include a term for the mass gained by loading.

For situations where more than two substances interact, additional equations could be written. Similarly an investigator might be interested in the levels of substances at various locations within the volume. The system can then be divided into subvolumes for which separate mass-balance equations would be developed. Additional transport terms could be included to account for the mass transfer between the subvolumes. This mathematical division of space and matter into compartments—termed ***segmentation***—is fundamental to the application of mass conservation to water-quality problems.

The remainder of the book will be devoted to using mass balance to develop various expressions of Eq. 1.11. In the course of doing this we will describe how transport and reactions can be formulated mathematically and how segmentation figures in the process. To place these efforts into proper perspective we will now briefly review the historical evolution of water-quality modeling.

1.4 HISTORICAL DEVELOPMENT OF WATER-QUALITY MODELS

Water-quality modeling has evolved appreciably since its innovation in the early years of the twentieth century. As depicted in Fig. 1.6, this evolution can be broken down into four major phases. These phases relate both to societal concerns and to the computational capabilities that were available during each of the periods.

Most of the early modeling work focused on the urban wasteload allocation problem (Fig. 1.1). The seminal work in the field was the model developed by Streeter and Phelps (1925) on the Ohio River. This and subsequent investigations provided a means to evaluate dissolved oxygen levels in streams and estuaries (e.g., Velz 1938, 1947; O'Connor 1960, 1967). In addition, bacteria models were also developed (O'Connor 1962).

Because of the nonavailability of computers, model solutions were closed form. This meant that applications were usually limited to linear kinetics, simple geometries, and steady-state receiving waters. Thus the scope of the problems that could be addressed was constrained by the available computational tools.

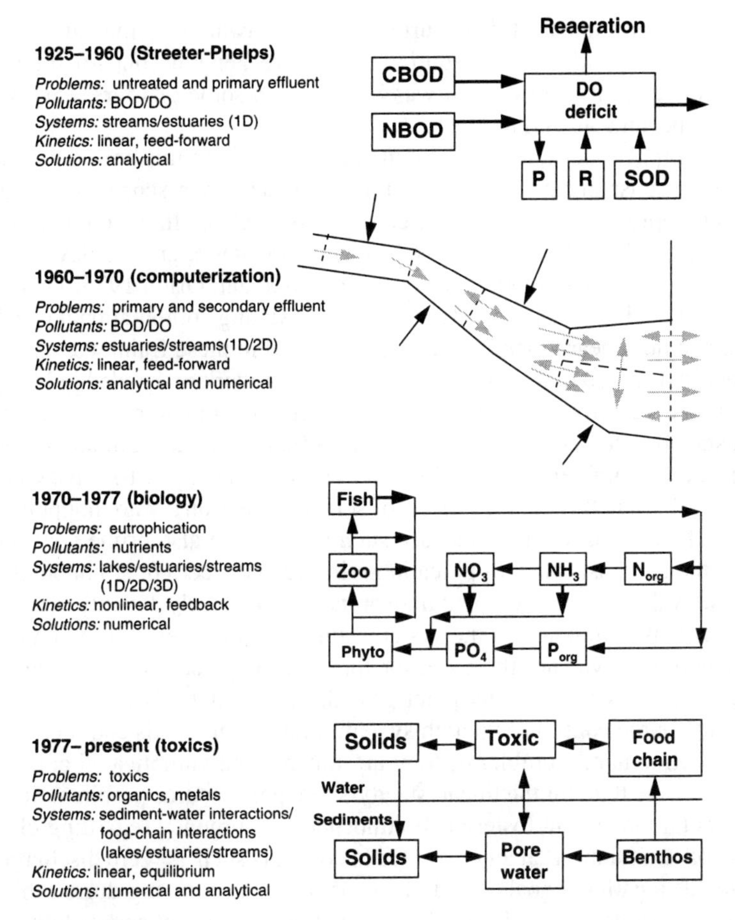

FIGURE 1.6
Four periods in the development of water-quality modeling.

In the 1960s digital computers became widely available. This led to major advances in both the models and the ways in which they could be applied. The first modeling advances involved numerical expressions of the analytical frameworks (e.g., Thomann 1963). Oxygen was still the focus, but the computer allowed analysts to address more complicated system geometries, kinetics, and time-variable simulations. In particular the models were extended to two-dimensional systems such as wide estuaries and bays.

The sixties also brought changes in the ways in which the models were applied. In particular the computer allowed a more comprehensive approach to water-quality problems. Rather than focusing on local effects of single point sources, one could view the drainage basin as a system. Tools developed originally in the field of operations research were coupled with the models to generate cost-effective treatment alternatives (e.g., Thomann and Sobal 1964, Deininger 1965, Ravelle et al. 1967). Although the focus was still on point sources, the computer allowed a more holistic perspective to be adopted.

In the 1970s another shift occurred. Societal awareness moved beyond dissolved oxygen and urban point sources to a more general concern for the environment. An ecological movement was born and, in some quarters, environmental remediation became an end in itself.

The principal water-quality problem addressed during this period was eutrophication. As a consequence, modelers broadened their own scope to include more mechanistic representations of biological processes. Capitalizing on oceanographic research (e.g., Riley 1946, Steele 1962), environmental engineers developed elaborate nutrient/food-chain models (Chen 1970; Chen and Orlob 1975; Di Toro et al. 1971, Canale et al. 1974, 1976). Because of the existing computational capabilities, feedback and nonlinear kinetics could be employed in these frameworks.

It should be noted that during this period, major work proceeded in bringing the urban point-source problem under control. In fact, most municipalities in the United States installed secondary treatment of their effluents. Aside from ameliorating the dissolved oxygen problem in many locales, for areas where point-source control was insufficient, this had the ancillary effect of shifting attention toward nonpoint sources of oxygen-demanding wastes. Because such sources are also prime contributors of nutrients, the emphasis on eutrophication reinforced concern over nonpoint inputs.

At face value the environmental awareness of the early seventies should have led to an increased reliance on the systems approach to water-quality management. Unfortunately this was not the case, for three primary reasons. First, because eutrophication deals with seasonal plant growth, it is a more dynamic problem than urban point-source control. Although systems analysis methods could be devised to optimize such dynamic problems, they are much more complicated and computationally intensive than for the linear, steady-state, point-source problem. Second, the environmental movement fostered an atmosphere of urgency regarding cleanup. A mentality of remediation "at any cost" led to concepts such as "zero discharge" being articulated as a national goal. Third, the economy was booming during this period. Therefore the economic feasibility of such a strategy was not seriously questioned. As a consequence the idea of balancing costs and benefits to devise an economical solution waned. Legalities supplanted sound engineering as the basis for most pollution control strategies. Although progress was made during this period, the unrealistic goals were never achieved.

The most recent stage of model development evolved in the wake of the energy crisis of the mid-seventies. Together with increased deficit spending, the energy crisis brought the pollution control effort back to economic reality. Unfortunately the initial response amounted to an overreaction to the excesses of the early seventies. Now, rather than an "at any cost" strategy, the public and their representatives had to be "sold" on the efficacy of environmental remediation. Consequently attention turned to problems such as toxic substances (and to a lesser extent acid rain) that, although they certainly represented a major threat to human and ecosystem health, could also be marketed effectively in the political arena.

The major modeling advance in this period has been to recognize the prominent role of solid matter in the transport and fate of toxicants (e.g., Thomann and Di Toro 1983, Chapra and Reckhow 1983, O'Connor, 1988). In particular the association of toxicants with settling and resuspending particles represents a major mechanism controlling their transport and fate in natural waters. Further, small organic particles, such as phytoplankton and detritus, can be ingested and passed along to higher organisms (Thomann 1981). Such food-chain interactions have led the modelers to view nature's organic carbon cycle as more than an end in itself. Rather, the food chain is viewed as a conveyer and concentrator of contaminants.

Today another shift is occurring in the development and application of water-quality modeling. As was the case in the late sixties and early seventies, there is a strong (and growing) recognition that environmental protection is critical to the maintenance of a high quality of life. However, added to this awareness are four factors that should make the coming decade different from the past:

1. Economic pressures are, if anything, more severe than during the late seventies. Thus incentives for cost-effective solutions are stronger than ever. This notion is reinforced by the fact that in the United States the least expensive point-source treatment options have already been implemented. As depicted in Fig. 1.7, treatment currently deals with the steepest part of the cost curve for point sources. Further, reductions of nonpoint or diffuse sources are typically more expensive than point-source controls. Today better models are needed to avoid the severe economic penalties associated with faulty decisions.
2. Developing countries around the world are beginning to recognize that environmental protection must be coupled with economic development. For these

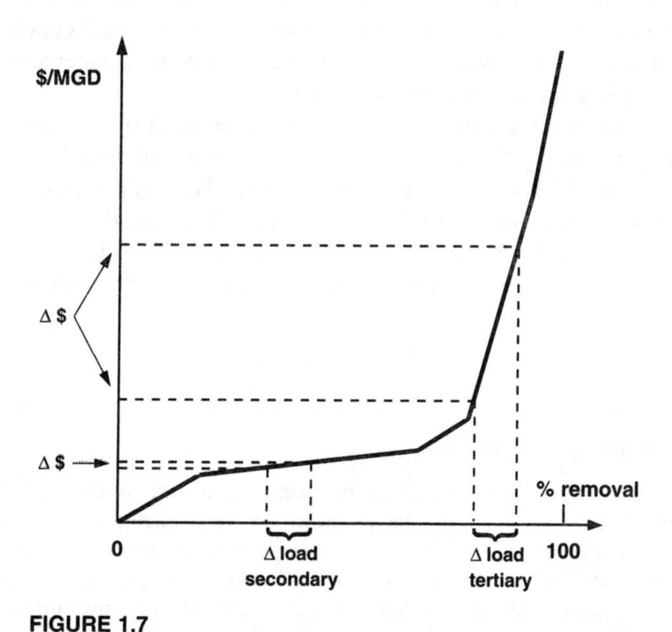

FIGURE 1.7
Capital construction costs versus degree of treatment for municipal wastewater treatment. Note that most decisions relating to tertiary waste treatment presently deal with high-percent removals. Consequently a faulty decision carries a much higher economic penalty today than in earlier years when primary and secondary waste treatment were dominant.

countries, cost-effective, model-based control strategies could provide a means to control pollution and sustain a high quality of life while maintaining economic growth.

3. Computer hardware and software have undergone a revolution over the past decade that rivals the initial advances made during the 1960s. In particular, graphical user interfaces and decision support systems are being developed that facilitate the generation and visualization of model output. Further, hardware advances are removing computational constraints that limited the scope of earlier models. Today two- and three-dimensional models with highly mechanistic kinetics can be simulated at a reasonable cost.
4. Finally significant research advances have occurred in the recent past. In particular, mechanistic characterizations of sediment-water interactions and hydrodynamics have advanced to the point that they can be effectively integrated into water-quality-modeling frameworks. Aside from the scientific advances involved in developing these mechanisms, their subsequent integration into usable frameworks is being made possible by the advances in computer technology.

In summary the evolution of water-quality modeling over the last 70 years has resulted in a unified theoretical framework that encompasses both conventional and toxic pollutants. In addition a variety of computer codes are available to implement the theory. Finally, although infrequently used, there are a number of systems analysis techniques that could be linked with the simulation models to provide cost-effective engineering solutions. Together with strong societal concern for the environment, these factors have provided a climate for a new management-oriented, computer-aided phase in water-quality modeling.

As already stated in the preface, there is a possible down side to this new phase of water-quality modeling; that is, widespread and easy use of models could lead to their being applied without insight as "black boxes." The main thesis of this book is that models must be applied with insight and with regard to their underlying assumptions, and I have gone to great lengths to maximize such a "clear box" approach to water-quality modeling. Now I'll describe how the book is organized to attain this objective.

BOX 1.2. Reading the Literature

Aside from providing some historical perspective, I had an ulterior motive in writing this section. If you really want to become an expert water-quality modeler, I would urge you to read the articles quoted in the above paragraphs (as well as many of the other references throughout the remainder of the book). Although you can learn a lot from textbooks, space constraints prevent an author from pursuing any particular topic in great depth. The published literature is a gold mine that presents knowledge in a detailed and comprehensive fashion, allows you to appreciate your heritage, and keeps you at the cutting edge. Thus I hope that you recognize that this book is merely a starting point—the tip of an iceberg represented by the wealth of insights found in the broader literature.

1.5 OVERVIEW OF THIS BOOK

The lectures are divided into seven parts that roughly follow the historical overview from the previous section. Parts I and II concentrate on modeling fundamentals. *Part I* is devoted to modeling well-mixed systems. This material includes an overview of analytical and computer-oriented solution techniques as well as an introduction to reaction kinetics. The analytical approaches emphasize the linear models that formed the basis of early water-quality modeling. In addition the computer-oriented methods provide the quantitative basis for extending the early approaches to address more complex systems. *Part II* uses a similar approach but for incompletely mixed systems. It also includes an overview of diffusion.

Part III provides an introduction to the environments that water-quality modelers usually address: streams, estuaries, lakes, and sediments. I provide general background information on these systems with special emphasis on how their transport regimes are quantified. In addition I also devote a lecture to the "modeling" environment, dealing with issues such as model calibration and verification.

Part IV looks at the first problems addressed by the pioneers of water-quality modeling: dissolved oxygen and bacteria. The material on dissolved oxygen starts with background on the problem. This is followed by a detailed discussion of the Streeter-Phelps model. Next more modern aspects of the problem such as nitrification, plant effects, and sediment oxygen demand are covered. This is followed by computer-oriented models, including a lecture on the QUAL2E software package.

Part V addresses the next major phase of water-quality modeling: eutrophication. After an overview of the problem and some simple approaches, the remainder of the section focuses on seasonal nutrient/food-chain models. Because of the importance of thermal stratification to this type of model, a couple of lectures are devoted to heat budgets and temperature modeling. This is followed by a detailed elaboration of phytoplankton and food-chain simulation.

The following parts focus on the major advances that have occurred in the 1980s. *Part VI* presents approaches to simulate the chemistry of natural waters. Then *Part VII* addresses the topic of toxic substance modeling.

It should be noted that each lecture includes detailed examples and is followed by a set of problems. These problems have been selected to test your understanding of the material and to extend the material to areas not covered in the text. I strongly urge you to work all these problems. There is an old proverb that states:

I hear . . . and I forget
I see . . . and I remember
I do . . . and I understand.

Nowhere is this saying more relevant than in the area of modeling. All too often, novice modelers believe that they "know" how to model because they understand and can write the underlying equations. Although this is obviously a necessary prerequisite, mechanics must be tempered and refined by the experience of combining the model with data. It is this experiential side that is difficult to convey in a text. I have attempted to provide a start in this direction through the problem sets. Together with additional practice and a strong scientific understanding of the system and the water-quality problem, they should prove useful in your efforts to gain proficiency in modeling.

PROBLEMS

1.1. How much mass (in g) is contained in 2.5 L of estuarine water that has a salt concentration of about 8.5 ppt?

1.2. Each individual in a city of 100,000 people contributes about 650 L capita^{-1} d^{-1} of wastewater and 135 g capita^{-1} d^{-1} of biochemical oxygen demand (BOD).
(*a*) Determine the flow rate (m^{3}s^{-1}) and the mass loading rate of BOD (mta) generated by such a population.
(*b*) Determine the BOD concentration of the wastewater (mg L^{-1}).

1.3. You are studying a 3-km stretch of stream that is about 35 m wide. A gaging station on the stream provides you with an estimate that the average flow rate during your study was 3 cubic meters per second (cms). You toss a float into the stream and observe that it takes about 2 hr to traverse the stretch. Calculate the average velocity (mps), cross-sectional area (m^2), and depth (m) for the stretch. To make these estimates, assume that the stretch can be idealized as a rectangular channel (Fig. P1.3).

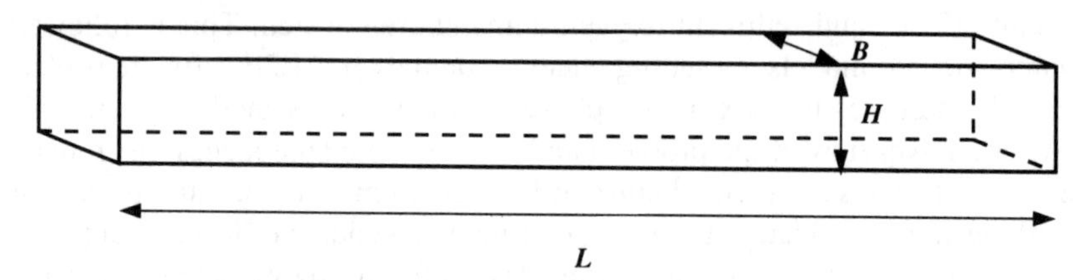

FIGURE P1.3
A stream stretch idealized as a rectangular channel: L = length, B = width, and H = mean depth.

1.4. In the early 1970s Lake Michigan had a total phosphorus loading of 6950 mta and an in-lake concentration of 8 μg L^{-1} (Chapra and Sonzogni 1979).
(*a*) Determine the lake's assimilation factor (km^3 yr^{-1}).
(*b*) What loading rate would be required to bring in-lake levels down to approximately 5μg L^{-1}?
(*c*) Express the results of (*b*) as a percent reduction, where

$$\%\ \text{reduction} = \frac{W_{\text{present}} - W_{\text{future}}}{W_{\text{present}}}(100\%)$$

1.5. A waste source enters a river as depicted in Fig. P1.5.
(*a*) What is the resulting flow rate in m^3 s^{-1} (cms)?
(*b*) If instantaneous mixing occurs, what is the resulting concentration in ppm?

1.6. You mix two volumes of water having the following characteristics:

	Volume 1	Volume 2
Volume	1 gal	2 L
Concentration	250 ppb	2000 mg m^{-3}

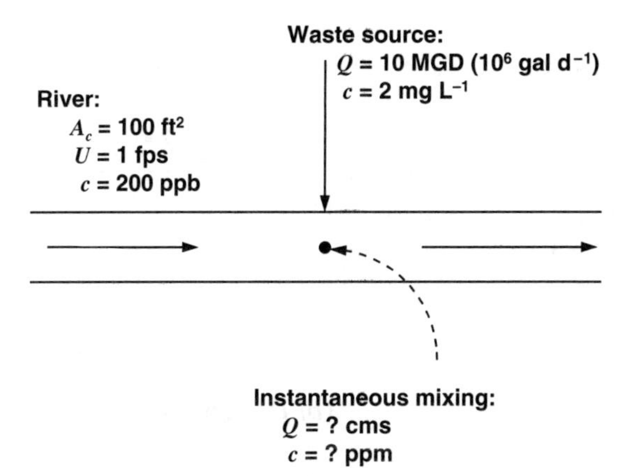

FIGURE P1.5

(*a*) Calculate the concentration (mg L^{-1}) for the mixture.
(*b*) Determine the mass in each volume and in the final mixture. Express your result in grams.

1.7. You require 4 m^3 s^{-1} of water with a salt content of 0.1 g L^{-1} for irrigation purposes. You have two reservoirs from which you can draw water (Fig. P1.7). Reservoir A has a concentration of 500 ppm, whereas reservoir B has 50 ppm. What flow rate must be pumped from each reservoir to meet the objective?

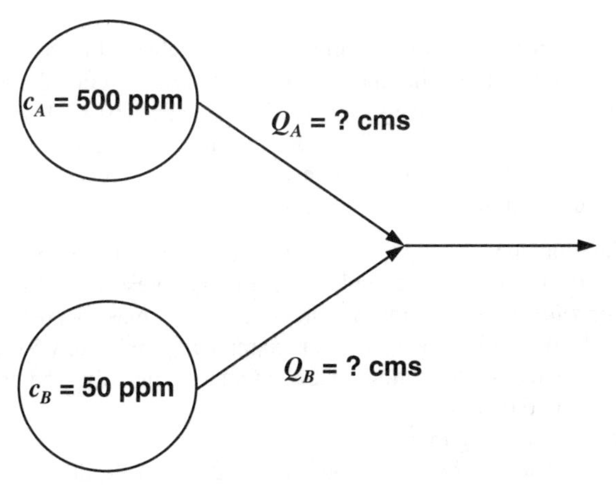

FIGURE P1.7

1.8. You add 10 mL of a glucose solution to a 300-mL bottle and then fill up the remainder of the bottle with distilled water. If the glucose solution has a concentration of 100 mg L^{-1},
(*a*) What is the concentration in the filled bottle?
(*b*) How many grams of glucose are in the bottle?

1.9. As depicted in Fig. P1.9, many lakes in temperate regions are thermally stratified, consisting of an upper layer (epilimnion) and a lower layer (hypolimnion).

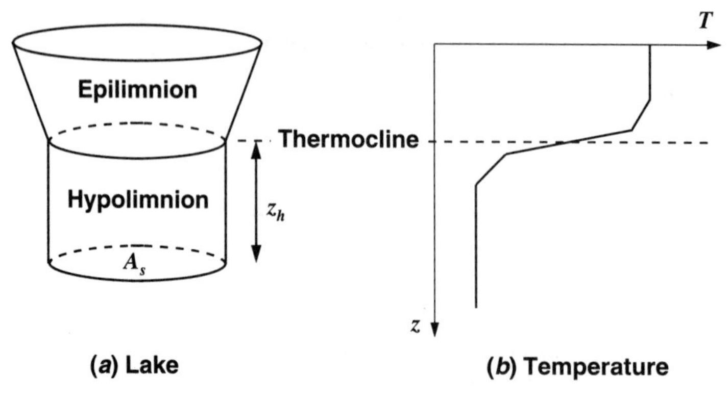

FIGURE P1.9

Onondaga Lake in Syracuse, New York, had the following characteristics at the end of a recent summer:

	Volume (m^3)	Dissolved oxygen concentration ($mg\ L^{-1}$)
Epilimnion	12×10^6	8.3
Hypolimnion	9×10^6	1.0

Compute the oxygen concentration following a severe storm that mixes the lake from top to bottom.

1.10. You must measure the flow in a small brook. Unfortunately the channel is so irregular and shallow that you cannot measure either the velocity or the cross-sectional area adequately. You therefore feed a conservative tracer with a concentration of 100 mg L^{-1} into the brook at a constant rate of 1 L min^{-1}. Note that the tracer does not occur naturally in the system. You then go downstream and measure a concentration of 5.5 mg L^{-1}. What is the original flow rate in the creek in $m^3\ s^{-1}$?

1.11. When solids enter a lake or impoundment, a portion settles and collects in an area called the ***deposition zone***. The rate of accumulation is often expressed as a flux that is called the lake's ***deposition rate***. This rate is calculated as the mass deposited per unit time normalized to the deposition zone area. For example approximately 5 million tonnes of sediment is deposited on the bottom of Lake Ontario annually. The area of the deposition zone is about 10,000 km^2.

(*a*) Calculate the deposition rate in g $m^{-2}\ yr^{-1}$.

(*b*) If the concentration of suspended solids in the water is 2.5 mg L^{-1}, determine the settling velocity from the water to the sediments, assuming that no resuspension occurs.

(*c*) Suppose that the sediment itself has a porosity of 0.90 (that is, 90% water by volume) and that the individual sediment particles have a density of 2.5 g cm^{-3}. Determine the sediment burial velocity (that is, the velocity at which the bottom is being filled in).

1.12. During warm periods, some lakes become thermally stratified (see Fig. P1.9). You measure total phosphorus concentration in the hypolimnion and find that the concentration increases from 20 $\mu g\ L^{-1}$ to 100 $\mu g\ L^{-1}$ over a 1-month period. If the bottom area is

1 km^2 and the average depth of the hypolimnion is 5 m, calculate the sediment flux of phosphorus needed to cause the observation. Express your results in mg m^{-2} d^{-1}. In your calculation assume that the thermocline forms an impermeable barrier between the upper and lower layers. (*Note:* Although the thermocline greatly diminishes exchange, some transport actually occurs. In later lectures, I show how this can be incorporated.)

1.13. The Boulder, Colorado, wastewater treatment plant enters Boulder Creek just above a USGS (U.S. Geological Survey) flow gaging station.

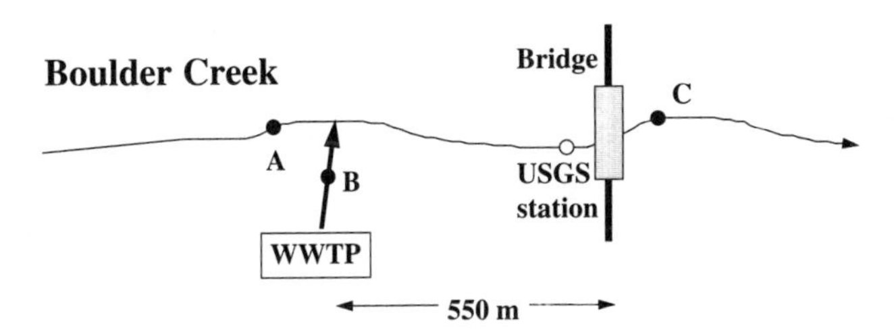

FIGURE P1.13

At 8:00 A.M. on December 29, 1994, conductivities of 170, 820, and 639 μmho cm^{-1} were measured at A, B, and C, respectively. (*Note:* Conductivity provides an estimate of the total dissolved solids in a solution by measuring its capability to carry an electrical current.) If the flow at the gaging station was 0.494 cms, estimate the flows for the treatment plant and the creek.

1.14. Sediment traps are small collecting devices that are suspended in the water column to measure the downward flux of settling solids.

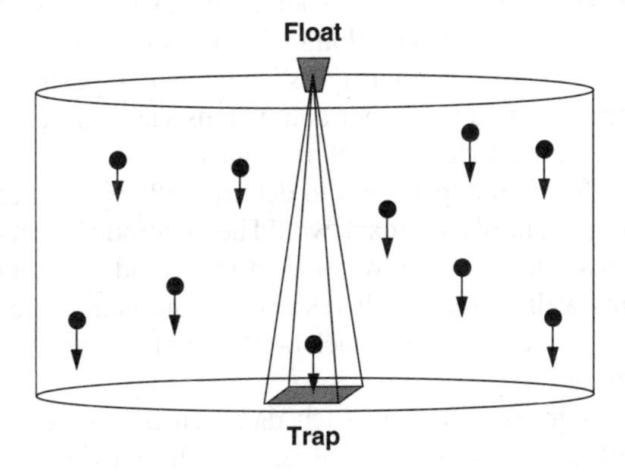

FIGURE P1.14

Suppose that you suspend a rectangular trap (1 m $\times$ 1 m) at the bottom of a layer of water. After 10 d, you remove the trap and determine that 20 g of organic carbon has collected on its surface.

(*a*) Determine the downward flux of organic carbon.

(*b*) If the concentration of organic carbon in the water layer is 1 mgC L^{-1}, determine the downward velocity of organic carbon.

(*c*) If the surface area at the bottom of the layer is 10^5 m^2, calculate how many kilograms of carbon are transported across the area over a 1-month period.

LECTURE 2

Reaction Kinetics

Lecture overview: I present some methods for characterizing reactions in natural waters. Then several graphical and computer methods are outlined to determine the order and rate of a reaction. In addition I review the reaction stoichiometry and the effect of temperature.

As described in Fig. 1.5, a number of things can happen to a pollutant once it enters a water body. Some of these relate to transport. For example it can be translated and dispersed by currents within the system. In addition the pollutant can exit the system by volatilization, by sedimentation, or by transport along with outflowing water. All these mechanisms affect the pollutant without altering its chemical composition. In contrast the pollutant might be transformed into other compounds via chemical and biochemical reactions. In this lecture we focus on such reactions.

Suppose that you want to perform an experiment to determine how a pollutant reacts after it enters a natural water. A simple approach would be to introduce some of the pollutant into a series of bottles filled with the water. A stirrer could be included in each bottle to keep the contents well mixed. Such vessels are commonly referred to as ***batch reactors***. By measuring concentration in each bottle over time, you would develop data for time and concentration (Fig. 2.1).

The purpose of this lecture is to explore how such data can be employed to characterize the reactions that affect the pollutant. That is, we will investigate how to quantitatively summarize (model) the reaction.

2.1 REACTION FUNDAMENTALS

Before discussing how reactions can be quantified, we must first develop some general definitions and nomenclature.

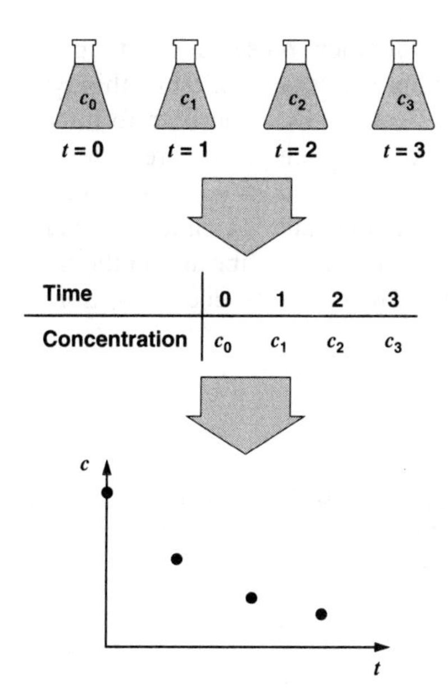

FIGURE 2.1
A simple experiment to collect rate data for a pollutant in a natural water.

2.1.1 Reaction Types

Heterogeneous reactions involve more than one phase, with the reaction usually occurring at the surfaces between phases. In contrast a ***homogeneous reaction*** involves a single phase (that is, liquid, gas, or solid). Because they are the most fundamental type of reaction employed in water-quality modeling, this lecture focuses on homogeneous reactions that take place in the liquid phase.

A ***reversible reaction*** can proceed in either direction, depending on the relative concentrations of the reactants and the products:

$$aA + bB \rightleftharpoons cC + dD \tag{2.1}$$

where the lowercase letters represent stoichiometric coefficients and the uppercase letters designate the reacting compounds. Such reactions tend to approach an equilibrium state where the forward and backward reactions are in balance. They are the basis for the area known as ***equilibrium chemistry***. We will return to these types of reactions when we address the topic of pH later in the book.

Although reversible reactions are important in water-quality modeling, more emphasis has been placed on ***irreversible reactions***. These proceed in a single direction and continue until the reactants are exhausted. For these cases, we are dealing with the determination of the rate of disappearance of one or more of the substances that is taking part in the reaction. For example for the irreversible reaction

$$aA + bB \rightarrow cC + dD \tag{2.2}$$

we might be interested in determining the rate at which substance A disappears.

A common example of an irreversible reaction is the decomposition of organic matter, which can be represented generally by

$$C_6H_{12}O_6 + 6O_2 \rightarrow 6CO_2 + 6H_2O \tag{2.3}$$

where $C_6H_{12}O_6$ is glucose, which can be taken as a simple representation of organic matter. When sewage is discharged into a receiving water, a reaction of this type takes place. The organic matter in the sewage is oxidized by bacteria to form carbon dioxide and water. Although photosynthesis (that is, plant growth) represents a reverse reaction that produces organic matter and oxygen, it does not usually occur in the same vicinity as the decomposition. In addition because decomposition and photosynthesis are relatively slow, they would not come to equilibrium on the time scales of interest in most water-quality problems. Therefore the decomposition is usually characterized as a one-way process.

2.1.2 Reaction Kinetics

The ***kinetics*** or rate of such reactions can be expressed quantitatively by the ***law of mass action***, which states that the rate is proportional to the concentration of the reactants. This rate can be represented generally as

$$\frac{dc_A}{dt} = -kf(c_A, c_B, \ldots) \tag{2.4}$$

This relationship is called a ***rate law***. It specifies that the rate of reaction is dependent on the product of a temperature-dependent constant k and a function of the concentrations of the reactants $f(c_A, c_B, \ldots)$.

The functional relationship $f(c_A, c_B, \ldots)$ is almost always determined experimentally. A common general form is

$$\frac{dc_A}{dt} = -kc_A^\alpha c_B^\beta \tag{2.5}$$

The powers to which the concentrations are raised are referred to as the ***reaction order.*** In Eq. 2.5 the reaction is α order with respect to reactant A and β order with respect to reactant B. The overall order of the reaction is

$$n = \alpha + \beta \tag{2.6}$$

The overall order of the reaction, or the order with respect to any individual component, does not have to be an integer. However, several of the most important reactions used in water-quality modeling exhibit integer orders.

In this lecture we focus on a single reactant. For this case Eq. 2.5 is often simplified as

$$\frac{dc}{dt} = -kc^n \tag{2.7}$$

where c = the concentration of the single reactant and n = the order.

2.1.3 Zero-, First-, and Second-Order Reactions

Although there are an infinite number of ways to characterize reactions, Eq. 2.7 with n = 0, 1, and 2 is most commonly employed in natural waters.

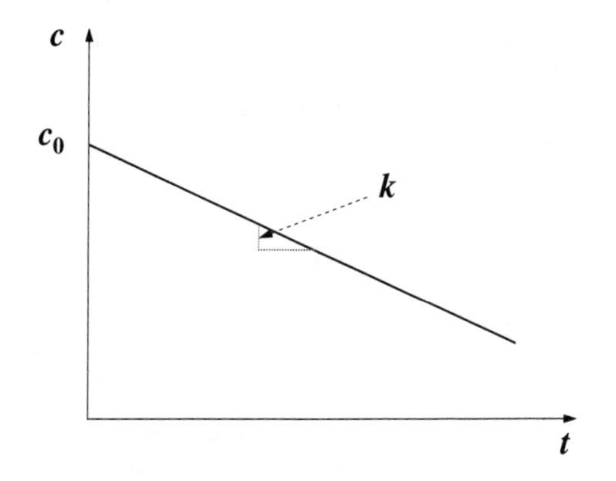

FIGURE 2.2
Plot of concentration versus time for a zero-order reaction.

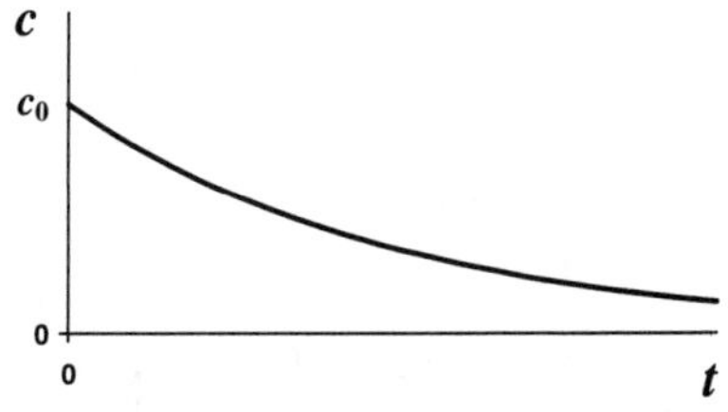

FIGURE 2.3
Plot of concentration versus time for a first-order reaction.

Zero-order. For the zero-order model ($n = 0$), the equation to integrate is

$$\frac{dc}{dt} = -k \tag{2.8}$$

where k has units of M L^{-3} T^{-1}. If $c = c_0$ at $t = 0$, then this equation can be integrated by separation of variables to yield

$$c = c_0 - kt \tag{2.9}$$

As denoted by this equation and the graph in Fig. 2.2, this model specifies a constant rate of depletion per unit time. Thus, if a plot of concentration versus time yields a straight line, we can infer that the reaction is zero-order.

First-order. For the first-order model the equation to integrate is

$$\frac{dc}{dt} = -kc \tag{2.10}$$

where k has units of T^{-1} (see Box 2.1). If $c = c_0$ at $t = 0$, then this equation can be integrated by separation of variables to yield

$$\ln c - \ln c_0 = -kt \tag{2.11}$$

Taking the exponential of both sides gives

$$c = c_0 e^{-kt} \tag{2.12}$$

As denoted by this equation, this model specifies an exponential depletion; that is, the concentration halves per unit time. Thus, as in Fig. 2.3, the concentration curve asymptotically approaches zero with time.

BOX 2.1. The "Meaning" of a First-Order Rate Constant

You may have noticed that the units of the reaction rate depend on the order of the reaction. For the zero-order reaction the rate and its units are easy to interpret. If someone states that a zero-order decay reaction has a rate of 0.2 mg L^{-1} d^{-1}, it simply means that the substance is disappearing at a rate of 0.2 mg L^{-1} every day.

In contrast a first-order rate of 0.1 yr^{-1} is not as straightforward. What does it "mean?" A way to gain insight is provided by the Maclaurin series approximation of the exponential function:

$$e^{-x} = 1 - x + \frac{x^2}{2!} - \frac{x^3}{3!} + \cdots$$

If the series is truncated after the first-order term, it is

$$e^{-x} \cong 1 - x$$

As depicted in Fig. B2.1, we see that the first-order approximation describes the rate of decrease well for small values of x. Below $x = 0.5$ the discrepancy is less than 20%. At higher values the approximation and the true value diverge.

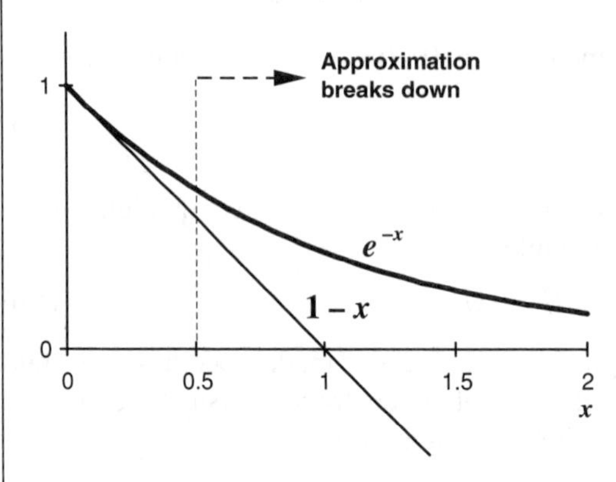

FIGURE B2.1
Plot of the exponential function along with the first-order Maclaurin-series approximation.

This leads us to the following interpretation of the "meaning" of the first-order rate constant. If its magnitude is less than 0.5, it can loosely be interpreted as the fraction of the pollutant that is lost per unit time. Thus a rate of 0.1 yr^{-1} means that 0.1 or 10% is lost in a year. If the magnitude of the rate is higher than 0.5, a change of the units can be used to interpret it. For example a rate of 6 d^{-1} clearly cannot be interpreted as meaning that 600% goes away per day. However, by converting it to an hourly rate,

$$k = 6\ \text{d}^{-1}\left(\frac{1\ \text{d}}{24\ \text{hr}}\right) = 0.25\ \text{hr}^{-1}$$

we can state that 25% goes away per hour.

The decay rate used in Eq. 2.12 is called a "base-e" rate, because it is used in conjunction with the exponential function to define the depletion of concentration with time. It should be noted that any base can be employed to describe the same trend. For example it should be recognized that the base-e or Naperian logarithm is related to the base-10 or common logarithm by

$$\log x = \frac{\ln x}{\ln 10} = \frac{\ln x}{2.3025} \tag{2.13}$$

This relationship can be substituted into Eq. 2.11 to give

$$\log c - \log c_0 = -k't \tag{2.14}$$

where $k' =$ a "base-10" rate that is related to the base-e rate by

$$k' = \frac{k}{2.3025} \tag{2.15}$$

Taking the inverse logarithm of Eq. 2.14 yields

$$c = c_0 10^{-k't} \tag{2.16}$$

This equation yields identical predictions to Eq. 2.12.

Although most first-order rates are written in terms of base-e, some are expressed in base-10. Therefore it is important to understand which base is being used. Misinterpretation would lead to using a rate that was incorrect by a factor of 2.3025 (Eq. 2.15).

Second-order. For the second-order model the equation to evaluate is

$$\frac{dc}{dt} = -kc^2 \tag{2.17}$$

where k has units of $L^3\ M^{-1}\ T^{-1}$. If $c = c_0$ at $t = 0$, then this equation can be integrated by separation of variables to yield

$$\frac{1}{c} = \frac{1}{c_0} + kt \tag{2.18}$$

Therefore if the reaction is second-order, a plot of $1/c$ versus t should yield a straight line. Equation 2.18 can also be expressed in terms of concentration as a function of time by inverting it to give

$$c = c_0 \frac{1}{1 + kc_0 t} \tag{2.19}$$

Thus, as was the case for the first-order reaction, the concentration approaches zero in a curved, asymptotic fashion.

Finally it should be obvious that a pattern is emerging that can be employed to model higher order rates. That is, for positive integer values of n, where $n \neq 1$,

$$\frac{1}{c^{n-1}} = \frac{1}{c_0^{n-1}} + (n-1)kt \tag{2.20}$$

or solving for c,

$$c = c_0 \frac{1}{\left[1 + (n-1)kc_0^{n-1}t\right]^{1/(n-1)}} \tag{2.21}$$

2.2 ANALYSIS OF RATE DATA

There are a variety of ways to analyze batch-reactor data of the type shown in Fig. 2.1. In the present section, we will review several methods. Although we will use Eq. 2.7 as the basis for illustrating these techniques, many of the general ideas apply to other rate models.

2.2.1 The Integral Method

The integral method consists of guessing n and integrating Eq. 2.7 to obtain a function, $c(t)$. Graphical methods are then employed to determine whether the model fits the data adequately.

The graphical approaches are based on linearized versions of the underlying models. For the zero-order reaction, merely plotting c versus t should yield a straight line (Eq. 2.9). For the first-order reaction, Eq. 2.11 suggests a semi-log plot. These and the other commonly applied models are summarized in Table 2.1.

TABLE 2.1
Summary of the plotting strategy used for applying the integral method to irreversible, unimolecular reactions

Order	Rate units	Dependent (y)	Independent (x)	Intercept	Slope
Zero ($n = 0$)	$M(L^3\ T)^{-1}$	c	t	c_0	$-k$
First ($n = 1$)	T^{-1}	$\ln c$	t	$\ln c_0$	$-k$
Second ($n = 2$)	$L^3(M\ T)^{-1}$	$1/c$	t	$1/c_0$	k
General ($n \neq 1$)	$(L^3\ M^{-1})^{n-1}T^{-1}$	c^{1-n}	t	c_0^{1-n}	$(n-1)k$

EXAMPLE 2.1. INTEGRAL METHOD. Employ the integral method to determine whether the following data is zero-, first-, or second-order:

t (d)	0	1	3	5	10	15	20
c (mg L^{-1})	12	10.7	9	7.1	4.6	2.5	1.8

If any of these models seem to hold, evaluate k and c_0.

Solution: Figure 2.4 shows plots to evaluate the order of the reaction. Each includes the data along with a best-fit line developed with linear regression. Clearly the plot of $\ln c$ versus t most closely approximates a straight line. The best-fit line for this case is

$$\ln c = 2.47 - 0.0972t \qquad (r^2 = 0.995)$$

Therefore the estimates of the two model parameters are

$$k = 0.0972\ \text{d}^{-1}$$
$$c_0 = e^{2.47} = 11.8\ \text{mg L}^{-1}$$

Thus the resulting model is

$$c = 11.8e^{-0.0972t}$$

The model could also be expressed to the base 10 by using Eq. 2.15 to calculate

$$k' = \frac{0.0972}{2.3025} = 0.0422$$

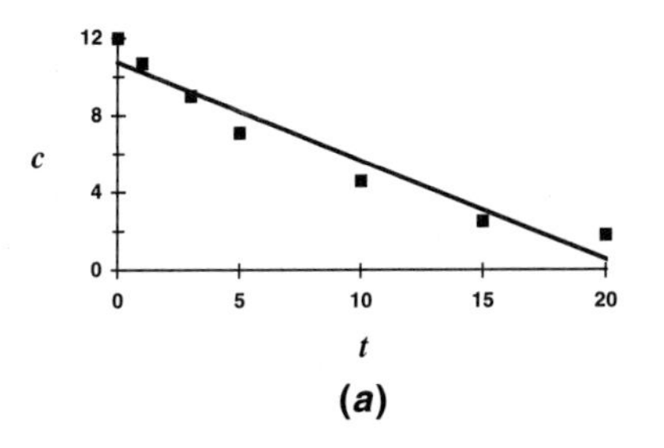

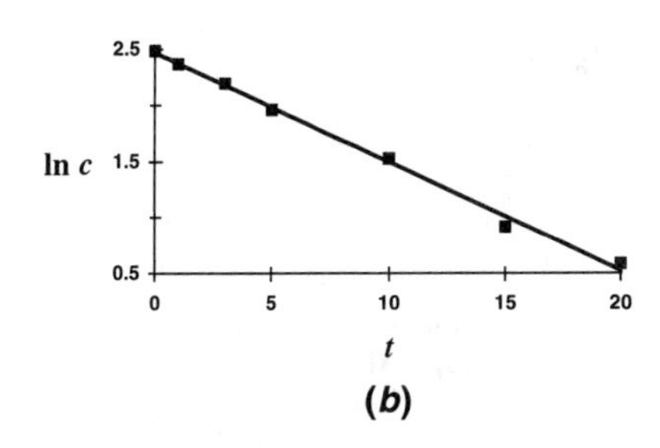

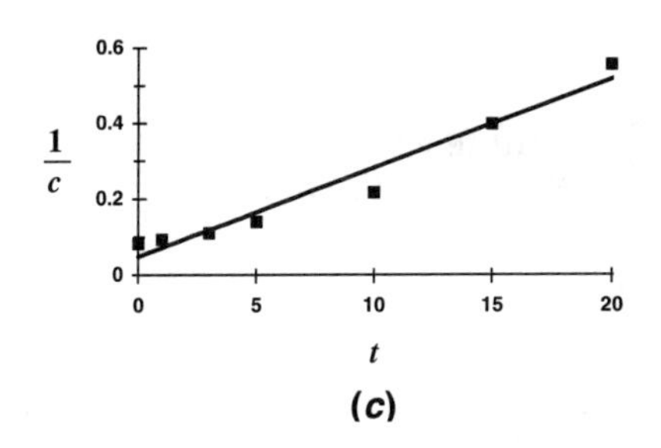

FIGURE 2.4
Plots to evaluate whether the reaction is (*a*) zero-order, (*b*) first-order, or (*c*) second-order.

which can be substituted into Eq. 2.16,

$$c = 11.8(10)^{-0.0422t}$$

The equivalence of the two expressions can be illustrated by computing c at the same value of time,

$$c = 11.8e^{-0.0972(5)} = 7.26$$

$$c = 11.8(10)^{-0.0422(5)} = 7.26$$

Thus they yield the same result.

2.2.2 The Differential Method

The differential method applies a logarithmic transform to Eq. 2.7 to give

$$\log\left(-\frac{dc}{dt}\right) = \log k + n \log c \tag{2.22}$$

Therefore if the general model (Eq. 2.7) holds, a plot of the $\log(-dc/dt)$ versus $\log c$ should yield a straight line with a slope of n and an intercept of $\log k$.

The differential approach has the advantage that it automatically provides an estimate of the order. It has the disadvantage that it hinges on obtaining a numerical estimate of the derivative. This can be done in several ways. One of the most common is based on numerical differentiation.

Numerical differentiation. Numerical differentiation uses finite-difference approximations to estimate derivatives (Chapra and Canale 1988). For example a

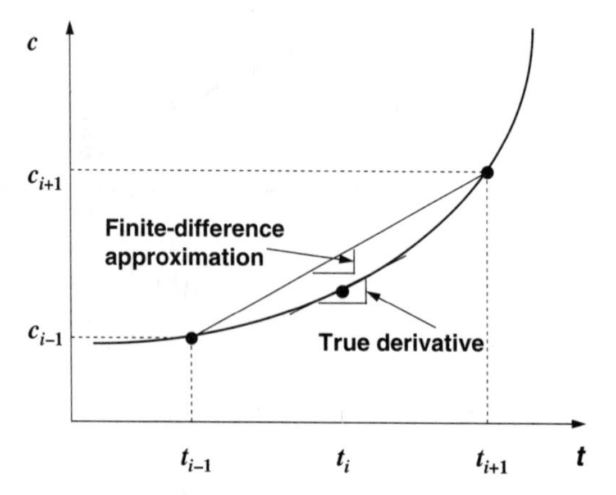

FIGURE 2.5
Numerical differentiation.

centered difference can be employed (Fig. 2.5):

$$\frac{dc_i}{dt} \cong \frac{\Delta c}{\Delta t} = \frac{c_{i+1} - c_{i-1}}{t_{i+1} - t_{i-1}} \tag{2.23}$$

Although this is certainly a valid approximation, numerical differentiation is an inherently ***unstable*** operation—that is, it amplifies errors. As depicted in Fig. 2.6, because the finite differences (Eq. 2.23) are subtractive, random positive and negative errors in the data are additive. As described in the following example, a technique known as equal-area differentiation can be used to moderate this problem (Fogler 1986).

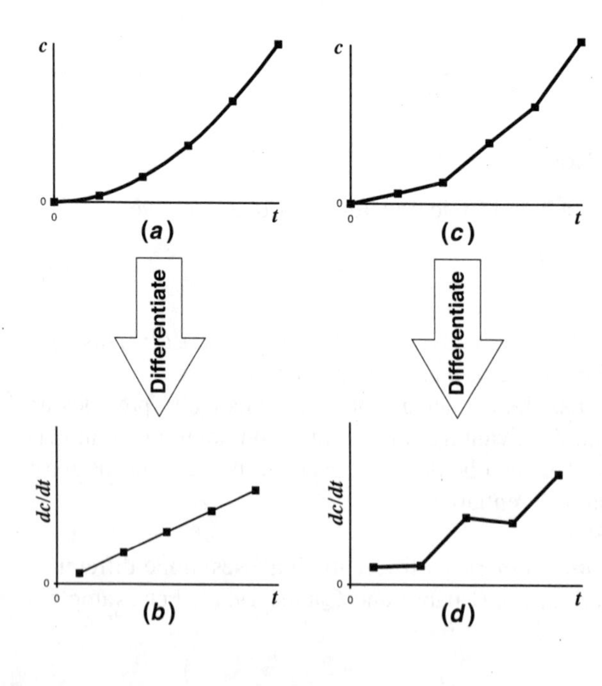

FIGURE 2.6
Illustration of how even small data errors are amplified by numerical differentiation. (*a*) Data with no error; (*b*) the resulting numerical differentiation; (*c*) data modified slightly; (*d*) the resulting differentiation manifesting increased variability (reprinted from Chapra and Canale 1988).

EXAMPLE 2.2. DIFFERENTIAL METHOD. Use the differential method to evaluate the order and the constant for the data from Example 2.1. Use equal-area differentiation to smooth the derivative estimates.

Solution: The data from Example 2.1 can be differentiated numerically to yield the estimates in Table 2.2. The derivative estimates can be graphed as a bar chart (Fig. 2.7). Then a smooth curve can be drawn that best approximates the area under the histogram. In other words try to balance out the histogram areas above and below the drawn curve. Then the derivative estimates at the data points can be read directly from the curve. These are listed in the last column of Table 2.2. Figure 2.8 shows a plot of the log of the negative derivative versus the log of concentration. The best-fit line for this case is

$$\log\left(-\frac{dc}{dt}\right) = -1.049 + 1.062 \log c \qquad (r^2 = 0.992)$$

TABLE 2.2
Data analysis to determine derivative estimates from time series of concentration

t (d)	c (mg L^{-1})	$-\frac{\Delta c}{\Delta t}$ (mg L^{-1} d^{-1})	$-\frac{dc}{dt}$ (mg L^{-1} d^{-1})
0	12.0		1.25
		1.3	
1	10.7		1.1
		0.85	
3	9.0		0.9
		0.95	
5	7.1		0.72
		0.50	
10	4.6		0.45
		0.42	
15	2.5		0.27
		0.14	
20	1.8		0.15

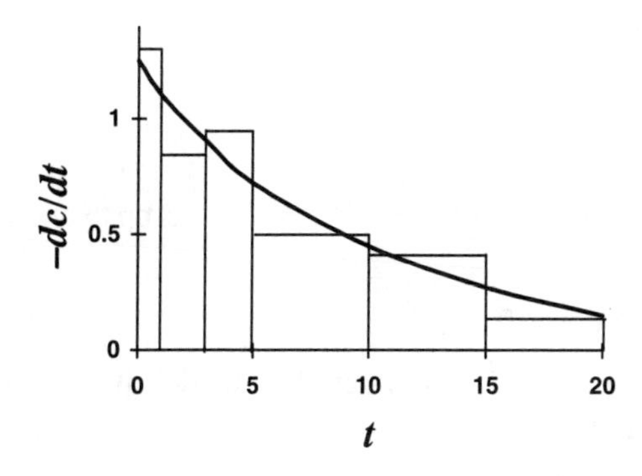

FIGURE 2.7
Equal-area differentiation.

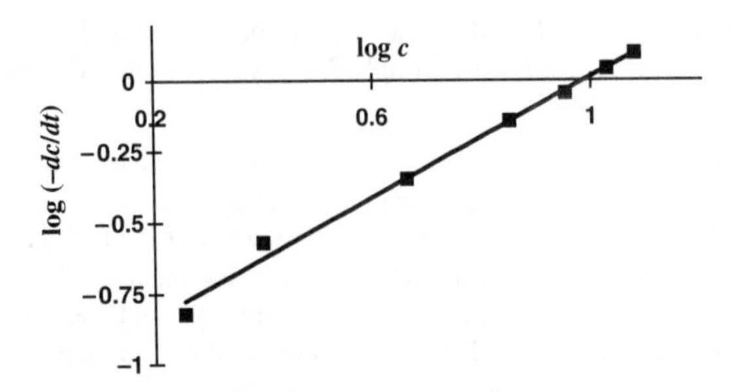

FIGURE 2.8
Plot of log($-dc/dt$) versus log c.

Therefore the estimates of the model parameters are

$$n = 1.062$$
$$k = 10^{-1.049} = 0.089\,\text{d}^{-1}$$

Thus the differential approach suggests that a first-order model is a valid approximation.

2.2.3 The Method of Initial Rates

There are cases where reactions occur in which complications arise over time. For example a significant reverse reaction might occur. Further some reactions are very slow and the time required for the complete experiment might be prohibitive. For such cases the method of initial rates uses data from the beginning stages of the experiment to determine the rate constant and order.

In this method a series of experiments is carried out at different initial concentrations c_0. For each experiment, the initial rate $-dc_0/dt$ is determined by differentiating the data and extrapolating to zero time. For the case where the rate law follows Eq. 2.7, the differential method [that is, a plot of $\log(-dc_0/dt)$ versus $\log c_0$] can be used to estimate k and n. How this is accomplished can be illustrated by taking the logarithm of the negative of Eq. 2.7:

$$\log\left(-\frac{dc_0}{dt}\right) = \log k + n \log c_0 \tag{2.24}$$

Thus the slope provides an estimate of the order, whereas the intercept provides an estimate of the logarithm of the rate.

2.2.4 The Method of Half-Lives

The ***half-life*** of a reaction is the time it takes for the concentration to drop to one-half of its initial value. In other words

$$c(t_{50}) = 0.5c_0 \tag{2.25}$$

where t_{50} = half-life. Again we use Eq. 2.7 as our rate law model. If $c = c_0$ at $t = 0$, Eq. 2.7 can be integrated to give

$$t = \frac{1}{kc_0^{n-1}(n-1)}\left[\left(\frac{c_0}{c}\right)^{n-1} - 1\right] \tag{2.26}$$

If Eq. 2.25 is combined with 2.26, the result is

$$t_{50} = \frac{2^{n-1} - 1}{k(n-1)}\frac{1}{c_0^{n-1}} \tag{2.27}$$

Taking the logarithm of this equation provides a linear relationship,

$$\log t_{50} = \log\frac{2^{n-1} - 1}{k(n-1)} + (1-n)\log c_0 \tag{2.28}$$

Thus a plot of the log of the half-life versus the log of the initial concentration will yield a straight line with a slope of $1 - n$ (providing, of course, that Eq. 2.7 holds). The estimate of n can then be used in conjunction with the intercept to evaluate k.

It should be noted that the choice of a half-life is arbitrary. In fact we could have picked any other response time t_ϕ, where ϕ is the percent reduction. For this general case, Eq. 2.27 becomes

$$t_\phi = \frac{[100/(100-\phi)]^{n-1} - 1}{k(n-1)}\frac{1}{c_0^{n-1}} \tag{2.29}$$

2.2.5 The Method of Excess

When a reaction involves many reactants, it is often possible to add excess quantities of all but one of the reactants. In such cases the reaction will depend solely on the single scarce reactant. For example several decomposition reactions for toxic substances (such as biodegradation and hydrolysis) can sometimes be represented by the reaction

$$\text{A} + \text{B} \rightarrow \text{products} \tag{2.30}$$

where A = the toxic compound and B = another quantity (such as bacteria or hydrogen ion) that participates in the reaction. The following simple rate expression is often employed to model the reaction:

$$\frac{dc_a}{dt} = -kc_ac_b \tag{2.31}$$

where c_a and c_b = concentrations of the two reactants. If the initial concentration of B (c_{b0}) is much greater than A (c_{a0}), the ensuing reaction can have a measurable effect on A whereas B will be affected minimally. Consequently the reaction can be reformulated as

$$\frac{dc_a}{dt} = -(kc_{b0})c_a = -k_{b2}c_a \tag{2.32}$$

where $k_{b2} = kc_{b0}$ = a pseudo-first-order reaction rate. The other techniques described in the previous sections can then be employed to evaluate the coefficients.

2.2.6 Numerical and Other Methods

Aside from the foregoing approaches, there are computer-oriented methods for evaluating rate data. The ***integral/least-squares method*** offers the benefits of both the integral and differential approaches in a single method. In this approach, values are assumed for the parameters (n and k) and Eq. 2.7 is solved for $c(t)$. However, rather than by calculus, the solution is obtained numerically. The solution consists of a table of predicted concentrations corresponding to the measured values. The sum of the squares of the residuals between the measured and predicted concentrations can be calculated. The assumed values of n and k are then adjusted until a minimum or least-squares condition is reached. This can be done by trial-and-error. However, modern software tools such as spreadsheets include nonlinear optimization algorithms that provide an automated way to accomplish the same goal.

The final parameter values represent the n and k that correspond to a best-fit of the data. Thus the technique has the advantage of the integral technique in the sense that it is not overly sensitive to data errors. Further it has the benefit of the differential approach in that no a priori assumption of reaction order is required.

EXAMPLE 2.3. INTEGRAL LEAST-SQUARES METHOD. Use the integral least-square method to analyze the data from Example 2.1. Use a spreadsheet to perform the calculation.

Solution: The solution to this problem is shown in Fig. 2.9. The Excel spreadsheet was used to perform the computation. Similar calculations can be implemented with other popular packages such as Quattro Pro and Lotus 123.

Initial guesses for the reaction rate and order are entered into cells B3 and B4, respectively, and the time step for the numerical calculation is typed into cell B5. For this case a column of calculation times is entered into column A starting at 0 (cell A7) and ending at 20 (cell A27). The k_1 through k_4 coefficients of the fourth-order RK method (see Lec. 7 for a description of this method) are then calculated in the block B7..E27. These are then used to determine the predicted concentrations (the c_p values) in column F. The measured values (c_m) are entered in column G adjacent to the corresponding predicted values. These are then used in conjunction with the predicted values to compute the squared residual in column H. These values are summed in cell H29.

At this point each of the spreadsheets determines the best fit in a slightly different way. At the time of this book's publication, the following menu selections would be made on Excel (v. 5.0), Quattro Pro (v. 4.5) and 123 for Windows (v. 4.0):

Excel or 123: t(ool) s(olver)   QP: t(ool) o(ptimizer)

Once you have accessed the solver or optimizer, you are prompted for a target or solution cell (H29), queried whether you want to maximize or minimize the target cell (minimize), and prompted for the cells that are to be varied (B3..B4). You then activate the algorithm [s(olve) or g(o)], and the results are as in Fig. 2.9. As shown, the values in cells B3..B4 minimize the sum of the squares of the residuals (SSR = 0.155) between the predicted and measured data. Note how these coefficient values differ from Examples 2.1 and 2.2. A plot of the fit along with the data is shown in Fig. 2.10.

	A	B	C	D	E	F	G	H
1	Fitting of reaction rate							
2	data with the integral/least-squares approach							
3	k	0.091528						
4	n	1.044425						
5	dt	1						
6	t	k1	k2	k3	k4	cp	cm	(cp-cm)^2
7	0	-1.22653	-1.16114	-1.16462	-1.10248	12	12	0
8	1	-1.10261	-1.04409	-1.04719	-0.99157	10.83658	10.7	0.018653
9	2	-0.99169	-0.93929	-0.94206	-0.89225	9.790448		
10	3	-0.89235	-0.84541	-0.84788	-0.80325	8.849344	9	0.022697
11	4	-0.80334	-0.76127	-0.76347	-0.72346	8.002317		
12	5	-0.72354	-0.68582	-0.68779	-0.65191	7.239604	7.1	0.019489
13	6	-0.65198	-0.61814	-0.61989	-0.5877	6.552494		
14	7	-0.58776	-0.55739	-0.55895	-0.53005	5.933207		
15	8	-0.53011	-0.50283	-0.50424	-0.47828	5.374791		
16	9	-0.47833	-0.45383	-0.45508	-0.43175	4.871037		
17	10	-0.4318	-0.40978	-0.4109	-0.38993	4.416389	4.6	0.033713
18	11	-0.38997	-0.37016	-0.37117	-0.35231	4.005877		
19	12	-0.35234	-0.33453	-0.33543	-0.31846	3.635053		
20	13	-0.31849	-0.30246	-0.30326	-0.28798	3.299934		
21	14	-0.28801	-0.27357	-0.2743	-0.26054	2.996949		
22	15	-0.26056	-0.24756	-0.24821	-0.23581	2.7229	2.5	0.049684
23	16	-0.23583	-0.22411	-0.22469	-0.21352	2.474917		
24	17	-0.21354	-0.20297	-0.20349	-0.19341	2.250426		
25	18	-0.19343	-0.18389	-0.18436	-0.17527	2.047117		
26	19	-0.17529	-0.16668	-0.16711	-0.1589	1.862914		
27	20	-0.15891	-0.15115	-0.15153	-0.14412	1.695953	1.8	0.010826
28								
29							SSR =	0.155062

FIGURE 2.9
The application of the integral least-squares method to determine the order and rate coefficient of reaction data. This application was performed with the Excel spreadsheet.

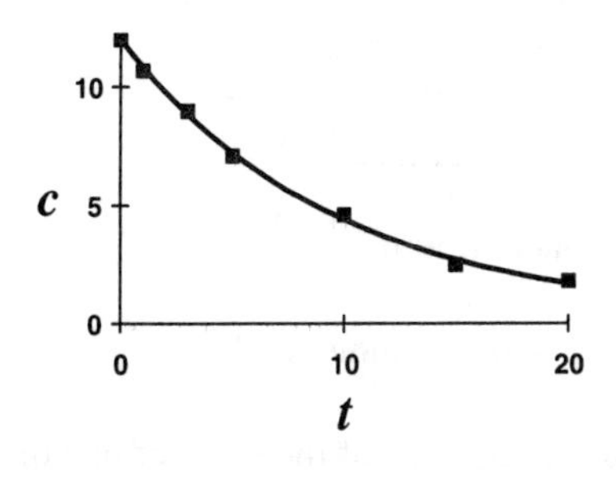

FIGURE 2.10
Plot of fit generated with the integral/least-squares approach.

There are a variety of other approaches for analyzing rate data beyond the ones described in this lecture. In addition there are other reactions aside from Eq. 2.7 that are used in water-quality modeling. The reader should consult the excellent presentations of Fogler (1986) and Grady and Lim (1980) for additional information. I will be reviewing some additional rate laws in later sections of this text. Methods for evaluating their rate constants will be reviewed in a "just-in-time" fashion as they are needed.

2.3 STOICHIOMETRY

In the previous lecture I introduced the notion of mass concentration as a means to quantify the strength of a single chemical compound in water. Now that we are dealing with reactions, several compounds may react to form other compounds. Therefore we might want to determine "how much" of a reactant or product is consumed or created as the reaction proceeds. The answer to this question resides in the ***stoichiometry***, or number of moles, taking part in a reaction.

For example the decomposition or oxidation of sugar is represented by (recall Eq. 2.3)

$$C_6H_{12}O_6 + 6O_2 \rightarrow 6CO_2 + 6H_2O \tag{2.33}$$

This equation specifies that 6 moles of oxygen will react with 1 mole of glucose to form 6 moles of carbon dioxide and 6 moles of water. In later lectures we directly use molar concentrations when we mathematically manipulate such equations to solve chemical equilibrium problems. For the time being, as outlined in the previous lecture, we must be able to interpret Eq. 2.33 from a mass-concentration perspective.

First, let's understand how the glucose in Eq. 2.33 would be expressed in mass units. This is usually done in two ways. The most direct way is to express the concentration on the basis of the whole molecule. For example we might say that a beaker contained 100 g m^{-3} as glucose. This is often abbreviated as 100 g-glucose m^{-3}. The number of moles of glucose in this solution can be determined with the gram molecular weight of glucose. The gram molecular weight can be calculated as

	Number of moles		Mass of one mole		
$6 \times C =$	6	$\times$	12 g	=	72 g
$12 \times H =$	12	$\times$	1 g	=	12 g
$6 \times O =$	6	$\times$	16 g	=	96 g
	Gram molecular weight			=	180 g

This result can be used to compute the molar concentration,

$$100\frac{\text{g-glucose}}{\text{m}^3}\left(\frac{1\text{ mole}}{180\text{ g-glucose}}\right) = 0.556\text{ mole m}^{-3} \tag{2.34}$$

An alternative is to express the concentration in terms of the mass of one of the components of glucose. Because it is an organic carbon compound, glucose could be expressed as g m^{-3} of carbon. For example

$$100\frac{\text{g-glucose}}{\text{m}^3}\left(\frac{6\text{ moles C} \times 12\text{ gC/mole C}}{180\text{ g-glucose}}\right) = 40\text{ gC m}^{-3} \tag{2.35}$$

Thus 100 g-glucose m^{-3} corresponds to 40 g m^{-3} of organic carbon, or 40 gC m^{-3}.

Such conversions are often expressed as stoichiometric ratios. For example the mass of carbon per mass of glucose can be expressed as

$$a_{cg} = \frac{6 \text{ moles C} \times 12 \text{ gC/mole C}}{180 \text{ g-glucose}} = 0.4 \text{ gC g-glucose}^{-1} \tag{2.36}$$

where a_{cg} = stoichiometric ratio of **carbon** to glucose. This ratio can be used to formulate Eq. 2.35 alternatively as

$$c_c = a_{cg} c_g = 0.4 \frac{\text{gC}}{\text{g-glucose}} \left(100 \frac{\text{g-glucose}}{\text{m}^3}\right) = 40 \text{ gC m}^{-3} \tag{2.37}$$

where the subscripts c and g designate carbon and glucose, respectively.

Aside from calculating how much individual element is contained in a molecule, stoichiometric conversions are often used to determine how much of a reactant or product is consumed or produced by a reaction. For example how much oxygen would be consumed if 40 gC m^{-3} of glucose reacted according to Eq. 2.33? First, we can calculate the mass of oxygen consumed per mass of glucose carbon decomposed,

$$r_{oc} = \frac{6 \text{ moles } O_2 \times 32 \text{ gO/mole } O_2}{6 \text{ moles C} \times 12 \text{ gC/mole C}} = 2.67 \text{ gO gC}^{-1} \tag{2.38}$$

where r_{oc} = mass of oxygen consumed per carbon decomposed. This ratio can be used to determine

$$2.67 \frac{\text{gO}}{\text{gC}} \left(40 \frac{\text{gC}}{\text{m}^3}\right) = 106.67 \text{ gO m}^{-3} \tag{2.39}$$

Thus if 40 gC m^{-3} of glucose (or 100 g-glucose m^{-3}) is decomposed, 106.67 gO m^{-3} will be consumed.

EXAMPLE 2.4. STOICHIOMETRIC RATIOS. Aside from the decomposition of organic carbon compounds such as glucose, other reactions consume oxygen in natural waters. One such process, called ***nitrification***, involves the conversion of ammonium (NH_4^+) to nitrate (NO_3^-). Although we will learn in Lec. 23 that it's a little more complicated, the nitrification reaction can be represented by

$$NH_4^+ + 2O_2 \rightarrow 2H^+ + H_2O + NO_3^-$$

Suppose you are told that a beaker contains 12 g-ammonium m^{-3} is nitrified according to the first-order reaction

$$\frac{dn_a}{dt} = -k_n n_a$$

where n_a = ammonium concentration and k_n = first-order rate constant for nitrification. (a) Convert the concentration to gN m^{-3}. (b) Determine how much oxygen is consumed if the nitrification reaction goes to completion. (c) Calculate the rate of oxygen consumption at the start of the process (k_n = 0.1 d^{-1}).

Solution: (*a*)

$$12\frac{\text{gNH}_4{}^+}{\text{m}^3}\left(\frac{1\times 14\text{ gN}}{1\times 14+4\times 1\text{ gNH}_4{}^+}\right)=9.33\text{ gN m}^{-3}$$

(*b*)

$$r_{on}=\frac{2\times 32}{1\times 14}=4.57\text{ gO gN}^{-1}$$

Therefore 4.57 gO are taken up for every gN that is nitrified. For our example,

$$9.33\frac{\text{gN}}{\text{m}^3}\left(4.57\frac{\text{gO}}{\text{gN}}\right)=42.67\text{ gO m}^{-3}$$

(*c*) At the onset of the experiment, the ammonium concentration will be at 9.33 gN m^{-3}. Using the oxygen-to-nitrogen ratio, the initial rate of oxygen consumption can be calculated as

$$\frac{do}{dt}=-r_{on}k_n n_a=-4.57\frac{\text{gO}}{\text{gN}}\left(0.1\text{ d}^{-1}\right)\left(9.33\frac{\text{gN}}{\text{m}^3}\right)=-4.264\text{ gO m}^{-3}\text{ d}^{-1}$$

2.4 TEMPERATURE EFFECTS

The rates of most reactions in natural waters increase with temperature. A general rule of thumb is that the rate will approximately double for a temperature rise of 10°C.

A more rigorous quantification of the temperature dependence is provided by the ***Arrhenius equation***,

$$k(T_a)=Ae^{\frac{-E}{RT_a}} \tag{2.40}$$

where A = a preexponential or frequency factor
E = activation energy (J mole^{-1})
R = the gas constant (8.314 J mole^{-1} K^{-1})
T_a = absolute temperature (K)

Equation 2.40 is often used to compare the reaction rate constant at two different temperatures. This can be done by expressing the ratio of the rates, as in

$$\frac{k(T_{a2})}{k(T_{a1})}=e^{\frac{E(T_{a2}-T_{a1})}{RT_{a2}T_{a1}}} \tag{2.41}$$

Equation 2.41 can be simplified by realizing that:

- Because temperatures in most water bodies vary over a rather narrow range (273 to 313 K), the product of T_{a1} and T_{a2} is relatively constant.
- The difference in temperature ($T_{a2}-T_{a1}$) is identical whether an absolute or a centigrade scale is used.

Consequently the following can be defined as a constant:

$$\theta\equiv e^{\frac{E}{RT_{a2}T_{a1}}} \tag{2.42}$$

and Eq. 2.41 can be reexpressed as

TABLE 2.3
Some typical values of θ used in water-quality modeling

θ	Q_{10}	Reaction
1.024	1.27	Oxygen reaeration
1.047	1.58	BOD decomposition
1.066	1.89	Phytoplankton growth
1.08	2.16	Sediment oxygen demand (SOD)

$$\frac{k(T_2)}{k(T_1)} = \theta^{T_2 - T_1} \tag{2.43}$$

where the temperature is expressed in °C.

In water-quality modeling, many reactions are reported at 20°C (see Prob. 2.16). Therefore, Eq. 2.43 is usually expressed as

$$k(T) = k(20)\theta^{T-20} \tag{2.44}$$

Table 2.3 summarizes some commonly used values for θ. Figure 2.11 illustrates the functional dependency on temperature across the range commonly encountered in natural waters.

The temperature dependence of biologically mediated reactions is often expressed as the quantity Q_{10}, which is defined as the ratio

$$Q_{10} = \frac{k(20)}{k(10)} \tag{2.45}$$

Substituting Eq. 2.44 yields

$$Q_{10} = \theta^{10} \tag{2.46}$$

Note that Eq. 2.46 can be used to compute that a Q_{10} of 2 (recall the heuristic at the beginning of this section) is equivalent to a θ of $2^{0.1} = 1.072$. Thus a $\theta = 1.072$ corresponds to a doubling of the rate for a temperature rise from 10 to 20°C.

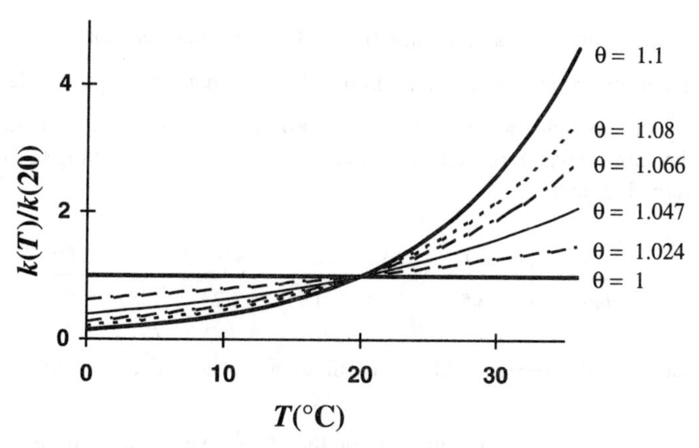

FIGURE 2.11
The effect of temperature on reaction rate for various values of θ.

EXAMPLE 2.5. EVALUATION OF TEMPERATURE DEPENDENCY OF REACTIONS. A laboratory provides you with the following results for a reaction:

$$T_1 = 4°\text{C} \qquad k_1 = 0.12\ \text{d}^{-1}$$
$$T_2 = 16°\text{C} \qquad k_2 = 0.20\ \text{d}^{-1}$$

(*a*) Evaluate θ for this reaction.
(*b*) Determine the rate at 20°C.

Solution: (*a*) To evaluate this information, we can take the logarithm of Eq. 2.43 and raise the result to a power of 10 to give

$$\theta = 10^{\frac{\log k(T_2) - \log k(T_1)}{T_2 - T_1}}$$

Substituting the data gives

$$\theta = 10^{\frac{\log 0.12 - \log 0.20}{4-16}} = 1.0435$$

(*b*) Equation 2.43 can then be used to compute

$$k(20) = 0.20 \times 1.0435^{20-16} = 0.237\ \text{d}^{-1}$$

Finally it should be noted that there are some reactions that do not follow the Arrhenius equation. For example certain biologically mediated reactions shut down at very high and very low temperatures. The formulations used in such situations are introduced in later lectures.

PROBLEMS

2.1. You perform a series of batch experiments and come up with the following data:

t (hr)	0	2	4	6	8	10
c (μg L^{-1})	10.5	5.1	3.1	2.8	2.2	1.9

Determine the order (n) and the rate (k) of the underlying reaction.

2.2. Derive a graphical approach to determine whether a reaction is third-order.

2.3. To study the photodegradation of aqueous bromine, we dissolved a small quantity of liquid bromine in water, placed it in a clear jar, and exposed it to sunlight. The following data were obtained:

t (min)	10	20	30	40	50	60
c (ppm)	3.52	2.48	1.75	1.23	0.87	0.61

Determine whether the reaction is zero-, first-, or second-order and estimate the reaction rate.

2.4. At a later date the laboratory informs you that they have a more complete data set than the two measurements used in Example 2.5:

T (°C)	4	8	12	16	20	24	28
k (d^{-1})	0.120	0.135	0.170	0.200	0.250	0.310	0.360

Use this data to estimate θ and k at 20°C

2.5. An article in a limnological[†] journal reports a Q_{10} for a phytoplankton growth rate of 1.9. If the growth rate is reported as 1.6 d^{-1} at 20°C, what is the rate at 30°C?

2.6. You set up a series of 300-mL bottles and add 10 mL of a glucose solution to each. Note that the glucose solution has a concentration of 100 mgC L^{-1}. To each bottle you add a small quantity (that is, with an insignificant amount of carbon compared to the glucose) of bacteria. You fill the remainder of their volumes up with water. Finally you seal each bottle and incubate them at 20°C. At various times you open one of the bottles and measure its oxygen content. The following data results:

t (d)	0	2	5	10	20	30	40	50	60	70
c ($mgO_2\ L^{-1}$)	10	8.4	6.5	4.4	2.3	1.6	1.3	1.2	1.1	1.1

(*a*) Develop a conceptual model for what is taking place inside the bottle.
(*b*) Using the information from this lecture, attempt to estimate the decay rate for the glucose.

2.7. In the fall of 1972 Larsen et al. (1979) measured the following concentrations of total phosphorus in Shagawa Lake, Minnesota:

Day	**mg m^{-3}**	**Day**	**mg m^{-3}**	**Day**	**mg m^{-3}**
250	97	270	72	290	62
254	90	275	51	295	55
264	86	280	57	300	46

It is known that the primary reason for the reduction in concentration during this period was the settling of particulate phosphorus. If the lake is assumed to act as a batch reactor and settling is assumed to follow a first-order process, determine the removal rate of total phosphorus for the lake. If the lake's mean depth is 5.5 m, calculate the settling velocity for total phosphorus.

2.8. Population dynamics is important in predicting how human development of a watershed might influence water quality. One of the simplest models incorporates the assumption that the rate of change of the population p is proportional to the existing population at any time t:

$$\frac{dp}{dt} = Gp \tag{P2.8}$$

[†]Limnology is the study of lakes. The terminology is derived from the Greek word for lake: *limnos.*

where G = the growth rate (yr^{-1}). Suppose that census data provides the following trend in population of a small town over a 20-yr period:

t	1970	1975	1980	1985	1990
p	100	212	448	949	2009

If the model (Eq. P2.8) holds, estimate G and the population in 1995.

2.9. The world took about 300 years to grow from about 0.5 billion to 4 billion people. Assuming first-order growth, determine the growth rate. Estimate the population over the next century if this rate continues.

2.10. Many lakes in temperate regions are thermally stratified in the summer, consisting of an upper layer (epilimnion) and a lower layer (hypolimnion). In general the surface layer has dissolved oxygen concentration near saturation. If it is productive (that is, has high plant growth), settling plant matter can collect in the hypolimnion. The decomposition of this matter can then lead to severe oxygen depletion in the bottom waters. When turnover occurs in the fall (that is, vertical mixing due to decreasing temperature and increasing winds), the mixing of the two layers can result in the lake's having an oxygen concentration well below saturation. The following data were collected for Onondaga Lake in Syracuse, New York:

Date	Sep. 30	Oct. 3	Oct. 6	Oct. 9	Oct. 12	Oct. 15	Oct. 18	Oct. 21
Oxygen conc. (mg L^{-1})	4.6	6.3	7.3	8.0	8.4	8.7	8.9	9.0

If the saturation concentration is 9.2, use this data to evaluate a first-order reaeration rate for the system (units of d^{-1}). Assume that the lake acts as an open batch reactor; that is, ignore inflows and outflows of oxygen except gas transfer across the lake's surface. Also, express the rate as a transfer velocity (units of m d^{-1}). Note that Onondaga Lake has a surface area of 11.7 km^2 and a mean depth of 12 m.

2.11. A reaction has a Q_{10} of 2.2. If the reaction rate at 25°C is 0.853 wk^{-1}, what is the reaction rate at 15°C?

2.12. A commonly used anesthetic is absorbed by human body organs at a rate proportional to its concentration in the bloodstream. Assume that a patient requires 10 mg of the anesthetic per kg of body weight to maintain an acceptable level of anesthesia for surgery. Compute how many mg must be administered to a 50-kg patient to maintain a proper level for a 2.5-hr operation. Assume that the anesthetic can be introduced into the patient's bloodstream as a pulse input and that it decays at a rate of 0.2% per minute.

2.13. Estimate the age of the fossil remains of a skeleton with 2.5% of its original carbon-14 content. Note that carbon-14 has a half-life of 5730 yr.

2.14. In 1828 Friedrich Wohler discovered that the inorganic salt ammonium cyanate (NH_4OCN) can be converted into the organic compound urea (NH_2CONH_2), as in

$$NH_4OCN\ (aq) \rightarrow NH_2CONH_2\ (aq)$$

The proof that this reaction occurred marked the beginning of modern organic and biochemistry. An investigator has reported the following data for an experiment initially containing a pure solution of ammonium cyanate:

Time (min)	0	20	50	65	150
NH_4OCN (mole L^{-1})	0.381	0.264	0.180	0.151	0.086

Determine the order and rate of the reaction.

2.15. You perform a batch experiment and develop the following data:

t	0	2	4	6	8	10
c	10.0	8.5	7.5	6.7	6.2	5.8

You know from experience that the reaction should be following a third-order reaction. Use this information and the integral method to determine a value for the reaction rate.

2.16. Suppose that the temperature dependence of a reaction rate is based on its value at 25°C (note that this is the convention in areas such as chemical engineering). For example

$$k(T) = 0.1(1.06)^{T-25}$$

Reexpress this relationship based on the rate's value at 20°C.

2.17. The following data for concentrations and times were developed for a series of batch experiments having different initial conditions:

t	c			
0	1.00	2.00	5.00	10.00
1	0.95	1.87	4.48	8.59
2	0.91	1.74	4.04	7.46

Assuming that Eq. 2.7 holds, use the method of initial rates to determine the order and rate of the reaction.

2.18. Assuming that Eq. 2.7 holds, use the method of half-lives to determine the reaction order and rate by evaluating the following half-lives and initial concentrations developed from a series of batch experiments:

c_0	1	2	5	10
t_{50}	16	11	7	5

2.19. Assuming that Eq. 2.7 holds, use the integral least-squares method to determine the reaction order and rate by evaluating the following data collected from a batch experiment:

t	0	2	4	6	8	10
c	10	7.5	5.8	4.6	3.8	3.1

2.20. The concentration of inorganic phosphorus in natural waters is usually expressed as phosphorus (P). However, it is sometimes expressed as phosphate (PO_4). When reading a scientific article, you see that an estuary has an inorganic phosphorus concentration of

10 mg m^{-3}. As is sometimes the case no guidance is given regarding how the concentration is expressed. How does the concentration change if it is actually $mgPO_4\ m^{-3}$? By what factor would you be off?

2.21. A more complete representation of the decomposition reaction is provided by

$$C_{106}H_{263}O_{110}N_{16}P_1 + 107O_2 + 14H^+ \rightarrow 106CO_2 + 16NH_4{}^+ + HPO_4{}^{2-} + 108H_2O$$

In contrast to the simplified version in Eq. 2.3, this reaction reflects that organic matter contains the nutrients nitrogen (N) and phosphorus (P). On the basis of this equation, given that 10 gC m^{-3} of organic matter is decomposed, calculate

(*a*) the stoichiometric ratio for the amount of oxygen consumed per carbon decomposed, r_{oc} (gO gC^{-1})

(*b*) the amount of oxygen consumed (gO m^{-3})

(*c*) the amount of ammonium released (expressed as mgN m^{-3})

LECTURE 3

Mass Balance, Steady-State Solution, and Response Time

LECTURE OVERVIEW: I introduce the primary organizing principle of water-quality modeling: the mass balance. It is used to derive a water-quality model for the simple case of a continuously stirred tank reactor, or CSTR. I develop and interpret a steady-state solution and present the notions of transfer functions and residence times. I also derive the general time-variable solution of the mass-balance model and calculate a response time to quantify temporal characteristics of the recovery of such systems.

Now that we have reviewed some fundamental concepts, let's tie them together and actually develop a water-quality model. Then we will solve the model to answer the two most commonly posed questions in water-quality modeling: If we institute a treatment program,

- How much will the water body improve?
- How long will it take for the improvement to occur?

3.1 MASS BALANCE FOR A WELL-MIXED LAKE

A completely mixed system, or ***continuously stirred tank reactor (CSTR)***, is among the simplest systems that can be used to model a natural water body. It is appropriate for a receiving water in which the contents are sufficiently well mixed as to be uniformly distributed. Such a characterization is often used to model natural lakes and some impoundments.

A hypothetical completely mixed system is depicted in Fig. 3.1. Note that I have included a number of sources and sinks that are typically encountered when

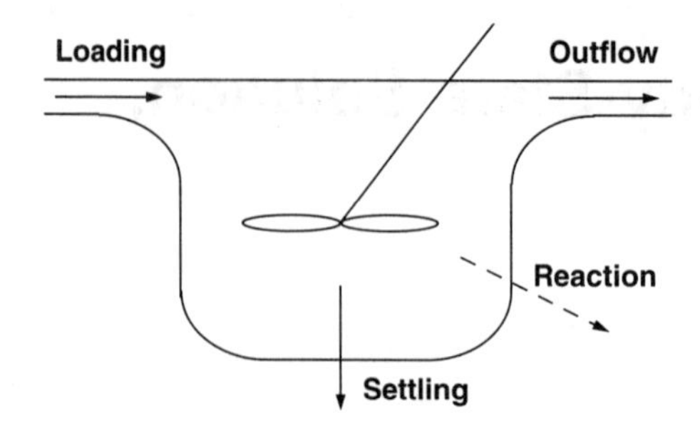

FIGURE 3.1
A mass balance for a well-mixed lake. The arrows represent the major sources and sinks of the pollutant. The dashed arrow for the reaction sink is meant to distinguish it from the other sources and sinks, which are transport mechanisms.

modeling water quality. For a finite time period the mass balance for the system can be expressed as

$$\text{Accumulation} = \text{loading} - \text{outflow} - \text{reaction} - \text{settling} \tag{3.1}$$

Thus there is a single source that contributes matter (loading) and three sinks that deplete matter (outflow, reaction, and settling) from the system. Note that other sources and sinks could have been included. For example volatilization losses (that is, transfer of the pollutant from the water to the atmosphere) could exit across the lake's surface. However, for simplicity, we limit ourselves to the sources and sinks depicted in Fig. 3.1.

Although Eq. 3.1 has descriptive value, it cannot be used to predict water quality. For this we must express each term as a function of measurable variables and parameters.

Accumulation. Accumulation represents the change of mass M in the system over time t:

$$\text{Accumulation} = \frac{\Delta M}{\Delta t} \tag{3.2}$$

Mass is related to concentration by (Eq. 1.1)

$$c = \frac{M}{V} \tag{3.3}$$

where V = volume of system (L^3). Equation 3.3 can be solved for

$$M = Vc \tag{3.4}$$

which can be substituted into Eq. 3.2 to give

$$\text{Accumulation} = \frac{\Delta Vc}{\Delta t} \tag{3.5}$$

In the present case we assume that the lake's volume is constant.† This assumption allows us to bring the term V outside the difference:

† Although this is a good assumption in many cases, it is not always true. For example many lakes and impoundments in the western United States are used for power production and water supply. Some of these can exhibit significant changes in their volumes over relatively short time periods. We will learn how to model such systems in Lec. 16.

$$\text{Accumulation} = V\frac{\Delta c}{\Delta t} \tag{3.6}$$

Finally Δt can be made very small and Eq. 3.6 reduces to

$$\text{Accumulation} = V\frac{dc}{dt} \tag{3.7}$$

Thus mass accumulates as concentration increases with time (positive dc/dt) and diminishes as concentration decreases with time (negative dc/dt). For the steady-state case, mass remains constant ($dc/dt = 0$). Note that the units of accumulation (as with all other terms in the balance) are mass per time (M T^{-1}).

Loading. Mass enters a lake from a variety of sources and in a number of different ways. For example mass carried by treatment plant effluents and tributary streams enters a lake at a point on its periphery. In contrast atmospheric sources, such as precipitation and dry fallout, are introduced in a distributed fashion across the air-water interface at the lake's surface. Whereas the position and manner of entry of loadings would have fundamental importance for incompletely mixed water bodies such as streams and estuaries, it is unimportant for our completely mixed system. This is because, by definition, all inputs are instantaneously distributed throughout the volume. Thus, for the present case, we lump all loadings into a single term, as in

$$\text{Loading} = W(t) \tag{3.8}$$

where $W(t)$ = rate of mass loading (M T^{-1}) and (t) signifies that loading is a function of time.

It should be noted that in a later part of this lecture we formulate loading in a slightly different fashion than in Eq. 3.8. Rather than as a single value $W(t)$, we will represent it as the product (recall Eq. 1.3)

$$\text{Loading} = Qc_{\text{in}}(t) \tag{3.9}$$

where Q = volumetric flow rate of all water sources entering the system ($\text{L}^3\ \text{T}^{-1}$) and $c_{\text{in}}(t)$ = average inflow concentration of these sources (M L^{-3}). Note that we have assumed that flow is constant and that all the temporal variations in loading are the result of temporal variations in the inflow concentration. Also recognize that average inflow concentration can be related to loading by equating Eqs. 3.8 and 3.9 and solving for

$$c_{\text{in}}(t) = \frac{W(t)}{Q} \tag{3.10}$$

Outflow. In our simple system (Fig. 3.1) mass is carried from the system by an outflow stream. The rate of mass transport can be quantified as the product of the volumetric flow rate Q and the outflow concentration c_{out} (M L^{-3}). But, because of our well-mixed assumption, the outflow concentration by definition equals the in-lake concentration $c_{\text{out}} = c$, and the outflow sink can be represented by

$$\text{Outflow} = Qc \tag{3.11}$$

Reaction. Although there are many different ways to formulate reactions that purge pollutants from natural waters, the most common by far is a first-order representation (recall Eq. 2.10)

$$\text{Reaction} = kM \tag{3.12}$$

where k = a first-order reaction coefficient (T^{-1}). Thus a linear proportionality is assumed between the rate at which the pollutant is purged and the mass of pollutant that is present. Equation 3.12 can be expressed in terms of concentration by substituting Eq. 3.4 into Eq. 3.12 to yield

$$\text{Reaction} = kVc \tag{3.13}$$

Settling. Settling losses can be formulated as a flux of mass across the surface area of the sediment-water interface (Fig. 3.2). Thus by multiplying the flux times area, a term for settling in the mass balance can be developed as

$$\text{Settling} = vA_s c \tag{3.14}$$

where v = apparent settling velocity ($L\,T^{-1}$) and A_s = surface area of the sediments (L^2). The settling velocity is called "apparent" because it represents the net effect of the various processes that act to deliver pollutant to the lake's sediments. For example some of the pollutant may be in dissolved form and hence not subject to settling. For such cases a "real" settling velocity cannot be used to represent the net effect of this mechanism.

Because volume is equal to the product of mean depth H and lake surface area A_s, Eq. 3.14 can also be formulated in a fashion similar to the first-order reaction, as in

$$\text{Settling} = k_s Vc \tag{3.15}$$

where k_s = a first-order settling rate constant = v/H. Notice that the ratio v/H has the same units (T^{-1}) as the reaction rate k. The validity of this representation is contingent on the assumption that the lake's surface area and the sediment area are equal. The format of Eq. 3.14 is preferable to Eq. 3.15 because the former more faithfully captures the mechanistic nature of settling, that is, as a mass transfer across a surface (see Box 3.1).

Total balance. The terms can now be combined into the following mass balance for a well-mixed lake:

$$V\frac{dc}{dt} = W(t) - Qc - kVc - vA_s c \tag{3.16}$$

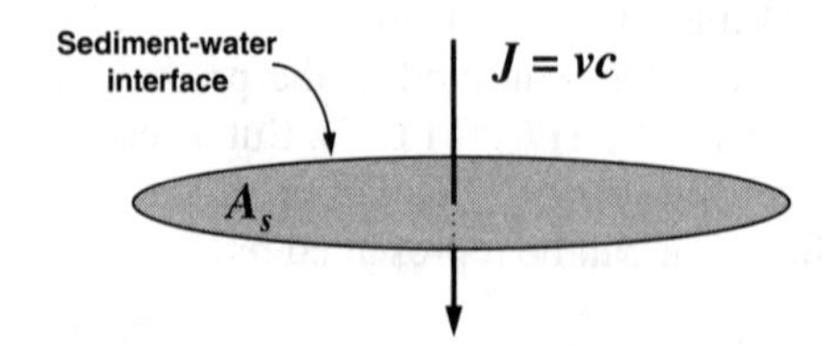

FIGURE 3.2
Settling losses formulated as a flux of mass across the sediment-water interface.

Before proceeding to solutions for Eq. 3.16 we should introduce some nomenclature. Concentration c and time t are the ***dependent*** and the ***independent variables***, respectively, because the model is designed to predict concentration as a function of time. The loading term $W(t)$ is referred to as the model's ***forcing function*** because it represents the way in which the external world influences or "forces" the system. Finally the quantities V, Q, k, v, and A_s are referred to as ***parameters*** or coefficients. Specification of these parameters will allow us to apply our model to particular lakes and pollutants.

BOX 3.1. Parameterization

As described in this section, settling losses can be parameterized as the product of settling velocity, surface area, and concentration vA_sc. However, as in Eq. 3.15, the settling mechanism can also be "parameterized" as a first-order rate. This is done by multiplying the settling-velocity version by H/H and collecting terms to yield

$$vA_sc\frac{H}{H} = \frac{v}{H}(AH)c = k_sVc$$

where k_s = a settling first-order rate constant (T^{-1}) that is equal to

$$k_s = \frac{v}{H}$$

This alternative formulation has been commonly employed in water-quality modeling.

Now the question arises, is either way superior? From a strictly mathematical standpoint they are identical. However, because it is more fundamental, the settling velocity parameterization is superior. By fundamental I mean that it more directly represents the process being modeled. That is, each term in vA_sc represents a characteristic of the process that can be measured independently. In contrast the k_s term confounds two independent properties: settling and depth.

Why is this a problem? First, the k_s version is system specific (because it implicitly includes a system-specific property, mean depth) and hence is awkward to extrapolate to other systems. If we measure a k_s in a particular system, we could use it only in other systems of the same depth. Thus, to extrapolate to a system with a different depth, we would have to revert to the settling velocity format anyway. Second, what if depth is changing? For this case the use of k_s clearly breaks down.

Now, where might confounding parameters be advantageous? For one thing, within a mathematical calculation for a particular system, we often find it useful to collect terms for mathematical convenience. Second, it is often of use to collect terms so that processes can be compared in commensurate units. For example the relative magnitudes of settling and a reaction could be assessed by comparing v/H versus k. Finally there are some instances where we might confound several parameters because one or more do not vary between systems and/or they are difficult to measure.

Throughout the remainder of this book the issue of proper parameterization will continuously arise as we attempt to quantify the processes observed in natural waters. When it arises we will discuss its further nuances.

3.2 STEADY-STATE SOLUTIONS

If the system is subject to a constant loading W for a sufficient time, it will attain a dynamic equilibrium condition called a ***steady-state***. In mathematical terms this means that accumulation is zero (that is, $dc/dt = 0$). For this case Eq. 3.16 can be solved for

$$c = \frac{W}{Q + kV + vA_s} \tag{3.17}$$

or using the format of Eq. 1.8,

$$c = \frac{1}{a}W \tag{3.18}$$

where the assimilation factor is defined as

$$a = Q + kV + vA_s \tag{3.19}$$

The steady-state solution provides our first illustration of the benefits of the mechanistic approach. That is, it has successfully yielded a formula that defines the assimilation factor in terms of measurable variables that reflect the system's physics, chemistry, and biology.

EXAMPLE 3.1. MASS BALANCE. A lake has the following characteristics:

Volume $= 50{,}000\ m^3$
Mean depth $= 2\ m$
Inflow $=$ outflow $= 7500\ m^3\ d^{-1}$
Temperature $= 25°C$

The lake receives the input of a pollutant from three sources: a factory discharge of $50\ kg\ d^{-1}$, a flux from the atmosphere of $0.6\ g\ m^{-2}\ d^{-1}$, and the inflow stream that has a concentration of $10\ mg\ L^{-1}$. If the pollutant decays at the rate of $0.25\ d^{-1}$ at 20°C ($\theta = 1.05$),

(*a*) Compute the assimilation factor.
(*b*) Determine the steady-state concentration.
(*c*) Calculate the mass per time for each term in the mass balance and display your results on a plot.

Solution: (*a*) The decay rate must first be corrected for temperature (Eq. 2.44):

$$k = 0.25 \times 1.05^{25-20} = 0.319\ d^{-1}$$

Then the assimilation factor can be calculated as

$$a = Q + kV = 7500 + 0.319(50{,}000) = 23{,}454\ m^3\ d^{-1}$$

Notice how the units look like flow (that is, volume per time). This is because the same mass units are used in the numerator and the denominator and they cancel, as in

$$\frac{g\ d^{-1}}{g\ m^{-3}} \rightarrow m^3\ d^{-1}$$

(*b*) The surface area of the lake is needed to calculate the atmospheric loading

$$A_s = \frac{V}{H} = \frac{50{,}000}{2} = 25{,}000 \text{ m}^2$$

The atmospheric load is then computed as

$$W_{\text{atmosphere}} = JA_s = 0.6(25{,}000) = 15{,}000 \text{ g d}^{-1}$$

The load from the inflow stream can be calculated as

$$W_{\text{inflow}} = 7500(10) = 75{,}000 \text{ g d}^{-1}$$

Therefore the total loading is

$$W = W_{\text{factory}} + W_{\text{atmosphere}} + W_{\text{inflow}} = 50{,}000 + 15{,}000 + 75{,}000 = 140{,}000 \text{ g d}^{-1}$$

and the concentration can be determined as (Eq. 3.18)

$$c = \frac{1}{a}W = \frac{1}{23{,}454}140{,}000 = 5.97 \text{ mg L}^{-1}$$

(*c*) The loss due to flushing through the outlet can be computed as

$$Qc = 7500(5.97) = 44{,}769 \text{ g d}^{-1}$$

and the loss due to reaction as

$$kVc = 0.319(50{,}000)5.97 = 95{,}231 \text{ g d}^{-1}$$

These results along with the loading can be displayed as in Fig. 3.3.

The representation in Fig. 3.3 can now be related back to the parable of the blind men and the elephant. Each arrow, representing a source or sink mechanism, is analogous to the individual parts of the elephant. It is only when they are tied

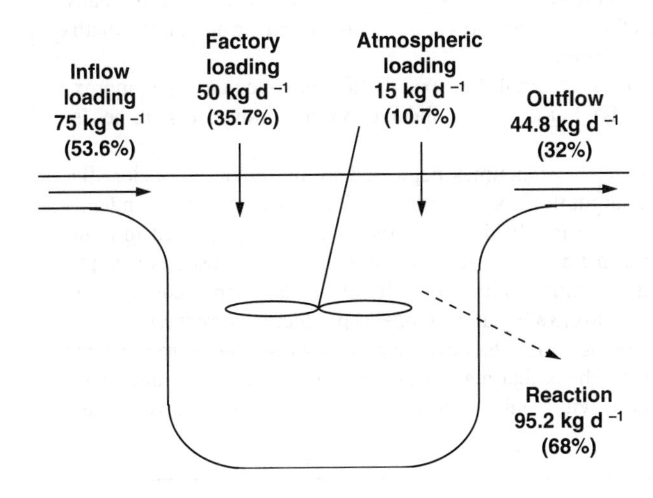

FIGURE 3.3
A mass balance for the well-mixed lake from Example 3.1. The arrows represent the major sources and sinks of the pollutant. The mass-transfer rates have also been included along with the percent of total mass inflow accounted for by each term.

together by the mass balance that we can assess their combined effect. Thus the model provides an integrated view of the system.

BOX 3.2. "Stream-of Consciousness" Versus "Cartoon" Modeling

"Stream of consciousness" is a psychological term, coined by the psychologist William James, that characterizes individual conscious experience as a succession of states constantly moving onward in time. The idea has been transferred to literature in the form of "stream-of-consciousness" writing. In its finest expression an individual's interior monologue is used to reveal character and comment on life. At its worst it amounts to a self-indulgent mind dump.

Unfortunately many creative exercises can be approached in the latter fashion. For example computer programs are often written without prior thought. Individuals sit down at a computer console and just begin typing. Invariably the final result (as well as the ultimate time investment) suffers from the haphazard approach and lack of design.

Mathematical models can also be developed in a stream-of-consciousness fashion. There is often the tendency to start writing mass balances without adequate forethought. As expected, the results are often incorrect or incomplete. In the best case a correct model results only after many time-consuming revisions.

Some simple steps can be applied to avoid such pitfalls:

- *Draw a diagram.* For the simple well-mixed models described up to now, this merely consists of sketching the major sources and sinks of the pollutant being modeled. Although this might seem trivial, the act of drawing forces you to delineate the mechanisms governing pollutant dynamics. In later lectures, as we deal with multiple pollutants in segmented systems, diagrams will become essential. Dr. Bob Broshears of the U.S. Geological Survey calls this "cartoon modeling." Although this terminology might sound flippant, it is not meant to be. Experienced modelers recognize that a well-thought-out schematic is critical to keeping track of all the variables and processes in a complicated model.
- *Write equations.* After a schematic is developed, it can be translated into model equations. For the simple case discussed so far, each arrow represents a term in the mass balance. In later lectures, there will be many variables (boxes) connected by many processes (arrows). Thus the schematic provides a guide for ensuring that the mathematical characterization is complete.
- *Obtain a solution.* This can be accomplished exactly (algebra or calculus) or approximately (numerical methods). For more complicated systems, computers are necessary.
- *Check results.* This last step is sometimes neglected by the novice modeler. Too many people trust model output if it "looks reasonable." Unfortunately this tendency increases when computers are involved. And if the results are displayed in high-resolution graphics in multiple colors, certain individuals lose any vestige of skepticism. Therefore, whether checking a homework solution or a large professional code, sufficient testing is required to ensure that the model is producing correct results. Beyond obvious and easily recognizable bloopers (e.g., a negative concentration), the simplest starting point is to check that mass is conserved. Beyond that, more complicated tests are required. I will touch on these when I review model development in Lec. 18.

3.2.1 Transfer Functions and Residence Time

Aside from the assimilation factor, there are a variety of other ways to summarize the ability of a steady-state system to assimilate pollutants.

Transfer function. An alternative way to formulate Eq. 3.17 is based on expressing the loading in the format of Eq. 3.9. For the steady-state case this is

$$W = Qc_{\text{in}} \tag{3.20}$$

Equation 3.20 can be substituted into Eq. 3.17, and both the numerator and denominator of the result can be divided by c_{in} to yield

$$\frac{c}{c_{\text{in}}} = \beta \tag{3.21}$$

where β = the transfer function

$$\beta = \frac{Q}{Q + kV + vA_s} \tag{3.22}$$

Equation 3.22 is called a ***transfer function***† because it specifies how the system input (as represented by c_{in}) is transformed or "transferred" to an output (as represented by c). Examination of Eq. 3.22 provides insight into how the model "works." If $\beta \ll 1$, then the lake's removal mechanisms will act to greatly reduce the level of pollutant in the lake; that is, such a lake has great assimilative capacity. Conversely if $\beta \to 1$, then the lake's removal mechanisms (the denominator) are weak relative to its supply mechanism (the numerator). For such cases the pollutant level will approach that of the inflow. In other words the lake's assimilative capacity is minimal.

Thus the lake's assimilative capacity can be evaluated by the dimensionless number β. Inspection of Eq. 3.22 indicates that for the simple model in Fig. 3.1, assimilation increases for large values of reaction rate, settling velocity, volume, and area. Note that flow which appears in both the numerator and the denominator acts to both increase and decrease assimilation. It increases assimilation as it reflects flushing of pollutant through the lake's outlet. It decreases assimilation as it reflects delivery of pollutant through the lake's inflow.

Residence time. The residence time τ_E of a substance E represents the mean amount of time that a molecule or particle of E would stay or "reside" in a system. It is defined for a steady-state, constant-volume system as (Stumm and Morgan 1981)

$$\tau_E = \frac{E}{|dE/dt|_{\pm}} \tag{3.23}$$

where E = quantity of E in the volume (either M or M L^{-3}) and $|dE/dt|_{\pm}$ = absolute value of either the sources or the sinks (either M T^{-1} or M L^{-3} T^{-1}).

One of the simpler applications of Eq. 3.23 is the determination of the residence time of water in a lake. Since the density of water is by definition approximately 1 g cm^3, the quantity of water in a lake is equivalent to its volume. In a similar

†The term "transfer function" has a related but more sophisticated definition in linear systems analysis.

sense the "sink" of water from a lake is measured by the magnitude of its outflow (assuming that evaporation = precipitation). Substituting these values into Eq. 3.23 yields the water residence time

$$\tau_w = \frac{V}{Q} \tag{3.24}$$

This relationship is useful for understanding the general notion of residence time since it has a straightforward physical interpretation—it is the amount of time that would be required for the outflow to replace the quantity of water in the lake. Thus it is a measure of the lake's flushing rate. If the volume is large and the flow is small the lake has a long residence time; that is, it is a slow flusher. Conversely lakes with short residence times (high flow and small volume) are referred to as fast flushers.

Equation 3.23 can also be used to compute a "pollutant residence time." For example for the system in Fig. 3.1, the sinks can be represented on a mass basis by

$$\left|\frac{dM}{dt}\right|_{\pm} = Qc + kVc + vA_s c \tag{3.25}$$

This equation, along with Eq. 3.4, can be substituted into Eq. 3.23 to give

$$\tau_c = \frac{V}{Q + kV + vA_s} \tag{3.26}$$

Note that Eqs. 3.24 and 3.26 are similar in form with the exception that the pollutant residence time is affected by reactions and settling in addition to the outflow.

EXAMPLE 3.2. TRANSFER FUNCTION AND RESIDENCE TIMES. For the lake in Example 3.1, determine the (*a*) inflow concentration, (*b*) transfer function, (*c*) water residence time, and (*d*) pollutant residence time.

Solution: (*a*) The inflow concentration is computed as

$$c_{in} = \frac{W}{Q} = \frac{140{,}000}{7500} = 18.67 \text{ mg L}^{-1}$$

(*b*) The transfer coefficient can now be determined as

$$\beta = \frac{c}{c_{in}} = \frac{Q}{Q + kV} = 0.32$$

Thus the removal processes act to create a lake concentration that is 32% of the inflow concentration.

(*c*) The residence time can be calculated as

$$\tau_w = \frac{V}{Q} = \frac{50{,}000}{7500} = 6.67 \text{ d}$$

(*d*) The pollutant residence time is

$$\tau_c = \frac{V}{Q + kV} = \frac{50{,}000}{7500 + 0.319(50{,}000)} = 2.13 \text{ d}$$

Because of the addition of the decay term, the residence time of a pollutant is about one-third the water residence time.

BOX 3.3. Estimating Reaction Kinetics with a Steady-State CSTR

Grady and Lim (1980) describe a method for evaluating reaction kinetics using experiments conducted with CSTRs. In this approach, which they call the ***algebraic method***, the reactors are run to a steady-state. At this point the mass balance can be written as

$$Qc_{\text{in}} - Qc - rV = 0 \tag{3.27}$$

where r = rate of consumption of the reactant ($\text{M L}^{-3}\ \text{T}^{-1}$). If all the other quantities are measured the balance can be solved for the consumption rate

$$r = \frac{Qc_{\text{in}} - Qc}{V} = \frac{1}{\tau_w}(c_{\text{in}} - c) \tag{3.28}$$

If we assume that Eq. 2.7 holds,

$$r = kc^n \tag{3.29}$$

The parameters k and n can be determined by taking the natural logarithm,

$$\ln r = \ln k + n \ln c \tag{3.30}$$

Thus if a plot of $\ln r$ versus ln c yields a straight line, Eq. 2.7 holds, and the slope and intercept can be used to calculate k and n. An exercise on the approach is presented in Prob. 3.5 at the end of this lecture.

3.3 TEMPORAL ASPECTS OF POLLUTANT REDUCTION

To this point we have focused on steady-state solutions. These provide an estimate of the average water quality that will result if loadings are held constant for a sufficiently long time period. In addition to steady-state predictions, water-quality managers are also interested in the temporal response of natural waters.

Suppose that a system is at steady-state. At a specific time a waste removal project is implemented. As depicted in Fig. 3.4, two interrelated questions arise:

- How long will it take for improved water quality to occur?
- What will the "shape" of the recovery look like?

To determine the correct trajectory, let's start with the mass-balance model (Eq. 3.16)

$$V\frac{dc}{dt} = W(t) - Qc - kVc - vA_s c \tag{3.31}$$

Before solving this equation, we can divide it by volume to yield

$$\frac{dc}{dt} = \frac{W(t)}{V} - \frac{Q}{V}c - kc - \frac{v}{H}c \tag{3.32}$$

Collecting terms gives

$$\frac{dc}{dt} + \lambda c = \frac{W(t)}{V} \tag{3.33}$$

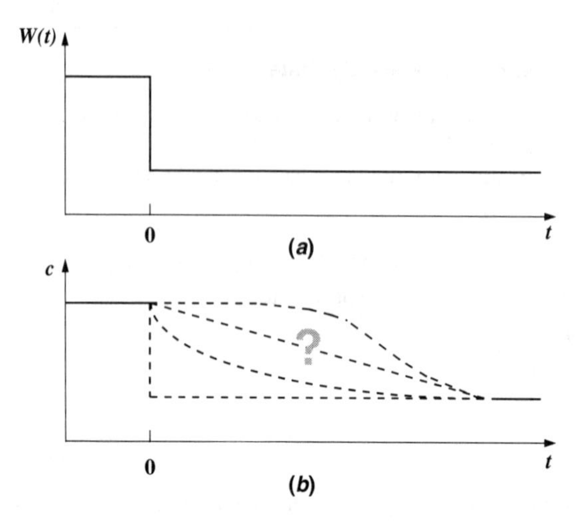

FIGURE 3.4
(*a*) A waste load reduction along with (*b*) four possible recovery scenarios for concentration.

where

$$\lambda = \frac{Q}{V} + k + \frac{v}{H} \tag{3.34}$$

in which λ is called an ***eigenvalue*** (that is, a characteristic value).

If all the parameters (Q, V, k, v, H) are constant, Eq. 3.33 is a nonhomogeneous, linear, first-order, ordinary differential equation. Its solution consists of two parts,

$$c = c_g + c_p \tag{3.35}$$

where c_g = general solution for the case $W(t) = 0$ and c_p = particular solution for specific forms of $W(t)$.

Because the general solution corresponds to the case where the loading is terminated, it is ideal for investigating a system's recovery time. As described next it will also provide us with insight into the shape of the recovery.

3.3.1 The General Solution

If $c = c_0$ at $t = 0$, Eq. 3.33 with $W(t) = 0$ can be solved by the separation of variables (recall solution of Eq. 2.10):

$$c = c_0 e^{-\lambda t} \tag{3.36}$$

Thus we have arrived at an equation that describes how the lake's concentration changes as a function of time following the termination of waste loading.

The behavior of Eq. 3.36 is clearly dictated by the exponential function. As in Fig. 3.5, for the case where the argument of the function (that is, the value to which e is raised: x) is zero, the exponential function's value is unity. Thereafter if the argument is positive, the function increases in an accelerated fashion; that is, it doubles its value at set intervals of x (= 0.693). In contrast if the argument is negative, the function asymptotically decreases toward zero by halving at the same set intervals.

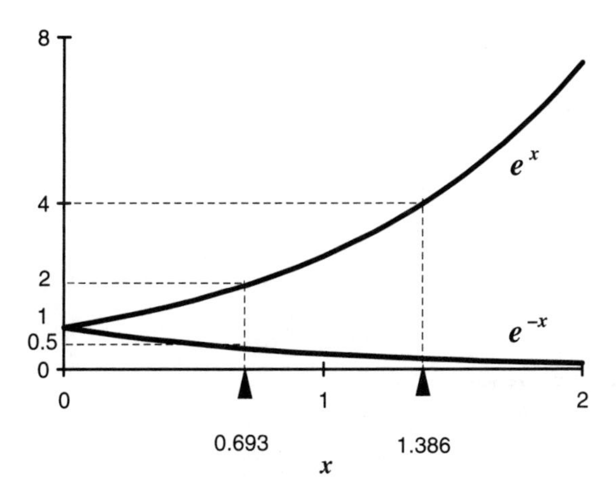

FIGURE 3.5
The exponential function.

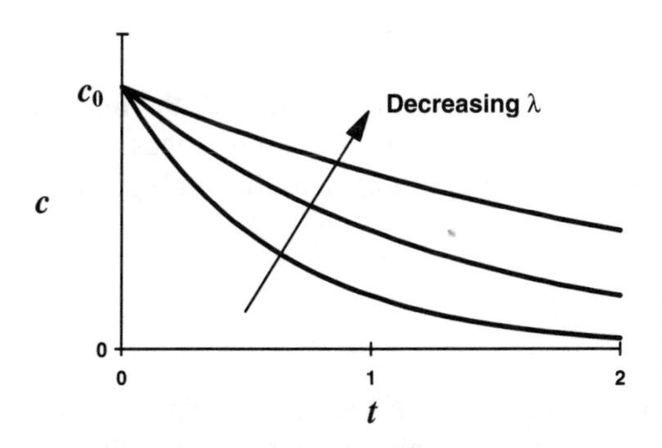

FIGURE 3.6
The temporal response of our well-mixed lake model following the termination of all loadings at $t = 0$.

Thus we can now interpret Eq. 3.36. As in Fig. 3.6 the negative value of the argument means that the concentration decreases and asymptotically approaches zero. Further the rate of decrease is dictated by the magnitude of the eigenvalue λ. If λ is large, the lake's concentration will decrease rapidly. If λ is small, the lake's response will be slow.

EXAMPLE 3.3. GENERAL SOLUTION. In Example 3.1 we determined the steady-state concentration for a lake having the following characteristics:

Volume $= 50{,}000 \text{ m}^3$	Temperature $= 25°\text{C}$
Mean depth $= 2 \text{ m}$	Waste loading $= 140{,}000 \text{ g d}^{-1}$
Inflow $=$ outflow $= 7500 \text{ m}^3 \text{ d}^{-1}$	Decay rate $= 0.319 \text{ d}^{-1}$

If the initial concentration is equal to the steady-state level (5.97 mg L^{-1}), determine the general solution.

Solution: The eigenvalue can be computed as

$$\lambda = \frac{Q}{V} + k = \frac{7500}{50{,}000} + 0.319 = 0.469 \text{ d}^{-1}$$

Thus the general solution is

$$c = 5.97e^{-0.469t}$$

which can be displayed graphically as

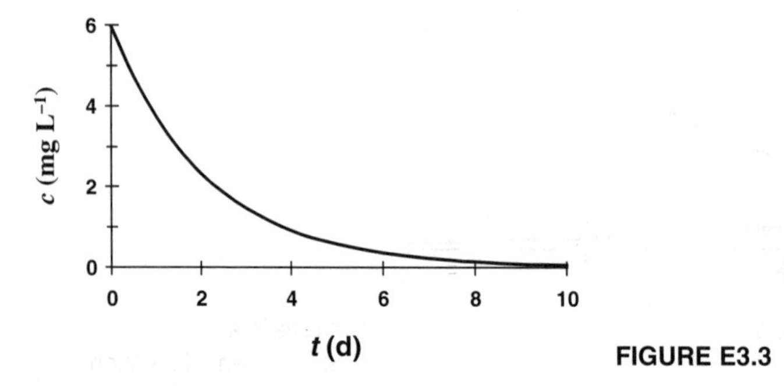

FIGURE E3.3

Note that by $t = 5$ d the concentration is reduced to less than 10% of its original value. By $t = 10$ d, for all intents and purposes, it has reached zero.

An interesting property of the general solution is that even though the loading is reduced to zero, the concentration will never reach zero. This introduces an element of ambiguity into the analysis. We now attempt to resolve this ambiguity by introducing the concept of response time.

3.3.2 Response Time

Although the parameter group λ clearly dictates the lake's temporal response characteristics, it has shortcomings for communicating with decision makers. First, it has the counterintuitive property that as it gets large the time for the lake to respond gets small. Second, as mentioned in the previous section, its interpretation is clouded by the fact that, from a strictly mathematical perspective, the underlying cleansing process never reaches completion. Try telling a politician that a cleanup would theoretically take forever! They tend to react very unfavorably to asymptotic solutions that extend beyond the next election.

Both these shortcomings can be rectified by using the general solution to derive a new parameter group. Called the ***response time***, this parameter group represents the time it takes for the lake to complete a fixed percentage of its recovery. Thus the problem of ambiguity is remedied by deciding "how much" of the recovery is judged as being "enough." For example we might assume that if the lake has experienced 95% of its recovery we would be satisfied that, for all practical purposes, the remedial measure is successful.

In terms of Eq. 3.36 a 50% response time means that the concentration is lowered to 50% of its initial value, or

$$0.50c_0 = c_0 e^{-\lambda t_{50}} \tag{3.37}$$

where t_{50} = 50% response time (T). Dividing by the exponential and $0.50c_0$ yields

$$e^{\lambda t_{50}} = 2 \tag{3.38}$$

TABLE 3.1
Response times

Response time	t_{50}	$t_{63.2}$	t_{75}	t_{90}	t_{95}	t_{99}
Formula	$0.693/\lambda$	$1/\lambda$	$1.39/\lambda$	$2.3/\lambda$	$3/\lambda$	$4.6/\lambda$

Taking the natural logarithm and solving for t_{50} gives

$$t_{50} = \frac{0.693}{\lambda} \tag{3.39}$$

Thus we can see that the 0.693 we observed previously (recall discussion of Fig. 3.5) is actually the natural logarithm of 2. Note that the quantity t_{50} is also commonly referred to as a half-life (recall Sec. 2.2.4).

The above derivation can be generalized to compute an arbitrary response time by the formula

$$t_\phi = \frac{1}{\lambda} \ln \frac{100}{100 - \phi} \tag{3.40}$$

where $t_\phi = \phi\%$ response time. For example if we are interested in determining how long it takes to reach 95% of its ultimate recovered level, we could compute

$$t_{95} = \frac{1}{\lambda} \ln \frac{100}{100 - 95} = \frac{3}{\lambda} \tag{3.41}$$

Table 3.1 and Fig. 3.7 show other response times. As would be expected, the higher the percentage of recovery, the longer the response time.

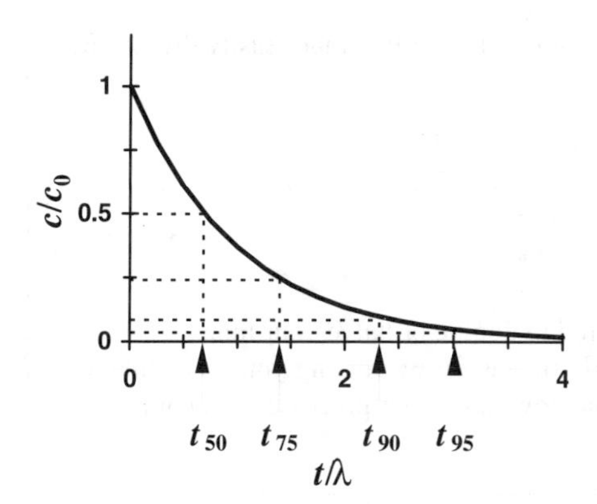

FIGURE 3.7
A plot of the general solution showing values of several response times.

EXAMPLE 3.4. RESPONSE TIME. Determine the 75%, 90%, 95%, and 99% response times for the lake in Example 3.3.

Solution: The 75% response time can be computed as

$$t_{75} = \frac{1.39}{0.469} = 2.96\,\text{d}$$

In a similar fashion we can compute $t_{90} = 3.9\,\text{d}$, $t_{95} = 6.4\,\text{d}$, and $t_{99} = 9.8\,\text{d}$.

As in Example 3.4 there is a certain amount of subjectivity involved in deciding "how much" of the recovery is judged as being "enough." In general I recommend using either t_{90} or t_{95}. They are neither too lenient nor too stringent and conform to what most individuals would deem an acceptable level of recovery.

BOX 3.4. The Rule of 72

Before the days of handheld calculators and computers, bankers and financiers needed a quick way to evaluate their investments. To do this they developed a heuristic that is called ***the rule of 72***. According to the rule, the time required to double your money can be estimated as

$$\text{Doubling time} \cong \frac{72}{\text{interest rate (\%)}}$$

For example if you invested some money at an annual interest rate of 6%, it would double in approximately 12 years. The same formula can also be employed to assess how the value of your money decreases due to inflation. For example if the inflation rate is 3%, the money hidden in your mattress would lose half its value in 24 years.

This formula is derived from the concept of the half-life. In fact a more accurate representation, based on Eq. 3.41, might be called "the rule of 69.3,"

$$\text{Doubling time} = \frac{69.3}{\text{interest rate (\%)}}$$

The reason that a numerator of 72 was chosen is that it is more easily divided by whole number interest rates. For example

$$\begin{array}{lll} 72/1 = 72 \text{ yr} & 72/5 \cong 14 \text{ yr} & 72/9 = 8 \text{ yr} \\ 72/2 = 36 \text{ yr} & 72/6 = 12 \text{ yr} & 72/10 \cong 7 \text{ yr} \\ 72/3 = 24 \text{ yr} & 72/7 \cong 10 \text{ yr} & \\ 72/4 = 18 \text{ yr} & 72/8 = 9 \text{ yr} & \end{array}$$

Thus you can quickly figure the time required to double or halve your money.

Aside from providing you with a handy means for evaluating your investments, we have included this discussion to illustrate how first-order processes and compound interest are based on similar mathematics.

PROBLEMS

3.1. A pond with a single inflow stream has the following characteristics:

Mean depth $= 3$ m
Surface area $= 2 \times 10^5$ m^2
Residence time $= 2$ weeks
Inflow BOD concentration $= 4$ mg L^{-1}

A subdivision housing 1000 people will discharge raw sewage into this system. Each individual contributes about 150 gal capita^{-1} d^{-1} of wastewater and 0.25 lb capita^{-1} d^{-1} of biochemical oxygen demand (BOD).

(*a*) Determine the BOD concentration of the wastewater in mg L^{-1}.

(*b*) If the BOD decays at a rate of 0.1 d^{-1} and settles at a rate of 0.1 m d^{-1}, calculate the assimilation factor for the pond prior to building the subdivision. Which of the purging mechanisms are most effective? List them in decreasing order of effectiveness.

(*c*) Calculate the transfer function factor after building the subdivision.

(*d*) Determine the steady-state concentration for the lake with and without the subdivision.

3.2. A lake with a single inflow stream has the following characteristics:

Mean depth = 5 m
Surface area = 11×10^6 m^2
Residence time = 4.6 yr

An industrial plant presently discharges malathion ($W = 2000 \times 10^6$ g yr^{-1}) to the lake. In addition the inflowing stream also contains malathion ($c_{in} = 15$ mg L^{-1}). Note that the volumetric rate of inflow and outflow are equal. Assuming that a first-order decay reaction can be used to characterize malathion decay ($k = 0.1$ yr^{-1}),

(*a*) Write a mass-balance equation for malathion for this system.

(*b*) If the lake is at steady-state, compute the in-lake malathion concentration.

(*c*) If the lake is at steady-state, what industrial plant loading rate must be maintained to lower the lake's concentration to 30 ppm? Express your result as a percent reduction.

(*d*) Evaluate each of the following engineering options to determine which is the most effective for lowering the steady-state concentration:

(i) Reduce the present loading rate of the industrial plant by building a waste treatment facility that will remove 50% of the malathion from the plant's effluent.

(ii) Double the lake's depth by dredging.

(iii) Double the lake's outflow rate Q by diverting malathion-free water from a nearby unpolluted stream into the lake.

(*e*) What other factors would need to be considered (aside from lowering concentration) when making a decision in (*d*) in the "real world"?

(*f*) Determine the 95% response times for each of the options in part (*d*).

3.3. Recall from Example 1.3 that in the early 1970s Lake Ontario had a total phosphorus loading of approximately 10,500 mta (metric tons per annum) and an in-lake concentration of 21 μg L^{-1} (Chapra and Sonzogni 1979). It is known that the only losses of total phosphorus for the lake are settling and flushing through the lake's outlet. Assume the outflow rate is 212 km^3 yr^{-1} and the sediment area is 10,500 km^2.

(*a*) Calculate the inflow concentration for the system. Assume that water inflow equals outflow.

(*b*) Use a mass balance to estimate the apparent settling velocity for total phosphorus in this lake.

3.4. A lake has the following characteristics:

Volume = 1×10^6 m^3
Surface Area = 1×10^5 m^2
Water residence time = 0.75 yr

A soluble pesticide is input to the lake at a rate of 10×10^6 mg yr^{-1}. The in-lake concentration is 0.8 μg L^{-1}.
(*a*) Determine the inflow concentration (assume inflow = outflow).
(*b*) Determine the transfer function.
(*c*) If the only removal mechanism (other than flushing) is volatilization, compute the flux of the pesticide out the lake's surface and to the atmosphere.
(*d*) Express the result of part (*c*) as a volatilization velocity.

3.5. A rate experiment is performed for a 1-L CSTR. The inflowing concentration is held steady at a level of 100 mg L^{-1}. The flow is varied and the resulting outflow concentrations are measured:

Q (L hr^{-1})	0.1	0.2	0.4	0.8	1.6
c (mg L^{-1})	23	31	41	52	64

Use the algebraic method to determine the rate and order of the reaction.

3.6. Derive Eq. 3.40.

3.7. A pond with a single inflow stream has the following characteristics:

Mean depth = 3 m
Surface area = 2×10^5 m^2
Residence time = 2 wk

A subdivision will discharge raw sewage into this system. If BOD decays at a rate of 0.1 d^{-1} and settles at a rate of 0.1 m d^{-1}, calculate the 75%, 90%, and 95% response times for the pond.

3.8. Determine a half-life for a batch reactor with a second-order decay reaction.

3.9. Compute the first-order reaction rates for the following substances:
(*a*) cesium-137 (half-life = 30 yr)
(*b*) iodine-131 (half-life = 8 d)
(*c*) tritium (half-life = 12.26 yr)

3.10. A lake (volume = 10×10^6 m^3, water residence time = 2 months) is located adjacent to a railway line that carries considerable traffic of chemicals. You are hired as a consultant to provide insight into potential spills into the lake. If the lake is assumed to be completely mixed, a spill would be distributed instantaneously throughout the volume. Therefore the resulting concentration would be $c_0 = m/V$, where m is the mass of pollutant that is spilled. Thereafter the lake's response would follow the general solution.
(*a*) Develop a plot of t_{75}, t_{95}, and t_{99} versus pollutant half life. Use logarithmic scales where you believe they would be helpful.
(*b*) Include a short "user's manual" for the plot to provide managers with guidance for its use and interpretation.

LECTURE 4

Particular Solutions

Lecture overview: I develop particular solutions for specific forms of the loading function: impulse, step, linear, exponential, and sinusoidal loadings. Then I place emphasis on summarizing each solution's behavior in terms of a few simple "shape parameters."

Up to this point we have derived steady-state and general solutions for the mass balance equation for a CSTR with first-order kinetics,

$$\frac{dc}{dt} + \lambda c = \frac{W(t)}{V} \tag{4.1}$$

where λ is an eigenvalue that, for the present model, is equal to

$$\lambda = \frac{Q}{V} + k + \frac{v}{H} \tag{4.2}$$

Now let's turn to particular solutions. These relate to closed-form solutions for specific forms of $W(t)$. As might be expected, solutions are available only for idealized forms of the loading function. Some of the more useful forms are summarized in Fig. 4.1 (*a* through *e*).

Although these forms might appear too idealized to be realistic, they have great utility in water-quality modeling for three reasons:

- First, there are a surprisingly large number of problem contexts where they provide an adequate approximation of loading trends.
- Second, models are frequently used to predict future impacts. For these cases we would not know the exact way in which future loads would vary. Consequently idealizations become extremely useful.

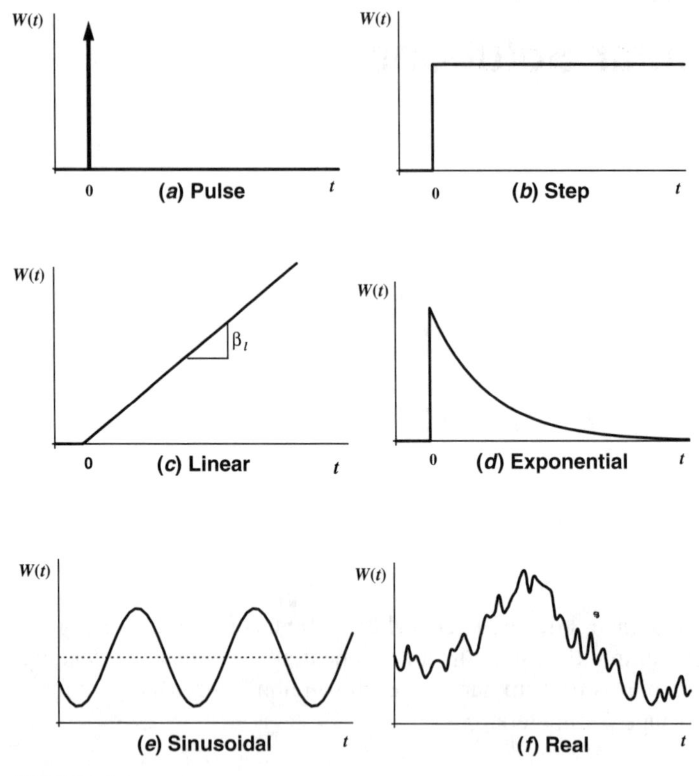

FIGURE 4.1
Loading functions $W(t)$ versus time t.

- Third, idealized loading functions allow us to gain a better understanding of how a model works. This is because realistic loading functions (see Fig. 4.1*f*) often make it difficult to discern model behavior. In contrast idealized solutions provide an uncomplicated view.

In the following sections I present the analytical solutions for each of these loading functions. Aside from the mathematical representation, I will illustrate that each solution has characteristic shape parameters that summarize its behavior. These parameters are akin to the response times that allowed us to neatly summarize the general solution in the previous lecture. In stressing these shape parameters, my ulterior motive is to help you build up mathematical insight regarding these simple solutions. It is hoped, by focusing on the underlying patterns, that you will develop a more intuitive sense of how natural waters respond to loadings.

4.1 IMPULSE LOADING (SPILL)

The most fundamental form of $W(t)$ is the impulse loading representing the discharge of waste over a relatively short time period. The accidental spill of a contaminant to a waterbody would be of this type. Mathematically the ***Dirac delta function*** (or ***impulse function***) $\delta(t)$ has been developed to represent such phenomena. The delta

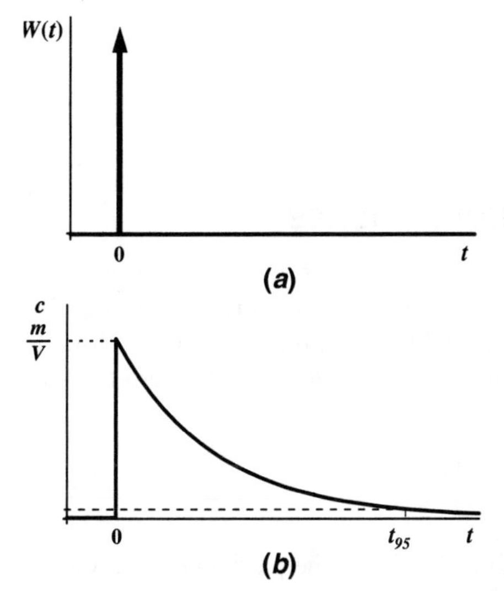

FIGURE 4.2
Plot of (*a*) loading and (*b*) response for impulse loading.

function can be visualized as an infinitely thin spike centered at $t = 0$ and having unit area. It has the following properties (Fig 4.2*a*):

$$\delta(t) = 0 \qquad t \neq 0 \tag{4.3}$$

and

$$\int_{-\infty}^{\infty} \delta(t)\, dt = 1 \tag{4.4}$$

The impulse load of mass to a water body can be represented in terms of the delta function as

$$W(t) = m\delta(t)$$

where m = quantity of pollutant mass discharged during a spill (M) and the delta function has units of T^{-1}. Substituting into Eq. 4.1 gives

$$\frac{dc}{dt} + \lambda c = \frac{m\delta(t)}{V} \tag{4.5}$$

The particular solution for this case is

$$c = \frac{m}{V} e^{-\lambda t} \tag{4.6}$$

This solution indicates that the spill is instantaneously distributed throughout the lake's volume, resulting in an initial concentration of m/V. Thereafter the result is identical to the general solution; that is, the concentration decreases exponentially at a rate dictated by the magnitude of λ (Fig. 4.2*b*).

Thus we can see that two shape parameters summarize the impulse response. The initial concentration defines the height of the response:

$$c_0 = \frac{m}{V} \tag{4.7}$$

and the response time defines its temporal extent. If a 95% response is arbitrarily chosen, this is

$$t_{95} = \frac{3}{\lambda} \tag{4.8}$$

4.2 STEP LOADING (NEW CONTINUOUS SOURCE)

If at $t = 0$ the lake's loading is changed to a new constant level, then the forcing function is called a ***step input***. Such behavior is given mathematical expression by the ***step function***. This is essentially an "on-off" function that has a jump discontinuity at $t = 0$. For the present case, where our loading jumps from zero to W at $t = 0$ (Fig. 4.3*a*), it can be represented as

$$\begin{aligned} W(t) &= 0 \qquad t < 0 \\ W(t) &= W \qquad t \geq 0 \end{aligned} \tag{4.9}$$

where $W =$ the new constant level of loading (M T^{-1}). The particular solution for this case is

$$c = \frac{W}{\lambda V}(1 - e^{-\lambda t}) \tag{4.10}$$

As depicted in Fig. 4.3*b* this solution starts at zero and then asymptotically converges on a new steady-state concentration. At $t = \infty$, Eq. 4.10 becomes equivalent to the steady-state solution (Eq. 3.17), since the exponential term becomes very small as time progresses.

Thus we can see that again two parameters summarize the step response. The steady-state concentration defines the height of the response:

$$\bar{c} = \frac{W}{\lambda V} \tag{4.11}$$

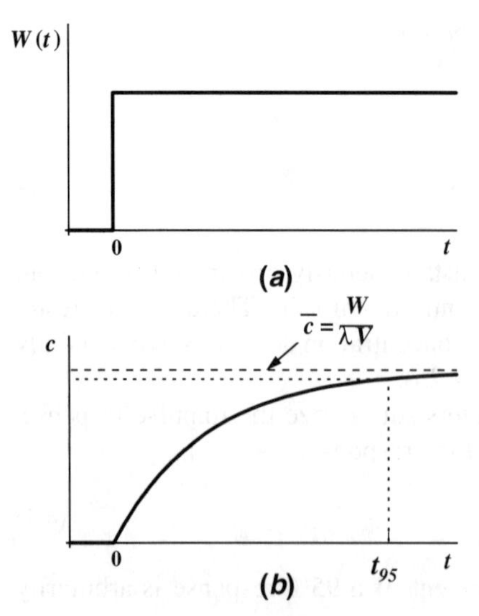

FIGURE 4.3
Plot of (*a*) loading and (*b*) response for step loading.

and the response time defines its temporal extent. If a 95% response is arbitrarily chosen, Eq. 4.8 holds.

EXAMPLE 4.1. STEP LOADING. At time zero, a sewage treatment plant began to discharge 10 MGD of wastewater with a concentration of 200 mg L^{-1} to a small detention basin (volume = 20×10^4 m^3). If the sewage decays at a rate of 0.1 d^{-1}, compute the concentration in the system during the first 2 wk of operation. Also determine the shape parameters to assess the ultimate effect of the plant.

Solution: The flow must be converted to the proper units:

$$10 \text{ MGD} \frac{1 \text{ m}^3 \text{ s}^{-1}}{22.8245 \text{ MGD}} \left(\frac{86{,}400 \text{ s}}{\text{d}} \right) = 37{,}854 \text{ m}^3 \text{ d}^{-1}$$

The eigenvalue can be determined as

$$\lambda = \frac{37{,}854}{20 \times 10^4} + 0.1 = 0.28927 \text{ d}^{-1}$$

Therefore the concentration can be computed with Eq. 4.10 as

$$c = \frac{W}{\lambda V}(1 - e^{-\lambda t}) = \frac{200(37{,}854)}{0.28927(20 \times 10^4)}(1 - e^{-0.28927t}) = 131(1 - e^{-0.28927t})$$

The results for the first 2 wk are

t (d)	0	2	4	6	8	10	12	14
c (mg L^{-1})	0	57.48	89.72	107.79	117.92	123.61	126.79	128.58

These values can also be displayed graphically as

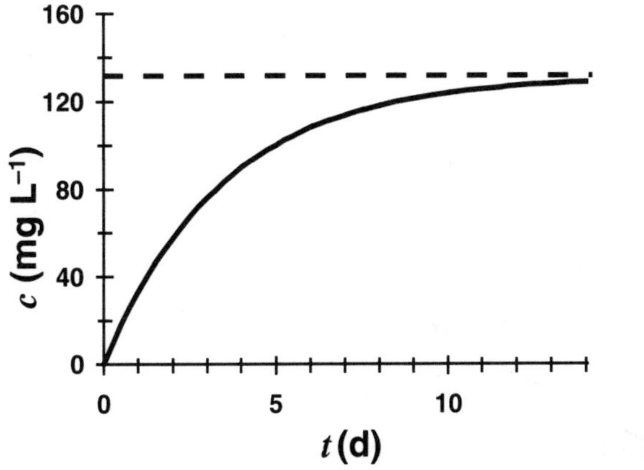

FIGURE E4.1

The shape parameters for this case are an ultimate steady-state concentration of 131 mg L^{-1}, of which 95% will be attained in 3/0.28927 ≅ 10.4 d.

4.3 LINEAR ("RAMP") LOADING

The simplest representation of a trend is a straight line with a nonzero slope. Waste inputs can often be represented in this way, as in

$$W(t) = \pm\beta_l t \tag{4.12}$$

where β_l = rate of change or slope of the trend (M T^{-2}). Notice that the trend can be either positive or negative. The particular solution for this case is

$$c = \pm\frac{\beta_l}{\lambda^2 V}\left(\lambda t - 1 + e^{-\lambda t}\right) \tag{4.13}$$

In water-quality applications, the positively increasing case (Fig. 4.4*a*) is more frequently used. Sometimes loadings that are proportional to population growth can be adequately approximated in this way. For such cases, after an initial start-up time (which can be defined by a 95% response time equal to $3/\lambda$), the solution becomes

$$c = \frac{\beta_l}{\lambda V}\left(t - \frac{1}{\lambda}\right) \tag{4.14}$$

Thus, as illustrated in Fig. 4.4*b*, the solution eventually increases at a constant slope of

$$\beta_l' = \frac{\beta_l}{\lambda V} \tag{4.15}$$

with response lagging the input by

$$t_l = \frac{1}{\lambda} \tag{4.16}$$

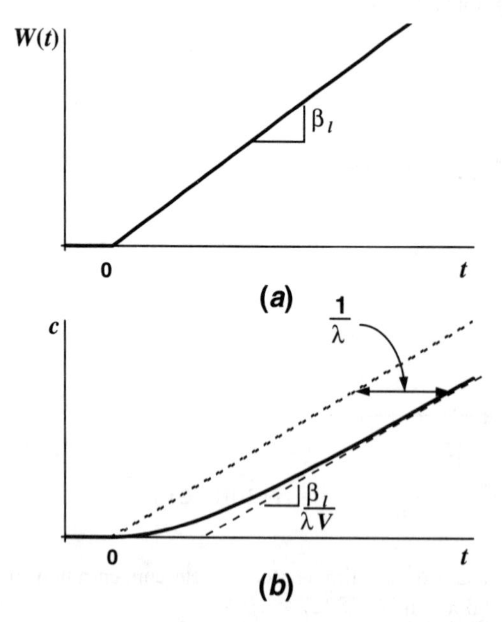

FIGURE 4.4
Plot of (*a*) loading and (*b*) response for a linearly increasing loading.

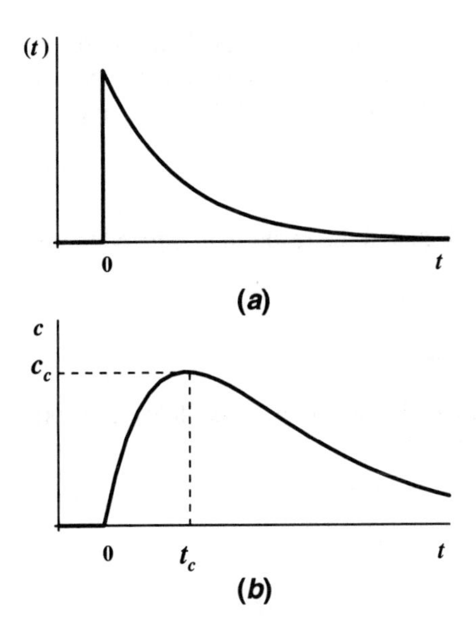

FIGURE 4.5
Plot of (a) loading and (b) response for exponentially decaying loading.

4.4 EXPONENTIAL LOADING

Another standard way to characterize loading trends is as an exponential function (Fig. 4.5*a*),

$$W(t) = W_e e^{\pm\beta_e t} \tag{4.17}$$

where W_e = a parameter that denotes the value at $t = 0$ (M T^{-1}) and β_e specifies the rate of growth (positive) or decay (negative) of the loading (T^{-1}).

Such a function characterizes waste sources that grow or decay in proportion to their own magnitude. The particular solution for this case is

$$c = \frac{W_e}{V(\lambda \pm \beta_e)}\left(e^{\pm\beta_e t} - e^{-\lambda t}\right) \tag{4.18}$$

This solution has many applications in water-quality modeling. The positive exponential has been used to model loadings that are dependent on population growth (see Example 4.4). The decaying exponential is more commonly encountered because it applies to series of first-order decaying processes (see Example 4.2 for such a case). Figure 4.5*b* shows the result for the decaying exponential load. Note how the solution rises to a peak and then begins to decline exponentially.

Thus we see that there are two shape parameters: the magnitude and the time of the peak. The time is easily determined by differentiating Eq. 4.18 and setting the result equal to zero. After some algebraic manipulation, the time of the peak can be determined as

$$t_c = \frac{\ln(\beta_e/\lambda)}{\beta_e - \lambda} \tag{4.19}$$

where the subscript c signifies that this is the time of the "critical" concentration. The magnitude can be determined by merely substituting Eq. 4.19 into Eq. 4.18. An alternative approach can be developed by recognizing that the peak concentration corresponds to Eq. 4.1, with the loading equal to Eq. 4.17 (with negative exponent) and $dc/dt = 0$. Substituting the critical time into Eq. 4.1 with $dc/dt = 0$ gives

$$c_c = \frac{W_e}{\lambda V} e^{-\beta_e t_c} \tag{4.20}$$

Further, Eq. 4.19 can be substituted into Eq. 4.20 and the result manipulated to give

$$c_c = \frac{W_e}{\lambda V} \left(\frac{\beta_e}{\lambda}\right)^{\frac{\beta_e}{\lambda - \beta_e}} \tag{4.21}$$

EXAMPLE 4.2. EXPONENTIAL FORCING FUNCTION. The following series of first-order reactions takes place in a batch reactor:

$$A \xrightarrow{k_l} B \xrightarrow{k_2}$$

Mass balance equations for these reactions can be written as

$$\frac{dc_A}{dt} = -k_1 c_A$$

and

$$\frac{dc_B}{dt} = k_1 c_A - k_2 c_B$$

Suppose that an experiment is conducted where $c_{A0} = 20$ and $c_{B0} = 0$ mg L^{-1}. If $k_1 = 0.1$ and $k_2 = 0.2$ d^{-1}, compute the concentration of reactant B as a function of time. Also, determine its shape parameters.

Solution: The concentration of reactant A can be determined by integrating the first differential equation to give

$$c_A = c_{A0} e^{-k_1 t}$$

This result can be substituted into the second differential equation to yield

$$\frac{dc_B}{dt} + k_2 c_B = k_1 c_{A0} e^{-k_1 t}$$

Thus the mass balance is now in the form of a first-order differential equation with an exponential forcing function. The solution is

$$c_B = \frac{k_l c_{A0}}{(k_2 - k_1)} \left(e^{-k_1 t} - e^{-k_2 t}\right)$$

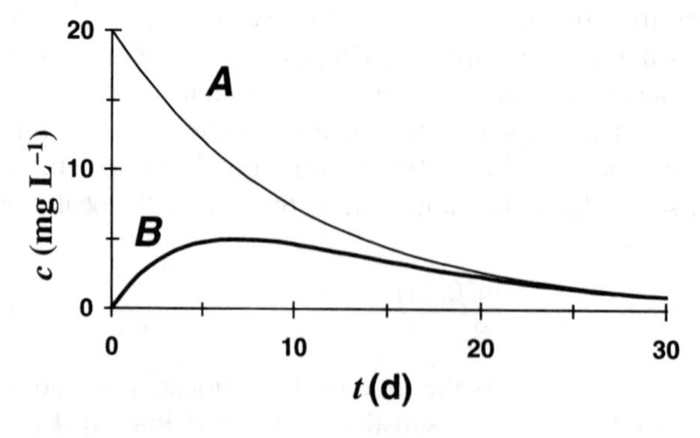

FIGURE E4.2

or substituting the parameter values,

$$c_B = 20\left(e^{-0.1t} - e^{-0.2t}\right)$$

The plot of this solution, along with the solution for c_A, is shown in Fig. E4.2.

The shape parameters for this case can be computed as

$$c_c = \frac{0.1(20)}{0.2}\left(\frac{0.1}{0.2}\right)^{\frac{0.1}{0.2-0.1}} = 5 \text{ mg L}^{-1}$$

which will occur at

$$t_c = \frac{\ln(0.1/0.2)}{0.1 - 0.2} = 6.93 \text{ d}$$

These values are consistent with the graph.

4.5 SINUSOIDAL LOADING

The simplest periodic input is the sinusoidal function (Fig. 4.6*a*), which can be represented mathematically as

$$W(t) = \overline{W} + W_a \sin(\omega t - \theta) \tag{4.22}$$

where $\overline{W}$ = mean loading (M T^{-1})
W_a = amplitude of the loading (M T^{-1})
θ = phase shift (radians)
ω = angular frequency of the oscillation (radians T^{-1}), which can be defined as

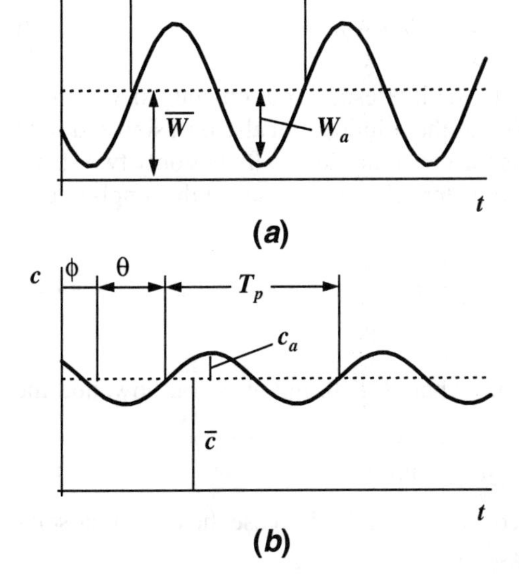

FIGURE 4.6
Plot of (*a*) loading and (*b*) response for the sinusoidal loading function. Note that a constant input is also shown in this illustration.

$$\omega = \frac{2\pi}{T_p} \tag{4.23}$$

where T_p = period of the oscillation (T). Note that the simple frequency of the oscillation can be computed as

$$f = \frac{1}{T_p} \tag{4.24}$$

where f has units of cycles T^{-1}.

Before proceeding to the particular solution, recognize that Eq. 4.22 itself has four shape parameters. These are ω (how frequently the sine wave oscillates), $\overline{W}$ (the mean value for the sine wave), W_a (how high the oscillation swings vertically), and θ (how much the function is shifted horizontally relative to the standard sine wave). Thus the parameters provide a lot of flexibility for characterizing a sinusoidal load.

Now the particular solution for this case is

$$c = \frac{\overline{W}}{\lambda V}\left(1 - e^{-\lambda t}\right) + \frac{W_a}{V\sqrt{\lambda^2 + \omega^2}}\sin[\omega t - \theta - \phi(\omega)] - \frac{W_a}{V\sqrt{\lambda^2 + \omega^2}}\sin[-\theta - \phi(\omega)]e^{-\lambda t} \tag{4.25}$$

where $\phi(\omega)$ = an additional phase shift (radians) that is a function of the frequency,

$$\phi(\omega) = \tan^{-1}\left(\frac{\omega}{\lambda}\right) \tag{4.26}$$

Inspection of Eq. 4.25 leads us to recognize that the first part of the solution approaches the steady-state level, and the last part of the solution will die away given sufficient time (which again, since it depends on an exponential decay, can be parameterized by a response time). Thus, after this initial start-up period, the solution will be

$$c = \frac{\overline{W}}{\lambda V} + \frac{W_a}{V\sqrt{\lambda^2 + \omega^2}}\sin[\omega t - \theta - \phi(\omega)] \tag{4.27}$$

Now inspection of this result leads to the interesting (but obvious, if you think about it) conclusion that the concentration of the solution will also be a sinusoid with the same frequency as the loading. Thus the concentration really has only two shape parameters that change as the model parameters change. These are the amplitude of the concentration response,

$$c_a = \frac{W_a}{V\sqrt{\lambda^2 + \omega^2}} \tag{4.28}$$

and the phase shift defined by Eq. 4.26.

An especially interesting case occurs when $\theta = 0$, inflow = outflow, and the loading is defined as

$$W(t) = Qc_{a,\text{in}}\sin(\omega t) \tag{4.29}$$

where $c_{a,\text{in}}$ = amplitude of the inflow concentration. In this case the particular solution becomes (after the transients have died out)

$$c = c_{a,\text{in}} A(\omega) \sin[\omega t - \phi(\omega)] \tag{4.30}$$

where $A(\omega)$ = an amplitude attenuation coefficient,

$$A(\omega) = \frac{\lambda}{\sqrt{\lambda^2 + \omega^2}} \tag{4.31}$$

Depending on the frequency of the oscillation (as reflected by the value of ω), the solution's amplitude will be attenuated (as reflected by A) and shifted (as reflected by ϕ). The implications of this solution are explored in the following example.

EXAMPLE 4.3. SINUSOIDAL FORCING FUNCTION. A completely mixed lake receives a conservative substance. Water inflow and outflow are equal, and the inflow concentration varies sinusoidally as

$$c_{\text{in}} = \bar{c}_{\text{in}} + c_{a,\text{in}} \sin(\omega t)$$

where $\bar{c}_{\text{in}}$ = average inflow concentration
$c_{a,\text{in}}$ = amplitude of the inflow concentration
ω = angular frequency ($= 2\pi/T_p$), in which T_p = period of the oscillation

If the lake has a volume of 2.5×10^6 m^3 and inflow = outflow = 9×10^6 m^3 yr^{-1}, determine its sensitivity to the sinusoidal component of the loading if the period of the oscillation is (*a*) 10 yr, (*b*) 1 yr, or (*c*) 0.1 yr.

Solution: The eigenvalue can be computed as

$$\lambda = \frac{Q}{V} = \frac{9 \times 10^6}{2.5 \times 10^6} = 3.6 \text{ yr}^{-1}$$

For a period of 10 yr, $\omega = 2\pi/10 = 0.628$ yr^{-1} and

$$\phi(0.628) = \tan^{-1}\left(\frac{0.628}{3.6}\right) = 0.1727 \text{ radian}\left(\frac{10 \text{ yr}}{2\pi \text{ radians}}\right) = 0.275 \text{ yr (100 d)}$$

and $$A(0.628) = \frac{3.6}{\sqrt{3.6^2 + 0.628^2}} = 0.985$$

Therefore the solution is almost identical to the forcing function. The amplitude is diminished only 1.5% and the phase shift is a mere 100 d (compared to the period of 10 yr). However, as shown by the following table, this correspondence breaks down as the frequency of the forcing function increases:

Period (yr)	***f* (cycles yr^{-1})**	***A*(ω)**	**$\phi(\omega)$ (d)**	**$[\phi(\omega)/T_p] \times 100\%$**
10	0.1	0.985	100	2.8%
1	1	0.573	61	16.7%
0.1	10	0.057	8.8	24.1%

As the frequency increases, the solution exhibits two effects. First, the amplitude diminishes and approaches zero. Second, as indicated by the normalized phase shift, the solution increasingly lags the forcing function. In essence both effects reflect the fact that the system is too sluggish (as manifested by its eigenvalue, λ) to "keep up" with the forcing function.

It should be noted that there is a formal way of presenting the information developed in this example. That is, both the amplitude and phase characteristics can be plotted versus frequency in what are called ***Bode diagrams***. These are conventionally constructed by plotting either the amplitude or the phase characteristic versus the angular frequency (ω). As shown in Fig. E4.3, we have chosen to plot the characteristics versus the period. We have also plotted the phase shift in degrees. In the present context we believe that both modifications make the plots somewhat easier to interpret.

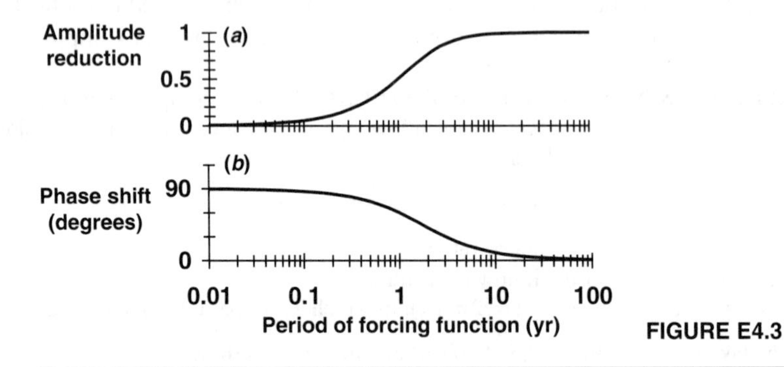

FIGURE E4.3

Although the phase shift certainly provides a measure of how much the solution is shifted temporally, an alternative shape parameter that represents the same effect is the time of the peak relative to time zero. This quantity can be computed by taking the derivative of Eq. 4.30 and setting it equal to zero to yield

$$t_{\max} = \frac{(\pi/2) + \phi}{\omega} \tag{4.32}$$

This format is useful because we are often concerned with maximum concentrations in water-quality modeling.

4.6 THE TOTAL SOLUTION: LINEARITY AND TIME SHIFTS

Figure 4.7 summarizes the analytical solutions we have developed. Although they are useful in their own right, they can also be employed in tandem to assess the impact of several loading trends simultaneously. That is, if we derive a number of particular solutions c_{pi}, and a_i are any constants, the total solution can be derived by simple addition of the general and particular solutions,

$$c = c_g + \sum_{i=1}^{n} a_i c_{pi} \tag{4.33}$$

This is possible only because the underlying model is linear.

In addition we can add loadings that are initiated at different times by merely shifting the solutions. If you have a function $c(t)$ and want it shifted a units of time into the future, you would merely subtract a from the argument, $c(t-a)$. For example suppose that we have two impulse loads, one that occurs at time zero and another at

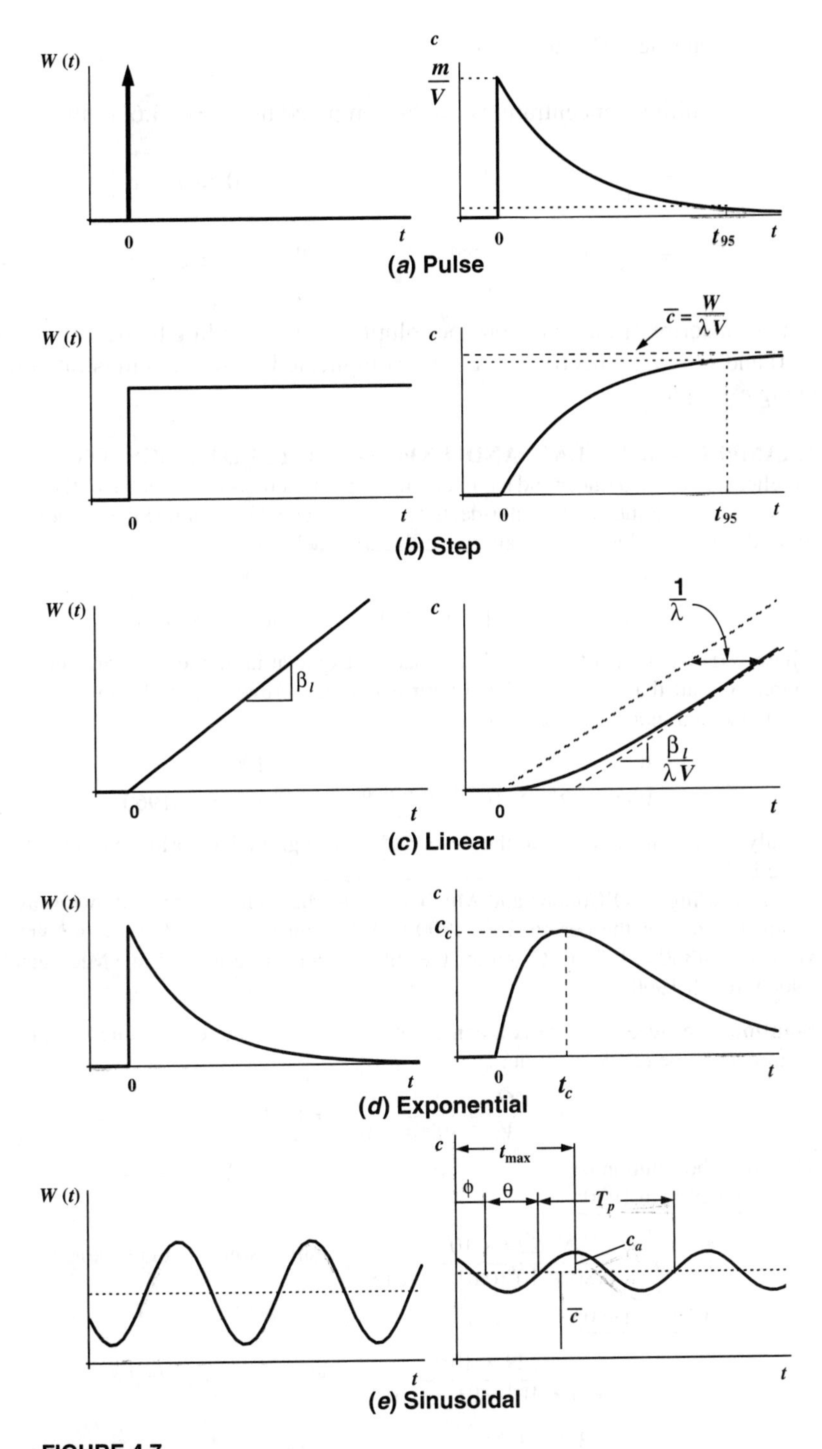

FIGURE 4.7
A summary of the solutions developed to this point for a substance reacting with first-order kinetics. Note that the shape parameters are shown on the concentration response plots.

time a. The resulting concentrations can be computed using Eq. 4.6, as in

$$c = \frac{m_1}{V} e^{-\lambda t} \qquad 0 \le t < a \tag{4.34}$$

$$c = \left(\frac{m_1}{V} e^{-\lambda a}\right) e^{-\lambda(t-a)} + \frac{m_2}{V} e^{-\lambda(t-a)} \qquad t \ge a \tag{4.35}$$

where the subscripts 1 and 2 denote the solutions corresponding to the impulse load at $t = 0$ and $t = a$, respectively. A more complicated problem is illustrated in the following example.

EXAMPLE 4.4. LINEAR AND EXPONENTIAL LOADINGS. O'Connor and Mueller (1970) used linear and exponential forcing functions to characterize the loadings of a conservative substance, chloride, to Lake Michigan. For example they characterized chloride loadings due to road salt by the linear model

$$W(t) = 0 \qquad t < 1930$$

$$W(t) = 13.2 \times 10^9 (t - 1930) \qquad 1930 \le t \le 1960$$

where $W(t)$ has units of g yr^{-1}. They used an exponential model to characterize other sources of salt that were correlated with population growth in the basin (for example municipal and industrial sources),

$$W(t) = 0 \qquad t < 1900$$

$$W(t) = 229 \times 10^9 e^{0.015(t-1900)} \qquad 1900 \le t \le 1960$$

Finally they considered that the lake had a background chloride concentration of 3 mg L^{-1}.

According to O'Connor and Mueller, Lake Michigan had the following average characteristics for the period from 1900 to 1960: outflow = 49.1×10^9 m^3 yr^{-1} and volume = 4880×10^9 m^3. Calculate the chloride concentration in Lake Michigan from 1900 through 1960.

Solution: Because chloride is a conservative substance, the eigenvalue is simply the reciprocal of the residence time:

$$\lambda = \frac{Q}{V} = \frac{49.1 \times 10^9}{4880 \times 10^9} = 0.01 \text{ yr}^{-1}$$

Therefore the solution is

From 1900 to 1930:

$$c = 3 + \frac{229 \times 10^9}{4880 \times 10^9 (0.01 + 0.015)} (e^{+0.015(t-1900)} - e^{-0.01(t-1900)})$$

From 1930 to 1960:

$$c = 3 + \frac{229 \times 10^9}{4880 \times 10^9 (0.01 + 0.015)} (e^{+0.015(t-1900)} - e^{-0.01(t-1900)})$$

$$- \frac{13.2 \times 10^9}{(0.01)^2 4880 \times 10^9} [1 - e^{-0.01(t-1930)} - 0.01(t - 1930)]$$

Note that initial conditions are not included explicitly because they are reflected in the constant background level of 3 mg L^{-1}. The results, along with data, are depicted in Fig. E4.4.

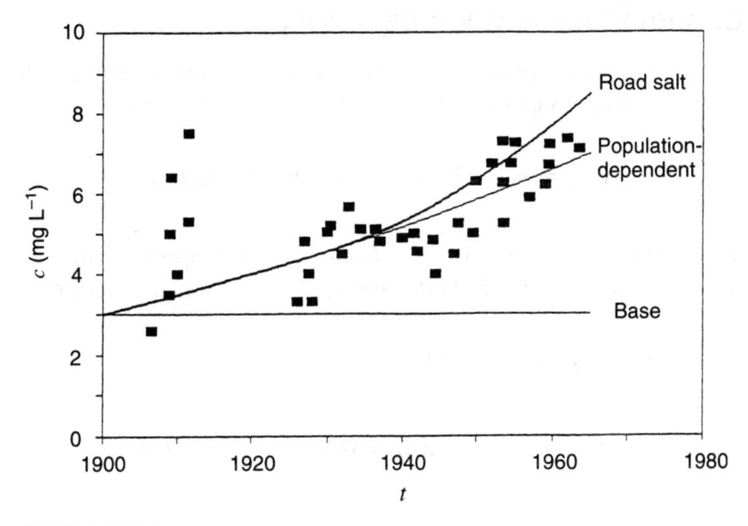

FIGURE E4.4

BOX 4.1. "Back-of-the-Envelope" Calculations

A substantial portion of water-quality modeling includes large, elaborate calculations involving digital computers. Although such computations can be extremely useful, most practitioners perform a significant amount of what is called "back-of-the-envelope" modeling.

As the name implies, this alternative approach involves sketching out a solution on a piece of paper no bigger than the back of an envelope. Although this might seem a tad "low-tech," a lot of useful modeling gets accomplished in this manner.

The past two lectures have been oriented toward such calculations. In Lec. 3, I presented the steady-state and general solutions for a well-mixed waterbody to provide quick estimates of the "how much?" and the "how long?" of load reductions. In this lecture the shape parameters provide a rapid means to extract the essential features of concentration response to a variety of loading shapes. In all cases a pencil, a pad, and a pocket calculator are all the computational firepower required to obtain rapid estimates.

In some studies such "quick-and-dirty" estimates might be all that's required (or affordable) to address a water-quality management question. Further, even when large computations are inevitable, back-of-the-envelope calculations can provide invaluable insights and checks.

Finally a back-of-the-envelope approach can sometimes be very effective for communicating results to decision makers. In the hands of a master modeler, the exercise can amount to an elegant distillation of the essential elements of a problem solution. As such it can serve as a vehicle for expressing complex model results in a clear and easily understandable fashion.

Much of the remainder of the text points toward large-scale numerical calculations. However, wherever possible, I will attempt to provide concise analytical methods to facilitate your own back-of-the-envelope calculations. In particular many of the end-of-lecture problems must necessarily be kept brief and pointed. As such they can provide good practice and experience in developing quick, broad estimates to complement your computer modeling.

4.7 FOURIER SERIES (ADVANCED TOPIC)

Aside from the sinusoid presented in Sec. 4.5, there are other periodic inputs (Fig. 4.8). Such functions can be represented by a Fourier series

$$W(t) = a_0 + \sum_{k=1}^{\infty} \left[a_k \cos(k\omega t) + b_k \sin(k\omega t)\right] \tag{4.36}$$

where $\omega = 2\pi/T_p$ is referred to as the ***fundamental frequency***, and its constant multiples $2\omega_0$, $3\omega_0$, etc., are called ***harmonics***. The coefficients can be evaluated by

$$a_0 = \frac{1}{T_p}\int_0^{T_p} W(t)\,dt \tag{4.37}$$

$$a_k = \frac{2}{T_p}\int_0^{T_p} W(t)\cos(k\omega t)\,dt \qquad \text{for } k = 1, 2, \ldots \tag{4.38}$$

and

$$b_k = \frac{2}{T_p}\int_0^{T_p} W(t)\sin(k\omega t)\,dt \qquad \text{for } k = 1, 2, \ldots \tag{4.39}$$

The Fourier series represents the loading as a summation of sinusoids. Then this summation can be substituted into Eq. 4.1. The individual solutions are then obtained (for example using Eq. 4.27 for the sinusoidal terms) and the results summed to determine the concentration response. The process is elaborated on in the following example.

EXAMPLE 4.5. HALF-SINUSOID FORCING FUNCTION. The half-sinusoid (Fig. 4.8*a*) can be employed in a number of water-quality-modeling contexts. For example it has been used successfully to simulate the affect of diurnal variations of plant photosynthesis on stream dissolved-oxygen levels.

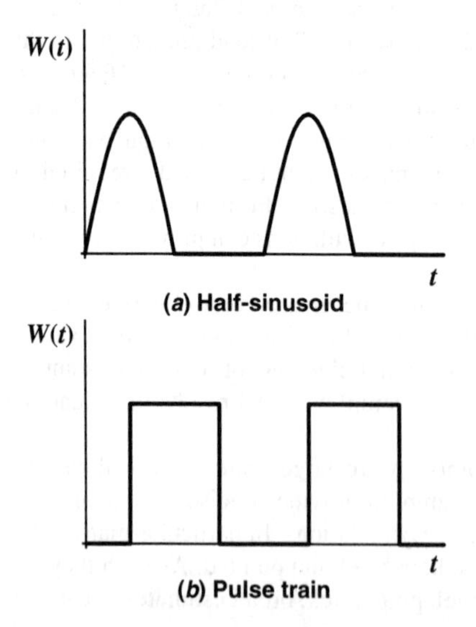

FIGURE 4.8
Two commonly used periodic inputs. (*a*) Half-sinusoid; (*b*) pulse train.

Another example relates to the design of a detention basin. Suppose that during the summer a company discharges a small flow of a highly concentrated contaminant into a tributary of a larger river. The company discharges from 6:00 P.M. until 6:00 A.M. Measurements at the mouth of the tributary indicate that a half-sine wave provides an adequate approximation of the concentration time series over the period

$$c_{\text{in}}(t) = c_a \sin(\omega t) \qquad 0 \le t \le \frac{T_p}{2}$$

$$c_{\text{in}}(t) = 0 \qquad \frac{T_p}{2} \le t \le T_p$$

where $c_{\text{in}}(t)$ = time series of the inflow concentration at the tributary mouth ($\mu\text{g L}^{-1}$) and c_a = amplitude of the half-sinusoid ($\mu\text{g L}^{-1}$). Note that time zero is assumed to be at 6:00 P.M.

It is proposed that the company build a detention basin at the tributary mouth to moderate the effect of the discharge on the river. The flow in the tributary (including the discharge) can be adequately characterized by a constant level of $Q = 1\times10^6 \text{ m}^3 \text{ d}^{-1}$. In addition the detention basin's volume is $1\times10^6 \text{ m}^3$ and the amplitude of the concentration is 10 $\mu\text{g L}^{-1}$. If no losses occur, determine the long-term concentration response using the well-mixed-lake model (Eq. 4.1) along with a Fourier-series approximation of the half-sinusoidal loading up to the second harmonic.

Solution: The mass balance for the detention basin can be written as

$$\frac{dc}{dt} + \lambda c = \lambda c_{\text{in}}(t)$$

where the eigenvalue for this example is $Q/V = 1$. The coefficients for the Fourier-series approximation of the half-sinusoid can be evaluated, as in

$$a_0 = c_a \frac{1}{1}\int_0^{1/2} \sin(2\pi t)\,dt = c_a \frac{1}{\pi}$$

$$a_1 = c_a \frac{2}{1}\int_0^{1/2} \sin(2\pi t)\cos(2\pi t)\,dt = 0$$

$$b_1 = c_a \frac{2}{1}\int_0^{1/2} \cos(2\pi t)\sin(2\pi t)\,dt = c_a\left(\frac{1}{2}\right)$$

$$a_2 = c_a \frac{2}{1}\int_0^{1/2} \sin(2\pi t)\cos(4\pi t)\,dt = c_a\left(-\frac{2}{3\pi}\right)$$

$$b_2 = c_a \frac{2}{1}\int_0^{1/2} \sin(2\pi t)\sin(4\pi t)\,dt = 0$$

Therefore the loading function can be approximated by (with $c_a = 10$)

$$c_{\text{in}}(t) = \frac{10}{\pi} + 5\sin(2\pi t) - \frac{20}{3\pi}\cos(4\pi t)$$

Note that the last term can be represented as a sine with a phase shift of $\pi/2$,

$$\sin\left(4\pi t + \frac{\pi}{2}\right) = \cos(4\pi t)$$

Therefore the loading function becomes

$$c_{\text{in}}(t) = \frac{10}{\pi} + 5\sin(2\pi t) - \frac{20}{3\pi}\sin\left(4\pi t + \frac{\pi}{2}\right)$$

This approximation is displayed in Fig. 4.9*a*. Notice that although it is not perfect (there are even some slightly negative values), the series provides a reasonable approximation of the half-sinusoid.

To evaluate concentration, the approximation is substituted into the mass balance to give

$$\frac{dc}{dt} + \lambda c = \frac{10\lambda}{\pi} + 5\lambda\sin(2\pi t) - \frac{20\lambda}{3\pi}\sin\left(4\pi t + \frac{\pi}{2}\right)$$

The response can be determined as

$$c = \frac{10}{\pi} + \frac{5\lambda}{\sqrt{\lambda^2 + (2\pi)^2}}\sin(2\pi t - \theta_1) - \frac{20\lambda}{3\pi\sqrt{\lambda^2 + (4\pi)^2}}\sin\left(4\pi t + \frac{\pi}{2} - \theta_2\right)$$

where

$$\theta_1 = \tan^{-1}\left(\frac{2\pi}{1}\right) = 1.413 \text{ radians } (= 5.5 \text{ hr})$$

$$\theta_2 = \tan^{-1}\left(\frac{4\pi}{1}\right) = 1.491 \text{ radians } (= 5.7 \text{ hr})$$

The results, displayed in Fig. 4.9*b*, indicate that the detention pond has a definite effect on the concentration discharged to the river. Whereas the inflow concentration swings between 0 and 10 $\mu\text{g L}^{-1}$, the pond concentration moves between about 2.4 and 4.1 $\mu\text{g L}^{-1}$. Further, the peak occurs at about 4:00 A.M. rather than at midnight.

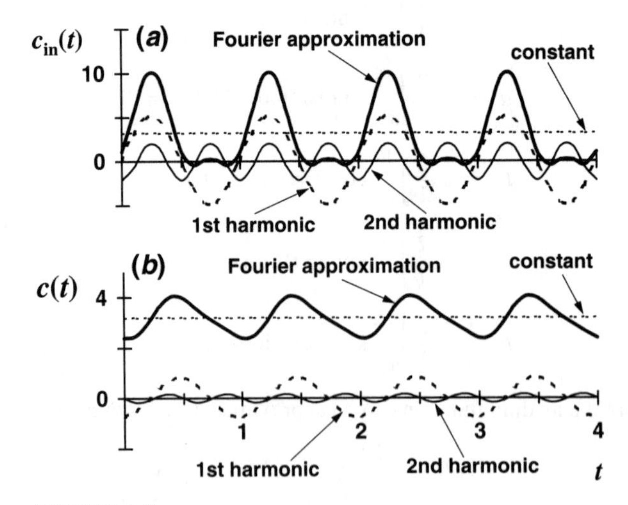

FIGURE 4.9
Loading and response for a detention basin. (*a*) Inflow concentration approximated by a half-sinusoid; (*b*) the resulting concentration in the basin.

BOX 4.2. Donald O'Connor and Closed-Form Solutions

You may be wondering why in this age of computers I've devoted this lecture (and several other parts of this book) to analytical or "closed-form" solutions. In fact prior to 1960 such solutions provided the only means to address complex mathematical problems (recall Fig. 1.6).

Way back then there were two primary reasons for developing such solutions. On the one hand certain mathematical manipulations can be expressly designed to generate answers in an efficient manner. An example was the use of "equivalent resistances" to simplify circuit analysis in electrical engineering. In water-quality modeling, linearization was often employed to make otherwise impossible or difficult problems tractable. Today because computing power is so accessible, some of these manipulations have become less utilized and more of historical interest.

The second reason for developing analytical solutions is to provide insight. Thus closed-form calculus can be employed to neatly summarize and illuminate the essential behavior of the systems being modeled. A classic example is the use of dimensionless numbers to characterize systems involving fluid dynamics. Such manipulations are far from outmoded and remain highly relevant in all areas of engineering.

In the field of water-quality modeling Prof. Donald O'Connor of Manhattan College has been the main innovator and proponent of such insight-oriented closed-form solutions. Although he certainly had predecessors (notably Harold W. Streeter, Gordon Fair, and Clarence J. Velz), Dr. O'Connor is generally considered the father of water-quality modeling. Starting with his doctoral work on stream reaeration in the 1950s (O'Connor and Dobbins 1958), he has published (and continues to publish) a series of significant papers over the past four decades that have advanced and guided the development of our field. Although all involve generating quantitative results, their hallmark is the fundamental insights they provide.

Beyond acknowledging Dr. O'Connor's contributions, I also wanted to emphasize the importance of closed-form solutions at this juncture. They are especially critical nowadays because you can implement just about any water-quality-modeling calculation on the computer. As will be stressed repeatedly throughout this text, use of computer models as "black boxes" can lead to blunders and erroneous interpretation of results. Insight-oriented analytical solutions are one important and effective way to build up modeling intuition and use computer-generated results wisely.

PROBLEMS

4.1. Use calculus to derive Eq. 4.6.

4.2. Use calculus to derive Eq. 4.18.

4.3. Use calculus to derive Eq. 4.27 for the case $\theta = 0$, $W = 0$, and the transients have died out. *Hint:* if you use Laplace transforms, recognize that the following partial fraction expansion applies:

$$\frac{(W/V)\omega}{(s^2 + \omega^2)(s + \lambda)} = \frac{As + B}{s^2 + \omega^2} + \frac{D}{s + \lambda}$$

4.4. A spill of 5 kg of a soluble pesticide takes place into a well-mixed pond. The pesticide is subject to volatilization that can be characterized by the following first-order flux:

$$J = v_v c$$

where v_v = a volatilization mass-transfer coefficient of 0.01 m d^{-1}. Other parameters for the lake are

Surface area = 0.1×10^6 m^2
Mean depth = 5 m
Outflow = 1×10^6 m^3 d^{-1}

(*a*) Predict the concentration in the lake as a function of time.
(*b*) Determine the 95% response time for the system.
(*c*) Calculate the time required for the concentration to be reduced to 0.1 μg L^{-1}.

4.5. A well-mixed lake has a steady-state concentration of 5 μg L^{-1} of total phosphorus. At the beginning of 1994 it received an additional loading of 500 kg yr^{-1} from a fertilizer processing plant. The lake has the following characteristics:

Inflow = outflow = 5×10^5 m^3 yr^{-1}
Volume = 4×10^7 m^3
Surface area = 5×10^6 m^2

If total phosphorus settles at a rate of 8 m yr^{-1}, compute the concentration in the lake from 1994 through 2010.

4.6. For the lake described in Prob. 4.5 suppose that, rather than a fertilizer plant, a small subdivision will be located on the lake in 1997. The population of this subdivision can be described by an exponential model:

$$p = 200e^{G_p t}$$

where G_p = first-order population growth rate = 0.2 yr^{-1}. If each person generates 0.5 kg yr^{-1} of total phosphorus, calculate the concentration in the lake from 1997 through 2010.

4.7. Determine the concentration in the lake described in Probs. 4.5 and 4.6 if both the fertilizer plant and the subdivision are allowed to discharge total phosphorus. Perform your computations from 1994 through 2010.

4.8. A lake has the following characteristics:

Inflow = outflow = 20×10^6 m^3 yr^{-1}
Mean depth = 10 m
Surface area = 10×10^6 m^2

A canning plant discharges a pollutant to the system that decays at a rate of 1.05 yr^{-1}. Because of the seasonal nature of the product being canned, peak waste discharges occur in the fall (October 1) and minimum levels in the spring (April 1). The discharge has a mean level of 30×10^6 g yr^{-1}, with a range between the extremes of 30×10^6 g yr^{-1}. If the initial condition is $c = 0$ at $t = 0$,
(*a*) Compute the concentration in the system from $t = 0$ to 10 yr.
(*b*) After sufficient time the concentration will reach a dynamic steady-state. At this point, on what day of the year will the in-lake concentration be at a maximum value?

4.9. Repeat Example 4.5 for the case where the contaminant reacts at a first-order rate of 0.1 d^{-1}. Interpret how the reaction affects the mean, amplitude, and lag of the response.

4.10. Derive Eq. 4.32.

4.11. The following data describe the seasonal cycle of solar energy incident on a small mountain lake:

Month	J	F	M	A	M	J	J	A	S	O	N	D
Solar (cal cm^{-2} d^{-1})	73	157	278	406	505	549	527	443	322	194	95	51

Fit the following function to this data (that is, estimate the parameters):

$$J(t) = \overline{J} + J_{\text{amp}} \sin(\omega t - \theta)$$

Assume a 360-d year.

4.12. A well-mixed lake is subject to the following pulse-train loading:

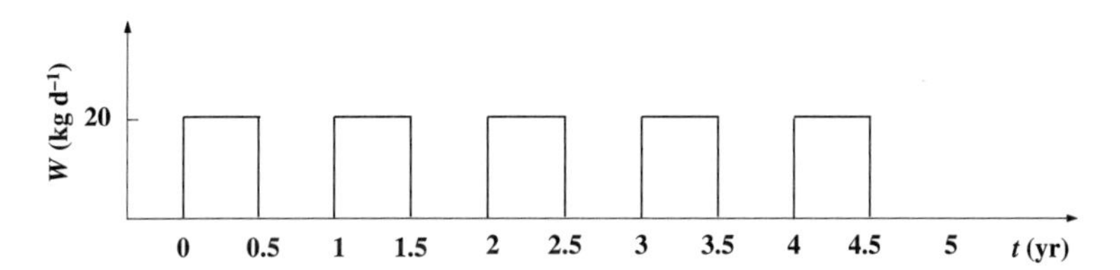

FIGURE P4.12

The concentration response of a CSTR to such a loading can be constructed using the step and the general solutions. Determine the response of a small pond to such a loading from $t = 0$ to 5 yr. Note that the initial condition is $c = 0$ at $t = 0$. The pond has a residence time of 2 yr, a volume of 100, 000 m^3, and a pollutant reaction rate of $k = 0.25$ yr^{-1}.

4.13. Repeat Prob. 4.12, but use the Fourier-series approach to obtain your solution. Note that the Fourier series for the pulse wave shown in Fig. P4.12 is

$$W(t) = 10 + \frac{40}{\pi}\left[\sin(2\pi t) + \frac{1}{3}\sin(6\pi t) + \frac{1}{5}\sin(10\pi t) + \cdots\right]$$

Use three terms for your solution and compare it with that obtained in Prob. 4.12.

4.14. A radioactive substance decays by first-order kinetics and forms a daughter substance that also decays in a first-order fashion:

$$c_1 \xrightarrow{k_1} c_2 \xrightarrow{k_2}$$

You perform an experiment in which you place some of the first compound in a batch reactor. The following data is collected over time:

t (d)	0	20	40	60	80	100
c_1 (nCi L^{-1})	10	6.3	4.0	2.5	1.6	1.0

In addition the peak concentration of the second compound occurs at 35 d. Use this information to determine the half-lives of the compounds.

LECTURE 5

Feedforward Systems of Reactors

LECTURE OVERVIEW: I develop steady-state and time-variable solutions for coupled reactors without feedback and apply these solutions to a series of lakes. In addition I extend the analysis to a series of first-order reactions within a single reactor.

To this point we have examined the steady-state and time-variable solutions for a single CSTR. We can now attempt to model more complicated systems by combining reactors.

Figure 5.1 shows two general ways in which completely mixed reactors can be connected: feedforward (Fig. 5.1*a*) and feedback (Fig. 5.1*b*). The difference between the two types of configurations is that in the feedforward setup, water never flows through the same reactor twice. This greatly simplifies both steady-state and time-variable solutions.

Aside from providing a model for systems of reactors, the feedforward framework is also useful for characterizing natural water systems. For example it can be employed to model a chain of lakes or a stream. Furthermore, it can be used to simulate serial reactions.

5.1 MASS BALANCE AND STEADY-STATE

The reactors in series model applies to a variety of problem contexts in environmental engineering. Among the most straightforward is a series of lakes connected by short rivers.

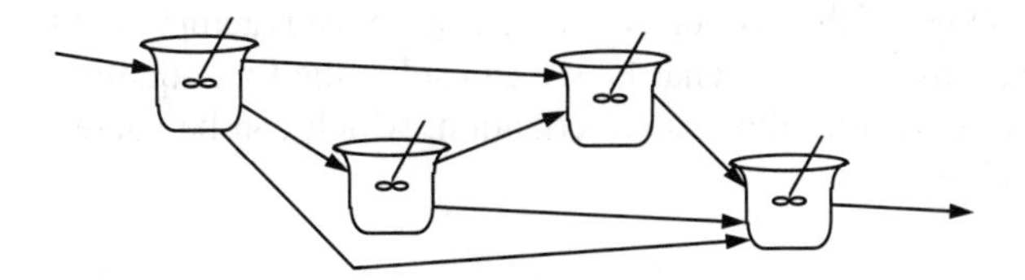

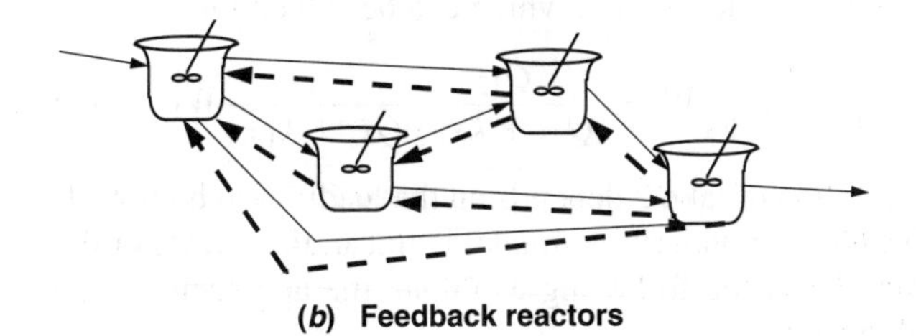

FIGURE 5.1
Two general ways in which reactors can be connected. The dashed arrows in (*b*) represent feedback.

We will keep the following derivation simple by limiting it to two lakes in series (Fig. 5.2). Mass balances for these reactors can be written as

$$V_1 \frac{dc_1}{dt} = W_1 - Q_{12}c_1 - k_1 V_1 c_1 \tag{5.1}$$

$$V_2 \frac{dc_2}{dt} = W_2 + Q_{12}c_1 - Q_{23}c_2 - k_2 V_2 c_2 \tag{5.2}$$

Note that both lakes are assumed to have constant volumes. For simplicity the inflow into the first reactor has not been shown explicitly.

The term $Q_{12}c_1$ representing the outflow of mass from the first lake appears in both balances. Thus, because mass flows from lake 1 to lake 2, the solution for lake 2 depends on the solution for lake 1.

At steady-state the terms can be collected and expressed as

$$a_{11}c_1 \qquad\qquad = W_1 \tag{5.3}$$

$$-a_{21}c_1 + a_{22}c_2 = W_2 \tag{5.4}$$

where

$$a_{11} = Q_{12} + k_1 V_1 \tag{5.5}$$

$$a_{21} = Q_{12} \tag{5.6}$$

$$a_{22} = Q_{23} + k_2 V_2 \tag{5.7}$$

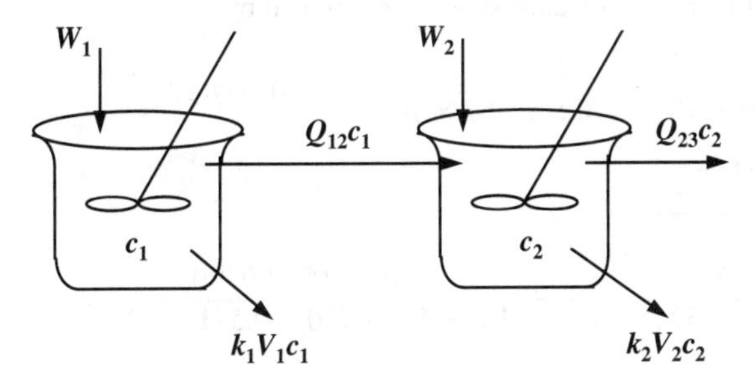

FIGURE 5.2
Lakes in series.

We can now see the advantage of this sort of system. Because the reactors are in series, we can solve the equations in series. That is, we can solve the first equation for c_1 and then substitute the result into the second equation, which can be solved for c_2. Thus Eq. 5.3 can be solved for

$$c_1 = \frac{1}{a_{11}} W_1 = \frac{1}{Q_{12} + k_1 V_1} W_1 \tag{5.8}$$

This result in turn can be substituted into Eq. 5.4, which can be solved for

$$c_2 = \frac{W_2 + a_{21} c_1}{a_{22}} = \frac{1}{Q_{23} + k_2 V_2} W_2 + \frac{Q_{12}}{Q_{23} + k_2 V_2} \frac{1}{Q_{12} + k_1 V_1} W_1 \tag{5.9}$$

We see clearly that the concentration of lake 2 depends on the loadings to both itself and the loading to lake 1. Because the underlying model is linear, the effects of the two loadings are independent. As in the following example, the approach can be easily extended to additional reactors.

EXAMPLE 5.1. LAKES IN SERIES. Suppose that three lakes connected in series have the following characteristics:

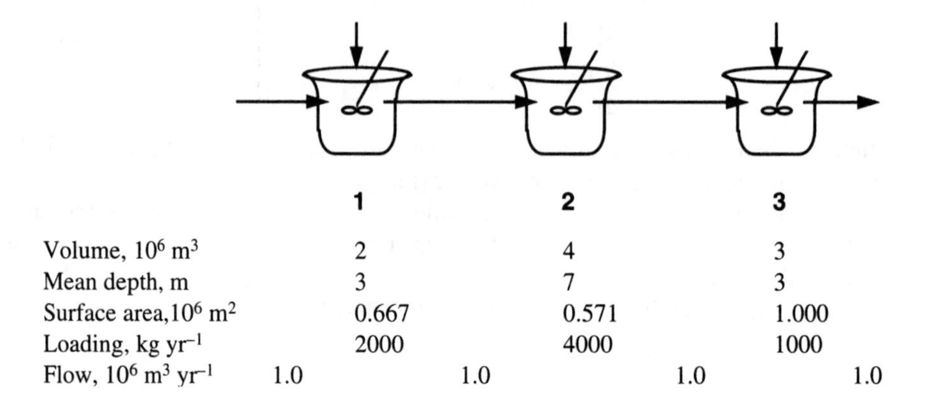

		1		2		3	
Volume, 10^6 m^3		2		4		3	
Mean depth, m		3		7		3	
Surface area, 10^6 m^2		0.667		0.571		1.000	
Loading, kg yr^{-1}		2000		4000		1000	
Flow, 10^6 m^3 yr^{-1}	1.0		1.0		1.0		1.0

If the pollutant settles at a rate of 10 m yr^{-1},

(*a*) Calculate the steady-state concentration in each of the reactors.

(*b*) Determine how much of the concentration in the third reactor is due to the loading to the second reactor.

Solution: (*a*) The concentration for the reactors can be determined by

$$c_1 = \frac{W_1}{Q_{12} + vA_1} = \frac{2 \times 10^9}{1.0 \times 10^6 + (10 \times 0.667 \times 10^6)} = 260.76\ \mu\text{g L}^{-1}$$

$$c_2 = \frac{W_2}{Q_{23} + vA_2} + \frac{Q_{12} c_1}{Q_{23} + vA_2}$$

$$= \frac{4 \times 10^9}{1.0 \times 10^6 + (10 \times 0.571 \times 10^6)} + \frac{1.0 \times 10^6 (260.76)}{1.0 \times 10^6 + (10 \times 0.571 \times 10^6)}$$

$$= 596.13 + 38.86 = 634.99\ \mu\text{g L}^{-1}$$

$$c_3 = \frac{W_3}{Q_{34} + vA_3} + \frac{Q_{23}c_2}{Q_{34} + vA_3}$$

$$= \frac{1 \times 10^9}{1 \times 10^6 + (10 \times 1 \times 10^6)} + \frac{1.0 \times 10^6(634.99)}{1 \times 10^6 + (10 \times 1 \times 10^6)}$$

$$= 148.64\ \mu\text{g L}^{-1}$$

(*b*) The determination of how much of the third reactor's concentration is due to the loading to the second reactor can be established by inspecting the solution for c_2 above. As can be seen, 596.13 μg L^{-1} of c_2 is due to direct loadings (that is, W_2), whereas 38.86 is due to the loadings to reactor 1. Therefore the effect of the second reactor on the third reactor can be calculated as

$$c_3(\text{due to loading to reactor 2}) = \frac{1.0 \times 10^6(596.13)}{1 \times 10^6 + (10 \times 1 \times 10^6)} = 54.19\ \mu\text{g L}^{-1}$$

5.1.1 Cascade Model

An interesting instance of reactors in series is the case where the reactors' size and flow are identical. This is sometimes referred to as a cascade of CSTRs (Fig. 5.3).

For this case the equations are simplified,

$$c_1 = \frac{Q}{Q + kV} c_0 \tag{5.10}$$

$$c_2 = \frac{Q}{Q + kV} c_1 = \frac{Q}{Q + kV} \frac{Q}{Q + kV} c_0 \tag{5.11}$$

$$\vdots$$

$$c_n = \left(\frac{Q}{Q + kV}\right)^n c_0 \tag{5.12}$$

Thus we can see that each reactor attenuates the concentration by a constant fraction because the parenthetical term will always be less than one.

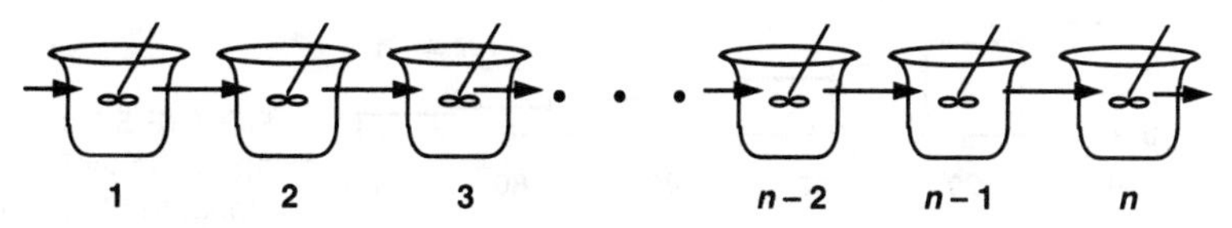

FIGURE 5.3
Cascade of CSTRs.

EXAMPLE 5.2. CASCADE MODEL OF AN ELONGATED TANK. Use the cascade model to simulate the steady-state distribution of concentration in an elongated tank.

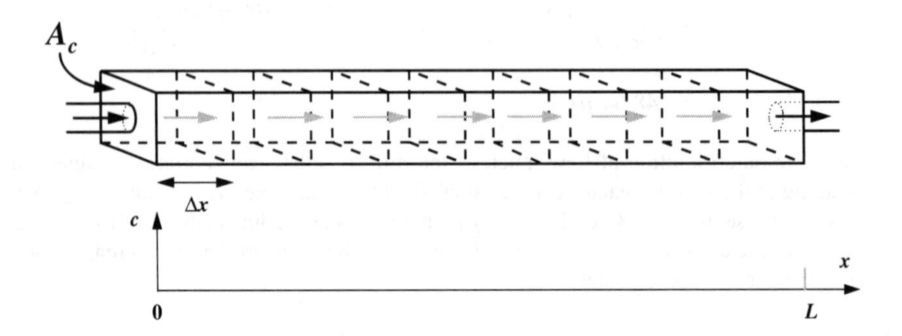

The tank has cross-sectional area $A_c = 10\ \text{m}^2$, length $L = 100$ m, velocity $U = 100$ m hr^{-1}, and first-order reaction rate $k = 2\ \text{hr}^{-1}$. The inflow concentration is 1 mg L^{-1}. Use $n = 1, 2, 4$, and 8 CSTRs to approximate the tank. Plot the results.

Solution: As in Eq. 5.12, the model for such a system is

$$c_n = \left(\frac{Q}{Q + kV}\right)^n c_0$$

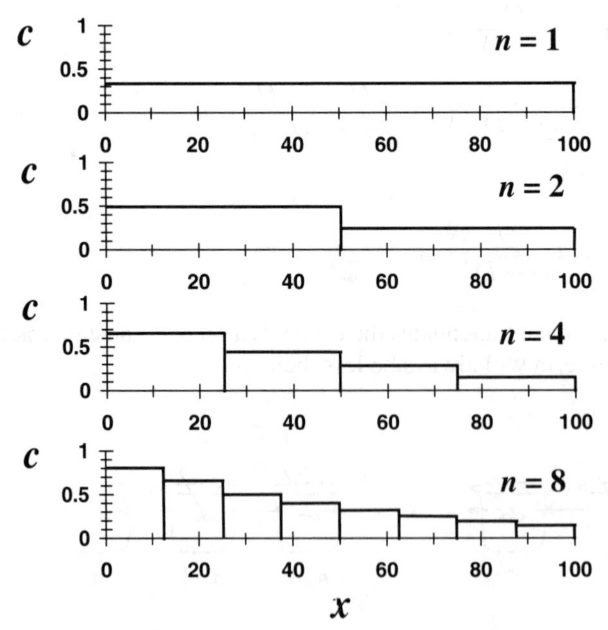

FIGURE 5.4
Approximation of an elongated tank by a series of n CSTRs.

Therefore for the single-segment approximation,

$$c(50) = \left[\frac{1000}{1000 + 2(1000)}\right]1 = 0.3333 \text{ mg L}^{-1}$$

where $Q = UA = 100(10) = 1000$. Similarly for $n = 2$,

$$c(25) = \left[\frac{1000}{1000 + 2(500)}\right]1 = 0.5 \text{ mg L}^{-1}$$

and

$$c(75) = \left[\frac{1000}{1000 + 2(500)}\right]^2 1 = 0.25 \text{ mg L}^{-1}$$

The other cases can be computed in similar fashion. The results, as summarized in Fig. 5.4, are interesting. As more (and smaller) reactors are used, the solution approaches a pattern that looks just like an exponential decay. We will explore this observation in more detail when we investigate models of elongated reactors in Lec. 9.

5.2 TIME VARIABLE

Equations 5.1 and 5.2, with zero loadings, can be written as

$$\frac{dc_1}{dt} = -\lambda_{11} c_1 \tag{5.13}$$

and

$$\frac{dc_2}{dt} = \lambda_{21} c_1 - \lambda_{22} c_2 \tag{5.14}$$

where

$$\lambda_{11} = \frac{Q_{12}}{V_1} + k_1 \tag{5.15}$$

$$\lambda_{21} = \frac{Q_{12}}{V_2} \tag{5.16}$$

$$\lambda_{22} = \frac{Q_{23}}{V_2} + k_2 \tag{5.17}$$

Again, as was the case for steady-state, the equations can be solved in sequence. If $c_1 = c_{10}$ and $c_2 = c_{20}$ at $t = 0$, the general solution for Eq. 5.13 can be developed as

$$c_1 = c_{10} e^{-\lambda_{11} t} \tag{5.18}$$

This result can be substituted into Eq. 5.14, which can then be solved for

$$c_2 = c_{20} e^{-\lambda_{22} t} + \frac{\lambda_{21} c_{10}}{\lambda_{22} - \lambda_{11}} \left(e^{-\lambda_{11} t} - e^{-\lambda_{22} t}\right) \tag{5.19}$$

Note also that this result is identical to the solution for a single lake with an exponential loading (Eq. 4.18).

Now suppose that there are a third and a fourth lake in series. For these cases solutions are (O'Connor and Mueller 1970, Di Toro 1972)

$$c_3 = c_{30}e^{-\lambda_{33}t} + \frac{\lambda_{32}c_{20}}{\lambda_{33}-\lambda_{22}}\left(e^{-\lambda_{22}t} - e^{-\lambda_{33}t}\right) + \frac{\lambda_{32}\lambda_{21}c_{10}}{\lambda_{22}-\lambda_{11}}\left(\frac{e^{-\lambda_{11}t} - e^{-\lambda_{33}t}}{\lambda_{33}-\lambda_{11}} - \frac{e^{-\lambda_{22}t} - e^{-\lambda_{33}t}}{\lambda_{33}-\lambda_{22}}\right) \tag{5.20}$$

and

$$\begin{aligned} c_4 = {} & c_{40}e^{-\lambda_{44}t} + \frac{\lambda_{43}c_{30}}{\lambda_{44}-\lambda_{33}}\left(e^{-\lambda_{33}t} - e^{-\lambda_{44}t}\right) \\ & + \frac{\lambda_{43}\lambda_{32}c_{20}}{\lambda_{33}-\lambda_{22}}\left(\frac{e^{-\lambda_{22}t} - e^{-\lambda_{44}t}}{\lambda_{44}-\lambda_{22}} - \frac{e^{-\lambda_{33}t} - e^{-\lambda_{44}t}}{\lambda_{44}-\lambda_{33}}\right) \\ & + \frac{\lambda_{43}\lambda_{32}\lambda_{21}c_{10}}{(\lambda_{22}-\lambda_{11})(\lambda_{33}-\lambda_{11})}\left(\frac{e^{-\lambda_{11}t} - e^{-\lambda_{44}t}}{\lambda_{44}-\lambda_{11}} - \frac{e^{-\lambda_{33}t} - e^{-\lambda_{44}t}}{\lambda_{44}-\lambda_{33}}\right) \\ & - \frac{\lambda_{43}\lambda_{32}\lambda_{21}c_{10}}{(\lambda_{22}-\lambda_{11})(\lambda_{33}-\lambda_{22})}\left(\frac{e^{-\lambda_{22}t} - e^{-\lambda_{44}t}}{\lambda_{44}-\lambda_{22}} - \frac{e^{-\lambda_{33}t} - e^{-\lambda_{44}t}}{\lambda_{44}-\lambda_{33}}\right) \end{aligned} \tag{5.21}$$

Inspection of Eqs. 5.18 through 5.21 suggests that a pattern is emerging. O'Connor and Mueller (1970) first noted this pattern and applied it to simulate lakes in series. Di Toro (1972) developed the insight further by recognizing that the pattern could be expressed as a recurrence relationship (Box 5.1).

BOX 5.1. Efficient Schemes to Compute Concentrations for Serial Systems

Inspection of Eqs. 5.18 to 5.21 suggests a pattern that Di Toro (1972) expressed as a recurrence relation. For example for the case where only the first lake has an initial condition c_{10}, the equations can be reformulated as

$$c_1(t, \lambda_{11}) = c_{10}e^{-\lambda_{11}t} \tag{5.22}$$

$$c_2(t, \lambda_{11}, \lambda_{22}) = \frac{\lambda_{21}}{\lambda_{22}-\lambda_{11}}\left[c_1(t, \lambda_{11}) - c_1(t, \lambda_{22})\right] \tag{5.23}$$

$$c_3(t, \lambda_{11}, \lambda_{22}, \lambda_{33}) = \frac{\lambda_{32}}{\lambda_{33}-\lambda_{22}}\left[c_2(t, \lambda_{11}, \lambda_{22}) - c_2(t, \lambda_{11}, \lambda_{33})\right] \tag{5.24}$$

$$c_4(t, \lambda_{11}, \lambda_{22}, \lambda_{33}, \lambda_{44}) = \frac{\lambda_{43}}{\lambda_{44}-\lambda_{33}}\left[c_3(t, \lambda_{11}, \lambda_{22}, \lambda_{33}) - c_3(t, \lambda_{11}, \lambda_{22}, \lambda_{44})\right] \tag{5.25}$$

The sequence can be generalized further for the case of the nth reactor,

$$c_n(t, \lambda_{11}, \ldots, \lambda_{n-1,n-1}, \lambda_{n,n}) = \prod_{j=1}^{n-1}\lambda_{j+1,j}\sum_{i=1}^{n}\frac{c_1(t, \lambda_{i,i})}{\prod_{j=1(j\neq i)}^{n}(\lambda_{j,j}-\lambda_{i,i})} \tag{5.26}$$

Di Toro (1972) further showed that this form requires 2^{n-1} evaluations of the c_1 function (Eq. 5.22). Since there are only n independent values of this function, Eq. 5.26 is inefficient. He then offered a more efficient version based on a partial-fraction expansion. The general form of this factored version is

$$c_n(t, \lambda_{11}, \ldots, \lambda_{n-1,n-1}, \lambda_{n,n}) = \prod_{j=1}^{n-1} \lambda_{j+1,j} \sum_{i=1}^{n} \frac{c_1(t, \lambda_{i,i})}{\prod_{j=1(j \neq i)}^{n} (\lambda_{j,j} - \lambda_{i,i})} \tag{5.27}$$

EXAMPLE 5.3. TEMPORAL RESPONSE OF LAKES IN SERIES. During the late 1950s and early 1960s, nuclear weapons testing introduced large quantities of radioactive substances into the atmosphere. As Fig. E5.3-1 shows, this resulted in a fallout flux of these substances to the surface of the earth.

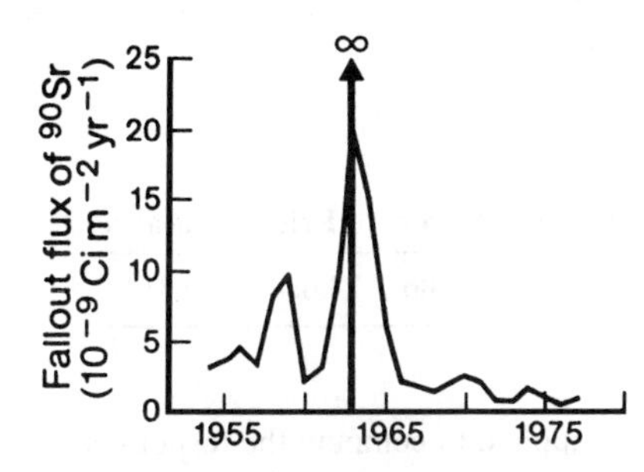

FIGURE E5.3-1
Fallout flux of ^{90}Sr to the Great Lakes (from Lerman 1972) along with the impulse load used to approximate the input.

Although the fallout has continued beyond the 1960s, the pronounced peak in 1963 allows idealization of the resulting load as an impulse function,

$$W(t) = J_{sr} A_s \delta(t - 1963)$$

where $J_{sr} = 70 \times 10^{-9}$ Ci m^{-2} (Ci denotes the radioactivity unit the curie)

A_s = lake surface area (m^2)

$\delta(t - 1963)$ = unit impulse function located at 1963

Predict the response of the Great Lakes to this flux if the half-life of ^{90}Sr is approximately 28.8 yr ($k = 0.0241$ yr^{-1}).

Solution: The Great Lakes can be represented as a series of reactors:

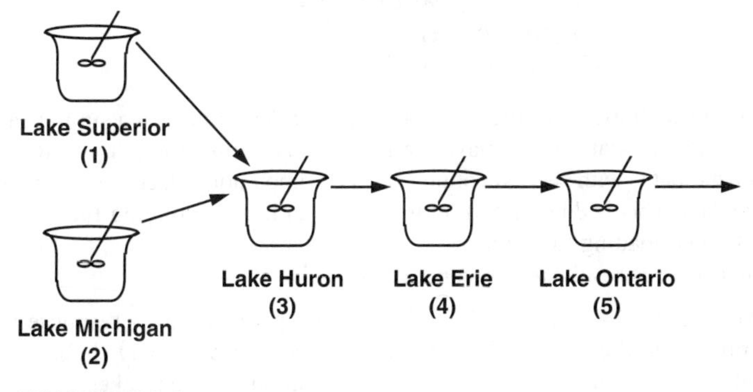

FIGURE E5.3-2

Both Lakes Superior and Michigan are "headwater" lakes. Their outflows feed into Lake Huron, which discharges to Lake Erie. Lake Ontario is the last lake in the chain. The parameters for the system are summarized as

Parameter	Units	Superior	Michigan	Huron	Erie	Ontario
Mean depth	m	146	85	59	19	86
Surface area	10^6 m^2	82,100	57,750	59,750	25,212	18,960
Volume	10^9 m^3	12,000	4,900	3,500	468	1,634
Outflow	10^9 m^3 yr^{-1}	67	36	161	182	212

The initial concentration in each lake in 1963 can be computed by

$$c_0 = \frac{J_{sr}}{H} \tag{E5.3.1}$$

where H = mean depth (m). The results are

	Units	Superior	Michigan	Huron	Erie	Ontario
c_0	10^{-9} Ci m^{-3}	0.479	0.824	1.186	3.684	0.814

Equations 5.18 through 5.21 can then be applied to compute the responses of each lake, and the total solution arrived at by summing the individual components. As in Eq. 5.18 the model for Lake Superior is

$$c_1 = 0.479e^{-0.02968t}$$

and for Lake Michigan is

$$c_2 = 0.824e^{-0.03145t}$$

Equation 5.18 is also used to predict how Lake Huron purges itself of its initial concentration. However, to compute the total response Eq. 5.19 is also employed to calculate the effect of Lakes Superior and Michigan on Huron's concentration:

$$c_3 = 1.186e^{-0.0701t} + \frac{0.01914(0.479)}{0.0701 - 0.02968}\left(e^{-0.02968t} - e^{-0.0701t}\right) + \frac{0.01029(0.824)}{0.0701 - 0.03145}\left(e^{-0.03145t} - e^{-0.0701t}\right)$$

The concentrations for the other lakes can be determined in a similar fashion. The results along with data are displayed in Fig. 5.5. The simulation duplicates the general trend of the data, with the exception that the computation decreases somewhat faster than the data. This is due, in part, to the use of an impulse forcing function to idealize the continuous loading function.

This analysis results in two conclusions:

- If two lakes receive an equal impulse flux of a pollutant, their response is inversely proportional to their depth (Eq. E5.3.1). This is the reason why shallow Lake Erie's initial concentration is about 4 times higher than for the other lakes.

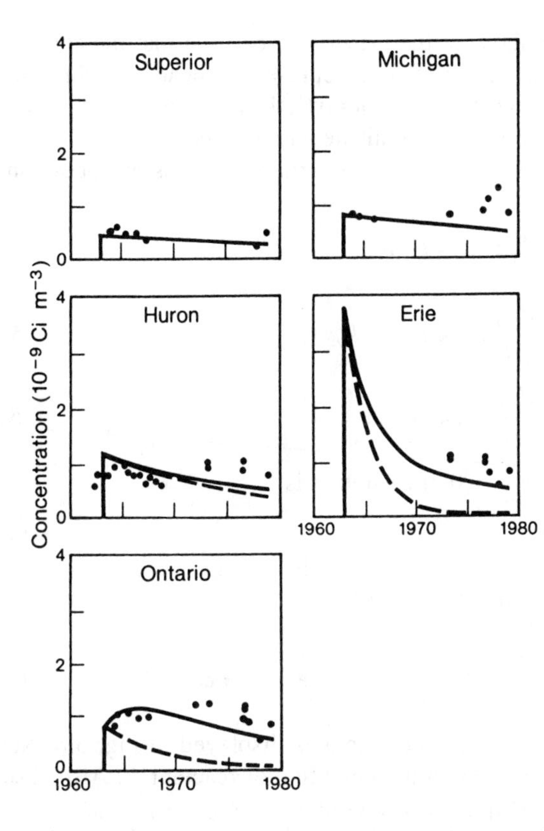

FIGURE 5.5
Response of the Great Lakes to an impulse loading of ^{90}Sr in 1963. Data (•) from Lerman (1972), Alberts and Wahlgren (1981), and International Joint Commission (1979). The dashed line represents the response of the lake to its own loading excluding the effect of upstream lakes.

- For a slowly decaying contaminant like ^{90}Sr, the upstream Great Lakes have a significant effect on the downstream lakes. The effect on Lake Ontario is so pronounced that its peak concentration does not occur in 1963 but lags 2 to 3 yr due to upstream effects.

5.3 FEEDFORWARD REACTIONS

Although we are emphasizing reactors in series, the models developed in this lecture can just as easily be applied to a series of reactions occurring within a single reactor. For example if the following series of reactions

$$A \rightarrow B \rightarrow C \rightarrow D \rightarrow \cdots \tag{5.28}$$

follows first-order kinetics, the mathematics described in the previous section are directly applicable.

Although environmental reactions do proceed as in Eq. 5.28, they do not involve an infinite chain of reactions. For example the following sequence is an interesting and relevant case:

$$A \xrightarrow{k_{ab}} B \xrightarrow{k_{bc}} C \tag{5.29}$$

This situation differs from those in the previous section in that the compound C does not undergo further reaction. In effect it represents a dead end. Such serial reactions appear in a number of water-quality contexts, the most noteworthy example being the nitrification of ammonium to form nitrite and nitrate.

If the transformations are assumed to be first-order, the mass balances can be written as

$$\frac{dc_a}{dt} = -k_{ab}c_a \tag{5.30}$$

$$\frac{dc_b}{dt} = k_{ab}c_a - k_{bc}c_b \tag{5.31}$$

$$\frac{dc_c}{dt} = k_{bc}c_b \tag{5.32}$$

If $c_a = c_{a0}$, $c_b = c_c = 0$ at $t = 0$, the solution is

$$c_a = c_{a0}e^{-k_{ab}t} \tag{5.33}$$

$$c_b = \frac{k_{ab}c_{a0}}{k_{ab} - k_{bc}}\left(e^{-k_{bc}t} - e^{-k_{ab}t}\right) \tag{5.34}$$

$$c_c = c_{a0} - c_{a0}e^{-k_{ab}t} - \frac{k_{ab}c_{a0}}{k_{ab} - k_{bc}}\left(e^{-k_{bc}t} - e^{-k_{ab}t}\right) \tag{5.35}$$

An example simulation using these equations is displayed in Fig. 5.6. Notice that the three species must sum to the initial condition of reactant A. Also observe how the solution for reactant C (Eq. 5.35) is merely $c_c = c_{a0} - c_a - c_b$.

Now an interesting question relates to the middle reactant. Since it is being gained and lost simultaneously, when will its concentration be significant? Conversely could there be cases where the reaction from B to C would be so much faster than from A to B that B would never reach significantly high levels and hence could be ignored?

The situation is depicted in the triangular or phase diagram in Fig. 5.7. Such representations are commonly used in fields like chemical engineering to map systems that manifest three states. In our case each vertex represents a pure solution

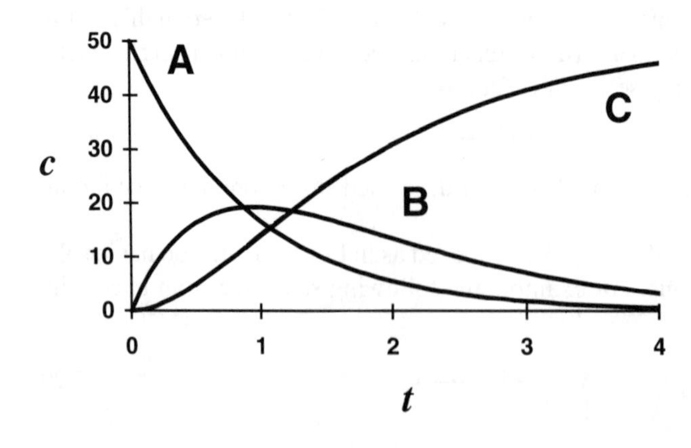

FIGURE 5.6
Concentrations of a series of reactants where the last is a " dead end."

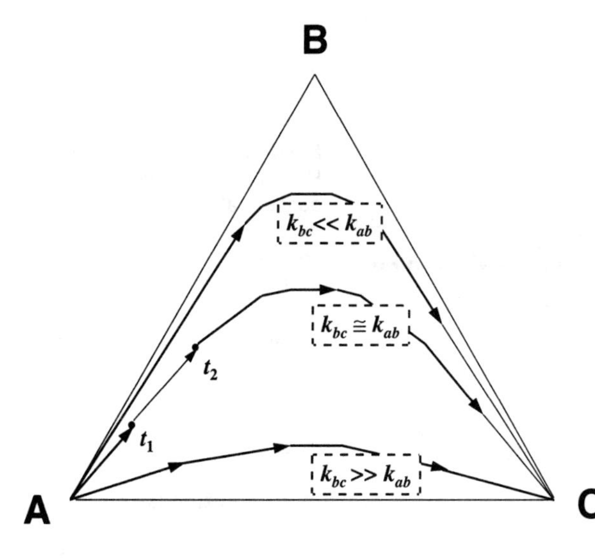

FIGURE 5.7
Concentrations of a series of three reactants mapped on a triangular space (redrawn from Fogler 1986).

of our three reactants, A, B, and C. If we set up our experiment with the initial conditions, $c_a = c_{a0}$, $c_b = c_c = 0$, the initial state would begin at vertex A. Then the reaction would move up and to the right as first B and then C are created as time progressed. After a certain time, B would pass its peak and begin to decrease. Thus the curve would move down toward the lower right vertex as the vessel approached a pure solution of C. Now the height of the curve, which indicates how much B is formed, depends on the relative values of k_{ab} and k_{bc}. If $k_{bc} \ll k_{ab}$, significant amounts of B will be formed as the second reaction acts as a bottle neck. Conversely if $k_{bc} \gg k_{ab}$, negligible amounts of B will be formed.

An alternative way to look at the same phenomena, but in the more familiar temporal context, is provided by a dimensionless analysis. In such analyses dimensionless groups are developed so that system parameters are incorporated into the variables. If done properly this generalizes the solution behavior as a function of a few dimensionless parameter groups.

For the present case this can be done by defining

$$t^* = k_{ab} t \tag{5.36}$$

$$c_a^* = \frac{c_a}{c_{a0}} \qquad c_b^* = \frac{c_b}{c_{a0}} \qquad c_c^* = \frac{c_c}{c_{a0}} \tag{5.37}$$

$$k^* = \frac{k_{bc}}{k_{ab}} \tag{5.38}$$

Equations 5.36 and 5.37 can be solved for nondimensionless time and concentrations and the results substituted into Eqs. 5.30 to 5.32 to yield

$$\frac{dc_a^*}{dt^*} = -c_a^* \tag{5.39}$$

$$\frac{dc_b^*}{dt^*} = c_a^* - k^* c_b^* \tag{5.40}$$

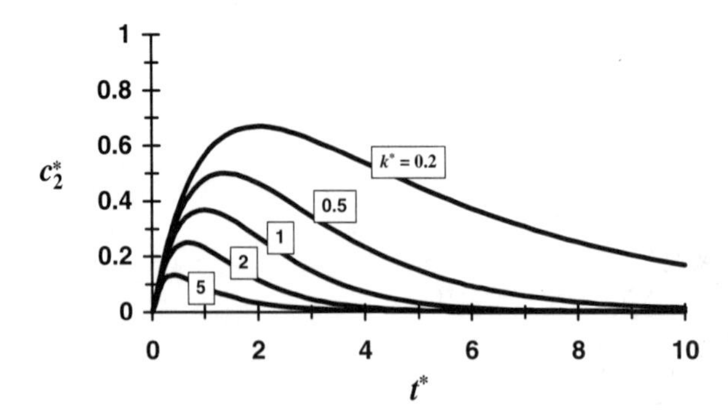

FIGURE 5.8
Plot of dimensionless concentration versus dimensionless time for the second of three reactants. The different lines represent different values of the dimensionless reaction rate.

$$\frac{dc_c^*}{dt^*} = k^* c_b^* \tag{5.41}$$

The initial conditions are now $c_a^* = 1$, $c_b^* = c_c^* = 0$, and the solution is

$$c_a^* = e^{-t^*} \tag{5.42}$$

$$c_b^* = \frac{1}{1 - k^*}\left(e^{-k^* t^*} - e^{-t^*}\right) \tag{5.43}$$

$$c_c^* = 1 - e^{-t^*} - \frac{1}{1 - k^*}\left(e^{-k^* t^*} - e^{-t^*}\right) \tag{5.44}$$

Now, aside from appearing much simpler (compare with Eqs. 5.33 to 5.35), the power of the dimensionless approach is illustrated in Fig. 5.8. This diagram, which shows c_b^* for various values of k^*, illustrates clearly that for $k^* > 5$ the concentration of the second reactant never reaches significantly high levels. In other words if the second reaction is 5 (or more) times faster than the first, the second reactant can be effectively ignored. Thus the analysis provides a way to determine when an analysis could be simplified by ignoring a state variable. That is, it provides an objective basis for deciding when a box (state variable) becomes an arrow (rate).

The conclusion of our present analysis is underscored by Fig. 5.9, which shows the dimensionless results for the cases where $k^* = 5$ and $k^* = \infty$. The latter amounts

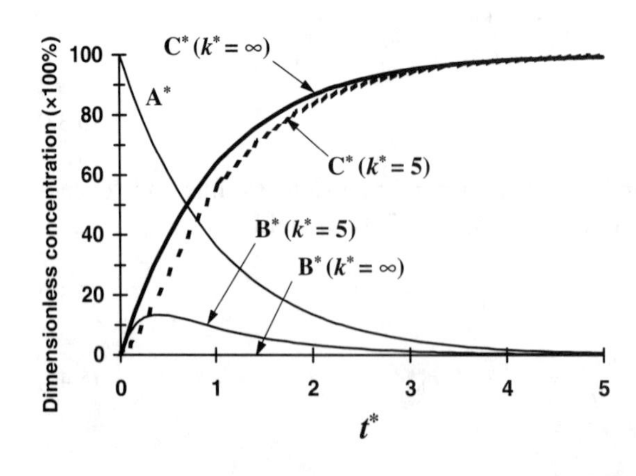

FIGURE 5.9
Dimensionless concentrations versus dimensionless time for a series of reactants where the last is a "dead end."

to omitting B from the analysis. The graph shows that results for C are quite close for the two cases, indicating that for all intents and purposes we would be justified in omitting B.

PROBLEMS

5.1. A spill of 5 kg of a soluble pesticide takes place in the first of two lakes in series. The pesticide is subject to volatilization that can be characterized by the first-order flux $J = v_v c$, where v_v = a volatilization mass-transfer coefficient of 0.01 m d^{-1}. Other parameters for the lakes are

	Lake 1	Lake 2
Surface area (m^2)	0.1×10^6	0.2×10^6
Mean depth (m)	5	3
Outflow (m^3 yr^{-1})	1×10^6	1×10^6

(*a*) Predict the concentration in both lakes as a function of time. Present your results as a plot.

(*b*) The second lake is used as a water supply reservoir. Therefore it is critical to determine the time required for the second lake to reach its maximum concentration. Derive an analytical formulation to make this determination. Verify your results by comparing it to the plot from part (*a*).

5.2. In 1970 total phosphorus loadings to the Great Lakes were (Chapra and Sonzogni 1979)

	Units	Superior	Michigan	Huron	Erie	Ontario
W	tonnes yr^{-1}	4000	6950	4575	18,150	6650

Using the parameters from Example 5.3 and assuming that total phosphorus settles at a rate of approximately 16 m yr^{-1},

(*a*) Calculate the steady-state concentration for each lake.

(*b*) Determine how much of each lake's concentration is due to upstream lakes.

5.3. Suppose that 5000 Ci of ^{137}Cs ($t_{50} \cong 30$ yr) was spilled into Lake Huron due to a nuclear accident. Calculate the response of Lakes Huron, Erie, and Ontario to the spill.

5.4. Duplicate Example 5.2, but do not limit yourself to $n \leq 8$. Plot your results on semi-logarithmic paper. As n increases, the plot should approach a straight line. Interpret the slope that results.

5.5. You are given the task of building detention ponds to remove settleable solids from a small stream prior to discharge into a lake. The ponds must be 2 m deep, the flow in the stream is 20 cfs, the solids settle at a rate of 0.2 m d^{-1}, and you must achieve a steady-state removal of 60% of the solids.

(*a*) Determine the size of a single CSTR to achieve the desired removal.

(*b*) Determine the sizes of a pair of identical CSTRs in series needed to achieve the desired removal.

(*c*) Which seems to be the best option? Why?

5.6. Develop a computer program to implement Di Toro's efficient scheme for implementing feedforward first-order systems. Test your framework by using it to duplicate the results of Example 5.3.

5.7. Everyone's heard about the five Great Lakes described in this lecture. However, not everyone knows that there is a sixth lake in the system, Lake St. Clair, that is located on the river connecting Lakes Huron and Erie. By the standards of most other lakes, it's more than adequate ($V = 6.6\ \text{km}^3$, $A_s = 1114\ \text{km}^2$, $H = 5.9\ \text{m}$, $Q = 170.5\ \text{km}^3\ \text{yr}^{-1}$). However, in the context of the Great Lakes, it's at best a "Very Good Lake."

(*a*) Determine the responses of Lakes Huron, St. Clair, and Erie to an impulse flux of strontium (as described in Example 5.3) to Lake Huron. That is, compute the response as if the fallout fell only on Huron.

(*b*) Use the dimensionless analysis described at the end of the lecture to justify omitting Lake St. Clair from the Great Lakes.

LECTURE 6

Feedback Systems of Reactors

LECTURE OVERVIEW: I develop steady-state and time-variable solutions for coupled reactors with feedback. For the steady-state case, matrices are offered as a means to concisely represent systems of coupled reactors. I introduce the matrix inverse as a means to sort out the interactions between the reactors. For the time-variable case, the general solution is derived and an eigenvalue approach used to gain insight into the system's dynamics. In addition to systems of reactors the lecture also deals with modeling coupled reactions within a single reactor.

In the previous lecture we developed models for reactors in series. Now let's add feedback flows to these systems. Although this greatly increases the range of application of such models, it also complicates their solution. In fact, for all but the simplest systems (that is, two or three reactors), it means that computers are essential.

6.1 STEADY-STATE FOR TWO REACTORS

Mass balances for two CSTRs with feedback (Fig. 6.1) can be written as

$$V_1 \frac{dc_1}{dt} = W_1 + Q_{01}c_0 - Q_{12}c_1 - k_1 V_1 c_1 + Q_{21}c_2 \tag{6.1}$$

$$V_2 \frac{dc_2}{dt} = W_2 + Q_{12}c_1 - Q_{21}c_2 - Q_{23}c_2 - k_2 V_2 c_2 \tag{6.2}$$

At steady-state the terms can be collected and expressed as

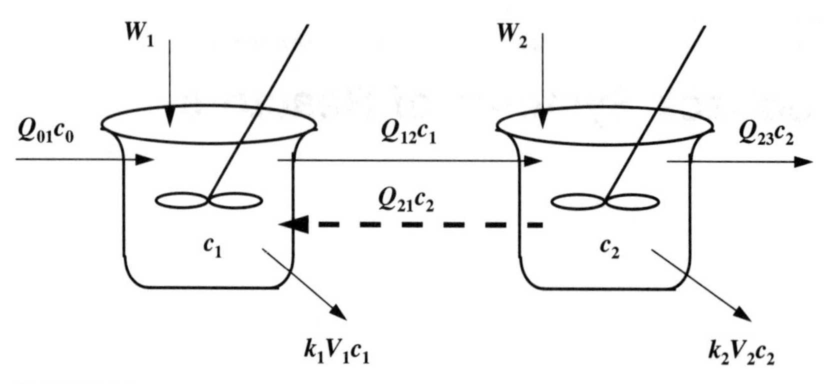

FIGURE 6.1
A system of reactors. The dashed arrow represents feedback.

$$a_{11}c_1 + a_{12}c_2 = W_1 \tag{6.3}$$

$$a_{21}c_1 + a_{22}c_2 = W_2 \tag{6.4}$$

where

$$a_{11} = Q_{12} + k_1V_1 \tag{6.5}$$

$$a_{12} = -Q_{21} \tag{6.6}$$

$$a_{21} = -Q_{12} \tag{6.7}$$

$$a_{22} = Q_{21} + Q_{23} + k_2V_2 \tag{6.8}$$

and the loading to the first reactor must be modified to include the inflow input

$$W_1 \leftarrow W_1 + Q_{01}c_0 \tag{6.9}$$

Thus we have two equations (6.3 and 6.4) with two unknowns. There are a variety of ways to obtain solutions. One convenient way, which is appropriate for small numbers of equations, is ***Cramer's rule.*** This rule states that each unknown in a system of linear algebraic equations may be expressed as a fraction of two determinants with the denominator as the system determinant. For example for a two equation system the denominator would be

$$D = \begin{vmatrix} a_{11} & a_{12} \\ a_{21} & a_{22} \end{vmatrix} = a_{11}a_{22} - a_{12}a_{21} \tag{6.10}$$

The numerator is obtained from D by replacing the column of coefficients of the unknown in question by the constants W. For example c_1 would be computed as

$$c_1 = \frac{\begin{vmatrix} W_1 & a_{12} \\ W_2 & a_{22} \end{vmatrix}}{D} = \frac{a_{22}W_1 - a_{12}W_2}{a_{11}a_{22} - a_{12}a_{21}} \tag{6.11}$$

Similarly the second unknown could be determined as

$$c_2 = \frac{\begin{vmatrix} a_{11} & W_1 \\ a_{21} & W_2 \end{vmatrix}}{D} = \frac{a_{11}W_2 - a_{21}W_1}{a_{11}a_{22} - a_{12}a_{21}} \tag{6.12}$$

Algebraic manipulation can also be employed to express these results as

$$c_1 = \frac{1}{a_{11} - (a_{21}a_{12}/a_{22})} W_1 + \frac{1}{a_{21} - (a_{11}a_{22}/a_{12})} W_2 \tag{6.13}$$

$$c_2 = \frac{1}{a_{12} - (a_{11}a_{22}/a_{21})} W_1 + \frac{1}{a_{22} - (a_{21}a_{12}/a_{11})} W_2 \tag{6.14}$$

Thus we see that the concentrations are expressed in the format of Eq. 1.8. However, these solutions are considerably more complicated than Eq. 1.8. For Eqs. 6.13 and 6.14 the assimilation factors are now composites of a number of parameter groups. But due to the linearity of the original model (Eqs. 6.3 and 6.4), these results can provide great insight into how the system operates. In particular notice how each equation consists of two independent parts—one that is solely dependent on the first reactor's loading, W_1, and the other dependent on the loading to the second, W_2. This fact, which is a product of the model's linearity, allows us to separately evaluate the impact of the individual loadings.

Note that for more than three equations Cramer's rule becomes impractical because, as the number of equations increases, the determinants become too time-consuming to evaluate by hand (or by computers if they're calculated by expanding minors!). Consequently, as described next, more efficient methods are used.

6.2 SOLVING LARGE SYSTEMS OF REACTORS

The extension of the mass balance to more than two reactors is straightforward. For example for three coupled reactors with feedback, a set of three equations with three unknowns would be generated,

$$a_{11}c_1 + a_{12}c_2 + a_{13}c_3 = W_1 \tag{6.15}$$

$$a_{21}c_1 + a_{22}c_2 + a_{23}c_3 = W_2 \tag{6.16}$$

$$a_{31}c_1 + a_{32}c_2 + a_{33}c_3 = W_3 \tag{6.17}$$

where a's = constants that reflect the system parameters (that is, Q, V, k, etc.)
W's = constant loadings
c's = unknown concentrations

Note that the unknowns are all raised to the first power and are multiplied by a constant. Equations with these characteristics are called ***linear algebraic equations.***

In the previous section we used Cramer's rule to solve two equations with two unknowns. For more than two equations, this technique becomes computationally burdensome. Therefore in such cases, computers and numerical methods must be used. Several numerical methods are available for this purpose (see App. E and Chapra and Canale 1988 for details).

Aside from the mechanics of obtaining solutions, the representation and manipulation of large numbers of equations is difficult using simple algebra. For this reason the next section is devoted to matrix algebra.

6.2.1 Matrix Algebra

A ***matrix*** consists of a rectangular array of elements represented by a single symbol. For example

$$[A] = \begin{bmatrix} a_{11} & a_{12} & a_{13} \\ a_{21} & a_{22} & a_{23} \\ a_{31} & a_{32} & a_{33} \end{bmatrix} \tag{6.18}$$

where $[A]$ is the shorthand notation for the entire matrix and a_{ij} designates an individual ***element*** of the matrix.

A horizontal set of elements is called a row and a vertical set is called a ***column.***† The first subscript i always designates the number of the row in which the element lies. The second subscript j designates the column. For example element a_{23} is in row two and column three.

A matrix can be characterized by its dimensions. It has n rows and m columns and is said to have a ***dimension*** of n by m (or $n \times m$). Such a matrix is commonly referred to as an ***n by m matrix***.

Matrices with column dimension $m = 1$ such as

$$\{B\} = \begin{Bmatrix} b_1 \\ b_2 \\ b_3 \end{Bmatrix} \tag{6.19}$$

are called ***column vectors.*** Note that for simplicity the second subscript of each element is dropped. Also it should be mentioned that it is often desirable to distinguish a column vector from other types of matrices. Consequently we use special brackets, { }, to enclose the values.

Matrices such as the one in Eq. 6.18, where $n = m$, are called ***square matrices.*** Square matrices are particularly important when solving sets of simultaneous linear equations. For such systems the number of equations (corresponding to the number of rows) and the number of unknowns (corresponding to the number of columns) must be equal for a unique solution to be possible. Consequently square matrices of coefficients are encountered when dealing with such systems.

There are a number of special types of square matrices. In the present context the most important is the ***identity matrix.*** For this case all elements are zero except for the diagonal that consists of ones. For the three by three case,

$$[I] = \begin{bmatrix} 1 & 0 & 0 \\ 0 & 1 & 0 \\ 0 & 0 & 1 \end{bmatrix} \tag{6.20}$$

This matrix has properties similar to unity. Just as $a \times 1 = a$ in simple algebra, so also does

$$[A] \times [I] = [A] \tag{6.21}$$

---

†A simple mnemonic is provided by the horizontal "rows" of a theater and the vertical "columns" of a temple.

Matrix algebra follows rules in the same fashion as algebraic manipulations of simple variables. In the present context the two most important involve matrix multiplication and matrix inversion.

Matrix multiplication. The ***product*** of two matrices is represented as $[C] = [A][B]$, where the elements of $[C]$ are defined as

$$c_{ij} = \sum_{k=1}^{n} a_{ik} b_{kj} \tag{6.22}$$

where n = the column dimension of $[A]$ and the row dimension of $[B]$. That is, the c_{ij} element is obtained by adding the product of individual elements from the ith row of the first matrix, in this case $[A]$, and the jth column of the second matrix $[B]$. Box 6.1 illustrates a simple way to visualize matrix multiplication.

BOX 6.1. A Simple Method for Multiplying Two Matrices

Although Eq. 6.22 is well-suited for implementation on a computer, it is not the simplest means for visualizing the mechanics of multiplying two matrices. What follows gives more tangible expression to the operation.

Suppose that we want to multiply $[X]$ by $[Y]$ to yield $[Z]$:

$$[Z] = [X][Y] = \begin{bmatrix} 3 & 1 \\ 8 & 6 \\ 0 & 4 \end{bmatrix} \begin{bmatrix} 5 & 9 \\ 7 & 2 \end{bmatrix}$$

A simple way to visualize the computation of $[Z]$ is to raise $[Y]$, as in

$$\begin{array}{ccc} & \Uparrow & \\ & \begin{bmatrix} 5 & 9 \\ 7 & 2 \end{bmatrix} & \leftarrow [Y] \\ [X] \rightarrow \begin{bmatrix} 3 & 1 \\ 8 & 6 \\ 0 & 4 \end{bmatrix} & \begin{bmatrix} & & \\ & ? & \\ & & \end{bmatrix} & \leftarrow [Z] \end{array}$$

Now the answer $[Z]$ can be computed in the space vacated by $[Y]$. This format has utility because it aligns the appropriate rows and columns that are to be multiplied. For example according to Eq. 6.22 the element z_{11} is obtained by multiplying the first row of $[X]$ by the first column of $[Y]$. This amounts to adding the product of x_{11} and y_{11} to the product of x_{12} and y_{21}, as in

$$\begin{array}{cc} & \begin{bmatrix} \mathbf{5} & 9 \\ \mathbf{7} & 2 \end{bmatrix} \\ & \downarrow \\ \begin{bmatrix} \mathbf{3} & \mathbf{1} \\ 8 & 6 \\ 0 & 4 \end{bmatrix} \rightarrow & \begin{bmatrix} \mathbf{3 \times 5 + 1 \times 7 = 22} \\ \\ \\ \end{bmatrix} \end{array}$$

Thus z_{11} is equal to 22. Element z_{21} can be computed in a similar fashion, as in

$$\begin{bmatrix} \mathbf{5} & 9 \\ \mathbf{7} & 2 \end{bmatrix}$$

$$\downarrow$$

$$\begin{bmatrix} 3 & 1 \\ \mathbf{8} & \mathbf{6} \\ 0 & 4 \end{bmatrix} \rightarrow \begin{bmatrix} 22 \\ \mathbf{8 \times 5 + 6 \times 7 = 82} \\ \end{bmatrix}$$

The computation can be continued in this way, following the alignment of the rows and columns, to yield the result

$$[Z] = \begin{bmatrix} 22 & 29 \\ 82 & 84 \\ 28 & 8 \end{bmatrix}$$

Note how this simple method makes it clear why it is impossible to multiply two matrices if the number of columns of the first matrix does not equal the number of rows in the second matrix. Also note how it demonstrates that the order of multiplication matters (that is, matrix multiplication is not commutative).

It should now be clear that matrices provide a succinct notation for representing simultaneous linear algebraic equations. For example Eqs. 6.15 to 6.17 can be expressed concisely as

$$[A]\{C\} = \{W\} \tag{6.23}$$

where the matrix $[A]$ contains the coefficients

$$[A] = \begin{bmatrix} a_{11} & a_{12} & a_{13} \\ a_{21} & a_{22} & a_{23} \\ a_{31} & a_{32} & a_{33} \end{bmatrix} \tag{6.24}$$

the vector $\{C\}$ contains the unknowns

$$\{C\} = \begin{Bmatrix} c_1 \\ c_2 \\ c_3 \end{Bmatrix} \tag{6.25}$$

and the vector $\{W\}$ contains the right-hand-side constants or forcing functions,

$$\{W\} = \begin{Bmatrix} W_1 \\ W_2 \\ W_3 \end{Bmatrix} \tag{6.26}$$

At this point you should apply the rule for matrix multiplication to convince yourself that Eqs. 6.23 and 6.15 to 6.17 are equivalent.

Matrix inversion. Although multiplication is possible, matrix division is not a defined operation. However, if a matrix $[A]$ is square, there is usually another matrix $[A]^{-1}$ called the ***inverse*** of $[A]$ for which

$$[A][A]^{-1} = [A]^{-1}[A] = [I] \tag{6.27}$$

Thus the multiplication of a matrix by its inverse is analogous to division, in the sense that a number divided by itself is equal to 1. That is, multiplication of a matrix by its inverse leads to the identity matrix.

The inverse of a two-dimensional square matrix can be calculated simply, as in

$$[A]^{-1} = \frac{1}{a_{11}a_{22} - a_{12}a_{21}} \begin{bmatrix} a_{22} & -a_{12} \\ -a_{21} & a_{11} \end{bmatrix} \tag{6.28}$$

Similar formulas for higher dimensional matrices are much more complicated. Consequently computer algorithms are usually used. One simple method involves standard computer solution algorithms such as Gauss elimination. For such cases each column j of the matrix inverse can be determined by using a unit vector (with a 1 in the jth row and 0 elsewhere) as the forcing-function vector (that is, the right-hand-side constants). Additional details can be found in App. E.

Once the inverse is obtained, a formal way to obtain a solution is to multiply each side of Eq. 6.23 by the inverse of $[A]$,

$$[A][A]^{-1}\{C\} = [A]^{-1}\{W\} \tag{6.29}$$

Because $[A]^{-1}[A]$ equals the identity matrix, the equation becomes

$$\{C\} = [A]^{-1}\{W\} \tag{6.30}$$

Therefore if we multiply the inverse of the coefficient matrix $[A]^{-1}$ by the matrix of constants $\{W\}$, we obtain the solution for the unknowns $\{C\}$. This is another example of how the inverse plays a role in matrix algebra similar to division. That is, just as $c = (1/a)W$ represents the steady-state solution for a single CSTR, Eq. 6.30 represents the solution for a system of CSTRs.

It should be noted that the matrix inverse is not a very efficient means to solve a system of equations. Thus other approaches, such as the elimination and iterative methods described in App. E, are used in numerical algorithms. However, as described next, the matrix inverse has great value in the engineering analysis of such systems.

6.3 STEADY-STATE SYSTEM RESPONSE MATRIX

Now that we have learned a little matrix algebra, we can explore its implications for steady-state solutions of coupled reactors with first-order kinetics. As described in the previous section, the matrix inverse can be used to obtain the solution for the steady-state case. In addition the terms in Eq. 6.30 have a definite physical interpretation. For example the elements of $\{C\}$ are the response of the system as reflected by the concentrations of the reactors. The right-hand-side vector $\{W\}$ contains the values of the system's stimuli or forcing functions—the loadings. Finally the matrix inverse $[A]^{-1}$ contains the parameters that express how the parts of the system are coupled and purge themselves of the pollutant. Consequently Eq. 6.30 might be reexpressed as

$$\{\text{Response}\} = [\text{interactions}]\{\text{stimuli}\} \tag{6.31}$$

In addition notice that Eq. 6.30 is the multidimensional manifestation of Eq. 1.8. This can be seen by applying the definition of matrix multiplication to Eq. 6.30. This gives

$$c_1 = a_{11}^{(-1)} W_1 + a_{12}^{(-1)} W_2 + a_{13}^{(-1)} W_3 \tag{6.32}$$

$$c_2 = a_{21}^{(-1)} W_1 + a_{22}^{(-1)} W_2 + a_{23}^{(-1)} W_3 \tag{6.33}$$

$$c_3 = a_{31}^{(-1)} W_1 + a_{32}^{(-1)} W_2 + a_{33}^{(-1)} W_3 \tag{6.34}$$

where $a_{ij}^{(-1)}$ denotes the element in the ith row and the jth column of the matrix inverse. Thus we find that the inverted matrix itself, aside from providing a solution, has extremely useful properties. That is, each of its elements represents the response of a single part of the system to a unit stimulus of any other part of the system.

EXAMPLE 6.1. STEADY-STATE SOLUTION FOR LAKES WITH FEEDBACK. Recall that in Example 5.1 we computed the steady-state distribution of a pollutant in three lakes connected in series. The same system is shown below, with the exception that we have recycled a fraction of the flow (α) from the third lake back to the first lake.

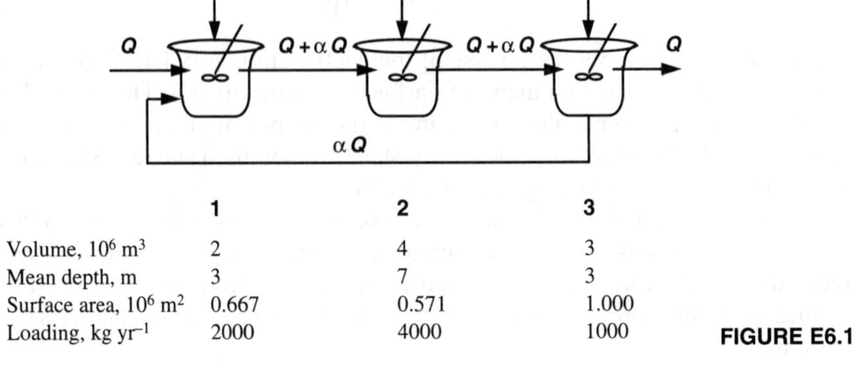

	1	2	3
Volume, 10^6 m^3	2	4	3
Mean depth, m	3	7	3
Surface area, 10^6 m^2	0.667	0.571	1.000
Loading, kg yr^{-1}	2000	4000	1000

FIGURE E6.1

(*a*) If $Q = 1 \times 10^6$ m^3 yr^{-1}, $\alpha = 0.5$, and the pollutant settles at a rate of 10 m yr^{-1}, calculate the concentration in each of the reactors.

(*b*) Use the matrix inverse to determine how much of the concentration in the third reactor is due to the loading to the second reactor.

(*c*) Determine the matrix inverse for the case where $\alpha = 0$.

Solution: (*a*) The steady-state mass balances for the three reactors can be written as

$$0 = W_1 - (Q + \alpha Q)c_1 - vA_1c_1 + \alpha Qc_3$$

$$0 = W_2 + (Q + \alpha Q)c_1 - (Q + \alpha Q)c_2 - vA_2c_2$$

$$0 = W_3 + (Q + \alpha Q)c_2 - (Q + \alpha Q)c_3 - vA_3c_3$$

Substituting the parameter values, the three simultaneous equations can be expressed in matrix form as

$$\begin{bmatrix} 8.17 \times 10^6 & 0 & -0.5 \times 10^6 \\ -1.5 \times 10^6 & 7.21 \times 10^6 & 0 \\ 0 & -1.5 \times 10^6 & 11.5 \times 10^6 \end{bmatrix} \begin{Bmatrix} c_1 \\ c_2 \\ c_3 \end{Bmatrix} = \begin{Bmatrix} 2 \times 10^9 \\ 4 \times 10^9 \\ 1 \times 10^9 \end{Bmatrix}$$

The matrix inverse can be determined as

$$\begin{bmatrix} 1.23 \times 10^{-7} & 1.11 \times 10^{-9} & 5.33 \times 10^{-9} \\ 2.55 \times 10^{-8} & 1.39 \times 10^{-7} & 1.11 \times 10^{-9} \\ 3.33 \times 10^{-9} & 1.81 \times 10^{-8} & 8.71 \times 10^{-8} \end{bmatrix}$$

which can be multiplied by $\{W\}$ to give

$$\begin{Bmatrix} c_1 \\ c_2 \\ c_3 \end{Bmatrix} = \begin{Bmatrix} 255 \\ 608 \\ 166 \end{Bmatrix}$$

(*b*) The effect of a load to segment two on the concentration of segment three can be based on $a_{32}^{(-1)} = 1.81 \times 10^{-8}\ \mu\text{g L}^{-1}/\text{mg yr}^{-1}$. The effect of the 4×10^9 mg yr^{-1} load to segment two on the concentration of segment three can be calculated as

$$c_3 \text{ (due to loadings to reactor 2)} = 1.81 \times 10^{-8}(4 \times 10^9) = 72.5\ \mu\text{g L}^{-1}$$

(*c*) For the case where $\alpha = 0$, the three simultaneous equations can be expressed in matrix form as

$$\begin{bmatrix} 7.67 \times 10^6 & 0 & 0 \\ -1 \times 10^6 & 6.71 \times 10^6 & 0 \\ 0 & -1 \times 10^6 & 11.0 \times 10^6 \end{bmatrix} \begin{Bmatrix} c_1 \\ c_2 \\ c_3 \end{Bmatrix} = \begin{Bmatrix} 2 \times 10^9 \\ 4 \times 10^9 \\ 1 \times 10^9 \end{Bmatrix}$$

The matrix inverse can be determined as

$$\begin{bmatrix} 1.30 \times 10^{-7} & 0 & 0 \\ 1.94 \times 10^{-8} & 1.49 \times 10^{-7} & 0 \\ 1.77 \times 10^{-9} & 1.35 \times 10^{-8} & 9.09 \times 10^{-8} \end{bmatrix}$$

Notice how the superdiagonal terms go to zero when feedback is omitted.

Before proceeding to time-variable solutions, we should comment a bit more on the structure of the matrix inverse. As depicted in Fig. 6.2, regions of the matrix

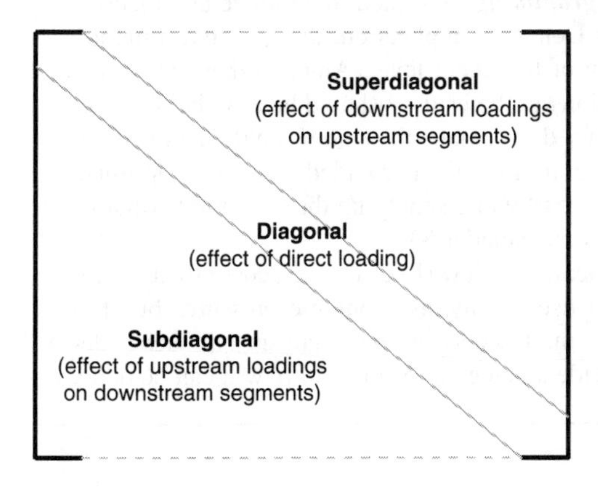

FIGURE 6.2
The elements of the matrix inverse $[A]^{-1}$ have a specific physical interpretation.

inverse $[A]^{-1}$ have a specific physical interpretation. Diagonal terms specify the response of the segments to direct loadings. The superdiagonal terms reflect the effect of downstream segments on upstream segments. The subdiagonal terms reflect the effect of upstream segments on downstream segments. Thus, as in part (*c*) of the previous example, a feedforward series system would have zero superdiagonal terms.

BOX 6.2. Input-Output Modeling and the Delaware Estuary Study

The steady-state system response matrix described in this lecture is an example of ***input-output modeling***. This approach was originally developed by W. W. Leontiff, the winner of the 1973 Nobel Prize for Economics. Leontiff derived linear systems of equations very similar to the ones in this lecture. However, rather than linking pollutant loadings with concentrations, he devised linear models that linked economic inputs (e.g., production of goods by sectors of the economy) with outputs (the consumption of the goods by other sectors).

The Delaware Estuary was the site of the first application of this approach in the water-quality domain. Located near Philadelphia, the Delaware Estuary is a critical water resource in the heavily populated Eastern United States. In the 1960s it was the site of the Delaware Estuary Comprehensive Study (FWPCA 1966), one of the first applications of computer-oriented water-quality modeling and systems analysis (Fig. 1.6). Despite having occurred over 30 yr ago, it remains one of the most comprehensive and innovative studies of its type (see Thomann 1972 for a nice summary).

Among the study's many contributions, the most relevant to water-quality modeling was the development of the control-volume approach described in this lecture. This framework (Thomann 1963) represented the doctoral research of the study's Technical Director, Bob Thomann, now a professor at Manhattan College. Along with the model, Dr. Thomann developed his own expression of the input-output approach in the form of the steady-state system response matrix.

In the present lectures I have emphasized how the matrix provides a concise way to link loadings with response. It should be noted that Dr. Thomann and his team (including Matt Sobal and Dave Marks, now professors at SUNY Stony Brook and at M.I.T., respectively) extended it well beyond this application. In particular they used the computer to develop treatment strategies that produced acceptable water quality at a minimum cost (Thomann and Sobal 1964). Specifically they employed a systems analysis approach known as ***linear programming***, designed to optimize an objective function subject to constraints. For the Delaware application, the objective function represented treatment cost as a function of treatment level. Among other factors the constraints reflected that the concentrations in the estuary should be at or below standards. The system response matrix was used to connect the loadings with the resulting concentrations in the estuary. Linear programming then provided a means to optimize the objective function (that is, minimize cost) while satisfying the constraints (that is, making sure the resulting concentrations met standards).

As already mentioned in our historical overview (Lec. 1), such cost-effective approaches have never found widespread use. Today, as economic pressures build, a revival may be at hand. The cause-effect models and systems analysis approaches developed 30 yr ago on the Delaware provide an elegant example of how it can be done.

6.4 TIME-VARIABLE RESPONSE FOR TWO REACTORS

Equations 6.1 and 6.2 (with no loads) can be written as

$$\frac{dc_1}{dt} = -\alpha_{11}c_1 + \alpha_{12}c_2 \tag{6.35}$$

$$\frac{dc_2}{dt} = \alpha_{21}c_1 - \alpha_{22}c_2 \tag{6.36}$$

where

$$\alpha_{11} = \frac{Q_{12}}{V_1} + k_1 \tag{6.37}$$

$$\alpha_{12} = \frac{Q_{21}}{V_1} \tag{6.38}$$

$$\alpha_{21} = \frac{Q_{12}}{V_2} \tag{6.39}$$

$$\alpha_{22} = \frac{Q_{23} + Q_{12}}{V_2} + k_2 \tag{6.40}$$

If $c_1 = c_{10}$ and $c_2 = c_{20}$, then the general solution can be developed as

$$c_1 = c_{1f}e^{-\lambda_f t} + c_{1s}e^{-\lambda_s t} \tag{6.41}$$

$$c_2 = c_{2f}e^{-\lambda_f t} + c_{2s}e^{-\lambda_s t} \tag{6.42}$$

where the λ's are eigenvalues that are defined as

$$\begin{matrix}\lambda_f \\ \lambda_s\end{matrix} = \frac{(\alpha_{11} + \alpha_{22}) \pm \sqrt{(\alpha_{11} + \alpha_{22})^2 - 4(\alpha_{11}\alpha_{22} - \alpha_{12}\alpha_{21})}}{2} \tag{6.43}$$

and the coefficients are

$$c_{1f} = \frac{(\lambda_f - \alpha_{22})c_{10} - \alpha_{12}c_{20}}{\lambda_f - \lambda_s} \tag{6.44}$$

$$c_{1s} = \frac{\alpha_{12}c_{20} - (\lambda_s - \alpha_{22})c_{10}}{\lambda_f - \lambda_s} \tag{6.45}$$

$$c_{2f} = \frac{-\alpha_{21}c_{10} + (\lambda_f - \alpha_{11})c_{20}}{\lambda_f - \lambda_s} \tag{6.46}$$

$$c_{2s} = \frac{-(\lambda_s - \alpha_{11})c_{20} + \alpha_{21}c_{10}}{\lambda_f - \lambda_s} \tag{6.47}$$

As we saw for the steady-state solution, the time-variable case is much more complicated than for a single completely mixed lake or for reactors in series. Aside from the complexity of the coefficients, the major difference is that the recovery of each segment now depends on two exponential decays.

The total recovery depends on the relative magnitudes of the eigenvalues. Note that λ_f will always be greater than λ_s. Consequently λ_f and λ_s are often referred to as the "fast" and the "slow" eigenvalues, respectively. This nomenclature derives from the rate at which they approach zero as time progresses. As in the following example, there are cases where the fast eigenvalue is much larger than the slow eigenvalue.

EXAMPLE 6.2. LAKES WITH FEEDBACK (TIME-VARIABLE). The following two lakes include feedback of a fraction of the flow (α) from the second lake back to the first lake:

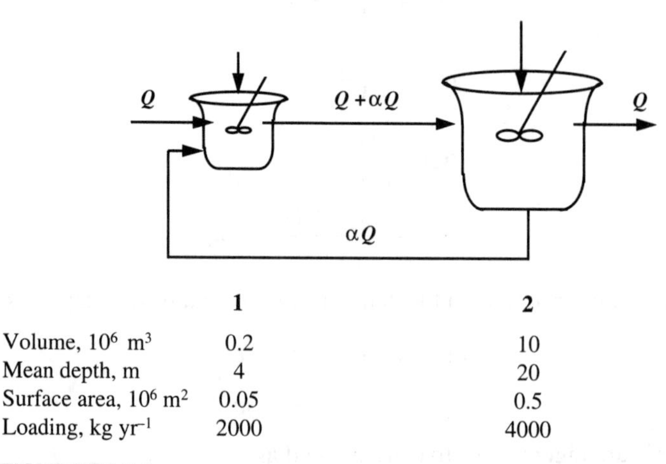

	1	2
Volume, 10^6 m^3	0.2	10
Mean depth, m	4	20
Surface area, 10^6 m^2	0.05	0.5
Loading, kg yr^{-1}	2000	4000

FIGURE E6.2-1

Notice that for this case the second lake is much larger than the first.
(*a*) If $Q = 1 \times 10^6$ m^3 yr^{-1}, $\alpha = 0.5$, and the pollutant settles at a rate of 10 m yr^{-1}, calculate the steady-state concentration in each of the reactors.
(*b*) Using the concentrations calculated in (*a*) as initial conditions, determine the response of each lake if their loadings are terminated at $t = 0$.

Solution: (*a*) Using the same approach as in Example 6.1, the steady-state concentrations for the two reactors can be calculated as

$$\{c\} = \begin{Bmatrix} 1224.5 \\ 898 \end{Bmatrix}$$

(*b*) The temporal response can be determined by substituting the parameter values into Eqs. 6.41 and 6.42 to yield

$$c_1 = 981.24e^{-10.04t} + 243.25e^{-0.61t}$$

$$c_2 = -15.67e^{-10.04t} + 913.63e^{-0.61t}$$

The results are plotted in Fig. E6.2-2. As depicted, the concentration of the first reactor initially drops precipitously because of the dominance of its fast eigenvalue. After this initial drop its recovery slows down as its response becomes tied to the more sluggish recovery of the second reactor.

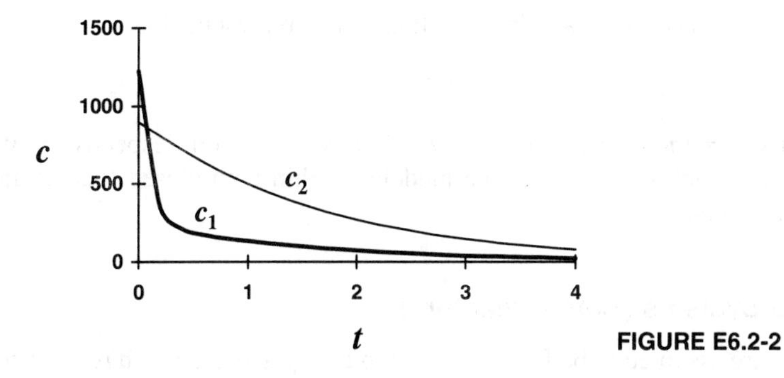

FIGURE E6.2-2

Although the foregoing analysis can be extended to many coupled reactors, most analyses of such systems are implemented numerically. We will review such techniques in the next lecture. Despite this fact the analytical, two-reactor case has general utility in water-quality analysis. This is because several important problem contexts can be modeled as two coupled reactors (Fig. 6.3). For example when we model toxic substances, a lake and its underlying sediments can be characterized in this way (Fig. 6.3*c*). In such cases the sediments typically respond much more slowly than the water. When loads are reduced, the lake's recovery will be retarded as sediment toxics seep back into the water. The eigenvalue approach outlined above can be useful in analyzing such cases.

6.5 REACTIONS WITH FEEDBACK

Aside from coupled reactors, the models developed in this lecture can be applied to reversible reactions occurring within a single reactor. For example, as previously

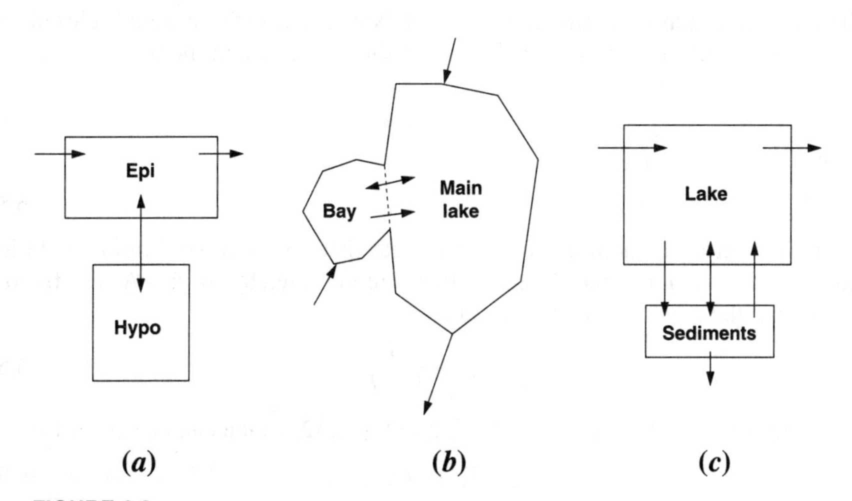

FIGURE 6.3
Three systems that can be characterized as a pair of coupled reactors. (*a*) The epilimnion (surface) and hypolimnion (bottom) of a stratified lake; (*b*) a lake with a major embayment; (*c*) a lake underlain by a sediment layer.

developed in Lec. 2, a simple reversible reaction can be represented as

$$\mathrm{A} \underset{k_{ba}}{\overset{k_{ab}}{\rightleftharpoons}} \mathrm{B} \tag{6.48}$$

where k_{ab} and k_{ba} = the forward and backward rates of reaction, respectively. We now examine how such reactions can be modeled within both closed (batch) and open (CSTR) reactors.

6.5.1 Closed Systems (Batch Reactors)

If the reactions are assumed to be first-order and to take place in a batch reactor, the following balances can be written:

$$\frac{dc_a}{dt} = -k_{ab}c_a + k_{ba}c_b \tag{6.49}$$

$$\frac{dc_b}{dt} = k_{ab}c_a - k_{ba}c_b \tag{6.50}$$

The steady-state solution of these equations can be obtained by setting the derivatives to zero and solving either equation for

$$\frac{c_b}{c_a} = \frac{k_{ab}}{k_{ba}} = K \tag{6.51}$$

where K = an equilibrium constant. Thus we arrive at the familiar result that, at equilibrium, the ratio of the products to the reactants of a chemical reaction will be constant.

To determine the magnitudes of c_a and c_b, we can define

$$c = c_a + c_b \tag{6.52}$$

where c = total mass of substances A and B. Now because the system is closed, this quantity is a constant. Thus Eq. 6.51 can be solved for c_b and the result substituted into Eq. 6.52 to give

$$c = c_a + Kc_a \tag{6.53}$$

which can be solved for

$$\bar{c}_a = F_a c \tag{6.54}$$

where the overbar is included to make it clear that this is an equilibrium solution, and F_a = the fraction of the total mass that is represented by species A. The fraction is related to the equilibrium coefficient by

$$F_a = \frac{1}{1 + K} \tag{6.55}$$

This result can then be substituted back into Eq. 6.52, which can be solved for

$$\bar{c}_b = F_b c \tag{6.56}$$

where F_b = the fraction of the total mass that is represented by species B,

$$F_b = 1 - F_a = \frac{K}{1 + K} \tag{6.57}$$

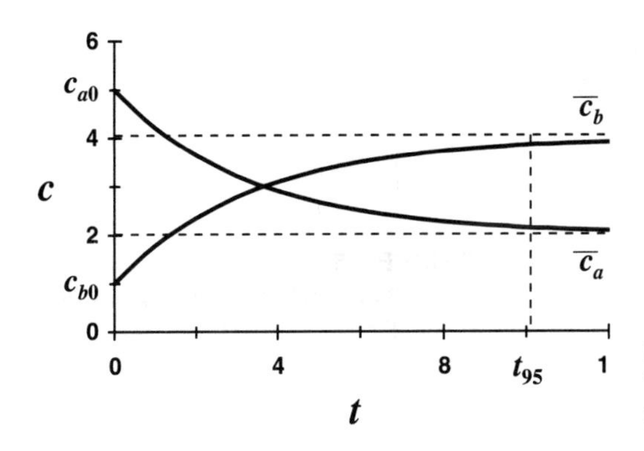

FIGURE 6.4
The progress of a reversible reaction within a batch reactor.

Thus, because we are at equilibrium, the concentrations of A and B are at fixed fractions of the total.

Next let us examine the time-variable solution. Although these equations can be solved using the methods described earlier in this lecture, an alternative approach involves solving Eq. 6.52 for c_b and substituting the result into Eq. 6.49 to give

$$\frac{dc_a}{dt} = -k_{ab}c_a + k_{ba}(c - c_a) \tag{6.58}$$

or collecting terms,

$$\frac{dc_a}{dt} + (k_{ab} + k_{ba})c_a = k_{ba}c \tag{6.59}$$

Thus the equation is a first-order ODE with a constant forcing function (recall Sec. 4.2). If $c = c_{a0}$ at $t = 0$, it can be solved for

$$c_a = c_{a0}e^{-(k_{ab}+k_{ba})t} + \overline{c}_a(1 - e^{-(k_{ab}+k_{ba})t}) \tag{6.60}$$

where $\overline{c}_a$ = the ultimate steady-state concentration of c_a, defined by Eq. 6.54.

The concentration of c_b can be determined by substituting Eq. 6.60 into 6.52. After some manipulation the result is

$$c_b = c_{b0}e^{-(k_{ab}+k_{ba})t} + \overline{c}_b(1 - e^{-(k_{ab}+k_{ba})t}) \tag{6.61}$$

where c_{b0} = the initial condition for B and $\overline{c}_b$ = the ultimate steady-state concentration defined by Eq. 6.56.

As depicted in Fig. 6.4, the solution asymptotically approaches the steady-state condition. The eigenvalue $k_{ab} + k_{ba}$ can be used to estimate that the reaction will have a t_{95} of $3/(k_{ab} + k_{ba})$.

The foregoing analysis has significance beyond merely obtaining solutions. In particular the eigenvalue provides insight into how the solutions would be obtained mathematically. If we are interested in time scales less than t_{95}, the differential equations would be solved directly to yield concentrations as a function of time. This is sometimes called a ***kinetic approach***, because we are interested in the time-variable or dynamic variations of the constituents. However, if we are not concerned with shorter time scales, the steady-state solutions (Eqs. 6.55 and 6.56) would be perfectly adequate. This is sometimes called an ***equilibrium approach***, because we are interested in the equilibrium states of the reactants rather than in how they got there.

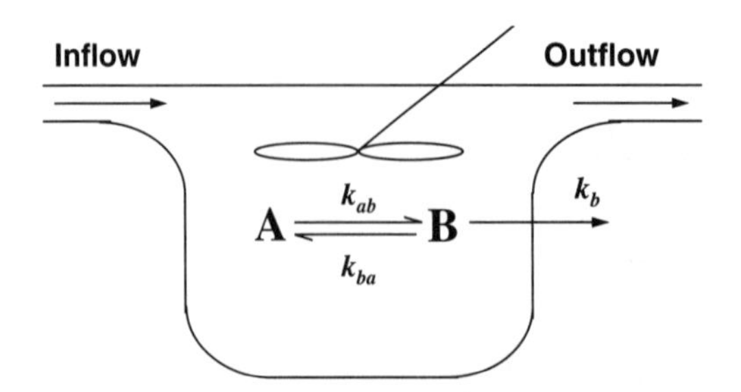

FIGURE 6.5
A mass balance for a coupled reaction taking place in a CSTR.

6.5.2 Open Systems (CSTRs)

Now let's explore how the foregoing analysis pertains to an open system. As in Fig. 6.5 suppose that the reversible reaction occurs in a CSTR. Both A and B are carried in and out of the reactor by flow. In addition assume that B is also subject to an additional reaction that removes it from the system. Mass balances can be written as

$$\frac{dc_a}{dt} = \frac{Q}{V}c_{a,\text{in}} - \frac{Q}{V}c_a - k_{ab}c_a + k_{ba}c_b \tag{6.62}$$

$$\frac{dc_b}{dt} = \frac{Q}{V}c_{b,\text{in}} - \frac{Q}{V}c_b + k_{ab}c_a - k_{ba}c_b - k_bc_b \tag{6.63}$$

These equations can be solved using the approaches described in earlier parts of this lecture. However, we will look at a special case. This relates to situations in which the coupling reaction rates k_{ab} and k_{ba} are much faster than the input-output or purging rates Q/V and k_b. For example if k_{ab} and k_{ba} were on the order of 1 hr^{-1} and the purging rates were 1 yr^{-1}, the coupling reactions would always be at a ***local equilibrium*** on an annual or even a daily time scale. In this case Di Toro (1976) has shown that Eqs. 6.62 and 6.63 can be added together to yield

$$\frac{dc}{dt} = \frac{Q}{V}c_{\text{in}} - \frac{Q}{V}c - k_bc_b \tag{6.64}$$

where c = total concentration = $c_a + c_b$ and c_{in} = total inflow concentration = $c_{a,\text{in}} + c_{b,\text{in}}$.

Notice that the terms representing the fast reactions were canceled by the addition of the equations. This is justified by the fact that if the reactions are at equilibrium, the terms $k_{ab}c_a$ and $k_{ba}c_b$ are equal and would cancel out.

Now at this point the mass balance, although simpler, cannot be solved because it is a single equation with two unknowns, c and c_b. However, the second can be eliminated by substituting Eq. 6.56 to give

$$\frac{dc}{dt} = \frac{Q}{V}c_{\text{in}} - \frac{Q}{V}c - k_bF_bc \tag{6.65}$$

Thus the original system of differential equations has been replaced by a single mass balance that can be solved for the total mass in the system as a function of time. Then at every time step, Eqs. 6.54 and 6.56 can be used to calculate the concentrations of each of the two interacting species.

PROBLEMS

6.1. Three matrices are defined as

$$[X] = \begin{bmatrix} 2 & 6 \\ 3 & 10 \\ 7 & 4 \end{bmatrix} \qquad [Y] = \begin{bmatrix} 6 & 0 \\ 1 & 4 \end{bmatrix} \qquad [Z] = \begin{bmatrix} 2 & 1 \\ 6 & 8 \end{bmatrix}$$

Perform all possible multiplications that can be computed between pairs of these matrices.

6.2. In 1970 the total phosphorus loadings to the Great Lakes were (Chapra and Sonzogni 1979)

	Units	Superior	Michigan	Huron	Erie	Ontario
W	tonnes yr^{-1}	4000	6950	4575	18,150	6650

(*a*) If total phosphorus settles at a rate of approximately 16 m yr^{-1}, calculate the steady-state concentration for each lake using the matrix inverse approach.
(*b*) Use the matrix inverse to determine how much of Lake Ontario's concentration is due to the loading to Lake Huron.
(*c*) Use the matrix inverse to determine how much Lake Ontario's concentration will change if Lake Erie's loading is reduced by 25% and Lake Huron's loading is halved.

6.3. Use Laplace transforms to solve Eqs. 6.35 and 6.36 for Eqs. 6.41 and 6.42.

6.4. As depicted in Fig. P6.4, a lake and its bottom segments can be modeled as two CSTRs with feedback.

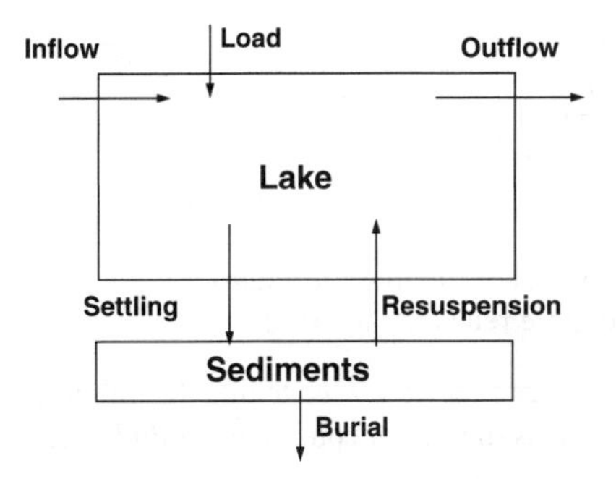

FIGURE P6.4

The following parameters are given:

Inflow = outflow = 20×10^6 m^3 yr^{-1}
Lake volume = 150×10^6 m^3
Settling velocity = 10 m yr^{-1}
Burial velocity = 2 mm yr^{-1}
Lake area = sediment area = 2.5×10^6 m^2
Sediment volume = 100×10^4 m^3
Resuspension velocity = 1 mm yr^{-1}

(*a*) If the lake receives a constant inflow concentration of 100 μg L^{-1} and the pollutant does not react, but does settle, compute the steady-state concentration in the sediments and the water. Use the matrix inverse to determine the concentrations.
(*b*) If a sediment concentration of 100,000 μg L^{-1} must be maintained, use the matrix inverse to come up with the necessary inflow concentration.

(*c*) Suppose that a spill of 20 kg of a contaminant occurs in the lake. Calculate the concentration in the lake and the sediments as a function of time in the following years.

6.5. Suppose that a spill of 5 kg of a pesticide takes place in the lake described in Prob. 6.4. Assume that 50% of the pesticide associates with solid matter and settles but does not volatilize (that is, it is not lost to the atmosphere). The other 50% is soluble and subject to volatilization (but not settling) that can be characterized by the first-order flux

$$J = v_v F_d c$$

where v_v = a volatilization mass-transfer coefficient of 0.01 m d^{-1} and F_d = the fraction of pesticide in dissolved form (in this case 0.5). Calculate the response of the system to the spill. When does the peak concentration occur in the sediments? Note that the lake is pesticide-free prior to the spill.

6.6. Suppose that the following first-order reactions take place in a batch reactor:

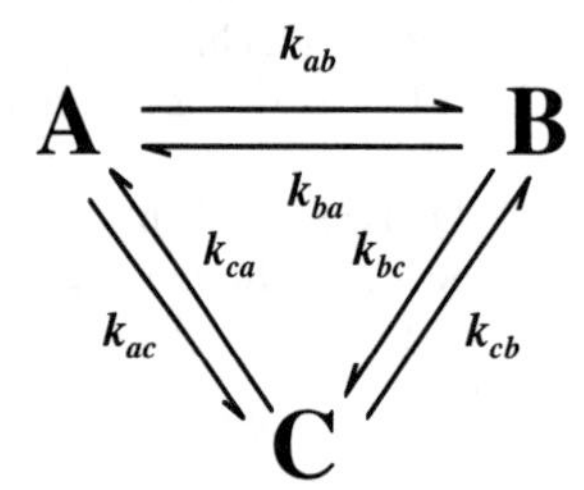

FIGURE P6.6

The following parameters and initial conditions apply:

$k_{ab} = 0.1\ d^{-1}$ $\qquad$ $k_{ba} = 0.2\ d^{-1}$
$k_{bc} = 0.4\ d^{-1}$ $\qquad$ $k_{cb} = 0.5\ d^{-1}$
$k_{ac} = 0.3\ s^{-1}$ $\qquad$ $k_{ca} = 0.6\ s^{-1}$
$c_{a0} = 10\ mg\ L^{-1}$ $\qquad$ $c_{b0} = 20\ mg\ L^{-1}$
$c_{c0} = 70\ mg\ L^{-1}$

(*a*) Calculate the steady-state concentrations of the three reactants.
(*b*) Calculate the concentrations of the three reactants for 10 d.

6.7. Repeat Prob. 6.6 for a CSTR with a residence time of 1 d. Note that the inflow concentrations of the three reactants are the same as the initial conditions in Prob. 6.6.

6.8. Figure P6.8 shows a simplified schematic of the Colorado River system, and characteristics of the impoundments are summarized in Table P6.8.
(*a*) Solve for the chloride concentration in each of the reservoirs.
(*b*) Determine the steady-state system response matrix.
(*c*) Use the result of part (*b*) to evaluate the new concentration in Lake Havasu if the loading to Lake Powell is halved.
(*d*) Suppose that global warming reduces all outflows by 20%. Recalculate the concentrations for this scenario.
(*e*) Calculate the effect on Lake Powell if 1 tonne (1000 kg) of a conservative contaminant is spilled into Flaming Gorge.
(*f*) Repeat part (*e*) but for a contaminant with a half-life of 2 yr.

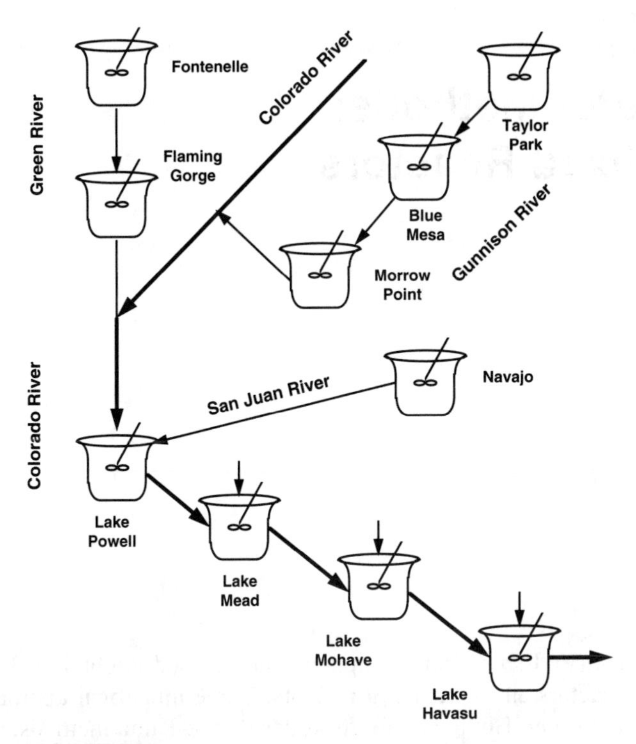

FIGURE P6.8
The Colorado River system.

TABLE P6.8
Characteristics of impoundments on the Colorado River system

Impoundment	Outflow (10^6 m^3 yr^{-1})	Volume (10^6 m^3)	Chloride loading (10^6 g yr^{-1})
Taylor Park	226	602	900
Blue Mesa	853	4,519	4,000
Morrow Point	914	851	400
Fontenelle	1,429	2,109	5,700
Flaming Gorge	1,518	19,581	20,000
Navajo	1,250	9,791	5,000
Powell	13,422	150,625	714,500
Mead	12,252	180,750	300,000
Mohave	12,377	12,050	102,000
Havasu	11,797	4,142	30,000

LECTURE 7

Computer Methods: Well-Mixed Reactors

LECTURE OVERVIEW: I show how computers can be used to obtain solutions for individual reactors and systems of reactors. Three numerical approaches are described: the Euler, Heun, and fourth-order Runge-Kutta methods.

To this point we have used calculus to develop solutions for a number of idealized cases. Although these solutions are extremely useful for obtaining a fundamental understanding of pollutant fate in natural waters, they are somewhat limited in their ability to characterize real problems. There are four reasons why the analytical approaches are limited:

- *Nonidealized loading functions.* To attain closed-form solutions, idealized loading functions must be used. For example the loading must be adequately represented by functions such as the impulse, step, linear, exponential, or sinusoidal forms described in Lec. 4. Although real loadings may sometimes be represented in this way, they are more often arbitrary, with no apparent underlying pattern (recall Fig. 4.1*f*).
- *Variable parameters.* To this point we have assumed that all the model parameters (that is, Q, V, k, v, etc.) are constant. In fact, they may vary temporally.
- *Multiple-segment systems.* As you have learned in the past two chapters, systems of more than two segments require computers for their efficient solution.
- *Nonlinear kinetics.* To this point we have emphasized first-order kinetics. This means that we limited our studies to linear algebraic and differential equations. Although first-order reactions are important, there are a variety of water-quality problems that require nonlinear reactions. In most of these cases analytical solutions cannot be obtained.

All the above problems can be addressed by using the computer and numerical methods. For these reasons you must now begin to learn how computers are used to solve differential equations.

7.1 EULER'S METHOD

Euler's method is the simplest numerical method for solving ordinary differential equations. One way to introduce Euler's method is to derive it to solve the completely mixed lake model,

$$\frac{dc}{dt} = \frac{W(t)}{V} - \lambda c \tag{7.1}$$

where

$$\lambda = \frac{Q}{V} + k + \frac{v}{H} \tag{7.2}$$

The fundamental approach for solving a mathematical problem with the computer is to reformulate the problem so that it can be solved by arithmetic operations. Our one stumbling block for solving Eq. 7.1 in this way is the derivative term dc/dt. However, as we have already shown in Lec. 2, difference approximations can be used to express derivatives in arithmetic terms. For example using a forward difference, we can approximate the first derivative of c with respect to t by

$$\frac{dc_i}{dt} \cong \frac{\Delta c}{\Delta t} = \frac{c_{i+1} - c_i}{t_{i+1} - t_i} \tag{7.3}$$

where c_i and c_{i+1} = concentrations at a present and a future time t_i and t_{i+1}, respectively. Substituting Eq. 7.3 into Eq. 7.1 yields

$$\frac{c_{i+1} - c_i}{t_{i+1} - t_i} = \frac{W(t)}{V} - \lambda c_i \tag{7.4}$$

which can be solved for

$$c_{i+1} = c_i + \left[\frac{W(t)}{V} - \lambda c_i\right](t_{i+1} - t_i) \tag{7.5}$$

Notice that the term in square brackets is the differential equation itself (Eq. 7.1), which provides a means to compute the rate of change or slope of c. Thus the differential equation has been transformed into an algebraic equation that can be used to determine the concentration at t_{i+1} using the slope and a previous value of c. If you are given an initial value for concentration at some time t_i, you can easily compute concentration at a later time t_{i+1}. This new value of c at t_{i+1} can in turn be used to extend the computation to t_{i+2}, and so on. Thus at any time along the way,

$$\text{New value} = \text{old value} + (\text{slope})\,(\text{step}) \tag{7.6}$$

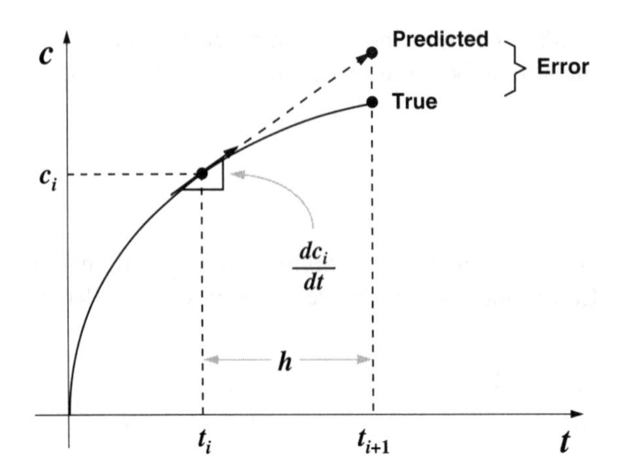

FIGURE 7.1
Graphical depiction of Euler's method.

This approach can be represented generally as

$$c_{i+1} = c_i + f(t_i, c_i)h \tag{7.7}$$

where $f(t_i, c_i) = dc_i/dt =$ the value of the differential equation evaluated at t_i and c_i and $h =$ step size ($= t_{i+1} - t_i$). This formula is referred to as ***Euler's*** (or the ***point-slope***) ***method*** (Fig. 7.1).

EXAMPLE 7.1. EULER'S METHOD. A well-mixed lake has the following characteristics:

$Q = 10^5 \text{ m}^3 \text{ yr}^{-1}$ $\quad V = 10^6 \text{ m}^3$
$z = 5 \text{ m}$ $\quad k = 0.2 \text{ yr}^{-1}$
$v = 0.25 \text{ m yr}^{-1}$

At $t = 0$ it receives a step loading of 50×10^6 g yr^{-1} and has an initial concentration of 15 mg L^{-1}. Use Euler's method to simulate the concentration from $t = 0$ to 20 yr using a time step of 1 yr. Compare the results with the analytical solution

$$c = c_0 e^{\lambda t} + \frac{W}{\lambda V}(1 - e^{-\lambda t})$$

Solution: First, the eigenvalue can be computed as

$$\lambda = \frac{10^5}{10^6} + 0.2 + \frac{0.25}{5} = 0.35 \text{ yr}^{-1}$$

At the start of the computation ($t_i = 0$) the concentration is 15 mg L^{-1} and the loading is 50×10^6 g yr^{-1}. Using this information and the parameter values, Eq. 7.5 can be employed to compute concentration at t_{i+1}:

$$c(1) = 15 + \left[\frac{50 \times 10^6}{10^6} - 0.35(15)\right]1.0 = 59.75 \text{ mg L}^{-1}$$

For the next interval (from $t = 1$ to 2 yr) the computation is repeated, with the result

$$c(2) = 59.75 + \left[\frac{50 \times 10^6}{10^6} - 0.35(59.75)\right]1.0 = 88.8375 \text{ mg L}^{-1}$$

The calculation is continued in a similar fashion to obtain additional values. The results, along with the analytical solution, are

	c (mg L^{-1})		t(yr)	c (mg L^{-1})	
t(yr)	**Numerical**	**Analytical**	t(yr)	**Numerical**	**Analytical**
0	15.00	15.00	6	133.21	127.20
1	59.75	52.75	7	136.59	131.82
2	88.84	79.37	8	138.78	135.08
3	107.74	98.12	9	140.21	137.38
4	120.03	111.33	10	141.14	139.00
5	128.02	120.64	∞	142.86	142.86

The numerical solution is plotted in Fig. 7.2 along with the analytical result. It can be seen that the numerical method accurately captures the major features of the exact solution. However, because we have used straight-line segments to approximate a continuously curving function, there is some discrepancy between the two results. One way to minimize such discrepancies is to use a smaller step size. For example applying Eq. 7.5 at 0.5-yr intervals results in a smaller error, as the straight-line segments track closer to the true solution. Using hand calculations the effort associated with using smaller and smaller step sizes would make such numerical solutions impractical. However, with the aid of the computer large numbers of computations can be performed easily. Thus you can accurately model the concentration without having to solve the differential equation exactly.

The Euler method is a first-order approach. Among other things this means that it would yield perfect results if the underlying solution of the differential equation were a linear polynomial or straight line. The accuracy of Euler's method can be improved by using a smaller time step. Other methods are also available that attain better accuracy by improving the slope estimate used for the extrapolation (see Chapra and Canale 1988 for a review). The following sections describe some of these methods.

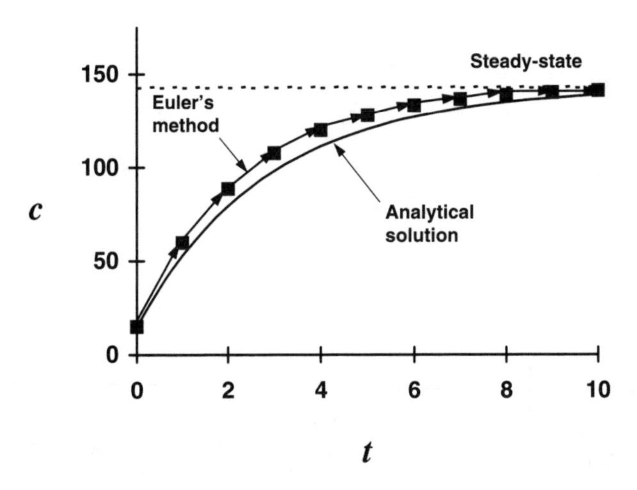

FIGURE 7.2
Comparison of analytical and Euler-method solutions for the well-mixed lake with a step loading.

7.2 HEUN'S METHOD

A fundamental source of error in Euler's method is that the derivative at the beginning of the interval is assumed to apply across the entire interval. One method to improve the estimate of the slope involves the determination of derivatives across the interval—one at the initial point and another at the end point. The two derivatives are then averaged to obtain an improved estimate of the slope for entire interval. This approach, called ***Heun's method***, is depicted graphically in Fig 7.3.

Recall that in Euler's method the slope at the beginning of an interval

$$\frac{dc_i}{dt} = f(t_i, c_i) \tag{7.8}$$

is used to extrapolate linearly to c_{i+1}:

$$c_{i+1}^0 = c_i + f(t_i, c_i)h \tag{7.9}$$

For the standard Euler method we would stop at this point. However, in Heun's method the c_{i+1}^0 calculated in Eq. 7.9 is not the final answer but an intermediate prediction. This is why we have distinguished it with a superscript 0. Equation 7.9 provides an estimate of c_{i+1} that allows the calculation of an estimated slope at the end of the interval:

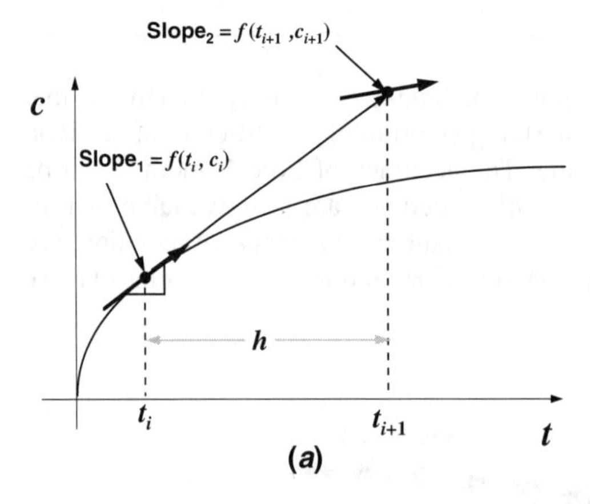

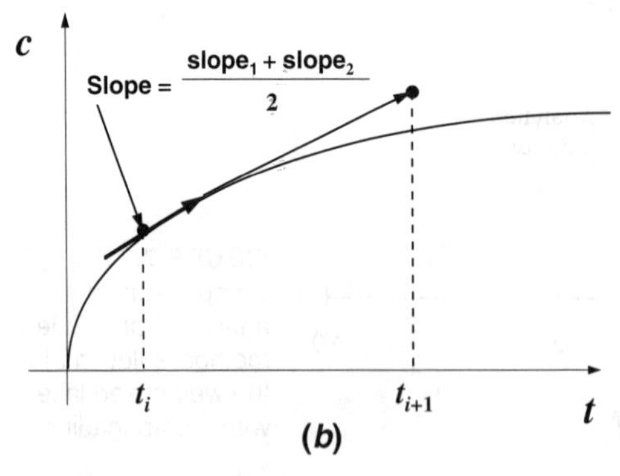

FIGURE 7.3
Graphical depiction of Heun's method. (*a*) Predictor; (*b*) corrector.

$$\frac{dc_{i+1}}{dt} = f(t_{i+1}, c_{i+1}^0) \tag{7.10}$$

Thus the two slopes (Eqs. 7.8 and 7.10) can be combined to obtain an average slope for the interval

$$\overline{\frac{dc}{dt}} = \frac{f(t_i, c_i) + f(t_{i+1}, c_{i+1}^0)}{2} \tag{7.11}$$

This average slope is then used to extrapolate linearly from c_i to c_{i+1},

$$c_{i+1} = c_i + \frac{f(t_i, c_i) + f(t_{i+1}, c_{i+1}^0)}{2} h \tag{7.12}$$

The Heun method is called a ***predictor-corrector method***. As derived above it can be expressed concisely as

Predictor: $$c_{i+1}^0 = c_i + f(t_i, c_i)h \tag{7.13}$$

Corrector: $$c_{i+1} = c_i + \frac{f(t_i, c_i) + f(t_{i+1}, c_{i+1}^0)}{2} h \tag{7.14}$$

Note that because Eq. 7.14 has c_{i+1} on both sides of the equal sign, it can be solved iteratively to refine the result. That is, an old estimate can be used repeatedly to provide an improved estimate of c_{i+1}. It should be understood that this iterative process will not necessarily converge on the true answer but will converge on an estimate with a finite truncation error. However, this truncation error will be smaller than for cruder approaches like Euler's method.

As with other iterative methods, a termination criterion for convergence of the corrector is provided by

$$\%\text{ error} = \left| \frac{c_{i+1}^j - c_{i+1}^{j-1}}{c_{i+1}^j} \right| (100\%) \tag{7.15}$$

where c_{i+1}^{j-1} and c_{i+1}^j are the results from the prior and the present iteration of the corrector, respectively.

EXAMPLE 7.2. THE HEUN METHOD. Use the Heun method to solve the same problem as in Example 7.1. For the present application do not iterate the corrector.

Solution: At the start of the computation ($t_i = 0$) the concentration is 15 mg L^{-1} and the loading is 50×10^6 g yr^{-1}. Using this information and the parameter values, Eq. 7.1 can be used to compute a slope estimate at t_i:

$$f(0, 15) = 50 - 0.35(15) = 44.75$$

which can then be used to calculate the concentration at the end of the interval

$$c(1) = 15 + (44.75)1.0 = 59.75 \text{ mg L}^{-1}$$

This value can in turn be employed to estimate a slope at the end of the interval

$$f(1, 59.75) = 50 - 0.35(59.75) = 29.0875$$

Then the two slopes can be input into the corrector to calculate the final result

$$c(1) = 15 + \tfrac{1}{2}(44.75 + 29.0875)1 = 51.91875 \text{ mg L}^{-1}$$

which is much closer to the true value than was obtained with Euler's method. The calculation is continued in a similar fashion to obtain additional values. The results, along with the analytical solution, are

	c (mg L⁻¹)			**c (mg L⁻¹)**	
t (yr)	**Numerical**	**Analytical**	**t (yr)**	**Numerical**	**Analytical**
0	15.00	15.00	6	126.30	127.20
1	51.92	52.75	7	131.08	131.82
2	78.18	79.37	8	134.48	135.08
3	96.85	98.12	9	136.90	137.38
4	110.14	111.33	10	138.62	139.00
5	119.58	120.64	∞	142.86	142.86

The Heun method is a second-order approach. Among other things this means that it would yield perfect results if the underlying solution of the differential equation were a quadratic polynomial. Thus it is superior to the Euler approach. However, the fact that two derivative evaluations must be made for each time step means that a computational price is paid for the gain in accuracy.

7.3 RUNGE-KUTTA METHODS

The Runge-Kutta (or RK) methods are a family of numerical methods that are used extensively in water-quality modeling. The RK methods all have the general form

$$c_{i+1} = c_i + \phi h \tag{7.16}$$

where ϕ = a slope estimate (formally called an ***increment function***). Comparison of Eq. 7.16 with Eq. 7.7 indicates that Euler's method is actually a first-order RK method with $\phi = f(t_i, c_i)$. In addition the Heun method (without corrector iteration) is a second-order RK algorithm (Chapra and Canale 1988).

The most commonly used RK method is the classical fourth-order method that has the form

$$c_{i+1} = c_i + \left[\tfrac{1}{6}(k_1 + 2k_2 + 2k_3 + k_4)\right]h \tag{7.17}$$

where

$$k_1 = f(t_i, c_i) \tag{7.18}$$

$$k_2 = f\left(t_i + \tfrac{1}{2}h, c_i + \tfrac{1}{2}hk_1\right) \tag{7.19}$$

$$k_3 = f\left(t_i + \tfrac{1}{2}h, c_i + \tfrac{1}{2}hk_2\right) \tag{7.20}$$

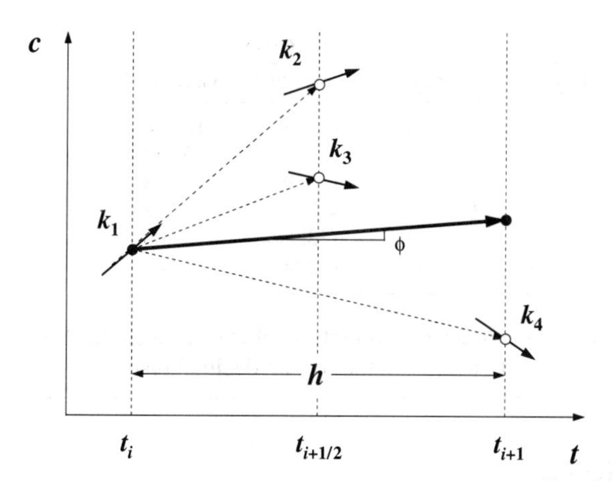

FIGURE 7.4
Graphical depiction of the fourth-order RK method.

$$k_4 = f(t_i + h,\ c_i + hk_3) \tag{7.21}$$

where the functions are merely the original differential equation evaluated at specific values of t and c. That is,

$$f(t, c) = \frac{dc}{dt}(t, c) \tag{7.22}$$

The fourth-order RK method is similar to the Heun approach in that multiple estimates of the slope are developed to come up with an improved average slope for the interval. As depicted in Figure 7.4, each of the k's represents a slope. Equation 7.17 then represents a weighted average of these to arrive at the improved slope.

EXAMPLE 7.3. THE FOURTH-ORDER RK METHOD. Use the classical fourth-order RK method to solve the same problem as in Example 7.1.

Solution: At the start of the computation ($t_i = 0$) the concentration is 15 mg L^{-1} and the loading is 50×10^6 g yr^{-1}. Using this information and the parameter values, Eqs. 7.17 to 7.21 can be used to compute concentration at t_{i+1}:

$$k_1 = f(0, 15) = 50 - 0.35(15) = 44.750$$

$$k_2 = f\left[0 + \tfrac{1}{2}(1), 15 + \tfrac{1}{2}(1)44.750\right] = f(0.5, 37.375) = 50 - 0.35(37.375) = 36.919$$

$$k_3 = f\left[0 + \tfrac{1}{2}(1), 15 + \tfrac{1}{2}(1)36.919\right] = f(0.5, 33.459) = 50 - 0.35(33.459) = 38.289$$

$$k_4 = f[0 + 1, 15 + 1(38.289)] = f(0.5, 53.289) = 50 - 0.35(53.289) = 31.349$$

$$c(1) = 15 + \left[\tfrac{1}{6}(44.750 + 2(36.919 + 38.289) + 31.349)\right](1) = 52.75 \text{ mg L}^{-1}$$

For the next interval ($t = 1$ to 2 yr) the computation is repeated, with the result

$$c(2) = 52.75 + \left[\tfrac{1}{6}(31.537 + 2(26.018 + 26.984) + 22.092)\right](1) = 79.36 \text{ mg L}^{-1}$$

The calculation is continued in a similar fashion to obtain additional values. The results, along with the analytical solution, are

	c (mg L^{-1})			c (mg L^{-1})	
t (yr)	Numerical	Analytical	t (yr)	Numerical	Analytical
0	15.00	15.00	6	127.19	127.20
1	52.75	52.75	7	131.82	131.82
2	79.36	79.37	8	135.08	135.08
3	98.11	98.12	9	137.38	137.38
4	111.32	111.33	10	138.99	139.00
5	120.63	120.64	∞	142.86	142.86

Thus, for this example, the numerical method very accurately follows the analytical solution. Discrepancies are limited to a unit quantity in the second decimal place.

7.4 SYSTEMS OF EQUATIONS

All the methods described previously can easily be adapted to simulate systems of differential equations of the form

$$\frac{dc_1}{dt} = f_1(c_1, c_2, \ldots, c_n) \tag{7.23}$$

$$\frac{dc_2}{dt} = f_2(c_1, c_2, \ldots, c_n) \tag{7.24}$$

$$\vdots$$

$$\frac{dc_n}{dt} = f_n(c_1, c_2, \ldots, c_n) \tag{7.25}$$

The solution of such a system requires that n initial conditions be known at the starting point. Then Eqs. 7.23 to 7.25 can be employed to develop slope estimates for each of the unknowns. These slope estimates are used to predict values of concentration at a new time. The process can then be repeated to project out another time step.

EXAMPLE 7.4. SPILL FOR TWO LAKES IN SERIES. A spill of 5 kg of a soluble pesticide takes place in the first of two lakes in series. Note that both lakes are completely mixed. The pesticide is subject to no losses except for flushing. Parameters for the lakes are

	Lake 1	Lake 2
Volume (m^3)	0.5×10^6	0.6×10^6
Outflow (m^3 yr^{-1})	1×10^6	1×10^6

Predict the concentration of both lakes as a function of time using Euler's method. Compare these results with the analytical solution. Present your results graphically.

Solution: First, we can develop the mass balances for the system:

$$\frac{dc_1}{dt} = -\lambda_{11}c_1 \qquad \frac{dc_2}{dt} = \lambda_{21}c_1 - \lambda_{22}c_2$$

where

$$\lambda_{11} = \frac{Q_{12}}{V_1} \qquad \lambda_{21} = \frac{Q_{12}}{V_2} \qquad \lambda_{22} = \frac{Q_{23}}{V_2}$$

Substituting the parameter values results in

$$\frac{dc_1}{dt} = -2c_1$$

$$\frac{dc_2}{dt} = 1.667c_1 - 1.667c_2$$

If $c_1 = 10\ \mu\text{g L}^{-1}$ and $c_2 = 0$, then the general solution can be developed as

$$c_1 = 10e^{-2t}$$

$$c_2 = \frac{1.667(10)}{1.667 - 2}(e^{-2t} - e^{-1.667t})$$

Now we can proceed with Euler's method. First, the differential equations are employed to calculate the slopes at $t = 0$,

$$\frac{dc_1}{dt}(0) = -2(10) = -20$$

$$\frac{dc_2}{dt}(0) = 1.667(10) - 1.667(0) = 16.667$$

These values can then be used to extrapolate out to $t = 0.1$ yr,

$$c_1(0.1) = 10 - 20(0.1) = 8\ \mu\text{g L}^{-1}$$

$$c_2(0.1) = 0 + 16.667(0.1) = 1.667\ \mu\text{g L}^{-1}$$

For the next interval ($t = 0.1$ to 0.2 yr) the computation is repeated. First, the slopes are determined at $t = 0.1$ yr:

$$\frac{dc_1}{dt}(0.1) = -2(8) = -16$$

$$\frac{dc_2}{dt}(0.1) = 1.667(8) - 1.667(1.667) = 10.556$$

These values can then be used to extrapolate out to $t = 0.2$ yr,

$$c_1(0.2) = 8 - 16(0.1) = 6.4\ \mu\text{g L}^{-1}$$

$$c_2(0.2) = 1.667 + 10.556(0.1) = 2.722\ \mu\text{g L}^{-1}$$

The calculation is continued in a similar fashion to obtain additional values. The results, along with the analytical solution, are

	c_1 (μg L^{-1})		c_2 (μg L^{-1})	
t (yr)	**Numerical**	**Analytical**	**Numerical**	**Analytical**
0.0	10.00	10.00	0.00	0.00
0.1	8.00	8.19	1.67	1.39
0.2	6.40	6.70	2.72	2.31
0.3	5.12	5.49	3.35	2.89
0.4	4.10	4.49	3.63	3.20
0.5	3.28	3.68	3.71	3.34
0.6	2.62	3.01	3.64	3.33
0.7	2.10	2.47	3.47	3.24
0.8	1.68	2.02	3.24	3.09
0.9	1.34	1.65	2.98	2.89
1.0	1.07	1.35	2.71	2.68

The results are plotted in Fig. 7.5 along with the analytical solution. As was the case for the single equation (Example 7.1), one way to minimize the discrepancies is to use a smaller step size.

The Heun and the fourth-order Runge-Kutta methods can also be used to simulate systems of ODEs. However, as illustrated by the following example, care should be taken to sequence the calculation of the slopes prior to computing new concentrations.

EXAMPLE 7.5. SYSTEMS OF ODES WITH THE FOURTH-ORDER RK METHOD. Use the classical fourth-order RK method to solve the same problem as in Example 7.4.

Solution: The important feature of this computation is that all the k's must be computed for the entire system of ODEs before computing the next set of concentrations. For example

$$k_{1,1} = -20 \qquad k_{2,1} = -18 \qquad k_{3,1} = -18.2 \qquad k_{4,1} = -16.36$$
$$k_{1,2} = 16.667 \qquad k_{2,2} = 13.611 \qquad k_{3,2} = 14.032 \qquad k_{4,2} = 11.295$$

These can then be used to compute the two increment functions

$$\phi_1 = \tfrac{1}{6}[-20 + 2(-18 - 18.2) - 16.36] = -18.127$$
$$\phi_2 = \tfrac{1}{6}[16.667 + 2(13.611 + 14.032) + 11.295] = 13.875$$

These can be used to extrapolate out in time,

$$c_1(0.1) = 10 - 18.127(0.1) = 8.19 \text{ mg L}^{-1}$$
$$c_2(0.1) = 0 + 13.875(0.1) = 1.39 \text{ mg L}^{-1}$$

The calculation is continued in a similar fashion to obtain additional values. The results match the analytical solution to two decimal places of accuracy. Thus, for this example, the numerical method very accurately follows the analytical solution.

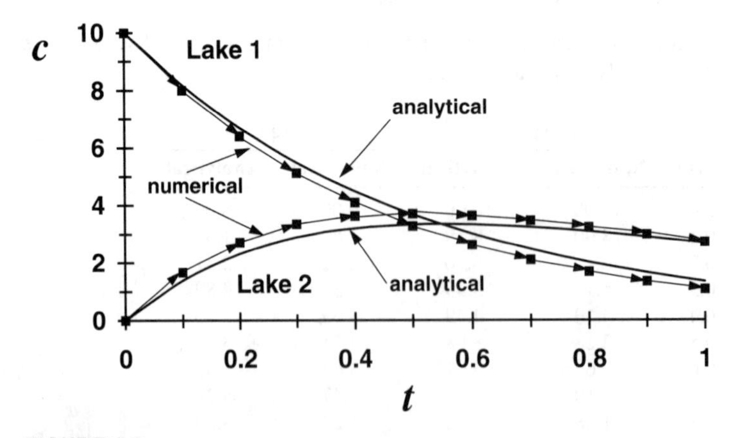

FIGURE 7.5
Comparison of analytical and Euler-method solutions for a system of well-mixed lakes.

PROBLEMS

7.1. Population-growth dynamics is important in a variety of engineering planning studies. One of the simplest models of such growth incorporates the assumption that the rate of change of the population p is proportional to the existing population at any time t:

$$\frac{dp}{dt} = Gp \tag{P7.1}$$

where G = a growth rate (yr^{-1}). This model makes intuitive sense because the greater the population, the greater the number of potential parents.

At time $t = 0$ an island has a population of 10,000 people. If $G = 0.075\ \text{yr}^{-1}$, use Euler's method to predict the population at $t = 20$ yr using a step size of 0.5 yr. Plot p versus t on standard and semi-log graph paper. Determine the slope of the line on the semi-log plot. Discuss your results.

7.2. Although the model in Prob. 7.1 works adequately when population growth is unlimited, it breaks down when factors such as food shortages, pollution, and lack of space inhibit growth. In such cases the growth rate itself can be thought of as being inversely proportional to population. One model of this relationship is

$$G = G'(p_{\text{max}} - p) \tag{P7.2}$$

where G' = a population-dependent growth rate [(people-yr)$^{-1}$] and p_{max} = the maximum sustainable population. Thus when population is small ($p \ll p_{\text{max}}$), the growth rate will be at a high, constant rate of $G'p_{\text{max}}$. For such cases growth is unlimited and Eq. P7.2 is essentially identical to Eq. P7.1. However, as population grows (that is, as p approaches p_{max}), G decreases until at $p = p_{\text{max}}$ it is zero. Thus the model predicts that when the population reaches the maximum sustainable level, growth is nonexistent, and the system is at a steady-state. Substituting Eq. P7.2 into Eq. P7.1 yields

$$\frac{dp}{dt} = G'(p_{\text{max}} - p)p$$

For the same island studied in Prob. 7.1 use Euler's method to predict the population at $t = 20$ yr using a step size of 0.5 yr. Use values of $G' = 10^{-5}$ (people-yr)$^{-1}$ and $p_{\text{max}} = 20{,}000$ people. At time $t = 0$ the island has a population of 10,000 people. Plot p versus t and interpret the shape of the curve.

7.3. Recall that in Prob. 4.8 we studied a lake having the following characteristics:

Inflow = outflow = $20 \times 10^6\ \text{m}^3$
Mean depth = 10 m
Surface area = $10 \times 10^6\ \text{m}^2$

A canning plant discharges a pollutant to the system that decays at a rate of $1.05\ \text{yr}^{-1}$. In Prob. 4.8 you were asked to approximate the seasonal loading from a cannery as a sinusoidal input. The following measurements provide a better estimate of the loading pattern over the course of a year.

Month	J	F	M	A	M	J	J	A	S	O	N	D
Load (mta)	29	26	11	0	0	9	23	43	44	64	53	50

If the initial condition is $c = 0$ at $t = 0$,

(*a*) Use the numerical method of your choice to compute the concentration in the system from $t = 0$ to 10 yr.
(*b*) After sufficient time the concentration will approach a dynamic steady-state. At this point, on what day of the year will the in-lake concentration be at a maximum value?

7.4. A small pond has the following characteristics:

Surface area = 0.2×10^6 m^2
Mean depth = 5 m
Outflow = 1×10^6 m^3 d^{-1}

The temperature in the pond varies diurnally as

Time	Midnight	2:00	4:00	6:00	8:00	10:00
Temp. (°C)	21	20	17	16	18	21
Time	Noon	2:00	4:00	6:00	8:00	10:00
Temp. (°C)	25	27	28	26	23	21

Determine the response of the system to a 2-kg spill of a pollutant that decays at a rate $k = 2$ d^{-1}. Calculate the response if the spill takes place (*a*) at midnight or (*b*) at noon. Plot the responses on the same graph. Note that the reaction has a temperature dependence characterized by $\theta = 1.08$.

7.5. Lake Washington, a beautiful lake located in Seattle, Washington, has the following general characteristics:

Volume = 2.9×10^9 m^3
Mean depth = 33 m
Outflow = 1.25×10^9 m^3 yr^{-1}

In the 1950s and 1960s it began to deteriorate because of increased loadings of the nutrient phosphorus. As a consequence, in the late 1960s, sewage inputs were greatly curtailed. The loading pattern from 1930 through the late 1970s can be summarized as in the following table:

Year	Load (mta)	Year	Load (mta)	Year	Load (mta)
1930	40	1961	137.4	1972	103.4
1940	40	1962	148.5	1973	42.9
1941	55	1963	156.5	1974	58.5
1949	55	1964	204.2	1975	99.3
1950	84.8	1965	142.8	1976	42.9
1951	81	1966	124.8	1977	60.3
1956	81	1967	54.3	1978	48.6
1957	93.2	1968	59.1	1979	60.5
1958	104.3	1969	48.2	≥ 1980	60.5
1959	115.3	1970	59.0		
1960	126.4	1971	53.8		

Total phosphorus settles at a rate of about 12 m yr^{-1}.
(*a*) Use the Heun method to compute the lake's response from 1930 through 1990. *Note:* Set the initial condition in 1930 at 17.4 μg L^{-1}.
(*b*) Compare your results with Euler's method.
(*c*) Compare your results with the fourth-order RK method.

7.6. A spill of 5 kg of a soluble pesticide takes place in the first of two lakes in series. The pesticide is subject to volatilization that can be characterized by first-order decay rates $k_1 = 0.002\ \text{d}^{-1}$ and $k_2 = 0.00333\ \text{d}^{-1}$. Other parameters for the lakes are

	Lake 1	**Lake 2**
Surface area (m^2)	0.1×10^6	0.2×10^6
Mean depth (m)	5	3
Outflow ($\text{m}^3\ \text{yr}^{-1}$)	1×10^6	1×10^6

Using the fourth-order RK method,

(*a*) Predict the concentration in both lakes as a function of time; present your results as a plot.

(*b*) Determine the time required for the second lake to reach its maximum concentration.

7.7. Aside from the Heun method, there is another second-order approach for solving ordinary differential equations called the ***midpoint*** or ***improved polygon method***. It can be represented mathematically by

$$c_{i+\frac{1}{2}} = c_i + f(t_i, c_i)\frac{h}{2}$$

$$c_{i+1} = c_i + f(t_{i+\frac{1}{2}}, c_{i+\frac{1}{2}})h$$

Thus the first equation uses Euler's method to make a prediction for c at the midpoint of the interval. This value is then used to generate a centered slope estimate that is applied to predict the value at the end of the interval with the second equation. Employ this approach to solve the ODE described in Example 7.2.

PART II

Incompletely Mixed Systems

Part II is designed to introduce you to the fundamentals of modeling incompletely mixed systems. *Lecture 8* provides an introduction to transport with emphasis on the important topic of diffusion. This process allows CSTRs to be applied to systems composed of segments linked by open boundaries. The case of an embayment connected with a large lake is used as an illustration.

The next two lectures are devoted to closed-form solutions for idealized elongated reactors.These are the basis for analytical solutions for streams (plug-flow reactor) and estuary (mixed-flow reactor) water-quality models. *Lecture 9* derives the mass balances and steady-state solutions for these systems. *Lecture 10* deals with time-variable solutions.

The remaining three lectures in this part focus on computer-oriented methods for incompletely mixed systems. First, a general steady-state method, the control-volume approach, is developed in *Lecture 11*. In addition to presenting the method, this lecture also explores the two constraints on the steady-state approach, model positivity, and numerical dispersion. The steady-state approach is extended to time-variable simulations in *Lecture 12*. The issues of accuracy and stability are introduced and discussed. Finally several advanced time-variable methods are described in *Lecture 13*.

PART

Incompletely Mixed Systems

LECTURE 8

Diffusion

Lecture overview: To extend our framework to incompletely mixed systems, I introduce the mechanism of diffusion, because of its role in the transfer of mass across open boundaries. Then I develop mass balances for a simple incompletely mixed case: a lake with an embayment. Both steady-state and time-variable solutions are derived for this system.

Recall that in Lec. 1, I stated that two major processes dictated how pollutants behaved in natural waters: kinetics and transport. I then devoted Lec. 2 to kinetics, outlining different types of reactions and how they can be measured.

In the ensuing lectures we treated the other process—mass transport—in a simple fashion. Because we were dealing almost exclusively with CSTRs, we had to contend only with flows between reactors. Such one-way mass transport through rivers or pipes could be handled as a simple product of flow and concentration.

In this lecture I extend our theoretical framework to encompass incompletely mixed systems. These include systems such as rivers, estuaries, embayments, and coastal zones that are not chemically homogeneous. Hence we now must deal with the issue of the movement of mass within systems, that is, across open boundaries.

8.1 ADVECTION AND DIFFUSION

Numerous types of water motion transport matter within natural waters. Wind energy and gravity impart motion to the water that leads to mass transport. In the present context within-system motion can be divided into two general categories: advection and diffusion.

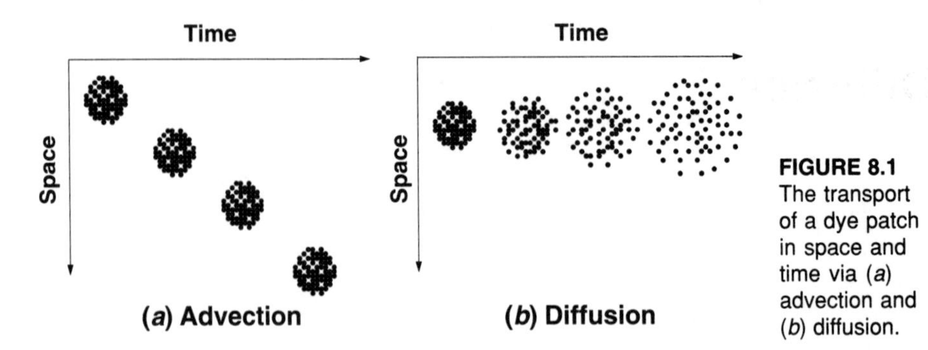

FIGURE 8.1 The transport of a dye patch in space and time via (*a*) advection and (*b*) diffusion.

Advection results from flow that is unidirectional and does not change the identity of the substance being transported. As in Figure 8.1*a* advection moves matter from one position in space to another. Simple examples of transport primarily of this type are the flow of water through a lake's outlet and the downstream transport due to flow in a river or estuary.

Diffusion refers to the movement of mass due to random water motion or mixing. Such transport causes the dye patch depicted in Figure 8.1*b* to spread out and dilute over time with negligible net movement of its center of mass. On a microscopic scale ***molecular diffusion*** results from the random Brownian motion of water molecules. A similar kind of random motion occurs on a larger scale due to eddies and is called ***turbulent diffusion***. Both have a tendency to minimize gradients (that is, differences in concentration) by moving mass from regions of high to low concentration.

The breakdown of the motion into the two idealized forms of advection and diffusion is influenced by the scales of the phenomena being modeled. For example the motion of water in an estuary might be perceived as primarily advective on a short time scale, since the rising or falling tide causes water to move unidirectionally into or out of the estuary. If the modeling problem concerns the effect of bacterial pollution from a short-term stormwater overflow episode, characterization of the transport as advection would be necessary. On a longer time scale, however, the tides would move water back and forth in a cyclical fashion and the motion might primarily be classified as diffusive. In many cases transport is considered a combination of the two modes, with the emphasis dependent on the scale of the problem.

Water motion within a natural water is a complicated process and, as with any idealization, the above concepts are intended as a brief introduction to some of its essential features. In the following sections we focus our attention on the simplest example of an incompletely mixed system—a lake with a single embayment. After using this context to introduce diffusion, a subsequent section will be devoted to some other introductory information regarding transport.

8.2 EXPERIMENT

Figure 8.2 illustrates an experiment that demonstrates what I mean by diffusive mass transport. A tank is divided in half by a removable partition. A number of tiny, neutrally buoyant particles are introduced into the left half of the tank. At the beginning

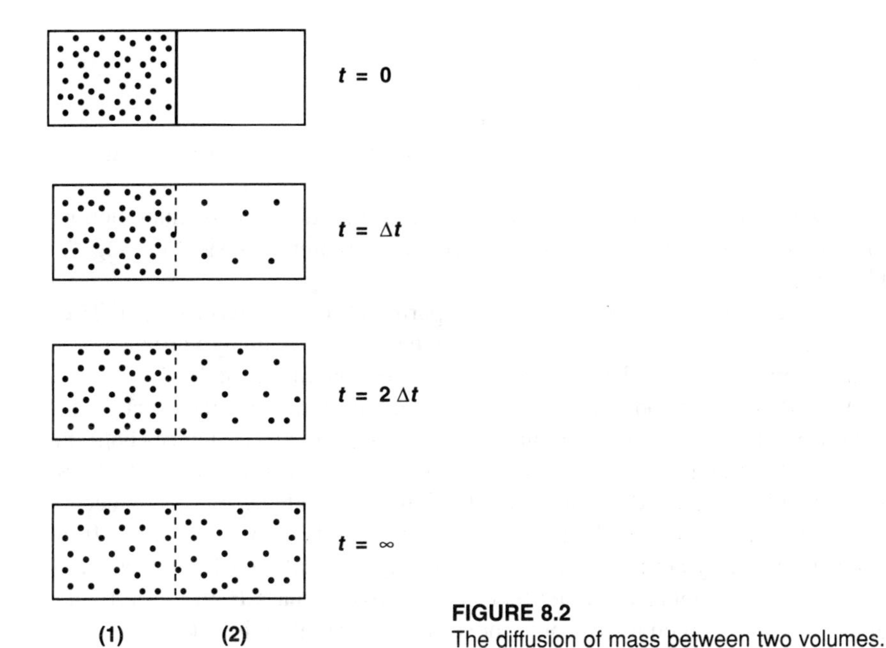

FIGURE 8.2
The diffusion of mass between two volumes.

of the experiment ($t = 0$) the partition is gently removed so that all the particles remain on the left side. If we return to the tank after a period of time ($t = \Delta t$), several of the particles will have wandered to the right side of the tank. Later ($t = 2\Delta t$) more would migrate to the right. Finally after a sufficiently long period, the concentrations for the two sides would be equal.

The speed of the process will be dictated by the strength of the mixing forces. If the tank were covered, Brownian motion would impart enough energy to the particles to move some to the right, albeit at a slow rate. The imposition of mechanical mixing, say, by blowing air randomly across the tank's surface, would accelerate the process. In any event, as time progressed, more particles would show up on the right until eventually they would be uniformly distributed throughout the tank.

As we already know, such movement of mass due to random water motion or mixing is referred to as ***diffusion***. We can quantify this mechanism by developing a mathematical model for the experiment from Fig. 8.2. If the left and right sides of the tank are designated 1 and 2, respectively, a mass balance for the left side can be written as

$$V_1 \frac{dc_1}{dt} = D'(c_2 - c_1) \tag{8.1}$$

where V_1 = volume of the left side
c_1 and c_2 = concentrations of the particles in the left and right sides, respectively
D' = diffusive flow ($m^3\ yr^{-1}$)

Thus diffusion is modeled as an equal, two-way flow connecting the volumes Fig. 8.3).

Notice that three factors contribute to the diffusive transport between the two sides of the tank. First, the mixing flow D' reflects the intensity of the mixing. Thus

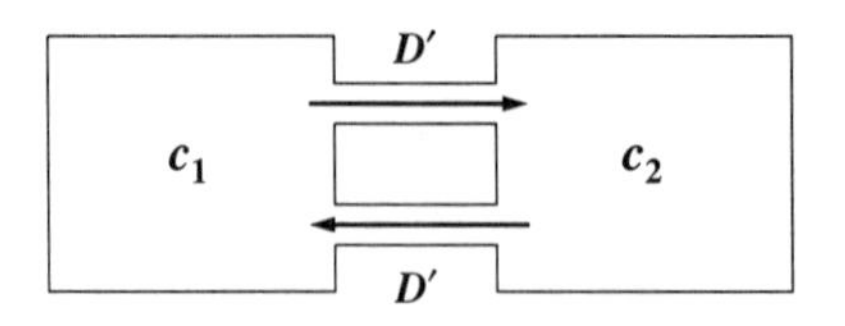

FIGURE 8.3
A two-way flow model of the diffusion of mass.

if the tank were subject to only weak mixing such as that due to Brownian motion, D' would be small. If, however, it were subjected to vigorous physical mixing, D' would be large.

Second, the mass transport is directly proportional to the interface area. Thus if the interface area were doubled, we would expect twice as many particles to be transported. This effect would also be reflected by the magnitude of D'.

Finally diffusive transport is proportional to the difference in concentration between the two sides. This concentration difference, or ***gradient***, influences both the direction and the magnitude of the transport. In the case of direction, if $c_1 > c_2$, then mass would be expected to move from the left to the right. Such motion occurs because more particles are available to move from the left than would move from the right. Consequently net motion is from left to right. Equation 8.1 represents this motion by yielding a negative value; that is, as a loss of mass from the left side. Conversely if $c_2 > c_1$, then mass would move from the right to the left.

In the case of magnitude, a large difference would mean that proportionately more particles are transported. If no difference existed ($c_1 = c_2$), then we would expect that leftward motion would balance rightward motion, giving a net transport of zero. Such zero net transport is reflected by the fact that Eq. 8.1 would yield a zero value.

EXAMPLE 8.1. DIFFUSION OF MASS BETWEEN TWO VOLUMES. Model the time required for the experiment shown in Fig. 8.2 to go to 95% of completion.

Solution: Recall that in Sec. 6.5.1 we simulated the temporal progression of two reactants undergoing a reversible reaction in a batch reactor. A similar approach can be used here. First, recognize that if the volumes of the two sides of the tank are identical, the balance for the left side can be expressed as

$$\frac{V}{2}\frac{dc_1}{dt} = D'(c_2 - c_1)$$

where V = total volume of the tank.

Further, again because the two sides have equal volume, the sum of the concentrations in each side must equal the initial concentration in the left side,

$$c_{10} = c_1 + c_2 \tag{E8.1.1}$$

This equation can be solved for c_2 and substituted into the mass balance to give

$$\frac{dc_1}{dt} + \frac{4D'}{V}c_1 = \frac{2D'}{V}c_{10}$$

which can be solved for

$$c_1 = c_{10}e^{-\frac{4D'}{V}t} + \frac{c_{10}}{2}\left(1 - e^{-\frac{4D'}{V}t}\right)$$

This result can then be substituted back into Eq. E8.1.1 to yield

$$c_2 = \frac{c_{10}}{2}\left(1 - e^{-\frac{4D'}{V}t}\right)$$

Thus, as depicted below, the sides asymptotically approach a concentration of $c_{10}/2$ with eigenvalues of $4D'/V$. Therefore, the time to get 95% complete mixing is $3V/(4D')$.

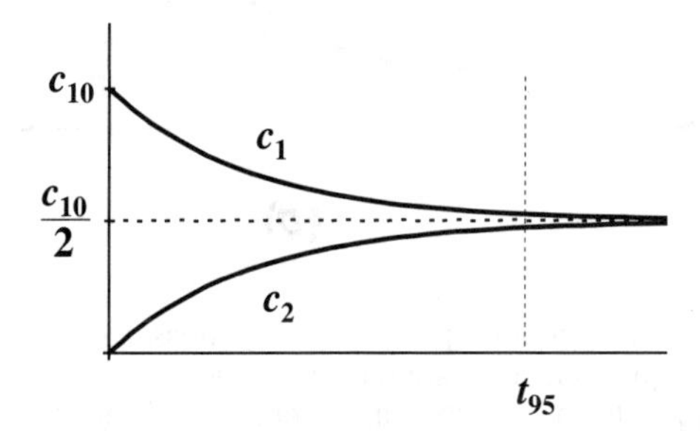

FIGURE E8.1

8.3 FICK'S FIRST LAW

Now that we have developed an intuitive and mathematical feel for diffusion, we will explore how scientists and engineers have formally characterized the process. In 1855 the physiologist Adolf Fick proposed the following model of diffusion:

$$J_x = -D\frac{dc}{dx} \tag{8.2}$$

where J_x = mass flux in the x direction (M L^{-2} T^{-1}) and D = a diffusion coefficient (L^2 T^{-1}). This model, which is called ***Fick's first law***, specifies that mass flux is proportional to the gradient (that is, the derivative or rate of change) of concentration. As depicted in Fig. 8.4 the negative sign is included to ensure that mass flux proceeds in the correct direction. Thus it is analogous to other constitutive laws such as Fourier's law for heat conduction or Ohm's law for electrical conduction. For example Fourier's law states that heat flows from regions of high to low temperature. In a similar way Fick's law states that mass flows from regions of high to low concentration.

The diffusion coefficient D is a parameter used to quantify the rate of the diffusive process. As in our discussion of the experiment in the previous section, if the tank were covered, the diffusion coefficient would be low to reflect the small rate of diffusion. If mechanical mixing were present it would be large.

Fick's law can now be used to model the situation depicted in Fig. 8.2. To do this we can write a mass balance for the left side of the reactor:

$$V_1\frac{dc_1}{dt} = -JA_c \tag{8.3}$$

where A_c = cross-sectional area of the interface between the two sides (m^2) and J = flux between the volumes.

A finite-difference approximation can be used to estimate the derivative at the interface between the two halves,

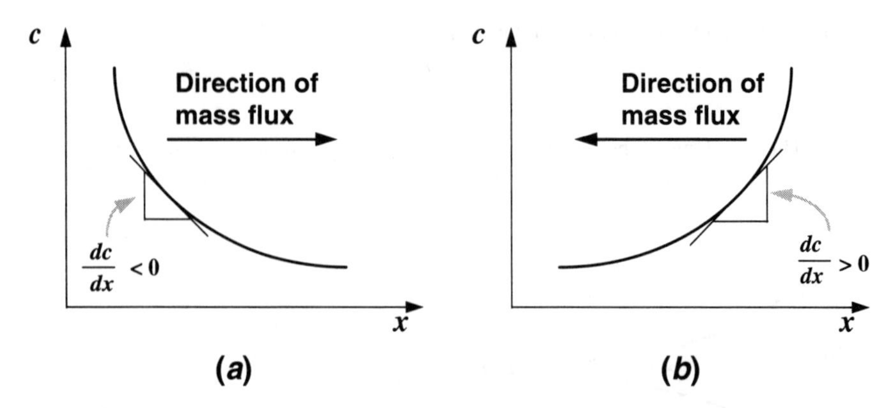

FIGURE 8.4
Graphical depiction of the effect of the concentration gradient on the mass flux. Because mass moves "downhill" from high to low concentrations, the flow in (*a*) is from left to right in the positive *x* direction. However, due to the orientation of cartesian coordinates, the slope is negative for this case. Thus a negative gradient leads to a positive flux. This is the origin of the negative sign in Fick's first law. The reverse case is depicted in (*b*), where the positive gradient leads to a negative mass flow from right to left.

$$\frac{dc}{dx} \cong \frac{c_2 - c_1}{\ell} \tag{8.4}$$

where ℓ = mixing length (L), which is the length over which the diffusive mixing takes place. Equations 8.2 and 8.4 can be substituted into Eq. 8.3 to give

$$V_1 \frac{dc_1}{dt} = \frac{DA_c}{\ell}(c_2 - c_1) \tag{8.5}$$

Comparison of Eq. 8.5 with Eq. 8.1 indicates that we have now defined the diffusion flow in terms of more fundamental parameters,

$$D' = \frac{DA_c}{\ell} \tag{8.6}$$

According to Fick's law, diffusion flow is made up of three components. The diffusion coefficient D reflects the vigor of the mixing process. The area A_c accounts for the fact that the mass transport should be directly proportional to the size of the interface across which the mixing occurs. Finally the mixing length ℓ defines the distance across which the mixing takes place.

The diffusion coefficient D (L^2 T^{-1}) serves as the fundamental parameter to quantify the diffusion process. It should be mentioned that alternative parameterizations and nomenclature are sometimes used and can lead to confusion.

For example a distinction between molecular diffusion and turbulent mixing is often made. Even though they take the same form mathematically, we use different nomenclature to designate them: D for molecular and E for turbulent diffusion.

Aside from nomenclature, parameters are often clustered for convenience. In cases where estimation of the mixing length is difficult or impossible, the length is often combined with the diffusion coefficient to develop a single parameter. For the molecular case the parameter is defined as

$$v_d = \frac{D}{\ell} \tag{8.7}$$

where v_d is called a diffusion mass-transfer coefficient (L T^{-1}).

Further the bulk diffusion coefficient D' or E' is frequently used for mathematical convenience. Recall that this parameterization (Eq. 8.6) confounds both the area and mixing length with the diffusion coefficient. For example for turbulent diffusion, the bulk coefficient

$$E' = \frac{EA_c}{\ell} \tag{8.8}$$

is often used for convenience in model formulations.

8.4 EMBAYMENT MODEL

As depicted in Fig. 8.5 a well-mixed bay connected to a large lake is a simple illustration of an incompletely mixed system (Chapra 1979). This is particularly instructive because it is directly comparable with the completely mixed lake model described in the previous lectures.

Mass balances for the main lake and the bay can be written as

$$V_1\frac{dc_1}{dt} = W_1 - Q_1c_1 - k_1V_1c_1 + Q_2c_2 + E'(c_2 - c_1) \tag{8.9}$$

and

$$V_2\frac{dc_2}{dt} = W_2 - Q_2c_2 - k_2V_2c_2 + E'(c_1 - c_2) \tag{8.10}$$

where the subscripts 1 and 2 refer to the main lake and the bay, respectively.

8.4.1 Estimation of Diffusion

The bay/lake model can be used to estimate the bulk diffusion coefficient for cases where there is a gradient of a conservative substance (that is, one with $k = 0$)

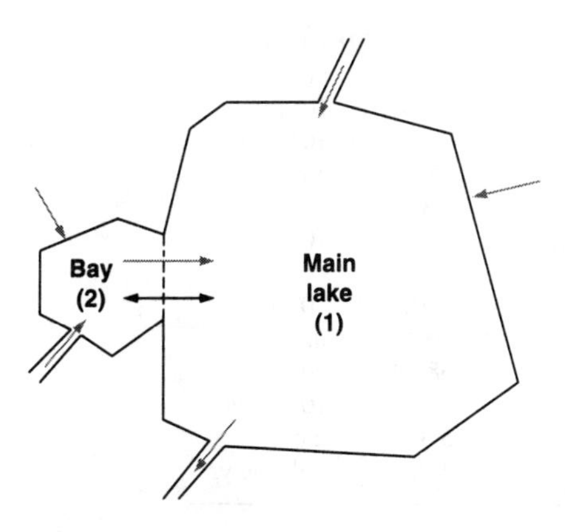

FIGURE 8.5
A lake/embayment system.

between the main lake and the bay. This is done by writing a mass balance for the conservative substance, as in

$$V_2 \frac{ds_2}{dt} = W_2 - Q_2 s_2 + E'(s_1 - s_2) \tag{8.11}$$

where s designates the concentration of the conservative substance ($M\ L^{-3}$). At steady-state, Eq. 8.11 can be solved for

$$E' = \frac{W_2 - Q_2 s_2}{s_2 - s_1} \tag{8.12}$$

EXAMPLE 8.2. USING NATURAL TRACERS TO ESTIMATE DIFFUSION. For Saginaw Bay the bulk diffusion coefficient can be estimated by using the gradient of the conservative substance chloride. Estimate the diffusion coefficient and the mass-transfer coefficient for the process.

Solution: Substituting values from Table 8.1 into Eq. 8.12 yields

$$E' = \frac{0.353 \times 10^{12} - [7 \times 10^9(15.2)]}{15.2 - 5.4} = 25.2 \times 10^9 \text{ m}^3 \text{ yr}^{-1}$$

Note that the rate of diffusion is considerably greater than the bay's outflow rate of 7×10^9 m^3 yr^{-1}.

The cross-sectional area between Saginaw Bay and Lake Huron is approximately 0.17×10^6 m^2. The mixing length is approximately 10 km. Therefore the mass-transfer coefficient can be computed as

$$v_d = \frac{E'}{A_c} = \frac{25.2 \times 10^9}{0.17 \times 10^6} = 1.48 \times 10^5 \text{ m yr}^{-1}$$

and the diffusion coefficient as

$$E = v_d \ell = 1.48 \times 10^5 (10 \times 10^3) = 1.48 \times 10^9 \text{ m}^2 \text{ yr}^{-1}$$

TABLE 8.1
Parameters for the Saginaw Bay/Lake Huron system

Parameter	Symbol	Value	Units
Saginaw Bay:			
Volume	V_2	8	10^9 m^3
Depth	H_2	5.81	m
Surface area	A_2	1,376	10^6 m^2
Outflow	Q_2	7	10^9 m^3 yr^{-1}
Chloride concentration	s_2	15.2	g m^{-3}
Chloride loading	W_{s2}	0.353	10^{12} g yr^{-1}
Phosphorus loading	W_{p2}	1.42	10^{12} mg yr^{-1}
Lake Huron:			
Volume	V_1	3507	10^9 m^3
Depth	H_1	60.3	m
Surface area	A_1	58,194	10^6 m^2
Outflow	Q_1	161	10^9 m^3 yr^{-1}
Chloride concentration	s_1	5.4	g m^{-3}
Phosphorus loading	W_{p1}	4.05	10^{12} mg yr^{-1}

or in more conventional units,

$$E = 1.48 \times 10^9 \text{ m}^2 \text{ yr}^{-1} \times \left(\frac{10^4 \text{ cm}^2}{\text{m}^2} \frac{\text{yr}}{365 \text{ d}} \frac{\text{d}}{86,400 \text{ s}} \right) = 4.7 \times 10^5 \text{ cm}^2 \text{ s}^{-1}$$

8.4.2 Steady-State Solution

At steady-state the concentration in the bay and the main lake can be determined using methods described in Lec. 6.

EXAMPLE 8.3. STEADY-STATE TOTAL P BUDGET FOR LAKE HURON AND SAGINAW BAY. Information for the Lake Huron/Saginaw Bay system is summarized in Table 8.1. The sedimentation loss of total phosphorus can be parameterized by an apparent settling velocity v of approximately 16 m yr^{-1} (Chapra 1977). Use the models described in Lec. 6 to determine (*a*) the inflow concentration and (*b*) the steady-state concentration.

Solution: (*a*) To calculate the inflow concentration for the main lake, we must determine its inflow rate. This is done by correcting its outflow rate to account for the flow from Saginaw Bay:

$$Q_{1,\text{in}} = Q_1 - Q_2 = 161 \times 10^9 - 7 \times 10^9 = 154 \times 10^9 \text{ m}^3 \text{ yr}^{-1}$$

The other values from Table 8.1 can then be used to compute

$$c_{1,\text{in}} = \frac{4.05 \times 10^{12}}{154 \times 10^9} = 26.3 \text{ } \mu\text{g L}^{-1}$$

$$c_{2,\text{in}} = \frac{1.42 \times 10^{12}}{7 \times 10^9} = 202.9 \text{ } \mu\text{g L}^{-1}$$

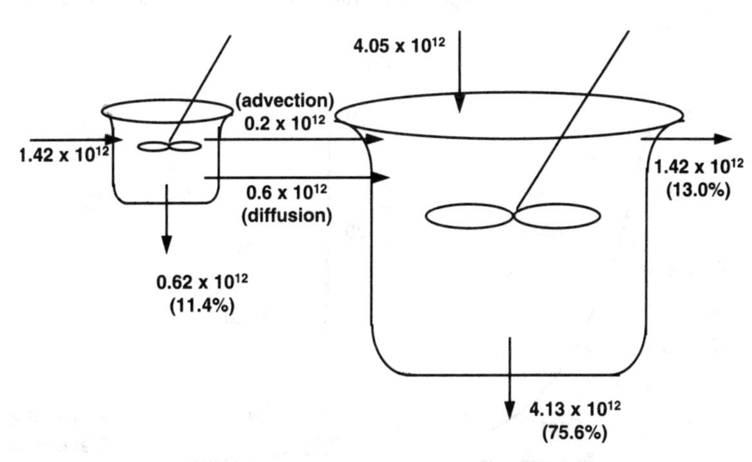

FIGURE 8.6
A total phosphorus budget for Lake Huron and Saginaw Bay computed in Example 8.2. Note that the percentages are based on the total loading to the system.

Thus the inflow concentration for Saginaw Bay is extremely high.

(*b*) The steady-state solution is (Eqs. 6.13 and 6.14)

$$c_1 = \frac{1}{1.102 \times 10^{12}} W_{p1} + \frac{1}{1.857 \times 10^{12}} W_{p2} = 3.671 + 0.768 = 4.44\ \mu\text{g L}^{-1}$$

$$c_2 = \frac{1}{2.373 \times 10^{12}} W_{p1} + \frac{1}{5.345 \times 10^{10}} W_{p2} = 1.705 + 26.658 = 28.36\ \mu\text{g L}^{-1}$$

Figure 8.6, which shows the magnitude of the individual terms of the steady-state budget, indicates the importance of the sedimentation mechanism. Note that about 87% of the total phosphorus load ultimately finds its way into the bottom sediments. In addition note that diffusion represents a significant mass-transfer mechanism. It accounts for approximately 42% of the total phosphorus that leaves Saginaw Bay.

8.4.3 Time-Variable Solution

The time-variable concentrations in the bay and the main lake can be determined using the methods described in Lec. 6.

EXAMPLE 8.4. TIME-VARIABLE SOLUTION FOR LAKE HURON AND SAGINAW BAY. Determine the temporal response of the Saginaw Bay/Lake Huron system following termination of loads. Assume that the system is initially at the steady-state concentrations determined in Example 8.3.

Solution: The eigenvalues can be computed as (Eq. 6.43)

$$\begin{matrix}\lambda_f \\ \lambda_s\end{matrix} = 3.545\left(1 \pm \sqrt{1 - 0.1693}\right) = \begin{matrix}6.777 \\ 0.3141\end{matrix}\ \text{yr}^{-1}$$

These, along with other values from Table 8.1, can be used to determine the general solutions

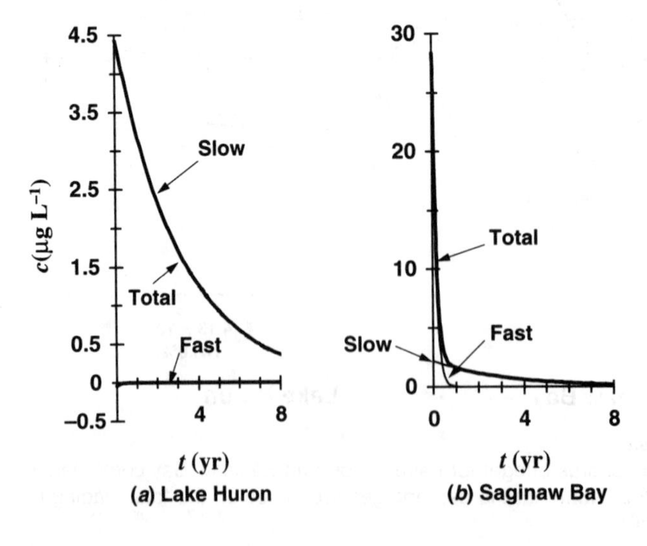

FIGURE 8.7
General solutions for (*a*) Lake Huron and (*b*) Saginaw Bay following the termination of loadings of total phosphorus.

$$c_1 = -0.037e^{-6.777t} + 4.476e^{-0.3141t}$$
$$c_2 = 26.18e^{-6.777t} + 2.18e^{-0.3141t}$$

Values of time can be substituted into these equations and the results displayed in Fig. 8.7. In addition a semi-logarithmic plot of the response is depicted in Fig. 8.8. The response of Lake Huron is dominated by its slow eigenvalue. This implies that its response is only slightly influenced by the response of Saginaw Bay. For Saginaw Bay the fast eigenvalue dominates initially. Thus the bay manifests a substantial improvement in less than 1 yr. However, after 1 yr the slow eigenvalue retards Saginaw Bay's recovery as its fate becomes dominated by the sluggish recovery of Lake Huron.

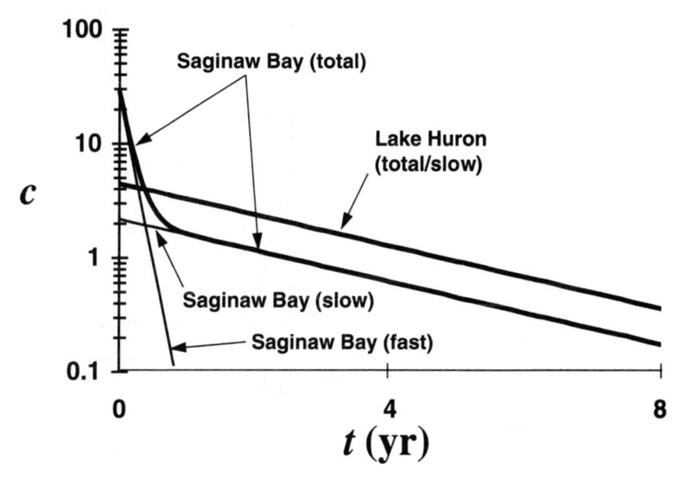

FIGURE 8.8
General solutions for Lake Huron and Saginaw Bay following the termination of loadings of total phosphorus. In this case the responses are displayed on a semi-logarithmic plot to make the eigenvalues obvious.

BOX 8.1. The Mixing Length (Confusion about Diffusion)

Estuaries were the first systems where the diffusion mechanism was incorporated into water-quality modeling. When the estuary-modeling theory was applied to other incompletely mixed systems, some confusion resulted regarding the proper choice of the mixing length.

The typical estuary modeling situation is depicted in Fig. B8.1*a*. As shown, such systems were usually divided into segments. Because many segments were employed, the gradients of pollutant between the segments were usually small. Consequently the segment length served as the proper choice for the mixing length. The validity of this approach can be affirmed by recalling that the purpose of the length is to provide an accurate finite-difference estimate of the derivative in Fick's law (recall Eq. 8.4).

In subsequent years, when modelers began to simulate bay/lake systems, the situation was usually quite different. As depicted in Fig. B8.1*b* the bay and the lake are treated as CSTRs separated by a constriction. Thus each has a relatively constant

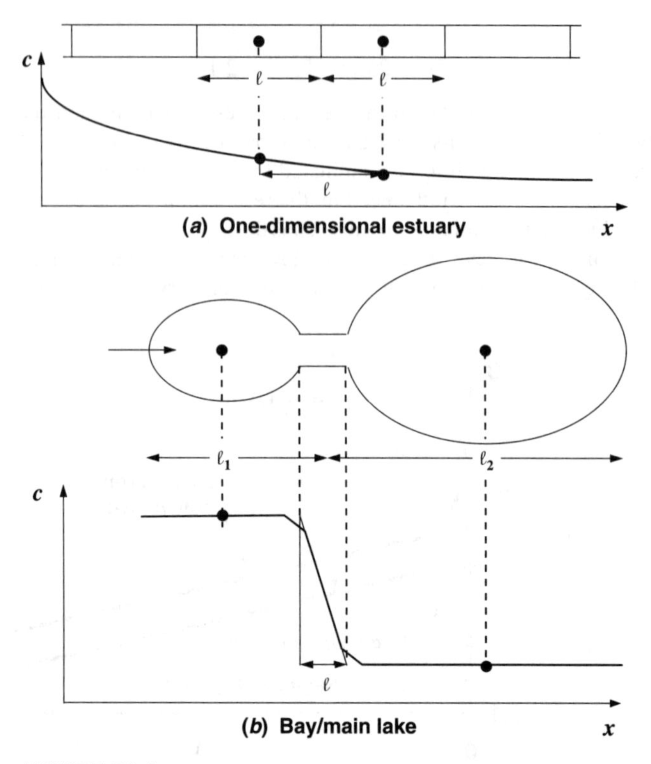

FIGURE B8.1

concentration, and the gradient or transition occurs in the narrow boundary region where diffusive mixing is minimized. Consequently the mixing length is best represented by the length of the boundary region rather than by the average length of the adjacent sections.

When we use either the bulk diffusion E' or the mass transfer coefficients v_d, the issue of mixing length is a moot point because the mixing length is part of the coefficient. However, when the analysis is based on the diffusion coefficient E, the choice of the length is critical. A related example will occur later in this text when we model vertical temperature profiles in stratified lakes.

In summary the incorporation of an embayment into our completely mixed model has certainly added to its complexity. If the lake had two bays, the analytical solution would be even more unwieldy (three equations with three unknowns). Consequently more complicated incompletely-mixed systems are almost invariably solved numerically with a computer (see Lecs. 11 through 13 for details).

Although the modeling of most incompletely-mixed systems involves computerized solution techniques, the conceptual basis for these models is straightforward. As in Fig. 8.9, an estuary is often treated as a series of coupled, well-mixed reactors. The derivation of the mass balances for these reactors is then handled in a similar fashion to the lake/embayment system addressed here. The equations are then solved simultaneously using numerical methods.

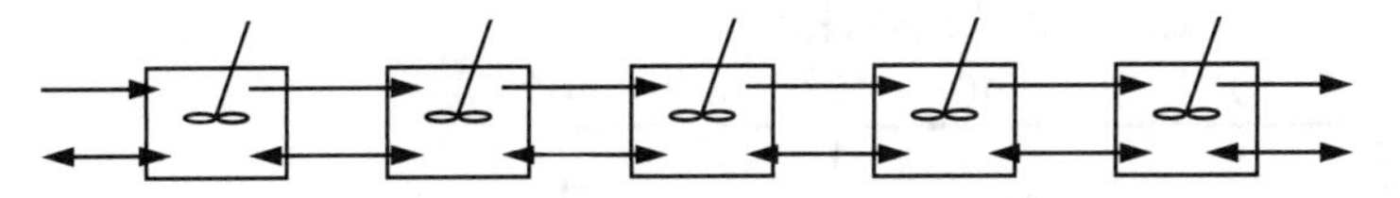

FIGURE 8.9
Complicated incompletely mixed systems such as estuaries can be represented as a series of coupled completely mixed systems.

8.5 ADDITIONAL TRANSPORT MECHANISMS

To this point we have provided a brief introduction to the diffusion mechanism. Now we would like to present some additional information about transport. First, we discuss the magnitudes of diffusion coefficients encountered in natural waters.

8.5.1 Turbulent Diffusion

Just as mass is transported via random molecular motion, matter is mixed by larger scale eddies or whirls in natural waters. If long enough observational time and space scales are employed, this motion can be viewed as random and treated mathematically as a diffusion process. Although equations of the kind developed for molecular motion can be used to characterize turbulent mass transport, two important differences should be noted.

First, because the eddies are much larger than the random motion of molecules, mixing by turbulent diffusion is much greater than by molecular diffusion. As shown in Fig. 8.10, turbulent diffusion coefficients are several orders of magnitude greater than those on a molecular scale. Note also that horizontal diffusion is generally much greater than vertical diffusion. Likewise, effective diffusion through porous media such as bottom sediments is less than molecular diffusion in free solution because, among other things, the solute must move around particles.

Second, in contrast to molecular diffusion where the motion is assumed to be identical in scale, turbulence is composed of a wide range of eddy sizes. Thus it

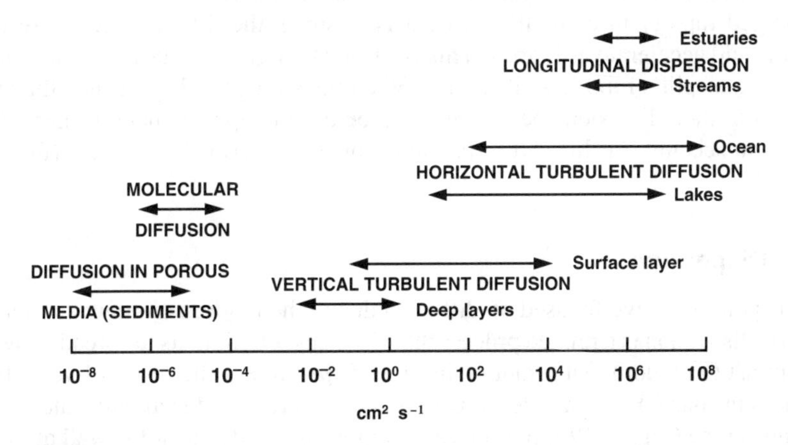

FIGURE 8.10
Typical ranges of the diffusion coefficient in natural waters and sediments.

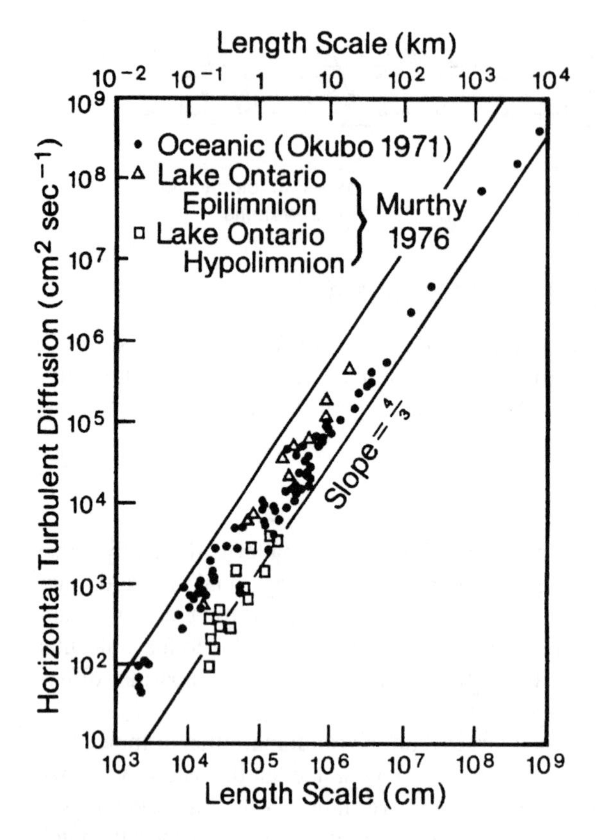

FIGURE 8.11
Relationship of horizontal diffusivity and scale length in the ocean and Lake Ontario. Lines define an envelope around the oceanic data with a slope of 4/3 (Okubo 1971, with additional data from Murthy 1976).

would be expected that the resulting transport is scale-dependent. It has been shown that the turbulent diffusion coefficient E varies with the $\frac{4}{3}$ power of the scale of the phenomenon (Richardson 1926). This relationship is supported by observations in both oceans and lakes (Fig. 8.11).

The scale dependence of eddy diffusivity can have great practical consequence for modeling pollutant transport in natural waters. For cases where a pollutant is discharged rapidly in a small area (that is, a spill), the diffusion of the resulting cloud would accelerate as it grew. This is caused by larger and larger eddies playing a role in the spill's mixing as it spread. When the spill grew bigger than the largest eddy, a constant diffusion coefficient could be used to approximate further mixing. Models to account for this sort of phenomenon are described elsewhere (Fischer et al. 1979).

8.5.2 Dispersion

To this point we have focused on diffusion due to the random motion of water (Fig. 8.12*a*). Dispersion is a related process that also causes pollutants to spread. However, in contrast to random water motion in time, ***dispersion*** is the result of velocity differences in space. For example suppose that dye were introduced into water flowing through a pipe (Fig. 8.12*b*). In such cases a molecule of dye near the wall of the pipe would move more slowly than a molecule near the center due to a velocity gradient

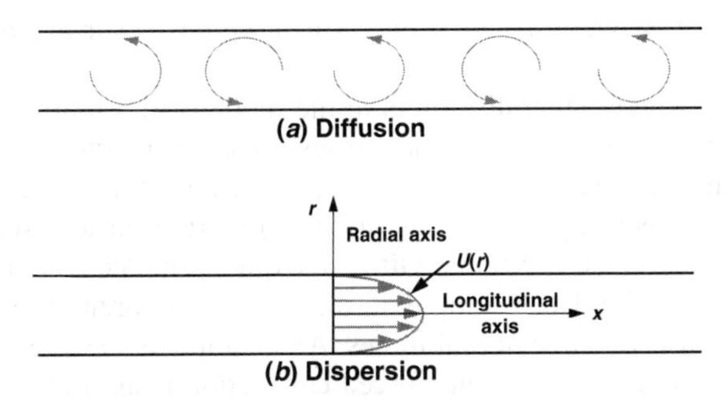

FIGURE 8.12
Contrast between diffusion and dispersion. Both tend to "spread out" pollutants. Diffusion is due to random motion of the water in time, whereas dispersion is due to differential movement of the water in space.

or shear. The net effect of these differences in mean velocity is to spread or mix the dye along the pipe's axis. Interestingly it can be shown (Taylor 1953, Fischer et al. 1979) that, coupled with random radial movement due to diffusion, this dispersion (given enough time) can be represented as a Fickian diffusion process.

In the environment, turbulent diffusion and dispersion can, individually or in concert, cause mixing of a substance. For example in rivers and estuaries, dispersion usually predominates because of the strong shears developed by the large mean flows (due to gravity flow and tides) and the constraining channel. Dispersion can also be of some importance in highly advective river-run reservoirs. Additionally, for all classes of water, dispersion is often important on smaller time and space scales. In particular, shears can be developed near system boundaries such as the shoreline and the bottom.

For lakes and wide bodies of water such as bays, diffusion tends to predominate. For these cases the wind is the primary agent imparting random motion to the water. In particular, for longer term simulations, wind mixing acts as a random process and can be adequately represented as turbulent diffusion.

In the remainder of this book we will use the term "dispersion" when dealing with narrow, flowing bodies of water such as one-dimensional streams and estuaries. For lakes, bays, and vertical transport we will usually employ "diffusion."

8.5.3 Conduction/Convection

Finally there are some additional mechanisms and terminology that are used in water-quality modeling and bear mention. Conduction and convection are two processes that originate from heat transfer and aerodynamics that are roughly analogous to diffusion and advection.

Conduction refers to the transfer of heat by molecular activity from one substance to another or through a substance. Because conduction is similar to molecular diffusion of mass, individuals will often use the two terms interchangeably. Some may even take the liberty of extending the term "conduction" to refer to turbulent diffusion. In this text we will use it only when we cover surface heat transfer in Lec. 30.

Convection, which generally refers to the motions in a fluid that result in the transport and mixing of the fluid's properties, takes two forms. ***Free convection*** refers to vertical atmospheric motions due to the buoyancy of heated or cooled fluid. For example in meteorology, the rising of heated surface air and sinking of cooler air aloft is called "free convection." Oftentimes the term "convection" is assumed to imply free convection [for example, in meteorology (Ahrens 1988)]. In contrast ***forced convection*** is due to external forces. An example is the lateral movement of heat or mass due to the wind. Thus forced convection is akin to advection. In the present context, we will use the term convection only when we are discussing free convection in relation to thermal stratification.

In summary water motion in natural waters is a complicated process and, as with any idealization, the above concepts are intended as a brief introduction to some of its essential features and terminology. In-depth discussions of the subtleties of mixing (Fischer et al. 1979) and lake hydrodynamics (Hutchinson 1957, Mortimer 1974, Boyce 1974) are presented elsewhere. In addition a more detailed discussion of transport processes in a number of natural waters is included in Lecs. 14 through 17.

BOX 8.2. Modeling Cocktail-Party Etiquette

Imagine you're at a cocktail party with a group of environmental engineers. As might be expected they're making small talk about pollution. In a loud voice you offhandedly mention that you are studying a lake with a horizontal eddy diffusion coefficient of about 10^{-6} cm^2 s^{-1}. Suddenly the room grows unnaturally quiet. One of the engineers rolls her eyes, while two glance at each other and snicker. Others avoid eye contact, as one by one they wander off. Before you know it you're standing alone. Upon reflection your heart plummets. You recognize that you have committed an environmental-modeling *faux pas*. Sweat trickles down the small of your back and your face burns as you recall (Figs. 8.10 and 8.11) that your stated diffusion coefficient is at least 6 to 8 orders of magnitude below a credible value. You slink toward the door, bereft, a pariah. No one says good-bye.

Clearly one mark of the experienced modeler is the capability to recognize reasonable values for common model parameters and variables. Aside from preventing social embarrassment, such knowledge can be invaluable in professional contexts.

For example you could be called upon as an expert to critique someone else's model. Recognizing odd parameters and variables is an important tool in such exercises. In addition, when developing your own models, you will often be unable to estimate all model parameters directly. In such cases educated guesses must be made. Finally sometimes a seemingly odd parameter estimate serves as a valuable clue in the model-development process. For example it could mean that the model is missing an important mechanism. Alternatively it could mean that the system being studied has atypical characteristics. Such evidence could result in the identification of research and data gaps.

In all these cases the ability to identify typical ranges of parameters and variables is an invaluable asset for the expert modeler. Throughout the remainder of this text I will endeavor to present such ranges to help you build this capability.

PROBLEMS

8.1. A well-mixed tidal bay is connected with the ocean by a channel.

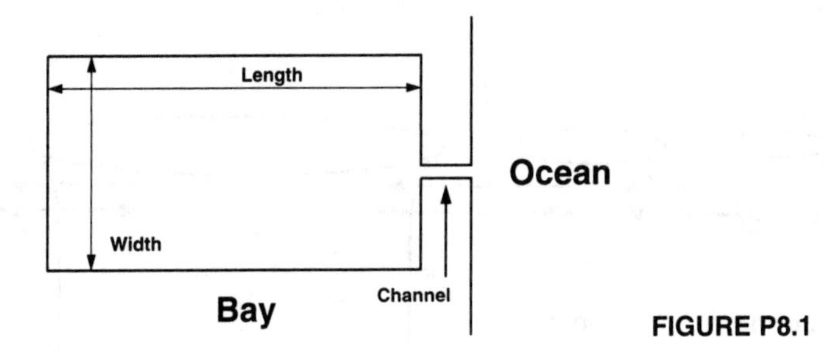

FIGURE P8.1

The bay and the channel have the following characteristics:

	Bay	Channel
Length (m)	1000	100
Width (m)	500	10
Depth (m)	5	2

At $t = 0$ d, 100 kg of a conservative dye is instantaneously released and distributed across the bay. The following measurements of dye concentration are made over the ensuing period:

t (d)	20	40	60	80	100	120
c (ppb)	32	29	23	20	17	16

(*a*) What is the diffusion coefficient for exchange between the bay and the ocean? Express your result in $m^2\,d^{-1}$.

(*b*) What mass loading of a reactive pollutant ($k = 0.01\ d^{-1}$) could be input to this system under steady-state conditions if the allowable concentration in the system were 1 ppm? Express your result in $g\,d^{-1}$. Note that both the conservative dye and the reactive pollutant have negligible concentration in the ocean. Also the residence time in the channel is negligible compared to the residence time of the bay.

8.2. A well-mixed tidal bay is adjacent to the ocean as shown in Fig. P8.2. The bay has a diameter of 1 km and a depth of 5 m. The short channel connecting the two waterbodies is 50 m long, 10 m wide, and 2 m deep. The transfer between the bay and the ocean

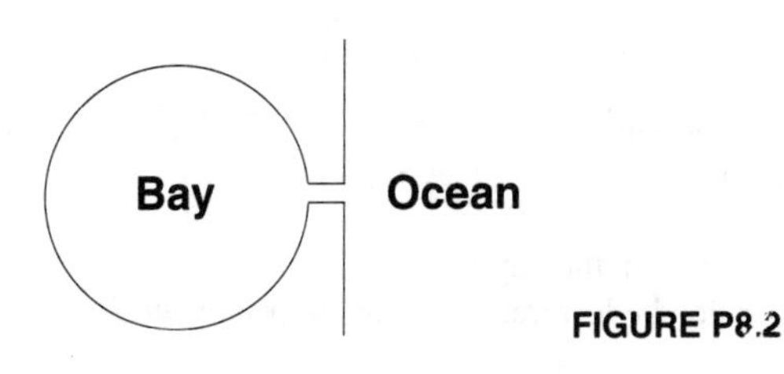

FIGURE P8.2

has been characterized by a turbulent diffusion coefficient of 10^5 m^2 d^{-1}. Determine the bay's 95% response time for a spill of a substance that settles at a rate of 0.1 m d^{-1}.

8.3. Green Bay is a major embayment connected with Lake Michigan. The Green Bay/Lake Michigan system can be modeled as three reactors,

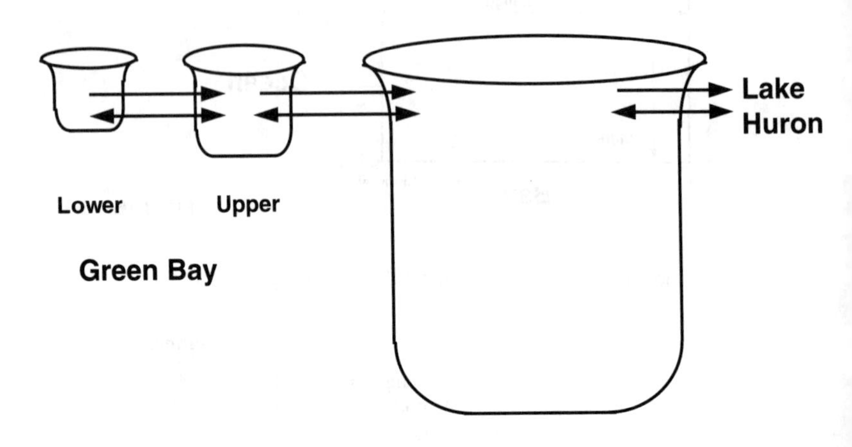

	Lower		Upper		Lake Michigan	
Flow, km^3 yr^{-1}		5.4		10.8		36
Bulk diffusion, km^3 yr^{-1}		20		30		70
Volume, km^3	7.5		55.4		4846	
Surface area, km^2	953		3260		53537	
Phosphorus loading, mta	1200		200		5500	

FIGURE P8.3

If phosphorus settles at a rate of approximately 12 m yr^{-1}, use the matrix inverse approach to compute the steady-state concentrations. Assume that Lake Huron has a constant concentration of 4.4 μg L^{-1}.

8.4. A bay is subject to a runoff event of a contaminant (see Fig. P8.4) that volatilizes at a rate of 1 m d^{-1}. Assume that inflow = outflow during the runoff event and that the main body of water has negligible concentration of this contaminant. The bay has the following characteristics:

Volume = 10×10^6 m^3
Mean depth = 3 m
Normal outflow = 0
Interface diffusion coefficient = 1×10^5 m^2 d^{-1}
Interface cross-sectional area = 1500 m^2
Interface mixing length = 100 m

(*a*) Determine the peak concentration in the bay.
(*b*) When will the bay concentration be lowered to 5% of the peak value?

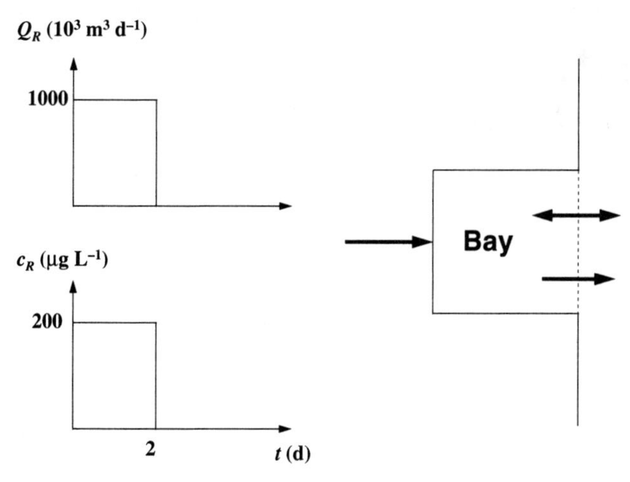

FIGURE P8.4

8.5. A well-mixed circular bay is connected to a large lake. The lake is so big relative to the bay that for all intents and purposes it can be assumed to be an "ocean"; that is, the bay has negligible impact on its state. At steady-state the bay and the lake have chloride concentrations of 30 and 7 mg L^{-1}, respectively. The bay has the following characteristics:

Radius = 0.25 km
Inflow = outflow to main lake = 5×10^6 m^3 yr^{-1}
Inflow chloride concentration = 70 mg L^{-1}
Mean depth = 5 m
Cross-sectional area of interface with main lake = 500 m^2
Mixing length of transition zone between the bay and lake = 0.1 km

(*a*) Determine the diffusion coefficient to parameterize turbulent exchange between the two water bodies. Express your result in cm^2 s^{-1}.

(*b*) How much loading of a highly concentrated waste (that is, its flow is negligible) could be added to the bay while maintaining a concentration of 10 mg L^{-1}? Note that the waste decays according to first-order kinetics at 1 yr^{-1}. Assume that the main lake will have zero concentration of this pollutant. Express your result in kg yr^{-1}.

(*c*) What is the bay's 95% response time for the waste in days?

LECTURE 9

Distributed Systems (Steady-State)

LECTURE OVERVIEW: We now turn to distributed systems. I introduce two idealized reactors that are used to model these systems: plug-flow and mixed-flow reactors. Then I show how these models can be used to simulate steady-state distributions of pollutants in streams and estuaries.

To this point I have focused on completely mixed reactors or CSTRs. In Lec. 8, I illustrated how incompletely mixed systems can be modeled by breaking them into a coupled set of CSTRs (recall Fig. 8.9). Such characterizations are formally referred to as ***lumped-parameter systems***, so called because each reactor has its own set of parameters. Thus even though the system is continuous, we approximate it as "lumps."

In contrast it is possible to model such systems while maintaining their continuous nature. Such characterizations are formally referred to as ***distributed-parameter systems***. In this lecture I will introduce you to such systems. To do this I focus on two idealized reactors: plug-flow and mixed-flow reactors. I also show how these models can be used to simulate steady-state distributions of pollutants in streams and estuaries.

9.1 IDEAL REACTORS

Both plug-flow and mixed-flow reactors are elongated rectangular basins, as depicted in Fig. 9.1. They are assumed to be well-mixed laterally (y) and vertically (z). Thus we are concerned only with variations in the longitudinal dimension (x).

As in Fig. 9.1, a mass balance is taken for a differential element of length Δx,

$$\Delta V \frac{\partial c}{\partial t} = J_{\text{in}} A_c - J_{\text{out}} A_c \pm \text{reaction} \tag{9.1}$$

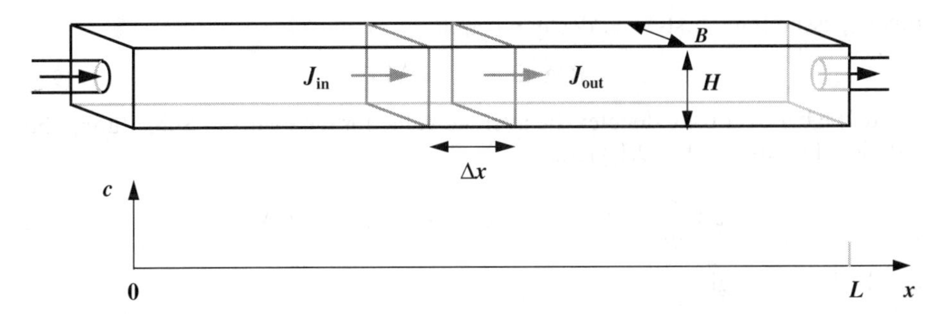

FIGURE 9.1
An elongated rectangular reactor.

where ΔV = volume of the element (L^3) = $A_c\,\Delta x$
A_c = cross-sectional area of the reactor (L^2) = BH
B = channel width (L)
H = depth (L)
J_{in} and J_{out} = flux of mass in and out of the element due to transport ($M\,L^{-2}\,T^{-1}$)
reaction = gain or loss of mass within the element due to reaction ($M\,T^{-1}$)

9.1.1 Plug-Flow Reactor (PFR)

The plug-flow reactor is one in which advection dominates. As in Fig. 9.2, this means that a "plug" of conservative dye introduced at one end will remain intact as it passes through the reactor. In other words substances are discharged in the same sequence as they enter the reactor.

For a plug-flow reactor, the flux into the element is defined as

$$J_{in} = Uc \tag{9.2}$$

where U = velocity ($L\,T^{-1}$) = Q/A_c. The flux out is estimated by a first-order Taylor-series expansion,

$$J_{out} = U\left(c + \frac{\partial c}{\partial x}\Delta x\right) \tag{9.3}$$

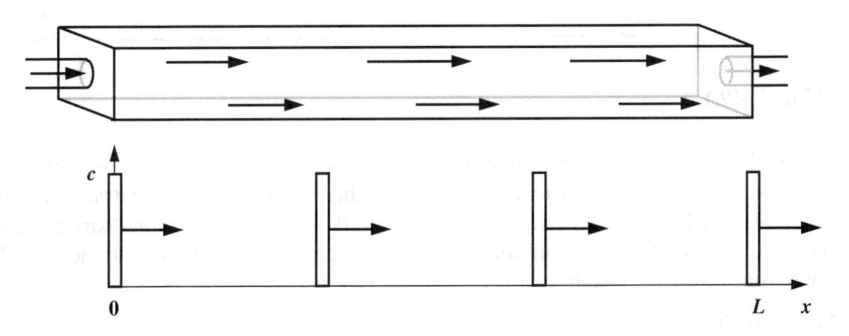

FIGURE 9.2
A plug-flow reactor.

Finally, assuming a first-order decay reaction,

$$\text{Reaction} = -k\Delta V\bar{c} \tag{9.4}$$

where the bar over the c denotes the average value for the element. Substituting the individual terms into Eq. 9.1 yields

$$\Delta V\frac{\partial c}{\partial t} = UA_c c - UA_c\left(c + \frac{\partial c}{\partial x}\,\Delta x\right) - k\,\Delta V\bar{c} \tag{9.5}$$

Combining terms gives

$$\Delta V\frac{\partial c}{\partial t} = -UA_c\frac{\partial c}{\partial x}\,\Delta x - k\,\Delta V\bar{c} \tag{9.6}$$

Dividing by $\Delta V = A_c\,\Delta x$ and taking the limit ($\Delta x \to 0$) yields

$$\frac{\partial c}{\partial t} = -U\frac{\partial c}{\partial x} - kc \tag{9.7}$$

At steady-state Eq. 9.7 becomes

$$0 = -U\frac{dc}{dx} - kc \tag{9.8}$$

which, if $c = c_0$ at $x = 0$, can be solved for

$$c = c_0 e^{-\frac{k}{U}x} \tag{9.9}$$

EXAMPLE 9.1. PLUG-FLOW REACTOR. In Example 5.2 we used a cascade model to simulate the steady-state distribution of concentration in an elongated tank.

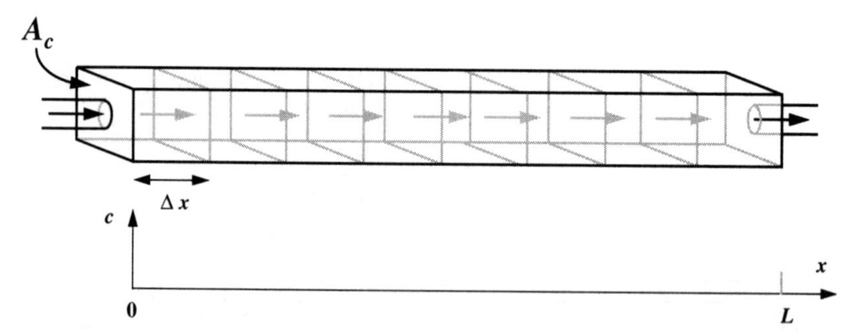

FIGURE E9.1-1

The tank has cross-sectional area $A_c = 10\text{ m}^2$, length $L = 100$ m, velocity $U = 100\text{ m hr}^{-1}$, and a first-order reaction rate $k = 2\text{ hr}^{-1}$. The inflow concentration is 1 mg L^{-1}. Use the plug-flow model to compute the steady-state concentration distribution for the tank. Display the concentrations along with the results for the $n = 4$ and 8 CSTR approximations from Example 5.2.

Solution: The plug-flow model is

$$c = 1e^{-\frac{2}{100}x} = 1e^{-0.02x}$$

Therefore this equation can be used to compute the distribution. The results along with those from Example 5.2 are shown below:

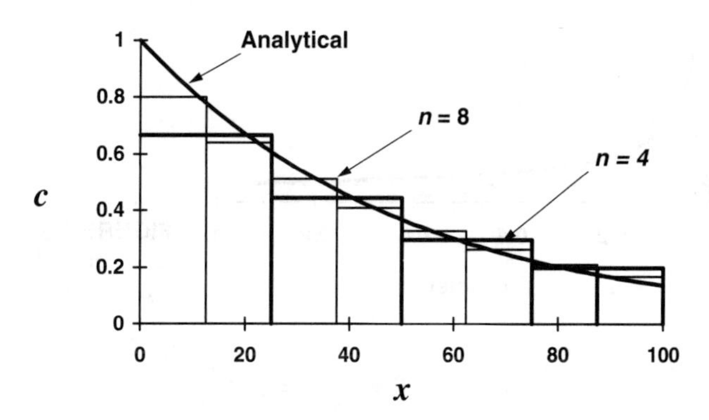

FIGURE E9.1-2

As the number of reactors becomes larger, the cascade model seems to be approaching the analytical solution. In fact it can be proved that the analytical solution is the limit of the cascade approximation.

9.1.2 Comparison of the CSTR and the PFR

Performance of the CSTR and the plug-flow reactor can now be compared. A nice way to do this is via the residence time required to achieve a certain level of performance efficiency. For example recall that the steady-state solution for the CSTR with first-order reaction can be represented as

$$c = c_{\text{in}} \frac{Q}{Q + kV} \tag{9.10}$$

The performance efficiency can be defined in terms of the transfer function (Eq. 3.22)

$$\beta \equiv \frac{c}{c_{\text{in}}} = \frac{Q}{Q + kV} \tag{9.11}$$

Dividing the numerator and the denominator by Q gives

$$\beta = \frac{1}{1 + k\tau_w} \tag{9.12}$$

where τ_w = water residence time = V/Q. This result can be solved for residence time,

$$\tau_w = \frac{1}{k} \frac{1 - \beta}{\beta} \tag{9.13}$$

In a similar fashion the efficiency of the PFR can be defined as

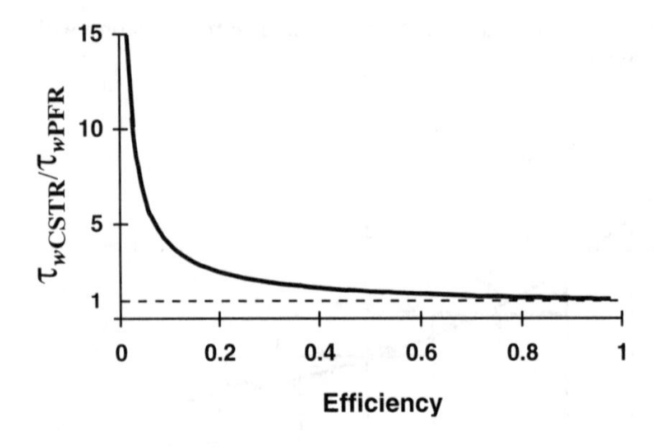

FIGURE 9.3
Ratio of CSTR to PFR residence times versus efficiency.

$$\beta \equiv \frac{c}{c_{\text{in}}} = e^{-\frac{k}{U}x} \tag{9.14}$$

Recognizing that the residence time of the PFR is equal to L/U,

$$\beta = e^{-k\tau_w} \tag{9.15}$$

This result can also be solved for residence time,

$$\tau_w = \frac{1}{k} \ln\left(\frac{1}{\beta}\right) \tag{9.16}$$

The two types of reactors can be compared by plotting the ratio of the two residence times versus efficiency. The result is shown in Fig. 9.3. This plot shows that the CSTR always requires a higher residence time to achieve a desired efficiency. This leads to the general conclusion that for first-order reactions the PFR is more efficient than the CSTR for removing pollutants.

9.1.3 Mixed-Flow Reactor (MFR)

The mixed-flow reactor is one in which both advection and diffusion/dispersion are important. As in Fig. 9.4, this means that a plug of conservative dye introduced at one end will "spread out" as it passes through the reactor.

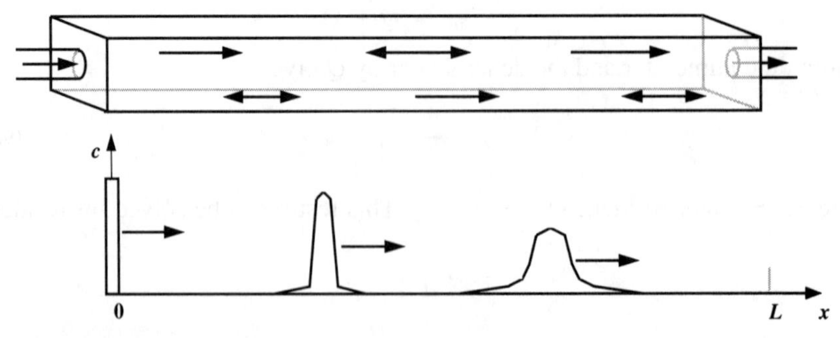

FIGURE 9.4
A mixed-flow reactor.

For the mixed-flow reactor, the flux in is defined as

$$J_{\text{in}} = Uc - E\frac{\partial c}{\partial x} \tag{9.17}$$

where E = turbulent diffusion. Note that the second term is Fick's first law (Eq. 8.2). The flux out is

$$J_{\text{out}} = U\left(c + \frac{\partial c}{\partial x}\Delta x\right) - E\left[\frac{\partial c}{\partial x} + \frac{\partial}{\partial x}\left(\frac{\partial c}{\partial x}\right)\Delta x\right] \tag{9.18}$$

Substituting the individual terms into Eq. 9.1 yields

$$\Delta V\frac{\partial c}{\partial t} = UA_c c - UA_c\left(c + \frac{\partial c}{\partial x}\Delta x\right) - EA_c\frac{\partial c}{\partial x} + EA_c\left[\frac{\partial c}{\partial x} + \frac{\partial}{\partial x}\left(\frac{\partial c}{\partial x}\right)\Delta x\right] - k\,\Delta V\bar{c} \tag{9.19}$$

Combining terms gives

$$\Delta V\frac{\partial c}{\partial t} = -UA_c\frac{\partial c}{\partial x}\Delta x + EA_c\frac{\partial^2 c}{\partial x^2}\Delta x - k\,\Delta V\bar{c} \tag{9.20}$$

Dividing by $\Delta V = A_c\,\Delta x$ and taking the limit ($\Delta x \to 0$) yields

$$\frac{\partial c}{\partial t} = -U\frac{\partial c}{\partial x} + E\frac{\partial^2 c}{\partial x^2} - kc \tag{9.21}$$

or at steady-state,

$$0 = -U\frac{dc}{dx} + E\frac{d^2 c}{dx^2} - kc \tag{9.22}$$

The general solution can be obtained in a variety of ways. A simple approach is to assume that the solution has the form

$$c = e^{\lambda x} \tag{9.23}$$

This solution can be substituted into Eq 9.22 to arrive at the characteristic equation

$$E\lambda^2 - U\lambda - k = 0 \tag{9.24}$$

which can be solved for

$$\begin{matrix}\lambda_1 \\ \lambda_2\end{matrix} = \frac{U}{2E}\left(1 \pm \sqrt{1 + \frac{4kE}{U^2}}\right) = \frac{U}{2E}(1 \pm \sqrt{1 + 4\eta}) \tag{9.25}$$

where $\eta = kE/U^2$.

Therefore the general solution is

$$c = Fe^{\lambda_1 x} + Ge^{\lambda_2 x} \tag{9.26}$$

where F and G = constants of integration.

The constants of integration can be evaluated by boundary conditions. For the tank in Fig. 9.1 a boundary condition can be developed by taking a mass balance at the inlet,

$$Qc_{\text{in}} = Qc(0) - EA_c\frac{dc}{dx}(0) \tag{9.27}$$

Dividing by A_c and substituting Eq. 9.26 gives

$$(U - E\lambda_1)F + (U - E\lambda_2)G = Uc_{\text{in}} \tag{9.28}$$

The second boundary condition relates to the fact that no diffusion of mass is assumed to occur through the outlet. Therefore no gradient should exist at the end of the tank,

$$\frac{dc}{dx}(L) = 0 \tag{9.29}$$

Using Eq. 9.26,

$$\left(\lambda_1 e^{\lambda_1 L}\right)F + \left(\lambda_2 e^{\lambda_2 L}\right)G = 0 \tag{9.30}$$

These two boundary conditions are commonly called ***Danckwerts boundary conditions***, after the chemical engineer P. V. Danckwerts, who proposed them originally (Danckwerts 1953).

Equations 9.28 and 9.30 now represent a system of two equations with two unknowns. They can be solved for

$$F = \frac{Uc_{\text{in}}\lambda_2 e^{\lambda_2 L}}{(U - E\lambda_1)\lambda_2 e^{\lambda_2 L} - (U - E\lambda_2)\lambda_1 e^{\lambda_1 L}} \tag{9.31}$$

and

$$G = \frac{Uc_{\text{in}}\lambda_1 e^{\lambda_1 L}}{(U - E\lambda_2)\lambda_1 e^{\lambda_1 L} - (U - E\lambda_1)\lambda_2 e^{\lambda_2 L}} \tag{9.32}$$

These constants can be substituted into Eq. 9.26, which can then be used to calculate concentration along the tank's length.

EXAMPLE 9.2. MIXED-FLOW REACTOR. For the same system as in Example 9.1, compute concentration using the mixed-flow model with diffusion coefficients of $E = 2000$ and 10,000 $\text{m}^2\ \text{hr}^{-1}$. Plot your results along with the PFR and CSTR models for the same tank.

Solution: The following results can be generated:

	$E = 0$(PFR)	$E = 2000$	$E = 10{,}000$	$E = \infty$(CSTR)
λ_1	∞	0.0653	0.02	0.0
λ_2	−0.02	−0.0153	−0.01	0.0
F	0	5.66×10^{-5}	0.0126	0.1667
G	1	0.7656	0.5063	0.1667
$x = 0$	1.0000	0.7656	0.5189	0.3333
$x = 20$	0.6703	0.5638	0.4333	0.3333
$x = 40$	0.4493	0.4157	0.3674	0.3333
$x = 60$	0.3012	0.3083	0.3197	0.3333
$x = 80$	0.2019	0.2354	0.2899	0.3333
$x = 100$	0.1353	0.2044	0.2794	0.3333

They can be displayed as

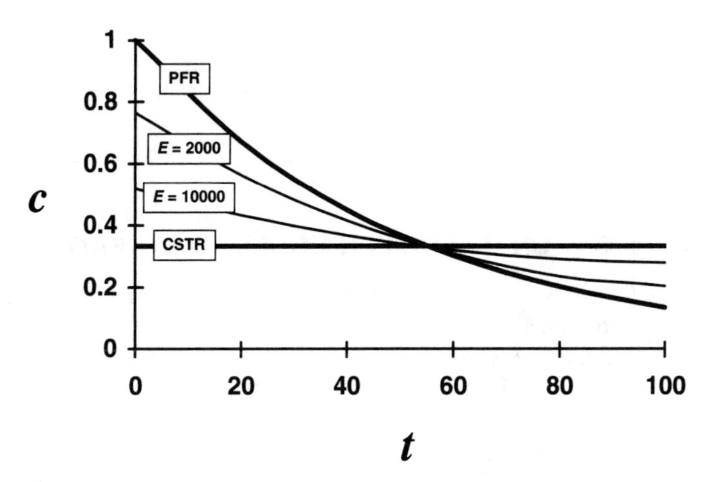

FIGURE E9.2

Thus at zero diffusion, the model converges on the PFR. As diffusion is increased it approaches the CSTR.

BOX 9.1. The Peclet Number

A dimensionless analysis of Eq. 9.22 can be used to gain insight into the results of Example 9.2. To do this, the following dimensionless parameter groups can be defined:

$$x^* = \frac{x}{L} \tag{9.33}$$

$$c^* = \frac{c}{c_{\text{in}}} \tag{9.34}$$

These can be solved for x and c and the result substituted into Eq. 9.22 to give

$$0 = -\frac{U}{L}\frac{dc^*}{dx^*} + \frac{E}{L^2}\frac{d^2c^*}{dx^{*2}} - kc^* \tag{9.35}$$

Multiplying this equation by L/U yields

$$0 = -\frac{dc^*}{dx^*} + \frac{E}{UL}\frac{d^2c^*}{dx^{*2}} - \frac{kL}{U}c^* \tag{9.36}$$

or

$$0 = -\frac{dc^*}{dx^*} + \frac{1}{P_e}\frac{d^2c^*}{dx^{*2}} - D_a c^* \tag{9.37}$$

where P_e is called the ***Peclet number***,

$$P_e = \frac{LU}{E} = \frac{\text{rate of advective transport}}{\text{rate of diffusive/dispersive transport}} \tag{9.38}$$

and D_a is called the ***Damkohler number***,

$$D_a = \frac{kL}{U} = \frac{\text{rate of consumption by decay}}{\text{rate of advective transport}} \tag{9.39}$$

Thus if P_e is high (> 10), the system approaches a PFR because the first-derivative term dominates the second-derivative term in Eq. 9.37. Conversely for low $P_e (< 0.1)$, longitudinal mixing dominates and the system becomes more like a CSTR. For intermediate values the MFR model is required.

9.2 APPLICATION OF THE PFR MODEL TO STREAMS

Just as the CSTR is the fundamental model for a lake, the PFR is the fundamental model for a stream. In this section we emphasize how a plug-flow model can be applied to analyze two types of sources: point and distributed loadings (Fig. 9.5).

9.2.1 Point Source

The following analysis focuses on the case where a point-source loading is injected into a channel having constant characteristics. For this case the solution would follow Eq. 9.9.

The initial concentration c_0 can be computed by taking a mass balance at the injection point. If it is assumed that complete mixing takes place in the lateral and vertical dimensions, the situation is as depicted in Fig. 9.6. A flow balance yields

$$Q = Q_w + Q_r \tag{9.40}$$

A mass balance can then be developed,

$$0 = Q_w c_w + Q_r c_r - (Q_w + Q_r) c_0 \tag{9.41}$$

that can be solved for

$$c_0 = \frac{Q_w c_w + Q_r c_r}{Q_w + Q_r} \tag{9.42}$$

or (because $Q_w c_w = W$)

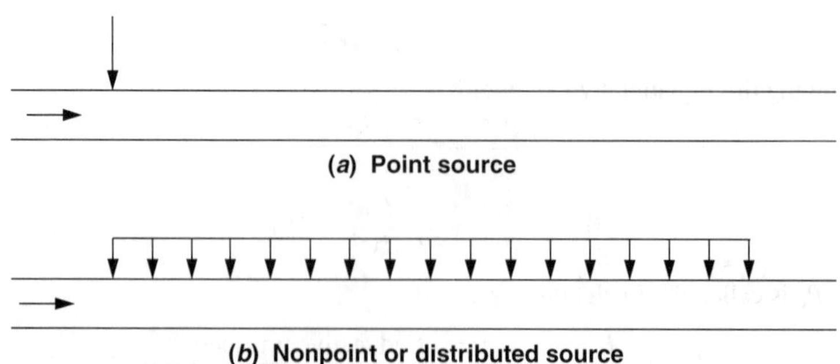

FIGURE 9.5
Point and nonpoint sources for one-dimensional systems.

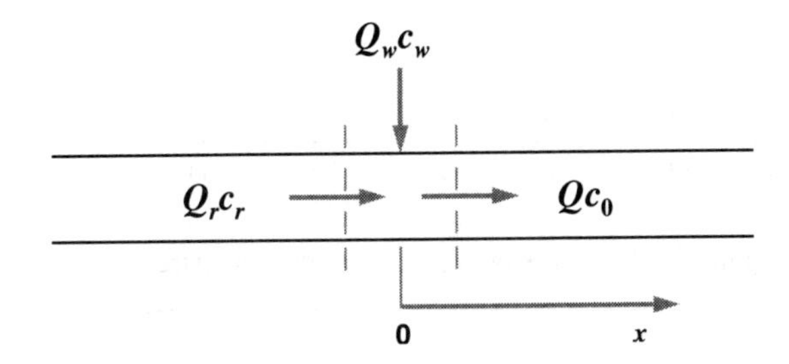

FIGURE 9.6
Mass balance for a point source discharged into a plug-flow system.

$$c_0 = \frac{W + Q_r c_r}{Q_w + Q_r} \tag{9.43}$$

Now, in some cases, the waste source has a relatively small flow and a high concentration, whereas the river has a relatively high flow and a small concentration. In other words,

$$Q_r \gg Q_w \tag{9.44}$$

and

$$c_r \ll c_w \tag{9.45}$$

For these cases, the following approximation holds:

$$c_0 = \frac{W}{Q} \tag{9.46}$$

EXAMPLE 9.3. POINT SOURCE FOR A PLUG-FLOW SYSTEM. A point source is discharged to a river having the following characteristics: $Q_r = 12 \times 10^6$ m^3 d^{-1}, $c_r = 1$ mg L^{-1}, $Q_w = 0.5 \times 10^6$ m^3 d^{-1}, and $c_w = 400$ mg L^{-1}.
(*a*) Determine the initial concentration assuming complete mixing vertically and laterally. Assess whether Eq. 9.46 is a good approximation for this case.
(*b*) Calculate the concentration of the pollutant for 8 km below the injection point. Note that the stream has a cross-sectional area of 2000 m^2 and the pollutant reacts with first-order decay ($k = 0.8$ d^{-1}).

Solution: (*a*) First, Eq. 9.42 can be used to compute the initial concentration,

$$c_0 = \frac{0.5 \times 10^6(400) + 12 \times 10^6(1)}{0.5 \times 10^6 + 12 \times 10^6} = 16.96 \text{ mg L}^{-1}$$

Equation 9.46 can also be used to make the same estimate, with the result

$$c_0 = \frac{0.5 \times 10^6(400)}{0.5 \times 10^6 + 12 \times 10^6} = 16 \text{ mg L}^{-1}$$

which represents a 5.7% error. This might be considered too large an error and the approximation deemed inadequate.
(*b*) The velocity in the stream can be computed as

$$U = \frac{Q}{A_c} = \frac{12.5 \times 10^6}{2000} = 6250 \text{ m d}^{-1}$$

Using Eq. 9.9,

$$c = 16.96e^{-\frac{0.8}{6250}x}$$

Substituting values of x gives

x (km)	0	1.6	3.2	4.8	6.4	8
c (mg L^{-1})	16.96	13.82	11.26	9.18	7.48	6.09

Thus the pollutant is decaying exponentially as it is being transported downstream.

9.2.2 Distributed Source

A ***distributed source*** is one that enters a water body in a diffuse manner.† In the context of a channel, it would enter along its length. The simplest type is the uniform distributed source, which is constant spatially (Fig. 9.7).

The distributed source can be incorporated into the steady-state mass balance,

$$0 = -U\frac{dc}{dx} - kc + S_d \tag{9.47}$$

If $c = c_0$ at $x = 0$, then

$$c = c_0 e^{-\frac{k}{U}x} + \frac{S_d}{k}\left(1 - e^{-\frac{k}{U}x}\right) \tag{9.48}$$

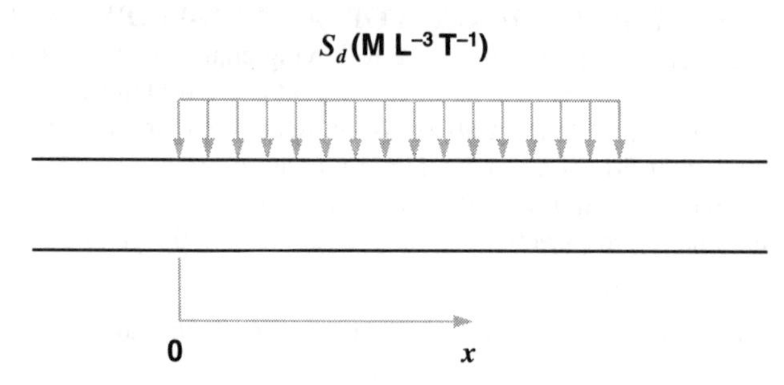

FIGURE 9.7
Uniform distributed source.

EXAMPLE 9.4. DISTRIBUTED SOURCE INTO A PLUG-FLOW SYSTEM. Suppose that for the same problem setting as in Example 9.3, a uniform diffuse source load with no flow contribution begins 8 km downstream from the point source. The diffuse source has a value of 15 g m^{-3} d^{-1} and continues for an additional 8 km. Thereafter all loadings terminate. Determine the concentration for the stretch 24 km below the point source. Note that the area increases to 3000 m^2 at the 8-km point.

---

†Note that the distributed sources described here contribute mass but no flow. In Lec. 22 we will model distributed sources that contribute both mass and flow.

Solution: The evaluation from Example 9.3 holds for the first 8 km and provides the boundary condition for the next reach. For km 8 to km 16 the velocity must be recalculated because of the increased area,

$$U = \frac{12.5 \times 10^6}{3000} = 4167 \text{ m d}^{-1}$$

Equation 9.48 can then be used to calculate the concentration in the second 8-km stretch,

$$c = 6.09e^{-\frac{0.8}{4167}(x-8000)} + \frac{15}{0.8}\left(1 - e^{-\frac{0.8}{4167}(x-8000)}\right)$$

Substituting values of x gives

x (km)	8	9.6	11.2	12.8	14.4	16
c (mg L^{-1})	6.09	9.44	11.90	13.71	15.05	16.03

Finally Eq. 9.9 can then be used to calculate the concentration in the third 8-km stretch,

$$c = 16.03e^{-\frac{0.8}{4167}(x-16{,}000)}$$

Substituting values of x gives

x (km)	16	17.6	19.2	20.8	22.6	24
c (mg L^{-1})	16.03	11.79	8.67	6.38	4.69	3.45

The entire solution can be displayed graphically as

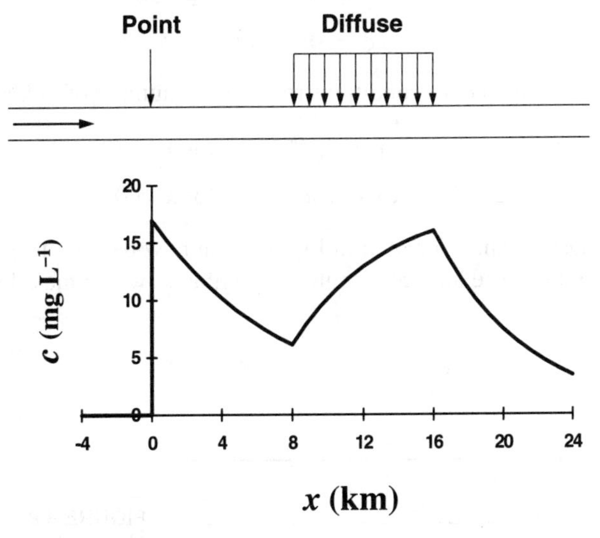

FIGURE E9.4

9.3 APPLICATION OF THE MFR MODEL TO ESTUARIES

The mixed-flow system is the fundamental model for one-dimensional estuaries. In this section I emphasize how the mixed-flow model can be applied to analyze point and distributed loadings.

9.3.1 Point Source

The following analysis focuses on the case where a point-source loading is injected into a channel having constant characteristics. The impact of dispersion on the initial concentration at a point source can be computed by taking a mass balance at the injection point. If it is assumed that complete mixing takes place in the lateral and vertical dimensions, the situation is as depicted in Fig. 9.8. For simplicity we assume that the loading does not add significant flow at the mixing point (that is, relative to the flow in the estuary). Problem 9.8 explores the case where the waste flow significantly elevates the estuary flow.

We use the same approach as was used to solve Eq. 9.22, and the general solution for this case is

$$c = Fe^{\lambda_1 x} + Ge^{\lambda_2 x} \tag{9.26}$$

where the λ's are

$$\begin{matrix}\lambda_1 \\ \lambda_2\end{matrix} = \frac{U}{2E}\left(1 \pm \sqrt{1 + 4\eta}\right) \tag{9.25}$$

The constants of integration can be evaluated by boundary conditions. For the present case these are

$$c = 0 \qquad @\ x = -\infty \tag{9.49}$$

$$c = 0 \qquad @\ x = \infty \tag{9.50}$$

Applying these boundary conditions reduces the number of unknowns to one,

$$c_1 = c_0 e^{\lambda_1 x} \qquad \text{for } x \le 0 \tag{9.51}$$

$$c_2 = c_0 e^{\lambda_2 x} \qquad \text{for } x > 0 \tag{9.52}$$

The mass balance depicted in Fig. 9.8 can now be used to evaluate c_0. For the case where the waste load contributes negligible flow, the mass balance is

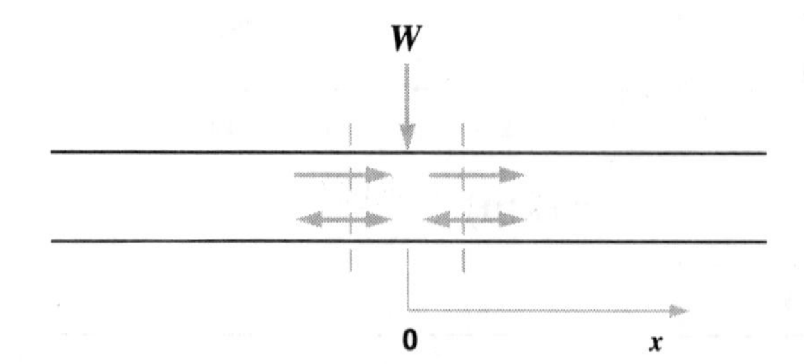

FIGURE 9.8
Mass balance for a point source discharged into a mixed-flow system.

$$W + UAc_1(0) - EA\frac{dc_1}{dx}(0) - UAc_2(0) + EA\frac{dc_2}{dx}(0) = 0 \tag{9.53}$$

Substituting Eqs. 9.51 and 9.52 yields

$$W + UAc_0 - EA\lambda_1 c_0 - UAc_0 + EA\lambda_2 c_0 = 0 \tag{9.54}$$

which can be solved for

$$c_0 = \frac{W}{Q}\frac{1}{\sqrt{1+4\eta}} \tag{9.55}$$

Therefore the final solution is

$$c = \frac{W}{Q\sqrt{1+4\eta}}e^{\frac{U}{2E}\left(1+\sqrt{1+4\eta}\right)x} \qquad x \leq 0 \tag{9.56}$$

$$c = \frac{W}{Q\sqrt{1+4\eta}}e^{\frac{U}{2E}\left(1-\sqrt{1+4\eta}\right)x} \qquad x \geq 0 \tag{9.57}$$

Note that when E approaches zero, this model converges on the plug-flow model (see Prob. 9.9). Also observe that Eq. 9.55 is in the format of Eq. 1.8. That is, it relates a critical concentration c_0 to the loading W. Thus as illustrated in the following example, it provides a means to perform an assimilative capacity calculation for an estuary.

EXAMPLE 9.5. POINT SOURCE TO A MIXED-FLOW SYSTEM. A point source is discharged into an estuary having the following characteristics:

	Value	Units
Dispersion coefficient	80×10^6	$m^2\ d^{-1}$
Flow	12×10^3	$m^3\ d^{-1}$
Width	0.5	m
Depth	8	m

The pollutant settles ($v = 1\ m\ d^{-1}$) and reacts via first-order kinetics ($k = 0.2\ d^{-1}$). What mass loading could be input to this system under steady-state conditions if the allowable concentration at the outfall is 10 ppm? Express your results in kg yr^{-1}. Assume complete lateral and vertical mixing at the outfall.

Solution: The velocity can be calculated,

$$U = \frac{12 \times 10^3}{0.5 \times 8} = 3000\ m\ d^{-1}$$

Equation 9.55 can be used to compute

$$W = Qc_0\sqrt{1 + \frac{4\left(k + \frac{v}{H}\right)E}{U^2}}$$

$$= 12 \times 10^3(10)\sqrt{1 + \frac{4\left(0.2 + \frac{1}{8}\right)80 \times 10^6}{(3000)^2}} = 425{,}206 \text{ g d}^{-1}$$

or converting to the proper units,

$$W = 425{,}206 \text{ g d}^{-1}\left(\frac{365 \text{ d}}{\text{yr}}\frac{\text{kg}}{10^3\text{g}}\right) = 155{,}200 \text{ kg yr}^{-1}$$

9.3.2 Distributed Source

Thomann and Mueller (1987) present the following solution for the case of a distributed load of a first-order decaying substance into an advective/dispersive system (see Fig. 9.9):

$$c = \frac{S_d}{k}\left(\frac{\sqrt{1+4\eta}-1}{2\sqrt{1+4\eta}}\right)\left(1 - e^{-\frac{U}{2E}\left(1+\sqrt{1+4\eta}\right)a}\right)e^{\frac{U}{2E}\left(1+\sqrt{1+4\eta}\right)x} \qquad x \le 0 \tag{9.58}$$

$$c = \frac{S_d}{k}\left[1 - \frac{\sqrt{1+4\eta}-1}{2\sqrt{1+4\eta}}e^{\frac{U}{2E}\left(1+\sqrt{1+4\eta}\right)(x-a)} - \frac{\sqrt{1+4\eta}+1}{2\sqrt{1+4\eta}}e^{\frac{U}{2E}\left(1-\sqrt{1+4\eta}\right)x}\right] \qquad 0 \le x \le a \tag{9.59}$$

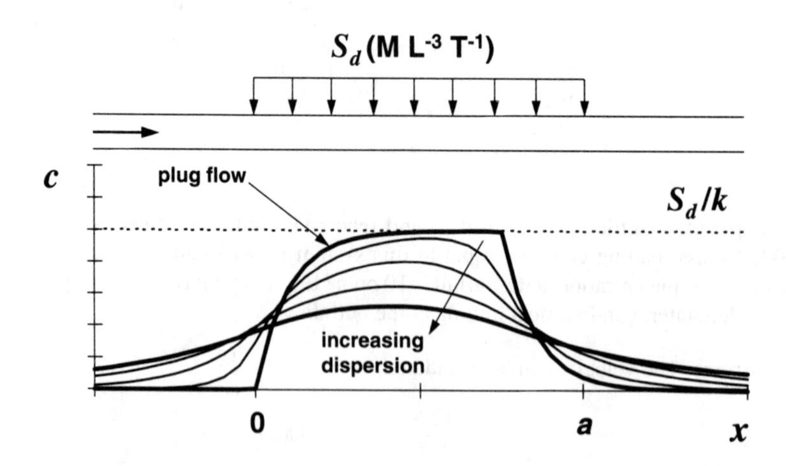

FIGURE 9.9
Uniform distributed source in a mixed-flow system. Several different levels of dispersion are displayed along with the plug-flow solution.

$$c = \frac{S_d}{k}\left(\frac{\sqrt{1+4\eta}+1}{2\sqrt{1+4\eta}}\right)\left(1 - e^{\frac{U}{2E}\left(1-\sqrt{1+4\eta}\right)a}\right)e^{\frac{U}{2E}\left(1-\sqrt{1+4\eta}\right)(x-a)} \qquad x \geq a \qquad (9.60)$$

As depicted in Fig. 9.9 the solution approaches the plug-flow case when E is low. As E increases, the solution eventually approaches a bell-shaped distribution centered on the midpoint of the distributed loading.

PROBLEMS

9.1. For the mixed-flow model notice that $c(0) \leq c_{in}$. That is, the concentration immediately inside the reactor will be less than or equal to the inflow concentration (Fig. E9.2). When will they be equal? Use mathematical and physical arguments to establish why they are different.

9.2. Determine the Peclet numbers for the four cases represented in Example 9.2.

9.3. At the end of Box 9.1 we specified bounds for the Peclet number. Verify these bounds by determining concentration profiles for the tank from Example 9.2 using Peclet numbers of 0.1 and 10 (by changing E). Display your results graphically along with the CSTR and PFR solutions.

9.4. Determine the steady-state distribution of a pollutant ($k = 0.1\ \text{d}^{-1}$) for $x = 0$ to 32 km in the following system. Note that the concentration in the river immediately upstream of the distributed load is 5 mg L^{-1}. How far downstream will the system return to the level of 5 mg L^{-1}?

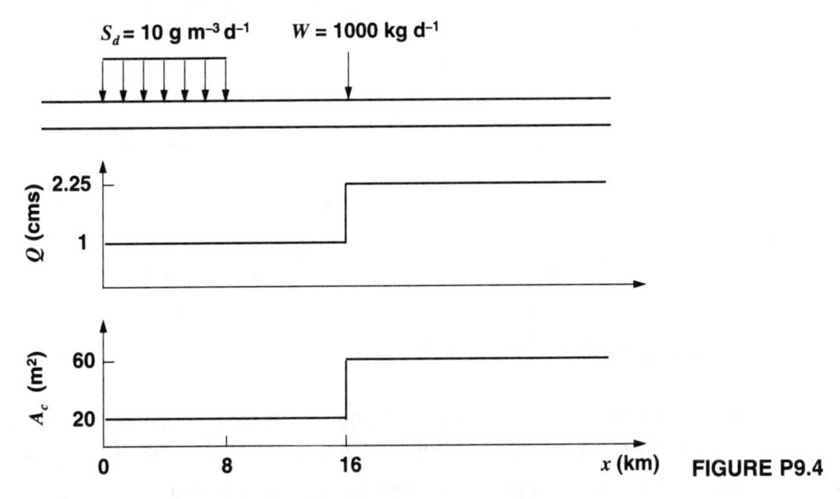

FIGURE P9.4

9.5. A point source is discharged into an estuary having the following characteristics:

	Value	Units
Dispersion coefficient	10^5	cm^2 s^{-1}
Flow	5×10^4	m^3 d^{-1}
Width	100	m
Depth	2	m

The pollutant settles ($v = 0.11$ m d^{-1}). What mass loading could be input to this system under steady-state conditions if the allowable concentration at the outfall is 10 ppb? Express your results in kg d^{-1}. Assume complete lateral and vertical mixing at the outfall.

9.6. A point source is discharged into an estuary having the following characteristics:

	Value	Units
Dispersion coefficient	10^6	m^2 d^{-1}
Flow	5×10^4	m^3 d^{-1}
Width	200	m
Depth	2	m

The pollutant decays at a rate of 0.2 d^{-1} at $T = 20°$C with $Q_{10} = 1.7$. The estuary has an ambient temperature of 27.5°C.

(*a*) Determine the reaction rate for the ambient temperature.

(*b*) What mass loading could be input to this system under steady-state conditions if the allowable concentration at the outfall is 20 ppb? Express your results in kg d^{-1}. Assume complete lateral and vertical mixing at the outfall.

9.7. A stream receives a point and a diffuse source as shown in Fig. P9.7:

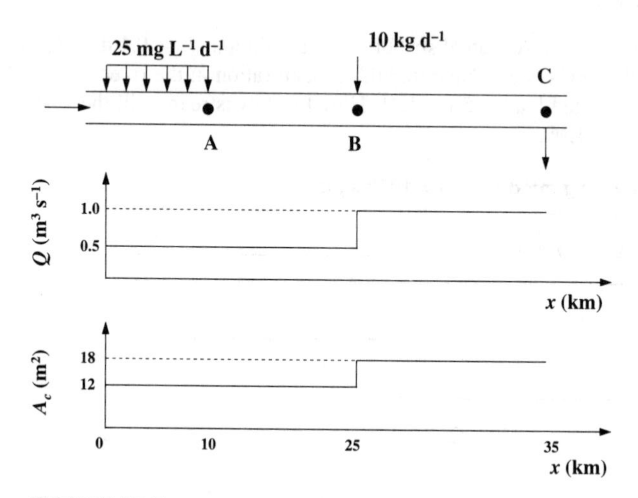

FIGURE P9.7

The boundary condition (at $x = 0$) is $c = 10$ mg L^{-1}. If the pollutant decays at a rate of 0.2 d^{-1}, determine the steady-state concentrations at points *A*, *B*, and *C*.

9.8. Rederive Eqs. 9.56 and 9.57 for the case where the loading contributes a significant flow to the estuary.

9.9. Rederive Eq. 9.57 so that it converges to the plug-flow solution at low dispersion. Note that one way to do this utilizes the following alternative formula for determining the roots of a quadratic (Chapra and Canale 1988):

$$x = \frac{-c}{b \pm \sqrt{b^2 - 4ac}} \tag{9.61}$$

LECTURE 10

Distributed Systems (Time-Variable)

Lecture overview: I continue our discussion of distributed systems by studying the temporal characteristics of plug-flow and mixed-flow systems, with focus on the instantaneous discharge into a one-dimensional channel. Such models are useful for modeling spills and for tracer studies in streams and estuaries.

I now describe some models for simulating time-variable changes in distributed systems. As in the previous lecture, the discussion is limited to one-dimensional systems with constant characteristics and to the instantaneous discharge of a substance into such channels.

The resulting models are extremely useful in environmental engineering. In particular they can be used to simulate an accidental spill in a stream or an estuary. In addition they can be used to analyze tracer studies, that is, for those cases where we deliberately introduce a tracer (such as a dye) into a water body to estimate some of its characteristics (for example velocity, dispersion, reaction rate).

10.1 PLUG FLOW

The time-variable mass balance for the plug-flow system can be written as (Fig. 10.1)

$$\frac{\partial c}{\partial t} = -U\frac{\partial c}{\partial x} - kc \qquad (10.1)$$

As derived in Box 10.1, if a spill causes a concentration c_0 at $t = x = 0$, the solution is

$$\begin{aligned} c &= c_0 e^{-kt} && \text{for } t = x/U \\ c &= 0 && \text{otherwise} \end{aligned} \qquad (10.2)$$

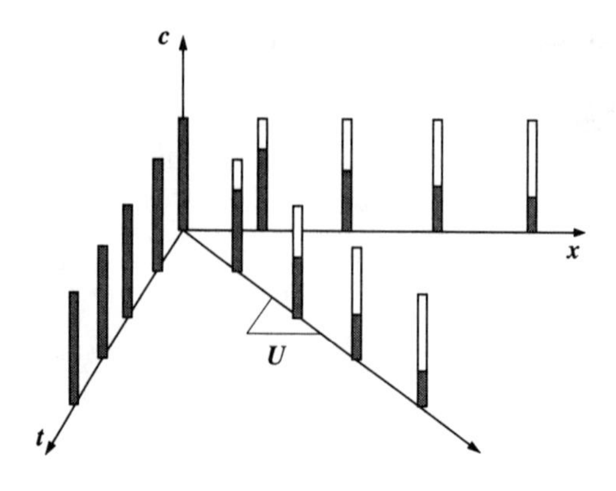

FIGURE 10.1
Depiction of the movement of dye in space and time for a plug-flow system. The whole plugs show the movement of a conservative substance. The shaded portions show a substance that reacts with first-order kinetics.

Further, the velocity establishes a direct relationship between time and space,

$$t = \frac{x}{U} \tag{10.3}$$

That is, the solution can also be written as

$$\begin{aligned} c &= c_0 e^{-\frac{k}{U}x} && \text{for } x = Ut \\ c &= 0 && \text{otherwise} \end{aligned} \tag{10.4}$$

BOX 10.1. Characteristics of the Plug-Flow Equation

The time-variable mass balance for the plug-flow system can be written as

$$\frac{\partial c}{\partial t} + U\frac{\partial c}{\partial x} = -kc \tag{10.5}$$

Now suppose that we want to find a solution to this equation along an arbitrary curve in the x-t plane (Fig. B10.1). The change of c, dc, from points A to B can be written as

$$dc = \frac{\partial c}{\partial t}dt + \frac{\partial c}{\partial x}dx \tag{10.6}$$

Dividing this equation by dt gives

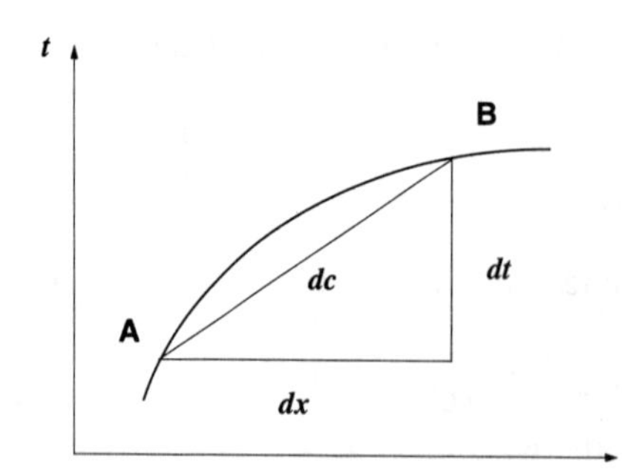

FIGURE B10.1

$$\frac{dc}{dt} = \frac{\partial c}{\partial t} + \frac{\partial c}{\partial x}\frac{dx}{dt} \tag{10.7}$$

where dx/dt is the slope of the curve AB in the x-t plane. The term dc/dt on the left-hand-side represents the rate of change as measured by a moving observer. The term $\partial c/\partial t$ represents the change of t at a fixed position. The term $(dx/dt)(\partial c/\partial x)$ represents the change due to the observer's moving into a region of possibly different c.

Now suppose that the curve is chosen so that its slope is equal to U. That is,

$$\frac{dx}{dt} = U \tag{10.8}$$

In other words suppose that the observer is moving at the same rate as the constant velocity U. If this is true Eq. 10.7 becomes

$$\frac{dc}{dt} = \frac{\partial c}{\partial t} + U\frac{\partial c}{\partial x} \tag{10.9}$$

Thus the right-hand side of Eq. 10.9 is equal to the left-hand side of Eq. 10.5. Consequently

$$\frac{dc}{dt} = -kc \tag{10.10}$$

This is a nice result. In essence we have converted the original partial differential equation into a pair of ordinary differential equations (10.8 and 10.10). The former (10.8) represents a curve in the x-t plane that is called the ***characteristic curve***. In the present case it is a straight line that has a slope of U. If we assume that $t = 0$ at $x = 0$ (that is, these are the coordinates where the observer starts to move downstream with the velocity), Eq. 10.8 can be integrated to yield

$$x = Ut \tag{10.11}$$

Thus the velocity defines a linear relationship between space and time. That is, at time t the observer will have moved a distance Ut downstream.

The concentration seen by the observer is then obtained by integrating Eq. 10.10 along this curve. For example if $c = c_0$ at $t = 0$,

$$c = c_0 e^{-kt} \tag{10.12}$$

Consequently, as depicted in Fig. 10.1, the solution to the original partial differential equation is an exponential decay that occurs along the line defined by Eq. 10.11.

EXAMPLE 10.1. SPILL INTO A PLUG-FLOW SYSTEM. Five kg of a conservative pollutant is spilled into a stream over a period of about 5 min. The stream has the following characteristics: flow = 2 $m^3\ s^{-1}$ and cross-sectional area = 10 m^2. Determine the concentration and the extent of the spill and how long it takes to reach a water intake located 6.48 km downstream.

Solution: The spill concentration can be estimated as follows:

$$W_{\text{spill}} = \frac{5\text{ kg}}{5\text{ min}}\left(\frac{1000\text{ g}}{\text{kg}}\frac{1440\text{ min}}{\text{d}}\right) = 1.44 \times 10^6\text{ g d}^{-1}$$

$$Q = 2\text{ m}^3\text{ s}^{-1}\left(\frac{86{,}400\text{ s}}{\text{d}}\right) = 0.1728 \times 10^6\text{ m}^3\text{ d}^{-1}$$

$$c = \frac{W_{\text{spill}}}{Q} = \frac{1.44 \times 10^6}{0.1728 \times 10^6} = 8.33 \text{ g m}^{-3}$$

The extent of the spill in this stretch is calculated as (note, $U = 2/10 = 0.2 \text{ m s}^{-1}$)

$$x_{\text{spill}} = 5 \text{ min}(0.2 \text{ m s}^{-1})\left(\frac{60 \text{ s}}{\text{min}}\right) = 60 \text{ m}$$

The front of the spill will reach the water intake at

$$t = \frac{6480 \text{ m}}{0.2 \text{ m s}^{-1}}\left(\frac{\text{hr}}{3600 \text{ s}}\right) = 9 \text{ hr}$$

from the start of the spill. Five minutes later it will have passed beyond the intake.

The preceding example took the perspective of an observer at a fixed location x downstream from the spill. This observer will see nothing until $t = x/U$. At this instant the spill will pass the observation point. For the case where the spill is decaying, the concentration will be reduced as specified by the exponential decay in Eq. 10.2.

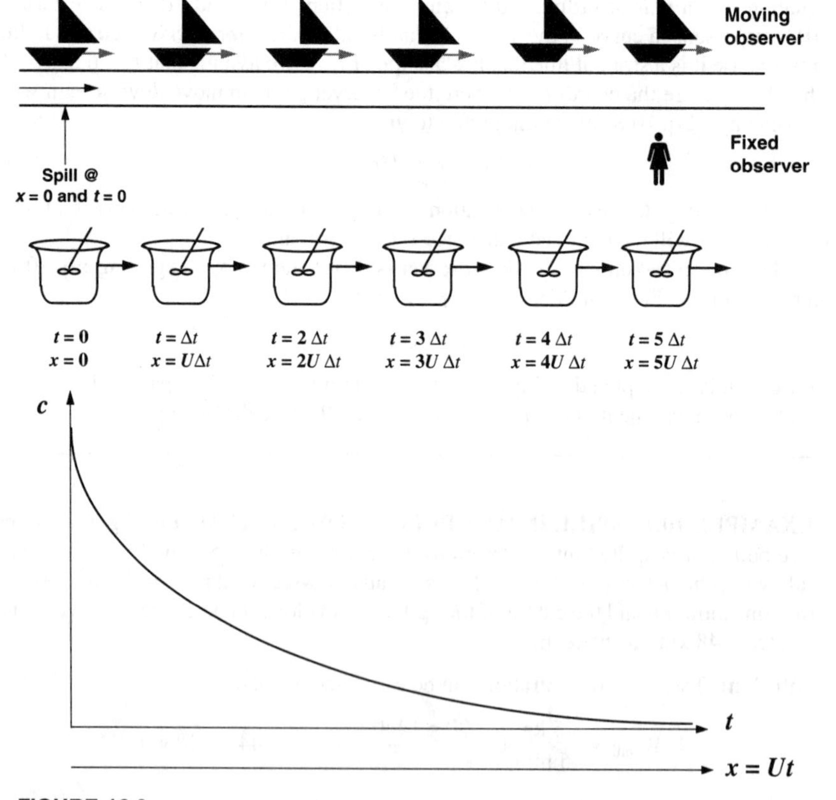

FIGURE 10.2
Two perspectives for viewing temporal and spatial changes in a plug-flow system.

An alternative perspective exists for a moving observer who travels downstream at the same velocity as the spill. As depicted in Fig. 10.2 this observer would watch the concentration decay exponentially as if the process were taking place in a batch reactor.

The interrelationship between space and time also applies to steady-state solutions. For example recall that the steady-state solution for a continuous load into a plug-flow system is

$$c = c_0 e^{-\frac{k}{U}x} \tag{10.13}$$

Now we should recognize that this equation could also be written as

$$c = c_0 e^{-kt} \tag{10.14}$$

where t represents travel time below the point source. The interchangeability of space and time will recur in our subsequent analyses of plug-flow systems throughout the remainder of this book.

Clearly the foregoing discussion is a highly idealized representation of stream transport. In even the most highly advective system, some dispersion will occur. Before showing how the dispersive process can be modeled, I'll first introduce the random walk model of the diffusion/dispersion process.

10.2 RANDOM (OR "DRUNKARD'S") WALK

The ***random*** or "***drunkard's***" ***walk*** is a term used to describe the type of random motion found in many diffusion processes. The name "drunkard's walk" stems from the similarity between this type of motion and the random stumbling around that a drunkard might exhibit.

Suppose that a population of particles is confined to motion along a one-dimensional line (Fig. 10.3). Assume that each particle has an equal likelihood of moving a small distance Δx to either the left or the right over a time interval Δt. At $t = 0$ all particles are grouped at $x = 0$ and are allowed to take one random step in either direction. After Δt has elapsed, approximately one-half of the particles would have stepped to the right (Δx) and the remainder would have stepped left ($-\Delta x$). After another time interval (that is, after $2\,\Delta t$ had elasped) approximately one-fourth would be at $-2\,\Delta x$, one-fourth at $2\,\Delta x$, and one-half would have stepped back to the origin.

With additional time the particles would spread out (Fig. 10.3). Note that the distribution of the population is not uniform but has a higher density at the origin and diminishes at the end. This is due to the fact that a particle would have to execute many successive moves in a single direction to reach the extremes. For example in Figure 10.3, after $4\,\Delta t$, a particle would have to execute 4 successive right steps to reach $4\,\Delta x$. Because there is a 50-50 chance of moving left or right in each time interval, it is more likely that a particle would stumble around in the vicinity of the origin. The net outcome is that the random walk of the individual particles results in a spreading bell-shaped distribution of the population. In addition note that this spreading tendency amounts to a general movement of particles from high to low concentrations.

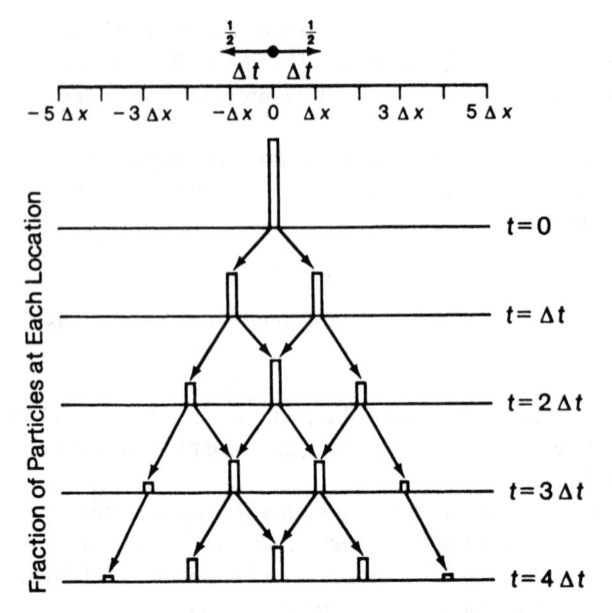

FIGURE 10.3
Graphic representation of a random walk. At time $t = 0$, all particles are grouped at the origin ($x = 0$). During each time step Δt, half the particles at each location move left and half move right. The result is that over time the particles spread out in a bell-shaped pattern.

The random-walk process can be expressed mathematically as (see Box 10.2 for details)

$$p(x, t) = \frac{1}{2\sqrt{\pi D t}} e^{-\frac{x^2}{4Dt}} \tag{10.15}$$

where $p(x, t)$ = the probability that a particle will be at x after an elapsed time t and D = a diffusion coefficient, defined as

$$D = \frac{\Delta x^2}{2\,\Delta t} \tag{10.16}$$

If the population is grouped at the origin at time zero, the number of individuals at position x at a subsequent time t would be proportional to the probability of an individual particle's being at x. Thus Eq. 10.15 can be expressed in terms of mass and concentration,

$$c(x, t) = \frac{m_p}{2\sqrt{\pi D t}} e^{-\frac{x^2}{4Dt}} \tag{10.17}$$

where m_p = total mass of the particles normalized to the cross-sectional area

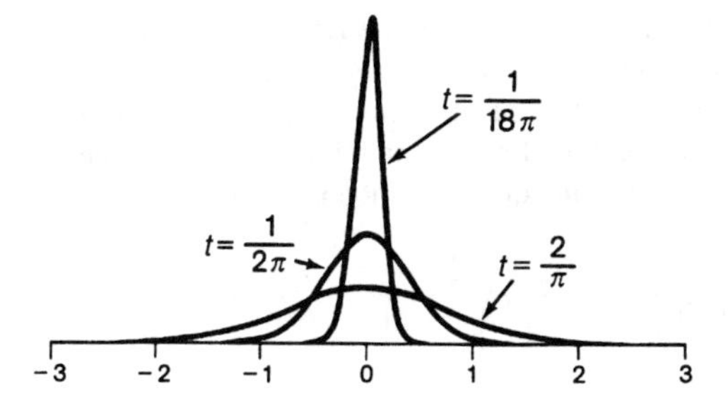

FIGURE 10.4
Representation of the random walk by a normal or "bell-shaped" distribution.

($M\,L^{-2}$).† Thus the distribution of the population of particles is described by a series of bell-shaped curves that spread out over time symmetrically about the origin (Fig. 10.4).

It should be noted that Eqs. 10.15 to 10.17 apply only if the number of time steps and spatial intervals are much greater than 1. In other words the observation time t and the observation space x must be much greater than the magnitudes of the duration Δt and size Δx of the individual steps. Because the random walk is an analog of the diffusion process, these conditions will have relevance to the application of diffusion models in the real world. That is, the process holds only when the random motions, whether molecular or due to eddies, have smaller time and space scales than do the phenomena being modeled.

BOX 10.2. Mathematics of the Random Walk

The random walk can be formulated mathematically by realizing that it can be represented as a binomial distribution. The ***binomial*** or ***Bernoulli's distribution*** results from our tendency to place observations in one or the other of two mutually exclusive categories. If p is the probability that an event will occur and $q = 1 - p$ is the probability that it will not, the probability that the event will happen exactly x times in n trials is given by

$$p(x, n) = \binom{n}{n-x} p^x q^{n-x}$$

where the parenthetical operation represents the number of possible ways or combinations by which the event can occur x times in n trials, where

$$\binom{n}{n-x} = \frac{n!}{x!(n-x)!}$$

and $n! = n(n-1)(n-2)\ldots 1$.

The random walk is this sort of process because the particle is limited to two modes of motion: left or right. Thus the binomial distribution can be used to determine the probability that, after n_t time steps, the particle will be n_x spatial intervals from the origin. To do this the total number of steps would be divided into n_r to the right and

†Note that m_p is formally called a ***plane source*** since it enters the system at a plane surface—the cross-sectional area.

$n_t - n_r$ to the left. To wind up at n_x, the difference between the right and left steps, $n_r - (n_t - n_r)$, would have to equal n_x. Therefore the number of right steps would have to be $n_r = (n_t + n_x)/2$ and left steps would have to be $(n_t - n_r) = (n_t - n_x)/2$. The probability that after n_t time steps the particle would be n_x spatial intervals from the origin can be represented as (remember there is an equal likelihood of moving left or right; that is, $p = q = 0.5$)

$$p(n_x, n_t) = \binom{n_t}{\frac{n_x + n_t}{2}} \left(\frac{1}{2}\right)^{\frac{n_t+n_x}{2}} \left(\frac{1}{2}\right)^{\frac{n_t-n_x}{2}}$$

or

$$p(n_x, n_t) = \left(\frac{1}{2}\right)^{n_t} \frac{n_t}{\left(\frac{n_t + n_x}{2}\right)!\left(\frac{n_t - n_x}{2}\right)!}$$

As the number of intervals becomes very large, the binomial distribution approaches the normal distribution. If we define continuous variables $x = n_x \Delta x$ and $t = n_t \Delta t$, the probability that the particle would be at distance x at time t can be formulated as a normal distribution with a mean of zero and a variance of $t(\Delta x)^2/\Delta t$, as in (see Pielou 1969)

$$p(x, t) = \frac{1}{2\sqrt{\pi D t}} e^{-\frac{x^2}{4Dt}}$$

where $p(x, t)$ = the probability that a particle will be at x after an elapsed time t and D = a diffusion coefficient, defined in the limit (as Δx and Δt become small) as

$$D = \frac{\Delta x^2}{2\,\Delta t}$$

10.3 SPILL MODELS

Now that we have reviewed the fundamentals of the random-walk model, we can integrate dispersion into models of instantaneous discharges into uniform, one-dimensional channels. First, we investigate the case where the spill occurs instantaneously. Then we present a model for a continuous input.

10.3.1 Instantaneous or "Impulse" Spills

For this case a significant input occurs over a very short time period at a point in space. Many spills can be approximated in this fashion. In the following paragraphs we build the solution by adding mechanisms on a term-by-term basis. First, we examine the situation where dispersion or turbulent diffusion is the only mechanism.

Diffusion/dispersion. As developed in Lec. 9, a mass balance for a substance that disperses in a one-dimensional channel can be written as (Eq. 9.21 with $k = U = 0$)

$$\frac{\partial c}{\partial t} = E\frac{\partial^2 c}{\partial x^2} \tag{10.18}$$

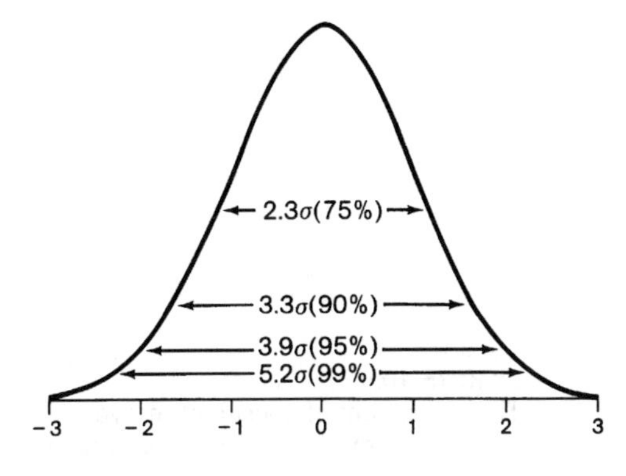

FIGURE 10.5
A standardized normal distribution showing the probability (expressed as percent) encompassed by various multiples of the standard deviation. For example 3.9σ encompasses 95% of the area under the curve.

This relationship is sometimes referred to as ***Fick's second law***. The solution for the case where the substance is initially concentrated at $x = 0$ is

$$c(x, t) = \frac{m_p}{2\sqrt{\pi E t}} e^{-\frac{x^2}{4Et}} \tag{10.19}$$

which is identical to the solution based on the random walk (Eq. 10.17). Thus the solution is a bell-shaped curve with mean at zero and a variance of $2Et$.

Because the variance is a measure of spread, this equation can be used as the basis of a simple engineering computation to assess the effect of dispersion on a pollutant spill. For example if a conservative substance were discharged in a lump sum to a water body, its tendency to spread outward from its center of mass could be represented by the standard deviation

$$\sigma = \sqrt{2Et} \tag{10.20}$$

or multiples of the standard deviation as depicted in Fig. 10.5. For example 95% and 99% spreads can be roughly approximated by 4σ and 5σ, respectively.

EXAMPLE 10.2. CONSERVATIVE SPILL IN A CHANNEL WITH NO FLOW. A barge releases a large quantity of a highly persistent contaminant in the center of a canal that is not flowing. If the dispersion coefficient is approximately 10^5 m^2 d^{-1}, how far will the contaminant spread in 1 d? In 2 d? Assume that a 95% band adequately approximates the extent of the spill.

Solution: Using Eq. 10.20 and the multiple of the standard deviation that encompasses 95% of the distribution (Fig. 10.5) yields

$$x(1\text{ d}) = 3.9\sqrt{2(10^5)1} = 1744\text{ m}$$

and

$$x(2\text{ d}) = 3.9\sqrt{2(10^5)2} = 2466\text{ m}$$

Dispersion/advection. Now we can add advection to the model. This relationship is sometimes referred to as the ***advection-diffusion*** (or ***advection-dispersion***) ***equation***

$$\frac{\partial c}{\partial t} = -U\frac{\partial c}{\partial x} + E\frac{\partial^2 c}{\partial x^2} \tag{10.21}$$

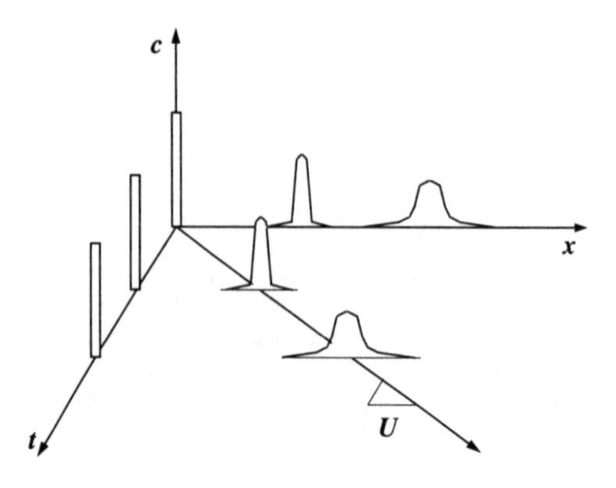

FIGURE 10.6
The movement of conservative dye in space and time for a mixed-flow system.

The solution for the case where the substance is initially concentrated at $x = 0$ is

$$c(x,t) = \frac{m_p}{2\sqrt{\pi E t}} e^{-\frac{(x-Ut)^2}{4Et}} \tag{10.22}$$

Note that in comparison with Eq. 10.19, the effect of advection is to "move" the dispersion solution intact downstream at velocity U (Fig. 10.6).

EXAMPLE 10.3. SPILL INTO A MIXED-FLOW SYSTEM. Evaluate the spill from Example 10.1, but include the effect of dispersion. Assume that a dispersion coefficient of 0.1 $m^2\ s^{-1}$ holds for the entire stretch. Also assume that the spill occurs instantaneously as a plane source.

Solution: The spill can be expressed as a plane source, as in

$$m_p = \frac{5 \times 10^3\ \text{g}}{10\ \text{m}^2} = 500\ \text{g m}^{-2}$$

The spill concentration in the stretch can then be estimated by

$$c(x,t) = \frac{500}{2\sqrt{\pi(0.1)t}} e^{-\frac{(x-0.2t)^2}{4(0.1)t}}$$

Results can be calculated and are displayed in Fig. E10.3. These results are very different from Example 10.1. Although the travel time to the water intake is the

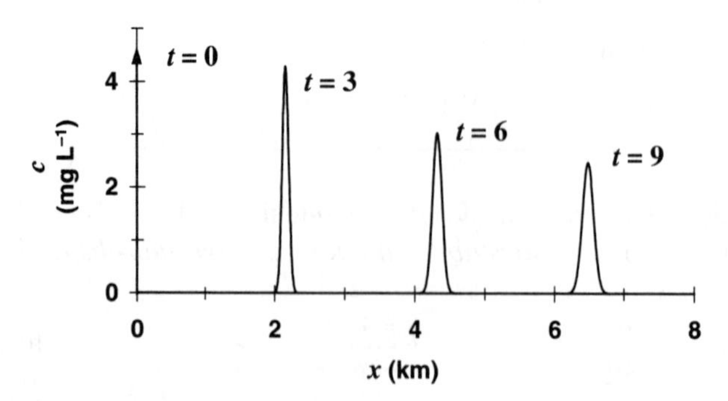

FIGURE E10.3

same, the peak progressively decays due to dispersion. In addition the spill spreads as time progresses. These results can be summarized as

	t = 3 hr	t = 6 hr	t = 9 hr
Extent of spill (m)†	181	256	314
Peak concentration (mg L^{-1})	4.29	3.03	2.48

† Defined as distance encompassing 95% of the mass.

Dispersion/advection/decay. Finally a first-order reaction can be added to the model,

$$\frac{\partial c}{\partial t} = -U\frac{\partial c}{\partial x} + E\frac{\partial^2 c}{\partial x^2} - kc \tag{10.23}$$

The solution for the case where the substance is initially concentrated at $x = 0$ is

$$c(x, t) = \frac{m_p}{2\sqrt{\pi E t}} e^{-\frac{(x-Ut)^2}{4Et} - kt} \tag{10.24}$$

In comparison with Eq. 10.22, the effect of decay reduces to the area under the bell-shaped curve as it moves downstream (Fig. 10.7).

Fixed versus global observer. Note that Eq. 10.24 is a function of two independent variables x and t. Thus it can be viewed from two perspectives. As in Example 10.3 we can compute the spatial distribution at a fixed time. Thus we can develop the global perspective manifested by the bell-shaped curves shown in Fig. 10.7.

Conversely we can compute the temporal distribution at a fixed point in space. This latter case relates to a static observer. For this case the view can be skewed because the bell-shaped curve continues to spread out as it is being observed (Fig. 10.8).

10.3.2 Continuous Spills

For some tracer studies, as well as some spills, the input jumps to a constant level. As depicted in Fig. 10.9, two idealized cases can be modeled. First, the concentration

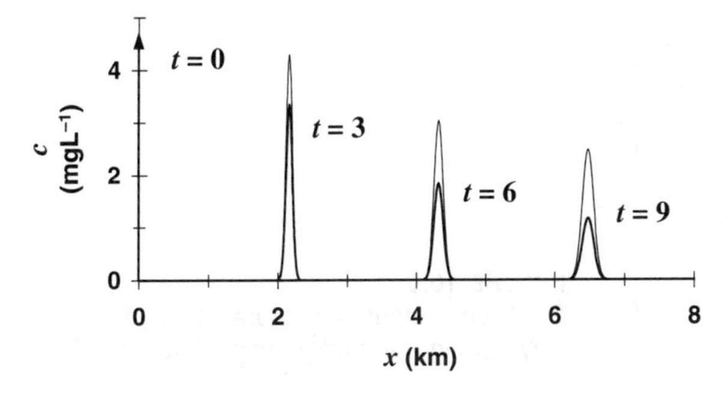

FIGURE 10.7
The effect of decay on the spill model. The thin lines are the same as calculated in Example 10.3. The heavy lines are for the same case but with k = 2 d^{-1}.

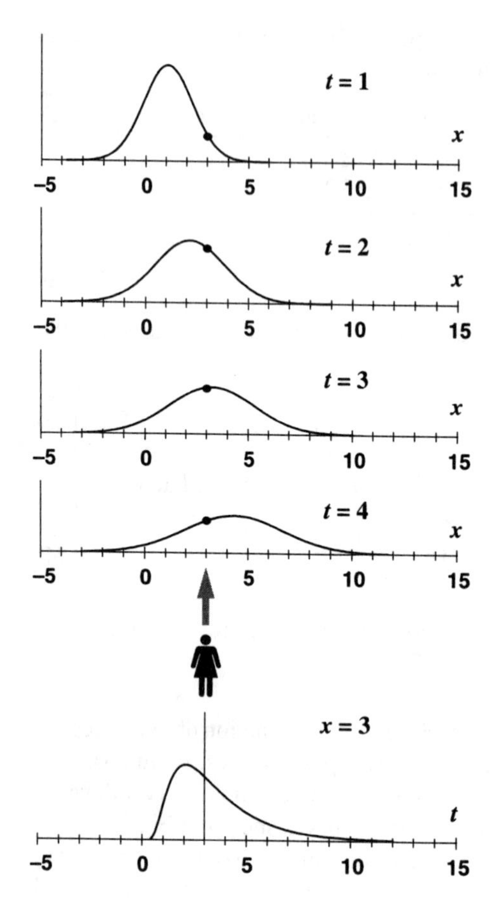

FIGURE 10.8
Although the distribution of a pollutant spill is bell-shaped in space, a fixed observer would "see" a skewed shape in time because the bell-shaped curve continues to spread out as it is being observed.

increase can be maintained for an infinite duration (Fig. 10.9*a*). For this case the solution of Eq. 10.23 with constant coefficients can be expressed as (O'Loughlin and Bowmer 1975)

$$c(x,t) = \frac{c_0}{2}\left[e^{\frac{Ux}{2E}(1-\Gamma)}\operatorname{erfc}\left(\frac{x-Ut\Gamma}{2\sqrt{Et}}\right) + e^{\frac{Ux}{2E}(1+\Gamma)}\operatorname{erfc}\left(\frac{x+Ut\Gamma}{2\sqrt{Et}}\right)\right] \quad (10.25)$$

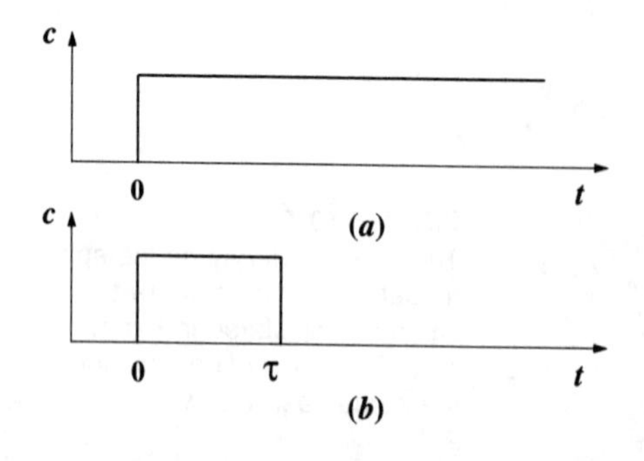

FIGURE 10.9
Continuous inputs are characterized in two ways: (*a*) infinite and (*b*) finite durations.

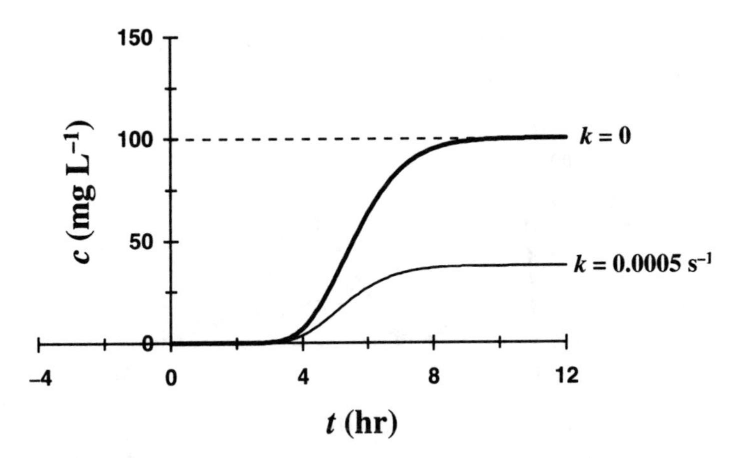

FIGURE 10.10
A simulation of a "breakthrough" curve. A step increase to a concentration of 100 mg L^{-1} is initiated at $x = 0$ at $t = 0$. Shown is the change in concentration at a sampling point 2000 m downstream for a conservative and a nonconservative release. This example used $U = 0.1$ m s^{-1} and $E = 5$ m^2 s^{-1}.

where

$$\Gamma = \sqrt{1 + 4\eta} \tag{10.26}$$

and

$$\eta = \frac{kE}{U^2} \tag{10.27}$$

The error function complement, erfc, is equal to one minus the error function: $1 - \text{erf}$. Also, $\text{erf}(-x) = -\text{erf}(x)$. The error function is simply the evaluation of the following definite integral,

$$\text{erf}(b) = \frac{2}{\sqrt{\pi}} \int_0^b e^{-\beta^2}\, d\beta \tag{10.28}$$

where β = a dummy variable. We have included selected values for the error function in App. G. It should also be noted that the error function is available on standard software libraries [e.g., the International Math and Statistics Library (IMSL), Numerical Recipes (Press et al. 1992, etc.), and as a function on many software packages (e.g., Excel, Mathematica, etc.)].

A simulation using Eq. 10.25 is shown in Fig. 10.10. Such curves are referred to as "breakthrough" curves. They are used extensively in both surface and groundwater problem contexts.

The second idealized application applies to cases in which the step input terminates after a finite time (Fig. 10.9*b*). For this situation the solution is presented in two parts. For $t < \tau$, Eq. 10.25 holds. Thereafter the following formula applies (Runkel 1996):

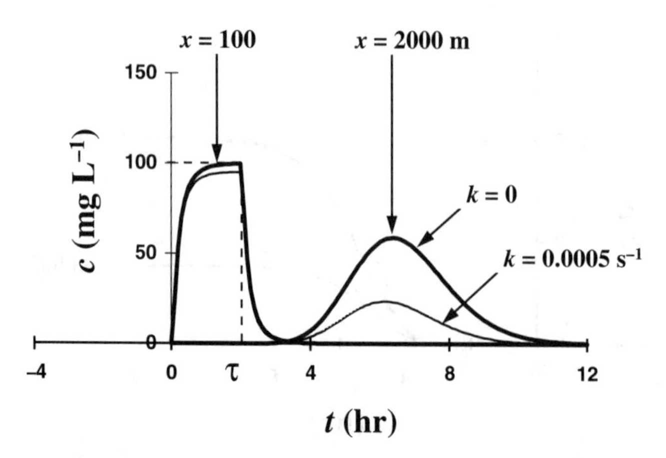

FIGURE 10.11
A simulation of a dye release or spill of finite duration—a "breakthrough" curve. A step increase to a concentration of 100 mg L^{-1} is initiated at $x = 0$ at $t = 0$ and lasts for $\tau = 2$ hr. Shown is the distribution at $x = 0$ (dashed line) along with curves at $x = 100$ and 2000 m for a conservative (bold line) and a nonconservative release (light line). This example used $U = 0.1$ m s^{-1} and $E = 5$ m^2 s^{-1}.

$$c(x,t) = \frac{c_0}{2}\left\{ e^{\frac{Ux}{2E}(1-\Gamma)}\left[\operatorname{erfc}\left(\frac{x - Ut\Gamma}{2\sqrt{Et}}\right) - \operatorname{erfc}\left(\frac{x - U(t-\tau)\Gamma}{2\sqrt{E(t-\tau)}}\right)\right]\right.$$
$$\left. + e^{\frac{Ux}{2E}(1+\Gamma)}\left[\operatorname{erfc}\left(\frac{x + Ut\Gamma}{2\sqrt{Et}}\right) - \operatorname{erfc}\left(\frac{x + U(t-\tau)\Gamma}{2\sqrt{E(t-\tau)}}\right)\right]\right\} \quad (10.29)$$

A simulation using Eq. 10.25 is shown in Fig. 10.11. Notice how the solution approaches a bell-shaped curve as it moves downstream.

10.4 TRACER STUDIES

Aside from accidental spills, the models determined above have utility when compounds are deliberately discharged as in a tracer study. In such cases the distribution downstream from the injection point can be used to determine key characteristics such as the velocity, the dispersion coefficient, and the decay rate.

To do this it is necessary to estimate some quantities from concentration data. Such data is usually measured at discrete points in time (Fig. 10.12).

For such cases the following formulas can be used:

Mean Concentration

$$\bar{c} = \frac{\sum_{i=0}^{n-1}(c_i + c_{i+1})(t_{i+1} - t_i)}{2(t_n - t_0)} \quad (10.30)$$

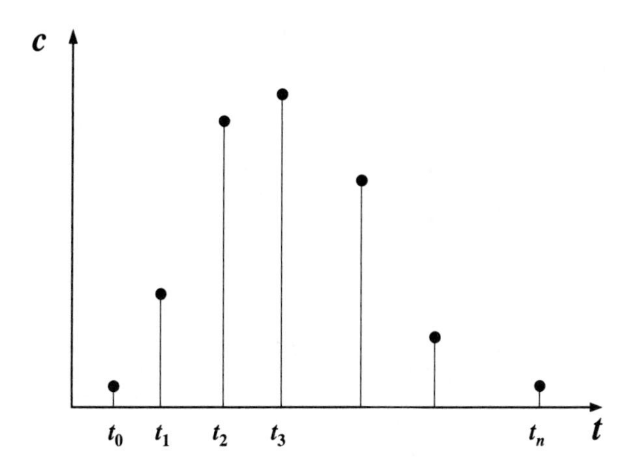

FIGURE 10.12
Concentration data sampled at a point in space to characterize the distribution of a tracer.

Mass

$$M = Q\bar{c}(t_n - t_0) \tag{10.31}$$

Travel Time

$$\bar{t} = \frac{\sum_{i=0}^{n-1} (c_i t_i + c_{i+1} t_{i+1})(t_{i+1} - t_i)}{\sum_{i=0}^{n-1} (c_i + c_{i+1})(t_{i+1} - t_i)} \tag{10.32}$$

Temporal Variance

$$s_t^2 = \frac{\sum_{i=0}^{n-1} \left(c_i t_i^2 + c_{i+1} t_{i+1}^2\right)(t_{i+1} - t_i)}{\sum_{i=0}^{n-1} (c_i + c_{i+1})(t_{i+1} - t_i)} - \left(\bar{t}\right)^2 \tag{10.33}$$

If data are available from two stations, located at x_1 and x_2, the mean velocity can be estimated by

$$U = \frac{x_2 - x_1}{\bar{t}_2 - \bar{t}_1} \tag{10.34}$$

The velocity estimate can, in turn, be used to calculate the dispersion coefficient by (Fischer 1968)

$$E = \frac{U^2\left(s_{t2}^2 - s_{t1}^2\right)}{2(\bar{t}_2 - \bar{t}_1)} \tag{10.35}$$

EXAMPLE 10.4. EVALUATION OF A TRACER STUDY. A tracer study is conducted in a stream with a flow of 3×10^5 m^3 d^{-1} and a width of 45 m. At $t = 0$, 5 kg of a conservative substance, lithium, is instantaneously injected at $x = 0$. Concentrations are measured at two downstream stations:

$x = 1$ km

t (min)	30	40	50	60	70	80	90	100	110	120
Lithium (μg L^{-1})	0	100	580	840	560	230	70	15	3	0

$x = 8$ km

t (min)	370	400	430	460	490	520	550	580	610
Lithium (μg L^{-1})	0	10	80	250	280	140	35	5	0

Determine (*a*) the velocity (m d^{-1}) and (*b*) the dispersion coefficient (cm^2 s^{-1}).

Solution: (*a*) The velocity can be evaluated by determining the travel time between the two sampling points,

$$\bar{t}_1 = \frac{2{,}982{,}600}{47{,}960} = 62.2 \text{ min}$$

$$\bar{t}_8 = \frac{23{,}133{,}000}{48{,}000} = 481.9 \text{ min}$$

Therefore the velocity can be computed as

$$U = \frac{8 - 1 \text{ km}}{481.9 - 62.2 \text{ min}}\left(\frac{1000 \text{ m}}{\text{km}}\right) = 16.67 \text{ m min}^{-1}\left(\frac{1440 \text{ min}}{\text{d}}\right) = 24{,}014 \text{ m d}^{-1}$$

(*b*) The dispersion coefficient can be estimated by calculating the time variances of the lithium concentration distributions,

$$s_{t1}^2 = \frac{1.92 \times 10^8}{47{,}960} - 62.2^2 = 137 \text{ min}^2$$

$$s_{t2}^2 = \frac{1.12 \times 10^{10}}{48{,}000} - 481.9^2 = 1043 \text{ min}^2$$

Then Eq. 10.35 can be used to compute

$$E = \frac{(16.67 \text{ m min}^{-1})^2(1043 \text{ min}^2 - 137 \text{ min}^2)}{2(481.9 \text{ min} - 62.2 \text{ min})}$$

$$= 300 \text{ m}^2 \text{ min}^{-1}\left(\frac{10^4 \text{ cm}^2}{\text{m}^2}\frac{\text{min}}{60 \text{ s}}\right) = 50{,}019 \text{ cm}^2 \text{ s}^{-1}$$

Dye studies can also be used to determine first-order reaction rates. For this case the mass under the concentration-time curve is determined at two positions. The rate can then be estimated by

$$k = \frac{1}{\bar{t}_2 - \bar{t}_1} \ln \frac{M_1}{M_2} \tag{10.36}$$

where the travel times and the masses are calculated according to Eqs. 10.32 and 10.31, respectively. We elaborate more on the use of tracers to estimate reaction rates when we discuss the measurement of reaeration later in the text (Lec. 20).

10.5 ESTUARY NUMBER

The relative importance of advection and dispersion can be assessed by a dimensionless analysis of Eq. 10.23:

$$\frac{\partial c}{\partial t} = -U\frac{\partial c}{\partial x} + E\frac{\partial^2 c}{\partial x^2} - kc \tag{10.23}$$

Three dimensionless parameter groups can be defined:

$$c^* = \frac{c}{c_0} \tag{10.37}$$

$$x^* = \frac{kx}{U} \tag{10.38}$$

$$t^* = kt \tag{10.39}$$

These equations can be solved for c, x, and t, respectively, and the results substituted into Eq. 10.23 to yield

$$\frac{\partial c^*}{\partial t^*} = \eta\frac{\partial^2 c^*}{\partial x^{*2}} - \frac{\partial c^*}{\partial x^*} - c^* \tag{10.40}$$

where η is called the ***estuary number***,

$$\eta = \frac{kE}{U^2} \tag{10.41}$$

From Eq. 10.41 it should be clear that the magnitude of η determines whether the second derivative term in the differential equation is significant relative to the other terms. Thus the following guidelines can be developed

		Suggested ranges
$\eta \gg 1$	Diffusion predominates	$\eta > 10$
$\eta \approx 1$	Advection/diffusion important	$0.1 < \eta < 10$
$\eta \ll 1$	Advection predominates	$\eta < 0.1$

EXAMPLE 10.5. ESTUARY NUMBER. Evaluate the estuary number for the stream from Example 10.4 for a nonconservative tracer with a half-life of 1 d.

Solution: The parameters must be expressed in common units,

$$E = 50{,}000 \text{ cm}^2\text{ s}^{-1}\left(\frac{\text{m}^2}{10{,}000 \text{ cm}^2}\frac{86{,}400 \text{ s}}{\text{d}}\right) = 432{,}000 \text{ m}^2\text{ d}^{-1}$$

$$U = 24{,}000 \text{ m d}^{-1}$$

$$k = \frac{0.693}{1\ \text{d}} = 0.693\ \text{d}^{-1}$$

The estuary number can be computed as

$$\eta = \frac{0.693(432{,}000)}{(24{,}000)^2} = 5.2 \times 10^{-4}$$

which clearly indicates a highly advective situation.

PROBLEMS

10.1. A one-dimensional estuary has constant dimensions and flow:

Width = 1000 ft
Depth = 10 ft
Flow = 500 cfs
Dispersion coefficient = 1×10^6 m^2 d^{-1}

(*a*) Calculate the estuary number for a herbicide that is subject to first-order decay (0.05 d^{-1}) and volatilization (0.3 m d^{-1}).
(*b*) Suppose that 10 kg of the herbicide is spilled at mile point 2. Plot the concentration distribution as a function of time at a water intake at mile point 10.

10.2. A tracer study is conducted in a stream with a flow of 3.7×10^5 m^3 d^{-1} and a width of 60 m. At $t = 0$, 50 kg of lithium is instantaneously injected at $x = 0$. Lithium concentration is measured at two downstream stations:

$x = 1$ km

t (min)	60	80	100	120	140	160	180	200	220	240	260
Lithium (μg L^{-1})	0	2	24	78	108	89	52	23	9	3	0

$x = 5$ km

t (min)	550	600	650	700	750	800	850	900	950
Lithium (μg L^{-1})	0	7	26	47	43	23	8	2	0

Determine

(*a*) velocity, m d^{-1}
(*b*) depth, m
(*c*) dispersion coefficient, cm^2 s^{-1}

10.3. At $t = 0$, 10 g of a conservative algicide is introduced into a trout hatchery pond having $V = 2500$ m^3. Flow in the channel and the river is plug flow and the hatchery pond is completely mixed. The dimensions of the channel are $B = 2$ m, $H = 1$ m, and $L = 0.5$ km. Flow in the river upstream of the diversion is 20,000 m^3 d^{-1}, and 2000 m^3 d^{-1} is diverted through the hatchery.

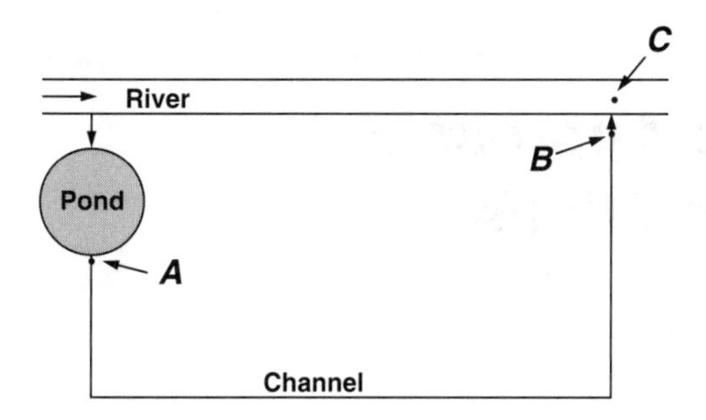

FIGURE P10.3

(*a*) Determine the concentration time series at points *A*, *B*, and *C*. Plot your results and express the concentration in ppb.

(*b*) How much algicide can be applied if the concentration in the river is never to exceed 10 ppb?

10.4. At $t = 0$, 10 g of a conservative algicide is introduced into a trout hatchery pond having $V = 2500$ m^3. Flow in the channel and the river is plug flow and the hatchery pond is completely mixed. The dimensions of the channel are $B = 2$ m, $H = 1$ m, and $L =$ 0.5 km. Flow in the river upstream of the diversion is 20,000 m^3 d^{-1}, and 2000 m^3 d^{-1} is diverted through the hatchery. Both the river and the pond flow into a well-mixed lake having $V = 10{,}000$ m^3. Determine the time when the maximum concentration occurs in the lake.

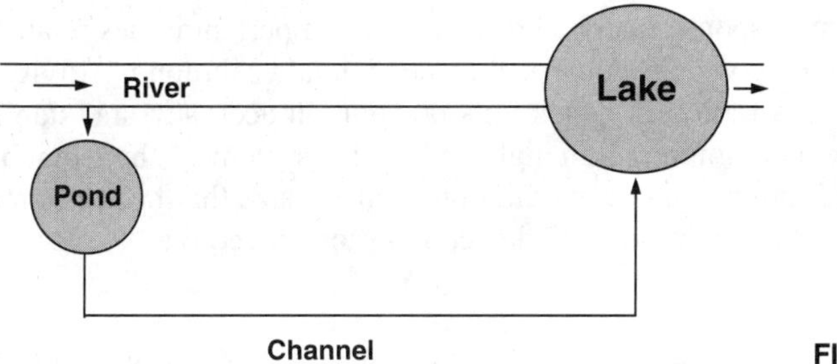

FIGURE P10.4

10.5. Suppose that a contaminant is injected into a sediment at a depth of 10 cm. Assuming that the diffusion coefficient is 10^{-6} cm^2 s^{-1},

(*a*) How long will it take the contaminant to reach the sediment water interface? Define the front as being the point that encompasses 99% of the contaminant.

(*b*) Determine the flux into the overlying water at this time.

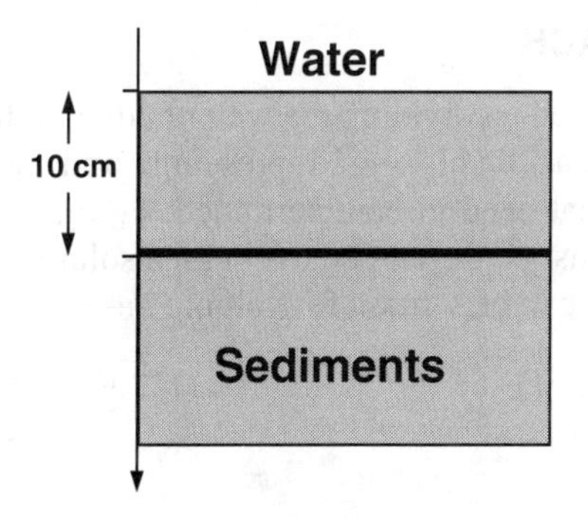

FIGURE P10.5

LECTURE 11

Control-Volume Approach: Steady-State Solutions

LECTURE OVERVIEW: I develop a control-volume method for simulating distributed systems with the computer. Then I derive a steady-state approach and present the system response matrix. I discuss some important issues related to the control-volume approach: numerical dispersion and solution positivity. In essence these issues represent constraints on solution accuracy and stability that become especially important in highly advective systems. I show that both can be subsumed into a single constraint on segment size that becomes more stringent as the system being modeled becomes more advective.

In the previous lectures you have learned how calculus can be used to obtain analytical solutions for distributed systems. Now I describe a method for modeling such systems with the computer. I take a straightforward approach by developing some simple computations to illustrate the method. Then I generalize the approach and explore its limitations.

11.1 CONTROL-VOLUME APPROACH

As depicted in Fig. 11.1 the approach is based on dividing the water body into finite segments or "control volumes." The segments 0 and $n + 1$ represent boundary segments. Therefore there are n unknowns that need to be determined: $c_1, c_2, \ldots, c_n$. Consequently n simultaneous equations must be developed to obtain a solution.

To do this a steady-state mass balance can be written for volume i as

$$0 = W_i + Q_{i-1,i}c_{i-1,i} - Q_{i,i+1}c_{i,i+1} + E'_{i-1,i}(c_{i-1} - c_i) + E'_{i,i+1}(c_{i+1} - c_i) - k_i V_i c_i \tag{11.1}$$

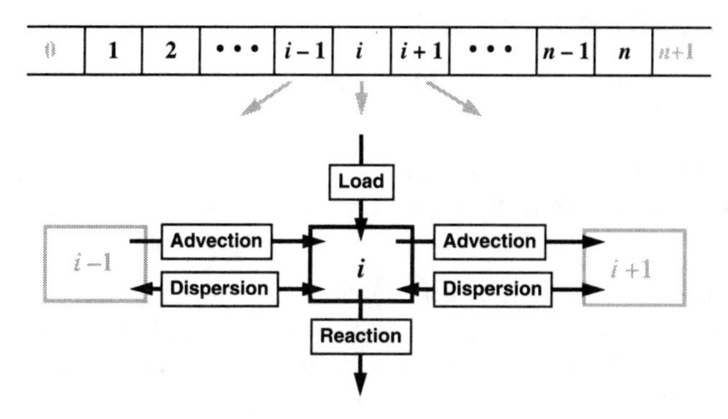

FIGURE 11.1
Mass balance around a control volume.

where double-subscripted terms refer to interfaces between volumes. For example the term $Q_{i-1,i}$ refers to the flow between volumes $i - 1$ and i.

Note that a problem with Eq. 11.1 relates to the terms, $c_{i-1,i}$ and $c_{i,i+1}$—that is, the concentrations at the interfaces between the segment and its upstream and downstream neighbors. Although it makes physical sense, inclusion of these concentrations introduces two additional unknowns to the equation. To eliminate these unknowns we must reexpress these interface concentrations in terms of the model unknowns. One approach for doing this is referred to as a ***backward*** or ***upstream difference***. Such terminology relates to the fact that the interface concentration is taken to be approximated by the concentration of the "upstream" segment. Using this approach, we can express Eq. 11.1 as

$$0 = W_i + Q_{i-1,i}c_{i-1} - Q_{i,i+1}c_i + E'_{i-1,i}(c_{i-1} - c_i) + E'_{i,i+1}(c_{i+1} - c_i) - k_i V_i c_i \tag{11.2}$$

Thus the model is now totally formulated in terms of the unknown concentrations. Collecting terms yields

$$-(Q_{i-1,i} + E'_{i-1,i})c_{i-1} + (Q_{i,i+1} + E'_{i-1,i} + E'_{i,i+1} + k_i V_i)c_i - (E'_{i,i+1})c_{i+1} = W_i \tag{11.3}$$

or

$$a_{i,i-1}c_{i-1} + a_{i,i}c_i + a_{i,i+1}c_{i+1} = W_i \tag{11.4}$$

where

$$a_{i,i-1} = -Q_{i-1,i} - E'_{i-1,i} \tag{11.5}$$

$$a_{i,i} = Q_{i,i+1} + E'_{i-1,i} + E'_{i,i+1} + k_i V_i \tag{11.6}$$

$$a_{i,i+1} = -E'_{i,i+1} \tag{11.7}$$

If Eq. 11.4 is written for volumes 1 through n in Fig. 11.1, there are n equations with $n + 2$ unknowns (c_0 through c_{n+1}). Therefore boundary conditions are needed to eliminate two of these unknowns.

11.2 BOUNDARY CONDITIONS

There are two fundamental types of boundary conditions used in water-quality modeling:

- ***Dirichlet boundary conditions*** specify the concentration at the boundary.
- ***Neumann boundary conditions*** specify the derivative of concentration at the boundary.

In addition there are occasions when a combination of these might be appropriate. In this section, because they are the most commonly used, I describe how Dirichlet conditions are treated. Some examples of Neumann conditions will be explored in the end-of-chapter problems and in later lectures.

Two types of Dirichlet conditions are illustrated by Fig 11.2. In Fig. 11.2*a* the boundaries are open, as they would be in a channel. For such cases Eq. 11.4 can be applied directly. For example for volume 1, the equation is

$$W_1 = a_{1,1}c_1 + a_{1,2}c_2 \tag{11.8}$$

where the loading is modified to reflect the fact that it includes the effect of inflow from volume 0,

$$W_1 \leftarrow W_1 - a_{1,0}c_0 \tag{11.9}$$

Similarly the balance for volume n reflects the fact that c_{n+1} is a known,

$$W_n = a_{n,n-1}c_{n-1} + a_{n,n}c_n \tag{11.10}$$

where the loading is modified to include the effect of volume $n + 1$,

$$W_n \leftarrow W_n - a_{n,n+1}c_{n+1} \tag{11.11}$$

For both cases the coefficients (the a's) are as defined in Eqs. 11.5 to 11.7.

When mass enters and leaves the tank via plug flow, the situation is more interesting. For the case of the inflow volume ($i = 1$), the original mass balance becomes

$$0 = W_1 + Q_{0,1}c_0 - Q_{1,2}c_1 + E'_{1,2}(c_2 - c_1) - k_1V_1c_1 \tag{11.12}$$

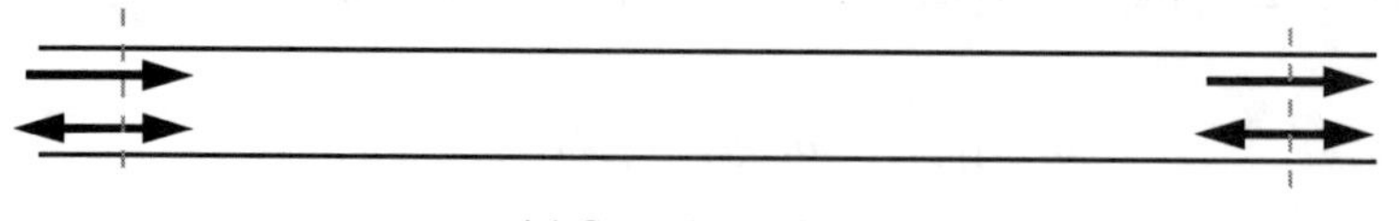

(*a*) Open boundaries

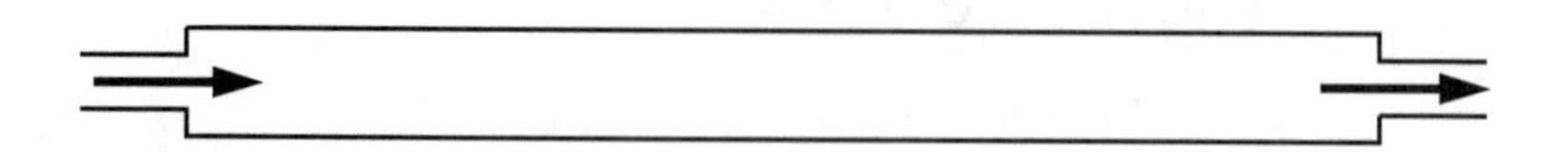

(*b*) Pipe boundaries

FIGURE 11.2
Two types of Dirichlet (fixed concentration) boundary conditions for one-dimensional systems. (*a*) Open advective/dispersive; (*b*) pipe advective boundaries.

Notice how we have merely omitted the dispersion term for the entry interface. When we collect terms the final equation looks like

$$W_1 = a_{1,1}c_1 + a_{1,2}c_2 \tag{11.13}$$

where

$$W_1 \leftarrow W_1 - a_{1,0}c_0 \tag{11.14}$$

$$a_{1,0} = -Q_{0,1} \tag{11.15}$$

$$a_{1,1} = Q_{1,2} + E'_{1,2} + k_1 V_1 \tag{11.16}$$

$$a_{1,2} = -E'_{1,2} \tag{11.17}$$

Thus the loading term is modified to include the inflow concentration, and the a's are adjusted to omit dispersion across the inflow interface.

A similar derivation at the outlet yields

$$W_n = a_{n,n-1}c_{n-1} + a_{n,n}c_n \tag{11.18}$$

where

$$a_{n,n-1} = -(Q_{n-1,n} + E'_{n-1,n}) \tag{11.19}$$

$$a_{n,n} = Q_{n,n+1} + E'_{n-1,n} + k_n V_n \tag{11.20}$$

Notice that because of the lack of dispersion, the loading term is not affected by c_{n+1}. Similarly the diagonal term $a_{n,n}$ does not include the term $E'_{n,n+1}$ because no dispersion occurs across the outlet interface.

11.3 STEADY-STATE SOLUTION

The complete set of equations to be solved is now specified as

$$a_{1,1}c_1 + a_{1,2}c_2 = W_1 \tag{11.21}$$

$$a_{2,1}c_1 + a_{2,2}c_2 + a_{2,3}c_3 = W_2 \tag{11.22}$$

$$a_{3,2}c_2 + a_{3,3}c_3 + a_{3,4}c_4 = W_3 \tag{11.23}$$

$$\vdots$$

$$a_{n-1,n-2}c_{n-2} + a_{n-1,n-1}c_{n-1} + a_{n-1,n}c_n = W_{n-1} \tag{11.24}$$

$$a_{n,n-1}c_{n-1} + a_{n,n}c_n = W_n \tag{11.25}$$

which can also be expressed in matrix form as

$$[A]\{c\} = \{W\} \tag{11.26}$$

where

$$[A] = \begin{bmatrix} a_{11} & a_{12} & 0 & \cdot & \cdot & \cdot & 0 \\ a_{21} & a_{22} & a_{23} & 0 & \cdot & \cdot & 0 \\ 0 & a_{32} & a_{33} & a_{34} & 0 & \cdot & 0 \\ \vdots & & & & & & \vdots \\ 0 & \cdot & \cdot & 0 & a_{n-1,n-2} & a_{n-1,n-1} & a_{n-1,n} \\ 0 & \cdot & \cdot & \cdot & 0 & a_{n,n-1} & a_{n,n} \end{bmatrix} \tag{11.27}$$

$$\{c\} = \begin{Bmatrix} c_1 \\ c_2 \\ c_3 \\ \vdots \\ c_{n-1} \\ c_n \end{Bmatrix} \qquad \{W\} = \begin{Bmatrix} W_1 \\ W_2 \\ W_3 \\ \vdots \\ W_{n-1} \\ W_n \end{Bmatrix} \tag{11.28}$$

Note that, as described previously in Lec. 6, a variety of methods are available for solving such equations. An application is provided by the following example.

EXAMPLE 11.1. APPLYING THE CONTROL-VOLUME APPROACH TO A MIXED-FLOW REACTOR. Recall that in Example 5.2 we used a cascade model to simulate the steady-state distribution of concentration in an elongated tank. In this example use the same system but add dispersive transport to the framework, as indicated in Fig. E11.1-1.

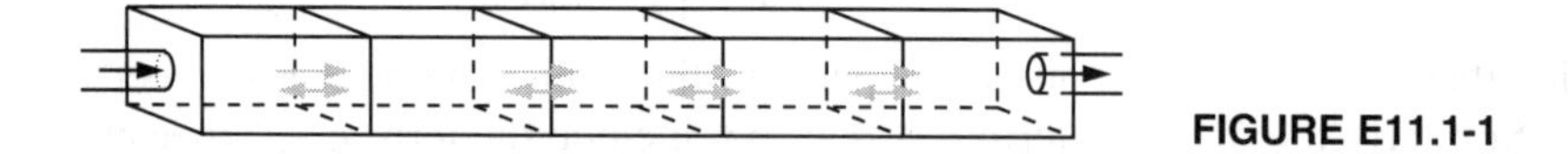

FIGURE E11.1-1

The tank has cross-sectional area $A_c = 10\ \text{m}^2$, length $L = 100$ m, velocity $U = 100\ \text{m hr}^{-1}$, first-order reaction rate $k = 2\ \text{hr}^{-1}$, and $E = 2000\ \text{m}^2\ \text{hr}^{-1}$. The inflow concentration is $1\ \text{mg L}^{-1}$. Use $n = 5$ and display the concentrations along with the analytical solution computed in Example 9.2.

Solution: The following matrix representation can be developed for the system:

$$\begin{bmatrix} 2400 & -1000 & 0 & 0 & 0 \\ -2000 & 3400 & -1000 & 0 & 0 \\ 0 & -2000 & 3400 & -1000 & 0 \\ 0 & 0 & -2000 & 3400 & -1000 \\ 0 & 0 & 0 & -2000 & 2400 \end{bmatrix} \begin{Bmatrix} c_1 \\ c_2 \\ c_3 \\ c_4 \\ c_5 \end{Bmatrix} = \begin{Bmatrix} 1000 \\ 0 \\ 0 \\ 0 \\ 0 \end{Bmatrix}$$

which can be solved for the concentrations. These results, along with the calculations from Example 9.2, are summarized in the following table and figure:

Distance	Analytical	Numerical
0	0.76563	
10	0.65700	0.60919
30	0.48401	0.46206
50	0.35753	0.35261
70	0.26760	0.27476
90	0.21319	0.22897
100	0.20441	

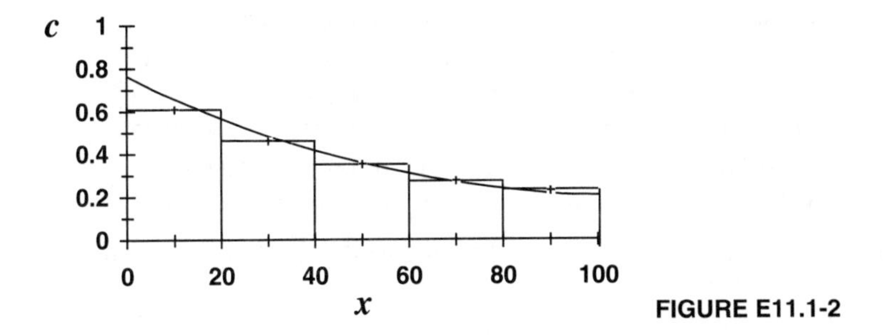

FIGURE E11.1-2

The numerical results generally follow the trend of the exact solution. If a finer segmentation were imposed (that is, a smaller Δx), the numerical outcome would approach the analytical case.

Despite the fact that the results of the foregoing example seem acceptable, closer inspection of Fig. E11.1-2 indicates that the discrepancies seem systematic. That is, at the head end of the tank the numerical results are lower than the analytical solution. At the tail end the numerical results are higher. In addition the discrepancy is most pronounced where the gradient is steepest, at the head of the tank. Such errors might occur if we had used too high a dispersion coefficient for the numerical calculation. In fact, in Sec. 11.6, I will show that there is a "numerical" dispersion that accompanies computations of the sort illustrated here, that is, those that use backward differences.

11.4 SYSTEM RESPONSE MATRIX

The solution for Example 11.1 can be obtained in a variety of ways. For example both iterative and elimination methods are available for accomplishing this task (App. E). In addition the matrix inverse could be used to generate the solution

$$\{c\} = [A]^{-1}\{W\} \tag{11.29}$$

Although this is not the most efficient means for obtaining a solution, aside from providing the solution vector, the matrix inverse allows the computation of the effect of loading to a particular segment on any other segment. As previously described in Sec. 6.3, the element of the matrix inverse $a_{i,j}^{-1}$ represents the change in concentration of segment i due to a unit loading change to segment j. This property is explored in further detail in the following example.

EXAMPLE 11.2. MATRIX INVERSE ANALYSIS OF A MIXED-FLOW REACTOR. Use the matrix inverse to evaluate the following loading scenarios for the mixed-flow reactor from Example 11.1:

(*a*) Determine how much the inflow concentration must be reduced to reduce the outflow concentration to 0.1 mg L^{-1}.

(*b*) Calculate the concentration distribution if 2000 g hr^{-1} is injected in the middle of the reactor and 1000 g hr^{-1} is injected at the end. Set the inflow concentration to zero.

Solution: (*a*) The matrix inverse for the system can be calculated as

$$\begin{bmatrix} 0.0006092 & 0.0002310 & 0.0000882 & 0.0000343 & 0.0000143 \\ 0.0004621 & 0.0005545 & 0.0002116 & 0.0000824 & 0.0000343 \\ 0.0003526 & 0.0004231 & 0.0005430 & 0.0002116 & 0.0000882 \\ 0.0002748 & 0.0003297 & 0.0004231 & 0.0005545 & 0.0002310 \\ 0.0002290 & 0.0002748 & 0.0003526 & 0.0004621 & 0.0006092 \end{bmatrix}$$

The load reduction needed to lower the outlet concentration to 0.1 mg L^{-1} can be calculated as

$$c_5 = a_{51}^{-1} W_1 = a_{51}^{-1} Q c_0$$

which can be solved for

$$c_0 = \frac{c_5}{a_{51}^{-1} Q} = \frac{0.1 \text{ g m}^{-3}}{0.0002290[\text{g m}^{-3}(\text{g hr}^{-1})^{-1}](1000 \text{ m}^3 \text{ hr}^{-1})} = 0.4367 \text{ g m}^{-3}$$

Therefore the inflow concentration must be reduced,

$$\% \text{ reduction} = \frac{1 - 0.4367}{1}(100\%) = 56.3\%$$

(*b*) The appropriate elements can be used to evaluate the effect of the two loadings on the entire reactor, as in

$$\begin{aligned}
c_1 &= 2000(0.0000882) + 1000(0.0000143) = 0.1764 + 0.0143 = 0.1907 \text{ mg L}^{-1} \\
c_2 &= 2000(0.0002116) + 1000(0.0000343) = 0.4232 + 0.0343 = 0.4575 \text{ mg L}^{-1} \\
c_3 &= 2000(0.0005430) + 1000(0.0000882) = 1.0860 + 0.0882 = 1.1742 \text{ mg L}^{-1} \\
c_4 &= 2000(0.0004231) + 1000(0.0002310) = 0.8462 + 0.2310 = 1.0772 \text{ mg L}^{-1} \\
c_5 &= 2000(0.0003526) + 1000(0.0006092) = 0.7052 + 0.6092 = 1.3144 \text{ mg L}^{-1}
\end{aligned}$$

These results are displayed below:

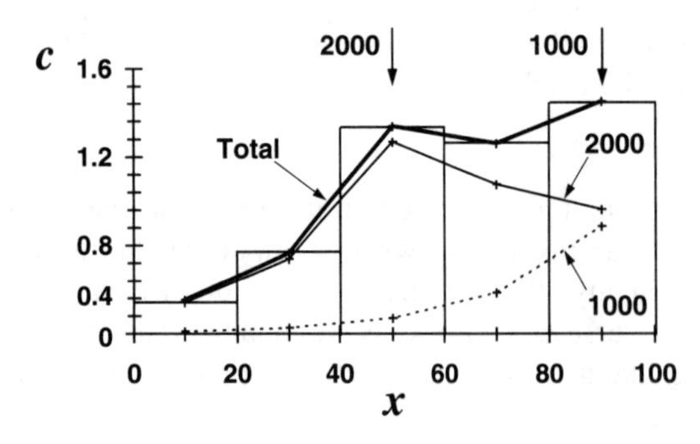

FIGURE E11.2

11.5 CENTERED-DIFFERENCE APPROACH

Note that the backward difference presented in the previous sections is but one way to represent the advective term in finite form. Another alternative is to specify the interface concentration as the average concentration of the two adjacent segments. For equal-sized volumes, the resulting balance is

$$0 = W_i + Q_{i-1,i}\left(\frac{c_{i-1} + c_i}{2}\right) - Q_{i,i+1}\left(\frac{c_i + c_{i+1}}{2}\right)$$
$$+ E'_{i-1,i}(c_{i-1} - c_i) + E'_{i,i+1}(c_{i+1} - c_i) - k_i V_i c_i \tag{11.30}$$

Collecting terms yields

$$a_{i,i-1}c_{i-1} + a_{i,i}c_i + a_{i,i+1}c_{i+1} = W_i \tag{11.31}$$

where

$$a_{i,i-1} = -\frac{Q_{i-1,i}}{2} - E'_{i-1,i} \tag{11.32}$$

$$a_{i,i} = E'_{i-1,i} + E'_{i,i+1} + k_i V_i \tag{11.33}$$

$$a_{i,i+1} = -E'_{i,i+1} + \frac{Q_{i,i+1}}{2} \tag{11.34}$$

EXAMPLE 11.3. CENTERED-DIFFERENCE MODEL. Repeat Example 11.1, but now use centered differences.

Solution: Using centered differences we can develop the following matrix representation for the system. Note that the inflow and outflow segments must be treated differently from what was done in Eqs. 11.13 and 11.18 to account for the centered differences (see Prob. 11.1),

$$\begin{bmatrix} 1900 & -500 & 0 & 0 & 0 \\ -1500 & 2400 & -500 & 0 & 0 \\ 0 & -1500 & 2400 & -500 & 0 \\ 0 & 0 & -1500 & 2400 & -500 \\ 0 & 0 & 0 & -1500 & 1900 \end{bmatrix} \begin{Bmatrix} c_1 \\ c_2 \\ c_3 \\ c_4 \\ c_5 \end{Bmatrix} = \begin{Bmatrix} 1000 \\ 0 \\ 0 \\ 0 \\ 0 \end{Bmatrix}$$

which can be solved for the concentrations. These results, along with the calculations from Example 9.2, are summarized in the following table and in Fig. E11.3:

Distance	Analytical	Numerical
0	0.76563	
10	0.65700	0.65338
30	0.48401	0.48284
50	0.35753	0.35748
70	0.26760	0.26741
90	0.21319	0.21111
100	0.20441	

The numerical results follow the trend of the exact solution much better than in Example 11.1.

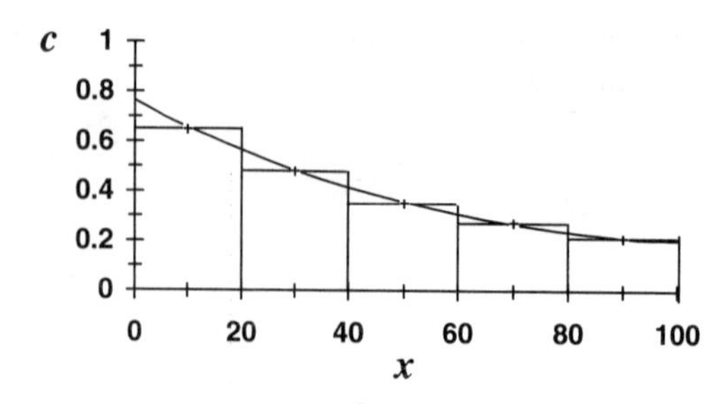

FIGURE E11.3

In Example 11.3 we observe that the centered difference does not exhibit the "numerical dispersion" that afflicted the backward difference in Example 11.1. Therefore, at this point, you might be wondering why we ever used the backward-difference approach in the first place. In brief, although the centered-difference method yields superior results for the present example, there are other cases where it yields physically unrealistic results. For such cases the backward difference provides an alternative. The next section deals with these issues.

In addition note that the centered and backward differences are actually two specific instances of a more general weighting scheme for handling the advective term in our model. This approach is outlined in Box 11.1.

BOX 11.1. A Weighted-Difference Formulation

Now that we have developed the backward and centered approximations, we can generalize them into a general weighted-difference approach. To do this we can formulate the concentration at the interface between segments by a weighted average,

$$c_{j,k} = \alpha_{j,k} c_j + \beta_{j,k} c_k$$

where

$$\alpha_{j,k} = \frac{\Delta x_k}{\Delta x_j + \Delta x_k}$$

and

$$\beta_{j,k} = 1 - \alpha_{j,k} = \frac{\Delta x_j}{\Delta x_j + \Delta x_k}$$

These can be substituted into Eq. 11.1 to give

$$0 = W_i + Q_{i-1,i}(\alpha_{i-1,i} c_{i-1} + \beta_{i-1,i} c_i) - Q_{i,i+1}(\alpha_{i,i+1} c_i + \beta_{i,i+1} c_{i+1})$$
$$+ E'_{i-1,i}(c_{i-1} - c_i) + E'_{i,i+1}(c_{i+1} - c_i) - k_i V_i c_i$$

The α's and β's serve two purposes. First, as depicted in Fig. B11.1, they provide a way to apply the centered-difference approximation for unequally spaced segments. In addition to accounting for unequal segment sizes, they also provide a means for shifting between backward and centered spatial approximations. Note that for $\alpha = 1$ and $\beta = 0$ the equation becomes the backward-difference representation. In contrast if $\alpha = \beta = 0.5$ (or for unequal-sized segments using the weights calculated above), the result amounts to a centered difference.

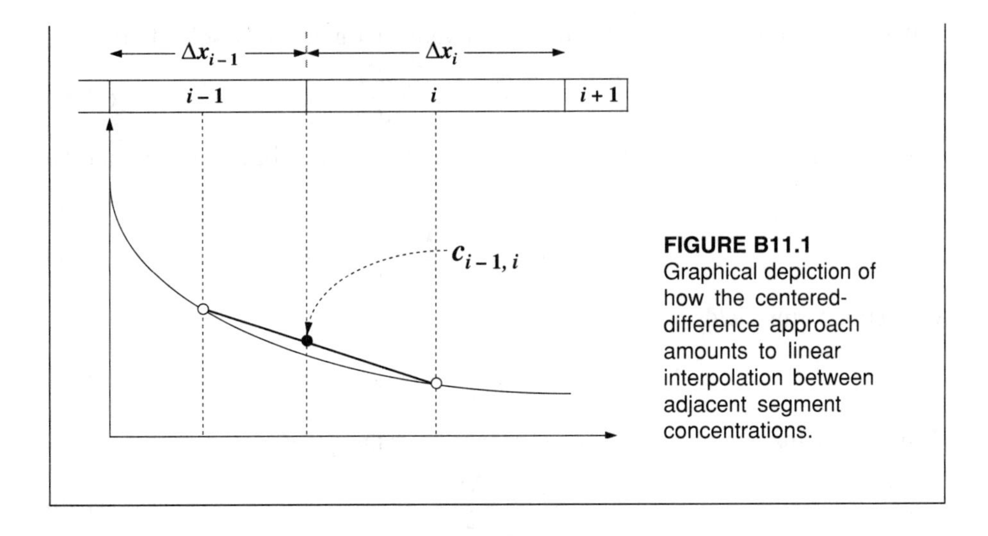

FIGURE B11.1
Graphical depiction of how the centered-difference approach amounts to linear interpolation between adjacent segment concentrations.

11.6 NUMERICAL DISPERSION, POSITIVITY, AND SEGMENT SIZE

In the previous sections we developed a straightforward numerical approach for modeling steady-state pollutant levels in distributed systems. In Example 11.1 we discovered a discrepancy between the numerical and analytical results for a backward-difference approach. Based on this experience, we take a more systematic look at two fundamental questions. First, does the control-volume approach always yield meaningful results? Second, when it works, how accurate are the calculations? We first look at the accuracy issue.

11.6.1 Numerical Dispersion

Now let us look at the problem we encountered in Example 11.1 when we developed a numerical approximation to the mass balance equation

$$0 = -U\frac{dc}{dx} + E\frac{d^2c}{dx^2} - kc \tag{11.35}$$

Recall that using a backward-difference approximation seemed to cause the computation to exhibit too much dispersion. To gain insight into what is happening, we can recognize that for constant-parameter systems, the control-volume approach is identical to a finite-difference approximation of the underlying partial differential equation (see Box 11.2 for details).

The finite-difference perspective allows the use of Taylor-series expansions to characterize the errors of the finite-difference approximations. For example we can expand concentration in a backward Taylor series,

$$c_{i-1} = c_i - \frac{dc}{dx}\Delta x + \frac{d^2c}{dx^2}\frac{\Delta x^2}{2!} - \cdots \tag{11.36}$$

If the series is truncated after the second-derivative term, it can be solved for

$$\frac{dc}{dx} \cong \frac{c_i - c_{i-1}}{\Delta x} + \frac{\Delta x}{2} \frac{d^2c}{dx^2} \tag{11.37}$$

The divided difference is substituted for the first derivative in Eq. 11.35, and the result is

$$0 = E\frac{d^2c}{dx^2} - U\left(\frac{dc}{dx} - \frac{\Delta x}{2}\frac{d^2c}{dx^2}\right) - kc \tag{11.38}$$

Collecting terms yields

$$0 = \left(E + \frac{\Delta x}{2}U\right)\frac{d^2c}{dx^2} - U\frac{dc}{dx} - kc \tag{11.39}$$

Thus we see that the real dispersion is enhanced by the quantity

$$E_n = \frac{\Delta x}{2}U \tag{11.40}$$

where E_n is called ***numerical dispersion***.

BOX 11.2. The Finite-Difference Approach

The finite-difference method offers a more conventional (though not necessarily superior) approach to deriving computer solutions for distributed systems. In this method we substitute finite-difference approximations for the derivatives in our fundamental equation

$$0 = -U\frac{dc}{dx} + E\frac{d^2c}{dx^2} - kc$$

For example the first derivative can be approximated with a backward difference,

$$\frac{dc}{dx} \cong \frac{c_i - c_{i-1}}{\Delta x}$$

The second derivative can be approximated with a centered difference,

$$\frac{d^2c}{dx^2} \cong \frac{\dfrac{c_{i+1} - c_i}{\Delta x} - \dfrac{c_i - c_{i-1}}{\Delta x}}{\Delta x} = \frac{c_{i+1} - 2c_i + c_{i-1}}{\Delta x^2}$$

Substituting these into the mass balance yields

$$0 = E\frac{c_{i+1} - 2c_i + c_{i-1}}{\Delta x^2} - U\frac{c_i - c_{i-1}}{\Delta x} - kc_i$$

Collecting terms and multiplying by $V_i = A_c\,\Delta x$ gives

$$-(Q + E')c_{i-1} + (Q + 2E' + kV_i)c_i - E'c_{i+1} = 0$$

which is of the same form as Eq. 11.3. Thus for constant-parameter systems, the control-volume and finite-difference approaches yield identical formulations.

By substituting the centered-difference approximation for the first derivative,

$$\frac{dc}{dx} \cong \frac{c_{i+1} - c_{i-1}}{2\,\Delta x}$$

a formulation can also be developed that is the equivalent of Eq. 11.30. Problem 11.4 deals with this case.

EXAMPLE 11.4. CORRECTING FOR NUMERICAL DISPERSION. Repeat Example 11.1, but calculate the numerical dispersion. Then reduce the dispersion coefficient by this amount and see what happens.

Solution: Equation 11.40 can be used to calculate numerical dispersion as

$$E_n = \frac{20\text{ m}}{2}(100\text{ m hr}^{-1}) = 1000\text{ m}^2\text{ hr}^{-1}$$

Therefore we can reduce the dispersion coefficient from 2000 to 1000 $\text{m}^2\text{ hr}^{-1}$. If this is done, the following matrix representation can be developed for the system:

$$\begin{bmatrix} 1900 & -500 & 0 & 0 & 0 \\ -1500 & 2400 & -500 & 0 & 0 \\ 0 & -1500 & 2400 & -500 & 0 \\ 0 & 0 & -1500 & 2400 & -500 \\ 0 & 0 & 0 & -1500 & 1900 \end{bmatrix} \begin{Bmatrix} c_1 \\ c_2 \\ c_3 \\ c_4 \\ c_5 \end{Bmatrix} = \begin{Bmatrix} 1000 \\ 0 \\ 0 \\ 0 \\ 0 \end{Bmatrix}$$

Note that this system is identical to the one derived using the centered-difference approach in Example 11.3. Consequently the computed results will be identical with the centered-difference characterization. Thus correcting the dispersion coefficients or using centered differences yield the same outcome!

On the basis of the foregoing example, you might now be wondering why we did not just use a centered difference to start with. The answer lies in an additional problem—solution positivity.

11.6.2 Positivity

It can be shown that as a system becomes relatively more advective, solutions can go negative (Hall and Porsching 1990). The following example illustrates this phenomenon.

EXAMPLE 11.5. NEGATIVE SOLUTIONS FOR HIGHLY ADVECTIVE SYSTEMS. Repeat Example 11.3, but increase the velocity in the tank to 400 m hr^{-1}. Also set the inflow concentration to zero and add a loading of 4000 g hr^{-1} to the middle of the tank. As in Example 11.3 use a centered-difference approximation.

Solution: The following matrix representation can be developed for the system:

$$\begin{bmatrix} 3400 & 1000 & 0 & 0 & 0 \\ -3000 & 2400 & 1000 & 0 & 0 \\ 0 & -3000 & 2400 & 1000 & 0 \\ 0 & 0 & -3000 & 2400 & 1000 \\ 0 & 0 & 0 & -3000 & 3400 \end{bmatrix} \begin{Bmatrix} c_1 \\ c_2 \\ c_3 \\ c_4 \\ c_5 \end{Bmatrix} = \begin{Bmatrix} 0 \\ 0 \\ 4000 \\ 0 \\ 0 \end{Bmatrix}$$

which can be solved for the concentrations. These results, along with calculations from an analytical solution, are summarized in the following table and figure:

Distance	Analytical	Numerical
0	0.00003	
10	0.00026	0.08477
30	0.01584	−0.28823
50	0.95346	0.94608
70	0.86483	0.86470
90	0.78665	0.76297
100	0.76479	

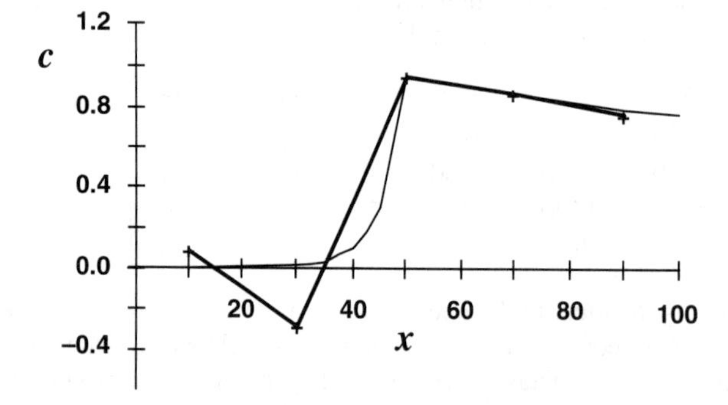

FIGURE E11.5

Although the results downstream from the load are adequate, the simulation above is clearly in error. In fact it yields the physically unrealistic outcome of negative concentrations.

Notice that the matrix in Example 11.5 differs from the others in this lecture. In all the other matrices the off-diagonal terms have a different sign from the terms on the main diagonal. That is, the main diagonal terms are positive, whereas the off-diagonal terms are negative. In the previous example the superdiagonal terms are positive. It can be shown (Hall and Porsching 1990) that this is the root cause of the negative results. To quantify this constraint, we can rephrase it as specifying that the superdiagonal term must be negative to obtain positive results. Recall the general formulation for the superdiagonal term (that is, the constants multiplying c_{i+1}),

$$-E' + 0.5Q < 0 \tag{11.41}$$

Substituting $E' = EA_c/\Delta x$ and $Q = A_c U$ yields

$$-\frac{EA_c}{\Delta x} + 0.5UA_c < 0 \tag{11.42}$$

which can be solved for

TABLE 11.1
Constraints due to numerical dispersion and positivity for solving the advection-dispersion-reaction equation with backward and centered finite-difference schemes

	Numerical dispersion	Positivity
Backward difference ($\alpha = 1; \beta = 0$)	$E = 0.5U\Delta x$	$\Delta x < \infty$
Centered difference ($\alpha = 0.5; \beta = 0.5$)	$E = 0$	$\Delta x < \dfrac{E}{0.5U}$

$$\Delta x < \frac{E}{0.5U} \tag{11.43}$$

Thus if centered differences are used, the spatial discretization must be below the level specified by this equation to maintain positivity. Note that this constraint is often formulated as

$$P_e = \frac{U\Delta x}{E} < 2 \tag{11.44}$$

where P_e is the cell Peclet or Reynolds number (recall Box 9.1).

11.6.3 What It All Means

At this point let's regroup and try to make sense out of what we've just learned. We now know that we are dealing with two constraints:

$$E_n = U\Delta x(\alpha - 0.5) \qquad \text{numerical dispersion} \tag{11.45}$$

and

$$\Delta x < \frac{E}{(1-\alpha)U} \qquad \text{positivity} \tag{11.46}$$

Note that these are now expressed in terms of the weighted-difference formulation of Box 11.1 (compare Eqs. 11.45 and 11.46 with Eqs. 11.40 and 11.43).

The situation is summarized in Table 11.1. If we use centered differences we do not have numerical dispersion. However, for systems where advection is high relative to dispersion, we are forced to use small segment sizes to obtain physically realistic solutions. In contrast if we use backward differences, our solutions are always positive, but they are contaminated with numerical dispersion.

What's the bottom line? Close inspection of Eqs. 11.45 and 11.46 show that the numerical dispersion and positivity constraints are structurally identical. To understand what this means, recognize that regardless of the difference scheme our goal should be

$$E_p = E_m + E_n \tag{11.47}$$

Real "physical" dispersion — The number you use in your "model" — "Numerical" dispersion induced by difference scheme

That is, we would like the model output to match the actual dispersion that occurs in the physical world.

Substitute Eq. 11.45 into 11.47 and rearrange,

$$E_m = E_p - U\Delta x(\alpha - 0.5) \tag{11.48}$$

and substitute this result into Eq. 11.46 to give

$$\Delta x < \frac{E_p - U\Delta x(\alpha - 0.5)}{U(1 - \alpha)} \tag{11.49}$$

The right side of this equation can be manipulated algebraically to give

$$\Delta x_c < \frac{E_p}{0.5U} \tag{11.50}$$

Thus the α's drop out and we see that both the numerical dispersion and the positivity constraints amount to a single constraint on the segment size. Consequently we have used the nomenclature Δx_c to connote that Eq. 11.50 defines a critical segment size.

In essence if backward differences are used, this critical length induces a numerical dispersion that is exactly equivalent to the physical dispersion. Thus we could set the model dispersion E_m equal to zero and obtain an accurate simulation because the backward difference would induce the proper dispersion numerically. Conversely this length also represents the point at which we could set the model dispersion to the level of the physical dispersion and obtain positive concentrations with a centered-difference scheme.

Now what is the significance of these results in real streams and estuaries? Figure 11.3 shows the critical length versus the width of a collection of streams and estuaries. This plot indicates that because they are predominantly dispersive systems, the estuaries have critical lengths of more than 1 km. This is also true for the larger rivers. However, for the smaller, more highly advective streams, the lengths fall below 0.1 km. This means that many more segments would be required to adequately simulate such systems.

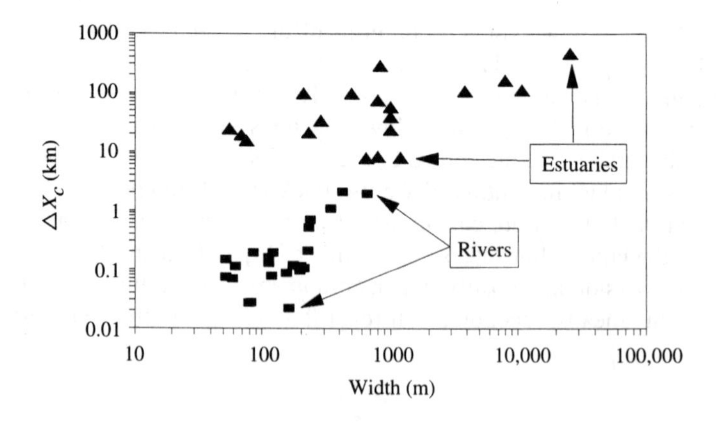

FIGURE 11.3
Plot of critical segment length versus width for a number of rivers and estuaries. The original data are taken from Hydroscience (1971) and Fischer et al. (1979).

In summary, for estuaries a centered difference can usually be used along with measured dispersion coefficients. If backward differences are used, the model dispersion should be reduced to account for numerical dispersion. For streams, smaller segment lengths are usually required to ensure positive solutions and to avoid excess numerical dispersion.

One additional practical point should be noted. Although smaller segment lengths are required to exactly match the physical dispersion in streams, the fact is that dispersion is not that important in such systems relative to advection. Therefore, particularly for steady-state solutions where gradients are not especially sharp, the constraint embodied in Eq. 11.50 can often be relaxed without introducing major inaccuracies into the solution. Test calculations against analytical solutions can often be performed to verify whether this is valid. However, it should be stressed that time-variable solutions for problems such as spills can be very sensitive to numerical dispersion. We'll explore this issue in Lecs. 12 and 13.

11.7 SEGMENTATION AROUND POINT SOURCES

Finer segmentation is typically used to better capture the sharper gradients around point sources. The problem is illustrated in Fig. 11.4. When a finer scheme is used (Fig. 11.4*a*), the error amounts to less than 5%. In contrast the coarse segmentation scheme in Fig. 11.4*b* results in a 30% underestimation of the concentration at the entry point of the point source. Mueller (1976, as described in Thomann and Mueller 1987) has developed some guidelines to show how segment sizes can be related to acceptable errors.

Based on a backward-difference scheme, he developed a set of formulas to compute segment size needed to achieve a desired error level in the peak concentration. For advective systems ($\eta < 1$),

$$\Delta x_0 = \frac{U}{k}\left[\sqrt{\frac{1+\eta}{(1-\varepsilon)^2}} - \eta - 1\right] \tag{11.51}$$

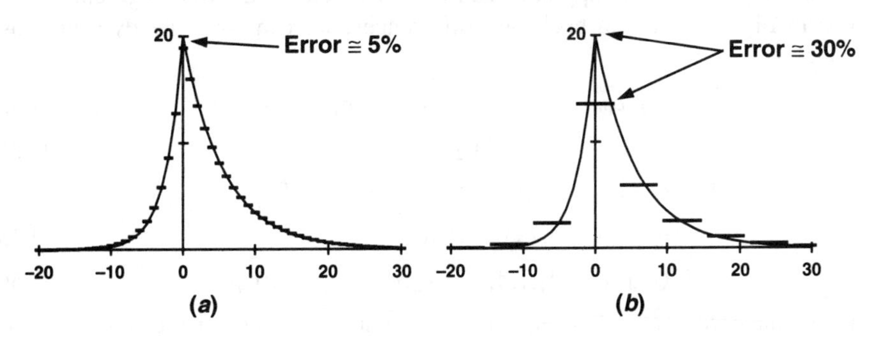

FIGURE 11.4
Comparison of numerical and analytical solutions for a point source into an estuary. The fine segmentation scheme in (*a*) yields much better resolution than the coarse version in (*b*).

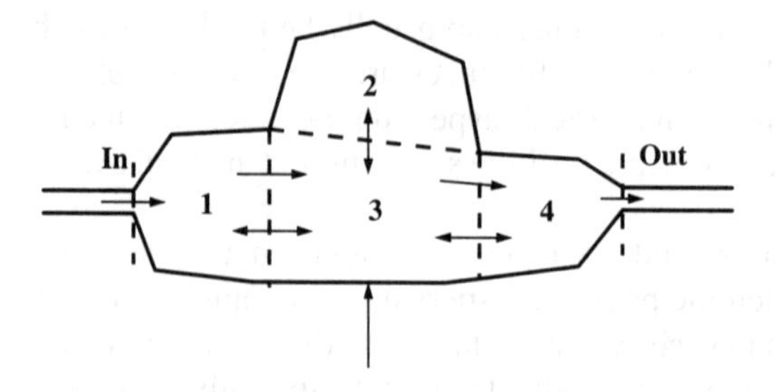

FIGURE 11.5
An elongated lake with a side bay is a simple example of a two-dimensional system.

where U = velocity
k = reaction rate
η = estuary number = kE/U^2
ε = the desired fractional relative error

For diffusive systems ($\eta > 1$), a comparable formula is

$$\Delta x_0 = \sqrt{\frac{4E\left[\sqrt{\frac{1+(1/\eta)}{(1-\varepsilon)^2}} - 1 - \frac{1}{\sqrt{\eta}}\right]^2}{k}} \tag{11.52}$$

These formulas provide useful starting points for choosing segment sizes around point sources. Additional information regarding their derivation can be found in Thomann and Mueller (1987).

11.8 TWO- AND THREE-DIMENSIONAL SYSTEMS

The extension of the methods described in this lecture to two- and three-dimensional systems is straightforward. In most cases backward-difference approximations of the advective terms can be used because any numerical dispersion that is introduced is usually dwarfed by physical dispersion. However, centered differences can also be simply applied for cases where numerical dispersion could have a detrimental effect on solution accuracy.

The simplicity of the approach can be illustrated for the four-segment system shown in Fig. 11.5. Using backward differences, we can write steady-state mass balances as

$$0 = Q_{0,1}c_{\text{in}} - Q_{1,3}c_1 + E'_{1,3}(c_3 - c_1) - kV_1c_1 \tag{11.53}$$

$$0 = E'_{2,3}(c_3 - c_2) - kV_2c_2 \tag{11.54}$$

$$0 = Q_{1,3}c_1 - Q_{3,4}c_3 + E'_{1,3}(c_1 - c_3) + E'_{2,3}(c_2 - c_3) + E'_{3,4}(c_4 - c_3) - kV_3c_3 \tag{11.55}$$

$$0 = Q_{3,4}c_3 - Q_{4,\text{out}}c_4 + E'_{3,4}(c_3 - c_4) - kV_4c_4 \tag{11.56}$$

Thus we have generated four equations that can be solved for the four unknown concentrations.

Three-dimensional problems can be handled in a similar fashion. For example suppose that the fourth segment were vertically stratified into an upper and a lower

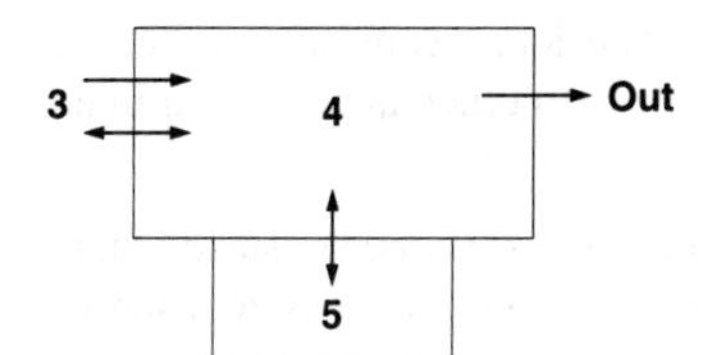

FIGURE 11.6
A two-dimensional vertical segmentation of the far-right segment from Fig. 11.5.

layer as depicted in Fig. 11.6. The fourth mass balance would be modified to give

$$0 = Q_{3,4}c_3 - Q_{4,\text{out}}c_4 + E'_{3,4}(c_3 - c_4) - kV_4c_4 + E'_{4,5}(c_5 - c_4) \quad (11.57)$$

and a fifth mass balance could be written for the lower layer,

$$0 = E'_{4,5}(c_4 - c_5) - kV_5c_5 \quad (11.58)$$

PROBLEMS

11.1. Develop equations for the entry ($i = 1$) and exit ($i = n$) segments of a reactor using centered finite-difference approximations for the derivatives in the advection-dispersion-reaction equation. Assume that the flow enters and leaves the reactor via pipes. Your equations should be presented in the format of Eqs. 11.13 to 11.20.

11.2. (*a*) Perform the same computations as in Example 11.1, but use a finer segmentation ($n = 10$).
(*b*) Compute percent relative errors for your results and for the coarser case ($n = 5$) as compared with the analytical results. Discuss your results.

11.3. Repeat part (*a*) of Example 11.2 using a segmentation scheme of $n = 10$. Discuss how and why the finer segmentation affects the reduction of the inflow concentration to meet the 0.1 mg L^{-1} goal.

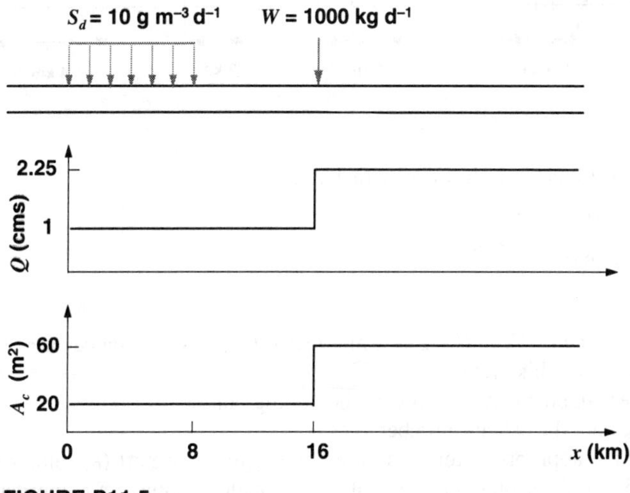

FIGURE P11.5

11.4. Derive the centered-difference version of the one-dimensional, constant-coefficient, advective-dispersive-reaction equation using the finite-difference approach outlined in Box 11.2.

11.5. (*a*) Use a backward-difference scheme ($\Delta x = 0.5$ km) to determine the steady-state distribution of a pollutant ($k = 0.1\ \text{d}^{-1}$) for $x = 0$ to 32 km in the system shown in Fig. P11.5. *Note:* The concentration in the river immediately upstream of the distributed load is 5 mg L^{-1}.
(*b*) Estimate the numerical dispersion for the calculation.
(*c*) Compare your results with the analytical solution from Prob. 9.4.

11.6. A one-dimensional estuary has the following characteristics:

	Value	Units
Dispersion coefficient	10^5	$\text{cm}^2\ \text{s}^{-1}$
Flow	5×10^4	$\text{m}^3\ \text{d}^{-1}$
Width	100	m
Depth	2	m

The pollutant settles ($v = 0.11\ \text{m d}^{-1}$). A mass loading of 2 kg d^{-1} is discharged into this estuary. Note that the boundary conditions are

$$c_0 = 0$$

$$\frac{dc_{18-19}}{dx} = 0$$

Use centered differences and the segmentation scheme in Fig. P11.6 to model the distribution of the pollutant in the estuary. Compare your results with the analytical solution.

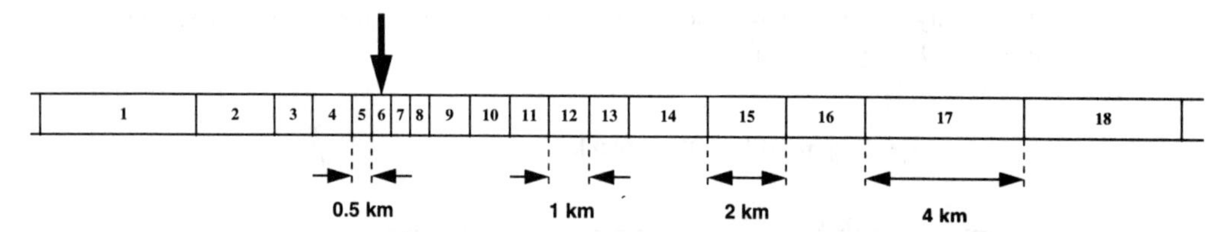

FIGURE P11.6

11.7. A tidal river has the following characteristics:

$$E = 1 \times 10^6\ \text{m}^2\ \text{d}^{-1}$$
$$U = 250\ \text{m d}^{-1}$$
$$A_c = 1000\ \text{m}^2$$

A point source of 12×10^6 g of a first-order decaying substance ($k = 0.075\ \text{d}^{-1}$) is introduced into this system.
(*a*) Determine the concentration at the mixing point.
(*b*) Calculate the estuary number.
(*c*) Use the appropriate formula based on the result of part (*b*) (either Eq. 11.51 or 11.52) to determine the segment length needed to obtain a numerical estimate of the concentration at the mixing point that is approximately 5% low ($\varepsilon = 0.05$).

(*d*) Verify your result by simulating the system with the backward-difference numerical scheme.

11.8. The lake in Fig. P11.8 receives pollutant inputs that settle at a rate of 20 m yr^{-1}. Using the data from Table P11.8,

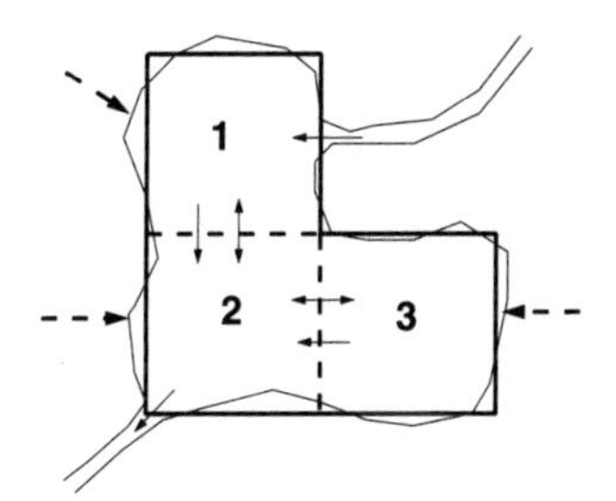

FIGURE P11.8

(*a*) Calculate the system's response to the loadings.
(*b*) Determine the improvement that would occur in segment 3 if the loading to segment 2 were reduced by 50%.

TABLE P11.8

Segment	A_s (km^2)	z (m)	V (km^3)	W (10^3 kg yr^{-1})	ℓ (km)	
1	280	4.3	1.2	500	16.7	
2	280	4.3	1.2	2000	16.7	
3	280	4.3	1.2	300	16.7	
Interface	Q (km^3 yr^{-1})	E (km^2 yr^{-1})	A_c (km^2)	**Width** (km)	E' (km^3 yr^{-1})	c_0 (μg L^{-1})
Inlet-1	170					10
1-2	170	1500	0.072	16.7	6.5	
2-3	10	1500	0.072	16.7	6.5	
2-outlet	180					

11.9. A mixed-flow reactor is 10 m long, 2 m wide, and 1 m deep and has the following hydraulic characteristics: $E = 2\ m^2\ hr^{-1}$ and $U = 1\ m\ hr^{-1}$. It is subject to a constant inflow concentration of a chemical, $c_{in} = 100\ mg\ L^{-1}$, that decays by first-order kinetics $k = 0.2\ hr^{-1}$. Determine the steady-state distribution of the chemical along the length of the reactor.

LECTURE 12

Simple Time-Variable Solutions

Lecture overview: After presenting a simple numerical scheme for obtaining time-variable solutions, I discuss the important issue of stability and describe how time-variable solutions generate numerical dispersion.

Now let's turn to time-variable computer solutions for distributed systems. Aside from describing numerical methods, I emphasize the issues of stability and accuracy. ***Stability*** connotes that errors are not amplified by the solution scheme. Hence an unstable solution is one where errors ultimately "swamp" the true solution. ***Accuracy*** deals with how well the solution scheme conforms to the true solution of the underlying differential equation.

12.1 AN EXPLICIT ALGORITHM

To simplify the analysis the time-variable approach is derived for the mass balance,

$$\frac{\partial c}{\partial t} = E\frac{\partial^2 c}{\partial x^2} \tag{12.1}$$

Thus we neglect advection and reaction and assume that the dispersion coefficient is constant.

A numerical solution for this equation can be developed by substituting finite-difference approximations for the derivatives. In contrast to the previous lectures on steady-state solutions, we must now discretize in time as well as space. Figure 12.1 shows a computational grid for one-dimensional spatial cases.

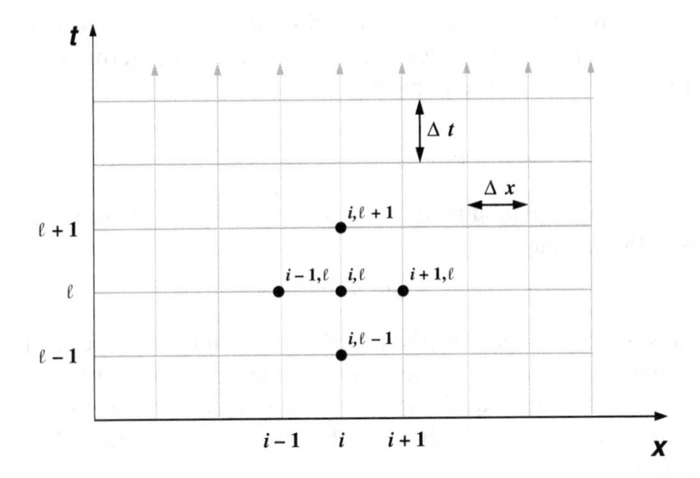

FIGURE 12.1
A computational grid used to characterize the spatial and temporal dimensions.

A simple approach for discretizing Eq. 12.1 for such a grid involves using a centered spatial derivative and a forward temporal derivative. That is,

$$\frac{\partial c}{\partial t} \cong \frac{c_i^{\ell+1} - c_i^\ell}{\Delta t} \tag{12.2}$$

and

$$\frac{\partial^2 c}{\partial x^2} \cong \frac{c_{i+1}^\ell - 2c_i^\ell + c_{i-1}^\ell}{\Delta x^2} \tag{12.3}$$

Notice the notation change in that superscripts are used to denote time. This is done so that a second subscript could be used to designate a second spatial dimension in the event that the approach is expanded to cases that are two-dimensional in space.

Substituting these differences into Eq. 12.1 gives

$$\frac{c_i^{\ell+1} - c_i^\ell}{\Delta t} = E\frac{c_{i+1}^\ell - 2c_i^\ell + c_{i-1}^\ell}{\Delta x^2} \tag{12.4}$$

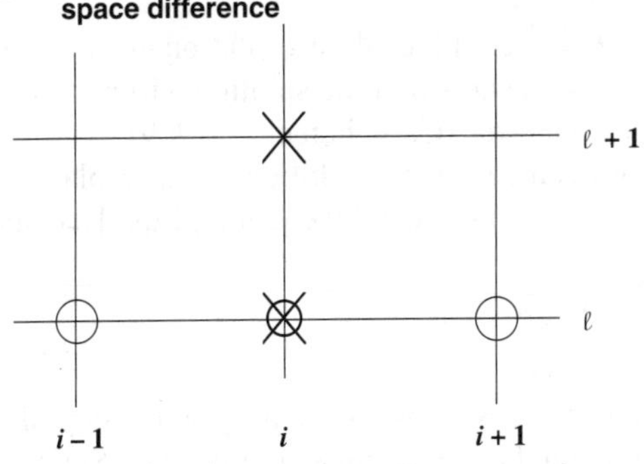

FIGURE 12.2
A computational molecule or stencil for the simple explicit method using FTCS difference approximations.

This approach is sometimes referred to as a ***forward-time/centered-space (FTCS)*** difference equation. It can be further simplified by collecting terms and solving for the concentration at the advanced time step,

$$c_i^{\ell+1} = c_i^\ell + E\frac{c_{i+1}^\ell - 2c_i^\ell + c_{i-1}^\ell}{\Delta x^2}\Delta t \tag{12.5}$$

Inspection of this equation reveals that it is actually a form of Euler's method. That is, it is in the format

$$\text{New} = \text{old} + \text{slope(step)} \tag{12.6}$$

This method is referred to as an ***explicit method***, so named because each difference equation has one unknown (the concentration at the next time step) that can be isolated on the left side of the equal sign. A computational molecule for this explicit method is depicted in Fig. 12.2, showing the nodes that constitute the spatial and temporal approximations.

We should also recognize that substituting finite differences for the right-hand side of Eq. 12.1 effectively transforms the partial differential equation into a system of ordinary differential equations. Thus the use of Euler's method to integrate the system of ODEs, which results from substituting Eq. 12.2 for the left-hand side, represents the most straightforward way to obtain solutions.

12.2 STABILITY

To investigate the stability of Eq. 12.5, we first collect terms to give

$$c_i^{\ell+1} = \lambda c_{i-1}^\ell + (1 - 2\lambda)c_i^\ell + \lambda c_{i+1}^\ell \tag{12.7}$$

where

$$\lambda = \frac{E\Delta t}{\Delta x^2} \tag{12.8}$$

and λ is formally called the ***diffusion number***.

Now we will examine the stability of Eq. 12.7. ***Stability*** connotes that errors are not amplified as the computation progresses. It can be shown that Eq. 12.7 will yield stable results if the diagonal term is positive; that is, $1 - 2\lambda > 0$. In other words,

$$\lambda < \tfrac{1}{2} \tag{12.9}$$

In addition it should be noted that setting $\lambda < \frac{1}{2}$ could result in a solution in which errors do not grow but oscillate. Setting $\lambda < \frac{1}{4}$ ensures that the solution will not oscillate, and $\lambda = \frac{1}{6}$ tends to minimize truncation error (Carnahan et al. 1969).

Although satisfaction of Eq. 12.9 guarantees stable solutions, it also places a strong limitation on the explicit method. This can be seen by equating Eqs. 12.8 and 12.9 and solving for

$$\Delta t < \frac{1}{2}\frac{\Delta x^2}{E} \tag{12.10}$$

Because it allows us to see how Δt changes as we vary the spatial step size Δx, this equation offers a nice perspective on the constraint's significance. Suppose that we are dissatisfied with our spatial resolution and double the number of spatial segments; that is, we cut Δx in half. According to Eq. 12.10 we must quarter the time step to maintain stability. Thus to perform comparable calculations, the number of time steps must be increased by a factor of 4. Furthermore the computation for each of these time steps will take twice as long because halving Δx doubles the number of segments for which equations must be written. Consequently for the one-dimensional case, halving Δx results in an 8-fold increase in the number of calculations. This is a serious computational penalty for a 2-fold increase in spatial resolution.

For two- and three-dimensional problems, the stability constraint is even more severe. For example for a two-dimensional problem, halving the spatial grid results in a 16-fold increase in computations. Although the simple explicit method is severely limited by the stability constraint, it is still widely used for water-quality modeling because it is extremely easy to program. This advantage is particularly important when dealing with nonlinear kinetics (we elaborate on this aspect in the next lecture).

12.3 THE CONTROL-VOLUME APPROACH

I'll now show how the general control-volume approach developed in the previous lecture for steady-state solutions can be extended to simulate time-variable results. Using a forward difference for time and a weighted difference for space, we can write a mass balance for segment i as

$$V_i \frac{c_i^{\ell+1} - c_i^\ell}{\Delta t} = W_i^\ell + Q_{i-1,i}\left(\alpha_{i-1,i}c_{i-1}^\ell + \beta_{i-1,i}c_i^\ell\right) - Q_{i,i+1}\left(\alpha_{i,i+1}c_i^\ell + \beta_{i,i+1}c_{i+1}^\ell\right)$$
$$+ E'_{i-1,i}\left(c_{i-1}^\ell - c_i^\ell\right) + E'_{i,i+1}\left(c_{i+1}^\ell - c_i^\ell\right) - k_i V_i c_i^\ell \tag{12.11}$$

Dividing both sides by V_i and rearranging yields

$$c_i^{\ell+1} = c_i^\ell + \frac{\Delta t}{V_i}\Big[W_i^\ell + Q_{i-1,i}\left(\alpha_{i-1,i}c_{i-1}^\ell + \beta_{i-1,i}c_i^\ell\right) - Q_{i,i+1}\left(\alpha_{i,i+1}c_i^\ell + \beta_{i,i+1}c_{i+1}^\ell\right)$$
$$+E'_{i-1,i}\left(c_{i-1}^\ell - c_i^\ell\right) + E'_{i,i+1}\left(c_{i+1}^\ell - c_i^\ell\right) - k_i V_i c_i^\ell\Big] \tag{12.12}$$

We collect terms to give

$$c_i^{\ell+1} = \frac{W_i^\ell}{V_i}\Delta t - \frac{\Delta t}{V_i}(-Q_{i-1,i}\alpha_{i-1,i} - E'_{i-1,i})c_{i-1}^\ell$$
$$+\left[1 - \frac{\Delta t}{V_i}\left(-Q_{i-1,i}\beta_{i-1,i} + Q_{i,i+1}\alpha_{i,i+1} + E'_{i-1,i} + E'_{i,i+1} + k_i V_i\right)\right]c_i^\ell$$
$$-\frac{\Delta t}{V_i}\left(Q_{i,i+1}\beta_{i,i+1} - E'_{i,i+1}\right)c_{i+1}^\ell \tag{12.13}$$

Depending on the value of α, this equation can represent either a forward-time/backward-space (FTBS) for $\alpha = 1$ or an FTCS for $\alpha = 0.5$.

As with the simple model described in the first section, solutions for this equation will be stable if the diagonal term is positive. For the case where all parameters are constant this is

$$1 - \frac{\Delta t}{V}(-Q\beta + Q\alpha + 2E' + kV) > 0 \tag{12.14}$$

which can be solved for

$$\Delta t < \frac{V}{Q(\alpha - \beta) + 2E' + kV} \tag{12.15}$$

or dividing the numerator and the denominator by the cross-sectional area,

$$\Delta t < \frac{(\Delta x)^2}{U\Delta x(\alpha - \beta) + 2E + k(\Delta x)^2} \tag{12.16}$$

Careful inspection of either Eq. 12.15 or 12.16 indicates that, beyond the fact that the matrix diagonal elements must be positive, the stability criterion actually has a physical meaning. That is, it says that the time step cannot exceed the pollutant residence time (recall Sec. 3.2.1) for the segment. This can be seen clearly for the case involving pure advection with a backward difference ($\alpha = 1, k = E = 0$). For this case the stability criterion becomes

$$\Delta t < \frac{V}{Q} \tag{12.17}$$

which is the cell residence time. Further we can also multiply the top and bottom of the equation by the cross-sectional area to yield

$$\Delta t < \frac{\Delta x}{U} \tag{12.18}$$

This result indicates that another interpretation of stability is that the time step cannot be so large that the water could move more than the segment length Δx for a given velocity U. In other words the stability criterion specifies that if the computation "gets ahead of itself" it will "blow up."

Note that Eq. 12.18 is formally referred to as the ***Courant condition***. It is often presented as $\gamma < 1$, where $\gamma = U\Delta t/\Delta x$ is called the ***advection number*** or the ***Courant number.***

12.4 NUMERICAL DISPERSION

In Lec. 11 we used a Taylor-series expansion to estimate the numerical dispersion created by approximating advection with backward spatial differences. Now we extend this analysis to determine the effect of forward temporal differences. It can be shown that the numerical dispersion can be represented as (see Box 12.1 for details)

$$E_n = U\Delta x\left[(\alpha - 0.5) - \frac{U\Delta t}{2\,\Delta x}\right] \tag{12.19}$$

BOX 12.1. Error Analysis to Estimate Numerical Dispersion

For simplicity we analyze the truncation error of the advection equation

$$\frac{\partial c}{\partial t} = -U\frac{\partial c}{\partial x} \tag{12.20}$$

Recall from Lec. 11 that a backward Taylor-series expansion can be truncated to yield (recall Eq. 11.37)

$$\frac{\partial c}{\partial x} \cong \frac{c_i^\ell - c_{i-1}^\ell}{\Delta x} + \frac{\Delta x}{2}\frac{\partial^2 c}{\partial x^2} \tag{12.21}$$

In similar fashion a forward series can be written in time and truncated to give

$$\frac{\partial c}{\partial t} \cong \frac{c_i^{\ell+1} - c_i^\ell}{\Delta t} - \frac{\Delta t}{2}\frac{\partial^2 c}{\partial t^2} \tag{12.22}$$

Equations 12.21 and 12.22 can be combined with 12.20 to yield

$$\frac{\partial c}{\partial t} = -U\frac{\partial c}{\partial x} + U\frac{\Delta x}{2}\frac{\partial^2 c}{\partial x^2} - \frac{\Delta t}{2}\frac{\partial^2 c}{\partial t^2} \tag{12.23}$$

The second-order time derivative can be converted to a space derivative by differentiating Eq. 12.20 with respect to time,

$$\frac{\partial^2 c}{\partial t^2} = -U\frac{\partial^2 c}{\partial t\,\partial x} \tag{12.24}$$

Next differentiate Eq. 12.20 with respect to space,

$$\frac{\partial^2 c}{\partial x\,\partial t} = -U\frac{\partial^2 c}{\partial x^2} \tag{12.25}$$

Therefore Eqs. 12.24 and 12.25 can be combined to give

$$\frac{\partial^2 c}{\partial t^2} = U^2\frac{\partial^2 c}{\partial x^2} \tag{12.26}$$

which can be substituted into Eq. 12.23 to yield

$$\frac{\partial c}{\partial t} = \left(\frac{\Delta x}{2}U - \frac{U^2 \Delta t}{2}\right)\frac{\partial^2 c}{\partial x^2} - U\frac{\partial c}{\partial x} \tag{12.27}$$

Thus the numerical dispersion is

$$E_n = \frac{\Delta x}{2}U - \frac{U^2 \Delta t}{2} \tag{12.28}$$

Close inspection of this formula indicates that there are two independent effects on numerical dispersion: spatial and temporal discretization effects,

$$E_n = \underbrace{U\Delta x(\alpha - 0.5)}_{\text{Spatial discretization}} - \underbrace{\frac{U^2 \Delta t}{2}}_{\text{Temporal discretization}} \tag{12.29}$$

An interesting aspect of this formula is that the time step has a negative effect upon the numerical dispersion. In other words as the time step is increased, the formula states that the numerical dispersion should decrease. In the extreme it might lead you to believe that at a sufficiently high value of the time step a negative dispersion could result. Since this is clearly physically unrealistic (it means that gradients would sharpen up, a clear violation of the second law of thermodynamics!), we must closely examine the formula to truly understand its implications.

To do this let's first determine the time step that would yield zero numerical dispersion. This can be done for the case of a backward difference ($\alpha = 1$) by setting Eq. 12.29 to zero and solving for

$$\Delta t = \frac{\Delta x}{U} \tag{12.30}$$

Comparison of this result with Eq. 12.18 indicates that this time step conforms to the point at which the stability condition would be violated. Thus because of the stability constraint, we cannot use a larger time step and hence it is impossible to create a negative dispersion.

Now the second implication of Eq. 12.29 is that the total numerical dispersion increases as the time step is decreased. Why would this occur? A graphical depiction of the situation for the FTBS difference is shown in Fig. 12.3. In Fig. 12.3*a* an initial concentration is set in a segment that is Δx wide. This concentration is advected downstream over a time step equal to the critical value calculated in Eq. 12.30. Thus as shown, the entire mass in the segment is displaced downstream with no mixing. As in Eq. 12.29 this occurs because the positive spatial numerical dispersion is canceled by the negative temporal dispersion.

In Fig. 12.3*b* the numerical experiment is repeated but using a time step that is one-half the critical value. In this case half the mass will be moved downstream while half remains in the original segment. Thus a net numerical dispersion is created since only part of the spatial dispersion is diminished by the negative temporal dispersion. In the limit ($\Delta t \to 0$) the temporal component will approach zero and the total dispersion will approach the spatial component.

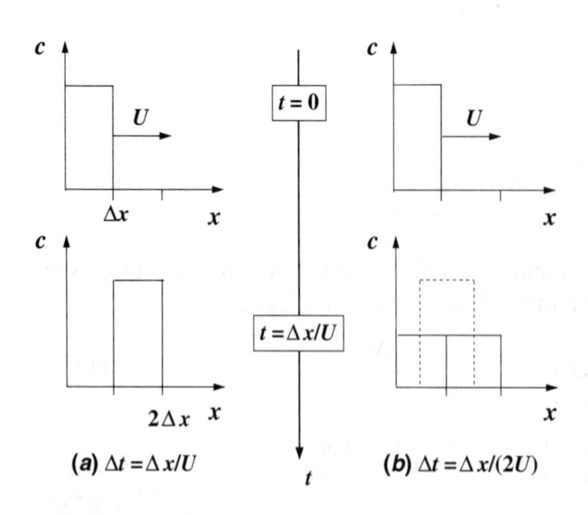

FIGURE 12.3
Depiction of the numerical dispersion created by the forward-time/backward-space (FTBS) difference approximations. In (*a*) the calculation uses the critical time step based on stability. For this case no numerical dispersion occurs. In (*b*) a smaller time step is used. For this case the concentration is dispersed.

Now let us turn to the FTCS approach, which exhibits no spatial numerical dispersion. For this case Eq. 12.29 reduces to

$$E_n = -\frac{U^2 \Delta t}{2} \tag{12.31}$$

Thus a negative numerical dispersion is created. In this case an additional dispersion equivalent to the absolute value of the quantity calculated in Eq. 12.31 would be added to the actual physical dispersion used in the model. This correction is illustrated in the following example.

EXAMPLE 12.1. NUMERICAL SIMULATION OF A SPILL. A river has the following characteristics:

Depth = 1 m	Flow = 144,000 m^3 hr^{-1}
Width = 60 m	Dispersion = 150,000 m^2 hr^{-1}
Velocity = 2400 m hr^{-1}	Cross-sectional area = 60 m^2

A spill of 5 kg of a conservative contaminant occurs about 0.5 km below a city. Use both (*a*) analytical and (*b*) centered-difference numerical models to simulate the distribution 0.2, 0.4, and 0.6 hr following the spill.

Solution: (*a*) Equation 10.22 can be used to generate the analytical solution

$$c(x, t) = \frac{5 \times 10^3/60}{2\sqrt{\pi(150{,}000)t}} e^{-\frac{(x-500-2400t)^2}{4(150{,}000)t}}$$

Values for time and space are substituted into this formula and used to generate the solid lines in Fig. 12.4. As expected, the spill is transported 2400 m hr^{-1} × 0.6 hr = 1440 m downstream and has spread

$$x_{99} = 5.2\sqrt{2(150{,}000)0.6} = 2206 \text{ m}$$

over the 0.6-hr period.

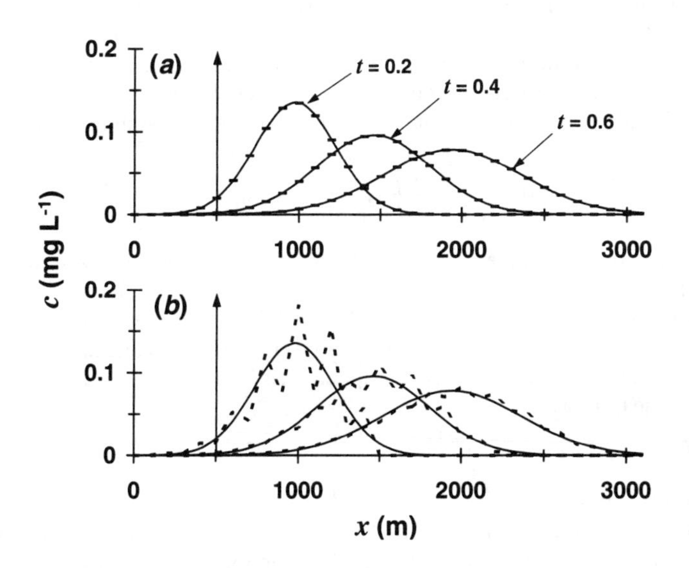

FIGURE 12.4
(*a*) Analytical (solid line) and numerical simulations (horizontal dashes) of a spill in a river obtained with a Δt = 0.01 hr. (*b*) Unstable results with a Δt = 0.0225 hr. Note that dashed lines are used for the numerical results to make the instability more evident.

(*b*) The first step in developing a numerical solution is to determine the critical segment length (Eq. 11.50),

$$\Delta x_c < \frac{150{,}000}{0.5(2400)} = 125 \text{ m}$$

To be on the safe side, we adopt a $\Delta x = 100$ m.

Next we determine the time step required to maintain stability using Eq. 12.16,

$$\Delta t < \frac{(100)^2}{2(150{,}000)} = 0.033 \text{ hr } (= 2 \text{ min})$$

Again to be on the safe side, we use a time step of 0.01 hr.

The numerical dispersion can be determined with Eq. 12.31,

$$E_n = -\frac{(2400)^2 0.01}{2} = -28{,}800 \text{ m}^2 \text{ hr}^{-1}$$

This means that a value of $150{,}000 + 28{,}800 = 178{,}800$ m^2 hr^{-1} will be used in the numerical model.

Now before performing the simulation, we must address the issue of an initial condition. The initial condition for the analytical solution is an impulse function. That is, the spill is assumed to occur instantaneously in space and time. Because our control-volume approach breaks time and space into discrete chunks, it is impossible to exactly represent such a function in our control-volume approach. An approximation involves distributing the mass across a single segment at the start of the computation. For the present case this means that the following concentration would be established in the segment where the spill occurred:

$$c(500) = \frac{m}{B\,\Delta x H} = \frac{5000}{60(100)1} = 0.833 \text{ mg L}^{-1}$$

A value of zero would be used as an initial condition for the other cells.

Using this approach we depict the results of the model run in Fig. 12.4*a*, along with the analytical solution. As manifested by a small reduction in the peak and small extension of the tails, the numerical values seem to exhibit slightly more dispersion than for the analytical case. However, the overall agreement is adequate. The numerical results generally follow the trend of the exact solution. If finer segmentation were imposed (that is, smaller Δt and Δx), the numerical outcome would approach the analytical case.

However, it should be noted that the total number of calculations required to obtain the solution amounts to

$$n = \frac{x_{\text{total}}}{\Delta x}\frac{t_{\text{total}}}{\Delta t} = \frac{3000}{100}\frac{0.6}{0.01} = 1800$$

If the space discretization were refined, the calculations would increase in the rapid fashion specified by Eq. 12.10.

Figure 12.4*b* shows the results for the case $\Delta t = 0.0225$ hr. As can be seen, the solution is beginning to become unstable. If larger steps were used, the calculation would begin to oscillate even more violently. Thus the explicit method's stability constrains our ability to lengthen the time step.

Now one more question needs to be addressed: For the centered-difference case is it possible to generate a negative dispersion? Such might be the case for a highly advective system; that is, one where the physical dispersion was low relative to advection. In such a case there might not be sufficient real dispersion to compensate for the numerical dispersion.

As in the previous example, the first step in answering this question is to calculate the critical segment length (Eq. 11.50),

$$\Delta x_c = \frac{2E_p}{U} \tag{12.32}$$

Next the stability criterion must be met. For the centered difference with no reaction this is (recall Eq. 12.16)

$$\Delta t < \frac{(\Delta x)^2}{2E_p} \tag{12.33}$$

Now substitute Eq. 12.32 into Eq. 12.33 to give

$$\Delta t < \frac{(2E_p/U)^2}{2E_p} = \frac{2E_p}{U^2} \tag{12.34}$$

Finally substitute this time step into the numerical dispersion estimate (Eq. 12.31),

$$E_n = -\frac{U^2\left(2E_p/U^2\right)}{2} = -E_p \tag{12.35}$$

Consequently for this case the stability point represents a boundary where a negative numerical dispersion exactly balances the physical dispersion. Thus because moving beyond this point with a larger Δt will yield an unstable solution, a physically unrealistic negative dispersion cannot be generated.

In summary we have successfully developed a simple approach for simulating the time-variable distribution of a pollutant in a distributed system. Because the scheme is so straightforward, it has utility for water-quality applications. However, because of its stability constraint, more refined techniques have been developed. The next lecture deals with some of these methods.

PROBLEMS

12.1. Suppose that for Example 12.1 you reduced the spatial scale to $\Delta x = 25$ m. Calculate the time step required to (*a*) maintain stability, (*b*) avoid oscillations, and (*c*) minimize truncation errors. For (*c*) determine exactly how many calculations would be required to perform the same calculation as in the example (that is, from $t = 0$ to 0.6 hr).

12.2. Boulder Creek has the following characteristics:

Parameter	Value
Width	12 m
Depth	0.3 m
Flow	1.7 $m^3\ s^{-1}$
Dispersion coefficient	4.5 $m^2\ s^{-1}$

If you are evaluating a pollutant that decays with first-order kinetics ($k = 1\ d^{-1}$),

(*a*) Calculate the estuary number. Interpret your result—is the system advective, dispersive, or advective/dispersive?

(*b*) For a steady-state calculation, determine the critical segment length.
(*c*) If you use this segment length, determine the time step needed to just maintain a stable solution with a centered-difference, explicit (that is, Euler) solution.
(*d*) What numerical dispersion would be manifested for part (*c*)?
(*e*) Calculate how a 2-kg spill of a conservative substance would move through a 20-km stretch of the system.

12.3. An estuary has the following characteristics:

Depth = 4 m	Flow = 100,000 m^3 hr^{-1}
Velocity = 250 m hr^{-1}	Cross-sectional area = 400 m^2
Width = 100 m	Dispersion = 1,000,000 m^2 hr^{-1}

A spill of 10 kg of a reactive contaminant ($k = 0.1\ d^{-1}$) occurs. Use both (*a*) an analytical and (*b*) a centered-difference numerical model to simulate the distribution of the contaminant in the estuary.

12.4. A mixed-flow reactor is 10 m long, 2 m wide, and 1 m deep and has the following hydraulic characteristics: $E = 2\ m^2\ hr^{-1}$ and $U = 1\ m\ hr^{-1}$. It initially contains pure water. At $t = 0$ it is subject to a step loading of a chemical, $c_{in} = 100\ mg\ L^{-1}$, that decays by first-order kinetics $k = 0.2\ hr^{-1}$. Determine the distribution of the chemical along the length of the reactor as a function of time using the explicit method and centered spatial differences.

12.5. Suppose that 20 kg of a conservative toxic compound were spilled in segment 1 of the lake described in Prob. 11.8. Use the explicit approach with backward spatial differences to calculate the time-variable response of the system to the spill.

LECTURE 13

Advanced Time-Variable Solutions

LECTURE OVERVIEW: I present two types of advanced methods for simulating time-variable distributions of pollutants in channels. The first type, called an implicit approach, is best known for its improved stability. I describe two such methods in this lecture: the backward-time/centered-space and the Crank-Nicolson methods. Then I present an advanced explicit approach—the MacCormack method. This algorithm, which is a predictor-corrector method, has the advantage that it is well-suited for solving nonlinear systems.

The simple explicit finite-difference formulation described in the last lecture has problems related to stability and numerical dispersion. The two types of algorithms described herein, implicit and predictor-corrector methods, attempt to obviate these deficiencies.

13.1 IMPLICIT APPROACHES

One source of the difficulties with the methods described to this point is that, as depicted in Fig. 13.1, they exclude information that has a bearing on the solution. That is, the value at the advanced time step is affected only by the values of the segment and its two adjacent neighbors at the previous step. Clearly it is also impacted by the concentrations at other segments in both the spatial and temporal domains.

Implicit methods overcome this difficulty at the expense of somewhat more complicated algorithms. In this section I describe two types of implicit approaches that are commonly used.

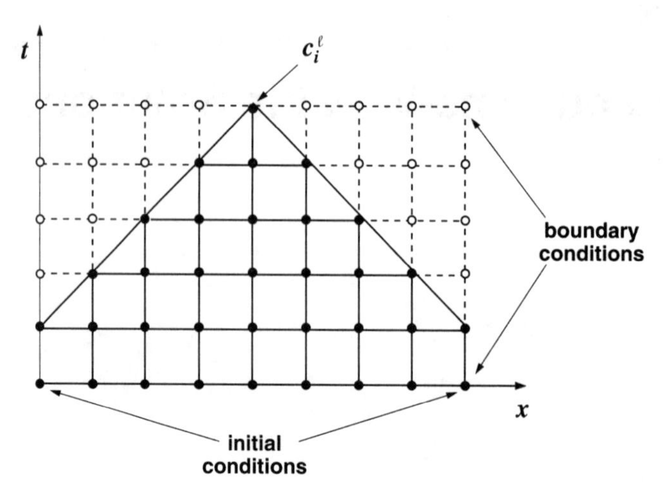

FIGURE 13.1
Representation of the effect of other nodes on the finite-difference approximation at node (i, ℓ) using an explicit finite-difference scheme. The shaded nodes have an influence on (i, ℓ), whereas the unshaded nodes, which in reality affect (i, ℓ), are excluded.

13.1.1 Simple Implicit or "Backward-Time" Algorithm

The fundamental difference between explicit and implicit approximations is depicted in Fig. 13.2. For the explicit form, we approximate the spatial derivative at time level ℓ (Fig. 13.2*a*). Recall that when we substituted this approximation into the partial differential equation, we obtained a difference equation with a single unknown. Thus we could solve "explicitly" for this unknown.

In implicit methods, the spatial derivative is approximated at an advanced time level. For example the first and second derivatives could be approximated by centered differences at time $\ell + 1$ by (Fig. 13.2*b*)

$$\frac{\partial c}{\partial x} \cong \frac{c_{i+1}^{\ell+1} - c_{i-1}^{\ell+1}}{2\,\Delta x} \tag{13.1}$$

$$\frac{\partial^2 c}{\partial x^2} \cong \frac{c_{i+1}^{\ell+1} - 2c_i^{\ell+1} + c_{i-1}^{\ell+1}}{\Delta x^2} \tag{13.2}$$

Note that because the spatial derivatives are evaluated at the advanced time step, this approach is sometimes referred to as a ***backward-time/centered-space (BTCS)*** implicit scheme.

When this relationship is substituted into the original PDE, the resulting difference equation contains several unknowns. Thus solutions cannot be obtained by simple algebraic rearrangement as was done in the explicit approach. Instead the entire system of equations must be solved simultaneously. This is possible because, along with the boundary conditions, the implicit formulations result in a set of linear algebraic equations with the same number of unknowns. The method therefore reduces to the solution of a set of simultaneous equations at each point in time.

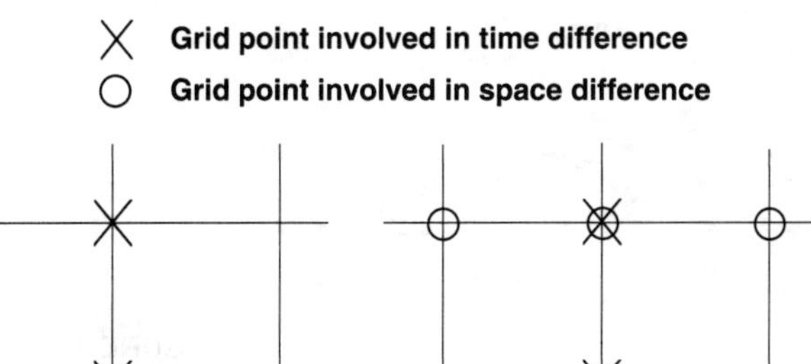

FIGURE 13.2
Computational molecules demonstrating the fundamental differences between (*a*) explicit and (*b*) implicit methods. The computational module in (*b*) is sometimes called a "backward-time/centered-space" (BTCS) scheme.

We now illustrate how the scheme is set up for the control-volume approach. Using a weighted difference for the advective terms,

$$V_i \frac{c_i^{\ell+1} - c_i^{\ell}}{\Delta t} = W_i^{\ell+1} + Q_{i-1,i}\left(\alpha_{i-1,i} c_{i-1}^{\ell+1} + \beta_{i-1,i} c_i^{\ell+1}\right) - Q_{i,i+1}\left(\alpha_{i,i+1} c_i^{\ell+1} + \beta_{i,i+1} c_{i+1}^{\ell+1}\right) + E'_{i-1,i}\left(c_{i-1}^{\ell+1} - c_i^{\ell+1}\right) + E'_{i,i+1}\left(c_{i+1}^{\ell+1} - c_i^{\ell+1}\right) - k_i V_i c_i^{\ell+1} \tag{13.3}$$

Collecting terms gives

$$c_i^{\ell} + \frac{W_i^{\ell+1}\Delta t}{V_i} = -\frac{\Delta t}{V_i}\left(Q_{i-1,i}\alpha_{i-1,i} + E'_{i-1,i}\right)c_{i-1}^{\ell+1} + \left[1 + \frac{\Delta t}{V_i}\left(-Q_{i-1,i}\beta_{i-1,i} + Q_{i,i+1}\alpha_{i,i+1} + E'_{i-1,i} + E'_{i,i+1} + k_i V_i\right)\right]c_i^{\ell+1} - \frac{\Delta t}{V_i}\left(-Q_{i,i+1}\beta_{i,i+1} + E'_{i,i+1}\right)c_{i+1}^{\ell+1} \tag{13.4}$$

The relationship can be simplified further for the centered-difference approximation ($\alpha = \beta = 0.5$) and constant parameters,

$$c_i^{\ell} + \frac{W_i^{\ell+1}\Delta t}{V_i} = \left[-\frac{U\Delta t}{2\,\Delta x} - \frac{E\Delta t}{(\Delta x)^2}\right]c_{i-1}^{\ell+1} + \left[1 + \frac{2E\Delta t}{(\Delta x)^2} + k\,\Delta t\right]c_i^{\ell+1} + \left[\frac{U\Delta t}{2\,\Delta x} - \frac{E\Delta t}{(\Delta x)^2}\right]c_{i+1}^{\ell+1} \tag{13.5}$$

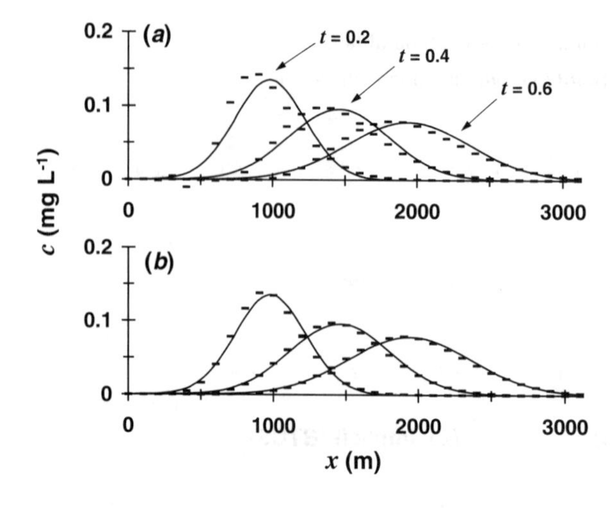

FIGURE 13.3
(*a*) Analytical (solid line) and numerical simulations (horizontal dashes) of a spill in a river obtained with the BTCS implicit approach with a $\Delta t = 0.025$ hr. (*b*) The same calculation is shown, but with the Crank-Nicolson algorithm.

This equation applies to all but the first and the last interior nodes, which must be modified to reflect the boundary conditions. These are handled in a fashion similar to the steady-state case (recall Sec. 11.2). When Eq. 13.5 is written for all the elements, the resulting set of n linear algebraic equations has n unknowns. In addition for the one-dimensional problems described here, the method has the added bonus that the system is tridiagonal. Thus we can utilize the efficient solution algorithms that are available for tridiagonal systems.

Using an analysis similar to the one in Box 12.1, it can be shown that the numerical dispersion for Eq. 13.5 is

$$E_n = U\Delta x\left[(\alpha - 0.5) + \frac{U\Delta t}{2\Delta x}\right] \tag{13.6}$$

Consequently this equation again shows that, when we use centered spatial differences, there is no spatial numerical dispersion. However, because the time derivative is biased (that is, backward), a temporal numerical dispersion is induced,

$$E_n = \frac{U^2\Delta t}{2} \tag{13.7}$$

Note that this result is opposite in sign to the simple explicit result (compare with Eq. 12.31). Therefore an additional dispersion equivalent to the quantity calculated in Eq. 13.7 would be subtracted from the actual physical dispersion used in the model.

EXAMPLE 13.1. SIMPLE IMPLICIT SIMULATION. Apply the backward-time/centered-space algorithm to the spill simulation described in Example 12.1. Use a temporal step size of 0.025 hr for the calculation.

Solution: The numerical dispersion can be determined with Eq. 13.7,

$$E_n = \frac{(2400)^2 0.025}{2} = 72{,}000 \text{ m}^2 \text{ hr}^{-1}$$

This means that a value of $150{,}000 - 72{,}000 = 78{,}000$ m^2 hr^{-1} will be used in the numerical model.

The results of applying the BTCS approach are shown in Fig. 13.3*a*. Observe how, although there is some inaccuracy, the numerical solution follows the general trend of the analytical solution and is much better than the unstable explicit technique from Fig. 12.4*b*.

Although the simple implicit method is unconditionally stable, there is an accuracy limit to the use of large time steps. Consequently it is not that much more efficient than the explicit approaches. Where it does shine is for steady-state problems.

Recall from Lec. 12 that matrix solution methods can be used to obtain steady-state solutions. An alternative approach would be to run a time-variable solution until it reached steady-state. In such cases, because inaccurate intermediate results are not an issue, implicit methods allow you to use larger time steps and hence can generate steady-state results in an efficient manner. In fact the simple implicit approach outlined here is particularly well-suited for this task since it converges quickly (for example it is superior to the Crank-Nicolson method in this regard). Consequently water-quality models such as QUAL2E (Brown and Barnwell 1987) use this algorithm to attain steady-state solutions.

13.1.2 Crank-Nicolson

In the last lecture we used an explicit scheme that was forward-time/centered-space. In the previous section we used an alternative algorithm that was backward-time/centered-space. Both approaches have the disadvantage that the spatial and temporal derivatives are out of synch—whereas the space differences are centered, the time differences are biased.

The ***Crank-Nicolson*** method remedies this situation by using a centered-time/centered-space strategy (Fig. 13.4). In essence, estimates of the spatial derivative are made at both the present *and* the future time step. These estimates are then averaged to obtain a spatial estimate that corresponds to the midpoint of the time step.

I'll now illustrate how the scheme is set up. Using constant parameters and centered spatial differences,

$$\frac{c_i^{\ell+1} - c_i^\ell}{\Delta t} = -U\left(\frac{\dfrac{c_{i+1}^\ell - c_{i-1}^\ell}{2\,\Delta x} + \dfrac{c_{i+1}^{\ell+1} - c_{i-1}^{\ell+1}}{2\,\Delta x}}{2}\right) + E\left[\frac{\dfrac{c_{i+1}^\ell - 2c_i^\ell + c_{i-1}^\ell}{(\Delta x)^2} + \dfrac{c_{i+1}^{\ell+1} - 2c_i^{\ell+1} + c_{i-1}^{\ell+1}}{(\Delta x)^2}}{2}\right] - k\left(\frac{c_i^\ell + c_i^{\ell+1}}{2}\right) \tag{13.8}$$

Collecting terms gives

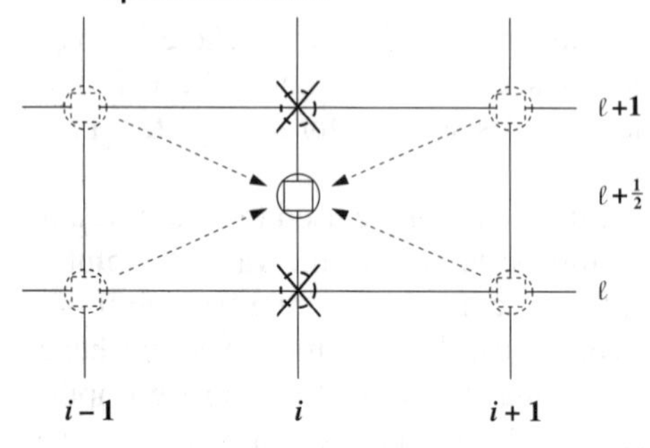

FIGURE 13.4
A computational molecule or stencil for the Crank-Nicolson method using a centered difference approximation for the advective term. The dashed lines are meant to show that the spatial derivative estimates at the present (ℓ) and future ($\ell + 1$) times are averaged to obtain the estimate at the intermediate time ($\ell + \frac{1}{2}$).

$$-(\lambda + \gamma)c_{i-1}^{\ell+1} + 2(1 + \lambda + k\,\Delta t)c_i^{\ell+1} - (\lambda - \gamma)c_{i+1}^{\ell+1}$$
$$= (\lambda + \gamma)c_{i-1}^{\ell} + 2(1 - \lambda - k\,\Delta t)c_i^{\ell} + (\lambda - \gamma)c_{i+1}^{\ell} \quad (13.9)$$

where

$$\lambda = \frac{E\Delta t}{(\Delta x)^2} \quad (13.10)$$

$$\gamma = \frac{U\Delta t}{2\,\Delta x} \quad (13.11)$$

This equation applies to all but the first and the last interior nodes, which must be modified to reflect the boundary conditions. These are handled in a fashion similar to the steady-state case (recall Sec. 11.2). When Eq. 13.9 is written for all the elements, the resulting set of n linear algebraic equations has n unknowns. In addition the method again has the added bonus that the system is tridiagonal. Thus we can utilize the efficient solution algorithms that are available for tridiagonal systems.

EXAMPLE 13.2. CRANK-NICOLSON SIMULATION. Apply the Crank-Nicolson algorithm to the spill simulation described in Example 12.1. Use a temporal step size of 0.025 hr for the calculation.

Solution: Because we are using centered differences in both space and time, numerical dispersion is not an issue for Crank-Nicolson. Consequently the actual dispersion can be used. Using the other parameters from Example 12.1, the results of applying the Crank-Nicolson approach are shown in Fig. 13.3*b*. Observe how, although there is some inaccuracy, the numerical solution follows the general trend of the analytical solution and is superior to the simple implicit approach outlined in Example 13.1. In addition the results are much better than the unstable explicit technique from Fig. 12.4*b*.

Although the Crank-Nicolson method is efficient for linear problems, it becomes less convenient when used for the nonlinear models that are sometimes encountered in water-quality modeling. For such cases the system of equations contains nonlinear terms and cannot be directly solved using matrix algebra approaches. Thus, as described next, an advanced explicit method can be used.

13.2 THE MacCORMACK METHOD

The MacCormack (1969) method is a predictor-corrector approach that is closely related to the Heun method described previously in Sec. 7.2. In its most conventional form it uses the stencil shown in Fig. 13.5. Thus it uses the following equation to make an initial prediction of the slope at the base point:

$$s_{1,i} = -U\frac{c_{i+1}^{\ell} - c_i^{\ell}}{\Delta x} + E\frac{c_{i+1}^{\ell} - 2c_i^{\ell} + c_{i-1}^{\ell}}{(\Delta x)^2} - kc_i^{\ell} \tag{13.12}$$

This slope is then used in conjunction with Euler's method to finish the predictor calculation,

$$c_i^{\ell+1} = c_i^{\ell} + s_{1,i}\,\Delta t \tag{13.13}$$

Notice that this is the first time we have used a *forward* spatial difference (that is, in the advection term).

Then the predictor is used to make a slope estimate at the end of the interval,

$$s_{2,i} = -U\frac{c_i^{\ell+1} - c_{i-1}^{\ell+1}}{\Delta x} + E\frac{c_{i+1}^{\ell+1} - 2c_i^{\ell+1} + c_{i-1}^{\ell+1}}{(\Delta x)^2} - kc_i^{\ell+1} \tag{13.14}$$

Observe that now we use a backward difference for the advection term.

Finally an average of the two slopes is used to obtain the final corrector equation,

$$c_i^{\ell+1} = c_i^{\ell} + \left(\frac{s_{1,i} + s_{2,i}}{2}\right)\Delta t \tag{13.15}$$

The MacCormack method has the advantage that it is more stable than the FTCS approach described in Lec. 12. In addition it is not plagued by numerical dispersion. Consequently, as described in the next example, it yields a good prediction at a reasonable computational price.

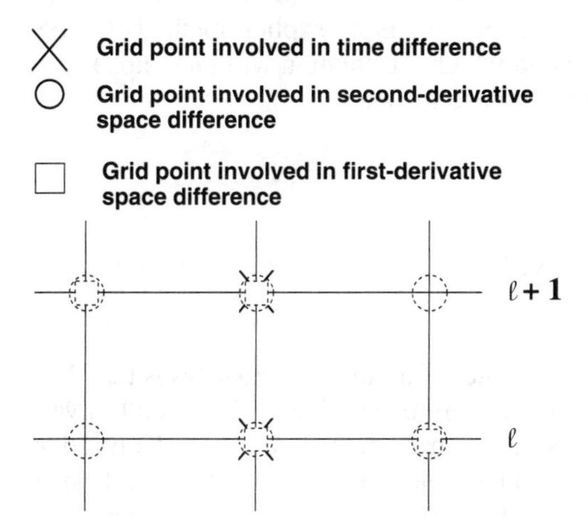

FIGURE 13.5
A computational molecule or stencil for the MacCormack method.

EXAMPLE 13.3. MACCORMACK SIMULATION. Apply the MacCormack algorithm to the spill simulation described in Example 12.1. Use a temporal step size of 0.025 hr for the calculation.

Solution: As was the case with Crank-Nicolson, since we are using centered differences in both space and time, numerical dispersion is not an issue for the MacCormack method. Consequently the actual dispersion can be used. Using the other parameters from Example 12.1, the results are shown in Fig. 13.6. Observe how, although there is some inaccuracy, the numerical solution follows the general trend of the analytical solution. It is comparable in performance to the Crank-Nicolson method and is superior to the simple implicit approach outlined in Example 13.1. In addition the results are much better than the unstable explicit technique from Fig. 12.4*b*.

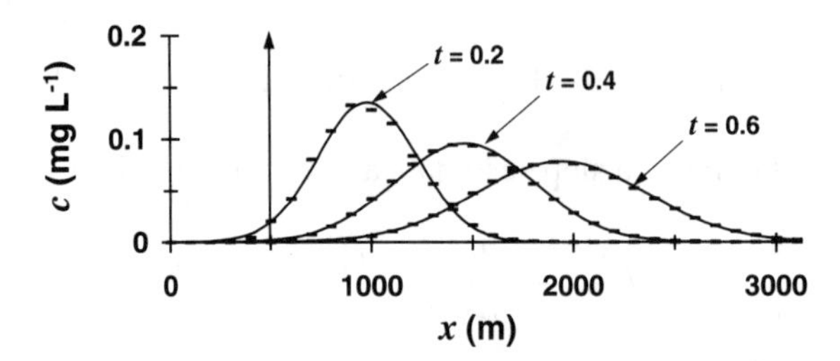

FIGURE 13.6
Analytical (solid line) and numerical simulations (horizontal dashes) of a spill in a river obtained with the MacCormack approach with a $\Delta t = 0.025$ hr.

It should be noted that variations of MacCormack's method use a backward difference for the predictor and a forward difference for the corrector. Further, for more diffusive systems, centered differences can be employed.

The MacCormack method is a good, general numerical approach for water-quality calculations. First, because it is a centered-time/centered-space approach, it has good accuracy. Second, because the method is explicit, it is relatively easy to program. In particular, since the MacCormack method does not involve simultaneous equations, it handles nonlinear problems in a straightforward fashion.

The down side of the approach is that it is not unconditionally stable. Although its stability constraint is more liberal than for the simple explicit methods (FTBS or FTCS), there is a time-step limit beyond which the solution will blow up. These and other trade-offs among the methods are described next.

13.3 SUMMARY

Table 13.1 summarizes all the methods we have described in the previous two lectures.

Explicit methods. The primary advantages of all these approaches is that they are relatively easy to code for both linear and nonlinear problems. The simplest versions are manifestations of Euler's method. Two treatments of the spatial advection derivative are usually used: backward (FTBS) and centered (FTCS). The former has both spatial and temporal numerical dispersion, whereas the latter has only

TABLE 13.1
Comparison of numerical methods for solving one-dimensional water-quality models

Method	Numerical Dispersion	Stability	Programming: Linear	Programming: Nonlinear
Explicit:				
FTBS (backward-space Euler)	$\frac{1}{2}U\Delta x - \frac{1}{2}U^2\Delta t$	Conditional	Easy	Easy
FTCS (centered-space Euler)	$-\frac{1}{2}U^2\Delta t$	Conditional	Easy	Easy
MacCormack (Heun)	None	Conditional	Easy/moderate	Easy/moderate
Implicit:				
BTCS (simple implicit)	$\frac{1}{2}U\Delta x + \frac{1}{2}U^2\Delta t$	Unconditional	Moderate	Difficult
Crank-Nicolson	None	Unconditional	Moderate	Difficult

temporal. Both types of numerical dispersion are problematic because they vary as a function of step size (both Δx and Δt) and velocity. Because these (particularly spatial step and velocity) are not usually constants, this means that a variable numerical dispersion is introduced. Beyond such accuracy considerations, the principal shortcoming of the Euler-based approaches relates to stability. In short they extract a harsh execution-time penalty for increases in spatial resolution.

By centering its time-derivative approximation, the MacCormack approach removes all second-order temporal numerical dispersion. In addition, although it is conditionally stable, its stability constraint is more liberal than for the simple explicit methods. This combined with its ease of programming for both linear and nonlinear models makes it an attractive alternative.

Implicit methods. Two implicit methods are commonly used. The simple implicit approach writes all spatial derivatives (the transport terms) at the advanced time step. Thus the resulting equations are implicit and solutions must be obtained by solving simultaneous equations. Although the implicit method is unconditionally stable, it requires small time steps to obtain accurate solutions for highly advective systems. In addition it generates time-step-dependent numerical dispersion. Although these represent serious deficiencies when generating time-variable solutions, they become less important when steady-state solutions are desired. Thus it is applied with a large time step for such cases, because inaccurate intermediate results are not a detriment. In addition the numerical dispersion drops out as the steady-state condition is reached. Finally second-order approaches such as Crank-Nicolson are available that do not manifest time-step-dependent numerical dispersion. However, they are limited to linear models.

Finally it should be noted that all the techniques in Table 13.1 exhibit third-order truncation errors. These tend to manifest themselves by inducing oscillations and skewness into model simulations. These shortcomings tend to be significant in the simulation of sharp fronts. Although they have not traditionally been emphasized in water-quality modeling, such fronts should receive future attention in problems involving spills and transient flows. Hoffman (1992) discusses the issue.

PROBLEMS

13.1. An estuary has the following characteristics:

Depth = 4 m	Flow = 100,000 m^3 hr^{-1}
Velocity = 250 m hr^{-1}	Cross-sectional area = 400 m^2
Width = 100 m	Dispersion = 1,000,000 m^2 hr^{-1}

A spill of 10 kg of a reactive contaminant ($k = 0.1\ d^{-1}$) occurs. Use the (*a*) simple implicit, (*b*) Crank-Nicolson, and (*c*) MacCormack methods to simulate the distribution of the contaminant in the estuary.

13.2. A mixed-flow reactor is 10 m long, 2 m wide, and 1 m deep and has the following hydraulic characteristics: $E = 2\ m^2\ hr^{-1}$ and $U = 1\ m\ hr^{-1}$. It initially contains pure water. At $t = 0$ it is subject to a step loading of a chemical, $c_{in} = 100\ mg\ L^{-1}$, that decays by first-order kinetics $k = 0.2\ hr^{-1}$. Determine the distribution of the chemical along the length of the reactor as a function of time using centered spatial differences and the

(*a*) simple implicit method
(*b*) Crank-Nicolson method
(*c*) MacCormack method

13.3. Calculate how a 2-kg spill of a conservative substance would move through a 20-km stretch of Boulder Creek as described in Prob. 12.2 using the

(*a*) simple implicit method
(*b*) Crank-Nicolson method
(*c*) MacCormack method

PART III

Water-Quality Environments

Part III is designed to introduce you to some of the environments commonly encountered in water-quality modeling. Most of the material deals with physical aspects of these environments. Just as Lec. 2 served as an introduction to reactions, this part provides information and modeling techniques that have relevance to the other major facet of water-quality modeling—transport.

The first three lectures describe the three major classes of natural waters encountered in water-quality modeling: rivers, estuaries, and lakes. *Lecture 14* deals with characterizing flows and mixing in rivers and streams. This includes using data to estimate model parameters and an introduction to how the hydraulics of such systems are simulated. *Lecture 15* performs a similar function for estuaries and tidal rivers. Emphasis is on characterizing mixing due to tidal motion. *Lecture 16* introduces aspects of lake and reservoir physics. Aside from describing lake water budgets, this lecture includes additional models to assess the distribution of pollutants in the near-shore zone. This latter material can also be applied to the other water bodies (that is, wide rivers and estuaries) as well as to pollutant discharge in the oceanic coastal environment.

Lecture 17 describes an additional environment that is common to most water-quality modeling contexts—sediments. Information is provided on both suspended and bottom sediments.

Finally a very different type of environment is explored in *Lecture 18*—the "modeling environment." This lecture deals with issues that are commonly encountered in applying models, and includes information on model calibration, verification, and sensitivity analysis.

LECTURE 14

Rivers and Streams

LECTURE OVERVIEW: After a brief description of river types, I address the selection of low-flow values and the determination of parameters such as river dispersion coefficients and velocities. In addition I provide a brief introduction to some elementary aspects of stream hydrodynamics.

As described in Lec. 1, water-quality modeling originally focused on river and stream pollution. Because they provided drinking water and navigation, rivers were the sites of many large cities and their associated waste discharges. In addition to bearing the brunt of urban point-source pollution, rivers also provide water for other uses such as agriculture, industry, and power supply. Each of these uses can potentially affect water quality.

Beyond their utility, clean rivers and streams are beautiful. Anyone who has ever meditated at the foot of a waterfall or strolled along the banks of an unpolluted creek can appreciate their aesthetic value and their contribution to the quality of life.

Consequently rivers and streams are still the site of great environmental concern. This lecture is designed to provide you with background information on these systems. In particular we focus on aspects of river transport that have a bearing on water-quality modeling.

14.1 RIVER TYPES

Rivers can be classified in many ways. In the present context we are primarily concerned with their hydrogeometric properties. That is, to model rivers, we must know their hydraulic (flow, velocity, dispersion) and geometric (depth, width, slope) parameters. Table 14.1 includes a size range of rivers and streams along with some of their hydrogeometric parameters.

TABLE 14.1
Hydrogeometric parameters for a range of rivers ordered by flow (Fischer et al. 1979)

River	Mean depth (m)	Width (m)	Slope	Velocity (mps)	Flow ($m^3 s^{-1}$)	Dispersion ($10^5 cm^2 s^{-1}$)
Missouri	2.70	200	0.00021	1.55	837.0	150.00
Sabine	3.40	116	0.00013	0.61	254.6	49.30
Windy/Big Horn	1.63	64	0.00135	1.22	144.1	10.10
Yadkin	3.10	71	0.00044	0.60	140.1	18.50
Clinch, Tennessee	1.68	53	0.00054	0.70	74.5	3.83
John Day	1.53	30	0.00239	0.92	41.8	3.95
Nooksack	0.76	64	0.00979	0.67	32.6	3.50
Coachella Canal, California	1.56	24	—	0.71	26.6	0.96
Bayou Anacoco	0.93	32	0.00050	0.37	10.9	3.60
Cinch, Virginia	0.58	36	—	0.21	4.4	0.81
Powell, Tennessee	0.85	34	—	0.15	4.3	0.95
Copper, Virginia	0.56	17	0.00130	0.32	3.6	1.51
Comite	0.43	16	0.00059	0.37	2.5	1.40

Aside from size, the other important aspect of stream physics is the temporal distribution of flow. This time series is quantified by a plot of flow rate versus time of year that is called the ***annual hydrograph***. Figure 14.1 shows four types.

The first (Fig. 14.1*a*) is for a stream in a temperate climate with a ***perennial*** flow regime. This type of flow pattern is typical of many of the streams in the eastern and midwestern United States, where water-quality modeling was first developed. Note that most of the peak runoff occurs during the spring when heavy precipitation and some snowmelt occur. More important, from the standpoint of early water-quality modeling, is the occurrence of a relatively stable low-flow period during the warm summer months. Such warm, low-flow periods are usually the most critical from a water-quality perspective. Thus the low-flow period provided a stable design context for early wasteload allocation studies.

Figure 14.1*b* shows a hydrograph for a perennial stream, but in a more humid climate typical of the southeastern United States. The spikes are due to rain storms, whereas the slowly varying underlying trend, called ***baseflow***, is due to groundwater contributions. For this case the difference between summer and spring is not as pronounced because of the more even distribution of precipitation over the year and the absence of snowmelt. Although there seems to be a low-flow summer period, it is more punctuated by storm events.

Figure 14.1*c* shows a hydrograph for a river with a significant amount of snowmelt. The bulk of the peak flow occurs because of heavy snowmelt in the early summer. The release of water from baseflow and residual slowly melting snow results in a steadier distribution over the remainder of the year.

Finally the hydrograph in Fig. 14.1*d* is an example of an ***ephemeral*** flow regime for a stream in an arid climate. For this case there are significant periods when the river is dry. The periods of flow are due to large storms over the basin.

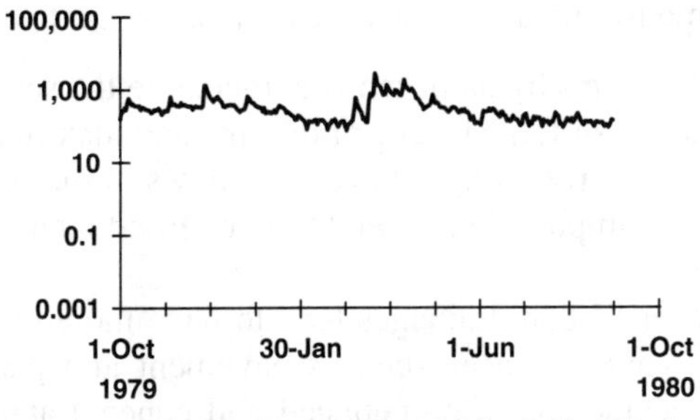

(*a*) Hudson River at Green Island, New York
(watershed area = 20,950 km²)

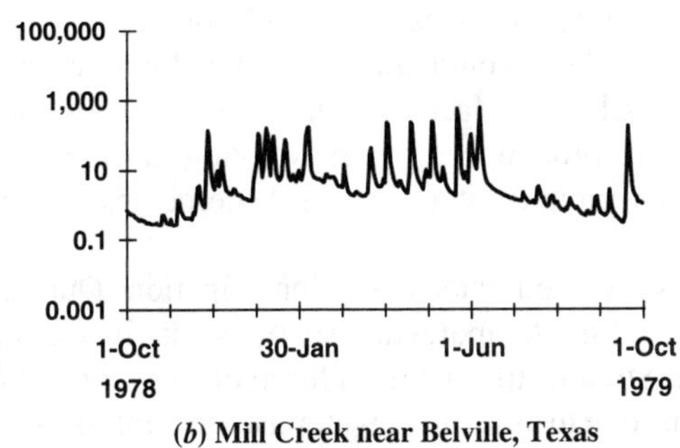

(*b*) Mill Creek near Belville, Texas
(watershed area = 974 km²)

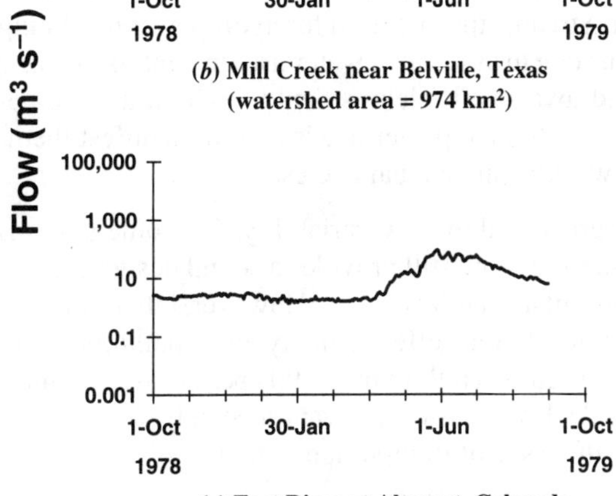

(*c*) East River at Altmont, Colorado
(watershed area = 748 km²)

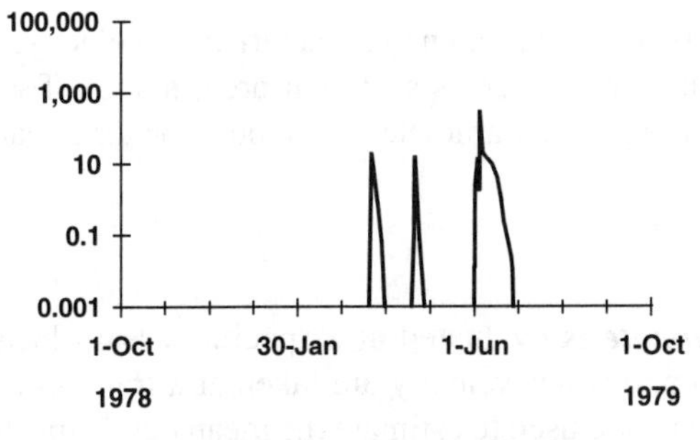

(*d*) Frio River near Uvalde, Texas
(watershed area = 1712 km²)

FIGURE 14.1
Annual streamflow hydrographs for (*a*) perennial temperate river, (*b*) perennial humid river, (*c*) snowfed river, and (*d*) ephemeral river.

Human influence can further modify hydrographs in several significant ways:

Impounding. The regulation of streams by dams tends to moderate the seasonal swings in flow. This also tends to shift the timing of minima and maxima. For example whereas the unregulated river might have peak flows in the spring, the peaks could be captured by impoundments and then released later in the summer for agricultural use.

Urbanization and channelization. Urban drainages tend to have more of their land area covered by impervious surfaces such as pavement and parking lots. Consequently runoff becomes more pronounced and concentrated because little rainfall will be absorbed by soil and vegetation. Further, storm sewers, which are constructed to convey the water off the streets rapidly, result in speedy delivery to receiving waters. In addition many streams have been channelized so that they would transmit high flows more efficiently and prevent flooding. All these factors contribute to making spikes in the annual hydrograph more pronounced. Moreover flow during the dry season is usually lowered as groundwater sources become less predominant.

Human water use. Many rivers are used extensively for irrigation. During the growing season this can lead to high temporal and spatial variability as farmers take and discharge water. Streams that are used for hydropower or cooling water also exhibit fluctuations due to associated water use. In contrast to other uses, power-associated withdrawals and releases tend to exhibit a more periodic nature. This is so because human power needs tend to manifest themselves on predictable daily, weekly, and annual cycles.

In summary streams exhibit a great deal of flow variability. For some systems the classical pattern of low, warm summer flow still provides a sound design context for evaluating pollution control, particularly point sources. However, other systems, because of their highly transient natures, human effects, and types of pollution problems (for example, nonpoint runoff), require a rethinking of this perspective. In most of the remainder of this lecture we deal with issues relevant to steady-flow modeling. In the final section we address the issue of time-variable flows.

14.2 STREAM HYDROGEOMETRY

A stream's ***hydrogeometry*** consists of its hydrologic characteristics (velocity, flow, dispersion) and its geometry (depth, width, cross-sectional area, slope). Two approaches are available for determining these parameters: at a point or for a reach.

14.2.1 Point Estimates

As the name implies, a ***point estimate*** is evaluated at a specific stream location. Typically, measurements of depth and water velocity are taken at a transect across the stream (Fig. 14.2). The depth data are used to estimate the mean depth and cross-sectional area. This is done via integration,

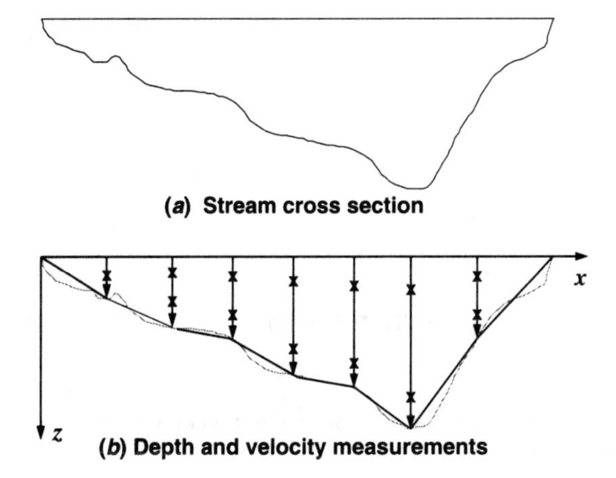

FIGURE 14.2
(*a*) A stream cross section along with (*b*) a transect showing depth and velocity measurements needed to calculate mean depth, flow, and other hydrogeometric parameters. Note that the velocity measurements (x) are taken at 60% depth for shallow water and at 20% and 80% for deep points.

$$A_c = \int_0^B z(x)\,dx \tag{14.1}$$

$$H = \frac{A_c}{B} \tag{14.2}$$

where A_c = cross-sectional area (m^2)
x = distance measured across the stream (m)
$z(x)$ = depth measured at location x (m)
H = mean depth (m)
B = stream width

At each depth measurement a corresponding velocity measurement can be taken. There are several ways to do this (Gupta 1989). The most common are:

1. For deeper waters, depth > 0.61 m (2 ft), the velocities taken at 20% and 80% of the total depth are averaged.
2. For shallower water, depth < 0.61 m, a single velocity measurement is taken at 60% of the depth.

It can be shown (Box 14.1) that these choices have a mathematical basis.

BOX 14.1. Mathematical Basis of the Two-Point and Six-Tenths-Depth Methods

A typical distribution of velocity with depth is depicted in Fig. B14.1-1. Dickinson (1967) determined that the distribution of velocity in streams could be fit by the simple power law

$$U(z) = U_0\left(\frac{z_b - z}{z_b - z_0}\right)^{1/m} \tag{14.3}$$

where $U(z)$ = velocity at depth z
U_0 = known velocity at a depth z_0 from the surface

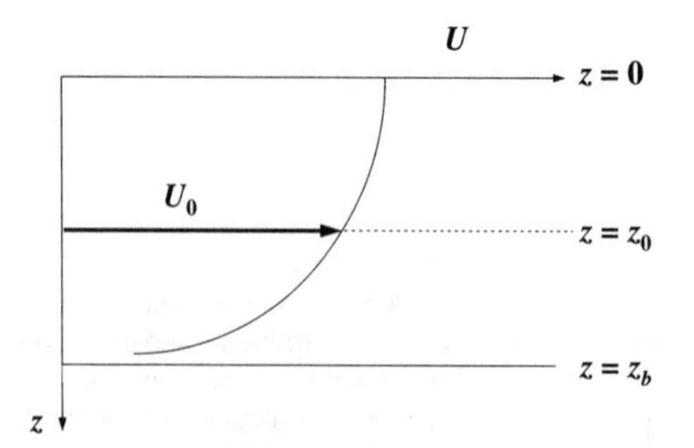

FIGURE B14.1-1

z_b = depth of the bottom
m = a constant that varies from 6 to 10, depending on the Reynolds number, and is approximately 7 (Daily and Harleman 1966)

This equation can be integrated to determine the average velocity,

$$\overline{U} = \frac{1}{z_b}\int_0^{z_b} U(z)\,dz = \frac{1}{z_b}\int_0^{z_b} U_0\left(\frac{z_b - z}{z_b - z_0}\right)^{1/m} dz = \frac{m}{m+1}U_0\left(\frac{z_b}{z_b - z_0}\right)^{1/m} \tag{14.4}$$

Now suppose that the mean velocity occurs at a distance Z from the bottom. Setting $U_0 = \overline{U}$ and $z_b - z = Z$ in Eq. 14.3 and equating the result with Eq. 14.4 gives

$$U_0\left(\frac{Z}{z_b - z_0}\right)^{1/m} = \frac{m}{m+1}U_0\left(\frac{z_b}{z_b - z_0}\right)^{1/m} \tag{14.5}$$

which can be solved for

$$Z = \left(\frac{m}{m+1}\right)^m z_b \tag{14.6}$$

For values of m between 6 and 10, this equation can be used to determine that $Z \cong 0.4\ z_b$. Thus a single measurement taken at about 60% of the total depth should provide a good approximation of the average velocity over the entire depth.

A similar analysis can be performed to determine a best two-point approach. For this case the mean is obtained by taking the average of the values at 20% and 80% of the total depth (Gupta 1989).

Interestingly, independent support for the two-point approach comes from the numerical integration method called ***Gauss quadrature.*** If the integration interval ranges from -1 to 1, it can be shown (Chapra and Canale 1988) that an optimal integral estimate can be determined by adding the values of the integrated variable at $1/\sqrt{3}$ and $-1/\sqrt{3}$. Thus an average value for the interval could be computed as

$$\overline{f} = \frac{\int_{-1}^{1} f(x)\,dx}{1-(-1)} = 0.5f\left(\frac{1}{\sqrt{3}}\right) + 0.5f\left(\frac{-1}{\sqrt{3}}\right) \tag{14.7}$$

Now by a simple change of variable, it can be recognized that $-1/\sqrt{3}$ on a domain from -1 to 1 corresponds to a value of approximately 0.22 z_b on a domain from 0 to z_b. Similarly $1/\sqrt{3}$ corresponds to 0.78 z_b (see Fig. B14.1-2). Hence the results support the strategy of averaging values taken at 20 and 80% of the total depth (Fig. B14.1-2).

A nice result of the Gauss-quadrature approach is that it is more general than the foregoing derivations. That is, it does not depend on having a power function of the form of Eq. 14.3. Consequently it should hold regardless of the velocity distribution.

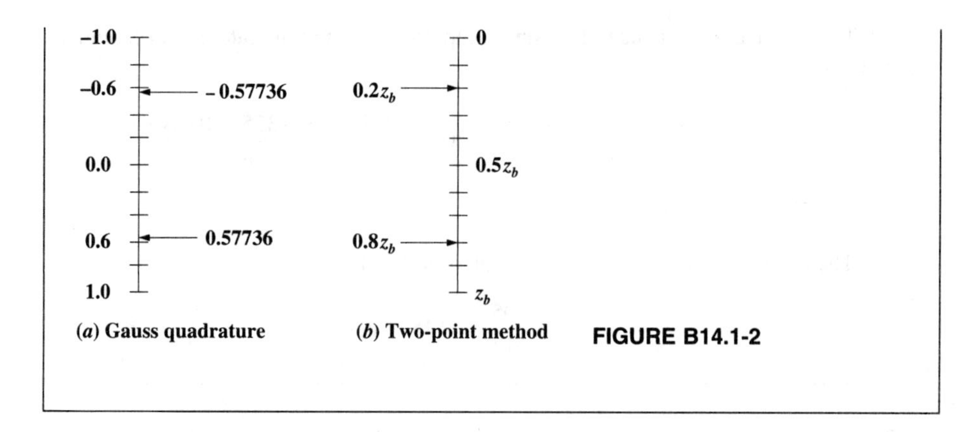

(*a*) Gauss quadrature (*b*) Two-point method **FIGURE B14.1-2**

Once the average velocity for each point on the transect is derived, $\overline{U}(x)$, the values can be integrated to arrive at the mean flow

$$Q = \int_0^B \overline{U}(x)z(x)\,dx \tag{14.8}$$

Finally the mean velocity can be calculated with the continuity equation

$$U = \frac{Q}{A_c} \tag{14.9}$$

The approach is illustrated in the following example.

EXAMPLE 14.1. POINT ESTIMATION. The following data were collected for a stream cross section:

x (m)	0	4	8	12	16	20
z (m)	0	0.4	1	1.5	0.2	0
$U(x)_{0.2}$ (mps)	0		0.2	0.3		0
$U(x)_{0.6}$ (mps)	0	0.05			0.07	0
$U(x)_{0.8}$ (mps)	0		0.12	0.2		0
$\overline{U}(x)$ (mps)	0	0.05	0.16	0.25	0.07	0
$\overline{U}(x)z$ (m^2 s^{-1})	0	0.02	0.16	0.375	0.014	0

Use this data to determine the (*a*) area, (*b*) mean depth, (*c*) flow, and (*d*) velocity.

Solution: (*a*) Because the sampling points are equally spaced, Simpson's rules (Chapra and Canale 1988) can be used to implement Eq. 14.1. The $\frac{1}{3}$ rule can be applied to the first two intervals and the $\frac{3}{8}$ rule to the last three,

$$A_c = (8 - 0)\frac{0 + 4 \times 0.4 + 1}{6} + (20 - 8)\frac{1 + 3(1.5 + 0.2) + 0}{8} = 12.617 \text{ m}^2$$

(*b*) Equation 14.2 can then be used to calculate the mean depth,

$$H = \frac{12.617}{20} = 0.6308 \text{ m}$$

(*c*) The flow can be evaluated by using Simpson's rules to integrate Eq. 14.8 numerically,

$$Q = (8 - 0)\frac{0 + 4 \times 0.02 + 0.16}{6} + (20 - 8)\frac{0.16 + 3(0.375 + 0.014) + 0}{8}$$

$$= 2.3105 \text{ m}^3 \text{ s}^{-1}$$

(*d*) The mean velocity can then be calculated as (Eq. 14.9)

$$U = \frac{2.3105}{12.617} = 0.1831 \text{ m s}^{-1}$$

The above procedure can be repeated for a number of transects along a stream. The data can then be plotted versus stream length and the average segment parameters determined (Fig. 14.3).

14.2.2 Reach Estimates

A ***reach*** approach provides an alternative to point estimates. This method is usually predicated on the generally valid assumption that the stream width is less variable than its depth. If this is true a stream reach is identified that has a relatively constant width. After measuring this width the flow at the end of the reach is determined by a point estimate. In addition gaging stations may also provide the same estimate.

Next a travel time is determined by injecting a tracer such as a dye at the head end and timing how long it takes to traverse the reach. The mean velocity for the reach can then be calculated as

$$U = \frac{x}{t} \tag{14.10}$$

where x = reach length and t = travel time.

The velocity and flow rate can then be used to determine the average cross-sectional area

$$A_c = \frac{Q}{U} \tag{14.11}$$

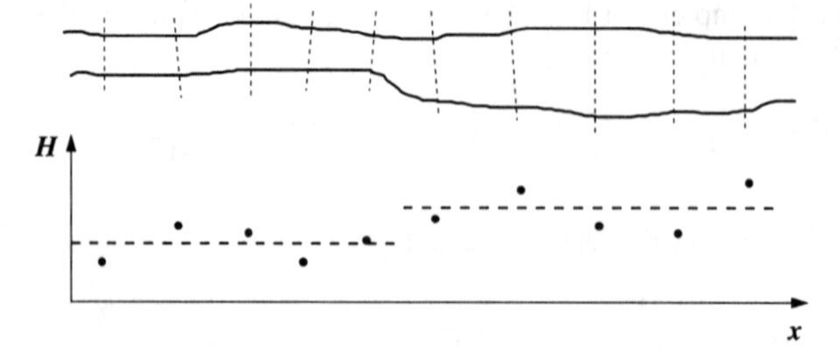

FIGURE 14.3
Data from individual transects can be plotted versus distance to develop average values of parameters for reaches.

which can then be used to determine the mean depth

$$H = \frac{A_c}{B} \tag{14.12}$$

EXAMPLE 14.2. REACH ESTIMATION OF VELOCITY AND MEAN DEPTH. Suppose that the point estimate calculated in Example 14.1 is at the downstream end of a 2-km reach with a mean width of 22 m. Recall that the point estimate of flow was 2.3105 $m^3\ s^{-1}$. You perform a dye study and determine that it takes 3.2 hr for the dye to traverse the 2 km. Use the reach approach to determine the velocity, cross-sectional area, and mean depth for the reach.

Solution: The mean velocity can be calculated as (Eq. 14.10)

$$U = \frac{2\ \text{km}}{3.2\ \text{hr}}\left(\frac{1\ \text{hr}}{3600\ \text{s}}\ \frac{1000\ \text{m}}{\text{km}}\right) = 0.1736\ \text{m s}^{-1}$$

The velocity and flow rate can then be used to determine the average cross-sectional area according to Eq. 14.11,

$$A_c = \frac{2.3105}{0.1736} = 13.3\ \text{m}^2$$

which can then be used to determine the mean depth (Eq. 14.12),

$$H = \frac{13.3}{22} = 0.605\ \text{m}$$

In general the reach approach is particularly effective for shallow systems. This is because the width is usually much less variable than the depth for such streams. In contrast the transect method is more commonly applied to deeper systems.

14.3 LOW-FLOW ANALYSIS

As mentioned in this lecture's introduction, traditional water-quality modeling has used the steady, low-flow summer period as its design condition. Therefore the question arises: How do you establish the low-flow condition?

The minimum 7-d flow that would be expected to occur every 10 yr is generally accepted as the standard design flow. Called the ***7Q10***, this value can be determined for the entire year, by month, or for a season. For the present discussion we limit ourselves to a hypothetical summer period consisting of the months of July through September.

The first step in computing the 7Q10 is to obtain a long-term flow record for the location being modeled. For our case we would examine the data for each year to determine the smallest flow that occurs for seven consecutive days in the summer period.

Next we would tabulate the n flows in ascending order and assign them a rank m. The cumulative probability of occurrence is given by

$$p = \frac{m}{N+1} \tag{14.13}$$

The recurrence interval is then defined as

$$T = \frac{1}{p} \tag{14.14}$$

Probability paper can be used to pick off the 7Q10 as depicted in Fig. 14.4 and described in the following example.

EXAMPLE 14.3. CALCULATION OF 7Q10. The following 7-d low flows were compiled for a river:

1971	1.72	1976	4.23	1981	4.48	1986	5.39
1972	3.03	1977	4.11	1982	3.03	1987	3.00
1973	2.76	1978	1.92	1983	2.84	1988	2.50
1974	1.65	1979	2.14	1984	3.66	1989	2.47
1975	2.00	1980	1.48	1985	1.87	1990	3.07

Use this data to determine the 7Q10.

Solution: The data can be ordered and tabulated along with the cumulative probability of occurrence and the recurrence interval:

Rank	Flow	Probability	Recurrence interval	Rank	Flow	Probability	Recurrence interval
1	1.48	4.76	21.00	11	2.84	52.38	1.91
2	1.65	9.52	10.50	12	3.00	57.14	1.75
3	1.72	14.29	7.00	13	3.03	61.90	1.62
4	1.87	19.05	5.25	14	3.03	66.67	1.50
5	1.92	23.81	4.20	15	3.07	71.43	1.40
6	2.00	28.57	3.50	16	3.66	76.19	1.31
7	2.14	33.33	3.00	17	4.11	80.95	1.24
8	2.47	38.10	2.63	18	4.23	85.71	1.17
9	2.50	42.86	2.33	19	4.48	90.48	1.11
10	2.76	47.62	2.10	20	5.39	95.24	1.05

This data can be plotted on normal probability paper. As in Fig. 14.4 a value of about 1.66 cms corresponds to the 7Q10.

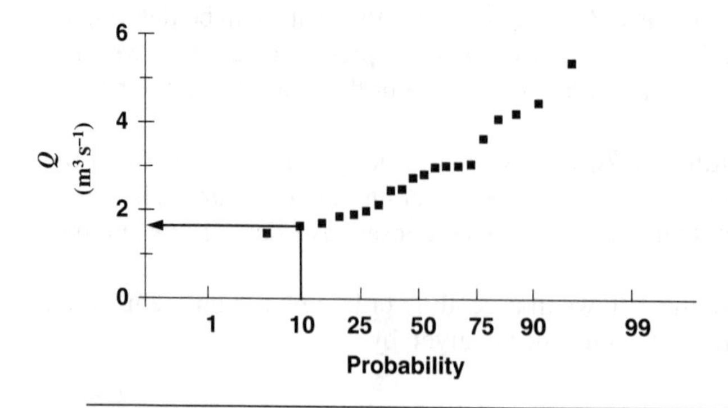

FIGURE 14.4
Frequency curve corresponding to the 7-d minimum flow for the river from Example 14.3.

14.4 DISPERSION AND MIXING

In rivers, we examine two mixing regimes. First, for one-dimensional models, we are concerned with mixing in the direction of flow or ***longitudinal mixing***. This process is parameterized by a dispersion coefficient. In addition we are also interested in the mixing across the stream or ***lateral mixing***. In this case we would like to assess our assumption that point sources are instantaneously mixed in this dimension. Therefore the goal here is to quantify the longitudinal flow length required to attain lateral mixing.

14.4.1 Longitudinal Dispersion

Several formulas are available to estimate the longitudinal dispersion coefficient for streams and rivers. Fischer et al. (1979) have developed the following:

$$E = 0.011 \frac{U^2 B^2}{H U^*} \tag{14.15}$$

where E has units of ($m^2\ s^{-1}$)
U = velocity ($m\ s^{-1}$)
B = width (m)
H = mean depth (m)
U^* = shear velocity ($m\ s^{-1}$), which is related to more fundamental characteristics by

$$U^* = \sqrt{gHS} \tag{14.16}$$

where g = acceleration due to gravity ($m\ s^{-2}$) and S = channel slope (dimensionless).

Table 14.1 contains measured values of longitudinal dispersion for several streams and rivers. Figure 14.5 shows a comparison between these measurements

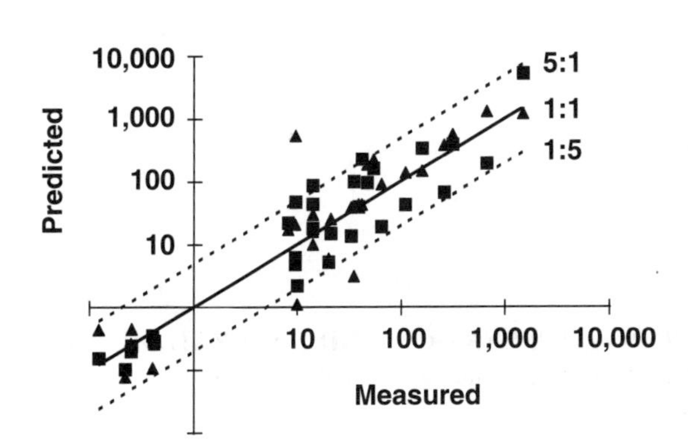

FIGURE 14.5
Comparison of measurements (Table 14.1) with predictions using Eqs. 14.15 (■) and 14.17 (▲). The solid line represents a perfect correlation between the predictions and the measurements, whereas the dashed lines correspond to a factor of 5 above and below the 1:1 correlation. Note that some additional data beyond Table 14.1, from some laboratory flumes and canals (Fischer et al. 1979), have been included on this plot.

and values predicted with Eq. 14.15. As can be seen the predictions are within a factor of 5 of the data.

McQuivey and Keefer (1974) proposed the alternative formula

$$E = 0.05937\frac{Q}{SB} \tag{14.17}$$

where Q = mean flow ($m^3\ s^{-1}$). They studied rivers with flows ranging from 35 to 33,000 cfs. They limit this formulation to systems with Froude numbers ($F = U/\sqrt{gH}$) less than 0.5. We have also included predictions from this model on Fig. 14.5. As with the Fischer formula (Eq. 14.15) the predictions are approximate to a range of plus or minus half an order of magnitude.

14.4.2 Lateral Mixing

Lateral mixing of point sources is the second facet of mixing relevant to one-dimensional stream water-quality modeling. Fischer et al. (1979) have developed the following formula to estimate the lateral or transverse dispersion coefficient for a stream:

$$E_{lat} = 0.6HU^* \tag{14.18}$$

where H = mean depth (m) and U^* = shear velocity (mps). This value can then be used to compute the length required to attain complete lateral mixing. For a side discharge, the formula is

$$L_m = 0.4U\frac{B^2}{E_{lat}} \tag{14.19}$$

For a discharge in the center of the channel,

$$L_m = 0.1U\frac{B^2}{E_{lat}} \tag{14.20}$$

Yotsukura (1968) has proposed the following alternative formula for a side discharge:

$$L_m = 8.52U\frac{B^2}{H} \tag{14.21}$$

where L_m = mixing length (m)
U = velocity (mps)
B = width (m)
H = depth (m)

EXAMPLE 14.4. LONGITUDINAL DISPERSION AND LATERAL MIXING. The following data applies to Boulder Creek, Colorado, immediately below the Boulder wastewater treatment plant discharge:

$U = 0.3$ mps
$B = 15$ m
$H = 0.4$ m
$S = 0.004$
$Q = 1.8$ cms

Use this data to determine the longitudinal dispersion coefficient and the length required to attain complete lateral mixing.

Solution: Substituting the values into Eqs. 14.16 and 14.15 yields

$$U^* = \sqrt{9.8(0.4)0.004} = 0.125 \text{ m s}^{-1}$$

$$E = 0.011 \frac{(0.3)^2(15)^2}{0.4(0.125)} = 4.45 \text{ m}^2 \text{ s}^{-1}$$

The McQuivey and Keefer formula (Eq. 14.17) can also be applied to calculate

$$E = 0.05937 \frac{1.8}{0.004(15)} = 1.78 \text{ m}^2 \text{ s}^{-1}$$

Thus for this case the estimates differ by about a factor of 2.5.

The mixing length can be determined as (Eqs. 14.18 and 14.19)

$$E_{lat} = 0.6(0.4)0.125 = 0.03 \text{ m}^2 \text{ s}^{-1}$$

$$L_m = 0.4(0.3)\frac{15^2}{0.03} = 898 \text{ m}$$

or with Eq. 14.21 as

$$L_m = 8.52(0.3)\frac{(15)^2}{0.4} = 1438 \text{ m}$$

Consequently these results indicate that it would require about 1 km of distance to reach complete mixing. In actuality a USGS gaging station is located just below the outfall. A weir at this point constricts the flow severely and effectively results in complete mixing at that point.

14.5 FLOW, DEPTH, AND VELOCITY

Suppose that you have a stretch of river that has constant dimensions and bottom slope. If a steady flow rate Q enters the upstream end for a sufficiently long period of time, the water will reach a steady, uniform-flow condition. That is, the flow will be constant in time and space.

Now what about the depth and velocity of the water? They will also be constant and are related to flow by the continuity equation

$$Q = UA_c \tag{14.22}$$

where A_c = cross-sectional area and U = mean velocity.

Equation 14.22 represents a dilemma. It says that the flow can manifest itself in two ways: as depth and as velocity. Clearly additional information is needed

TABLE 14.2
Average values and ranges of exponents in hydrogeometric correlations

Correlation	Exponent	Value	Range
Velocity-flow	b	0.45	0.3–0.7
Depth-flow	β	0.4	0.1–0.6
Width-flow	f	0.15	0.05–0.25

to determine both quantities. As described next this information is handled in two ways: discharge coefficients and Manning's equation.

14.5.1 Discharge Coefficients

Power equations can be used to relate mean velocity, depth, and width to flow (Leopold and Maddock 1953),

$$U = aQ^b \tag{14.23}$$

$$H = \alpha Q^\beta \tag{14.24}$$

$$B = cQ^f \tag{14.25}$$

where H = mean depth and $a, b, \alpha, \beta, c, f$ are empirical constants that are determined from stage-discharge rating curves (log-log plots between the parameters).

Because these parameters are interrelated, the coefficients are not totally independent. For example because $Q = UA_c$ and $A_c = BH$, the sum of the exponents should be equal to unity, $b + \beta + f = 1$. Table 14.2 outlines values and ranges for the exponents.

14.5.2 The Manning Equation

The ***Manning equation***, which is derived from a momentum balance for the channel, provides a means to relate velocity to channel characteristics,

$$U = \frac{C_o}{n} R^{2/3} S_e^{1/2} \tag{14.26}$$

where C_o = a constant (= 1.0 for metric and 1.486 for English units)
n = Manning's roughness coefficient (Table 14.3)
R = channel's hydraulic radius (m or ft) = A_c/P
P = wetted perimeter (m or ft)
S_e = slope of the channel's energy grade line† (dimensionless)

†Note that since we have assumed that flow is steady and the cross section is constant, the energy slope is equal to the channel slope.

TABLE 14.3
The Manning roughness coefficient for various open channel surfaces (Chow 1959)

Material	n
Artificial channels:	
Concrete	0.012
Gravel bottom with sides:	
Concrete	0.020
Mortared stone	0.023
Riprap	0.033
Natural stream channels:	
Clean, straight	0.030
Clean, winding	0.040
Weeds and pools, winding	0.050
Heavy brush, timber	0.100

The Manning formula can also be substituted into the continuity equation to compute flow,

$$Q = \frac{C_o}{n} A_c R^{2/3} S_e^{1/2} \tag{14.27}$$

If flow is given and the area and hydraulic radius can be expressed as functions of depth, Eq. 14.27 has one unknown—depth. Consequently it can be solved as a roots problem. Depth can then be used to determine the cross-sectional area. Finally depth can be used in conjunction with flow to calculate velocity with the continuity equation.

To illustrate this method, suppose that the geometry can be represented as a trapezoid (Fig. 14.6) with constant dimensions and bottom slope.

Because we are dealing with a trapezoid, the cross-sectional area and the hydraulic radius can be expressed as a function of depth,

$$A_c = (B_0 + sy)y \tag{14.28}$$

$$P = B_0 + 2y\sqrt{s^2 + 1} \tag{14.29}$$

$$R = \frac{A_c}{P} = \frac{(B_0 + sy)y}{B_0 + 2y\sqrt{s^2 + 1}} \tag{14.30}$$

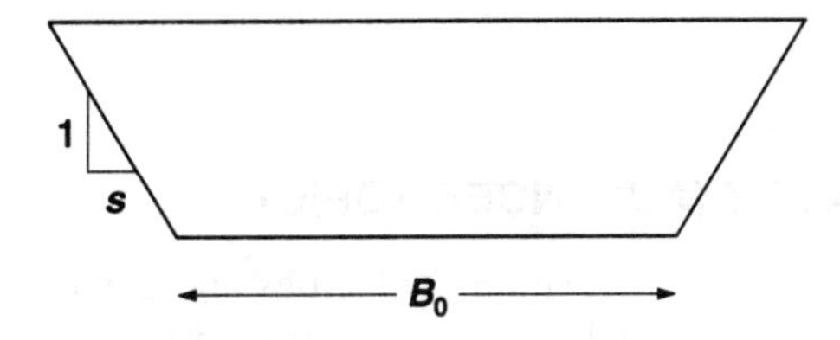

FIGURE 14.6
A cross section of a trapezoidal channel showing the parameters needed to uniquely define the geometry: B_0 = bottom width, s = side slope.

Equations 14.28 and 14.30 can be substituted into Eq. 14.28 to yield

$$Q = \frac{1}{n} \frac{\left[(B_0 + sy)y\right]^{5/3}}{\left(B_0 + 2y\sqrt{s^2+1}\right)^{2/3}} S_e^{1/2} \tag{14.31}$$

If Q is given, Eq. 14.31 is a nonlinear equation with one unknown, y. Thus this equation can be reformulated as a roots problem,

$$f(y) = \frac{1}{n} \frac{\left[(B_0 + sy)y\right]^{5/3}}{\left(B_0 + 2y\sqrt{s^2+1}\right)^{2/3}} S_e^{1/2} - Q \tag{14.32}$$

The root (that is, the value of y that makes this equation zero) is the reach depth.

EXAMPLE 14.5. THE MANNING EQUATION. A channel has the following characteristics:

Flow = 6.25 cms
Channel slope = 0.0002
Bottom width = 10 m
Side slope = 2
Roughness = 0.035

Determine the depth, cross-sectional area, and velocity for the channel.

Solution: Substituting values into Eq. 14.32 yields

$$f(y) = 0.4041 \frac{\left[(10 + 2y)y\right]^{5/3}}{(10 + 4.472y)^{2/3}} - 6.25$$

which can be solved for $y = 1.24$ m. This result can be substituted into Eq. 14.28 to compute

$$A_c = \left[10 + 2(1.24)\right] 1.24 = 15.5 \text{ m}^2$$

which, along with the flow rate, can be substituted into the continuity equation to determine

$$U = \frac{Q}{A_c} = \frac{6.25}{15.5} = 0.403 \text{ m s}^{-1}$$

Computer packages, such as EPA's stream model QUAL2E, use the Manning equation to model steady flow in streams. To do this the stream is broken into reaches that consist of constant hydrogeometric properties. Then Manning's equation is used to determine the velocity and depth for each reach. This information is then used in mass balances for individual pollutants.

14.6 ROUTING AND WATER QUALITY (ADVANCED TOPIC)

Although the Manning approach is adequate for determining velocities and depths for streams with steady flows, there are other cases where flow changes temporally. For example in agricultural regions, farmers withdraw and release irrigation water on a frequent basis during the growing season. Another relevant example might be the tailwater below a power-supply reservoir. In such cases flows often vary on an hourly basis due to changes in demand for electricity. Transport could be an extremely significant determinant of the water quality for such systems.

14.6.1 Routing Water

There is a spectrum of models that are available to simulate dynamic water movement for streams. All are based on the ***St. Venant equations***. These result from taking mass and momentum balances on a one-dimensional channel. For the case where lateral inflow, wind shear, and eddy losses are omitted, the results are

Continuity equation (mass balance):

$$\frac{\partial Q}{\partial x} + \frac{\partial A_c}{\partial t} = 0 \tag{14.33}$$

Momentum equation (momentum balance):

$$\frac{1}{A_c}\frac{\partial Q}{\partial t} + \frac{1}{A_c}\frac{\partial}{\partial x}\left(\frac{Q^2}{A_c}\right) + g\frac{\partial y}{\partial x} - g(S_o - S_f) = 0 \tag{14.34}$$

As depicted in Fig. 14.7 the momentum balance can be simplified by dropping terms. In this section, we develop a solution for the simplest form, the kinematic wave. For this case the pressure and acceleration force terms are dropped and the following pair of differential equations result:

$$\frac{\partial Q}{\partial x} + \frac{\partial A_c}{\partial t} = 0 \tag{14.35}$$

$$S_o = S_f \tag{14.36}$$

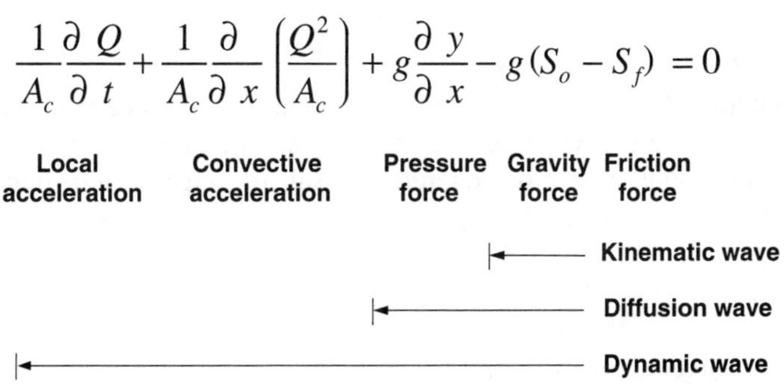

$$\frac{1}{A_c}\frac{\partial Q}{\partial t} + \frac{1}{A_c}\frac{\partial}{\partial x}\left(\frac{Q^2}{A_c}\right) + g\frac{\partial y}{\partial x} - g(S_o - S_f) = 0$$

Local acceleration	Convective acceleration	Pressure force	Gravity force	Friction force

Kinematic wave (gravity and friction force terms)

Diffusion wave (pressure, gravity and friction force terms)

Dynamic wave (all terms)

FIGURE 14.7
The St. Venant equation for momentum can be simplified by dropping terms as shown (redrawn from Chow et al. 1988).

The continuity equation merely expresses the idea that if the area is increasing temporally at a point in space, then the flow rate passing this point at that time must be decreasing spatially to account for the water that is stored in the increasing cross section. The momentum equation states that the accelerating effect of slope is exactly balanced by the retarding effect of friction.

Now Manning's equation can be incorporated into the scheme by substituting $S_o = S_f$ and $R = A_c/P$,

$$Q = \frac{1}{n}\frac{A_c^{5/3}}{P^{2/3}}S_o^{1/2} \tag{14.37}$$

which can be solved for

$$A_c = \alpha Q^{\beta} \tag{14.38}$$

where $\beta = \frac{3}{5}$ and

$$\alpha = \left(\frac{nP^{2/3}}{\sqrt{S_o}}\right)^{3/5} \tag{14.39}$$

If the channel has uniform, ideal geometry Eq. 14.39 can be simplified further. For example for a rectangular channel that is much wider than it is deep, it becomes

$$\alpha = \left(\frac{nB^{2/3}}{\sqrt{S_o}}\right)^{3/5} \tag{14.40}$$

Thus for this simple case, α is a constant.

Next Eq. 14.38 can be differentiated with respect to time,

$$\frac{\partial A_c}{\partial t} = \alpha\beta Q^{\beta-1}\frac{\partial Q}{\partial t} \tag{14.41}$$

which can be substituted into Eq. 14.35 to yield

$$\frac{\partial Q}{\partial x} + \alpha\beta Q^{\beta-1}\frac{\partial Q}{\partial t} = 0 \tag{14.42}$$

Thus these manipulations have generated a single differential equation with one unknown, Q, that is the equivalent of the original two equations (14.35 and 14.36).

A numerical solution can be generated by substituting forward-time/backward-space differences into Eq. 14.42 to give

$$\frac{Q_i^\ell - Q_{i-1}^\ell}{\Delta x} + \alpha\beta(Q_i^\ell)^{\beta-1}\frac{Q_i^{\ell+1} - Q_i^\ell}{\Delta t} \tag{14.43}$$

which can be solved explicitly for

$$Q_i^{\ell+1} = Q_i^\ell + \left\{\left[\frac{(Q_i^\ell)^{1-\beta}}{\alpha\beta\Delta x}\right](Q_{i-1}^\ell - Q_i^\ell)\right\}\Delta t \tag{14.44}$$

EXAMPLE 14.6. KINEMATIC WAVE. Suppose that we have a rectangular channel that has the following characteristics

Flow = 2.5 cms
Channel slope = 0.004
Bottom width = 15 m
Roughness = 0.07

For simplicity we assume that the width is much greater than the depth, and therefore Eq. 14.40 holds. At $x = 0$ the flow increases and then decreases due to a power-plant release. The excess flow Q_e can be approximated by the half-sinusoid,

$$Q_e = 2.5 \sin \omega t \qquad 0 \leq t \leq 0.25 \text{ d}$$
$$Q_e = 0 \qquad t > 0.25 \text{ d}$$

where $\omega = 2\pi(0.5 \text{ d})^{-1}$.

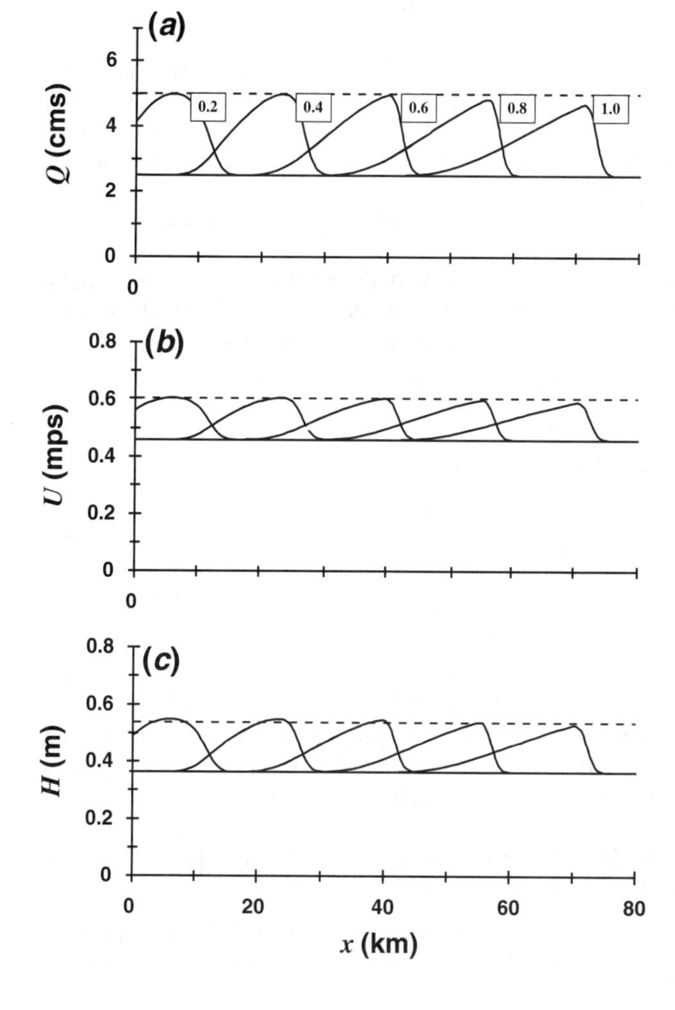

FIGURE 14.8
Simulation of a wave moving through a channel. The boxed numbers represent the amount of time (d) following the onset of the wave. Also the dashed lines indicate the peak values.

(*a*) Determine the steady-state depth and velocity prior to the power-plant release.
(*b*) Then use the explicit numerical method to predict how the depth and the velocity change as the power-plant flow propagates down the channel.

Solution: (*a*) Substituting values into Eq. 14.40 yields

$$\alpha = \left[\frac{0.07(15)^{2/3}}{\sqrt{0.004}}\right]^{3/5} = 3.14$$

Next Eq. 14.38 can be solved for the steady-state area,

$$A_c = 3.14(2.5)^{0.6} = 5.44 \text{ m}^2$$

which can be used to calculate depth and velocity:

$$U = \frac{2.5 \text{ m}^3 \text{ s}^{-1}}{5.44 \text{ m}^2} = 0.46 \text{ m s}^{-1}$$

$$H = \frac{5.44 \text{ m}^2}{15 \text{ m}} = 0.3627 \text{ m}$$

(*b*) Using time and space scales of 0.007 d and 1 km, respectively, we can use Eq. 14.44 to generate the solution shown in Fig. 14.8. The solution in Fig. 14.8 exhibits some numerical dispersion. This would not be unexpected because of our use of the forward-time/backward-space scheme. More refined numerical methods can be used to mitigate this effect.

Before using the foregoing model to simulate pollutant transport, I want to point out an important feature of the kinematic wave model. To do this recall that the advection mass-balance equation for a conservative pollutant can be written as

$$\frac{\partial c}{\partial t} = -U\frac{\partial c}{\partial x} \tag{14.45}$$

We can rearrange Eq. 14.42 so that it is in the same format,

$$\frac{\partial Q}{\partial t} = -U_c\frac{\partial Q}{\partial x} \tag{14.46}$$

where

$$U_c = \frac{Q^{1-\beta}}{\alpha\beta} = \frac{\partial Q}{\partial A_c} \tag{14.47}$$

Comparison indicates that just as pollutant mass is propagated down a channel at a velocity U (Eq. 14.45), a flow wave moves at a velocity U_c (Eq. 14.46). The velocity U_c is formally termed the kinematic wave's ***celerity.***

EXAMPLE 14.7. CELERITY. Determine the celerity for Example 14.6 and use Fig. 14.8 to verify that the wave moves at this velocity.

Solution: In Example 14.6 the flow in the channel varies between 2.5 and 5 cms. Thus the celerity would also vary. Substituting values into Eq. 14.47 for each bound yields

$$U_c = \frac{2.5^{1-0.06}}{3.14(0.6)} = 0.766 \text{ m s}^{-1}$$

$$U_c = \frac{5^{1-0.6}}{3.14(0.6)} = 1.01 \text{ m s}^{-1}$$

Now suppose that we approximate the wave as a half-sinusoid. The average velocity can be calculated as

$$\overline{U}_c = \frac{\int_0^{0.25}[0.766 + (1.01 - 0.766)\sin(\omega t)]\,dt}{0.25} = 0.92 \text{ m s}^{-1}$$

Inspection of Fig. 14.8 indicates that the peak flow is at 6500 and 71,500 m at t = 0.2 and 1.0 d, respectively. These values can be used to determine an average celerity of

$$U_c = \frac{(71{,}500 - 6500)\text{ m}}{(1.0 - 0.2)\text{ d}} = 81{,}250 \text{ m d}^{-1}\left(\frac{1\text{ d}}{86{,}400\text{ s}}\right) = 0.94 \text{ m s}^{-1}$$

which corresponds closely to the computed average. Also note that the water velocity ranges between about 0.46 and 0.6 m s^{-1} (Fig. 14.8). Thus the celerity is approximately 1.67 times larger than the water velocity.

The foregoing example shows that the wave and water velocities are different. Their relative magnitudes can be evaluated for the simple case of a wide, shallow rectangular channel. For such a channel, the hydraulic radius can be approximated by the mean depth

$$R = \frac{A_c}{P} = \frac{By}{2y + B} \cong \frac{By}{B} = y \tag{14.48}$$

Then the Manning equation can be used to represent the velocity and flow as

$$U = \frac{1}{n}y^{2/3}S^{1/2} \tag{14.49}$$

and

$$Q = \frac{1}{n}By^{5/3}S^{1/2} \tag{14.50}$$

Next Eq. 14.50 can be differentiated to give

$$\frac{\partial Q}{\partial y} = \frac{5}{3}\left(\frac{1}{n}By^{2/3}S^{1/2}\right) \tag{14.51}$$

Recognize that for a rectangular channel,

$$\frac{\partial Q}{\partial A_c} = \frac{1}{B}\frac{\partial Q}{\partial y} \tag{14.52}$$

Thus the celerity is equal to

$$U_c = \tfrac{5}{3}U \tag{14.53}$$

In other words the wave velocity propagates approximately 1.6 times faster than the water velocity. This result confirms the calculations in Example 14.7. Although this result was derived for a simple case, the fact that the kinematic wave celerity is faster than the water velocity holds for other geometries.

14.6.2 Routing Pollutants

Now we will investigate how the varying flows can be used to model water quality. A mass balance for a conservative substance with time-varying flow can be written as

$$\frac{\partial(A_c c)}{\partial t} = -\frac{\partial(Qc)}{\partial x} \tag{14.54}$$

A forward-time/backward-space scheme can be used to approximate the derivatives

$$\frac{(A_c c)_i^{\ell+1} - (A_c c)_i^{\ell}}{\Delta t} = -\frac{(Qc)_i^{\ell} - (Qc)_{i-1}^{\ell}}{\Delta x} \tag{14.55}$$

To simplify we can multiply both sides by Δx to give

$$\frac{(Vc)_i^{\ell+1} - (Vc)_i^{\ell}}{\Delta t} = (Qc)_{i-1}^{\ell} - (Qc)_i^{\ell} \tag{14.56}$$

Now it can be recognized that the term $(Vc)_i^{\ell+1}$ can be evaluated as

$$(Vc)_i^{\ell+1} = \left[V_i^{\ell} + \left(Q_{i-1}^{\ell} - Q_i^{\ell}\right)\Delta t\right] c_i^{\ell+1} \tag{14.57}$$

This result can then be substituted into Eq. 14.56. Because there is only one term at the advanced time point $\ell + 1$, the resulting equation can be solved explicitly for

$$c_i^{\ell+1} = \frac{V_i^{\ell} c_i^{\ell} + \left(Q_{i-1}^{\ell} c_{i-1}^{\ell} - Q_i^{\ell} c_i^{\ell}\right)\Delta t}{V_i^{\ell} + \left(Q_{i-1}^{\ell} - Q_i^{\ell}\right)\Delta t} \tag{14.58}$$

This equation has a simple interpretation. The numerator represents an Euler-method prediction of the mass in segment i at the next time step. The denominator is an Euler prediction of its volume. The ratio therefore provides an estimate of the new concentration.

EXAMPLE 14.8. WATER-QUALITY WITH THE KINEMATIC WAVE MODEL. Suppose that the system from Example 14.7 initially has a steady concentration of 100 mg L^{-1} of a conservative pollutant. If the sinusoidal flow carries no concentration, determine the transient concentration of the pollutant in the channel.

Solution: The results of applying Eq. 14.58 are displayed in Fig. 14.9. We have included the flow for comparative purposes. Notice that the concentration shows an inverse pattern to the flow. Thus a dilution "wave" moves through the river.

Although the shapes are similar, notice that the concentration wave moves slower than the flow wave. In fact the former moves at about 60% the velocity of the latter. This is due to the difference between the celerity of the wave and the velocity of the water. Consequently the evolution of the concentration changes proceeds at a slower rate than those that influence the hydrodynamics.

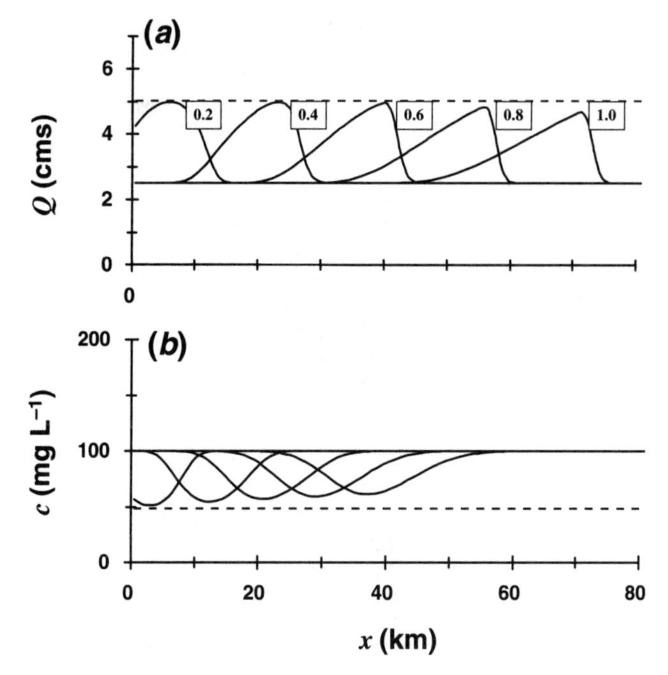

FIGURE 14.9
Simulation of (*a*) a hydrograph and (*b*) a dilution wave or "dilutograph" moving through a channel.

The foregoing example demonstrates that pollution and water waves travel at different velocities in one-dimensional streams. For the kinematic wave studied here, the pollutograph lags the hydrograph. A practical implication relates to water withdrawals during wave events. Ignorance of the lag might lead a water user to begin withdrawals after the passage of the crest of the water wave, only to find that the maximum pollution level would occur some time later.

It should be noted that the foregoing description of river hydraulics is a very simplified representation. It was intended to introduce you to how such computations are performed and some of the issues involved in their implementation. The expansion of the water-quality modeling paradigm to explicitly include hydrodynamics is currently an important area of research. You should consult other sources such as texts on hydrology (Ponce 1989, Chow et al. 1988, Bras 1990) or open-channel flow (Chaudry 1993) for additional information on routing.

PROBLEMS

14.1. The data in Table P14.1 were collected on Boulder Creek, Colorado, on December 23, 1994. The slope at the sampling location is approximately 0.004. Use the data along with appropriate formulas to determine (*a*) cross-sectional area, (*b*) mean depth,

TABLE P14.1

x (ft)	H (ft)	$U_{60\%}$ (fps)	$U_{20\%}$ (fps)	$U_{80\%}$ (fps)
0	0			
3	1.3	0.3		
6	1.25	0.35		
9	0.85	0.48		
12	1.2	0.46		
15	1.85	0.58		
18	1.85	0.66		
21	1.75	0.68		
24	1.9	0.67		
27	2.25		0.4	0.68
30	2.4		0.4	0.62
33	2.1		0.29	0.5
36	1.5	0.37		
39	1	0.25		
42	1.3	0.09		
45	1.2	0.13		
48	1	0.14		
51	0.8	0.08		
54	0	0		

(*c*) flow, (*d*) mean velocity, (*e*) longitudinal dispersion coefficient, and (*f*) distance needed to attain lateral mixing.

14.2. Suppose a stream channel has the following characteristics:

Bottom width = 10 m
Side slope = 1.5
Channel slope = 0.004
Roughness = 0.07

Develop log-log plots and power equations of depth and velocity versus flow based on the Manning equation. Use a range of flows from 1 to 20 cms.

14.3. The following data represent 7-d minimum flows in cms:

1977	0.023	1982	0.311	1987	0.091	1992	0.142
1978	0.105	1983	1.331	1988	0.127	1993	0.340
1979	0.024	1984	0.051	1989	0.089	1994	0.651
1980	0.187	1985	0.065	1990	0.453	1995	0.017
1981	0.010	1986	0.062	1991	0.849		

Determine the 7Q10 for this stream.

14.4. An industrial plant discharges at the side of a stream having the following characteristics: width = 30 m, depth = 0.9 m, velocity = 0.9 mps, and slope = 0.0002.

(*a*) Determine the length of stream needed for complete mixing.

(*b*) If Manning's equation holds and the channel is roughly rectangular, what is the stream's roughness?

(*c*) Determine the longitudinal dispersion coefficient.

14.5. The data in Table P14.5 were collected on Boulder Creek on December 29, 1994. Use the kinematic wave model to simulate this data. Assume the following values for the creek's hydrogeometric parameters: $B = 12$ m, $n = 0.04$, $S = 0.0042$, and $E = 7 \times 10^4$ cm^2 s^{-1}. Note that your simulation results for KP 5.5 might differ from the data.

TABLE P14.5

KP = 0.5			KP = 5.5					
Time (hr)	Q_1 **(cms)**	c_1	**Time (hr)**	Q_2 **(cms)**	c_2	**Time (hr)**	Q_2 **(cms)**	c_2
5.08	0.544	522	5.40	0.716	450	12.75	1.016	
6.88	0.479	597	7.00	0.544	500	13.00	0.994	692
7.87	0.479	639	7.50	0.532	500	13.25	0.929	703
8.00	0.500		7.75	0.524	508	13.50	0.917	710
8.42	0.544	665	8.00	0.507	514	13.75	0.906	709
8.95	0.642	693	8.25	0.490	518	14.00	0.906	707
9.42	0.726	708	8.50	0.481	522	14.25	0.883	700
10.00	0.819	696	8.75	0.470	529	14.50	0.872	694
10.57	0.923	672	9.00	0.453	536	14.75	0.861	684
10.70	0.942		9.25	0.453	543	15.00	0.841	678
11.15	0.999	626	9.50	0.470	550	15.25	0.818	674
11.67	0.923	657	9.75	0.481	557	15.50	0.818	678
12.12	0.887		10.00	0.507	566	15.75	0.818	685
12.37	0.853	706	10.25	0.524	575	16.00	0.830	699
12.75	0.836	712	10.50	0.544	583	16.25	0.818	711
13.43	0.819	717	10.75	0.617	593	16.50	0.810	720
14.00	0.787	723	11.00	0.696	598	16.75	0.810	727
14.67	0.787	723	11.25	0.716	608	17.00	0.798	733
14.80	0.787	723	11.50	0.779	618	17.25	0.798	736
15.23	0.772	723	11.75	0.818	621	17.50	0.798	738
16.17	0.756	722	12.00	0.872	628			
16.80	0.726	720	12.25	0.929	642			
17.23	0.726		12.50	0.982	656			

14.6. A rectangular channel has the following characteristics:

Flow = 20 cms — Channel slope = 0.0005
Width = 30 m — Roughness = 0.025

Determine the depth, cross-sectional area, and velocity for the channel.

14.7. Repeat Prob. 14.6, but for a trapezoidal channel with a bottom width = 30 m and a side slope = 1.5.

LECTURE 15

Estuaries

Lecture overview: After a brief description of estuary types, I focus on dispersion. In particular I show how salinity gradients can be used to determine dispersion coefficients, and compare estuary dispersion with turbulence in streams and lakes.

Situated at the interface between rivers and the sea, estuaries are the site of tremendous human development and biotic diversity. From a human perspective the role of estuaries as shipping ports made them the locales of some of the world's greatest cities. From a biotic perspective, estuaries receive large quantities of nutrients from their inflowing rivers. As a consequence the typical estuarine ecosystem teems with life.

This lecture is designed to provide you with background information on these systems. In particular we focus on aspects of estuarine transport that have a bearing on water-quality modeling.

15.1 ESTUARY TRANSPORT

An ***estuary*** is the region where a free-flowing river meets the ocean. As depicted in Fig. 15.1 it can be treated as a number of zones based on the interactions of advection, dispersion, and salinity. At one boundary is the river, which is devoid of sea salt and in which tidal motion is negligible. Next is the tidal river, where the effect of tides begins to affect transport, but the water is still relatively fresh. This is followed by the estuary itself. This region is marked by full tidal reversals. The water here is called ***brackish,*** meaning that it is a mixture of salt and fresh waters. Oftentimes the estuary will widen into a bay. The bay has a higher salt concentration due to

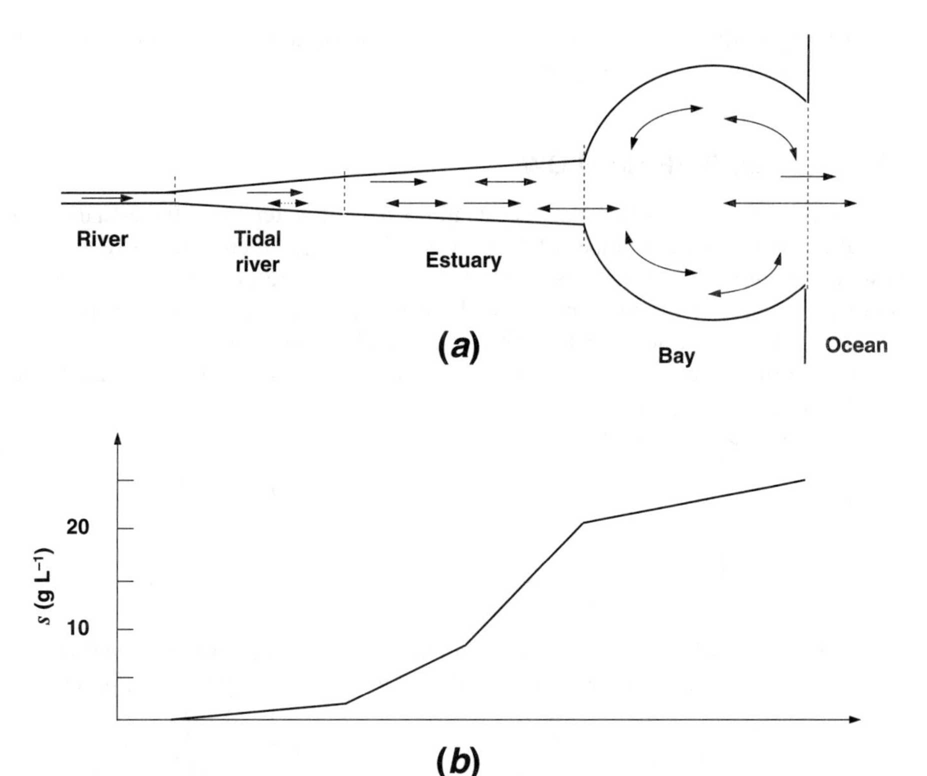

FIGURE 15.1
(*a*) A schematic of the various zones in an estuarine system. (*b*) Corresponding salinity concentration.

its proximity to the ocean. In addition it is distinguished from the estuary in that it is wider. Thus it must usually be modeled as a two-dimensional horizontal system. Also it can sometimes be deep enough that it must be treated as a three-dimensional body. In the bay, the freshwater flow is dwarfed by the action of tides and winds. Finally the estuarine system ends at the ocean.

There are several aspects of estuarine transport that have direct relevance to water-quality modeling. Foremost is the effect of tidal action. Due to the attraction of both the moon and the sun, tidal motion causes water to move in and out of the estuary in a periodic fashion. Thus the flow, velocity, and depth change dynamically.

Depending on the scale of the problem being addressed, the tidal motion can be perceived as being either advective or dispersive. For short-scale problems such as the discharge of highly reactive substances or spills, the motion would be perceived primarily as advection. On a longer time scale, however, the tides would move water back and forth in a cyclical fashion and the motion might be classified as dispersive.

In this lecture we limit ourselves primarily to the long-term perspective. Thus we focus on the steady-state condition averaged over a number of tidal cycles. The shorter scale or "real time" aspects of estuarine transport are beyond the scope of this text. The reader can consult other references for additional information on estuarine hydrodynamics (for example Dailey and Harleman 1972; Officer 1976, 1983; Fischer et al. 1979; Lung 1994; etc.).

15.2 NET ESTUARINE FLOW

Net estuarine flow refers to the advective movement of water out of the estuary over a tidal cycle or a given number of tidal cycles. For cases where the upstream river flow dwarfs other water sources, the net flow would correspond to the river flow. More commonly, where point inflows and diffuse inflows such as surface runoff and groundwater are significant, a water balance must be developed.

Where the advective flow is large, the tidal flows can sometimes be analyzed to determine the net flow. As depicted in Fig. 15.2 the tidal flow variations can be idealized as a pair of half-sinusoids,

$$Q = q_e \sin\left[\frac{\pi(t-\theta)}{T_e}\right] \qquad \theta \le t \le \theta + T_e \tag{15.1}$$

$$Q = q_f \sin\left[\pi + \frac{\pi(t-\theta-T_e)}{T_f}\right] \qquad \theta + T_e \le t \le \theta + T_e + T_f \tag{15.2}$$

where the subscripts e and f denote ebb and flood, respectively. An average flow can be developed by integrating Eqs. 15.1 and 15.2 and dividing by the tidal period,

$$Q_n = \frac{\displaystyle\int_{\theta}^{\theta+T_e} q_e \sin\left[\frac{\pi(t-\theta)}{T_e}\right] dt + \int_{\theta+T_e}^{\theta+T_e+T_f} q_f \sin\left[\frac{\pi(t-\theta-T_e)}{T_f}\right] dt}{T_e + T_f} \tag{15.3}$$

which can be evaluated to yield

$$Q_n = \frac{2}{\pi}\,\frac{q_e T_e - q_f T_f}{T_e + T_f} \tag{15.4}$$

Although Eq. 15.4 can be useful, it should be employed with caution for systems where the net flow is small. For such cases the numerator would consist of the difference between two very similar, large numbers. Such differences are notoriously sensitive to data error.

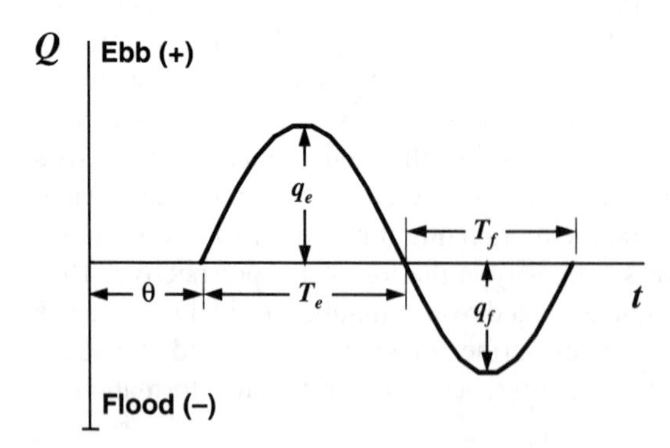

FIGURE 15.2
Simple sinusoidal approximation of the tidal flow can be used to estimate the net estuarine flow.

15.3 ESTUARY DISPERSION COEFFICIENT

The natural salt gradient can be employed to estimate the dispersion coefficient in an estuary. The following sections show how this can be done. In the first section the estuary is assumed to have constant hydrogeometric characteristics. Although this is not usually the case, the analysis provides insight into the method. This is followed by a more general, finite-difference approach.

15.3.1 Constant-Parameter Analysis

A constant-parameter estuary can be represented as in Fig. 15.3. The steady-state mass balance for this case is

$$0 = -U\frac{ds}{dx} + E\frac{d^2s}{dx^2} \tag{15.5}$$

Using the same approach as was used to solve Eq. 9.26, we obtain the solution for this case as

$$c = Fe^{\lambda_1 x} + Ge^{\lambda_2 x} \tag{15.6}$$

where the λ's are

$$\lambda_1 = \frac{U}{E} \tag{15.7}$$

$$\lambda_2 = 0 \tag{15.8}$$

The constants of integration can be evaluated by the boundary conditions,

$$s = s_0 \qquad @\ x = 0 \tag{15.9}$$

$$s = 0 \qquad @\ x = -\infty \tag{15.10}$$

Applying these boundary conditions leads to the solution

$$s = s_0 e^{\frac{U}{E}x} \qquad \text{for } x \le 0 \tag{15.11}$$

$$s = s_0 \qquad \text{for } x > 0 \tag{15.12}$$

Equation 15.11 provides a model for evaluating the dispersion coefficient. Taking its natural logarithm yields

$$\ln\frac{s}{s_0} = \frac{U}{E}x \tag{15.13}$$

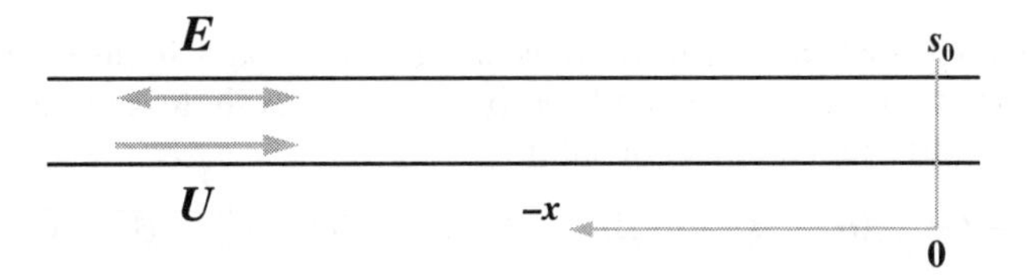

FIGURE 15.3
Representation of a constant dimension estuary as a mixed-flow system with an ocean boundary condition.

Therefore if Eq. 15.11 holds, a plot of the natural log of s/s_0 versus distance should yield a straight line with a slope of U/E. If we know U, we can use this slope to estimate E.

EXAMPLE 15.1. USING THE SALINITY GRADIENT TO ESTIMATE DISPERSION. A one-dimensional estuary has constant dimensions and flow: width = 300 m, depth = 3 m, and flow = 15 cms. You measure the following chloride concentrations along its length (note that kilometer points increase toward the ocean):

Kilometer point	0	3	6	9	12	15	18	21
Chloride (ppt)	0.3	0.5	0.8	1.4	2.2	3.7	6.0	10.0

Determine the dispersion coefficient for the estuary in $\text{cm}^2\ \text{s}^{-1}$.

Solution: The velocity for the system can be computed as

$$U = \frac{Q}{A_c} = \frac{15}{300(3)} = 0.016667\ \text{m s}^{-1}\left(\frac{86{,}400\ \text{s}}{\text{d}}\right) = 1440\ \text{m d}^{-1}$$

A semi-log plot of s/s_0 yields

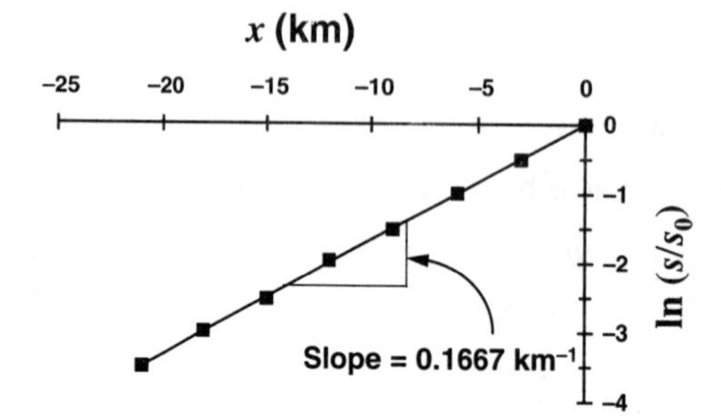

FIGURE E15.1

The slope can be converted to $1.667 \times 10^{-4}\ \text{m}^{-1}$ and then used to estimate the dispersion coefficient

$$E = \frac{U}{\text{slope}} = \frac{1440\ \text{m d}^{-1}}{1.667 \times 10^{-4}\ \text{m}^{-1}}\left(\frac{10^4\ \text{cm}^2}{\text{m}^2}\ \frac{1\ \text{d}}{86{,}400\ \text{s}}\right) = 1 \times 10^6\ \text{cm}^2\text{s}^{-1}$$

15.3.2 Finite-Difference Analysis

Although estuary reaches can have constant geometry and hydrology, it is more likely that they will vary. To extend the previous analysis to such cases, we can develop a finite-difference model of salinity,

$$0 = Q_{i-1,i}(\alpha_{i-1,i}s_{i-1} + \beta_{i-1,i}s_i) - Q_{i,i+1}(\alpha_{i,i+1}s_i + \beta_{i,i+1}s_{i+1}) + E'_{i-1,i}(s_{i-1} - s_i) + E'_{i,i+1}(s_{i+1} - s_i) \tag{15.14}$$

where

$$\alpha_{j,k} = \frac{\Delta x_k}{\Delta x_j + \Delta x_k} \tag{15.15}$$

and

$$\beta_{j,k} = 1 - \alpha_{j,k} = \frac{\Delta x_j}{\Delta x_j + \Delta x_k} \tag{15.16}$$

If salinities and flows are known, and this equation is written for an n-segment estuary, there will be n equations and $n + 1$ unknown dispersion coefficients. Consequently one of the coefficients must be specified. In many cases the most natural choice is to extend the analysis up the estuary to the point where dispersion becomes negligible. Thus the dispersion across the first interface is set to zero and the equations can be solved.

To simplify the analysis we assume that we are dealing with equal-sized segments; that is, $\alpha = \beta = 0.5$. The steady-state salinity mass balance for the first segment can be written as

$$0 = Q_{0,1}(0.5s_0 + 0.5s_1) - Q_{1,2}(0.5s_1 + 0.5s_2) + E'_{1,2}(s_2 - s_1) \tag{15.17}$$

which can be solved for

$$E'_{1,2} = \frac{Q_{0,1}(s_0 + s_1) - Q_{1,2}(s_1 + s_2)}{2(s_1 - s_2)} \tag{15.18}$$

Subsequent values can then be computed by solving Eq. 15.14 for

$$E'_{i,i+1} = \frac{Q_{i-1,1}(s_{i-1} + s_i) - Q_{i,i+1}(s_i + s_{i+1}) + 2E'_{i-1,i}(s_{i-1} - s_i)}{2(s_i - s_{i+1})} \tag{15.19}$$

EXAMPLE 15.2. USING THE SALINITY GRADIENT TO ESTIMATE DISPERSION. Thomann and Mueller (1987) present the following data from the Wicomico Estuary, a small tributary of Chesapeake Bay. Use this data to determine the dispersion coefficients for the system.

Distance (km)	Segment	Interface	Area (m²)	Flow (m³ s⁻¹)	Salinity (ppt)
1.61	Bay				11.3
0		7,bay	2601	2.83	10.1
−1.61	7				9.0
−3.22		6,7	2322	2.72	8.0
−4.83	6				7.0
−6.44		5,6	1672	2.66	6.1
−8.05	5				5.2
−9.66		4,5	1393	2.60	4.35
−11.26	4				3.5
−12.87		3,4	1208	2.55	2.75
−14.48	3				2.0
−16.09		2,3	929	2.41	1.5
−17.70	2				1.0
−19.31		1,2	836	2.35	0.6
−20.92	1				0.2
−22.53		0,1	650	2.29	0.2
−24.14	0				0.2

Solution: Equation 15.18 can be employed to compute the bulk dispersion coefficient for interface (1,2),

$$E'_{1,2} = \frac{2.29(0.2 + 0.2) - 2.35(0.2 + 1)}{2(0.2 - 1)} = 1.19 \text{ m}^3 \text{ s}^{-1}$$

which can be used to calculate a dispersion coefficient of

$$E_{1,2} = E'\frac{\Delta x}{A_c} = 1.19 \text{ m}^3 \text{ s}^{-1} \left(\frac{3.22 \text{ km}}{836 \text{ m}^2} \frac{1000 \text{ m}}{\text{km}} \frac{10{,}000 \text{ cm}^2}{\text{m}^2} \right) = 4.58 \times 10^4 \text{ cm}^2 \text{ s}^{-1}$$

Equation 15.19 can be employed to compute the bulk dispersion coefficient for interface (2,3),

$$E'_{2,3} = \frac{2.35(0.2 + 1) - 2.41(1 + 2) + 2(1.189)(0.2 - 1)}{2(1 - 2)} = 3.157 \text{ m}^3 \text{ s}^{-1}$$

which can be used to calculate a dispersion coefficient of

$$E_{2,3} = 3.157 \text{ m}^3 \text{ s}^{-1} \left(\frac{3.22 \text{ km}}{929 \text{ m}^2} \frac{1000 \text{ m}}{\text{km}} \frac{10{,}000 \text{ cm}^2}{\text{m}^2} \right) = 1.09 \times 10^5 \text{ cm}^2 \text{ s}^{-1}$$

The remainder of the values can be calculated and tabulated, as below. They can also be displayed graphically (Fig. 15.4).

Distance (km)	Interface	E' ($m^3 s^{-1}$)	E ($cm^2 s^{-1}$)
0	7,8	12.29	1.52×10^5
3.2	6,7	10.65	1.48×10^5
6.4	5,6	8.76	1.69×10^5
9.7	4,5	6.38	1.48×10^5
12.9	3,4	4.37	1.16×10^5
16.1	2,3	3.16	1.09×10^5
19.3	1,2	1.19	0.46×10^5

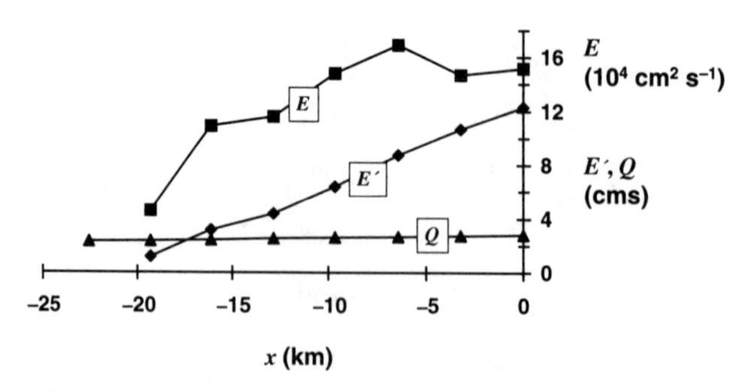

FIGURE 15.4
Plot of flow, bulk dispersion, and dispersion versus distance as calculated in Example 15.2.

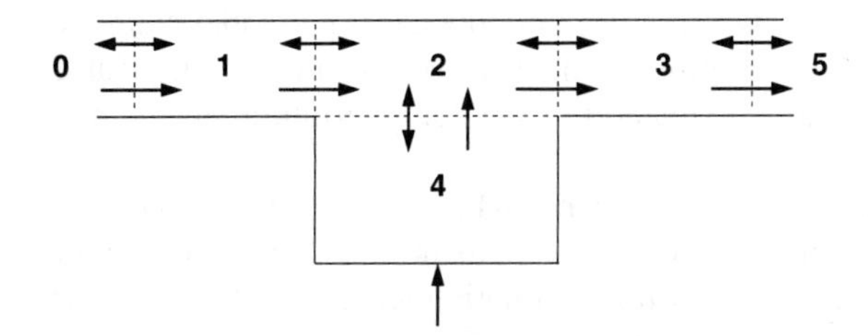

FIGURE 15.5
An overview of a simple two-dimensional estuary.

The foregoing analysis can be extended to two- and three-dimensional systems using matrix techniques. An example of a simple two-dimensional system is shown in Fig. 15.5. If dispersion across the first interface is zero, then mass balances of a conservative substance can be written as

$$0 = Q_{0,1}(\alpha_{0,1}s_0 + \beta_{0,1}s_1) - Q_{1,2}(\alpha_{1,2}s_1 + \beta_{1,2}s_2) + E'_{1,2}(s_2 - s_1) \tag{15.20}$$

$$0 = Q_{1,2}(\alpha_{1,2}s_1 + \beta_{1,2}s_2) - Q_{2,3}(\alpha_{2,3}s_2 + \beta_{2,3}s_3) + Q_{2,4}(\alpha_{2,4}s_4 + \beta_{2,4}s_2) + E'_{1,2}(s_1 - s_2) + E'_{2,3}(s_3 - s_2) + E'_{2,4}(s_4 - s_2) \tag{15.21}$$

$$0 = Q_{2,3}(\alpha_{2,3}s_2 + \beta_{2,3}s_3) - Q_{3,5}(\alpha_{3,5}s_3 + \beta_{3,5}s_5) + E'_{2,3}(s_2 - s_3) + E'_{3,5}(s_5 - s_3) \tag{15.22}$$

$$0 = W_4 + E'_{2,4}(s_2 - s_4) - Q_{2,4}(\alpha_{2,4}s_4 + \beta_{2,4}s_2) \tag{15.23}$$

These are four equations with four unknowns. However, in contrast to previous simultaneous equations (e.g., Lec. 6), the unknowns are not concentrations but bulk dispersion coefficients. Collecting terms gives

$$[\Delta S]\{E'\} = \{Qs\} \tag{15.24}$$

where $[\Delta S]$ = a matrix of concentration differences,

$$\Delta S = \begin{bmatrix} s_2 - s_1 & 0 & 0 & 0 \\ -(s_2 - s_1) & s_3 - s_2 & 0 & -(s_2 - s_4) \\ 0 & -(s_3 - s_2) & s_5 - s_3 & 0 \\ 0 & 0 & 0 & s_2 - s_4 \end{bmatrix} \tag{15.25}$$

and $\{Qs\}$ = a vector of advection terms and sources,

$$\{Q_s\} = \begin{Bmatrix} -Q_{0,1}(\alpha_{0,1}s_0 + \beta_{0,1}s_1) + Q_{1,2}(\alpha_{1,2}s_1 + \beta_{1,2}s_2) \\ -Q_{1,2}(\alpha_{1,2}s_1 + \beta_{1,2}s_2) + Q_{2,3}(\alpha_{2,3}s_2 + \beta_{2,3}s_3) - Q_{2,4}(\alpha_{2,4}s_2 + \beta_{2,4}s_4) \\ -Q_{2,3}(\alpha_{2,3}s_2 + \beta_{2,3}s_3) + Q_{3,5}(\alpha_{3,5}s_3 + \beta_{3,5}s_5) \\ -W_4 + Q_{2,4}(\alpha_{2,4}s_2 + \beta_{2,4}s_4) \end{Bmatrix} \tag{15.26}$$

and $\{E'\}$ = a vector of unknown dispersion coefficients,

$$\{E'\} = \begin{Bmatrix} E'_{1,2} \\ E'_{2,3} \\ E'_{3,5} \\ E'_{2,4} \end{Bmatrix} \tag{15.27}$$

Because both Eqs. 15.20 and 15.23 have only one unknown, this particular example can be solved simply with algebraic manipulation. However, for more complicated cases, computer methods provide a more general means for determining the dispersion coefficients.

Notice that segment 4 has a source term, which can be either a tributary (Qs) or a waste loading (W). In essence all upstream terminal segments must have such source terms for the successful implementation of this estimation technique. For example if segment 4 had no source term its mass balance would be

$$0 = E'_{2,4}(s_2 - s_4) \tag{15.28}$$

which specifies that there is no gradient; that is, $s_2 = s_4$. If such were the case, segments 2 and 4 should be consolidated into a single well-mixed cell.

The foregoing approach should be applied with caution because, as with all data-based approaches, its results are dependent on the quality of the underlying data. Further, because the calculation is based on gradients (that is, differences), the results are particularly sensitive to data errors (recall Fig. 2.6). However, if these reservations are considered, it can sometimes provide order-of-magnitude estimates of dispersion. Several problems at the end of the lecture are devoted to the approach.

15.3.3 Slack-Tide Sampling

The presence of tides frequently makes estuarine sampling a case of "hitting a moving target." The issue is especially relevant to the steady-state dispersion approach used in the previous sections. For the steady-state approach, we are trying to define an average condition for a system that is continuously changing.

Slack-tide sampling provides one way to address this dilemma. The term ***slack tide*** refers to the point at which the tides reverse direction, that is, when the flow goes to zero. In some systems this state moves progressively up or down the estuary. Thus if the sampling team has a fast enough boat (approximately 40 km hr^{-1}), they can collect samples at the slack tide.

Because there is both a high-water (HWS) and a low-water (LWS) slack tide, such surveys would yield two "snapshots" of water quality: one biased up the estuary (HWS) and the other down (LWS). Therefore an average would have to be developed (Fig. 15.6) to be consistent with the long-term approach described in the previous section.

15.3.4 Comparison with Streams, Lakes, and the Ocean

Recall that in Lec. 8, I presented a diffusion diagram (Fig. 8.11) that related the diffusion coefficient in lakes and the open ocean to length scales. At this point it would be interesting to relate these values to the dispersion coefficients in streams and estuaries.

Figure 15.7 reproduces Fig. 8.11 but adds dispersion coefficients from a number of streams and estuaries. The channel width is taken as the length scale for these waters.

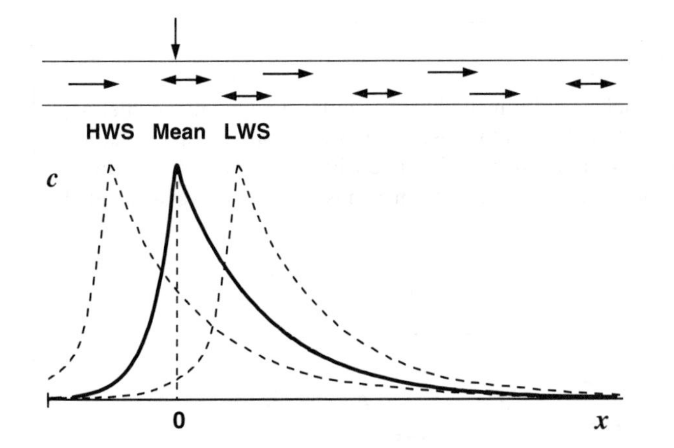

FIGURE 15.6
Samples around a point discharge of a substance that decays with first-order kinetics. Both high- and low-water slacks are shown along with a mean concentration profile.

The plot indicates that, given the same length scale, mixing is more intense in streams and estuaries than in lakes and the ocean. This should not be surprising since streams and estuaries are subject to the same wind forces that are the primary cause of horizontal turbulence in lakes and the ocean. However, additional forces supplement the wind and lead to higher total mixing. In streams, gravity-induced flow leads to shear that causes dispersion. In estuaries the tidal motion creates the same sort of enhancement.

Because dispersion and diffusion are different processes, we have mixed "apples and oranges" in developing Fig. 15.7. However, it is useful in contrasting the magnitudes of horizontal mixing for the three types of waters.

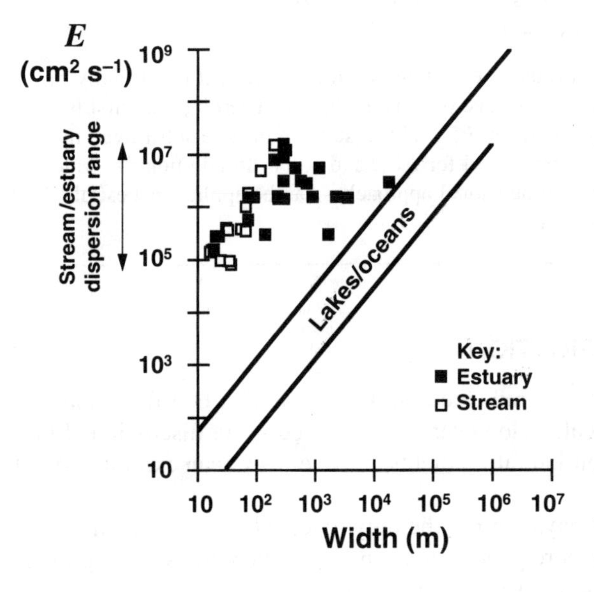

FIGURE 15.7
Okubo's diffusion diagram (recall Fig. 8.11) along with dispersion values from a number of streams and estuaries.

BOX 15.1. Freshwater Bays

Freshwater systems such as lakes and impoundments can have embayments that act like estuaries. We have already addressed the case where such systems can be characterized as a single well-mixed system when we discussed diffusion in Lec. 8. However, as depicted in Fig. B15.1, some are elongated and hence must be treated as a distributed system.

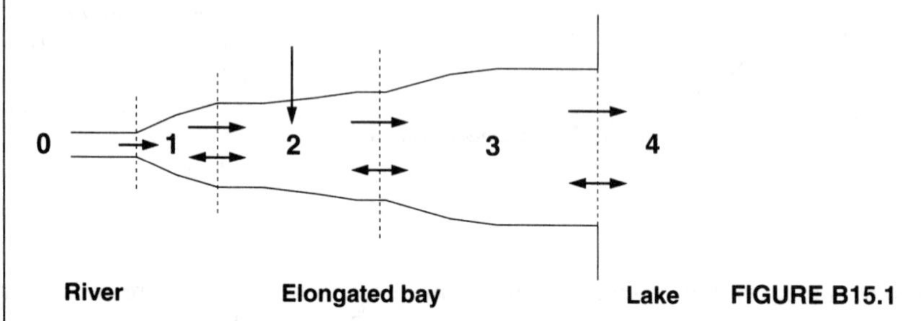

FIGURE B15.1

Because such systems are often fed by polluted rivers, gradients of conservative substances often occur. As with saltwater systems, such gradients can be used to estimate turbulent diffusion coefficients. However, in contrast to estuaries where the concentration increases toward the sea, a polluted bay's concentration usually decreases toward the main body of the lake.

Mass balances of a conservative substance can be written as

$$0 = Q_{0,1}s_0 - Q_{1,2}(\alpha_{1,2}s_1 + \beta_{1,2}s_2) + E'_{1,2}(s_2 - s_1) \tag{15.29}$$

$$\begin{aligned} 0 = {} & W_2 + Q_{1,2}(\alpha_{1,2}s_1 + \beta_{1,2}s_2) - Q_{2,3}(\alpha_{2,3}s_2 + \beta_{2,3}s_3) \\ & + E'_{1,2}(s_1 - s_2) + E'_{2,3}(s_3 - s_2) \end{aligned} \tag{15.30}$$

$$\begin{aligned} 0 = {} & Q_{2,3}(\alpha_{2,3}s_2 + \beta_{2,3}s_3) - Q_{3,4}(\alpha_{3,4}s_3 + \beta_{3,4}s_4) \\ & + E'_{2,3}(s_2 - s_3) + E'_{3,4}(s_4 - s_3) \end{aligned} \tag{15.31}$$

Now it can be seen that because the system is fed by a river, it can usually be assumed that diffusion across the interface (0,1) is negligible. Therefore the first balance can be solved for its only unknown, $E'_{1,2}$. This result can then be substituted into the second equation, which can be solved for $E'_{2,3}$, and so on. In addition for wider and deeper bays, two- and three-dimensional approaches can be applied as described previously at the end of Sec. 15.3.2.

15.4 VERTICAL STRATIFICATION

The bulk of this lecture has been predicated on the assumption that the estuary is well-mixed laterally and vertically. However, as mentioned in our discussion of Fig. 15.1, wide estuaries can exhibit lateral gradients and deep bays can exhibit vertical stratification.

It must also be noted that bays are not the only place where vertical stratification takes place. In addition a more generally occurring vertical transport regime is established due to the interaction of salinity and heat.

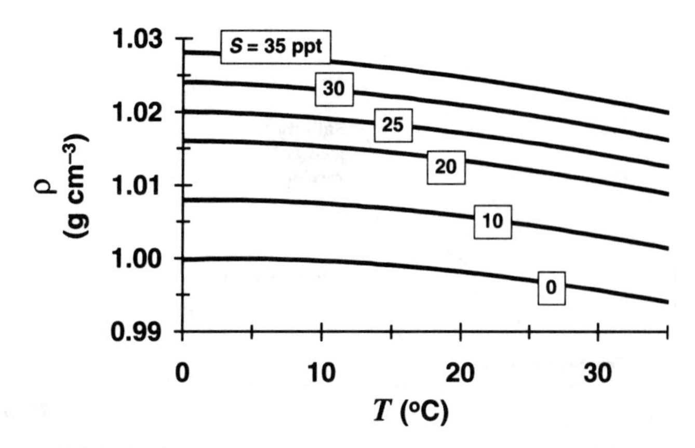

FIGURE 15.8
Density as a function of temperature and salinity.

To understand this regime, the following equations of state quantify the relationship of density to salinity and temperature (Millero and Poisson 1981):

$$\rho = \rho_o + AS + BS^{3/2} + CS^2 \tag{15.32}$$

where ρ = density (g L^{-1}), S = salinity (ppt), and

$$A = 8.24493 \times 10^{-1} - 4.0899 \times 10^{-3}T + 7.6438 \times 10^{-5}T^2 - 8.2467 \times 10^{-7}T^3 + 5.3875 \times 10^{-9}T^4 \tag{15.33}$$

$$B = -5.72466 \times 10^{-3} + 1.0227 \times 10^{-4}T - 1.6546 \times 10^{-6}T^2 \tag{15.34}$$

$$C = 4.8314 \times 10^{-4} \tag{15.35}$$

in which T = temperature (°C) and ρ_o = density of fresh water,

$$\rho_o = 999.842594 + 6.793952 \times 10^{-2}T - 9.095290 \times 10^{-3}T^2 + 1.001685 \times 10^{-4}T^3 - 1.120083 \times 10^{-6}T^4 + 6.536332 \times 10^{-9}T^5 \tag{15.36}$$

As displayed in Fig. 15.8, density increases with salinity and decreases with temperature. Note that a peak density occurs at 4°C for the freshwater case.

Because of its high salinity, sea water is denser than fresh water. Further, in many cases, ocean water is often denser because it is frequently colder than river

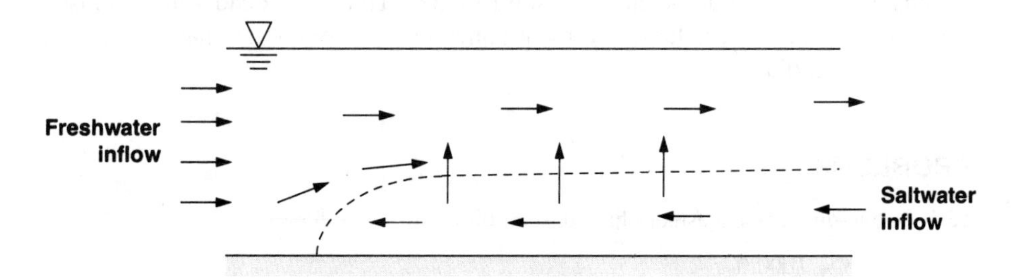

FIGURE 15.9
Two-dimensional vertical transport in a chemically stratifi ed estuary.

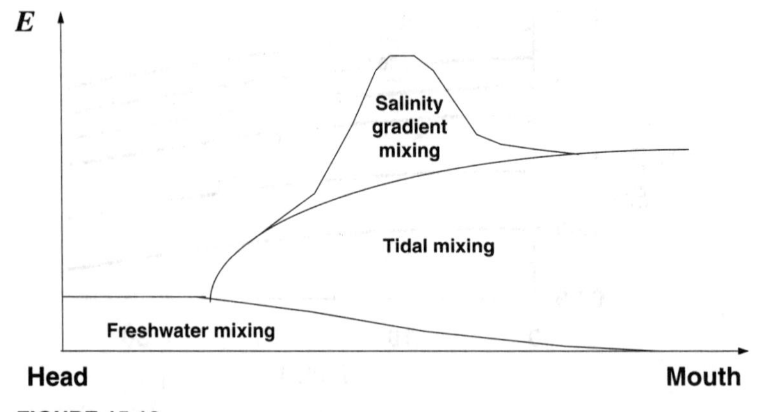

FIGURE 15.10
Factors contributing to the tidally averaged dispersion coefficient in a chemically stratified estuary.

water. As a consequence, as shown in Fig. 15.9, denser sea water can form a salt wedge at the bottom of the estuary. Less dense freshwater river water flows over the top of the wedge. This seaward transport is supplemented by salt water moving upward. Thus the surface layer is brackish.

The net result is that longitudinal mixing is enhanced in the region of the wedge. As depicted in Fig. 15.10, estuarine longitudinal dispersion is made up of three components. In the freshwater region the system is dominated by riverine dispersion due to the shear flow. In the estuarine section this mechanism is enhanced by the tidal oscillations. And in the area of the wedge the effect of salinity leads to a region of maximum mixing.

Aside from dispersion, estuarine hydraulics have additional effects on water quality through the impact of flow and mixing on suspended solids. This is due to the fact that there is a point in the estuary called the ***dead zone*** where the downstream river flow and the tidal motion tend to cancel. This convergence zone is the site of enhanced sedimentation.

The solids themselves are carried into the estuary with the freshwater flow and are also generated by plant growth. These particles settle to the bottom saline layer where they are transported back upstream to the dead zone. This concentration of particles is important in most water-quality contexts. For example many toxic compounds preferentially associate with solid matter. Thus the "dead zone" is also a "deposition zone" where large sediment concentrations of contaminants and nutrients can accumulate.

PROBLEMS

15.1. A one-dimensional estuary has constant dimensions and flow:

Width = 1000 ft
Depth = 10 ft
Flow = 500 cfs

You measure the following chloride concentrations along its length (note that mile points increase toward the ocean):

Mile point	0	2	4	6	8	10	12	14
Chloride (ppt)	0.3	0.5	0.8	1.4	2.2	3.7	6.0	10.0

Determine the dispersion coefficient for the estuary in $cm^2\ s^{-1}$.

15.2. A tidal estuary has the following characteristics:

Depth = 5 m
Cross-sectional area = 100 m^2
Flow = 0.3 $m^3\ s^{-1}$

x (miles from mouth)	−30	−24	−18	−12	−6	0
Chloride (ppt)	0.4	0.7	1.4	2.7	5.2	10

A pollutant that settles at a rate of 0.2 m d^{-1} is to be discharged at kilometer point 18.
(*a*) What is the dispersion coefficient?
(*b*) What is the estuary number?
(*c*) What loading can be discharged if the maximum allowable concentration is 1 ppm?

15.3. Determine the net estuarine flow for the following case:

Start of ebb = 0700	End of ebb = 2000	Peak ebb flow = 2.5 cms
Start of flood = 2000	End of flood = 3100	Peak flood flow = 2.1 cms

15.4. Thomann and Mueller (1987) report the following salinity data for the Back River, an estuary of Chesapeake Bay:

Distance from mouth (km)	0	5.5	8.6
Salinity (ppt)	3.9	2.5	0.32

If the net nontidal velocity is 3 cm s^{-1}, determine the dispersion coefficient in $cm^2\ s^{-1}$.

15.5. Suppose that the following data are available for a vertically stratified estuary (Fig. P15.5):

$Q_{01} = 4 \times 10^6\ m^3/s$ $\quad Q_{32} = 2.5 \times 10^6\ m^3/s$ $\quad E'_{01} = 0$ $\quad s_1 = 8.5$ ppt

$s_2 = 14.5$ ppt $\quad s_3 = 15$ ppt $\quad s_0 = 0$ ppt

(a) Determine the bulk dispersion coefficients, E'_{12}, E'_{13}, and E'_{23}. Assume that the latter two are equal and use backward differences for the advective terms.
(b) Suppose that the inflowing river has a concentration of 5 mg L^{-1} of a conservative pollutant. Assuming that segment 3 contains negligible concentration of the substance, determine the concentration in segments 1 and 2.

15.6. The following data are available for the system depicted in Fig. P15.5:

$Q_{01} = 1 \times 10^6\ m^3\ d^{-1}$ $\quad s_0 = 0$ ppt $\quad s_5 = 30$ ppt

$Q_{12} = 1 \times 10^6\ m^3\ d^{-1}$ $\quad s_1 = 1.74$ ppt $\quad W_4 = 0$

$Q_{23} = 1.5 \times 10^6 \text{ m}^3 \text{ d}^{-1}$ $\quad s_2 = 10.47$ ppt

$Q_{35} = 1.5 \times 10^6 \text{ m}^3 \text{ d}^{-1}$ $\quad s_3 = 22.33$ ppt

$Q_{24} = 0.5 \times 10^6 \text{ m}^3 \text{ d}^{-1}$ $\quad s_4 = 7.48$ ppt

Determine the bulk dispersion coefficients for the interfaces in this system.

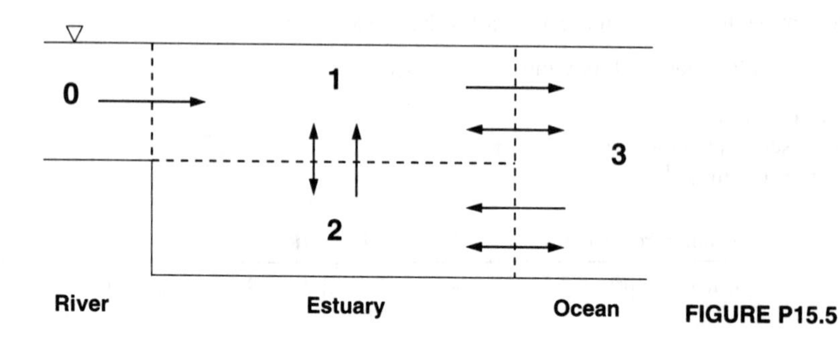

FIGURE P15.5

15.7. A vertically stratified water-supply reservoir has a small embayment (Fig. P15.7). The water utility typically fluoridates the water at a constant level with the result that all water in the lake has a constant fluoride concentration of 1 mg L^{-1}. To study the lake's mixing characteristics, you talk the utility into abruptly stopping fluoridation. In the ensuing weeks you measure the following concentrations as the lake purges itself of fluoride:

Times, d	0	20	40	60	80	100	120	140	160
Epilimnion	1	0.49	0.26	0.15	0.10	0.08	0.05	0.04	0.04
Hypolimnion	1	0.98	0.92	0.85	0.78	0.73	0.65	0.60	0.54
Bay	1	0.93	0.77	0.61	0.47	0.38	0.27	0.20	0.16

The parts of the lake have the following volumes:

$V_e = 3 \times 10^7 \text{ m}^3$

$V_h = 8 \times 10^6 \text{ m}^3$

$V_b = 3 \times 10^6 \text{ m}^3$

The inflow = outflow = $1 \times 10^6 \text{ m}^3 \text{ d}^{-1}$, all of which flows through the epilimnion. Use this information to determine the bulk diffusion coefficients across the thermocline and between the bay and the epilimnion.

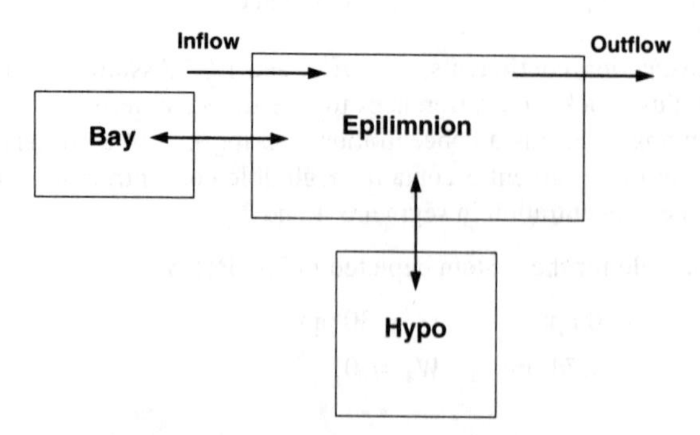

FIGURE P15.7

15.8. Two tributaries enter an elongated impoundment along its length (Fig. P15.8). Flow and chloride concentration for these tributaries and the inflow river are

	Flow ($m^3\ yr^{-1}$)	Chloride concentration ($mg\ L^{-1}$)	Phosphorus concentration ($\mu g\ L^{-1}$)
Inflow	10×10^6	150	30
Tributary A	7.5×10^6	400	100
Tributary B	5×10^6	100	50

The resulting steady-state chloride concentrations are shown at the bottom of the plot.

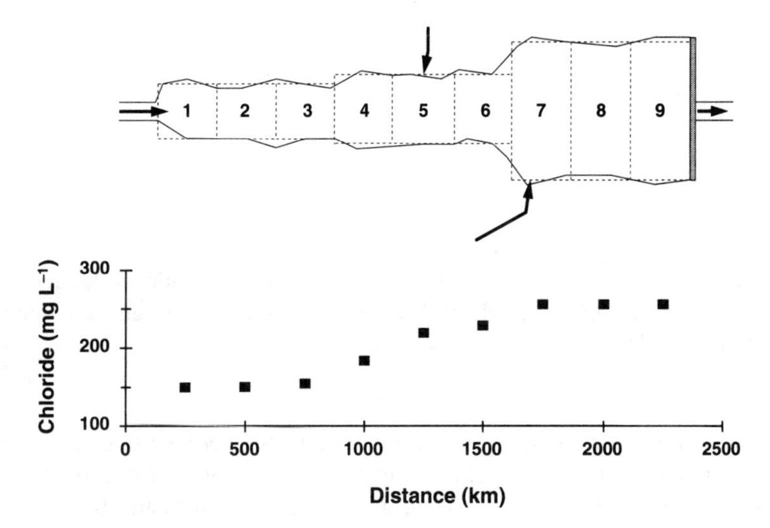

FIGURE P15.8

The following segment and interface data are available:

Segment	Length (m)	Width (m)	Depth (m)	Chloride ($mg\ L^{-1}$)	Interface	Area (m^2)
1	500	250	1	150	1–2	250
2	500	250	1	151	2–3	250
3	500	250	1	157	3–4	250
4	500	300	2	200	4–5	600
5	500	300	2	253	5–6	600
6	500	300	2	245	6–7	600
7	500	1000	5	222	7–8	5000
8	500	1000	5	222	8–9	5000
9	500	1000	5	222		

(*a*) Employ this data to estimate the longitudinal turbulent diffusion coefficients for the interfaces in this system.

(*b*) Determine the total phosphorus concentration profile if TP settles at a rate of $10\ m\ yr^{-1}$.

LECTURE 16

Lakes and Impoundments

LECTURE OVERVIEW: After a brief overview of lake types, I address the determination of lake morphometry. Then I show how its water budget is modeled, and review models of the near-shore zone.

In contrast to flowing waters, lakes and impoundments were not emphasized in the early years of water-quality modeling. This is because, with the exception of large navigable systems like the Great Lakes, they have not historically been the major focus of urban development.

Starting in the 1970s, however, it was recognized that natural and man-made lakes are equally, if not more, important than estuaries and rivers from a recreational standpoint. Further their use for water supply, hydropower, and flood control also contribute to their significance.

This lecture is designed to provide you with background information on these systems. After a brief overview of lake types, I describe how lake morphometry and hydrology can be quantified. Then I describe how pollutants advect and diffuse in the near-shore zone.

16.1 STANDING WATERS

This lecture focuses on standing waters. Such water bodies range from small detention ponds to huge systems like the Great Lakes and Lake Baikal. Lake scientists (or ***limnologists***) classify lakes in many ways (see Hutchinson 1957 and Wetzel 1975 for reviews). From the standpoint of water-quality modeling there are three major features of lakes that relate to our efforts to simulate their transport and fate:

Origin. By this I mean whether the water body is natural (a lake) or artificial (an impoundment or a reservoir). Although there is a tremendous amount of variability within these two categories, there are some generalizations that typify each in a general sense. In particular impoundments often have controlled outflow. In contrast natural lakes are uncontrolled. In addition, as described next, they usually have different shapes.

Shape. Artificial impoundments are almost always created by damming a river. Consequently they tend to be elongated or dendritic since they consist of drowned river valleys. In contrast natural lakes tend to be less elongated and more circular. Of course there are also some dendritic and elongated natural lakes and circular impoundments, but the reverse situation is more likely.

Size. The two aspects of size that have a strong bearing on water quality are residence time and depth. In general, lakes are divided into short ($\tau_w < 1$ yr) and long ($\tau_w > 1$ yr) residence time systems. Further they are classified as shallow ($H < 7$ m) and deep ($H > 7$ m). The latter classification is significant because deep lakes are often subject to thermal stratification during certain times of the year.

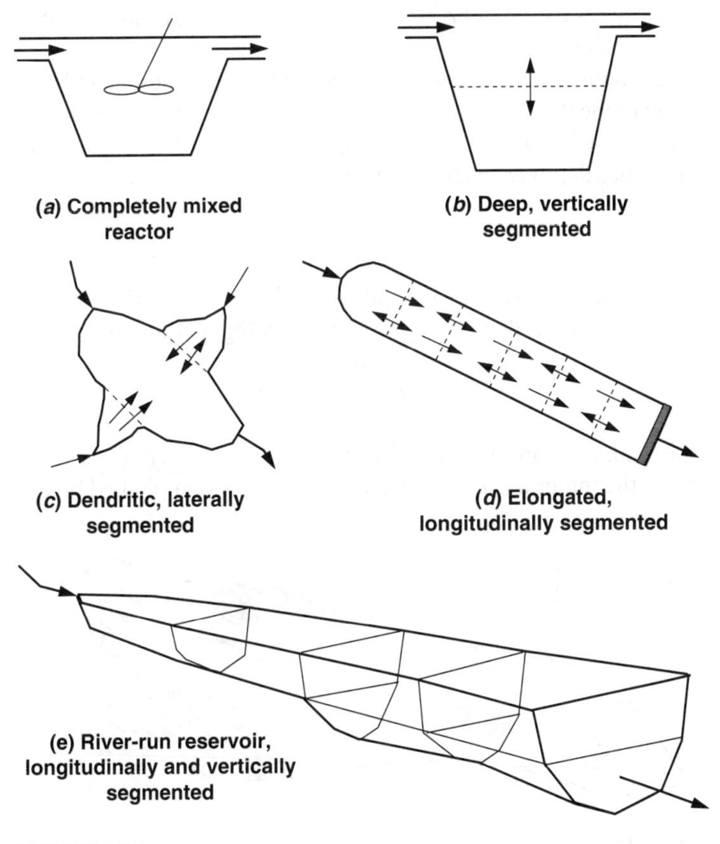

FIGURE 16.1
Typical segmentation schemes used for lakes and impoundments.

As we will see throughout the remainder of this text, these attributes have an impact on how lakes and impoundments are modeled. For the time being the primary implication relates to model segmentation. As depicted in Fig. 16.1*a*, CSTRs are often used to model standing waters. However, such systems are also segmented vertically (Fig. 16.1*b*), laterally (Fig. 16.1*c*), and longitudinally (Fig. 16.1*d*). Further, as in the deep river-run reservoir in Fig. 16.1*e*, the system can be divided in several dimensions. In Lec. 18 we revisit the topic of model segmentation in a more general sense.

16.2 LAKE MORPHOMETRY

The first step in characterizing any lake or impoundment is to determine its geometry or, as it is formally called, ***morphometry***. To do this the lake's bathymetry must be mapped. ***Bathymetry*** refers to a topographic map showing depth contour lines (Fig. 16.2).

The analyst can then determine how much area is encompassed by each depth contour. This can be done using a mechanical device called a ***planimeter***. If a planimeter is unavailable, an alternative is to superimpose a grid over the bathymetric map. The areas can then be estimated by summing the areas of the grid cells falling within each contour.

As depicted in Fig. 16.3 the resulting areas and corresponding depths represent a tabular function of the dependence of area on depth, $A(z)$. The volume for the system can be determined by integration. For example the volume from the surface ($z = 0$) down to a particular depth ($z = H$) can be calculated as

$$V(H) = \int_0^H A(z)\,dz \tag{16.1}$$

In addition the volume between two depths can be evaluated. For example the volume between two adjacent depths can be represented generally as

$$V_{i,i+1} = \int_{H_i}^{H_{i+1}} A(z)\,dz \tag{16.2}$$

Clearly, because we are dealing with tabular data, numerical methods must be used to evaluate the integrals. This is most commonly accomplished by applying the

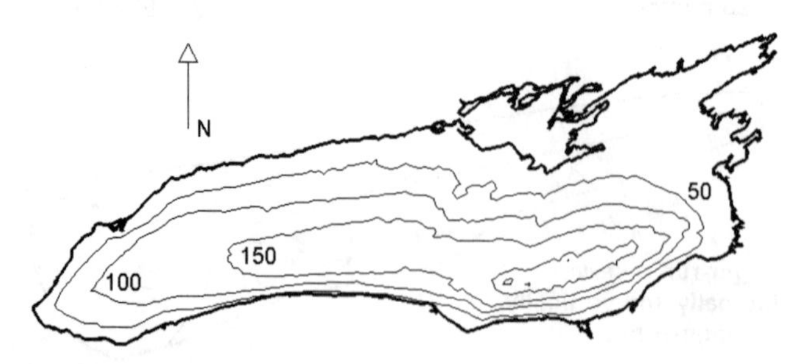

FIGURE 16.2
A bathymetric map of Lake Ontario (courtesy of Mike McCormick, GLERL/NOAA).

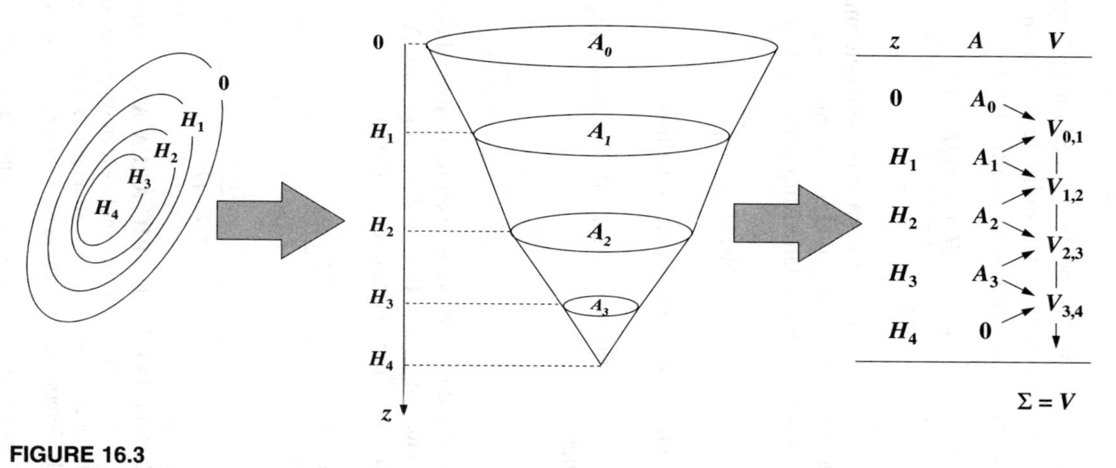

FIGURE 16.3
The process of calculating lake and reservoir morphometry consists of determining areas from a bathymetric map. These areas are then tabulated and used to determine volumes by numerical integration.

trapezoidal rule to Eq. 16.2,

$$V_{i,i+1} = \left[\frac{A(H_i) + A(H_{i+1})}{2}\right](H_{i+1} - H_i) \tag{16.3}$$

In this way the volume can be determined for each layer starting at the surface and working down to the bottom. The individual layer volumes can be accumulated to determine the total volume down to a particular depth,

$$V_{i+1} = \sum_{j=0}^{i}\left[\frac{A(H_j) + A(H_{j+1})}{2}\right](H_{j+1} - H_j) \tag{16.4}$$

where $i + 1 =$ depth at which the volume is to be determined.

Not only can volume be computed from areas, but area can be determined from volumes. Based on the inverse relationship between differentiation and integration,

$$A(H) = \frac{dV(H)}{dz} \tag{16.5}$$

Thus the rate of change of the cumulative volume at a point provides an estimate of the area.

Again numerical methods are required. A common approach is to use a centered divided difference,

$$A_i = \frac{dV_i}{dz} \cong \frac{V_{i+1} - V_{i-1}}{z_{i+1} - z_{i-1}} \tag{16.6}$$

However, this approach is flawed for unequally spaced depths. In addition it cannot be used to evaluate the area of the top-most segment. A formula that takes care of both shortcomings is (Chapra and Canale 1988)

$$A(z) = \frac{dV(z)}{dz} \cong V_{i-1}\frac{2z - z_i - z_{i+1}}{(z_{i-1} - z_i)(z_{i-1} - z_{i+1})} + V_i\frac{2z - z_{i-1} - z_{i+1}}{(z_i - z_{i-1})(z_i - z_{i+1})}$$
$$+ V_{i+1}\frac{2z - z_{i-1} - z_i}{(z_{i+1} - z_{i-1})(z_{i+1} - z_i)} \tag{16.7}$$

EXAMPLE 16.1. BATHYMETRIC AREA AND VOLUME CALCULATIONS. Bathymetric maps of the Tolt Reservoir (a water-supply impoundment for Seattle, Washington) can be used to determine the areas and depths in the second and third columns of Table 16.1. Determine the volume if the reservoir level is at $z = 0$.

Solution: Equation 16.3 can be used to compute the volume of the top segment,

$$V_{0,1} = \left(\frac{5.180 \times 10^6 + 4.573 \times 10^6}{2}\right)(6.10 - 0.00) = 29.73 \times 10^6\ \text{m}^3$$

Similar calculations can be performed for the other segments (Table 16.1) and the results substituted into Eq. 16.4 to compute the total volume,

$$V_{11} = \sum 29.73 \times 10^6 + 13.44 \times 10^6 + \cdots + 0.493 \times 10^6 = 120.3 \times 10^8\ \text{m}^3$$

The cumulative volume is listed in the last column of Table 16.1.

TABLE 16.1
Summary of bathymetric data and morphometric calculations for the Tolt Reservoir, Washington

Index	Depth (m)	Area (10^6m^2)	Volume (10^6m^3)	Cumulative volume (10^6m^3)
0	0	5.180	29.73	0
1	6.10	4.573	13.44	29.73
2	9.14	4.249	12.64	43.17
3	12.19	4.047	11.78	55.81
4	15.24	3.683	10.67	67.59
5	18.29	3.318	16.90	78.26
6	24.38	2.226	11.22	95.16
7	30.48	1.457	7.401	106.4
8	36.57	0.971	4.255	113.8
9	42.67	0.425	1.789	118.0
10	48.77	0.162	0.493	119.8
11	54.86	0	0	120.3

Now Eq. 16.6 can be used to calculate the area at interface 1,

$$A_1 = \frac{dV_1}{dz} \cong \frac{V_2 - V_0}{z_2 - z_0} = -\frac{(43.17 \times 10^6 - 0)\ \text{m}^3}{(9.14 - 0.00)\ \text{m}} = 4.72 \times 10^6\ \text{m}^2$$

Observe the discrepancy between this result and the true area (4.573×10^6) in Table 16.1. The disparity is due in part to the inadequacy of Eq. 16.6. The alternative formulation (Eq. 16.7) provides a better estimate,

$$\frac{dV(6.10)}{dz} \cong 0 + 29.73 \times 10^6 \frac{2(6.10) - 0 - 9.14}{(6.10 - 0)(6.10 - 9.14)} + 43.17 \times 10^6 \frac{2(6.10) - 0 - 6.10}{(9.14 - 0)(9.14 - 6.10)} = 4.572 \times 10^6\ \text{m}^2$$

In addition, Eq. 16.7 can be used to determine the top surface area,

$$\frac{dV(0)}{dz} \cong 0 + 29.73 \times 10^6 \frac{2(0) - 0 - 9.14}{(6.10 - 0)(6.10 - 9.14)} + 43.17 \times 10^6 \frac{2(0) - 0 - 6.10}{(9.14 - 0)(9.14 - 6.10)} = 5.176 \times 10^6\ \text{m}^2$$

which also compares well with the measured area (5.180×10^6).

At the end of the previous example we saw how Eq. 16.7 provides superior estimates of the area based on the derivative of the volume. However, because numerical differentiation tends to amplify data errors, it is not recommended that area be determined in this fashion. Rather the procedure outlined in Fig. 16.3 (bathymetry → area → volume) is the proper approach.

It should be noted that the results showed in Table 16.1 are often presented in the graphical format of area-depth (Fig. 16.4*a*) and volume-depth (Fig. 16.4*b*) plots. As described above, the area-depth version is more fundamental, with volumes being generated through integration of the area-depth curve.

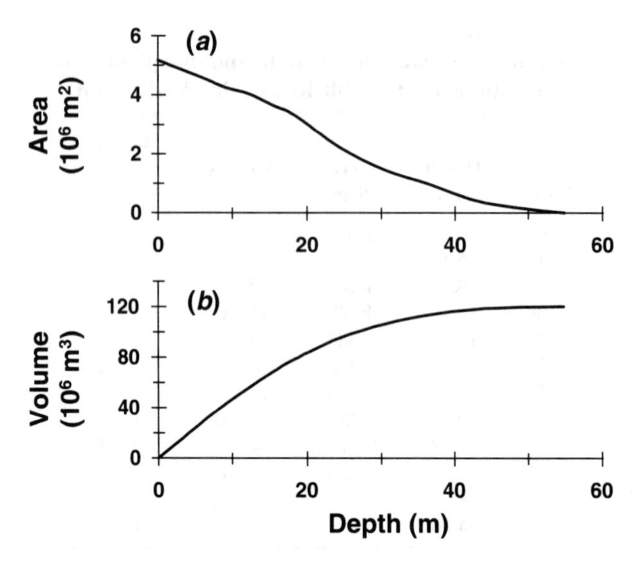

FIGURE 16.4
(*a*) Area-depth and (*b*) volume-depth curves for the Tolt Reservoir, Washington.

16.3 WATER BALANCE

A water balance for a well-mixed lake can be written as

$$S = \frac{dV}{dt} = Q_{\text{in}} - Q_{\text{out}} + G + PA_s - EA_s \tag{16.8}$$

where S = storage (m^3 d^{-1})
V = volume (m^3)
t = time (d)
Q_{in} = inflow (m^3 d^{-1})
Q_{out} = outflow (m^3 d^{-1})
G = groundwater flow (m^3 d^{-1})
P = precipitation (m d^{-1})
E = evaporation (m d^{-1})
A_s = surface area (m^2)

Equation 16.8 can be applied in two general fashions: steady-state and time-variable. The following sections treat both these applications.

16.3.1 Steady-State

In many instances lakes and impoundments (especially large ones) do not experience drastic volume changes for the time periods addressed by water-quality models. For such cases Eq. 16.8 simplifies to

$$0 = Q_{\text{in}} - Q_{\text{out}} + G + PA_s - EA_s \tag{16.9}$$

For most cases it is much easier to measure inflow and outflow than the other quantities in the balance. Thus for many water-quality modeling applications, the other terms are neglected. This simplification is based on the assumption that precipitation balances evaporation and groundwater flow is negligible.

Although this simplification is often invoked for expediency, it should be checked for validity. Unfortunately although precipitation measurements can sometimes be obtained (usually contingent on access to a nearby weather station), direct measurements of evaporation and groundwater are much more difficult to obtain. The following example illustrates how an inaccurate water budget can impact modeling calculations.

EXAMPLE 16.2. EFFECT OF WATER BUDGET ON QUALITY MODELING. A lake has the following characteristics:

Volume $= 1 \times 10^7$ m^3
River inflow $= 1 \times 10^6$ m^3 d^{-1}
River outflow $= 0.8 \times 10^6$ m^3 d^{-1}

Suppose that a first-order decaying (0.2 yr^{-1}) dissolved pollutant is discharged to this system at a constant rate of mass loading of 1×10^7 g yr^{-1}. Calculate the lake concentration for two cases, if the discrepancy between inflow and outflow is due to (*a*) a groundwater loss or (*b*) an evaporation loss.

Solution: (*a*) For this case the loss will carry the pollutant into the groundwater. Therefore the total outflow is equal to 1×10^6 m^3 d^{-1} and the concentration can be calculated as

$$c = \frac{W}{Q + kV} = \frac{1 \times 10^7}{1 \times 10^6 + 0.1(1 \times 10^7)} = 5 \text{ mg L}^{-1}$$

(*b*) For this case the loss will not carry the pollutant into the atmosphere. Therefore the total outflow is equal to 0.8×10^6 m^3 d^{-1} and the concentration can be calculated as

$$c = \frac{1 \times 10^7}{0.8 \times 10^6 + 0.1(1 \times 10^7)} = 5.556 \text{ mg L}^{-1}$$

Therefore a discrepancy of about 10% results, depending on your assumption regarding the discrepancy between the river inflow and outflow.

Of course if either evaporation or groundwater flow can be deemed unimportant, the situation is greatly simplified. For example if the lake overlies impermeable bedrock, the assumption could be made that groundwater is unimportant. In such instances if inflow, outflow, and precipitation are known, Eq. 16.9 can be used to estimate the evaporation rate.

16.3.2 Evaporation

Although evaporation can be estimated by difference, there are many systems where this is impossible. In such cases direct measurements and model equations provide alternatives.

The most common direct method involves deploying a pan of water. The pan can be either floated on the water surface or positioned on land close to the water body. The amount of water lost over time is monitored daily (with corrections for precipitation) and is expressed as the pan evaporation rate E_p(cm d^{-1}). It is then adjusted by a correction factor called the pan coefficient k_p to extrapolate the result to the natural water. It can then be used to compute the quantity of flow lost via evaporation, as in

$$Q_e = 0.01 k_p E_p A_s \tag{16.10}$$

where Q_e = evaporative water flow (m^3 d^{-1}), A_s = surface area (m^2), k_p averages about 0.70 for the United States with a range from 0.64 to 0.81, and the 0.01 is included to convert cm to m.

There are also some equations available to calculate evaporation based on meteorological and lake conditions. For example the energy flux due to evaporation can be computed as

$$H_e = f(U_w)(e_s - e_{\text{air}}) \tag{16.11}$$

where $f(Uw)$ = a function reflecting the effect of wind on evaporation and e_s and e_{air} = the vapor pressure corresponding to the water and the dew-point temperatures (mmHg), respectively. The heat transfer can then be converted to a water flow rate by

$$Q_e = 0.01 \frac{f(U_w)(e_s - e_{\text{air}})}{L_e \rho_w} A_s \tag{16.12}$$

where L_e = the latent heat of vaporization (cal g^{-1}) and ρ_w = water density (g cm^{-3}). The 0.01 is included so that flow is in m^3 s^{-1}. The latent heat and the wind function can be computed with

$$L_e = 597.3 - 0.57T \tag{16.13}$$

and (Brady, Graves, and Geyer 1969)

$$f(U_w) = 19.0 + 0.95U_w^2 \tag{16.14}$$

where T = temperature (°C) and U_w = the wind speed measured in m s^{-1} at a height of 7 m above the water surface. The vapor pressures can be determined by the formula (Raudkivi 1979)

$$e = 4.596 e^{\frac{17.27T}{237.3+T}} \tag{16.15}$$

where the surface water and the dew-point temperature are used to generate e_s and e_{air}, respectively.

These formulas can then be used to estimate the flow lost via evaporation. This value, along with inflow, outflow, and precipitation flows, can then be substituted into Eq. 16.9 to estimate the groundwater flow.

EXAMPLE 16.3. EVAPORATION CALCULATION. A lake has the following characteristics:

Surface area = 1×10^6 m^3
Wind speed = 2 mps
Water temperature = 25°C
Dew-point temperature = 20°C

Compute the evaporation flow.

Solution: The required parameters can be determined as

$$L_e = 597.3 - 0.57(25) = 583.05 \text{ cal g}^{-1}$$

$$f(U_w) = 19.0 + 0.95(2)^2 = 22.8$$

$$e_s = 4.596e^{\frac{17.27(25)}{237.3+(25)}} = 23.84$$

$$e_{air} = 4.596e^{\frac{17.27(20)}{237.3+(20)}} = 17.59$$

These values can be substituted into Eq. 16.12 to give the flow rate,

$$Q_e = 0.01\frac{22.8(17.59 - 23.84)}{585.05(1)}1 \times 10^6 = 2440.8 \text{ m}^3 \text{ d}^{-1}$$

Note that the positive sign connotes that the flow is lost from the lake.

It should be noted that there are many other predictive approaches for calculating evaporation. Any good book on hydrology (e.g., Chow et al. 1988, Ponce 1989, Bras 1990) will contain an extensive review on the subject. In addition we provide additional information on evaporation in our discussion of thermal modeling in Lec. 30.

16.3.3 Time-Variable

For simplicity we assume that precipitation and evaporation are approximately equal and that groundwater flow is negligible. For this case Eq. 16.8 reduces to

$$\frac{dV}{dt} = Q_{in} - Q_{out} \tag{16.16}$$

If the initial volume and the inflow time series are known, the solution of Eq. 16.16 depends on characterizing the outflow. Of course if the outflow is regulated or measured, the solution is simple. In many cases, however, it is not given. For these situations the relationship between outflow and volume must be determined. This is done by establishing a functional relationship between outflow and head (volume) in the reservoir.

For certain spillway structures, equations are available to establish the relationship. These formulas are usually of the general form

$$Q_{out} = CLH^a \tag{16.17}$$

where C and a = coefficients
L = length of the spillway
H = the total head or surface-water elevation

For cases where an equation is not given, relationships can be established by measurements of flow and elevation. In either case Eq. 16.16 can be written as

$$\frac{dV}{dt} = Q_{in}(t) - Q_{out}(H) \tag{16.18}$$

The process is further complicated by the fact that the relationship of depth and volume must be established. In other words

$$dV = A(H)\,dH \tag{16.19}$$

Thus an area-stage relationship $A(H)$ must be established (recall Fig. 16.4*a*). If this is done Eq. 16.19 can be substituted into Eq. 16.16 to give

$$\frac{dH}{dt} = \frac{Q_{\text{in}}(t) - Q_{\text{out}}(H)}{A(H)} \tag{16.20}$$

This approach is sometimes called the level-pool routing technique.

EXAMPLE 16.4. LEVEL-POOL RESERVOIR FLOW ROUTING. A small detention pond has a surface area of 2 ha and vertical sides. The discharge-head relationship has been measured:

Elevation (m)	Outflow ($m^3\ s^{-1}$)
0.0	0.0
0.5	0.0
1.0	0.0
1.5	1.7
2.0	5.0
2.5	9.0
3.0	14.0
3.5	20.0
4.0	26.0

Notice that no outflow occurs when the water level is below 1 m. This volume is called ***dead storage***. A storm creates the inflow hydrograph in Fig. 16.5. Use the level-pool model to route this flow through the detention basin. Assume that initially the pond has a depth of 1 m.

Solution: The results of applying the level-pool routing technique are displayed in Fig. 16.5. Notice how the pond tends to diminish and spread the flow. The peak discharge is diminished from 10 cms at $t = 50$ min for the inflow to approximately 4.5 cms at about $t = 78$ min for the outflow.

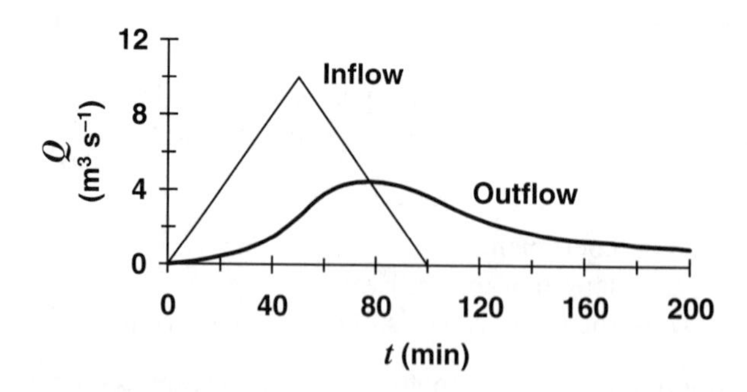

FIGURE 16.5
Inflow and outflow hydrographs for a small detention pond. The outflow hydrograph was calculated with the level-pool routing method.

Now that we have learned how to route water through a pond, we can model pollutant transport and fate for such a system. To do this we can write a general mass balance for a pollutant that reacts with first-order kinetics as

$$\underset{\text{Accumulation}}{\frac{dM}{dt}} = \underset{\text{Inflow}}{Q_{\text{in}}(t)c_{\text{in}}(t)} - \underset{\text{Outflow}}{Q_{\text{out}}(H)\frac{M}{V}} - \underset{\text{Reaction}}{kM} \tag{16.21}$$

Observe how this equation has been written in terms of mass rather than concentration because the volume is changing. Thus to solve this problem, we must integrate Eqs. 16.20 and 16.21 simultaneously to calculate mass and volume at each time step. Then, if we would like to determine the pond's concentration, we merely use

$$c = \frac{M}{V} \tag{16.22}$$

EXAMPLE 16.5. LEVEL-POOL RESERVOIR POLLUTANT ROUTING. Suppose that for the detention pond from Example 16.4, the inflow has a constant concentration of 100 mg L^{-1} of a pollutant that settles at a rate of 1 m d^{-1}. Calculate the concentration and the rate of mass outflow. Assume that $c = 0$ at $t = 0$.

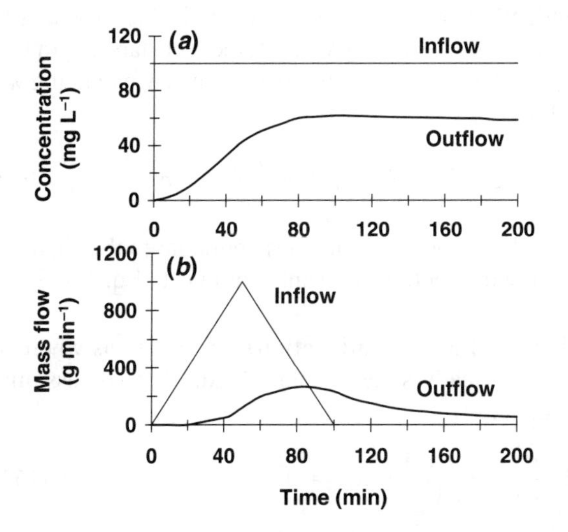

FIGURE 16.6
Inflow and outflow (*a*) concentration and (*b*) mass for a small detention pond.

Solution: The results of applying the level-pool routing technique are displayed in Fig. 16.6. Notice how the concentration increases rapidly as the inflow carries the pollutant into the system. After the peak flow has passed, the concentration diminishes slowly as settling becomes the primary removal mechanism. In contrast the mass outflow rate more closely follows the outflow hydrograph (Fig. 16.5), which for this case is determined mainly by the outflow rate.

16.4 NEAR-SHORE MODELS (ADVANCED TOPIC)

Since many pollutants enter lakes (and other waterbodies for that matter) at their peripheries, another important water-quality problem is the distribution of contaminants in the vicinity of a waste discharge or a river. Such areas are important in a

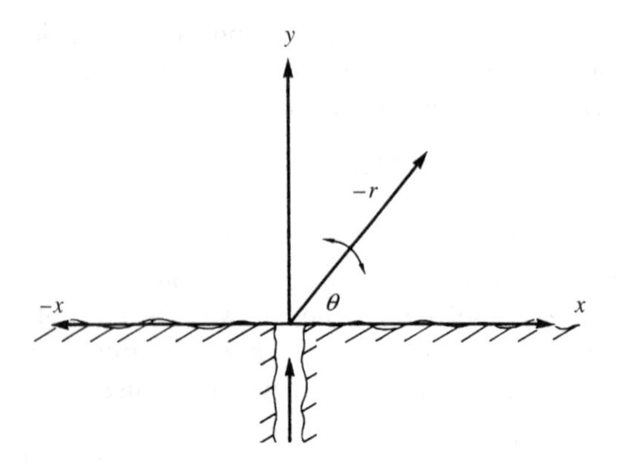

FIGURE 16.7
Cartesian (x, y) and radial (r) coordinates used for coastal zone models.

management context since human use and perception of a lake are typically intense in the coastal or near-shore zone, where beaches and other recreational areas are located.

After an initial period of mixing due to the turbulence of the discharge jet (see Fischer et al. 1979 for a discussion of modeling jets and plumes), the concentration of a near-shore input is dependent on the transport processes in the lake and the pollutant's reaction characteristics. For a vertically well-mixed, constant-depth layer such as the epilimnion of a stratified lake, the distribution of a substance reacting with first-order decay is represented by

$$\frac{\partial c}{\partial t} = -U_x \frac{\partial c}{\partial x} - U_y \frac{\partial c}{\partial y} + E_x \frac{\partial^2 c}{\partial x^2} + E_y \frac{\partial^2 c}{\partial y^2} - kc \tag{16.23}$$

where the x and y axes are defined to be parallel and perpendicular to the shoreline, respectively (Fig. 16.7). The following sections present solutions to Eq. 16.23.

Steady-state case in an infinite fluid (no advection). In situations where advective currents are negligible, the steady-state version of Eq. 16.23 is (assuming that diffusion is equal in all directions)

$$E\left(\frac{\partial^2 c}{\partial x^2} + \frac{\partial^2 c}{\partial y^2}\right) - kc = 0 \tag{16.24}$$

O'Connor (1962) transformed and solved Eq. 16.24 for polar coordinates, as depicted in Fig. 16.7. If r is the radial axis, Eq. 16.24 becomes

$$\frac{\partial^2 c}{\partial r^2} + \frac{1}{r}\frac{\partial c}{\partial r} + \frac{1}{r^2}\frac{\partial^2 c}{\partial \theta^2} - \frac{k}{E}c = 0 \tag{16.25}$$

If we assume that c is constant for a given r, then $\partial c/\partial \theta$ and $\partial^2 c/\partial \theta^2$ are zero and Eq. 16.25 reduces to a Bessel equation[†] of order zero,

[†] Differential equations of a particular form (e.g., Eq. 16.26) are called Bessel equations. The solutions for these equations are called Bessel functions. These functions are tabulated in numerous mathematical reference volumes and handbooks. Additionally many computer systems carry easy-to-use library functions to compute their value.

$$\frac{d^2c}{dr^2} + \frac{1}{r}\frac{dc}{dr} - \frac{k}{E}c = 0 \tag{16.26}$$

the solution of which is

$$c = BI_0\sqrt{\frac{kr^2}{E}} + CK_0\sqrt{\frac{kr^2}{E}} \tag{16.27}$$

where B and C are constants of integration and I_0 and K_0 are modified Bessel functions of the first and second kind, respectively. By invoking the following boundary conditions:

$$c(r_0) = c_0 \tag{16.28}$$

and

$$c(\infty) = 0 \tag{16.29}$$

O'Connor solved Eq. 16.27 for

$$\frac{c}{c_0} = \frac{K_0\sqrt{\dfrac{kr^2}{E}}}{K_0\sqrt{\dfrac{kr_0^2}{E}}} \tag{16.30}$$

O'Connor's first boundary condition (Eq. 16.28) sets the concentration equal to a constant at a distance r_0 from the origin. This distance can be thought of as the periphery of a mixing zone. Although this formulation has the disadvantage of not being directly related to the waste source at $r = 0$, it is useful for circumventing the problem that the solution (Eq. 16.30) actually approaches infinity as r approaches zero. This can be seen from an alternative solution to Eq. 16.26 that uses a boundary condition at $r = 0$,

$$c = \frac{W}{\pi HE}K_0\sqrt{\frac{kr^2}{E}} \tag{16.31}$$

where H = the water depth. This solution goes to infinity as r approaches zero since, mathematically, the waste source emanates from a point (or more accurately a line since the source enters over the depth of the layer) of infinitely small thickness at the shoreline. Di Toro (1972) has presented an analysis of the problem along with a tabulated solution that is finite at the origin. While the modeler should consult Di Toro's paper, Eqs. 16.30 and 16.31 provide adequate approximations in many cases and will be used in subsequent examples.

EXAMPLE 16.6. RADIAL BACTERIAL MODEL. O'Connor (1962) used Eq. 16.30 to solve for bacterial distribution in the vicinity of Indiana Harbor in Lake Michigan. The radius of the mixing zone was taken to be 45.7 m, which is one-half the approximate width of the harbor outlet. A diffusion coefficient of $2.6 \times 10^6\ m^2\ d^{-1}$ ($3 \times 10^5\ cm^2\ s^{-1}$) was used to characterize turbulent mixing. Two decay rates, 0.5 and 3.0 d^{-1}, were used to estimate bounds for bacterial die-off under summer temperature conditions.

Solution: The results of applying O'Connor's model (Fig. 16.8) indicate general agreement between model results and 3-month averages (June, July, and August) of bacterial concentration.

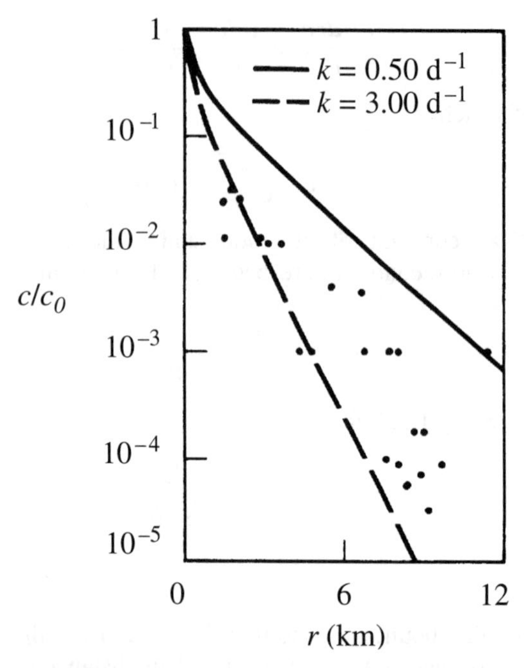

FIGURE 16.8
Profiles of bacterial concentration (normalized to concentration at the edge of the mixing zone) versus distance r (km) from the edge the mixing zone in the vicinity of Indiana Harbor, Lake Michigan, as originally computed by O'Connor (1962). The lines represent model calculations based on two estimates of the bacterial die-off rate. The data are 3-month averages for June through August.

Steady-state case in an infinite fluid (advection along the shoreline). Although the preceding is appropriate where turbulent mixing is the primary transport mechanism, some lakes have persistent unidirectional currents that can advect pollutants along the shoreline. For this case Eq. 16.23 becomes, at steady-state,

$$U_x \frac{\partial c}{\partial x} = E\left(\frac{\partial^2 c}{\partial x^2} + \frac{\partial^2 c}{\partial y^2}\right) - kc \tag{16.32}$$

where U_x = current velocity along the shoreline. Boyce and Hamblin (1975) have presented the following solution:

$$c = \frac{W}{\pi HE} e^{\frac{U_x x}{2E}} K_0\left[r\sqrt{\frac{k}{E} + \left(\frac{U_x}{2E}\right)^2}\right] \tag{16.33}$$

where

$$r = \sqrt{x^2 + y^2} \tag{16.34}$$

Notice that for $U_x = 0$, Eq. 16.33 reduces to Eq. 16.31.

Steady-state case in a bounded fluid. The preceding solutions are appropriate for situations where the reaction is rapid and/or the transport weak and/or the lake wide enough that the opposite shoreline of the lake has no effect on the solution. For example Eqs. 16.31 and 16.33 would be valid for the analysis of discharge of coliform bacteria into one of the Great Lakes since the bacteria would die off within a few kilometers of shore and such lakes are on the order of 100 km wide. However, for narrow lakes or for substances that react slowly or are conservative, the effect of the opposite shoreline must be considered. In that case Eq. 16.33 is expressed directly in cartesian coordinates as

$$c(x, y) = \frac{W}{\pi H E} e^{\frac{U_x x}{2E}} K_0 \sqrt{(x^2 + y^2)\left[\frac{k}{E} + \left(\frac{U_x}{2E}\right)^2\right]} \tag{16.35}$$

Then for a lake of width Y, the solution is obtained by the iterative formula (Boyce and Hamblin 1975),

$$c = c(x, y) + \sum_{n=1}^{\infty} [c(x, y + 2nY) + c(x, y - 2nY)] \tag{16.36}$$

In this solution the confining effect of the opposite shoreline is accounted for by the infinite series of additions in the second half of Eq. 16.36. In practice only a small number ($\cong$ 2 or 3) are needed to characterize the effect.

EXAMPLE 16.7. BACTERIAL MODEL WITH ADVECTION AND DIFFUSION. A municipality discharges sewage at a rate of 4×10^4 m^3 d^{-1} at a point on a lake's shoreline. The wastewater is laden with coliform bacteria (30×10^6 number per 100 mL) that die at a rate of 1.0 d^{-1}. Calculate concentration profiles in the lake's vertically well-mixed surface layer (H = 20 m) for the following cases:

(*a*) The lake is infinitely long and wide and horizontal diffusion ($E_x = E_y = 5 \times 10^6$ m^2 d^{-1}) is the sole transport mechanism.
(*b*) The same conditions exist as in (*a*), but with a current of 0.5×10^4 m d^{-1} moving from east to west.
(c) The same conditions exist as in (*b*), but with a finite width of 4 km.

Solution: (a) The waste load W is calculated as the product of flow and concentration of the discharge, as in

$$W = (4 \times 10^4 \text{ m}^3 \text{ d}^{-1})(30 \times 10^6 \text{ number per 100 mL})$$

Equation 16.31 can then be used to determine concentrations in the lake,

$$c = \frac{(4 \times 10^4)(30 \times 10^6)}{\pi(20)5 \times 10^6} K_0 \sqrt{\frac{r^2}{5 \times 10^6}}$$

The results of this computation for a section of the lake are displayed Fig. 16.9*a*.
(*b*) In this case Eq. 16.33 is used:

$$c = \frac{(4 \times 10^4)(30 \times 10^6)}{\pi(20)5 \times 10^6} e^{\frac{(0.5\times 10^4)x}{2(5\times 10^6)}} K_0 \left\{ r\sqrt{\frac{1}{5 \times 10^6} + \left[\frac{(0.5 \times 10^4)x}{2(5 \times 10^6)}\right]^2} \right\}$$

The results are shown in Figure 16.9*b*.

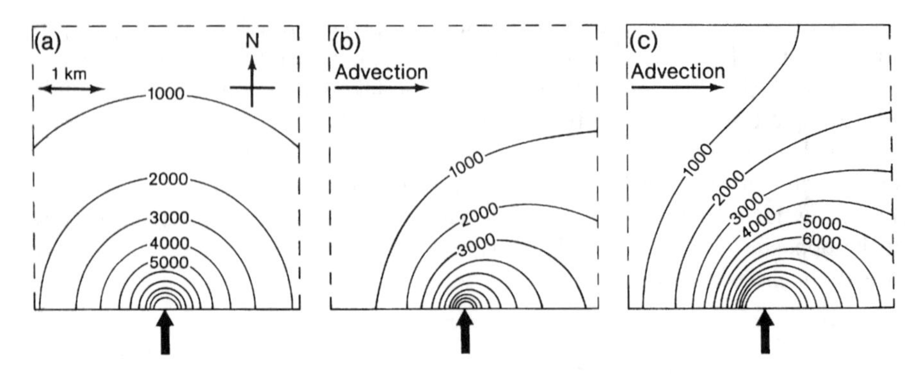

FIGURE 16.9
Countours of coliform bacteria (number per 100 mL) as a result of a waste discharge at a point on a lake's shoreline for three cases: (*a*) unbounded fluid with diffusion, (*b*) unbounded fluid with diffusion and advection, and (*c*) bounded fluid with diffusion and advection.

(*c*) In this case Eqs. 16.35 and 16.36 are used with $n = 2$. The results are displayed in Fig. 16.9*c*.

EXAMPLE 16.8. BACTERIAL MODEL FOR LAKE ERIE. Boyce and Hamblin (1975) developed Eq. 16.35 and used it to simulate steady-state chloride profiles in the central basin of Lake Erie (Fig. 16.10). They assumed that the controlling dimension of the lake (that is, its narrowest width) defines the largest energy-bearing eddy and specifies the point at which the acceleration of the diffusion process comes to a halt. Thus for long-term computations, Fig. 8.11 with scale length specified by the controlling width provides a first estimate of the order of magnitude of the diffusion coefficient. For example for central Lake Erie, the minimum width is approximately 80 km. Figure 8.11 indicates that the diffusion coefficient for this scale length ranges from approximately 5×10^5 to 8×10^6 cm^2 s^{-1}. Figure 16.10*b* shows Boyce and Hamblin's (1975) simulation using Eq. 16.35 and values of $E = 1.5 \times 10^6$ cm^2 s^{-1} and $n = 2$.

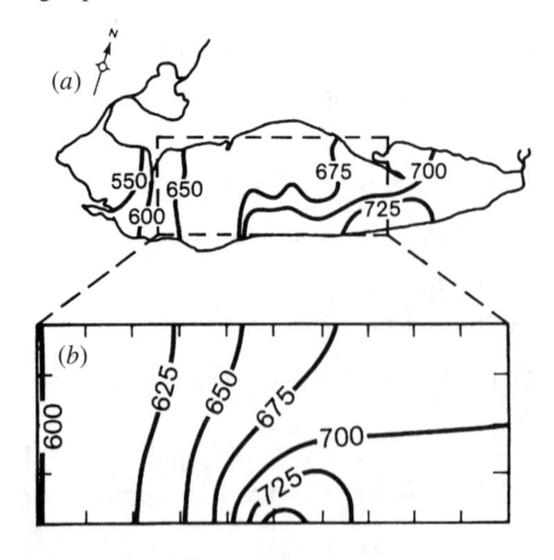

FIGURE 16.10
Chloride model of Lake Erie from Boyce and Hamblin (1975). (*a*) Chloride contours in μM for Lake Erie during 1970; (*b*) model results using Eq. 16.35.

As in Example 16.8, Fig. 8.11 provides a first estimate of the expected range of the horizontal eddy diffusion coefficient for such long-term computations. However, as with any default coefficient, additional measurements, such as calibration with conservative substances, are advisable. Boyce and Hamblin's calibration of their model to chloride contours is a good example of such an approach.

PROBLEMS

16.1. The following depth-area data is available for Lake Youngs in the State of Washington:

Depth (m)	0	6	14	28	31
Area (m^2)	2,830,000	2,405,500	1,188,600	424,500	0

(*a*) Determine the volume of the lake when it is filled.
(*b*) If the lake is vertically stratified into two layers at a depth of 10 m, determine the volume of the surface (epilimnion) and bottom (hypolimnion) layers.

16.2. The following polynomial defines the area of Cascade Reservoir, Idaho:

$$A(z) = 0.1675333(z - 1453)^4 - 7.446(z - 1453)^3 + 115.5717(z - 1453)^2 - 101.05(z - 1453)$$

where $A(z)$ = surface area (10^4 m^2) and z = depth (m). Note that z = 1453 m is the bottom of the reservoir and that depth increases upward.
(*a*) Determine the volume of the lake when it is filled (z = 1473 m).
(*b*) If the lake is vertically stratified into two layers at a depth of 10 m from the bottom (z = 1463 m), determine the volume of the surface (epilimnion) and bottom (hypolimnion) layers.
(*c*) Develop a polynomial to predict volume as a function of depth.

16.3. The following data have been reported for Shagawa Lake, Minnesota:

Depth (m)	0	2	4	6	8	10	12
Volume (%)	100	68	41	19.5	4.45	1	0

If the total volume is 53×10^6 m^3, use this information to determine the area-depth curve for the lake.

16.4. You are studying a small pond during the summer. It has the following characteristics:

Volume = 1×10^5 m^3 Surface area = 1×10^5 m^2
Outflow = 3×10^3 m^3 d^{-1} Precipitation = 0.1 cm d^{-1}
Inflow = 1×10^3 m^3 d^{-1}

You measure a pan evaporation rate of 0.25 cm d^{-1}. Assuming a pan coefficient of 0.7, determine the flow due to groundwater.

16.5. Compute the evaporation flow for the lake in Prob. 16.4 if

Wind speed = 1 mps
Water temperature = 30°C
Dew-point temperature = 25°C

16.6. A conical mining pit is shown in Fig. P16.6. The pit is located in an area where precipitation is negligible. The ratio of the radius to the depth is 20:1. What is its ultimate depth if it is being filled by a constant groundwater inflow at a rate of 1×10^4 m^3 d^{-1}? Note that it is subject to the following meteorological conditions:

Wind speed = 1 mps
Water temperature = 28°C
Dew-point temperature = 22°C

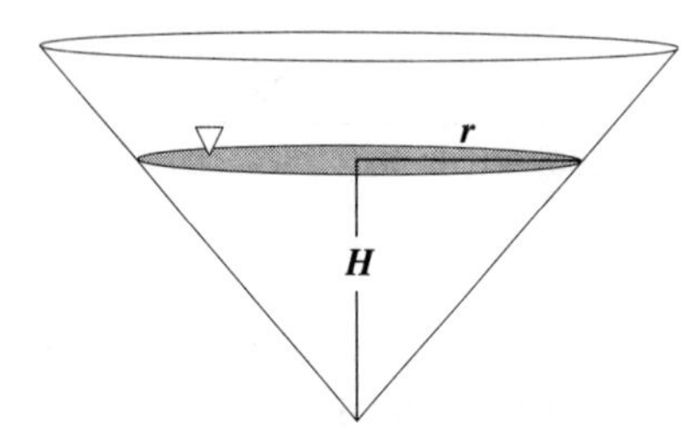

FIGURE P16.6

16.7. Route the following hydrograph and pollutograph through the system in Examples 16.4 and 16.5. Note that the pollutant is conservative.

Time (min)	0	20	40	60	80	100	120	140
Inflow (m^3 s^{-1})	0	4	10	10	4	2	1	0
Concentration (mg L^{-1})	0	10	20	10	0	0	0	0

16.8. A first-order decaying pollutant ($k = 0.1$ d^{-1}) is discharged at a continuous rate into a large lake. If the radius of the mixing zone is 100 m and advection is negligible, determine the radius at which the concentration drops to 5% of its initial value. Note that the diffusion coefficient is 10^5 cm^2 s^{-1}.

16.9. A municipality discharges sewage at a rate of 1×10^5 m^3 d^{-1} at a point on a lake's shoreline. The wastewater contains coliform bacteria (200×10^6 number per 100 mL) that die at a rate of 0.5 d^{-1}. Calculate concentration profiles in the lake's vertically well-mixed surface layer ($H = 10$ m) for the following cases:

(*a*) The lake is infinitely long and wide and horizontal diffusion ($E_x = E_y = 10^6$ m^2 d^{-1}) is the sole transport mechanism.

(*b*) The same conditions exist as in (*a*), but with a current of 10^4 m d^{-1} moving from east to west.

(*c*) The same conditions exist as in (*b*), but with a finite width of 5 km.

LECTURE 17

Sediments

Lecture overview: Following a general discussion of sediment transport and suspended solids, we present an overview of the mechanism of settling. This includes a review of both observed values and the Stokes' law model. After providing background on the bottom sediments, I develop a simple budget for a well-mixed lake underlain by a well-mixed surface sediment layer. This is followed by a description of models that characterize the sediments as a distributed system, that is, as a continuum. The lecture ends with an introduction to the topic of sediment resuspension.

For most of its history water-quality modeling has been concerned with how solutes moved through aquatic environments. Consequently adequate characterization of hydrodynamics and water budgets have been an essential component of our field.

Over the past 20 yr there has been increased attention directed to how water pollutants interact with solid matter. The interest originally focused on the simulation of the settling of organic particles produced by eutrophication and their subsequent decomposition in the bottom sediments. Today modeling of both suspended and bottom sediments has been heightened by the recognition that many toxic substances associate with solid matter. Finally in some systems, solids are considered a pollutant in their own right. For example the reproduction of some endangered fish can be affected by sediment deposition on spawning beds.

Thus bottom sediments and suspended solids represent an "environment" that must be understood to adequately simulate the transport and fate of many pollutants. This lecture provides background information on this environment.

17.1 SEDIMENT TRANSPORT OVERVIEW

Once suspended solids are introduced into a natural water they are transported and transformed by a number of different mechanisms. A portion of the organic solids will be lost by decomposition. The residual organic particles along with inorganic solids are subject to a number of transport processes.

The particles will be carried laterally by water currents. At the same time they will settle differentially, depending on their size and density. Although a portion will remain permanently on the bottom, solids can be reintroduced into the water by turbulence. Such resuspension tends to occur due to strong currents and in shallower areas due to wind mixing.

Although the situation is very complex, some general patterns have been observed. In particular, fine-grained sediments will tend to collect in low-energy areas. This has some interesting ramifications for the water environments we studied in the previous three lectures.

For streams (Fig. 17.1*a*), deposition zones tend to form at low-energy regions such as pools and the inside of bends in meandering rivers. In estuaries (Fig. 17.1*b*), a ***null zone*** is formed where inflowing river flow and tidal action tend to cancel. This convergence region is typified by higher particle concentration in the water and large sediment accumulation.

In lakes (Fig. 17.1*c*), wind- and current-induced turbulence leads to the accumulation of coarser solid matter in shallow water and finer particles at depth. This process, referred to as ***focusing,*** means that a fine-grained-solids deposition zone will be formed at the center of a lake.

Impoundments exhibit two effects (Fig. 17.1*d*). First, a delta can form at the point where the turbulent river enters the much lower energy impoundment. Second, very fine sediments can be carried past the delta by density currents that hug the bottom. Such advective transport can eventually lead to large accumulation of fine-grained solids at the pool above the dam.

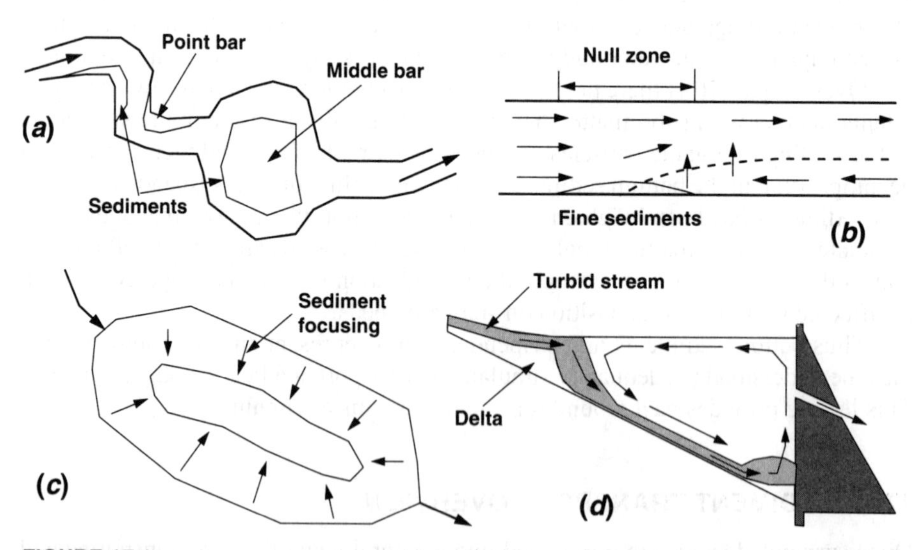

FIGURE 17.1
Patterns of deposition of fine sediments in natural waters. (*a*) Top view of river; (*b*) side view of estuary; (*c*) top view of lake; and (*d*) side view of impoundment.

Finally all these deposition patterns can be modified by the effects of humans and extreme natural events. For example sediment beds are typically found in the vicinity of waste outfalls. Further, floods and high winds can scour and redistribute bottom sediments.

The level of modeling detail required to simulate the patterns depicted in Fig. 17.1 is beyond the scope of this book. In the following sections we cover introductory information and some simple models related to both suspended and bottom solids that has direct relevance to water-quality modeling.

17.2 SUSPENDED SOLIDS

Let's start our discussion with suspended solids. The following section provides information on suspended solids properties such as typical concentration, size, and density. Then I introduce Stokes' law as an idealized model of the settling mechanism.

17.2.1 Suspended Solids Properties

The suspended solids concentration of natural waters is reported on a dry-weight basis. Values range from below 1 mg L^{-1} for extremely clean waters to over 100 mg L^{-1} for highly turbid systems. Levels for sewage and for storm-water overflows can be even higher. Table 17.1 lists some typical values.

Although suspended solids are expressed on a dry-weight basis (that is, dry weight of solids per volume of water), their dynamics require a more in-depth characterization of their composition. One aspect of this characterization relates to their origin.

TABLE 17.1
Suspended solids concentrations encountered in natural waters and sewage (from Di Toro et al. 1971, O'Connor 1988*c*, Lung 1994, Thomann and Mueller 1987)

System	Suspended solids (mg L^{-1})
Great Lakes:	
Superior/Huron	0.5
Saginaw Bay	8.0
Western Lake Erie	20.0
Flint River, Michigan	8–12
Rapid Creek, South Dakota	158
Clinton River, Michigan	10–120
Hudson River, NY	10–60
Potomac Estuary	5–30
James Estuary, Virginia	10–50
Sacramento–San Joaquin Delta, California	50–175
Raw sewage	300

Solids in natural waters have two primary origins: the drainage basin and the photosynthesis process. Particles from these two sources are formally referred to as ***allochthonous*** and ***autochthonous*** solids, respectively. They are also informally dubbed "brown" and "green" solids, respectively, based on their general color.

The two types of solids differ in several ways:

Organic carbon content. Autochthonous solids are organic matter and hence have high organic carbon content. In contrast allochthonous solids usually originate from soil erosion. Therefore allochthonous solids have much lower organic carbon content than fresh photosynthetically-derived solids because allochthonous solids (1) include inorganic solids from weathering of rocks that are low in organic carbon, and (2) their organic content has been reduced by decomposition. Further, the make up of the organic carbon differs. Freshly produced plants are composed of a variety of organic carbon compounds. After they die and begin decomposition, the more easily decomposable portion is lost first. Thus with time, the fraction remaining becomes more ***refractory*** (that is, resistant to decomposition). Consequently the organic carbon in the allochthonous solids is usually less reactive.

Density. The density is the mass of a particle normalized to its volume. Thus density has the same units as concentration. For example the density of water is approximately 1 g cm^{-3}. Autochthonous solids tend to be much less dense than allochthonous particles because inorganic minerals tend to be more dense than organic carbon. In addition fresh organic matter has a high water content. Consequently, as listed in Table 17.2, organic matter such as bacteria or plant matter has a density that is very close to that of water.

Size. A general classification scheme is shown in Fig. 17.2. Allochthonous solids tend to cover a wide spectrum of sizes. Although organic particles also range in size, the principle type of autochthonous solids (bacteria and floating unicellular plants called phytoplankton) tend to reside at the lower end of the range. However, it should be noted that both inorganic and organic particles can agglomerate to form larger assemblages called flocs and colonies. Such aggregation will have an effect on the particle's settling characteristics.

Finally it should be noted that chemical precipitation can enhance settling. In particular, hard-water systems experience calcium-carbonate or ***calcite precipitation***

TABLE 17.2
Densities of water and particulate matter

Substance	Density (g cm^{-3})
Water	1
Organic matter	
Wet-weight basis	1.02–1.1
Dry-weight basis	1.27
Siliceous minerals	2.65
Garnet sands	4

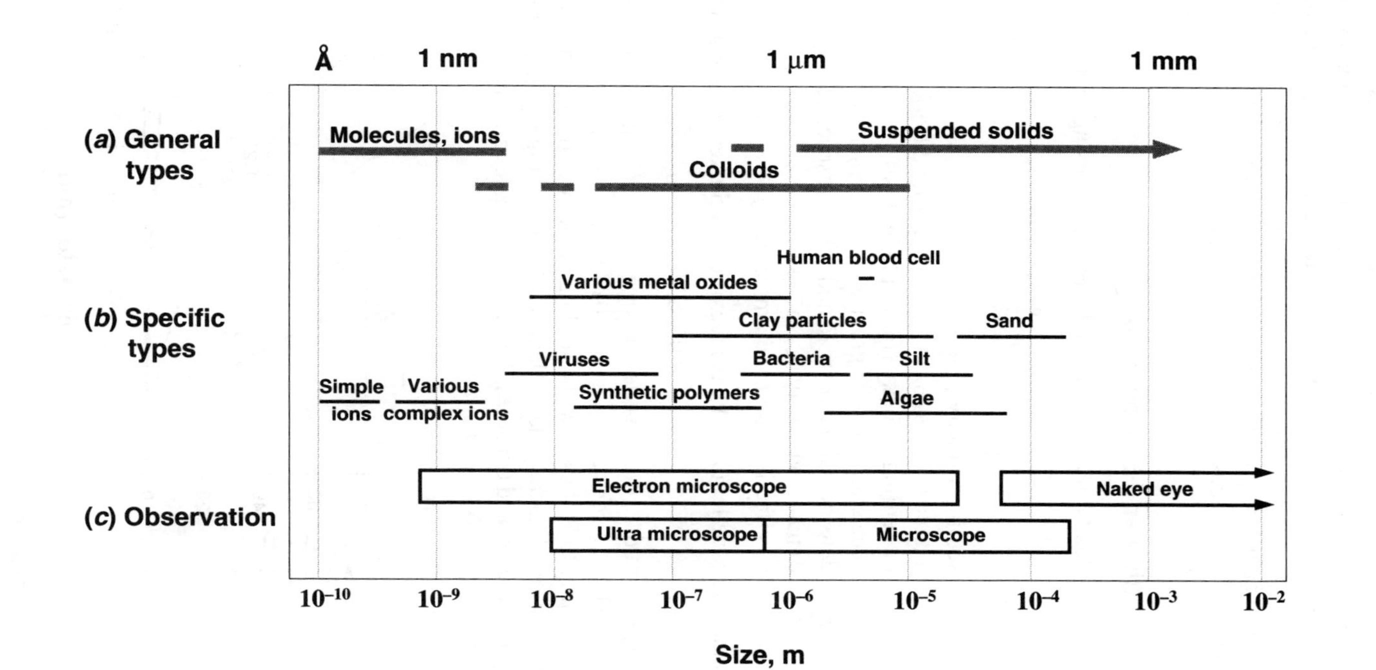

FIGURE 17.2
Sizes of particles in natural waters.

when plant productivity reduces carbon dioxide and raises pH. Such events sometimes produce so much calcium carbonate that they are commonly called "whitings" due to the resulting milky appearance of the water. Organic compounds can sorb to these particles and phytoplankton can serve as a nucleus for calcium carbonate formation. Hence as the particles settle, they carry organic matter with them. Wetzel (1975) provides a discussion of the mechanism.

17.2.2 Settling and Stokes' Law

The settling velocity of a particle can be estimated using Stokes' law,

$$v_s = \alpha \frac{g}{18}\left(\frac{\rho_s - \rho_w}{\mu}\right)d^2 \tag{17.1}$$

where v_s = settling velocity (cm s^{-1})
α = a dimensionless form factor reflecting the effect of the particle's shape on settling velocity (for a sphere it is 1.0)
g = acceleration due to gravity (= 981 cm s^{-2})
ρ_s and ρ_w = densities of the particle and water, respectively (g cm^{-3})
μ = dynamic viscosity (g cm^{-1} s^{-1})
d = an effective particle diameter (cm)

Thomann and Mueller (1987) have reexpressed Stokes' law in a convenient form,

$$v_s = 0.033634\alpha(\rho_s - \rho_w)d^2 \tag{17.2}$$

where v_s is in m d^{-1}, the densities are in g cm^{-3}, d is in μm, and water viscosity is assumed to have a constant value of 0.014 g cm^{-1} s^{-1}.

According to Stokes' law the settling velocity is linearly dependent on particle density and quadratically dependent on diameter. Figure 17.3 shows values for a range of diameters and densities. These dependencies are illustrated in the following example.

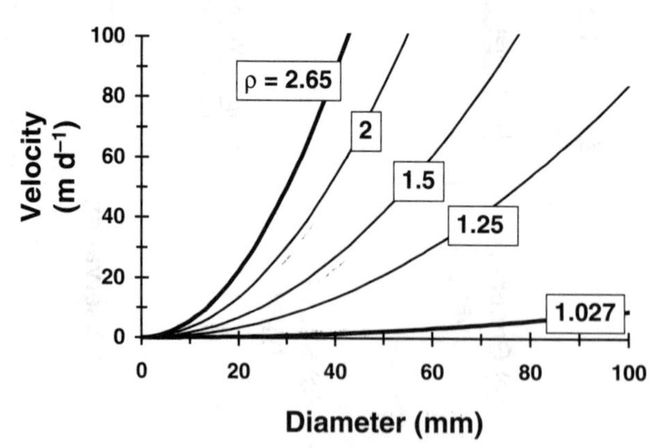

FIGURE 17.3
Plot of settling velocity versus diameter for various levels of particle density. This figure assumes that the particles are perfect spheres (α = 1.0 and d = diameter).

EXAMPLE 17.1. STOKES' LAW. Determine the settling velocities for (*a*) phytoplankton (ρ_s = 1.027 g cm^{-3}) and (*b*) silt (ρ_s = 2.65 g cm^{-3}) for the cases of diameters of 10 and 20 μm. Assume that all particles are perfect spheres (α = 1.0 and d = diameter).

Solution: Stokes' law (Eq. 17.2) can be used to calculate the settling velocity for the 10-μm phytoplankton,

$$v_s = 0.033634(1.027 - 1)10^2 = 0.09 \text{ m d}^{-1}$$

The other cases can be computed and tabulated

	d = 10 μm	d = 20 μm
Phytoplankton (ρ_s = 1.027 g cm^{-3})	0.09	0.36
Silt (ρ_s = 2.65 g cm^{-3})	5.55	22.20

Note how increasing diameter and density lead to much larger settling velocities.

In general, particles in natural waters have complex shapes (which leads to values of α lower than 1). Thus the results for both phytoplankton and other particles would be lower than for Example 17.1. As summarized in Table 17.3 measured values for phytoplankton and organic solids average about 0.25 m d^{-1}, with a range of 0.1 to 1 m d^{-1} (Table 17.3). For systems with calcite precipitation, the average value can rise to levels on the order of 0.5 m d^{-1}.

TABLE 17.3
Settling velocities of particles found in natural waters (Wetzel 1975, Burns and Rosa 1980)

Particle type	Diameter (μm)	Settling velocity (m d^{-1})
Phytoplankton:		
Cyclotella meneghiniana	2	0.08(0.24)[†]
Thalassiosira nana	4.3–5.2	0.1–0.28
Scenedesmus quadricauda	8.4	0.27(0.89)
Asterionella formosa	25	0.2(1.48)
Thalassiosira rotula	19–34	0.39–2.1
Coscinodiscus lineatus	50	1.9(6.8)
Melosira agassizii	54.8	0.67(1.87)
Rhizosolenia robusta	84	1.1(4.7)
Particulate organic carbon	1–10	0.2
	10–64	1.5
	> 64	2.3
Clay	2–4	0.3–1
Silt	10–20	3–30

[†] Parenthetical numbers are for the stationary phase (see Lec. 32 for an explanation of the different phases of microbial growth).

In practice few models use Stokes' law to directly determine settling velocities of organic matter. For one thing Stokes' law is predicated on the assumption that flow is laminar, whereas flow in most natural systems is turbulent. Further, living particles such as phytoplankton can become buoyant due to internal gas vacuoles. In addition their settling velocities may vary with their physiological state. For example as in Table 17.3, they may exhibit differing settling velocities, depending on whether or not they are actively growing. Finally the formation of flocs and precipitates complicates the direct application of Stokes' law.

The settling velocity also depends on the model segmentation (Lec. 18). Coarse, one-dimensional vertical models (e.g., those that treat the water column as two layers) typically use lower values than two- and three-dimensional characterizations. This has been attributed to the fact that the one-dimensional approaches do not adequately characterize hydrodynamic phenomena such as upwelling and entrainment that tend to reduce the effect of settling (Scavia and Bennett 1980). Thus a lower settling velocity is used to compensate for the omission of the mechanisms.

Because of all the above factors, the settling velocity is usually determined by direct measurement or calibration in most water-quality models. Nevertheless Stokes' law provides a useful theoretical reference, particularly in assessing the relative effects of density, diameter, and shape on particle settling. In addition it could become more directly useful in the future as both the hydrodynamic characterization are refined and water-quality modeling becomes increasingly involved with problems involving inorganic solids.

17.3 THE BOTTOM SEDIMENTS

Some of the suspended solids are eventually deposited and become part of a water body's bottom sediments. This section deals with how this environment can be quantified.

17.3.1 Porosity

In all the models described in the previous lectures, we dealt exclusively with the water column. Because natural waters are typically dilute (Fig. 17.4*a*), it made sense to represent concentration by normalizing the quantity of a chemical or solid to the water's volume. We now turn to a system, the bottom sediments, where the situation is not quite as clear cut. This is because a significant fraction of the sediment

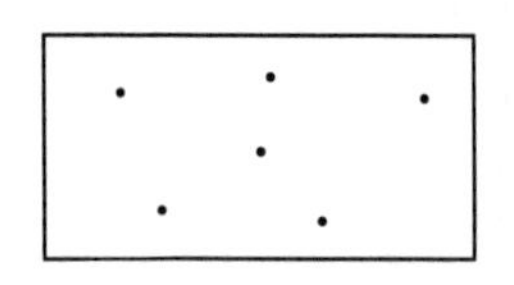

(*a*) Dilute aqueous suspension

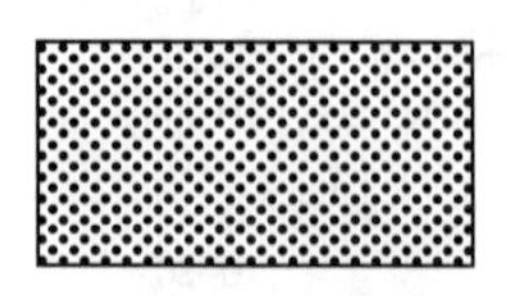

(*b*) Porous media

FIGURE 17.4
The contrast between (*a*) dilute natural water and (*b*) bottom sediments. The significant amount of solid matter present in the latter case must be accounted for when developing mass balances for sediment-water systems.

volume is solid (Fig. 17.4*b*). Such systems are referred to as ***porous media.*** They require the definition of several new parameters.

The ***porosity*** refers to the volume of the sediments that is in the liquid phase and is interconnected (Engelhardt 1977). Strictly speaking this excludes isolated pore space that is considered part of the solid phase. However, because such isolated pores are rarely found in fine-grained sediments (Berner 1980), the porosity ϕ is operationally defined as the fraction of the total volume that is in the liquid phase,

$$\phi = \frac{V_\ell}{V_2} \tag{17.3}$$

where V_ℓ = volume of the liquid part of the sediment layer (m^3) and V_2 = total volume of the sediment layer (m^3). Note that the subscript 2 is used for the sediments because we will shortly use this nomenclature when we model sediment-water interactions (look ahead to Fig. 17.5).

The fraction of the sediment that is in the solid phase immediately follows as

$$1 - \phi = \frac{V_p}{V_2} \tag{17.4}$$

where V_p = volume of the solid or particulate phase of the sediment (m^3).

17.3.2 Density and Sediment Solids Concentration

Another quantity that is used in modeling porous media is the density, which can be represented in terms of more fundamental parameters as

$$\rho = \frac{M_2}{V_p} \tag{17.5}$$

where ρ = density (g m^{-3}) and M_2 = mass of the solid phase in the sediments (g).

We can now use the above quantities to define a number of parameters that are needed to model sediment-water interactions. First, recall that the suspended solids concentration is the critical metric of the solids content of the water. A "suspended solids" concentration in the sediments, m_2, can be defined as

$$m_2 = \frac{M_2}{V_2} \tag{17.6}$$

Equation 17.5 can be solved for

$$M_2 = \rho V_p \tag{17.7}$$

Equation 17.4 can be solved for

$$V_2 = \frac{V_p}{1 - \phi} \tag{17.8}$$

Equations 17.7 and 17.8 can be substituted into Eq. 17.6 to yield

$$m_2 = (1 - \phi)\rho \tag{17.9}$$

Thus we have reexpressed the sediment solids concentration in terms of parameters that are conventionally used to measure porous media. We now use this expression to develop a solids budget for a sediment-water system. In the following discussions, it will be useful to recognize that the terms $(1 - \phi)\rho$ represent the "suspended solids" concentration of the bottom sediments.

17.4 SIMPLE SOLIDS BUDGETS

Now that we know something about suspended and bottom sediments, we can develop a solids model. For simplicity the model will be developed for allochthonous solids in a well-mixed lake. As in Fig. 17.5 two cases will be examined. In the first a one-way loss to the sediments is used. Then we couple the sediments and water by adding sediment resuspension.

For Fig. 17.5*a* the following mass balance can be written for the water:

$$V\frac{dm}{dt} = Qm_{\text{in}} - Qm - v_s A_s m \tag{17.10}$$

where v_s = settling velocity (m yr^{-1}) and A_s = area of the sediment-water interface (m^2). At steady-state Eq. 17.10 can be solved for

$$m = \frac{Qm_{\text{in}}}{Q + v_s A_s} \tag{17.11}$$

Now we can add a sediment layer to the model (Fig. 17.5*b*). Mass balances for solids in the water and the sediment layer can be written as

$$V_1\frac{dm_1}{dt} = Qm_{\text{in}} - Qm_1 - v_s A_s m_1 + v_r A_s m_2 \tag{17.12}$$

and

$$V_2\frac{dm_2}{dt} = v_s A_s m_1 - v_r A_s m_2 - v_b A_s m_2 \tag{17.13}$$

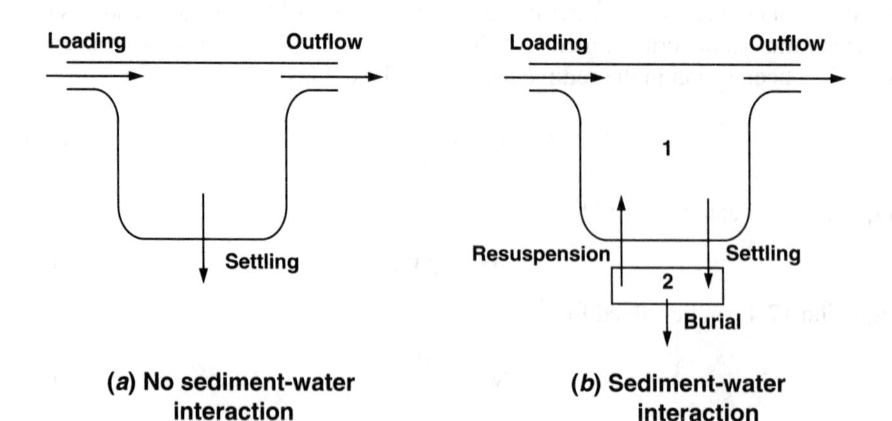

FIGURE 17.5
Schematic of a solids budget for a well-mixed lake (*a*) without and (*b*) with sediment feedback.

where v_r = resuspension velocity (m yr^{-1}) and v_b = burial velocity (m yr^{-1}). Equation 17.9 can be used to express sediment suspended solids m_2 in terms of sediment porosity and density. In addition the subscript of m_1 can be dropped. At steady-state, the resulting solid balance equations are

$$0 = Qm_{\text{in}} - Qm - v_s A_s m + v_r A_s(1 - \phi)\rho \tag{17.14}$$

and

$$0 = v_s A_s m - v_r A_s(1 - \phi)\rho - v_b A_s(1 - \phi)\rho \tag{17.15}$$

Next Eq. 17.15 can be solved for

$$(1 - \phi)\rho = \frac{v_s}{v_r + v_b} m \tag{17.16}$$

which can be substituted into Eq. 17.14 and the result solved for

$$m = \frac{Qm_{\text{in}}}{Q + v_s A_s(1 - F_r)} \tag{17.17}$$

where F_r is a resuspension factor that is defined as

$$F_r = \frac{v_r}{v_r + v_b} \tag{17.18}$$

Note that Eq. 17.17 is very similar to the well-mixed-lake models described previously. The effect of adding the sediment layer is isolated in the dimensionless parameter group F_r. This group represents the balance between the resuspension rate and the total rate at which the sediment purges itself of solids (that is, both burial and resuspension). Thus if burial dominates resuspension ($v_b \gg v_r$), $F_r \sim 0$ and Eq. 17.17 reduces to a well-mixed model with no sediments. In contrast, if resuspension dominates burial ($v_b \ll v_r$), $F_r \sim 1$ and Eq. 17.17 reduces to $m = m_{\text{in}}$. In other words when resuspension is relatively dominant, the water concentration will approach the inflow concentration because everything that settles is immediately resuspended.

The above solutions are in the simulation mode, where all the parameters are known. Although the solids model can be used in this way, it is more conventional for the model to be employed to estimate some of the parameters. In the next section we show how this can be done.

17.4.1 Parameter Estimation

The parameters in the model are ρ, ϕ, m, m_{in}, Q, A_s, v_s, v_r, and v_b. For the steady-state case, Eqs. 17.14 and 17.15 represent a pair of simultaneous algebraic equations. Thus given seven of the parameters, they generally provide us with a means of estimating the remaining two. Although an algorithm can be developed for this general problem, we take a different tack. We will try to assess which of the parameters are least likely to be available and then show how they can be estimated with the model.

Of the nine parameters, we assume that ρ and ϕ are known. Typical values for fine-grained sediments are ρ = 2.4–2.7 $\times$ 10^6 g m^{-3} and ϕ = 0.8–0.95. We also assume that the flow and area, Q and A_s, are given.

Therefore we are left with five unknown parameters: m, m_{in}, v_s, v_r, and v_b. Now among these, one is extremely difficult to measure: v_r. This is the focus of our parameter estimation. There are two situations that usually occur.

First, there is the case where m and m_{in} have been measured. In addition the settling velocity v_s may have been measured directly or estimated from literature values. For example a value of 2.5 m d^{-1} represents a typical value for allochthonous fine-grained particles (O'Connor 1988). Equations 17.14 and 17.15 can be added to give

$$0 = Qm_{in} - Qm - v_b A_s(1 - \phi)\rho \tag{17.19}$$

which can be used to estimate v_b,

$$v_b = \frac{Q}{A_s}\frac{m_{in} - m}{(1 - \phi)\rho} \tag{17.20}$$

Second, the burial velocity is sometimes measured directly. This is often accomplished using sediment-dating techniques (see Box 17.1)

Regardless of whether it is measured directly or computed, once v_b has been approximated the resuspension velocity can then be estimated by solving the steady-state version of Eq. 17.15 for

$$v_r = v_s\frac{m}{(1 - \phi)\rho} - v_b \tag{17.21}$$

EXAMPLE 17.2. SOLIDS BUDGET. Thomann and Di Toro (1983) presented the following data related to the solids budget for Lake Ontario:

Volume = 1666×10^9 m^3
Solids loading = 4.46×10^{12} g yr^{-1}
Area = $19{,}485 \times 10^6$ m^2
Suspended solids concentration = 0.5 mg L^{-1}
Flow = 212×10^9 m^3 yr^{-1}

They assumed that the solids settle at a rate of 2.5 m d^{-1} (912.5 m yr^{-1}) and that the sediments have $\rho = 2.4$ g cm^{-3} and $\phi = 0.9$. Determine the burial and resuspension velocities by a mass balance approach.

Solution: First, an inflow concentration can be determined as

$$m_{in} = \frac{4.46 \times 10^{12}}{212 \times 10^9} = 21 \text{ mg L}^{-1}$$

Next Eq. 17.20 can be used to calculate

$$v_b = \frac{212 \times 10^9}{19{,}485 \times 10^6}\frac{21 - 0.5}{(1 - 0.9)2.4 \times 10^6} = 0.000929 \text{ m yr}^{-1} = 0.929 \text{ mm yr}^{-1}$$

This result can be substituted into Eq. 17.21 to determine

$$v_r = 912.5\frac{0.5}{(1 - 0.9)2.4 \times 10^6} - 0.000929 = 0.000972 \text{ m yr}^{-1} = 0.972 \text{ mm yr}^{-1}$$

The simple budgets described above have been used in conjunction with contaminant balances to model toxic substance dynamics in lakes (see Lec. 40). However, it is recognized that they represent a simplification of the dynamics of solids in such systems. In particular it is understood that sediment resuspension is not a steady-state process. Rather, it occurs episodically—usually due to high-wind events in lakes and high-current events in rivers. In the last section of this lecture we introduce a model that simulates resuspension episodes in lakes.

17.5 BOTTOM SEDIMENTS AS A DISTRIBUTED SYSTEM

In the previous section we characterized the bottom sediments as a single layer. Although such lumped models have proved useful in areas such as toxicant modeling, sediments can also be characterized as distributed systems. As depicted in Fig. 17.6 the simplest such approach views the bottom sediments as a one-dimensional continuum in the vertical.

Three processes are shown in Fig. 17.6. The substance being modeled is subject to simple first-order decay. We also assume that it diffuses within the pore water. Finally because solid matter rains down from the overlying water, substances in the sediment are buried. Thus although a layer of sediment does not move physically, its distance from the sediment-water interface increases with time as matter accumulates on the bottom. In essence the sediment-water interface is advecting upward. From our modeling perspective, however, it is convenient to conceptualize the process as if the interface were static and the sediments were advecting downward.

For the case of a dissolved contaminant the three mechanisms can be combined into the following mass balance:

$$\frac{\partial c}{\partial t} = -v_b \frac{\partial c}{\partial z} + \phi D \frac{\partial^2 c}{\partial z^2} - kc \tag{17.22}$$

where c = concentration of a dissolved contaminant (mg L^{-1})
v_b = the burial velocity (m yr^{-1})
D = an effective diffusion coefficient through the sediment pore waters (m^2 yr^{-1})

We have assumed constant parameters for Eq. 17.22. It should be noted that this is not necessarily true. With respect to transport, sediments are subject to

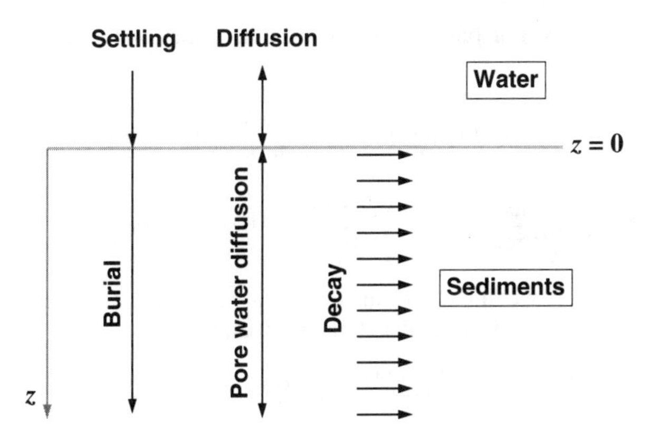

FIGURE 17.6
Schematic of a sediment viewed as a vertical distributed system.

compaction as the weight of overlying sediments presses down on deeper layers. In the simplest sense such a process means that both the velocity and the porosity vary with depth. Although in most water-quality contexts the effect is assumed to be negligible, modifications of Eq. 17.22 are available to account for the phenomenon (Robbins and Edgington 1975, Chapra and Reckhow 1983).

Equation 17.22 is of the same form as Eq. 10.23 for a mixed-flow reactor or an estuary. In fact we are treating the sediments as if they were an estuary on its side. Consequently all the previous analytical and numerical solutions developed in this text for such systems are applicable to the sediments.

17.5.1 Steady-State Distributions

Suppose that the pore water at the sediment-water interface is held at a constant level c_0 for a sufficiently long time so that the sediments come to a steady-state. For this case, Eq. 17.22 becomes

$$0 = -v_b \frac{dc}{dz} + \phi D \frac{d^2c}{dz^2} - kc \tag{17.23}$$

with boundary conditions

$$\begin{aligned} c(0, t) &= c_0 \\ c(\infty, t) &= 0 \end{aligned} \tag{17.24}$$

Using the techniques described in Lec. 9, the solution is

$$c = c_0 e^{\lambda z} \tag{17.25}$$

where

$$\lambda = \frac{v_b}{2\phi D}\left(1 - \sqrt{1 + \frac{4k\phi D}{v_b^2}}\right) \tag{17.26}$$

EXAMPLE 17.3. STEADY-STATE SEDIMENT DISTRIBUTION. A contaminant has a pore water concentration of 10 μg L^{-1} at the sediment-water interface. If it has a half-life of 10 yr, how far will it penetrate into the sediments if $\phi D = 0.9 \times 10^{-5}$ cm^2 s^{-1} and $v_b = 2$ mm yr^{-1}?

Solution: First, let's calculate the reaction rate and express the other parameters in commensurate units,

$$k = \frac{0.693}{10 \text{ yr}} = 0.0693 \text{ yr}^{-1} \qquad v_b = 2 \text{ mm yr}^{-1}\left(\frac{\text{m}}{1000 \text{ mm}}\right) = 0.002 \text{ m yr}^{-1}$$

$$\phi D = 0.9 \times 10^{-5} \text{ cm}^2 \text{ s}^{-1}\left(\frac{\text{m}^2}{10^4\text{cm}^2} \frac{86{,}400\text{s}}{\text{d}} \frac{365\text{d}}{\text{yr}}\right) = 0.02838 \text{ m}^2 \text{ yr}^{-1}$$

Before calculating the concentration, we first determine the estuary number (recall Sec. 10.5) to ascertain which mechanism dominates, advection or diffusion,

$$\eta = \frac{k\phi D}{v_b^2} = \frac{0.0693(0.02838)}{0.002^2} = 492$$

Thus diffusion overwhelms advection. To calculate the distribution in the sediments, the eigenvalue can be determined as

$$\lambda = \frac{0.002}{2(0.02838)}\left[1 - \sqrt{1 + 4(492)}\right] = -1.528 \text{ m}^{-1}$$

This value can be substituted into Eq. 17.25,

$$c = 10e^{-1.528z}$$

which can be used to calculate the profile displayed in Fig. E17.3.

A 95% penetration depth can be computed as

$$x_{95} = \frac{3}{1.528} = 1.96 \text{ m}$$

Thus as depicted in the graph, the pollutant penetrates about 200 cm into the sediments.

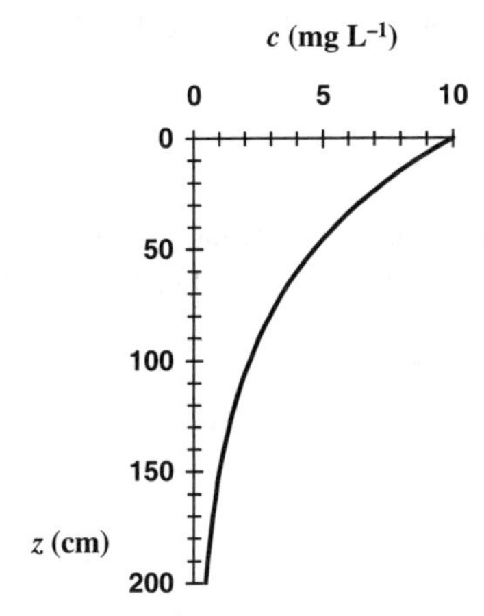

FIGURE E17.3

17.5.2 Time-Variable Distributions

Although analytical solutions are available for a variety of idealized cases (Carslaw and Jaeger 1959), numerical methods provide a more general approach. Equation 17.22 can be integrated using any of the finite-difference methods described in Lecs. 12 and 13. For problems where advection is negligible, the simple explicit FTCS approach (Lec. 12) can be used to express it as

$$c_i^{\ell+1} = c_i^\ell + \left(\phi D \frac{c_{i+1}^\ell - 2c_i^\ell + c_{i-1}^\ell}{\Delta z^2} - kc_i^\ell\right)\Delta t \tag{17.27}$$

Boundary conditions can be incorporated by developing finite-difference equations for the top and bottom segments. For the top segment (that is, the one at the sediment-water interface), the simplest case is the Dirichlet boundary condition. For this case the difference equation would be

$$c_1^{\ell+1} = c_1^\ell + \left(\phi D \frac{c_2^\ell - 2c_1^\ell + c_0^\ell}{\Delta z^2} - kc_1^\ell\right)\Delta t \tag{17.28}$$

where the subscript 0 designates the concentration of the overlying water.

For the bottom segment (subscript $= n$) a flat gradient can be assumed. This means the net transport across the lower boundary is zero,

$$c_n^{\ell+1} = c_n^\ell + \left(\phi D \frac{c_{n-1}^\ell - c_n^\ell}{\Delta z^2} - kc_n^\ell\right)\Delta t \tag{17.29}$$

EXAMPLE 17.4. TIME-VARIABLE SEDIMENT DISTRIBUTION. Use the same system as Example 17.3. Assume that at $t = 0$ the sediments are devoid of the contaminant and that the boundary concentration is immediately brought up to the level of 10 mg L^{-1}. Use the simple explicit approach to compute sediment profiles during the build-up in the sediments.

Solution: First, let's determine the time step to achieve stability. Using a spatial scale of 2.5 cm, Eq.12.16 can be employed to calculate

$$\Delta t = \frac{0.25^2}{2(0.02838) + 0.0693(0.25)^2} = 1.1 \text{ yr}$$

Therefore we will use a time step of 0.25 yr. The resulting calculation is shown in Fig. E17.4. As can be seen the sediment profile takes over a decade to build up to the steady-state previously computed in Example 17.3.

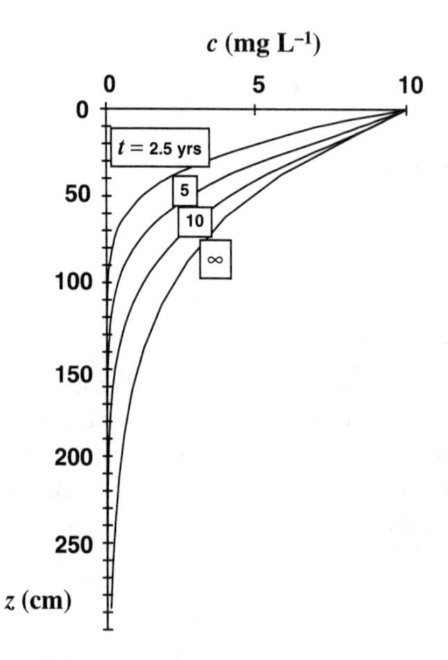

FIGURE E17.4

In this section we have limited ourselves to a single, dissolved compound. It should be noted that many problems involve several compounds. When we discuss sediment oxygen demand in Lec. 25, we will deal with multiple reactants in a sediment system.

Further, many contaminants sorb onto particulate matter. Consequently the diffusion mechanism can be reduced by this effect. In some cases, such as the radionuclide lead-210, sorption is so strong that diffusion can become negligible (see Box 17.1). We will describe this sorption effect in detail when we discuss the modeling of toxic substances in the sediments in Part VII.

BOX 17.1. Sediment Dating with Lead-210

Lead-210 or ^{210}Pb is a naturally occurring ***radionuclide*** (an unstable isotope) that is delivered to the sediment-water interface at a constant flux rate. Because it sorbs strongly and has a long (but not too long) half-life of 19.4 yr, it is an effective tool for quantifying the rate of sedimentation for lakes and other standing waters.

For a substance that associates exclusively with solid matter, Eq. 17.23 can be expressed as

$$0 = -v_b \frac{d\nu}{dz} - kc \tag{17.30}$$

where ν = amount of contaminant per mass of solids. When the contaminant is measured on a mass basis the units of ν might be something like $\mu g\ g^{-1}$. For a radionuclide like ^{210}Pb, where the amount is measured in curies (see Lec. 43 for additional details on this and other units of radioactivity), the units would be $Ci\ g^{-1}$. Aside from the change in the expression for concentration, notice that the diffusion term has dropped out of the model. Thus because the substance is not free to move in the pore water, the model becomes plug flow.

If the concentration at the sediment-water interface is ν_0 and the burial velocity and decay rate are constants, Eq. 17.30 can be solved for

$$\nu = \nu_0 e^{-\frac{k}{v_b} z} \tag{17.31}$$

According to this equation the concentration should decrease exponentially with depth into the sediments. Consequently if the model holds, a semi-log plot of concentration versus depth should yield a straight line with an intercept of ν_0 and a slope of $-k/v_b$. Because the half-life of ^{210}Pb is known, the slope can be used to evaluate the burial velocity. Problem 17.5 at the end of the lecture deals with this calculation.

Aside from estimating the burial velocity, the ^{210}Pb techniques can be used to date sediments. Once the burial velocity and the decay rate are known, each stratum can be assigned a time horizon below the sediment-water interface by recognizing that time and depth are related by $v_b = x/t$. Problem 17.5 also deals with this estimate.

Finally it should be mentioned that other approaches are available to date sediments and hence estimate the burial rate. A group of techniques are based on identifying the remains of aquatic and terrestrial organisms. Thus the approach is similar to archaeological dating based on fossil remains.

For example pollen in a watershed can change abruptly due to agricultural clearing of land, disease, or the introduction of exotic plants. For example farming in many parts of North America was accompanied by the sudden rise of *Ambrosia,* or common ragweed. If the time of agriculturalization can be pinpointed, the sediment strata where the pollen first shows up is associated with that time. Wetzel (1975) provides a nice discussion of this and other methods.

17.6 RESUSPENSION (ADVANCED TOPIC)

As depicted in Fig. 17.7 the resuspension of sediments is a function of several factors. The process begins with the energy delivered to the water surface via the wind. The amount of energy is dependent on both the wind velocity and fetch. The latter refers to the length of exposed water surface in the direction that the wind blows.

This wind energy creates waves. In general the greater the wind velocity and fetch, the greater the height and period of the resulting waves. Under the surface the water moves in circular eddies, the energy of which dissipates with depth. The orbital velocity of these gyres (large eddies) exerts a shear stress at the bottom. Scour or resuspension of the sediments can then occur, depending on the magnitude of the stress and the type of bottom sediments.

Engineers and scientists have developed a variety of formulas to quantify the process. The following development uses the approach outlined by Kang et al. (1982) to simulate wave action and bottom shear stresses in Lake Erie.

Wave height, period, and length. Ijima and Tang (1966) have developed formulas to predict wave height and period for shallow waters. These equations are dependent on mean depth, wind velocity, and fetch,

$$\frac{gH_s}{U^2} = 0.283\tanh\left[0.53\left(\frac{gH}{U^2}\right)^{0.75}\right]\tanh\left\{\frac{0.0125\left(\frac{gF}{U^2}\right)^{0.42}}{\tanh\left[0.53\left(\frac{gH}{U^2}\right)^{0.75}\right]}\right\} \tag{17.32}$$

$$\frac{gT_s}{2\pi U} = 1.2\tanh\left[0.833\left(\frac{gH}{U^2}\right)^{0.375}\right]\tanh\left\{\frac{0.077\left(\frac{gF}{U^2}\right)^{0.25}}{\tanh\left[0.833\left(\frac{gH}{U^2}\right)^{0.375}\right]}\right\} \tag{17.33}$$

where H_s = significant wave height (m)
T_s = significant wave period (s)
U = wind velocity (m s^{-1})
H = mean depth (m)
F = fetch (m)

Note that the hyperbolic tangent is available as an intrinsic function in Fortran as well as on other software such as some spreadsheets. It can be also be computed with the exponential function

$$\tanh(x) = \frac{e^x - e^{-x}}{e^x + e^{-x}} \tag{17.34}$$

The wavelength can also be determined by solving the following equation iteratively,

$$L = L_0 \tanh\left(\frac{2\pi H}{L}\right) \tag{17.35}$$

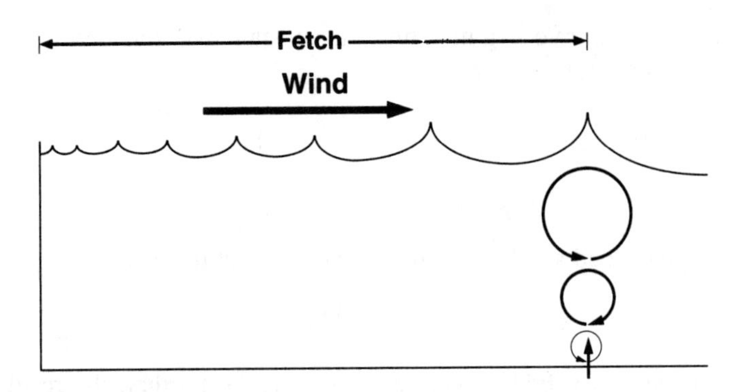

FIGURE 17.7
A simple model of how wind induces wave action that leads to sediment resuspension.

where L = wavelength (m) and L_0 = deep-water wavelength (m), which can be estimated by

$$L_0 = \frac{gT_s^2}{2\pi} \tag{17.36}$$

Orbital velocity. The orbital velocity created by the wave can be determined by

$$\overline{U} = \frac{\pi H_s}{T_s} \frac{100}{\sinh(2\pi H/L)} \tag{17.37}$$

where $\overline{U}$ = orbital velocity (cm s^{-1}). The hyperbolic sine can be computed as

$$\sinh(x) = \frac{e^x - e^{-x}}{2} \tag{17.38}$$

Shear stress. For shallow lakes where currents are generally small, the shear stress can be approximated by

$$\tau = 0.003\overline{U}^2 \tag{17.39}$$

where τ = shear stress (dyne cm^{-2}).

Bottom scour. The amount of mass of sediments scoured from the bottom can be calculated with

$$\begin{aligned} \varepsilon &= 0 & \tau &\le \tau_c \\ \varepsilon &= \frac{\alpha_0}{t_d^2}(\tau - \tau_c)^3 & \tau &> \tau_c \end{aligned} \tag{17.40}$$

where ε = mass of sediments scoured (g m^{-2})
α_0 = 0.008
t_d = 7
τ_c = critical shear stress (dyne cm^{-2})

Suspended solids concentration and entrainment rate. Experiments have shown that the entrainment process takes on the order of an hour at a constant shear stress. The entrainment rate E(g m^{-2} hr^{-1}) is constant over this period and considered negligible thereafter,

$$E = \frac{\varepsilon}{1\text{ hr}} \qquad t \le 1\text{ hr}$$
$$E = 0 \qquad t > 1\text{ hr} \tag{17.41}$$

The resulting suspended solids concentration can be estimated by

$$c = 10{,}000\frac{\varepsilon}{H} \tag{17.42}$$

where c = suspended solids concentration (mg L^{-1}). Further, the results can be expressed as a resuspension velocity by

$$v_r = \frac{E}{(1-\phi)\rho} \tag{17.43}$$

Inspection of Eqs. 17.39 and 17.40 indicates that below a critical shear stress no resuspension will occur. Above the critical level, however, resuspension is dependent on the orbital velocity raised to the sixth power. Thus scour will be particularly critical during extreme storm events.

EXAMPLE 17.5. CALCULATION OF SEDIMENT RESUSPENSION. A uniform wind blows over a rectangular basin as shown in Fig. E17.5:

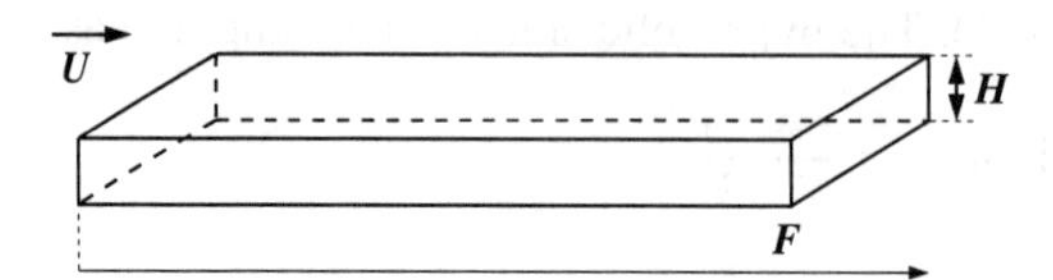

FIGURE E17.5

The critical shear stress is equal to 1 dyne cm^{-2}. If the wind velocity and the depth are equal to 20 mps and 5 m, respectively, determine the sediment resuspension at a point F = 10 km down the basin.

Solution: First, Eqs. 17.32 and 17.33 can be used to determine the significant wave height and period,

$$H_s = \frac{20^2}{9.8}0.283\tanh\left[0.53\left(\frac{9.8\times 5}{20^2}\right)^{0.75}\right]\tanh\left\{\frac{0.0125\left(\frac{9.8\times 10{,}000}{20^2}\right)^{0.42}}{\tanh\left[0.53\left(\frac{9.8\times 5}{20^2}\right)^{0.75}\right]}\right\} = 1.03\text{ m}$$

$$T_s = \frac{2\pi(20)^2}{9.8}1.2\tanh\left[0.833\left(\frac{9.8\times 5}{20^2}\right)^{0.375}\right]\tanh\left\{\frac{0.077\left(\frac{9.8\times 10{,}000}{20^2}\right)^{0.25}}{\tanh\left[0.833\left(\frac{9.8\times 5}{20^2}\right)^{0.375}\right]}\right\}$$
$$= 3.82\text{ s}$$

Substituting the parameters for the present problem into Eq. 17.35 gives,

$$L = 22.81025 \tanh\left(\frac{31.416}{L}\right)$$

Using L_0 as the initial guess for L, we can solve the equation iteratively for

Iteration	L	% error
0	22.810	
1	20.081	13.59
2	20.897	3.91
3	20.660	1.14
4	20.730	0.336
5	20.710	0.0983
6	20.716	0.0278
7	20.714	0.00843

This result can be substituted into Eq. 17.37 to calculate the orbital velocity

$$\overline{U} = \frac{\pi(1.03)}{3.82} \frac{100}{\sinh[2\pi(5)/20.714]} = 39.15 \text{ cm s}^{-1}$$

The orbital velocity can then be employed to determine the shear stress,

$$\tau = 0.003\overline{U}^2 = 0.003(39.15)^2 = 4.6 \text{ dyne cm}^{-2}$$

which is used to calculate the sediment erosion rate

$$\varepsilon = \frac{8 \times 10^{-3}}{49}(4.6 - 1)^3 = 0.0076 \text{ g cm}^{-2}$$

This result can be normalized to depth to yield a concentration in the water,

$$c = 10{,}000\frac{0.0076}{5} = 15.22 \text{ mg L}^{-1}$$

It can also be expressed as a resuspension velocity,

$$v_r = \frac{0.0076 \text{ g cm}^{-2} \text{ hr}^{-1}}{(1 - 0.9)2.5 \text{ g cm}^{-3}} = 0.0304 \text{ cm hr}^{-1}$$

PROBLEMS

17.1. Use Stokes' law to calculate the settling velocities for the phytoplankton and the inorganic solids (clay and silt) in Table 17.3. Assume that the form factor is 1. Develop a plot of the measured velocities versus your calculated values. Plot both the growth and stationary values for the phytoplankton. Use different symbols on the plot to discriminate among inorganic solids, growth-phase phytoplankton, and stationary-phase phytoplankton. Discuss the plot.

17.2. Repeat Example 17.1, but use the area of the deposition zone (10,000 km^2) rather than the lake surface area as the actual interface across which the vertical transfers take place.

17.3. A lake is fed by two rivers as shown in Fig. P17.3. The inflowing rivers deliver solids at rates of 0.1×10^9 and 0.2×10^9 g yr^{-1}, respectively, and the outlet carries 0.05×10^9 g yr^{-1}. The lake's deposition area is 10^6 m^2.

(*a*) If the sediment has $\rho = 2.5 \times 10^6$ g m^{-3} and $\phi = 0.9$, determine the burial velocity.

(*b*) Sediment traps (area = 250 cm^2) are deployed in the lake. These yield an annual average accumulation rate of 20 g yr^{-1} of solid matter. In addition the average suspended solids concentration in the water is measured as 5 mg L^{-1}. Determine the settling velocity.

(*c*) Estimate the resuspension velocity.

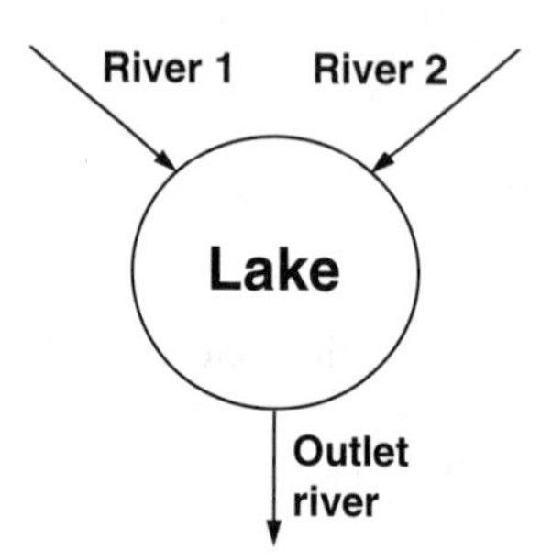

FIGURE P17.3

17.4. A lake has a steady-state concentration of 100 μg L^{-1} of a dissolved substance that has a half-life of 28 yr. Assuming that the sediments are also at a steady-state and that the burial velocity is negligible, determine the molecular diffusion coefficient based on the following sediment data ($\rho = 2.5 \times 10^6$ g m^{-3} and $\phi = 0.9$):

Depth (cm)	0	50	100	150	200
Concentration (μg L^{-1})	100	60	40	20	15

17.5. Determine the burial velocity and sediment deposition flux for Lake Michigan based on the data below for ^{210}Pb (assume that $\rho = 2.4 \times 10^6$ g m^{-3} and $\phi = 0.85$). Also if the sediment core were taken in 1975, determine the date at which the layer at 3 cm was deposited.

Depth (cm)	0	0.5	1.5	2.5	3.5	4.5	5.5
Concentration (10^{-12} Ci g^{-1})	8.1	7.2	5.4	3.6	1.5	0.6	0.5

17.6. Suppose that 5 Ci of a dissolved radionuclide (half-life = 20 yr) is spilled into a 5-m-deep lake that has a residence time of 10 yr and a surface area of 2×10^6 m^2. Calculate the profile in the sediments over the first 10 yr following the spill. Assume that the burial velocity is negligible and the diffusion coefficient is 5×10^{-6} cm^2 s^{-1}.

17.7. Repeat Example 17.5 for a range of wind velocities from 0 to 20 m s^{-1}. Plot the resulting suspended solids concentration versus wind velocity. Discuss your results.

LECTURE 18

The "Modeling" Environment

LECTURE OVERVIEW: I devote the bulk of the lecture to the various steps that constitute the modeling process. This is followed by descriptions of techniques that can be employed for two model-development tasks: sensitivity analysis and calibration. Finally I conclude the lecture with an overview of the interrelationship of model time, space, and kinetic scales.

Because their primary intent is to provide the reader with an organized body of knowledge, many textbooks give the impression that the world is a very neat place. In fact the "real world" is fraught with ambiguities and does not usually yield to our efforts in the same "black-and-white" fashion as in the end-of-chapter problems.

Water-quality modeling seems to be especially prone to such oversimplification because, before you can even develop realistic models, you must first assimilate a considerable body of technical information dealing with theory, mathematics, and computers. As a consequence the neophyte can come away from the topic thinking that it's all about writing mass balances and solving differential equations.

In reality the mathematical aspects are but one facet (albeit an important one) of how models are developed and applied to solve water-quality-modeling problems. It is the task of this lecture to provide a larger picture of the modeling process.

18.1 THE WATER-QUALITY-MODELING PROCESS

A water-quality model is merely one linkage in a larger management process (recall Fig. 1.2). Further the water-quality model itself does not arise out of thin air. Rather it results from a multistep model-development and application process (Fig. 18.1).

This lecture is designed to give you a glimpse of the larger picture by providing background information on the model-development process. In so doing I introduce tasks such as model calibration and confirmation that must be implemented prior to management decisions. I also cover some techniques such as sensitivity analysis that arise in model application.

18.1.1 Problem Specification: Getting Started

Before a single equation is written, as depicted in Fig. 18.1, the modeling process starts with a *problem specification* phase. In other words the water-quality engineer must be provided with a clear delineation of the objectives of the customer. This can be an individual or a decision-making entity such as a corporation, a municipality, or a regulatory agency. Oftentimes, because he or she usually has broad expertise related to water quality, the modeler is included in this process. This is particularly true when the decision-making entity requires additional technical expertise to clarify its ultimate goals.

Note that two primary information sources feed this phase. The first are management objectives, control options, and constraints, which might include physical constraints as well as legal, regulatory, and economic information.

The second source is data related to the physics, chemistry, and biology of the water body and its drainage basin. Oftentimes such information is sparse or nonexistent. In such cases some preliminary premodeling data collection could be in order.

When this phase is complete, the modeler should have a clear idea of the problem objectives and the accompanying water-quality variables required to assess whether the objectives are attainable. In addition, based on the available data, she or he should have a preliminary idea of how the system presently functions. In particular the modeler should understand the temporal, spatial, and kinetic resolution that will be compatible with the problem needs. I'll elaborate on the topic of model resolution at the end of this lecture.

Finally it should be noted that modelers often provide simple back-of-the-envelope calculations during this phase (recall Box 4.1). These are usually helpful in bounding the problem and providing rough estimates of system behavior.

18.1.2 Model Selection

Once you know what you're dealing with, the next step is to obtain a model. In some instances this can be accomplished by choosing an existing software package. This is obviously the preferred route for two reasons. First, someone else has done the work for you. Second, there are several models that have been widely used and, as a consequence, have a certain cachet in legal and regulatory contexts. Thus they have credibility in the eyes of judges and other decision makers.

Unfortunately there are many water-quality problems for which canned packages are inadequate or unavailable. These reflect both kinetics (for instance, they don't simulate the right pollutants) and the time-space scale considerations. For example current interest in nonpoint source pollution has prompted the need for water-quality models with built-in hydrodynamics. Unfortunately these are not available

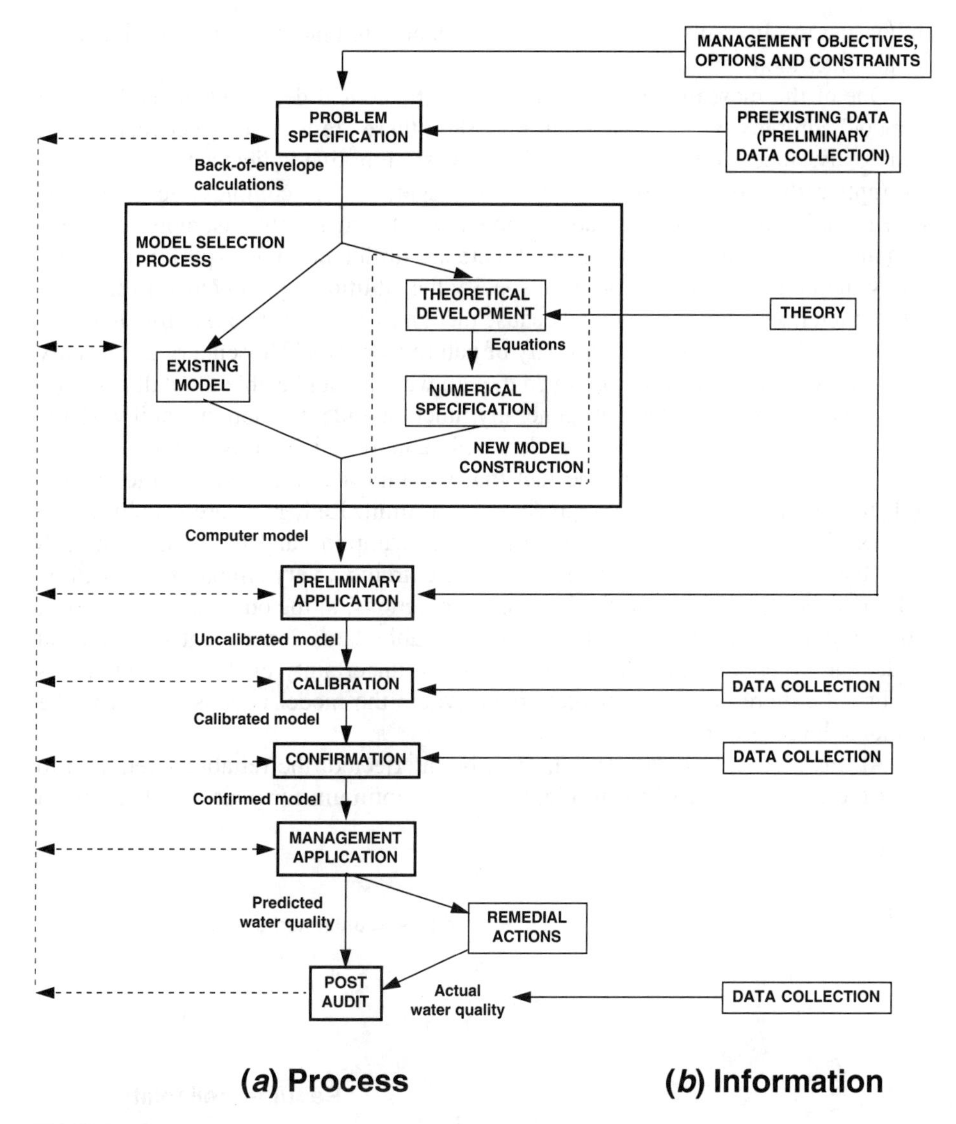

FIGURE 18.1
(*a*) The water-quality-modeling process along with (*b*) the necessary information needed for its effective implementation.

for the entire range of pollutants encountered in natural waters. In addition you may be confronting a new problem that has never been modeled before. In such cases you must develop your own model from scratch. This consists of two phases: theoretical development, and numerical specification and validation.

Theoretical development. First, you must provide a theoretical development that specifies the required variables and parameters along with the associated continuity equations. The continuity equations are most frequently written for mass and/or energy. In cases where hydrodynamics are simulated, momentum balances are also included.

One of the most important aspects of the theoretical design phase is the issue of model complexity. As depicted in Fig. 18.2, there are trade-offs between model complexity, uncertainty, and information. The straight line indicates the underlying assumption that, if we have an unlimited budget, a more complex model will be more reliable. In essence as we add complexity to the model (that is, more equations with more parameters), we assume that there are sufficient funds to perform the necessary field and laboratory studies to specify the additional parameters. In fact this assumption itself may not be true, because there may be limits to our ability to mathematically characterize the complexity of nature (sort of a "Heisenberg uncertainty principle" of ecology). However, even though we may not be able to totally characterize a natural water system, we generally function under the notion that if we have more and better information, our models will be more reliable predictors.

The real rub comes from the fact that we almost never have a large enough budget to even come close to approaching that limit. Rather we are usually in the position that we must make due with a limited sampling and lab budget. In such cases there are two extreme outcomes: at one extreme a very simple model will be so unrealistic that it will not yield reliable predictions; at the other, a very complex model will be so detailed that it outpaces available data. Consequently the second model can be equally unreliable because of the uncertainty of the parameters. As in Fig. 18.2 there is an intermediate point where the model is consistent with the available level of information.

At this point a third dimension must be interjected: the reliability required to solve the problem. Clearly you might be at the optimum for your data but have a

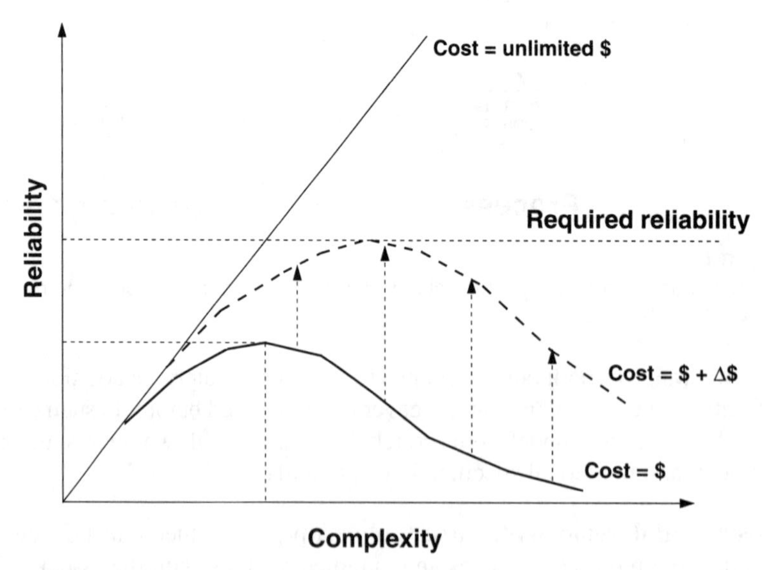

FIGURE 18.2
The trade-off between model reliability and complexity.

model that gives inadequate predictions for the problem being addressed. Consequently additional information must be collected to bring the model up to the required level.

All this is related to the fact that you should always strive to develop the simplest model that is consistent with the data and the problem requirements. This dictum is sometimes referred to as applying Ockham's razor† to the model. The idea was also expressed by Einstein. To paraphrase his words, "every model should be as simple as possible, but no simpler."

Numerical specification and validation. After we develop the theoretical specifications, the equations are implemented on the computer. This phase consists of several substeps, including algorithm design (that is, specification of data structure and numerical solution techniques), coding in a computer language, debugging, testing, and documentation. The numerical methods are chosen to generate sufficiently stable and accurate solutions for the minimum computational effort.

Before proceeding we should elaborate a bit on the testing phase of program development. One of the most common mistakes made by novice programmers is to assume that if the model yields "reasonable" results, it is adequately tested. Thus if results are positive (for variables such as concentration, which shouldn't go negative), relatively smooth, and have expected magnitudes, the computation is deemed successful.

In fact such criteria are no guarantee that the results are correct. For example it is relatively simple to violate the conservation of mass and still obtain smooth, positive values that are in the right ballpark. Consequently the model should be subjected to a battery of tests to ensure that it is mathematically correct. This is sometimes referred to as numerical validation (Oreskes et al. 1994). The following are some specific suggestions for establishing that the model calculation is valid:

- *Mass balances.* The model should be checked to ensure that mass is balanced. That is, over the period of the calculation the change of accumulation should equal the integral (that is, the summation) of the sources minus the sinks of mass. Some water-quality models do this as a part of their algorithms. For example the Corps of Engineers two-dimensional reservoir model, CE-QUAL-W2 (Cole and Buchak 1995), includes a mass-balance check as an automatic part of each run.
- *Simplified solutions.* The model output can be compared with simplified cases for which closed-form solutions are possible. For example time-variable models can be run to steady-state. The results can then be checked against closed-form steady-state solutions. Similarly analytical or closed-form solutions are sometimes available for idealized geometries and linear reaction kinetics. The model can be evaluated to see how well it matches such solutions.
- *Range of conditions.* Although the foregoing suggestions can help to identify major model flaws, they will not ensure that the model operates under a wide range of conditions. Therefore the model should be tested for robustness by applying it to

†Named after the monk William of Ockham.

different types of systems under differing initial conditions, boundary conditions, and loading scenarios.

- *Graphical results.* Although it does not address specific problems per se, a graphical interface is invaluable in assessing model performance and identifying model flaws. For example unless negative results occur, mild instabilities (that is, numerically generated oscillations) are extremely difficult to identify by perusing numerical output in the form of tables. However, they are readily visible as graphical output.
- *Benchmarking.* Finally benchmarking or "beta testing" involves having a wide number of users implement the model. Not only does this process result in identification of errors, but it also provides guidance on how to modify the model so that it more adequately meets the user's needs.

18.1.3 Preliminary Application

The next step in the process is to perform some preliminary simulations for the system being modeled. Although necessary data will often be lacking, the exercise is tremendously useful for identifying data deficiencies and theoretical gaps. In particular it can provide an important context for designing the field and laboratory studies required to fill the gaps.

Aside from identifying obvious information needs, this phase can also be useful in identifying which model parameters are most important. When working under a limited budget, it is important to recognize which model parameters have the greatest impact on model predictions. If this can be done, more of the available resources can be directed toward determining these parameters.

One way to identify the important parameters is with a model sensitivity analysis. In its simplest form a ***sensitivity analysis*** consists of merely varying each of the parameters by a set percent and observing how the predictions vary. A later section in this lecture provides details on how this can be done.

18.1.4 Calibration

The next step in the process is to calibrate or "tune" the model to fit a data set. This consists of varying the model parameters to obtain an optimal agreement between the model calculations and the data set.

The calibration data set should be as similar as possible to the design condition for the problem being studied. For example if the model is being used to analyze a wastewater discharge into a stream, the design condition would usually be for the summer low flow (that is, summer 7Q10 as described previously in Lec. 14). Consequently a data set collected during the summer under low-flow conditions would be most appropriate, whereas calibrating the model for a data set collected during the spring runoff would not be as effective.

After the data set is collected, there will be many parameters that can be varied. However, rather than arbitrarily trying to vary all of these, there is a systematic way to attain the best fit (Fig. 18.3). To understand how this is done, it should first be recognized that there are several types of information that must be fed into the model.

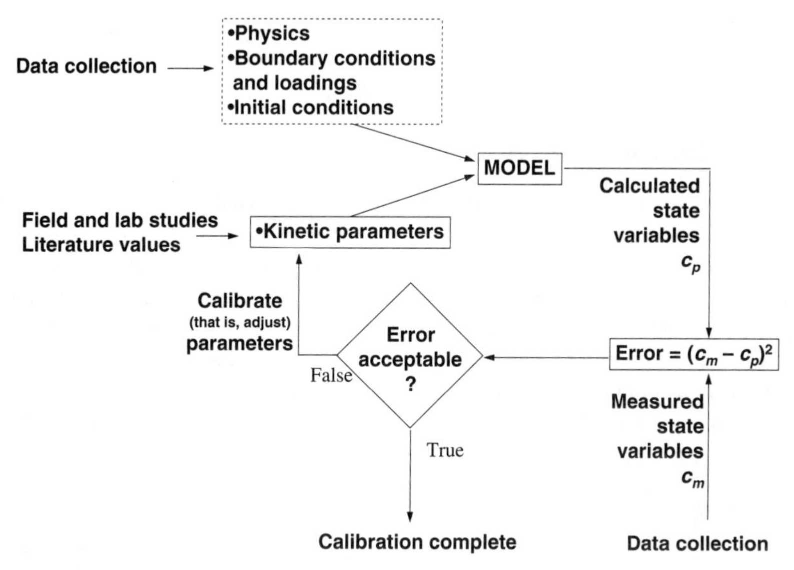

FIGURE 18.3
Schematic diagram of the model calibration process.

These include

- Forcing functions and physical parameters:
  - Boundary conditions and loads
  - Initial conditions
  - Physics
- Calibration parameters:
  - Kinetics

Notice how we have separated the kinetic parameters from the rest of the information. In essence we are assuming that sufficient information is available to adequately define the other parameters. Thus before beginning the calibration process, we must measure the system's loads, boundary conditions, and initial conditions with sufficient precision that they are not considered to be a significant source of uncertainty. Similarly an accurate characterization of the system's geometry and hydraulics must be developed. Once this is done, the physical parameters should not be varied.

When this is accomplished, the focus of the calibration becomes the adjustment of the kinetic parameters. This process also should be approached in a systematic fashion. For example if certain kinetic parameters have been directly measured in an adequate fashion, they should be fixed. Further some parameters can be estimated from the measured values of the state variables. Finally the remaining parameters can be varied within reasonable ranges until a best fit is attained. That is, adjustments are made until the simulation matches the data in some optimal sense.

There are several approaches for making the adjustments. At one extreme, tuning can proceed in a trial-by-error fashion. At the other, automated techniques such as least-squares regression can be implemented. Both are illustrated in an example later in this lecture.

18.1.5 Confirmation and Robustness

Once you have developed the calibrated model, all you really know is that the model fits a single data set. But before the model can be used with confidence to make management predictions, it must be confirmed. To do this the calibrated model should be run for a new data set (or ideally, several data sets), with the physical parameters and the forcing functions changed to reflect the new conditions. In contrast the kinetic coefficients must now be kept fixed at the values derived during the original calibration. If the new model simulations match the new data, the model has been confirmed as an effective prediction tool for the range of conditions defined by the original calibration and the confirmation data sets. If there is no match, the model can often be analyzed to determine possible reasons for the discrepancy(s). This may lead to additional mechanism characterization and model refinement.

Even when the model is deemed adequately confirmed, the simulation will not exactly match the confirmation data set. Thus some investigators fine tune the model after confirmation so that it optimally matches both the calibration and confirmation data.

It should be noted that in the past, water-quality modelers referred to this phase as "verification." This jargon implied that once the model successfully simulated an independent data set, it represented an establishment of "truth"; that is, it represented an accurate representation of physical reality. In fact this can never really be proved absolutely. At best all that can be concluded is that our testing has not proved the model wrong (Oreskes et al. 1994).

This then brings us to the notion of model robustness. If we can never totally verify a model, we are left with the quality or rigor of its confirmation. As put by Oreskes et al. (1994), "the greater the number and diversity of confirming observations, the more probable it is that the conceptualization embodied in the model is not flawed." Such a model is said to be ***robust***. Therefore the actual goal of the confirmation process should be to establish the model's robustness.

18.1.6 Management Applications

Many modeling studies result in remedial actions. For example waste treatment plants might be built or upgraded, or environmental modifications such as aeration or dredging could be implemented. The effectiveness of these actions can be evaluated by modifying the model parameters and forcing functions and running the model to predict the effects of the changes on the state variables.

18.1.7 Post-Audit

After implementation of remedial actions, a check can be made to learn whether the model predictions were valid. In a very few cases such model post-audits have shown excellent agreement between the predictions and the resulting water quality. However, in other instances, such is not the case and differences occur. Although they can be disturbing to the original modelers, such discrepancies can be extremely useful since they often provide clues to missing mechanisms and information. Thus they provide a further way for model developers to improve the robustness of their frameworks.

BOX 18.1. Great Lakes *Cladophora* Model

Over the past 40 years the Great Lakes have been threatened by a number of water-quality problems. One of the most visible is the growth of nuisance filamentous algae such as *Cladophora glomerata* on rocks, breakwalls, buoys, and other structures intersecting the water line. Profuse growth of such organisms takes place in areas enriched with nutrients. The problem is particularly vexing because it occurs near shore, where people come into the most frequent contact with the water body.

In the late 1970s and early 1980s Professors Ray Canale (University of Michigan) and Marty Auer (Michigan Technological University) conducted a major observational and modeling study of these organisms. Although they mapped *Cladophora* distribution in all the Great Lakes, their modeling study focused on the Harbor Beach area of Lake Huron.

Aside from representing a landmark in the modeling of large aquatic plants, this study was also significant from the perspective of model development. This was due to two factors.

First, after the model equations were developed, a traditional model calibration was not performed. Rather Canale and Auer directly measured all the model rates in the laboratory or in the field. For example to determine the plant photosynthesis and respiration rates for the model, they collaborated with Dr. Jim Graham of the University of Wisconsin to grow the organism under a controlled environment in a cross-gradient room called a BIOTRON (Graham et al. 1982). This facility allows plants to be grown and measurements to be carried out simultaneously over a matrix of light and temperature conditions. The resulting measurements were then used to develop model parameters in a fashion similar to the kinetic analyses previously described in Lec. 2. By measuring all rates independently, Canale and Auer eliminated calibration parameters from the resulting model. Thus for better or worse, it yielded predictions with no tuning.

Second, Canale and Auer then used the model to simulate their system in a "before and after" sense. In other words they first employed the model to simulate polluted conditions in 1979 and unpolluted conditions in 1980. The pollution reduction was due to the installation of phosphorus removal by Harbor Beach between their two sampling periods. In both cases the model parameters were set at the values determined in the laboratory and field measurements. Thus only the loading and physical parameters such as light and temperature were changed to reflect the differences between the 2 years.

The results of the model simulations are shown in Fig. B18.1. The model did a good job in capturing the improvement in soluble reactive phosphorus levels in the water. Although the response of the *Cladophora* standing crop (biomass) does not appear as accurate, it must be considered that the biomass measurements are much less precise than for a chemical measurement like phosphorus. If this is taken into consideration, the model generally captured the improvement in biomass, particularly in the vicinity of the outfall (distance = 0.0), where the problem was greatest.

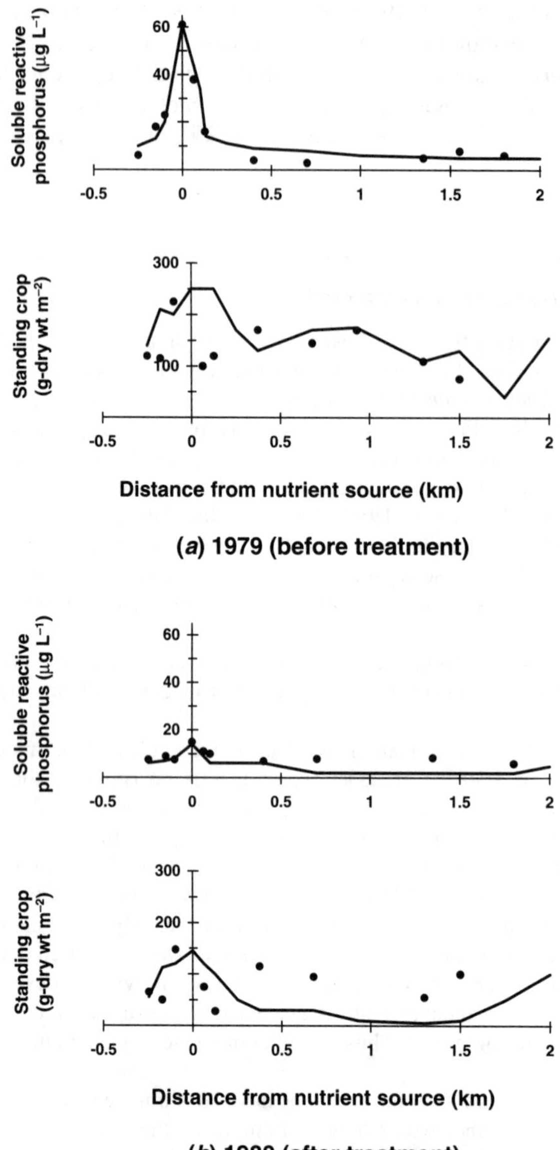

FIGURE B18.1
Plots of SRP and biomass versus distance from the outfall at Harbor Beach, Lake Huron (redrawn from Canale and Auer 1982). The plots represent the condition before and after phosphorus reduction technology was installed at the Harbor Beach wastewater treatment plant.

In summary Canale and Auer's Lake Huron *Cladophora* study is unique in coupling direct mechanism measurement with modeling. Professor Auer has edited a special issue of the *Journal of Great Lakes Research* (Auer 1982) devoted to the study. Aside from representing a seminal contribution in the area of algal growth modeling, it provides an interesting case study related to the model development and application issues described in this lecture.

18.2 MODEL SENSITIVITY

Before modeling, it is advisable to understand the general behavior of a water-quality model. One way to do this is called a ***sensitivity analysis***. There are a variety of ways to implement such an analysis. Two of the most common are simple parameter perturbation and first-order sensitivity analysis. In addition Monte Carlo approaches provide a more general approach. We describe each in the following sections.

18.2.1 Parameter Perturbation

For simplicity we illustrate both parameter perturbation and first-order sensitivity analysis for a simple mass-balance equation for a well-mixed lake,

$$V\frac{dc}{dt} = Qc_{\text{in}} - Qc - kVc \tag{18.1}$$

which, at steady-state, can be solved for

$$c = \frac{Q}{Q + kV}c_{\text{in}} \tag{18.2}$$

Recognize that according to this formula c is a function of each of the model parameters and forcing functions; that is, $c = f(Q, k, V, c_{\text{in}})$. Thus one way of visualizing the dependence of the solution on one of the parameters (for example k) is by developing a plot of c versus k (Fig. 18.4).

As the name implies, ***parameter perturbation*** consists of varying each of the model parameters (for example raised and lowered a fixed percent) while holding all other terms constant. The corresponding variations of the state variables reflect the sensitivity of the solution to the varied parameter. Thus if we desired to quantify the sensitivity of Eq. 18.2 to a parameter such as the reaction rate k, we would merely substitute $k - \Delta k$ and $k + \Delta k$ into Eq. 18.2 to calculate $c(k - \Delta k)$ and $c(k + \Delta k)$. The error of the prediction could then be calculated as

$$\Delta c = \frac{c(k + \Delta k) - c(k - \Delta k)}{2} \tag{18.3}$$

A graphical depiction of the approach is shown in Fig. 18.4*a*.

18.2.2 First-Order Sensitivity Analysis

An alternative technique that yields similar information is based on a ***first-order sensitivity analysis***. This approach uses the derivative of the function with respect to the parameter as an estimate of the sensitivity. One way to derive it is to employ first-order Taylor-series expansions of the model (Eq. 18.2) around the value of the parameter. For example forward and backward expansions can be written as

$$c(k + \Delta k) = c(k) + \frac{\partial c(k)}{\partial k}\Delta k \tag{18.4}$$

$$c(k - \Delta k) = c(k) - \frac{\partial c(k)}{\partial k}\Delta k \tag{18.5}$$

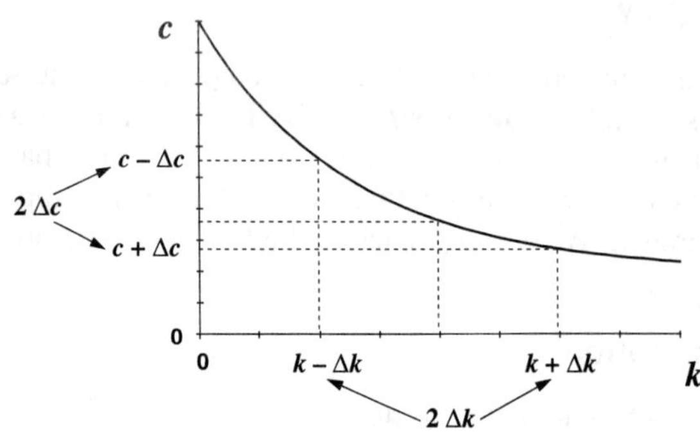

(a) Parameter perturbations

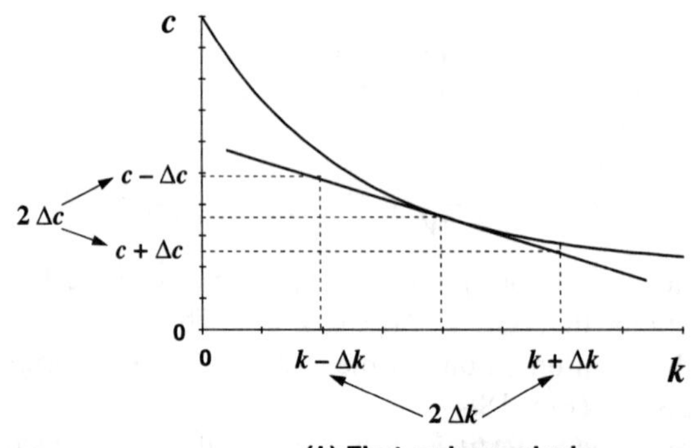

(b) First-order analysis

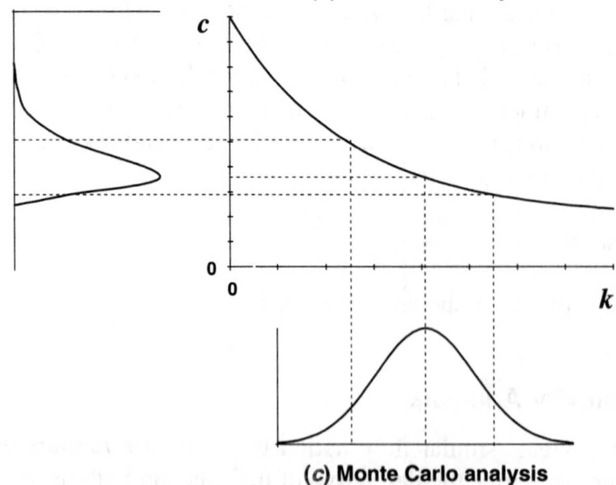

(c) Monte Carlo analysis

FIGURE 18.4 Graphical depictions of three methods for assessing model sensitivity.

Equation 18.5 can be subtracted from 18.4 and the result solved for

$$\Delta c = \frac{c(k + \Delta k) - c(k - \Delta k)}{2} = \frac{\partial c(k)}{\partial k}\Delta k \tag{18.6}$$

Note that for this formula the sign of the derivative indicates whether a positive variation of the parameter results in a positive or negative variation of the prediction. As depicted in Fig. 18.4*b*, because the slope of the straight-line approximation is negative, an increase in *k* results in a decrease in *c*.

A nice refinement on both the perturbation and first-order sensitivity analyses is to express the results as ***condition numbers*** (Chapra and Canale 1988). For the first-order sensitivity analysis such a number can be derived by dividing both sides of Eq. 18.6 by *c*. Then the right-hand-side can be multiplied by k/k and the result solved for

$$\frac{\Delta c}{c} = CN_k \frac{\Delta k}{k} \tag{18.7}$$

where CN_k is the condition number for the parameter *k*,

$$CN_k = \frac{k}{c}\frac{\partial c}{\partial k} \tag{18.8}$$

Thus Eq. 18.7 indicates that the condition number provides a transfer function to propagate the relative error of the parameter into the relative error of the prediction. Note that although the derivation is for the first-order sensitivity analysis, it can be used for the perturbation analysis. However, for this case a discrete form is used for the derivative,

$$CN_k = \frac{k}{c}\frac{\Delta c}{\Delta k} \tag{18.9}$$

EXAMPLE 18.1. SENSITIVITY ANALYSIS. As depicted below, two chemical species react within a lake. Mass balances for the two reactants can be written as

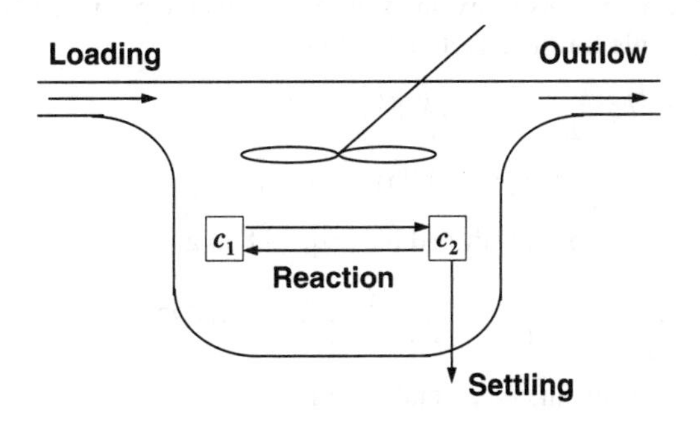

FIGURE E18.1

$$V\frac{dc_1}{dt} = Qc_{\text{in}} - Qc_1 - k_{12}Vc_1 + k_{21}Vc_2$$

$$V\frac{dc_2}{dt} = -Qc_2 + k_{12}Vc_1 - k_{21}Vc_2 - v_sA_sc_2$$

where V = volume of lake = 50,000 m^3
Q = inflow = outflow = 50,000 $m^3\ d^{-1}$
A_s = surface area of lake sediments = 16,667 m^2
k_{12} = first-order conversion rate of c_1 to c_2 = 1 to 4 d^{-1}
k_{21} = first-order conversion rate of c_2 to c_1 = 0.5 to 0.7 d^{-1}
v_s = settling velocity of c_2 = 0 to 1 m d^{-1}

The ranges for the parameters connote literature values. The term c_{in} represents the concentration of the inflowing stream (mg L^{-1}), which during the period of study has a mean value of approximately 10 mg L^{-1}. Use this information to estimate the sensitivity of the model to the three parameters k_{12}, k_{21}, and v_s. Employ first-order sensitivity analysis and express your results as a table of condition numbers.

Solution: First, we can develop the steady-state solution for each of the unknowns as

$$c_1 = \frac{k_q^2 + k_q k_{21} + k_q k_s}{k_q^2 + k_q k_{21} + k_q k_s + k_q k_{12} + k_{12} k_s} c_{in}$$

$$c_2 = \frac{k_q k_{12}}{k_q^2 + k_q k_{21} + k_q k_s + k_q k_{12} + k_{12} k_s} c_{in}$$

where $k_q = Q/V$
$k_s = v_s/H$
H = mean depth = V/A_s

Because we have no basis for anything better (at this point), we will take the average value of each parameter as the most likely value. For example for k_{12} we would use $(1 + 4)/2 = 2.5\ d^{-1}$. This value, along with the average values for the other parameters ($k_{21} = 0.6\ d^{-1}$ and $k_s = 0.5/3 = 0.1667\ d^{-1}$) can be used to compute our mean predictions,

$$c_1 = \frac{1^2 + 1(0.6) + 1(0.1667)}{1^2 + 1(0.6) + 1(0.1667) + 1(2.5) + 2.5(0.1667)} 10 = 3.772 \text{ mg L}^{-1}$$

$$c_2 = \frac{1(2.5)}{1^2 + 1(0.6) + 1(0.1667) + 1(2.5) + 2.5(0.1667)} 10 = 5.338 \text{ mg L}^{-1}$$

We can also determine the total concentration as $c_T = c_1 + c_2 = 9.110$ mg L^{-1}.

Now we can perform a first-order sensitivity analysis. To do this the partial derivatives must be analyzed. For example for the effect of k_{12} on c_1,

$$\frac{\partial c_1}{\partial k_{12}} = \frac{-\left(k_q^2 + k_q k_{21} + k_q k_s\right)\left(k_q + k_s\right)}{\left(k_q^2 + k_q k_{21} + k_q k_s + k_q k_{12} + k_{12} k_s\right)^2} c_{in}$$

which along with the other values can be substituted into Eq. 18.8 to give

$$CN_{1,k_{12}} = \frac{-k_{12}\left(k_q + k_s\right)}{k_q^2 + k_q k_{21} + k_q k_s + k_q k_{12} + k_{12} k_s} = -0.623$$

The other condition numbers can be evaluated and tabulated:

	k_{12}	k_{21}	k_s
c_1	−0.623	0.212	−0.030
c_2	0.377	−0.128	−0.125
c_T	−0.037	0.013	−0.086

These results lead to the following general conclusions:

- The transfer rate k_{12} has the greatest effect on both variables. However, it has an opposite impact on the two state variables. Whereas c_1 is inversely dependent on the rate, c_2 is directly dependent on the rate.
- The impact of k_{21} is opposite to that of k_{12}. That is, a rise in k_{21} causes c_1 to rise and c_2 to fall. However, the magnitude of the impact is less than for k_{12}.
- The settling rate causes both concentrations to fall, but has about 4 times more impact on c_2 than on c_1.
- The effect of the parameters on c_T is much less than on the individual constituents because flushing is the primary input-output mechanism (see Prob. 18.1). Of all the parameters, the total concentration is most sensitive to the only one that directly affects input-output: v_s.

A problem with the foregoing example is that it did not consider the uncertainty of the parameter estimates. By using constant perturbations we ascribe the same uncertainty to each. However, based on the ranges provided in the problem, they exhibit quite different uncertainties. For example k_{21} is known with a much greater precision than either k_{12} or the settling velocity. One way of rectifying this problem involves propagating the range through either the perturbation or first-order procedures. This can done simply by using Eq. 18.7 to propagate the relative error in each parameter into the resulting error in the prediction.

EXAMPLE 18.2. UNCERTAINTY ANALYSIS. For the case from Example 18.1, determine the expected uncertainties of the concentrations due to the uncertainties of the parameters.

Solution: The relative error in the parameter k_{12} can be calculated as

$$\frac{\Delta k_{12}}{k_{12}} = \frac{4 - 1}{2(2.5)} = 0.6$$

which can be propagated through the condition number to estimate the expected uncertainty in c_1,

$$\frac{\Delta c_1}{c_1} = -0.623(0.6) = -0.374$$

The other uncertainties can be evaluated and tabulated:

	k_{12}	k_{21}	k_s
c_1	−0.374	0.035	−0.030
c_2	0.226	−0.021	−0.125
c_T	−0.022	0.002	−0.086

These results indicate that:

- The uncertainty in k_{12} is a large contributing factor to the uncertainty in the relative values of c_1 and c_2. However, because it merely shifts mass between the two forms, the uncertainty does not strongly impact the total quantity, c_T.
- The uncertainty in k_{21} has little impact on the uncertainty of the components or the total concentration.

- The settling parameter has a larger impact on the uncertainty of c_2. It also makes a significant contribution to the uncertainty of the total prediction.

On the basis of this simple analysis, it can be concluded that effort should be expended in refining the estimates of k_{12} and k_s.

18.2.3 Monte Carlo Analysis

As the name implies, Monte Carlo analysis is related to games of chance. Rather than prescribing a range for the parameters, the analysis characterizes their distributions.

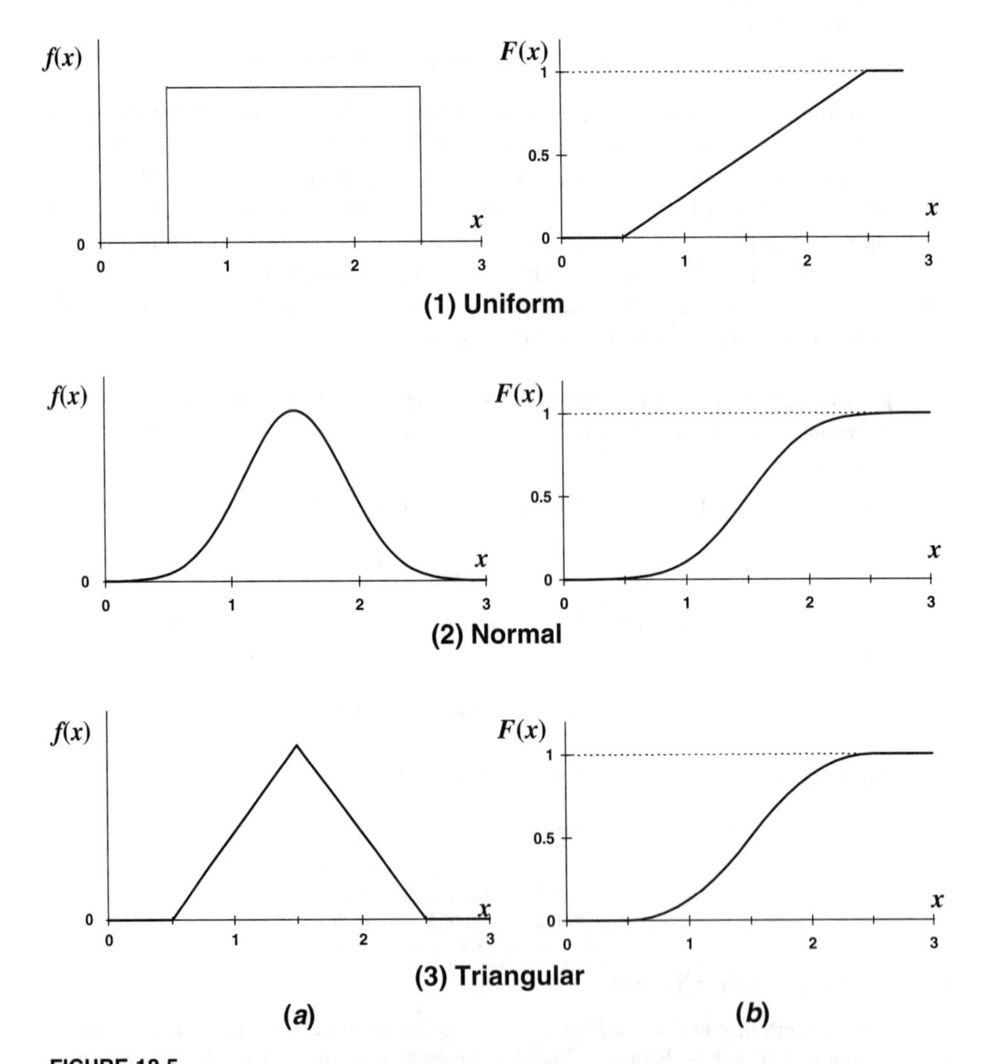

FIGURE 18.5
(*a*) Probability density and (*b*) cumulative distribution functions for three distributions commonly used to characterize uncertainty of water-quality modeling parameters: (1) uniform, (2) normal, and (3) triangular distributions.

Random numbers are then used to generate a series of parameter estimates that follow the distribution. Each of these estimates is used to compute model simulations. The outcome is a distribution of concentrations (Fig. 18.4*c*).

Figure 18.5*a* shows a few ideal probability distributions that are commonly used to describe parameter variability in water-quality modeling. The probability density functions or p.d.f. (Fig. 18.5*a*) represent plots of frequency of occurrence $f(x)$ versus the parameter value x. Thus the areas under the probability density functions are equal to 1, connoting that they encompass the total probability of occurrence.

The ***uniform distribution*** (Fig. 18.5.1) assumes that there is an equal likelihood of occurrence between two bounds. This distribution is often used when all that is known is a range for the parameter. In contrast the ***normal distribution*** assumes that the parameter follows a symmetrical bell-shaped curve. Thus there are more occurrences at a most likely value in the center and diminishing occurrences at the extremes (Fig. 18.5.2). Because of its flexibility, a ***triangular distribution*** has utility in describing a range of shapes. The symmetrical version shown in Fig. 18.5.3 can be used to approximate the normal distribution. However, because the peak does not have to be centered between the extremes, it can also be employed to represent distributions that are off-centered or "skewed."

The cumulative distribution function or c.d.f. represents the integral of the p.d.f.,

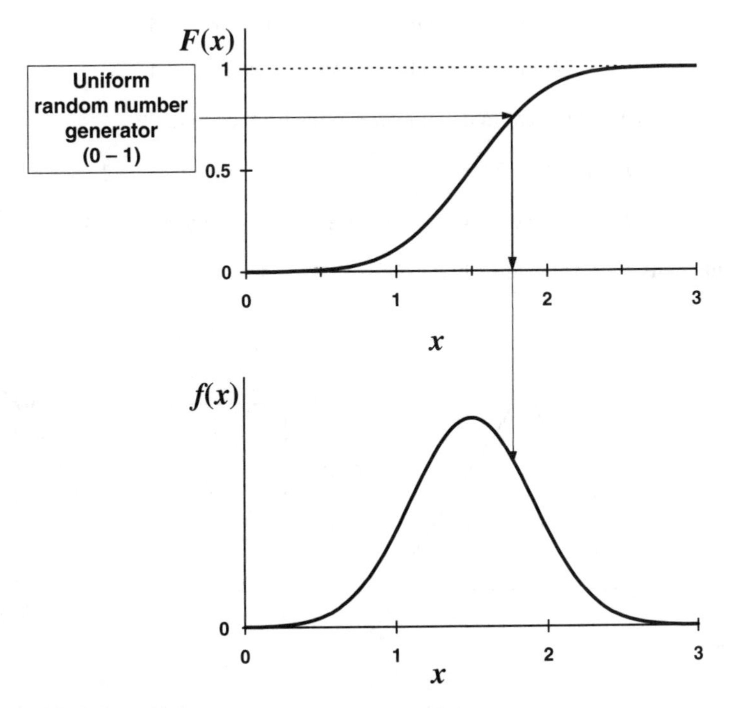

FIGURE 18.6
Graphical depiction of how the c.d.f. is used in conjunction with random numbers to generate values of model coefficients in Monte Carlo simulation.

$$F(x) = \int_{-\infty}^{x} f(x)\,dx \tag{18.10}$$

This integral specifies the probability that the parameter will be less than x. The two distributions are also related inversely by differentiation,

$$f(x) = \frac{dF(x)}{dx} \tag{18.11}$$

In other words the frequency of occurrence is equal to the rate of change of the c.d.f. with respect to the rate of change of the parameter.

As depicted in Fig. 18.6 the c.d.f. provides the key tool for implementing Monte Carlo analysis. A random uniformly distributed number in the range 0 to 1 can be generated with the computer. This value is then propagated through the c.d.f. to generate a value of the parameter x. By repeating the process, a series of parameters will be generated that together are distributed according the underlying p.d.f. These values are then used in a water-quality model to simulate distributions of concentration.

EXAMPLE 18.3. MONTE CARLO ANALYSIS. For the case from Example 18.1, determine the expected uncertainties of the concentrations due to the uncertainty of the parameter k_{12}. Assume that the parameter follows the triangular distribution

$$f(x) = 0.4444(x - 1) \qquad 1 \le x < 2.5$$

$$f(x) = -0.4444(x - 4) \qquad 2.5 \le x < 4$$

These can be integrated to develop equations for the c.d.f.,

$$F(x) = 0.2222(x - 1)^2 \qquad 1 \le x \le 2.5$$

$$F(x) = -0.4444(0.5x^2 - 4x + 5.75) \qquad 2.5 \le x \le 4$$

Both functions are displayed below:

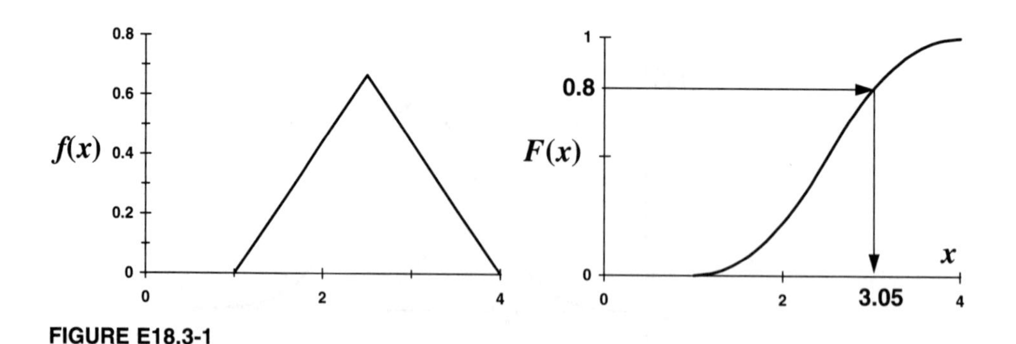

FIGURE E18.3-1

Random numbers between 0 and 1 can be generated. For example a value of 0.8 might be generated and substituted into the c.d.f. to give

$$0.8 = -0.4444(0.5x^2 - 4x + 5.75)$$

or collecting terms,

$$0 = -0.2222x^2 + 1.7776x - 3.3553$$

which yields the root 3.0515. (The quadratic formula also gives another root, 4.9485, which is disregarded because it's outside the parameter range, 1 to 4.) This value can then be substituted into the model (recall Example 18.1) along with the mean values for the other parameters to give $c_1 = 3.317$ and $c_2 = 5.729$. The process can be repeated for other random numbers. The resulting distributions are displayed below for the case where 400 random outcomes were generated. Notice that although the mean values are similar to those from Example 18.1, the distributions are skewed.

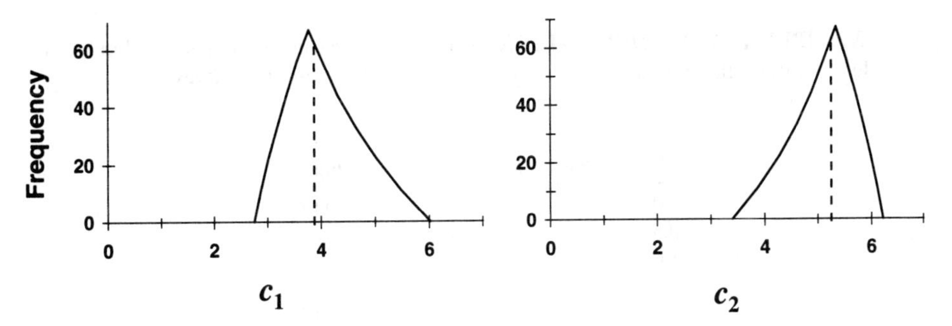

FIGURE E18.3-2

18.3 ASSESSING MODEL PERFORMANCE

As described previously in Sec. 18.1.4 the aim of calibration is to adjust the parameters so that model solutions fit observations in an optimal fashion. Now we must define what we mean by a "best fit." There are two general approaches for assessing the quality of a calibration: subjective and objective. Subjective assessment is based on a visual comparison of the simulation with the data. This usually consists of individual time series plots for each of the state variables. The modeler adjusts parameters until he or she deems that the agreement between the model output and data "looks" adequate.

In contrast objective approaches hinge on developing some quantitative measure of the quality of the fit (usually some measure of error). Once such a measure is adopted, the parameters are adjusted until a best (usually a minimum) value is attained.

There are a number of measures that can be developed to assess a fit. In the present discussion we will focus on minimizing the sum of the squares of the residuals,

$$S_r = \sum_{i=1}^{n} (c_{p,i} - c_{m,i})^2 \tag{18.12}$$

where $c_{p,i}$ = ith model prediction of concentration and $c_{m,i}$ = ith measured concentration.

The sum of the squares provides a total value that should be a minimum when the predictions and measurements are in agreement. Thus it provides a single score that can be monitored as you adjust parameters.

It should be noted that the parameter adjustment process for the objective approach can also be implemented in two fashions. First, as with the subjective method, it can be approached in a trial-and-error fashion. However, along with a visual assessment, the best-fit measure (Eq. 18.12) provides an additional means of judging the effects of parameter adjustment. Second, numerical optimization methods can be used to adjust parameters in an automatic fashion. Both are illustrated in the following example.

EXAMPLE 18.4. MODEL CALIBRATION. For the same system studied in Example 18.1, the following data were collected during a 5-d sampling survey:

t (d)	c_1 (mg L^{-1})	c_2 (mg L^{-1})	c_T (mg L^{-1})
0.0	7.0	4.0	11.0
0.5	7.3	3.5	10.8
1.0	6.6	4.7	11.3
1.5	8.9	4.0	12.9
2.0	6.6	5.0	11.7
2.5	6.2	3.1	9.3
3.0	3.3	4.3	7.6
3.5	4.7	2.2	6.8
4.0	3.75	3.9	7.7
4.5	6.1	1.3	7.4
5.0	6.0	4.0	10.0
Mean	6.03	3.64	9.67

The concentration of the inflowing stream c_{in} during the period of study can be represented by the sine function,

$$c_{in} = 10 + 5 \sin\left(\frac{\pi}{2}t\right)$$

In addition in situ experiments were performed at two times during the study to estimate k_{12} directly. These yielded estimates of 1.05 d^{-1} and 1.55 d^{-1} for that parameter. Use this information (and any other techniques at your disposal) to estimate the three parameters k_{12}, k_{21}, and v_s.

Solution: As in Example 18.1, in the absence of additional information, the average values for the parameters are the most likely: $k_{12} = 2.5$ d^{-1}, $k_{21} = 0.6$ d^{-1}, and $v_s =$ 0.5 m d^{-1}. Using these parameters along with the other information results in the simulation shown in Fig. 18.7*a*. Notice that the model does an adequate job of matching the total concentration. However, it underpredicts c_1 and overpredicts c_2. The differences between the data and model are large as reflected by the total sum of the squares (calculated based on both c_1 and c_2), which is 97.69.

The fact that the simulation misses c_1 low and c_2 high suggests that k_{12} might be a good candidate for adjustment (recall the analysis at the end of Example 18.1). Thus we can repeat the simulation using the average value of the direct measurement of k_{12}, $(1.05 + 1.55)/2 = 1.3$ d^{-1}. This value along with the average values for the other

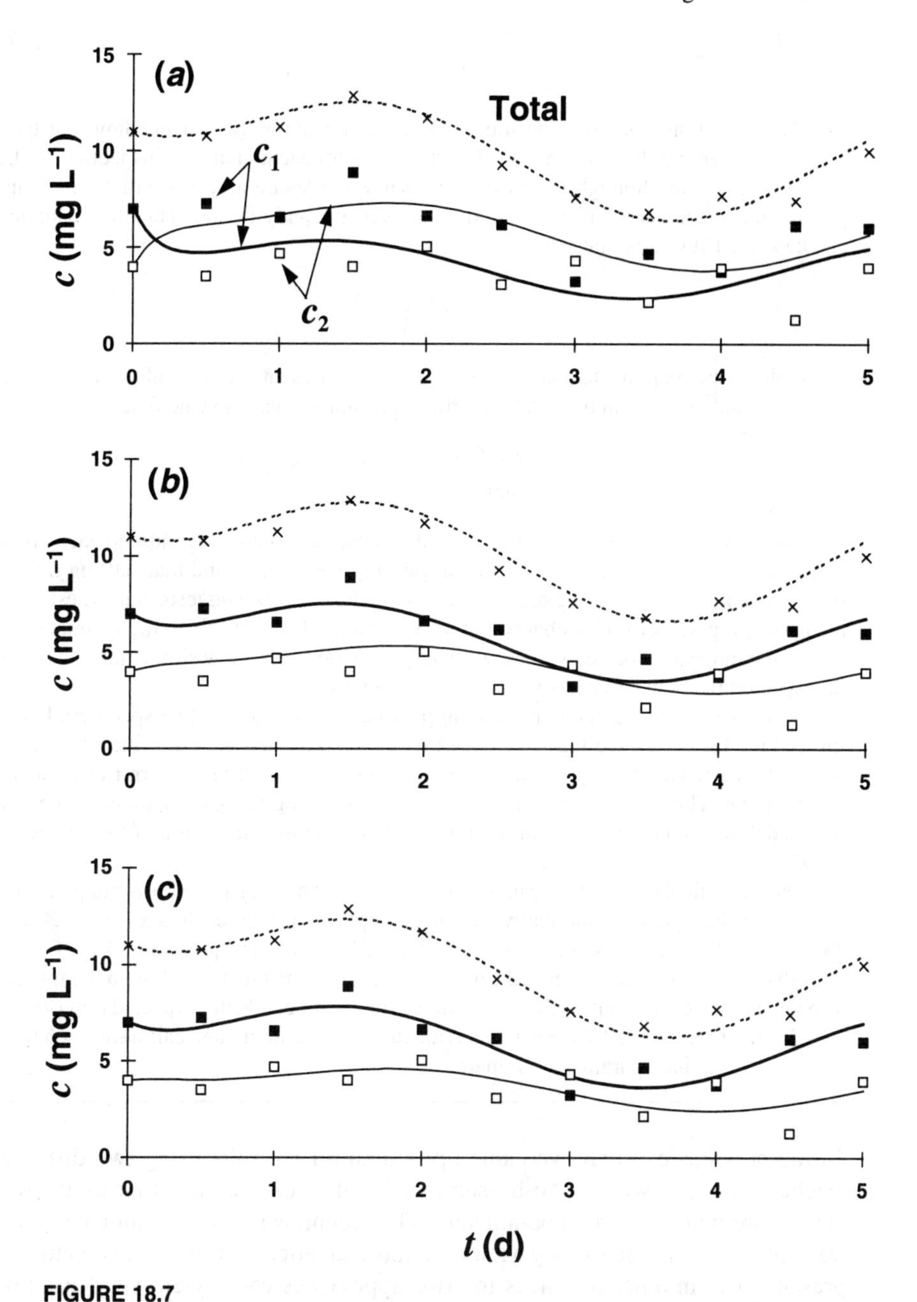

FIGURE 18.7
Calibration. (*a*) The original fit using average values for the parameters; (*b*) improved fit after using the measured value for k_{12}; (*c*) final calibration.

parameters can be employed to compute another prediction. As in Fig. 18.7*b* this adjustment brings a great improvement (SSR reduces to 18.35).

Next we can try to improve our estimate of v_s. A rough estimate of this parameter can be obtained by recognizing that the two original differential equations can be added to yield

$$V\frac{dc_T}{dt} = Qc_{\text{in}} - Qc_T - v_s A_s c_2$$

This equation describes the input-output relationships that govern how the total mass varies temporally. Now although it describes transient changes, inspection of the data indicates that although the two dependent variables c_T and c_2 oscillate, they approximately return to their initial conditions over the study period. Thus this equation can be solved at steady-state for

$$v_s = \frac{Q\bar{c}_{\text{in}} - Q\bar{c}_T}{A_s \bar{c}_2}$$

where the overbars indicate that the concentrations represent average values for the simulation period. These can be developed from the data and used to calculate

$$v_s = \frac{50{,}000(10.67 - 9.67)}{16{,}667(3.64)} = 0.82 \text{ m d}^{-1}$$

Notice how the average inflow and total concentrations are very close to each other. This implies that flushing is the dominant purging mechanism and that settling traps a relatively small fraction of the total influent in the lake. It also suggests that the estimate of the settling velocity is probably not very certain. However, since the estimate lies within the range of expected values, we can use it to develop a new simulation. As might be expected the new value has very little effect on the results.

Thus we are left with one remaining parameter, k_{21}, that must be specified. Looking at Fig. 18.7*b*, we see that c_2 is still being overpredicted. Thus an increase in k_{21} seems in order. This adjustment can now be made by either trial-and-error or by using an optimizer. The result is a value of 0.7 d^{-1}. Substituting this parameter estimate into the model yields the final calibration in Fig. 18.7*c*. The final total sum of the squares is 14.95.

Now as a final exercise we can repeat the calibration using an optimizer to minimize the sum of the squares numerically (recall Example 2.3). The result is $k_{12} = 1.24 \text{ d}^{-1}$, $k_{21} = 0.74 \text{ d}^{-1}$, and $v_s = 0.93 \text{ m d}^{-1}$, with a total sum of the squares of 14.23. Consequently for this case the automated calibration yields results that are close to the manual version. However, also notice that the estimate for k_{21} is outside the expected range of 0.5 to 0.7 noted in Example 18.1. Such behavior can occur in automated calibration schemes using unconstrained parameter estimates.

In the previous example we came up with similar results using two different approaches. The first was a step-by-step, trial-and-error approach that used some direct measurements of certain parameters. The second was a fully automated least-squares approach. Although they yielded almost identical parameter estimates for the present case, in other instances the two approaches could yield very different parameter estimates. Such would be the case when some of the key state variables were highly uncertain.

At face value it might seem that the objective method is superior. However, because the first approach uses additional information, it is preferable. Further the subjective, visual approach also has value since it employs the modeler's intuition in the process. Modern interactive computer graphics coupled with optimization software could provide a means to integrate both approaches in future model interfaces.

18.4 SEGMENTATION AND MODEL RESOLUTION

The final concepts that must be explored before developing mechanistic water-quality models are segmentation and model resolution. As stated previously, ***segmentation*** is the process of dividing space and matter into increments. In space a water body can be divided into volumes for which mass-balance equations are written. Within each volume, matter, in turn, may be divided into different chemical and biological forms for which separate equations would again be written. Thus, if a water body were divided into n segments in space and matter were divided into m substances, $m \times n$ mass-balance equations would be used to define the system.

Aside from space and matter, there is a temporal aspect to segmentation related to the fact that the mass balance defines changes in the water body *over a finite period of time.* Although this concept will become clearer in the subsequent development of actual models, the key point is that, just as a model describes additional spatial and material detail by using more segments, the model also has a temporal focus that can be increased by using a shorter "finite period" or ***time step*** for the mass-balance computation. For example the water body may be undergoing a significant yet gradual change due to longer term phenomena such as increases in nutrient loads (Fig. 18.8*a*). At the same time, as in Fig. 18.8*b*, phytoplankton[†] undergo seasonal cycles due to, among other things, fluctuating light and temperature levels over the course of the year. By adjusting the time step of the computation, we can design the model to resolve part or all of this variability. In a sense this can be thought of as a segmentation in time.

The degree to which space, time, and matter are segmented is called ***model resolution***. This concept is analogous to photography, where the camera's lens is

[†] Microscopic floating plants.

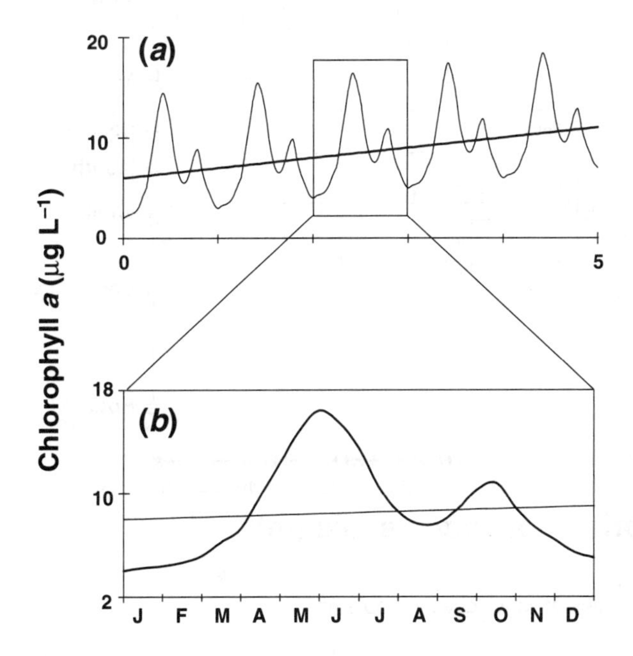

FIGURE 18.8
Plot of chlorophyll *a* concentration (μg L^{-1}) versus time. (*a*) Depiction of the long-term trend (straight line) along with the underlying seasonal cycle (wavy line); (*b*) depiction of the seasonal cycle (wavy line) along with the underlying long-term trend (straight line).

adjusted to bring different parts of the field of view into focus. At times the foreground is important; at other times distant details might be of interest.

In the present context there are two basic ways in which the "focus" or resolution of a model has significance for water-quality analysis. First, the fine-scale phenomenon may have a direct, causative influence on predictions made on the coarser scale. This is primarily a function of the substance's properties and the physical characteristics of the system. For example certain pollutants (e.g., enteric bacteria) die rapidly upon entering a water body. Therefore they typically are at high levels near a sewage discharge and then decrease rapidly, so that in the open waters of lakes and estuaries they are often at "background" levels. Thus a model of near-shore bacterial pollution would require relatively fine spatial and temporal segmentation around sewage outfalls. In contrast conservative or slowly reacting substances often can be modeled with coarser schemes that treat the whole water body as a single well-mixed segment.

The physical characteristics of the system also can dictate the required level of segmentation. For example dendritic lakes have numerous embayments that make treatment of the system as a single segment unrealistic. In addition problems such

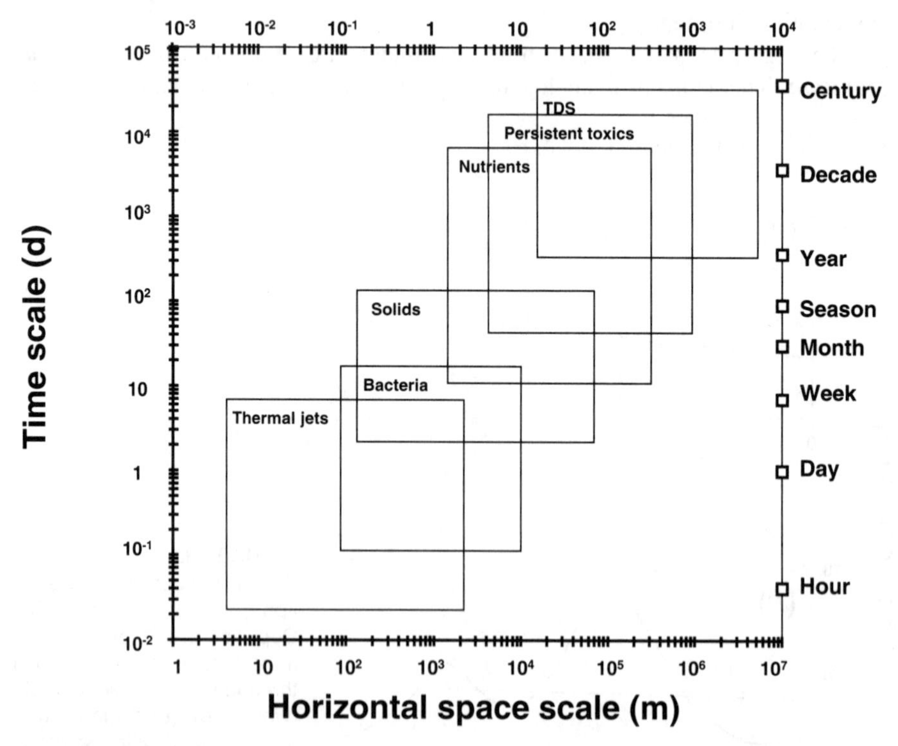

FIGURE 18.9
Approximate time and space scales of water-quality problems.

as eutrophication[†] that are strongly influenced by thermal stratification may need multiple vertical segments for adequate characterization.

The second way in which model resolution has significance is the influence of the problem context on the choice of scales. For example although phytoplankton go through growth cycles on seasonal time scales and exhibit patchiness on small space scales, a water-quality planner might not have the funds to develop models to simulate such short-term variability. This would often occur when large numbers of small lakes were being evaluated. Alternatively the issue of concern to the planner may be such that spatial and temporal aggregation results in a negligible loss of relevant planning information. In these cases the manager might opt for a coarser scale model. In other situations, as with bacterial contamination of beaches, a finer scale approach would be necessary to handle the problem effectively.

The temporal, spatial, and kinetic scales of a problem often are interrelated. As depicted in Fig. 18.9, fast kinetic processes such as jet mixing of thermal effluents or bacterial die-off tend to manifest themselves on local (that is, small) time and space scales. In contrast problems determined by slow reactions such as the decay of persistent contaminants are more likely to be important on a whole-system, long-term basis.

[†] Profuse growth of plants due to overfertilization.

PROBLEMS

18.1. Develop condition numbers for Example 18.1 for flow and inflow concentration.

18.2. The bay/main lake system depicted in Fig. P18.2 has the following characteristics:

$$V_1 = 2 \times 10^7 \text{ m}^3 \qquad V_2 = 5 \times 10^6 \text{ m}^3$$

$$A_{s,1} = 2 \times 10^6 \text{ m}^2 \qquad A_{s,2} = 1 \times 10^6 \text{ m}^2$$

The outflow from the main lake is 1×10^7 m^3 yr^{-1} and the flow from the bay to the main lake is 0.2×10^7 m^3 yr^{-1}. The main lake has an inflow concentration of 10 μg L^{-1} of a substance that reacts with first-order kinetics (half-life = 1 yr). An industrial plant is to be located on the bay with a flow of 0.1×10^7 m^3 yr^{-1} and a concentration of 100 μg L^{-1}. The interface between the main lake and the bay has a cross-sectional area of 250 m^2 and a mixing length of 200 m.

(*a*) Use Okubo's diffusion diagram (Fig. 8.11) to estimate the range of turbulent diffusion coefficients expected for such a system. Use a value of 500 m for the length

FIGURE P18.2

scale. In addition to the range, estimate a most likely value as the geometric mean of the minimum and maximum values obtained from the plot. The geometric mean can be computed as

$$GM = \sqrt{E_{\min} E_{\max}}$$

where $E_{\min}$ and $E_{\max}$ are the minimum and maximum diffusion coefficients from Okubo's plot. Note that the geometric mean is the appropriate statistic for central tendency when the parameter is distributed log normally (that is, the logarithms of the values follow a normal distribution).

(*b*) Use the mean diffusion coefficient along with the other data to estimate the expected concentration in the main lake and in the bay.

(*c*) Use a first-order sensitivity analysis to estimate the uncertainties in these predictions.

(*d*) Use a triangular distribution to approximate the log-normal distribution of the diffusion coefficient and perform a Monte Carlo analysis to compute the expected distribution of the bay's concentration.

18.3. A total phosphorus mass balance for a lake can be solved at steady-state for

$$p = \frac{W}{Q + vA_s}$$

where p = total phosphorus concentration (mg m^{-3})
W = total phosphorus loading (mg yr^{-1})
Q = outflow (m^3 yr^{-1})
v = an apparent settling velocity (m yr^{-1})
A_s = lake-surface area

(*a*) Apply this model to calculate the total phosphorus concentration of Lake Ontario in the early 1970s ($Q = 212 \times 109$ m^3 yr^{-1}, $V = 1634 \times 10^9$ m^3, $A_s = 19{,}000 \times 10^6$ m^2, $W = 10{,}000$ mta). Assume that the settling velocity is 12.4 m yr^{-1}.

(*b*) Suppose you have additional data available indicating that the settling velocity can range from about 4 to 40 m yr^{-1}. Use this information and a first-order sensitivity analysis to translate this range into an uncertainty estimate for the total phosphorus prediction.

18.4. The same kinetic reactions as in Example 18.1 occur in a pond having the following characteristics: $V = 20{,}000$ m^3, $Q = 50{,}000$ m^3 d^{-1}, and $A_s = 50{,}000$ m^2. The following data were collected during a 10-d sampling survey:

t (d)	c_1 (mg L^{-1})	c_2 (mg L^{-1})
0	1.7	7.3
1	3.5	6.9
2	2.5	10.5
3	3.9	10.2
4	2.6	13.6
5	4.6	12.8
6	2.8	13.6
7	3.3	11.0
8	1.3	11.9
9	2.9	7.2
10	0.7	8.8

The concentration of the inflowing stream, c_{in}, during the period of study can be represented by the sine function,

$$c_{\text{in}} = 15 + 10 \sin\left(\frac{\pi}{6}t\right)$$

Unfortunately none of the parameters were estimated directly. Use the observations to estimate the three parameters k_{12}, k_{21}, and v_s.

18.5. A general representation of the triangle function is depicted in Fig. P18.5.

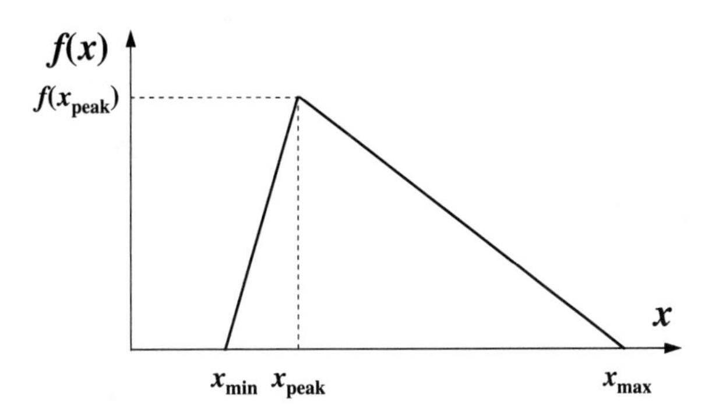

FIGURE P18.5

(*a*) Determine $f(x_{\text{peak}})$ so that the area under the function is equal to 1.
(*b*) Develop the equations describing $f(x)$.
(*c*) Develop the equations describing $F(x)$.

PART IV

Dissolved Oxygen and Pathogens

Part IV is designed to introduce you to our first water-quality-modeling problem context: dissolved oxygen. In addition it provides information on another classic water-quality problem: bacterial pollution.

The first two lectures describe the fundamental building blocks that form the foundation of oxygen modeling. *Lecture 19* deals with biochemical oxygen demand or BOD. In addition the subject of oxygen saturation is covered. *Lecture 20* is devoted to the general topic of gas transfer, with special emphasis on oxygen reaeration.

After this review of fundamentals, the next two lectures present the Streeter-Phelps oxygen model. *Lecture 21* describes the point-source version, whereas *Lecture 22* deals with diffuse sources.

The succeeding lectures outline refinements that have been made to the original Streeter-Phelps framework. These cover the inclusion of nitrification (*Lecture 23*), plant photosynthesis and respiration (*Lecture 24*), and sediment oxygen demand or SOD (*Lecture 25*).

Next, computer aspects of oxygen modeling are discussed. The first part of *Lecture 26* deals with general computer approaches, with particular emphasis on estuary models. Then the remainder of the lecture is devoted to a description of oxygen simulation with the QUAL2E software package.

Finally the subject of pathogen modeling is explored in *Lecture 27*. This lecture focuses on classic first-order frameworks designed to simulate the transport and fate of enteric bacteria in natural waters and sediments. In addition new models for protozoa are explored.

LECTURE 19

BOD and Oxygen Saturation

LECTURE OVERVIEW: After a brief introduction to the dissolved oxygen problem, I describe how BOD serves as a means to quantify the oxygen demand of a wastewater. Simple mass balances are developed for the BOD of a batch system. Following a brief review of Henry's law, I also present a general overview of the saturation of oxygen in water.

In Lec. 1, I stated that engineers originally became involved in water-quality modeling to assess the impact of sewage on receiving waters. We will now study the models that were developed to solve this problem. To place these models in context, I'll first describe the cycle of organic production and decomposition that occurs in the biosphere.

19.1 THE ORGANIC PRODUCTION/DECOMPOSITION CYCLE

As depicted in Fig. 19.1 the biosphere can be viewed as a cycle of life and death. Powered by the sun, autotrophic† organisms (primarily plants) convert simple inorganic nutrients into more complex organic molecules. In ***photosynthesis***, solar energy is stored as chemical energy in the organic molecules. In addition oxygen is liberated and carbon dioxide is consumed.

The organic matter then serves as an energy source for heterotrophic organisms (bacteria and animals) in the reverse processes of ***respiration*** and ***decomposition***.

† The term ***autotrophic*** refers to organisms like plants that do not depend on other organisms for nutrition. In contrast ***heterotrophic*** organisms consist of animals and most bacteria that subsist on organic matter.

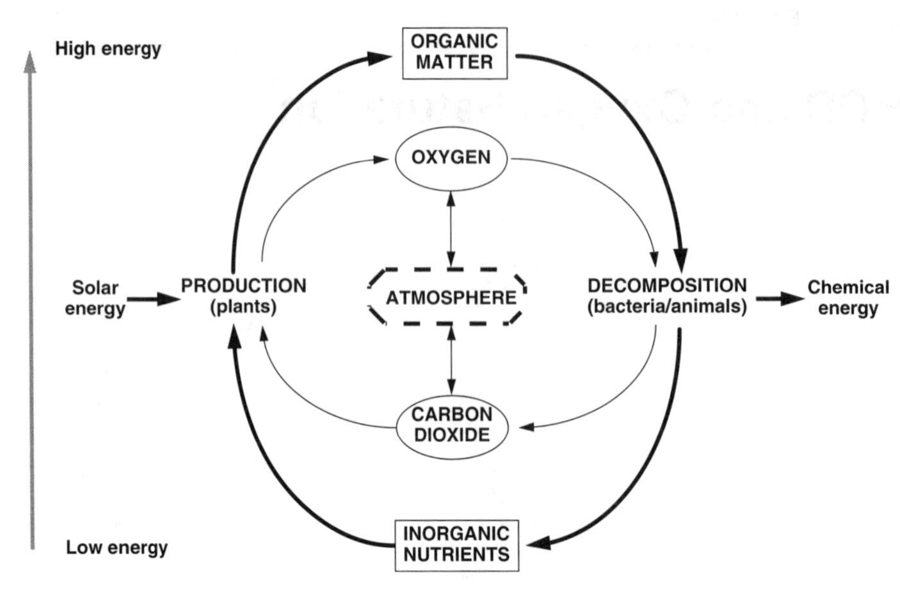

FIGURE 19.1
The natural cycle of organic production and decomposition.

These return the organic matter to the simpler inorganic state. During breakdown, oxygen is consumed and carbon dioxide is liberated.

The cycle can be represented in chemical terms by the following simple expression:

$$\underset{\text{Carbon dioxide}}{6CO_2} + \underset{\text{Water}}{6H_2O} \underset{\text{respiration}}{\overset{\text{photosynthesis}}{\rightleftharpoons}} C_6H_{12}O_6 + \underset{\text{Oxygen}}{6O_2} \tag{19.1}$$

Sugar

According to this reversible reaction, carbon dioxide and water are used to synthesize organic matter (the sugar glucose) and to create oxygen in the forward photosynthesis reaction. Conversely the organic matter is broken down and oxygen is consumed in the reverse respiration and decomposition reactions.

The chemistry of the production/decomposition cycle is far more complicated than Eq. 19.1. For example many different organic compounds are created and broken down in the process. In addition other elements beyond carbon, hydrogen, and oxygen are involved. In later lectures I'll present a more complete representation. However, Eq. 19.1 provides a starting point for our efforts to quantify the process.

19.2 THE DISSOLVED OXYGEN SAG

Now that we have the big picture, let's link the life/death cycle with the environment in a stream below a wastewater discharge (Fig. 19.2). If the stream is originally unpolluted, dissolved oxygen levels above the discharge will be near saturation. The introduction of the untreated sewage will elevate the levels of both dissolved and

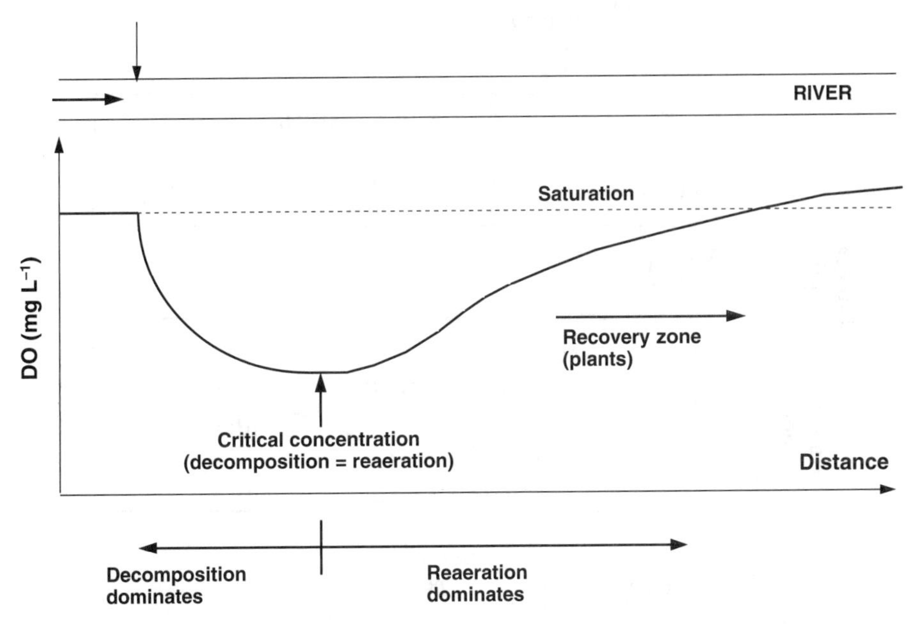

FIGURE 19.2
The dissolved oxygen "sag" that occurs below sewage discharges into streams.

solid organic matter. This has two impacts. First, the solid matter makes the water turbid. Thus light cannot penetrate and plant growth is suppressed. Some of the solids settle downstream from the sewage outfall and create sludge beds that can emit noxious odors (Fig. 19.3*a*). Second, the organic matter provides food for heterotrophic organisms. Consequently the right side of the cycle in Fig. 19.1 becomes dominant. Large populations of decomposer organisms break down the organic matter in the water and in the process deplete the dissolved oxygen. In addition decomposition of the organic matter takes place in the sludge bed and a sediment oxygen demand supplements the decay in the water.

As oxygen levels drop, atmospheric oxygen enters the water to compensate for the oxygen deficit. At first the oxygen consumption in the water and to the sediments dwarfs this ***reaeration***. However, as the organic matter is assimilated and the oxygen levels drop, there will come a point at which the depletion and the reaeration will be in balance. At this point the lowest or "critical" level of oxygen will be reached. Beyond this point reaeration dominates and oxygen levels begin to rise. In this recovery zone the water becomes clearer because much of the solid matter from the discharge will have settled. In addition inorganic nutrients liberated during the decomposition process will be high. Consequently the recovery zone will often be dominated by the growth of plants. Thus the left side of the cycle from Fig. 19.1 can become overemphasized.

Beyond the chemical changes, sewage also leads to significant effects on the biota. As in Fig. 19.3*b* and *c*, molds and bacteria dominate near the discharge. In addition the bacteria themselves provide a food source for a succession of organisms

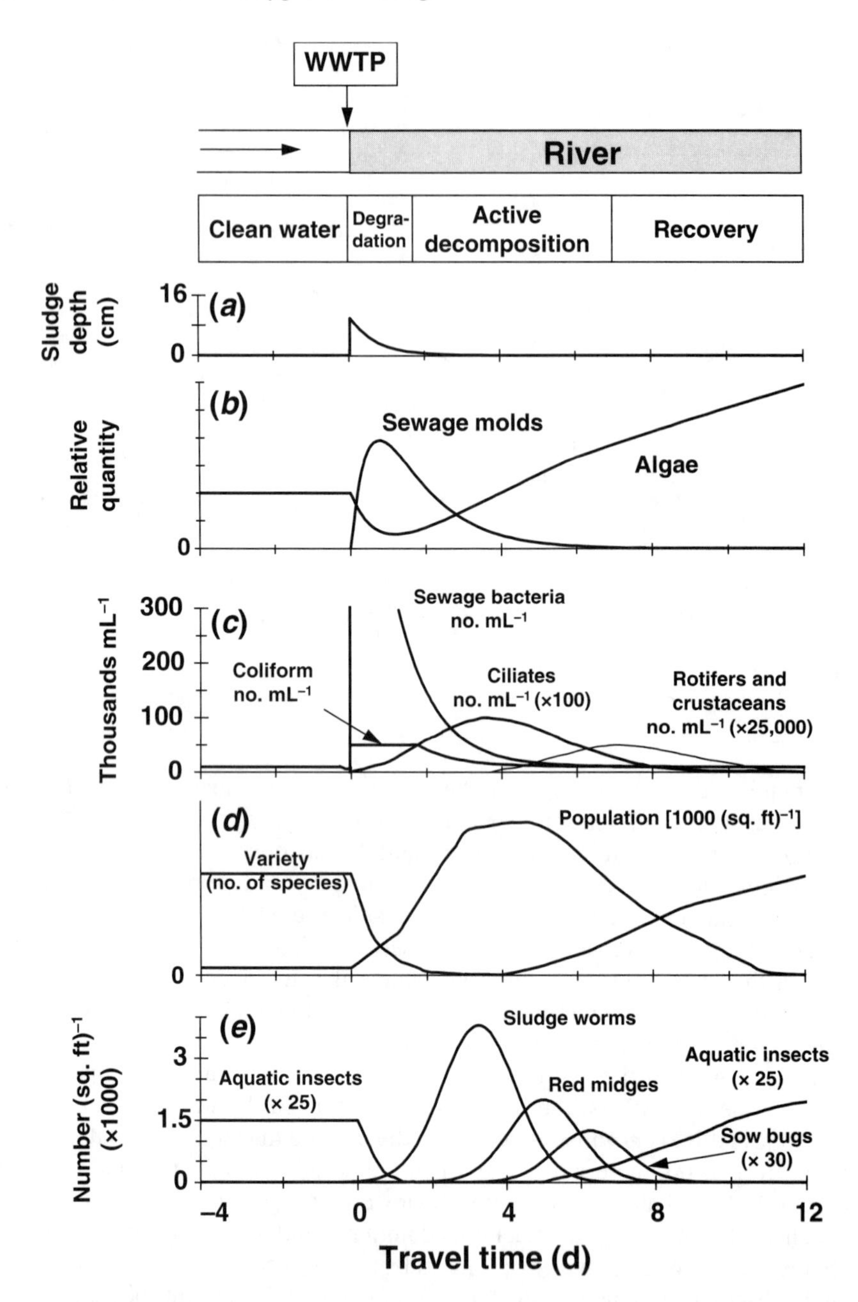

FIGURE 19.3
The changes in the biota below a sewage treatment plant effluent (redrawn from Bartsch and Ingram 1977).

consisting of ciliates, rotifers, and crustaceans. The diversity of higher organisms decreases drastically in the degradation and active decomposition zones below the discharge. At the same time the total number of organisms increases (Fig. 19.3*d* and *e*). As recovery ensues, these trends are reversed.

All the chemical interactions form the characteristic dissolved oxygen "sag" shown in Fig. 19.2. The key feature of the sag is the critical or minimum concentration. The location and magnitude of this critical concentration depends on a number of factors, including the size of the loading, the stream's flow and morphometry, water temperature, etc. The goal of the models formulated below will be to simulate the sag as a function of these factors.

19.3 EXPERIMENT

The first step in modeling the DO sag will be to characterize the strength of the wastewater. To do this we will focus on the respiration/decomposition portion of the life/death cycle. In terms of the general chemical representation from Eq. 19.1, this is

$$C_6H_{12}O_6 + 6O_2 \xrightarrow{\text{respiration}} 6CO_2 + 6H_2O \tag{19.2}$$

Now let's imagine a closed-batch experiment to investigate how this reaction affects a simple environment (recall Lec. 2). Suppose that you placed a quantity of sugar into a bottle of water with an initial oxygen content of o_0. You also add a small amount of bacteria and stopper the bottle. Assuming that decomposition proceeds as a first-order reaction, a mass balance for the glucose can be written as

$$V\frac{dg}{dt} = -k_1 V g \tag{19.3}$$

where g = glucose concentration (mg-glucose L^{-1}) and k_1 = the decomposition rate in the bottle (d^{-1}). If the initial level of glucose is g_0, this equation can be solved for

$$g = g_0 e^{-k_1 t} \tag{19.4}$$

Next a mass balance can be written for oxygen,

$$V\frac{do}{dt} = -r_{og} k_1 V g \tag{19.5}$$

where o = oxygen concentration (mgO L^{-1}) and r_{og} = the stoichiometric ratio of oxygen consumed to glucose decomposed (mgO mg-glucose^{-1}). Equation 19.4 can be substituted into Eq. 19.5 to give

$$V\frac{do}{dt} = -r_{og} k_1 V g_0 e^{-k_1 t} \tag{19.6}$$

If the initial level of oxygen is o_0, this equation can be solved for

$$o = o_0 - r_{og} g_0 (1 - e^{-k_1 t}) \tag{19.7}$$

According to this equation, the bottle will originally have a dissolved oxygen level of o_0. Thereafter the oxygen will decrease exponentially and asymptotically approach a level of

$$o \rightarrow o_0 - r_{og} g_0 \tag{19.8}$$

Before proceeding further, let's work an example.

EXAMPLE 19.1. OXYGEN DEPLETION IN A CLOSED BATCH SYSTEM. You place 2 mg of glucose in a 250-mL bottle. After adding a small quantity of bacteria, you fill the remainder of the volume with water and stopper the bottle. The initial concentration of oxygen is 10 mg L^{-1}. If glucose decomposes at a rate of 0.1 d^{-1}, determine the oxygen concentration as a function of time in this closed batch system.

Solution: First, we must determine the initial concentration of glucose,

$$g_0 = \frac{2 \text{ mg}}{250 \text{ mL}}\left(\frac{1000 \text{ mL}}{\text{L}}\right) = 8 \text{ mg L}^{-1}$$

Next we must determine the ratio of mass of oxygen consumed per mass of glucose decomposed. Using the stoichiometry of Eq. 19.2 we can calculate

$$r_{og} = \frac{6(32)}{6 \times 12 + 1 \times 12 + 6 \times 16} = 1.0667 \text{ mgO mg-glucose}^{-1}$$

Therefore total decomposition of the glucose would consume the following amount of oxygen:

$$r_{og} g_0 = 1.0667 \times 8 = 8.5333 \text{ mg L}^{-1}$$

Consequently the oxygen level in the bottle will ultimately approach (Eq. 19.8)

$$o \rightarrow 10 - 8.5333 = 1.4667 \text{ mg L}^{-1}$$

Equation 19.7 can then be used to compute the oxygen level as a function of time. Some results, along with the glucose levels (expressed as oxygen equivalents) computed with Eq. 19.4 are displayed below:

Time (d)	Glucose (mgO L^{-1})	Oxygen (mgO L^{-1})
0	8.5333	10.0000
4	5.7201	7.1867
8	3.8343	5.3009
12	2.5702	4.0369
16	1.7229	3.1895
20	1.1549	2.6215

A graph of the results can also be generated:

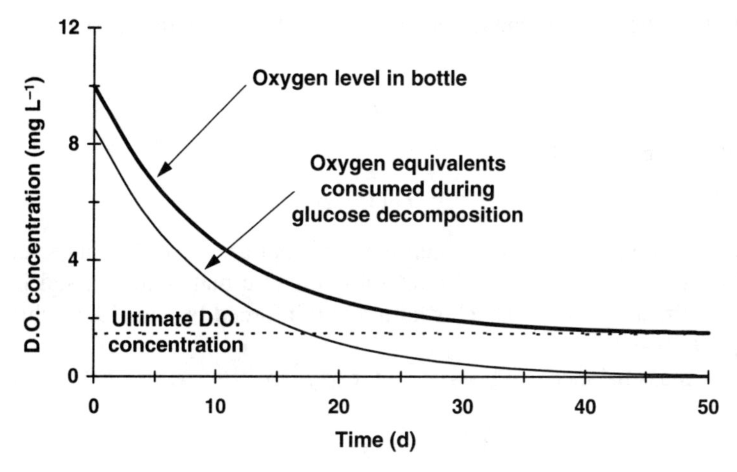

FIGURE E19.1

19.4 BIOCHEMICAL OXYGEN DEMAND

The experiment outlined at the end of the last section was intended to show how the decomposition process could be modeled for a simple batch system. At face value a similar approach could be used to model how sewage would affect oxygen levels. However, such an approach would be problematic because, as mentioned previously, sewage is not composed of simple sugar. Thus to rigorously apply the approach, we would have to characterize the concentrations of the myriad organic compounds in each sewage sample. We would further need to determine the stoichiometry of the decomposition for each reaction. Finally each of the compounds could decompose at a different rate. Obviously such a rigorous approach would be impractical.

In the early days of modeling, the constraints on such an approach were even more severe because of the limitations of the technology for characterizing organic compounds. As a consequence the first water-quality analysts took an empirical approach and simply disregarded the composition of the sewage. As in our simple experiment from Example 19.1, the analysts introduced some sewage into a batch reactor and merely measured how much oxygen was consumed. The resulting quantity was dubbed ***biochemical oxygen demand*** or ***BOD***.

In terms of our simple model we can now define a new variable L (mgO L^{-1}) that is the amount of oxidizable organic matter remaining in the bottle expressed as oxygen equivalents. A mass balance for L for the batch system can be written as

$$V\frac{dL}{dt} = -k_1 VL \tag{19.9}$$

If the initial level is L_0, this equation can be solved for

$$L = L_0 e^{-k_1 t} \tag{19.10}$$

Note that the oxygen consumed during the decomposition process can be defined as

$$y = L_0 - L \tag{19.11}$$

or substituting Eq. 19.10,

$$y = L_0(1 - e^{-k_1 t}) \tag{19.12}$$

where y = BOD (mgO L^{-1}). We can now see that the value L_0 can be defined as either the initial concentration of oxidizable organic matter (expressed in oxygen units) or as the ultimate BOD. This notion is reinforced by Fig. 19.4, which shows both Eq. 19.10 and 19.12.

Next a mass balance can be written for oxygen,

$$V\frac{do}{dt} = -k_1 V L_0 e^{-k_1 t} \tag{19.13}$$

If the initial level of oxygen is o_0, this equation can be solved for

$$o = o_0 - L_0(1 - e^{-k_1 t}) \tag{19.14}$$

According to this equation the bottle will originally have a dissolved oxygen level of o_0. Thereafter the oxygen will decrease exponentially and asymptotically approach a level of

$$o \to o_0 - L_0 \tag{19.15}$$

Notice that this development is identical to the glucose experiment. In fact the glucose experiment could be modeled in terms of BOD by substituting

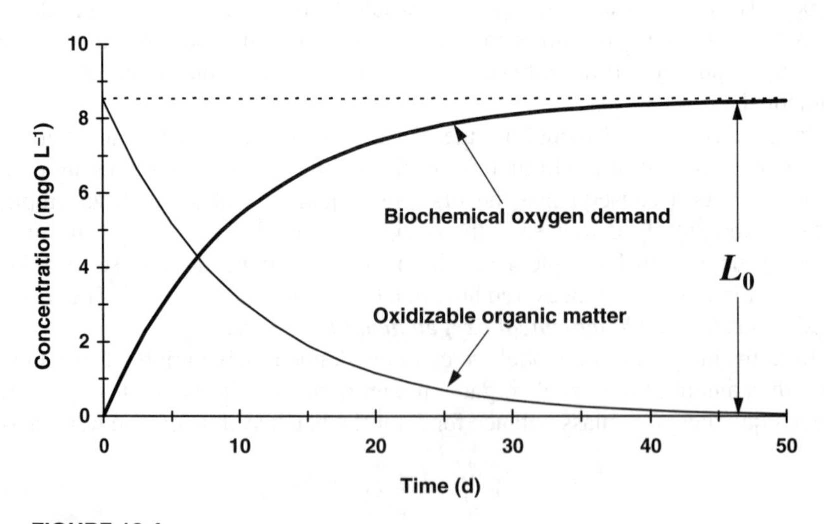

FIGURE 19.4
The value L_0 can be defined as either the initial concentration of oxidizable organic matter or as the ultimate BOD.

$$L_0 = r_{og} g_0 \tag{19.16}$$

Thus by disregarding the exact composition of the organic matter, we avoid the necessity of characterizing the organic matter and its stoichiometry relative to oxygen.

Finally it should be noted that although the exact composition of the organic matter in wastewater is not characterized, the organic carbon content can be measured directly. In these cases Eq. 19.16 can be reformulated to estimate the BOD on the basis of organic carbon content, as in

$$L_0 = r_{oc} C_{\text{org}} \tag{19.17}$$

where C_{org} = organic carbon concentration of the wastewater (mgC L^{-1}) and r_{oc} = ratio of mass of oxygen consumed per mass of carbon assimilated (mgO mgC^{-1}). Again, Eq. 19.2 provides a basis for estimating the ratio

$$r_{oc} = \frac{6(32)}{6(12)} = 2.67 \text{ mgO mgC}^{-1} \tag{19.18}$$

Finally it should be noted that along with the decomposition of carbonaceous matter, an additional oxygen demand is exerted due to the oxidation of ammonia to nitrate in the process called ***nitrification***. The oxygen demand due to nitrification is sometimes referred to as ***nitrogenous BOD*** or ***NBOD*** to distinguish it from the carbonaceous BOD or CBOD described above. In Lec. 23 we will describe nitrification and NBOD in detail.

19.5 BOD MODEL FOR A STREAM

Now we can take our model and apply it to a stream below a sewage treatment plant. When we do this we must now consider that, in addition to decomposition, the BOD can also be removed by sedimentation. Therefore the mass balance for a constant-flow, constant-geometry channel can be written as

$$\frac{\partial L}{\partial t} = -U \frac{\partial L}{\partial x} - k_r L \tag{19.19}$$

where k_r = total removal rate (d^{-1}), which is composed of both decomposition and settling,

$$k_r = k_d + k_s \tag{19.20}$$

where k_d = decomposition rate in the stream (d^{-1}) and k_s = settling removal rate (d^{-1}). Recognize that the decomposition rate represents the same type of process as the k_1 from the bottle experiment (recall Eqs. 19.3 and 19.9). We have used a different subscript in Eq. 19.20 to highlight that decomposition in a natural environment such as a river will generally be different from that in a bottle. In addition note that the settling rate is related to more fundamental parameters by

$$k_s = \frac{v_s}{H} \tag{19.21}$$

where v_s = BOD settling velocity (m d^{-1}) and H = water depth (m).

At steady-state, Eq. 19.19 becomes

$$0 = -U\frac{dL}{dx} - k_r L \tag{19.22}$$

If complete mixing is assumed at the location of the discharge, an initial concentration can be calculated as the flow-weighted average (recall Eq. 9.42) of the loading (subscript w) and the BOD in the river upstream of the discharge (subscript r),

$$L_0 = \frac{Q_w L_w + Q_r L_r}{Q_w + Q_r} \tag{19.23}$$

Using this value as an initial condition, we can solve Eq. 19.22 for

$$L = L_0 e^{-\frac{k_r}{U}x} \tag{19.24}$$

Thus the BOD is reduced by decomposition and settling as it is carried downstream.

Aside from modeling the distribution of BOD below a point source in a plug-flow stream with constant parameters, Eq. 19.24 provides a framework for estimating the removal rate in such systems. To do this, the natural logarithm can be taken,

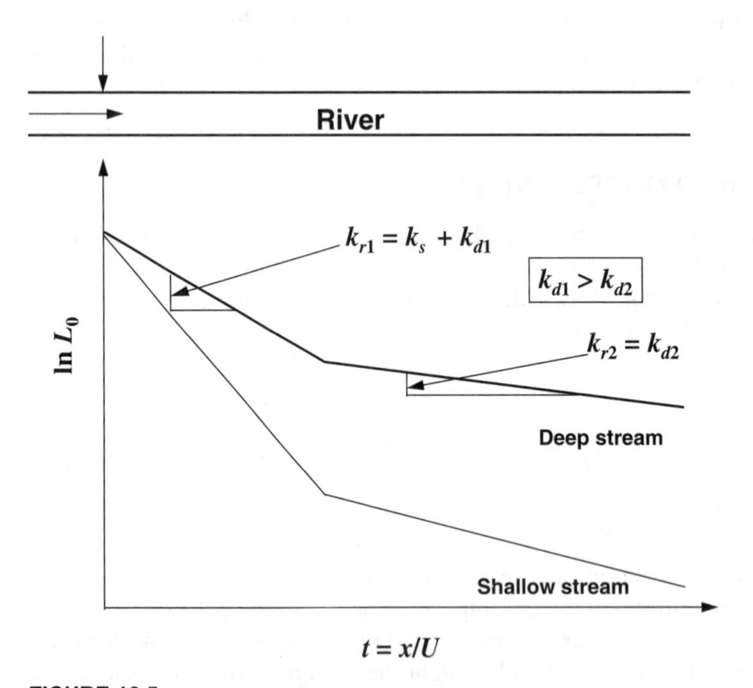

FIGURE 19.5
Plot of BOD downstream from a point source of untreated sewage into a plug-flow river having constant hydrogeometric characteristics. In the initial stretch, high BOD removal rates will occur due to settling and fast decomposition of easily degradable organic matter. Further downstream the lower removal rates will occur as the more refractory organic matter degrades at a slower rate.

$$\ln L = \ln L_0 - \frac{k_r}{U}x \tag{19.25}$$

Consequently if the simple model holds, a plot of $\ln L$ versus x/U (that is, travel time) should yield a straight line with a slope of k_r.

Figure 19.5 shows a typical pattern that might be observed in a stream receiving untreated sewage. Note how the rate is higher immediately below the discharge. Several factors underlie such a pattern. In particular the higher rates are usually caused by the fast degradation of readily decomposable organics and the settling of sewage particulates.

19.6 BOD LOADINGS, CONCENTRATIONS, AND RATES

Before proceeding to other aspects of DO modeling, let's review some of the parameters that relate to biochemical oxygen demand.

19.6.1 BOD_5 (5-Day BOD)

As in Table 19.1, typical values for the BOD bottle decay rate range from 0.05 to 0.5 d^{-1}, with a geometric mean of about 0.15 d^{-1}. This information can be used to estimate a 95% response time for the bottle test as $t_{95} = 3/0.15 = 20$ d. Because such a long measurement period is unacceptable, water-quality analysts early on adopted a 5-day BOD test.

Although shortening the incubation time to 5 d makes the test practical, we must then have a means to extrapolate the 5-d result to the ultimate BOD level. This is usually done by performing a long-term BOD to estimate the decay rate. If the first-order decomposition model holds, Eq. 19.12 can be used to compute

$$L_0 = \frac{y_5}{1 - e^{-k_1(5)}} \tag{19.26}$$

where y_5 = 5-day BOD. Table 19.1 includes typical values for the ratio of 5-day to ultimate BOD.

19.6.2 BOD Loadings and Concentrations

Table 19.2 provides typical values of flow rate and BOD for raw sewage from both the United States and developing countries. In general the flow rate for the United

TABLE 19.1
Typical values of the BOD bottle decomposition rate for various levels of treatment. BOD_u is the ultimate BOD. Values here are for CBOD

Treatment	k_1(20°C)	BOD_5/BOD_u
Untreated	0.35 (0.20–0.50)	0.83
Primary	0.20 (0.10–0.30)	0.63
Activated sludge	0.075 (0.05–0.10)	0.31

TABLE 19.2
Typical loading rates for untreated domestic sewage

	Per-capita flow rate (m^3 capita^{-1} d^{-1})	Per-capita CBOD (m^3 capita^{-1} d^{-1})	CBOD concentration (mg L^{-1})
United States	0.57 (150)†	125 (0.275)‡	220
Developing countries	0.19 (50)†	60 (0.132)‡	320

†Gallons capita^{-1} day^{-1}; ‡pounds capita^{-1} day^{-1}.

States is higher because higher water use typically accompanies a higher standard of living. Per capita generation rate of BOD is also higher because of garbage disposals and other accouterments of a developed economy like that of the United States. The average concentration of the developing countries is generally higher because the lower water use in these countries outweighs the higher per capita BOD contribution for the United States.

19.6.3 BOD Removal Rates

The bottle BOD decomposition rate provides a first estimate of the removal rate in natural waters. As listed in Table 19.1, the rates depend on the degree of treatment of the sewage prior to discharge. Raw sewage is a mixture of compounds ranging from easily decomposable sugars to refractory substances that take longer to break down. Because waste treatment tends to selectively remove the former, BOD bottle rates tend to be lower for treated sewage.

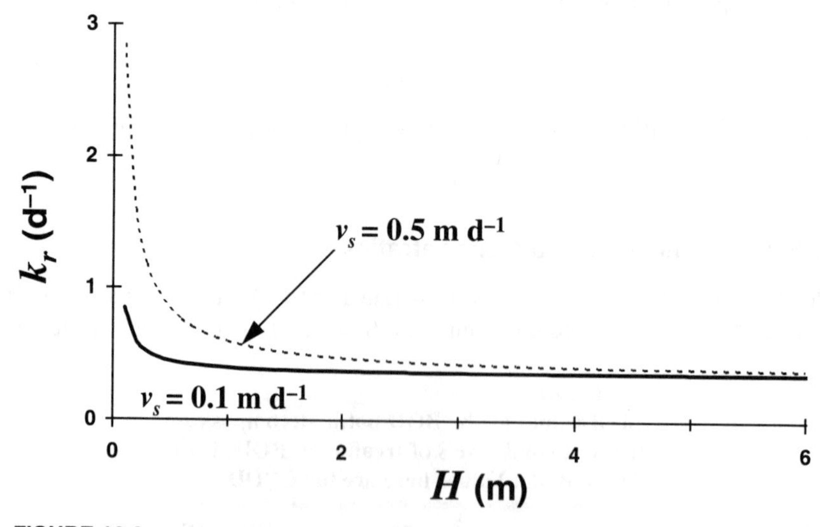

FIGURE 19.6
Plot of total removal rate versus stream depth for BOD that is 50% in settleable form. A range of settling velocities is depicted. Note that a decomposition rate of 0.35 d^{-1} is used.

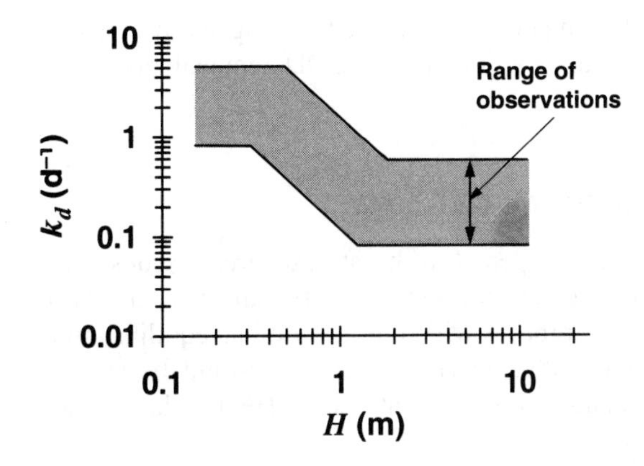

FIGURE 19.7
In-stream decomposition rate versus depth (Bowie et al. 1985).

As might be expected, the bottle rate can rarely be directly applied to rivers because the bottle environment is not a good representation of the river. In fact only in deep, slow rivers would the two converge. In most other rivers, environmental factors tend to make removal higher than for the bottle. The primary causes of this increase are settling and bed effects.

Settling. Settling effects relate to the fact that for sewage with a significant fraction of organic solids, the total removal rate in streams is a combination of settling and decomposition (recall Eqs. 19.20 and 19.21),

$$k_r = k_d + \frac{v_s}{H} \tag{19.27}$$

Using some typical settling velocities, we can see (Fig. 19.6) that the settling effect can be particularly significant for raw sewage in shallow streams (that is, < 1 m).

Bed effects. All other things equal, attached bacteria generally are more effective decomposers than free-floating bacteria. Bottom decomposition can be parameterized as a mass-transfer flux of BOD. Thus in a way similar to settling, bottom decomposition becomes more pronounced in shallower systems because the effect becomes more significant relative to the volumetric decomposition in the water.

This trend, which is displayed in Fig. 19.7, has been fit by the equations (Hydroscience 1971)

$$\begin{aligned} k_d &= 0.3\left(\frac{H}{8}\right)^{-0.434} && 0 \le H \le 8 \text{ ft} \\ k_d &= 0.3 && H > 8 \text{ ft} \end{aligned} \tag{19.28}$$

Thus up to about 2.4 m (8 ft) the rate decreases with depth. Above 2.4 m the rate approaches a constant value that is typical of bottle rates.

Finally, BOD decomposition rates can be extrapolated to other temperatures using Eq. 2.44 ($k = k_{20}\theta^{T-20}$), with $\theta \cong 1.047$.

In summary BOD removal rates tend to increase with temperature and tend to be higher immediately downstream from point sources. The latter effect is more pronounced for untreated wastewater. In addition enhanced settling and bed effects means that shallower systems typically exhibit higher BOD removal rates than deeper waters.

19.7 HENRY'S LAW AND THE IDEAL GAS LAW

If a beaker of gas-free distilled water is opened to the atmosphere, gaseous compounds such as oxygen, carbon dioxide, and nitrogen cross the air-water interface and enter into solution (Fig. 19.8). The process will continue until an equilibrium is established between the partial pressure of the gas in the atmosphere and the concentration in the aqueous phase. This equilibrium is quantified by Henry's law, which can be represented as

$$H_e = \frac{p}{c} \tag{19.29}$$

where H_e = Henry's constant (atm m^3 $mole^{-1}$)
p = partial pressure (atm)
c = water concentration (mole m^{-3})

Henry's constants for some gases commonly encountered in water-quality modeling are summarized in Table 19.3.

Equation 19.29 can also be represented in dimensionless form by invoking the ideal gas law

$$c = \frac{p}{RT_a} \tag{19.30}$$

where R = universal gas constant [8.206×10^{-5} atm m^3 (K mole)$^{-1}$] and T_a = absolute temperature (K). Thus the ideal gas law provides a means to express the partial pressure in concentration units of moles m^{-3}. Substituting Eq. 19.30 into 19.29 and rearranging yields the dimensionless Henry's constant

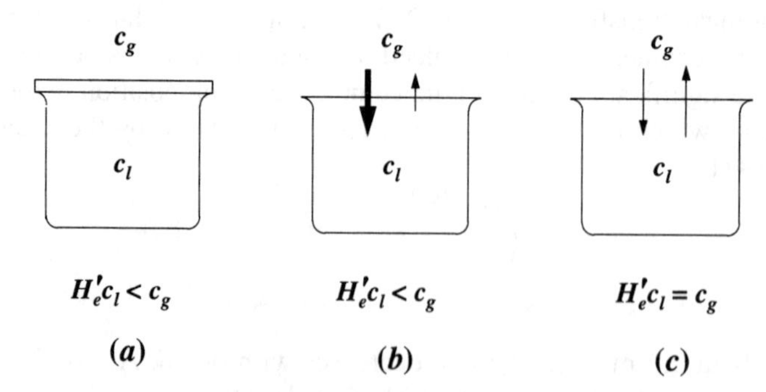

FIGURE 19.8
The closed system in (*a*) is undersaturated with oxygen. When it is opened to the atmosphere (*b*), oxygen comes into solution until an equilibrium (*c*) is reached. Henry's law provides the means to quantify this equilibrium condition.

TABLE 19.3
Henry's constants for some gases commonly encountered in water-quality modeling (modified from Kavanaugh and Trussell 1980)

		Henry's constant (20°C)	
Compound	**Formula**	**(Dimensionless)**	**(atm m^3 mole^{-1})**
Methane	CH_4	64.4	1.55×10^0
Oxygen	O_2	32.2	7.74×10^{-1}
Nitrogen	N_2	28.4	6.84×10^{-1}
Carbon dioxide	CO_2	1.13	2.72×10^{-2}
Hydrogen sulfide	H_2S	0.386	9.27×10^{-3}
Sulfur dioxide	SO_2	0.0284	6.84×10^{-4}
Ammonia	NH_3	0.000569	1.37×10^{-5}

$$H_e' = \frac{H_e}{RT_a} = \frac{c_g}{c_l} \tag{19.31}$$

where c_g and c_l are the gas and liquid concentrations, respectively (mole m^{-3}).

Notice that Henry's law specifies that at equilibrium the ratio of the gaseous to the water concentration will be maintained at a constant value. The water concentration for a particular gaseous level is referred to as the ***saturation concentration***. In the following example we determine the saturation concentration for oxygen.

EXAMPLE 19.2. HENRY'S LAW AND OXYGEN SATURATION. Determine the saturation concentration of oxygen in water at 20°C. Note that clean, dry air near sea level is composed of approximately 20.95% oxygen by volume.

Solution: Assuming that Dalton's law holds, the partial pressure of oxygen can be computed as

$$p = 0.2095(1 \text{ atm}) = 0.2095 \text{ atm}$$

This value can then be substituted into Eq. 19.29 along with the value of Henry's constant from Table 19.3 to yield

$$c = \frac{p}{H_e} = \frac{0.2095}{0.774} = 0.2707 \text{ mole m}^{-3}$$

or in mass units,

$$c = 0.2707 \text{ mole m}^{-3}\left(\frac{32 \text{ g-oxygen}}{\text{mole}}\right) = 8.66 \text{ mg L}^{-1}$$

19.8 DISSOLVED OXYGEN SATURATION

As calculated in the previous example, the saturation concentration of oxygen in a natural water is on the order of 10 mg L^{-1}. In general several environmental factors can affect this value. From the perspective of water-quality modeling, the most important of these are:

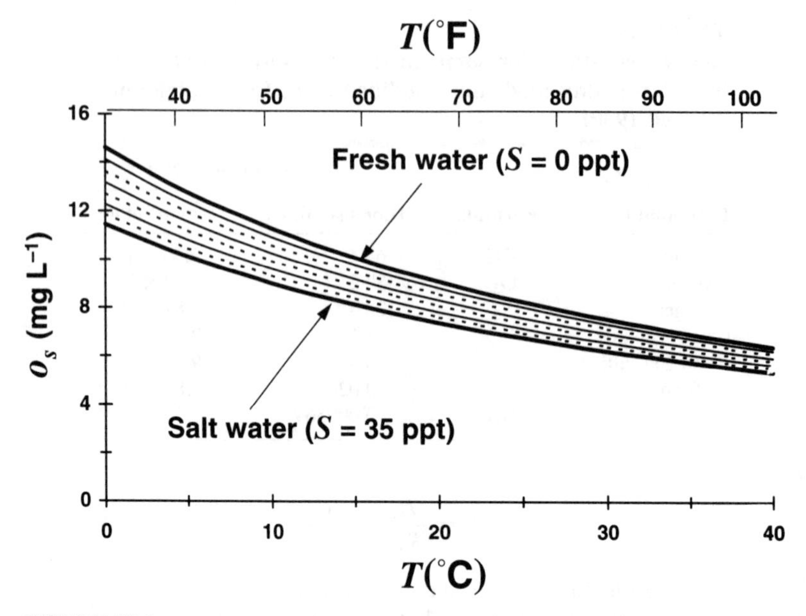

FIGURE 19.9
Relationship of oxygen saturation in water to temperature and salinity.

- Temperature
- Salinity
- Partial pressure variations due to elevation

Several empirically derived equations have been developed to predict how these factors influence saturation. These are reviewed in the following sections.

19.8.1 Temperature Effect

The following equation can be used to establish the dependence of oxygen saturation on temperature (APHA 1992):

$$\ln o_{sf} = -139.34411 + \frac{1.575701 \times 10^5}{T_a} - \frac{6.642308 \times 10^7}{T_a^2} + \frac{1.243800 \times 10^{10}}{T_a^3} - \frac{8.621949 \times 10^{11}}{T_a^4} \quad (19.32)$$

where o_{sf} = saturation concentration of dissolved oxygen in fresh water at 1 atm (mg L^{-1}) and T_a = absolute temperature (K). Remember that

$$T_a = T + 273.15 \quad (19.33)$$

where T = temperature (°C). According to this equation, saturation decreases with increasing temperature. As displayed in Fig. 19.9, freshwater concentration ranges from about 14.6 mg L^{-1} at 0°C to 7.6 mg L^{-1} at 30°C.

19.8.2 Salinity Effect

The following equation can be used to establish the dependence of saturation on salinity (APHA 1992):

$$\ln o_{ss} = \ln o_{sf} - S\left(1.7674 \times 10^{-2} - \frac{1.0754 \times 10^1}{T_a} + \frac{2.1407 \times 10^3}{T_a^2}\right) \tag{19.34}$$

where o_{ss} = saturation concentration of dissolved oxygen in saltwater at 1 atm (mg L^{-1}) and S = salinity (g L^{-1} = parts per thousand, ppt, sometimes given as ‰).

Salinity can be related to chloride concentration by the following approximation:

$$S = 1.80655 \times \text{Chlor} \tag{19.35}$$

where Chlor = chloride concentration (ppt). The higher the salinity, the less oxygen can be held by water (Fig. 19.9).

EXAMPLE 19.3. OXYGEN SATURATION FOR AN ESTUARY. Determine the saturation for an estuary with a temperature of 20°C and a salinity of 25 ppt.

Solution: Equation 19.32 can be used to compute

$$\ln o_{sf} = -139.34411 + \frac{1.575701 \times 10^5}{293.15} - \frac{6.642308 \times 10^7}{293.15^2} + \frac{1.243800 \times 10^{10}}{293.15^3} - \frac{8.621949 \times 10^{11}}{293.15^4} = 2.207$$

This corresponds to a freshwater saturation value of

$$o_{sf} = e^{2.207} = 9.092 \text{ mg L}^{-1}$$

Equation 19.34 can be used to correct for salinity,

$$\ln o_{ss} = 2.207 - 25\left(1.7674 \times 10^{-2} - \frac{1.0754 \times 10^1}{293.15} + \frac{2.1407 \times 10^3}{293.15^2}\right) = 2.060$$

which corresponds to a saltwater saturation value of

$$o_{ss} = e^{2.060} = 7.846 \text{ mg L}^{-1}$$

Thus the saltwater value is about 86% of the freshwater value.

19.8.3 Pressure Effect

The following equation can be used to establish the dependence of saturation on pressure (APHA 1992):

$$o_{sp} = o_{s1} p \left[\frac{\left(1 - \frac{p_{wv}}{p}\right)(1 - \theta p)}{(1 - p_{wv})(1 - \theta)}\right] \tag{19.36}$$

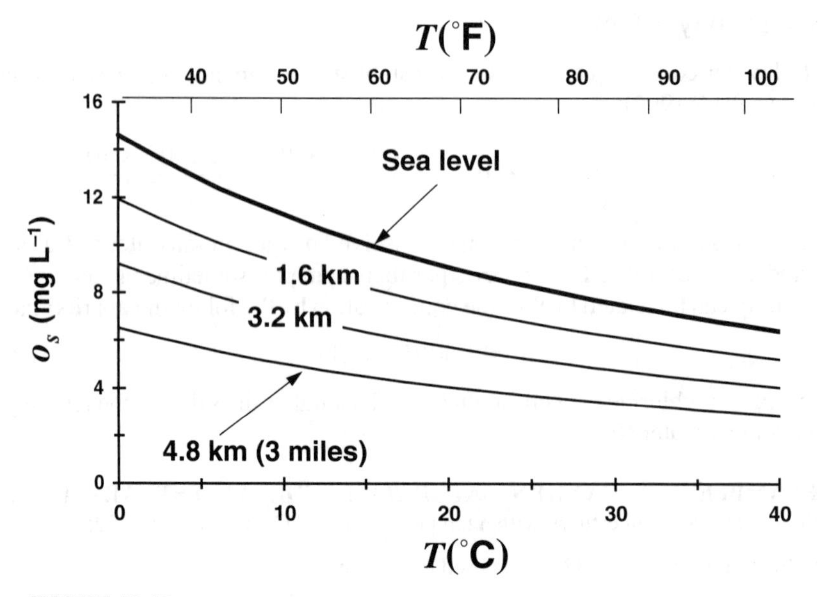

FIGURE 19.10
Relationship of oxygen saturation in water to temperature and elevation above sea level.

where p = atmospheric pressure (atm)
o_{sp} = saturation concentration of dissolved oxygen at p (mg L^{-1})
o_{s1} = saturation concentration of dissolved oxygen at 1 atm (mg L^{-1})
p_{wv} = partial pressure of water vapor (atm)

p_{wv} can be calculated by

$$\ln p_{wv} = 11.8571 - \frac{3840.70}{T_a} - \frac{216{,}961}{T_a^2} \tag{19.37}$$

The parameter θ can be computed as

$$\theta = 0.000975 - 1.426 \times 10^{-5}T + 6.436 \times 10^{-8}T^2 \tag{19.38}$$

Notice that this formula is written in terms of temperature in degrees Celsius rather than Kelvin.

Zison et al. (1978) have developed a handy approximation based on elevation,

$$o_{sp} = o_{s1}[1 - 0.1148 \times \text{elev(km)}] \tag{19.39}$$

or in English units,

$$o_{sp} = o_{s1}[1 - 0.000035 \times \text{elev(ft)}] \tag{19.40}$$

where elev = elevation above sea level. As displayed in Fig. 19.10 this relationship indicates that as pressure decreases at higher elevations, the saturation drops.

PROBLEMS

19.1. Use separation of variables to solve Eq. 19.6 for Eq. 19.7.

19.2. A tanker truck careens off the road and dumps 30,000 L of glucose syrup into a small mountain lake. The concentration of the syrup is 100 g-glucose L^{-1}.
(*a*) Compute the grams of CBOD spilled.
(*b*) Determine the lake's saturation concentration of oxygen (T = 10°C; elev = 11,000 ft.).

19.3. A lake with a bay has the following characteristics:

	Lake	**Bay**
Mean depth	8 m	3 m
Surface area	1.6×10^6 m^2	0.4×10^6 m^2
Inflow	50,000 m^3 d^{-1}	4800 m^3 d^{-1}
Inflow BOD concentration	0 mg L^{-1}	57 mg L^{-1}

A subdivision housing 1000 people is planned that will discharge raw sewage into the bay. Each individual contributes about 0.568 m^3 capita^{-1} d^{-1} of wastewater and 113.4 g capita^{-1} d^{-1} of carbonaceous biochemical oxygen demand.
(*a*) The bay inflow has a chloride concentrations of 50 mg L^{-1}. The lake and the bay have chloride concentrations of 5 and 10 mg L^{-1}, respectively. Determine the bulk diffusion coefficient between the lake and the bay.
(*b*) If the BOD decays at a rate of 0.1 d^{-1} and settles at a rate of 0.1 m d^{-1}, determine the steady-state BOD concentration of the lake and bay in mg L^{-1}. Make this determination with and without the subdivision.

19.4. The following data have been collected for dissolved and total BOD below a point source of untreated sewage into a stream. Use this data to estimate the BOD removal rates (k_r, k_s, and k_d) for the river. The velocity and depth are 6600 m d^{-1} and 2 m, respectively.

x (km)	0	5	10	15	20	25	30	35
Dissolved (mg L^{-1})	20.0	17.0	14.7	12.9	11.6	10.5	9.6	8.9
Total (mg L^{-1})	40.0	29.4	22.5	17.8	14.6	12.3	10.7	9.6
x (km)	40	45	50	60	70	80	90	100
Dissolved (mg L^{-1})	8.3	7.8	7.3	6.6	6.0	5.5	5.1	4.7
Total (mg L^{-1})	8.7	8.0	7.5	6.7	6.0	5.5	5.1	4.7

19.5. Determine the saturation concentrations at 20°C for (*a*) nitrogen gas (78.1% by volume in the atmosphere) and (*b*) carbon dioxide (0.0314%).

19.6. A flow of 2 cms with a 5-d BOD of 10 mg L^{-1} is discharged from an activated-sludge treatment plant to a stream with a flow of 5 cms and zero BOD. Stream characteristics are $k_{r,20}$ = 0.2 d^{-1}, cross-sectional area = 25 m^2, and T = 28°C.
(*a*) What is the concentration of BOD at the mixing point?
(*b*) How far below the effluent will the stream BOD concentration fall to 5% of its original value?

19.7. The following temperatures, salinities, and oxygen concentrations are measured in an estuary:

Distance from ocean, km	30	20	10
Temperature, °C	25	22	18
Salinity, ppt	5	10	20
Dissolved oxygen	5	6.5	7.5

Calculate the percent saturation of oxygen at the three locations.

19.8. What is the oxygen saturation concentration of a saline lake (mostly sodium chloride) that is located at an elevation of 1 km, and has a salinity of 10 ppt, and is at a temperature of 25°C?

LECTURE 20

Gas Transfer and Oxygen Reaeration

Lecture overview: I describe two theories of gas transfer: the two-film model and the surface renewal model, and show how they can be used to model oxygen transfer in natural waters. In addition I review formulas to predict reaeration in streams.

Suppose we fill an open bottle with oxygen-free distilled water. We know from the previous lecture that, given sufficient time, atmospheric oxygen will enter until the solution reaches the saturation level defined by Henry's law. Similarly if we had a bottle of water that was supersaturated, over time oxygen would leave the solution until the saturation value was reached.

The key question is, "How long will it take?" In other words we would like to assess the rate of the process. Now let's imagine an experiment to quantitatively answer this question.

We have our open bottle filled with oxygen-free distilled water. As depicted in Fig. 20.1, we place a mixing device in the bottle. This device keeps the volume well-mixed except in the bottle's narrow neck, where molecular diffusion governs transport.

To model this system, assume that the water at the air-water interface is at the saturation concentration. Under this assumption, a mass balance for the bottle can be written as

$$V\frac{do}{dt} = DA\frac{o_s - o}{\Delta H} \tag{20.1}$$

where D = molecular diffusion coefficient of oxygen in water ($m^2\ d^{-1}$)
A = cross-sectional area of the bottle neck (m^2)

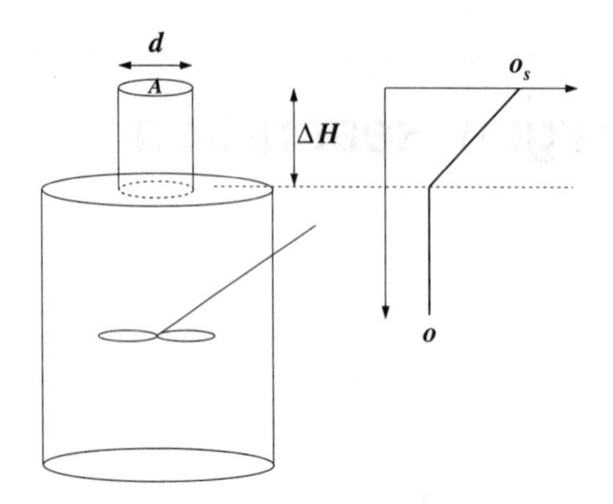

FIGURE 20.1
An open well-mixed bottle with a narrow neck in which molecular diffusion governs transport.

o_s = oxygen saturation concentration (mg L^{-1})
o = oxygen concentration in the bottle (mg L^{-1})
ΔH = length of the neck (m)

The model can also be expressed as

$$V\frac{do}{dt} = K_L A(o_s - o) \tag{20.2}$$

where K_L = oxygen mass-transfer velocity (m d^{-1}), which is equal to

$$K_L = \frac{D}{\Delta H} \tag{20.3}$$

Dividing both sides of Eq. 20.2 by the volume and rearranging yields

$$\frac{do}{dt} + k_a o = k_a o_s \tag{20.4}$$

where k_a = reaeration rate (d^{-1}), which is equal to $K_L A/V$. Together with the initial condition that $o = 0$ at $t = 0$, Eq. 20.4 can be solved for

$$o = o_s(1 - e^{-k_a t}) \tag{20.5}$$

EXAMPLE 20.1. OXYGEN TRANSFER FOR A BOTTLE. You fill the 300-mL bottle shown in Fig. 20.1 with oxygen-free water. Calculate the oxygen concentration as a function of time if $D = 2.09 \times 10^{-5}$ cm^2 s^{-1}, $d = 2$ cm, and $\Delta H = 2.6$ cm. Assume that the system is at a temperature of 20°C and the saturation concentration is 9.1 mg L^{-1}.

Solution: First, we must determine the mass-transfer velocity,

$$K_L = \frac{2.09 \times 10^{-5}\ \text{cm}^2\ \text{s}^{-1}}{2.6\ \text{cm}} \left(\frac{1\ \text{m}}{100\ \text{cm}} \frac{86{,}400\ \text{s}}{\text{d}}\right) = 0.006945\ \text{m d}^{-1}$$

Next we can calculate the reaeration coefficient,

$$k_a = \frac{0.006945\ \text{m d}^{-1}[\pi(0.01)^2]\ \text{m}^2}{300\ \text{mL}} \left(\frac{10^6\ \text{mL}}{\text{m}^3}\right) = 0.007273\ \text{d}^{-1}$$

The parameters can be substituted into Eq. 20.5,

$$c = 9.1(1 - e^{-0.007273t})$$

which can be used to determined the following values as a function of time:

Time	0	80	160	240	320	400
Oxygen	0.00	4.01	6.25	7.50	8.21	8.60

A graph of the results can also be generated:

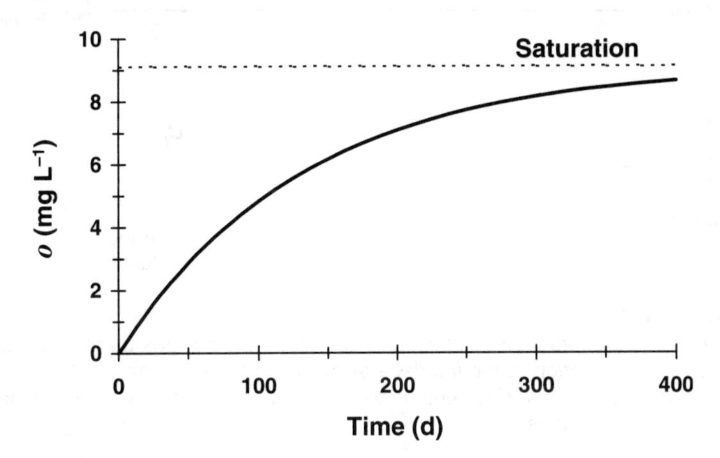

FIGURE E20.1.1

Over a long period of time the concentration approaches the saturation value. The time can be quantified by determining a 95% response time,

$$t_{95} = \frac{3}{0.007273} = 412 \text{ d}$$

Thus according to the model, it would take over 1 yr to reach 95% of saturation.

In the foregoing example we calculated that it would take over 1 yr for a bottle of water to reoxygenate. Although the situation in nature is not as slow, gas transfer in natural waters involves many of the principles we used to model the bottle.

20.1 GAS TRANSFER THEORIES

We will now describe two theories that are widely used to describe gas transfer in natural waters. Although both are used in streams, estuaries, and lakes, the stagnant-film theory is more widely used in standing waters such as lakes, whereas the surface-renewal model is more commonly used in flowing waters such as streams.

20.1.1 Whitman's Two-Film Theory

A simple model of gas exchange is provided by Whitman's two-film or two-resistance model (Whitman 1923, Lewis and Whitman 1924).

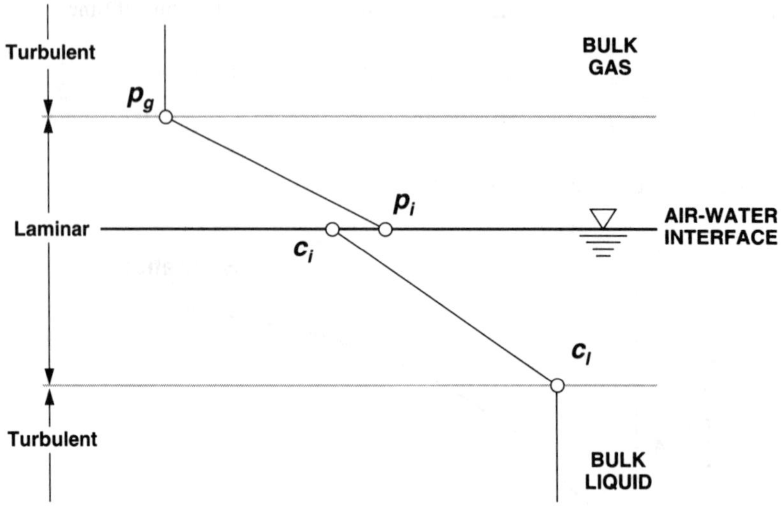

FIGURE 20.2
Schematic representation of Whitman's two-film theory of gas transfer. Liquid and gas concentration at the interface are assumed to be at an equilibrium as defined by Henry's law. Gradients in the films control the rate of gas transfer between the bulk fluids.

As depicted in Fig. 20.2, the bulk or main body of the gaseous and liquid phases are assumed to be turbulently well-mixed and homogeneous. The two-film theory assumes that a substance moving between the phases encounters maximum resistance in two laminar boundary layers where mass transfer is via molecular diffusion. The mass transfer through the individual films would be a function of a mass-transfer velocity and the gradient between the concentrations at the interface and in the bulk fluid. For example transfer through the liquid film can be represented by

$$J_l = K_l(c_i - c_l) \tag{20.6}$$

where J_l = mass flux from the bulk liquid to the interface (mole m^{-2} d^{-1})
K_l = mass-transfer velocity in the liquid laminar layer (m d^{-1})
c_i and c_l = liquid concentrations at the air-water interface and in the bulk liquid, respectively (mole m^{-3}).

Similarly transfer through the gaseous film can be represented by

$$J_g = \frac{K_g}{RT_a}(p_g - p_i) \tag{20.7}$$

where J_g = mass flux from the interface to the bulk gas (mole m^{-2} d^{-1}),
K_g = mass-transfer velocity in the gaseous laminar layer (m d^{-1})
p_g and p_i = the gas pressures in the bulk gas and at the air-water interface, respectively (atm)

Notice that for both Eqs. 20.6 and 20.7, a positive flux represents a gain to the water.

The transfer coefficients can be related to more fundamental parameters by

$$K_l = \frac{D_l}{z_l} \tag{20.8}$$

and

$$K_g = \frac{D_g}{z_g} \tag{20.9}$$

where D_l = liquid molecular diffusion coefficient (m^2 d^{-1})
D_g = gas molecular diffusion coefficient (m^2 d^{-1})
z_l = thickness of the liquid film (m)
z_g = thickness of the gas film (m)

A key assumption of the two-film theory is that an equilibrium exists at the air-water interface. In other words Henry's law (recall Eq. 19.29) holds:

$$p_i = H_e c_i \tag{20.10}$$

Equation 20.10 can be substituted into Eq. 20.6, which can be solved for

$$p_i = H_e\left(\frac{J_l}{K_l} + c_l\right) \tag{20.11}$$

Equation 20.7 can be solved for

$$p_i = p_g - \frac{RT_a J_g}{K_g} \tag{20.12}$$

Equations 20.11 and 20.12 can be equated and solved for flux,

$$J = v_v\left(\frac{p_g}{H_e} - c_l\right) \tag{20.13}$$

where v_v = net transfer velocity across the air-water interface (m d^{-1}), which can be computed by

$$\frac{1}{v_v} = \frac{1}{K_l} + \frac{RT_a}{H_e K_g} \tag{20.14}$$

Equation 20.13 now provides a means to compute mass transfer as a function of the gradient between the bulk levels in the gaseous and the liquid phases. In addition it yields a net transfer velocity (Eq. 20.14) that is a function of environmental characteristics K_l and K_g and the gas-specific parameter H_e. Note that Eq. 20.14 can be inverted to calculate the mass-transfer velocity directly,

$$v_v = K_l \frac{H_e}{H_e + RT_a(K_l/K_g)} \tag{20.15}$$

Notice that I have modified the nomenclature slightly by using a "*v*" rather than a "*K*" for the net transfer velocity. I did this to make the coefficient's nomenclature consistent with its units—that is, a velocity. The subscript *v* is intended to signify that the coefficient is a volatilization mass-transfer velocity.

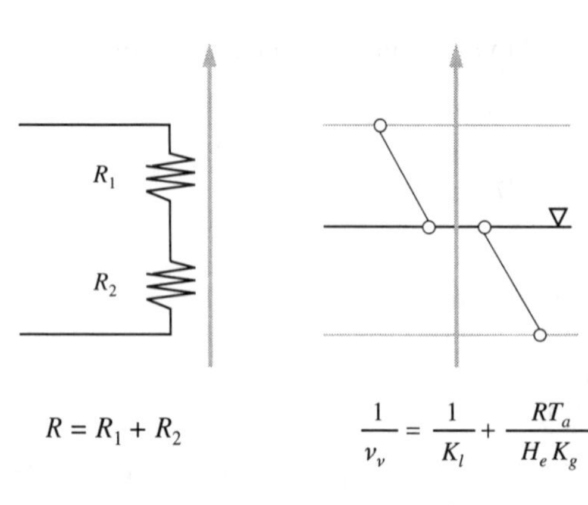

$R = R_1 + R_2$

(*a*) Two resistors in series

$$\frac{1}{v_v} = \frac{1}{K_l} + \frac{RT_a}{H_e K_g}$$

(*b*) Two films in series

FIGURE 20.3
The two-film theory of gas transfer is analogous to the formulation for two resistors in series in an electrical circuit.

Also observe that Eq. 20.14 seems to be analogous to the formulation used to determine the effect of two resistors in parallel in an electrical circuit:

$$\frac{1}{R} = \frac{1}{R_1} + \frac{1}{R_2} \tag{20.16}$$

Although Eq. 20.14 might superficially be in this format, it can be recognized that the resistance in each film is actually the reciprocal of its mass-transfer velocity. Consequently Eq. 20.14 is actually analogous to the formulation used to determine the effect of two resistors in series in an electrical circuit (Fig. 20.3).

As in Eq. 20.15 the total resistance to gas transfer is a function of the individual resistances in the liquid and the gaseous boundary layers. The liquid, the gas, or both layers can be the controlling or limiting factor depending on the values of the three coefficients K_l, K_g, and H_e. This can be quantified by using Eq. 20.15 to develop (Mackay 1977)

$$R_l = \frac{H_e}{H_e + RT_a(K_l/K_g)} \tag{20.17}$$

where R_l = ratio of the liquid-layer resistance to the total resistance. For lakes, K_g varies from approximately 100 to 12,000 m d^{-1} and K_l from 0.1 to 10 m d^{-1} (Liss 1975, Emerson 1975). The ratio of K_l to K_g generally ranges from 0.001 to 0.01, with the higher values in small lakes due primarily to lower K_g because of sheltering from wind. A plot of R_l versus H_e (Fig. 20.4) indicates where the liquid, gas, or both films govern transport for contaminants of differing solubility. In general the higher the Henry's constant, the more the control shifts to the liquid film. Also note that smaller lakes tend to be more gas-film controlled than larger lakes.

As mentioned previously the two-film theory usually represents a good approximation for standing waters such as lakes. Next we turn to another theory, one that extends the two-film theory to systems such as streams that have strong advective flow.

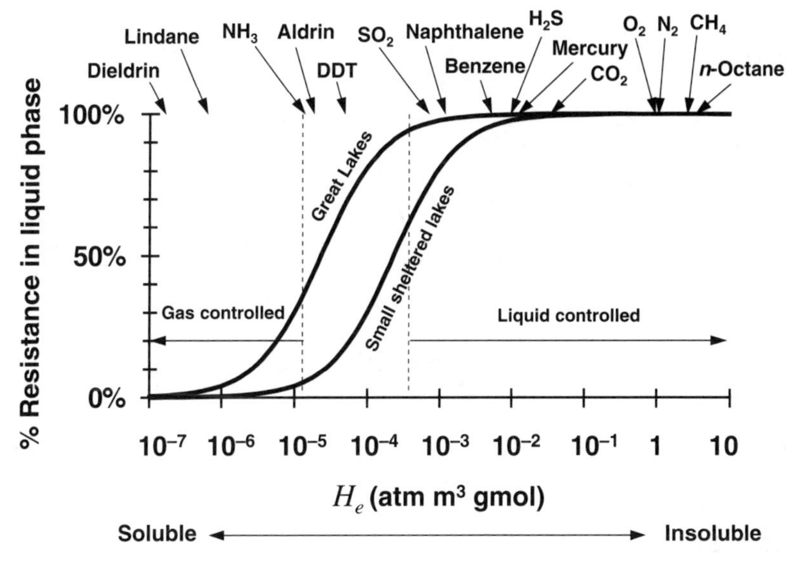

FIGURE 20.4
Percent resistance to gas transfer in the liquid phase as a function of H_e, the Henry's constant for lakes. Values of H_e for some environmentally important gases and toxic substances are indicated (modified from Mackay 1977).

20.1.2 Surface Renewal Model

We now turn to a model that takes a different approach to gas transfer from the two-film theory described in the previous section. Rather than as a stagnant film, the system is conceptualized as consisting of parcels of water that are brought to the surface for a period of time. While at the surface, exchange takes place. Then the parcels are moved away from the surface and mixed with the bulk liquid (Fig. 20.5).

Higbie (1935) suggested that when the liquid and gas are first brought into contact, the liquid film will be at the concentration of the bulk liquid. Thus, prior to the situation envisioned by the two-film theory (Fig. 20.2), the dissolved gas must penetrate the film. Hence it was dubbed the ***penetration theory***. The evolution of this penetration is depicted by the succession of dashed lines in Fig. 20.6. If the

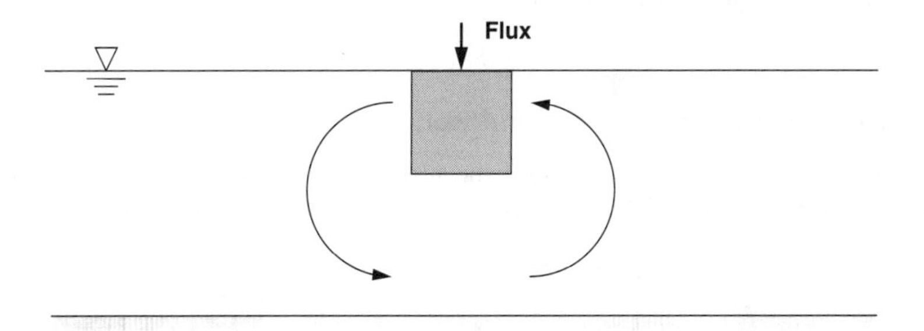

FIGURE 20.5
Depiction of surface renewal model of gas exchange.

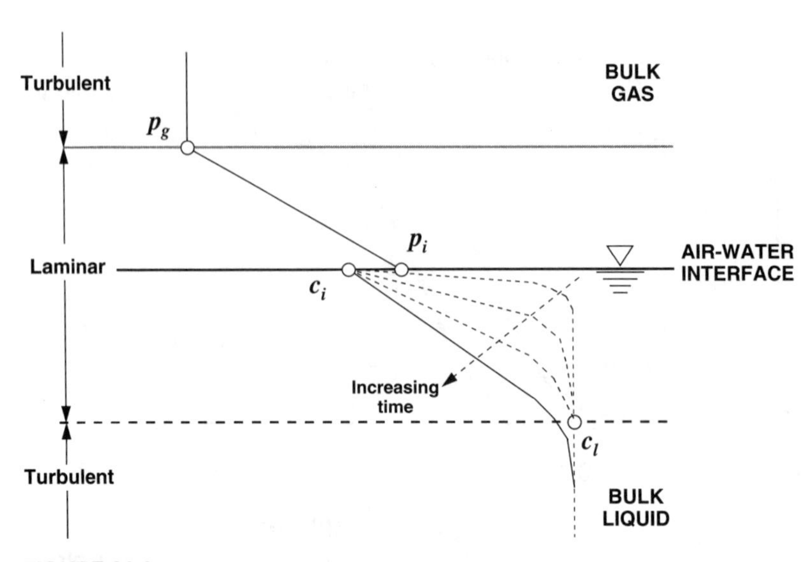

FIGURE 20.6
The temporal evolution of the liquid fi lm immediately after it is brought into contact with the gas.

process is not interrupted, the Whitman two-film condition (the solid line) will be attained.

As described in Box 20.1, the penetration theory can be used to estimate the flux of gas across the air-water interface as

$$J = \sqrt{\frac{D_l}{\pi t^*}}(c_s - c_l) \tag{20.18}$$

where D_l = liquid diffusion coefficient
c_s = concentration at the air-water interface
c_l = concentration in the bulk water
t^* = average contact time of the fluid parcel at the interface

This equation is of little value in itself because the average contact time at the interface is difficult to measure. However, Eq. 20.18 yields the valuable insight that if the penetration theory holds, the mass-transfer velocity is proportional to the square root of the gas's molecular diffusivity.

BOX 20.1. Derivation of Penetration Theory

Suppose that a parcel of water moves to the air-water interface (Fig. 20.5). The parcel can be idealized as a one-dimensional semi-infinite medium described by the equation

$$\frac{\partial c}{\partial t} = D_l \frac{\partial^2 c}{\partial z^2} \tag{20.19}$$

subject to the initial and boundary conditions

$c(z, 0) = c_l$ initial condition

$$c(0, t) = c_s \qquad \text{boundary condition at air-water interface} \tag{20.20}$$

$$c(\infty, t) = c_l \qquad \text{bottom boundary condition}$$

where D_l = liquid diffusion coefficient
c_s = concentration at the air-water interface
c_l = concentration in the bulk water

Applying these conditions, we can solve Eq. 20.19 for

$$c(z, t) = (c_s - c_l)\,\text{erfc}\left(\frac{z}{2\sqrt{D_l t}}\right) \tag{20.21}$$

where erfc is the error function complement, equal to 1 − erf, where erf is the error function (recall Sec. 10.3.2 and App. G),

$$\text{erf}\phi = \frac{2}{\sqrt{\pi}}\int_0^{\phi} e^{-\xi^2}\,d\xi \tag{20.22}$$

The flux across the air-water interface can be computed by applying Fick's first law at the interface ($z = 0$),

$$J(0, t) = -D_l\frac{\partial c(0, t)}{\partial z} \tag{20.23}$$

and the average flux is determined by

$$J = \frac{\int_0^{t^*} J(0, t)\,dt}{t^*} \tag{20.24}$$

where t^* = average contact time of the fluid parcel at the interface. Equation 20.21 can be differentiated and substituted into Eqs. 20.23 and 20.24 and solved for

$$J = \sqrt{\frac{D_l}{\pi t^*}}(c_s - c_l) \tag{20.25}$$

One of Higbie's underlying assumptions was that all packets of water have the same contact time at the interface. Danckwerts (1951) modified the approach by assuming that the fluid elements reach and leave the interface randomly. That is, their exposure is described by a statistical distribution. This approach, which is called the ***surface renewal theory***, was used to derive

$$J = \sqrt{D_l r_l}(c_s - c_l) \tag{20.26}$$

where r_l = liquid surface renewal rate, which has units of T^{-1}.

The surface renewal theory can also be applied to the gaseous side of the interface by assuming that packets of gas are brought into contact with the air-water interface in a random fashion. The transfer velocities for the liquid and gaseous phases can thus be written as

$$K_l = \sqrt{r_l D_l} \tag{20.27}$$

and

$$K_g = \sqrt{r_g D_g} \tag{20.28}$$

These relationships can be substituted into either Eqs. 20.14 or 20.15 to estimate a total transfer velocity for the interface.

We can now see that a major difference between two-film and the surface renewal theories relates to how the liquid and gas-film exchange velocities are formulated. In particular for the two-film theory, the velocities are proportional to D (Eqs. 20.8 and 20.9), whereas for the surface renewal theory they are proportional to the square root of D (Eqs. 20.27 and 20.28).

We return to the topic of gas transfer when we cover toxic substances later in this text. At that time, I provide additional information on Henry's constant and exchange coefficients for organic toxicants. For the time being let's narrow our focus and concentrate on the problem at hand: oxygen transfer.

20.2 OXYGEN REAERATION

At this point we have a general equation for the flux of any gas (Eq. 20.13),

$$J = v_v \left(\frac{p_g}{H_e} - c_l \right) \tag{20.29}$$

Now let's apply it to oxygen reaeration. Because of its high Henry's constant (≈ 0.8 atm m^3 mole^{-1}), oxygen is overwhelmingly liquid-film controlled. Consequently $v_v = K_l$ and Eq. 20.29 becomes

$$J = K_l \left(\frac{p_g}{H_e} - o \right) \tag{20.30}$$

where o = oxygen concentration in the water. Further, because oxygen is so abundant in the atmosphere, the partial pressure is constant and therefore

$$J = K_l(o_s - o) \tag{20.31}$$

where o_s = saturation concentration of oxygen.

Next the mole flux can be converted to a mass flux, and the liquid concentrations can be reexpressed in mass rather than mole units by multiplying both sides of Eq. 20.31 by the molecular weight of oxygen (32 g mole^{-1}). The equation can also be transformed from a flux to units of mass per time by multiplying it by the surface area of the liquid exposed to the atmosphere. Thus for a well-mixed open batch reactor, a mass balance for oxygen can be written as

$$V\frac{do}{dt} = K_l A_s(o_s - o) \tag{20.32}$$

where A_s = surface area of the water body.

Finally there are many cases (especially streams and rivers) where the transfer velocity is expressed as a first-order rate. In cases where the air-water interface is not constricted (as was not the case for the bottle from Fig. 20.1), the volume is

$$V = A_s H \tag{20.33}$$

where H = mean depth. If this is true Eq. 20.32 can be expressed as

$$V\frac{do}{dt} = k_a V(o_s - o) \tag{20.34}$$

where k_a = reaeration rate, which is equivalent to

$$k_a = \frac{K_l}{H} \tag{20.35}$$

Regardless of how the oxygen transfer rate is parameterized, Eq. 20.32 or 20.34 provides insight into how the mechanism of oxygen reaeration operates. The direction and magnitude of the mass transfer depends partially on the difference between the saturation value and the actual value in the water. If the water is undersaturated ($o < o_s$), then transfer will be positive (a gain) as oxygen moves from the atmosphere into the water to try to bring the water back to the equilibrium state of saturation. Conversely if the water is supersaturated ($o > o_s$), then transfer will be negative (a loss) as oxygen is purged from the system.

Oxygen reaeration rates can be extrapolated to other temperatures by

$$k_{a,T} = k_{a,20}\theta^{T-20} \tag{20.36}$$

where $\theta \cong 1.024$.

20.3 REAERATION FORMULAS

Many investigators have developed formulas for predicting reaeration in streams and rivers. Comprehensive reviews can be found elsewhere (Bowie et al. 1985). In this section we describe some of the more commonly used formulas that have been developed for natural waters.

20.3.1 Rivers and Streams

Numerous formulas have been proposed to model stream reaeration. Among these, three are very commonly used: the O'Connor-Dobbins, Churchill, and Owens-Gibbs formulas.

O'Connor-Dobbins. In Sec. 20.1.2 we developed the surface renewal model. For oxygen this model can be used to formulate the transfer velocity as

$$K_l = \sqrt{r_l D_l} \tag{20.37}$$

O'Connor and Dobbins (1956) hypothesized that the surface renewal rate could be approximated by the ratio of the average stream velocity to depth,

$$r_l = \frac{U}{H} \tag{20.38}$$

This hypothesis was backed up by experimental measurements. Substituting this value into Eq. 20.37 yields

$$K_l = \sqrt{\frac{D_l U}{H}} \tag{20.39}$$

The relationship is usually expressed as a reaeration rate,

$$k_a = \sqrt{D_l}\frac{U^{0.5}}{H^{1.5}} \tag{20.40}$$

The diffusivity of oxygen in natural waters is approximately 2.09×10^{-5} cm^2 s^{-1}. Therefore the ***O'Connor-Dobbins formula*** can be expressed as

Metric:

$$k_a = 3.93\frac{U^{0.5}}{H^{1.5}}$$

Units: k_a(d^{-1}), U(mps), H(m)

English:

$$k_a = 12.9\frac{U^{0.5}}{H^{1.5}} \tag{20.41}$$

Units: k_a(d^{-1}), U(fps), H(ft)

Churchill. Churchill et al. (1962) used a more empirical approach than that used by O'Connor and Dobbins. They exploited the fact that the water leaving some of the reservoirs in the Tennessee River valley were undersaturated with oxygen. They therefore measured oxygen levels in the stretches below these dams and calculated associated reaeration rates. They then correlated their results with depth and velocity to obtain

Metric:

$$k_a = 5.026\frac{U}{H^{1.67}}$$

English:

$$k_a = 11.6\frac{U}{H^{1.67}} \tag{20.42}$$

Same units as Eq. 20.41.

Owens and Gibbs. Owens et al. (1964) also used an empirical approach, but they induced oxygen depletion by adding sulfite to several streams in Great Britain. They combined their results with the data from the Tennessee River and fit the following formulas:

Metric:

$$k_a = 5.32\frac{U^{0.67}}{H^{1.85}}$$

English:

$$k_a = 21.6\frac{U^{0.67}}{H^{1.85}} \tag{20.43}$$

Same units as Eq. 20.41.

Comparison among formulas. As summarized in Table 20.1, the O'Connor-Dobbins, Churchill, and Owens-Gibbs formulas were developed for different types of streams. Covar (1976) found that they could be used jointly to predict reaeration for ranges of depths and velocity combinations (Zison et al. 1978). According to Fig. 20.7, O'Connor-Dobbins has the widest applicability being appropriate for moderate to deep streams with moderate to low velocities. The Churchill formula applies for similar depths but for faster streams. Finally the Owens-Gibbs relationship is used for shallower systems.

TABLE 20.1
Ranges of depth and velocity used to develop the O'Connor-Dobbins, Churchill, and Owens-Gibbs formulas for stream reaeration

Parameter	O'Connor-Dobbins	Churchill	Owens-Gibbs
Depth, m	0.30–9.14	0.61–3.35	0.12–0.73
ft	1–30	2–11	0.4–2.4
Velocity, mps	0.15–0.49	0.55–1.52	0.03–0.55
fps	0.5–1.6	1.8–5	0.1–1.8

Notice that the O'Connor-Dobbins formula generally gives lower values than the Churchill and Owens-Gibbs formulas. One possible explanation is that the slower, deeper channels for which O'Connor-Dobbins performs best are more idealized (i.e., more like a flume) than faster, shallower streams where drop structures and riffles may enhance reaeration.

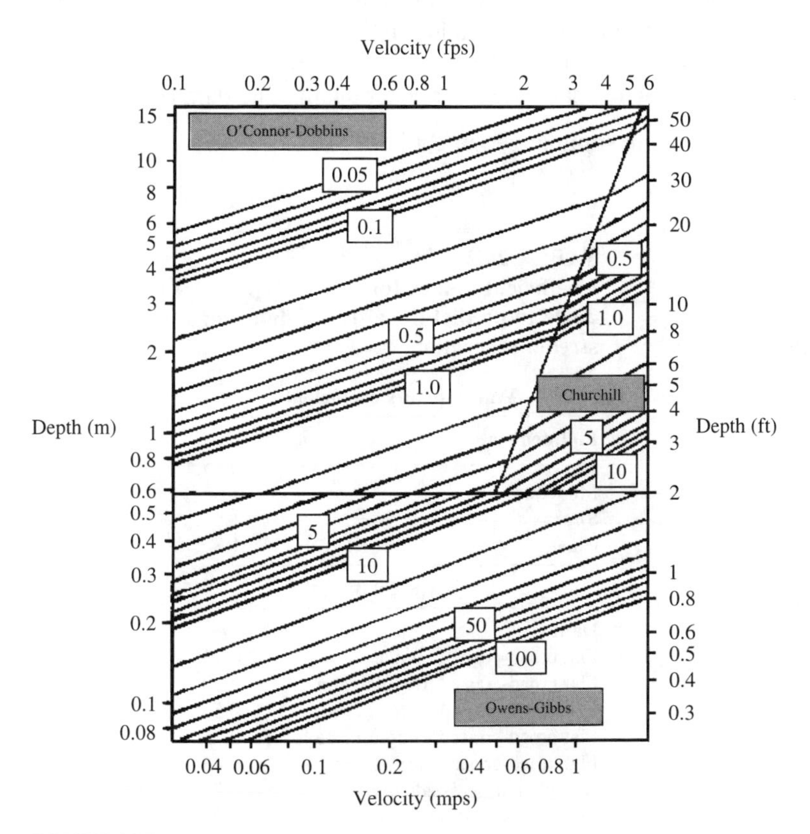

FIGURE 20.7
Reaeration rate (d^{-1}) versus velocity and depth (Covar 1976 and Zison et al. 1978).

Other formulas. There are many other reaeration equations beyond the O'Connor-Dobbins, Churchill, and Owens-Gibbs formulas. Bowie et al. (1985) provide an extensive compilation of many formulas along with references to major critiques and intercomparisons that have been performed.

In addition, along with allowing the user to specify reaeration values directly, software packages such as EPA's QUAL2E model also provide the option to automatically compute reaeration rates according to formulas. I will present these formulas when I describe the QUAL2E model in Lec. 26.

20.3.2 Waterfalls and Dams

Oxygen transfer in streams can be significantly influenced by the presence of waterfalls and dams. Butts and Evans (1983) have reviewed efforts to characterize this transfer and have suggested the following formula:

$$r = 1 + 0.38abH(1 - 0.11H)(1 + 0.046T) \tag{20.44}$$

where r = ratio of the deficit above and below the dam
H = difference in water elevation (m)
T = water temperature (°C)
a and b = coefficients that correct for water quality and dam type

Values of a and b are summarized in Table 20.2.

TABLE 20.2
Coefficient values for use in Eq. 20.44 to predict the effect of dams on stream reaeration

Water-quality coefficient	
Polluted state	a
Gross	0.65
Moderate	1.0
Slight	1.6
Clean	1.8

Dam-type coefficient	
Dam type	b
Flat broad-crested regular step	0.70
Flat broad-crested irregular step	0.80
Flat broad-crested vertical face	0.60
Flat broad-crested straight-slope face	0.75
Flat broad-crested curved face	0.45
Round broad-crested curved face	0.75
Sharp-crested straight-slope face	1.00
Sharp-crested vertical face	0.80
Sluice gates	0.05

20.3.3 Standing Waters and Estuaries

For standing waters, such as lakes, impoundments, and wide estuaries, wind becomes the predominant factor in causing reaeration.

Lakes. The oxygen-transfer coefficient itself can be estimated as a function of wind speed by a number of formulas. Some, such as the following relationship developed by Broecker et al. (1978), indicate a linear dependence,

$$K_l = 0.864 U_w \tag{20.45}$$

where K_l = oxygen mass-transfer coefficient (m d^{-1}) and U_w = wind speed measured 10 m above the water surface (m s^{-1}).

Others use various wind dependencies to attempt to characterize the different turbulence regimes that result at the air-water interface as wind velocity increases. For example the following is a widely used formula of this type (Banks 1975, Banks and Herrera 1977):

$$K_l = 0.728 U_w^{0.5} - 0.317 U_w + 0.0372 U_w^2 \tag{20.46}$$

Thus at high wind velocities, the relationship becomes dominated by the second-order term, as shown in Fig. 20.8.

As with stream reaeration formulas, lake oxygen-transfer formulas have both empirical and theoretical bases. For example Wanninkhof et al. (1991) used gas tracer experiments in lakes to develop the following formula:

$$K_l = 0.108 U_w^{1.64} \left(\frac{Sc}{600}\right)^{0.5} \tag{20.47}$$

where Sc = Schmidt number, which for oxygen in water is approximately 500. If this value is adopted the Wanninkhof formula reduces to $K_l = 0.0986 U_w^{1.64}$.

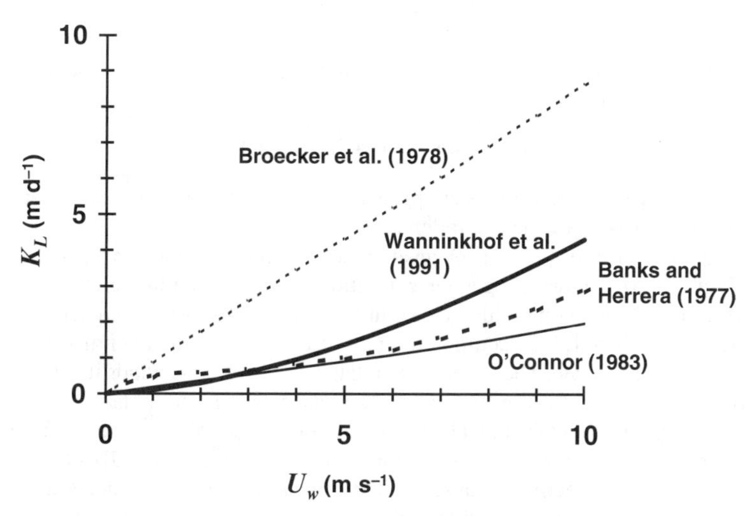

FIGURE 20.8
Comparison of wind-dependent reaeration formulas.

Finally O'Connor (1983) has developed a theoretically based set of formulas to compute transfer for low-solubility gases as a function of wind. His scheme can be applied to oxygen (Fig. 20.8).

It should be noted that there are many more formulas for calculating oxygen transfer as a function of wind. Many of these are summarized in general references such as Bowie et al. (1985). As evident from Fig. 20.8 these formulas yield a wide range of predictions. Consequently it is advisable to obtain system-specific measurements to check the validity of the formulas before using them in model calculations. As was the case for Wanninkhof et al. (1991) this can be done with artificial tracers (see Sec. 20.4). In addition natural oxygen-depletion events can sometimes be exploited to obtain direct measurements (Box 20.2).

BOX 20.2. Direct Measurement of Reaeration in Lakes

Many lakes in temperate regions are thermally stratified in the summer, consisting of an upper layer (epilimnion) and a lower layer (hypolimnion). In general the surface layer has dissolved oxygen concentration near saturation. If it is productive (that is, has high plant growth), settling plant matter can collect in the hypolimnion. The decomposition of this matter can then lead to severe oxygen depletion in the bottom waters. When turnover (that is, vertical mixing due to dropping temperature and increasing winds) occurs in the fall, the mixing of the two layers sometimes results in the lake's having an oxygen concentration well below saturation.

In certain cases the lake can be assumed to act as an open batch reactor: that is, we can ignore inflows and outflows of oxygen, except gas transfer across the lake's surface. If any additional sources and sinks of oxygen (such as sediment oxygen demand) are negligible, a mass balance for oxygen can be written for the lake in the period following overturn as

$$V\frac{do}{dt} = k_a(o_s - o) \tag{20.48}$$

If the saturation value is constant over the ensuing period, this equation can be solved for (with $o = o_i$ at $t = 0$)

$$o = o_i e^{-k_a t} + o_s(1 - e^{-k_a t}) \tag{20.49}$$

Thus if the oxygen concentrations are measured as a function of time, this model provides a means to estimate the reaeration rate.

Gelda et al. (1996) applied such an approach to Onondaga Lake in Syracuse, New York. Figure B20.2 shows oxygen concentrations that occurred in the lake following fall overturn in 1990, along with a curve fit with Eq. 20.49 using a reaeration rate of approximately 0.055 d^{-1}. In addition the plot also shows a simulation using a wind-dependent reaeration rate. The superior fit obtained with variable winds illustrates how important accurate wind estimates are for simulating gas transfer in lakes.

The approach of Gelda et al. (1996) is particularly appealing because it is nonobtrusive; that is, it does not depend on the introduction of tracers and dyes to the environment. It is also attractive because it directly measures oxygen concentration on a whole-lake basis. Wherever possible such direct measurements are preferable to indirect formula estimation.

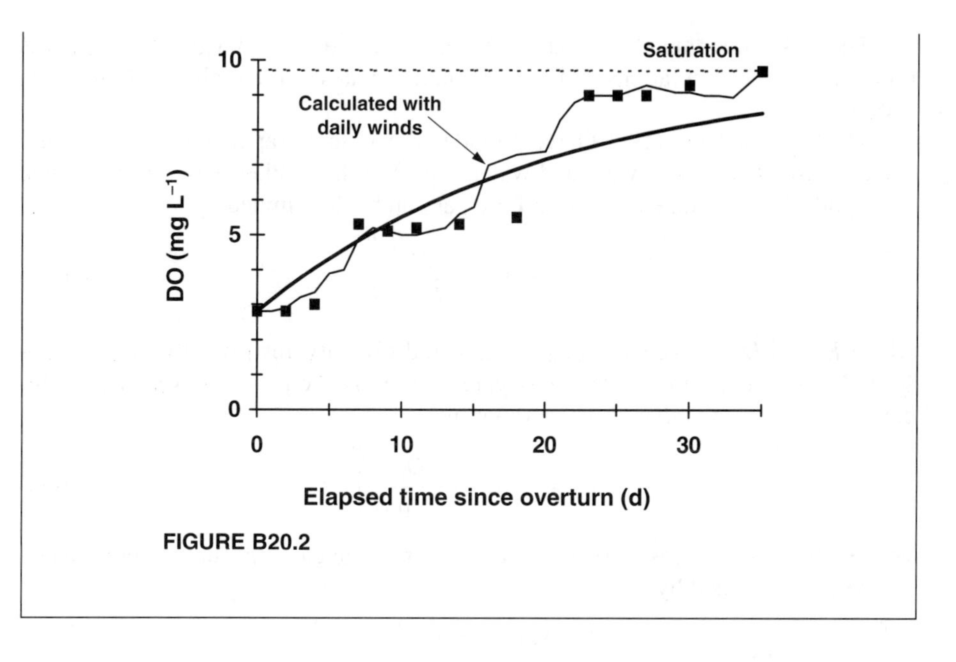

FIGURE B20.2

Estuaries. Because estuary gas transfer can be affected by both water and wind velocity, efforts to determine reaeration in estuaries combines elements of current and wind-driven approaches.

The water velocity effects are typically computed with the O'Connor-Dobbins formula (Eq. 20.40),

$$k_a = \frac{\sqrt{D_l U_o}}{H^{3/2}} \tag{20.50}$$

where U_o = mean tidal velocity over a complete tidal cycle.

The wind effects can be computed with any of the formulas developed for standing waters in the previous paragraphs. For example Eq. 20.46 can be expressed as a reaeration rate, as in

$$k_a = \frac{0.728U_w^{0.5} - 0.317U_w + 0.0372U_w^2}{H} \tag{20.51}$$

Thomann and Fitzpatrick (1982) have combined the two approaches for estuaries affected by both tidal velocities and wind,

$$k_a = 3.93\frac{\sqrt{U_o}}{H^{3/2}} + \frac{0.728U_w^{0.5} - 0.317U_w + 0.0372U_w^2}{H} \tag{20.52}$$

20.3.4 Extrapolating Reaeration to Other Gases

As stated earlier, we are going to return to the topic of gas transfer when we model toxic organics later in this text. However, beyond toxics there are a few other common gases that are of interest in environmental engineering.

The most important of these are carbon dioxide and ammonia gas. The former is important in pH calculations, whereas the latter relates to the problem of ammonia toxicity.

Mackay and Yeun (1983) have provided a way to extrapolate from commonly studied gases (such as oxygen and water vapor) to these other gases. For example the liquid-film exchange coefficient for a gas can be determined by

$$K_l = K_{l,O_2}\left(\frac{D_l}{D_{l,O_2}}\right)^{0.5} \tag{20.53}$$

where K_l and D_l = exchange coefficient and diffusivity, respectively, and the subscript O_2 designates the values for oxygen. Similarly the gas-film exchange coefficient can be scaled to that of water vapor by

$$K_g = K_{g,H_2O}\left(\frac{D_g}{D_{g,H_2O}}\right)^{0.67} \tag{20.54}$$

where it has been suggested (Mills et al. 1982) that the gas-film coefficient for water can be approximated by

$$K_{g,H_2O} = 168U_w \tag{20.55}$$

where K_{g,H_2O} has units of m d^{-1} and U_w = wind speed (m s^{-1}).

Schwarzenbach et al. (1993) have correlated diffusion coefficients with molecular weight. For a temperature of 25°C, the resulting equations are

$$D_l = \frac{2.7 \times 10^{-4}}{M^{0.71}} \tag{20.56}$$

and

$$D_g = \frac{1.55}{M^{0.65}} \tag{20.57}$$

Finally some investigators have combined relationships such as Eqs. 20.53 to 20.56 to directly calculate the exchange coefficients as a function of molecular weight. Using this approach Mills et al. (1982) have come up with

$$K_l = K_{l,O_2}\left(\frac{32}{M}\right)^{0.25} \tag{20.58}$$

and

$$K_g = K_{g,H_2O}\left(\frac{18}{M}\right)^{0.25} \tag{20.59}$$

20.4 MEASUREMENT OF REAERATION WITH TRACERS

Aside from formulas, reaeration can be measured directly in the field. Four methods are commonly used. The first three consist of techniques that back-calculate reaeration based on a mass balance model and field measurements of oxygen. These are:

- *Steady-state oxygen balance.* If all the other factors governing an oxygen sag (that is, deoxygenation rate, sediment oxygen demand, etc.) can be determined independently, the only unknown governing the sag will be the reaeration coefficient.

Unfortunately, because the other factors are difficult to measure accurately, estimates obtained in this fashion are usually highly uncertain. However, as in Churchill's studies on the Tennessee River (recall discussion of Eq. 20.42), there are certain problem settings where the approach works nicely. Box 20.2 outlines such a case that sometimes occurs in lakes.

- *Deoxygenation with sodium sulfite.* As in Owens' studies of British streams, oxygen levels can be artificially lowered by adding sodium sulfite to the stream. This method would be particularly attractive for relatively clean systems, where other effects would be negligible.
- *Diurnal oxygen swings.* In some streams, plant growth can induce diurnal swings in oxygen level. Chapra and Di Toro (1991) have illustrated how such data can be used to obtain reaeration estimates. I will describe this approach when I discuss the impact of photosynthesis on oxygen in Lec. 24.

The fourth method for measuring reaeration in the field takes a decidedly different tack. Rather than oxygen, a different volatile substance is injected into the system. Such substances are chosen because (1) they volatilize in an analogous fashion to oxygen, (2) they do not react, and (3) their concentrations can be measured at a relatively reasonable cost. Most commonly, radioactive (e.g., krypton-85), hydrocarbon (e.g., ethylene, propane, methyl chloride, etc.), and inorganic tracers (e.g., sulfur hexafluoride) are used. These tracers are usually discharged together with a conservative, nonvolatile tracer (tritium, lithium) to determine dispersion (recall Sec. 10.4).

Either continuous or pulse experiments are normally conducted. For the continuous case the tracers are injected at a constant rate until steady concentration levels are attained at two downstream locations. The first-order gas-transfer rate can then be estimated by Eq. 10.36,

$$k = \frac{l}{\bar{t}_2 - \bar{t}_1} \ln \frac{M_1}{M_2} \tag{20.60}$$

where the subscripts 1 and 2 represent the upstream and the downstream locations, the $\bar{t}$'s are the average travel times to the two locations, and the M's are the masses of the tracer. Because the experiment is continuous, the masses should be equal to the flow times concentration at each point. Therefore the equation can be expressed as (assuming constant flow)

$$k = \frac{1}{\bar{t}_2 - \bar{t}_1} \ln \frac{c_1}{c_2} \tag{20.61}$$

A similar approach can be used for the pulse experiment, except that the masses in Eq. 20.60 would be determined by integration as described previously in Lec. 10.

Once the first-order gas-transfer rate is estimated, the result must be extrapolated to oxygen. One way to do this would be by using the empirically derived correlations such as Eqs. 20.58. Although this can be done, the developers of tracer methods have directly estimated the scaling as represented by

$$k_a = Rk \tag{20.62}$$

where R = scale factor to relate the tracer exchange rate to reaeration (Table 20.3).

TABLE 20.3
Factors for scaling gas-tracer exchange rates to oxygen reaeration rates

Tracer	R	Reference
Ethylene	1.15	Rathbun et al. (1978)
Propane	1.39	Rathbun et al. (1978)
Methyl chloride	1.4	Wilcox (1984*a, b*)
Sulfur hexafluoride	1.38	Canale et al. (1995)
Krypton	1.2	Tsivoglou and Wallace (1972)

PROBLEMS

20.1. A tanker truck careens off the road and dumps 30,000 L of glucose syrup into a small mountain lake. The concentration of the syrup is 100 g-glucose L^{-1}. The lake has the following characteristics in the period immediately following the spill: residence time = 30 d, depth = 5 m, area = 5×10^4 m^2, altitude = 11,000 ft, wind speed = 2.235 m s^{-1}, and temperature = 10°C. Note that the lake is assumed to be completely mixed, and has zero BOD and is at saturation prior to the spill. Also recognize that the inflow river is at saturation.

(*a*) Compute the grams of CBOD spilled.
(*b*) Compute how the CBOD and oxygen in the lake change after the spill.
(*c*) Determine the time of the worst oxygen level in the system.

20.2. A lake in the United States has a surface area of 5×10^5 m^2, a mean depth of 5 m, and a residence time of 1 wk. How large a community could discharge to the system during the summer (wind speed = 0.89 mps, temperature = 30°C, and elevation = 1 km) if the BOD decay rate is 0.1 d^{-1} and the desired oxygen level is 6 mg L^{-1}? Assume that the sewage has zero dissolved oxygen concentration and does not settle. Also express your result as an equivalent inflow concentration.

20.3. Suppose that the 300-mL bottle described in Example 20.1 had an open top as illustrated in Fig P20.3. Repeat the example for this case. As with the bottle neck, assume that transfer through the thin liquid film takes place by molecular diffusion.

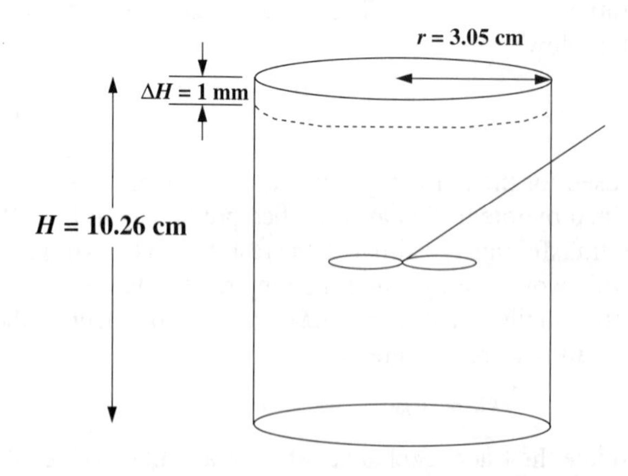

FIGURE P20.3

20.4. A flat broad-crested regular step dam with a drop of 2 m is situated on a grossly polluted stretch of river below a sewage outfall plant. The river is located at an elevation of 2 km. Determine the oxygen concentration below the dam if the water upstream has a concentration of 2 mg L^{-1} and a temperature of 26°C.

20.5. Derive a relationship of the form of Eq. 20.58 by combining Eq. 20.56 with Eq. 20.53.

20.6. A dye study yields the following data for time and concentration of ethylene:

Station 1 (6 km)

t (min)	0	10	20	30	40	50	60	70	80	90	100
c (μg L^{-1})	0	9	69	81	78	74	71	80	80	80	0

Station 2 (13.5 km)

t (min)	9	9.5	10	10.5	11	11.5	12	12.5
c (μg L^{-1})	0.0	0.0	1.4	2.9	3.4	4.6	3.4	3.0
t (min)	13	13.5	14	14.5	15	15.5		
c (μg L^{-1})	2.9	2.3	2.1	1.1	1.1	0.6		

An accompanying tracer study using rhodamine dye and lithium has yielded estimates of 0.5 d and 8.3×10^4 cm^2 s^{-1} for travel time and dispersion, respectively. The river has the following characteristics: Q = 3.7 cms, B = 46 m, and T = 21°C.

(*a*) Estimate the reaeration and compare your results with appropriate reaeration formulas.

(*b*) Use Eq. 10.24 to compute the continuous distribution of ethylene at the second station. Plot the data on the same graph for comparison.

20.7. You continuously discharge sulfur hexafluoride into a stream having constant hydrogeometric characteristics. You measure concentrations of 400 and 150 pptr at locations 0.5 and 4 km downstream from the injection point, respectively. Use this data to estimate the reaeration rate if the velocity over the stretch is 0.2 m s^{-1}.

20.8. The following data are measured for a polluted lake with a mean depth of 12 m following overturn:

Time (d)	0	4	8	12	16	20	24	28	32
DO (mg L^{-1})	5	6.4	6.8	7.8	8	8.5	8.5	8.5	8.8

Determine the reaeration rate and the oxygen mass-transfer coefficient if the saturation during the sampling period was 9 mg L^{-1}.

20.9. A stream has a velocity of 0.4 mps, a depth of 0.3 m, and a temperature of 23°C. Estimate (*a*) the reaeration rate and (*b*) the comparable rate for carbon dioxide.

20.10. A stream with a rectangular channel has the following characteristics: S = 0.001, B = 20 m, n = 0.03, Q = 1 cms, and T = 10°C. Determine the reaeration rate following the introduction of a new point source that will discharge an additional 0.5 cms (T = 25°C) to the channel.

20.11. The epilimnion of a lake ($H = 7$ m) has a temperature of 16°C and is subject to a constant wind of 2 mps. (*a*) Estimate the oxygen reaeration rate in d^{-1}. (*b*) Estimate the mass transfer velocity for ammonia gas in m d^{-1}. Assume Eq. 20.46 holds.

20.12. A wide river ($B = 200$ m) has a flow of 800 cms and a mean depth of 2.7 m. It is subject to a wind of 1.5 mps and its mean temperature is 25°C.
(*a*) Determine its reaeration rate in d^{-1}.
(*b*) Determine a mass-transfer rate for ammonia in m d^{-1}.
Assume Eq. 20.47 holds.

LECTURE 21

Streeter-Phelps: Point Sources

LECTURE OVERVIEW: I derive the classic Streeter-Phelps equation for a single point source of carbonaceous BOD and provide some guidance on how it can be calibrated. Next I show how it can be used to model multiple point sources and illustrate how care must be taken when making an oxygen balance for a point source. I then show how the model is affected by anaerobic conditions. Finally I contrast the plug-flow model with the mixed-flow version that is appropriate in systems such as estuaries, where dispersion is significant.

The Streeter-Phelps model ties together the two primary mechanisms governing dissolved oxygen in a stream receiving sewage: decomposition of organic matter and oxygen reaeration. As such it provides an analytical framework for predicting the effect of both point and nonpoint sources of organic wastewater on stream and estuary dissolved oxygen. In this lecture I focus on point sources.

21.1 EXPERIMENT

Now that we have explored ways to characterize gas transfer, let us return to the experiment that we developed in Lec. 19. Recall that we placed a quantity of organic matter in a sealed bottle and then simulated how oxygen was depleted. Now we can perform the same simulation, but with the bottle open to the atmosphere.

The mass balances for BOD and dissolved oxygen can be written as

$$V\frac{dL}{dt} = -k_d V L \tag{21.1}$$

and
$$V\frac{do}{dt} = -k_d VL + k_a V(o_s - o) \tag{21.2}$$

Now before proceeding with the solution, we make a transformation that simplifies the oxygen balance. To do this we introduce a new variable,

$$D = o_s - o \tag{21.3}$$

where D is called the ***dissolved oxygen deficit***. Equation 21.3 can be differentiated to give

$$\frac{dD}{dt} = -\frac{do}{dt} \tag{21.4}$$

Equations 21.3 and 21.4 can be substituted into Eq. 21.2 to give

$$V\frac{dD}{dt} = k_d VL - k_a VD \tag{21.5}$$

Thus the use of deficit simplifies the differential equation.

If $L = L_0$ and $D = 0$ at $t = 0$, then Eqs. 21.1 and 21.5 can be solved for

$$L = L_0 e^{-k_d t} \tag{21.6}$$

and
$$D = \frac{k_d L_0}{k_a - k_d}\left(e^{-k_d t} - e^{-k_a t}\right) \tag{21.7}$$

As depicted in Fig. 21.1, opening the lid means that oxygen first decreases, but it then recovers as reaeration replenishes the oxygen.

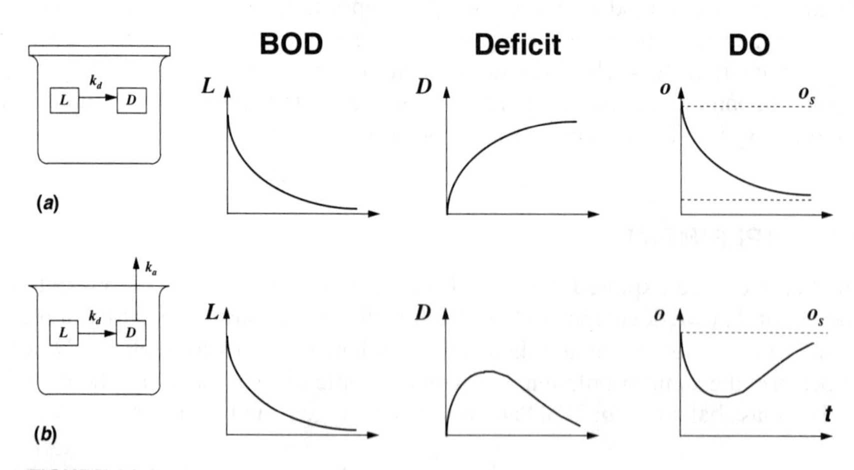

FIGURE 21.1
How (*a*) closed and (*b*) open systems are affected by BOD decomposition.

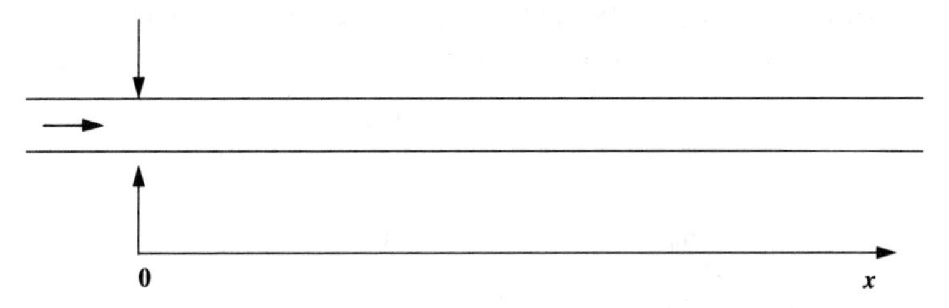

FIGURE 21.2
A point source discharged into a stream reach of constant geometry and hydrology.

21.2 POINT-SOURCE STREETER-PHELPS EQUATION

Now we can move the foregoing theory into a natural water. Specifically we will model a stream with a single point source of BOD. As depicted in Fig. 21.2 the reach is at steady-state and is characterized by plug flow with constant hydrology and geometry. This is the simplest manifestation of the classic Streeter-Phelps model.

Mass balances can be written as

$$0 = -U\frac{dL}{dx} - k_r L \tag{21.8}$$

and

$$0 = -U\frac{dD}{dx} + k_d L - k_a D \tag{21.9}$$

where $k_r = k_d + k_s$.

If $L = L_0$ and $D = D_0$ at $t = 0$, then these equations can be solved for

$$L = L_0 e^{-\frac{k_r}{U}x} \tag{21.10}$$

and

$$D = D_0 e^{-\frac{k_a}{U}x} + \frac{k_d L_0}{k_a - k_r}\left(e^{-\frac{k_r}{U}x} - e^{-\frac{k_a}{U}x}\right) \tag{21.11}$$

These equations constitute the classic "Streeter-Phelps" model and are the equations behind the oxygen "sag" we discussed in Fig. 19.2.

Aside from being written for a plug-flow system, these equations differ from Eqs. 21.6 and 21.7 in two important ways. First, only part of the removal has an effect on the deficit. Second, the deficit has an initial value.

21.3 DEFICIT BALANCE AT THE DISCHARGE POINT

We introduced the dissolved oxygen deficit to simplify our mathematical model. Unfortunately it also complicates the determination of the boundary concentration at the mixing point because most waste inflows exhibit a significantly different temperature from the receiving water. Consequently their oxygen saturations will differ and, hence, a simple deficit balance will be erroneous. This is best illustrated by example.

EXAMPLE 21.1. OXYGEN BALANCE AT A DISCHARGE POINT. A point source and a receiving stream at sea level have the following characteristics:

Values	Point source	River
Flow ($m^3\ s^{-1}$)	0.463	5.787
Temperature (°C)	28	20
DO ($mg\ L^{-1}$)	2	7.5
DO saturation ($mg\ L^{-1}$)	7.827	9.092
DO deficit ($mg\ L^{-1}$)	5.827	1.592

Perform mass balances for temperature and oxygen assuming complete mixing.

Solution: If we assume that the density and heat capacity of water are relatively constant, a heat balance for temperature can be developed in a similar fashion to a mass balance,

$$T_0 = \frac{5.787(20) + 0.463(28)}{5.787 + 0.463} = 20.59°\text{C}$$

A mass balance for oxygen can be calculated simply as

$$o_0 = \frac{5.787(7.5) + 0.463(2)}{5.787 + 0.463} = 7.093 \text{ mg L}^{-1}$$

The saturation value for 20.59°C is 8.987 mg L^{-1}. Therefore the deficit at the mixing point can be determined as $D_0 = 8.987 - 7.093 = 1.894$ mg L^{-1}. This is the correct value.

Now let us see what happens when we try to balance the deficit directly:

$$D_0 = \frac{5.787(1.592) + 0.463(5.827)}{5.787 + 0.463} = 1.906 \text{ mg L}^{-1}$$

which represents a discrepancy of $1.894 - 1.906 = -0.012$ mg L^{-1}. Why does this error occur? It results from the fact that temperature and oxygen saturation are related in a nonlinear fashion by Eq. 19.32.

It should also be mentioned that the type of errors illustrated in the previous section can also be incurred for systems where saturation changes longitudinally due to temperature, elevation, or salinity variations. Examples include systems with high elevation changes (upland streams) or estuaries. For such systems, deficit balances at junctions can introduce slight discrepancies if not handled correctly. Several end-of-chapter problems illustrate the errors.

Fortunately the discrepancy incurred by using deficit is usually not great because the relationship of oxygen to temperature, elevation, and salinity is not highly nonlinear over the ranges commonly encountered (recall Figs. 19.9 and 19.10). Therefore the errors incurred are usually not significant. In addition, now that computers are ubiquitous, dissolved oxygen (rather than deficit) is simulated directly, and the problems associated with deficit balances are becoming a moot point.

21.4 MULTIPLE POINT SOURCES

Now that we have an understanding of how to perform mass balances at discharge points, we can investigate how multiple point sources can be simulated. In general the approach consists of treating the river as a series of uniform reaches linked by boundary conditions. The computation starts at the farthest upstream point where boundary concentrations are specified. Then the model equations (Eqs. 21.10 and 21.11) are used to calculate concentrations in the downstream direction.

The calculation is continued until a boundary is encountered. Two types of boundaries occur. The first represents the case where the system parameters change. For example the bottom slope might change, which would lead to a change in the channel's velocity, depth, reaeration rate, etc. In such cases the concentrations at the end of the upstream reach directly serve as the initial concentrations for the model equations with revised parameters. Second, for situations where a point source enters the system, mass balances are used to establish the starting concentrations for the new reach. The approach is described in the following example.

EXAMPLE 21.2. MULTIPLE SOURCES. Figure 21.3 shows a river that receives a sewage treatment plant effluent at kilometer point 100 (KP 100) and a tributary inflow at KP 60. Note that the channel is trapezoidal with the characteristics shown. The deoxygenation rate for CBOD is equal to 0.5 d^{-1} at 20°C. For 20 km downstream from the treatment plant, there is a CBOD settling removal rate of 0.25 d^{-1}.

Assuming that the O'Connor-Dobbins reaeration formula holds and that the stream is at sea level, compute the concentration of dissolved oxygen in the system. To simplify the calculation, we have independently determined the heat balances. In addition we have computed the system's hydrogeometric parameters and reaction kinetics. This information is included in the following tables:

Parameter	Units	KP > 100	KP 100–60	KP < 60
Depth	m	1.19	1.24	1.41
	(ft)	(3.90)	(4.07)	(4.62)
Area	m^2	14.71	15.5	18.05
Flow	$m^3\ s^{-1}$	5.787	6.250	7.407
	$m^3\ d^{-1}$	500,000	540,000	640,000
	(cfs)	(204)	(221)	(262)
Velocity	$m\ s^{-1}$	0.393	0.403	0.410
	$m\ d^{-1}$	33,955	34,819	35,424
	(fps)	(1.29)	(1.32)	(1.35)

Parameter	KP > 100	KP 100–80	KP 80–60	KP < 60
T (°C)	20	20.59	20.59	19.72
o_s (mg L^{-1})	9.092	8.987	8.987	9.143
k_a (d^{-1})	1.902	1.842	1.842	1.494
k_r (d^{-1})	0.50	0.764	0.514	0.494
k_d (d^{-1})	0.50	0.514	0.514	0.494

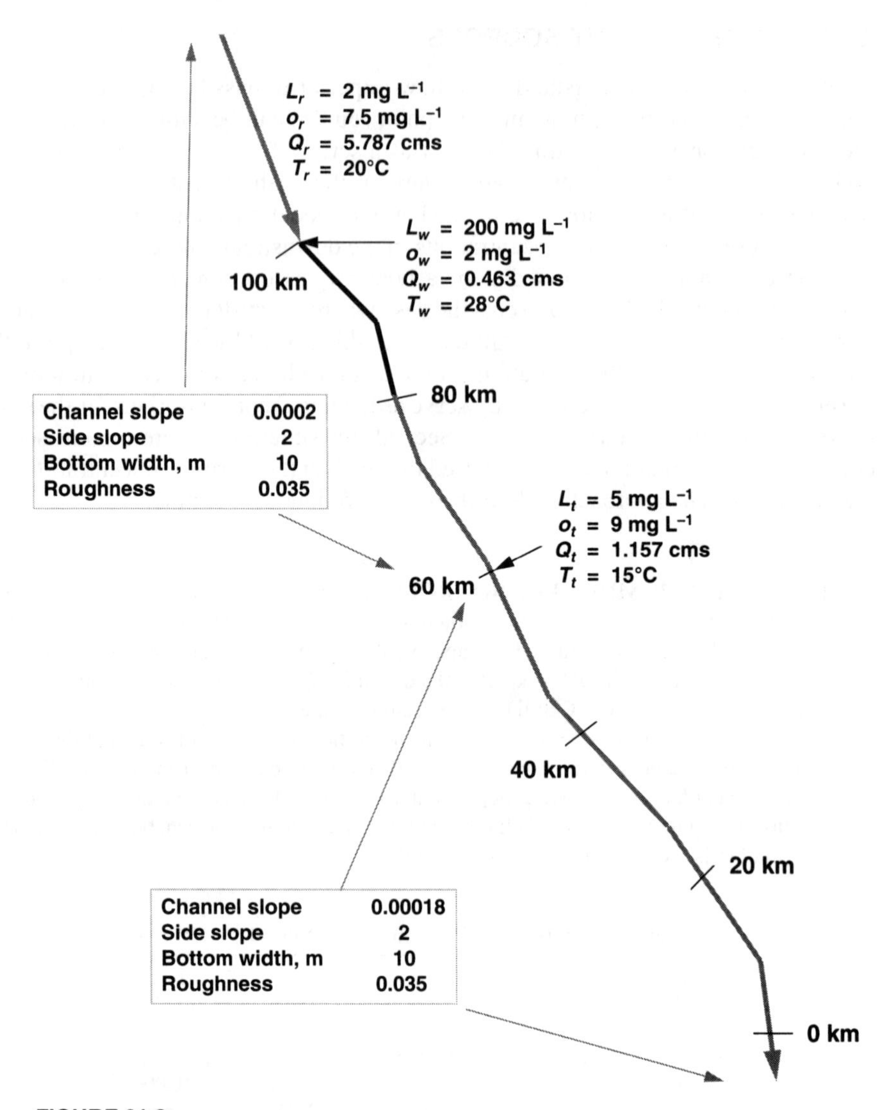

FIGURE 21.3
A stretch of stream with two point sources and changing hydrogeometric characteristics.

Solution: First, the concentration of carbonaceous BOD can be calculated for each stretch. A CBOD mass balance is developed at KP 100,

$$L_0 = \frac{40{,}000(200) + 500{,}000(2)}{540{,}000} = 16.667 \text{ mg L}^{-1}$$

which then decays downstream according to

$$L = 16.667e^{-\frac{0.514+0.25}{34{,}819}x}$$

At KP 80 (that is, 20 km downstream from the point source) the value of BOD will have dropped to

$$L = 16.667e^{-\frac{0.764}{34,819}20,000} = 10.75 \text{ mg L}^{-1}$$

In the next reach the BOD does not settle, but continues to decay. At KP 60 the value of BOD will have dropped to

$$L = 10.75e^{-\frac{0.514}{34,819}(40,000-20,000)} = 8 \text{ mg L}^{-1}$$

Now another mass balance must be taken to account for the inflowing tributary,

$$L_0 = \frac{540,000(8) + 100,000(5)}{640,000} = 7.53 \text{ mg L}^{-1}$$

which then decays downstream. Thus by KP 0, the value of BOD will have dropped to

$$L = 7.53e^{-\frac{0.494}{35,424}(100,000-40,000)} = 3.26 \text{ mg L}^{-1}$$

For dissolved oxygen a mass balance is developed at KP 100,

$$o_0 = \frac{40,000(2) + 500,000(7.5)}{540,000} = 7.093 \text{ mg L}^{-1}$$

which represents an initial deficit of

$$D_0 = 8.987 - 7.093 = 1.894 \text{ mg L}^{-1}$$

The deficit in the next downstream reach (KP 100 to KP 80) can be computed with the Streeter-Phelps formula

$$D = 1.894e^{-\frac{1.842}{34,819}x} + \frac{0.514(16.667)}{0.764 - 1.842}\left(e^{-\frac{1.842}{34,819}x} - e^{-\frac{0.764}{34,819}x}\right)$$

For example, at KP 80, the deficit would be

$$D = 1.894e^{-\frac{1.842}{34,819}20,000} + \frac{0.514(16.667)}{0.764 - 1.842}\left(e^{-\frac{1.842}{34,819}20,000} - e^{-\frac{0.764}{34,819}20,000}\right) = 3.02 \text{ mg L}^{-1}$$

which corresponds to an oxygen concentration of

$$o = 8.987 - 3.02 = 5.97 \text{ mg L}^{-1}$$

The deficit in the next downstream reach (KP 80 to KP 60) can be computed as

$$D = 5.97e^{-\frac{1.842}{34,819}(x-20,000)} + \frac{0.514(10.75)}{0.514 - 1.842}\left(e^{-\frac{1.842}{34,819}(x-20,000)} - e^{-\frac{0.514}{34,819}(x-20,000)}\right)$$

At KP 60 (that is, 40 km downstream from the point source), the deficit would be

$$\begin{aligned} D &= 5.97e^{-\frac{1.842}{34,819}(40,000-20,000)} \\ &\quad + \frac{0.514(10.75)}{0.514 - 1.842}\left(e^{-\frac{1.842}{34,819}(40,000-20,000)} - e^{-\frac{0.514}{34,819}(40,000-20,000)}\right) \\ &= 2.70 \text{ mg L}^{-1} \end{aligned}$$

which corresponds to an oxygen concentration of

$$o = 8.987 - 2.70 = 6.29 \text{ mg L}^{-1}$$

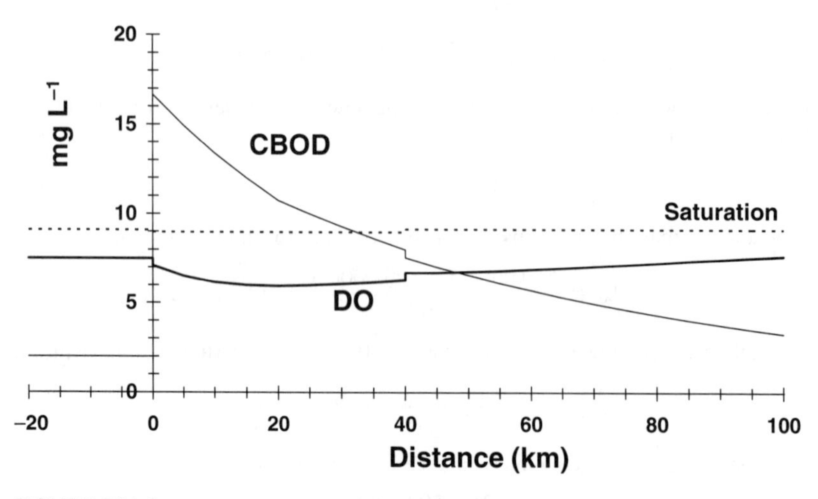

FIGURE E21.2

Now another mass balance is taken to account for the inflowing tributary,

$$L_0 = \frac{540{,}000(6.29) + 100{,}000(5)}{640{,}000} = 2.43 \text{ mg L}^{-1}$$

This value, along with the mass balance for CBOD, serves as the boundary condition for the next reach. The final results are summarized in Fig. E21.2 and the following table:

KP	CBOD (mg L^{-1})	Deficit (mg L^{-1})	DO (mg L^{-1})
−10	2.000	1.592	7.500
0	16.667	1.894	7.093
10	13.384	2.814	6.173
20	10.748	3.022	5.965
30	9.274	2.918	5.997
40	7.532	2.433	6.710
50	6.553	2.391	6.752
60	5.700	2.260	6.883
70	4.959	2.085	7.059
80	4.314	1.891	7.252
90	3.753	1.696	7.447
100	3.265	1.509	7.635

21.5 ANALYSIS OF THE STREETER-PHELPS MODEL

In this section we scrutinize the simple Streeter-Phelps model to try to gain some insight into its behavior. This is extremely important for model calibration. For simplicity the following analysis is written in terms of travel time. Recall that the basic equation is

$$D = D_0 e^{-k_a t} + \frac{k_d L_0}{k_a - k_r}\left(e^{-k_r t} - e^{-k_a t}\right) \tag{21.12}$$

The critical travel time can be determined by differentiating Eq. 21.12, setting the result equal to zero, and solving for

$$t_c = \frac{1}{k_a - k_r} \ln\left\{\frac{k_a}{k_r}\left[1 - \frac{D_0(k_a - k_r)}{k_d L_0}\right]\right\} \tag{21.13}$$

The critical deficit can be determined as

$$D_c = \frac{k_d L_0}{k_a}\left\{\frac{k_a}{k_r}\left[1 - \frac{D_0(k_a - k_r)}{k_d L_0}\right]\right\}^{-\frac{k_r}{k_a - k_r}} \tag{21.14}$$

These results indicate that the presence of an initial deficit causes the critical deficit to be larger and farther downstream from the discharge. To gain further insight, let us now simplify the analysis by assuming that D_0 is zero. For this case Eqs. 21.13 and 21.14 reduce to

$$t_c = \frac{1}{k_a - k_r} \ln \frac{k_a}{k_r} \tag{21.15}$$

and

$$D_c = \frac{k_d L_0}{k_a}\left(\frac{k_a}{k_r}\right)^{-\frac{k_r}{k_a - k_r}} \tag{21.16}$$

Now some additional general characteristics emerge. First, notice that the critical time depends only on the removal and the reaeration rates. To understand this relationship, note the plot of t_c versus k_r for various values of k_a in Fig. 21.4. This leads to the general conclusion that an increase of either one or both of these parameters means that the critical point on the sag will move closer to the source.

Second, inspection of Eq. 21.16 shows that the magnitude of the critical deficit is linearly related to k_d and L_0. This makes sense, because each of these parameters contributes linearly to the amount of BOD that is available for decomposition.

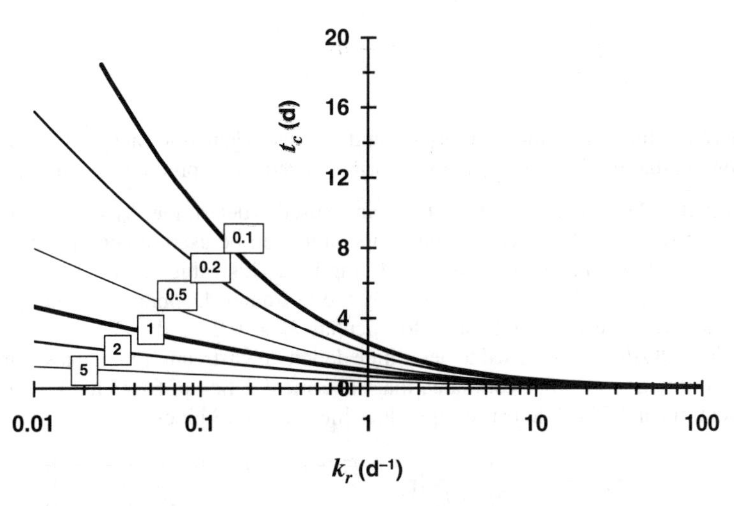

FIGURE 21.4
A plot of t_c versus k_r for various values of k_a (boxed numbers).

21.6 CALIBRATION

The relationships developed in the previous section can be extremely useful in model calibration for the case where the critical time and deficit are known. For example if the initial BOD and reaeration are predetermined, Eqs 21.15 and 21.16 can be used to calculate values for k_d and k_r.

EXAMPLE 21.3. PARAMETER ESTIMATION. A point source and a receiving stream at an altitude of 1524 m (5000 ft) above sea level have the following characteristics:

Values	Point source	River
Flow ($m^3\ s^{-1}$)	0.7078	3.398
Temperature (°C)	22.7	14.6
BOD (mg L^{-1})	40	5
DO (mg L^{-1})	2	8

For 32 km below the discharge point, the stream has a relatively constant width of 61 m and a mean depth of 1.83 m. The following DO measurements are available:

Downstream distance (km)	DO (mg L^{-1})	Downstream distance (km)	DO (mg L^{-1})
1.61	2.9	16.09	4.8
3.22	1.7	19.31	5.6
4.83	1.5	22.53	6.2
6.44	1.7	25.75	6.7
8.05	2.1	28.97	7.0
9.66	2.7	32.19	7.3
12.87	3.8		

Determine the total removal rate (k_r) and the deoxygenation rate (k_d) from the data, assuming that the O'Connor-Dobbins formula adequately predicts reaeration.

Solution: Mass balances at the outfall can be used to determine: $L_0 = 11$ mg L^{-1}, $o_0 =$ 6.97 mg L^{-1}, and $T_0 = 16$°C. The temperature can be used in conjunction with Eqs. 19.32 and 19.39 to compute $o_{so} = 8.143$ mg L^{-1}. This value, in turn, can be employed to calculate $D_0 = 1.173$ mg L^{-1}. Finally the O'Connor-Dobbins formula along with a temperature correction can be used to calculate $k_a = 0.2775$ d^{-1}.

The DO data can be used to determine deficit downstream from the discharge. From this information it can be estimated that the critical deficit of 6.643 mg L^{-1} occurs at a travel time of 1.517 d. Substituting values into Eq. 21.13 gives

$$1.517 = \frac{1}{0.2775 - k_r} \ln\left\{\frac{0.2775}{k_r}\left[1 - \frac{1.173(0.2775 - k_r)}{k_d(11)}\right]\right\}$$

and into Eq. 21.14,

$$6.643 = \frac{k_d(11)}{0.2775}\left\{\frac{0.2775}{k_r}\left[1 - \frac{1.173(0.2775 - k_r)}{k_d(11)}\right]\right\}^{-\frac{k_r}{0.2775-k_r}}$$

These two nonlinear equations can be solved simultaneously for k_r and $k_d = 1.159$ and $0.97\ d^{-1}$. These values, along with the other model parameters, can then be used in conjunction with the Streeter-Phelps model to compute the fit shown in Fig. E21.3.

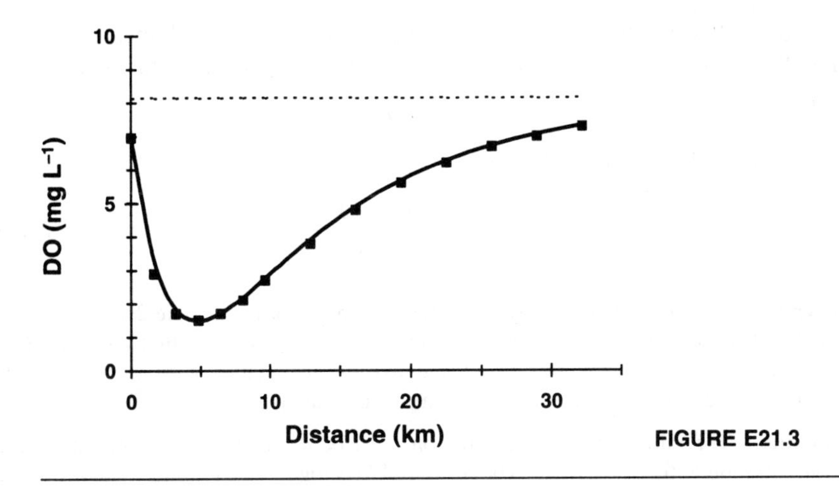

FIGURE E21.3

21.7 ANAEROBIC CONDITION

It is possible that the magnitude of a BOD discharge is so great that a stream will be devoid of oxygen. For such cases the Streeter-Phelps model must be modified. Gundelach and Castillo (1970) have developed a nice analysis for such cases.

For simplicity we again express the equations in terms of travel time. In addition we assume that $k_d = k_r$ (that is, there are no settling losses). For this situation the Streeter-Phelps model is

$$L = L_0 e^{-k_d t} \tag{21.17}$$

and

$$D = D_0 e^{-k_a t} + \frac{k_d L_0}{k_a - k_d}\left(e^{-k_d t} - e^{k_a t}\right) \tag{21.18}$$

The point where the system goes anaerobic can be determined by solving Eq. 21.18 for t with $D = o_s$. This must be done numerically. In other words, the solution amounts to finding the smallest root of

$$f(t) = D_0 e^{-k_a t} + \frac{k_d L_0}{k_a - k_d}\left(e^{-k_d t} - e^{k_a t}\right) - o_s \tag{21.19}$$

When this point is reached, oxygen depletion can no longer proceed at the rate of $k_d L$. Rather it will be limited by the rate at which oxygen passes across the air-water interface via reaeration. That is,

$$\frac{dL}{dt} = -k_a o_s \tag{21.20}$$

Thus once the oxygen is totally depleted, the reaction becomes zero-order. Using the initial condition that $L = L_i$ at $t = t_i$, we can calculate the BOD in the stretch from t_i to t_f as

$$L = L_i - k_a o_s (t - t_i) \tag{21.21}$$

Consequently as with all zero-order reactions, the BOD is reduced linearly.

Finally we must determine where the anaerobic zone ends. At this point,

$$k_a o_s = k_d L_f \tag{21.22}$$

Combining Eqs. 21.21 and 21.22 yields

$$t_f = t_i + \frac{1}{k_d} \frac{k_d L_i - k_a o_s}{k_a o_s} \tag{21.23}$$

EXAMPLE 21.4. ANAEROBIC CONDITIONS. Repeat Example 21.3, but double the BOD concentration of the treatment plant. That is, increase the point source of BOD to 80 mg L^{-1}. Also assume that no settling losses occur (that is, $k_r = k_d = 0.97\ d^{-1}$). All other values are assumed to be the same as for Example 21.3.

Solution: The initial CBOD concentration must be recalculated to reflect the higher concentration of the wastewater. The result is 17.93 mg L^{-1}. This value, along with the other parameters, can be substituted into Eq. 21.18 to give

$$D = 1.173 e^{-\frac{0.2775}{3182} x} + \frac{0.97(17.93)}{0.2775 - 0.97} \left(e^{-\frac{0.97}{3182} x} - e^{-\frac{0.2775}{3182} x} \right)$$

When we set $D = o_s$, this equation can be used to estimate that the oxygen concentration will reach zero at a travel time of 0.589 d, which corresponds to a distance of $x = 1.87$ km. At this point the BOD will have decomposed to a level of 10.13 mg L^{-1}. This serves as the value of L_i to be used in conjunction with Eqs. 21.21 and 21.23.

The length of the anoxic stretch can be determined via Eq. 21.23:

$$t_f = 0.589 + \frac{1}{0.97} \frac{0.97(10.13) - 0.2775(8.143)}{0.2775(8.143)} = 4.04 \text{ d}$$

which corresponds to $x = 12.85$ km. The BOD at this point can be calculated with Eq. 21.21,

$$L = 10.13 - 0.2775(8.143)(4.04 - 0.589) = 2.33 \text{ mg L}^{-1}$$

This then provides a boundary condition to calculate the remainder of the concentrations using the normal Streeter-Phelps model. The final results for both BOD and oxygen are shown in Fig. 21.5. Notice how for BOD, after an initial steep exponential drop, the degradation slows to a linear decrease in the anoxic region. After oxygen moves above zero, the more rapid exponential decrease returns.

For DO, along with the simulation corrected for anoxia, we also display the original result for the low-BOD load as well as a high-BOD-load simulation where we make no correction for anoxia (that is, we allow the model to go negative). Note how the correction causes the zone of low oxygen to expand because of the depressed decomposition that occurs.

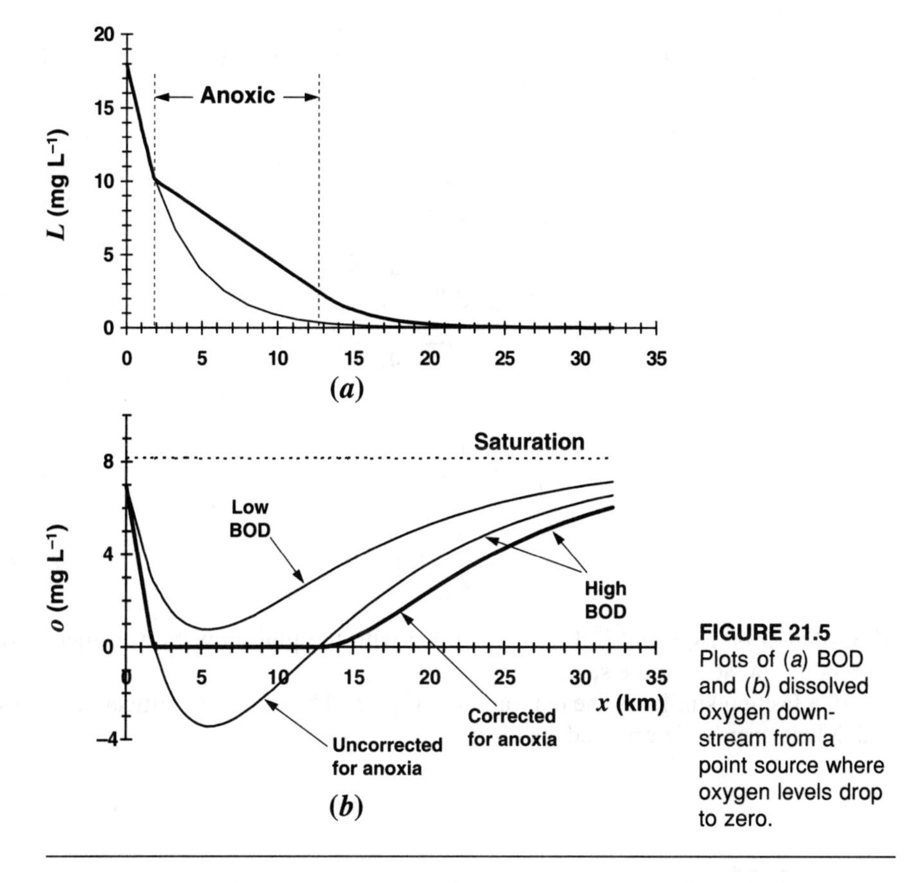

FIGURE 21.5
Plots of (*a*) BOD and (*b*) dissolved oxygen downstream from a point source where oxygen levels drop to zero.

21.8 ESTUARY STREETER-PHELPS

For a dispersive system such as an estuary, the Streeter-Phelps equation can be written as

$$0 = E\frac{d^2L}{dx^2} - U\frac{dL}{dx} - k_r L \tag{21.24}$$

and

$$0 = E\frac{d^2D}{dx^2} - U\frac{dD}{dx} + k_d L - k_a D \tag{21.25}$$

The solution for BOD is

$$L = L_0 e^{j_{1r}x} \qquad x \le 0 \tag{21.26}$$

$$L = L_0 e^{j_{2r}x} \qquad x \ge 0 \tag{21.27}$$

and for oxygen deficit is

$$D = \frac{k_d}{k_a - k_r}\frac{W}{Q}\left(\frac{e^{j_{1r}x}}{\alpha_r} - \frac{e^{j_{1a}x}}{\alpha_a}\right) \qquad x \le 0 \tag{21.28}$$

$$D = \frac{k_d}{k_a - k_r} \frac{W}{Q} \left(\frac{e^{j_{2r}x}}{\alpha_r} - \frac{e^{j_{2a}x}}{\alpha_a} \right) \qquad x \geq 0 \tag{21.29}$$

where

$$L_0 = \frac{W}{\alpha_r Q} \tag{21.30}$$

$$\alpha_r = \sqrt{1 + \frac{4k_r E}{U^2}} \tag{21.31}$$

$$\alpha_a = \sqrt{1 + \frac{4k_a E}{U^2}} \tag{21.32}$$

$$\begin{matrix} j_{1r} \\ j_{2r} \end{matrix} = \frac{U}{2E}(1 \pm \alpha_r) \tag{21.33}$$

$$\begin{matrix} j_{1a} \\ j_{2a} \end{matrix} = \frac{U}{2E}(1 \pm \alpha_a) \tag{21.34}$$

The solution is displayed in Fig. 21.6, along with the plug-flow case. Notice how dispersion spreads out the sag.

In a fashion similar to the derivation of Eqs. 21.15 and 21.16, critical distances and deficits can be determined, as in

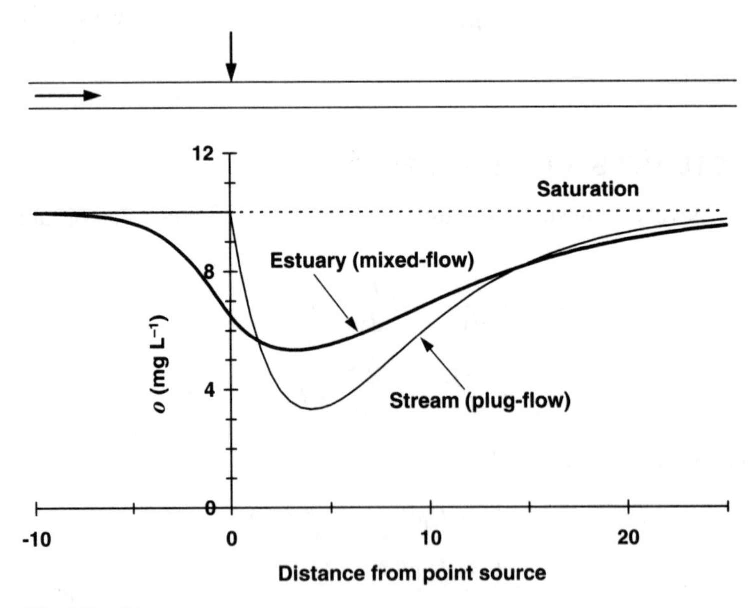

FIGURE 21.6
Plot of dissolved oxygen for a point source into an estuary. The solution for the plug-flow case is shown for comparison.

$$x_c = \frac{\ln\left(\dfrac{\alpha_r}{\alpha_a}\dfrac{j_{2a}}{j_{2r}}\right)}{j_{2r} - j_{2a}} \tag{21.35}$$

and

$$D_c = \frac{k_d}{k_a - k_r}\frac{W}{Q}\left(\frac{e^{j_{2r}x_c}}{\alpha_r} - \frac{e^{j_{2a}x_c}}{\alpha_a}\right) \tag{21.36}$$

Observe that, as with the plug-flow case, the critical deficit is linearly related to the loading and the deoxygenation rate. In addition the location is solely dependent on the removal and reaeration rates. As with the plug-flow model, higher removal and reaeration rates tend to move the critical deficit closer to the outfall.

PROBLEMS

21.1. A tanker truck careens off the road and dumps 30,000 L of glucose syrup into a small mountain lake. The concentration of the syrup is 100 g-glucose L^{-1}. The lake has the following characteristics in the period immediately following the spill: residence time = 30 d, depth = 5 m, area = 5×10^4 m^2, altitude = 11,000 ft, wind speed = 2.235 m s^{-1}, and temperature = 10°C. Note that the lake is assumed to be completely mixed, and has zero BOD and is at saturation prior to the spill. Compute the oxygen level in the lake during the period following the spill.

21.2. Some organic matter is added to a closed reactor so that its initial concentration is 4 mg L^{-1} of TOC. As the organic matter decays, you know (from a previous study) that it will deoxygenate the water at a rate of 0.2 d^{-1} at a temperature of 20°C.

(*a*) If the present experiment is conducted at 15°C, predict the concentration of BOD versus time in the reactor. Plot your results.

(*b*) Determine the dissolved oxygen and the deficit versus time within the closed reactor. Plot your results.

(*c*) Open the reactor to the atmosphere and repeat parts (*a*) and (*b*). Note that your laboratory is located at an altitude of 2 km and there is a gentle breeze of 1 mps blowing over the reactor during the experiment.

21.3. You make the following measurements below a sewage plant outfall into a stream at sea level:

Travel time (d)	0	2	4	6	8	10	12	16	20
CBOD (mg L^{-1})	50	38.9	30.0	23.0	17.7	13.9	11.4	6.7	3.7
DO (mg L^{-1})	10.0	5.3	2.7	1.6	1.3	1.6	2.2	4.0	5.7

If the water temperature is 10°C, determine the reaeration and deoxygenation rates.

21.4. A fresh stream having the characteristics Q_f = 7 cms, S_f = 1 ppt, o_f = 12 mg L^{-1}, and T_f = 10°C discharges into a saline stream (Q_s = 2 cms, S_s = 15 ppt, o_s = 3 mg L^{-1}, and T_s = 18°C). Determine the oxygen and the oxygen deficit assuming complete mixing at the confluence of the two streams. Note that the elevation is 1.5 km.

21.5. A waste source (Q_w = 1 cms, L_w = 25 mg L^{-1}, o_w = 2 mg L^{-1}, T_w = 25°C) discharges into a stream (Q_r = 10 cms, L_r = 2 mg L^{-1}, o_r = 10 mg L^{-1}, and T_r = 15°C). Downstream, the velocity is 0.3 mps and the depth is 0.3 m. Calculate the profiles of both BOD and oxygen downstream, assuming that the stream is located at 5500 ft elevation and that the BOD decays at a rate of 1 d^{-1}. Determine the value and the location of the maximum deficit.

21.6. Determine the profiles of BOD and dissolved oxygen for the following sea-level stream:

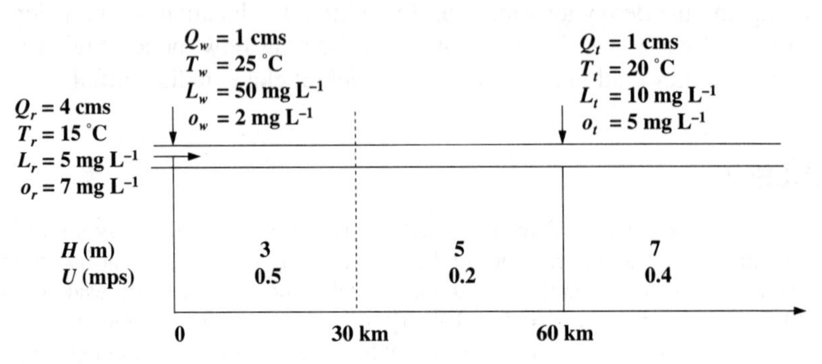

FIGURE P21.6

LECTURE 22

Streeter-Phelps: Distributed Sources

LECTURE OVERVIEW: I focus on the effect of distributed sources of BOD and deficit on the Streeter-Phelps model. The initial sections describe cases in which the distributed sources contribute pollutant mass but do not add significant quantities of flow. I follow this with a description of models of flow-contributing sources.

To this juncture I have focused on point sources. Now let's look at nonpoint or diffuse sources. As the name implies, these inputs enter the system in a diffuse manner. For one-dimensional streams and estuaries this means that the loadings enter the system along its length.

In this lecture I focus on two types of diffuse sources: those that do not contribute flow and those that do. Before doing that I'll briefly describe how the rates of diffuse source contributions are parameterized.

22.1 PARAMETERIZATION OF DISTRIBUTED SOURCES

Recall that in Lec. 9, I showed how a steady-state mass balance for a distributed source (Fig. 22.1*a*) could be written for a plug-flow system with uniform hydrology and geometry,

$$0 = -\frac{dc}{dt} - kc + S_d \tag{22.1}$$

where S_d = rate of the distributed source (g m^{-3} d^{-1}) and t = travel time for a plug-flow system. If $c = 0$ at $t = 0$, then

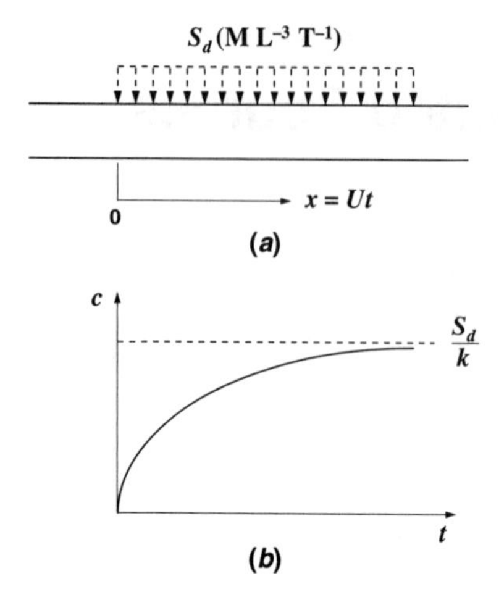

FIGURE 22.1
(*a*) Uniform distributed source and (*b*) resulting response if the substance decays with first-order kinetics.

$$c = \frac{S_d}{k}(1 - e^{-kt}) \tag{22.2}$$

Thus the solution starts at zero and asymptotically approaches a level of S_d/k as travel time increases (Fig. 22.1*b*).

Now although the proper units for the rate are M L^{-3} T^{-1}, we should note that such loads enter in three common ways. The manner in which the loadings enter the system influence how they are parameterized. As in Fig. 22.2*a*, some distributed loads enter as a line source along the side of the stream. Examples might be loads connected with bank erosion or with leaching from landfills. The best way to

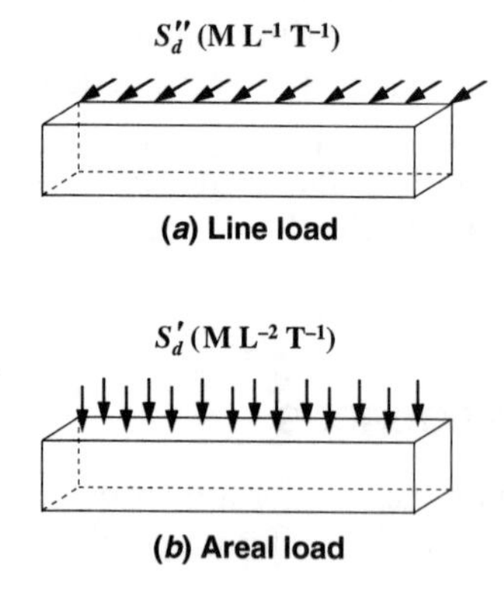

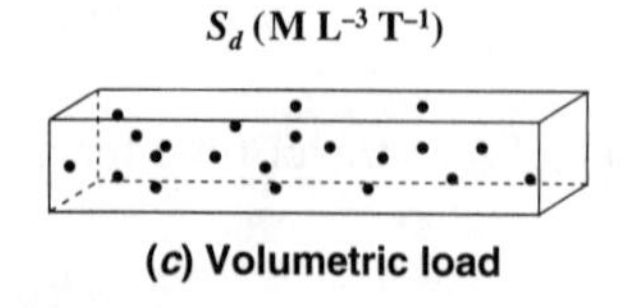

FIGURE 22.2
Ways in which distributed loads enter one-dimensional water bodies.

parameterize such inputs would be per length of stream, S_d'' (M L^{-1} T^{-1}). Such a rate could be converted to the proper volumetric units for Eq. 22.1 by

$$S_d = S_d'' \frac{L}{V} = \frac{S_d''}{A_c} \tag{22.3}$$

where L = total length
V = volume
A_c = cross-sectional area of the reach being loaded

Thus normalization to stream cross-sectional area would convert the per-length rate into the proper volumetric units for the mass balance.

As in Fig. 22.2*b* other distributed loads enter as an areal source across the bottom or the top surface areas. Some examples include sediment and atmospheric sources and oxygen fluxes from rooted plants. The best way to parameterize such inputs would be as a flux S_d' (M L^{-2} T^{-1}). Such areal units could be converted to the proper units for the mass balance by

$$S_d = S_d' \frac{A_s}{V} = \frac{S_d'}{H} \tag{22.4}$$

where H = stream depth.

Finally, as in Fig. 22.2*c*, some distributed loads enter in a volumetric fashion. An example is the oxygen produced by floating plants (phytoplankton). Because the best way to parameterize such inputs is on a volumetric basis, they would naturally be in the proper format S_d (M L^{-3} T^{-1}).

22.2 NO-FLOW SOURCES

Let's now focus on diffuse sources that deliver mass but not water to rivers and streams. Although these are primarily used to characterize oxygen, as described first, they can also be applied to BOD.

22.2.1 BOD

The primary examples of distributed sources that introduce BOD into a system without adding significant flow are interactions with porous media. For example a polluted bottom sediment can contain very high concentration of dissolved organic matter in its pore water. If the pore-water BOD is much greater than the stream BOD, the diffusion can be characterized as a zero-order source of mass that does not contribute water. For such cases the mass balance can be written in a fashion similar to Eq. 22.1,

$$0 = -\frac{dL}{dt} - k_r L + S_L \tag{22.5}$$

where S_L = rate of the BOD distributed source (g m^{-3} d^{-1}) and k_r = rate of BOD removal (d^{-1}). Note the use of k_r rather than k_d. This is done for completeness, although in reality a sediment or leaching source would be in dissolved form and not subject to settling losses. The solution, if $L = 0$ at $t = 0$, would be

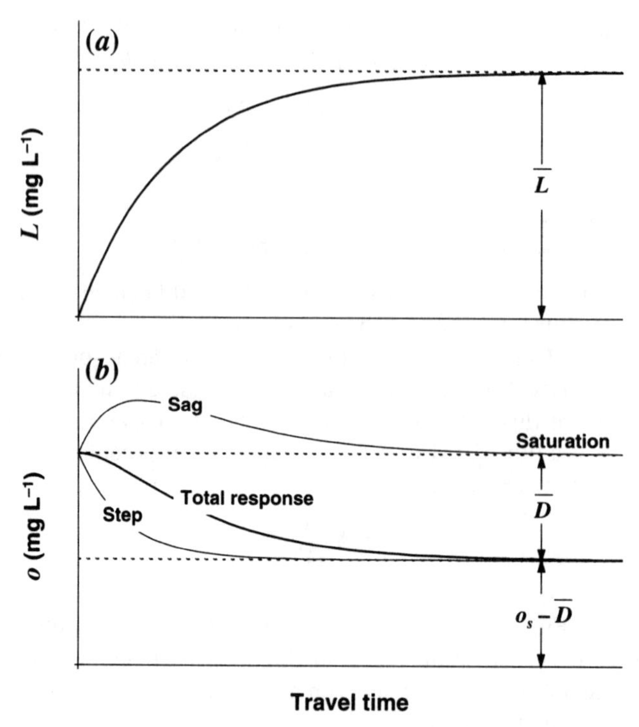

FIGURE 22.3
Plot of the (*a*) BOD and (*b*) oxygen responses due to a distributed BOD loading.

$$L = \frac{S_L}{k_r}(1 - e^{-k_r t}) \tag{22.6}$$

Thus the BOD increases in a step-response fashion and asymptotically approaches a steady-state level (Fig. 22.3*a*) $\overline{L}$, where the BOD gain is balanced by removal,

$$\overline{L} = \frac{S_L}{k_r} \tag{22.7}$$

A mass balance for oxygen deficit can be written as

$$0 = -\frac{dD}{dt} - k_a D + \frac{k_d}{k_r} S_L (1 - e^{-k_r t}) \tag{22.8}$$

Thus the BOD would act as a forcing function contribution to the deficit. For the case $D = 0$ at $t = 0$ this balance can be solved for

$$D = \frac{k_d S_L}{k_r k_a}(1 - e^{-k_a t}) - \frac{k_d S_L}{k_r (k_a - k_r)}(e^{-k_r t} - e^{-k_a t}) \tag{22.9}$$

This is an interesting solution since it contains both a step and a sag response. Thus as depicted in Fig. 22.3*b*, the initial response is s-shaped due to the sag term. Thereafter the deficit increases in a step-response fashion and asymptotically approaches a steady-state level,

$$\overline{D} = \frac{k_d S_L}{k_r k_a} \tag{22.10}$$

22.2.2 Dissolved Oxygen

The no-flow DO sources and sinks have traditionally been the most commonly employed distributed loads. It has been used to simulate the effect of plants and sediment oxygen demand. The mass balance can be written as

$$0 = -\frac{dD}{dt} - k_a D - P + R + \frac{S_B'}{H} \tag{22.11}$$

where P and R = volumetric rates of plant photosynthesis and respiration, respectively (g m^{-3} d^{-1})
S_B' = areal rate of sediment oxygen demand (g m^{-2} d^{-1})
H = depth (m)

The solution, if $D = 0$ at $t = 0$, would be

$$\overline{D} = \frac{-P + R + (S_B'/H)}{k_a}(1 - e^{-k_a t}) \tag{22.12}$$

Thus the deficit increases in a step-response fashion and asymptotically approaches a steady-state level $\overline{D}$, where the net deficit gain is balanced by reaeration losses (Fig. 22.4),

$$\overline{D} = \frac{-P + R + (S_B'/H)}{k_a} \tag{22.13}$$

22.2.3 Total Streeter-Phelps Model

We have now developed formulations for handling both point and nonpoint sources of BOD and oxygen. These can be combined into a comprehensive Streeter-Phelps model to handle both types of sources. For a stream stretch with a point source at its upstream end and diffuse sources along its length, the resulting equations for BOD

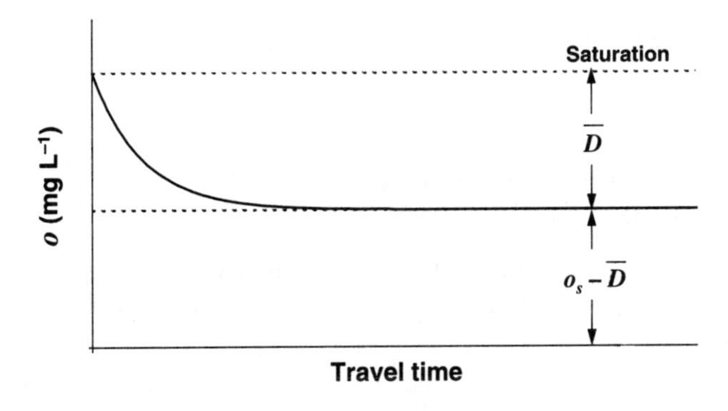

FIGURE 22.4
Plot of the oxygen response due to a distributed oxygen loading.

and oxygen deficit are

$$L = L_0 e^{-k_r t} + \frac{S_L}{k_r}(1 - e^{-k_r t}) \tag{22.14}$$

Point Distributed

$$D = D_0 e^{-k_r t} + \frac{k_d L_0}{k_a - k_r}(e^{-k_r t} - e^{-k_a t}) + \frac{-P + R + (S_B'/H)}{k_a}(1 - e^{-k_a t})$$

Point deficit Point BOD Distributed deficit

$$+ \frac{k_d S_L}{k_r k_a}(1 - e^{-k_a t}) - \frac{k_d S_L}{k_r(k_a - k_r)}(e^{-k_r t} - e^{-k_a t}) \tag{22.15}$$

Distributed BOD

These equations now provide a framework for simulating the steady-state effects of multiple-point and diffuse sources in a fashion similar to Sec. 21.4.

22.3 DIFFUSE SOURCES WITH FLOW

To this point we have investigated the response of a stream to diffuse sources that do not contribute significant flow. Although this has been the standard approach in traditional stream-oxygen modeling, recent concern over nonpoint-source pollution has directed attention to distributed sources that contribute flow.

In this section I'll first present closed-form solutions for some idealized steady-state cases to demonstrate the behavior of pollutants affected by such sources. Then I'll describe a numerical approach that has more general applicability.

22.3.1 Analytical Solutions

A mass balance for a diffuse source that contributes both flow and mass can be written as (Thomann 1972, Thomann and Mueller 1987)

$$\frac{\partial(A_c c)}{\partial t} + \frac{\partial(Qc)}{\partial x} = \frac{\partial Q}{\partial x} c_d - k A_c c \tag{22.16}$$

where A_c = cross-sectional area (m^2)
c = concentration ($mg\ L^{-1}$)
t = time (s)
Q = flow rate ($m^3\ s^{-1}$)
x = distance (m)
c_d = concentration of the diffuse source ($mg\ L^{-1}$)
k = first-order decay rate (s^{-1})

(Note that we are using seconds for our time unit to be consistent with subsequent parts of this section. In actuality day or hour units are more commonly used.) At steady-state, Eq. 22.16 becomes

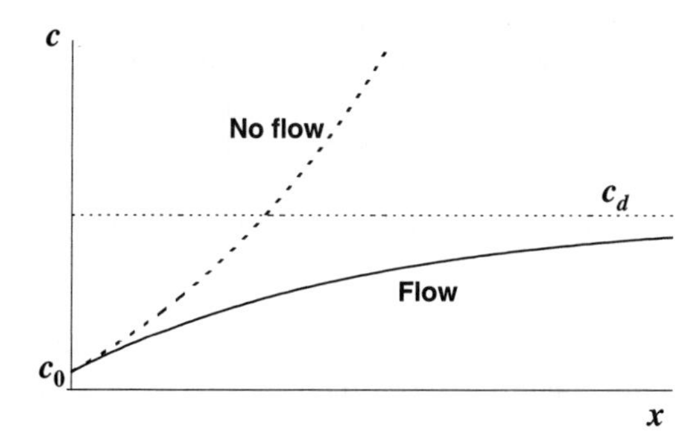

FIGURE 22.5
Plot of concentration versus distance where nonpoint sources add flow and a dissolved conservative pollutant to a stream. For this case the in-stream concentration asymptotically approaches the concentration of the diffuse source. In addition we also show how the concentration would increase exponentially for the case where the diffuse flow contributes mass but not flow.

$$0 = -\frac{d(Qc)}{dx} + \frac{dQ}{dx}c_d - kA_c c \tag{22.17}$$

O'Connor (1976) developed both steady-state and time-variable solutions for dissolved solids ($k = 0$). In his analysis O'Connor idealized flow increases as an exponential function,

$$Q = Q_0 e^{q'x} \tag{22.18}$$

where Q_0 = flow at $x = 0$ (m^3 d^{-1}) and q' = exponential rate of flow increase (m^{-1}). Equation 22.18 can be substituted into Eq. 22.17 to yield

$$\frac{dc}{dx} + q'c = q'c_d \tag{22.19}$$

By applying the boundary condition $c = c_0$ at $x = 0$, O'Connor obtained the solution

$$c = c_0 e^{-q'x} + c_d(1 - e^{-q'x}) \tag{22.20}$$

As in Fig. 22.5 this model predicts that stream concentration will asymptotically approach the nonpoint concentration. Thus, because both mass and flow are introduced at the same rate, the concentration of the conservative substance stabilizes at the nonpoint concentration.

This situation can be contrasted with the case where the pollutant mass enters in an exponentially increasing fashion, but without an associated flow increase. For this case, which is also depicted in Fig. 22.5, the conservative concentration increases exponentially because the new mass is not being diluted with new flow.

Now that we have an idea how conservative diffuse sources affect streams, we can turn to nonconservative substances. Unfortunately the addition of a reaction term complicates the analysis because we must now consider how the diffuse flow affects the stream's hydrogeometric parameters. In particular we must account for the fact that flow additions will impact both velocity and cross-sectional area (Fig. 22.6).

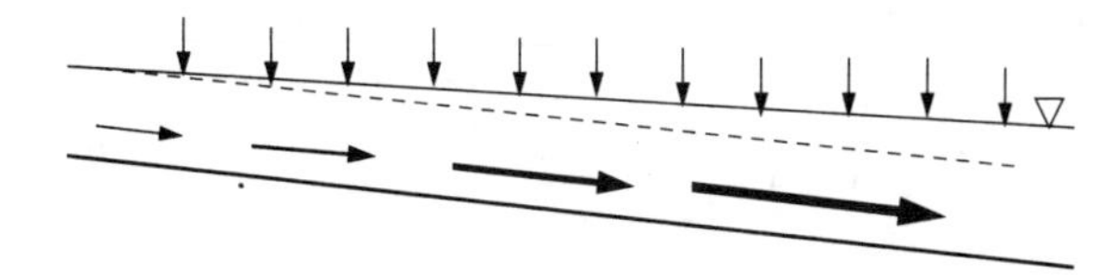

FIGURE 22.6
Distributed sources of flow cause water in a channel to both move faster and run deeper.

Early attempts to model diffuse sources did not account for these impacts. In most cases it was assumed that either the area or the velocity remained constant while the other was allowed to vary (for example Thomann and Mueller 1987). In fact as we have already seen in Lec. 14, both the velocity and area should increase as diffuse flow is added.

To incorporate such effects into our model we employ a linear function to represent the flow increases,

$$Q = Q_0 + qx \tag{22.21}$$

where q = a constant parameterizing the rate of the linear increase ($m^2\ d^{-1}$). Substituting Eq. 22.21 into Eq. 22.17 gives

$$0 = -(Q_0 + qx)\frac{dc}{dx} - (q + kA_c)c + qc_d \tag{22.22}$$

I have chosen the linear characterization (in place of the exponential model of Eq. 22.18) for two reasons. First, although inflow can exhibit an exponential increase over large distances, the increase can just as well be characterized by a linear relationship. This would particularly be the case over shorter distances that are typical when dealing with nonconservative pollutants. Second, the numerical approach described in the next section assumes linearly increasing flow.

As a prelude to incorporating realistic hydraulics, we first develop two solutions to Eq. 22.22 that correspond to extreme cases. The first assumes that the cross-sectional area is constant and all distributed flow manifests itself in increased velocity. For this case the solution is

$$c = c_0\left(\frac{U_0}{U_0 + \nu x}\right)^{\frac{k+\nu}{\nu}} + c_d\frac{\nu}{\nu + k}\left[1 - \left(\frac{U_0}{U_0 + \nu x}\right)^{\frac{k+\nu}{\nu}}\right] \tag{22.23}$$

where

$$\nu = \frac{Q}{A_{c0}} \tag{22.24}$$

in which A_{c0} = initial area (m^2).

At the other extreme the velocity is assumed to be constant and all distributed flow manifests itself in increased cross-sectional area. For this case the steady-state result is

$$c = \left[c_0 e^{-\frac{k}{U_0}x} + c_d\frac{\nu}{k}\left(1 - e^{-\frac{k}{U_0}x}\right)\right]\frac{U_0}{U_0 + \nu x} \tag{22.25}$$

Both solutions are shown in Fig. 22.7. As can be seen, the constant velocity result eventually goes to zero. In contrast the constant-area solution asymptotically approaches a constant concentration

$$c = c_d\frac{\nu}{\nu + kA_{c0}} \tag{22.26}$$

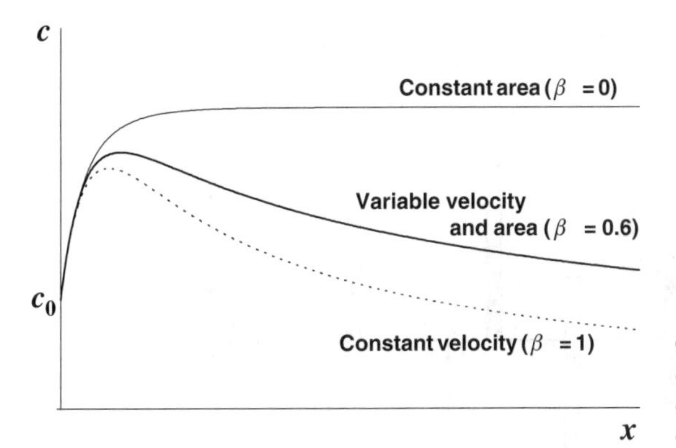

FIGURE 22.7
Plot of concentration versus distance for three cases: constant area, area and velocity varying according to Manning's equation, and constant velocity.

Because of the decay the ultimate result will always be less than the diffuse concentration. However, the solution does not go to zero. In fact it approaches this constant fraction of the diffuse concentration that represents a balance between the delivery (ν) and the decay (k) rates.

Clearly the correct result for realistic cases lies between these two extremes. To quantify this intermediate result we assume that the Manning equation provides an adequate representation of the momentum balance for steady-flow in the channel,

$$Q = \frac{C_o}{n} A_c R^{2/3} S_e^{1/2} \tag{22.27}$$

where C_o = a constant (= 1.0 for metric and 1.486 for English units)
n = Manning's roughness coefficient
R = channel's hydraulic radius (m or ft) = A_c/P
P = wetted perimeter (m or ft)
S_e = slope of the channel's energy grade line (dimensionless)

For steady flow the last quantity is approximated by the channel slope S_o.

Assuming metric units, we can solve Eq. 22.27 for

$$A_c = \alpha Q^\beta \tag{22.28}$$

where $\beta = \frac{3}{5}$ and, for the case of a wide, shallow rectangular channel,

$$\alpha = \left(\frac{nB^{2/3}}{\sqrt{S_o}}\right)^{3/5} \tag{22.29}$$

where B = channel width. Thus for this simple case, because α is a constant, a simple relationship between flow and area is established.

Note also that the parameter β specifies the relationship between area and velocity. The extreme cases of constant area and constant velocity correspond to values of β of 0 and 1, respectively.

We apply the method of characteristics, and the original partial differential equation (22.16) can be simplified by expressing it as a coupled pair of ordinary differential equations,

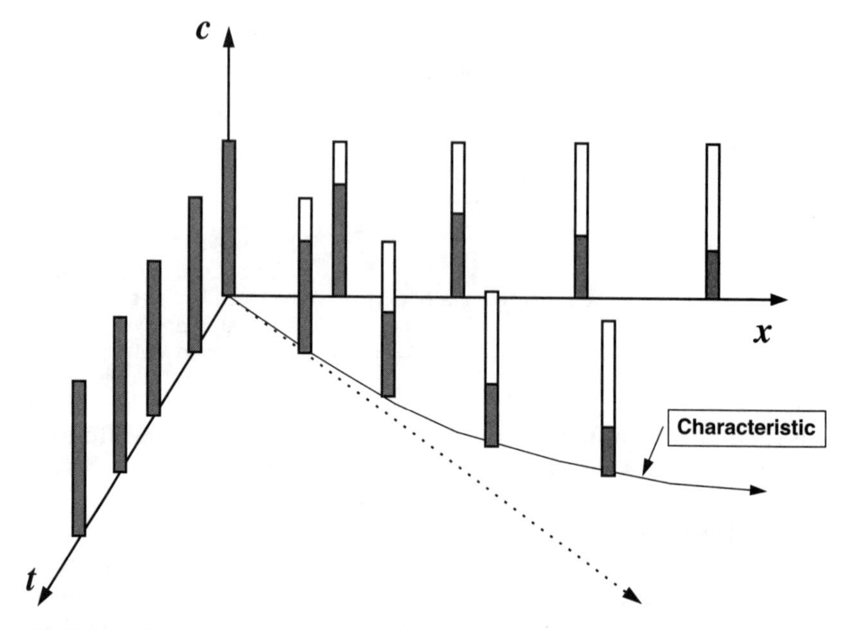

FIGURE 22.8
Graphical depiction of the method of characteristics for a plug-flow system where flow increases with distance.

$$\frac{dx}{dt} = U \tag{22.30}$$

$$\frac{dc}{dt} = -kc + \frac{q}{A_c}c_d - \frac{q}{A_c}c \tag{22.31}$$

As depicted in Fig. 22.8, the first equation describes the solution's characteristic trajectory, that is, how time and space are related via velocity. The second equation shows how the pollutant changes temporally as it moves along the ***characteristic***. In other words the first equation represents the position in space and time of an observer moving at the stream's velocity, and the second equation indicates the concentration changes that would be perceived by the moving observer as a function of time.

Substituting Eqs. 22.21 and 22.28 into Eqs. 22.30 and 22.31 yields

$$\frac{dx}{dt} = \frac{1}{\alpha}(Q_0 + qx)^{(1-\beta)} \tag{22.32}$$

$$\frac{dc}{dt} = \frac{q(c_d - c)}{\alpha(Q_0 + qx)^\beta} - kc \tag{22.33}$$

Applying appropriate initial conditions (at $t = 0$, $x = 0$, and $Q = Q_0$), we can solve Eq. 22.32 for

$$x = \left[\left(\frac{\beta q}{\alpha}t + Q_0^\beta\right)^{1/\beta} - Q_0\right]\frac{1}{q} \tag{22.34}$$

Unfortunately a closed-form solution for Eq. 22.33 cannot be obtained. However, it can be easily integrated numerically. The results are displayed in Fig. 22.7. As can be seen the solution lies between the constant area and velocity cases.

Notice that the solution eventually approaches zero. For the conservative case ($k = 0$) the concentration would eventually approach the diffuse source concentration as the rate of delivery of the diffuse source would eventually balance the reduction in concentration by dilution. In contrast the presence of the increasing area in the reaction term means that the denominator will continuously increase. Thus for the nonconservative case, the concentration would eventually approach zero.

22.3.2 Numerical Method

The analytical approach described previously has limited applicability. Consequently a computer-oriented numerical method is better suited for more general applications.

The following description is similar to the approach employed in the QUAL2E model. It is based on dividing the river into reaches with uniform geometry. Each reach is further divided into equal-length computational elements. These are the same as the control volumes described previously in Lec. 11.

Once this segmentation scheme is adopted, the transport through the system must be established. Because of our interest in diffuse sources we use the approach depicted in Fig. 22.9. For this situation a total diffuse flow Q_d is established for a reach. This flow is then divided equally among the segments. For an n-segment reach, each element would receive Q_d/n flow. Consequently such a scheme is the discrete form of the linearly increasing load described in the previous section (recall Eq. 22.21). In fact if the length of each segment is Δx, then Q_d and q are related by

$$Q_e = \frac{Q_d}{n} = q\,\Delta x \tag{22.35}$$

where Q_e = incremental flow for each element (m^3 d^{-1}).

The transport calculation is then performed on an element-by-element basis starting at the element farthest upstream. First, simple flow balances are computed

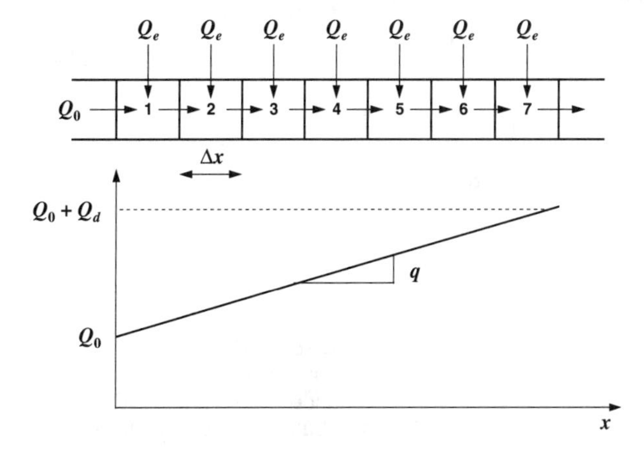

FIGURE 22.9
Plot of a finite-difference scheme accounting for a uniformly distributed load.

for the cell. For example for the first element,

$$Q_1 = Q_0 + Q_e \tag{22.36}$$

where Q_1 = outflow from element 1.

Next, for a wide rectangular channel, the flow can be used to determine the element volume in a similar fashion to the derivation of Eq. 22.28,

$$V = \Delta x A_c = \Delta x \alpha Q^{\beta} \tag{22.37}$$

Once the flow and volume are determined, the mass balance for a pollutant can be developed using backward differences. For example balances for BOD and oxygen can be written as

$$0 = Q_{i-1}L_{i-1} - Q_i L_i + Q_e L_{d,i} - k_{r,i}V_i L_i \tag{22.38}$$

$$0 = Q_{i-1}o_{i-1} - Q_i o_i + Q_e o_{d,i} - k_{d,i}V_i L_i + k_{a,i}V_i(o_{s,i} - o_i) + P_i - R_i - \frac{S'_{b,i}}{H_i} \tag{22.39}$$

Because we are using backward differences, the balances can be solved in sequence,

$$L_i = \frac{Q_{i-1}L_{i-1} + Q_e L_{d,i}}{Q_i + k_{r,i}V_i} \tag{22.40}$$

$$o_i = \frac{Q_{i-1}o_{i-1} + Q_e o_{d,i} - k_{d,i}V_i L_i + k_{a,i}V_i o_{s,i} + P_i - R_i - (S'_{b,i}/H_i)}{Q_i + k_{a,i}V_i} \tag{22.41}$$

The results of applying this model are displayed in Fig. 22.10. For this case we omitted photosynthesis, respiration, and benthic (sediment) oxygen demand. In addition we used boundary conditions of $L_0 = 0$ and $o_0 = 10 \text{ mg L}^{-1}$. The distributed BOD and oxygen loads had concentrations of 80 mg L^{-1} and 0 mg L^{-1}, respectively. As can be seen the BOD solution conforms closely to the analytical result obtained previously. The resulting oxygen profile has a similar shape to the classic DO sag. However, because it is the result of distributed sources, the sag is much more drawn out.

It should be noted that this solution technique will manifest a level of numerical dispersion that is approximately equal to (Eq. 11.40)

$$E_n = \frac{\Delta x}{2}U \tag{22.42}$$

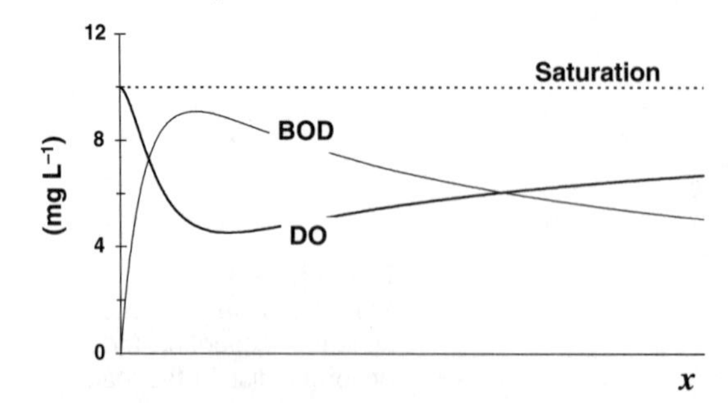

FIGURE 22.10
Plot of concentration versus distance for BOD and DO for distributed sources with flow.

Real streams exhibit dispersion (Sec. 14.4.1). Consequently, given a physical dispersion, Eq. 22.42 provides an upper bound on the spatial step,

$$\Delta x = \frac{2E}{U} \tag{22.43}$$

If this value is used the backward difference will generate a numerical dispersion equal to the real dispersion. Unfortunately, because velocity is changing, this means that each cell would require a different Δx to attain the desired result. However, because advection dominates dispersion in steady-state streams, using a constant Δx will not lead to significant error.

Finally real systems never exhibit uniform hydrogeometric properties for the distances used in the examples in this lecture. Thus the algorithm outlined above would usually be implemented in a piecewise fashion. This would be done by dividing the system into reaches with constant properties. The solutions for the individual reaches could then be linked by mass balances at their boundaries.

PROBLEMS

22.1. A small uncontaminated stream flows past a landfill for 10 km. At the end of the stretch, the stream has a concentration of 2 μg L^{-1} of a contaminant with a known half-life of 5 d. If the stream's velocity is 0.1 m s^{-1}, determine the rate of the distributed source in mg m(stream length)$^{-1}$ d^{-1}.

22.2. Determine the oxygen concentration for a 10-km stretch of stream flowing over a sludge bed that exerts a sediment oxygen demand of 5 g m^{-2} d^{-1}. Note that the stream has an oxygen saturation of 8 mg L^{-1}, a velocity of 0.2 m s^{-1}, a width of 50 m, a slope of 0.0002, and a roughness coefficient of 0.03. The stream's cross section can be approximated by a rectangular channel.

22.3. An excess of 2 mg L^{-1} of oxygen develops in a river at the end of a 20-km section with a high concentration of attached plants. Determine the net photosynthesis rate of these plants in g m^{-2} d^{-1}. The river has the following characteristics: U = 0.2 m d^{-1}, H = 0.5 m, and k_a = 2 d^{-1}. Assume that the river is at saturation when it starts to flow over the plants.

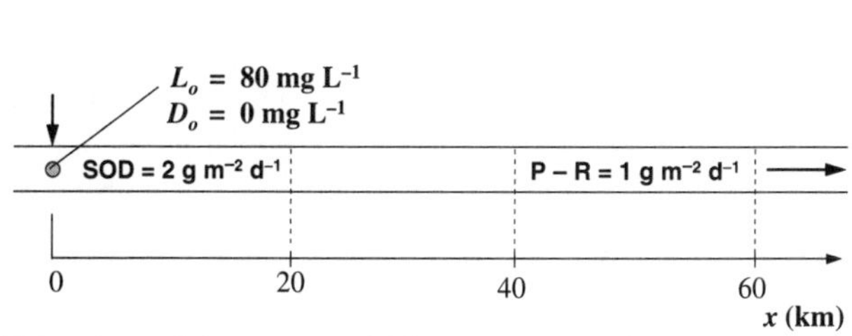

L_o = 80 mg L^{-1}
D_o = 0 mg L^{-1}

SOD = 2 g m^{-2} d^{-1} P – R = 1 g m^{-2} d^{-1}

0   20   40   60
x (km)

	0–20	20–40	40–60
U (mps)	0.1	0.15	0.1
H (m)	0.8	1	1
k_r (d^{-1})	0.2	0.1	0.1
k_d (d^{-1})	0.1	0.1	0.1
k_a (d^{-1})	1	1.2	1.2
o_s (mg L^{-1})	10	9	8

FIGURE P22.4

22.4. A river is subject to a point source of BOD and diffuse sources of oxygen deficit, as shown in Fig. P22.4. Compute the profiles of BOD and oxygen in this stretch.

22.5. Twenty km of a river in a developing country are being populated by squatters. At the head end of the river stretch the flow is 2 cms and the water is very clean and saturated with dissolved oxygen. For the stretch,

$$B = 20 \text{ m} \quad S = 0.001 \quad n = 0.035 \quad q = 0.00006 \text{ m}^2 \text{ s}^{-1} \quad L_d = 0 \quad o_d = o_s$$

The government is interested in evaluating the sustainable population for the area. This quantity is to be defined as the number of people who can populate the stretch if the minimum acceptable oxygen concentration is 2 mg L^{-1}. Assume the basin is located at sea level and that the temperature (20°C) is unaffected by the additional flow and loadings generated by the population. Employ the O'Connor-Dobbins formula as your reaeration model, and assume that all the BOD and flow generated by the population is input to the stream. Further assume that the wasteflow added by the populace has zero dissolved oxygen.

22.6. A stream flows past a landfill which leaches water at a rate of 0.001 m^2 s^{-1} with carbonaceous BOD and DO concentrations of 50,000 mg L^{-1} and 0 mg L^{-1}, respectively. The landfill runs for 5 km of stream length. Determine the BOD and DO of the stream for 20 km downstream from the source.

(*a*) Assuming that the leachate flow does not significantly affect the stream depth and velocity.

(*b*) Assuming that the leachate flow does significantly affect the stream depth and velocity.

The stream has a rectangular channel. Assume the following values for other pertinent parameters:

$$o_s = 10 \text{ mg L}^{-1} \quad o_0 = 10 \text{ mg L}^{-1} \quad L_0 = 0 \text{ mg L}^{-1} \quad T = 20°\text{C}$$

$$k_d = 0.1 \text{ d}^{-1} \quad B = 20 \text{ m} \quad S = 0.0005 \quad n = 0.03$$

LECTURE 23

Nitrogen

LECTURE OVERVIEW: I describe and then simulate the effect of nitrogen on stream-dissolved oxygen. After an overview of nitrogen and water quality, I provide a description of the process whereby nitrogen affects dissolved oxygen: nitrification. Then I present two frameworks that have been used to simulate the phenomenon: nitrogenous BOD and nitrification models. Finally I provide a brief introduction to ammonia toxicity.

At the beginning of this book I presented the parable of the "Blind Men and the Elephant." This story has great relevance to the impact of nitrogen on water quality in that no less than four specific, but interconnected, water-quality problems are associated with this nutrient. Although this lecture focuses on dissolved oxygen, I'll start by presenting an overview of nitrogen as a pollutant.

23.1 NITROGEN AND WATER QUALITY

Figure 23.1 depicts the nitrogen cycle in natural waters. As can be seen the cycle affects the water's oxygen level. In addition several other water-quality problems occur. As described next these problems can be grouped into two categories. In the first group are nitrification/denitrification and eutrophication. For these problems nitrogen serves as the cause of the problem rather than as a problem in itself. In the second group are nitrate pollution and ammonia toxicity. In these cases nitrogen species are the actual pollutants. Although we can divide the problems in this way, as described next, all the problems are interconnected.

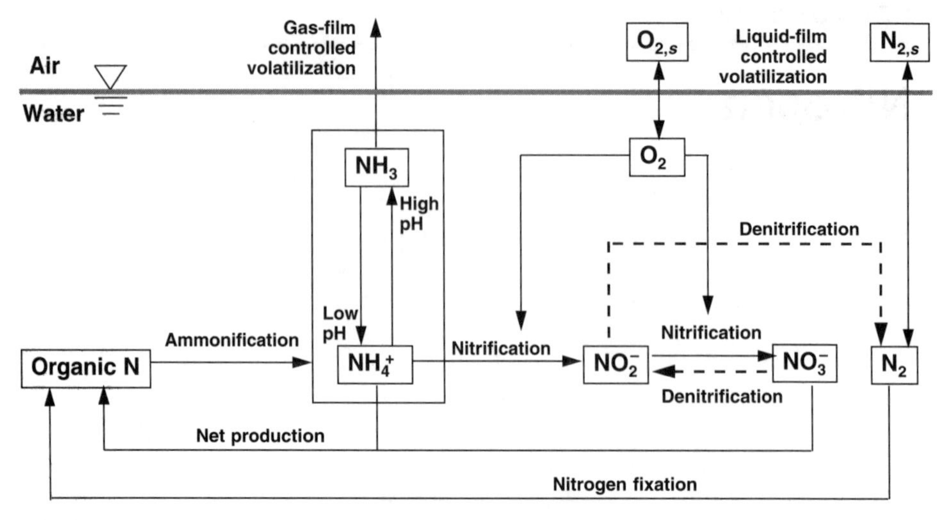

FIGURE 23.1
The nitrogen cycle in natural waters. The dashed arrows indicate that the denitrification reactions take place under anaerobic conditions. Note that, although it is not depicted on this diagram, the net production of organic N can have an impact on a water body's oxygen level.

- *Nitrification/denitrification.* As depicted in Fig. 23.1, ammonia due to direct loadings and to the decomposition of organic nitrogen is oxidized in a two-step process to form nitrite (NO_2^-) and nitrate (NO_3^-). The process consumes oxygen and, thus, can seriously deplete the water body's oxygen levels. If conditions go anaerobic the nitrate can be reduced to nitrite and the nitrite converted to free nitrogen by denitrification. Because free nitrogen is in gaseous form, this mechanism can result in the loss of nitrogen to the atmosphere. In addition free nitrogen can be utilized by certain nitrogen-fixing algae and bacteria.
- *Eutrophication.* Aside from its other characteristics, nitrogen serves as an essential nutrient for plant growth. Thus it acts as a fertilizer that can overstimulate plant growth in the process called eutrophication. This excess growth can impact water quality directly (e.g., unsightly scums, clogging of the water course, etc.) or indirectly by exacerbating other problems (e.g., oxygen, ammonia toxicity, etc.).
- *Nitrate pollution.* As depicted in Fig. 23.1, the ultimate result of the nitrification process is nitrate. In sufficiently high concentrations, nitrate in drinking water can have serious and occasionally fatal effects in infants. The problem can become especially critical in agricultural regions where nonpoint sources of nitrate from fertilization supplement high levels due to nitrification from point sources.
- *Ammonia toxicity.* Ammonia exists in two forms in natural waters: ammonium ion (NH_4^+) and ammonia gas (NH_3). Whereas the former is innocuous at the levels encountered in most natural waters, the un-ionized form is toxic to fish. The equilibrium relationship between the two forms is governed primarily by pH. At high pH (and to a lesser extent at high temperatures), ammonia exists principally in the toxic, un-ionized form. For example at moderate temperatures ($\approx 20°C$) and pH levels above 9, upward of 20% of the total ammonia will be in the un-ionized form. Such conditions can occur in the recovery zone below a secondary

sewage discharge into a shallow stream. For such cases suspended solids will be low and ammonia will be high. Consequently profuse plant growth can occur. In the late afternoon such plant photosynthesis can deplete carbon dioxide from the water, and for weakly buffered waters it can induce a large increase in pH. The late afternoon is also coincidentally the time of day when temperatures are highest. Consequently high pH and temperature can result in elevated levels of un-ionized ammonia.

In summary the nitrogen problem is interconnected and multifaceted. First, ammonia can cause oxygen depletion via nitrification. If this occurs, one of the byproducts is nitrate, which itself is a pollutant. Further, depending on temperature and pH, the ammonia can manifest itself in an un-ionized form that is toxic to aquatic organisms. Finally both ammonia and nitrate are essential nutrients for photosynthesis. Thus they can stimulate excessive plant growth, which constitutes a water-quality problem in its own right, and exacerbate some of the other problems.

With this overview as background the remainder of this lecture focuses on nitrification and, to a lesser extent, nitrate pollution and ammonia toxicity. In later lectures I'll describe eutrophication. Although this approach runs the risk of missing the "whole elephant," I have done this because of our present emphasis on dissolved oxygen and because nitrification was the first context in which modelers addressed the nitrogen cycle in natural waters. I hope that, after covering all relevant sections, you will appreciate the big picture.

23.2 NITRIFICATION

In addition to carbonaceous BOD, nitrogen compounds in wastewater also have an impact on a river's oxygen resources (Fig. 23.2). Sewage nitrogen can be broadly broken down into organic nitrogen compounds (for example proteins, urea, etc.) and ammonia. With time the organic nitrogen compounds are hydrolyzed to create additional ammonia. Autotrophic bacteria then assimilate the ammonia and create nitrite (NO_2^-) and nitrate (NO_3^-).

The conversion of ammonia to nitrate is collectively called ***nitrification.*** It can be represented by a series of reactions (Gaudy and Gaudy 1980). In the first, bacteria of the genus *Nitrosomonas* convert ammonium ion (NH_4^+) to nitrite,

$$NH_4^+ + 1.5O_2 \rightarrow 2H^+ + H_2O + NO_2^- \tag{23.1}$$

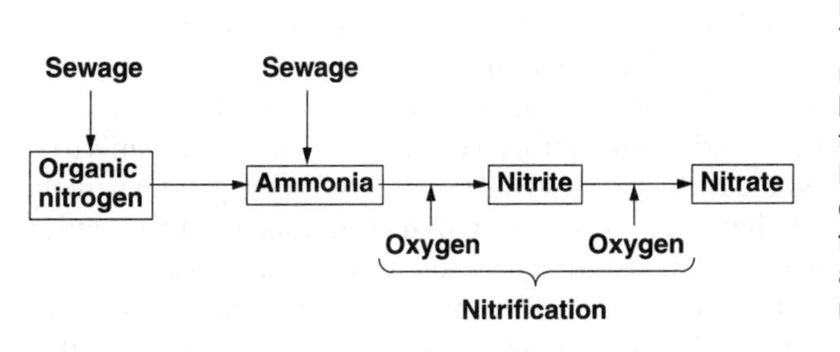

FIGURE 23.2
The decomposition of nitrogen compounds below a wastewater treatment plant discharge. Models have been developed to simulate this process based on a series of first-order reactions.

In the second, bacteria of the genus *Nitrobacter* convert nitrite to nitrate,

$$NO_2^- + 0.5O_2 \rightarrow NO_3^- \tag{23.2}$$

The growth rates of the two bacteria are generally slower than for heterotrophic bacteria (d^{-1} as contrasted with hr^{-1} rates). In addition the conversion from nitrite to nitrate is faster than the conversion of ammonium to nitrite. This reflects the fact that about 3 times the amount of substrate is consumed in the second step to derive the same amount of energy.

The oxygen consumed in the two stages can be computed as

$$r_{oa} = \frac{1.5(32)}{14} = 3.43 \text{ gO gN}^{-1} \tag{23.3}$$

$$r_{oi} = \frac{0.5(32)}{14} = 1.14 \text{ gO gN}^{-1} \tag{23.4}$$

where r_{oa} and r_{oi} represent the amount of oxygen consumed due to the nitrification of ammonium and nitrite, respectively. The entire process can be represented as

$$r_{on} = r_{oa} + r_{oi} = 4.57 \text{ gO gN}^{-1} \tag{23.5}$$

where r_{on} = amount of oxygen consumed per unit mass of nitrogen oxidized in the total process of nitrification. It should be noted that some of the ammonium will be used for bacterial cell production. As a consequence Gaudy and Gaudy (1980) suggest that the oxygen demand actually approaches 4.2 g of oxygen per g of nitrogen oxidized.

The occurrence of nitrification depends on additional factors besides the presence of ammonium. The most important of these cofactors are

1. The presence of adequate numbers of nitrifying bacteria. This depends on both the level of treatment and the stream type. If the sewage has been subject to secondary treatment, there is the possibility that some nitrification will have begun before discharge. As for stream type, shallower streams with rocky bottoms provide a substrate upon which bacteria can grow more effectively.
2. Alkaline pH levels (optimally at about 8). This tends to neutralize the acid that is produced.
3. Sufficient oxygen (greater than 1 to 2 mg L^{-1}).

The net effect of these cofactors is to inhibit nitrification immediately downstream from point-source effluents. In this region oxygen and pH will be depressed. In addition for deeper streams, the establishment of sufficient numbers of bacteria is hindered by their relatively slow growth rate and the lack of substrate. Thus nitrification and its accompanying oxygen demand would generally be exerted farther downstream than the oxygen demand due to oxidation of organic carbon.

As with the decomposition of carbonaceous matter, nitrification will cause a dissolved oxygen sag in the stream. The entire process, as depicted in Fig. 23.3, shows the collective effect of organic carbon and nitrogen decomposition on a stream's oxygen resources. In Fig. 23.3*a* the carbonaceous BOD is elevated at the point where the sewage enters the stream. It then decreases as the CBOD is decomposed and settles. The effect of decomposition in tandem with reaeration results in an oxygen sag.

In Fig. 23.3*b* nitrogen also induces a sag. However, because of the sequential nature of the nitrification reactions, the sag's minimum occurs farther downstream than

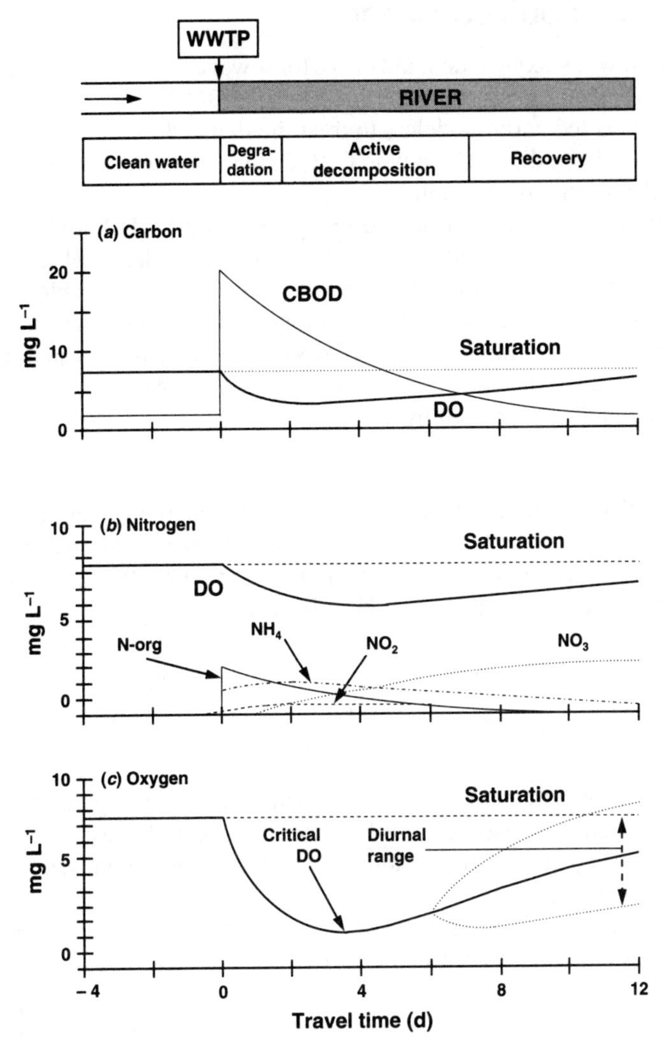

FIGURE 23.3
The trends of carbon, nitrogen, and oxygen below a wastewater treatment plant discharge into a river.

for the carbon induced depletion in Fig. 23.3*a*. In addition the cofactors described earlier can cause the nitrification sag to lag even farther downstream.

Figure 23.3*c* shows the total depletion that results from the combined effect of carbon and nitrogen. Note that the resulting DO profile is worse than the individual sags, and the critical DO level occurs at a distance somewhere between the carbon and the nitrogen cases.

Early attempts to simulate the impact of nitrification used a nitrogenous BOD (NBOD) to characterize the process. In recent years a more mechanistic approach that attempted to model organic nitrogen, ammonia, and nitrate explicitly has been developed. Both are described next.

23.3 NITROGENOUS BOD MODEL

Figure 23.4 shows an oxygen demand curve for sewage that has a significant amount of organic carbon and nitrogen. Note that a second BOD demand is exerted due to nitrification. This led early modelers to treat the demand as a nitrogenous BOD or NBOD. Thus, using a direct analogy to CBOD, mass balances could be written and solved for NBOD and oxygen deficit.

The NBOD could be measured directly from the BOD bottle measurement. In addition it could be measured indirectly through knowledge of the nitrification process. First, it was recognized that the total amount of oxidizable nitrogen was equal to the sum of organic and ammonia nitrogen. In fact there is an analytical measurement called total Kjeldahl nitrogen (TKN) that represents this quantity. Further, according to Eq. 23.5, the oxidation would consume 4.57 g of oxygen per g of TKN. Thus NBOD could be estimated by

$$L_N = 4.57\ \text{TKN} \tag{23.6}$$

where L_N = NBOD (mg L^{-1}) and TKN = total Kjeldahl nitrogen (mgN L^{-1}).

For a plug-flow system, mass balances for NBOD and oxygen deficit can be written in a similar fashion as for CBOD. For a point source, if $L_N = L_{N0}$ and $D = D_0$ at $t = 0$, then these equations can be solved for

$$L_N = L_{No} e^{-\frac{k_n}{U}x} \tag{23.7}$$

and

$$D = D_0 e^{-\frac{k_a}{U}x} + \frac{k_n L_o}{k_a - k_n}\left(e^{-\frac{k_n}{U}x} - e^{-\frac{k_a}{U}x}\right) \tag{23.8}$$

where k_n = rate of NBOD oxidation (d^{-1}). Thus an oxygen sag due to nitrification could be simulated. Because the mass balances are linear, the results for CBOD and NBOD could be added to derive the total deficit due to the combined effect of carbon and nitrogen oxidation.

Values of k_n generally range from 0.1 to 0.5 d^{-1} for deeper waters. For shallower streams, values of greater than 1 d^{-1} are often encountered.

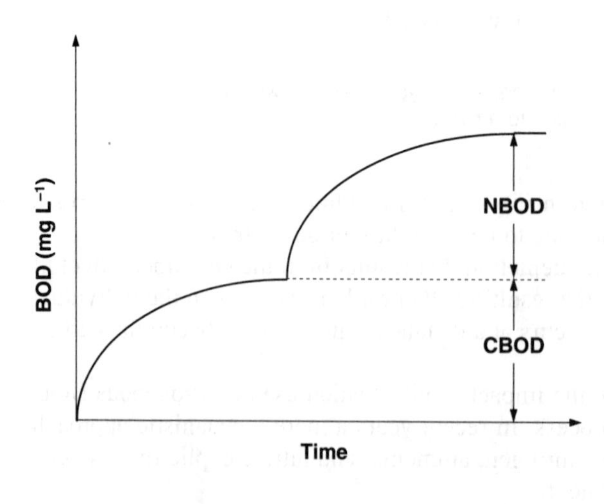

FIGURE 23.4
Oxygen demand curve for sewage that has a significant amount of organic carbon and nitrogen.

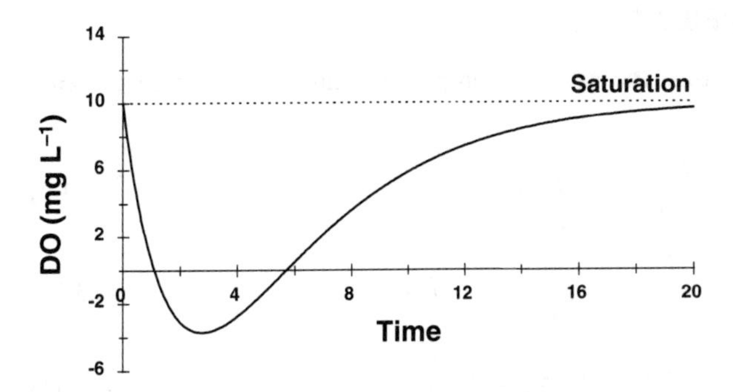

FIGURE 23.5
NBOD simulation.

Figure 23.5 shows the results of a simulation where the initial condition is 50% organic N and 50% ammonium. Oxygen saturation is set at 10 mg L^{-1}, the NBOD decay rate is set at 0.25 d^{-1}, and the reaeration rate is set at 0.5 d^{-1}. Although it is physically unrealistic, we have let the oxygen go negative. Notice that the maximum deficit is -3 mg L^{-1} and occurs at a travel time of 2.5 d from the outfall.

Although the NBOD framework provided a first way to incorporate nitrification into the Streeter-Phelps framework, it had some serious shortcomings. In particular it could not adequately simulate the time lags observed in the BOD bottle measurements and in natural systems. This can be better understood by scrutinizing the assumptions that are implicit in the adoption of NBOD as an adequate surrogate for the actual nitrification process:

1. If we lump the oxygen demands of organic nitrogen and ammonium into a single quantity (NBOD), the conversion rate of organic N to ammonium N is effectively ignored. Thus when organic nitrogen is abundant, a time lag would be erroneously omitted.
2. If we use a single, first-order NBOD decomposition rate, the two-step transformation of ammonium to nitrite to nitrate is ignored. Although the relative rapidity of the second step (nitrite to nitrate) mitigates this shortcoming, another time lag is omitted.
3. Finally, and probably most important, the NBOD perspective does not directly address the other cofactors (presence of bacteria, pH, and oxygen) that are required for nitrification to occur.

All three of these shortcomings mean that direct use of NBOD would result in simulated oxygen depletion occurring too close to the effluent. Analysts could circumvent this problem in two contrived ways. First, they could use slower rates for NBOD decomposition. Second, they could artificially add dead time to their models. Both remedies are inadequate: the first because it involves a distortion of reality and the second because, although it might adequately represent the observed lag under present conditions, it cannot address how the lag might change once treatment is implemented. As a consequence the NBOD approach has been discarded and, as described next, a more mechanistic characterization of the nitrification process developed.

23.4 MODELING NITRIFICATION

If we assume first-order kinetics, the nitrification process can be written as a series of first-order reactions,

$$\frac{dN_o}{dt} = -k_{oa}N_o \tag{23.9}$$

$$\frac{dN_a}{dt} = k_{oa}N_o - k_{ai}N_a \tag{23.10}$$

$$\frac{dN_i}{dt} = k_{ai}N_a - k_{in}N_i \tag{23.11}$$

$$\frac{dN_n}{dt} = k_{in}N_i \tag{23.12}$$

where the subscripts o, a, i, and n denote organic, ammonium, nitrite, and nitrate, respectively. An oxygen deficit balance can be written as

$$\frac{dD}{dt} = r_{oa}k_{ai}N_a + r_{oi}k_{in}N_i - k_aD \tag{23.13}$$

Because Eqs. 23.9 to 23.12 are sequential, they can be solved analytically using the techniques elaborated previously in Secs. 5.2 and 5.3 for feedforward reactors and reactions. For the case where $N_o = N_{o0}$ and $N_a = N_{a0}$ at $t = 0$, the solutions for the nitrogen species are

$$N_o = N_{o0}e^{-k_{oa}t} \tag{23.14}$$

$$N_a = N_{a0}e^{-k_{ai}t} + \frac{k_{oa}N_{o0}}{k_{ai} - k_{oa}}(e^{-k_{oa}t} - e^{-k_{ai}t}) \tag{23.15}$$

$$\begin{aligned} N_i = {} & \frac{k_{ai}N_{a0}}{k_{in} - k_{ai}}(e^{-k_{ai}t} - e^{-k_{in}t}) \\ & + \frac{k_{ai}k_{oa}N_{o0}}{k_{ai} - k_{oa}}\left(\frac{e^{-k_{oa}t} - e^{-k_{in}t}}{k_{in} - k_{oa}} - \frac{e^{-k_{ai}t} - e^{-k_{in}t}}{k_{in} - k_{ai}}\right) \end{aligned} \tag{23.16}$$

$$\begin{aligned} N_n = {} & N_{o0} + N_{a0} - N_{o0}e^{-k_{oa}t} - N_{a0}e^{-k_{ai}t} \\ & - \frac{k_{oa}N_{o0}}{k_{ai} - k_{oa}}(e^{-k_{oa}t} - e^{-k_{ai}t}) - \frac{k_{ai}N_{a0}}{k_{in} - k_{ai}}\left(e^{-k_{ai}t} - e^{-k_{in}t}\right) \\ & - \frac{k_{ai}k_{oa}N_{o0}}{k_{ai} - k_{oa}}\left(\frac{e^{-k_{oa}t} - e^{-k_{in}t}}{k_{in} - k_{oa}} - \frac{e^{-k_{ai}t} - e^{-k_{in}t}}{k_{in} - k_{ai}}\right) \end{aligned} \tag{23.17}$$

The deficit can be determined in a similar fashion (see Prob. 23.3). In addition the differential equations can be integrated numerically.

Figure 23.6 shows the results of a simulation where the initial condition is 50% organic N and 50% ammonium. Oxygen saturation is set at 10 mg L^{-1} and all rates are 0.25 d^{-1}, with the exceptions of k_{in} (at 0.75 d^{-1}) and reaeration (at 0.50 d^{-1}). In contrast with Fig. 23.5 the maximum deficit has been reduced from −3 mg L^{-1} to

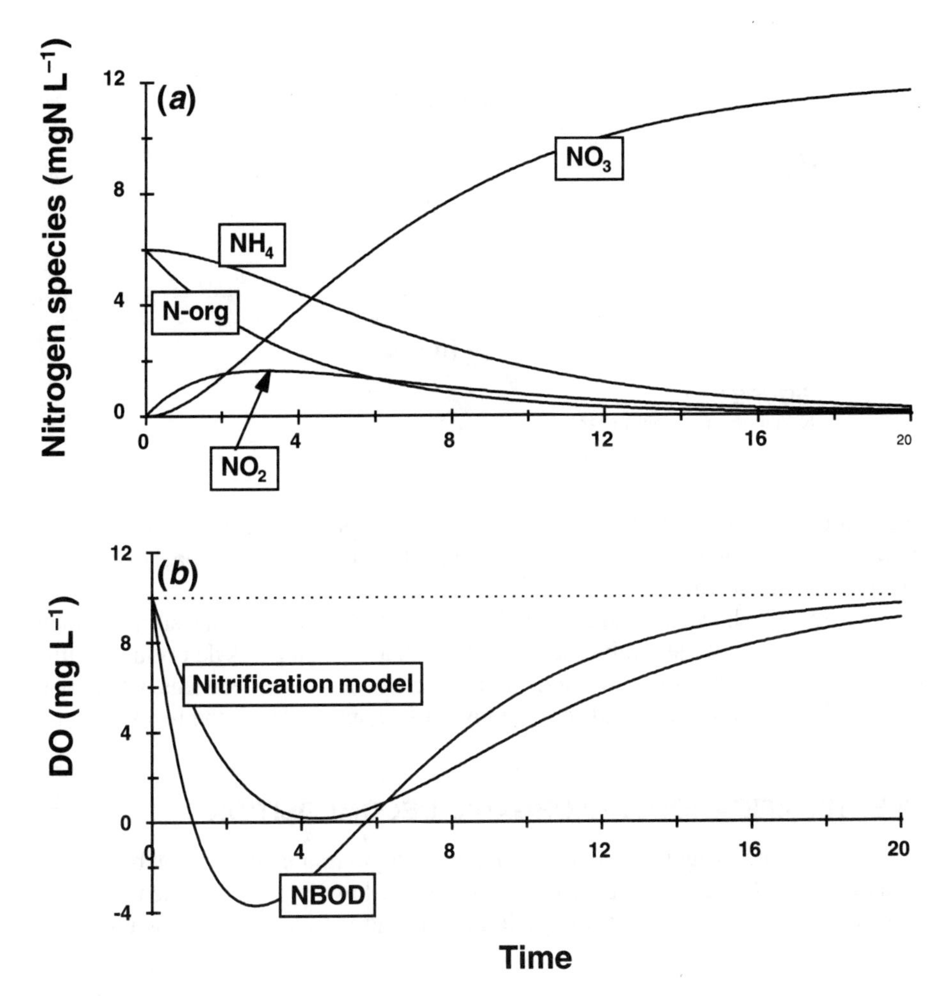

FIGURE 23.6
Nitrification simulation.

about 0 mg L^{-1} and has moved to a little over 4 d travel time below the source. Thus the use of sequential reactions has spread out and delayed the impact on oxygen.

23.4.1 Nitrification Inhibition

Although the explicit simulation of nitrification is more realistic than using NBOD, it still has some deficiencies. In particular it does not consider the limiting cofactors. One attempt to address this shortcoming involves the inhibition of the nitrification reaction due to depressed oxygen levels. The following factor is multiplied by each of the nitrification rates k_{ai} and k_{in} (Brown and Barnwell 1987):

$$f_{nitr} = 1 - e^{-k_{nitr} o} \tag{23.18}$$

where k_{nitr} = first-order nitrification inhibition coefficient (≈ 0.6 L mg^{-1}). As plotted in Fig. 23.7 the factor is close to 1 for dissolved oxygen concentrations greater

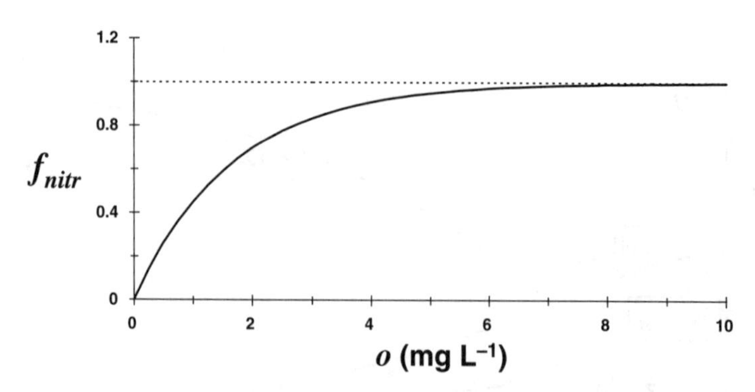

FIGURE 23.7
Nitrification inhibition factor.

than about 3 mg L^{-1}. At lower levels the factor approaches a linear relationship, which at zero oxygen goes to total inhibition. Thus as oxygen approaches zero, nitrification shuts down.

Figure 23.8 shows the results from Fig. 23.6, along with the same simulation but with the inclusion of the inhibition factor. Note that the latter was determined with a numerical solution (fourth-order RK). For this case the oxygen bottoms out at about 2 mg L^{-1}. In addition it spreads out the lowest DO and prolongs the recovery.

23.5 NITRIFICATION AND ORGANIC DECOMPOSITION

In addition to the oxygen yield, it is also interesting to relate nitrification back to the organic matter being decomposed. Recall that we have previously used the sugar glucose, $C_6H_{12}O_6$, to represent organic matter, and we used the following simple

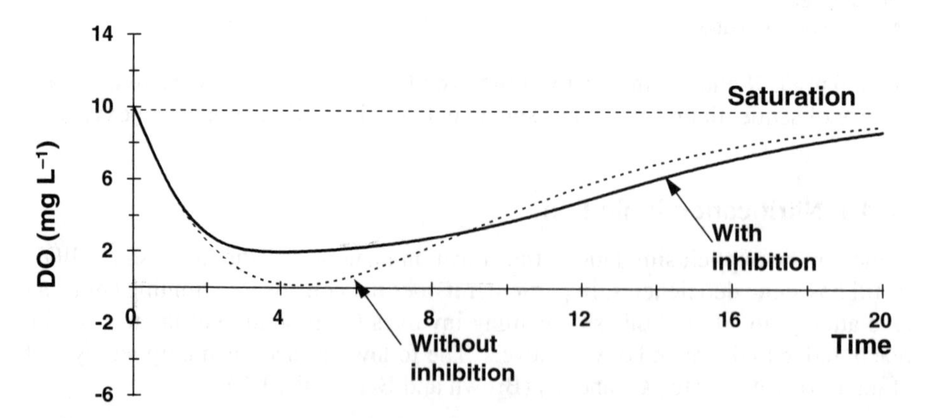

FIGURE 23.8
Nitrification simulation with inhibition at low dissolved oxygen levels.

representation for decomposition (Eq. 19.2):

$$C_6H_{12}O_6 + 6O_2 \xrightarrow{\text{respiration}} 6CO_2 + 6H_2O \tag{23.19}$$

Now that we are beginning to deal with nutrients, a more detailed representation is needed. Such a representation was provided by Redfield et al. (1963), who added phosphate and ammonium to develop a more complete stoichiometry for organic matter,

$$(C_1H_2O_1)_{106}(NH_3)_{16}(H_3PO_4) = C_{106}H_{263}O_{110}N_{16}P_1 \tag{23.20}$$

Using this formula a more comprehensive chemical representation of respiration can be expressed as (Stumm and Morgan 1981)

$$C_{106}H_{263}O_{110}N_{16}P_1 + 107O_2 + 14H^+ \xrightarrow{\text{respiration}}$$
$$106CO_2 + 16NH_4{}^+ + HPO_4{}^{2-} + 108H_2O \tag{23.21}$$

Note that this formula provides a similar result for the oxygen demand per unit mass of organic carbon as was obtained previously with Eq. 23.19 (recall Eq. 19.18),

$$r_{oc} = \frac{107(32)}{106(12)} = 2.69 \text{ gO gC}^{-1} \tag{23.22}$$

Although ammonia can originate from a variety of sources (for example fertilizers), let's limit ourselves to that which originates purely from the decomposition of organic matter. According to Eq. 23.21 the stoichiometric yield of nitrogen from the decomposition of a gram of organic carbon should be

$$a_{nc} = \frac{16(14)}{106(12)} = 0.176 \text{ gN gC}^{-1} \tag{23.23}$$

This result can now be linked with Eq. 23.5 to compute an oxygen demand generated per gram of carbon decomposed due to nitrification,

$$r_{on}a_{nc} = 4.57 \text{ gO gN}^{-1} (0.176 \text{ gN gC}^{-1}) = 0.804 \text{ gO gC}^{-1} \tag{23.24}$$

Thus the oxygen consumed in nitrification is about 30% of the oxygen consumed in carbonaceous oxidation of pure organic matter.

Finally let's look at wastewater. The nitrogen concentration of untreated wastewater in the United States is approximately 44 mgN L^{-1}, which breaks down into 17.5 and 26.5 mgN L^{-1} of organic and ammonia nitrogen, respectively (Metcalf and Eddy 1991, Thomann and Mueller 1987). Multiplying 44 mgN L^{-1} by 4.57 mgO mgN^{-1} yields an NBOD estimate of about 200 mgO L^{-1}. Thus the NBOD of raw sewage should be roughly equivalent to its CBOD ($\cong$ 220 mgO L^{-1} as in Table 19.2).

Now if sewage were pure organic matter (Eq. 23.20), we would expect that NBOD would be about 30% of CBOD. Several factors could explain the discrepancy. For example not all the organic matter might be decomposable under the conditions of the BOD test. Further, the sewage might be enriched in nitrogen from sources other than organic wastes. Regardless these simple calculations suggest that (*a*) NBOD can be significant and comparable to CBOD, and (*b*) wastewater is not simply composed of readily decomposable organic matter.

23.6 NITRATE AND AMMONIA TOXICITY

As described at the beginning of this lecture, forms of nitrogen are considered water pollutants in their own right. Both nitrate and ammonia fall into this category.

23.6.1 Nitrate Pollution

The presence of high levels of nitrate in drinking water appears to be the cause of ***methemoglobinemia*** or "blue babies." This disease primarily affects infants (< 6 months) but may impact children up to age 6. Levels above 10 mgN L^{-1} of nitrate are believed to lead to the condition (Salvato 1982).

Because nitrate is the terminal point of the aerobic nitrogen cycle and because raw sewage contains about 40 mg L^{-1} of total nitrogen, the potential exists for nitrate to become a problem below point sources without nitrogen removal. Further, the presence of nitrogen in agricultural fertilizers means that nonpoint sources could compound the problem.

The nitrification models described above provide a starting point for the analysis of nitrate pollution. However, more advanced frameworks should also consider other sources and sinks of nitrate beyond nitrification. For example plants use nitrate as a nutrient source and give off nitrogen when they decay. Further, the process of denitrification can act as a sink of nitrate under anoxic conditions. This process can occur in the water when oxygen levels drop to zero or near-zero levels. However, even when the water contains oxygen, denitrification can occur in anoxic sediments and biofilms. We describe such processes when we discuss sediment oxygen demand in Lec. 25.

23.6.2 Ammonia Toxicity

Ammonia nitrogen can exist in two states in natural waters: ammonium ion ($NH_4{}^+$) and un-ionized ammonia (NH_3),

$$[NH_3]_T = [NH_4{}^+] + [NH_3] \tag{23.25}$$

where the brackets connote molar concentrations and $[NH_3]_T$ = total ammonia. At sufficiently high levels (on the order of 0.01 to 0.1 mgN L^{-1}), the un-ionized form can be toxic to fish. Hence the term "ammonia toxicity" is associated with this effect.

The amounts of the two forms can be represented by the equilibrium dissociation reaction

$$NH_4{}^+ \rightleftharpoons NH_3 + H^+ \tag{23.26}$$

where the ratio of the reactants to the products is specified by an equilibrium coefficient

$$K = \frac{[NH_3][H^+]}{[NH_4{}^+]} \tag{23.27}$$

The equilibrium coefficient for the reaction is related to temperature by (Emerson et al. 1975)

$$pK = 0.09018 + \frac{2729.92}{T_a} \tag{23.28}$$

where T is in kelvins and $pK = -\log_{10}(K)$.

To gain insight into the partitioning, we can solve Eq. 23.27 for the ammonium ion and substitute the result into Eq. 23.25, which in turn can be solved for (see Prob. 23.4)

$$[NH_3] = F_u[NH_3]_T \tag{23.29}$$

where F_u = fraction of total ammonia in un-ionized form,

$$F_u = \frac{1}{1 + ([H^+]/K)} \tag{23.30}$$

Thus we can now see that the fraction un-ionized is a function of the system's pH and temperature. Although Eq. 23.29 is written in molar terms, it applies directly to mass concentrations expressed as nitrogen since both ammonia and ammonium have the same masses of nitrogen per mole.

EXAMPLE 23.1. CALCULATION OF UN-IONIZED AMMONIA. Determine the concentration of un-ionized ammonia at a point in a stream that has a total ammonia concentration of 1 mgN L^{-1}, a pH of 8.5, and a temperature of 30°C.

Solution: The temperature can be used to determine the equilibrium coefficient,

$$pK = 0.09018 + \frac{2729.92}{303.15} = 9.095$$

$$K = 10^{-pK} = 10^{-9.095} = 8.03 \times 10^{-10}$$

The hydrogen ion concentration can be determined from the pH,

$$[H^+] = 10^{-pH} = 10^{-8.5} = 3.16 \times 10^{-9}$$

These values can then be substituted into Eq. 23.30,

$$F_u = \frac{1}{1 + [(3.16 \times 10^{-9})/(8.03 \times 10^{-10})]} = 0.202$$

which can be used to compute

$$NH_3 = 0.202(1) = 0.202 \text{ mgN L}^{-1}$$

Figure 23.9 summarizes values for percent un-ionized ammonia for a range of temperatures and pH's. As can be seen, systems with high pH and temperature have the highest percentages.

As was the case with nitrate pollution, the nitrification model developed in Sec. 23.4 provides a starting point for modeling the un-ionized ammonia problem. In the simplest sense the nitrification model provides a means to simulate total ammonia as a function of loads and transport. Then Eq. 23.29 provides a means to determine how much of the total ammonia is in un-ionized form. Although this is a simplified view of the phenomenon, it provides a means to develop first-cut predictions to meet un-ionized ammonia standards in natural waters.

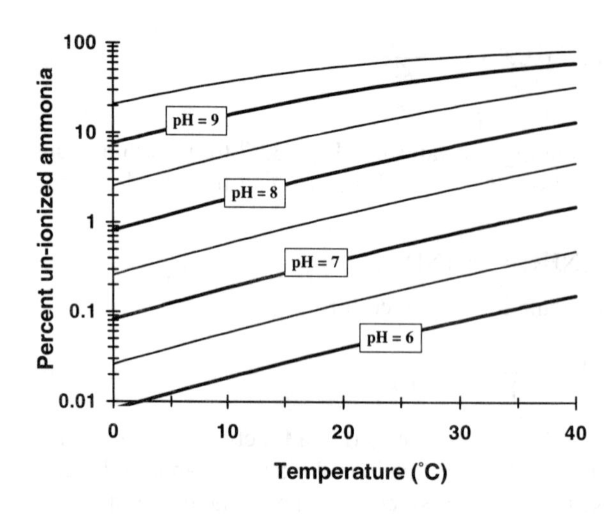

FIGURE 23.9
Percent un-ionized ammonia versus temperature for various levels of pH.

PROBLEMS

23.1. A point source has the following characteristics:

$N_o = 6\text{ mg L}^{-1}$ $\quad N_a = 6\text{ mg L}^{-1}$

$Q_w = 2\text{ cms}$ $\quad k_{oa} = 0.05\text{ d}^{-1}$

$k_{ai} = 0.075\text{ d}^{-1}$ $\quad k_{in} = 0.2\text{ d}^{-1}$

If the receiving water has a flow of 5 cms and negligible nitrogen, compute the profiles of the nitrogen species below the plant. Note that $n = 0.03$, $S = 0.001$, and $B = 20$ m. Disregard temperature.

23.2. A waste treatment plant discharges into a stream so that at the mixing point the concentration of organic nitrogen and ammonium are 5 mgN L^{-1} and 1 mgN L^{-1}, respectively, and the deficit is 2 mgO L^{-1}. Downstream the river has the following characteristics: $U = 0.2\text{ m d}^{-1}$, $H = 0.5$ m, $o_s = 10\text{ mg L}^{-1}$, and $k_a = 2\text{ d}^{-1}$.

(*a*) Compute the oxygen sag in the stretch given the following decay rates: $k_{oa} = 0.25\text{ d}^{-1}$, $k_{ai} = 0.2\text{ d}^{-1}$, and $k_{in} = 0.6\text{ d}^{-1}$.

(*b*) If the pH and temperature in the stretch are 8 and 15°C, respectively, compute the maximum level of un-ionized ammonia that occurs.

23.3. Develop a closed-form solution for deficit as a function of time by integrating Eq. 23.13 with $D = D_0$ at $t = 0$.

23.4. Perform the algebraic manipulations described in Sec. 23.6.2 to solve for the fraction un-ionized (Eq. 23.30) as a function of hydrogen ion concentration and the equilibrium coefficient. In addition solve for a fraction ionized, F_i, where $F_u + F_i = 1$.

23.5. A small detention pond (depth = 1 m, area = $1 \times 10^5\text{ m}^2$, and residence time = 2 wk) receives inflow concentrations of 5 and 2 mgN L^{-1} of organic and ammonium nitrogen, respectively. Determine the steady-state levels of all species of nitrogen if $k_{oa} = 0.1\text{ d}^{-1}$, $k_{ai} = 0.075\text{ d}^{-1}$, and $k_{in} = 0.25\text{ d}^{-1}$.

LECTURE 24

Photosynthesis/Respiration

LECTURE OVERVIEW: I focus on how photosynthesis and respiration rates are measured in streams and other natural waters. First, I review introductory material dealing with plant types, units, and light. Then I describe three approaches to measuring photosynthesis and respiration rates: direct measurement (light bottle/dark bottle), biomass, and the delta method.

The photosynthesis and respiration of plants can add and deplete significant quantities of oxygen from natural waters. As described in the previous lecture, such effects can be incorporated into the Streeter-Phelps framework as a zero-order distributed source. For such cases the resulting deficit can be calculated as

$$D = \frac{R - P}{k_a}(1 - e^{-k_a t}) \tag{24.1}$$

Such a zero-order characterization was deemed adequate when water-quality models were primarily used to assess the impact of treating raw sewage. However, as treatment became more refined and concern for nonpoint sources grew, such characterizations have been discarded.

In a later part of this book I will describe a mechanistic theory that predicts plant growth and death (and hence its effect on oxygen) as a function of nutrients, light, and other environmental factors. For the time being, we focus on quantitatively estimating photosynthesis and respiration rates and their impact on oxygen.

24.1 FUNDAMENTALS

Before describing measurement methods, we must cover some fundamental information. First, I'll provide a brief description of plant activity in streams. Then I will describe measurement units and parameterization of photosynthesis and respiration rates. Finally I'll show how light, the principal determining factor of photosynthesis, is quantified.

24.1.1 Stream Plant Activity

In Lec. 19, I described the environment in a stream below an untreated sewage outfall. Recall that for the stretch immediately below the discharge, biological activity is primarily heterotrophic. That is, it is dominated by organisms such as bacteria that obtain their energy by consuming organic matter, and in the process, deplete oxygen. Plant growth in this zone is suppressed because of a number of factors, including light extinction due to turbidity.

Further downstream, as the stream begins to recover, levels of nutrients such as nitrogen and phosphorus will be high. In addition most particulate solids will have settled and the water clears. Consequently autotrophic plant growth usually becomes dominant.

Because photosynthesis creates oxygen and respiration depletes oxygen, the plants will have an impact on the stream's oxygen resources. Because photosynthesis is light-dependent, this effect can have both seasonal and diurnal manifestations.

From a seasonal perspective, photosynthesis will tend to dominate during the growing season, whereas respiration and plant decomposition will prevail during the nongrowing period. Thus from a mean-daily perspective, plant activity will tend to ameliorate or exacerbate the oxygen problem depending on the time of year.

In the growing season, swings in oxygen can be induced by diurnal light variations. Thus stream oxygen levels could be supersaturated during the afternoon and severely depleted just before dawn.

Finally, depending on the depth, two types of plants would tend to grow. For deeper rivers, floating plants called phytoplankton would dominate. For shallower streams, where light can reach the lower depths, bottom plants would tend to make up most of the plant biomass. These can include rooted and attached macrophytes, as well as microfloral growth called periphyton.

Fixed plants tend to have a greater impact on stream oxygen resources than floating plants due to two factors:

1. Because they are usually situated in shallower water, fixed plants will tend to have a greater effect on oxygen levels. That is, for an equal growth or respiration rate, the impact on a shallower system will be greater.
2. Because they are fixed in space, they tend to be more concentrated longitudinally.

In addition, as described next, whether the plants are floating or attached will have a bearing on how their rates are parameterized.

24.1.2 Units and Parameterization

Recall that at the beginning of our description of oxygen modeling, we introduced the following simple chemical representation of the production and decomposition of organic matter in the environment,

$$6CO_2 + 6H_2O \rightleftharpoons C_6H_{12}O_6 + 6O_2 \tag{24.2}$$

We then used this representation to determine how much oxygen would be consumed in the decomposition of a unit mass of organic carbon,

$$r_{oc} = \frac{6(32)}{6(12)} = 2.67 \text{ gO gC}^{-1} \tag{24.3}$$

In the present context we can now see that just as this factor allows us to characterize the breakdown of organic matter, it also allows us to determine how much oxygen is produced when a quantity of organic carbon is created via photosynthesis.

This one-to-one correspondence between organic matter and oxygen brings up the point that photosynthesis and respiration rates can be expressed in either oxygen or carbon units. The choice is usually dictated by the problem context. Early oxygen modelers adopted the former, whereas those interested in the later problem of eutrophication tended to choose carbon. If a subscript c is used to designate the carbon version, the two expressions can be simply related by using the stoichiometric conversion of Eq. 24.3. Thus, for photosynthesis,

$$P = r_{oc} P_c \tag{24.4}$$

The second issue that must be addressed relates to how the rates should be normalized. This depends on whether the plants are free-floating or fixed. In the former case a volumetric expression, M L^{-3} T^{-1}, is proper, whereas in the latter an areal approach, M L^{-2} T^{-1}, is preferred. As described in the previous lecture, the two quantities can be related by depth. Thus for photosynthesis,

$$P = \frac{P'}{H} \tag{24.5}$$

where the prime denotes the areal rate. Note that the above conversions and parameterizations also apply to respiration.

24.1.3 Light and Photosynthesis

In most water-quality models the rate of photosynthesis is assumed to be directly proportional to light energy,

$$P(t) \propto I(t) \tag{24.6}$$

where $I(t)$ = available light (langleys d^{-1}), where 1 langley (ly) is 1 cal cm^{-2}.

One aspect of this relationship that has relevance to the present discussion is the light variation over the diurnal cycle. This variation can be idealized by a half-sinusoid function. If we assume that Eq. 24.6 holds, the photosynthesis rate can then also be represented by a half-sinusoid (Fig. 24.1),

$$\begin{aligned} P(t) &= P_m \sin[\omega(t - t_r)] \qquad && t_r < t < t_s \\ P(t) &= 0 && \text{otherwise} \end{aligned} \tag{24.7}$$

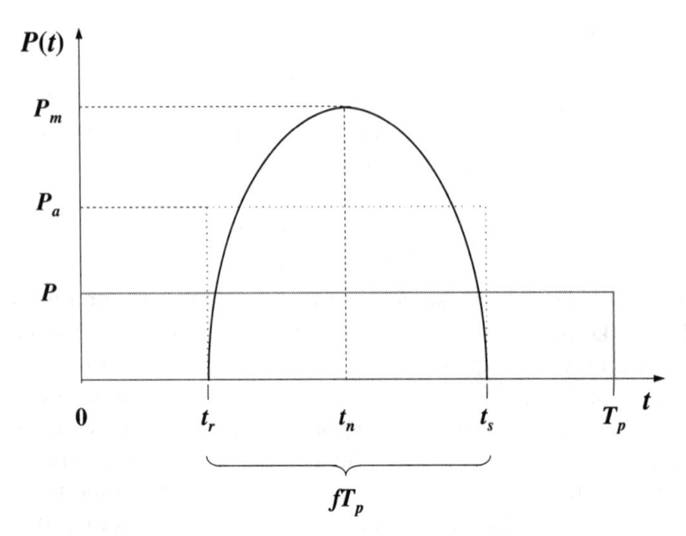

FIGURE 24.1
The variation in photosynthesis over the diurnal cycle due to a half-sinusoid representation of light intensity.

where P_m = maximum rate (g m^{-3} d^{-1})
ω = angular frequency [$= \pi/(fT_p)$]
t_r = time of sunrise (d)
t_s = time of sunset (d)
f = fraction of day subject to sunlight (the ***photoperiod***)
T_p = daily period (= 1 d, 24 hr, etc., depending on the units of time)

Note that the photoperiod is related to other parameters by

$$f = \frac{t_s - t_r}{T_p} \tag{24.8}$$

and solar noon can be calculated as

$$t_n = \frac{t_r + t_s}{2} \tag{24.9}$$

Finally in addition to its maximum value, we are often interested in the average value of Eq. 24.7. Two such averages can be developed. The average daily rate can be calculated by integration as

$$P = \frac{\int_0^{T_p} P(t)\,dt}{T_p} = P_m \frac{2f}{\pi} \tag{24.10}$$

The average daylight rate can be determined by

$$P_a = \frac{\int_0^{T_p} P(t)\,dt}{fT_p} = P_m \frac{2}{\pi} \tag{24.11}$$

Thus the average daylight rate is 63.7% of the maximum rate regardless of the photoperiod. The average daily rate is about 32% of the maximum for a 12-hr photoperiod, with a range of from 20 to 40% for 8-hr and 16-hr photoperiods.

EXAMPLE 24.1. PHOTOSYNTHESIS RATES AND STOICHIOMETRY. O'Connor and Di Toro (1970) present the following data for the Grand River, Michigan, on August 30, 1960:

Depth (m)	0.58
Sunrise (hr)	0700
Photoperiod	0.5417(13 hr)
Solar noon (hr)	1330
Average daily photosynthesis rate (mg-oxygen L^{-1} d^{-1})	17.3

Determine time of sunset, and the maximum and the average daylight photosynthesis rates. Express the average daily rate on an areal basis in carbon units.

Solution: The time of sunset can be calculated from Eq. 24.9 as

$$t_s = 2t_n - t_r = 2(13.5) - 7 = 20$$

Thus the sun sets at 2000 or 8:00 P.M.

The maximum rate can be calculated from Eq. 24.10,

$$P_m = 17.3\left[\frac{\pi}{2(0.5417)}\right] = 50.17 \text{ g m}^{-3} \text{ d}^{-1}$$

and the average daylight rate from Eq. 24.11,

$$P_a = 50.17\frac{2}{\pi} = 31.94 \text{ g m}^{-3} \text{ d}^{-1}$$

Finally the average daily rate can be converted to areal and carbon units by using Eqs. 24.4 and 24.5 to compute

$$P_c' = \frac{P}{r_{oc}}H = \frac{17.3 \text{ gO m}^{-3} \text{ d}^{-1}}{2.67 \text{ gO gC}^{-1}}0.58 \text{ m} = 3.76 \text{ gC m}^{-2} \text{ d}^{-1}$$

In general, values of the average daily areal production rate range from about 0.3 to 3 gO m^{-2} d^{-1} for moderately productive streams. Highly productive systems can range from 3 to 20 gO m^{-2} d^{-1}. In the following section we explore some of the ways in which such values are obtained.

24.2 MEASUREMENT METHODS

Now that you have some feel for the quantities involved in photosynthesis and respiration, I will describe techniques that are used to measure these rates. We first look at estimates that are made ***in situ***— directly in the environment. Then we examine two indirect approaches: the first uses biomass estimates in conjunction with a theoretical model of phytoplankton growth; the second uses diurnal oxygen data.

24.2.1 Light-Bottle/Dark-Bottle Technique

The first approach, called the light-bottle/dark-bottle technique, is very similar in spirit to the BOD bottle measurement described in Lec. 19. That is, a microcosm or bottle environment is used to simulate the chemical environment in the water body.

The method consists of placing a sample of a natural water in two bottles. One is a clear or "light" bottle, where both photosynthesis and respiration can occur. The other, which does not allow light penetration and is hence called "dark," should exhibit only respiration. After determining the oxygen concentration in each bottle, we cap them and place them in the water. Some time later we open the bottles and determine the oxygen concentrations. The difference between the oxygen levels can then be used to compute the photosynthesis and respiration rates, as described next.

Suppose that the experiment is conducted over a time t, and the initial and final concentrations in the light bottle are designated o_{li} and o_{lf}. Because we are dealing with a reaction that is assumed to be zero-order, the net photosynthesis rate can be determined simply by

$$P_{\text{net}} = \frac{o_{lf} - o_{li}}{t} \tag{24.12}$$

where P_{net} is defined as

$$P_{\text{net}} = P_b - R_{cm} \tag{24.13}$$

where the subscript b designates that this is the "bottle" rate, and the respiration rate is subscripted with "cm" to signify that it is a "community" rate which not only reflects plant respiration but also other oxygen-consuming reactions such as bacterial respiration.

The respiration in the dark bottle (in other words, the community respiration) can be determined in a similar fashion. If the initial and final concentrations in the dark bottle are designated o_{di} and o_{df} and we assume that the incubation time is the same,

$$R_{cm} = \frac{o_{di} - o_{df}}{t} \tag{24.14}$$

This value can then be used in conjunction with Eq. 24.13 to calculate P_b.

The community respiration can be corrected to compute plant respiration by determining the filtered BOD, L_f, that will be exerted over the incubation time. This value can be used to calculate

$$R_b = R_{cm} - \frac{L_f}{t} \tag{24.15}$$

where the correction can be measured directly or calculated from the bottle decomposition rate k_1 as $L_f/t = k_1 L_{fi}$, where L_{fi} is the initial filtered BOD in the bottle.

The results are average rates for the bottle over the period of the study. However, the photosynthesis rate varies according to the half-sinusoid function (Eq. 24.7). Thus further manipulation is required to determine maximum and average photosynthesis rates. The following example illustrates how this is done.

EXAMPLE 24.2. LIGHT BOTTLE/DARK BOTTLE. A light-bottle/dark-bottle experiment conducted over an 8-hr period starting at 9 A.M. yields the following data:

$$o_{li} = 7 \text{ mg L}^{-1} \qquad o_{lf} = 12 \text{ mg L}^{-1} \qquad o_{di} = 7 \text{ mg L}^{-1} \qquad o_{df} = 6.5 \text{ mg L}^{-1}$$

Note that sunrise and sunset occur at 6:00 A.M. and 6:00 P.M., respectively. The filtered BOD of the water is 2 mg L^{-1} and the BOD deoxygenation rate is 0.05 d^{-1}. Use this data to compute the stream's photosynthesis and respiration rate.

Solution: Equations 24.12 and 24.14 can be used to compute

$$P_{\text{net}} = \frac{(12 - 7) \text{ mg L}^{-1}}{0.33 \text{ d}} = 15 \text{ g m}^{-3} \text{ d}^{-1}$$

and

$$R_{cm} = \frac{(7 - 6.5) \text{ mg L}^{-1}}{0.33 \text{ d}} = 1.5 \text{ g m}^{-3} \text{ d}^{-1}$$

These values can then be substituted into Eq. 24.13 to give

$$P_b = 15 + 1.5 = 16.5 \text{ g m}^{-3} \text{ d}^{-1}$$

The plant respiration can be determined as

$$R_b = 1.5 - 0.05(2) = 1.4 \text{ g m}^{-3} \text{ d}^{-1}$$

Next the maximum photosynthesis rate can be determined by integrating Eq. 24.7 between the beginning and the end of the test, t_1 and t_2, and setting the result equal to $P_b(t_2 - t_1)$,

$$\int_{t_1}^{t_2} P_m \sin\left[\frac{\pi}{fT_p}(t - t_r)\right] dt = P_b(t_2 - t_1)$$

The left side can be integrated and the resulting expression solved for

$$P_m = P_b(t_2 - t_1)\frac{fT_p}{\pi\left\{\cos\left[\frac{\pi}{fT_p}(t_1 - t_r)\right] - \cos\left[\frac{\pi}{fT_p}(t_2 - t_r)\right]\right\}} = 20.66 \text{ g m}^{-3} \text{ d}^{-1}$$

Finally Eq. 24.10 can be used to determine the average daily rate,

$$P = P_m\frac{2f}{\pi} = 20.66\frac{2(0.5)}{\pi} = 6.6 \text{ g m}^{-3} \text{ d}^{-1}$$

There are various ways in which the light-bottle/dark-bottle approach can be implemented. For shallower systems, where bottom plants dominate, light and dark sediment chambers must be used. As depicted in Fig. 24.2*a*, probes and automated recorders are typically used to generate continuous oxygen readings within the chambers. For deep systems, it is necessary to deploy arrays of bottles (Fig. 24.2*b*) to determine the variation in photosynthesis due to light extinction with depth. Integration is used to translate the resulting information into a single depth-averaged value.

24.2.2 Biomass Estimates

As I stated at the beginning of this lecture, a theory of phytoplankton growth will be developed in a later lecture. For the time being, I present some results of this

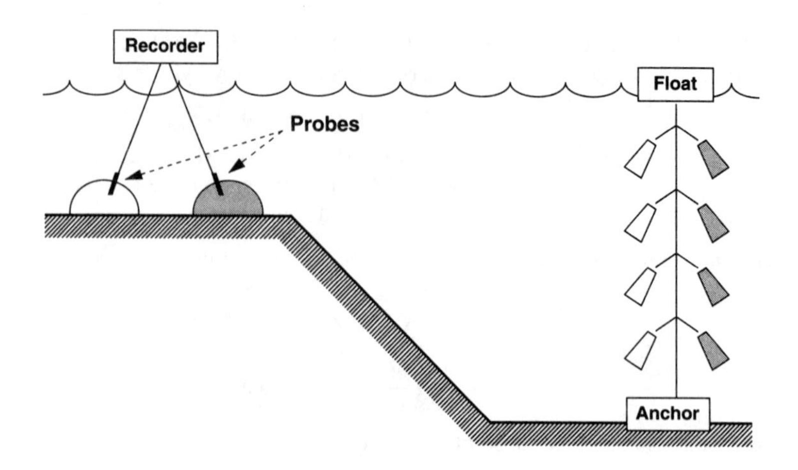

FIGURE 24.2
Variations of the *in situ* light-bottle/dark-bottle method. The benthic chambers in (*a*) are used to measure rates for bottom plants. The vertical array of bottles in (*b*) is employed to measure the effect of phytoplankton in deeper waters.

theory that can be employed to estimate photosynthesis from plant biomass measurements.

If nutrients are not limiting, which is often the case below sewage treatment plant outfalls, the daily average plant photosynthesis rate can be computed by

$$P = r_{oa} G_{\max} 1.066^{T-20} \phi_l a \tag{24.16}$$

where r_{oa} = oxygen generated per unit mass of plant biomass produced (range = 0.1 to 0.3 gO mg-Chla^{-1})
$G_{\max}$ = maximum plant growth rate for optimal light conditions and excess nutrients (range = 1.5 to 3.0 d^{-1})
T = water temperature (°C)
a = concentration of plant biomass (mg-Chl$a$ m^{-3})
ϕ_l = attenuation of growth due to light

The light attenuation factor for a layer of depth H is calculated as (complete derivations of the following relationships are presented in Lec. 33)

$$\phi_l = \frac{2.718 f}{k_e H}(e^{-\alpha_1} - e^{-\alpha_0}) \tag{24.17}$$

where

$$\alpha_1 = \frac{I_a}{I_s} e^{-k_e H} \tag{24.18}$$

and

$$\alpha_0 = \frac{I_a}{I_s} \tag{24.19}$$

where I_a = average daylight light intensity (ly d^{-1})
I_s = light intensity at which plant growth is optimal (range = 250 to 500 ly d^{-1})
k_e = a light extinction coefficient (m^{-1})

Note that extinction can be related to other parameters, such as Secchi-disk depth (m), by formulas. For example

$$k_e = \frac{1.8}{SD} \tag{24.20}$$

Equation 24.17 generates a number ranging between 0 and 1, which reflects the effect of light on plant growth. That is, a value of 0 indicates total suppression of productivity. If it is set to 1, Eq. 24.16 represents the oxygen production at optimal light levels,

$$P_s = r_{oa} G_{\max} 1.066^{T-20} a \tag{24.21}$$

where the subscript "s" denotes "saturating" light intensity. Thomann and Mueller (1987) have noted that, for T = 20°C, r_{oa} = 0.125 g mg^{-1}, and $G_{\max}$ = 2 d^{-1}, Eq. 24.20 reduces to

$$P_s = 0.25a \tag{24.22}$$

which is an oft-cited heuristic or "rule of thumb."

The plant respiration rate can be estimated by

$$R = r_{oa} k_{ra} 1.08^{T-20} a \tag{24.23}$$

where k_{ra} = respiration rate of the plants (range = 0.05 to 0.25 d^{-1}). Again, Thomann and Mueller (1987) point out that another commonly used rule of thumb,

$$R = 0.025a \tag{24.24}$$

corresponds to Eq. 24.23, with r_{oa} = 0.25 g mg^{-1} and T = 20°C.

EXAMPLE 24.3. BIOMASS ESTIMATE OF PHOTOSYNTHESIS AND RESPIRATION. A stream has a phytoplankton population of about 20 g dry weight m^{-2}. The dry weight is about 40% carbon. Employ this biomass estimate to determine the average daily areal production and respiration rates on a clear day in September. Note that the average temperature is 16°C, the photoperiod is about 12 hr, and the mean daily solar radiation is 485 ly d^{-1} at this time of year. The stream is about 1 m deep and has an extinction coefficient of approximately 0.5 m^{-1}. Use values of I_s = 300 ly d^{-1}, r_{oa} = 0.15 g mg^{-1}, and $G_{\max}$ = 2 d^{-1}.

Solution: The mean daylight light intensity can be calculated as

$$I_a = 485 \frac{24}{12} = 970 \text{ ly d}^{-1}$$

The parameter values can be substituted into Eq. 24.20 to compute the light attenuation

$$\phi_l = \frac{2.718(0.5)}{0.5(1)}(e^{-1.954} - e^{-3.233}) = 0.276$$

The biomass can be converted to organic carbon by 0.4(20) = 8 gC m^{-2}. Because we are dealing with carbon units we must use r_{oc} (Eq. 24.3) in place of r_{oa} in Eq. 24.16,

$$P' = 2.67 \text{ gO gC}^{-1}(2 \text{ d}^{-1})1.066^{16-20}(0.276)8 \text{ gC m}^{-2} = 9.1 \text{ gO m}^{-2} \text{ d}^{-1}$$

24.2.3 The Delta Method

Over 40 years ago, Odum (1956) suggested that photosynthetic production could be estimated using a mass-balance model in conjunction with diurnal oxygen measurements. Over the years a number of investigators (for example O'Connor and Di Toro 1970; Hornberger and Kelly 1972; Erdmann 1979*a*, 1979*b*; Schurr and Ruchti 1975, 1977, etc.) have extended and refined the idea. Recently Chapra and Di Toro (1993) have presented a graphical expression of Odum's idea that is easy to implement. In addition they have shown how the scheme can be extended to provide an estimate of stream reaeration. Their approach, called the ***delta method***, is the subject of the following paragraphs.

The basic approach is appealing because it is based on direct observations and hence is superior to the biomass approach described in the previous section. Further it is relatively inexpensive and can be applied to very shallow systems. Thus it also has advantages over the light-/dark-bottle technique.

As described by O'Connor and Di Toro (1970), a mass balance for dissolved oxygen deficit in a stream can be written as

$$\frac{\partial D}{\partial t} = -U\frac{\partial D}{\partial x} - k_a D + R - P(t) \tag{24.25}$$

where D = oxygen deficit (mg L^{-1})
t = time (d)
U = stream velocity (m d^{-1})
x = distance (m)
k_a = reaeration rate (d^{-1})
R = respiration (mg L^{-1} d^{-1})
$P(t)$ = primary production (mg L^{-1} d^{-1})

For situations where the plants are uniformly distributed for a sufficiently long distance ($> 3U/k_a$), deficit does not vary spatially ($\partial D/\partial x \cong 0$), and Eq. 24.25 simplifies to

$$\frac{dD}{dt} + k_a D = R - P(t) \tag{24.26}$$

The plant production rate can be approximated by a half-sinusoid (Eq. 24.7 and Fig. 24.3*a*). Note that for the present derivation, we have assumed that time zero corresponds to sunrise. O'Connor and Di Toro (1970) proposed that Eq. 24.7 can be expressed as a Fourier series,

$$P(t) = P_m\left\{\frac{2f}{\pi} + \sum_{n=1}^{\infty} b_n \cos\left[\frac{2\pi n}{T_p}\left(t - \frac{fT_p}{2}\right)\right]\right\} \tag{24.27}$$

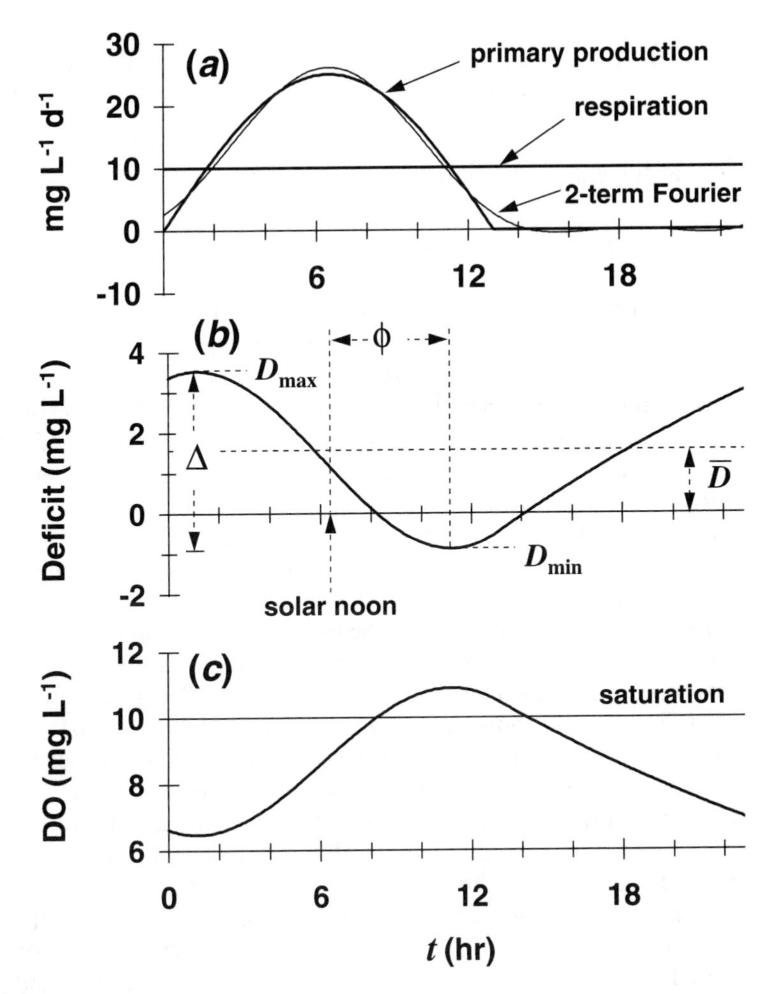

FIGURE 24.3
(*a*) Plant primary production and respiration (mg L^{-1} d^{-1}); (*b*) dissolved oxygen deficit (mg L^{-1}); and (*c*) dissolved oxygen concentration versus time (hr). Note that (*a*) shows the half-sinusoidal photosynthesis function along with the two-term Fourier series approximation (redrawn with permission from ASCE).

where

$$b_n = \cos(n\pi f)\frac{4\pi/f}{(\pi/f)^2 - (2\pi n)^2} \tag{24.28}$$

The two-term approximation is superimposed on Fig. 24.3*a*.

The endogenous respiration R is assumed to be constant. This is the traditional supposition of previous models [e.g., Odum (1956)]. In light of presently existing information (Wetzel 1983), it is considered a sound first approximation.

Assuming that the effects of initial conditions have decayed to zero, the solution to 24.26 with 24.27 as the forcing function, can be determined as

$$D(t) = \overline{D} - P_m \left\{ \sum_{n=1}^{\infty} \frac{b_n}{\sqrt{k_a^2 + (2\pi n/T_p)^2}} \cos\left[\frac{2\pi n}{T_p}\left(t - \frac{fT_p}{2}\right) - \tan^{-1}\left(\frac{2\pi n}{k_a T_p}\right)\right]\right\} \tag{24.29}$$

where $\overline{D}$ = average daily deficit (mg L^{-1}),

$$\overline{D} = \frac{R - P}{k_a} \tag{24.30}$$

Examples of the solution are shown in Fig. 24.4. To ease interpretation of this figure, I have plotted oxygen concentration above saturation rather than deficit. In addition the respiration is set equal to zero.

At large values of the reaeration coefficient k_a, the solution does not diverge much from saturation. Also the shape of the solution is similar to that of the forcing function $P(t)$, with the maximum oxygen concentration (minimum deficit) occurring shortly after solar noon. However, as k_a decreases, the solution moves farther from saturation. In addition the shape departs from that of the forcing function and becomes more sinusoidal. The maximum concentration occurs later in the afternoon, and a clearly defined minimum occurs just after dawn. In effect the small k_a retards the rate of reaeration transfer, allowing the effect of the previous day to carry over into the present day. This effect accounts for the minimum dissolved oxygen concentration (maximum deficit) occurring slightly after dawn. The decay toward equilibrium is slow during the evenings and is not completed at sunrise. Thus the dissolved oxygen is still above saturation as dawn arrives, and a slight decay occurs until the source rises to a value sufficient to start the solution increasing again.

Inspection of Eq. 24.29 and Fig. 24.4 leads to some generalizations regarding the model. As summarized in Fig. 24.3*b*, it can be concluded that the solution has three fundamental characteristics: mean vertical displacement, vertical range, and horizontal displacement. The mean vertical displacement of the entire curve $\overline{D}$ is governed by the interplay of reaeration, primary production, and respiration as specified by Eq. 24.30. The vertical range Δ is dictated by reaeration and the tem-

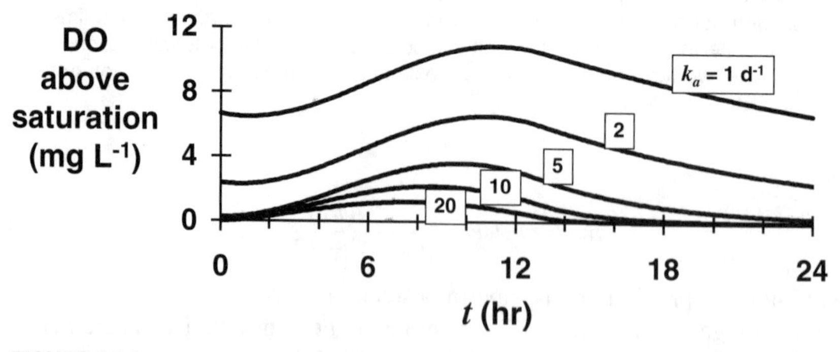

FIGURE 24.4
Dissolved oxygen concentration above saturation versus time as computed by Eq. 24.28; P_m = 25 mg L^{-1} d^{-1}, f = 13 hr, R = 0 mg L^{-1} d^{-1}. Boxed numbers indicate different values of the reaeration rate (redrawn with permission from ASCE).

poral variation of $P(t)$. Finally the time lag ϕ between solar noon and the time of minimum deficit is strictly a function of the reaeration rate. These observations can be used as the basis of a simple method for estimating reaeration, primary production, and respiration in a sequential fashion.

Reaeration. Because the solution's phase shift is dictated solely by reaeration, there must be a unique functional relationship linking k_a and ϕ. Unfortunately an analytical expression for this relationship is difficult to obtain because it involves the solution of a transcendental equation. However, a numerical solution is straightforward.

The numerical solution is shown in Fig. 24.5, where the reaeration rate is expressed as a function of the time lag ϕ. Figure 24.5 shows that lower k_a's are associated with longer time lags. It also suggests that for time lags greater than about 3 hr, the reaeration prediction is particularly sensitive to the photoperiod.

Note that Fig. 24.5 is extremely sensitive to data errors for cases with short time lags (that is, streams with high reaeration). Consequently uncertainty in the time lag can translate into a sizable uncertainty in the reaeration estimate. For such cases other methods might be required to obtain a sufficiently accurate estimate of reaeration. For example the rate can also be determined by direct measurement or by using predictive equations (Sec. 20.3).

Primary production. As with the reaeration estimate, an analytical expression for Δ as a function of primary production is difficult to obtain. However, a numerical solution can be easily developed.

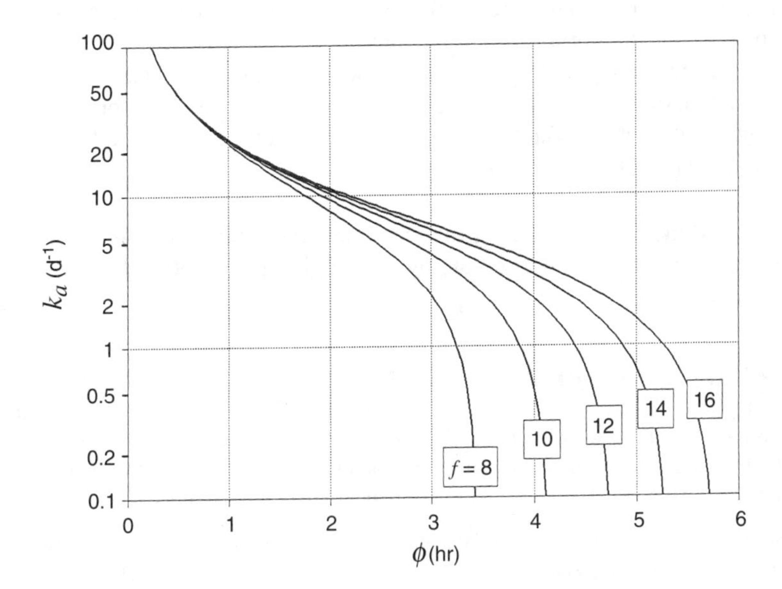

FIGURE 24.5
Plot of k_a (d^{-1}) versus ϕ (hr) for range of photoperiods (hr). Note that we have expressed photoperiod in hours for this plot (redrawn with permission from ASCE).

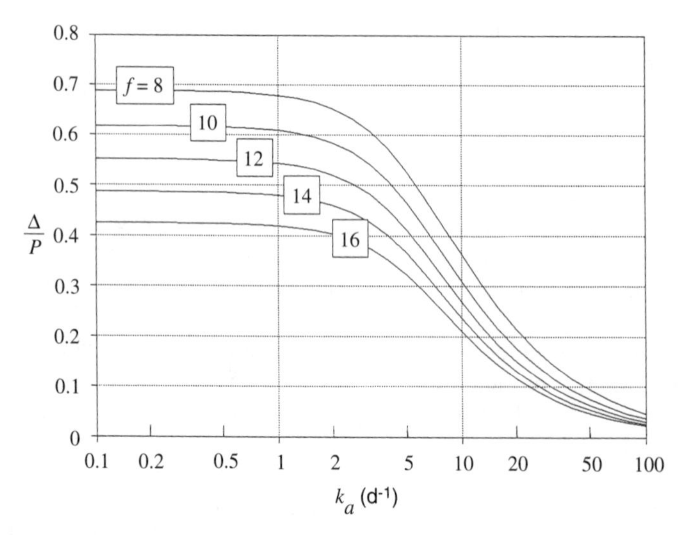

FIGURE 24.6
Plot of Δ/P versus k_a (d^{-1}) for range of photoperiods (hr). Note that we have expressed photoperiod in hours for this plot (redrawn with permission from ASCE).

As shown in Fig. 24.6, the numerical solution indicates that the ratio Δ/P is dependent on the photoperiod and the reaeration rate. At low reaeration levels (that is, $k_a < 1.0\ d^{-1}$) the ratio Δ/P becomes solely dependent on the photoperiod. In contrast, for streams with high reaeration the photoperiod is less important, since the ratio becomes dictated primarily by reaeration.

It should be noted that if f and k_a are known, this plot can be used in two ways. If diurnal oxygen data are available, Fig. 24.6 can be used to estimate primary production. Conversely for a production rate calculated by other means (light/dark bottle, etc.), the graph can be employed to estimate Δ.

Respiration. After obtaining the reaeration and primary production estimates, we can determine the respiration rate by rearranging Eq. 24.30 to give

$$R = P + k_a\overline{D} \tag{24.31}$$

EXAMPLE 24.4. THE DELTA METHOD. Table 24.1 summarizes O'Connor and Di Toro's (1970) results for a station on the Grand River, Michigan, that is marked by profuse growth of aquatic plants. The station is also sufficiently far downstream from the Lansing sewage treatment plant that the assumption of negligible spatial gradients ($\partial D/\partial x \cong 0$) holds. Use this data to implement the delta method.

Solution: The diurnal oxygen data for this case are shown in Fig. 24.7. Visual inspection of the plot suggests a mean oxygen level of about 8 mg L^{-1} with an amplitude of approximately 4 mg L^{-1}. In addition, the trend seems to peak between 5:00 P.M. and 6:00 P.M., which amounts to a time lag of about 4 hr.

Although a visual inspection can be used to estimate the trend parameters, regression techniques provide an alternative that is somewhat less subjective. For example

TABLE 24.1
Data for Grand River, Michigan (38 km below the Lansing sewage treatment plant) on August 10, 1960 as reported by O'Connor and Di Toro (1970).

Quantity	Symbol	Value	Units
	(*a*) Measured		
Depth	H	0.58 (1.9)	m (ft)
Flow	Q	8.35 (295)	$m^3\ s^{-1}$ (cfs)
Water temperature	T_w	25.9	°C
Oxygen saturation	o_s	8.13	mg L^{-1}
Sunrise	t_r	0700	hr
Photoperiod	f	0.54 (13)	d (hr)
Solar noon	t_n	1330	hr
	(*b*) Calculated		
Reaeration†	k_a	5.5	d^{-1}
Primary production			
Maximum	P_m	50	mg $L^{-1}\ d^{-1}$
Average	P	17.3	mg $L^{-1}\ d^{-1}$
Respiration	R	16	mg $L^{-1}\ d^{-1}$

†Computed with the O'Connor and Dobbins (1958) formula.

least-squares regression can be used to fit a sinusoid to the data (Chapra and Canale 1988). The resulting estimates are $\overline{D} = o_s - o = 8.128 - 8.4 = -0.272$ mg L^{-1}, $\Delta =$ 8.4 mg L^{-1}, and $\phi = 3.93$ hr.

Using Fig. 24.5 with $f = 13$ hr, the value of $\phi = 3.93$ hr can be employed to estimate $k_a = 2.7\ d^{-1}$. The reaeration estimate can then be employed in conjunction with Fig. 24.6 to determine $\Delta/P = 0.47$. This value can, in turn, be used to compute

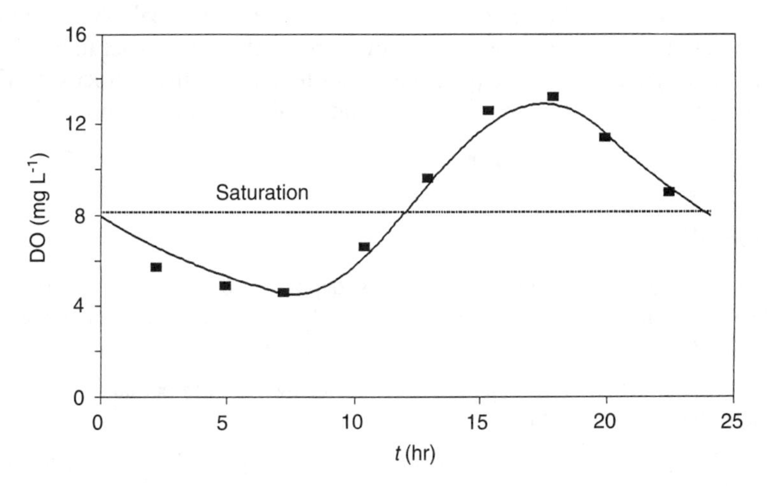

FIGURE 24.7
Plot of dissolved oxygen (mg L^{-1}) for Grand River, on August 10, 1960; superimposed curve was generated using Eq. 24.29 with parameter values estimated in Example 24.4 (redrawn with permission from ASCE).
(■ observations; ______ fitted curve).

$P = 17.9$ mg L^{-1} d^{-1}. Finally the primary production and the reaeration rate can be substituted into Eq. 24.31 to yield $R = 17.2$ mg L^{-1} d^{-1}.

Our estimates compare favorably with those obtained by O'Connor and Di Toro (Table 24.1) using their more complicated model. The reaeration rate manifests the greatest discrepancy: 2.7 versus 5.5 d^{-1}. It should be noted that O'Connor and Di Toro used a formula (O'Connor and Dobbins 1958) to obtain their rate. However, it can also be seen that if the phase lag had been estimated as 3.5 hr, we would have obtained a reaeration rate of about 5 d^{-1} from Fig. 24.5.

The delta method has three deficiencies: one specific to systems with high reaeration and two of a more general nature. As shown by Chapra and Di Toro (1991) the reaeration computation is particularly sensitive at low time lags. Thus there is considerable uncertainty associated with the resulting reaeration estimate. Consequently predictive equations or direct measurements might be necessary for such cases.

A more general difficulty pertains to the serial nature of the parameter evaluations; that is, the calculation of each parameter depends on the previous estimates. This could represent a problem in any application. However, it could be especially significant in highly aerated systems where the most uncertain estimate, reaeration, is the first to be computed. Consequently error in k_a would propagate to P and then R. An alternative approach might be to use nonlinear regression to approximate all parameters simultaneously.

A further limitation of the approach relates to our assumption that photosynthesis can be adequately represented as a half-sinusoid. This assumption precludes application of the delta method to periods (overcast days) or systems (canyon streams) that are shaded. If such effects can be quantified, some of the other more complicated expressions of the Odum framework can be employed. For example Schurr and Ruchti (1975) proposed a way to incorporate cloud cover into the framework.

Despite these deficiencies, because of its ease of application, the delta method has utility for water-quality modeling. For cases where physical or economic limitations preclude adequate measurement of plant activity and reaeration, the delta method provides a quick and inexpensive means to quantify these processes. For situations where acceptable sampling can be conducted, it offers a simple, independent check.

PROBLEMS

24.1. A stream has the following characteristics: depth = 0.8 m, sunrise = 0600, noon = 1300, and maximum photosynthesis rate = 10 gC m^{-2} d^{-1}. Determine the average daylight and daily photosynthesis rates in gO m^{-3} d^{-1}.

24.2. A light-bottle/dark-bottle experiment conducted over a 10-hr period starting at 8 A.M. yields the following data:

$$o_{li} = 8 \text{ mg L}^{-1} \qquad o_{lf} = 14 \text{ mg L}^{-1} \qquad o_{di} = 8 \text{ mg L}^{-1} \qquad o_{df} = 5.5 \text{ mg L}^{-1}$$

Note that sunrise and sunset occur at 5:00 A.M. and 7:00 P.M., respectively. The filtered BOD of the water is 4 mg L^{-1} and the BOD deoxygenation rate is 0.05 d^{-1}. Use this data to compute the stream's photosynthesis (average and maximum) and respiration rate.

24.3. The epilimnion of a lake has a phytoplankton population of 5 μg-chlorophyll L^{-1}. The carbon to chlorophyll ratio of the plankton is about 40:1. Use this biomass estimate to determine the average daily areal production and respiration rates on a clear day in September. Note that the average temperature is 22°C, the photoperiod is about 12 hr, and the mean daily solar radiation is 500 ly d^{-1} at this time of year. The epilimnion is about 7 m deep and has a Secchi-disk depth of 3.6 m. Use values of $I_s = 350$ ly d^{-1}, $r_{oa} = 0.15$ g mg^{-1}, and $G_{max} = 2\ d^{-1}$.

24.4. Temperature and dissolved oxygen were measured at a station on Boulder Creek (about 8.5 miles below the Boulder WWTP) on September 21 and 22, 1987:

Time	T (°C)	DO (mg L^{-1})	Time	T (°C)	DO (mg L^{-1})	Time	T (°C)	DO (mg L^{-1})
0400	12.4	6.0	1200	17.2	11.0	2000	18.5	6.7
0500	12.6	6.2	1300	19.2	11.2	2100	17.4	6.8
0600	12.6	6.2	1400	20.8	11.2	2200	16.9	5.4
0700	12.6	7.4	1500	21.5	10.6	2300	16.2	5.1
0800	12.8	8.2	1600	21.6	9.2	2400	15.6	5.4
0900	12.4	8.8	1700	21.3	8.8	2500	14.9	5.2
1000	12.4	10.8	1800	21.0	8.2	2600	14.0	5.3
1100	15.6	11.0	1900	19.2	6.9	2700	14.0	5.4

The elevation at the station is approximately 5500 feet above sea level. The stream's velocity above the station was about 0.8 fps and the depth was about 0.75 ft. Use the delta method to determine the reaeration rate and the maximum and the mean daily productivity and respiration at the station. Express the productivity and respiration in g m^{-2} d^{-1}. Note that the photoperiod is approximately 13 hr and solar noon is at 1:00 P.M.

24.5. A stream has a reaeration rate of 1 d^{-1}, DO saturation of 10 mg L^{-1}, an average daily photosynthesis rate P of 6 mg L^{-1} d^{-1}, and a respiration rate R of 5 mg L^{-1} d^{-1}. If the photoperiod is 10 hr and dawn is at 7:00 A.M., determine the value and the time of the maximum dissolved oxygen concentration.

24.6. A series of light and dark bottles are suspended in the 8-m-thick epilimnion of a lake. The resulting average daily rates of photosynthesis and respiration were calculated as

Depth (m)	0	2	4	6	8
Surface area (m^2)	1×10^6	0.95×10^6	0.9×10^6	0.83×10^6	0.7×10^6
Photosynthesis (g m^{-3} d^{-1})	1	6	10	8	5
Respiration (g m^{-3} d^{-1})	0.5	1	1	4	10

Use this data to determine the epilimnion's photosynthesis and respiration rates.

LECTURE 25

Sediment Oxygen Demand

LECTURE OVERVIEW: I review typical SOD values observed in natural waters and then outline mechanistic approaches for modeling the phenomenon. I first derive a simple model that is consistent with the classic Streeter-Phelps paradigm. Then I describe a more realistic framework that accounts for phenomena such as gas formation and hence provides a superior approach for simulating SOD in natural waters.

Sediment oxygen demand or SOD is due to the oxidation of organic matter in bottom sediments. These benthic deposits or "sludge beds" derive from several sources. Wastewater particulates, as well as other allochthonous particulates (leaf litter and eroded organic-rich soils), can result in sediments with high organic content. In addition, in highly productive environments such as eutrophic lakes, estuaries, and rivers, photosynthetically produced plant matter can settle and accumulate on the bottom. Regardless of the source, oxidation of the accumulated organic matter will result in a sediment oxygen demand.

As mentioned previously in Lec. 22, the first attempts to model SOD used a zero-order or constant source term. Such a characterization was deemed adequate when water-quality models were primarily used to assess the impact of treating raw sewage. However, as treatment becomes more refined and concern for nonpoint sources grows, such characterizations become increasingly inadequate. This is due to the fact that there is no satisfactory way to decide how the SOD changes following treatment.

The two most common approaches were to (1) leave the SOD unchanged or (2) assume linearity and lower SOD in direct proportion to the load reduction. Such arbitrary methods were usually justified under the assumption that a truly mechanistic understanding of the process was not necessary when making crude assessments of the effects of primary and secondary treatment. Although this assumption is probably

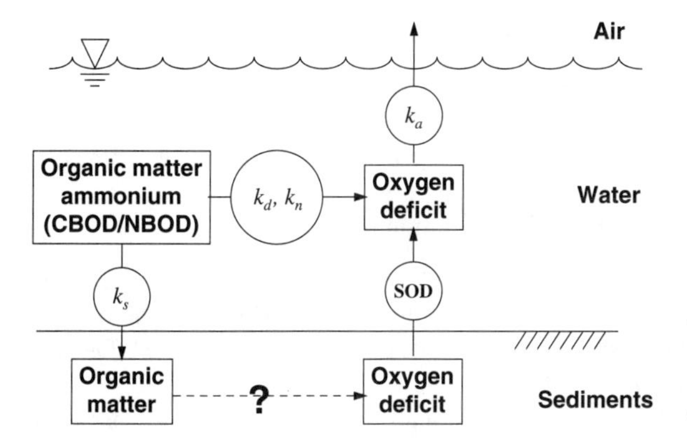

FIGURE 25.1
The "missing link" between BOD settling losses and SOD in classical Streeter-Phelps theory.

questionable in itself, it becomes even more tenuous as models are used to evaluate advanced waste treatment.

Such shortcomings arise because the zero-order approach treats SOD as a model input rather than as a calculated variable. This problem is illustrated in Fig. 25.1. Recall that in our earlier discussion of the Streeter-Phelps model, we included a parameter, k_s, to simulate that particulate sewage can be removed by settling. Later, we expanded the model to include the zero-order SOD to simulate the oxygen demand due to the decomposition of the settled sewage in the bottom sediments. As illustrated in Fig. 25.1, there is a missing piece that was left out of the process. That is, the use of the zero-order SOD neglects the mechanism whereby the sediment organic matter is converted into oxygen demand.

The present lecture is designed to show how this missing link can be modeled. To do this, I will first review data in order to delineate the levels that occur in natural waters and the major factors that influence SOD. Then I will construct a simple "Streeter-Phelps" SOD model which, although it is unrealistic, provides a context for our subsequent discussions. Finally I describe a more recently developed framework for calculating SOD in a mechanistic fashion.

25.1 OBSERVATIONS

SOD is typically measured in three ways: two approaches based on modeling observed oxygen levels and one based on direct measurement. In the first model-based approach, a DO model is developed for the water body and all rates except the SOD are determined. The SOD can then be estimated by adjusting the SOD rate until the model predictions match the observed DO levels. Although this calibration method was widespread in the early years of modeling, it is flawed because it assumes that all other model parameters (reaeration, deoxygenation rates, etc.) are known with confidence. Because such is rarely the case, values obtained with this method are no better than order-of-magnitude estimates.

The second model-based approach is expressly designed for stratified lakes. For such systems the deeper waters (or hypolimnion) can be idealized as a closed system, and an areal hypolimnetic oxygen demand can be determined as

$$\text{AHOD} = \frac{o_2 - o_1}{t_2 - t_1} H_h \tag{25.1}$$

where AHOD = areal hypolimnetic oxygen demand (gO m^{-2} d^{-1})
o_1 and o_2 = hypolimnetic oxygen levels (mg L^{-1}) measured at two times, t_1 and t_2 (d), during the stratified period
H_h = mean thickness of the hypolimnion (m)

If it is assumed that the primary cause of oxygen depletion in this layer is decomposition of organic matter in or at the surface of the bottom sediments, the oxygen depletion over the summer-stratified period provides an estimate of the SOD. That is, we assume that $S'_B \cong$ AHOD.

This approach has a number of deficiencies that relate to its underlying assumptions. First, it does not distinguish between decomposition in the water column and in the sediments. Whereas the former is not usually as important as the latter, in deeper systems the assumption might be violated. Second, although strong thermal gradients diminish mass transfer between the surface and bottom water layers, some transport does occur. It should be noted that corrections can be made to account for thermocline transport (Chapra and Reckhow 1983). However, since such corrections have rarely been made, most literature AHOD values usually represent underestimates of the actual depletion.

The final estimation method involves direct measurement. This is done by enclosing the sediments and some overlying water in a chamber and measuring the dissolved oxygen concentration in the water as a function of time. This approach is used in both the laboratory and the field. As with all microcosm measurements (e.g., the BOD or light-bottle/dark-bottle tests), this test has the drawback that enclosing the water in a chamber makes the system different from the natural environment.

Typical values of SOD are listed in Table 25.1. In general, values from about 1 to 10 g m^{-2} d^{-1} are considered indicative of enriched sediments. Also notice that the organism *Sphaerolitus* is included in the table to indicate that benthic organisms other than bacteria can process organic matter from the overlying water and exert an SOD (see Box 25.1).

TABLE 25.1
Sediment oxygen demand values (from Thomann 1972 and Rast and Lee 1978)

	$S'_{B,20}$ (g m^{-2} d^{-1})	
Bottom type and location	**Average value**	**Range**
Sphaerolitus (10 g-dry wt m^{-2})	7	—
Municipal sewage sludge:		
Outfall vicinity	4	2–10
Downstream of outfall, "aged"	1.5	1–2
Estuarine mud	1.5	1–2
Sandy bottom	0.5	0.2–1
Mineral soils	0.07	0.05–1
Areal hypolimnetic oxygen demand (AHOD)—lakes		0.06–2

The effect of temperature on SOD can be represented by

$$S'_B = S'_{B,20}\theta^{T-20} \tag{25.2}$$

where $S'_{B,20}$ = areal SOD rate at 20°C and θ = temperature coefficient. Zison et al. (1978) have reported a range of 1.04 to 1.13 for θ. A value of 1.065 is commonly employed. In addition the SOD for temperatures below 10°C typically declines faster than indicated by Eq. 25.2. In the range from 0 to 5°C, it approaches zero.

BOX 25.1. ZOD or "Zebra Mussel Oxygen Demand"

You may have noticed that along with typical environmental settings, Table 25.1 contains the organism *Sphaerolitus*. These organisms are attached, filamentous higher bacteria that are commonly and erroneously called "sewage fungus." They typically grow in channels dominated by raw sewage, and as noted, their respiration results in a high oxygen demand.

Other bottom organisms can have a similar effect. Among the most notorious are the freshwater mollusks commonly called zebra mussels (*Dreissena polymorpha*). Originally native to the Caspian Sea/Black Sea region, they were probably introduced into the Great Lakes in the late 1980s when a ship from Europe released its ballast water into Lake St. Clair (Ludyanskiy et al. 1993, Mackie 1991). This freshwater mollusk attaches to lake and river bottoms, channels, and pipe walls. To date, the biggest problem connected with the organism is clogging at drinking and cooling water intake pipes. However, recent evidence suggests that they can also have a significant effect on water quality.

One such case occurs in the Seneca River, N.Y., a tributary of Lake Ontario. In 1993 the Upstate Freshwater Institute (U.F.I.) in Syracuse, N.Y., conducted a study of a 16-km section of the river that was infested with the mussels. Steve Effler, Clifford Siegfried, and Ray Canale measured (Effler and Siegfried 1994) and modeled (Canale and Effler 1995) the section, and their results document the impact of the organism.

They measured an average mussel density over the stretch of 6000 individuals m^{-2}. They then independently measured an organism respiration rate of 0.9 mgO individual^{-1} d^{-1}. The product of these numbers yielded an oxygen demand, or as dubbed by Prof. Canale, a "ZOD," of 5.4 gO m^{-2} d^{-1}. Together with other measured parameters for the section, they used a simple oxygen model of the form of Eq. 22.12 to determine that oxygen levels should drop to about 3.9 mg L^{-1} at the end of the stretch. As depicted in Fig. B25.1, this prediction conforms nicely to measurements. Beyond that the model predicts that 80% of the oxygen demand is due to zebra mussel respiration.

The zebra mussels also have a dramatic effect on the nutrients in the system. The mussels in the reach filter water at a rate of 60 m^3 s^{-1}. This organism "flow rate" is 2 times the flow rate in the river! As they filter, they take in and consume floating phytoplankton. As a consequence, chlorophyll *a* levels in the stretch drop from 45 to 5 μg L^{-1}. At the same time, the mussels excrete and egest soluble phosphorus and ammonium, leading to increases of 48 μgP L^{-1} and 0.4 mgN L^{-1}, respectively. Although the phosphorus increase is consistent with model results, the ammonium observations are less than predicted. However, this discrepancy can be attributed to nitrification.

In summary, aside from its other nuisance impacts, the zebra mussel can also directly impact water chemistry and quality through their metabolism and processing of

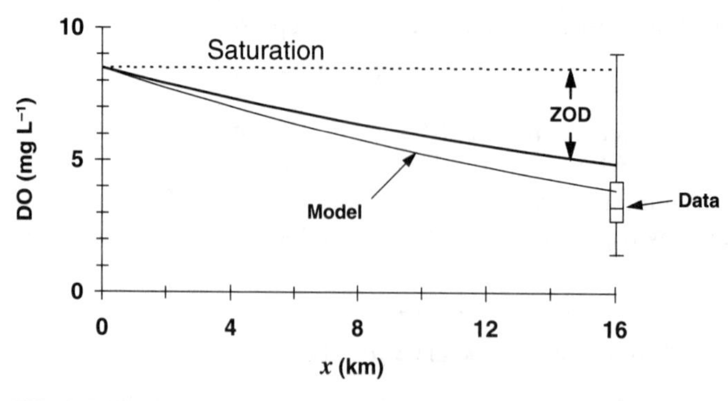

FIGURE B25.1

organic matter. The more general theme of such biological impacts will be elaborated on in great detail when we describe eutrophication in the next part of this book.

Aside from temperature, the two other factors that affect SOD are the organic content of the sediments and the oxygen concentration of the overlying waters. Baity (1938) and Fair et al. (1941) provided the first evidence of how the organic content of the sediments affected SOD. Their work suggests a square-root relationship between the SOD and the sediment volatile solids. A similar effect is provided by Fig. 25.2*a*, which correlates SOD with sediment COD.

AHOD also seems to be related to lake total phosphorus concentration by a square-root dependency (Fig. 25.2*b*). Because lakes with high phosphorus tend to have high productivity, such systems would be expected to have elevated sediment organic-carbon content. Thus Fig. 25.2*b* seems to imply the same square-root relationship between SOD and sediment carbon. In summary all the data displayed in Fig. 25.2 lead to the general conclusion that there is less SOD per organic carbon as the sediments become more enriched.

Oxygen is the other factor that affects SOD. Clearly if water oxygen concentration goes to zero the SOD will cease. Conversely above a certain level, it is usually assumed that the SOD is independent of the oxygen concentration in the overlying

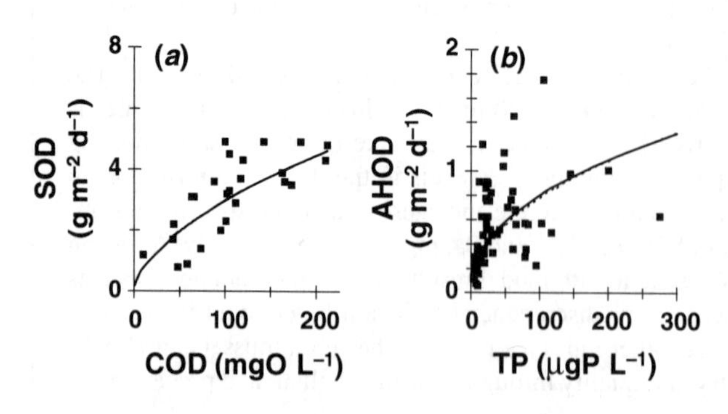

FIGURE 25.2
Two sets of observations that indicate a square-root relationship between SOD and organic content of sediments. (*a*) SOD versus surface sediment COD (Gardiner et al. 1984); (*b*) AHOD versus total phosphorus concentration (Rast and Lee 1978, Chapra and Canale 1991).

waters. Baity (1938) concluded that this was the case for oxygen levels greater than 2 mg L^{-1}. Several investigators have tried to represent the dependence by a saturating relationship,

$$S'_B(o) = \frac{o}{k_{so} + o} S'_B \tag{25.3}$$

where $S'_B(o)$ = oxygen-dependent SOD
o = oxygen concentration of the overlying water (mg L^{-1})
k_{so} = half-saturation value for the dependence (mg L^{-1})

Lam et al. (1984) have suggested a value for k_{so} of 1.4 mg L^{-1}. Thomann and Mueller (1987) fit the data of Fillos and Molof (1972) with a value of k_{so} = 0.7 mg L^{-1}. This latter result corresponds to independence at oxygen concentrations greater than about 3 mg L^{-1}.

25.2 A "NAIVE" STREETER-PHELPS SOD MODEL

Initial efforts to develop a mechanistic representation of SOD took the logical step of simulating the sediment compartment in a similar fashion to the water column. The segmentation scheme for such a model is depicted in Fig. 25.3. Notice that the subscripts w and 2 will be used for water and sediments, respectively. In addition the water concentrations will not be simulated as state variables. Rather they will serve as boundary conditions for the sediments. Finally we will assume that the surface area of the water is equal to the surface area of the sediments and that the temperatures of the two segments are identical.

We can now develop sediment CBOD and NBOD mass balances using first-order kinetics. For CBOD the equation would be

$$V_2 \frac{dL_2}{dt} = k_s V_w L_w - k_{d2} V_2 L_2 \tag{25.4}$$

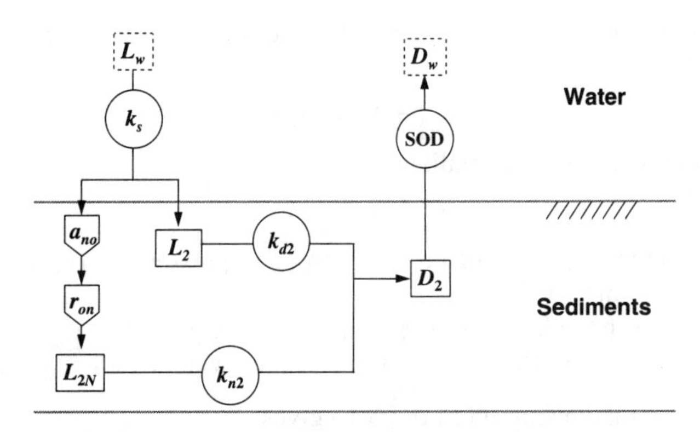

FIGURE 25.3
A "Streeter-Phelps" model for SOD. Note that the water concentrations of BOD and deficit are considered as boundary conditions, rather than state variables.

where k_{d2} = decomposition rate in the sediments (d^{-1}). At steady-state this balance can be solved for

$$L_2 = \frac{k_s H_w}{k_{d2} H_2} L_w \tag{25.5}$$

where H = layer thickness (m).

A similar balance for NBOD can be written as

$$V_2 \frac{dL_{2n}}{dt} = a_{no} r_{on} k_s V_w L_w - k_{n2} V_2 L_2 \tag{25.6}$$

where a_{no} = stoichiometric yield of nitrogen from the decomposition of settling BOD

r_{on} = oxygen demand to nitrogen ratio due to nitrification = 4.57 gBOD gN^{-1}

k_{n2} = nitrification rate in the sediments (d^{-1})

Note that, according to the stoichiometric relationships from Eq. 23.21, the nitrogen yield can be calculated as

$$a_{no} = \frac{16(14)}{107(32)} = 0.0654 \text{ gN gBOD}^{-1} \tag{25.7}$$

At steady-state, Eq. 25.6 can be solved for

$$L_{2n} = 0.3 \frac{k_s H_w}{k_{n2} H_2} L_w \tag{25.8}$$

where we have substituted and combined the values for a_{no} and r_{on}. Thus we see that according to our simple model, the settling particulate CBOD carries an additional 30% of oxygen demand due to nitrification.

Finally the oxygen deficit balance can be written as

$$V_2 \frac{dD_2}{dt} = k_{d2} V_2 L_2 + k_{n2} V_2 L_{2n} - S'_B A_s \tag{25.9}$$

where A_s = area of the sediment water interface (m^2). At steady-state this balance can be solved for

$$S'_B = 1.3 k_s H_w L_w \tag{25.10}$$

It can further be simplified by realizing that

$$J_{C^*} = k_s H_w L_w = v_s L_{pw} \tag{25.11}$$

where J_{C^*} = downward flux of organic carbon expressed as oxygen equivalents ($gO\ m^{-2}\ d^{-1}$)

v_s = settling velocity of particulate organics ($m\ d^{-1}$)

L_{pw} = particulate BOD concentration in the water ($mg\ L^{-1}$)

Substituting this relationship into Eq. 25.10 gives

$$S'_B = 1.3 J_{C^*} \tag{25.12}$$

Consequently the model yields the rather simple result that the steady-state SOD should be equal to about 130% of the downward flux of ultimate BOD. Table 25.2

TABLE 25.2
SOD in g m^{-2} d^{-1} calculated with the "naive" Streeter-Phelps SOD model

Particulate CBOD	Settling velocity (m d^{-1})		
(mg L^{-1})	0.1	0.2	0.5
0.5	0.065	0.13	0.325
1	0.13	0.26	0.65
5	0.65	1.3	3.25
10	1.3	2.6	6.5
50	6.5	13	32.5
100	13	26	65

shows a matrix that applies this result with a range of organic settling velocities and particulate BODs that are typically encountered in natural waters. Thus our simple framework would lead us to conclude that SODs in natural waters should range between about 0.05 to 65 g m^{-2} d^{-1}. The upper end of this range is well beyond the typical values in Table 25.1. This overprediction occurs because Eq. 25.12 specifies a linear relationship between SOD and the downward flux of organics. This contradicts the square-root relationships depicted in Fig. 25.2. Thus we can conclude that our "naive" Streeter-Phelps SOD model is not capable of capturing the diminished oxygen demand for highly productive systems. At best it provides an upper bound on the process. Now we turn to a more sophisticated framework that is capable of simulating the square-root dependency.

25.3 AEROBIC AND ANAEROBIC SEDIMENT DIAGENESIS

In a landmark paper, Di Toro et al. (1990) developed a model of the SOD process that mechanistically arrives at the same square-root relationship exhibited by Fig. 25.2. Before plunging into its mathematics, we would first like to provide a more realistic description of how organic carbon decomposition (or diagenesis) leads to SOD.

As depicted in Fig. 25.4 the sediments are divided into an aerobic (oxidizing) and an anaerobic (reducing) zone. Within these zones both organic carbon and nitrogen undergo transformations that ultimately create SOD. We will describe them separately.

Carbon. Particulate organic matter (POM) is delivered to the sediments by settling. Within the anaerobic sediments, the organic carbon decomposes to yield dissolved methane,

$$CH_2O \rightarrow \tfrac{1}{2}CO_2 + \tfrac{1}{2}CH_4 \tag{25.13}$$

where CH_2O is a simplified representation of organic matter. The methane diffuses upward to the aerobic zone where it is oxidized,

$$\tfrac{1}{2}CH_4 + O_2 \rightarrow \tfrac{1}{2}CO_2 + H_2O \tag{25.14}$$

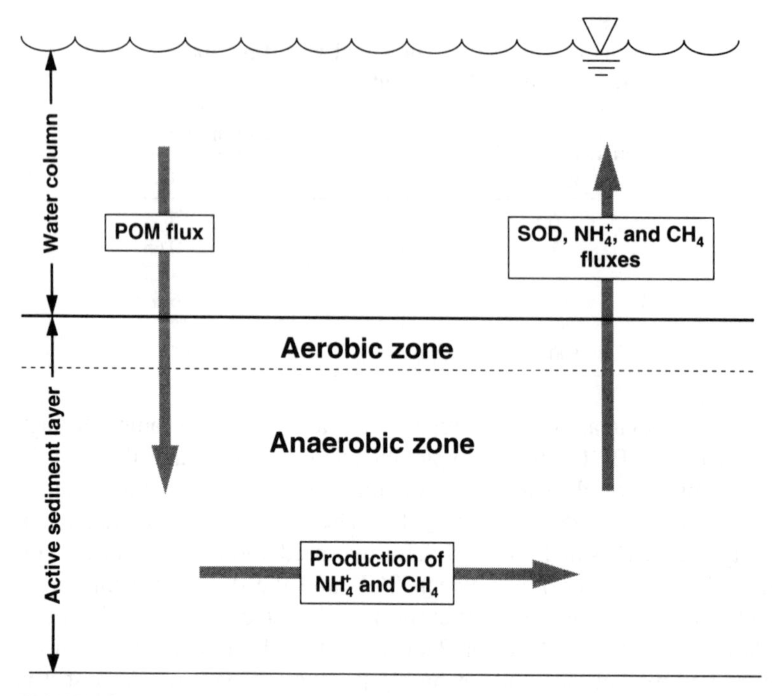

FIGURE 25.4
An overview of the mechanistic SOD framework (redrawn from Di Toro et al. 1990).

In the process an SOD is generated. Any residual methane that is not oxidized in the aerobic layer is diffused back into the water where additional oxidation can take place.

A problem with modeling this scheme is that we are dealing with several chemical species: CH_2O, CH_4, and O_2. Thus according to Eq. 25.13, 1 gC of organic matter would yield 0.5 gC of methane in the anaerobic sediments. Then, according to Eq. 25.14, the 0.5 gC of methane would consume 2.67 gO in the aerobic zone.

These conversions can be avoided by expressing all the species as ***oxygen equivalents***, that is, as the oxygen required for complete oxidation. Di Toro uses the nomenclature O* to distinguish oxygen equivalents from the actual oxygen concentration O_2. If this is done, 1 gO* of organic matter yields 1 gO* of methane that consumes 1 gO* of oxygen. Thus the need for stoichiometric conversions is unnecessary. It should be noted that the idea of oxygen equivalents should not be new to you. In fact we have already used them in the previous section when we used CBOD and NBOD to represent the oxygen equivalents of organic carbon and nitrogen.

Nitrogen. The two-zone scheme depicted in Fig. 25.4 also affects nitrogen. In the anaerobic zone, ammonium is generated via ammonification of organic N. Thus a gradient is created that feeds ammonium to the aerobic layer. In the aerobic zone, nitrification of the ammonium serves as an oxygen-demanding reaction that exerts an NSOD. Ammonium that is not converted to nitrate diffuses back into the water by diffusion.

The oxidation of ammonium in the aerobic layer can be represented by

$$NH_3 + 2O_2 \rightarrow HNO_3 + H_2O \tag{25.15}$$

This reaction consumes 4.57 gO gN^{-1}, which together with the nitrogen yield from organic matter decomposition (0.0654 gN gO^{-1}) leads to the nitrogenous oxygen demand calculated previously in Eq. 25.8.

Now, in contrast to the simple model described in the previous section, nitrification is not the last step in the process. In fact much of the nitrate is denitrified to form nitrogen gas. If it is assumed that the carbon source for denitrification is methane, this reaction can be represented as

$$\tfrac{5}{8}CH_4 + HNO_3 \rightarrow \tfrac{5}{8}CO_2 + \tfrac{1}{2}N_2 + \tfrac{7}{4}H_2O \tag{25.16}$$

The entire process can be compressed into a single reaction by combining Eqs. 25.14 to 25.16 to give

$$NH_3 + \tfrac{3}{4}O_2 \rightarrow \tfrac{1}{2}N_2 + \tfrac{6}{4}H_2O \tag{25.17}$$

Thus the oxidation of the nitrogenous matter yields

$$r'_{on} = \frac{0.75(32)}{14} = 1.714 \text{ gO gN}^{-1} \tag{25.18}$$

where r'_{on} = oxygen demand for nitrification, corrected for denitrification.

This is an extremely important result since it indicates that the oxygen consumption would be much lower than the 4.57 gO gN^{-1} usually expected from the nitrification reaction. This reduction occurs because methane is consumed in the denitrification process. Consequently Eq. 25.12 should be adjusted to

$$S'_B = 1.11 J_{C^*} \tag{25.19}$$

Thus, rather than the 30% additional SOD due to nitrification (Eq. 25.12), the inclusion of denitrification lowers the increase to about 11%.

In summary the foregoing description does not lead to serious modifications to the simple Streeter-Phelps approaches described in the previous section. However, it does suggest that denitrification mitigates the NSOD effect. Next we see that modeling the vertical distribution of reactants in the sediments along with gas formation will have a much more significant impact on our calculations.

25.4 SOD MODELING (ANALYTICAL)

Di Toro and his colleagues used the framework described in the previous section to explain the square-root relationship of SOD to sediment organic carbon content. Their theory is depicted in Fig. 25.5. In essence when the organic carbon content of the sediment gets high, the amount of methane produced in the anaerobic sediments can exceed its solubility and bubbles form. These bubbles advect upward due to their buoyancy and consequently represent a loss of organic carbon that does not

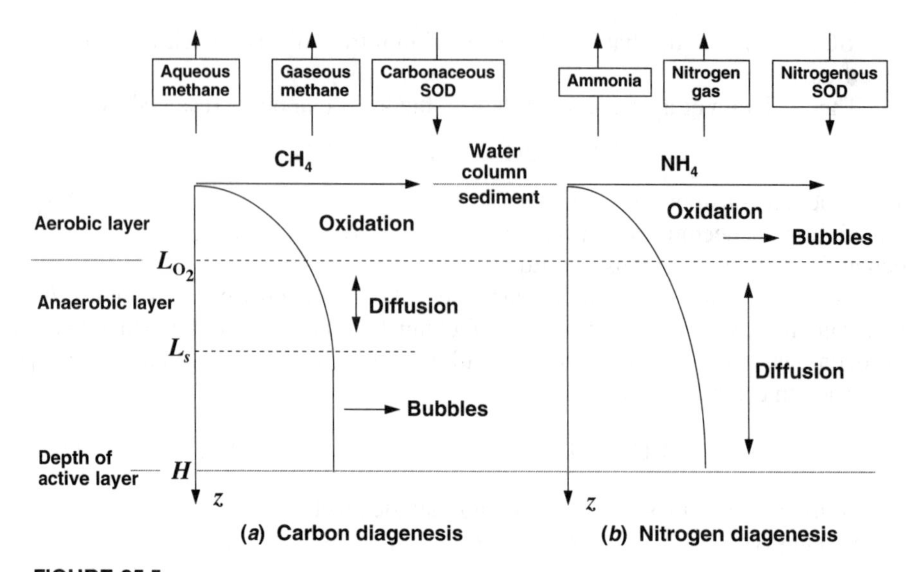

FIGURE 25.5
A schematic diagram of Di Toro's SOD model (redrawn from Di Toro et al. 1990).

exert a sediment oxygen demand. This loss is one component of the observed square-root relationship between SOD and sediment organic carbon content.

The following sections are intended to describe Di Toro's approach. Because the framework is somewhat complicated, I describe it in several parts. It is hoped that this will help you to understand it more thoroughly and, in so doing, appreciate its elegance.

25.4.1 A Simple Methane Balance

Rather than presenting the complete framework, we can develop a simplified version to gain insight into Di Toro's approach. To do this we initially concentrate on carbon. Then in a later section we broaden the framework to include nitrogen.

The active sediments can be idealized as a one-dimensional vertical system with a thickness H (Fig. 25.6). A mass balance for dissolved methane in the sediment

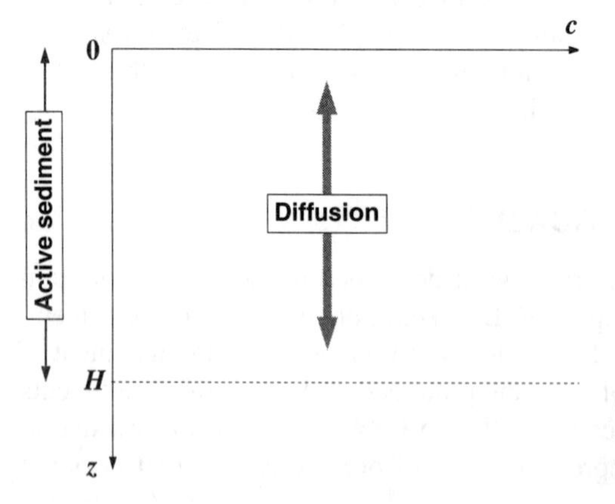

FIGURE 25.6
One-dimensional scheme for characterizing the vertical distribution of constituents in an active sediment layer.

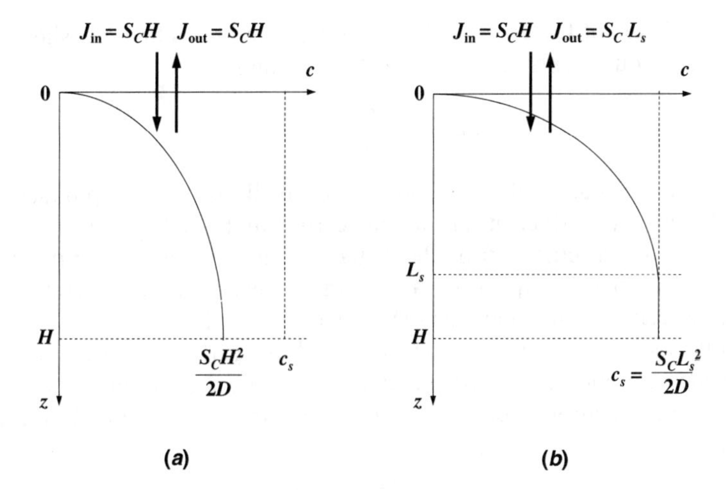

FIGURE 25.7
Vertical profile when methane in the pore waters (*a*) does not exceed and (*b*) does exceed saturation.

interstitial waters can be written as

$$0 = D\frac{d^2c}{dz^2} + S_C \tag{25.20}$$

where D = diffusion coefficient of methane in the sediment interstitial waters ($m^2\,d^{-1}$)
c = concentration of methane as measured in oxygen equivalents ($mgO\,L^{-1}$)
z = depth into the sediments (m)
S_C = a constant source of methane ($mgO\,L^{-1}\,d^{-1}$)

We can integrate this equation for two cases. In the first case we assume that the methane never exceeds saturation. Thus all methane is in dissolved form and no bubbles form. For this case the following boundary conditions can be assumed:

$$c(0) = 0 \tag{25.21}$$

$$\left.\frac{dc}{dz}\right|_{z=H} = 0 \tag{25.22}$$

The first condition specifies that the methane concentration of the overlying water is zero. The second stipulates that transport across the bottom of the active layer is zero.

Using these conditions, we can solve Eq. 25.20 for

$$c = \frac{S_C H z}{D} - \frac{S_C z^2}{2D} \tag{25.23}$$

As displayed in Fig. 25.7*a*, the solution is a parabola that is zero at the surface and increases to a maximum concentration at the bottom of the active sediments,

$$c(H) = \frac{S_C H^2}{2D} \tag{25.24}$$

Differentiation of Eq. 25.23 and substitution into Fick's first law shows that the flux of oxygen equivalents back to the water column is

$$J_{\text{out}} = D\frac{dc}{dz} = S_C H \tag{25.25}$$

As would be expected, the flux out corresponds to all the methane production in the active sediment layer. Since all the organic carbon in the sediment is assumed to be due to settling of particulates from the water column, this means that the flux out is equal to the flux in. Consequently if the methane never exceeds saturation, the linear model described in the previous section would be valid.

Next we look at a second case where the organic carbon levels are high enough that methane saturation is exceeded and, hence, bubbles form. For this case, along with the surface condition (Eq. 25.21), the following bottom boundary conditions hold:

$$c(L_s) = c_s \tag{25.26}$$

$$\left.\frac{dc}{dz}\right|_{z=L_s} = 0 \tag{25.27}$$

The first condition specifies that there is a depth L_s at which the methane concentration reaches saturation, c_s. The second condition stipulates that net transport across this depth is zero.

Employing these conditions, we can solve Eq. 25.20 for

$$c = c_s - \frac{S_C}{2D}(L_s - z)^2 \qquad z \le L_s \tag{25.28a}$$

$$c = c_s \qquad z > L_s \tag{25.28b}$$

As displayed in Fig. 25.7*b*, the solution is again parabolic. However, now it reaches saturation at a level L_s that is above the bottom of the active layer. This depth can be determined by substituting $c = 0$ and $z = 0$ into Eq. 25.28*a*, which can be solved for

$$L_s = \sqrt{\frac{2Dc_s}{S_c}} \tag{25.29}$$

In addition the flux at the sediment-water interface can be determined as

$$J_{\text{out}} = S_C L_s \tag{25.30}$$

Because $L_s < H$, the SOD for this case would be less than for Eq. 25.25 because some of the organic carbon is lost as methane bubbles. The gas flux loss can also be quantified as

$$J_{\text{gas}} = S_C(H - L_s) \tag{25.31}$$

Finally more useful forms of the flux equations can be developed by substituting Eq. 25.29 into Eqs. 25.30 and 25.31 to yield

$$J_{\text{out}} = \sqrt{2Dc_s S_C} \tag{25.32}$$

and

$$J_{\text{gas}} = S_C H - \sqrt{2Dc_s S_C} \tag{25.33}$$

Two further modifications can be made by recognizing that the rate of methane production can be related to the incoming carbon flux by

$$S_C = \frac{J_{C^*}}{H} \tag{25.34}$$

(where both S_C and J_{C^*} are measured in oxygen equivalents), and the diffusion coefficient can be expressed as a mass-transfer coefficient,

$$\kappa_D = \frac{D}{H} \tag{25.35}$$

If we assume that all the methane is completely oxidized in a small aerobic layer at the sediment-water interface, we can then express the maximum carbonaceous SOD as

$$\text{CSOD}_{\text{max}} = \sqrt{2\kappa_D c_s J_{C^*}} \tag{25.36}$$

EXAMPLE 25.1. MAXIMUM CSOD. Use the model developed above to determine the downward flux of organic carbon that corresponds to the point at which methane becomes saturated for a 10-cm-thick active sediment. Then develop a plot of CSOD_{max} versus downward flux for the same sediment. Note that the diffusion mass-transfer coefficient for methane is 0.00139 m d^{-1}, and $c_s \cong 100$ mgO L^{-1}.

Solution: First, we can combine Eqs. 25.24 [with $c(H) = c_s$], 25.34, and 25.35 and solve for

$$J_{C^*} = 2\kappa_D c_s = 2(0.00139)(100) = 0.278 \text{ gO m}^{-2}\text{ d}^{-1}$$

Thus, for carbon fluxes lower than this value, the linear SOD model holds. The square-root relationship of Eq. 25.36 can be used to determine the CSOD for higher values. The results are displayed below.

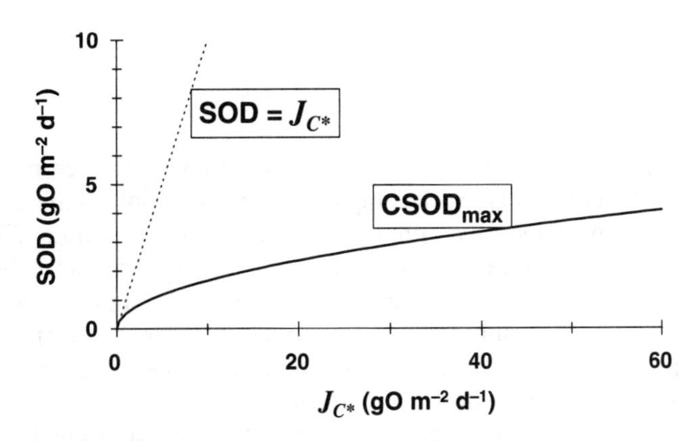

FIGURE E25.1

The foregoing analysis is very important because it provides an explanation for the square-root dependence of SOD on organic carbon. That is, it indicates that the

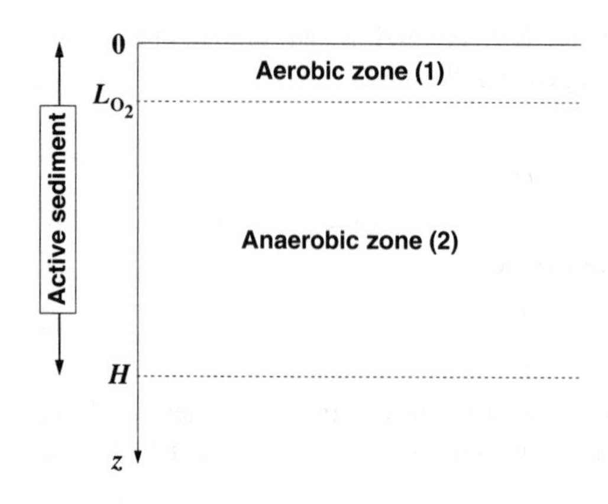

FIGURE 25.8
Segmentation scheme for an active sediment layer composed of aerobic and anaerobic zones.

square-root relationship is caused by carbon losses due to methane bubble formation at high organic carbon levels.

25.4.2 Oxidation in the Aerobic Zone

In the previous section we assumed that all the dissolved methane would be completely oxidized in a thin aerobic layer at the sediment surface. Depending on the relative magnitude of the oxidation reaction and diffusive transport, this might not be true. For example if the oxidation reaction were much slower than the transport, much of the methane would merely pass through the aerobic layer. If such were the case the resulting SOD would be less than computed by Eq. 25.36.

To quantify this effect, methane oxidation can be added to our model. To do this we can write mass balances for each zone in Fig. 25.8,

$$0 = D\frac{d^2c_1}{dz^2} - k_C c_1 \tag{25.37}$$

and

$$0 = D\frac{d^2c_2}{dz^2} + S_C \tag{25.38}$$

where c_1 and c_2 = dissolved methane concentrations in the aerobic and anaerobic zones, respectively, and k_C = decomposition rate of dissolved methane in the aerobic layer. Note that we have omitted the source term (S_C) from Eq. 25.37. Although this assumption is not necessary to obtain a solution, it greatly simplifies the analysis. The omission of the source term can be justified by the fact that, as we will discover, the thickness of the aerobic layer will be much smaller than that of the anaerobic layer. Consequently the methane production in this layer would be dwarfed by the magnitude of the methane flux from the anaerobic zone.

These differential equations are subject to four boundary conditions:

$$c_1(0) = 0 \quad \text{zero methane at the sediment-water interface} \tag{25.39}$$

$$c_1(L_{O_2}) = c_2(L_{O_2}) \quad \text{concentration continuity at aerobic-anaerobic interface} \tag{25.40}$$

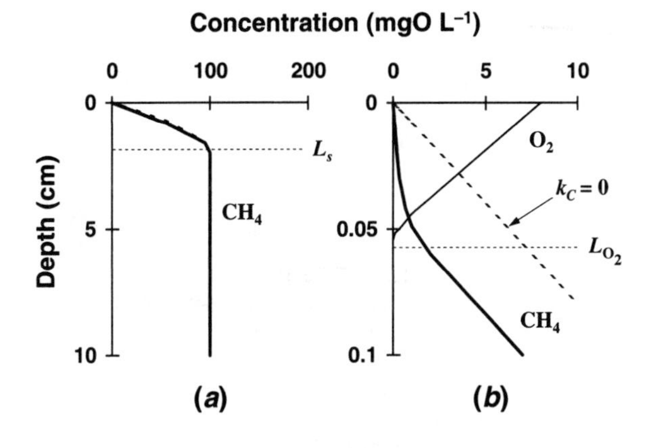

FIGURE 25.9 Profiles of (*a*) methane and (*b*) methane and oxygen for Di Toro's SOD model. The dashed line shows the limiting case where there is no oxidation in the aerobic zone (redrawn from Di Toro et al. 1991).

$$-D\left.\frac{dc_1}{dz}\right|_{z=L_{O2}} = -D\left.\frac{dc_2}{dz}\right|_{z=L_{O2}}$$ flux continuity at anaerobic-aerobic interface (25.41)

$$c_2(L_s) = c_s$$ saturation concentration at saturation depth (25.42)

A numerical solution for the active sediment is depicted in Fig. 25.9*a*. This graph shows two cases—with (solid) and without (dashed) oxidation—indicating that the inclusion of oxidation has negligible effect on the profile. However, a blow-up of the aerobic layer in Fig. 25.9*b* shows that oxidation has a major impact within the aerobic layer. Further, the oxygen profile for the aerobic layer, shown in Fig. 25.9*b*, indicates that (1) the aerobic layer is much smaller (on the order of 1 mm) than either the active layer (10 cm) or the undersaturated layer (2 cm), and (2) the oxygen profile is approximately linear.

Although the profiles shown in Fig. 25.9 are instructive, we are presently more interested in the impact of oxidation on the SOD. To quantify this impact, Di Toro et al. (1990) have also developed an analytical solution for the same system. Their solution can be differentiated and substituted into Fick's first law to compute a formula for the flux of dissolved methane at the sediment-water interface,

$$J_{\text{out}} = \sqrt{2\kappa_D c_s J_{C^*}}\,\text{sech}(\lambda_C L_{O_2}) \tag{25.43}$$

where

$$\lambda_C = \sqrt{\frac{k_C}{D}} \tag{25.44}$$

and the hyperbolic secant is computed as

$$\text{sech}(x) = \frac{2}{e^x + e^{-x}} \tag{25.45}$$

In addition they computed the flux of methane gas,

$$J_{\text{gas}} = J_{C^*} - \sqrt{2\kappa_D c_s J_{C^*}} \tag{25.46}$$

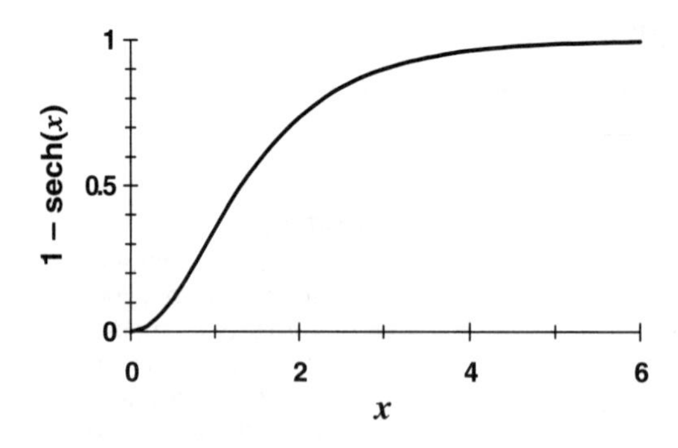

FIGURE 25.10
A plot showing the behavior of 1 − sech(x) versus x.

and the CSOD,

$$\text{CSOD} = \sqrt{2\kappa_D c_s J_{C^*}}[1 - \text{sech}(\lambda_C L_{O_2})] \tag{25.47}$$

Now inspection of Eq. 25.47 indicates that the incorporation of oxidation modifies the former solution (Eq. 25.36) by the term 1 − sech(x). As in Fig. 25.10, this term starts at zero and approaches 1 as x becomes large. In our case Eq. 25.47 will approach Eq. 25.36 when the product of λ_C and L_{O_2} becomes large. The former makes sense because λ_C is directly proportional to k_C and inversely proportional to D. Thus we would expect Eq. 25.47 to approach maximum levels as the oxidation rate increases or the diffusion coefficient decreases. Similarly the latter is reasonable because more oxidation would take place in a larger aerobic zone.

25.4.3 CSOD

The final step in the analysis involves the determination of the thickness of the aerobic layer, L_{O_2}. This can be accomplished by recognizing that the total SOD is the gradient of the pore-water oxygen concentration at the sediment-water interface. Using Fick's first law,

$$\text{SOD} = -D_o \left.\frac{do}{dz}\right|_{z=0} \tag{25.48}$$

where D_o = diffusion coefficient for oxygen in water. The derivative can be approximated by a first-order finite-divided difference (that is, by a straight line),

$$\left.\frac{do}{dz}\right|_{z=0} \cong \frac{o(0) - o(L_{O_2})}{0 - L_{O_2}} = -\frac{o(0)}{L_{O_2}} \tag{25.49}$$

Equation 25.49 can be substituted into Eq. 25.48, which can then be solved for

$$L_{O_2} = D_o \frac{o(0)}{\text{SOD}} \tag{25.50}$$

This result, along with Eq. 25.44, can be used to reformulate the argument of the hyperbolic secant term in Eq. 25.47 as,

$$\lambda_C L_{O_2} = \kappa_C \frac{o(0)}{\text{SOD}} \tag{25.51}$$

where

$$\kappa_C = \sqrt{k_C \frac{D_o^2}{D_c}} \tag{25.52}$$

where we have used the subscript c to distinguish the methane diffusion coefficient (D_c) from that for oxygen (D_o).

Finally Eq. 25.51 can be substituted into Eq. 25.47 to yield the final model of CSOD,

$$\text{CSOD} = \sqrt{2\kappa_D c_s J_{C^*}} \left\{ 1 - \text{sech}\left[\kappa_C \frac{o(0)}{\text{CSOD}} \right] \right\} \tag{25.53}$$

Note that for low carbon flux ($J_{C^*} < 2\kappa_D c_s$), the square root term is replaced by J_{C^*},

$$\text{CSOD} = J_{C^*} \left\{ 1 - \text{sech}\left[\kappa_C \frac{o(0)}{\text{CSOD}} \right] \right\} \tag{25.54}$$

Because CSOD is on both sides of these equations, this equation must be solved numerically as a roots problem,

$$f(\text{CSOD}) = J_{C^*} \left\{ 1 - \text{sech}\left[\kappa_C \frac{o(0)}{\text{CSOD}} \right] \right\} - \text{CSOD} = 0 \tag{25.55}$$

This numerical approach is described in the following example.

EXAMPLE 25.2. CSOD. Develop a plot of CSOD versus downward carbon flux for the same sediment as in Example 25.1. Also, compute the thickness of the aerobic layer for this case. Assume that the oxygen concentration in the overlying water is $o(0) = 4 \text{ mg L}^{-1}$ and that the reaction coefficient is $\kappa_C = 0.575 \text{ m d}^{-1}$. The diffusion coefficient for oxygen is approximately $2.1 \times 10^{-5} \text{ cm}^2 \text{ s}^{-1}$.

Solution: As an example suppose that the downward flux is $10 \text{ gO m}^{-2} \text{ d}^{-1}$. This value along with the other model parameters can be substituted into Eq. 25.55 to give

$$f(\text{CSOD}) = 10\left[1 - \text{sech}\left(0.575 \frac{4}{\text{CSOD}}\right)\right] - \text{CSOD}$$

The root of this equation can be determined in a number of ways. In this example we use the modified secant method,

$$\text{CSOD} = \text{CSOD} - \frac{\epsilon \text{CSOD} f(\text{CSOD})}{f(\text{CSOD} + \epsilon \text{CSOD}) - f(\text{CSOD})}$$

where ϵ = a small perturbation fraction. An excellent initial guess is provided by Eq. 25.36. For the present case this is

$$\text{CSOD}_{\text{max}} = \sqrt{2(0.00139)100(10)} = 1.6673$$

This value can be substituted into the formula for the modified secant method. If we use a value of $\epsilon = 0.1$, the resulting iterations are

Iteration	CSOD	Approximate error (%)
0	1.6673	
1	1.1557	44.27
2	1.1928	3.1
3	1.1926	0.015

The thickness of the aerobic layer can be calculated as

$$L_{O_2} = 2.1 \times 10^{-5}\ \text{cm}^2\ \text{s}^{-1}\left(\frac{4\ \text{g m}^{-3}}{1.1926\ \text{g m}^{-2}\ \text{d}^{-1}}\frac{\text{m}}{100\ \text{cm}}\frac{10\ \text{mm}}{\text{cm}}\frac{86{,}400\ \text{s}}{\text{d}}\right) = 0.608\ \text{mm}$$

Results for other levels of downward flux are displayed below along with the maximum CSOD values determined previously in Example 25.1. Notice how the inclusion of oxidation reduces the CSOD because some dissolved methane diffuses to the overlying waters before it can be oxidized.

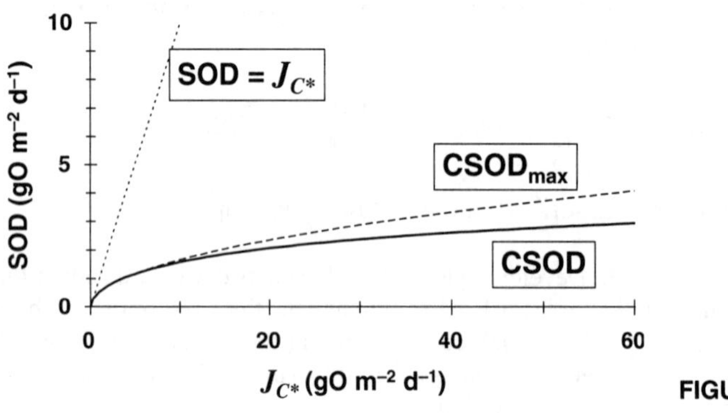

FIGURE E25.2

25.4.4 Nitrogen and Total SOD

Using a similar analysis as applied to carbon, Di Toro et al. (1990) also evaluated the effect of nitrification on SOD. As depicted in Fig. 25.5*b* nitrogen differs from carbon in that the ammonium formed from organic nitrogen does not become saturated and form bubbles as was the case for the methane from organic carbon.

This difference is accounted for by the following equation for the total SOD due to both carbon and nitrogen,

$$\text{SOD} = \underbrace{\sqrt{2\kappa_D c_s J_{C^*}}\left\{1 - \text{sech}\left[\kappa_C \frac{o(0)}{\text{SOD}}\right]\right\}}_{\text{CSOD}} + \underbrace{r'_{on} a_{no} J_{C^*}\left\{1 - \text{sech}\left[\kappa_N \frac{o(0)}{\text{SOD}}\right]\right\}}_{\text{NSOD}} \tag{25.56}$$

where

$$\kappa_N = \sqrt{k_N \frac{D_o^2}{D_n}} \tag{25.57}$$

where k_N = oxidation rate of ammonium to nitrogen gas and D_n = diffusion coefficient for ammonium ion in water. Note that again for low carbon flux ($J_{C^*} < 2\kappa_D c_s$), the square-root term is replaced by J_{C^*},

$$\text{SOD} = J_{C^*}\left\{1 - \text{sech}\left[\kappa_C \frac{o(0)}{\text{SOD}}\right]\right\} + r'_{on} a_{no} J_{C^*}\left\{1 - \text{sech}\left[\kappa_N \frac{o(0)}{\text{SOD}}\right]\right\} \tag{25.58}$$

Inspection of Eq. 25.56 shows that the NSOD is mathematically similar to the CSOD. The differences relate to the conversion of the carbon flux into nitrogenous oxygen demand ($r'_{on} a_{no}$) and the absence of the square-root term because of the lack of gas formation in the anaerobic layer. However, as with CSOD, the hyperbolic secant term accounts for both the diffusion/reaction competition in the aerobic zone and the shutdown of SOD at low oxygen levels. As was the case for Eqs. 25.53 and 25.54, because SOD appears on both sides of Eqs. 25.56 and 25.58, they must be solved numerically as a roots problem.

EXAMPLE 25.3. TOTAL SOD. Develop a plot of total SOD versus downward carbon flux for the same sediment as was studied in Examples 25.1 and 25.2. Assume that the parameters have the following values: $\kappa_N = 0.897$ m d^{-1}, $r'_{on} = 1.714$ gO gN^{-1}, and $a_{no} = 0.0654$ gN gO^{-1}.

Solution: As with Example 25.2, suppose that the downward flux is 10 gO m^{-2} d^{-1}. This value along with the other model parameters can be substituted into Eq. 25.56 to give

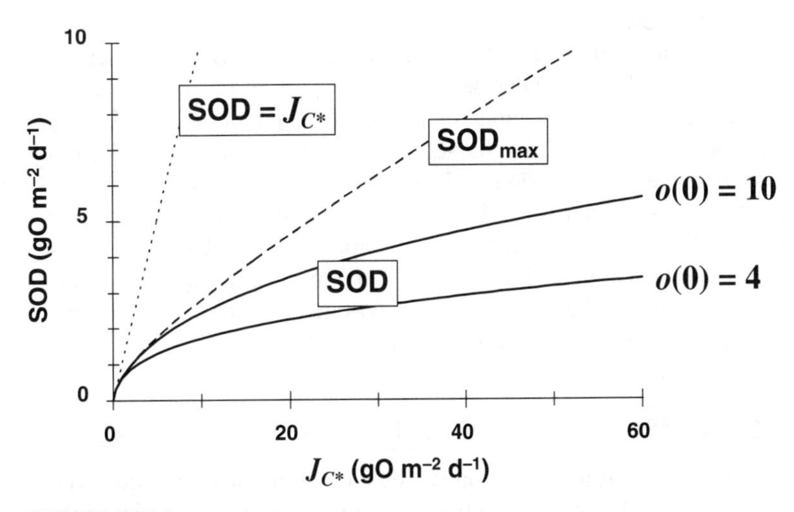

FIGURE E25.3

$$f(\text{SOD}) = \sqrt{2(0.00139)100(10)}\left[1 - \text{sech}\left(0.575\frac{4}{\text{SOD}}\right)\right]$$
$$+ 1.714(0.0654)(10)\left[1 - \text{sech}\left(0.897\frac{4}{\text{SOD}}\right)\right] - \text{SOD}$$

The root of this equation can be determined with the modified secant method, with an initial guess provided by

$$\text{CSOD}_{\text{max}} = \sqrt{2\kappa_D c_s J_{C^*}} + r'_{on} a_{no} J_{C^*}$$

This value can be substituted into the formula for the modified secant method. The results along with SODs for other flux rates are displayed in Fig. E25.3. In addition the plot shows results when the oxygen in the water is raised to 10 mg L^{-1}. As expected this leads to higher SODs.

Before proceeding, let's regroup and consider what the analytical model tells us about SOD:

1. If there is oxygen in the overlying water, oxidation of methane and ammonium takes place in a very thin aerobic layer at the sediment-water interface.
2. In the thin aerobic layer a competition exists between the oxidation reaction and diffusion. The 1 − sech terms in Eq. 25.58 account for this competition as well as the suppression of SOD when oxygen in the overlying water gets low.
3. Methane can exceed its saturation concentration in the anaerobic sediments. Consequently bubbles can be lost from the system, as reflected by the square-root term in Eq. 25.58. Because ammonium does not form a gaseous phase at the pHs encountered in sediments, the relationship of NSOD to downward oxygen-equivalent flux in Eq. 25.58 is linear.

By providing a means to calculate SOD as a function of external loadings, the Di Toro framework is a major advance over prior schemes. Predictions should be more realistic than those obtained with the two alternatives used by earlier modelers: fixed SOD or a linear decrease in proportion to load reductions. The former is overly conservative whereas the latter is overly optimistic. By allowing a more physically realistic middle course, the Di Toro approach should provide a basis for better management decisions. Most importantly it allows SOD to be computed internally in a mechanistic fashion rather than in a prescribed or empirical manner. To date Di Toro's framework has been applied to the Chesapeake Bay (Cerco and Cole 1993, Di Toro and Fitzpatrick 1993). In the coming years numerical versions of the approach will undoubtedly be incorporated into most water-quality models to provide a more rigorous characterization of SOD.

25.5 NUMERICAL SOD MODEL

Although the analytical solution is elegant and instructive, numerical solution techniques provide greater flexibility and broader application. The following models are similar in spirit to the approach outlined by Di Toro and Fitzpatrick (1993) for estuarine systems. However, in this section we limit our discussion to the freshwater case. At the end of the discussion we will describe how estuarine systems differ from freshwater environments.

In this section we first describe a simplified SOD model that focuses on ammonium creation and oxidation. Later we will incorporate carbon and other elements into the framework.

As in Fig. 25.11, we can represent the active sediments as two well-mixed layers. Notice that in contrast to the analytical model, we are representing the downward flux of organic matter in carbon units (rather than oxygen equivalents).

A mass balance can be written for organic carbon in the lower sediment layer as

$$V_2 \frac{dc_2}{dt} = J_C A_s - k_{c2} V_2 c_2 \tag{25.59}$$

where c_2 = concentration of organic carbon (gC m^{-3})
J_C = flux of organic carbon settling from the overlying water (gC m^{-2} d^{-1})
A_s = surface area of the sediment-water interface (m^2)
k_{c2} = first-order diagenesis rate of organic carbon (d^{-1})

Now because the surface area between all the model segments is equal to A_s and $V_2 = A_s H_2$, we can divide both sides of the balance by A_s to yield

$$H_2 \frac{dc_2}{dt} = J_C - k_{c2} H_2 c_2 \tag{25.60}$$

Thus we have expressed the mass balance in terms of flux (M L^{-2} T^{-1}) rather than rate of mass transfer (M T^{-1}). Similar flux balances can be written for ammonium as

$$H_2 \frac{dn_2}{dt} = a_{nc} k_{c2} H_2 c_2 + v_{dn12}(n_1 - n_2) \tag{25.61}$$

and

$$H_1 \frac{dn_1}{dt} = v_{dn12}(n_2 - n_1) - v_{dnw1} n_1 - k_{n1} H_1 n_1 \tag{25.62}$$

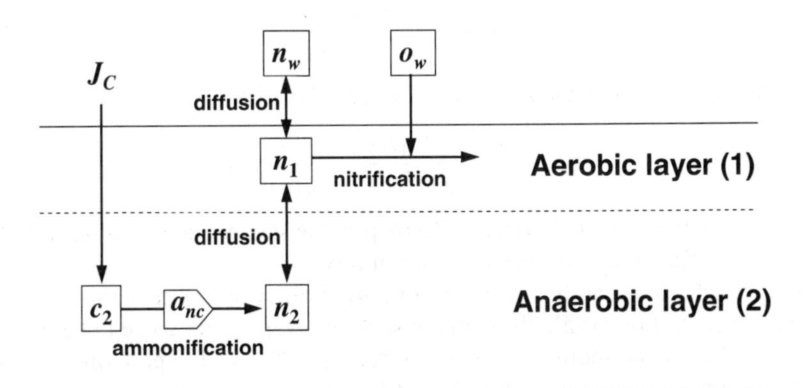

FIGURE 25.11
Schematic of a control-volume approach for modeling NSOD.

where n = ammonium concentration expressed as nitrogen (gN m^{-3})
v_{dn12} = diffusive mass-transfer coefficient between the two layers (m d^{-1})
v_{dnw1} = diffusive mass-transfer coefficient between the surface layer and the water (m d^{-1})
k_{n1} = nitrification rate of ammonium in the surface layer (d^{-1})

Notice that Eq. 25.62 assumes that the ammonium concentration of the overlying water is negligible. Although this will not necessarily be true, we have adopted this assumption to be consistent with the analytical framework. Also recognize that the mass-transfer coefficients can be related to more fundamental parameters by

$$v_{dn12} = \frac{D_n}{H_{12}} \qquad v_{dnw1} = \frac{D_n}{H_{w1}} \tag{25.63}$$

where D_n = diffusion coefficient for ammonium in water (m^2 d^{-1}) and H_{12} and H_{w1} = mixing lengths for the two diffusive transfer processes (m).

As described later these balances can be solved numerically to simulate the concentrations as a function of time. Before doing this it is instructive to first solve them at steady-state. Along with insight into how the model functions, such solutions allow direct comparison between our lumped, numerical characterization and Di Toro's distributed, analytical solution.

After setting Eqs. 25.60 to 25.62 to steady-state, Eq. 25.60 can be solved for

$$c_2 = \frac{J_C}{k_{c2}H_2} \tag{25.64}$$

which in turn can be substituted into Eq. 25.61,

$$0 = a_{nc}J_C + v_{dn12}(n_1 - n_2) \tag{25.65}$$

Next Eqs. 25.65 and 25.62 can be added to eliminate n_2,

$$0 = a_{nc}J_C - k_{n1}H_1 n_1 - v_{dnw1}n_1 \tag{25.66}$$

This equation can then be solved for

$$n_1 = \frac{a_{nc}J_C}{k_{n1}H_1 + v_{dnw1}} \tag{25.67}$$

which can then be substituted back into Eq. 25.62 to give,

$$n_2 = \frac{v_{dn12} + k_{n1}H_1 + v_{dnw1}}{v_{dn12}(k_{n1}H_1 + v_{dnw1})} a_{nc}J_C \tag{25.68}$$

Therefore Eqs. 25.64, 25.67, and 25.68 provide steady-state solutions for all the variables as a function of the organic carbon flux.

Oxygen demand can be integrated into the framework by recognizing that the oxygen concentration profile through the surface layer is very close to being linear. Thus the SOD at the sediment-water interface can be nicely approximated by the following simple expression of Fick's first law (recall Eq. 25.50),

$$\text{SOD} = \frac{D_o}{H_1} o_w \tag{25.69}$$

where D_o = oxygen diffusion coefficient ($m^2\ d^{-1}$). Now it can be recognized that the SOD is also equivalent to the rate at which ammonium is being oxidized,

$$\text{SOD} = r'_{on} k_{n1} H_1 n_1 \tag{25.70}$$

where r'_{on} = oxygen demand for nitrification/denitrification (according to Eq. 25.18, 1.714 gO gN^{-1}). Assuming that $H_{w1} = H_1$, Eqs. 25.67 and 25.69 can be substituted into Eq. 25.70 and the result manipulated to give

$$\text{NSOD} = r'_{on} a_{nc} J_C \frac{1}{1 + \dfrac{D_n \text{NSOD}^2}{D_o^2 o_w^2 k_{n1}}} \tag{25.71}$$

where we have changed SOD to NSOD to acknowledge that we are presently limiting ourselves to ammonium oxidation. This equation can be solved numerically for NSOD. This result can then be used in conjunction with Eq. 25.69 to solve for the thickness of the aerobic layer.

Thus Eq. 25.71 is a simple finite-difference representation of the analytical approach which for NSOD is the second part of Eq. 25.56,

$$\text{NSOD} = r'_{on} a_{nc} J_C \left[1 - \text{sech}\left(\sqrt{k_{n1} \frac{D_o^2}{D_n}} \frac{o_w}{\text{NSOD}}\right)\right] \tag{25.72}$$

EXAMPLE 25.4. COMPARISON OF LUMPED AND DISTRIBUTED NSOD. Develop a plot of NSOD versus downward carbon flux for the same sediment as was studied in Examples 25.1 through 25.3 using both the distributed and lumped NSOD models. Assume that the parameters have the following values: o_w = 10 mg L^{-1}, κ_N = 0.897 m d^{-1}, r'_{on} = 1.714 gO gN^{-1}, $D_o = 1.73 \times 10^{-4}\ m^2\ d^{-1}$, $D_n = 8.47 \times 10^{-5}\ m^2\ d^{-1}$, and a_{nc} = 0.176 gN gC^{-1}.

Solution: First, we must determine the nitrification rate from Eq. 25.57,

$$k_{n1} = \kappa_N^2 \frac{D_n}{D_o^2} = 0.897^2 \frac{8.47 \times 10^{-5}}{(1.73 \times 10^{-4})^2} = 2282\ d^{-1}$$

Next the parameter values can be substituted into Eq. 25.72 to obtain the distributed solution. For example for the case where J_{C*} = 10 gO $m^{-2}\ d^{-1}$ (which corresponds to J_C = 3.75 gC $m^{-2}\ d^{-1}$),

$$\text{NSOD} = 1.714(0.176)3.75 \left\{1 - \text{sech}\left[\sqrt{2282 \frac{(1.73 \times 10^{-4})^2}{8.47 \times 10^{-5}}} \frac{10}{\text{NSOD}}\right]\right\}$$

which can be solved iteratively for NSOD = 1.130 gO $m^{-2}\ d^{-1}$.

In a similar fashion the lumped case can be set up for Eq. 25.71,

$$\text{NSOD} = 1.714(0.176)3.75 \frac{1}{1 + \dfrac{8.47 \times 10^{-5} \text{NSOD}^2}{(1.73 \times 10^{-4})^2 (10)^2 2282}}$$

which can be solved for 1.113 gO $m^{-2}\ d^{-1}$. Thus the approaches yield similar results. Additional solutions are displayed in the plot shown in Figure E25.4:

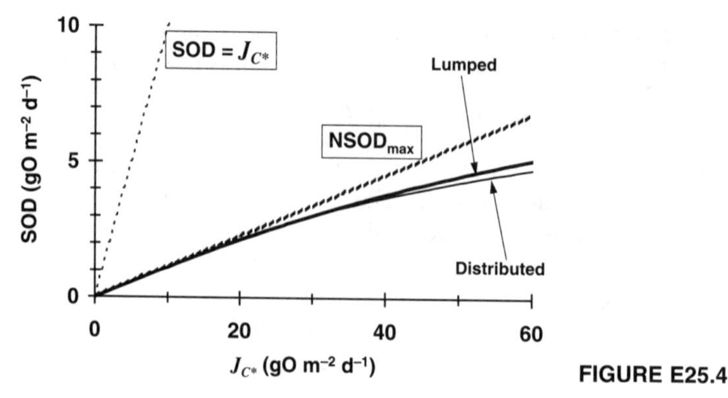

FIGURE E25.4

25.6 OTHER SOD MODELING ISSUES (ADVANCED TOPIC)

At the time of this book's publication, SOD modeling is an active and rapidly evolving research area. Consequently there are a number of relevant issues that are currently unresolved and that were not included in the prior sections. Because these topics are highly relevant to how SOD will be computed in the coming years, they are the focus of this section.

25.6.1 Methane Bubble Formation

Although the foregoing section shows how numerical, finite-difference methods can be developed, it did not address the issue of methane bubble formation in the anaerobic sediments. The simplest approach is merely to model methane in a fashion similar to ammonium, as in

$$H_2 \frac{dc_2}{dt} = J_C - k_{c2} H_2 c_2 \tag{25.73}$$

$$H_2 \frac{dm_2}{dt} = k_{c2} H_2 c_2 + v_{dm12}(m_1 - m_2) \tag{25.74}$$

$$H_1 \frac{dm_1}{dt} = v_{dm12}(m_2 - m_1) - v_{dmw1} m_1 - k_{m1} H_1 m_1 \tag{25.75}$$

Once the deep aerobic layer methane exceeds saturation, it can be kept fixed at the saturation level. This means that Eq. 25.75 becomes

$$H_1 \frac{dm_1}{dt} = v_{dm12}(c_s - m_1) - v_{dmw1} m_1 - k_{m1} H_1 m_1 \tag{25.76}$$

At this point this equation can be directly solved for the methane concentration and SOD in a fashion similar to the derivation of Eq. 25.71.

Although this approach may seem appealing, it has one major flaw. That is, once the saturation is exceeded, the CSOD becomes constant. Thus the square root

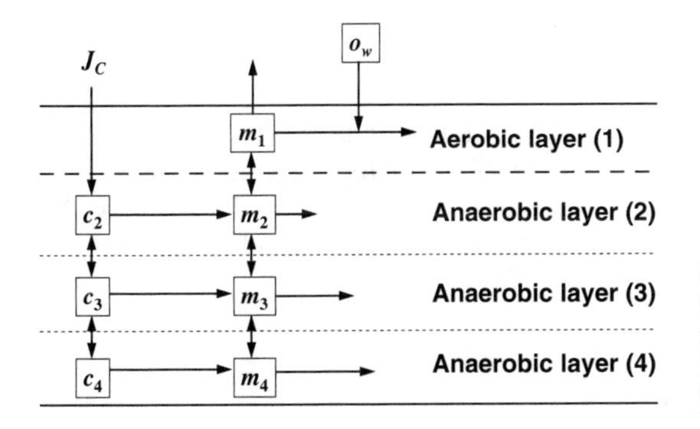

FIGURE 25.12
Schematic of a control-volume approach for modeling CSOD in freshwater systems with methane formation.

relationship manifested by the analytical solution does not occur. This problem is compounded because the saturation is usually exceeded at low organic carbon flux. Thus not only does the CSOD hit a ceiling, but it does so at an unrealistically low level.

A remedy for this problem is depicted in Fig. 25.12. By dividing the anaerobic zone into layers, the saturation phenomena will come into play in stages rather than at a single instant. In other words, the lowest layer will saturate first at which point its level becomes fixed. Thereafter, the middle and then the top layer will saturate. Once all layers saturate, the CSOD will be fixed. However, the sequential saturation up to that point will follow the analytical solution nicely. In addition it extends the validity of the approximation to higher levels than would be the case for the single layer.

25.6.2 Time-Variable SOD Calculations

To this point we have focused on steady-state computations. For most water-quality problems, time-variable solutions would be of equal, if not greater, importance. The most straightforward approach would be to merely integrate the equations numerically using either explicit or implicit numerical algorithms as described previously in Lecs. 7 and 11 through 13. For example Eqs. 25.60 to 25.62 could be integrated to determine time-variable NSOD.

Although this is a viable option, a problem arises in that the thin surface layer requires much finer temporal resolution (that is, a much smaller time step) to attain stable solutions than does the thicker anaerobic layer. In numerical methods this is referred to as a "stiff" solution.

This apparent dilemma leads to a beneficial insight. As recognized by Di Toro and Fitzpatrick (1993), because it is much thinner, the surface layer will always be at a steady-state relative to the time frame needed to resolve variations in the thicker anaerobic layer.

Consequently Eq. 25.62 can be solved at steady-state for

$$n_1 = R_{12} n_2 = \frac{v_{dn12}}{v_{dn12} + v_{dnw1} + k_{n1} H_1} n_2 \tag{25.77}$$

where R_{12} = ratio of the ammonium concentration in the aerobic (1) to the anaerobic layer (2). This value can then be substituted into Eq. 25.61 to give

$$H_2 \frac{dn_2}{dt} = a_{nc} k_{c2} H_2 c_2 + v_{dn12}(R_{12} - 1)n_2 \tag{25.78}$$

Thus instead of integrating three differential equations (one of which is less stable than the other two), we integrate the two more robust ODEs (Eqs. 25.60 and 25.78) and determine the third unknown, n_1, algebraically with Eq. 25.77. Note that this trick applies equally well to methane and CSOD as well as to other elements involved in sediment calculations.

25.6.3 Water Boundary Layer

In Example 25.4 a very high nitrification rate is needed so that the model as originally formulated yields realistic NSOD values. A similar conclusion can be also made for CSOD. In general, biologically mediated reactions such as nitrification and methane oxidation generally proceed at rates on the order of 0.01 to 5 d^{-1} rather that the magnitude of 1000 to 10,000 d^{-1} used in Example 25.4.

One reason for the high rates is that sediments have much higher bacterial biomass than the water. A second explanation lies in Di Toro's assumption that all the decomposition takes place in a thin aerobic sediment layer. Although this assumption was certainly valid from Di Toro's perspective and does not negate his conclusion, it must now be reexamined as we begin to integrate the mechanistic SOD model into large water-quality-modeling frameworks.

As depicted in Fig. 25.13, along with the sediment layers, a laminar boundary layer can exist in the water. The size of the two layers would be determined by different processes. The aerobic sediment layer thickness would be dictated by the balance between oxygen consumption and diffusion as expressed by Eqs. 25.69 and 25.70. In contrast the water boundary layer is an artifact of the turbulence in

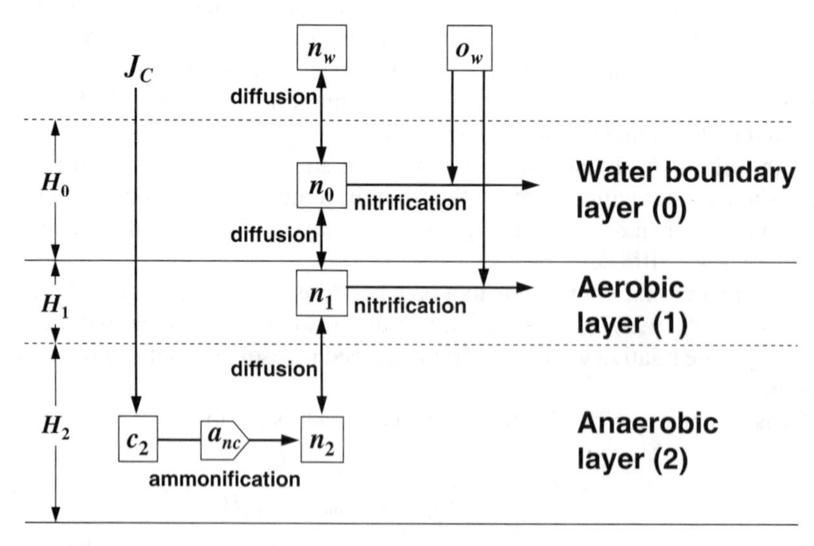

FIGURE 25.13
An NSOD model including a water laminar boundary layer.

the water. Thus for deeper waters with relatively low flows (for example, deeper lakes and estuaries), such a boundary layer can be significant. Consequently it could serve as a compartment where additional oxidation could take place.

The foregoing finite-difference models can be extended to account for such a layer. For example mass balances can be written for ammonium as

$$H_2 \frac{dn_2}{dt} = a_{nc} k_{c2} H_2 c_2 + v_{dn12}(n_1 - n_2) \tag{25.79}$$

$$H_1 \frac{dn_1}{dt} = v_{dn12}(n_2 - n_1) + v_{dn01}(n_0 - n_1) - k_{n1} H_1 n_1 \tag{25.80}$$

$$H_0 \frac{dn_0}{dt} = v_{dn01}(n_1 - n_0) + v_{dnw0}(n_w - n_0) - k_{n0} H_0 n_0 \tag{25.81}$$

Notice how we have now included the ammonium concentration in the overlying water, n_w.

At steady-state these equations can be solved simultaneously (along with Eq. 25.60) for the organic carbon and the ammonium concentrations. The values of ammonium for the upper two layers can then be substituted into the following relationship for SOD:

$$\text{SOD} = r'_{on}(k_{n1} H_1 n_1 + k_{n0} H_0 n_0) \tag{25.82}$$

This formula can be combined with Eq. 25.69 in order to develop an iterative method for determining the SOD.

Although the foregoing analysis provides an indication of the importance of the water boundary layer, it suffers in that the thickness of the boundary layer must be given. A superior approach would link the transfer in this layer to the flow velocity in the overlying water. Recent efforts are being made to develop such relationships (Nakamura and Stefan 1994).

25.6.4 Nitrate and Phosphate

Because of our present emphasis on oxygen, all the other compounds discussed to this point (ammonium, methane) directly impact SOD. However, beyond these compounds there are a variety of others that are important and are significantly affected by oxygen.

Nitrate. Nitrate is an important plant nutrient that is released during nitrification. To this point we have assumed that it is totally reduced to nitrogen gas by denitrification. Although this is a reasonable first assumption that works well for many systems, a more refined viewpoint would consider that only part of the generated nitrate flux would be denitrified and part released to the overlying water. Di Toro and Fitzpatrick (1993) show how this can be done by slightly modifying the SOD model described previously.

First, nitrate is modeled explicitly. For the two-layer model, balances would be written as

$$H_2 \frac{di_2}{dt} = v_{dn12}(i_1 - i_2) - k_{i2} H_2 i_2 \tag{25.83}$$

and
$$H_1 \frac{di_1}{dt} = v_{dn12}(i_2 - i_1) - v_{dnw1}(i_0 - i_1) + k_{n1} H_1 n_1 - k_{i1} H_1 i_1 \tag{25.84}$$

where i = nitrate concentration (mgN L^{-1}) and k_i = denitrification rate (d^{-1}). For the steady-state case these equations can be solved for i_1 and i_2. These values, in turn, can be used to compute the flux of nitrogen gas that would be created and lost from the sediments,

$$J_N = k_{i1} H_1 i_1 + k_{i2} H_2 i_2 \tag{25.85}$$

This value is then used to correct the carbon flux to acknowledge that methane is utilized during denitrification. The final model for conditions below methane saturation would look like

$$\text{SOD} = r_{oc}(J_C - a_{cn} J_N) \frac{1}{1 + \dfrac{D_c \text{SOD}^2}{D_o^2 o_w^2 k_{m1}}} + r_{on} a_{nc} J_C \frac{1}{1 + \dfrac{D_n \text{SOD}^2}{D_o^2 o_w^2 k_{n1}}} \tag{25.86}$$

where a_{cn} = methane loss due to denitrification = 1.07 gC gN^{-1}. Also, since we are directly discounting the CSOD, the true nitrification oxygen demand r_{on} = 4.57 gO gN^{-1} is used for the NSOD.

Phosphate. Phosphorus is another important plant nutrient that is greatly affected by sediment oxygen levels. However, in contrast to nitrate its accurate modeling involves some major modifications. In particular the dynamics of phosphorus are significantly dictated by its association with particulate matter in the sediments.

A number of investigators starting with Mortimer (1941, 1942) have theorized that phosphorus in the aerobic surface layer is sorbed to precipitated iron hydroxides. Such attachment then impedes its transfer back into the water. When oxygen levels drop, the iron hydroxides are dissolved as they become reduced. This leads to an associated release of dissolved phosphorus into the sediment pore waters where it is free to diffuse into the overlying water.

These effects can be integrated into the SOD framework by writing balances for phosphate that account for the sorption mechanism,

$$H_2 \frac{dp_2}{dt} = a_{pc} k_{c2} H_2 c_2 + v_{dp12}(F_{d1} p_1 - F_{d2} p_2) + v_{dp12}(F_{p1} p_1 - F_{p2} p_2) + v_b(p_1 - p_2) \tag{25.87}$$

$$H_1 \frac{dp_1}{dt} = v_{dp12}(F_{d2} p_2 - F_{d1} p_1) + v_{dp12}(F_{p2} p_2 - F_{p2} p_1) + v_{dmw1}(F_{dw} p_2 - F_{d1} p_1) \tag{25.88}$$

where F_d and F_p are the fractions associated with dissolved and particulate forms in the two layers. Later in this book when we discuss toxic substances, we will describe how they are derived. For the time being they can be calculated by

$$F_d = \frac{1}{\phi + K_{dp}(1 - \phi)\rho} \qquad F_p = \frac{K_{dp}(1 - \phi)\rho}{\phi + K_{dp}(1 - \phi)\rho} \tag{25.89}$$

where ϕ = sediment porosity ($\cong$ 0.8 to 0.95)
ρ = sediment density ($\cong 2.5 \times 10^6$ g m^{-3})
K_{dp} = phosphorus sorption coefficient (m^3 g^{-1})

The sorption coefficient is the key to the phosphorus model. At high values ($\cong$ 0.01 to 0.001 m^3 g^{-1}), the dissolved concentration will be low and diffusion in the pore water will be low. Conversely at low values of K_{dp} ($\cong$ 0.00005 to 0.0005 m^3 g^{-1}), pore water concentration and diffusion become high.

Aside from sorption, three additional features of Eqs. 25.87 and 25.88 bear mention. First, notice that I have included terms for particle mixing. Such transport is due to burrowing and other activities of benthic organisms. Second, I have included an advective term to account for the movement of matter away from the sediment-water interface due to the accumulation of settling solids in the sediments. Finally I should mention that whereas we have used the steady-state case for all previous constituents, it is not meaningful for a substance like phosphorus that is not continuously subject to a constant diffusion mechanism. Thus, unless the overlying water is always aerobic (in which case the phosphorus flux would probably be negligible anyway), phosphorus in the anaerobic layer must be simulated dynamically.

25.6.5 Estuaries

Because of the dominant role of sulfur, estuaries tend to exhibit a different carbon chemistry from freshwater systems like lakes and streams. In essence the SOD is tied to the reduction of sulfate (SO_4) to produce sulfide (H_2S). The key reaction is (Barnes and Goldberg 1976)

$$2CH_2O + H_2SO_4 \rightarrow 2CO_2 + H_2S + 2H_2O \qquad (25.90)$$

In the anaerobic zone, sulfide is produced. A portion of it reacts with iron to form particulate iron sulfide [$FeS(s)$]. The remainder diffuses into the aerobic zone where a part is oxidized. In addition particle mixing can move some of the particulate iron sulfide into the aerobic zone where it can be oxidized to form ferric oxyhydroxide [$Fe_2O_3(s)$], again consuming oxygen in the process.

Thus the model would be quite similar in format to the balances for phosphorus, with the exception that

- Sulfide is modeled in place of phosphorus.
- Sulfide undergoes oxidation in the surface layer.
- Precipitation reactions would not depend on oxygen in the way that phosphate sorption does.

Di Toro and Fitzpatrick (1993) provide a very thorough description of this sulfide model along with all the other elements of SOD modeling discussed in this lecture.

Finally, it should be noted that sulfur plays an important role in many freshwater sediments. In the future, sulfur dynamics must be incorporated into SOD frameworks developed for such systems.

PROBLEMS

25.1. A small pond located at 10,000 ft has the following characteristics:

Residence time = 1 wk
Area = 1×10^5 m^2
Mean depth = 2 m

The pond has an inflow CBOD concentration of 50 mg L^{-1}. The CBOD does not settle but decays at a rate of 0.05 d^{-1} at 20°C. The temperature of the pond is 15°C. The wind speed over the pond is 2 m s^{-1}. The inflow has a temperature of 10°C and a deficit of 3 mg L^{-1}. Half of the sediments exert no SOD whereas the other half exert an SOD of 0.2 g m^{-2} d^{-1} at 20°C. Compute the following (don't forget temperature correction factors):

(*a*) The CBOD concentration of the system
(*b*) The oxygen saturation concentration for the pond (use of table in App. B is acceptable)
(*c*) The reaeration rate for the pond at 15°C (units of d^{-1})
(*d*) The oxygen concentration of the pond

25.2. The epilimnion of a lake has a particulate organic carbon concentration of 1 mg L^{-1} and a dissolved oxygen concentration of 9 mg L^{-1}. The carbon settles at a rate of 0.05 m d^{-1} and is oxidized in the hypolimnion at a rate of 0.05 d^{-1}. The bulk diffusion coefficient across the thermocline is 0.05 m d^{-1} and the hypolimnion thickness is 5 m. The situation is depicted in the following diagram:

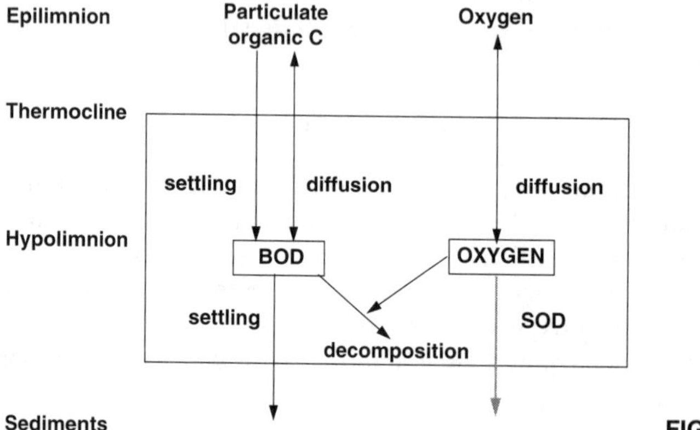

FIGURE P25.2

(*a*) Compute the steady-state concentration of carbonaceous BOD and dissolved oxygen in the hypolimnion. Assume that the epilimnion levels are constant and that once the particulate matter settles to the bottom it does not exert an SOD.
(*b*) For the case described in (*a*) compute the SOD that would be exerted by the particulate matter that settles from the hypolimnion to the sediments using the naive Streeter-Phelps SOD model described in Sec. 25.2. Assume that the particulate matter is completely oxidized in the sediments. Express the resulting SOD in units of g m^{-2} d^{-1}.

25.3. Repeat Example 25.2, but for a downward flux of 20 gO m^{-2} d^{-1} and a water oxygen concentration of 3 mg L^{-1}.

25.4. A lake has a particulate organic carbon concentration of 5 mg L^{-1} that settles at a rate of 0.25 m d^{-1}. Determine the steady-state SOD and the thickness of the aerobic layer for the system if the overlying water has an oxygen concentration of 4 mg L^{-1}. The active sediment layer is 10 cm thick and assume that all the nitrate produced by nitrification is denitrified to nitrogen gas. Assume that there is no water boundary layer.

(*a*) Use the analytical solution to obtain your answer.

(*b*) Use a numerical approach with the anaerobic sediments divided into four segments.

25.5. Repeat Prob. 25.4, but add a 1-cm-thick boundary layer in the water. Assume that diffusion and reaction in both the sediment aerobic and water boundary layers are the same.

LECTURE 26

Computer Methods

Lecture overview: I review some computer-oriented methods for solving oxygen balances with the computer. First I develop the steady-state system response matrix for the Streeter-Phelps framework. Then I provide an introduction to the QUAL2E model. After some historical background, I describe its underlying structure. Then I illustrate its use by applying it to simulate oxygen levels in a stream, with a focus on oxygen depletion due to BOD and SOD.

Although we have touched on some computer methods, to this point our discussion of oxygen modeling has focused on analytical, closed-form solutions. Now we broaden our perspective to show how numerical methods allow us to implement such calculations with the computer. First I describe a matrix approach that has great utility for developing steady-state solutions for linear DO models. Then I provide an introduction to EPA's QUAL2E software package.

26.1 STEADY-STATE SYSTEM RESPONSE MATRIX

In Lecs. 6 and 11 we developed the steady-state-system response matrix as a means to summarize the interactions of loadings and responses for linear water-quality models. Recall that the steady-state mass balances for such systems could be expressed as a set of linear algebraic equations. These equations could then be solved as

$$\{C\} = [A]^{-1}\{W\} \tag{26.1}$$

where $\{C\}$ = vector of unknown concentrations
$\{W\}$ = vector of loadings
$[A]^{-1}$ = matrix inverse or the steady-state-system response matrix

We showed how each element of $[A]^{-1}$ provides the change in concentration in element i due to a unit load change to segment j.

We now illustrate how this idea can be expanded to BOD/oxygen modeling. To do this we limit our analysis to a one-dimensional system with carbonaceous BOD and oxygen. The approach can be easily extended to multidimensional systems, NBOD, or linear representations of nitrification. A steady-state mass balance for CBOD in segment i can be written as

$$\begin{aligned} 0 = {} & W_{Li} + Q_{i-1,i}(\alpha_{i-1,i}L_{i-1} + \beta_{i-1,i}L_i) - Q(\alpha_{i,i+1}L_i + \beta_{i,i+1}L_{i+1}) \\ & + E'_{i-1,i}(L_{i-1} - L_i) + E'_{i,i+1}(L_{i+1} - L_i) - k_{ri}V_iL_i \end{aligned} \tag{26.2}$$

Writing this equation for an n-segment system with appropriate upstream and downstream boundary conditions gives

$$[A]\{L\} = \{W_L\} \tag{26.3}$$

with a solution

$$\{L\} = [A]^{-1}\{W_L\} \tag{26.4}$$

Now the coefficients of $[A]$ bear scrutiny. They are

$$a_{i,i-1} = \underbrace{-\alpha_{i-1,i}Q_{i-1,i} - E'_{i-1,i}}_{\text{Transport}} \tag{26.5}$$

$$a_{i,i} = \underbrace{\alpha_{i,i+1}Q_{i,i+1} - \beta_{i-1,i}Q_{i-1,i} + E'_{i-1,i} + E'_{i,i+1}}_{\text{Transport}} + \underbrace{k_{ri}V_i}_{\text{Kinetics}} \tag{26.6}$$

$$a_{i,i+1} = \underbrace{\beta_{i,i+1}Q_{i,i+1} - E'_{i,i+1}}_{\text{Transport}} \tag{26.7}$$

All the terms marked "transport" would be identical regardless of the pollutant. Thus the $[A]$ matrix can be divided into two parts, as in

$$[A] = [T] + [k_rV] \tag{26.8}$$

where $[T]$ is a "transport" matrix identical to matrix $[A]$ but containing only the transport terms, and $[k_rV]$ is a square diagonal matrix containing the terms $k_{ri}V_i$ on the diagonal and 0 elsewhere.

Using a similar approach, we can write mass balances for deficit (see Thomann and Mueller 1987). However, because such an approach can lead to errors for systems where saturation is variable (recall our discussion in Lec. 21), we write the mass balance for oxygen,

$$\begin{aligned} 0 = {} & W_{oi} + Q_{i-1,i}(\alpha_{i-1,i}o_{i-1} + \beta_{i-1,i}o_i) - Q_{i,i+1}(\alpha_{i,i+1}o_i + \beta_{i,i+1}o_{i+1}) \\ & + E'_{i-1,i}(o_{i-1} - o_i) + E'_{i,i+1}(o_{i+1} - o_i) - k_{di}V_iL_i \\ & + k_{ai}V_i(o_{si} - o_i) + P_iV_i - R_iV_i - S'_BA_{si} \end{aligned} \tag{26.9}$$

Writing this equation for an n-segment system with appropriate upstream and downstream boundary conditions gives

$$[B]\{o\} = \{W_o\} + \{PV\} - \{RV\} - \{S'_BA_s\} + \{k_aVo_s\} - \{k_dVL\} \tag{26.10}$$

where

$$[B] = [T] + [k_a V] \tag{26.11}$$

By collecting terms, we can write the resulting system of equations as

$$[B]\{o\} = \{W_o'\} - [k_d V]\{L\} \tag{26.12}$$

where $\{W_o'\}$ is a matrix containing all the external oxygen sources and sinks,

$$\underset{\text{External sources}}{\{W_o'\}} = \underset{\text{Direct loading}}{\{W_o\}} + \underset{\text{Photosynthesis gain}}{\{PV\}} - \underset{\text{Respiration loss}}{\{RV\}} - \underset{\text{SOD loss}}{\{S_B'A\}} + \underset{\text{Reaeration gain}}{\{k_a V o_s\}} \tag{26.13}$$

Multiplying both sides by the matrix inverse of $[B]$ yields

$$\{o\} = [B]^{-1}\{W_o'\} - [B]^{-1}[k_d V][A]^{-1}\{W_L\} \tag{26.14}$$

or

$$\{o\} = [B]^{-1}\{W_o'\} - [C]^{-1}\{W_L'\} \tag{26.15}$$

where $[C]^{-1}$ is a system response matrix that relates oxygen concentration to BOD loading,

$$[C]^{-1} = [B]^{-1}[k_d V][A]^{-1} \tag{26.16}$$

EXAMPLE 26.1. MATRIX APPROACH FOR OXYGEN. A one-dimensional estuary has the following characteristics:

Flow $= 1 \times 10^7$ m^3 d^{-1} BOD decay $= 0.2$ d^{-1}
Width $= 1500$ m   Reaeration $= 0.25$ d^{-1}
Depth $= 5$ m   Saturation $= 8$ mg L^{-1}
Dispersion $= 1 \times 10^7$ m^2 d^{-1}

The estuary is 100 km long. The boundary conditions at both the upstream and downstream ends are $L = 0$ and $o = o_s$. Loadings of BOD and oxygen of 300,000 and 100,000 kg d^{-1}, respectively, enter the estuary at KP 35. Centered differences were used to approximate space.

(*a*) Calculate the distribution of BOD and oxygen in the estuary using segment sizes of 10 km.

(*b*) Determine the BOD loading reduction needed to raise the minimum oxygen concentration in the estuary to 5 mg L^{-1}.

Solution: (*a*) Using the parameter values, we can write and solve the mass balance equations. The results are displayed in Fig. E26.1.

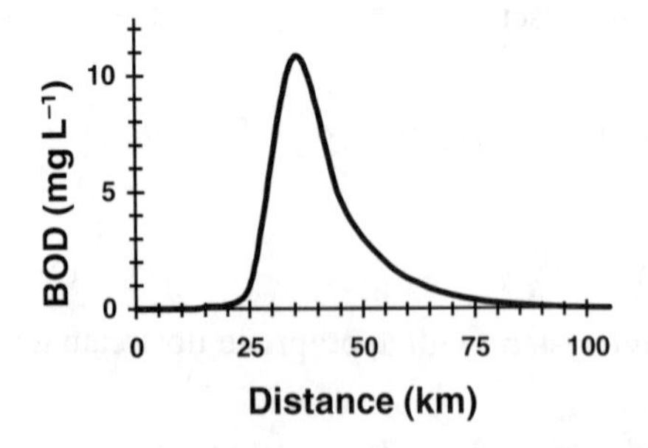

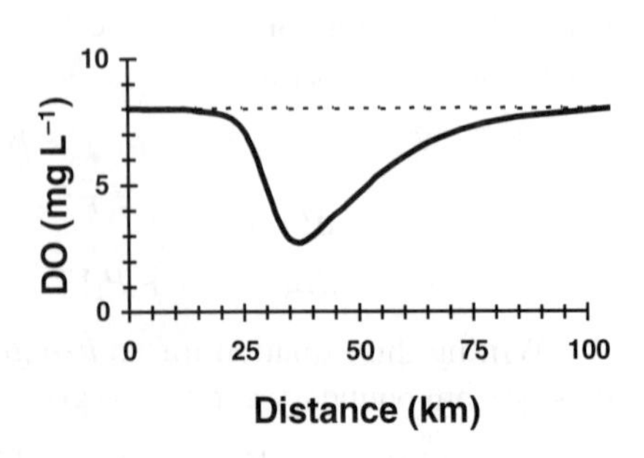

FIGURE E26.1

(*b*) The minimum DO is 2.84 mg L^{-1}. Thus, to bring this value up to 5 mg L^{-1}, we must increase the DO by $\Delta o = 5 - 2.84 = 2.16$ mg L^{-1}. This value can be translated into a load change by using the appropriate value from the matrix inverse,

$$[C]^{-1} = \begin{bmatrix} 1.64\text{E}-08 & 2.66\text{E}-09 & 3.25\text{E}-10 & 3.53\text{E}-11 & 3.61\text{E}-12 & 3.54\text{E}-13 & 3.38\text{E}-14 & 3.17\text{E}-15 & 2.92\text{E}-16 & 2.56\text{E}-17 \\ 1.33\text{E}-08 & 1.80\text{E}-08 & 2.84\text{E}-09 & 3.43\text{E}-10 & 3.71\text{E}-11 & 3.78\text{E}-12 & 3.70\text{E}-13 & 3.53\text{E}-14 & 3.29\text{E}-15 & 2.92\text{E}-16 \\ 8.12\text{E}-09 & 1.42\text{E}-08 & 1.81\text{E}-08 & 2.85\text{E}-09 & 3.44\text{E}-10 & 3.72\text{E}-11 & 3.78\text{E}-12 & 3.70\text{E}-13 & 3.53\text{E}-14 & 3.17\text{E}-15 \\ 4.42\text{E}-09 & 8.58\text{E}-09 & 1.42\text{E}-08 & 1.81\text{E}-08 & 2.85\text{E}-09 & 3.44\text{E}-10 & 3.72\text{E}-11 & 3.78\text{E}-12 & 3.70\text{E}-13 & 3.38\text{E}-14 \\ 2.25\text{E}-09 & 4.64\text{E}-09 & 8.60\text{E}-09 & 1.42\text{E}-08 & 1.81\text{E}-08 & 2.85\text{E}-09 & 3.44\text{E}-10 & 3.72\text{E}-11 & 3.78\text{E}-12 & 3.54\text{E}-13 \\ 1.11\text{E}-09 & 2.36\text{E}-09 & 4.65\text{E}-09 & 8.60\text{E}-09 & 1.42\text{E}-08 & 1.81\text{E}-08 & 2.85\text{E}-09 & 3.44\text{E}-10 & 3.71\text{E}-11 & 3.61\text{E}-12 \\ 5.28\text{E}-10 & 1.16\text{E}-09 & 2.36\text{E}-09 & 4.65\text{E}-09 & 8.60\text{E}-09 & 1.42\text{E}-08 & 1.81\text{E}-08 & 2.85\text{E}-09 & 3.43\text{E}-10 & 3.53\text{E}-11 \\ 2.47\text{E}-10 & 5.51\text{E}-10 & 1.16\text{E}-09 & 2.36\text{E}-09 & 4.65\text{E}-09 & 8.60\text{E}-09 & 1.42\text{E}-08 & 1.81\text{E}-08 & 2.84\text{E}-09 & 3.25\text{E}-10 \\ 1.14\text{E}-10 & 2.57\text{E}-10 & 5.51\text{E}-10 & 1.16\text{E}-09 & 2.36\text{E}-09 & 4.64\text{E}-09 & 8.58\text{E}-09 & 1.42\text{E}-08 & 1.80\text{E}-08 & 2.66\text{E}-09 \\ 5.00\text{E}-11 & 1.14\text{E}-10 & 2.47\text{E}-10 & 5.28\text{E}-10 & 1.11\text{E}-09 & 2.25\text{E}-09 & 4.42\text{E}-09 & 8.12\text{E}-09 & 1.33\text{E}-08 & 1.64\text{E}-08 \end{bmatrix}$$

The needed value is in the fourth row and the fourth column: $c_{44}^{-1} = 1.81 \times 10^{-8}$ (mg L^{-1})/(g d^{-1}). (Remember, the column location is dictated by the position of the loading and the row by the location of the resulting concentration.) This value can be used to compute

$$\Delta W_L = \frac{2.16 \text{ mg L}^{-1}}{1.81 \times 10^{-8} \dfrac{\text{mg L}^{-1}}{\text{g d}^{-1}}} \left(\frac{\text{kg}}{1000 \text{ g}}\right) = 119{,}337 \text{ kg d}^{-1}$$

Thus the BOD load must be reduced by 40% to meet the oxygen goal.

The matrix approach is predicated on having a linear model. Although this is true for the classic Streeter-Phelps framework, more recent oxygen models include nonlinear terms that preclude the matrix approach. Consequently it might seem that the method outlined in the present section is obsolete. There are three reasons why this isn't necessarily true:

1. Even though nonlinearities appear in models, some of the new mechanisms are not highly nonlinear. Consequently some mechanisms are amenable to linearization via a Taylor-series expansion. For such cases the foregoing approach could be employed without incurring major errors.

2. Although the Streeter-Phelps model may seem passé, it is still useful where primary and secondary treatment are to be evaluated. Such management contexts still occur in developing countries.
3. The general approach outlined here is useful for other problem contexts beyond oxygen. For example many of the modeling frameworks used to assess toxic substances are linear. Hence the matrix solution techniques are possible and useful.

26.2 THE QUAL2E MODEL

The QUAL2E software package is presently the most widely used computer model for simulating stream-water quality. It is capable of simulating up to 15 water-quality constituents (Table 26.1) in dendritic streams that are well-mixed laterally and vertically. Among its many capabilities it allows for multiple waste discharges, withdrawals, tributary flows, and incremental (that is, distributed) inflows and outflows.

QUAL2E has its roots in the QUAL-I model developed by F. D. Masch and Associates and the Texas Water Development Board (1970). In 1972 Water Resources Engineers, Inc. (now Camp, Dresser and McKee), under contract with the U.S. Environmental Protection Agency, modified and extended QUAL-I to create the first version of QUAL-II. Over the ensuing years the model was upgraded several times (for example Roesner et al. 1981*a*, 1981*b*). The present version (Brown and Barnwell 1987) is known as the "enhanced QUAL-II model" or QUAL2E for short. It is currently maintained by the EPA's Center for Water Quality Modeling in Athens, Georgia.

It should be noted that QUAL2E was originally developed using punch cards as its input media. As the software moved to time-sharing systems and then personal computers, the input files to run the model maintained the data structure of the punch-card input. Recently a user-friendly interface for entering the input file and viewing the results of QUAL2E simulations has been developed and is being distributed (Lahlou et al. 1995). This interface should further enhance the utility and widespread application of this model. Although the interface employs easy-to-use, spreadsheet-like input, it still requires precisely the same information as for the flat-file version. Thus this lecture (as well as later descriptions of QUAL2E applications in Lec. 36) uses the flat-file format. So whether you use the original version or the new interface, our input files provide a concise representation of all the necessary information for performing model simulations with QUAL2E.

TABLE 26.1
The 15 constituents that can be simulated by QUAL2E

Dissolved oxygen	Ammonia as N	Coliform bacteria
Biochemical oxygen demand	Nitrite as N	Arbitrary nonconservative constituent
Temperature	Nitrate as N	Conservative constituent I
Algae as chlorophyll *a*	Organic phosphorus as P	Conservative constituent II
Organic nitrogen as N	Dissolved phosphorus as P	Conservative constituent III

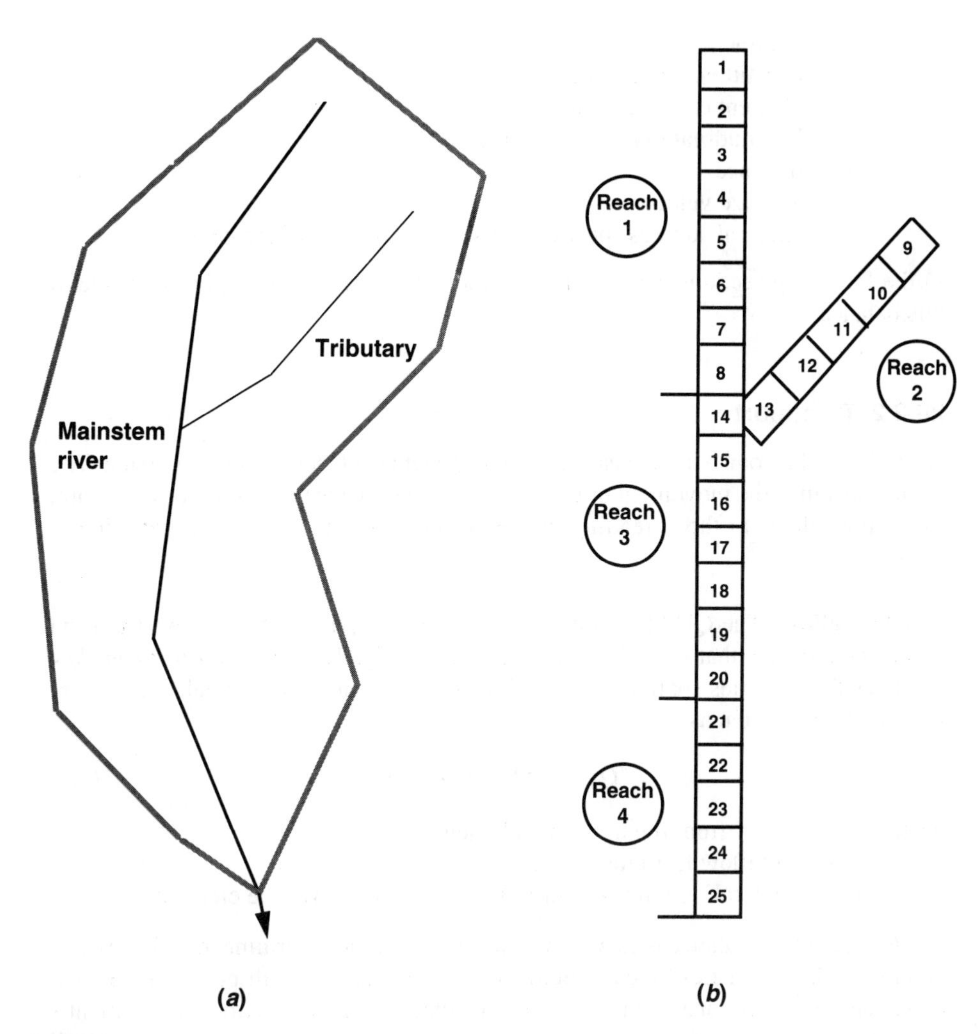

FIGURE 26.1
(*a*) A river basin and (*b*) a QUAL2E representation as reaches and elements.

26.2.1 Spatial Discretization and Model Overview

As depicted in Fig. 26.1, QUAL2E treats a river as a collection of reaches, each having homogeneous hydrogeometric properties. Each reach, in turn, is divided into a series of equal-length computational elements or control volumes.

A mass balance is used to keep track of the water-quality constituents listed in Table 26.1. This balance can be written generally as

$$\underbrace{V\frac{\partial c}{\partial t}}_{\text{Accumulation}} = \underbrace{\underbrace{\frac{\partial\left(A_c E \frac{\partial c}{\partial x}\right)}{\partial x} dx}_{\text{Dispersion}} - \underbrace{\frac{\partial(A_c U c)}{\partial x} dx}_{\text{Advection}}}_{\text{Transport}} + \underbrace{V\frac{dc}{dt}}_{\text{Kinetics}} + \underbrace{s}_{\text{External sources/sinks}} \tag{26.17}$$

where V = volume
c = constituent concentration
A_c = element cross-sectional area
E = longitudinal dispersion coefficient
x = distance
U = average velocity
s = external sources (positive) or sinks (negative) of the constituent

In the following sections we elaborate on the three terms on the right-hand side of this balance.

26.2.2 Transport

As in Eq. 26.17, transport consists of two components: advection and dispersion. The former specifies the movement of the constituent with water as it flows downstream. The latter relates to the spreading of the constituent that occurs primarily due to shear.

Advection. The QUAL2E model assumes steady, nonuniform flow. The term ***steady flow*** means that flow does not vary temporally. The term ***nonuniform flow*** connotes that it varies spatially. For such a characterization a flow balance for element i can be written as

$$Q_{i-1} \pm Q_{x,i} - Q_i = 0 \tag{26.18}$$

where Q_{i-1} = flow from the upstream element
Q_i = outflow from the element
$Q_{x,i}$ = lateral flow into (positive) or out of (negative) the element

Once the flow balance is established, it is necessary to determine the other hydrogeometric characteristics for each element. In particular the resulting water velocity, depth, and cross-sectional area must be established. The relationship of the element's other hydrogeometric characteristics to its flow rate is handled in two ways (recall Lec. 14):

1. Power equations can be used to relate mean velocity and depth to flow,

$$U = aQ^b \tag{26.19}$$

and

$$H = \alpha Q^\beta \tag{26.20}$$

where H = mean depth and a, b, α, and β are empirical constants that are determined from stage-discharge rating curves. Once velocity has been determined, the cross-sectional area can be calculated from the continuity equation

$$A_c = \frac{Q}{U} \tag{26.21}$$

2. The Manning equation provides a means to relate channel characteristics and flow. In metric units,

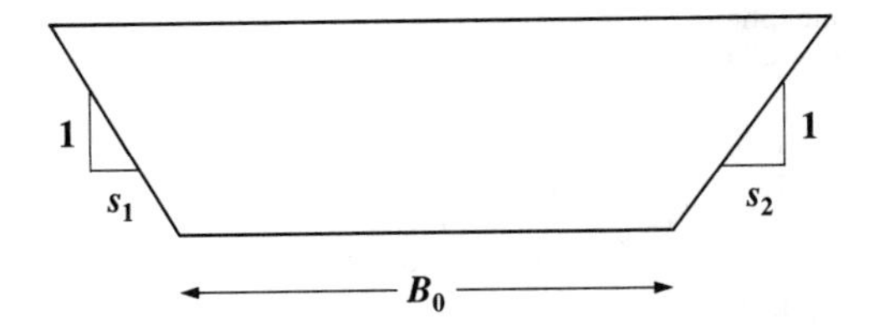

FIGURE 26.2
A trapezoidal channel showing the three parameters needed to uniquely defie the geometry. B_0 = bottom width; s_1 and s_2 = side slopes.

$$Q = \frac{1}{n} A_c R^{2/3} S_e^{1/2} \tag{26.22}$$

where R = channel's hydraulic radius (m) and S_e = slope of the channel's energy grade line (dimensionless). Note that since we have assumed that flow is steady and cross sections are constant, the energy slope is equal to the channel slope. QUAL2E assumes that the channel has a trapezoidal cross section (Fig. 26.2), so the cross-sectional area and hydraulic radius can be expressed as a function of depth. If Q is given, this means that Eq. 26.22 is a nonlinear equation that can be solved for depth numerically. Depth can then be employed to determine the area, which can be used to compute velocity with Eq. 26.21 (recall Example 14.5).

Dispersion. The QUAL2E model uses the following relationship to compute dispersion as a function of the channel's characteristics (Fischer et al. 1979),

$$E = 3.11 K n U H^{5/6} \tag{26.23}$$

where E = longitudinal dispersion coefficient $(\mathrm{m^2\ s^{-1}})$
n = channel's roughness coefficient (dimensionless)
U = mean velocity (mps)
H = mean depth (m)
K = a dispersion parameter (dimensionless)

defined as

$$K = \frac{E}{HU^*} \tag{26.24}$$

where U^* = shear velocity (m s^{-1}). Equations 26.23 and 26.24 represent circular reasoning. Thus they differ from Eqs. 14.15 and 14.17, which calculate dispersion purely as a function of channel parameters. However, once K is established it provides a formula to compute dispersion as a function of nonuniform flow conditions. It is in this way that K is used in QUAL2E.

26.2.3 Kinetics

To keep this lecture simple we limit the discussion to two constituents: carbonaceous BOD (CBOD) and dissolved oxygen. The kinetics for these constituents are displayed in Fig. 26.3 and can be represented mathematically by

$$\frac{dL}{dt} = -K_1 L - K_3 L \tag{26.25}$$

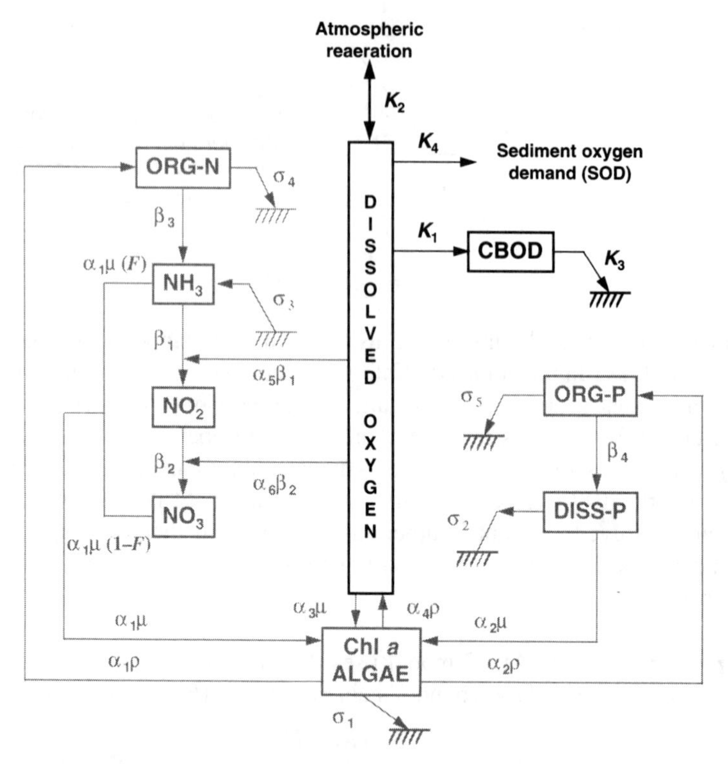

FIGURE 26.3
QUAL2E kinetics. Note that the highlighted constituents and processes are described and modeled in this lecture. We will extend our discussion to the remainder of the diagram in Lec. 36.

and
$$\frac{do}{dt} = K_2(o_s - o) - K_1L - \frac{K_4}{H} \tag{26.26}$$

where L = carbonaceous BOD (mg L^{-1})
K_1 = BOD decomposition rate (d^{-1})
K_3 = BOD settling rate (d^{-1})
o = dissolved oxygen concentration (mg L^{-1})
K_2 = reaeration rate (d^{-1})
o_s = dissolved oxygen saturation concentration (mg L^{-1})
K_4 = sediment oxygen demand (g m^{-2} d^{-1})

Note that all the rates (the K's) are corrected for temperature by

$$K = K_{20}\theta^{T-20} \tag{26.27}$$

where K = rate at temperature T
K_{20} = rate at 20°C
θ = a temperature correction factor

All the rates in Eqs. 26.25 and 26.26 can be entered directly to QUAL2E. However, the reaeration rate can also be internally calculated using eight different formulas. These formulas are summarized in Table 26.2.

TABLE 26.2
Reaeration formulas in QUAL2E. Note that some of these formulas (2, 3, and 4) differ slightly from the versions in Lec. 20 due to rounding errors

Option	Author(s)	K_2 (d^{-1} at 20°C)	Units
1	User-specified value		
2	Churchill et al. (1962)	$5.03\frac{U^{0.969}}{H^{1.673}}$	$\overline{U}$ (m s^{-1}) H (m)
3	O'Connor and Dobbins (1958)	$3.95\frac{U^{0.5}}{H^{1.5}}$	$\overline{U}$ (m s^{-1}) H (m)
4	Owens et al. (1964)	$5.34\frac{U^{0.67}}{H^{1.85}}$	$\overline{U}$ (m s^{-1}) H (m)
5	Thackston and Krenkel (1966)	$24.9\frac{(1+\sqrt{F})u^*}{H}$ where F is the Froude number, $F = \frac{u^*}{\sqrt{gH}}$ and u^* is the shear velocity, $u^* = \sqrt{HS_e g} = \frac{Un\sqrt{g}}{H^{1.67}}$	F (dimensionless) u^* (m s^{-1}) H (m) $\overline{U}$ (m s^{-1})
6	Langbien and Durum (1967)	$5.13\frac{U}{H^{1.33}}$	H (m) $\overline{U}$ (m s^{-1})
7	User-specified power function	aQ^b	Q (cms)
8	Tsivoglou and Wallace (1972); Tsivoglou and Neal (1976)	$c\frac{\Delta H}{t_f}$ where ΔH is change in water-surface elevation in the element, t_f is the flow time in the element, and c is a flow-dependent escape coefficient: $c = 0.36$ for $0.028 \le Q \le 0.28$ cms $c = 0.177$ for $0.708 \le Q \le 85$ cms	c (m^{-1}) ΔH (m) t_f (d)

26.2.4 Numerical Algorithm

Now that we have an understanding of the major components of QUAL2E, we can discuss how the model obtains solutions numerically. To do this, Eq. 26.17 can be divided by volume and written as

$$\frac{\partial c}{\partial t} = \frac{\partial\left(A_c E \frac{\partial c}{\partial x}\right)}{A_c \partial x} - \frac{\partial(A_c U c)}{A_c \partial x} + rc + p + \frac{s}{V} \tag{26.28}$$

Observe that we have divided the kinetics into two separate terms,

$$\frac{dc}{dt} = rc + p \tag{26.29}$$

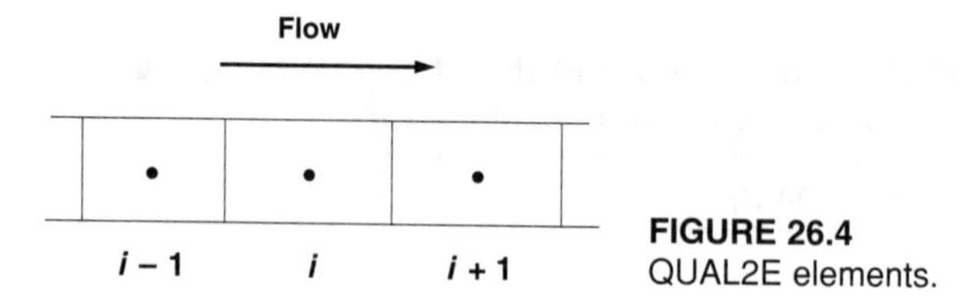

FIGURE 26.4
QUAL2E elements.

The first term on the right denotes those reactions that are linearly dependent on concentration, and the second term denotes internal constituent sources and sinks (for example, benthic sources, nutrient loss from algal growth, etc.). Although some of the latter are constants, others are nonlinear functions of constituent concentrations.

A general representation of the QUAL2E element scheme is shown in Fig. 26.4. Equation 26.28 can be written for element i by using a backward difference, as in

$$\underbrace{\frac{\partial c_i}{\partial t}}_{\text{Acc}} = \underbrace{\frac{-\left(A_c E \frac{\partial c}{\partial x}\right)_{i-1} + \left(A_c E \frac{\partial c}{\partial x}\right)_i}{V_i}}_{\text{Dispersion (In, Out)}} + \underbrace{\frac{(A_c U c)_{i-1} - (A_c U c)_i}{V_i}}_{\text{Advection (In, Out)}}$$

$$+ \underbrace{r_i c_i}_{\substack{\text{First-order}\\ \text{reactions}}} + \underbrace{p_i}_{\substack{\text{Internal}\\ \text{sources/sinks}}} + \underbrace{\frac{s_i}{V_i}}_{\substack{\text{External}\\ \text{sources/sinks}}} \tag{26.30}$$

Then backward differences can be used to approximate the remaining spatial derivatives,

$$\frac{\partial c_i}{\partial t} = \frac{(A_c E)(c_{i+1} - c_i)}{V_i \, \Delta x_i} + \frac{(A_c E)(c_{i-1} - c_i)}{V_i \, \Delta x_i} + \frac{Q_{i-1} c_{i-1} - Q_i c_i}{V_i} + r_i c_i + p_i + \frac{s_i}{V_i} \tag{26.31}$$

Finally a backward difference can be applied in time to yield

$$\frac{c_i^{\ell+1} - c_i^{\ell}}{\Delta t} = \frac{(A_c E)_{i,i+1}(c_{i+1}^{\ell+1} - c_i^{\ell+1})}{V_i \, \Delta x_i} + \frac{(A_c E)_{i-1,i}(c_{i-1}^{\ell+1} - c_i^{\ell+1})}{V_i \, \Delta x_i}$$

$$+ \frac{Q_{i-1} c_{i-1}^{\ell+1} - Q_i c_i^{\ell+1}}{V_i} + r_i c_i^{\ell+1} + p_i + \frac{s_i}{V_i} \tag{26.32}$$

Equation 26.32 can now be reexpressed by collecting terms to yield a linear system

$$e_i c_{i-1}^{n+1} + f_i c_i^{n+1} + g_i c_{i+1}^{n+1} = z_i \tag{26.33}$$

where

$$e_i = -\left[(A_c E)_{i-1} \frac{\Delta t}{V_i \, \Delta x_i} + \frac{Q_{i-1} \, \Delta t}{V_i}\right] \tag{26.34}$$

$$f_i = 1 + [(A_c E)_{i-1} + (A_c E)_i] \frac{\Delta t}{V_i \, \Delta x_i} + \frac{Q_i \, \Delta t}{V_i} - r_i \, \Delta t \tag{26.35}$$

$$g_i = -\left[(A_xE)_i \frac{\Delta t}{V_i \Delta x_i}\right] \tag{26.36}$$

$$z_i = c_i^n + \frac{s_i \Delta t}{V_i} + p_i \Delta t \tag{26.37}$$

These equations form a tridiagonal system that can be solved efficiently for concentration as a function of time. Note that the external sources and sinks (the s_i terms) are treated as constants in this formulation. As noted previously, some of these are nonlinear functions of other constituent concentrations. Thus the QUAL2E solution algorithm handles nonlinear terms by treating them as constant contributions to the forcing function that are updated at each time step.

Two forms of the solution are implemented in QUAL2E:

Steady-state. The model is run until it reaches a steady-state. This is the conventional implementation mode for QUAL2E. Thus the time-variable algorithm is a means to an end—the steady-state result (recall our discussion of the backward-time implicit approach in Lec. 13).

Time-variable. The model can also be run in a normal time-variable mode. At present this type of implementation is limited to diurnal simulations.

Now that we have covered the basics of the QUAL2E software package, I'll provide an overview of how it is implemented. However, before doing this I'll introduce a problem context where we will apply the model.

26.2.5 QUAL2E Application

Figure 26.5 shows a river that receives a sewage treatment plant effluent at kilometer point 100 (KP 100) and a tributary inflow at KP 60. Note that the channel is trapezoidal with the characteristics shown. The deoxygenation rate for CBOD is equal to 0.5 d^{-1} at 20°C. For 20 km downstream from the treatment plant there is a CBOD settling removal rate of 0.25 d^{-1}. In addition there is a sediment oxygen demand for this reach of 5 $g\,m^{-2}\,d^{-1}$. Assume that the O'Connor-Dobbins reaeration formula holds and that the stream is at sea level. Values for the channel's geometry and hydraulics are listed in Table 26.3. Kinetic parameters and temperatures are summarized in Table 26.4.

TABLE 26.3
Hydrogeometric parameters

Parameter	Units	KP > 100	KP 100–60	KP < 60
Depth	m	1.19	1.24	1.41
	(ft)	(3.90)	(4.07)	(4.62)
Area	m^2	14.71	15.5	18.05
Flow	$m^3\,s^{-1}$	5.787	6.250	7.407
	$m^3\,d^{-1}$	500,000	540,000	640,000
	(cfs)	(204)	(221)	(262)
Velocity	$m\,s^{-1}$	0.393	0.403	0.410
	$m\,d^{-1}$	33,955	34,819	35,424
	(fps)	(1.29)	(1.32)	(1.35)

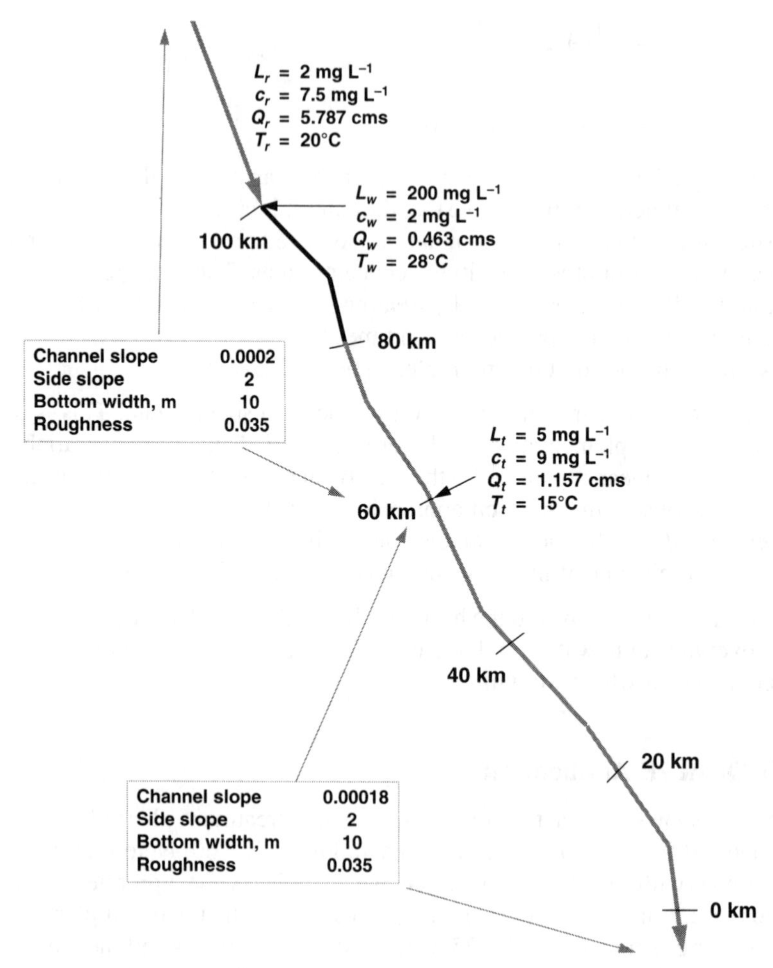

FIGURE 26.5
A stream receiving BOD loadings from a point source and a tributary.

TABLE 26.4
Temperatures and kinetic coefficients

Parameter	KP > 100	KP 100–80	KP 80–60	KP < 60
T (°C)	20	20.59	20.59	19.72
o_s (mg L^{-1})	9.092	8.987	8.987	9.143
k_a (d^{-1})	1.902	1.842	1.842	1.494
k_r (d^{-1})	0.500	0.767	0.514	0.494
k_d (d^{-1})	0.500	0.514	0.514	0.494
k_s (d^{-1})	0.000	0.254	0.000	0.000
SOD (g m^{-2} d^{-1})	0.000	5.175	0.000	0.000

The first step in using QUAL2E is to develop the spatial segmentation scheme for the system being modeled. This involves dividing the system into reaches of constant hydrogeometric characteristics. These reaches, in turn, consist of equal-length computational elements.

A segmentation scheme used for the present case is depicted in Fig. 26.6. Observe that we have divided the system into six reaches that we will call:

1. MS-HEAD
2. MS100-MS080
3. MS080-MS060
4. MS060-MS040
5. MS040-MS020
6. MS020-MS000

The names for each reach are at the user's discretion and merely serve to identify each reach. We have chosen to use the abbreviation "MS" to designate that we are simulating a "main stem" of a river with no tributaries modeled explicitly. Also notice that we have used distance in kilometers to identify the extent of each reach. However, we have also reversed the sense of the distances. That is, we have measured distance upstream from the downstream end of the system, rather than downstream as in the original problem statement. We have made this modification to make our scheme consistent with the way in which distances must be entered into QUAL2E.

Next, inspection of Fig. 26.6 shows that our scheme consists of 2-km elements. The elements that make up the reach must be numbered in order from the headwater to the most downstream point in the system. In addition the type of each element must be designated. There are seven element types:

1. Headwater element
2. Standard element
3. Element just upstream from a junction
4. Junction element
5. Last element in system
6. Input element
7. Withdrawal element

In the present example we use only four of these types. The first (1) and last (51) elements are type 1 and 5, respectively. Elements 2 and 22 are type 6 because they both receive point inflows. The remainder of the elements are the standard type 2.

Once the system segmentation is defined, we can create a data file to run QUAL2E. The file that conforms to the present problem is shown in Fig. 26.7. Note that the data in this file must be typed exactly as shown (minus lines and shading, of course). This is because QUAL2E is written in FORTRAN 77. Thus each line of input corresponds to an 80-column input-card format that derives from the punched cards used in early FORTRAN programs.

It should be noted that at the time of this book's printing, a user-friendly interface for entering the input file and viewing the results became available (Lahlou et al. 1995). The interface requires precisely the same information as in Fig. 26.7. Thus, whether you employ the original version or the new interface, Fig. 26.7 can serve as a guide for performing the QUAL2E simulation outlined in this lecture.

Reach name	Reach no.	Element no.	Element type
MS-HEAD	1	1	1
MS100-MS080	2	2	6
		3	2
		4	2
		5	2
		6	2
		7	2
		8	2
		9	2
		10	2
		11	2
MS080-MS060	3	12	2
		13	2
		14	2
		15	2
		16	2
		17	2
		18	2
		19	2
		20	2
		21	2
MS060-MS040	4	22	6
		23	2
		24	2
		25	2
		26	2
		27	2
		28	2
		29	2
		30	2
		31	2
MS040-MS020	5	32	2
		33	2
		34	2
		35	2
		36	2
		37	2
		38	2
		39	2
		40	2
		41	2
MS020-MS000	6	42	2
		43	2
		44	2
		45	2
		46	2
		47	2
		48	2
		49	2
		50	2
		51	5

FIGURE 26.6
The QUAL2E segmentation scheme conforming to the river shown in Fig. 26.5.

Columns — Data types

0 1 2 3 4 5 6 7 8
12345678901234567890123456789012345678901234567890123456789012345678901234567890

TITLE01 EXERCISE 1, QUAL-2EU WORKSHOP: SIMPLE CBOD/SOD
TITLE02 Steve Chapra, May 18, 1994
TITLE03 NO CONSERVATIVE MINERAL I
TITLE04 NO CONSERVATIVE MINERAL II
TITLE05 NO CONSERVATIVE MINERAL III
TITLE06 NO TEMPERATURE
TITLE07 YES BIOCHEMICAL OXYGEN DEMAND
TITLE08 NO ALGAE AS CHL-A IN UG/L
TITLE09 NO PHOSPHORUS CYCLE AS P IN MG/L — Title data
TITLE10 (ORGANIC-P; DISSOLVED-P)
TITLE11 NO NITROGEN CYCLE AS N IN MG/L
TITLE12 (ORGANIC-N; AMMONIA-N; NITRITE-N;' NITRATE-N)
TITLE13 YES DISSOLVED OXYGEN IN MG/L
TITLE14 NO FECAL COLIFORM IN NO./100 ML
TITLE15 NO ARBITRARY NON-CONSERVATIVE
ENDTITLE

NO LIST DATA INPUT — Program control (data type 1)
NO WRITE OPTIONAL SUMMARY
NO FLOW AUGMENTATION
STEADY STATE
TRAPEZOIDAL CHANNELS
NO PRINT LCD/SOLAR DATA
NO PLOT DO AND BOD
FIXED DNSTM CONC (YES=1)= 0. 5D-ULT BOD CONV K COEF = 0.25
INPUT METRIC = 1. OUTPUT METRIC = 1.
NUMBER OF REACHES = 6. NUMBER OF JUNCTIONS = 0.
NUM OF HEADWATERS = 1. NUMBER OF POINT LOADS = 2.
TIME STEP (HOURS) = 0. LNTH. COMP. ELEMENT (KM)= 2.
MAXIMUM ROUTE TIME (HRS)= 30. TIME INC. FOR RPT2 (HRS)=
LATITUDE OF BASIN (DEG) = 00. LONGITUDE OF BASIN (DEG)= 00.
STANDARD MERIDIAN (DEG) = 00. DAY OF YEAR START TIME = 0.
EVAP. COEF.,(AE) = 0.0000000 EVAP. COEF.,(BE) = .0000000
ELEV. OF BASIN (METERS) = 0. DUST ATTENUATION COEF. = 0.00
ENDATA1
ENDATA1A
ENDATA1B

STREAM REACH 1. RCH= MS-HEAD FROM 102.0 TO 100.0 — Reach identification and river Mile/kilometer data (data type 2)
STREAM REACH 2. RCH= MS100-MS080 FROM 100.0 TO 80.0
STREAM REACH 3. RCH= MS080-MS060 FROM 80.0 TO 60.0
STREAM REACH 4. RCH= MS060-MS040 FROM 60.0 TO 40.0
STREAM REACH 5. RCH= MS040-MS020 FROM 40.0 TO 20.0
STREAM REACH 6. RCH= MS020-MS000 FROM 20.0 TO 0.0
ENDATA2
ENDATA3

FLAG FIELD RCH= 1. 1. 1. — Computational elements Flag field data (data type 4)
FLAG FIELD RCH= 2. 10. 6.2.2.2.2.2.2.2.2.2.
FLAG FIELD RCH= 3. 10. 2.2.2.2.2.2.2.2.2.2.
FLAG FIELD RCH= 4. 10. 6.2.2.2.2.2.2.2.2.2.
FLAG FIELD RCH= 5. 10. 2.2.2.2.2.2.2.2.2.2.
FLAG FIELD RCH= 6. 10. 2.2.2.2.2.2.2.2.2.5.
ENDATA4

HYDRAULICS RCH= 1. 0.00 2.0 2.0 10. .0002 .035 — Hydraulics data (data type 5)
HYDRAULICS RCH= 2. 0.00 2.0 2.0 10. .0002 .035
HYDRAULICS RCH= 3. 0.00 2.0 2.0 10. .0002 .035
HYDRAULICS RCH= 4. 0.00 2.0 2.0 10. .00018 .035
HYDRAULICS RCH= 5. 0.00 2.0 2.0 10. .00018 .035
HYDRAULICS RCH= 6. 0.00 2.0 2.0 10. .00018 .035
ENDATA5
ENDATA5A

REACT COEF RCH= 1. 0.00 0.000 0.000 1. 0.000 0.0000 0.0000 — BOD and DO reaction Rate constants (data type 6)
REACT COEF RCH= 2. 0.50 0.250 5.000 3. 0.000 0.0000 0.0000
REACT COEF RCH= 3. 0.50 0.000 0.000 3. 0.000 0.0000 0.0000
REACT COEF RCH= 4. 0.50 0.000 0.000 3. 0.000 0.0000 0.0000
REACT COEF RCH= 5. 0.50 0.000 0.000 3. 0.000 0.0000 0.0000
REACT COEF RCH= 6. 0.50 0.000 0.000 3. 0.000 0.0000 0.0000
ENDATA6A
ENDATA6B

INITIAL COND-1 RCH= 1. 22.00 8.11 0.0 0.00 0.00 0.00 0.000 0.0 — Initial conditions-1 (data type 7)
INITIAL COND-1 RCH= 2. 20.59 8.11 0.0 0.00 0.00 0.00 0.000 0.0
INITIAL COND-1 RCH= 3. 20.59 8.11 0.0 0.00 0.00 0.00 0.000 0.0
INITIAL COND-1 RCH= 4. 19.72 8.11 0.0 0.00 0.00 0.00 0.000 0.0
INITIAL COND-1 RCH= 5. 19.72 8.11 0.0 0.00 0.00 0.00 0.000 0.0
INITIAL COND-1 RCH= 6. 19.72 8.11 0.0 0.00 0.00 0.00 0.000 0.0
ENDATA7
ENDATA7A

INCR INFLOW-1 RCH= 1. 0.000 00.00 0.0 0.0 0.0 0.0 0.0 0.0 0. — Incremental inflow-1 (data type 8)
INCR INFLOW-1 RCH= 2. 0.000 00.00 0.0 0.0 0.0 0.0 0.0 0.0 0.
INCR INFLOW-1 RCH= 3. 0.000 00.00 0.0 0.0 0.0 0.0 0.0 0.0 0.
INCR INFLOW-1 RCH= 4. 0.000 00.00 0.0 0.0 0.0 0.0 0.0 0.0 0.
INCR INFLOW-1 RCH= 5. 0.000 00.00 0.0 0.0 0.0 0.0 0.0 0.0 0.
INCR INFLOW-1 RCH= 6. 0.000 00.00 0.0 0.0 0.0 0.0 0.0 0.0 0.
ENDATA8
ENDATA8A
ENDATA9

HEADWTR-1 HDW= 1 UPSTREAM 5.7870 20.0 7.50 2.0 00.0 0.0 0.0 — Headwater sources-1 (data type 10)
ENDATA10
ENDATA10A

POINTLD-1 PTL= 1. MS0 0.00 0.463 28.0 2.00 200.0 0.0 0.0 0.0 — Point load-1 (data type 11)
POINTLD-1 PTL= 2. MS60 0.00 1.157 15.0 9.00 5.0 0.0 0.0 0.0
ENDATA11
ENDATA11A
ENDATA12
ENDATA13
ENDATA13A

FIGURE 26.7
QUAL2E input file.

Title data. These specify identification information for the run and set the constituents that are to be simulated. Notice that we have typed "YES" for BOD and dissolved oxygen.

Program control (data type 1). These consist of two parts. The first defines the program control options. The second sets the characteristics of the stream system configuration as well as some of the geographical/meteorological conditions for modeling temperature. Most of the information is self-explanatory. If data is not necessary it is omitted or set to zero. For example latitude is not needed because we are not simulating temperature. The "MAXIMUM ROUTE TIME (HRS)" entry bears additional explanation. When performing a steady-state computation, this entry sets the maximum number of iterations of the numerical method. This way, if the solution does not converge it will be halted. I have found that a value of 30.0 works well for the applications I have developed with QUAL2E. However, for other systems some adjustment of this parameter might be necessary.

Reach identification and river mile/kilometer data (data type 2). The cards in this group identify the stream reach system by name and river mile/kilometer. The latter is done by listing the stream reaches from the most upstream point to the most downstream point in the system. Observe that the river mile/kilometer data must be in descending order.

Computational elements flag field data (data type 4). This group of cards identifies the type of each computational element in each reach.

Hydraulics data (data type 5). Because we specified "TRAPEZOIDAL CHANNELS" in card 5 of data type 1, we use the Manning formula to determine the hydrogeometric characteristics of each reach. Consequently these cards include the parameters necessary for the Manning coefficient calculation (that is, channel slope, side slope, roughness, etc.).

BOD and DO reaction rate constants (data type 6). This group of cards includes reach information on the BOD decay rate coefficient and settling rate, sediment oxygen demand, and the method of computing the reaeration coefficient.

Initial conditions–1 (data type 7). This card group, one card per reach, establishes the initial values of the system for temperature, dissolved oxygen, BOD, and the three conservative minerals. Initial conditions for temperature must be specified whether it is simulated or not. For the present case, when it is not simulated, the initial conditions are used to determine the temperature correction factors for the rate constants. Other values can be set at zero for steady-state applications.

Incremental inflow–1 (data type 8). Even though we will not simulate incremental inflows, these cards must be included. As in our example, all the values can be set to zero.

Headwater sources–1 (data type 10). This card group, one card per headwater, defines the flow, temperature, dissolved oxygen, BOD, and conservative mineral concentrations of the headwater. Note that the headwater numbers are not the same

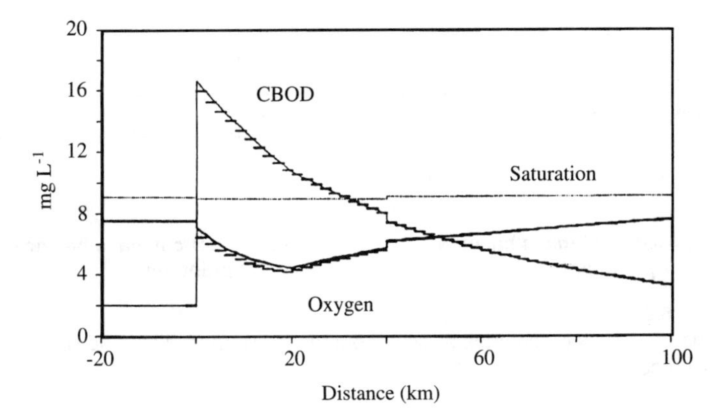

FIGURE 26.8
Comparison of QUAL2E output with analytical solution.

as either the reach or element numbers. Rather the headwaters are numbered consecutively (starting at 1) from the farthest upstream headwater.

Point load–1 (data type 11). This card group, one card per point source or withdrawal, defines the percent treatment, inflow or withdrawal, temperature, dissolved oxygen, BOD, and conservative mineral concentrations of each point load or withdrawal. Note that the point load numbers are not the same as either the reach or element numbers. Rather the point sources or withdrawals are numbered consecutively (starting at 1) from the most upstream to the most downstream.

Model output. The output for this run consists of

- Hydraulics summary
- Reaction coefficient summary
- Water-quality variables
- Dissolved oxygen data

A plot of the QUAL2E output for CBOD and oxygen is shown in Fig. 26.8. The QUAL2E results generally follow the analytical solution.

However, notice that there is a discrepancy at KP 100. This is due to the finite-difference approximation used in QUAL2E. Let us look at CBOD to understand what is happening. Recall that in the analytical solution, a mass balance is made at the mixing point,

$$Q_r L_r + Q_w L_w - (Q_r + Q_w) L_0 = 0$$

which can be used to calculate (Fig. 26.9*a*)

$$L_0 = \frac{40{,}000(200) + 500{,}000(2)}{540{,}000} = 16.667 \text{ mg L}^{-1}$$

which then decays downstream according to

$$L = 16.667 e^{-\frac{0.514+0.254}{34{,}819}x}$$

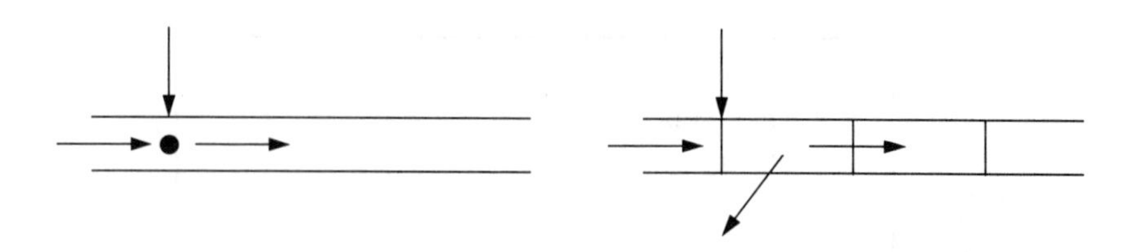

FIGURE 26.9
Comparison of mixing schemes at point sources for (*a*) analytical and (*b*) QUAL2E mass balances.

Thus at 1 km the value of BOD will have dropped to

$$L = 16.667e^{-\frac{0.767}{34,819}1000} = 16.304 \text{ mg L}^{-1}$$

In contrast, for QUAL2E, the loading is introduced into a well-mixed element (Fig. 26.9*b*). Therefore a mass balance must now include the sources and sinks for the volume,

$$Q_r L_r + Q_w L_w - (Q_r + Q_w)L_0 - (K_1 + K_3)VL = 0$$

which can be used to calculate

$$L_0 = \frac{40,000(200) + 500,000(2)}{540,000 + (0.76)31.01 \times 10^3} = 15.98 \text{ mg L}^{-1}$$

Thus there is a discrepancy because of the finite discretization at the mixing point. As discussed in more general terms in Sec. 11.7, care should be taken to ensure that the element size is small enough that such errors do not have a significant impact on the application of the model for decision making.

In summary the QUAL2E model provides a convenient tool to implement oxygen balances in stream networks. We will revisit the software in later lectures when we turn to the eutrophication problem in Part V.

PROBLEMS

26.1. Suppose that the following data are available for a vertically stratified estuary (Fig. P26.1):

$Q_{01} = 4 \times 10^6 \text{ m}^3 \text{ s}^{-1}$ $\quad Q_{32} = 2.5 \times 10^6$ $\quad Q_{21} = 2.5 \times 10^6$ $\quad Q_{13} = 6.5 \times 10^6$

$E'_{01} = 0 \text{ m}^3 \text{ s}^{-1}$ $\quad E'_{32} = 2.5 \times 10^6$ $\quad E'_{21} = 1.2 \times 10^6$ $\quad E'_{13} = 2.5 \times 10^6$

Determine the steady-state system response matrices for BOD and oxygen if the deoxygenation and reaeration rates are 0.1 and 0.2 d^{-1}, respectively. Use the matrices to determine how much BOD loading can be discharged to segment 1 if a level of

4 mg L^{-1} must be maintained in all segments. Assume that segments 0 and 3 contain negligible BOD and are at saturated levels of DO (o_s = 10 mg L^{-1}). In addition segment 2 has an SOD of 0.1 g m^{-2} d^{-1} and a surface area of 10,000 m^2.

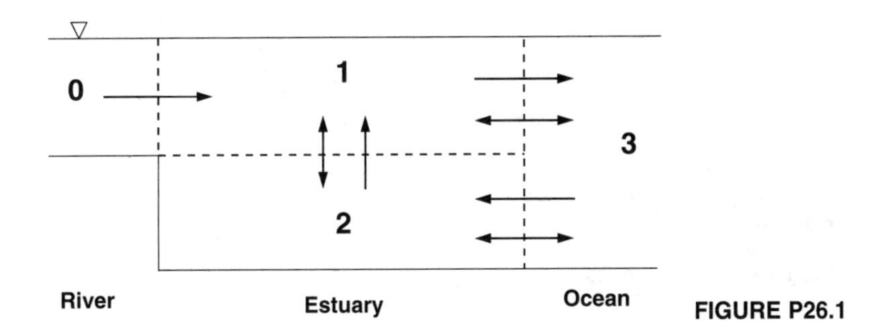

FIGURE P26.1

26.2. A waste source (Q_w = 1 cms, L_w = 25 mg L^{-1}, o_w = 2 mg L^{-1}, T_w = 25°C) discharges into a stream (Q_r = 10 cms, L_r = 2 mg L^{-1}, o_r = 10 mg L^{-1}, T_r = 15°C). Downstream the stream flows through a rectangular channel with roughness = 0.03, channel slope = 0.0005, bottom width = 20 m, and a side slope of 3. Use QUAL2E to calculate the profiles of both BOD and oxygen downstream assuming the BOD decays at a rate of 1 d^{-1}. Determine the value and the location of the maximum deficit.

26.3. Use QUAL2E to determine the profiles of BOD and dissolved oxygen for the following sea-level stream:

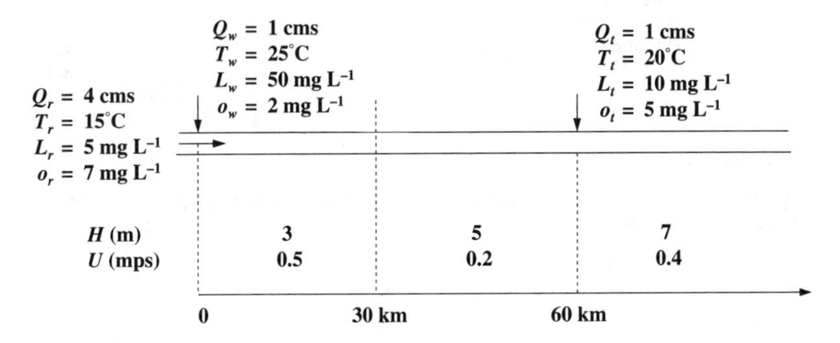

FIGURE P26.3

26.4. Figure P26.4 shows a river that receives a sewage treatment plant effluent at KP 0 and a withdrawal at KP 70. At a distance of 150 km, the stream enters a much larger river. Note that the channel is trapezoidal with the characteristics shown. The deoxygenation rate for CBOD is equal to 1 d^{-1} at 20°C. For 20 km downstream from the treatment plant, there is a CBOD settling removal rate of 2 d^{-1}. In addition there is a sediment oxygen demand for this reach of 4 g m^{-2} d^{-1}. Assume that the Churchill reaeration formula holds, that the stream is at sea level, and that the dispersion constant is zero. Use QUAL2E to simulate the levels of CBOD and oxygen for this case.

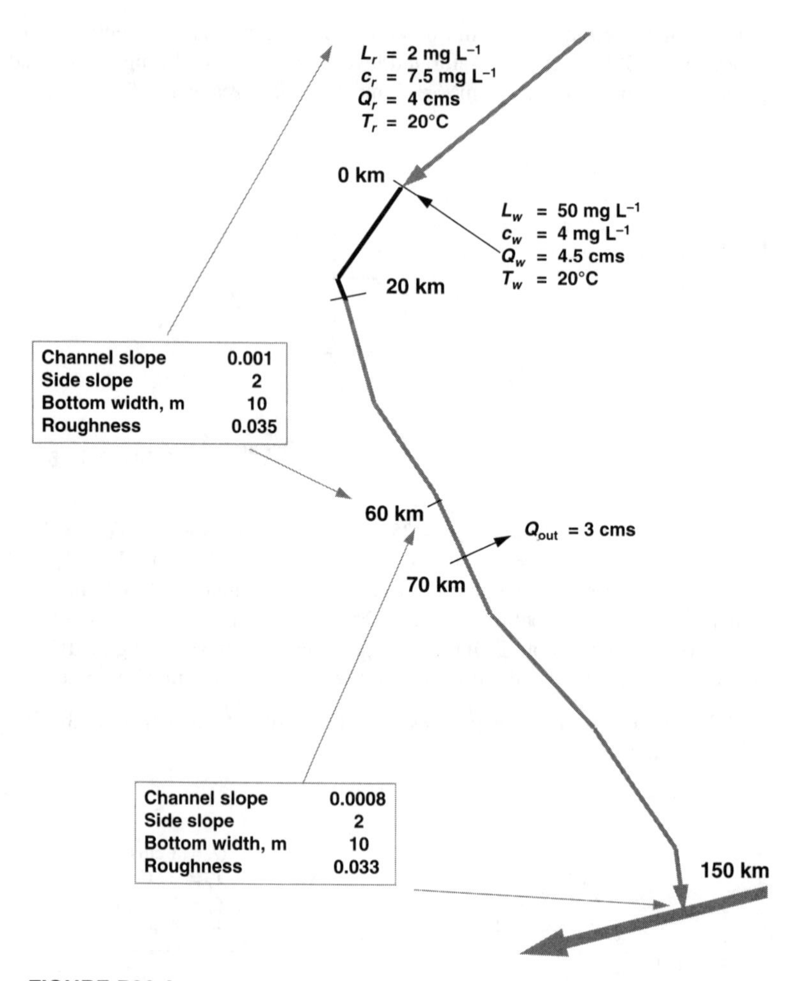

FIGURE P26.4
A stream receiving BOD loading from a point source and a loss of water through a withdrawal.

LECTURE 27

Pathogens

LECTURE OVERVIEW: I provide information on pathogens, with special emphasis on calculating the loss rate for bacteria. Then I use this information to develop an example for a small pond, including the impact of sediment-water interactions. I also describe models for pathogenic protozoa.

The initial impetus for water-quality control arose because of concerns over waterborne diseases. In developed countries, much has been done to diminish the threat posed by disease-carrying organisms in natural waters. However, bacteria, protozoa, and viruses still pose problems related to disease transmission and interference with uses of water for recreation. In addition most undeveloped and developing countries still experience great problems related to pathogens.

27.1 PATHOGENS

Contaminated water is responsible for the spread of many contagious diseases. The primary agents of these diseases are called ***pathogens***. These are disease-producing organisms that grow and multiply within the host. Some pathogens enter the human body through the skin. More commonly they are ingested along with drinking water.

Pathogens can be divided into categories. The most common groups associated with water pollution are summarized in Table 27.1.

Although the organisms listed are the cause of waterborne diseases, their concentrations are often very difficult to measure. Consequently, as described next, indicator organisms are used to track and manage pathogen problems.

TABLE 27.1
Some waterborne pathogenic organisms

Category	Description	Species and groups
Bacteria	Microscopic, unicellular organisms that lack a fully-defined nucleus and contain no chlorophyll	*Vibrio cholerae* *Salmonella* *Shigella* *Legionella*
Viruses	A large group of submicroscopic (10 to 25 nm) infectious agents. They are composed of a protein sheath surrounding a nucleic acid core and, thus, contain all the information required for their own reproduction. However, they require a host in which to live	Hepatitis A Enteroviruses Polioviruses Echoviruses Coxsackieviruses Rotaviruses
Protozoa	Unicellular animals that reproduce by fission	*Giardia lambia* *Entamoeba histolytica* *Cryptosporidium* *Naegleria fowleri*
Helminths (intestinal worms)	Intestinal worms and wormlike parasites	Nematodes *Schistosoma haematobium*
Algae	Large group of nonvascular plants. Certain species produce toxins that if consumed in large quantities may be harmful	*Anabaena flos-aquae* *Microcystis aeruginosa* *Aphanizomenon flos-aquae*

27.2 INDICATOR ORGANISMS

Because individual pathogens are usually difficult and/or expensive to measure directly, classical water-quality management and modeling has focused on the levels of indicator organisms. These are groups of organisms that are convenient to measure and that are abundant in human and animal waste. If they are present it is assumed that pathogens may also be present.

27.2.1 Types

There are three major types of indicator bacteria:

1. *Total coliform (TC).* A large group of anaerobic, gram-negative, nonspore-forming, rod-shaped bacteria that ferment lactose with gas formation within 48 hr at 35°C. They exist in both polluted and unpolluted soils and occur in the feces of warm-blooded animals. *Escherichia coli* (or *E. coli*) and *Aerobacter aerogenes* are common members of the group that occur in organisms and soils, respectively.
2. *Fecal coliform (FC).* A subset of TC that come from the intestines of warm-blooded animals. Thus, because they do not include soil organisms, they are

preferable to TC. They are measured by running the standard total coliform test at an elevated temperature (44.5°C). As a general rule-of-thumb the FC is about 20% of TC (Kenner 1978). However, there is a wide spread in the ratio.

3. *Fecal streptococci (FS).* These include several varieties of streptococci that originate from humans (*Streptococcus faecalis*) as well as from domesticated animals such as cattle (*Streptococcus bovis*) and horses (*Streptococcus equinus*).

Although the TC measurement has traditionally been the most widely used indicator of contamination, its use is problematic because of the presence of nonfecal coliform bacteria. Consequently emphasis is shifting more to fecal coliforms and fecal streptococci.

Further, the ratio of FC to FS (FC/FS) has been used to determine whether contamination is due to human or animal sources. In general an FC/FS > 4 is often taken to indicate human contamination whereas FC/FS < 1 is interpreted as originating from other warm-blooded animals. However, the ratio should be used with care because of differential die-off of FC and FS. Thus, as the distance from a sewage outfall increases, the ratio can change and simplistic interpretation of the ratio could be misleading.

27.2.2 Concentrations

Per capita generation rates of indicator organisms are summarized in Table 27.2. Loading concentrations depend on the water use. For example in the United States, where per capita water use is high, the concentration of raw sewage is on the order of 20×10^6 TC/100 mL. In contrast for a country like Brazil, where water consumption is lower, concentrations of 200×10^6 have been measured.

Diffuse sources can also be contaminated. For example the geometric mean of concentrations for urban runoff is about 0.3×10^6 TC/100 mL. Because the source of such pollution is animal waste (rodents, dogs, cats, etc.), the FC/FS ratio tends to be low (< 0.7) for such sources. In contrast combined sewer overflows (CSOs) carry both human wastewater and storm water. Consequently their FC/FS ratio is higher (> 4). In addition, because of the presence of wastewater, their concentration tends to be higher [geometric mean $\cong 6 \times 10^6$ TC/100 mL].

TABLE 27.2
Per capita generation rates of intestinal bacteria for warm-blooded animals (Metcalf and Eddy 1991)

	TC	FC	FS	
Animal	**(10^6 number capita^{-1} d^{-1})**			**FC/FS**
Human	100,000–400,000	2,000	450	4.4
Chicken		240	620	0.4
Cow		5,400	31,000	0.2
Duck		11,000	18,000	0.6
Pig		8,900	230,000	0.04
Sheep		18,000	43,000	0.4
Turkey		130	1,300	0.1

TABLE 27.3
Concentration standards for total and fecal coliform

Use	TC (No. / 100 mL)	FC (No. / 100 mL)
Drinking water	0	0
Shellfish	70	14
Fishing	1000–5000	100–1000
Contact recreation	1000–5000	100–1000

Finally Table 27.3 summarizes concentration standards for several water uses. Recognize that these values are much smaller than the typical concentrations of sewage. For example the least stringent standard in the table (1000 to 5000 number/100 mL for fishing and contact recreation) is about 4 orders of magnitude less than the average United States sewage concentration of 20×10^6.

27.3 BACTERIAL LOSS RATE

The loss rate for total coliform bacteria can be represented as

$$k_b' = k_{b1} + k_{bi} + k_{bs} \tag{27.1}$$

where k_b' = total loss rate (d^{-1})
k_{b1} = base mortality rate (d^{-1})
k_{bi} = loss rate due to solar radiation (d^{-1})
k_{bs} = settling loss rate (d^{-1})

Note that we have marked the loss rate with a prime because it includes a transport mechanism: settling. Later in this lecture, when we model sediment-water interactions, we will separate that effect from the actual mortality.

27.3.1 Natural Mortality and Salinity

The following equation can be used to calculate a base mortality rate for total coliforms (Mancini 1978, Thomann and Mueller 1987):

$$k_{b1} = (0.8 + 0.006P_s)1.07^{T-20} \tag{27.2}$$

where P_s = percent sea water. Thus this formulation assumes a freshwater loss rate of 0.8 d^{-1}. This freshwater loss is supplemented by a saltwater loss that is linearly dependent on salinity. Consequently the total loss rate ranges from 0.8 d^{-1} for fresh water to 1.4 d^{-1} for salt water. The total loss is then modified to account for temperature. Recall from Lec. 2 that a value of 1.07 denotes a strong temperature dependence (that is, a doubling of the rate for a 10°C temperature rise).

If sea water is assumed to have a salinity of 30 to 35 ppt, Eq. 27.2 can also be written as

$$k_{b1} = (0.8 + 0.02S)1.07^{T-20} \tag{27.3}$$

where S = salinity (ppt or g L^{-1}).

27.3.2 Light

The bacterial loss due to the effect of light can be represented as (Thomann and Mueller 1987)

$$k_{bi} = \alpha \bar{I} \tag{27.4}$$

where k_{bi} = light decay rate (d^{-1})
α = a proportionality constant
$\bar{I}$ = average light energy (ly hr^{-1})

Using data from Gameson and Gould (1974), Thomann and Mueller (1987) concluded that α is approximately unity.

As depicted in Fig. 27.1, light extinction for a well-mixed layer of water can be modeled as an exponential decay represented by the Beer-Lambert law

$$I(z) = I_0 e^{-k_e z} \tag{27.5}$$

where $I(z)$ = light energy (ly hr^{-1})
I_0 = surface light energy (ly hr^{-1})
k_e = an extinction coefficient (m^{-1})
z = depth (m)

Note that the extinction coefficient is a function of the quantity of particulate matter and the color of the water. It can be related to Secchi-disk depth (m) by

$$k_e = \frac{1.8}{SD} \tag{27.6}$$

or to suspended solids, m (mg L^{-1}), by (Di Toro et al. 1981)

$$k_e = 0.55m \tag{27.7}$$

The average light for the layer can then be estimated by integrating over the depth, as in

$$\bar{I} = \frac{\int_0^H I_0 e^{-k_e z}\, dz}{H} = \frac{I_0}{k_e H}\left(1 - e^{-k_e H}\right) \tag{27.8}$$

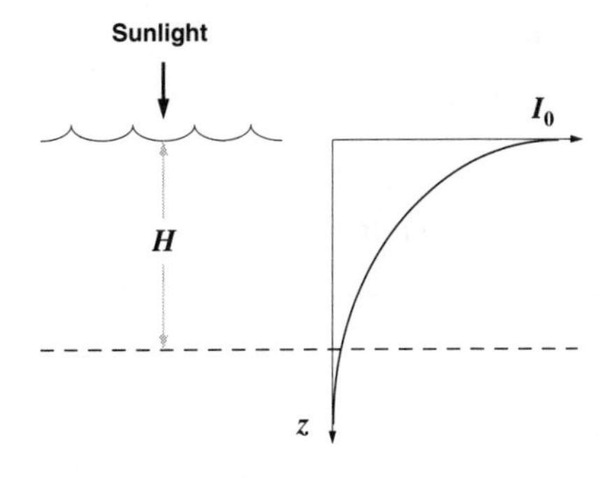

FIGURE 27.1
The exponential extinction of light for a well-mixed layer.

Equation 27.8 can be substituted into Eq. 27.4 to give

$$k_{bi} = \frac{\alpha I_0}{k_e H}\left(1 - e^{-k_e H}\right) \tag{27.9}$$

27.3.3 Settling

Settling losses depend on how many of the organisms are attached to particles. To model this process we must first discriminate between the quantity of bacteria that are free-floating and attached,

$$N = N_w + N_p \tag{27.10}$$

where N_w = concentration of bacteria that are free-floating (number/100 mL) and N_p = concentration of bacteria that are attached to particles (number/100 mL).

The quantity of bacteria on the particles is often expressed as a mass-specific concentration, r (number/g). Thus the volume-specific concentration on the particles can be expressed as

$$N_p = 10^{-4} rm \tag{27.11}$$

where m = suspended solids concentration (mg L^{-1}) and 10^{-4} is included to convert volume to 100 mL.

The tendency of the bacteria to attach to particles can also be represented by a linear partition coefficient,

$$K_d = 10^{-4} \frac{r}{N_w} \tag{27.12}$$

where K_d = a partition coefficient (m^3 g^{-1}) and 10^{-4} is again included to convert volume to 100 mL.

If the rate at which the bacteria adsorb and desorb from the particles is fast, a local equilibrium is assumed to occur. Equations 27.11 and 27.12 can be substituted into Eq. 27.10 to give

$$N = N_w + K_d m N_w \tag{27.13}$$

which can be solved for

$$N_w = F_w N \tag{27.14}$$

where F_w = fraction of the bacteria that are free floating,

$$F_w = \frac{1}{1 + K_d m} \tag{27.15}$$

Equation 27.14 can be substituted into Eq. 27.10 and solved for

$$N_p = F_p N \tag{27.16}$$

where F_p = fraction of the bacteria that are attached,

$$F_p = \frac{K_d m}{1 + K_d m} \tag{27.17}$$

If the settling velocity of the particles is v_s (m d^{-1}), the loss due to settling can then be represented as

$$k_{bs} = F_p \frac{v_s}{H} \tag{27.18}$$

27.3.4 Total Loss Rate

The total loss rate can now be formed by substituting Eqs. 27.3, 27.9, and 27.18 into 27.1 to give

$$k_b' = \underbrace{(0.8 + 0.02S)1.07^{T-20}}_{\text{Natural mortality}} + \underbrace{\frac{\alpha I_0}{k_e H}(1 - e^{-k_e H})}_{\text{Light}} + \underbrace{F_p \frac{v_s}{H}}_{\text{Settling}} \tag{27.19}$$

EXAMPLE 27.1. BACTERIA MODEL FOR A WELL-MIXED POND. A community of 10,000 people is being developed alongside a stream. The development's wastewater has a per capita flow rate of 0.5 m^3 capita^{-1} d^{-1}, a coliform generation rate of 1×10^{11} number capita^{-1} d^{-1}, and a suspended solids contribution of 100 g capita^{-1} d^{-1}. A sample of the raw wastewater is centrifuged and it is determined that $K_d = 0.05$ m^3 g^{-1}.

It is proposed that the sewage be passed through a small artificial pond prior to discharge to the stream. The pond has the following dimensions: $V = 2.4 \times 10^4$ m^3, $H = 4$ m, and $A_s = 6 \times 10^3$ m^2.

Determine how much the pond will reduce both the suspended solids and the coliform levels during calm periods. Assume that the solids settle at a rate of 0.4 m d^{-1} and that there is no resuspension of solids when the weather is calm. Also assume that the pond's temperature is 25°C and the mean daily solar radiation is 350 ly d^{-1}.

Solution: First, we can determine the flow that will be contributed by the development,

$$10{,}000 \text{ capita} \times 0.5 \text{ m}^3 \text{ capita}^{-1} \text{ d}^{-1} = 5000 \text{ m}^3 \text{ d}^{-1}$$

which means that the pond's residence time will be 24,000/5000 = 4.8 d.

The inflow's and the pond's suspended solids can be calculated as

$$m_{\text{in}} = \frac{10{,}000 \text{ capita}(100 \text{ g capita}^{-1} \text{ d}^{-1})}{5000 \text{ m}^3 \text{ d}^{-1}} = 200 \text{ mg L}^{-1}$$

$$m = \frac{5000}{5000 + 0.4(6000)} 200 = (0.67)200 = 135 \text{ mg L}^{-1}$$

Thus the pond removes 33% of the solids.

The extinction coefficient can be determined as $k_e = 0.55(135) = 74.3$ m^{-1}, and the fraction of the bacteria associated with the solids as

$$F_p = \frac{0.05(135)}{1 + 0.05(135)} = 0.871$$

This value, along with the other parameters, can be substituted into Eq. 27.19 to calculate

$$k_b' = (0.8)1.07^{25-20} + \frac{350/24}{74.3(4)}(1 - e^{-74.3(4)}) + 0.87\frac{0.4}{4}$$

$$= 1.122 + 0.049 + 0.087 = 1.258 \text{ d}^{-1}$$

The inflow and the pond bacteria concentrations can then be calculated as

$$N_{\text{in}} = \frac{10{,}000 \text{ capita}(1 \times 10^{11} \text{ number capita}^{-1} \text{ d}^{-1})}{5000 \text{ m}^3 \text{ d}^{-1}} \left(\frac{10^{-4} \text{ m}^3}{100 \text{ mL}}\right)$$

$$= 20 \times 10^6 \text{ number/100 mL}$$

$$N = \frac{5000}{5000 + 1.258(24{,}000)} 20 \times 10^6 = 2.84 \times 10^6 \text{ number/100 mL}$$

Thus although about 85% of the bacteria would be removed, the loading to the river would still be very large.

27.4 SEDIMENT-WATER INTERACTIONS

In the foregoing analysis, bacteria were lost to the sediments via settling. For shallow systems the bacteria can be reintroduced into the surface water via storm events. To gain insight into this mechanism, we must first reformulate Eq. 27.19 so that settling is separated from mortality,

$$k_b' = k_b + F_p \frac{v_s}{H} \tag{27.20}$$

where k_b = mortality rate,

$$k_b = (0.8 + 0.02S)1.07^{T-20} + \frac{\alpha I_0}{k_e H}(1 - e^{-k_e H}) \tag{27.21}$$

Once this is done we can write mass balances for the water and sediments as (see Fig. 27.2)

$$V_1 \frac{dN_1}{dt} = QN_{\text{in}} - QN_1 - F_p v_s A_s N_1 - k_b V_1 N_1 + v_r A_s N_2 \tag{27.22}$$

$$V_2 \frac{dN_2}{dt} = F_p v_s A_s N_1 - k_{b2} V_2 N_2 - v_r A_s N_2 - v_b A_s N_2 \tag{27.23}$$

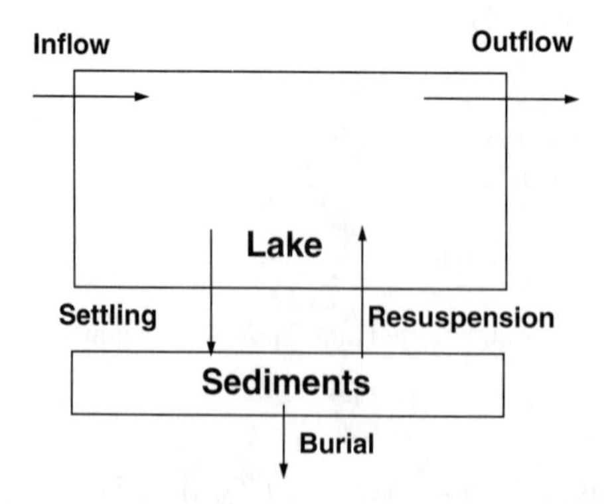

FIGURE 27.2
Lake and underlying sediment layer.

where the subscripts 1 and 2 designate the lake and sediments, respectively, and v_r and v_b = resuspension and burial velocities, respectively.

At steady-state, Eq. 27.23 can be solved for

$$N_2 = R_{21}N_1 \tag{27.24}$$

where R_{21} = ratio of sediment to water concentration,

$$R_{21} = \frac{F_p v_s}{v_r + v_b + k_{b2}H_2} \tag{27.25}$$

This can be substituted into Eq. 27.22, which can be solved for

$$N_1 = \frac{q_s}{q_s + F_p v_s + k_{b1}H_1 - v_r R_{21}} N_{\text{in}} \tag{27.26}$$

where $q_s = Q/A_s$.

EXAMPLE 27.2. BACTERIA MODEL FOR A WELL-MIXED POND AND SEDIMENTS. For the pond described in Example 27.1, (*a*) calculate the concentration in the sediments during calm periods (that is, no resuspension). Assume that the bacteria are confined to a thin surface sediment layer that is 5 mm thick. Also assume that burial losses are negligible, the temperature of the surface layer is 15°C, and negligible light reaches the sediment-water interface. (*b*) Determine the water concentration of bacteria immediately after a storm that resuspends the top 40% of the layer (2 mm).

Solution: (*a*) Bacterial death in the sediment layer can be estimated with Eq. 27.21,

$$k_b = (0.8)1.07^{15-20} + 0 = 0.57\ \text{d}^{-1}$$

The ratio of sediment to water concentration can be calculated as

$$R_{21} = \frac{0.87(0.4)}{0 + 0 + 0.57(0.005)} = 122$$

This ratio, along with $N_1 = 2.84 \times 10^6$ number/100 mL from Example 27.1, can be substituted into Eq. 27.24 to give

$$N_2 = 122(2.84 \times 10^6) = 347 \times 10^6 \text{ number/100 mL}$$

Thus the sediments have a much higher concentration of bacteria than the water.

(*b*) The storm loading to the water can be idealized as an impulse loading. The quantity of bacteria added to the water can be computed as

$$V_2N_2 = 30 \times 10^6 \text{ mL}(347 \times 10^6 \text{ number/100 mL})0.4 = 4.17 \times 10^{13}$$

which can be normalized to the pond's volume to determine the resulting increase in concentration,

$$\Delta N_1 = \frac{V_2N_2}{V_1} = \frac{4.17 \times 10^{13} \text{ number}}{24{,}000 \times 10^6 \text{ mL}} \frac{100 \text{ mL}}{(100 \text{ mL})} = 0.174 \times 10^6 \text{ number/100 mL}$$

27.5 PROTOZOANS: *GIARDIA* AND *CRYPTOSPORIDIUM*

As reflected by this chapter, bacteria have been the focus of most past efforts to manage pathogens. However, as outlined in Table 27.1, there are other organisms that can create waterborne health problems. In particular two parasitic protozoa, *Giardia* and *Cryptosporidium*, have received great attention as drinking-water pathogens.

Giardia was first found by Leeuwenhoek in 1681, who discovered it in his own stools (Schmidt and Roberts 1977). At present, *Giardia* is the most common disease-inducing intestinal parasite in the United States (Smith and Wolfe 1980). When ingested the organism causes ***giardiasis***, a disease marked by a number of symptoms including diarrhea, nausea, cramps, dehydration, and headaches.

Cryptosporidium was first discovered around the turn of the century. Species names are associated with the host organism. For example *Cryptosporidium parvum* is the name given to the species found in the domestic mouse. This is also the major species responsible for illness in humans. Currently there is a disputed number of distinct species because an organism from one animal host can infect another, including humans. In general the disease caused by the organism, ***cryptosporidiosis***, has similar symptoms to giardiasis and is self-limiting in individuals with healthy immune systems. However, for individuals with impaired immune systems (such as those infected with the AIDS virus) no cure is currently available.

Both *Giardia* and *Cryptosporidium* exist in the environment in resting stages called ***cysts*** and ***oocysts***, respectively. When ingested, these can hatch in the gastrointestinal tract to produce their growing stages, which can multiply and lead to the associated illnesses. Both cysts and oocysts range from spherical to ovoid in shape. LeChevallier and Norton (1995) report that *Giardia* has an 8.6-μm width (6.6 to 11.9) and a 12.3-μm length (8.6 to 16.5). These dimensions are equivalent to an effective diameter of about 10 μm (7.5 to 14). *Cryptosporidium* oocysts are smaller, with effective diameters on the order of 5 μm (3 to 7). Both have specific gravities in the range from 1.05 to 1.1.

27.5.1 Loadings

In general it is believed that elevated *Giardia* levels are mostly due to the introduction of sewage effluents, whereas *Cryptosporidium* is more closely associated with nonpoint sources (LeChevallier et al. 1991). However, both can originate from either point or nonpoint sources.

Sykora et al. (1991) reported that annual average geometric means of *Giardia* in raw sewage were about 1500 cysts L^{-1} (with a range from about 650 to 3000). Their data also indicated a significant seasonal trend with values from October through January being higher ($\approx$ 2500 cysts L^{-1}) than for the other months of the year ($\approx$ 1000 cysts L^{-1}). They also suggest that secondary treatment by activated sludge is superior (90 to 100% removal) to trickling filters (40 to 60%). Gassmann and Schwartzbrod (1991) reported a range from 800 to 14,000 cysts L^{-1} in sewage. Again, higher values were detected in the winter and early spring months.

Rose (1988) reported average values of *Cryptosporidium* in raw and treated sewage in the western United States as 28.4 and 17 oocysts L^{-1}, respectively. High

TABLE 27.4
Export coefficients for *Cryptosporidium* (Hansen and Ongerth 1991)

Basin	Land use	Drainage area (ha)	Export coefficient (oocysts ha^{-1} d^{-1})
Cedar	Protected; mountainous forest land	31,077	7.72×10^3
Snoqualmie (upper)	Unprotected; mountainous forest land; heavy recreational use	97,115	1.04×10^5
Snoqualmie (lower)	Unprotected; dairy farming	59,046	$1.20 \times 10^{6\dagger}$ $5.41 \times 10^{6\ddagger}$

† Entire study period.
‡ During high runoff.

values were obtained in Arizona, which had raw values of 1732 and treated values of 489 oocysts L^{-1}. Madore et al. (1987) reported higher values of 5180 oocysts L^{-1} (850 to 13,700) and 1063 oocysts L^{-1} (4 to 3960) for raw and treated effluents, respectively.

Hansen and Ongerth (1991) studied two watersheds in Washington that were not impacted by sewage. As in Table 27.4, they found a large difference in *Cryptosporidium* export coefficients between the protected Cedar River and the unprotected Snoqualmie River. In addition, within the Snoqualmie watershed they detected significant increases due to land use and seasonal effects.

27.5.2 Concentrations in Natural Waters

LeChevallier et al. (1991) have measured raw drinking-water levels of surface water treatment plants in the United States and Canada. The geometric mean of detectable (69 out of 85 samples) *Giardia* was 2.77 cysts L^{-1}, with levels ranging from 0.04 to 66. If the nondetectables are assigned the value of 0.04 at the lower end of the range, the geometric mean would be lowered to about 1 cyst L^{-1}. This is consistent with a value of about 0.9 cyst L^{-1} reported by Rose in Western U.S. streams and lakes.

LeChevallier et al. (1991) also reported a geometric mean of detectable (74 out of 85 samples) *Cryptosporidium* as 2.7 oocysts L^{-1}, with levels ranging from 0.07 to 484. If the nondetectables are assigned the value of 0.07 at the lower end of the range, the geometric mean would be lowered to about 1.7 oocyst L^{-1}. They suggest that their results compare favorably with a range of 0.002 to 112 oocysts L^{-1} reported by other investigators.

In summary, environmental concentrations for both *Cryptosporidium* and *Giardia* are on the order of 1 cyst or oocyst L^{-1}, ranging from about 0.05 for pristine waters to 100 for polluted systems. In general the *Cryptosporidium* levels tend to be somewhat higher. LeChavallier states that *Cryptosporidium* levels are typically about 1.5 times as high as for *Giardia.*

Before proceeding I should note that cysts and oocysts can become nonviable in the environment. Thus only a fraction of the total concentration may lead to infection. In general, lower viability tends to occur at high temperatures (Wickramanayake et al. 1985, deRegnier 1989).

27.5.3 Drinking-Water Treatment and Acceptable Risk Levels

The EPA (1989) has ruled that an acceptable risk for giardiasis is 10^{-4}, that is, one infection per year per 10,000 people. To achieve such a risk it has been calculated that potable water should not contain more than 7×10^{-6} cyst L^{-1} for *Giardia* on the basis of the geometric mean for 1 year. To achieve such a risk for *Cryptosporidium*, it has been calculated that potable water should not contain more than 3×10^{-5} cyst L^{-1} on the basis of the geometric mean for one year (Rose et al. 1991, Regli et al. 1991, LeChevallier and Norton 1995).

Figure 27.3, which summarizes all the information compiled above, indicates that significant levels of treatment will be necessary to meet these objectives. Present conventional filtration technology attains about 99 to 99.9% removal (2-log to 3-log reductions). Thus raw drinking waters subject to sewage and nonpoint runoff would require additional treatment. The problem is further complicated because *Giardia* can be treated effectively with chlorine whereas *Cryptosporidium* cannot. Thus, to mitigate the high cost of treatment, water-quality management may be needed in order to improve the quality of raw water supplies. Models of the sort described next may be useful in guiding such management.

27.5.4 *Giardia/Cryptosporidium* Model

A simple model for *Giardia* and *Cryptosporidium* can be developed for a stratified lake. In the following discussion we will use the term cyst to refer both to the *Giardia*

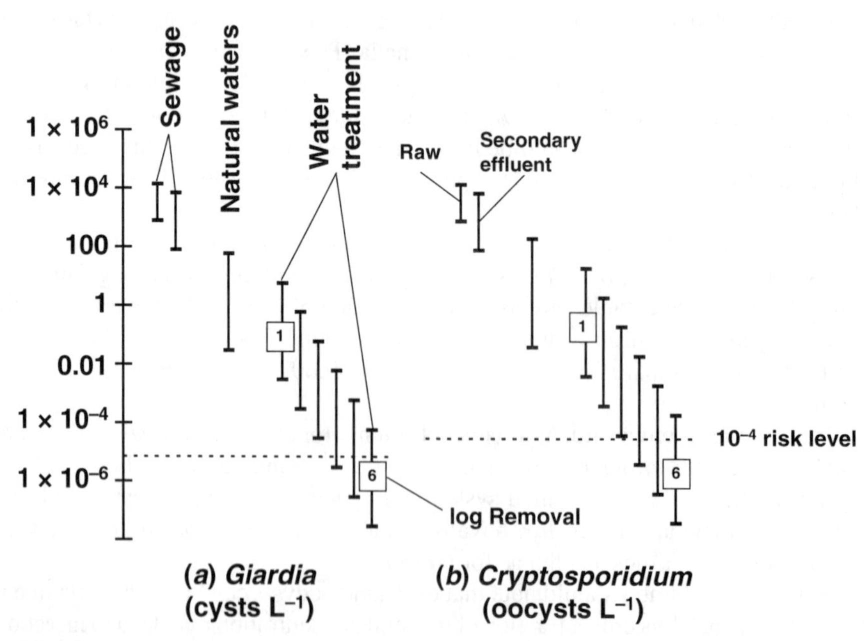

FIGURE 27.3
Summary of *Giardia* and *Crytosporidium* levels in waste, natural, and drinking waters.

cyst and the *Cryptosporidium* oocyst. Mass balances can be written for the surface and bottom layers of a thermally stratified lake as

$$V_e \frac{dc_e}{dt} = Qc_{\text{in}} - Qc_e - v_{s,e}A_t c_e + E_t'(c_h - c_e) \tag{27.27}$$

$$V_h \frac{dc_h}{dt} = v_{s,e}A_t c_e - v_{s,h}A_t c_h + E_t'(c_e - c_h) \tag{27.28}$$

where the subscripts e and h represent epilimnion (surface) and hypolimnion (bottom), respectively, V = volume (m^3), t = time (d), c = concentration (cysts L^{-1}), A_t = area of the thermocline separating the upper and lower layers (m^2), and E_t' = bulk diffusion coefficient for turbulent mixing across the thermocline (m^3 d^{-1}). Note that different settling velocities are used for the surface and bottom layers. This is done to reflect that settling should be higher in the surface layer because of the lower viscosity of warmer surface water.

EXAMPLE 27.3. *GIARDIA* AND *CRYPTOSPORIDIUM* SIMULATIONS. A water supply reservoir has surface and bottom volumes of $V_e = 3 \times 10^7$ and $V_h = 1 \times 10^7$ m^3. Simulate the lake's response given the following seasonally varying parameters:

Parameter	Unstratified ($t \le 120$ d)	Runoff ($120 < t \le 150$ d)	Stratified ($120 < t \le 150$ d)	Unstratified ($150 < t \le 360$ d)
T_e	4	6	20	6
T_h	4	6	8	6
E_t'	∞	∞	1.2×10^4	∞
Q	3×10^5	15×10^5	3×10^5	3×10^5
$c_{v,\text{in}}$	10	50	10	10
$c_{n,\text{in}}$	10	50	10	10
$v_{s,e}$–*Crypto.*	0.08	0.08	0.12	0.08
$v_{s,h}$–*Crypto.*	0.08	0.08	0.08	0.08
$v_{s,e}$–*Giardia*	0.3	0.3	0.475	0.3
$v_{s,h}$–*Giardia*	0.3	0.3	0.3	0.3

Note that the settling velocities are based on Stokes' law.

Solution: Equations 27.27 and 27.28 can be integrated to give the time series displayed in Fig. 27.4. In general the responses of the two organisms are similar in that both (1) exhibit heightened levels during runoff and (2) have higher hypolimnion concentrations during summer stratification. They differ in that *Cryptosporidium* exhibits somewhat higher concentrations. In addition the impact of spring runoff is more prolonged for *Cryptosporidium.* Both these results are due to *Cryptosporidium's* lower settling velocity.

The previous example points toward some possible management strategies for meeting drinking-water standards for protozoan pollution. First, it suggests that during stratified periods, water should be drawn from epilimnetic waters that are subject to both flushing and higher settling velocities due to higher temperatures. Second, it suggests that treatment might be managed from a seasonal perspective. That is, increased treatment might be required during and after high runoff events.

The foregoing analysis should be tempered because other drinking-water requirements such as toxic disinfection byproducts (e.g., trihalomethane formation due to chlorination) and taste and odor problems must be concurrently managed along with pathogen problems. In some cases these other problems might be minimized by adopting similar strategies as for pathogens. However, different and contradictory strategies might just as likely be required, e.g., taking water from the lake's hypolimnion to obtain cooler water to minimize trihalomethane formation.

In summary the management of pathogenic protozoans is complicated by a number of factors. In particular they are difficult to measure (especially their viability) and the concentration range from sources to finished water spans many orders of magnitude. However, sufficient information has been compiled to allow the development of simple, order-of-magnitude models of their levels in natural waters.

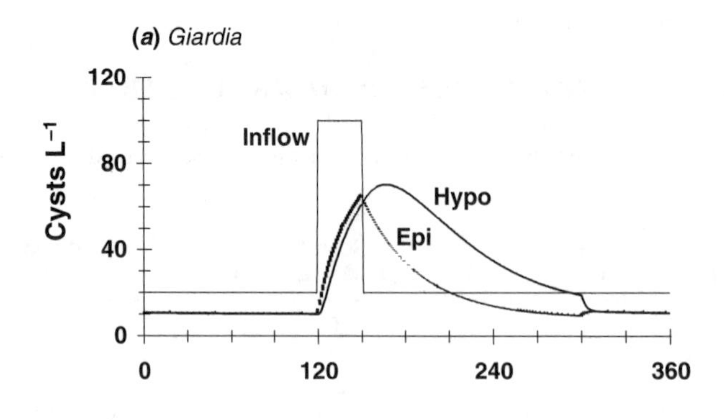

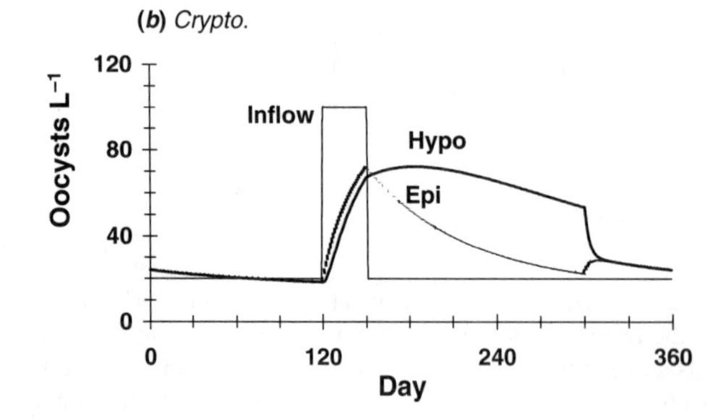

FIGURE 27.4

PROBLEMS

27.1. A stream receives wastewater from a city with a population of 100,000 people. The per capita loading rate of total coliform bacteria for the city is 1.5×10^{11} number capita^{-1} d^{-1} and the per capita flow rate is 0.5 m^3 capita^{-1} d^{-1}. The wastewater and the stream (above the discharge) have temperatures of 25 and 15°C, respectively. After mixing takes place, assume that the temperature in the stream remains constant. The peak solar radiation is 650 cal cm^{-2} d^{-1} and the photoperiod is 13 hr. The stream above the waste discharge has a flow of 100,000 $m^3 d^{-1}$ and negligible coliform bacteria. There are negligible suspended solids in the stream (the extinction coefficient is 0.5 m^{-1}).

(*a*) Assuming instantaneous mixing, compute the temperature at the discharge point.

(*b*) Assuming instantaneous mixing, compute the coliform concentration at the discharge point.

(*c*) Compute the velocity in the stream below the mixing point. Note that the stream is 20 m wide and 0.5 m deep.

(*d*) Compute the base mortality rate (k_{b1}) of the bacteria in the stream below the mixing point.

(*e*) Compute the light decay rate (k_{bi}) of the bacteria in the stream below the mixing point.

(*f*) Compute the concentration of the bacteria at a beach located 10 km downstream from the discharge.

27.2. A point source of bacteria enters an estuary channel having the following characteristics:

	Value	Units
Dispersion coefficient	10^6	$m^2 d^{-1}$
Flow	5×10^4	$m^3 d^{-1}$
Width	200	m
Depth	2	m

The estuary has an ambient temperature of 27.5°C. It also has a mean daily light intensity of 200 cal $cm^{-2} d^{-1}$ and an extinction coefficient of 0.2 m^{-1}. At the point of the input the estuary is 50% sea water. If the bacteria do not settle:

(*a*) Determine the bacterial die-off rate.

(*b*) What mass loading could be input to this system under steady-state conditions if the allowable concentration at the outfall is 1000 number/mL? Express your results in number d^{-1}. Assume complete lateral and vertical mixing at the outfall.

27.3. A lake with a residence time of 2 months (inflow = outflow) has a depth of 7 m and a surface area of 5×10^5 m^2. The average surface solar radiation is 200 ly d^{-1} and the temperature is 22°C. The lake has a suspended solids concentration of 2 mg L^{-1} that settle at a rate of 0.3 m d^{-1}. It also has an extinction coefficient of 0.4 m^{-1}. Note that the bacteria associate with the solids with a linear partition coefficient of 0.005 $m^3 g^{-1}$. If the total concentration of bacteria in the system cannot exceed 1000/100 mL, determine the acceptable loading rate for the lake.

27.4. For the same lake as in Prob. 27.3, compute the loading needed to maintain a sediment concentration of 100,000/100 mL. Assume that the sediment layer is 10 cm thick, that resuspension and burial are negligible, and that light has a negligible effect on bacterial die-off in the sediments.

27.5. Estimate the mean and range of settling velocities for both *Giardia* and *Cryptosporidium* using Stokes' law, along with the size and density information from Sec. 27.5. Base your estimate on the density and viscosity of fresh water at a temperature of 20°C.

27.6. (*a*) Calculate the rate of *Giardia* that could be loaded to a small lake (residence time = 1 month, volume = 10×10^6 m^3, and mean depth = 2 m), if the only loss mechanisms are flushing and settling (v_s = 0.4 m d^{-1}) and the desired concentration is 0.1 cyst L^{-1}. Express your result as an inflow concentration.

(*b*) How much additional treatment (express as a log removal) would be required to meet the 10^{-4} risk level for drinking water?

PART V

Eutrophication and Temperature

We now turn from decomposition processes to a more complete representation of the life/death cycle in the aquatic environment. The problem of eutrophication will serve as the focus for the following set of lectures. In addition we will also cover the rudiments of heat budget models that are so important in thermally stratified systems.

Lecture 28 provides an introduction to the eutrophication problem along with information on some of the nutrients that stimulate plant growth in natural waters. *Lecture 29* describes the phosphorus loading concept that was developed to provide simple models of eutrophication in lakes.

The succeeding two lectures deal with temperature modeling because of the great impact of thermal stratification on eutrophication. *Lecture 30* describes how heat enters and leaves the surface of a water body due to atmospheric interactions. Then *Lecture 31* illustrates how heat budgets and thermal stratification are simulated.

Next the interactions between nutrients and organisms are explored. *Lecture 32* provides an introduction to modeling microbial growth kinetics. This material relates to algae as well as to other organisms such as bacteria. Then the microbial growth model is applied to algae (*Lecture 33*) with special emphasis placed on the role of light in photosynthesis. *Lecture 34* deals with predator-prey interactions between organisms.

Finally a total nutrient/food-chain framework is developed and applied to lakes (*Lecture 35*) and streams (*Lecture 36*). The latter lecture includes information on how the QUAL2E model can be used to simulate both heat balances and stream eutrophication.

PART V

Eutrophication and Temperature

LECTURE 28

The Eutrophication Problem and Nutrients

Lecture overview: I provide background information on the eutrophication problem. After an overview of the symptoms of eutrophication, I discuss the primary nutrients that stimulate the process. I also describe plant stoichiometry and the nitrogen:phosphorus ratio.

Several times a year I fertilize my lawn and garden so I can grow green grass and lots of fresh vegetables. In a similar way the addition of nutrients to a natural water stimulates plant growth. In small quantities this can be a good thing. For example because of the higher human population of its drainage basin, Lake Michigan receives more nutrients than Lake Superior. Consequently Lake Michigan has more plant growth, which ultimately allows it to support more game fish.

However, as with most everything in life, there can always be too much of a good thing. When lakes, streams, and estuaries are overfertilized, the resulting excessive plant growth can become a serious water-quality problem.

This "overfertilization" phenomenon is generally referred to as ***eutrophication***. This terminology was originally coined to describe the natural aging process whereby a lake is transformed from a lake to a marsh to a meadow. This process can take thousands of years to occur naturally. However, the process can be greatly quickened by excess nutrients from human activities. This accelerated process is sometimes called ***cultural eutrophication***.

Water bodies are often classified as to their trophic state. The general terms are

- Oligotrophic (poorly nourished)
- Mesotrophic (moderately nourished)
- Eutrophic (well-nourished)
- Hypereutrophic (overnourished)

Although such terms were originally developed for and are most commonly applied to lakes, they are also appropriate descriptors of streams and estuaries.

28.1 THE EUTROPHICATION PROBLEM

In general, eutrophication can have a number of deleterious effects on water bodies. These include:

- *Quantity.* The profuse growth of floating plants decreases water clarity, and some species form unsightly scums. Further, certain floating plants can clog filters at water treatment plants, and overgrowth of rooted plants can hinder navigation and recreation by clogging waterways.
- *Chemistry.* Plant growth and respiration can affect the system's water chemistry. Most notably, oxygen and carbon dioxide levels are directly impacted by plant activity. Oxygen has implications related to the survival of organisms such as fish. In particular the bottom waters of thermally stratified systems can become totally devoid of oxygen due to the decomposition of dead plants. Carbon dioxide can impact pH.
- *Biology.* Eutrophication can alter the species composition of an ecosystem. Native biota may be displaced as the environment becomes more productive. Certain species of algae cause taste and odor problems in drinking water. Further, certain blue-green algae can be toxic when consumed by animals. Many of these problems become prominent as the water body becomes more eutrophic.

Now that we have a feeling for the types of problems caused by eutrophication, we must figure out how it works. The first step in this process involves the inorganic nutrients that serve as the raw materials from which plant biomass is synthesized.

28.2 NUTRIENTS

Inorganic nutrients provide chemical building blocks for life in aquatic systems. Some are required in large quantities for cell development and hence are called ***macronutrients***. These are carbon, oxygen, nitrogen, phosphorus, sulfur, silica, and iron. Smaller quantities of ***micronutrients***, such as manganese, copper, and zinc, are also necessary. Water-quality modeling has focused on four macronutrients: phosphorus, nitrogen, carbon, and silica.

28.2.1 Phosphorus

Phosphorus is essential to all life. Among other functions, it has a critical role in genetic systems and in the storage and transfer of cell energy.

From a water-quality perspective, phosphorus is important because it is usually in short supply relative to the other macronutrients. This scarcity is due to three primary factors:

1. It is not abundant in the earth's crust. Further, the phosphate minerals that do exist are not very soluble.
2. It does not exist in a gaseous form. Thus, in contrast to carbon and nitrogen, there is no gaseous atmospheric source.
3. Finally phosphate tends to sorb strongly to fine-grained particles. The settling of these particles, along with sedimentation of organic particles containing

phosphorus, serves to remove phosphorus from the water to the bottom sediments. For cases where the water in contact with the sediments contains oxygen, such sediment phosphorus becomes chemically trapped.

Although phosphorus is naturally scarce, many human activities result in phosphorus discharge to natural waters. Human and animal wastes both contain substantial amounts of phosphorus. In the recent past the former has been supplemented by detergent phosphorus. In addition nonpoint sources from agricultural and urban land both contribute excess phosphorus. Part of the enhancement of diffuse sources is due to fertilizers and other phosphorus-containing chemicals associated with human land use. Moreover, human uses lead to soil erosion, which also enhances phosphorus transport into waters.

Phosphorus in natural waters can be subdivided in several ways. One scheme, which stems from conventional measurement techniques and modeling necessity, is (Fig. 28.1)

- *Soluble reactive phosphorus (SRP).* Also called orthophosphate or soluble inorganic P, this is the form that is readily available to plants. It consists of the species $H_2PO_4^-$, HPO_4^{2-}, and PO_4^{3-}.
- *Particulate organic P.* This form mainly consists of living plants, animals, and bacteria as well as organic detritus.
- *Nonparticulate organic P.* These are dissolved or colloidal organic compounds containing phosphorus. Their primary origin is the decomposition of particulate organic P.
- *Particulate inorganic P.* This category consists of phosphate minerals (e.g., apatite phosphorus), sorbed orthophosphate (e.g., on clays), and phosphate complexed with solid matter (e.g., calcium carbonate precipitates or iron hydroxides).
- *Nonparticulate inorganic.* This group includes condensed phosphates such as those found in detergents.

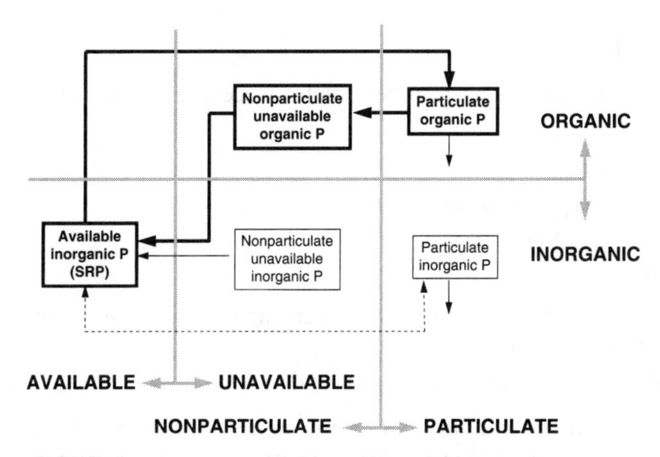

FIGURE 28.1
Forms of phosphorus found in natural waters. The principal forms involved in the production/decomposition life cycle are shown in bold.

Phosphorus is measured in several ways. Soluble reactive phosphorus (SRP) is measured by adding ammonium molybdate, which forms a colored complex with the phosphate. Organic phosphorus is not detected by the test unless it is first hydrolyzed, that is, converted to orthophosphate by digestion with heat and a strong acid. In addition filtering can be used to divide between particulate and nonparticulate components. Finally application of the SRP method to a digested, unfiltered sample provides a means to measure all the phosphorus in a sample. As we will see later in this lecture and the next, such a ***total phosphorus (TP)*** measurement has been used widely to quantify eutrophication.

As mentioned above, the partitioning in Fig. 28.1 is based on available measurement techniques and modeling necessity. The distinction between particulate and nonparticulate forms is made so that the former can selectively be removed by settling. The division of available phosphorus from the other species is made because it is the only form that is directly available for plant growth. It should be understood that the other forms are not absolutely "unavailable." Rather, they must first be converted to SRP before they can be consumed by plants.

It also should be noted that Fig. 28.1 is not the last word regarding phosphorus segmentation. In fact from a strictly scientific basis, each of the compartments could be broken down into finer detail. Conversely from the perspective of water-quality modeling, some of the compartments are often consolidated whereas others are broken down into finer segments. For example the distinction between inorganic and organic unavailable forms is not usually made. That is, they are lumped into unavailable particulate P and unavailable nonparticulate groupings. An example of more refinement is that living particulate organic P is often distinguished from nonliving forms. In other words, groups of organisms (such as phytoplankton, zooplankton, etc.) are modeled separately. In such cases the amount of phosphorus contained in these groups would have to be subtracted from the particulate P compartment. We return to this topic when we build models in a later lecture.

28.2.2 Nitrogen

I have already described the nitrogen cycle in Lec. 23. Recall that the primary forms are

- Free nitrogen (N_2)
- Ammonium (NH_4^+)/ammonia (NH_3)
- Nitrite (NO_2^-)/nitrate (NO_3^-)
- Organic nitrogen

As depicted in Fig. 28.2, the organic nitrogen can be broken down further into particulate and dissolved components.

Some of the major processes governing the dynamics of these groups are

- *Ammonia and nitrate assimilation.* This includes the uptake of inorganic nitrogen by phytoplankton. Although phytoplankton utilize both ammonia and nitrate, their preference for the former has been demonstrated (Harvey 1955, Walsh and Dugdale 1972, Bates 1976).

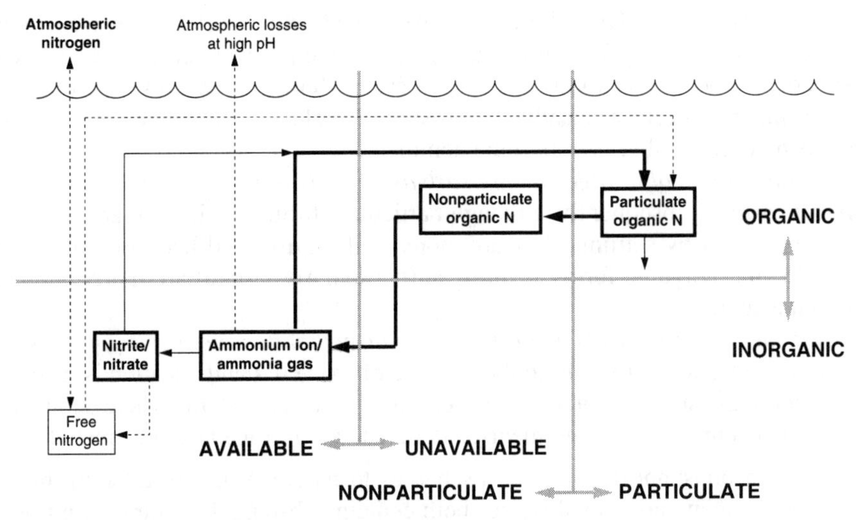

FIGURE 28.2
Forms of nitrogen found in natural waters. The principal forms involved in the production/decomposition life cycle are shown in bold.

- *Ammonification.* This is the transformation of organic nitrogen to ammonia. This is a complicated process involving several mechanisms, including bacterial decomposition, zooplankton excretion, and direct autolysis after cell death.
- *Nitrification.* This is the oxidation of ammonia to nitrite and nitrite to nitrate via the action of a select group of aerobic bacteria. This process utilizes oxygen and typically is represented by first-order reactions. The fact that the transformation of nitrite to nitrate is relatively fast supports the use of a single functional group to encompass both nitrate and nitrite in some nutrient/food-chain models. For these cases the nitrification process would be represented kinetically by the oxidation of ammonia.
- *Denitrification.* Under anaerobic conditions—for example in the sediments and the anoxic hypolimnia of some lakes—nitrate can serve as an electron acceptor for certain bacteria. Nitrite is formed as an intermediate, with the principal end product being free nitrogen.
- *Nitrogen fixation.* A number of organisms can fix elemental nitrogen. An important group from the standpoint of nutrient/food-chain modeling is blue-green algae possessing heterocysts. These organisms are important because, for lakes with high phosphorus loadings, phytoplankton growth can depress nitrogen levels to the point where nonfixing algae will become nitrogen-limited. The ability of the blue-green algae to utilize free nitrogen gives them a competitive advantage in such situations. The resulting dominance by the blue-greens has implications to water quality since many species have objectionable characteristics; for example they form floating scums.

Although nitrogen is just as necessary for life as phosphorus, they differ in three ways:

- *Nitrogen has a gas phase.* Further, as mentioned above, certain blue-green algae are capable of fixing free nitrogen. This gives them a competitive advantage in situations where other forms of nitrogen are in short supply. This state of affairs can sometimes occur when advanced treatment includes nitrogen removal. In such cases blue-green algae can become dominant.
- *Inorganic forms of nitrogen do not sorb as strongly to particulate matter as does phosphorus.* Consequently, although particulate forms of nitrogen are carried to the sediments by settling, they are more easily introduced back into the water. In addition inorganic forms of nitrogen (particularly nitrate) are more mobile in groundwater.
- *Denitrification represents a purging mechanism that does not occur for phosphorus.* Because it occurs only in the absence of oxygen, denitrification is insignificant for many surface waters. However, for productive systems where denitrification can occur in anoxic sediments, a deficiency of nitrogen can be created.

As with phosphorus, nitrogen discharges to natural water result from human activities. Human and animal wastes both contain substantial amounts of nitrogen. In addition nonpoint sources from agricultural and urban land both contribute excess nitrogen. As mentioned above, because forms such as nitrate do not associate strongly with solid matter, they can be easily transmitted to surface waters along with groundwater flow.

Most of the aforementioned characteristics mean that phosphorus has usually been identified as the primary controllable nutrient governing the eutrophication process in fresh waters. However, productive estuaries can tend to be nitrogen limited.

28.2.3 Carbon

Carbon can play three roles in water-quality modeling:

- *Nutrient.* In the same way as phosphorus and nitrogen, carbon can be thought of as a nutrient. However, in most cases modelers usually assume that carbon cannot limit algal growth, although some models (Chen 1970, Chen and Orlob 1975) do allow for potential limitation. Studies dealing with carbon-limited algal cultures raise questions regarding the form of carbon needed for photosynthesis by different algae (Goldman et al. 1974, King and Novak 1974). It also has been suggested that the relative abilities of green and blue-green algae to use various forms of inorganic carbon could partially explain the succession from greens to blue-greens in enriched natural waters (King 1972, Shapiro 1973). These and other developments (see references in Goldman et al. 1974) indicate that control of primary production by carbon limitation could be important in certain systems.
- *Biomass.* Because it usually constitutes a large component of organic compounds, carbon is often used as a measure of biomass.
- *Pollutant.* Finally carbon is an important factor in other pollution problems beyond eutrophication. First, as already discussed in great detail in Part IV, the decomposition of organic carbon can greatly affect a system's oxygen concentration. Second, it is known that many toxicants preferentially associate with organic matter. Thus the dynamics of toxics in the environment is intimately connected with

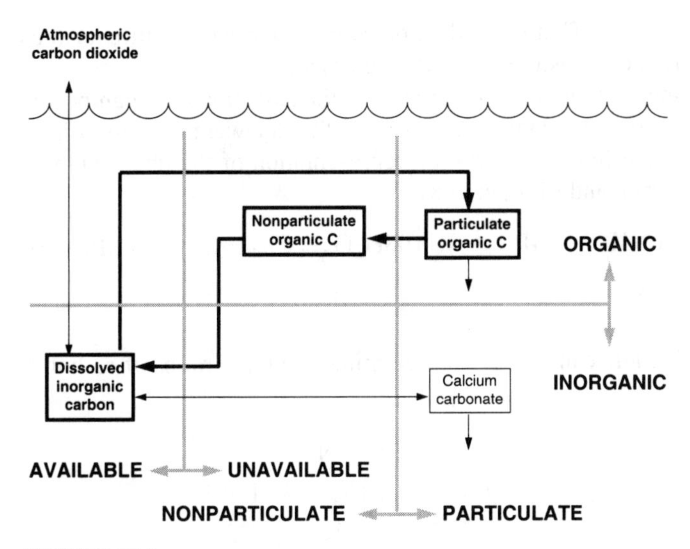

FIGURE 28.3
Forms of carbon found in natural waters. The principal forms involved in the production/decomposition life cycle are shown in bold.

the generation, transport, and fate of organic carbon. Finally naturally produced organic carbon can, itself, be transformed into a toxic compound. For example chlorine can react with organic compounds to form toxic trihalomethanes.

Figure 28.3 is an effort to represent the carbon cycle in natural waters. Note that the dissolved inorganic carbon compartment actually consists of several species: carbon dioxide (CO_2), bicarbonate (HCO_3^-), and carbonate (CO_3^-).

28.2.4 Silicon

Although it might be considered a minor nutrient, silicon has significance in the dynamics of phytoplankton because of its importance as a major structural element in the cells of an important phytoplankton group—the diatoms. These organisms use dissolved reactive silicon [mainly as $Si(OH)_4$] to build a ***frustule*** or "glass wall" that surrounds the cell. Silicon in frustules is not available to other diatoms, and ambient concentrations of available silicon can therefore become low enough to limit further growth of these algae.

In most modeling work to date, silicon has not been simulated. Where it has been included, silicon has been treated as one compartment, such as dissolved inorganic silicon (Lehman et al. 1975), or two compartments, such as available silicon and unavailable silicon (Scavia 1980).

28.3 PLANT STOICHIOMETRY

Aside from nutrients, the other key part of the eutrophication process is the food chain. As depicted previously in Fig. 19.1, the exchange between the two components represents a cycle. That is, production converts inorganic nutrients into organic matter whereas decomposition reverses the process.

An important factor in the process is the stoichiometric composition of organic matter. Although the composition varies, the dry-weight[†] composition can be idealized as in the following detailed representation of the photosynthesis/respiration process (Stumm and Morgan 1981):[‡]

$$106CO_2 + 16NH_4^+ + HPO_4^{2-} + 108H_2O \rightleftharpoons \underset{\text{“Algae”}}{C_{106}H_{263}O_{110}N_{16}P_1} + 107O_2 + 14H^+ \tag{28.1}$$

This formula can be used to determine the mass ratios of carbon to nitrogen to phosphorus:

$$\begin{array}{ccccc} C & : & N & : & P \\ 106 \times 12 & : & 16 \times 14 & : & 1 \times 31 \\ 1272 & : & 224 & : & 31 \end{array} \tag{28.2}$$

It is also known that plant protoplasm is about 1% phosphorus on a dry-weight basis. Therefore we can normalize the ratios to the mass of phosphorus and express the results as percentages of dry weight,

$$\begin{array}{ccccc} C & : & N & : & P \\ 40\% & : & 7.2\% & : & 1\% \end{array} \tag{28.3}$$

Thus a gram dry weight of organic matter contains approximately 10 mg of phosphorus, 72 mg of nitrogen, and 400 mg of carbon. Finally it should be noted that dry-weight biomass has a density of about 1.27 g cm^{-3} and wet-weight biomass is about 90% water. Figure 28.4 summarizes the fundamental information regarding cell stoichiometry. Additional information, particularly regarding the variability of these numbers, can be found in references such as Bowie et al. (1985).

Although the foregoing information provides a way to break down biomass into its individual components, additional information is required because phytoplankton are often measured in units other than dry weight. In cases where organic carbon is used, the foregoing stoichiometric information can be employed directly. More commonly the phytoplankton are measured as chlorophyll *a*. In general the chlorophyll-to-carbon ratio ranges from 10 to 50 μgChl mgC^{-1}. The lower value is usually typical of well-illuminated waters such as oligotrophic systems. For such systems less chlorophyll is required because of the high solar radiation. In contrast, waters with less light, such as eutrophic and turbid systems, would tend to have phytoplankton with a higher chlorophyll content.

[†]Dry weight means the weight of the organic matter after it has been dehydrated.

[‡]This formula holds when ammonium is the source of inorganic nitrogen. For the case where nitrate is the source, the reaction is modified,

$$106CO_2 + 16NO_3^- + HPO_4^{2-} + 122H_2O + 18H^+ \rightleftharpoons C_{106}H_{263}O_{110}N_{16}P_1 + 138O_2$$

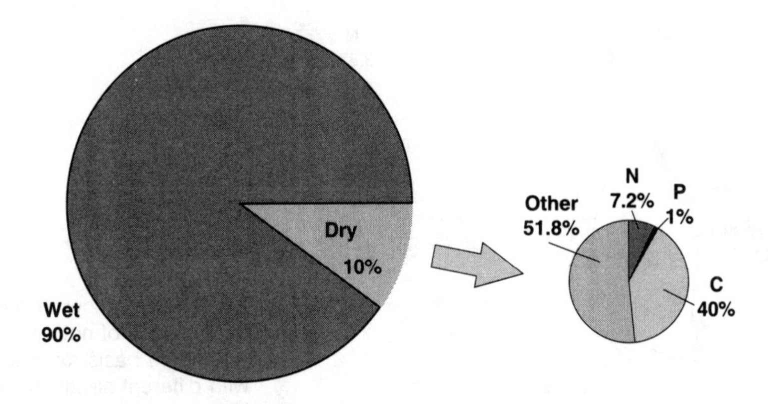

FIGURE 28.4
Pie diagram showing the percentages of nutrients and water constituting the average phytoplankton biomass.

EXAMPLE 28.1. PHYTOPLANKTON STOICHIOMETRY. Suppose that a lake has a volume of 1×10^6 m^3 and a phytoplankton concentration of 10 μg L^{-1} of chlorophyll *a*. If the carbon-to-chlorophyll ratio is 25 μgChl mgC^{-1} and all the other stoichiometry follows Fig. 28.4: (*a*) Reexpress the phytoplankton concentration as organic carbon. (*b*) If the phytoplankton are decomposing at a rate of 0.1 d^{-1}, what is the resulting rate of oxygen demand in g m^{-3} d^{-1}? (*c*) What is the rate of release of nitrogen and phosphorus in g d^{-1}?

Solution:

(*a*) The phytoplankton concentration as organic carbon can be calculated as

$$10\frac{\text{mgChl}a}{\text{m}^3}\left(\frac{\text{gC}}{25\ \text{mgChl}a}\right) = 0.40\ \text{gC m}^{-3}$$

(*b*) The decomposition of a gram of organic carbon utilizes 2.67 g of oxygen (recall Eq. 19.18). Thus the amount of oxygen consumed can be calculated as

$$r_{oc}k_d c = 2.67\frac{\text{gO}}{\text{gC}}\left(\frac{0.1}{\text{d}}\right)0.40\frac{\text{gC}}{\text{m}^3} = 0.1068\ \text{gO m}^{-3}\ \text{d}^{-1}$$

(*c*) The phosphorus generation rate can be calculated by

$$a_{pa}k_d V a = 1\frac{\text{mgP}}{\text{mgChl}a}\left(\frac{0.1}{\text{d}}\right)1 \times 10^6\ \text{m}^3\left(\frac{\text{mgChl}a}{10\ \text{m}^3}\right)\left(\frac{1\ \text{gP}}{1000\ \text{mgP}}\right) = 1000\ \text{gP d}^{-1}$$

The nitrogen-to-chlorophyll ratio can be computed as

$$a_{na} = 1\frac{\text{mgP}}{\text{mgChl}a}\left(7.2\frac{\text{mgN}}{\text{mgP}}\right) = 7.2\frac{\text{mgN}}{\text{mgChl}a}$$

which can then be employed to determine the nitrogen generation rate,

$$a_{na}k_d V a = 7.2\frac{\text{mgN}}{\text{mgChl}a}\left(\frac{0.1}{\text{d}}\right)1 \times 10^6\ \text{m}^3\left(10\frac{\text{mgChl}a}{\text{m}^3}\right)\left(\frac{1\ \text{gN}}{1000\ \text{mgN}}\right) = 7200\ \text{gN d}^{-1}$$

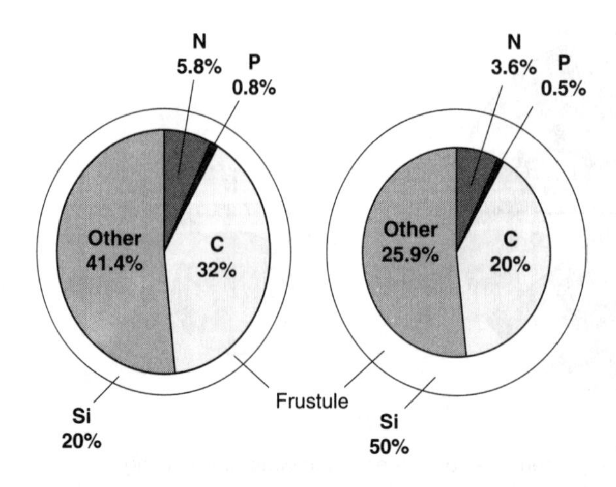

FIGURE 28.5
Pie diagram showing the percentages of nutrients on a dry-weight basis for diatoms with different silicon content in their "glass" cell walls or frustules.

Diatoms differ from other forms of phytoplankton in that silicon makes up a large fraction of their biomass. The percentage of dry weight ranges from 20 to 50% depending on the structure of their frustule. It is usually assumed that the remainder of their dry-weight biomass follows the proportions in Eq. 28.4. Figure 28.5 shows how the dry-weight percentages would decrease as more of the total dry weight is dedicated to silica.

28.4 NITROGEN AND PHOSPHORUS

Because they are the primary controllable nutrients, nitrogen and phosphorus have been the focus of most efforts to control eutrophication.

28.4.1 Nitrogen and Phosphorus Sources

Point sources of nitrogen and phosphorus are summarized in Table 28.1. In general the use of phosphate detergents has had a great impact on the amount of phosphorus in wastewater.

Some typical nonpoint sources are summarized in Table 28.2. Notice that urban and agricultural use greatly increase the export of both nitrogen and phosphorus from the land.

28.4.2 N:P Ratio

We now know that as plants grow they take up inorganic nutrients from the water in proportion to their stoichiometry. In eutrophication management it is often important to identify which of the several nutrients used for plant nutrition ultimately controls the level of plants in the water body. A first cut at identifying this "limiting nutrient" is to compare the levels of the nutrients in the water with the cell stoichiometry.

This is most commonly done for nitrogen and phosphorus. A rough rule of thumb for assessing which nutrient is limiting relates to the nitrogen-to-phosphorus ratio. Recall that the ratio of nitrogen to phosphorus in biomass is approximately

TABLE 28.1
Amounts of nitrogen and phosphorus in untreated domestic sewage in the United States. Numbers in parentheses represent ranges (from Metcalf and Eddy 1991, Thomann and Mueller 1987)

Nutrient	Concentration ($mg\ L^{-1}$)	Per-capita loading rate ($g\ capita^{-1}\ d^{-1}$)†
Nitrogen	40 (20–85)	23
Organic N	15 (8–35)	8.5
Free ammonia	25 (12–50)	14.2
Phosphorus (with detergents)	8 (4–15)	4.5
Organic P	3 (1–5)	1.7
Inorganic P	5 (3–10)	2.8
Phosphorus (without detergents)	4	2.3

† Based on a per-capita flow rate of 0.57 $m^3\ capita^{-1}\ d^{-1}$ (150 gal $capita^{-1}\ d^{-1}$).

TABLE 28.2
Nitrogen and phosphorus export rates ($kg\ ha^{-1}\ yr^{-1}$) generated from various nonpoint sources in the United States (numbers in parentheses represent ranges)

Nutrient	Forest	Agricultural	Urban	Atmospheric
Nitrogen	3 (1.3–10.2)	5 (0.5–50)	5 (1–20)	24
Phosphorus	0.4 (0.01–0.9)	0.5 (0.1–5)	1 (0.1–10)	1 (0.05–5)

7.2. Hence an N:P ratio in the water that is less than 7.2 suggests that nitrogen is limiting. Conversely, higher levels imply that phosphorus will limit plant growth. The rationale behind the ratio is derived in the following example.

EXAMPLE 28.2. N:P RATIOS. Suppose that a batch reactor has an initial concentration of algae of 1 $\mu gChl\ L^{-1}$. If the plants are growing according to first-order kinetics ($k_g = 1\ d^{-1}$), determine how both plant and nutrient concentrations evolve for initial nutrient levels of (*a*) $n_0 = 100\ \mu gN\ L^{-1}$ and $p_0 = 10\ \mu gP\ L^{-1}$, and (*b*) $n_0 = 72\ \mu gN\ L^{-1}$ and $p_0 = 36\ \mu gP\ L^{-1}$.

Solution: Mass balances for algae, phosphorus, and nitrogen can be written as

$$\frac{da}{dt} = k_g a \qquad \frac{dp}{dt} = -a_{pa} k_g a \qquad \frac{dn}{dt} = -a_{na} k_g a$$

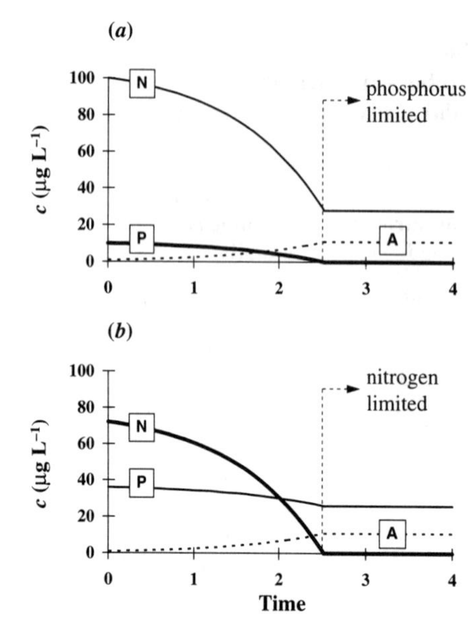

FIGURE 28.6
Algal growth in a batch reactor starting with initial N:P ratios that lead to (*a*) phosphorus limitation (initial N:P = 10) and (*b*) nitrogen limitation (initial N:P = 2).

which can be solved for

$$a = a_0 e^{k_g t} \qquad p = p_0 - a_{pa} a_0 (e^{k_g t} - 1) \qquad n = n_0 - a_{na} a_0 (e^{k_g t} - 1)$$

(*a*) The solutions for the first case are depicted in Fig. 28.6*a*. Because it is in relatively short supply (N:P = 10 > 7.2), the phosphorus runs out first at approximately $t = 2.5$ d. At this point the algae can no longer grow and excess nitrogen remains in the water.

(*b*) The second case, where nitrogen is in short supply (N:P = 2 < 7.2), is depicted in Fig. 28.6*b*. For this case the nitrogen runs out first. Again, at this point, the algae can no longer grow. However, now excess phosphorus remains in the water.

As displayed in Table 28.3, sewage is generally enriched in phosphorus. Therefore water bodies dominated by wastewater effluents tend to be nitrogen limited.

TABLE 28.3
N:P ratios for point, nonpoint, and marine waters (data from Thomann and Mueller 1987, Omernik 1977, and Goldman et al. 1973)

Source type	TN/TP†	IN/IP‡	Limiting nutrient
Raw sewage	4	3.6	Nitrogen
Activated sludge	3.4	4.4	Nitrogen
Activated sludge plus nitrification	3.7	4.4	Nitrogen
Activated sludge plus phosphorus removal	27.0	22.0	Phosphorus
Activated sludge plus nitrogen removal	0.4	0.4	Nitrogen
Activated sludge plus nitrogen and phosphorus removal	3.0	2.0	Nitrogen
Nonpoint sources	28	25	Phosphorus
Marine waters	—	2	Nitrogen

†TN/TP = total nitrogen/total phosphorus; ‡IN/IP = inorganic nitrogen/inorganic phorphorus.

Similarly estuaries tend to be deficient in nitrogen and hence are usually nitrogen limited. In contrast those systems subject to phosphorus removal and nonpoint-source input are generally phosphorus limited.

Although Table 28.3 provides general patterns, individual natural waters must be assessed on a case-by-case basis. Consequently mathematical models have been developed to simulate the eutrophication process.

PROBLEMS

28.1. A sewage treatment pond has the following characteristics:

Residence time = 3 wk
Area = 1×10^5 m^2
Mean depth = 2 m

The inflow concentrations of nitrogen and phosphorus are 50 and 5 mg L^{-1}, respectively. These inputs result in a level of 200 mgChl*a* m^{-3} of phytoplankton in the pond. The phytoplankton are removed by settling at a rate of 0.5 m d^{-1}. Also, the nitrogen is removed by denitrification at a rate of 0.05 d^{-1}. Compute the following:
(*a*) The steady-state nitrogen and phosphorus concentrations in the pond
(*b*) Based on the nitrogen-to-phosphorus ratio, which is the limiting nutrient?

Note that the chlorophyll-to-phosphorus ratio in the phytoplankton is 1.

28.2. A stretch of a river (U = 1 m s^{-1}) has a uniform, net photosynthesis rate of 10 gO m^{-2} d^{-1}. The boundary conditions of inorganic nitrogen and phosphorus at the head end of the stretch are 20 mgN L^{-1} and 4 mgP L^{-1}, respectively. If plant activity is the only significant source or sink of nutrients, determine and plot the N:P ratio for the stretch until one of the nutrients runs out.

28.3. Right after thermal stratification is established, a lake has available phosphorus and available nitrogen (mostly nitrate) concentrations of 10 μgP L^{-1} and 100 μgN L^{-1}. If the chlorophyll-to-carbon ratio is 25 μgChl mg^{-1}: (*a*) How much biomass could potentially be produced by photosynthesis in μgChl L^{-1}? (*b*) How much carbon in mgC L^{-1}? (*c*) How much oxygen would be required in mgO L^{-1} to nitrify the organic nitrogen produced by photosynthesis?

28.4. A town of 20,000 people discharges raw sewage into a river with a flow of 1 cms. If the river has 10 μgP L^{-1} and 100 μgN L^{-1} prior to the addition of the sewage, what is the N:P ratio at the mixing point? What is the limiting nutrient above and below the discharge?

LECTURE 29

Phosphorus Loading Concept

LECTURE OVERVIEW: I first describe the phosphorus loading plots that provide an easy-to-use method for obtaining order-of-magnitude estimates of lake eutrophication. Then I develop a simple mass-balance framework to perform similar predictions and illustrate how the plots and the mass balances are related. Next I review some of the empirical correlations that have been devised to predict eutrophication symptoms (chlorophyll a, hypolimnetic oxygen demand, etc.) as a function of loadings and in-lake concentrations. Finally I end the lecture with two mechanistic models based on the phosphorus loading approach—a sediment-water model for total phosphorus and a water-column model that differentiates between different forms of phosphorus.

The phosphorus loading concept is based on the premise that phosphorus is the primary, controllable limiting nutrient of lake and reservoir eutrophication. A number of simple empirical models have been developed to predict lake eutrophication on the basis of this premise. The earliest and most fundamental were developed by Richard Vollenweider. As a consequence they are often termed "Vollenweider plots."

29.1 VOLLENWEIDER LOADING PLOTS

The first loading plot was developed by Vollenweider (1968). This model was based on Rawson's (1955) insight that deeper lakes are less susceptible to eutrophication than shallower systems. Vollenweider compiled areal loadings of total phosphorus L_p (mgP m^{-2} yr^{-1}) and mean depth H (m) from north temperate lakes from around the world. He used these points to locate the lakes on a space defined by log L_p versus log H (Fig. 29.1). He then labeled each lake as to its trophic status (oligotrophic, mesotrophic, eutrophic). Finally he superimposed lines dividing the various categories of lakes.

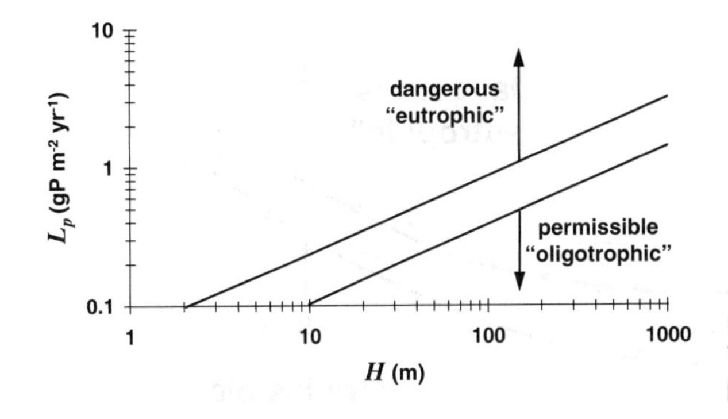

FIGURE 29.1
Vollenweider's (1968) loading plot.

As displayed in Fig. 29.1, the plot provides an easy-to-use model to perform both simulation and wasteload allocation calculations for lakes. For a simulation calculation the modeler would use loading and depth data to predict trophic state. For a wasteload allocation calculation the modeler would determine the loading required to attain a desired trophic state for a lake of a particular mean depth.

In a subsequent paper Vollenweider (1975) added a second determining factor into the loading plot framework. He recognized that not only depth but also residence time had an impact on eutrophication. In essence he observed that faster flushing lakes seemed to be less susceptible to eutrophication than lakes with long residence times. He incorporated this effect into the plot by adding the inverse residence time to the abscissa.

As in Fig. 29.2, he again plotted lakes on this space and superimposed lines. However, along with straight lines, he suggested that curves represented a superior fit of the data. Again, as with the earlier version (Fig. 29.1), the plot can be used for both simulation and wasteload allocation predictions.

It should be noted that one problem with using mean depth and residence time is that they are not independent. In fact, the abscissa, H/τ_w, can be shown to be independent of depth,

$$\frac{H}{\tau_w} = \frac{HQ}{V} = \frac{HQ}{HA_s} = \frac{Q}{A_s} \equiv q_s \tag{29.1}$$

where q_s is called the hydraulic overflow rate (m yr^{-1}). Note that engineers involved in water and waste treatment have historically correlated sedimentation in treatment reactors with the overflow rate (Reynolds 1982).

A final refinement was developed independently by Vollenweider (1976) and Larsen and Mercier (1976). Their models corresponded to plots of the logarithm of L_p versus the logarithm of $q_s(1 + \sqrt{\tau_w})$. Again, curves were used to divide the loading levels on the plot.

Other investigators, notably Rast and Lee (1978), have applied Vollenweider's (1976) approach to larger data bases and have extended it to predict trophic status. We review these extensions in Sec. 29.3.

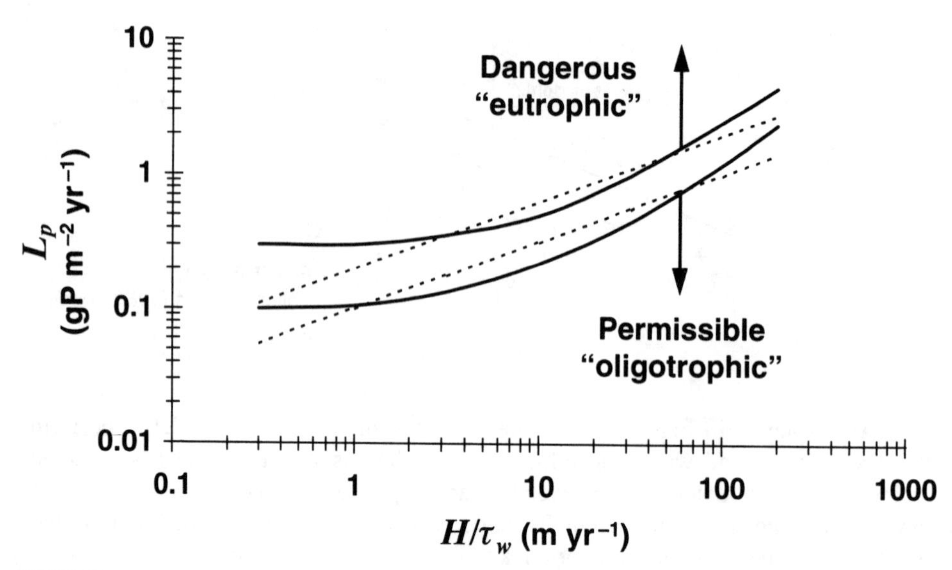

FIGURE 29.2
Vollenweider's (1975) loading plot.

29.2 BUDGET MODELS

Early in the development of phosphorus loading models, it was recognized that simple mass-balance models could provide the same predictions as loading plots. In fact Vollenweider (1976) wrote one of the first phosphorus mass-balance models for a well-mixed lake as

$$V\frac{dp}{dt} = W - Qp - k_s V p \tag{29.2}$$

where V = volume (m^3)
p = total phosphorus concentration (mg m^{-3})
t = time (yr)
W = total P loading rate (mg yr^{-1})
Q = outflow (m^3 yr^{-1})
k_s = a first-order settling loss rate (yr^{-1})

At steady-state this equation can be solved for

$$p = \frac{W}{Q + k_s V} \tag{29.3}$$

Based on phosphorus budget data (that is, inputs, outputs, and concentration of phosphorus), the loss rate can be determined as

$$k_s = \frac{W - Qp}{Vp} = \frac{W}{Vp} - \frac{1}{\tau_w} \tag{29.4}$$

On the basis of such budget calculations, Vollenweider concluded that the loss rate could be approximated by

$$k_s = \frac{10}{H} \tag{29.5}$$

Chapra (1975) suggested that because the loss of phosphorus was due to settling of particulate phosphorus, the loss term should be represented as

$$V\frac{dp}{dt} = W - Qp - vA_s p \tag{29.6}$$

where v is the apparent settling velocity (m yr^{-1}). If this is true, at steady-state Eq. 29.6 can be solved for

$$p = \frac{W}{Q + vA_s} \tag{29.7}$$

Thus we can see that Vollenweider's estimate that $k_s = 10/H$ supports the settling velocity approach. In fact Eq. 29.5 is equivalent to using Eqs. 29.6 and 29.7 with $v =$ 10 m yr^{-1}. Data analysis by a number of individuals (e.g., Chapra 1975, Dillon and Rigler 1975, Thomann and Mueller 1987) has determined that the settling velocity most commonly takes on values in the range from about 5 to 20 m yr^{-1}. However, values have been reported from less than 1 to over 200 m yr^{-1}.

The unity between loading plots and budget models can be illustrated (Chapra and Tarapchak 1976) by dividing the numerator and denominator of Eq. 29.7 by the surface area to give

$$p = \frac{L}{q_s + v} \tag{29.8}$$

or

$$L = p(q_s + v) \tag{29.9}$$

Taking the logarithm of Eq. 29.9 gives

$$\log L = \log p + \log(q_s + v) \tag{29.10}$$

It is assumed that for phosphorus-limited systems, trophic state is correlated with phosphorus concentrations. Vollenweider and others have suggested that mesotrophy is bounded by total phosphorus concentrations of 10 and 20 μgP L^{-1} (Table 29.1). If this is true, Eq. 29.10 can be used to draw lines on a graph of $\log L_p$ versus $\log q_s$. As in Fig. 29.3, the result is a loading plot that is quite similar to Vollenweider's (1975) model.

Beyond showing the unity of the two approaches, the budget approach is useful in elucidating the mechanisms that underlie the shape of the loading plot. Inspecting Eq. 29.10 indicates that there are two asymptotes. At one extreme (low flushing lakes; small q_s) Eq. 29.10 reduces to

$$\log L = \log p + \log v = \text{constant} \tag{29.11}$$

Thus as assimilation becomes solely dependent on sedimentation, the curves flatten out on the left side of Fig. 29.3. Conversely for high flushing lakes (high q_s), Eq. 29.10 approaches

TABLE 29.1
Trophic-state classification based on total phosphorus concentration as well as on other variables reflective of eutrophication

Variable	Oligotrophic	Mesotrophic	Eutrophic
Total phosphorus (μgP L^{-1})	< 10	10–20	> 20
Chlorophyll *a* (μgChl*a* L^{-1})	< 4	4–10	> 10
Secchi-disk depth (m)	> 4	2–4	< 2
Hypolimnion oxygen (% saturation)	> 80	10–80	< 10

$$\log L = \log p + \log q_s \tag{29.12}$$

Consequently, as assimilation becomes solely dependent on flushing, the curves approach straight lines with a slope of one.

Beyond the Vollenweider (1975) plot, the above theoretical development can be employed to gain insight into his 1976 model, which can be formulated as

$$p = \frac{L}{q_s(1 + \sqrt{\tau_w})} \tag{29.13}$$

Comparing Eqs. 29.8 and 29.13 leads to the conclusion that the 1976 model has a settling velocity

$$v = q_s \sqrt{\tau_w} = \frac{H}{\sqrt{\tau_w}} \tag{29.14}$$

or a first-order rate of

$$k_s = \frac{1}{\sqrt{\tau_w}} \tag{29.15}$$

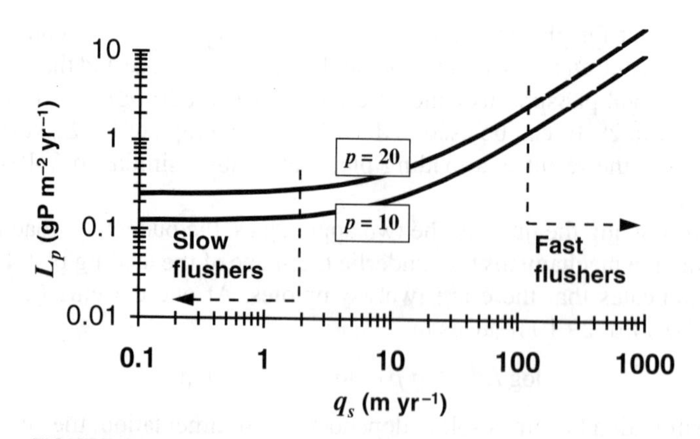

FIGURE 29.3
Loading plot derived from a phosphorus budget model (Eq. 29.10 with v = 12.4 m yr^{-1}).

29.3 TROPHIC-STATE CORRELATIONS

In the previous sections we calculated total phosphorus concentrations and interpreted the resulting levels as indicators of trophic status. Another approach is to use phosphorus concentration (or in some cases loadings) to predict other trophic-state variables that more directly reflect the deleterious effects of eutrophication.

As in Table 29.1, these other variables provide measures of eutrophication. In fact because they are more directly reflective of the adverse effects of eutrophication, they are preferable to total phosphorus concentration.

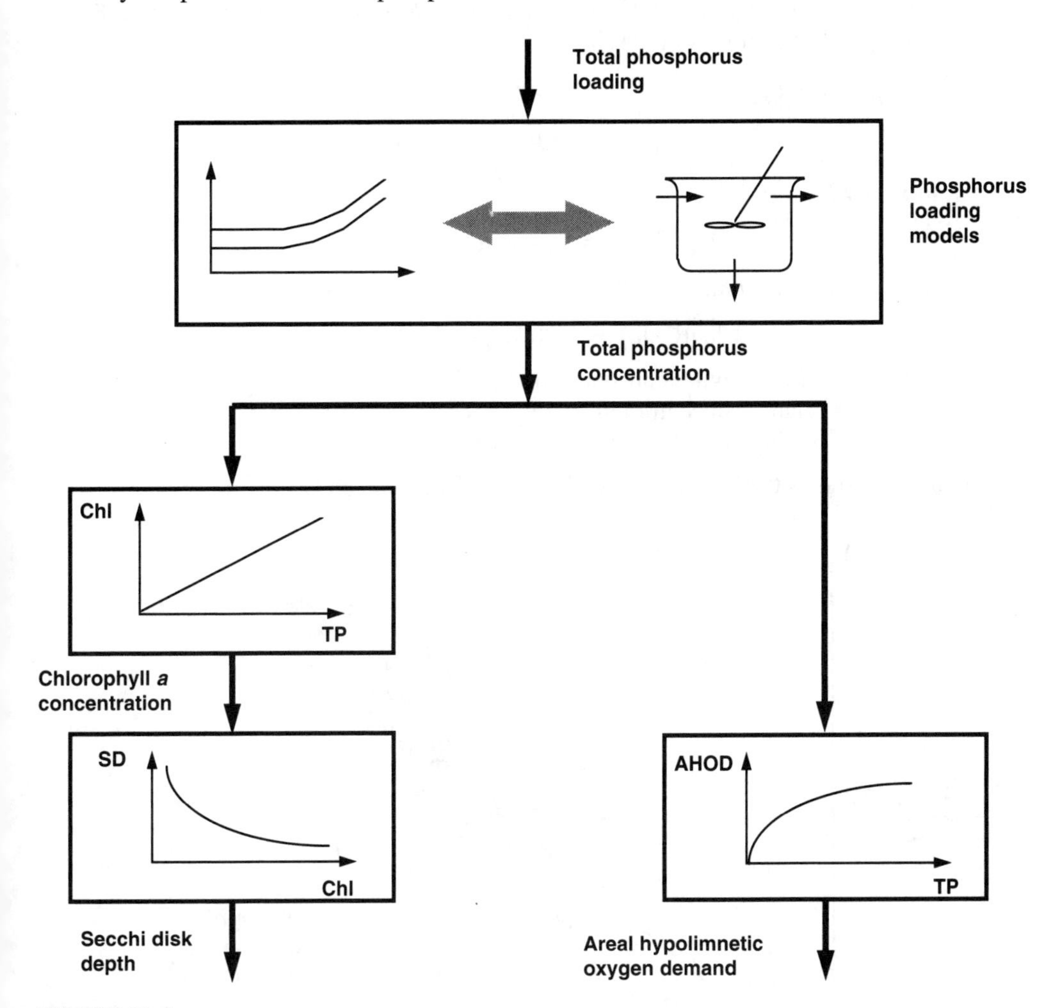

FIGURE 29.4
Schematic of approach used by Chapra (1980) to predict trophic state variables based on phosphorus loading model predictions. The approach consists of a number of submodels and correlations that form a hypothesized causal chain that starts with total P concentration predictions based on budget models or loading plots. This concentration is used in conjunction with a series of correlation plots to estimate symptoms of eutrophication such as chlorophyll *a* concentration, Secchi-disk depth, and hypolimnetic oxygen demand.

One such scheme is outlined in Fig. 29.4. The approach consists of relating predicted total phosphorus concentrations to symptoms of eutrophication such as chlorophyll *a* concentration, Secchi-disk depth, and hypolimnetic oxygen demand. Each step is dictated by empirically derived correlations, as described next.

29.3.1 Phosphorus-Chlorophyll Correlations

Initial attempts to extend phosphorus loading models attempted to calculate chlorophyll *a* levels as a function of total P concentration. Most of these are based on a fit of a log-log plot. Several examples are

Dillon and Rigler (1974):

$$\log(\text{Chl}a) = 1.449\log(p_v) - 1.136 \tag{29.16}$$

Rast and Lee (1978):

$$\log(\text{Chl}a) = 0.76\log(p) - 0.259 \tag{29.17}$$

Bartsch and Gakstatter (1978):

$$\log(\text{Chl}a) = 0.807\log(p) - 0.194 \tag{29.18}$$

where Chl$a$ = chlorophyll a concentration (μg L^{-1})
p = total P concentration (μg L^{-1})
p_v = spring total P concentration (μg L^{-1})

Figure 29.5 shows the Bartsch and Gakstatter (1978) version.

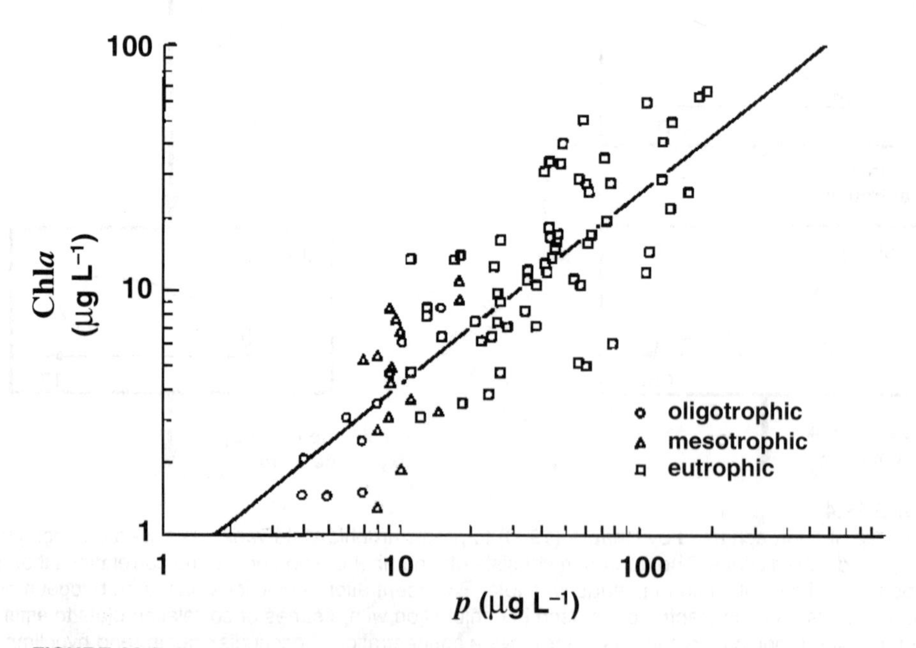

FIGURE 29.5
The relationship between chlorophyll *a* and phosphorus in some United States lakes and reservoirs (from Bartsch and Gakstatter 1978).

All these models show an increase of chlorophyll with increasing phosphorus. In all cases the relationship is nonlinear. However, the Dillon and Rigler model differs in that its exponent is greater than 1 (slope of log-log plot = 1.449), connoting that more polluted lakes exhibit proportionately higher chlorophyll than less polluted lakes. The other correlations show less chlorophyll per phosphorus for more productive systems.

In addition all the models are assumed to be appropriate only for phosphorus-limited systems. Smith and Shapiro (1981) have presented a modified correlation that attempts to account for potential nitrogen limitation,

$$\log(\text{Chl}a) = 1.55 \log(p) - b \tag{29.19}$$

where

$$b = 1.55 \log\left[\frac{6.404}{0.0204(\text{TN:TP}) + 0.334}\right] \tag{29.20}$$

in which TN:TP = total nitrogen to phosphorus ratio.

29.3.2 Chlorophyll–Secchi-Disk Depth Correlations

Attempts to relate Secchi-disk depth to chlorophyll levels have again usually started with log-log plots. One such graph is shown in Fig. 29.6, and it can be described by the equation

$$\log(SD) = -0.473 \log(\text{Chl}a) + 0.803 \tag{29.21}$$

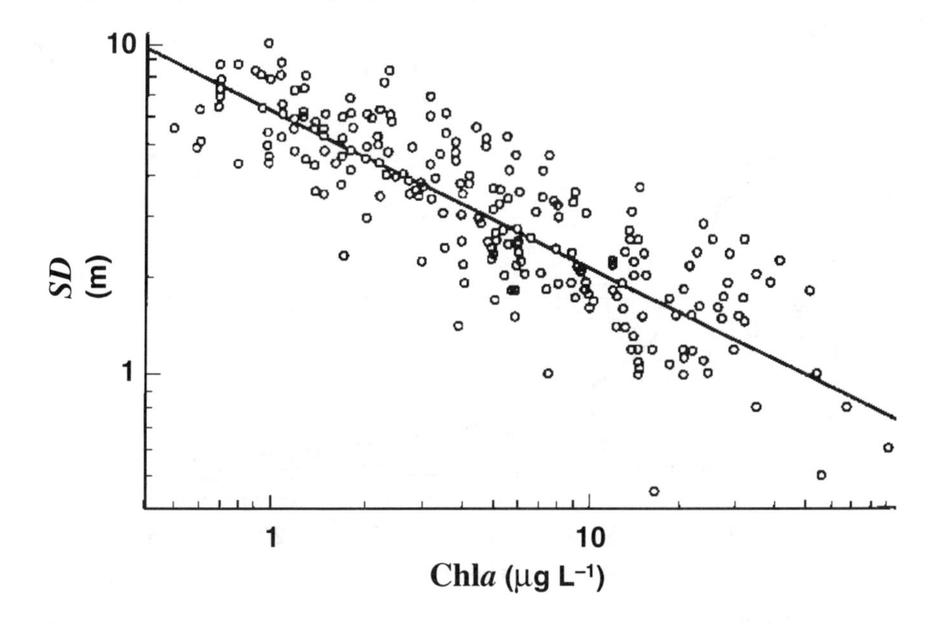

FIGURE 29.6
The relationship between Secchi-disk depth and chlorophyll *a* (from Rast and Lee 1978).

where SD = Secchi-disk depth (m).

When this equation is transformed into normal coordinates, it becomes

$$SD = 6.35\text{Chl}\,a^{-0.473} \tag{29.22}$$

Thus the fit traces out a hyperbolic shape that, as expected, exhibits large Secchi-disk depths at low chlorophyll concentrations and small Secchi-disk depths at high chlorophyll concentrations.

The shape can be related to more fundamental measurements by recognizing that light extinction in natural waters is often described by the Beer-Lambert law

$$I = I_0 e^{-k_e H} \tag{29.23}$$

where I = light at depth H
I_0 = light at the surface
k_e = extinction coefficient of the water

A number of investigators have related Secchi-disk depth to light extinction. For example a rough rule of thumb is that the Secchi-disk depth corresponds to the depth at which about 85% of the surface light is extinguished (Sverdrup et al. 1942, Beeton 1958). When we assume this level, Eq. 29.23 becomes

$$0.15 = e^{-k_e SD} \tag{29.24}$$

Further, the extinction coefficient is often related to chlorophyll levels. One common model is a linear proportionality,

$$k_e = k_{wc} + \alpha\text{Chl}a \tag{29.25}$$

where k_{wc} = extinction due to water, color, and nonalgal particles (m^{-1}) and α = a coefficient (L μg^{-1} m^{-1}). Substituting this relationship and taking the natural logarithm gives

$$\ln 0.15 = -(k_{wc} + \alpha\text{Chl}a)SD \tag{29.26}$$

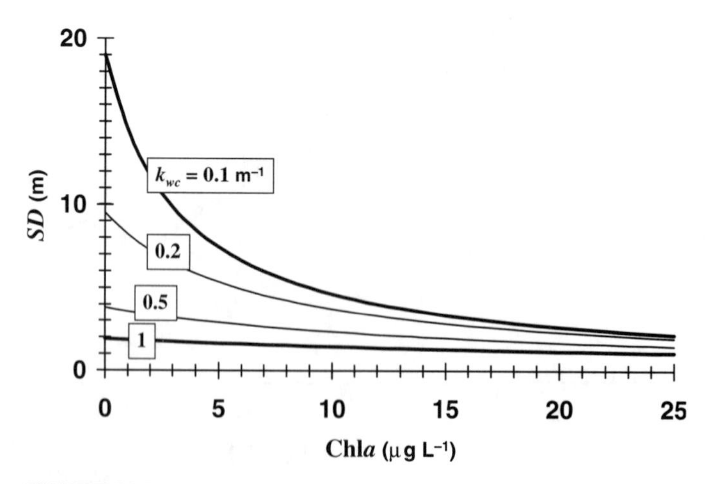

FIGURE 29.7
The relationship between Secchi-disk depth and chlorophyll derived from the Beer-Lambert law and light extinction relationships.

which can be manipulated to give

$$SD = \frac{1}{1 + \mu \text{Chl}a} SD_{\max} \tag{29.27}$$

where $\mu = \alpha / k_{wc}$ and $SD_{\max} = 1.9 / k_{wc}$. Thus the relationship starts at an initial Secchi-disk depth corresponding to particle-free water. Then, as chlorophyll fosters light extinction, the Secchi-disk depth is reduced to zero. Figure 29.7 illustrates this pattern.

29.3.3 Areal Hypolimnetic Oxygen Demand

Rast and Lee (1978) presented the following correlation to predict the areal hypolimnetic oxygen demand in lakes (Fig. 29.8),

$$\log \text{AHOD} = 0.467 \log \left[\frac{L}{q_s(1 + \sqrt{\tau_w})} \right] - 1.07 \tag{29.28}$$

where AHOD = areal hypolimnetic oxygen demand $(\text{gO m}^{-2}\,\text{d}^{-1})$, or taking the antilog,

$$\text{AHOD} = 0.0851 \left[\frac{L}{q_s(1 + \sqrt{\tau_w})} \right]^{0.467} \tag{29.29}$$

Thus even though the equation seems to represent a correlation with loading, it actually correlates AHOD with in-lake total P concentration (recall Eq. 29.13).

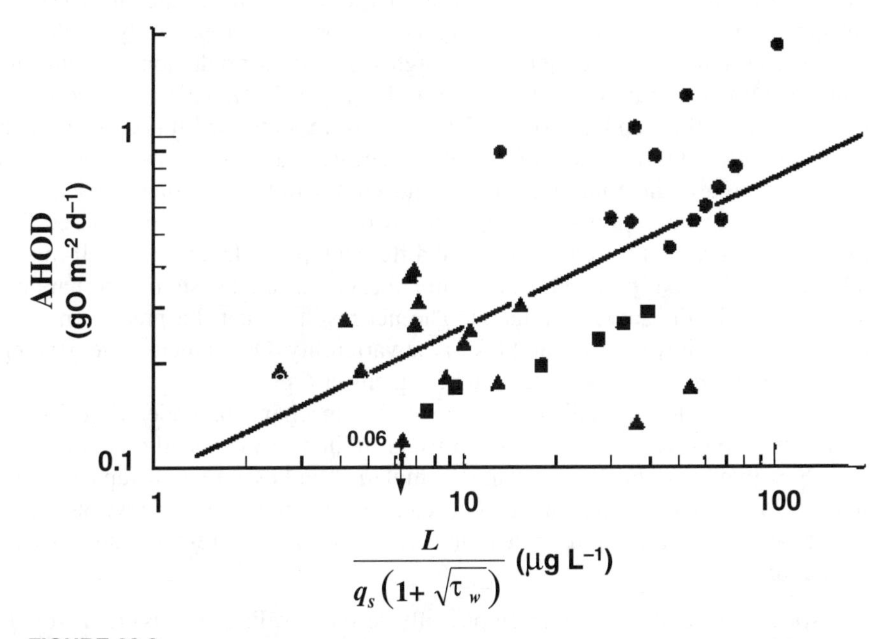

FIGURE 29.8
The relationship between areal hypolimnetic oxygen demand and phosphorus loading (Rast and Lee 1978).

Chapra and Canale (1991) realized this and reanalyzed Rast and Lee's data to determine a direct correlation with total P concentration,

$$\text{AHOD} = 0.086p^{0.478} \tag{29.30}$$

where p = mean total P concentration of the lake (μgP L^{-1}).

When this equation is plotted on untransformed scales (recall Fig. 25.2*b*), it traces a saturating curve that suggests that as lakes become more productive, the amount of AHOD does not increase as rapidly as the phosphorus increases. This is consistent with Fair's observation (Fair et al. 1941) that sediment oxygen demand increases as the square of the organic carbon content of a sediment. In addition it conforms to Di Toro et al.'s (1990) SOD model. If it is assumed that lakes with higher total P have higher sediment organic content, Eq. 29.30 and Fair's observation and Di Toro's model show some consistency.

29.3.4 Summary

The models and correlations constituting the phosphorus loading concept have been widely used, in part because they are so easy to apply. They also have the intrinsic appeal of any empirical approach. That is, they directly reflect observations—"what you see, is what you get."

Unfortunately these models and correlations also have shortcomings:

- Because the plots are all log-log and exhibit large scatter, the prediction errors are substantial. In the forms presented here, these errors are not explicitly displayed. Thus the user might naively employ a highly uncertain prediction with unwarranted confidence in its validity. Reckhow (Reckhow 1977, 1979; Reckhow and Chapra 1979, 1983) and Walker (1977, 1980) have addressed this deficiency in detail and offered remedies. Unfortunately because the loading models and plots are so easy to use, uncertainty is rarely connected with the resulting predictions.
- The models are commonly developed from widely heterogeneous data bases. For example lakes from different regions and different types of lakes (e.g., well-mixed lakes and elongated impoundments, some nitrogen-limited systems, etc.) are often included in the same correlation. The net result is that the prediction error becomes inflated by regional and lake-type variability. One remedy is to develop customized correlations for specific lake regions or types.
- They provide little mechanistic insight into the mechanisms underlying the eutrophication process. Further, their utility is limited to the specific applications for which they were intended—that is, simulation and assimilation capacity estimation. In contrast, mechanistic models can be extended to assess environmental modifications (e.g., dredging, reaeration, etc.) and to guide research and experimentation.

In spite of these shortcomings, empirically derived loading models often provide useful order-of-magnitude estimates. As such they provide a quick means to "see the big picture." In other words they offer a way to discern how eutrophication in a particular lake relates to how lakes generally behave.

29.4 SEDIMENT-WATER INTERACTIONS

Bottom sediments have long been acknowledged as a potential source of phosphorus to the overlying waters of lakes and impoundments. As such, sediment feedback could have a significant impact on the recovery of such systems. This would be particularly true in shallow lakes or those with anaerobic hypolimnia.

We have already alluded to sediment feedback of nutrients in Sec. 25.6.4. Such mechanistic frameworks provide one means to simulate the process. In this section we develop an alternative approach that is more akin to the phosphorus loading models described herein. That is, semiempirical formulations are used to simulate sediment feedback in conjunction with simple total phosphorus budgets for the water and the sediments. Such a simplified approach is sometimes more consistent with typical data collection programs for many water bodies.

29.4.1 Sediment-Water Model

In this section we develop a simple modeling framework to address this problem for stratified lakes. The framework, which is expressly designed for management applications, includes two components: a total phosphorus budget and a model of hypolimnetic oxygen deficit. Each is described briefly in the following paragraphs.

Total phosphorus model. A sediment-water model for total phosphorus in a lake and its underlying sediments (Fig. 29.9) can be written as

$$V_1 \frac{dp_1}{dt} = W - Qp_{\text{in}} - v_s A_s p_1 + v_r A_s p_2 \tag{29.31}$$

$$V_2 \frac{dp_2}{dt} = v_s A_s p_1 - v_r A_s p_2 - v_b A_s p_2 \tag{29.32}$$

where the subscripts 1 and 2 designate the water and the enriched surface sediment layer, respectively, v_s = settling velocity of phosphorus from the water to the sediments (m yr^{-1}), A_s = surface area of the deposition zone (m^2), v_r = recycle mass-transfer coefficient from the sediments to the water (m yr^{-1}), and v_b = a burial mass-transfer coefficient from the enriched surface layer to the deep sediments (m yr^{-1}).

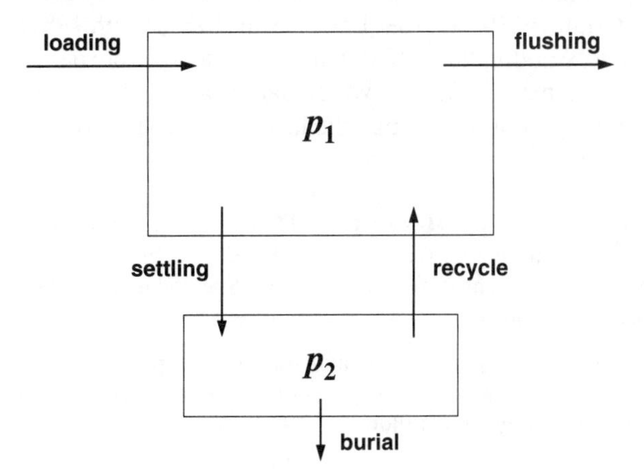

FIGURE 29.9
Schematic diagram of a phosphorus budget model for a lake underlain by sediments.

Hypolimnetic oxygen model. A zero-order model is employed here to simulate hypolimnetic oxygen during periods when the lake is stratified,

$$o_h = o_i - \frac{\text{AHOD}}{H_h}(t - t_s) \tag{29.33}$$

where o_h = hypolimnetic dissolved oxygen level (g m^{-3})
o_i = initial oxygen concentration at the onset of stratification (g m^{-3})
AHOD = areal hypolimnetic oxygen demand (g m^{-2} d^{-1})
H_h = average hypolimnion thickness (m)
t = time (d)
t_s = time of onset of stratification (d)

Although frameworks to predict AHOD as a function of the quality of the overlying water are currently under development (Lec. 25), empirical approaches provide a much simpler means to estimate this quantity. For example, Eq. 29.30 can be used.

For dimictic lakes, an AHOD can also be exerted during winter inverse stratification. To simulate depletion for these systems Eq. 29.30 can be extrapolated on the basis of temperature using

$$\text{AHOD}_w = \text{AHOD}_s 1.08^{T_w - T_s} \tag{29.34}$$

where T_s is the temperature (°C) at which the summer AHOD$_s$ is measured and T_w is the temperature (°C) corresponding to the desired winter AHOD$_w$. Note that a similar temperature correction factor is used to scale the recycle velocity.

29.4.2 Application: Shagawa Lake

Shagawa Lake, Minnesota, is an ideal setting to illustrate the model. This is because it is one of the first systems where sediment feedback was recognized as being important. In addition it has been the subject of numerous studies and, hence, there is a large quantity of information available for model calibration. Finally after being polluted for many years, it was the subject of significant nutrient load reductions in the early 1970s.

Model calibration is based on the extensive data reported by Larsen, Malueg, and colleagues (e.g. Larsen and Malueg, 1976, 1981; Larsen et al. 1975, 1979, 1981; Malueg et al., 1975; Bradbury and Waddington, 1973). Parameter values were determined by calibrating for the pretreatment period for which data is available (1967–1972). As in the following example, it is assumed that the lake was at a steady-state during this time frame.

EXAMPLE 29.1. CALIBRATION OF SEDIMENT-WATER TOTAL P MODEL. Data for Shagawa Lake from 1967 through 1972 are summarized in Tables 29.2 and 29.3. Use this data to calibrate the sediment-water total P model. Specifically, estimate (*a*) the burial and (*b*) the recycle velocities.

Solution: If the system is at steady-state, the sources and sinks of phosphorus should balance. As depicted in Fig. E29.1-1, the amount of phosphorus buried must be equal to the difference between phosphorus inflow and outflow.

TABLE 29.2
Data for Shagawa Lake (1967 to 1972)

Parameter	Symbol	Value	Units
Volume	V_1	53×10^6	m^3
Surface area	A_1	9.6×10^6	m^2
Mean depth	H_1	5.5	m
Hypolimnion thickness	H_h	2.2	m
Deposition zone area	A_2	4.8×10^6	m^2
Surface sediment thickness	H_2	10	cm
Total P loading	W_{in}	6692×10^6	mg yr^{-1}
Total P outflow	W_{out}	4763×10^6	mg yr^{-1}
Mean water P concentration	p_1	56.3	mg m^{-3}
Mean sediment P concentration	p_2	500,000	mg m^{-3}
Hypolimnion temperature—summer	$T_{h,s}$	15	°C
Hypolimnion temperature—winter	$T_{h,w}$	4	°C
Total P settling velocity	v_s	42.2	m yr^{-1}
Summer initial hypolimnetic DO	$DO_{i,s}$	8	mg L^{-1}
Winter initial hypolimnetic DO	$DO_{i,w}$	8	mg L^{-1}

TABLE 29.3
Stratification data for Shagawa Lake (1967 to 1972)

Event	Day
Start of spring mixed period	120
Start of summer stratification	150
Start of fall mixed period	255
Start of winter stratification	320

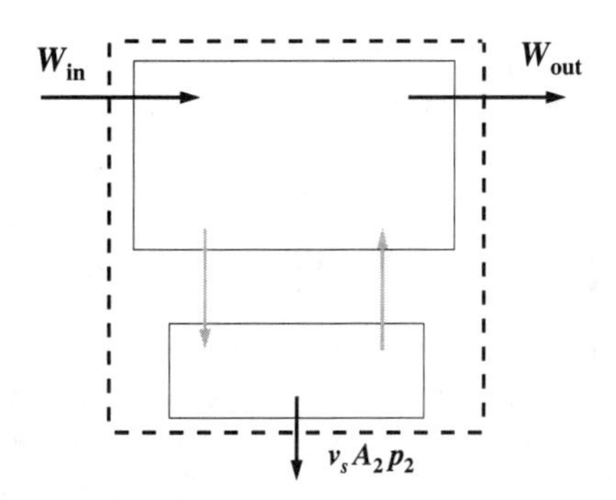

FIGURE E29.1-1

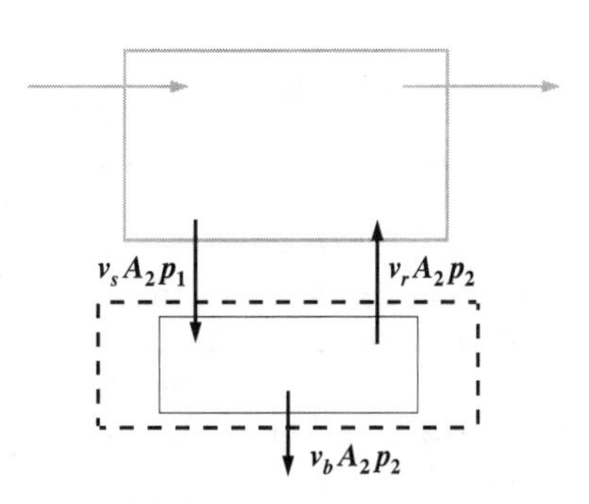

FIGURE E29.1-2

Thus the burial velocity can be computed as

$$v_b = \frac{W_{\text{in}} - W_{\text{out}}}{A_2 p_2} = \frac{6692 \times 10^6 - 4763 \times 10^6}{4.8 \times 10^6(500{,}000)} = 8.03 \times 10^{-4} \text{ m yr}^{-1}$$

(*b*) The recycle velocity can be estimated by taking a mass balance around the sediments to determine how much phosphorus is recycled from the sediments to the water on an annual basis (Fig. E29.1-2). The balance can be solved for the amount of phosphorus recycled per year,

$$v_r A_2 p_2 = v_s A_2 p_1 - v_b A_2 p_2 = 11{,}410 \times 10^6 - 1928 \times 10^6 = 9476 \times 10^6 \text{ mg yr}^{-1}$$

Now, this value must be distributed over the summer and winter anoxic periods. To do this we must determine the AHOD rates. For the summer,

$$\text{AHOD} = 0.086(56.3)^{0.478} = 0.5905 \text{ g m}^{-2} \text{ d}^{-1}$$

This value along with other parameters can be substituted into Eq. 29.33 to determine how long after stratification the lake would go anoxic (that is, below approximately 1.5 mg L^{-1}),

$$(t - t_s) = \frac{(o_i - o_{\text{anoxic}})(H_h)}{\text{AHOD}} = \frac{(8 - 1.5)2.2}{0.5905} = 24.2 \text{ d}$$

which means that in the summer the lake will be anoxic for $105 - 24.2 = 80.8$ d. A similar calculation (with the temperature correction from Eq. 29.34) can be used to determine that the lake would be anoxic for 108.5 d during the winter.

The total amount of recycled phosphorus can be set equal to the terms in the model accounting for recycle in the summer and winter anoxic periods,

$$W_{\text{recycle}} = F_{a,s} v_r 1.08^{T - T_{h,s}} A_2 p_2 + F_{a,w} v_r 1.08^{T - T_{h,w}} A_2 p_2$$

where $F_{a,s}$ and $F_{a,w}$ = fractions of the year when the hypolimnion is anoxic during summer and winter, respectively. This equation can then be solved for

$$v_r = \frac{W_{\text{recycle}}}{A_2 p_2 (\Delta t_{a,s} 1.08^{T_{h,s} - 20} + \Delta t_{a,w} 1.08^{T_{h,w} - 20})}$$

$$= \frac{9476 \times 10^6}{4.6 \times 10^6 (500{,}000) \left[\frac{80.8}{365}(1.08^{15-20}) + \frac{108.5}{365}(1.08^{4-20}) \right]} = 0.01663 \text{ m yr}^{-1}$$

Figure 29.10 shows simulation results for a single year during the steady-state calibration period. Data shown are for 1972 (Larsen et al. 1979). Notice how the phosphorus increases due to the heightened release rate when the oxygen level falls below 1.5 mg L^{-1}. A better fit could have been accomplished by additional tuning of the parameters or by allowing the model parameters to vary seasonally. However, considering the simplicity of the present calibration process, Fig. 29.10 is judged to be an adequate approximation of the general trend of the data.

To obtain some perspective on the long-term dynamics of Shagawa Lake, a simulation was performed for the period from 1880 to 2000. Actual flows were used for the period 1967 to 1979. An average flow is used for all other years.

Because measurements were not made prior to 1967, an idealized long-term loading scenario was developed (Fig. 29.11). Measured loadings are used for the

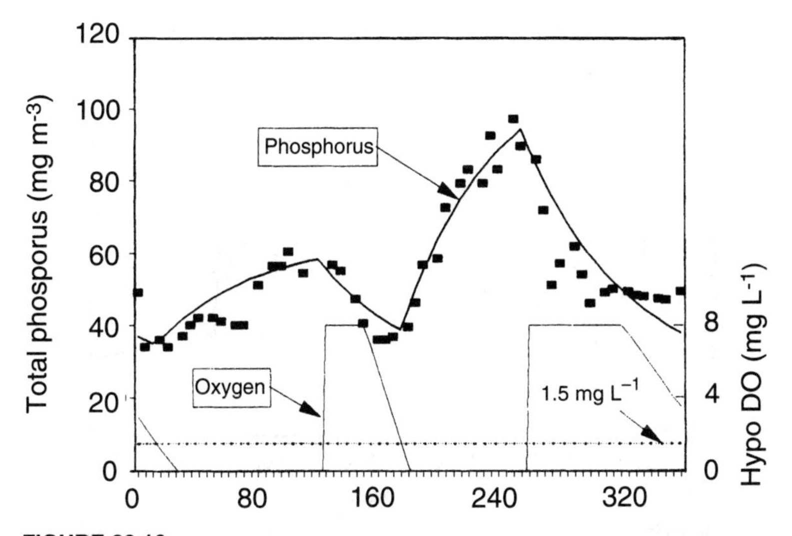

FIGURE 29.10
Plot of phosphorus and oxygen in the pretreatment period (1967–1972) for Shagawa Lake. The data are for 1972 (Larsen et al. 1979).

period 1967 to 1979. Before 1890 and after 1979, an average "natural" loading of 1311 kg yr^{-1} (Larsen et al. 1975) is assumed to apply.

In 1890 the town of Ely was established in the lake watershed. From 1890 to the present, the town's population has been relatively stable. An idealized scenario is used to characterize the town's contribution. The scenario assumes that immediately upon Ely's establishment, the loadings increased stepwise to the high average levels of the late 1960s. Although the actual loading was undoubtedly different from this idealization, in the absence of direct measurements it is considered to be an adequate first approximation.

The results of the long-term simulation along with measured data are shown in Fig. 29.11. The plot indicates that after the 1973 load reduction, the lake experienced an immediate quick response. However, by 1974 the lake's recovery slowed significantly due to feedback of phosphorus from the sediments.

Aside from its simplicity, the use of a step increase to characterize the historical loading scenario allows a clearer visualization of the model's response characteristics. As in Fig. 29.11, the model shows an immediate sharp increase in 1890 following the step load increase. Then, as sediment feedback begins to dominate, the remainder of the increase proceeds at an extremely slow rate (about 0.2 mg m^{-3} yr^{-1}). By the early 1960s, the lake was just beginning to approach a steady-state. Although it had not totally reached a steady-state at that time, it seems to have been close enough to make our calibration acceptable.

The simulation results and data for the recent past are depicted in Fig. 29.12. The adequacy of the simulation can be assessed by three features of the response. First, the model approximately matches the prediversion levels of 50 to 60 mg m^{-3}. Second, the initial rapid drop of about 25 mg m^{-3} is close to the observed drop. Finally, the rate of the subsequent retarded recovery is quite close in the data. Considering that the model parameters are untuned, these results are encouraging.

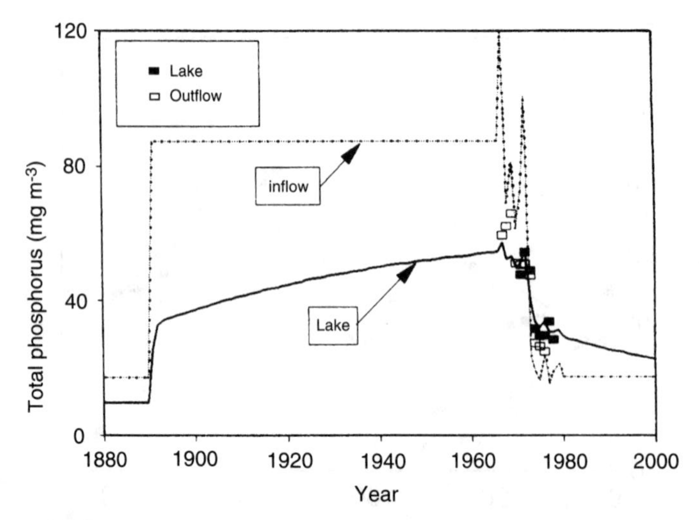

FIGURE 29.11
Long-term total phosphorus concentration for Shagawa Lake as simulated by the phosphorus-oxygen model (thick line). A plot of inflow concentration is superimposed (thin line) for comparison.

Figure 29.13 is a plot of external and internal loadings as calculated by the model. Notice how the internal sediment feedback load takes many decades to build up to the high levels observed during the 1960s. Also notice how after advanced treatment was installed in 1973, the sediment feedback experienced a small abrupt drop followed by a very gradual decrease. It also indicates that, although the constant feedback model would be adequate for short-term projections (that is, less than a decade), long-term predictions would have to account for the gradual decline in sediment release.

In summary the foregoing provides a preliminary framework for assessing the impact of sediment feedback of phosphorus on long-term lake recovery, which because it depends on a few parameter values, offers a simple means to assess the long-term impact of nutrient loadings on lake eutrophication.

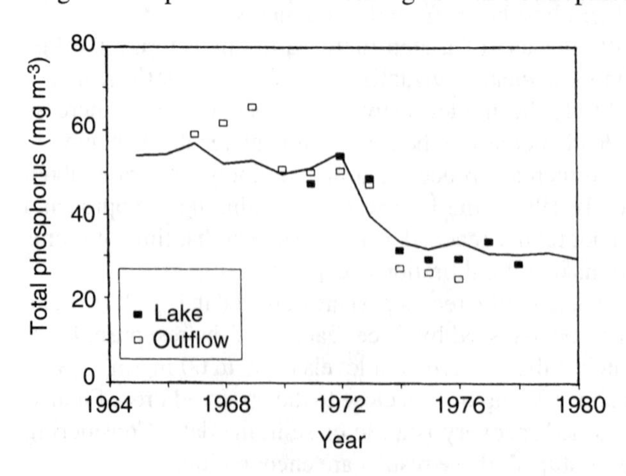

FIGURE 29.12
Recent total phosphorus concentration for Shagawa Lake as simulated by the phosphorus-oxygen model.

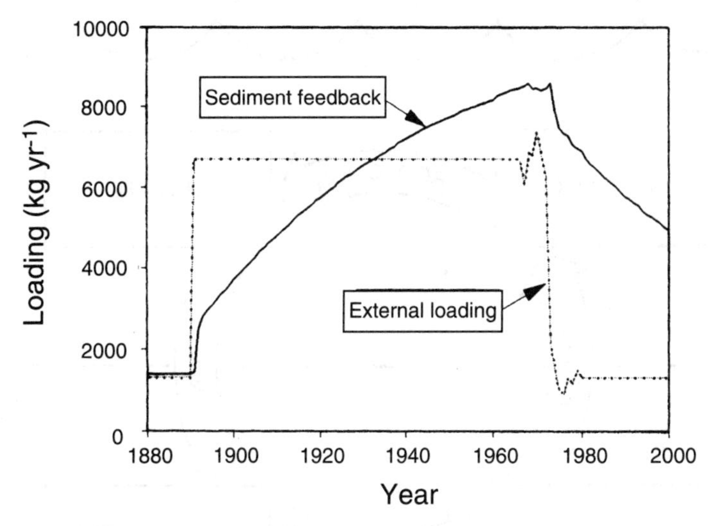

FIGURE 29.13
Long-term trends of sediment recycle (thick line) of phosphorus for Shagawa Lake as simulated by the phosphorus-oxygen model. The external loading is superimposed (thin line) for comparison.

29.5 SIMPLEST SEASONAL APPROACH

Temperature changes have a profound effect on mass cycling within the water column. The presence of a strong thermocline essentially divides the lake into two vertical layers with markedly different characteristics (Fig. 29.14).

The surface layer (epilimnion) is warm and well-illuminated. Consequently algal photosynthesis leads to transformation of dissolved nutrients into particulate organic matter. Although the thermocline greatly reduces vertical mixing, some of this particulate matter settles and diffuses to the bottom layer (hypolimnion), where it decomposes and eventually returns to a soluble form. Mixing across the thermocline then reintroduces some of the dissolved nutrient to the surface water, where it is again taken up by the phytoplankton. Because many contaminants in lakes are associated with particulate organic matter, this cycle has significance to their transport and fate.

This section describes an approach to simulating basic features of this cycle. The key characteristic of the approach is that it partitions the substance being modeled into two fractions. Although the model is specifically developed for phosphorus, this partitioning is basic to many other substances (such as certain toxic compounds), and the approach could serve as a preliminary framework for analysis of the seasonal dynamics of these contaminants. In addition to its use in simulating mass cycling, the model can also be employed to simulate oxygen concentration in the water column.

29.5.1 Model Formulation

Most of the models described in earlier lectures deal with the dynamics of a single substance in a vertically well-mixed water body. As previously mentioned, many

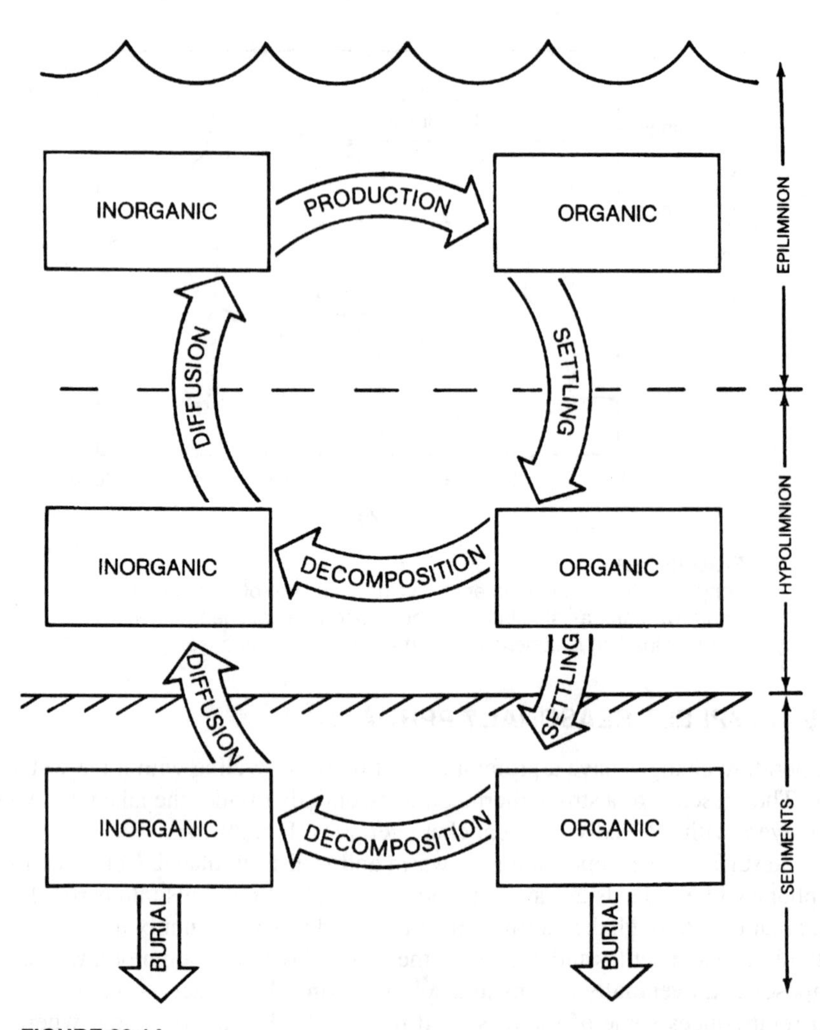

FIGURE 29.14
Idealized representation of the cycle of production and decomposition that plays a critical role in determining the vertical distribution of matter in stratified lakes.

constituents occur in various chemical and/or biological forms that are subject to transformations due to thermal stratification of the water column. O'Melia (1972), Imboden (1974), and Snodgrass (1974; and O'Melia 1975) have provided a modeling framework that makes a first attempt to simulate some of these changes for the nutrient phosphorus. Simons and Lam (1980) have dubbed this framework the "simplest seasonal approach" or SSA (Fig. 29.15). Although each of their approaches has unique features (for which the reader can refer to the original publications), in general all the models have the following characteristics:

1. Phosphorus is separated into two components. These conventionally refer to soluble reactive phosphorus (SRP) and phosphorus that is not soluble reactive

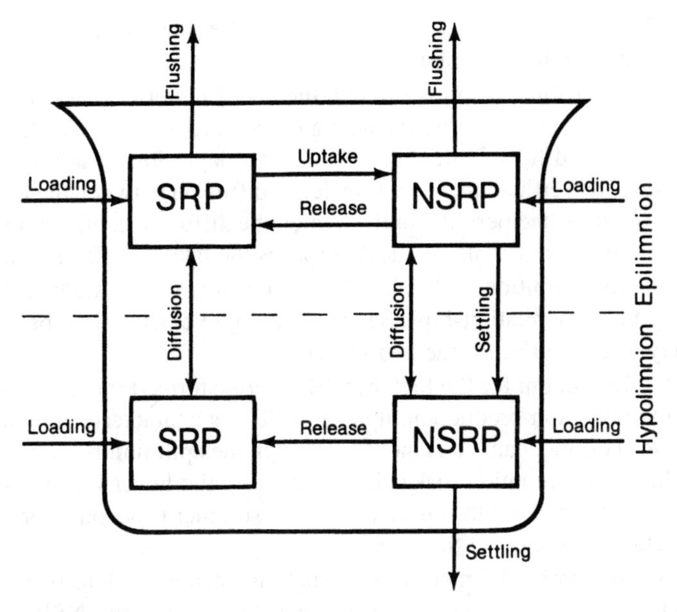

FIGURE 29.15
Schematic representation of the "simplest seasonal approach" developed by O'Melia, Imboden, and Snodgrass.

phosphorus (NSRP).[†] This breakdown has an operational basis because the two most common phosphorus measurements are for SRP and total phosphorus (TP). NSRP can therefore be estimated by the difference of the two quantities as in NSRP $=$ TP $-$ SRP. In the model itself a further distinction is made in that NSRP is subject to settling losses whereas SRP is not. This is based on the assumption that a significant portion of the NSRP is in particulate form. However, because NSRP also includes nonsettleable dissolved organic phosphorus (DOP), the distinction is not precise. We return to the subject of the definition and interpretation of the phosphorus fractions and the general question of kinetic segmentation at the end of this section.

2. The lake is segmented spatially into well-mixed upper and lower layers of constant thicknesses.
3. The year is divided into two seasons representing a summer, stratified period during which turbulent exchange between the layers is minimal and a winter, circulation period when turbulent transport is intense and the lake is well-mixed vertically.
4. Linear first-order differential equations are used to characterize the transport and kinetics representing the mass exchange between the components. As schematized in Fig. 29.15, these exchanges include the following:

[†]Note that Imboden (1974) treats phosphorus as dissolved P available for bioproduction and particulate P. He therefore does not explicitly account for dissolved organic phosphorus. The question of kinetic segmentation is explored in further detail at the end of this section.

- *Waste inputs.* Mass loadings of SRP and NSRP fractions can be input to both the epi- and the hypolimnion.
- *Transport.* Flushing of mass through the lake's outlet is characterized mathematically in a similar fashion to a single CSTR (that is, as flow times concentration), yet it differs in that only the surface layer loses mass in this way. A small level of vertical turbulent transfer or diffusion between the layers is used during the stratified period. For the winter, the diffusion coefficient is increased to the point where for all practical purposes the lake is well-mixed.
- *Settling.* By definition, only NSRP is removed from the water column using settling velocity relationships. Separate settling velocities can be input for the two layers as well as for the two seasons.
- *Uptake.* To account for the fact that SRP is transformed into particulate matter via phytoplankton production, uptake of SRP is characterized by a first-order reaction. This mechanism is included only in the epilimnion under the assumption that light limitation makes it negligible in the lower layers. In addition a much higher uptake rate is used during the summer to account for the fact that production is greatest at that time.
- *Release.* A number of mechanisms such as decomposition, respiration, zooplankton grazing, etc., act to return phosphorus from the NSRP to the SRP pool. A first-order reaction dependent on the concentration of NSRP is used to approximate this phenomenon.

To this point the kinetics of most models in this volume have been characterized as simple first-order decay of a single pollutant. With the SSA we begin to introduce more elaborate representations of substance interactions. As in Fig. 29.15, the phosphorus forms are coupled by uptake and release reactions. For example in the epilimnion, phosphorus is lost from the NSRP compartment or "pool" by a release reaction. This loss, in turn, represents a gain for the SRP pool. Therefore, when we write the mass balance for a particular pool, each arrow in Fig. 29.15 represents a term in the resulting differential equation. For example for the SRP pool in the epilimnion,

$$\underset{\text{Accumulation}}{V_e \frac{dp_{s,e}}{dt}} = \underset{\text{Loading}}{W_{s,e}} - \underset{\text{Flushing}}{Qp_{s,e}} + \underset{\text{Diffusion}}{v_t A_t (p_{s,h} - p_{s,e})} - \underset{\text{Uptake}}{k_{u,e} V_e p_{s,e}} + \underset{\text{Release}}{k_{r,e} V_e p_{n,e}} \tag{29.35}$$

where the subscripts n and s designate NSRP and SRP fractions, respectively, and the subscripts e and h designate epilimnion and hypolimnion, respectively.

Although the reactions in Eq. 29.35 are all first-order, they could just as easily be more complicated formulations. For example in Lec. 33, nonlinear relationships are used to characterize phytoplankton growth. In addition if other substances (such as additional forms of phosphorus or other nutrients) were to be included in the model, compartments (each representing a differential equation) and arrows (each representing a term in the differential equation) could be included. The point is that regardless of the complexity of the situation, the conservation of mass, as reflected by the set of differential equations, is a simple bookkeeping exercise to account for how, when, and where mass moves within the system.

TABLE 29.4
Typical ranges of parameters for the simplest seasonal approach for modeling phosphorus

Parameter	Season	Symbol	Range[†]	Units
Epilimnetic uptake rate	Summer	$k_{u,e}$	0.1–5.0	d^{-1}
	Winter	$k_{u,e}$	0.01–0.5	d^{-1}
Epilimnetic release rate	Summer	$k_{r,e}$	0.01–0.1	d^{-1}
	Winter	$k_{r,e}$	0.003–0.07	d^{-1}
Hypolimnetic release rate	Summer	$k_{r,h}$	0.003–0.07	d^{-1}
	Winter	$k_{r,h}$	0.003–0.07	d^{-1}
Settling velocity	Annual	v_e, v_h	0.05–0.6	$m\ d^{-1}$

[†] Values are taken primarily from Imboden (1974) and Snodgrass (1974).

The mass balances for the three remaining pools are

$$V_e \frac{dp_{n,e}}{dt} = W_{n,e} - Qp_{n,e} + v_t A_t(p_{n,h} - p_{n,e}) + k_{u,e} V_e p_{s,e} - k_{r,e} V_e p_{n,e} - v_e A_t p_{n,e} \tag{29.36}$$

$$V_h \frac{dp_{s,h}}{dt} = W_{s,h} + v_t A_t(p_{s,e} - p_{s,h}) + k_{r,h} V_h p_{n,h} \tag{29.37}$$

$$V_h \frac{dp_{n,h}}{dt} = W_{n,h} + v_t A_t(p_{n,e} - p_{n,h}) - k_{r,h} V_h p_{n,h} + v_e A_t p_{n,e} - v_h A_t p_{n,h} \tag{29.38}$$

Definitions and typical values of the parameters are contained in Table 29.4.

29.5.2 Application to Lake Ontario

Parameters for Lake Ontario are summarized in Tables 29.5 and 29.6. In addition the heat exchange coefficient across the thermocline is 0.0744 $m\ d^{-1}$ during the summer stratified period, and the water column is well-mixed vertically during other times of the year.

Equations 29.35 to 29.38 can be integrated numerically with a method such as the fourth-order Runge-Kutta technique. The results are displayed in Fig. 29.16. As can be seen, the primary feature in the epilimnion is the shift of mass from the SRP and the NSRP fraction during summer due to the large uptake rate. Because of the lack of production, the hypolimnion is generally a more stable system, with concentrations maintained at fairly constant levels throughout the year.

Although the above framework captures many of the essential features of the seasonal cycle, it has several limitations. In particular, its use of constant coefficients and first-order kinetics limits its general applicability. Many of the processes governing substance interactions in the water column are nonlinear and dependent on factors not accounted for in this model. For example the epilimnetic uptake rate depends, among other things, on light intensity, temperature, and levels of both the phytoplankton and the dissolved nutrient. In addition the dependence on the nutrient is best described by a nonlinear relationship. Thus some efforts to refine Equations

TABLE 29.5
Information on Lake Ontario in the early 1970s

Parameter	Symbol	Value	Units
Area			
Surface	A_s	19,000	10^6 m^2
Thermocline	A_t	18,500	10^6 m^2
Mean depth			
Whole lake	H	86	m
Epilimnion	H_e	15	m
Hypolimnion	H_h	71	m
Volume			
Whole lake	V	1634	10^9 m^3
Epilimnion	V_e	254	10^9 m^3
Hypolimnion	V_h	1380	10^9 m^3
Outflow	Q	212	10^9 m^3 yr^{-1}
SRP load			
Epilimnion	$W_{s,e}$	4000	10^9 mg yr^{-1}
Hypolimnion	$W_{s,h}$	0	10^9 mg yr^{-1}
NSRP load			
Epilimnion	$W_{n,e}$	8000	10^9 mg yr^{-1}
Hypolimnion	$W_{n,h}$	0	10^9 mg yr^{-1}

29.35 to 29.38 have focused on more mechanistic characterizations of the kinetic interactions (Imboden and Gachter 1978). Later lectures review nutrient/food-chain models that incorporate sufficient refinements as to constitute an alternative method.

Other ways in which the approach can be modified would be to divide the hypolimnion into several layers to try to resolve vertical gradients in the bottom waters (Imboden and Gachter 1978). In addition the metalimnion can be modeled as a third segment.

Likewise, phosphorus can be subdivided into additional components to more realistically define its dynamics. For example a distinction can be made between living particulate phosphorus and detrital phosphorus or between dissolved organic and dissolved inorganic forms. Figure 29.17 depicts a number of possible kinetic segmentation schemes for phosphorus.

TABLE 29.6
Kinetic parameters for Lake Ontario in the early 1970s

Parameter	Season	Value	Units
$k_{u,e}$	Summer	0.36	d^{-1}
	Winter	0.045	d^{-1}
$k_{r,e}$	Summer	0.068	d^{-1}
	Winter	0.005	d^{-1}
$k_{r,h}$	Summer	0.005	d^{-1}
	Winter	0.005	d^{-1}
v_e, v_h	Annual	0.103	m d^{-1}

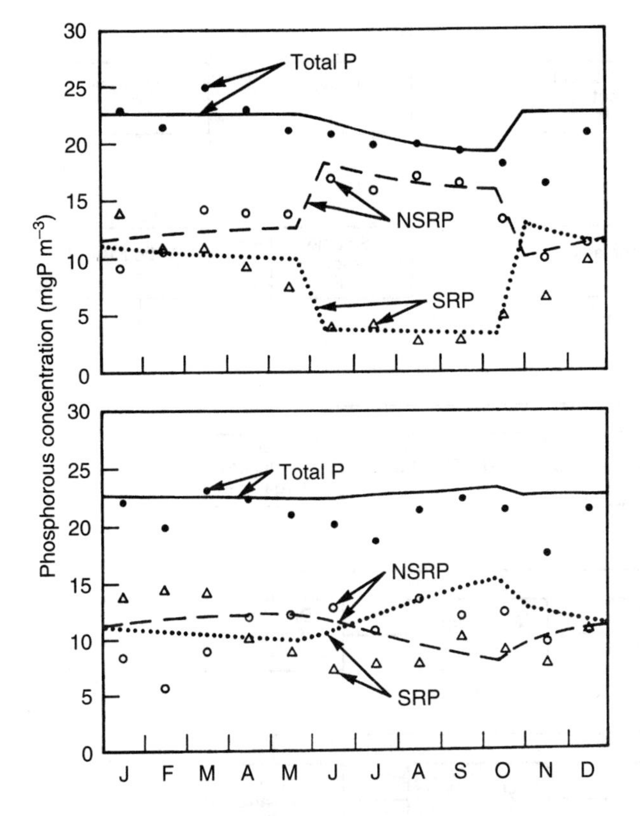

FIGURE 29.16
Data and simulation results using the simplest seasonal approach for total phosphorus in Lake Ontario. (Top—epilimnion; Bottom—hypolimnion).

29.5.3 Kinetic Segmentation

There are three basic rationales underlying kinetic segmentation:

- First, the division of matter can be based on measurement techniques as in the case of the SRP/NSRP scheme. Similarly a dissolved/particulate split (Fig. 29.17*b*) is, in part, based on the use of filtration to discriminate between these pools.
- Second, the segmentation can have a mechanistic basis. For example the breakdown of matter into pools with similar kinetic characteristics facilitates derivation and measurement of the input-output terms and coupling mechanisms between components. The division of the particulate phosphorus into phytoplankton and detrital components (Fig. 29.17*h*) is illustrative of this rationale since the settling rates of these two pools are different and can be measured separately.
- Finally the segmentation scheme can have a management basis. For example the explicit formulation of a phytoplankton pool has informational value for the planner trying to assess the deleterious effects of eutrophication.

A general advantage of adding compartments is that the transport and reaction processes can usually be formulated in a more mechanistic manner based on measurements. A disadvantage is that additional effort must be expended to obtain these measurements. In addition the more "sophisticated" representations usually require more effort to program and run on the computer, and they are more difficult to

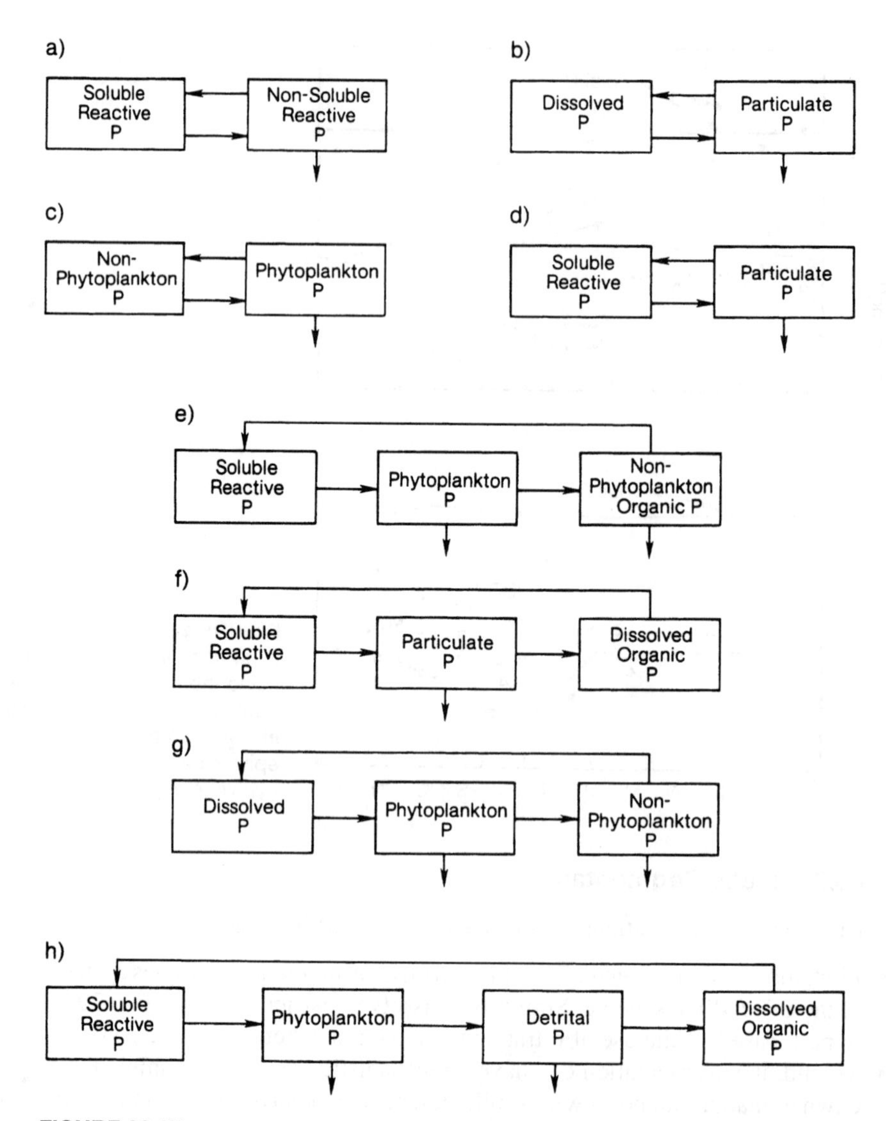

FIGURE 29.17
Alternative kinetic segmentation schemes to model seasonal phosphorus dynamics.

interpret. Further, prediction reliability may actually decrease with additional compartments, because the least understood pools would typically be incorporated last (recall Fig. 18.2). Therefore the decision to expand on a basic framework such as Fig. 29.15 must be made only after considering these factors.

PROBLEMS

29.1. A gravel company excavates a 405-ha gravel pit that is 1.52 m deep. The pit is filled with groundwater at a rate of 100 $m^3\ d^{-1}$. The groundwater is devoid of nutrients. The pit has a small drainage area of 607 ha. Rainfall on this drainage area yields runoff of

25 cm yr^{-1} with a phosphorus content of 40 kg km^{-2} yr^{-1}. Atmospheric sources of total phosphorus are 24 kg km^{-2} yr^{-1}. If total phosphorus settles at a rate of 12 m yr^{-1}, calculate the pit's total phosphorus concentration. Note that evaporation exactly equals direct precipitation.

29.2. A lake (surface area = 10^6 m^3, mean depth = 5 m, residence time = 2 yr) has a total phosphorus loading of 2.5×10^8 mg yr^{-1}. Note that the lake has a k_{wc} = 0.15 m^{-1}. Determine the following:
(*a*) Total phosphorus inflow and lake concentrations
(*b*) Chlorophyll *a* concentration
(*c*) Secchi-disk depth
(*d*) Loading required to maintain the lake at the border between oligotrophy and mesotrophy

29.3. A stratified lake (surface area = 1.5×10^6 m^2, thermocline area = 1×10^6 m^2, epilimnion volume = 1×10^7 m^3, hypolimnion volume = 0.8×10^7 m^3, residence time = 3 yr) has a total phosphorus loading of 7×10^8 mg yr^{-1}. Determine its total P and chlorophyll *a* concentrations and its Secchi-disk depth. Also, calculate its AHOD and determine the concentration of oxygen in its hypolimnion over the summer stratified period of 3 months.

29.4. Lake Sammamish, located in Washington State, has the following characteristics:

Lake surface area = 119.8 km^2
Sediment surface area = 13.068 km^2
Water volume = 3.5×10^8 m^3
Hypolimnion volume = 9.8×10^7 m^3
Hypolimnion thickness = 7.5 m
Flow = 2.03×10^8 m^3 yr^{-1}
Active sediment thickness = 10 cm
Sediment porosity = 0.9
Sediment density = 2.5×10^6 g m^{-3}

The lake was heavily polluted in the 1960s when the total P concentration in the lake was approximately 33 mg m^{-3}. At this time the lake's inflow concentration was about 100 mg m^{-3}. The sediment P concentration at the same time was about 0.12% P.

The lake is monomictic with a stratified period from day 135 to day 315. The oxygen at the beginning of the stratified period is about 8 mg L^{-1} and the summer hypolimnetic temperature is approximately 10°C.

In 1969 the inflow concentration was abruptly dropped to 65 mg m^{-3}. The following data are available for lake total P (in mg m^{-3}) before and after the diversion:

Year	1964	1965	1966	1971	1972	1973	1974
Total P	32	35	32	27	29	32.5	25.5
Year	1975	1979	1981	1982	1983	1984	
Total P	20	14	22.5	18.5	16.5	18.5	

If a value of 46 m yr^{-1} is assumed for the settling velocity, calibrate and simulate total P in this lake from 1960 through the year 2000. Present your results graphically. In addition investigate the response if the inflow concentration had been dropped to 45 or 25 mg m^{-3} in 1969.

29.5. Use the simplest seasonal approach to determine the concentrations in Lake Ontario over the annual cycle if the loadings in the early 1970s were doubled.

LECTURE 30

Heat Budgets

Lecture overview: I outline how a heat budget can be developed for a well-mixed system. Emphasis is placed on characterizing the surface heat exchange between the water and the atmosphere.

We ended the previous lecture with a model of a thermally stratified lake. As we saw, thermal stratification can have a pronounced effect on water quality. Consequently, understanding heat and temperature modeling is an important facet of water-quality modeling.

The mathematical modeling of the transport and fate of heat in natural waters has been the subject of extensive study. Edinger et al. (1974) provide an excellent and comprehensive report of this research. Thomann and Mueller (1987) have summarized the fundamental approach as it relates to water-quality modeling.

Most of this work has been oriented toward evaluating cooling-water discharges and has dealt with systems such as cooling ponds and large rivers. Today there is a broadening of interest in temperature modeling that goes beyond the effects of point sources of heat such as power plants. Several problem contexts that involve temperature and heat are:

- Physical processes (such as thermal stratification) along with biological and chemical transformations are sensitive to temperature. Therefore to adequately characterize these other problems, there may be cases where an accompanying analysis of heat would be needed.
- There is growing interest in the diurnal temperature variations of shallow, turbulent streams that are commonly found in upland regions. Although these systems are sometimes subjected to anthropogenic heat loads, their response to natural forcing functions is also of interest. Physical modifications, such as channelization and riparian zone denudation, can have a pronounced effect on their thermal regimes.

Remedial measures have been proposed to reverse these effects (Oswald and Roth 1988). Mathematical models could prove useful in the evaluation of these modifications. In addition diurnal temperature variations are relevant to the modeling of the fate of pollutants in such systems. For example water-quality problems such as ammonia toxicity are sensitive to diurnal temperature variations.

- Temperature effects on biota, some of which are threatened or endangered, are directing attention toward heat management. Aside from the drainage-basin modifications noted above, reservoir releases can also have an impact on temperatures in tail waters.

For all these problems, the extensive body of theory concerning the fate of heat provides a basis for temperature models of streams, estuaries, and lakes. In this lecture I'll describe the part of this theory that relates to surface heat exchange.

30.1 HEAT AND TEMPERATURE

To this point we have focused on mass balances of substances in natural waters. In particular we have used concentration to provide a measure of the intensity of a pollutant. For a volume of water V, the concentration c is related to the mass m by

$$c = \frac{m}{V} \tag{30.1}$$

Recall from our discussion in Lec. 1 that mass is an extensive (that is, size-dependent) property whereas concentration is intensive (size-independent).

A similar relationship can be developed for heat,

$$T = \frac{H}{\rho C_p V} \tag{30.2}$$

where T = temperature
H = heat
ρ = density
C_p = specific heat

Thus heat is the extensive quantity whereas temperature is intensive (Fig. 30.1).

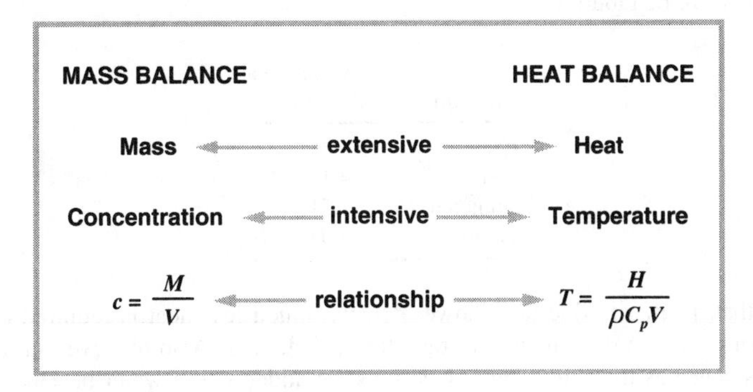

FIGURE 30.1
The analogy between mass and heat.

TABLE 30.1
Thermal units

	Symbol	MKS	CGS	English
Temperature	T	°C or K	°C or K	°F or °R
Heat	H	joules	calories	Btu
Density	ρ	kg m^{-3}	g cm^{-3}	lb$_m$ ft^{-3}
Volume	V	m^3	cm^3	ft^3
Specific heat	C_p	J (kg °C)$^{-1}$	cal (g °C)$^{-1}$	Btu (lb$_m$ °F)$^{-1}$

Note that several systems of units can be used (Table 30.1) for the quantities in Eq. 30.2.

Observe that two parameters, density and specific heat, are included in Eq. 30.2 to reflect the type of substance that is being described. As in the following example, these parameters influence the temperature change that results from adding heat to different substances.

EXAMPLE 30.1. TEMPERATURE AND HEAT. Determine how much heat needs to be added to 1 m^3 of air, water, brick, and iron to induce a 1°C rise in temperature. Thermal properties of these substances (Kreith and Bohn 1986) are listed below:

Substance	Density (kg m^{-3})	Specific heat [J(kg °C)$^{-1}$]
Dry air	1.164	1012
Water	998.2	4182
Common brick	1800	840
Cast iron	7272	420

Note: all values are for 20°C.

Solution: For air, Eq. 30.2 can be rearranged to compute

$$H = \rho C_p V T = 1.164 \frac{\text{kg}}{\text{m}^3}\left(1012 \frac{\text{J}}{\text{kg °C}}\right)(1 \text{ m}^3)(1°\text{C}) = 1178 \text{ J}$$

The amount of heat for the other substances can be computed in a similar fashion. All the results can be tabulated as

Substance	Added heat (J)
Dry air	1178
Water	4.17×10^6
Common brick	1.51×10^6
Cast iron	3.05×10^6

Notice that, primarily due to its lower density, much less heat is required to induce a temperature rise in dry air than in the other substances. Also observe that although it is not as dense as the solids, more heat must be added to the water because of its high specific heat.

30.2 SIMPLE HEAT BALANCE

Just as a mass balance can be written for a volume of water, a ***heat balance*** can also be developed. As was the case for mass, the heat balance states that for a finite volume of water over a unit time period,

$$\text{Accumulation} = \text{sources} - \text{sinks} \tag{30.3}$$

A hypothetical completely mixed system is depicted in Fig. 30.2. For a finite time period, the heat balance for the system can be expressed as

$$\text{Accumulation} = \text{inflow} - \text{outflow} \pm \text{surface heat exchange} \tag{30.4}$$

In this balance there is a single inflow source that contributes heat. Although this is labeled "inflow" it represents both heat entering through tributary streams as well as point discharges. A loss is included due to heat leaving the system through the lake's outlet. Finally surface heat exchange represents the heat gained across the air-water interface due to interactions with the atmosphere. As indicated by the plus or minus sign, this term can be either a source or a sink, depending on the state of the lake and the atmosphere.

Other sources and sinks could have been included in the heat balance. For example exchange of energy with the sediments can be an important term in very shallow systems. However, since sediment-water exchange is not significant for most natural water bodies, we limit ourselves to the terms outlined in Fig. 30.2 and Eq. 30.4.

Although Eq. 30.4 has descriptive value, it cannot be used to predict water quality. To do this we now express each term as a function of measurable variables and parameters.

Accumulation. Accumulation represents the change of heat H in the system over time t,

$$\text{Accumulation} = \frac{\Delta H}{\Delta t} \tag{30.5}$$

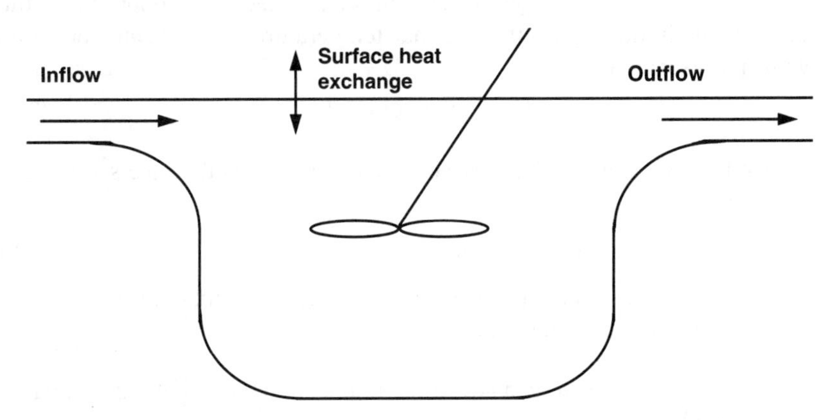

FIGURE 30.2
A heat balance for a well-mixed lake. The arrows represent the major sources and sinks of heat.

Substituting Eq. 30.2 into 30.5 yields

$$\text{Accumulation} = \frac{\Delta \rho C_p V T}{\Delta t} \tag{30.6}$$

In the present case we assume that the density and specific heat of water are relatively constant over the range of temperatures for the lake. We also assume that the lake's volume is constant. These assumptions allow us to bring the parameters outside the difference. Finally when Δt is made very small, Eq. 30.6 reduces to

$$\text{Accumulation} = \rho C_p V \frac{dT}{dt} \tag{30.7}$$

Thus heat accumulates as temperature increases with time (positive dT/dt) and diminishes as it decreases with time (negative dT/dt). For the steady-state case, heat remains constant ($dT/dt = 0$). Note that the units of accumulation (as with all other terms in the balance) are heat per time (J d^{-1}). This can be seen by inspecting the units of the individual terms in Eq. 30.7,

$$\rho C_p V \frac{dT}{dt} = \frac{\text{kg}}{\text{m}^3}\left(\frac{\text{J}}{\text{kg °C}}\right)(\text{m}^3)\left(\frac{\text{°C}}{\text{d}}\right) = \frac{\text{J}}{\text{d}} \tag{30.8}$$

Inflow. As was the case with mass, we lump all point and nonpoint sources of heat entering the lake's periphery into a single term,

$$\text{Inflow} = Q\rho C_p T_{\text{in}}(t) \tag{30.9}$$

where Q = volumetric flow rate of all water sources entering the system and $T_{\text{in}}(t)$ = average inflow temperature of these sources. For the present case we have assumed that flow is constant and that all the temporal variations in heat inputs are due to temporal variations in the inflow temperature.

Outflow. In our simple system (Fig. 30.2) heat is carried from the system by an outflow stream. The rate of heat transport can be quantified as a function of the outflow temperature T_o. However, because of our well-mixed assumption, the outflow temperature by definition equals the mid-lake temperature, $T_o = T$, and the outflow sink can be represented by

$$\text{Outflow} = Q\rho C_p T \tag{30.10}$$

Surface heat exchange. The net heat exchange across the lake's surface can be represented as a flux,

$$\text{Surface heat exchange} = A_s J \tag{30.11}$$

where A_s = lake's surface area (m^2) and J = surface heat flux (J m^{-2} d^{-1}), with a positive flux representing a heat gain.

Total balance. The terms can now be combined into the following heat balance for a well-mixed lake:

$$V\rho C_p \frac{dT}{dt} = Q\rho C_p T_{\text{in}}(t) - Q\rho C_p T + A_s J \tag{30.12}$$

EXAMPLE 30.2. HEAT BALANCE FOR A WELL-MIXED POND. A pond has the following characteristics:

Volume = 50,000 m^3
Surface area = 25,000 m^2
Mean depth = 2 m
Inflow = outflow = 7500 m^3 d^{-1}

The pond's inflow has a temperature of 20°C. The net heat gain from the atmosphere is 250 cal cm^{-2} d^{-1}. If there is no other heat exchange, calculate the steady-state temperature.

Solution: First, the surface heat flux can be converted to the proper units,

$$J = 250 \frac{\text{cal}}{\text{cm}^2\ \text{d}} \left(\frac{10{,}000\ \text{cm}^2}{\text{m}^2}\right)\left(\frac{\text{J}}{0.2388\ \text{cal}}\right) = 1.047 \times 10^7\ \text{J m}^{-2}\ \text{d}^{-1}$$

Then, Eq. 30.12 can be solved for

$$T = T_{\text{in}} + \frac{A_s J}{Q \rho C_p} = 20 + \frac{25{,}000(1.047 \times 10^7)}{7500(998.2)(4182)} = 20 + 8.36 = 28.36°\text{C}$$

Thus the temperature is increased 8.36°C by the surface heat flux.

In the previous example, we treated atmospheric heat exchange as a single entity. In fact, as described next, it consists of a number of different mechanisms.

30.3 SURFACE HEAT EXCHANGE

As depicted in Fig. 30.3, surface heat exchange can be modeled as a combination of five processes. Note that the processes can be clustered in two ways. First, we can distinguish between radiative and nonradiative mechanisms. ***Radiation*** refers to energy that is transmitted in the form of electromagnetic waves and thus does not

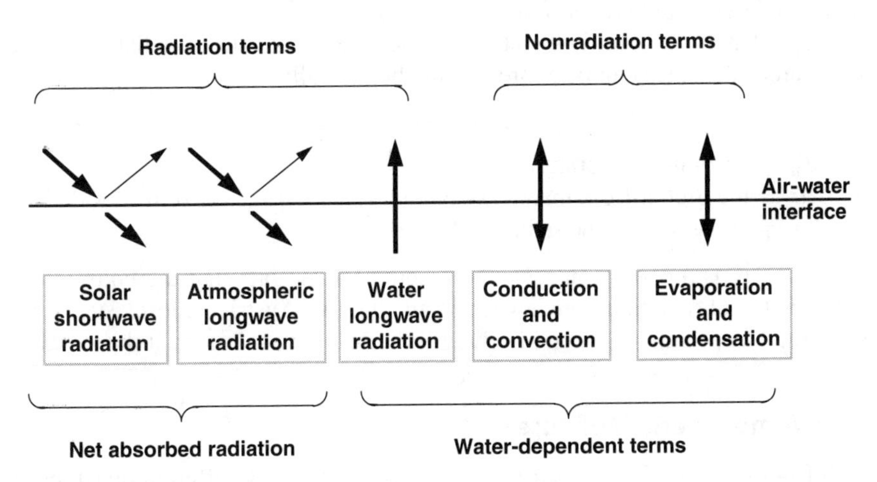

FIGURE 30.3
The components of surface heat exchange.

depend on matter for its transmission. In contrast, processes such as conduction and evaporation depend on the motion of molecules.

The second way in which the mechanisms can be divided is based on whether they are dependent on the temperature of the water. This division relates to whether the process acts as a model forcing function or includes the dependent variable: water temperature. Processes such as solar radiation and atmospheric longwave radiation are independent of the condition of the water. Hence these terms act as forcing functions. As labeled in Fig. 30.3, they constitute the net radiation absorbed by the water. In contrast, although they are dependent on external conditions (air temperature, moisture, wind speed, etc.), processes such as evaporation are also a function of water temperature.

The total surface heat flux can be represented as

$$J = \underbrace{J_{sn} + J_{an}}_{\text{Net absorbed radiation}} - \underbrace{(J_{br} + J_c + J_e)}_{\text{Water-dependent terms}} \tag{30.13}$$

where J_{sn} = net solar shortwave radiation
J_{an} = net atmospheric longwave radiation
J_{br} = longwave back radiation from the water
J_c = conduction
J_e = evaporation

In the following sections we outline how these terms are quantified. Before doing this we must provide some background information on two subjects that have a strong bearing on some of the terms: the Stefan-Boltzmann law and atmospheric moisture.

30.3.1 Stefan-Boltzmann Law

All objects with a temperature above absolute zero emit radiation. The higher the object's temperature, the shorter the wavelength of the emission and the greater the quantity of energy emitted per unit of surface area.

The Stefan-Boltzmann law states that the maximum rate of radiation emitted per unit area of surface can be expressed mathematically as

$$J_{\text{rad}} = \epsilon\sigma T_a^4 \tag{30.14}$$

where T_a = absolute temperature (K)
σ = the Stefan-Boltzmann constant [$= 11.7 \times 10^{-8}$ cal (cm^2 d K^4)$^{-1}$]
ϵ = emissivity of the radiating body

The emissivity is a correction factor to account for the fact that the body is not a perfect emitter of radiation. The law is named after two Austrians, Josef Stefan, who discovered it experimentally, and Ludwig Boltzmann, who derived it theoretically.

30.3.2 Atmospheric Moisture

Recall from Example 30.1 that dry air and water have very different densities and specific heats. Consequently something must be done to account for the fact that air can have a wide range of water content. The range of water content can be represented by the percent relative humidity R_h, which is the ratio of the air's actual water content to its maximum possible level at the same temperature,

$$R_h = 100\frac{e_{air}}{e_{sat}} \tag{30.15}$$

where e_{air} = vapor pressure of the air (mmHg) and e_{sat} = saturation vapor pressure (mmHg). The latter quantity is a function of temperature and can be calculated by (modified for units from Raudkivi 1979)

$$e_{sat} = 4.596e^{\frac{17.27T}{237.3+T}} \tag{30.16}$$

Figure 30.4 shows the resulting dependency of saturation on temperature.

Finally the ***dew-point temperature*** is the temperature at which a mass of air just becomes saturated when cooled at a constant pressure and water content. The name makes sense if you think about the situation that occurs at night. At sundown, if there is no precipitation, the air will likely have a water content below saturation. After the sun goes down, the air usually cools until it reaches its lowest value just before sunrise. As the air cools (assuming that the pressure and the moisture content remain relatively constant), the saturation level decreases according to Eq. 30.16. If the process continues, there will come a point at which the moisture content will equal the saturation value. At this point the air cannot hold additional water and dew forms.

Inspection of Fig. 30.4 indicates that the air vapor pressure is equivalent to the saturation pressure corresponding to the dew-point temperature,

$$e_{air} = 4.596e^{\frac{17.27T_d}{237.3+T_d}} \tag{30.17}$$

These concepts are important since they relate to quantifying the heat loss due to evaporation. For example Dalton (1802) proposed that evaporation was proportional to the difference between the vapor pressure of the air and the saturated vapor pressure calculated at the temperature of the water surface T_s:

$$J_e = -f(U_w)(e_s - e_{air}) \tag{30.18}$$

where e_s = saturation vapor pressure corresponding to the surface temperature of

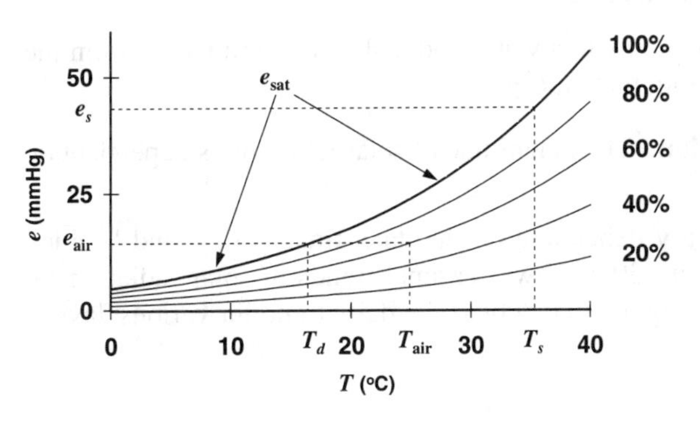

FIGURE 30.4
The dependency of saturation vapor pressure on temperature. Lines of constant relative humidity are displayed. Also shown are the vapor pressures corresponding to the water, air, and dew-point temperature from Example 30.3.

the water (°C) and $f(U_w)$ = a transfer coefficient that depends on the wind speed measured a fixed distance above the water surface.

It is unlikely that you will be provided with a value for the vapor pressure in the air above a water body. It is more likely that you will be able to obtain estimates for air temperature, dew-point temperature, and/or relative humidity. As illustrated by the following example, all you require is two of these quantities to calculate the remaining value as well as the vapor pressure of the air.

EXAMPLE 30.3. RELATIVE HUMIDITY, DEW POINT, AND AIR TEMPERATURE. The atmosphere above a lake has an air temperature of 25°C, a relative humidity of 60%, and the surface water temperature is 35°C. Use this information to determine (*a*) the air vapor pressure and the dew-point temperature and (*b*) whether evaporation or condensation will occur.

Solution: (*a*) Equation 30.16 can be used to compute the saturation vapor pressure,

$$e_{sat} = 4.596e^{\frac{17.27(25)}{237.3+25}} = 23.84 \text{ mmHg}$$

Therefore Eq. 30.15 can be used to determine the air vapor pressure,

$$e_{air} = \frac{R_h e_s}{100} = \frac{60(23.84)}{100} = 14.3 \text{ mmHg}$$

Then this value can be substituted into Eq. 30.17,

$$14.3 = 4.596e^{\frac{17.27T_d}{237.3+T_d}}$$

which can be solved for

$$T_d = \frac{237.3}{\frac{17.27}{\ln(14.3/4.596)} - 1} = 16.7°\text{C}$$

(*b*) The saturation vapor pressure corresponding to the surface temperature of the water can be computed as

$$e_s = 4.596e^{\frac{17.27(35)}{237.3+35}} = 31.93 \text{ mmHg}$$

Because $e_s > e_{air}$, Dalton's law indicates that heat (and water) will be lost from the lake by evaporation.

30.3.3 Net Absorbed Radiation

Two sources of radiation add energy to a water body: shortwave radiation from the sun and longwave radiation from the atmosphere.

Solar shortwave radiation. The magnitude of solar radiation is dependent on several factors:

- *Solar altitude.* This will vary depending on the date, time of day, and location on the earth's surface. Figure 30.5 shows seasonal trends of solar radiation for different values of latitude. Figure 30.6 depicts the daily trend for various days of the year.

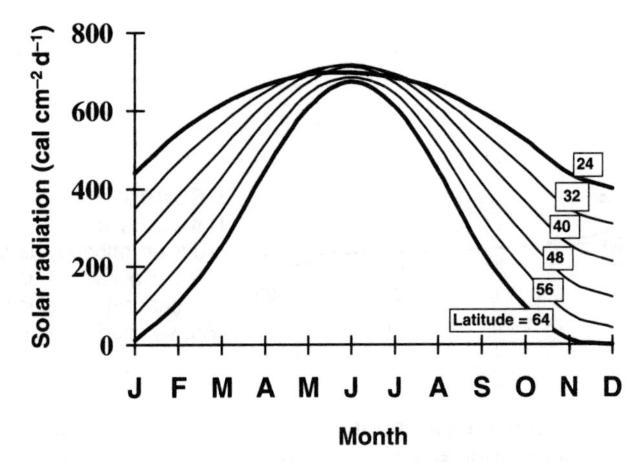

FIGURE 30.5
Daily totals of solar radiation for different latitudes as a function of time of year (from Kreider 1982).

- *Scattering and absorption.* Once sunlight enters the atmosphere it is scattered by dust, reflected by clouds, and absorbed by atmospheric gases.
- *Reflection.* Upon reaching the water, a portion of the radiation will be scattered by reflection at the surface. The percent of radiation returning from a surface compared to that which strikes it is formally referred to as the surface's ***albedo***. As depicted in Fig. 30.7, for a flat water surface the fraction reflected is significant only when the sun is low in the sky. In addition the condition of the sky (clear/overcast) and the water surface (flat/waves) can affect reflection.
- *Shading.* Some streams are located in deep canyons or are lined by tall trees. In such cases the resulting shading can greatly decrease solar radiation.

Solar radiation is usually obtained from direct measurements or equations. The latter are a function of factors such as time, position, and cloud cover. Some of the commonly used algorithms are summarized in a number of references (Eagleson 1970, Bras 1990, Brown and Barnwell 1987, etc.)

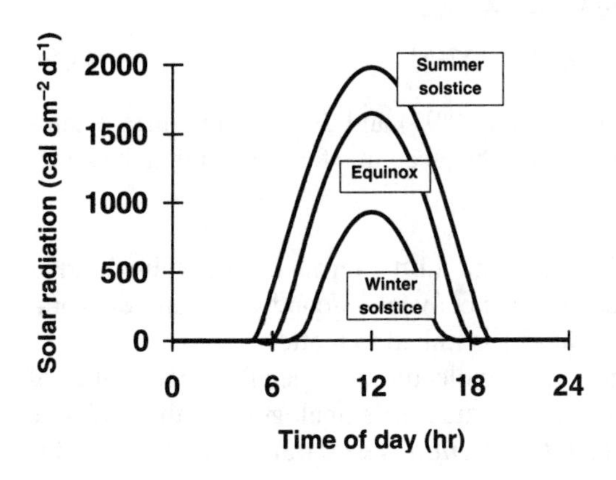

FIGURE 30.6
Diurnal trends of clear-sky solar radiation on a horizontal surface at latitude 40°N on the summer solstice (June 21), the winter solstice (Dec. 21), the vernal equinox (Mar. 21), and the autumnal equinox (Sep. 21) (from Kreider 1982).

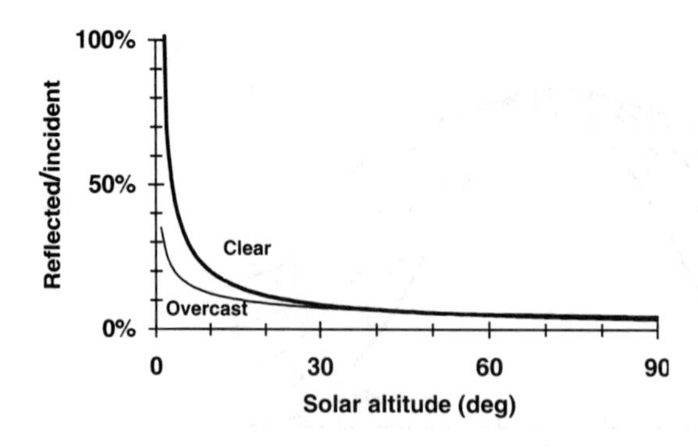

FIGURE 30.7
Ratio of reflected to incident solar radiation (albedo) as a function of solar altitude. Both clear and overcast sky conditions are shown (Brown and Barnwell 1987).

Atmospheric longwave radiation. The atmosphere itself emits longwave radiation. This gain can be represented as a modification of the Stefan-Boltzmann law,

$$J_{an} = \underbrace{\sigma(T_{\text{air}} + 273)^4}_{\text{Stefan-Boltzmann law}} \underbrace{\left(A + 0.031\sqrt{e_{\text{air}}}\right)}_{\text{Atmospheric attenuation}} \underbrace{(1 - R_L)}_{\text{Reflection}} \tag{30.19}$$

where σ = the Stefan-Boltzmann constant = 4.9×10^{-3} J (m² d K⁴)⁻¹ or 11.7×10^{-8} cal (cm² d K⁴)⁻¹

T_{air} = air temperature (°C)

A = a coefficient (0.5 to 0.7)

e_{air} = air vapor pressure (mmHg)

R_L = reflection coefficient

The reflection coefficient is generally small ($\cong 0.03$).

30.3.4 Water-Dependent Terms

Water longwave radiation. The back radiation from the water surface can also be represented by the Stefan-Boltzmann law,

$$J_{br} = \epsilon\sigma(T_s + 273)^4 \tag{30.20}$$

where ϵ = emissivity of water (approximately 0.97) and T_s = water surface temperature. The emissivity is a correction factor that accounts for the fact that the water is not a perfect emitter of radiation.

Conduction and convection. To this point all the terms have involved radiation. We now turn to the last two mechanisms: convection/conduction and evaporation/condensation. In both cases heat transfer is linked to matter.

Conduction is the transfer of heat from molecule to molecule when matter of different temperatures come into contact. As such, it is analogous to the diffusive transport previously described in Lec. 8. ***Convection*** is heat transfer that occurs due to mass movement of fluids. Both can occur at the air-water interface and can be described by

$$J_c = c_1 f(U_w)(T_s - T_{air}) \tag{30.21}$$

where c_1 = Bowen's coefficient ($\cong$ 0.47 mmHg °C^{-1}). The term $f(U_w)$ defines the dependence of the transfer on wind velocity over the water surface, where U_w is the wind speed measured a fixed distance above the water surface. Many relationships exist to define the wind dependence. Bras (1990) provides a review. Edinger et al. (1974) have suggested the relationship proposed by Brady, Graves, and Geyer (1969),

$$f(U_w) = 19.0 + 0.95U_w^2 \tag{30.22}$$

where the wind speed is measured in m s^{-1} at a height of 7 m above the water surface.

Evaporation and condensation. The heat loss due to evaporation can be represented by Dalton's law,

$$J_e = f(U_w)(e_s - e_{air}) \tag{30.23}$$

where e_s = saturation vapor pressure at the water surface and e_{air} = vapor pressure in the overlying air (mmHg).

30.3.5 Total Heat Budget

The individual terms developed in the previous paragraphs can now be incorporated into Eq. 30.13,

$$J = \underbrace{J_{sn}}_{\text{Net solar}} + \underbrace{\sigma(T_{air} + 273)^4 \left(A + 0.031\sqrt{e_{air}}\right)(1 - R_L)}_{\text{Atmospheric longwave}}$$

$$- \underbrace{\epsilon\sigma(T_s + 273)^4}_{\text{Water longwave}} - \underbrace{c_1 f(U_w)(T_s - T_{air})}_{\text{Conduction}} - \underbrace{f(U_w)(e_s - e_{air})}_{\text{Evaporation}} \tag{30.24}$$

This surface heat flux can then be substituted into the heat balance for the well-mixed lake (Eq. 30.12). As described next, the resulting equation then provides a basis for predicting water temperature as a function of heat loads and atmospheric conditions.

30.4 TEMPERATURE MODELING

Now that we have developed a mechanistic model of surface heat exchange, the total heat balance can be used to predict the temperature of a natural water. For simplicity the following discussion focuses on the simple completely mixed lake introduced at the beginning of this lecture. It should be noted that the approach can be easily extended to other bodies of water such as streams and estuaries.

30.4.1 Steady-State

At steady-state a heat balance for the well-mixed lake in Fig. 30.2 can be written as

$$0 = \frac{\rho C_p Q T_{in}}{A_s} + J_{sn} + \sigma(T_{air} + 273)^4 \left(A + 0.031\sqrt{e_{air}}\right)(1 - R_L) - \frac{\rho C_p Q T_s}{A_s}$$
$$- \epsilon\sigma(T_s + 273)^4 - c_1 f(U_w)(T_s - T_{air}) - f(U_w)(e_s - e_{air}) \tag{30.25}$$

EXAMPLE 30.4. STEADY-STATE HEAT BALANCE FOR A WELL-MIXED LAKE. A pond has the following characteristics:

Volume = 250,000 m^3
Surface area = 25,000 m^2
Inflow = outflow = 7500 $m^3\ d^{-1}$

The pond's inflow has a temperature of 10°C. In addition, it is subject to the following meteorological conditions:

Net solar radiation = 300 cal $cm^{-2}\ d^{-1}$
Air temperature = 25°C
Dew-point temperature = 16.7°C
Wind speed = 3 m s^{-1}
Relative humidity = 60%

Calculate the steady-state temperature.

Solution: Each term in the steady-state heat balance (Eq. 30.25) can be dealt with separately.

Inflow:

$$\frac{\rho C_p Q T_{in}}{A_s} = \frac{1(1)7500 \times 10^6(10)}{250 \times 10^6} = 300 \text{ cal cm}^{-2}\text{ d}^{-1}$$

Atmospheric longwave: This term requires estimates of A and e_{air}. We will arbitrarily choose a value of $A = 0.6$ as the midpoint of the range suggested previously. The vapor pressure of the air can be computed as

$$e_{air} = 0.6\left(4.596e^{\frac{17.27(25)}{237.3+25}}\right) = 14.3 \text{ mmHg}$$

These values can be substituted along with other parameters to compute

$$\sigma(T_{air} + 273)^4 \left(A + 0.031\sqrt{e_{air}}\right)(1 - R_L)$$
$$= 11.7 \times 10^{-8}(25 + 273\text{K})^4 \left(0.6 + 0.031\sqrt{14.3}\right)(1 - 0.03) = 642 \text{ cal cm}^{-2}\text{ d}^{-1}$$

Outflow:

$$-\frac{\rho C_p Q T_s}{A_s} = -\frac{0.9982(1)7500 \times 10^6}{250 \times 10^6} T_s = -30T_s$$

Water longwave:

$$-\epsilon\sigma(T_s + 273)^4 = -0.97(11.70 \times 10^{-8})(T_s + 273)^4 = -11.35 \times 10^{-8}(T_s + 273)^4$$

Conduction: First, we must compute the wind effect (Eq. 30.22),

$$f(U_w) = 19.0 + 0.95(3)^2 = 27.55$$

This result can be substituted along with other values to compute the conduction losses,

$$-c_1 f(U_w)(T_s - T_{\text{air}}) = -0.47(27.55)(T_s - 25) = -12.95(T_s - 25)$$

Evaporation: We must first determine the vapor pressure for the water,

$$e_s = 4.596e^{\frac{17.27T_s}{237.3+T_s}}$$

This result, along with other parameters, can be substituted into Eq. 30.18,

$$-f(U_w)(e_s - e_{\text{air}}) = -27.55\left(4.596e^{\frac{17.27T_s}{237.3+T_s}} - 14.3\right)$$

The individual terms can now be consolidated into the total heat budget,

$$0 = 300 + 300 + 642 - 30T_s - 11.35 \times 10^{-8}(T_s + 273)^4$$

$$- 12.95(T_s - 25) - 27.55\left(4.596e^{\frac{17.27T_s}{237.3+T_s}} - 14.3\right)$$

This nonlinear equation can be solved numerically for $T_s = 17.3°\text{C}$. This value can then be substituted back into the heat budget to assess the relative magnitude of the individual terms. The results are shown below:

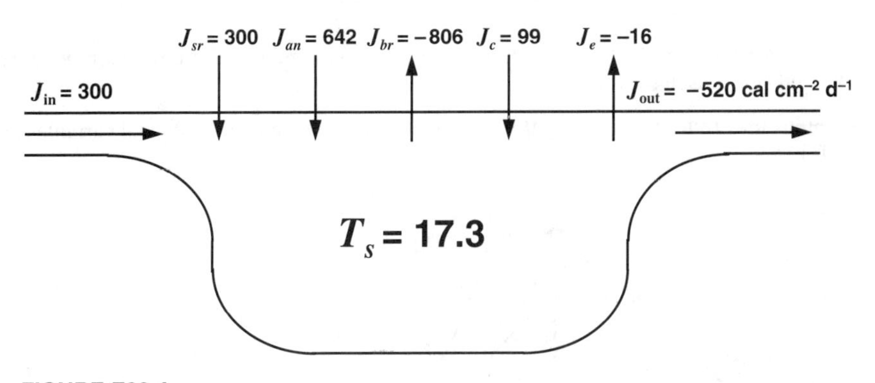

FIGURE E30.4

30.4.2 Time Variable

The surface heat exchange (Eq. 30.24) can also be incorporated into Eq. 30.12 to derive a time-variable heat budget,

$$\frac{dT_s}{dt} = \frac{Q}{V}T_{\text{in}} + \frac{J_{sn}}{\rho C_p H} + \frac{\sigma(T_{\text{air}} + 273)^4\left(A + 0.031\sqrt{e_{\text{air}}}\right)(1 - R_L)}{\rho C_p H}$$

$$- \frac{Q}{V}T_s - \frac{\epsilon\sigma(T_s + 273)^4}{\rho C_p H} - \frac{c_1 f(U_w)(T_s - T_{\text{air}})}{\rho C_p H} - \frac{f(U_w)(e_s - e_{\text{air}})}{\rho C_p H} \tag{30.26}$$

EXAMPLE 30.5. TIME-VARIABLE HEAT BALANCE FOR A WELL-MIXED LAKE. Compute the annual heat budget for the same pond used in Example 30.4. Values for meteorological variables have been provided (Table E30.5):

TABLE E30.5
Thermal units

Month	Solar radiation (cal cm^{-2} d^{-1})	Air temperature (°C)	Dew-point temperature (°C)	Wind speed (km hr^{-1})
Jan	169	8.3	2.8	11.6
Feb	274	9.0	3.3	11.7
Mar	414	13.5	4.9	16.4
Apr	552	13.9	4.0	15.6
May	651	21.8	5.3	16.6
Jun	684	24.7	7.8	16.7
Jul	642	29.4	11.8	12.7
Aug	537	26.6	11.5	11.7
Sep	397	24.9	7.7	14
Oct	259	15.0	6.8	12.9
Nov	160	9.7	6.5	14.8
Dec	127	6.6	2.4	11.6

Assume that the volume and flows are constant and that the inflow temperature is fixed at a constant level of 10°C.

Solution: Before developing the solution, we can plot the meteorological variables (Fig. E30.5-1):

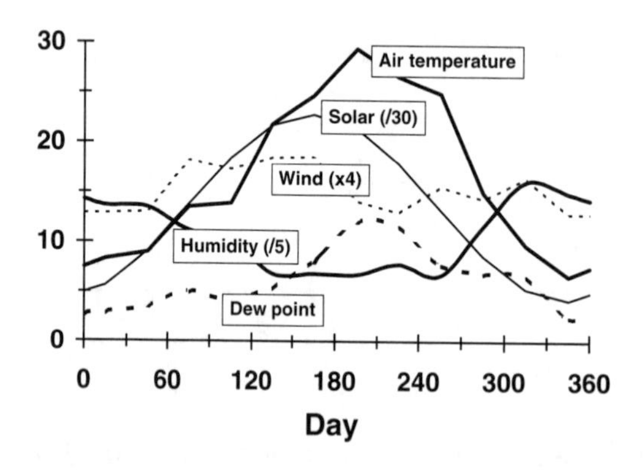

FIGURE E30.5-1
Meteorological variables.

Notice that we have computed and displayed the relative humidity along with the other meteorological variables. The peak solar radiation occurs on day 167 (mid-June) whereas the peak air and dew-point temperatures fall about a month later on day 198 (mid-July). High winds blow in the spring and to a lesser extent in the fall. Finally the highest relative humidities take place in the winter months.

Equation 30.26 can now be integrated with a method such as the fourth-order Runge-Kutta approach described in Lec. 7. The results are displayed in Fig. E30.5-2. We have included the solar radiation, as well as the inflow and the air and dew-point temperatures, on the plot for comparative purposes. Notice how the peak water temperature occurs at about the same time of year as the peak air temperature. In contrast the peak solar radiation occurs about a month earlier.

Finally Fig. E30.5-3 shows the individual terms of the heat balance in flux units. Observe that the largest gains and losses are due to atmospheric and water radiation.

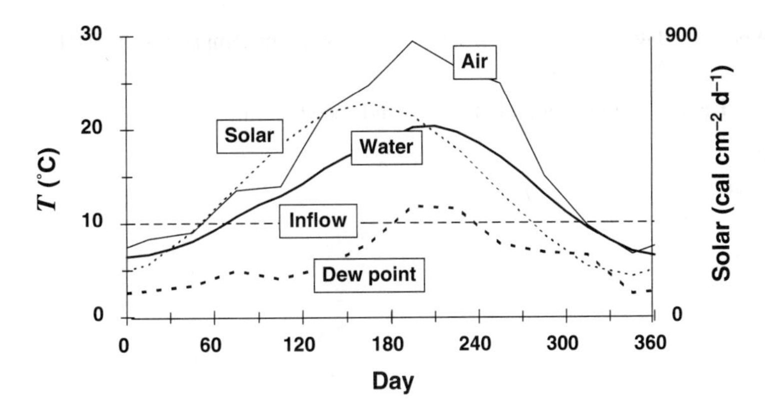

FIGURE E30.5-2
Simulation results.

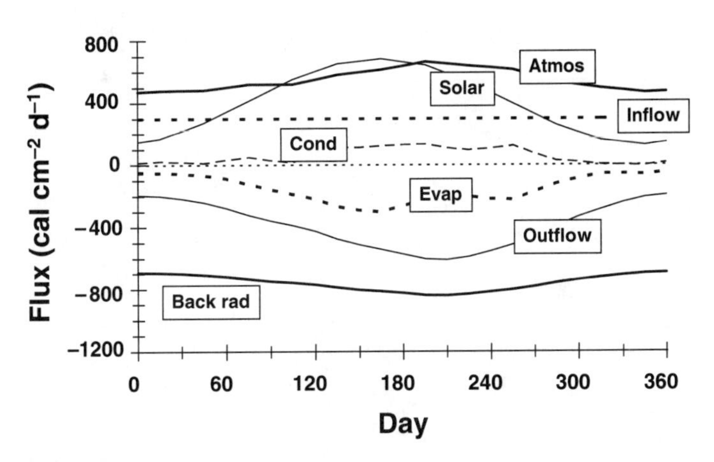

FIGURE E30.5-3
Components of the heat balance.

PROBLEMS

30.1. A $30 \times 10 \times 10$ cm brick is placed in a 2-m^3 container along with 1 m^3 of water. The container is perfectly insulated and is immediately sealed after the brick is added. If the brick, the water, and the air in the container are originally at 50, 10, and 20°C, respectively, determine the equilibrium temperature of the system. How much heat will be contained in the brick, the water, and the air before and after equilibrium? Express your results in kilocalories (kcal).

30.2. A heated discharge has a flow of 7 $m^3\ s^{-1}$ and a temperature of 40°C. A completely mixed cooling pond is to be built. If it is assumed that $J = -200$ cal $(cm^2\ d)^{-1}$, determine the surface area so that the pond's exit temperature is 25°C.

30.3. Determine the net atmospheric surface heat flux so that the lake from Example 30.2 has a temperature of 30°C.

30.4. Calculate the air temperature if the dew-point temperature is 21°C and the relative humidity is 30%.

30.5 Repeat Example 30.4, but use a wind speed of 6 $m\ s^{-1}$.

LECTURE 31

Thermal Stratification

LECTURE OVERVIEW: I describe the vertical distribution of temperature in a lake. Then I illustrate how vertical temperature distributions can be employed to quantify vertical transport in freshwater systems such as lakes.

A lake's vertical thermal regime has dual significance for the water-quality modeler. As mentioned throughout this text, temperature has a strong influence on the rates of chemical and biological reactions. However, it also has additional significance as a tracer of transport in the water column. In fact, heat balances are a primary tool for estimating vertical mixing rates in freshwater systems. Before describing how this is done, I will briefly describe the seasonal temperature changes in the water column of a lake.

31.1 THERMAL REGIMES IN TEMPERATE LAKES

Hutchinson (1957) defines temperate lakes as those "with temperature above 4°C in winter, thermal gradients large, two circulation periods in spring and late autumn." Although other lake types can be severely polluted,[†] the present discussion focuses on temperate lakes because many of the world's developed areas are in temperate

[†] For example lakes that never mix (*amictic*) or mix incompletely (*meromictic*) are extremely sensitive to pollutant inputs. In addition, because of growing urbanization and industrialization in the world's tropical regions, many nontemperate lakes are being subjected to severe water-quality stress. Although some of the approaches in this section might serve as a starting point for modeling these systems, additional research is needed for nontemperate lakes.

climates and, consequently, many lakes in these climates have been subject to pollution. Thus most engineering models have been developed for temperate systems.

The thermal regime of temperate lakes is primarily the result of the interplay of two processes: (1) heat and momentum transfer across the lake's surface and (2) the force of gravity acting on density differences within the lake. Depending on the season of the year, heat transfer tends to either raise or lower the temperature at the lake's surface as a consequence of a number of factors, including the magnitude of solar radiation, air temperature, relative humidity, wind speed, and cloud cover as described in the previous lecture. Winds blowing over the lake's surface tend to mix the surface waters and transfer heat and momentum down through the water column. The extent of this mixing is, in turn, inhibited by buoyancy effects. This relates to the fact that the density of water varies over the range of temperatures encountered in lakes (Fig. 31.1). Therefore, denser waters accumulate at the lake's bottom and are overlaid with lighter waters.

For example in Lake Ontario in August (Fig. 31.2), surface waters of 18°C with a density of 0.9986 g cm^{-3} overlay deep waters of 4°C with a density of 1.0000 g cm^{-3}. Although these density differences may seem small, considerable work must be expended to mix the entire column (that is, lift the heavier bottom waters against the force of gravity to mix them with the lighter surface waters). The energy to do this work comes from the wind. The result is that buoyancy works against and mitigates wind-induced turbulence. The interplay between these factors can be expressed quantitatively by a dimensionless parameter, the ***Richardson number,*** that represents the ratio of buoyancy to shear forces, as in

$$R_i = \frac{\text{buoyancy}}{\text{shear}} = \frac{(g/\rho)(\partial \rho/\partial z)}{(\partial u/\partial z)^2} \tag{31.1}$$

where z = depth (L), which is positive in the downward direction
g = acceleration due to gravity (L T^{-2})
ρ = density of the fluid (M L^{-3})
$\partial \rho/\partial z$ = gradient of density with depth (M L^{-4})
$\partial u/\partial z$ = gradient of horizontal velocity with depth (T^{-1}) or shear

If R_i is significantly greater than a critical level ($\sim$ 0.25), a stable regime results. If R_i is significantly less than 0.25, connoting strong shear relative to stratification, then shear-induced turbulence is generated.

Figure 31.2 depicts the seasonal changes in the vertical temperature distribution of Lake Ontario. Although this lake is very deep, its thermal regime has many of the characteristic features of smaller temperate lakes. Some time after the disappearance of any ice cover in spring, temperatures throughout the water column rise to, or within a few degrees of, the maximum density of water (4°C). At this temperature, heating or cooling of the lake's surface results in very small density differences (Fig. 31.1*b*) and, consequently, only a small amount of wind stress is required to keep the water column well-mixed. In terms of the Richardson number, buoyancy forces (the numerator) are small, and therefore shear forces (the denominator) of small magnitude are sufficient to keep R_i below the critical level.

As spring progresses, solar radiation increases, air temperatures rise, and thermal stratification will be established in the near-surface waters. However, density

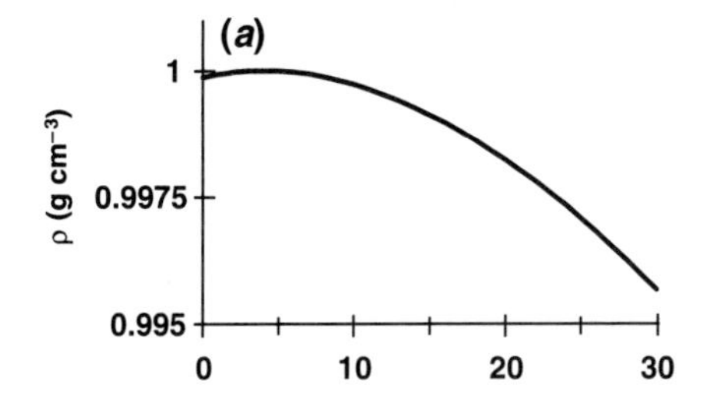

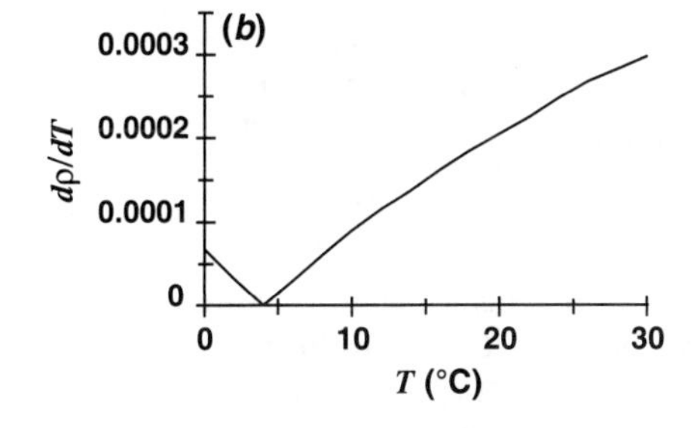

FIGURE 31.1
Plots of (*a*) density (g cm^{-3}) and (*b*) rate of change of density per rate of change of temperature (g cm^{-3} $°C^{-1}$) versus temperature (°C). Note that the maximum density of water occurs at 4°C.

gradients are neither large enough nor deep enough to prevent mixing of the water column by major storms. At the end of spring and the beginning of summer, surface heating increases to the point that mixing is confined to an upper layer. In terms of the bulk Richardson number, the fluid has reached the point where the density gradient is sharp enough that even large storms do not reduce R_i below the stability criterion. At this stage the lake is said to be stably stratified. The attainment of persistent stratification in a lake leads to the existence of three regimes: the upper (epilimnion) and lower (hypolimnion) layers separated by a narrow region of sharp temperature change—the ***thermocline*** or ***metalimnion.***[†]

During midsummer the net daily heat flux at the surface is low and, although the thermocline deepens gradually, the density gradient between the epi- and the hypolimnion remains strong and stable. Although transport of heat and energy across the thermocline occurs, it is at a low level and exchange between the upper and lower layers is at a minimum.

In late summer and fall, loss of heat due largely to falling air temperatures results in a net heat loss from the lake. As surface waters cool, they become more dense than underlying epilimnetic water. Since this is an unstable situation, strong vertical mixing called ***convection*** occurs. Together with increased winds during fall, the process erodes the metalimnion from above, giving the impression of a sinking thermocline. As the lake cools further, a point is reached at which the deepening surface layer becomes denser than the bottom layer, and complete mixing of the column occurs. This episode is called ***fall overturn.***

[†]The terms "thermocline" and "metalimnion" often are used interchangeably to designate the layer of sharp temperature change. Although the metalimnion is really a layer, the thermocline is actually the plane passing through the point of maximum decrease in temperature with depth.

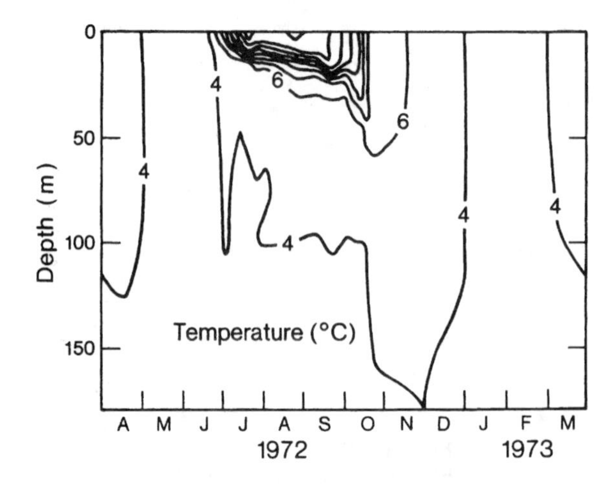

FIGURE 31.2
Depth-time diagram of isotherms (°C) at a mid-lake station in Lake Ontario, 1972– 1973.

The lake continues to be well-mixed and to lose heat as temperatures drop in winter. In some cases the surface water cools below 4°C and an inverse stratification results owing to the low density of water below 4°C (Fig. 31.1*a*). This inverse stratification is strengthened if ice forms at the lake's surface.

In summary the seasonal changes in a temperate lake can be idealized in both time and space. Temporally the cycle consists of two stages: a summer period of strong stratification and a nonstratified period of intense vertical mixing. Spatially the summer stratified period can be treated as consisting of two layers separated by an interface of minimal vertical mixing. These idealizations form the basis of some efforts to develop engineering models of vertical heat and mass distribution in temperate lakes.

31.2 ESTIMATION OF VERTICAL TRANSPORT

In previous lectures (for example Lecs. 8 and 15) we showed how gradients of conservative substances, such as chloride, could be used to estimate horizontal diffusion. Because of the large seasonal temperature gradients in temperate lakes, heat balances serve an analogous role in vertical models. This implies that water motion has an identical effect on determining heat and mass transport in the water column. Although this analogy does not hold strictly for large particles or dense solutions, it is a good first approximation for most pollutants.

We can now write a heat budget in a similar fashion to those developed in the previous lecture. Consider the simplest case of a temperate lake during the mid-summer stratified period. At this time (Fig. 31.2) the temperatures in the epi- and hypolimnion are fairly uniform, with the major gradient occurring at the thermocline. Although the thermocline deepens during this period, we assume that the descent is, at most, very gradual. A simple model for this case consists of two well-mixed layers of constant thickness separated by an interface across which diffusive transport occurs (Fig. 31.3). Therefore the metalimnion is not modeled explicitly. Rather the thermocline acts as the interface between the surface and the bottom layers. For such a system, heat balances can be written for each of the layers as

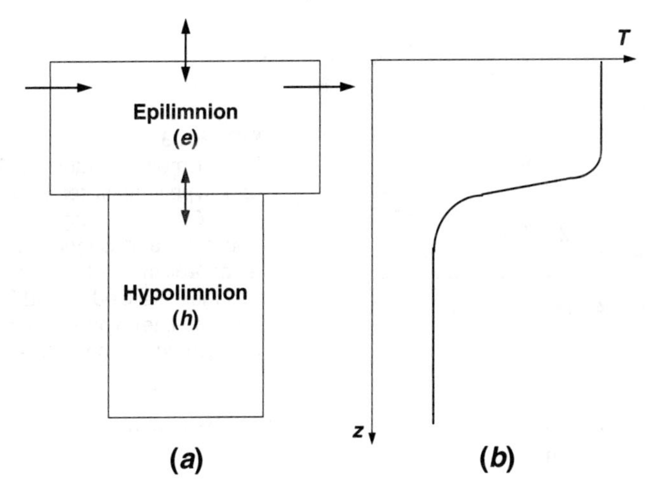

FIGURE 31.3
Idealization of summer stratified period in a temperate lake. (*a*) The segmentation of the water column into a well-mixed epilimnion and hypolimnion corresponding to the vertical temperature distribution in (*b*).

$$V_e \rho C_p \frac{dT_e}{dt} = Q\rho C_p T_{\text{in}}(t) - Q\rho C_p T_e \pm JA_s + v_t A_t \rho C_p (T_h - T_e) \tag{31.2}$$

$$V_h \rho C_p \frac{dT_h}{dt} = v_t A_t \rho C_p (T_e - T_h) \tag{31.3}$$

where the subscripts e and h designate the epilimnion and hypolimnion, respectively, T = temperature (°C), ρ = density (g cm^{-3}), C_p = specific heat (cal g^{-1} °C^{-1}), Q = volumetric flow rate of all water sources entering the system (g cm^{-3}), $T_{\text{in}}(t)$ = average inflow temperature of these sources (°C), A_s = lake's surface area (cm^2), J = surface heat flux (cal cm^{-2} d^{-1}), v_t = thermocline heat transfer coefficient (cm d^{-1}), and A_t = thermocline area (cm^2).

It should be noted that exchange across the thermocline also can be parameterized as a vertical diffusion coefficient E_t, where E_t has units of cm^2 d^{-1} and is related to the heat exchange coefficient by

$$E_t = v_t H_t \tag{31.4}$$

where H_t = thermocline thickness (cm). This alternative expression is useful in comparing the magnitude of vertical mixing with other turbulent processes. However, it has the disadvantage that it requires specification of the thermocline thickness to parameterize vertical mixing.

Equation 31.3 can now be used to estimate the thermocline heat exchange coefficient. To do this we assume that during the summer stratified period, the epilimnion temperature is constant (see Fig. 31.4). If this is true, Eq. 31.3 can be written as

$$\frac{dT_h}{dt} + \lambda_h T_h = \lambda_h \overline{T}_e \tag{31.5}$$

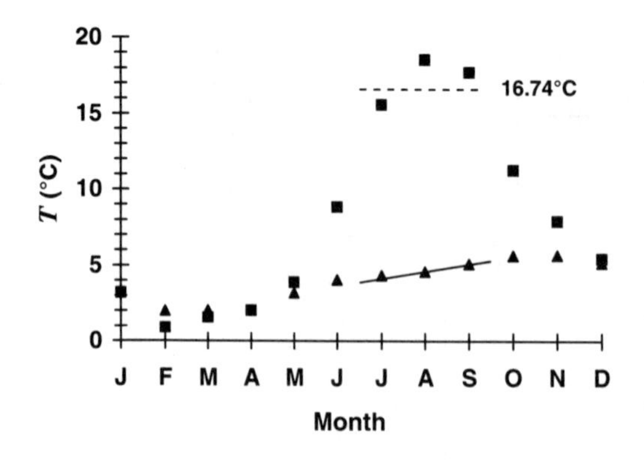

FIGURE 31.4
Plot of mean monthly epilimnetic and hypolimnetic temperatures for Lake Ontario during 1967 from Thomann and Segna (1980). The mean epilimnion temperature for the summer stratified period is displayed (dashed line) along with the simulated hypolimnetic temperature (solid line).

where $\overline{T}_e$ designates that the epilimnion temperature is constant and λ_h is the eigenvalue for the hypolimnion,

$$\lambda_h = \frac{v_t A_t}{V_h} \tag{31.6}$$

Equation 31.5 is identical in form to the completely mixed lake model, with step input described previously in Lec. 4. If the hypolimnion temperature at the beginning of the summer period is $T_{h,i}$, Eq. 31.5 can be solved for

$$T_h = T_{h,i}e^{-\lambda_h t} + \overline{T}_e(1 - e^{-\lambda_h t}) \tag{31.7}$$

Equation 31.7 can then be arranged to estimate the heat exchange coefficient across the thermocline (Chapra 1980),

$$v_t = \frac{V_h}{A_t t_s} \ln\left(\frac{\overline{T}_e - T_{h,i}}{\overline{T}_e - T_{h,s}}\right) \tag{31.8}$$

where t_s = time after the onset of stratification at which the hypolimnion temperature $T_{h,s}$ is measured.

EXAMPLE 31.1. THERMOCLINE TRANSFER COEFFICIENT. Lake Ontario is strongly stratified during the months of July, August, and September. During this time the thermocline is at a depth of approximately 15 m. The average epilimnetic temperature over the period is approximately 16.74°C and the surface area of the thermocline is approximately $15{,}000 \times 10^6$ m^2. The temperature in the hypolimnion (volume $= 1380 \times 10^9$ m^3) rises from 4.20°C in the beginning of July to 5.38°C at the end of September (Table 31.1). Use Eq. 31.8 to estimate the thermocline transfer coefficient. In addition compute the thermocline diffusion coefficient if $H_t = 7$ m.

Solution: Equation 31.8 can be used to compute

$$v_t = \frac{1380 \times 10^{15}\ \text{cm}^3}{15{,}000 \times 10^{10}\ \text{cm}^2(90\ \text{d})} \ln\left(\frac{16.74 - 4.20}{16.74 - 5.38}\right) = 10.1\ \text{cm d}^{-1}$$

which in turn can be used to compute a diffusion coefficient

$$E_t = 10.1\ \text{cm d}^{-1}(700\ \text{cm})\left(\frac{\text{d}}{86{,}400\ \text{s}}\right) = 0.082\ \text{cm}^2\ \text{s}^{-1}$$

TABLE 31.1
Mean monthly epilimnetic (0 to 15 m) and hypolimnetic (15 m to bottom) temperatures for Lake Ontario (Chapra and Reckhow 1983).

Month	T_e (°C)	T_h (°C)	Month	T_e (°C)	T_h (°C)	Month	T_e (°C)	T_h (°C)
Dec	5.47	5.18	May	3.91	3.20	Oct	11.29	5.64
Jan	3.18	3.30	Jun	8.86	4.05	Nov	7.92	5.68
Feb	0.89	2.00	Jul	15.58	4.35	Dec	5.47	5.18
Mar	1.55	2.04	Aug	18.56	4.61	Jan	3.18	3.30
Apr	2.06	2.04	Sep	17.74	5.12			

These coefficients can now be employed in conjunction with Eq. 31.7 to calculate the hypolimnetic temperature and response time. First, the eigenvalue can be computed as (Eq. 31.6)

$$\lambda_h = \frac{10.1(15{,}000 \times 10^{10})}{1380 \times 10^{15}} = 0.001087 \text{ d}^{-1} \ (= 0.397 \text{ yr}^{-1})$$

Then Eq. 31.7 can be used to calculate

$$T_h = 4.2e^{-0.001087t} + 16.75(1 - e^{-0.001087t})$$

Values calculated with this formula are plotted in Fig. 31.4. As can be seen, the fit seems linear because the response time of the hypolimnion is so long,

$$t_{95} = \frac{3}{\lambda_h} = \frac{3}{0.397} = 7.6 \text{ yr}$$

The transfer coefficient can also be used to compute mass transfer of substances across the thermocline. For example in Lake Ontario during the summer, the concentrations of soluble reactive phosphorus in the epilimnion and hypolimnion are approximately 3.1 and 8.6 μg L^{-1}, respectively. The mass-transfer coefficient from the hypolimnion to the epilimnion can be estimated as

$$10.1 \text{ cm d}^{-1}(15{,}000 \times 10^{10} \text{ cm}^2)(8.6 - 3.1) \text{ mg m}^{-3}\left(\frac{\text{m}^3}{10^6 \text{ cm}^3}\right)\left(\frac{365 \text{ d}}{\text{yr}}\right)\left(\frac{\text{mta}}{10^9 \text{ mg yr}^{-1}}\right)$$
$$= 3001 \text{ tonnes yr}^{-1}$$

This estimate can be put into perspective by comparing it to the 12,000 metric tons per year (mta) of total phosphorus that enters the lake from external sources. The fact that diffusive transport to the epilimnion across the thermocline amounts to about 25% of the external loading suggests its relative importance in regulating surface production of phytoplankton during the growing season.

Snodgrass (1974) has summarized estimates of the thermocline diffusion coefficient E_t for a number of lakes. The values are positively correlated with mean depth (Fig. 31.5) and range over several orders of magnitude from 0.003 to 2.4 cm^2 s^{-1}. Note that the following formula can be derived from the plot:

$$E_t = 7.07 \times 10^{-4} H^{1.1505} \tag{31.9}$$

where E_t is in cm^2 s^{-1} and H = mean depth (m). With knowledge of the

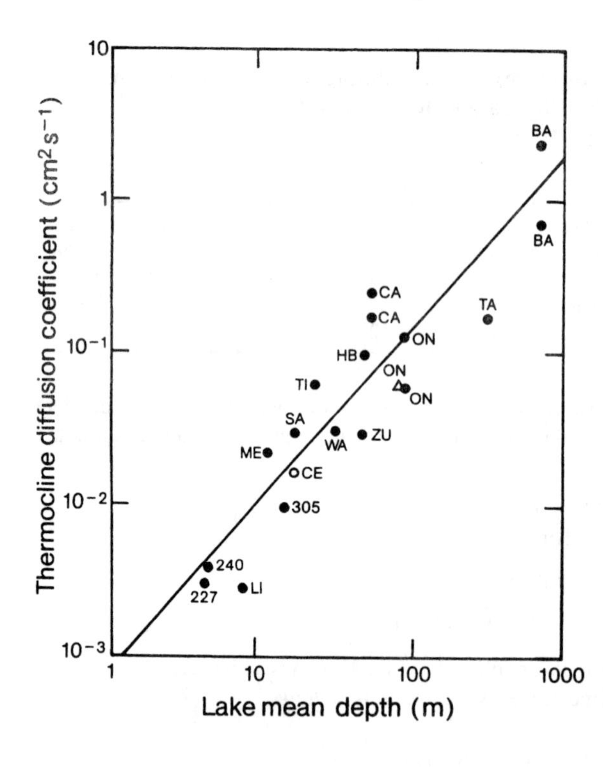

FIGURE 31.5
Thermocline diffusion coefficient ($cm^2\ s^{-1}$) versus lake mean depth for a number of temperate lakes. Redrawn from Snodgrass (1974) with new data from Lake Ontario (Δ) and the central basin of Lake Erie (O).

approximate thickness of the metalimnion, the plot or the equation can be used to make a first estimate of the thermocline exchange coefficient for a lake.

Now that we have estimated the thermocline exchange coefficient, it is relatively straightforward to combine the two-layer lake model with the surface heat exchange model described in the previous lecture.

EXAMPLE 31.2. TIME-VARIABLE HEAT BALANCE FOR A STRATIFIED LAKE. Compute the annual heat budget for the same pond used in Examples 30.4 and 30.5. Note that the following additional information is available:

$$V_e = 175{,}000\ \text{m}^3 \qquad A_t = 11{,}000\ \text{m}^2 \qquad A_s = 25{,}000\ \text{m}^2$$
$$V_h = 75{,}000\ \text{m}^3 \qquad H_t = 3\ \text{m}$$

Assume that all atmospheric interactions are limited to the epilimnion and use Eq. 31.9 to determine the thermocline transfer coefficient for the summer stratified period (April 1 through September 30). Also assume that the lake does not stratify during the winter.

Solution: First, we must determine the thermocline transfer coefficient. To do this the pond's mean depth can be computed as

$$H = \frac{V}{A_s} = \frac{250{,}000}{25{,}000} = 10\ \text{m}$$

Then Eq. 31.9 can be used to compute

$$E_t = 7.07 \times 10^{-4}(10)^{1.1505} = 0.01\ \text{cm}^2\ \text{s}^{-1}$$

The transfer coefficient can be determined as

$$v_t = \frac{E_t}{H_t} = \frac{0.01 \text{ cm}^2 \text{ s}^{-1}}{3 \text{ m}} \left(\frac{\text{m}^2}{10{,}000 \text{ cm}^2} \right) \left(\frac{86{,}400 \text{ s}}{\text{d}} \right) = 0.0288 \text{ m d}^{-1}$$

For the unstratified period, a large enough coefficient is used to simulate complete mixing. The magnitude of this exchange coefficient is estimated by trial and error.

Equations 31.2 and 31.3 can now be integrated with a method such as the fourth-order Runge-Kutta approach described in Lec. 7. The results are displayed in Fig. E31.2. We have included the single-layer model from Example 30.5 for comparison. Observe how the single-layer-model temperature falls between the results for the two-layer case. As expected, because it is subject to meteorological influences, the epilimnion temperature is warmer during the stratified period.

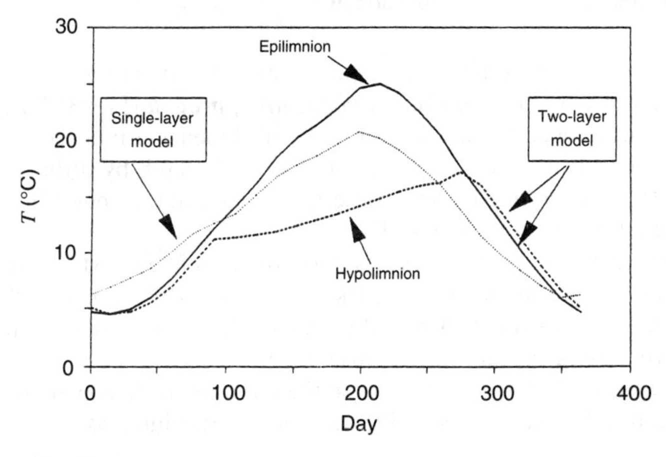

FIGURE E31.2

31.3 MULTILAYER HEAT BALANCES (ADVANCED TOPIC)

Although the lumped approach of the previous section is widely used, one-dimensional distributed approaches are also employed. The primary reason for using such approaches relates to the fact that the hypolimnion is not completely mixed. In fact it is more accurately characterized as a one-dimensional system where heat transport occurs by diffusion.

Such a distributed approach may be less important for unproductive or very deep systems (Fig. 31.6*a*). For these cases, vertical gradients of constituents may not be significant in the hypolimnion. Thus a well-mixed bottom layer serves as an adequate approximation.

However, for shallower, productive systems, significant hypolimnetic gradients can occur during stratification (Fig. 31.6*b*). In most cases such gradients are created by sediment release of constituents. Because the hypolimnion is governed by diffusion, concentrations rise because of the sediment sources. Thus, higher concentrations first develop near the sediment-water interface and slowly diffuse up through the water column. A distributed characterization is required to simulate such gradients and the associated mass transport.

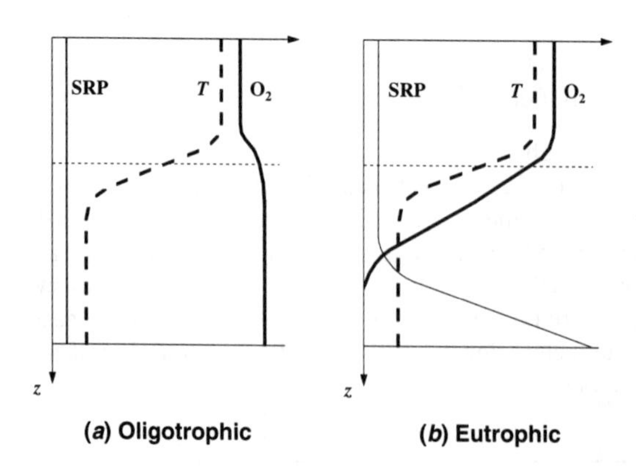

FIGURE 31.6
Typical vertical profiles of temperature, oxygen, and soluble reactive phosphorus in (*a*) an oligotrophic and (*b*) a eutrophic lake. For the oligotrophic system, sediment sources are not strong enough to induce vertical gradients in the bottom water. In contrast eutrophic systems typically have productive sediment that take up oxygen and release nutrients. Consequently, strong vertical gradients usually occur.

Two approaches have been developed to model vertical temperature distributions. The first is based on a ***turbulent diffusion approach*** (Fig. 31.7*a*). Numerical solutions are developed by dividing the vertical dimension into segments. Heat is supplied to the surface and then distributed through depth by diffusion in a fashion similar to the distributed models described earlier in the book (for example the elongated reactors described in Part II).

The second method, called a ***mixed-layer approach*** (Fig. 31.7*b*), uses a mechanical energy balance to predict the thickness of the epilimnion. This thick surface layer is then modeled as a well-mixed segment. The hypolimnion is modeled as a series of layers using the turbulent diffusion approach.

Because of its simplicity we focus on the turbulent diffusion approach. A one-dimensional heat balance can be written for the water column as

$$\frac{\partial T}{\partial t} = \frac{\partial}{\partial z}\left[E(z)\frac{\partial T}{\partial z}\right] + \frac{J_{sn}(z)A_s}{\rho C_p} \tag{31.10}$$

where T = temperature
z = depth, measured downward from the surface
$E(z)$ = a vertical turbulent diffusion coefficient
$J_{sn}(z)$ = net solar radiation distributed by depth
A_s = surface area
ρ = water density
C_p = water specific heat

Note that the solar radiation is distributed by depth according to the formula

$$J_{sn}(z) = (1 - F_a)J_{sn}e^{-k_e z} \tag{31.11}$$

where F_a = fraction of the solar radiation that is absorbed immediately at the surface
J_{sn} = net solar radiation delivered to the surface
k_e = extinction coefficient

The following boundary conditions are required at the water surface and the bottom:

$$-E(0)\frac{\partial T}{\partial z} = \frac{J_{sn} - (1 - F_a)J_{sn}}{\rho C_p} \tag{31.12}$$

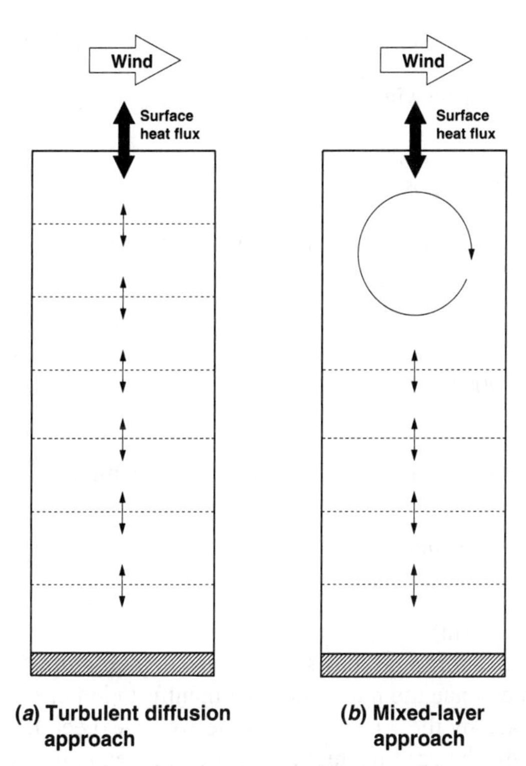

FIGURE 31.7
Contrast between the two primary models used to distribute heat vertically in lakes and reservoirs.

and
$$\frac{\partial T}{\partial z}(H) = 0 \tag{31.13}$$

The first condition specifies the heat flux at the air-water interface. The second condition sets an insulated condition at the bottom.

Also observe how a portion of the solar radiation is absorbed within a few centimeters of the surface. The remainder is absorbed in an exponential fashion as it radiates down through the water column.

At this point the model formulated above merely distributes heat through the water column by mixing. The key to simulating stratification relates to how the vertical diffusion coefficient varies with depth. A variety of approaches have been proposed. One of the first was that of Munk and Anderson (1948). This model assumed that the diffusion coefficient was a function of the Richardson number,

$$E(z) = \frac{E_0}{(1 + aR_i)^{3/2}} \tag{31.14}$$

where E_0 = the diffusion coefficient at neutral stability
R_i = Richardson number
a = a constant

A variety of formulations have been proposed to relate E_0 to depth and wind velocity. We will use the simplest formulation,

$$E_0 = c\omega_\theta \tag{31.15}$$

where c = an empirical constant and ω_θ = wind shear velocity,

$$\omega_\theta = \sqrt{\frac{\tau_s}{\rho_w}} \tag{31.16}$$

in which ρ_w = water density and τ_s = shear stress at the air-water interface,

$$\tau_s = \rho_{\text{air}} C_d U_w^2 \tag{31.17}$$

where ρ_{air} = air density
C_d = a drag coefficient = $0.00052 U_w^{0.44}$
U_w = wind speed

Aside from Eq. 31.1, alternative formulations are available to compute the Richardson number. For example

$$R_i = \frac{-(g/p)(\partial \rho/\partial x)}{\omega_0/(z_s - z)^2} \tag{3.18}$$

where z_s = the water surface elevation (m).

Although the foregoing sequence of formulas may seem involved, their ultimate effect is embodied in Eq. 31.14. In essence the numerator is a manifestation of the kinetic energy delivered to the surface by the wind. Thus, as we would expect, the vertical diffusion within the lake will increase for higher winds. The denominator decreases the diffusion as the Richardson number increases. Thus, as the density gradient increases or as the velocity gradient decreases, the diffusion coefficient decreases and stratification can occur.

The foregoing is an introduction to vertical temperature modeling in lakes. It reviews one of many formulations that have been proposed to model this phenomenon. Ford and Johnson (1986) and Henderson-Sellers (1984) can be consulted for additional information.

PROBLEMS

31.1. A pond has the following characteristics:

Epilimnion volume = 150,000 m^3 Thermocline area = 10,000 m^2
Hypolimnion volume = 50,000 m^3 Inflow = outflow = 5000 m^3 d^{-1}
Surface area = 25,000 m^2 Thermocline thickness = 3 m

The pond's inflow has a temperature of 10°C and enters and leaves the hypolimnion. In addition it is subject to the following meteorological conditions:

Net solar radiation = 250 cal cm^{-2} d^{-1} Dew-point temperature = 15°C
Air temperature = 20°C Wind speed = 2 m s^{-1}

Calculate the steady-state temperatures of the epilimnion and hypolimnion.

31.2. The Central Basin of Lake Erie had the following general characteristics for the summer stratified period from 1967 through 1972:

Mean depth = 17.8 m	Hypolimnion thickness = 4 m
Volume hypolimnion = 40 km^3	Epilimnetic temperature = 19.94°C
Area thermocline = 10,000 km^2	Epilimnion oxygen = 9.66 mg L^{-1}
Metalimnion thickness = 7 m	Total P concentration = 20 μg L^{-1}

In addition, the following data can be extracted from temperature and oxygen time series:

	June 1	September 2
Hypolimnion temperature (°C)	7.77	12.5
Hypolimnion oxygen (mg L^{-1})	10.72	0.60

(*a*) Determine the thermocline diffusion coefficient in $cm^2\ s^{-1}$ and compare it with the value from the Snodgrass plot.

(*b*) Determine:

(*i*) The apparent areal hypolimnetic oxygen demand in g $m^{-2}\ d^{-1}$

(*ii*) The actual areal hypolimnetic oxygen demand in g $m^{-2}\ d^{-1}$ corrected for diffusion

(*iii*) The empirical areal hypolimnetic oxygen demand in g $m^{-2}\ d^{-1}$ as a function of total P concentration (Eq. 29.30)

(*c*) Compare and discuss the results of (*b*).

31.3 Use the fourth-order RK method to compute the hypolimnetic temperature for a lake having the characteristics given in Table P31.3. Use a value of 0.13 $cm^2\ s^{-1}$ for the thermocline diffusion coefficient for the summer stratified period. Also assume that the lake does not stratify during the winter. Use the same epilimnion temperatures as in Table 31.1.

TABLE P31.3
Lake characteristics

Parameter	Value	Units
Lake surface area	25,000	m^2
Thermocline area	11,000	m^2
Epilimnion volume	175,000	m^3
Hypolimnion volume	75,000	m^3
Thermocline thickness	3	m
Epilimnion thickness	7	m
Hypolimnion thickness	6.8	m
Inflow = outflow	274×10^4	$m^3\ yr^{-1}$
Start of summer stratification	120	d
End of stratification	290	d

LECTURE 32

Microbe/Substrate Modeling

LECTURE OVERVIEW: I describe how microbial growth can be modeled as a function of substrate level. After a brief review of bacterial growth and biological oxidation, I present the Michaelis-Menten rate formulation and show how its parameters can be estimated from data. Then I illustrate how the Michaelis-Menten formulation can be used to model microbial growth in both batch reactors and CSTRs. In the latter I introduce the important idea of a washout rate. Finally I show how the theory can be applied to algae growing on a limiting nutrient.

A large body of theory has been developed to simulate how microorganisms such as bacteria break down organic carbon. Although this theory was largely developed to model waste treatment processes, it has great relevance to the modeling of natural waters. In particular it is pertinent to the algal models discussed in later lectures.

Beyond its connection with algal modeling, microbial kinetics has several other applications in water-quality modeling. First, it provides an alternative way to model bacteria from the traditional first-order decay approach described previously in Lec. 27. In particular it offers a means to incorporate growth, and in so doing, could lead to improved pathogen models. Second, it provides a theory to characterize the microbial-mediated decomposition mechanisms in models of both conventional and toxic organic pollutants.

32.1 BACTERIAL GROWTH

Suppose that a batch reactor initially contains a large amount of biodegradable organic carbon and a small amount of bacteria. As depicted in Fig. 32.1, the numbers of bacteria will typically manifest four phases over time:

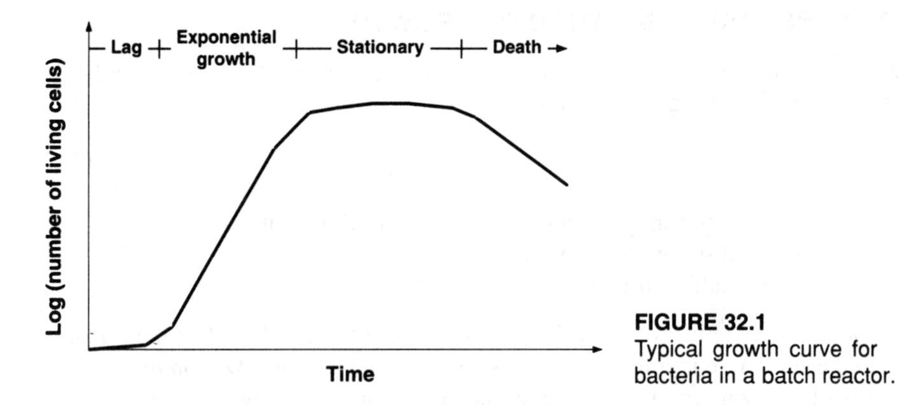

FIGURE 32.1
Typical growth curve for bacteria in a batch reactor.

- *Lag phase.* When bacteria are added to the reactor, time is often required for the bacteria to acclimate to the new environment.
- *Exponential- (or log-) growth phase.* After the cells are acclimated, because of the abundance of substrate (food), they typically grow at a maximum rate limited only by their ability to process the substrate. Because they grow by binary fission, the resulting increase follows the exponential growth model.
- *Stationary phase.* Growth eventually levels off as the bacteria exhaust their food supply. Thus the growth rate of new cells comes into balance with the death of old cells, and net growth goes to zero. The leveling off may be hastened by environmental factors such as the generation of toxic metabolic by-products by the bacteria.
- *Death phase.* If incubation is continued, cell death will eventually outpace growth, and a decline will set in.

The following mass balance represents a simple starting point for modeling Fig. 32.1:

$$\frac{dX}{dt} = (k_g - k_d)X \tag{32.1}$$

where X = bacterial concentration (cells L^{-1})
k_g = bacterial growth rate (hr^{-1})
k_d = bacterial death rate (hr^{-1})

Thus the status of the bacterial population is dictated by the balance of growth and death.

It is important to recognize that the growth and death rates should not necessarily be considered as constants in this simple representation. However, if the growth rate is constant and the death rate is negligible, Eq. 32.1 would aptly model the exponential growth phase. Similarly if a constant death rate grew much larger than growth, the model could approximate the death phase. Further, making growth dependent on the food supply could allow the model to simulate the progression from exponential growth to the stationary phase as the substrate is depleted.

In the following sections we will be concerned primarily with modeling this shift from exponential growth to substrate-limited growth. To do this we now discuss how substrate limitation of growth is modeled.

32.2 SUBSTRATE LIMITATION OF GROWTH

The relation between the growth rate and the concentration of substrate can be described by the following empirical model:

$$k_g = k_{g,\max} \frac{S}{k_s + S} \tag{32.2}$$

where $k_{g,\max}$ = maximum growth rate when food is abundant (hr^{-1})
S = substrate concentration (mg L^{-1})
k_s = a half-saturation constant (mg L^{-1})

This model is sometimes called the ***Michaelis-Menten model***, after the equation used in the study of enzyme kinetics. It is also referred to as the ***Monod model***, after the Nobel Prize-winning microbiologist Jacques Monod, who was the first to apply it to model microbial growth.

As depicted in Fig. 32.2, the half-saturation constant represents the substrate concentration at which growth is half the maximum. Thus it provides the parameter that dictates at what level the substrate becomes limiting. At low food levels ($S \ll k_s$), the growth rate becomes directly proportional to the food supply,

$$k_g \cong \frac{k_{g,\max}}{k_s} S \tag{32.3}$$

In the context of Eq. 32.1 the growth process becomes second-order.

At high food levels ($S \gg k_s$) the growth rate becomes constant and independent of the food supply,

$$k_g \cong k_{g,\max} \tag{32.4}$$

In the context of Eq. 32.1 the growth process becomes first-order. As in Fig. 32.2, this level is approximately 5 times the half-saturation constant. Thus Eq. 32.2 adjusts the rate as a function of whether food is abundant or scarce.

Before proceeding we should pause and consider how the Michaelis-Menten formulation is so radically different from the way in which we have characterized reactions to this point. Aside from the mention of zero- and second-order reactions

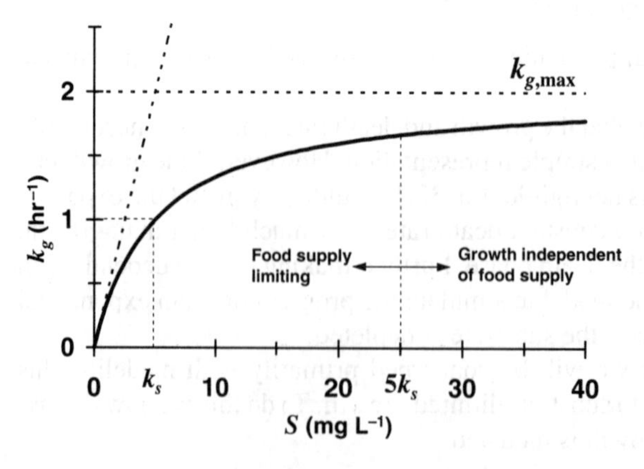

FIGURE 32.2
Michaelis-Menten or Monod curve used to characterize the microbial growth rate as the function of substrate concentration. For this plot, $k_{g,\max}$ = 2 hr^{-1} and k_s = 5 mg L^{-1}.

in Lec. 2, almost all the reactions until now have been first-order. This was a valid approximation for modeling the gross decomposition of sewage in oxygen modeling.

However, now that we must deal with biological growth as an issue unto itself, the boundless nature of first-order growth begins to become inadequate. Hence the Michaelis-Menten formulation allows us to incorporate a limit to growth (food supply) into model kinetics. The introduction of such limits is characteristic of the eutrophication models in the coming lectures and is reflective of the fact that unbridled growth is checked and moderated by finite resources in nature.

32.2.1 Parameter Estimation

Recall that in Lec. 2 we presented a variety of methods to estimate rate parameters for simple reactions. Michaelis-Menten parameters can be evaluated in a similar fashion. Many of the approaches involve transforming Eq. 32.2 so that it plots as a straight line. One of the earliest methods involved inverting it to give

$$\frac{1}{k_g} = \frac{k_s}{k_{g,\max}} \frac{1}{S} + \frac{1}{k_{g,\max}} \tag{32.5}$$

Thus, if the Michaelis-Menten model holds, a plot of $1/k_g$ versus $1/S$ should yield a straight line with an intercept of $1/k_{g,\max}$ and a slope of $k_s/k_{g,\max}$. A linear regression can then be used to determine estimates of the coefficients. This approach is called a ***Lineweaver-Burk plot.***

Another linearized version is obtained by multiplying Eq. 32.5 by the substrate concentration to give

$$\frac{S}{k_g} = \frac{1}{k_{g,\max}} S + \frac{k_s}{k_{g,\max}} \tag{32.6}$$

For this case, which is called a ***Hanes equation***, a plot of S/k_g versus S should yield a straight line with an intercept of $k_s/k_{g,\max}$ and a slope of $1/k_{g,\max}$.

A third option is obtained by multiplying Eq. 32.2 by $(k_s + S)/S$ to give

$$k_g = k_{g,\max} - k_s\left(\frac{k_g}{S}\right) \tag{32.7}$$

For this case, which is called a ***Hofstee equation***, a plot of k_g versus k_g/S should yield a straight line with an intercept of $k_{g,\max}$ and a slope of $-k_s$.

Finally, aside from linearization, the integral/least-squares method (recall Sec. 2.2.6) can be used to estimate the coefficients numerically. All the approaches are applied in the following example.

EXAMPLE 32.1. PARAMETER ESTIMATION FOR THE MICHAELIS-MENTEN MODEL. The following data were generated from Eq. 32.2 with values of $k_{g,\max} = 2$ and $k_s = 5$. The resulting values of k_g were then modified by adding and subtracting an error of 0.1 from alternate values:

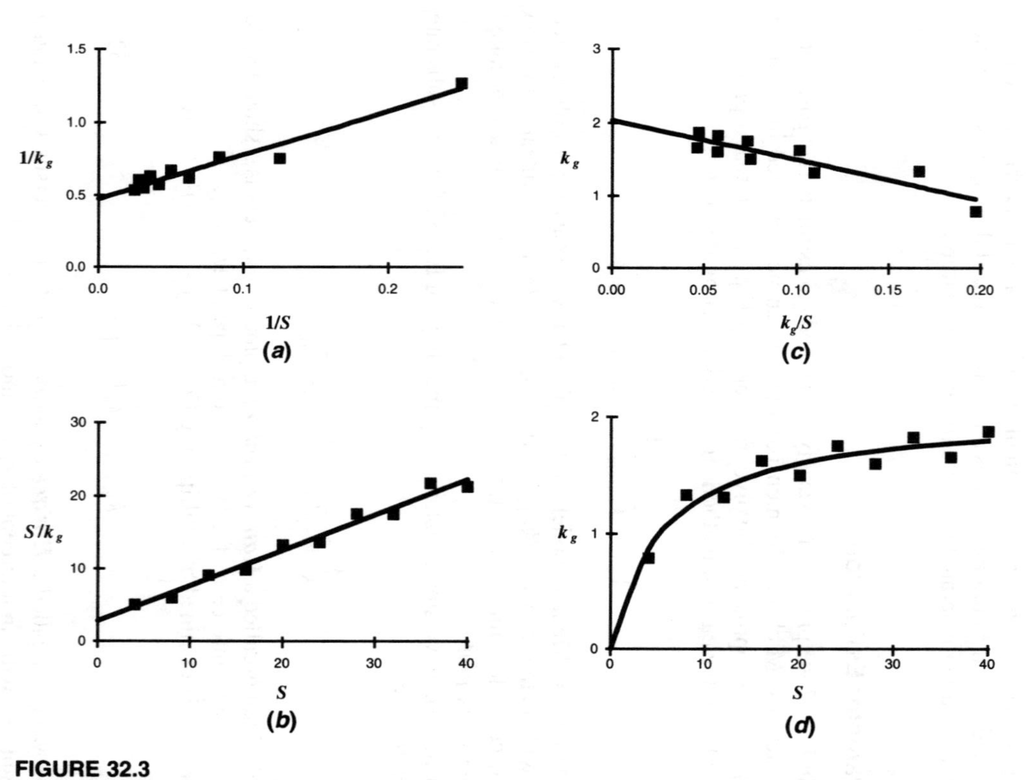

FIGURE 32.3
Best-fit lines for four methods for estimating the parameters of the Michaelis-Menten model. (*a*) Lineweaver-Burk, (*b*) Hanes, (*c*) Hofstee, and (*d*) integral/least-squares.

S	k_g	S	k_g	S	k_g
4	0.7889	20	1.5000	36	1.6561
8	1.3308	24	1.7552	40	1.8778
12	1.3118	28	1.5970		
16	1.6238	32	1.8297		

Estimate the parameters of Eq. 32.2 with the Lineweaver-Burk, Hanes, and Hofstee transformations and the integral/least-squares approach.

Solution: The Lineweaver-Burk approach can be implemented by plotting $1/k_g$ versus $1/S$. As displayed in Fig. 32.3*a*, a straight line can be fit to the data with linear regression. The best-fit line is

$$\frac{1}{k_g} = 3.031\frac{1}{S} + 0.472$$

which can be used to determine

$$k_{g,\max} = \frac{1}{0.472} = 2.117$$

$$k_s = 2.117(3.031) = 6.418$$

The other transformation approaches can be implemented in a similar fashion. Transformation plots and fits are displayed in Fig. 32.3*b* and *c*. The integral/least-squares approach can be applied in a similar fashion to Example 2.3. The fit line is shown in Fig. 32.3*d*. The parameter estimates of all methods are summarized in Table 32.1.

The results of the previous example provide some insights into the strengths and weaknesses of the various approaches. Of the four methods the Lineweaver-Burk performs the poorest. This is because the most stable points (that is, those at higher levels of S) tend to cluster together near the origin, whereas those that usually manifest the most variability are spread out far from the origin and hence will have an inordinately large influence on the slope. Thus the value of k_s would be expected to be particularly poor.

The Hanes transformation provides a better slope estimate by spreading out the data along the abscissa. However, the intercept tends to be close to the origin and

TABLE 32.1
Results of four methods for obtaining parameter estimates for the Michaelis-Menten model. The data were generated with perturbed values based on $k_{g,\max} = 2\ \text{hr}^{-1}$ and $k_s = 5\ \text{mg L}^{-1}$

Method	$k_{g,\max}$	k_s
Lineweaver-Burk	2.117	6.418
Hanes	2.054	5.767
Hofstee	2.040	5.521
Integral/least-squares	2.048	5.551

hence becomes less reliable. The value of k_s tends to be poor because, for the Hanes model, it depends on the intercept.

The Hofstee approach rectifies both problems by spreading the data out and generating a better intercept. However, because the dependent variable (k_g) appears in both the ordinate and the abscissa, least-squares regression is not strictly applicable.[†] As in Table 32.1, the result nevertheless often provides good results.

Finally, the integral/least-squares approach also provides decent results. In addition, it has the advantage that the equation is fit directly rather than via a transformation.

32.3 MICROBIAL KINETICS IN A BATCH REACTOR

Michaelis-Menten kinetics can now be used to simulate the dynamics of bacteria and substrate in a batch reactor (Fig. 32.4). For simplicity, we will use organic carbon as the substrate and as the measure of bacterial biomass.

As in Fig. 32.4, the bacteria grow by utilizing the substrate. However, not all of the carbon that is utilized becomes new bacterial cells. A significant part is converted into carbon dioxide and water. This is quantified by the cell yield,

$$Y = \frac{\text{gC cells}}{\text{gC substrate}} \tag{32.8}$$

Thus, for this example, because the cells and the substrate are measured in commensurate units, the cell yield represents the efficiency of the conversion of organic carbon into cell carbon.

Returning back to Fig. 32.4, the cells are degraded by two processes: death and decay. Death represents losses whereby the cell organic carbon is released back into

[†] Least-squares regression is predicated on the assumption that the independent variable (the "x" axis) is measured precisely and that all the measurement error is posited in the dependent variable (see Chapra and Canale 1988 for additional details regarding regression).

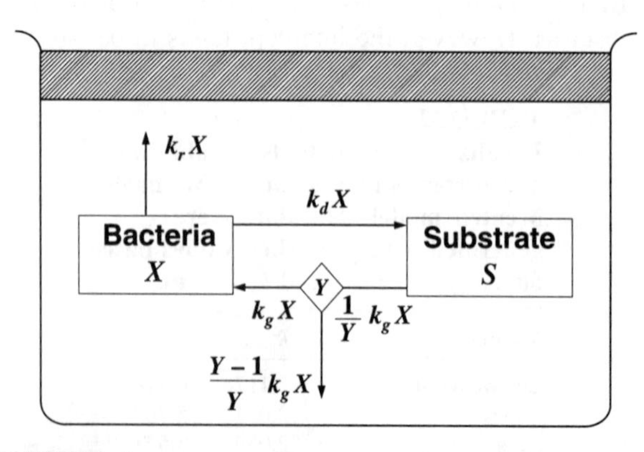

FIGURE 32.4
Flow diagram for kinetic interactions between bacteria and organic carbon substrate in a batch reactor.

the substrate pool. In contrast, decay represents losses of bacterial biomass that neither synthesize new cells nor release organic carbon (for example, respiration). Mass balances for the bacteria and the substrate can be written,

$$\frac{dX}{dt} = \left(k_{g,\max}\frac{S}{k_s + S} - k_d - k_r\right)X \tag{32.9}$$

$$\frac{dS}{dt} = -\frac{1}{Y}k_{g,\max}\frac{S}{k_s + S}X + k_d X \tag{32.10}$$

Notice how we have used the Michaelis-Menten formulation to quantify the dependence of bacterial growth on substrate concentration.

EXAMPLE 32.2. BATCH REACTOR. A batch reactor has the following characteristics:

$X_0 = 2$ mgC L^{-1} $\quad$ $S_0 = 998$ mgC L^{-1}
$k_{g,\max} = 0.2$ hr^{-1} $\quad$ $k_s = 150$ mgC L^{-1}
$k_d = k_r = 0.01$ hr^{-1} $\quad$ $Y = 0.5$ gC cells (gC substrate)$^{-1}$

Simulate how the substrate, bacteria, and total organic carbon change over time in this reactor.

Solution: The parameters can be substituted into Eqs. 32.9 and 32.10 and solved numerically with the fourth-order Runge-Kutta method. The results are displayed graphically in Fig. 32.5. Notice how the bacteria first grow exponentially because the substrate concentration far exceeds the half-saturation constant. However, at about 40 days the substrate concentration becomes low enough that it limits growth and the bacteria level off.

On the basis of the yield coefficient, we would expect that the original 998 mgC L^{-1} of substrate would be converted to 449 mgC L^{-1} of bacteria. However, because of the decay loss, a peak of only about 433 mgC L^{-1} is reached. Thereafter the bacterial population declines gradually as both decay and assimilation slowly deplete the system's carbon resources.

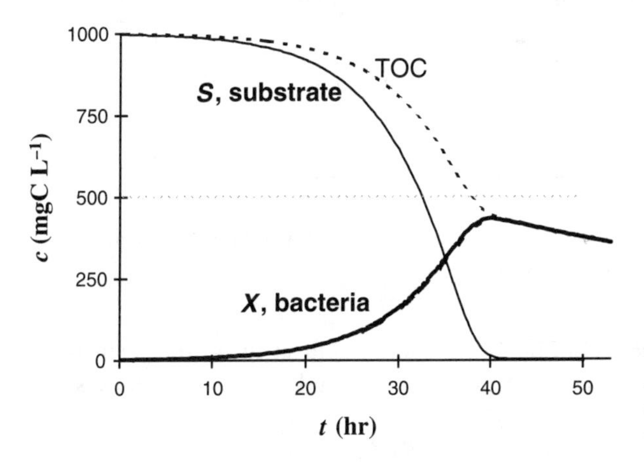

FIGURE 32.5
Growth of bacteria as a function of substrate concentration in a batch reactor.

32.4 MICROBIAL KINETICS IN A CSTR

Although the batch reactor is informative, natural systems are not closed. Therefore we now examine how the bacterial growth behaves in a continuously stirred tank reactor or CSTR (Fig. 32.6). Adding inflow and outflow terms to Eqs. 32.9 and 32.10 gives

$$\frac{dX}{dt} = \left(k_{g,\max}\frac{S}{k_s + S} - k_d - k_r - \frac{Q}{V}\right)X \tag{32.11}$$

$$\frac{dS}{dt} = -\frac{1}{Y}k_{g,\max}\frac{S}{k_s + S}X + k_d X + \frac{Q}{V}(S_{\text{in}} - S) \tag{32.12}$$

EXAMPLE 32.3. CSTR REACTOR. A CSTR has the same parameters as the batch reactor from Example 32.2. In addition it has the following characteristics:

$V = 10\text{ L}$
$S_{\text{in}} = 1000\text{ mgC L}^{-1}$
$S_0 = 0\text{ mgC L}^{-1}$

Simulate how the substrate, bacteria, and total organic carbon change over time in this reactor for three residence times: (*a*) $\tau_w = 20$ hr, (*b*) $\tau_w = 10$ hr, and (*c*) $\tau_w = 5$ hr.

Solution: The parameters can be substituted into Eqs. 32.11 and 32.12 and solved numerically with the fourth-order Runge-Kutta method. The results are displayed graphically in Fig. 32.7. For the longest residence time, case (*a*), the results are qualitatively similar to the batch reactor. As the residence time is lowered, less bacteria are grown. For the shortest case, (*c*), the bacteria go extinct! That is, they do not grow fast enough to counterbalance the rate at which they are being flushed.

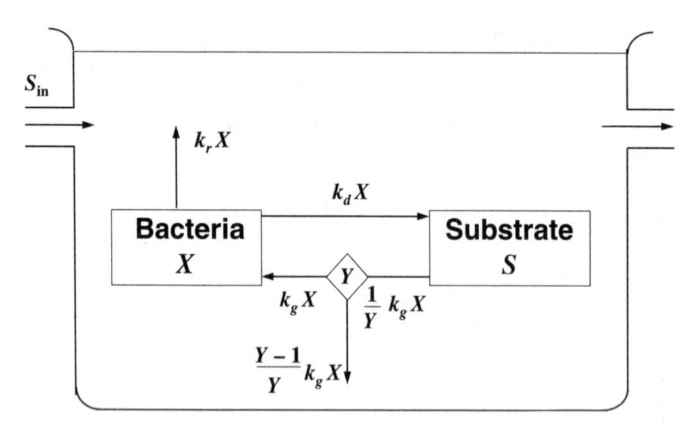

FIGURE 32.6
Flow diagram for kinetic interactions between bacteria and organic carbon substrate in a CSTR.

The condition depicted in Fig. 32.7*c* is referred to as "wash out." The residence time at which it occurs can be determined by solving Eq. 32.11 at steady-state for

$$k_{g,\max}\frac{S}{k_s + S} = k_d + k_r + \frac{Q}{V} \tag{32.13}$$

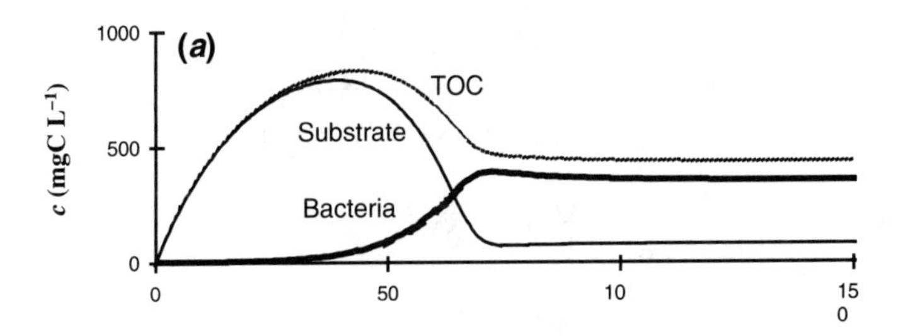

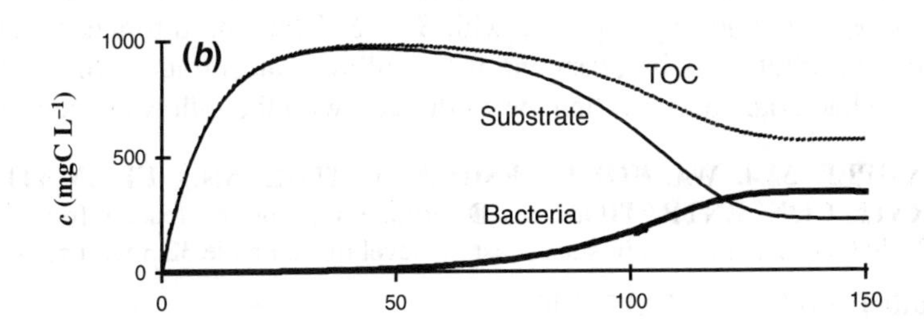

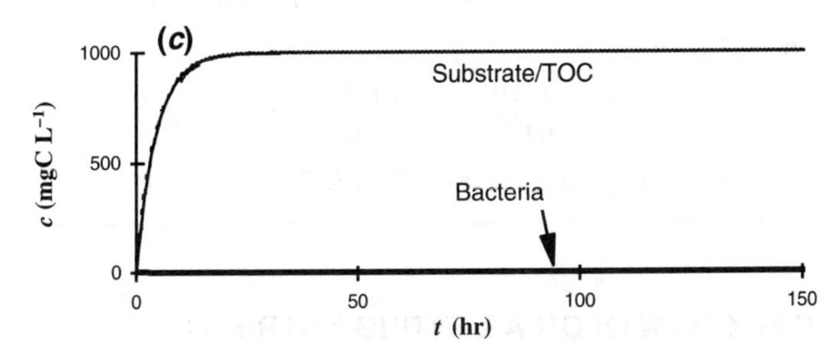

FIGURE 32.7
Growth of bacteria as a function of substrate concentration in a CSTR for three residence times. (*a*) $\tau_w = 20$ hr, (*b*) $\tau_w = 10$ hr, and (*c*) $\tau_w = 5$ hr.

Now it can be recognized that the maximum possible growth rate would occur if $S = S_{\text{in}}$. Therefore Eq. 32.13 can be reformulated as

$$k_{g,\max} \frac{S_{\text{in}}}{k_s + S_{\text{in}}} = k_d + k_r + \frac{1}{\tau_{\min}} \tag{32.14}$$

where $\tau_{\min}$ = residence time at which washout occurs. Equation 32.14 can be solved for

$$\tau_{\min} = \frac{k_s + S_{\text{in}}}{(k_{g,\max} - k_d - k_r)S_{\text{in}} - (k_d + k_r)k_s} \tag{32.15}$$

Another interesting feature of the model can be illustrated by solving the steady-state versions of Eqs. 32.11 and 32.12 for

$$\overline{S} = \frac{\left(k_d + k_r + \frac{Q}{V}\right)k_s}{k_{g,\max} - \left(k_d + k_r + \frac{Q}{V}\right)} \tag{32.16}$$

and

$$\overline{X} = \frac{\frac{Q}{V}(S_{\text{in}} - \overline{S})}{\frac{1}{Y}k_g - k_d} \tag{32.17}$$

where $\overline{S}$ and $\overline{X}$ = steady-state concentrations of the substrate and bacteria, respectively, and k_g is defined by Eq. 32.2 with $S = \overline{S}$. Thus the ultimate steady-state substrate concentration is independent of the inflow concentration. In contrast the steady-state bacterial concentration varies directly with the inflow concentration.

EXAMPLE 32.4. WASHOUT RESIDENCE TIME AND ULTIMATE SUBSTRATE CONCENTRATION. (*a*) Determine $\tau_{\min}$ for the reactor from Example 32.3. (*b*) Calculate the steady-state substrate level for Example 32.3 with $\tau_w = 20$ hr.

Solution: (*a*) Equation 32.15 yields

$$\tau_{\min} = \frac{150 + 1000}{(0.2 - 0.01 - 0.01)1000 - (0.01 + 0.01)150} = 6.5 \text{ hr}$$

(*b*) Equation 32.16 gives

$$\overline{S} = \frac{(0.01 + 0.01 + 0.05)1.50}{0.2 - (0.01 + 0.01 + 0.05)} = 80.77 \text{ mgC L}^{-1}$$

which corresponds to the numerical result (Fig. 32.7*a*).

32.5 ALGAL GROWTH ON A LIMITING NUTRIENT

The theory of the previous sections can be easily extended to the growth of algae as a function of a limiting nutrient. A CSTR is depicted in Fig. 32.8 for the case of algae and the nutrient phosphorus. Mass balances can be written as

$$\frac{da}{dt} = \left(k_{g,\max}\frac{p}{k_{sp} + p} - k_d - \frac{Q}{V}\right)a \tag{32.18}$$

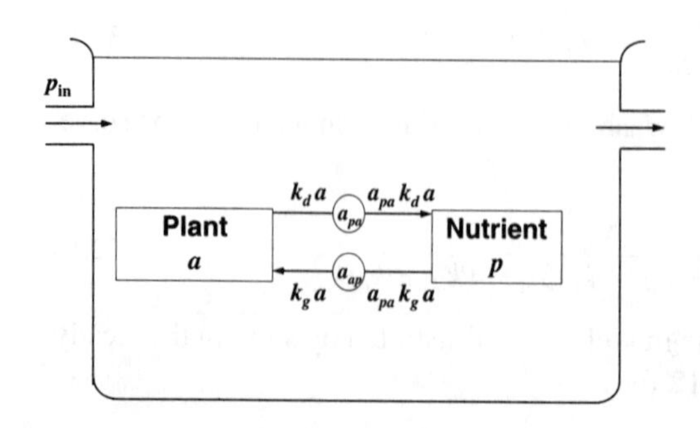

FIGURE 32.8
Flow diagram for kinetic interactions between plants and a limiting nutrient in a CSTR.

$$\frac{dp}{dt} = -a_{pa}k_{g,\max}\frac{p}{k_{sp}+p}a + a_{pa}k_d a + \frac{Q}{V}(p_{\text{in}} - p) \tag{32.19}$$

where a and p = concentration of plants ("a" stands for "algae") (mgChl$a$ m^{-3}) and phosphorus (mgP m^3), respectively

k_{sp} = half-saturation constant for phosphorus (mgP m^{-3})

a_{pa} = ratio of phosphorus to chlorophyll a in algae (mgP mgChla^{-1})

Two important distinctions can be made between this scheme and the one for microbial decomposition from Fig. 32.6:

- Because the plants and the nutrient are measured on a different basis, stoichiometric conversions must be applied to transfers between the compartments.
- There is no loss of mass due to degradation processes as was the case for carbon. Thus the stoichiometric conversions do not involve a loss of mass, merely a change of units.

EXAMPLE 32.5. ALGAL/NUTRIENT INTERACTIONS. The epilimnion of a lake has the following characteristics at the beginning of the stratified period:

a_0 = 0.5 mgChl$a$ m^{-3} $\quad$ p_0 = 9.5 mgP m^{-3}

p_{in} = 10 mgP m^{-3} $\quad$ a_{pa} = 1.5 mgP mgChla^{-1}

$k_{g,\max}$ = 1 d^{-1} $\quad$ k_s = 2 mgP m^{-3}

k_d = 0.1 d^{-1} $\quad$ τ_w = 30 d

Assuming that mass transfer across the thermocline due to diffusion and settling is negligible, (*a*) simulate how the algae and phosphorus change over time, (*b*) calculate the ultimate, steady-state level of phosphorus, and (*c*) determine $\tau_{\min}$.

Solution: (*a*) The parameters can be substituted into Eqs. 32.18 and 32.19 and solved numerically with the fourth-order Runge-Kutta method. The results are displayed graphically in Fig. 32.9. The algae grow to a peak of 6.6 mg m^{-3} in about 4 d.

(*b*)

$$\bar{S} = \frac{[k_d + (Q/V)]k_s}{k_{g,\max} - [k_d + (Q/V)]} = \frac{[0.1 + 1/30]2}{1 - [0.1 + (1/30)]} = 0.308 \text{ mgP m}^{-3}$$

(*c*)

$$\tau_{\min} = \frac{k_s + p_{\text{in}}}{(k_{g,\max} - k_d)p_{\text{in}} - k_d(k_s)} = \frac{2 + 10}{(1 - 0.1)10 - 0.1(2)} = 1.36 \text{ d}$$

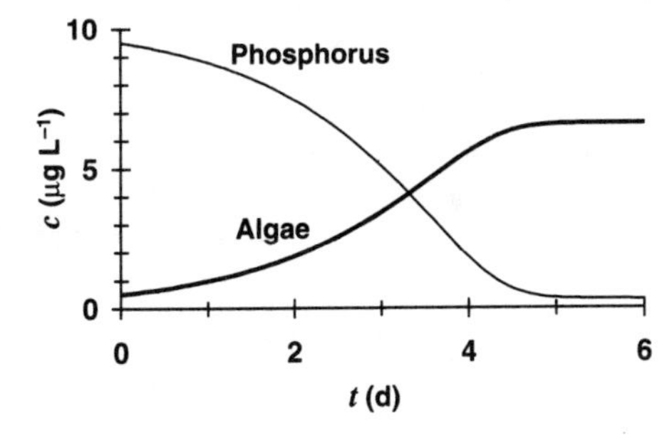

FIGURE 32.9
Flow diagram for kinetic interactions between plants and a limiting nutrient in a CSTR.

PROBLEMS

32.1. The following data were collected for algal growth rate as a function of the inorganic phosphorus concentration:

SRP (μgP L^{-1})	0.5	2.0	3.5	5.0	6.5	8.0	9.5	11.0
k_g (d^{-1})	0.33	0.52	1.15	1.02	1.09	1.28	1.27	1.55

Use (*a*) the Lineweaver-Burk and (*b*) the integral/least-squares methods to estimate k_s and $k_{g,\max}$.

32.2. Suppose that during the springtime the algae in a lake's epilimnion die at a rate of 0.025 d^{-1}. Thermocline diffusion is negligible, but the phytoplankton settle at a rate of 0.1 m d^{-1} and the epilimnion thickness is 5 m. The following conditions hold:

a_0 = 0.25 mgChl*a* m^{-3} p_0 = 20 mgP m^{-3}
p_{in} = 50 mgP m^{-3} a_{pa} = 1.5 mgP mgChl*a*$^{-1}$
k_s = 2 mgP m^{-3} τ_w = 30 d

All other parameters are as in Example 32.5. Calculate (*a*) the time required to reach equilibrium, (*b*) the ultimate concentrations of phosphorus and algae, and (*c*) the washout rate.

32.3. Repeat Examples 32.3 and 32.4, but with an S_{in} of 500 mgC L^{-1}.

LECTURE 33

Plant Growth and Nonpredatory Losses

LECTURE OVERVIEW: I review the factors that contribute to phytoplankton growth: temperature, nutrients, and light. Then I show how the resulting growth rate can be used to compute mass rates such as primary production and plant oxygen generation. Finally I conclude with a brief introduction to recently developed algal growth models employing variable carbon-to-chlorophyll ratios.

In the previous lecture we learned how to model the dynamics of microorganisms. At the end of our discussion we focused in on a specific group of microorganisms: algae. Now we will refine our plant growth framework by developing a more complete model of an important group of algae: the free-floating microorganisms called ***phytoplankton.*** In particular, along with nutrient limitation, we will expand the framework to include the effects of light and temperature on phytoplankton growth.

33.1 LIMITS TO PHYTOPLANKTON GROWTH

Just as we used first-order reactions as the starting point for characterizing the decay of organic matter in Lec. 19, we can use a similar approach to model growth. For a batch system, a mass balance can be written for algae as

$$\frac{da}{dt} = k_g a \tag{33.1}$$

where a = concentration of algae (mgChl$a$ m^{-3}) and k_g = first-order growth rate (d^{-1}). If the initial condition is $a_0 = 0$ at $t = 0$, then Eq. 33.1 can be solved for

$$a = a_0 e^{k_g t} \tag{33.2}$$

It is known that the phytoplankton growth rate is on the order of 2 d^{-1}. If an initial condition of 1 mg m^{-3} were used, the following values can be computed with Eq. 33.2:

t (d)	0	1	10	100
a (mg m^{-3})	1	7.8	4.85×10^8	7.2×10^{86}

Thus in only 100 d we would have a ridiculously large quantity of plants.

In nature such levels are never reached because, along with growth, there are a number of loss processes. Some of these are transport-related such as settling and diffusion/dispersion. Others are kinetic such as respiration, excretion, and death by predation. Further, the growth rate itself is not a simple constant, but varies in response to environmental factors such as temperature, nutrients, and light. At low levels, and in some instances at high levels, these factors can limit growth.

To incorporate these effects, Eq. 33.1 can more realistically be written as

$$\frac{da}{dt} = k_g(T, N, I)a - k_d a \tag{33.3}$$

where $k_g(T, n, I)$ = growth rate as a function of temperature T, nutrients N, and light I and k_d = loss rate. Later in this lecture we learn that the loss rate is also not a simple constant. However, for now, we will focus on how the growth rate is formulated.

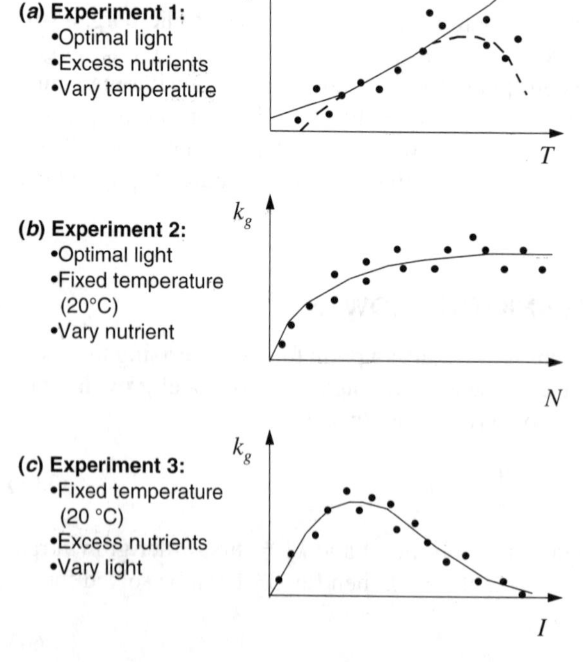

FIGURE 33.1
The results of three experiments to determine the phytoplankton growth rate as a function of (*a*) temperature, (*b*) a nutrient, and (*c*) light. The lines are intended to indicate the general pattern exhibited by the points.

The growth rate from Eq. 33.3 can be represented by

$$k_g(T, N, I) = k_{g,T}\phi_N\phi_L \tag{33.4}$$

where $k_{g,T}$ = maximum growth rate (that is, under optimal light and excess nutrients) at a particular temperature and ϕ_N and ϕ_L = attenuation factors for nutrient and light limitation, respectively. The attenuation factors take on values between 0 and 1, specifying complete limitation (0) and no limitation (1).

The shapes of the individual terms in Eq. 33.4 can be explored by conducting experiments. As displayed in Fig. 33.1, the effect of temperature, nutrients, and light can be determined by holding two of these characteristics fixed while varying the third. Such experiments indicate that in general all the factors seem to stimulate growth. However, light and temperature show a suppression of growth at high levels. We will now look at how these effects can be formulated mathematically.

33.2 TEMPERATURE

A variety of formulations have been developed to represent the effect of temperature on plant growth (Fig 33.2). The simplest is a linear model with some minimum temperature below which growth does not occur,

$$\begin{aligned} k_{g,T} &= 0 && T \le T_{min} \\ k_{g,T} &= k_{g,ref}\frac{T - T_{min}}{T_{ref} - T_{min}} && T > T_{min} \end{aligned} \tag{33.5}$$

where $k_{g,T}$ = growth rate (d^{-1}) at temperature T(°C)
$k_{g,ref}$ = growth rate (d^{-1}) at the reference temperature T_{ref}(°C)
T_{min} = temperature below which growth ceases

A more commonly used form is the theta model previously described in Lec. 2,

$$k_{g,T} = k_{g,20}\theta^{T-20} \tag{33.6}$$

Eppley (1972) proposed a value of θ = 1.066 based on a large number of studies involving many species of phytoplankton. Recall from Lec. 2 that this coefficient value roughly corresponds to a doubling of the rate for a 10°C rise in temperature.

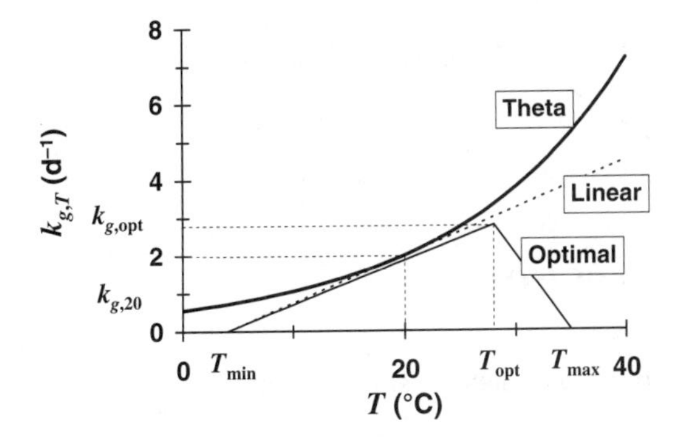

FIGURE 33.2
Some of the models used to characterize the effect of temperature on phytoplankton growth.

Several models have been proposed to represent a temperature dependence that is zero at a minimum temperature, increases to a peak growth rate at an optimal temperature, and then decreases at higher temperatures. The simplest approach is a linear representation,

$$\begin{aligned} k_{g,T} &= 0 && T \leq T_{\min} \\ k_{g,T} &= k_{g,\text{opt}} \frac{T - T_{\min}}{T_{\text{opt}} - T_{\min}} && T_{\min} \leq T \leq T_{\text{opt}} \\ k_{g,T} &= k_{g,\text{opt}} \frac{T_{\max} - T}{T_{\max} - T_{\text{opt}}} && T > T_{\text{opt}} \end{aligned} \tag{33.7}$$

Other investigators have suggested various functions to fit such a shape smoothly (e.g., Shugart et al. 1974, Lehman et al. 1975, Thornton and Lessem 1978, etc.). For example Cerco and Cole (1994) have used the following formulation based on the normal or bell-shaped distribution:

$$\begin{aligned} k_{g,T} &= k_{g,\text{opt}} e^{-\kappa_1 (T - T_{\text{opt}})^2} && T \leq T_{\text{opt}} \\ k_{g,T} &= k_{g,\text{opt}} e^{-\kappa_2 (T_{\text{opt}} - T)^2} && T > T_{\text{opt}} \end{aligned} \tag{33.8}$$

where κ_1 and κ_2 are parameters that determine the shape of the relationship of growth to temperature below and above the optimal temperature, respectively.

Which approach is best? Equation 33.6 is typically used when phytoplankton are simulated as a single state variable. This usually implies that there will always be species that grow at any particular temperature. Further it can also be used when the simulations are performed for temperatures lower than the optimal temperature for a particular species.

When several individual species or groups are modeled, the level of detail provided by Eqs. 33.7 or 33.8 may become necessary. For example Canale and

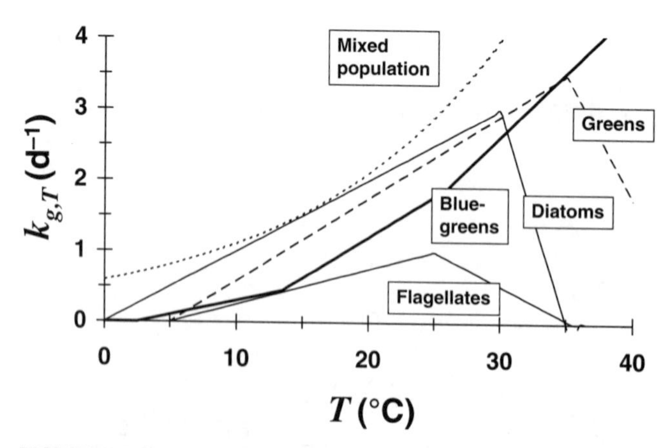

FIGURE 33.3
Temperature dependence of growth rates for several algal groups (Canale and Vogel 1974).

TABLE 33.1
Half-saturation constants for nutrient limitation of phytoplankton growth

Nutrient	k_s
Phosphorus	1–5 μgP L^{-1}
Nitrogen	5–20 μgN L^{-1}
Silica (diatoms)	20–80 μgSi L^{-1}

Vogel (1974) showed that major algal groups had differing sensitivity to temperature (Fig. 33.3). Their work illustrated that groups such as diatoms grew at low temperatures. In contrast blue-green algae grew better at high temperatures. Inclusion of such effects in water-quality models is necessary to determine their influence on competition among groups.

33.3 NUTRIENTS

As previously described in Sec. 32.2, the Michaelis-Menten equation provides the most common approach for handling nutrient limitation,

$$\phi_N = \frac{N}{k_{sN} + N} \tag{33.9}$$

where N = concentration of the limiting nutrient and k_{sN} = half-saturation constant. This equation is linearly proportional to concentration at low nutrient levels and approaches a constant value of one at high levels (Fig. 33.4). Values of the half-saturation constant are summarized in Table 33.1. Note that the half-saturation constant may also vary depending on the form of the nutrient that is limiting (e.g., SRP

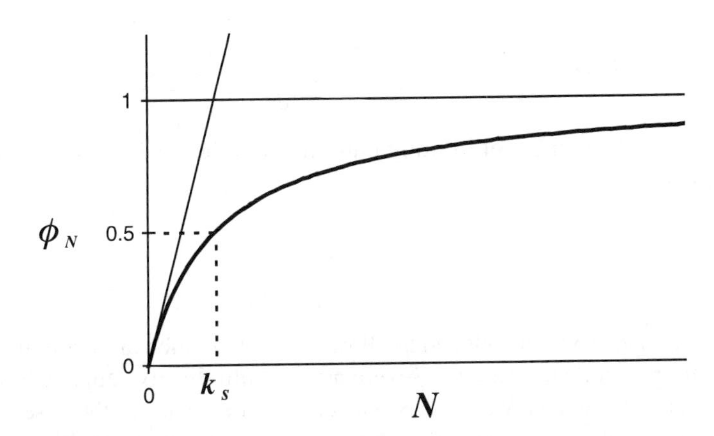

FIGURE 33.4
The Michaelis-Menten (or Monod) model of nutrient limitation.

vs. dissolved P for phosphorus; ammonium vs. nitrate vs. total inorganic nitrogen for nitrogen).

Multiple nutrients. There are several ways in which the nutrient limitation term can be refined. The most important relates to how multiple nutrients are handled. The most common case involves phosphorus and nitrogen. For such cases, separate limitation terms would be developed for each nutrient,

$$\phi_p = \frac{p}{k_{sp} + p} \tag{33.10}$$

and

$$\phi_n = \frac{n}{k_{sn} + n} \tag{33.11}$$

where p and n are concentrations of available phosphorus and nitrogen, respectively.

Three primary approaches are used to determine the combined effect of the nutrients:

1. *Multiplicative.* In this approach the two limitation terms are multiplied, as in

$$\phi_N = \phi_p \phi_n \tag{33.12}$$

Thus it is assumed that the nutrients have a synergistic effect. That is, several nutrients in short supply will more severely limit growth than a single limiting nutrient. The multiplicative form has been criticized as being excessively low when several nutrients are limiting. It also suffers in that, as more nutrients are included, it becomes more restrictive.

2. *Minimum.* At the other extreme is the case where the nutrient in shortest supply controls growth,

$$\phi_N = \min\{\phi_p, \phi_n\} \tag{33.13}$$

This type of approach, which is similar in spirit to ***Liebig's law of the minimum,*** is the most commonly accepted formulation.

3. *Harmonic mean.* The reciprocals of the limitation terms are combined, as in

$$\phi_N = \frac{m}{\sum_{j=1}^{m} \frac{1}{\phi_j}} \tag{33.14}$$

where m = the number of limiting nutrients. For the case of nitrogen and phosphorus, it would be

$$\phi_N = \frac{2}{\frac{1}{\phi_p} + \frac{1}{\phi_n}} \tag{33.15}$$

This formulation was developed to allow some interaction among multiple limiting nutrients while not being as severe as the multiplicative approach. However, it has been criticized (Walker 1983) as being unrealistic for the case where one nutrient is limiting. For example suppose that phosphorus is limiting (near zero) and nitrogen is in excess (near one). For this case Eq. 33.15 approaches $2\phi_p$ rather than ϕ_p as expected.

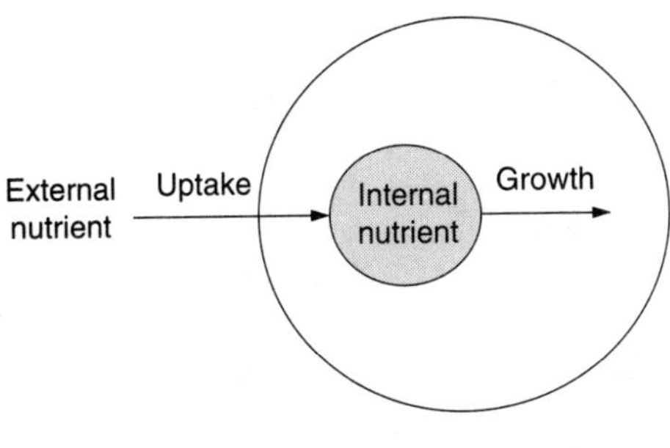

FIGURE 33.5
Internal nutrient pool phytoplankton model.

Luxury uptake and variable stoichiometry. Most water-quality models assume a fixed stoichiometry. However, it is known that phytoplankton stoichiometry varies.

A variety of models have been developed to simulate this phenomenon. They are all based on the notion that nutrient growth limitation is a function of the phytoplankton's internal concentration rather than the nutrient concentration in the water. This necessitates a separation of the mechanisms of uptake and growth (Fig. 33.5).

The first attempts to model the phenomenon used the Michaelis-Menten formulation to model uptake,

$$v = \frac{v_{\max} N}{k_{su} + N} \tag{33.16}$$

where v = uptake rate
$v_{\max}$ = maximum uptake rate
k_{su} = half-saturation rate for uptake

The growth limitation term was then computed as (Droop 1974)

$$\phi_N = 1 - \frac{q_0}{q} \tag{33.17}$$

where q = internal nutrient concentration, mass of nutrient per dry weight of algae, and q_0 = minimum internal concentration. Subsequent investigators (e.g., Auer 1979, Gotham and Rhee 1981, etc.) have refined this basic approach.

The internal nutrient pool models have not been widely used (noteworthy exceptions in engineering-oriented water-quality modeling are Grenney et al. 1973, Bierman 1976, and Canale and Auer 1982). However, models are now being developed to acknowledge that cell chlorophyll content varies. These models also include variable nutrient content. We will describe such approaches at the end of this lecture.

33.4 LIGHT

The effect of light on phytoplankton growth is complicated by the fact that several factors have to be integrated to come up with the total effect. As summarized in Fig. 33.6, these factors are diurnal surface-light variation, light attenuation with depth, and dependence of the growth rate on light.

The dependence of the growth rate on light can be quantified by experiment. As depicted in Fig. 33.6*c*, the growth rate peaks at an optimal light level. A variety

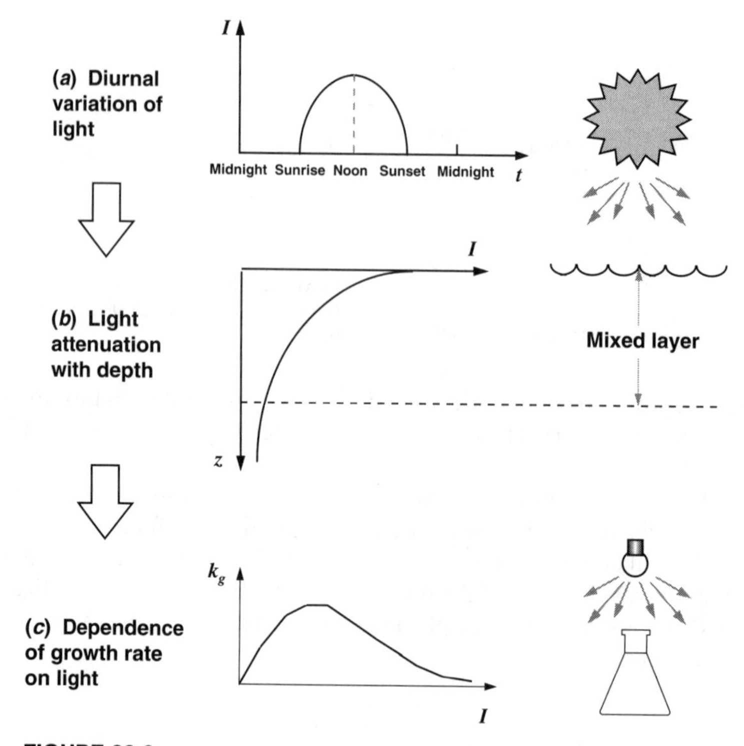

FIGURE 33.6
Incorporating light into the phytoplankton growth model involves integrating three separate factors: (*a*) the temporal variation of solar radiation at the water surface, (*b*) the attenuation of light as it passes through the layer of water being modeled, and (*c*) the effect of the light on plankton growth.

of models have been developed to fit light dependence. For example a Michaelis-Menten formulation is sometimes used,

$$F(I) = \frac{I}{k_{si} + I} \tag{33.18}$$

where k_{si} = a half-saturation constant for light (ly d^{-1}). Thus, as for nutrient limitation, this relationship is linearly proportional to light intensity at low light levels and constant at high levels. Consequently it does not capture growth attenuation at high light intensity. As was the case for Eq. 33.6 for temperature, such a formulation might be justified when the phytoplankton was simulated as a single group. In such cases it would be implicitly assumed that (1) some species of phytoplankton would grow optimally at any particular range of light intensity and (2) increasing light always leads to increasing growth. It would also be appropriate for a single species when the experiments were conducted below the optimal light intensity.

A somewhat different approach was taken by Steele (1965), whose model acknowledges that growth is inhibited at high light levels,

$$F(I) = \frac{I}{I_s} e^{-\frac{I}{I_s}+1} \tag{33.19}$$

where I = light level and I_s = optimal light level. Note that I_s ranges from about 100 to 400 ly d^{-1}. Lower values are appropriate for low-light-adapted species, whereas high values are for those that are adapted to high light intensity.

The temporal variation in light can be characterized by a half-sinusoid. For such cases the average light over the daylight hours can be computed as (Eq. 24.11)

$$I_a = I_m\left(\frac{2}{\pi}\right) \tag{33.20}$$

where I_m = maximum light intensity and f = photoperiod (sunlight fraction of day). Thus the average daylight value is about $\frac{2}{3}$ of the maximum.

The spatial variation of light down through the water column can be modeled by the Beer-Lambert law,

$$I(z) = I_0 e^{-k_e z} \tag{33.21}$$

where I_0 = solar radiation at the surface and k_e = extinction coefficient. The latter can be related to more fundamental quantities by (Riley 1956)

$$k_e = k_e' + 0.0088a + 0.054a^{2/3} \tag{33.22}$$

where k_e' = light extinction due to factors other than phytoplankton, which can be either measured directly or calculated via (Di Toro 1978),

$$k_e' = k_{ew} + 0.052N + 0.174D \tag{33.23}$$

where k_{ew} = light extinction due to particle-free water and color (m^{-1})
N = nonvolatile suspended solids (mg L^{-1})
D = detritus (nonliving organic suspended solids) (mg L^{-1})

All the preceding formulas can be applied to compute the mean light limitation for a well-mixed layer (Fig. 33.7). For example using the Steele model, we first substitute Eq. 33.21 into Eq. 33.19 to give an equation for the growth limitation at depth z,

$$F(I) = \frac{I_a e^{-k_e z}}{I_s} e^{-\frac{I_a e^{-k_e z}}{I_s}+1} \tag{33.24}$$

This function can then be integrated over depth and time to develop the mean value

$$\phi_L = \frac{1}{H}\int_{H_1}^{H_2} \frac{1}{T_p}\int_0^{fT_p} \frac{I_a e^{-k_e z}}{I_s} e^{-\frac{I_a e^{-k_e z}}{I_s}+1}\, dt\, dz \tag{33.25}$$

Evaluating this double integral results in

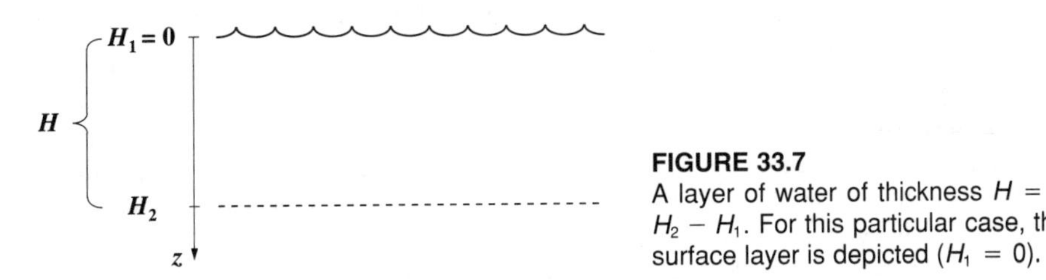

FIGURE 33.7
A layer of water of thickness $H = H_2 - H_1$. For this particular case, the surface layer is depicted ($H_1 = 0$).

$$\phi_L = \frac{2.718f}{k_e H}\left(e^{-\alpha_1} - e^{-\alpha_0}\right) \tag{33.26}$$

where

$$\alpha_0 = \frac{I_a}{I_s} e^{-k_e H_1} \tag{33.27}$$

and

$$\alpha_1 = \frac{I_a}{I_s} e^{-k_e H_2} \tag{33.28}$$

It should be noted that the light values used in all the preceding equations is visible, photosynthetically available light. This value is typically about 40 to 50% of the energy in the complete standard spectrum used for such calculations as heat budgets (Lec. 30) or photolysis (Lec. 42). For example Bannister (1974) and Stefan et al. (1983) suggest values from 44 to 46%. It should be noted that almost all the radiation outside the visible range is absorbed in the first meter below the surface (Orlob 1977).

33.5 THE GROWTH-RATE MODEL

The complete model of phytoplankton growth can now be developed as

$$k_g = \underbrace{k_{g,20} 1.066^{T-20}}_{\text{Temperature}} \underbrace{\left[\frac{2.718f}{k_e H}\left(e^{-\alpha_1} - e^{-\alpha_0}\right)\right]}_{\text{Light}} \underbrace{\min\left(\frac{n}{k_{sn} + n}, \frac{p}{k_{sp} + p}\right)}_{\text{Nutrients}} \tag{33.29}$$

This formula has a variety of applications. In a later lecture we use it to calculate phytoplankton growth as part of a nutrient/food-chain model. In addition it can be employed to calculate some other quantities of interest to water-quality modelers. For example if the quantity of phytoplankton is known, it can be used to compute the primary production,

$$Pr = a_{ca} k_g H a$$

with units of gC m^{-2} d^{-1}, or the oxygen produced by photosynthesis,

$$P = r_{oc} a_{ca} k_g a$$

with units of gO m^{-2} d^{-1}. Finally it can be used to gain insight into which factors are limiting phytoplankton growth. All these applications are explored in the following example.

EXAMPLE 33.1. PHYTOPLANKTON GROWTH RATE. The epilimnion of a lake has the following parameters:

$T = 20°$C
$I_s = 300$ ly d^{-1}
Available P concentration = 3 mg m^{-3}
Available N concentration = 20 mg m^{-3}
Chlorophyll *a* concentration = 4 mg m^{-3}
$k_{g,20} = 2$ d^{-1}

$I_a = 500$ ly d^{-1}
$k'_e = 0.3$ m^{-1}
P half-saturation constant = 2 mg m^{-3}
N half-saturation constant = 10 mg m^{-3}
$f = 0.5$
$H = 5$ m

(*a*) Compute the phytoplankton growth rate. Assume that the suspended solids concentration (other than phytoplankton) is negligible.
(*b*) Determine the primary production rate in g m^{-2} d^{-1} if the chlorophyll-to-carbon ratio is 20 μgChl*a* mgC^{-1}.

Solution: (*a*) The extinction coefficient can be determined by

$$k_e = 0.3 + 0.0088(4) + 0.54(4)^{2/3} = 0.471 \text{ m}^{-1}$$

which can used in conjunction with the light data to calculate

$$\alpha_0 = \frac{500}{300} e^{-0.471(0)} = 1.667$$

and

$$\alpha_1 = \frac{500}{300} e^{-0.471(5)} = 0.158$$

which can be substituted into Eq. 33.29 to give

$$k_g = 2 \cdot 1.066^{25-20} \left[\frac{2.718(0.5)}{0.471(5)} \left(e^{-1.667} - e^{-0.158} \right) \right] \min \left(\frac{20}{10+20}, \frac{3}{2+3} \right)$$

$$2.753 \quad \times \quad 0.3838 \quad \times \quad 0.6 \quad = 0.634 \text{ d}^{-1}$$

Thus we can see that the temperature sets the maximum rate, whereas the light and nutrients diminish the ultimate value. Further, the equation indicates that phosphorus is the limiting nutrient.
(*b*) The rate can be translated into a daily primary production rate by

$$Pr = a_{ca} k_g H a$$

Substituting values gives

$$Pr = \frac{1 \text{ mgC}}{20\ \mu\text{gChl}a} \left(\frac{0.634}{\text{d}} \right) (5\text{m}) \left(4 \frac{\mu\text{gChl}a}{L} \right) \left(\frac{1 \text{ g}}{1000 \text{ mg}} \right) \left(\frac{1000 \text{ L}}{\text{m}^3} \right) = 0.634 \text{ gC m}^{-2} \text{ d}^{-1}$$

33.6 NONPREDATORY LOSSES

A number of processes contribute to the loss rate of phytoplankton in Eq. 33.3. In water-quality modeling, three losses are emphasized:

- *Respiration.* This refers to the process opposite to photosynthesis, where the plant utilizes oxygen and releases carbon dioxide.
- *Excretion.* This process has traditionally focused on the release of nutrients. However, algae can also release organic carbon as extracellular byproducts.
- *Predatory losses.* Death of algae due to grazing by zooplankton.

Because the first two processes are difficult to measure separately, they have usually been modeled as a single first-order decay. Thus the death rate from Eq. 33.3 is usually expanded, as in

$$k_d = k_{ra} + k_{gz} \tag{33.30}$$

where k_{ra} = loss due to the combined effects of respiration and excretion (d^{-1}), and k_{gz} = grazing losses (d^{-1}). Values for k_{ra} range between 0.01 and 0.5 d^{-1}, with typical values on the order of 0.1 to 0.2 d^{-1}. A theta model is usually used to correct the respiration/excretion rate for temperature. A value of $\theta = 1.08$, connoting a strong temperature effect, is conventional.

It should be noted that although they are often treated as a single process, the division between respiration and excretion should not be considered a trivial distinction. This is particularly true as nutrient/food-chain models evolve toward a more accurate representation of the organic carbon cycle. In such cases, processes that tend to generate carbon dioxide and liberate available nutrients (respiration) should be separated from processes that liberate organic forms of carbon and nutrients (excretion).

At this point we can now integrate the growth and decay mechanisms into our modeling framework. In an analogous fashion to Eqs. 32.18 and 32.19, we can develop the following mass-balance equations for a limiting nutrient and algae for a CSTR:

$$\frac{da}{dt} = \left[k_{g,T}\frac{p}{k_{sp}+p}\frac{2.718f}{k_eH}\left(e^{-\alpha_1} - e^{-\alpha_0}\right) - k_{ra} - \frac{Q}{V}\right]a \tag{33.31}$$

$$\frac{dp}{dt} = -a_{pa}k_{g,T}\frac{p}{k_{sp}+p}\frac{2.718f}{k_eH}\left(e^{-\alpha_1} - e^{-\alpha_0}\right)a + a_{pa}k_{r,a}a + \frac{Q}{V}\left(p_{\text{in}} - p\right) \tag{33.32}$$

Notice that we have omitted grazing losses, which will be described in more detail in the following lecture.

EXAMPLE 33.2. ALGAL/NUTRIENT INTERACTIONS WITH LIGHT LIMITATION. Perform the same calculation as in Example 32.5, but now include the effect of light limitation. Note that the epilimnion has the following characteristics at the beginning of the stratified period:

$a_0 = 0.5$ mgChl$a$ m^{-3}	$p_0 = 9.5$ mgP m^{-3}	$p_{\text{in}} = 10$ mgP m^{-3}
$a_{pa} = 1.5$ mgP mgChla^{-1}	$k_{g,T} = 1$ d^{-1}	$k_{sp} = 2$ mgP m^{-3}
$k_{ra} = 0.1$ d^{-1}	$\tau_w = 30$ d^{-1}	$f = 0.5$
$I_a = 400$ ly d^{-1}	$I_s = 250$ ly d^{-1}	$H = 10$ m
$k_e' = 0.1$ m^{-1}		

Assuming that mass transfer across the thermocline due to diffusion and settling is negligible, simulate how the algae and phosphorus change over time.

Solution: Using Eqs. 33.22 and 33.26 to quantify the light effect, we display the results graphically in Fig. 33.8. In contrast to Fig. 32.9, two major differences are evident. First, the final algae level is lower and the phosphorus higher for the present case. Second, the time to steady-state is longer when light limitation is included. Both are due to the reduction of algal growth from light limitation. Thus, because the growth rate is smaller, the final partitioning between plants and nutrient is less extreme. Similarly the response time is lengthened.

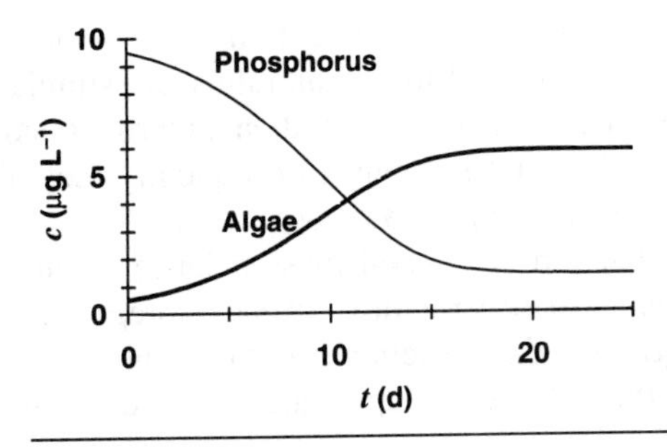

FIGURE 33.8
Flow diagram for kinetic interactions between plants and a limiting nutrient in a CSTR.

33.7 VARIABLE CHLOROPHYLL MODELS (ADVANCED TOPIC)

As described to this point, most water-quality models have used a constant stoichiometry to characterize algal concentration. An easy-to-measure quantity such as chlorophyll *a* was usually adopted as the measure of algal biomass. Transfers between phytoplankton and other model state variables such as nutrient pools were then handled by simple stoichiometric conversion factors.

Although this has proved to be a good first approximation, it has long been understood by ***phycologists*** (scientists who study algae) that cell stoichiometry is not constant. We have already alluded to this in our earlier discussion of luxury uptake.

Beyond nutrients, a more fundamental issue relates to carbon and chlorophyll. In particular, the chlorophyll-to-carbon ratio is not constant, but varies in response to light levels and the physiological state of the cells. Reviews of the published literature (e.g., Bowie et al. 1985) indicate that the chlorophyll-to-carbon ratio varies between about 10 and 100 μgChl*a* mgC^{-1}.

There are two reasons why such variability could have significance in the coming years:

- The focus of modeling is shifting from nutrient/food-chain interactions to an organic carbon cycle characterization. This is due to the application of water-quality frameworks beyond eutrophication to encompass problems such as toxic pollution, sediment-water interactions, and the impact of disinfection byproducts on drinking water.
- Water-quality models are being increasingly used to analyze cleaner systems than were studied in the past. Many water bodies are cleaner because their loadings have been reduced through waste treatment. Further, new problems such as drinking-water quality are being addressed. Because heavily polluted systems are more turbid and dimly illuminated, they tend to have a more constant chlorophyll-to-carbon ratio. In contrast clean water bodies typically exhibit more variable light levels because significant light can penetrate beyond well-mixed surface layers. The existence of deep chlorophyll layers in some estuarine systems and clear lakes provides circumstantial evidence for such effects.

Over the past decade a number of models have been developed to account for variable chlorophyll content in plants. Today the models are beginning to be integrated into water-quality frameworks. The purpose of this section is to illustrate how such integration can be done.

Laws and Chalup (1990) have developed such a model. In their approach, growth and respiration rates, along with chlorophyll-to-carbon ratios, are simulated as a function of light and nutrient levels. Because they focused on a marine diatom, Laws and Chalup used nitrogen as their limiting nutrient. Consequently they also computed a nitrogen-to-carbon ratio for phytoplankton biomass.

To derive their model, Laws and Chalup developed mass balances around a phytoplankton cell for four types of cell carbon: structural carbon, storage carbon, and carbon associated with both the light and dark reactions of photosynthesis. The resulting model can be represented by the following coupled algebraic equations:

$$\mu_s = \frac{K_e(1 - r_g)(1 - \mathrm{S/C})I}{K_e/f_{p0} + I[1 + k_e/(I_s f_{p0})]} - \frac{r_0}{C} \tag{33.33}$$

$$r/\mathrm{C} = (r_0/C + r_g\mu)(1 - r_g) \tag{33.34}$$

$$\mathrm{N/C} = \frac{F + (1 - F)(1 - \mu/\mu_s)}{W_N} \tag{33.35}$$

$$\mathrm{Chl}a/\mathrm{C} = \frac{1 - (1 - F)(1 - \mu/\mu_s) - \mathrm{S/C} - \dfrac{\mu + r_0/C}{(1 - r_g)K_e}}{W_{\mathrm{Chl}}} \tag{33.36}$$

where I = incident irradiance (mol quanta $\mathrm{m}^{-2}\ \mathrm{d}^{-1}$)
μ = algal growth rate (d^{-1})
μ_s = nutrient-saturated algal growth rate (d^{-1})
r = respiration rate per cell (gC $\mathrm{cell}^{-1}\ \mathrm{d}^{-1}$)

The remaining model parameters along with typical values are listed in Table 33.2.

Insight into the model can be gained by substituting the values from Table 33.2 into the model equations. For example we can first substitute the parameters into Eq. 33.33 to give

$$\mu_s = \mu_{\max}\frac{I}{k_{si} + I} \tag{33.37}$$

where $\mu_{\max}$ = maximum growth rate at saturated light and nutrient levels and k_{si} = light half-saturation constant. For the parameters in Table 33.2, $\mu_{\max}$ = 1.94 d^{-1} and k_{si} = 10.7 mol quanta $\mathrm{m}^{-2}\ \mathrm{d}^{-1}$. The latter value can be converted to more familiar units by multiplying it by a conversion factor of about 5 ly (mol quanta $\mathrm{m}^2)^{-1}$ to give 53.5 ly d^{-1}. Thus the model specifies that nutrient-saturated growth is related to light level by the Michaelis-Menten model specified earlier in this lecture (Eq. 33.18).

If the Michaelis-Menten model is also assumed to apply to nutrient limitation (as we did previously in Eq. 33.9), the ratio of the actual to the saturated growth rate would be specified by

$$\frac{\mu}{\mu_s} = \frac{n}{k_{sn} + n} \tag{33.38}$$

and the actual growth rate by the product of Eq. 33.37 and 33.38,

TABLE 33.2
Parameters developed by Laws and Chalup (1990) for the application of their algal growth model to the marine diatom *Pavlova lutheri*

Parameter	Symbol	Value	Range	Units
Rate of change of r per unit change in gross photosynthesis rate per cell	r_g	0.28	0.03–0.43	gC cell^{-1} d^{-1}
Basal respiration rate	r_0/C	0.03	−0.11–0.16	d^{-1}
Ratio of total cell C to nutrient in structural, light/dark reactions of C	W_N	6.9	6.2–7.7	gC gN^{-1}
Quotient of nutrient-limited N/C ratios at relative growth rates of 0 and 1	F	0.22	0.17–0.27	
Gross photosynthesis rate per unit quantity of dark reaction carbon	K_e	3.6	2.7–4.2	d^{-1}
Ratio of total cell C to chlorophyll a in the light reaction component of C	W_{Chl}	17	13–34	gC gChl^{-1}
Structural C per total C	S/C	0.1	−0.2–0.2	
Value of gross rate of photosynthesis per unit light reaction carbon per unit light intensity in the limit of zero irradiance	f_{p0}	0.28	0.16–0.51	m^2 mol quanta^{-1}
Half-saturation irradiance	I_s	64	10–122	mol quanta m^{-2} d^{-1}

$$\mu = \mu_{\max} \frac{I}{k_{si} + I} \frac{n}{k_{sn} + n} \tag{33.39}$$

Consequently we now can see that growth is related to light and nutrients in a similar fashion to Eq. 33.4. That is, it corresponds to a maximum rate times light and nutrient attenuation coefficients.

Next, we see that respiration is a simple linear function of growth,

$$r/\mathrm{C} = 0.042 + 0.389\mu \tag{33.40}$$

In other words, beyond a small basal metabolism rate of 0.042 d^{-1}, growth-related respiration will be about 39% of the actual growth rate.

The nitrogen-to-carbon ratio is linearly related to the cells nutrient condition, as in

$$\mathrm{N/C} = 32 + 113\frac{\mu}{\mu_s} \tag{33.41}$$

where we have changed the expression of the ratio from Laws and Chalup's gN gC^{-1} to the mgN gC^{-1} that is more consistent with the units used in previous sections. Thus in plants grown in nutrient-poor environments ($\mu/\mu_s = 0$), the nitrogen to carbon ratio is 32 mgN gC^{-1}. For nutrient-rich waters ($\mu/\mu_s = 1$), it approaches 145 mgN gC^{-1}. Recall that in Lec. 28 we developed an average N/C ratio for plant matter of 7200 mgN/40 gC = 180 mgN gC^{-1}, which corresponds to the upper bound of the range. Thus the model implies that phytoplankton are at the upper bound when they are in a nutrient-rich environment, and drop to about 20% of their maximum level when nutrients are severely depleted.

Last but not least, let's look at the chlorophyll-to-carbon ratio

$$\text{Chl}a/\text{C} = 6.378 + 45.882\frac{\mu}{\mu_s} - 22.7\mu \tag{33.42}$$

where again I have changed the units from gChl$a$ gC^{-1} to mgChl$a$ gC^{-1}.

This relationship indicates that nutrient enrichment tends to raise the ratio whereas light tends to lower it. It can be simplified further by substituting Eqs. 33.37 and 33.38 to give

$$\text{Chl}a/\text{C} = 6.378 + 45.9\frac{n}{k_{sn} + n} - 44\frac{n}{k_{sn} + n}1 - \frac{I}{k_{si} + I} \tag{33.43}$$

or collecting terms and assuming that 45.9 and 44 can both be approximated by 45,

$$\text{Chl}a/\text{C} = 6.4 + 45\frac{n}{k_{sn} + n}\frac{k_{si}}{k_{si} + I} \tag{33.44}$$

Now this relationship provides additional insight into the model. The last term indicates that the negative effect of light on the ratio has a hyperbolic shape that is the mirror image of the Michaelis-Menten curve. That is, Eq. 33.44 can be expressed as

$$\text{Chl}a/\text{C} = 6.4 + 45\phi_n(1 - \phi_L) \tag{33.45}$$

The result, as displayed in Fig. 33.9, illustrates that higher chlorophyll levels should occur in darker waters and when the cells have plenty of nutrients. Conversely they would have low chlorophyll in highly illuminated waters or when they were nutrient limited.

Now it can be seen how this model can be integrated into our larger algal modeling framework. For the simple case of phytoplankton and a single limiting nutrient, mass balances for a CSTR can be written as

$$\frac{dc_a}{dt} = 0.389\left(1.94\frac{I_{av}}{53.5 + I_{av}}\frac{n}{k_{sn} + n}\right)c_a - 0.042c_a - \frac{Q}{V}c_a \tag{33.46}$$

$$\frac{dn}{dt} = -(\text{N/C})\left[0.389\left(1.94\frac{I_{av}}{53.5 + I_{av}}\frac{n}{k_{sn} + n}\right)c_a - 0.042c_a\right] + \frac{Q}{V}(n_{in} - n) \tag{33.47}$$

where n = concentration of available nitrogen (mgN L^{-1})
c_a = algal carbon (mgC L^{-1})
I_{av} = average irradiance for the layer (ly d^{-1})

For a constant temporal light source, the average irradiance would be related to depth and light extinction by

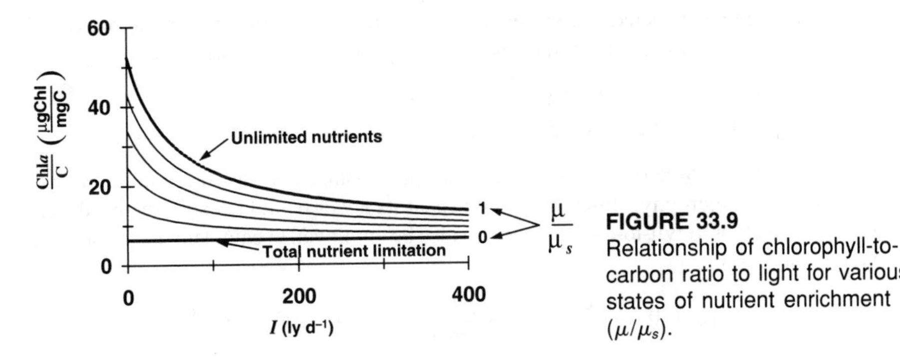

FIGURE 33.9
Relationship of chlorophyll-to-carbon ratio to light for various states of nutrient enrichment (μ/μ_s).

$$I_{av} = \frac{I_0}{k_e H}\left(1 - e^{-k_e H}\right) \tag{33.48}$$

where I_0 = surface light (ly d^{-1}).

Thus the first term in Eq. 33.46 is net growth (photosynthesis minus growth-related respiration) and the second term is basal respiration. Note how the N/C ratio is employed to translate the plant photosynthesis and respiration sources and sinks into sinks and sources in the nitrogen balance.

Finally where does Chl*a*/C fit in? In the present framework it would be employed to translate algal carbon into algal chlorophyll for use in determining the extinction coefficient as in Eq. 33.22. It could also be used to make the conversion in order to compare model output in carbon units with measured chlorophyll concentrations.

EXAMPLE 33.3. ALGAL/NUTRIENT INTERACTIONS WITH VARIABLE CHLOROPHYLL. A batch system has the following characteristics:

a_0 = 0.5 mgChl*a* m^{-3}	n_0 = 700 mgN m^{-3}
$k_{g,T}$ = 1 d^{-1}	k_{sp} = 15 mgN m^{-3}
$k_{d,r}$ = 0.1 d^{-1}	τ_w = 30 d
f = 0.5	I_{av} = 400 ly d^{-1}
H = 10 m	k_e' = 0.1 m^{-1}

Determine how the algae and phosphorus change over time.

Solution: Using Eq. 33.22, the initial extinction coefficient can be computed as

$$k_e = 0.1 + 0.0088(0.5) + 0.054(0.5)^{2/3} = 0.138 \text{ m}^{-1}$$

which can be used in conjunction with Eq. 33.48 to determine the average irradiance in the reactor,

$$I_{av} = \frac{400}{0.138(10)}\left(1 - e^{-0.138(10)}\right) = 216.6 \text{ ly d}^{-1}$$

The initial nutrient saturation state and growth rate can be computed as

$$\frac{\mu}{\mu_s} = \frac{700}{15 + 700} = 0.979$$

$$\mu = 1.94\frac{216.6}{53.5 + 216.6}\frac{700}{15 + 700} = 1.94(0.802)0.979 = 1.523 \text{ d}^{-1}$$

Thus, as might be expected, the system has very little light or nutrient limitation. Initial N/C and Chl$a$/C ratios can be determined with Eqs. 33.41 and 33.42,

$$\text{N/C} = 32 + 113(0.979) = 143 \text{ mgN gC}^{-1}$$

$$\text{Chl}a\text{/C} = 6.378 + 45.882(0.979) - 22.7(1.523) = 16.72 \text{ mgChl gC}^{-1}$$

The nitrogen-to-carbon ratio is near its upper bound because there are abundant nutrients. The chlorophyll-to-carbon ratio is near its lower bound because the irradiance is high. The chlorophyll-to-carbon ratio can also be employed to calculate the initial algal biomass as carbon,

$$c_a = 0.5\frac{\text{mgChl}a}{\text{m}^3}\left(\frac{\text{g C}}{16.72 \text{ mgChl}a}\right) = 0.0299 \text{ gC m}^{-3}$$

Now, using these initial conditions, Eqs. 33.46 and 33.47 can be integrated (with $Q/V = 0$ to simulate the batch system). At each time step, all of the ratios are recomputed. The results are shown in Fig. E33.3.

The results are interesting. The nitrogen and the algal carbon fall and rise as sigmoid (that is, S-shaped) curves as would be expected from our previous models of batch

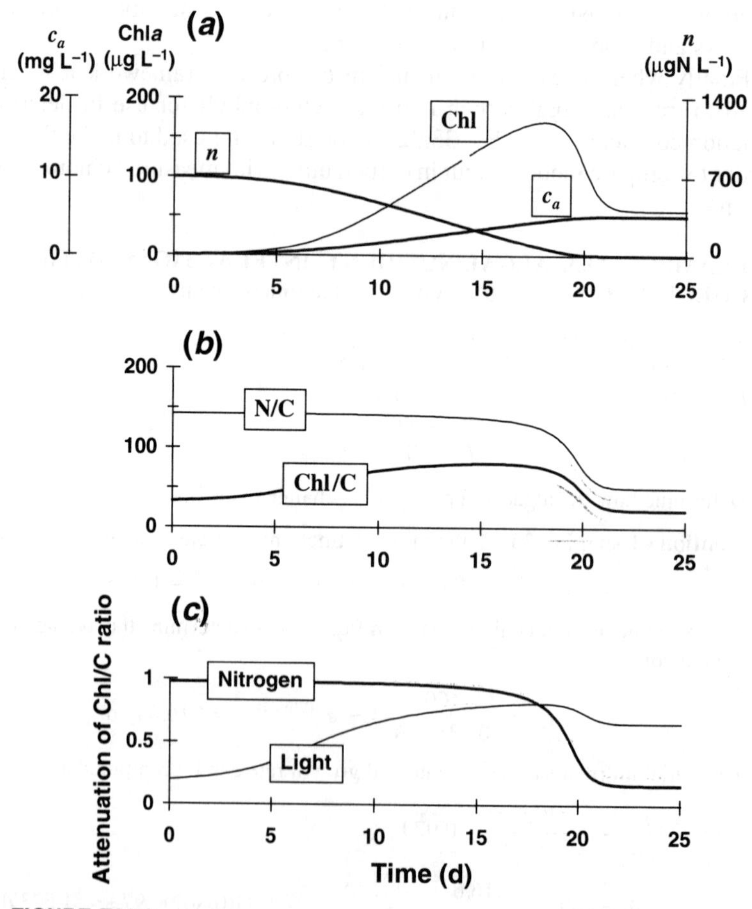

FIGURE E33.3

systems. However, the algal chlorophyll rises to a peak and then declines. The reason for this behavior is illustrated by Fig. 33.3*c*, which shows how the chlorophyll-to-carbon ratio is affected by light and nutrients. As the simulation progresses, light is diminished because of shading by the increasing algal biomass. This leads to an increased Chl*a*/C ratio. The resulting peak is then rapidly diminished as the ratio drops when the nutrients are eventually used up.

The foregoing example is intended to illustrate how variable carbon can be integrated into water-quality frameworks. It also suggests a number of implications of the mechanism. According to the model,

- Because it varies, chlorophyll is a deceptive measure of true biomass. This is somewhat unfortunate since the chlorophyll *a* measurement provides such a convenient and economical way to discriminate plant biomass from other particulate matter in natural waters. However, if the foregoing model is an adequate descriptor of reality, the calculated ratio provides a means to translate chlorophyll into carbon.
- Because of the opposing effects of nutrients and light on the ratio, the mechanism seems to sharpen up growth events. This behavior might allow water-quality models to better simulate both temporal and spatial gradients of carbon and chlorophyll in natural waters. Examples of such phenomena are spring blooms, deep chlorophyll layers, and horizontal gradients in waste plumes.

PROBLEMS

33.1. A very clear body of water has the following conditions:

$T = 10°C$	Available P concentration in hypolimnion
$I_a = 700 \text{ ly d}^{-1}$	$= 4 \text{ mg m}^{-3}$
$I_s = 250 \text{ ly d}^{-1}$	P half-saturation constant $= 2 \text{ mg m}^{-3}$
$k_e' = 0.2 \text{ m}^{-1}$	$f = 0.5$
(same in epi and hypo)	$k_{g,20} = 2 \text{ d}^{-1}$

Calculate the phytoplankton growth rate in the hypolimnion. Note that the surface layer has a thickness of 4 m and the hypolimnion has a thickness of 3 m. Assume that phosphorus is the limiting nutrient.

33.2. Use (*a*) Eq. 33.7 and (*b*) Eq. 33.8 to fit the effect of temperature on the growth of the flagellates, as in Fig. 33.3.

33.3. Develop the equation for light limitation (comparable to Eq. 33.26) for the Michaelis-Menten form of Eq. 33.18.

33.4. Repeat Example 33.2, but include a phytoplankton settling velocity of 0.2 m d^{-1}.

33.5. Perform a sensitivity analysis (recall Lec. 18) on Example 33.2 for the following parameters: a_0, p_0, p_{in}, $k_{g,T}$, k_{sp}, k_{ra}, and I_s.

33.6. Duplicate the computation from Example 33.3, but for a CSTR with a residence time of 10 d. Discuss how flushing modifies the shapes of the curves in Fig. E33.3.

LECTURE 34

Predator-Prey and Nutrient/Food-Chain Interactions

LECTURE OVERVIEW: I now describe the predator-prey interactions that dictate an important component of plant death. To do this I introduce you to the mathematics of predator-prey interactions. Then I apply this theory to phytoplankton and their key predators: zooplankton. Finally I integrate predator-prey interactions with a nutrient balance to form a simple nutrient/food-chain model.

In the previous lecture I outlined some external physical and chemical factors, such as temperature, light, and nutrients, that limit phytoplankton growth in natural waters. I also described losses such as respiration and excretion. Now let's turn to factors that limit algal populations by causing their death. In particular we'll focus on the action of predators such as zooplankton that use algae as a food source. To provide background for these phytoplankton-zooplankton interactions, I'll first describe general mathematical models that have been developed to simulate predator-prey interactions.

34.1 LOTKA-VOLTERRA EQUATIONS

Aside from phytoplankton and zooplankton, there are other cases involving a pair of organisms in which one serves as the primary food source for the other. For example in southern oceans, the shrimplike Antarctic krill serves as the principal food source for baleen whales. Moose provide the primary food source for wolves on isolated island ecosystems such as Isle Royale in Lake Superior.

In 1926 the Italian biologist Humberto D'Ancona estimated the populations of predator and prey species in the upper Adriatic Sea based on the numbers sold in fish markets from 1910 to 1923. On the basis of the data, he hypothesized that lack of fishing during World War I led to a higher proportion of predators. He communicated his findings to his father-in-law, the noted mathematician Vito Volterra. Over the following year, Volterra developed a number of mathematical models to simulate the interactions among two or more species. Independently the American biologist A. J. Lotka came up with many of the same models.

There are a variety of these models, which are now commonly called ***Lotka-Volterra equations.*** In this section we describe the simplest version. To do this we can first write a growth equation for a single prey species in an isolated environment (that is, no predators and plenty of food). For such a case a simple first-order model can be written as

$$\frac{dx}{dt} = ax \tag{34.1}$$

where x = number of prey and a = a first-order growth rate.

Next an equation can be written for a single predator y in the absence of its sole food source x,

$$\frac{dy}{dt} = -cy \tag{34.2}$$

where c = a first-order death rate.

Now the interaction between the two species should depend on both their magnitudes. If there are either too few predators or too few prey, the interactions would decrease. A simple way to represent this interaction is as a product xy. Thus, because the interaction spells death for the prey, it can be added as a loss to Eq. 34.1,

$$\frac{dx}{dt} = ax - bxy \tag{34.3}$$

where b = a parameter that quantifies the impact of the interaction on prey mortality. Conversely for the predator, the interaction represents a gain,

$$\frac{dy}{dt} = -cy + dxy \tag{34.4}$$

where d quantifies the impact on predator growth.

The results of integrating these equations is shown in Fig. 34.1. Note that a cyclical pattern emerges. Thus, because predator population is initially small, the

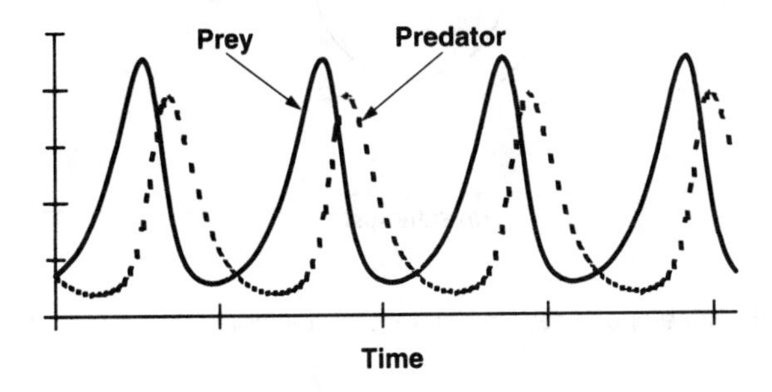

FIGURE 34.1
Time-domain representation of numbers of predators and prey for the Lotka-Volterra model.

prey grows exponentially. At a certain point the prey become so numerous, that the predator population begins to grow. Eventually the increased predators cause the prey to decline. This decrease, in turn, leads to a decrease of the predators. Eventually the process repeats. Notice that, as expected, the predator peak lags the prey. Also observe that the process has a fixed period; that is, it repeats in a set time.

Now if the parameters used to simulate Fig. 34.1 were changed, although the general pattern would remain the same, the magnitudes of the peaks, lags, and period would change. Thus there are an infinite number of cycles that could occur. The state-space perspective (Box 34.1) provides a nice way to see the underlying structure of these cycles.

BOX 34.1. State-Space Representation of Dynamic Systems

I can introduce you to the state-space approach with a very simple model,

$$\frac{dx}{dt} = -y \qquad \frac{dy}{dt} = x \tag{34.5}$$

If $y = a$ and $x = 0$ at $t = 0$, the solution is

$$x = a \cos t \qquad y = a \sin t \tag{34.6}$$

As shown in Fig. 34.2*a*, the variables trace out sinusoids that are $\pi/2$ radians out of phase. Because the curves evolve in time, this is called the ***time-domain representation.***

Now, just as with predator prey interactions, different curves would result depending on the value of the parameter a. An effective way to summarize all the different results is provided by the state-space representation. This can be developed by dividing the two original differential equations to yield

$$\frac{dy/dt}{dx/dt} = \frac{dy}{dx} = -\frac{x}{y} \tag{34.7}$$

Thus the division eliminates time from the model, and we are left with a single differential equation that describes how the two state variables co-vary. It can be solved by

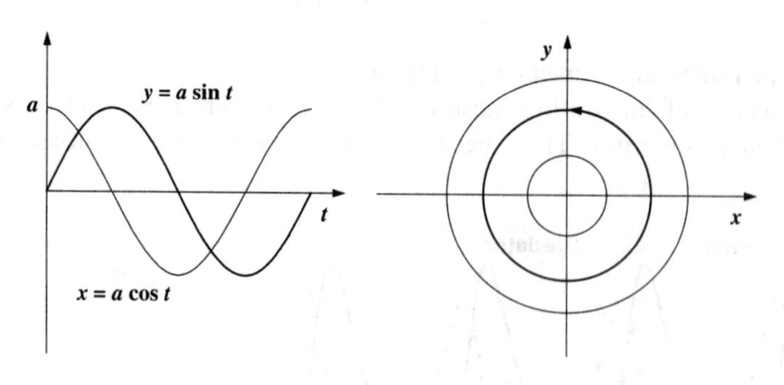

FIGURE 34.2
(a) Time-domain and (b) state-space representations of a simple periodic model.

the separation of variables for

$$x^2 + y^2 = 2a \tag{34.8}$$

which is the equation for a circle of radius a. This equation can be displayed on a two-dimensional y versus x plot (Fig. 34.2*b*). Because the plot illustrates the way that the state variables interact, it is referred to as a ***state-space representation.*** By showing plots for different values of a, we have effectively displayed all solutions on a single graph. Although the results are not tremendously illuminating because of the simplicity of the present example, the representation is often a powerful tool for examining solutions in more complex cases. This is the advantage of the approach.

A similar approach can be applied to the Lotka-Volterra equations. First, the ratio of Eqs. 34.4 and 34.3 can be developed,

$$\frac{dy}{dx} = \frac{-cy + dxy}{ax - bxy} \tag{34.9}$$

This equation is separable and solvable for

$$a \ln y - by + c \ln x - dx = K' \tag{34.10}$$

where $K' =$ a constant of integration. The solution can be simplified further by exponentiation,

$$(y^a e^{-by})(x^c e^{-dx}) = K \tag{34.11}$$

where $K = e^{K'}$. Equation 34.11 defines counterclockwise orbits of the type displayed on Fig. 34.4*b*.

The state-space approach can be applied to the Lotka-Volterra model. First, we can examine the "rest points" of the model. These are the values that result in no change for the variables. To determine these points, we set the derivatives in Eqs. 34.3 and 34.4 to zero,

$$ax - bxy = 0 \tag{34.12}$$

$$-cy + dxy = 0 \tag{34.13}$$

which can be solved for

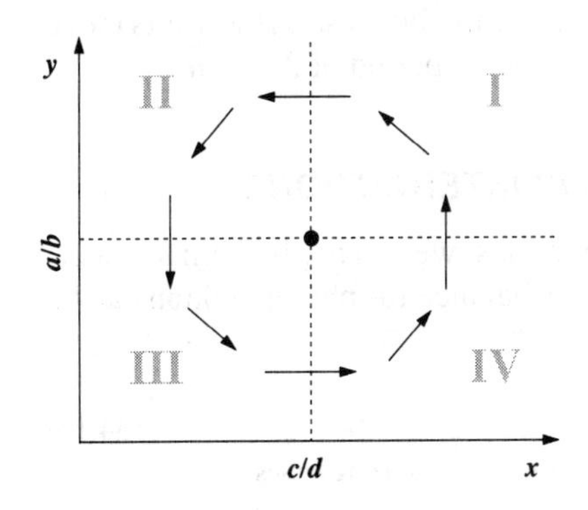

FIGURE 34.3
Rest points and trajectories in a state-space representation of the Lotka-Volterra model.

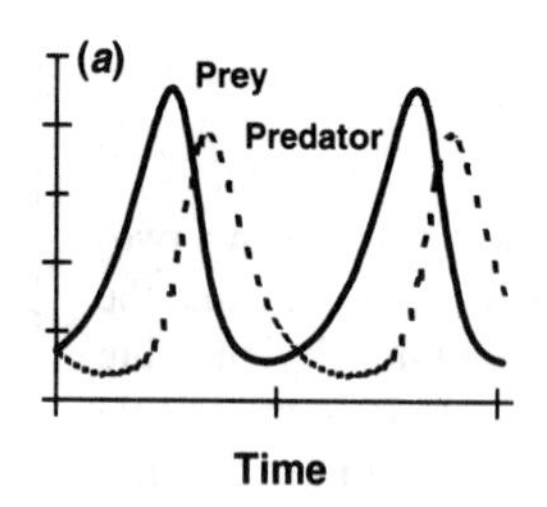

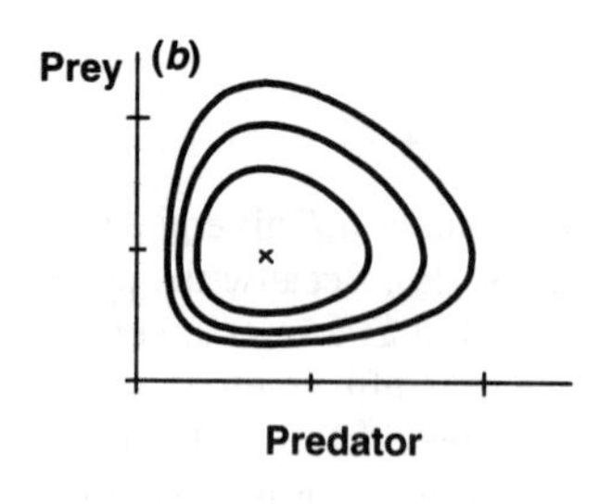

FIGURE 34.4
(a) Time-domain representation of a single instance of the Lotka-Volterra model. (b) State-space representation of all cycles for a particular parameter set.

$$(x, y) = (0, 0)$$
$$(x, y) = \left(\frac{c}{d}, \frac{a}{b}\right) \tag{34.14}$$

Thus one of the solutions is the trivial result that if we have neither predators nor prey, nothing will happen. The more interesting result is that if $x = c/d$ and $y = a/b$, the derivatives will be zero and the populations will remain constant. This rest point can be plotted on a state-space plot (Fig. 34.3).

Next the ratio of Eqs. 34.4 and 34.3 can be developed,

$$\frac{dy}{dx} = \frac{-cy + dxy}{ax - bxy} \tag{34.15}$$

Four quadrants can be demarcated by drawing a vertical and a horizontal line through the nontrivial rest point in Fig. 34.3. By inspecting Eq. 34.15 it can then be recognized that, depending on the initial values of x and y, the trajectories (that is, dy/dx) in each of these quadrants will have the same sign. For example in quadrant I ($x > c/d$ and $y > a/b$) the slope will always be negative. The other slopes are shown on Fig. 34.3. Thus it appears that the result will be a counterclockwise cycle.

This insight can be refined by solving Eq. 34.15 and sketching the resulting solution on the space shown in Fig. 34.4*b*. One of the time-domain realizations is shown for comparison (Fig. 34.4*a*). Analysis of the solution leads to several conclusions regarding the model:

- The orbits are closed curves that cycle around the nontrivial rest point. Thus every orbit represents a periodic cycle that repeats infinitely. In other words neither the predator nor the prey will become extinct.
- The peak for the predators always lags behind the prey.
- The periods are not constant. In fact they monotonically increase as the orbits move out from the rest point. The smaller orbits approach a period of $2\pi/\sqrt{ac}$.

34.2 PHYTOPLANKTON-ZOOPLANKTON INTERACTIONS

Now that we have reviewed predator-prey interactions, we can apply them to water-quality modeling. Recall from Lec. 33 that a mass balance for phytoplankton can be written as

$$\frac{da}{dt} = \underbrace{k_g(T, n, I)a}_{\text{Growth}} - \underbrace{k_{ra}a}_{\text{Respiration/excretion}} - \underbrace{k_{gz}a}_{\text{Grazing losses}} \tag{34.16}$$

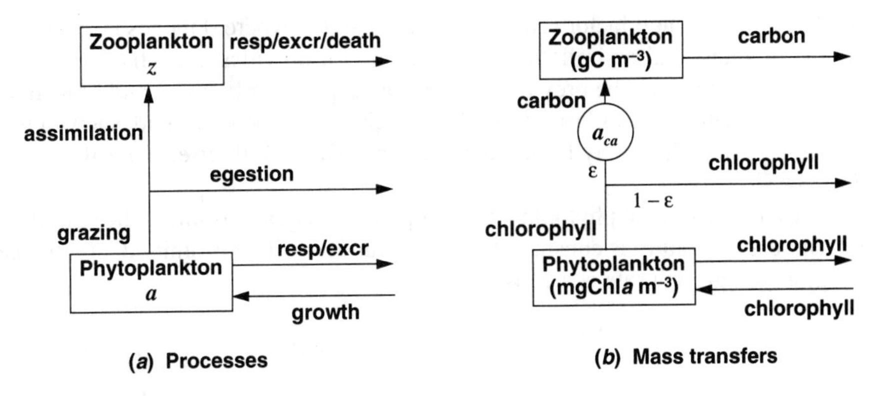

FIGURE 34.5
A simple representation of the interactions simulated in a food-chain model.

The grazing loss rate can be represented simply as

$$k_{gz} = C_{gz} z \theta_{gz}^{T-20} \tag{34.17}$$

where C_{gz} = zooplankton grazing rate (m^3 gC^{-1} d^{-1})
θ_{gz} = temperature correction factor
z = zooplankton concentration (gC m^{-3})

Recognize that the product $C_{gz}z$ has units of d^{-1}. Thus when Eq. 34.17 is substituted into Eq. 34.16, the grazing loss term will have units of μgChl L^{-1} d^{-1}.

Note that there have been a variety of refinements and enhancements made to this representation. For example some investigators have included a Michaelis-Menten term to account for the fact that at high levels of phytoplankton, zooplankton grazing levels off,

$$k_{gz} = \frac{a}{k_{sa} + a} C_{gz} z \theta_{gz}^{T-20} \tag{34.18}$$

where k_{sa} = the half-saturation constant for zooplankton grazing on algae (μgChl L^{-1}). Other formulations are summarized in Bowie et al. (1985).

Incorporating Eq. 34.18 into Eq. 34.16 yields the final balance for the algae,

$$\frac{da}{dt} = k_g(T, n, I)a - k_{ra}a - \frac{a}{k_{sa} + a} C_{gz} z \theta_{gz}^{T-20} a \tag{34.19}$$

Next we can turn to the zooplankton. As depicted in Fig. 34.5, the zooplankton gain biomass by assimilating phytoplankton and lose biomass by respiration, excretion, and death. These processes can be incorporated into a zooplankton balance,

$$\frac{dz}{dt} = a_{ca}\varepsilon \frac{a}{k_{sa} + a} C_{gz} \theta_{gz}^{T-20} za - k_{dz}z \tag{34.20}$$

where a_{ca} = ratio of carbon to chlorophyll a in the phytoplankton biomass (gC mgChla^{-1})
ε = a grazing efficiency factor
k_{dz} = a first-order loss rate for respiration, excretion, and death (d^{-1}).

Note that the efficiency factor ranges between 0 and 1, where 0 means no assimilation and 1 means total assimilation. Consequently it provides a mechanism for establishing how much of the prey biomass becomes predator biomass and how much is released as detritus. Similarly the carbon-to-chlorophyll ratio provides a means to translate ingested chlorophyll into zooplankton carbon. Both effects are illustrated in Fig. 34.5*b*.

Now let's look at the phytoplankton-zooplankton equations under the simplified case where temperature is constant, the half-saturation effect is omitted, and the net phytoplankton growth rate is a constant,

$$\frac{da}{dt} = (k_g - k_{ra})a - C_{gz}za \tag{34.21}$$

$$\frac{dz}{dt} = (a_{ca}\varepsilon C_{gz})za - k_{dz}z \tag{34.22}$$

Because all the parameters are constants, these represent Lotka-Volterra equations.

EXAMPLE 34.1. ALGAL-ZOOPLANKTON INTERACTIONS. A batch system containing phytoplankton and zooplankton has the following characteristics:

$a_0 = 1$ mgChl*a* m^{-3} $\quad$ $z_0 = 0.05$ gC m^{-3}

$a_{ca} = 0.04$ gC mgChl^{-1} $\quad$ $C_{gz} = 1.5$ m^3 gC^{-1} d^{-1}

$\varepsilon = 0.6$ $\quad$ $k_g - k_{ra} = 0.3$ d^{-1}

$k_{dz} = 0.1$ d^{-1}

Calculate a few cycles of both organisms.

Solution: The results of integrating Eqs. 34.21 and 34.22 with the fourth-order RK method are displayed in Fig. E34.1. Note that we have expressed the phytoplankton concentration in carbon units so that the biomass values are directly comparable. We have also included the total of zooplankton and phytoplankton on the plot.

The biota peak about every 45 d. The zooplankton peak about 5 d after the phytoplankton reach a maximum. Note that the behavior exhibited by this plot is influenced

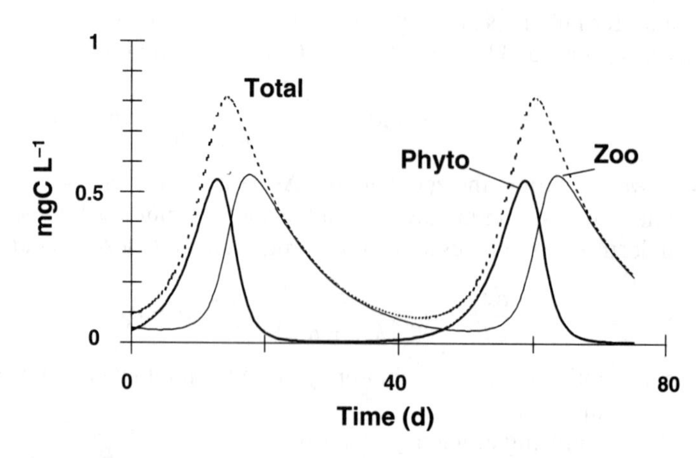

FIGURE E34.1

by our choice of parameters and the fact that we are using a batch system. Other parameters, including transport, such as flushing and settling, can alter the size and phasing of the peaks. You will be provided some opportunity to explore such sensitivity in the end-of-lecture problems.

34.3 ZOOPLANKTON PARAMETERS

Grazing rates generally vary between 0.5 and 5, with values most commonly in the range from 1 to 2 $m^3\ gC^{-1}\ d^{-1}$. The temperature correction factor is generally on the high side, with a value of 1.08 being commonly used. The half-saturation constant varies between 2 and 25, with values most commonly between 5 and 15 μgChl L^{-1}. Grazing efficiencies commonly vary between 0.4 and 0.8.

Because it is a composite of several factors, the zooplankton loss rate must be broken down into its component parts. As with phytoplankton, a common approach is to break it into nonpredatory and predatory losses,

$$k_{dz} = k_{rz} + k_{gzc} \tag{34.23}$$

where k_{rz} = nonpredation loss rate (d^{-1}) and k_{gzc} = predation loss rate (d^{-1}), in which the subscript c connotes that the predation is carnivorous. It should also be recognized that there are omnivorous zooplankton that eat both animals and plants.

The nonpredatory losses consist primarily of respiration and excretion. Because these are difficult to measure separately, they are lumped together. This rate is commonly between 0.01 and 0.05 d^{-1}, with reported values ranging between 0.001 and 0.1 d^{-1}.

The grazing loss is typically handled in two fashions. Where upper levels of the food chain are not modeled explicitly, it is often treated as a constant with a temperature correction. Conversely when higher predators such as carnivores are simulated, it is modeled in a similar fashion to Eq. 34.17. We will illustrate how such "food-chain" modeling is implemented in the following lecture.

Summaries of data for all the aforementioned zooplankton grazing parameters can be found in Bowie et al. (1985). This publication also summarizes alternative formulations and approaches for modeling algal-zooplankton interactions.

34.4 NUTRIENT/FOOD-CHAIN INTERACTIONS

Now we integrate a limiting nutrient into our phytoplankton-zooplankton scheme to complete our fundamental picture of how nutrient/food-chain interactions are modeled. Following the simple kinetic scheme in Fig. 34.6, the resulting equations could be written as

$$\frac{da}{dt} = (k_g - k_{ra})a - C_{gz}za \tag{34.24}$$

$$\frac{dz}{dt} = (a_{ca}\varepsilon C_{gz})za - k_{rz}z \tag{34.25}$$

$$\frac{dp}{dt} = a_{pa}(1 - \varepsilon)C_{gz}za + a_{pc}k_{rz}z - a_{pa}(k_g - k_{ra})a \tag{34.26}$$

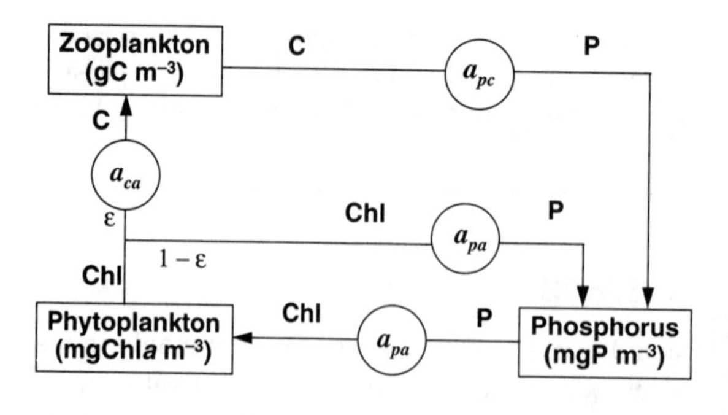

FIGURE 34.6
A simple representation of the interactions simulated in a nutrient/food-chain model.

The equations now form a completely closed system where nutrients are turned into biomass by net production and recycled by respiration and death. The following example illustrates the resulting dynamics of the three components.

EXAMPLE 34.2. NUTRIENT/FOOD-CHAIN INTERACTIONS. A batch system containing phytoplankton, zooplankton, and available phosphorus has the following characteristics:

$a_0 = 1$ mgChl*a* m^{-3} $\quad z_0 = 0.05$ gC m^{-3}
$a_{ca} = 0.04$ gC mgChl^{-1} $\quad a_{pa} = 1$ mgP mgChl^{-1}
$C_{gz} = 1.5$ m^3 gC^{-1} d^{-1} $\quad \varepsilon = 0.6$
$k_g - k_{ra} = 0.3$ d^{-1} $\quad k_{dz} = 0.1$ d^{-1}
$p_0 = 20$ μgP L^{-1}

Calculate a few cycles of both organisms and the nutrient. Include a nutrient limitation effect in the phytoplankton growth term. Use a value of 2 μgP L^{-1} for the half-saturation constant.

Solution: The nutrient limitation term is included by modifying the growth term, as in

$$k_g = k_{g,m} \frac{p}{k_{sp} + p}$$

where $k_{g,m}$ = a constant growth rate at saturated nutrient levels. The results of substituting this relationship into the system and integrating with the fourth-order RK method are displayed in Fig. E34.2. Note that we have expressed the phytoplankton and phosphorus

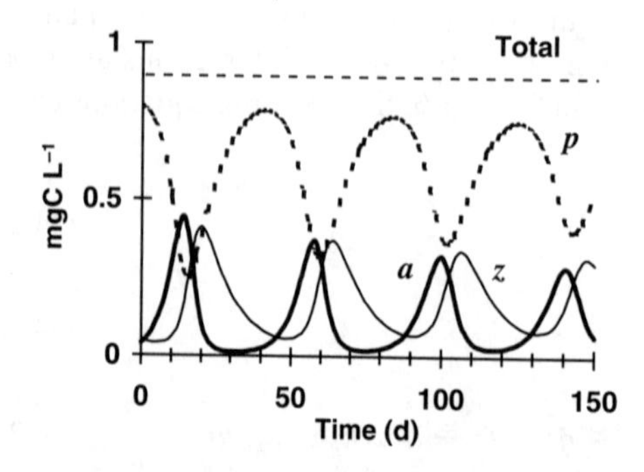

FIGURE E34.2

in equivalent carbon units so that the values of the three forms are directly comparable. We have also included the total of the three components on the plot.

The biota peak about every 42 d. The zooplankton peak about 1 wk after the phytoplankton reach a maximum. Notice that the plants recede because of the grazing effect rather than the nutrient limitation (the phosphorus never falls below 6 μgP L^{-1} during the simulation). In other cases the nutrients could be depleted first. One value of nutrient/food-chain modeling is the capability to simulate and discriminate between such effects. As with Example 34.1 the behavior exhibited by this plot is influenced by our choice of parameters and the fact that we are using a batch system.

This lecture has been designed to illustrate how predator-prey kinetics work and to show how they can be integrated into a more comprehensive nutrient/food-chain computation. For simplicity we have limited our exposition to batch systems. In the next lecture we expand the framework and implement it in an open, natural water body: a stratified lake.

PROBLEMS

34.1. Integrate Eqs. 34.3 and 34.4 numerically using fourth-order RK and parameter values of $a = 1, b = 0.1, c = 0.5$, and $d = 0.02$ and initial conditions of $x = 20$ and $y = 5$.

34.2. Perform a sensitivity analysis on Example 34.1 as described in Lec. 18. Vary the following parameters in your analysis: net growth $(k_g - k_{ra})$, C_{gz}, ε, and k_{dz}.

34.3. Repeat Example 34.1, but use the following parameter values:

$a_0 = 2$ mgChl$a$ m^{-3} $z_0 = 0.03$ gC m^{-3}

$a_{ca} = 0.04$ gC mgChl^{-1} $C_{gz} = 2$ m^3 gC^{-1} d^{-1}

$\varepsilon = 0.5$ $k_g - k_{ra} = 0.5$ d^{-1}

$k_{dz} = 0.075$ d^{-1}

34.4. Modify Example 34.1 by including a carnivorous zooplankter that eats the herbivorous zooplankton. Also, modify the model by incorporating a saturation term for herbivore grazing on algae. Along with the parameters from the example, you will require the following additional values:

$z_{co} = 0.02$ gC m^{-3} $C_{gzc} = 3$ m^3 gC^{-1} d^{-1}

$\varepsilon_c = 0.6$ $k_{dzc} = 0.1$ d^{-1}

$k_{sa} = 15$ μgChl m^{-3}

where the subscript c denotes the carnivores. Also, remove the lumped respiration/excretion/death term for the herbivores and model respiration/excretion explicitly ($k_{rzh} = 0.01$ d^{-1}) along with the carnivore grazing effect.

34.5. Modify Example 34.2 so that the system is a CSTR with a residence time of 20 d and inflow concentrations of $a_{in} = 0.5$ mgChl L^{-1} and $p_{\text{in}} = 10$ mgP L^{-1}.

34.6. Isle Royale National Park is a 210-square-mile archipelago composed of a single large island and many small islands in Lake Superior. Moose arrived around 1900 and by 1930 their population approached 3000, thus ravaging vegetation. In 1949, wolves crossed an ice bridge from the Canadian Province of Ontario. Since the late 1950s the numbers of the moose and wolves have been tracked (Allen 1973, Peterson et al. 1984).

Year	Moose	Wolves	Year	Moose	Wolves
1960	700	22	1972	836	23
1961	—	22	1973	802	24
1962	—	23	1974	815	30
1963	—	20	1975	778	41
1964	—	25	1976	641	43
1965	—	28	1977	507	33
1966	881	24	1978	543	40
1967	—	22	1979	675	42
1968	1000	22	1980	577	50
1969	1150	17	1981	570	30
1970	966	18	1982	590	13
1971	674	20	1983	811	23

(*a*) Integrate the Lotka-Volterra equations from 1960 through 2020 with the fourth-order RK method. Determine the coefficient values that yield an optimal fit. Compare your simulation with the data using a time-series approach, and comment on the results.

(*b*) Plot the simulation of (*a*), but use a state-space approach.

(*c*) After 1993 suppose that the wildlife managers trap one wolf per year and transport it off the island. Predict how the populations of both wolves and moose would evolve to the year 2020. Present your results as both time series and state-space plots. For this case [as well as for (*d*)], use the following coefficients: $a = 0.3$, $b = 0.01111$, $c = 0.2106$, and $d = 0.0002632$.

(*d*) Suppose that in 1993 some poachers sneak onto the island and kill 50% of the moose. Predict how the populations of both wolves and moose would evolve to the year 2020. Present your results as both time-series and state-space plots.

LECTURE 35

Nutrient/Food-Chain Modeling

LECTURE OVERVIEW: I combine the material from the preceding two lectures into a complete framework to compute nutrient/food-chain interactions for a stratified lake.

In this lecture a simple model of a two-layer lake is used to illustrate how the nutrient/food-chain framework developed at the end of the previous lecture can be applied to a natural water. It should be emphasized that the framework described below is but one of a variety of formulations proposed to model nutrient/food-chain dynamics in natural waters. The reader also can consult Scavia's (1979) review of the subject, as well as critical articles by Riley (1963), Steele (1965), and Mortimer (1975*b*).

35.1 SPATIAL SEGMENTATION AND PHYSICS

The present framework handles physical segmentation (two vertical layers), loadings, and transport in the same way as the two-component model (SSA) described in Sec. 29.5. Mass balances for a substance in the epilimnion (1) and the hypolimnion (2) can be written as (Fig. 35.1)

$$V_1 \frac{dc_1}{dt} = W(t) - Qc_1 + v_t A_t (c_2 - c_1) + S_1 \tag{35.1}$$

and

$$V_2 \frac{dc_2}{dt} = v_t A_t (c_1 - c_2) + S_2 \tag{35.2}$$

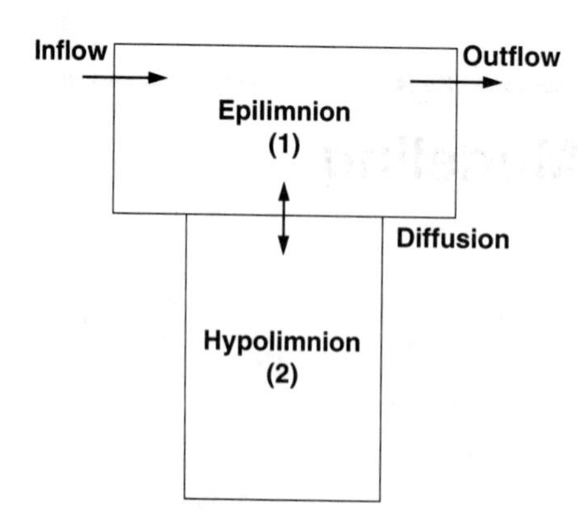

FIGURE 35.1
Physical segmentation scheme and transport representation.

where V = volume
c = concentration
t = time
$W(t)$ = loading
Q = outflow
v_t = thermocline transfer coefficient
A_t = thermocline area
S = sources and sinks

35.2 KINETIC SEGMENTATION

As depicted in Fig. 35.2, the model consists of eight state-variables. These can be divided into three major groups (Table 35.1). Mass-balance equations are written for each of the state-variables for each of the layers. The source and sink terms for each state variable are described in this section.

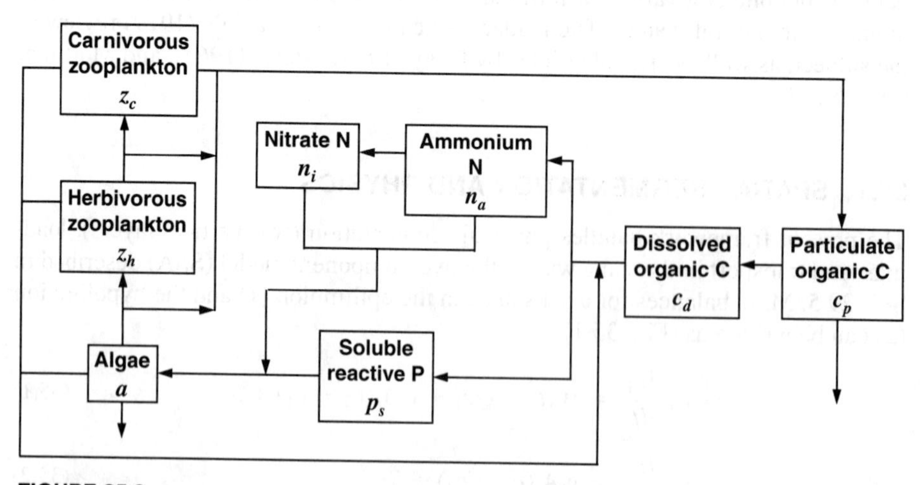

FIGURE 35.2
Kinetic segmentation.

TABLE 35.1
Model state-variables

State-variable	Symbol	Units
Food chain:		
1. Algae	a	mgChl$a$ m^{-3}
2. Herbivorous zooplankton	z_h	gC m^{-3}
3. Carnivorous zooplankton	z_c	gC m^{-3}
Nonliving organic carbon:		
4. Particulate	c_p	gC m^{-3}
5. Dissolved	c_d	gC m^{-3}
Nutrients:		
6. Ammonium nitrogen	n_a	mgN m^{-3}
7. Nitrate nitrogen	n_i	mgN m^{-3}
8. Soluble reactive phosphorus	p_s	mgP m^{-3}

All the equations in Secs. 35.2.1 and 35.2.2 are written for a generic layer. Thus they can be applied to either the epilimnion or the hypolimnion. The only cases where the equations for the layers would differ is for the variables subject to settling (algae and particulate organic carbon). For these variables the following equations are written for a layer that is overlain by another layer. As such they apply strictly to the hypolimnion, which receives settling gains from the epilimnion. When applied to the surface layer, these settling gains are merely omitted from the equation.

35.2.1 Food Chain

The food chain consists of a single plant group along with two zooplankton groups.

Algae. As described in Lec. 33, algae grow as a function of temperature, nutrients, and solar radiation. For the hypolimnion, algae are also gained by settling from the surface layer. Sinks include respiration/excretion, grazing, and settling losses:

$$V\frac{da}{dt} = k_g(T, n_t, p_s, I)Va - k_{ra}(T)Va - C_{gh}(T, a, z_h)Va + v_a A_t a_u - v_a A_t a \tag{35.3}$$

where $k_g(T, n_t, p_s, I)$ = algal growth rate (d^{-1})
$k_{ra}(T)$ = losses due to respiration and excretion (d^{-1})
$C_{gh}(T, a, z_h)$ = grazing losses (d^{-1})
v_a = phytoplankton settling velocity (m d^{-1})

and the subscript u designates an upper layer.

The growth and the grazing rates are dependent on environmental factors, as in (recall Eqs. 33.29 and 34.18)

$$k_g(T, n_t, p_s, I) = k_{g,20}1.066^{T-20}\left[\frac{2.718f}{k_e H}\left(e^{-\alpha_1} - e^{-\alpha_0}\right)\right]\min\left(\frac{n_t}{k_{sn} + n_t}, \frac{p_s}{k_{sp} + p_s}\right) \tag{35.4}$$

and
$$C_{gh}(T, a, z_h) = \frac{a}{k_{sa} + a} C_{gh} \theta^{T-20} z_h \tag{35.5}$$

where n_t = total inorganic nitrogen = $n_a + n_i$ and other parameters are defined in Lecs. 33 and 34.

In addition the respiration rate is corrected for temperature with the theta model,

$$k_{ra}(T) = k_{ra,20} \theta^{T-20} \tag{35.6}$$

All other temperature-dependencies are corrected in this fashion.

Herbivorous zooplankton. Part of the consumed algae is converted into herbivorous zooplankton. The herbivores are depleted by carnivore grazing and respiration/excretion losses:

$$V \frac{dz_h}{dt} = a_{ca} \varepsilon_h C_{gh}(T, a, z_h) V a - C_{gc}(T, z_c) V z_h - k_{rh}(T) V z_h \tag{35.7}$$

where a_{ca} = the stoichiometric coefficient for the conversion of algal chlorophyll a to zooplankton carbon (gC mgChla^{-1}).

Carnivorous zooplankton. Part of the consumed herbivorous zooplankton is converted into carnivorous zooplankton. The carnivores are depleted by respiration/excretion losses and a first-order death due to grazing by organisms higher on the food chain (primarily fish),

$$V \frac{dz_c}{dt} = \varepsilon_c C_{gc}(T, z_c) V z_h - k_{rc}(T) V z_c - k_{dc}(T) V z_c \tag{35.8}$$

35.2.2 Nonliving Organic Carbon

The nonliving organic carbon is divided into particulate and dissolved fractions in order to distinguish between settleable and nonsettleable forms.

Particulate. Inefficient grazing (egestion) along with carnivore death result in gains to the particulate nonliving organic carbon pool. For the hypolimnion, POC is also gained by settling from the surface layer. Sinks include a first-order dissolution reaction and settling losses:

$$V \frac{dc_p}{dt} = a_{ca}(1 - \varepsilon_h) C_{gh}(T, a, z_h) V a + (1 - \varepsilon_c) C_{gc}(T, z_c) V z_h + k_{dc}(T) V z_c$$
$$- k_p(T) V c_p + v_p A_t c_{pu} - v_p A_t c_p \tag{35.9}$$

Dissolved. DOC is gained via the first-order dissolution reaction and lost by hydrolysis:

$$V \frac{dc_d}{dt} = k_p(T) V c_p - k_h(T) V c_d \tag{35.10}$$

35.2.3 Nutrients

The nutrients are divided into inorganic nitrogen and phosphorus. The former is also split into ammonium and nitrate nitrogen.

Ammonium nitrogen. Ammonium ion is gained due to hydrolysis of dissolved organic carbon and from food-chain respiration. It is lost via plant uptake and nitrification:

$$V\frac{dn_a}{dt} = a_{nc}k_h(T)Vc_d + a_{na}k_{ra}(T)Va + a_{nc}k_{rh}(T)Vz_h + a_{nc}k_{rc}(T)Vz_c - F_{am}a_{na}k_g(T, n_t, p_s, I)Va - k_n(T)Vn_a \tag{35.11}$$

where F_{am} = fraction of inorganic nitrogen that is taken from the ammonium pool by plant uptake,

$$F_{am} = \frac{n_a}{k_{am} + n_a} \tag{35.12}$$

in which k_{am} = a half-saturation constant for ammonium preference, and a_{nc} and a_{na} = the ratios of nitrogen to carbon and chlorophyll, respectively.

Nitrate nitrogen. Nitrate is gained due to nitrification and it is lost via plant uptake:

$$V\frac{dn_i}{dt} = k_n(T)Vn_a - (1 - F_{am})a_{na}k_g(T, n_t, p_s, I)Va \tag{35.13}$$

Soluble reactive phosphorus. SRP is gained due to hydrolysis of dissolved organic carbon and from food-chain respiration. It is lost via plant uptake:

$$V\frac{dp_s}{dt} = a_{pc}k_h(T)Vc_d + a_{pa}k_{ra}(T)Va + a_{pc}k_{rh}(T)Vz_h + a_{pc}k_{rc}(T)Vz_c - a_{pa}k_g(T, n_t, p_s, I)Va \tag{35.14}$$

where a_{pc} and a_{pa} = the ratios of phosphorus to carbon and chlorophyll, respectively.

35.3 SIMULATION OF THE SEASONAL CYCLE

By using the kinetic interactions from Eqs. 35.3 through 35.14, we can write Eqs. 35.1 and 35.2 for each of the eight state-variables. The resulting 16 ordinary differential equations can be integrated simultaneously using a numerical method such as the fourth-order Runge-Kutta method.

EXAMPLE 35.1. NUTRIENT/FOOD-CHAIN MODEL FOR LAKE ONTARIO. Physical parameters for Lake Ontario during the early 1970s are summarized in Table 35.2 and Fig. 35.3. Note that the hypolimnetic temperatures are simulated with a heat-balance model (recall Lec. 31). Loads and initial conditions are summarized in Table 35.3, and model parameters are listed in Table 35.4.

TABLE 35.2
Hydrogeometric parameters for Lake Ontario

Parameter	Symbol	Value	Units
Surface area	A_s	$19{,}000 \times 10^6$	m^2
Thermocline area	A_t	$10{,}000 \times 10^6$	m^2
Epilimnion volume	V_e	254×10^9	m^3
Hypolimnion volume	V_h	$1{,}380 \times 10^9$	m^3
Thermocline thickness	H_t	7	m
Epilimnion thickness	H_e	17	m
Hypolimnion thickness	H_h	69	m
Outflow	Q	212×10^9	$m^3\ yr^{-1}$
Thermocline diffusion	v_t		$cm^2\ s^{-1}$
Summer-stratified		0.13	
Winter-mixed		13	
Start of summer stratification		100	d
Time to establish stratification		58	d
Onset of end of stratification		315	d
End of stratification		20	d

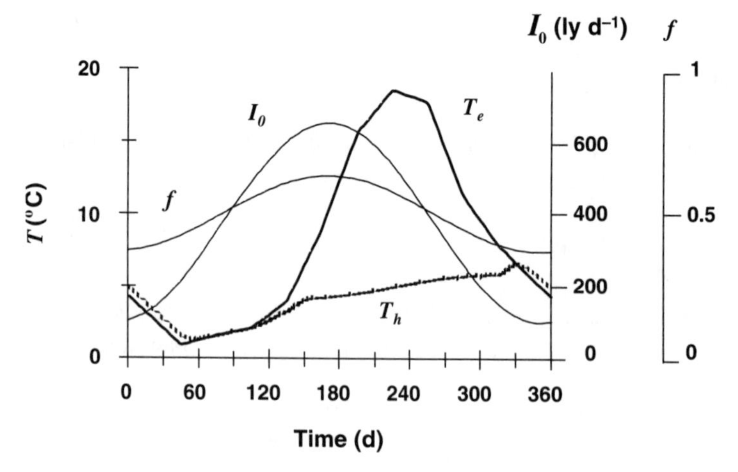

FIGURE 35.3
Temperature and light data for Lake Ontario.

TABLE 35.3
Boundary (loading) and initial conditions for Lake Ontario in the early 1970s

Variable	Units	Loading†	Initial conditions‡
Algae	μgChl*a* L^{-1}	1	1
Herbivorous zooplankton	mgC L^{-1}	0	0.005
Carnivorous zooplankton	mgC L^{-1}	0	0.005
Particulate organic carbon	mgC L^{-1}	0.8	0.12
Dissolved organic carbon	mgC L^{-1}	0.8	0.12
Ammonium	μgN L^{-1}	15	15
Nitrate	μgN L^{-1}	220	250
Soluble reactive phosphorus	μgP L^{-1}	14.3	12

† Multiply by outflow to convert to mass-loading rate.
‡ Same for epilimnion and hypolimnion.

TABLE 35.4
Model parameter values used for Lake Ontario simulation in the early 1970s

Parameter	Symbol	Value	Units
Algae:			
Growth rate	$k_{g,20}$	2	d^{-1}
Temperature factor	θ_a	1.066	
Respiration rate	k_{ra}	0.025	d^{-1}
Temperature factor	θ_{ra}	1.08	
Settling velocity	v_a	0.2	$m\ d^{-1}$
Optimal light	I_s	350	$ly\ d^{-1}$
P half-saturation	k_{sp}	2	$\mu gP\ L^{-1}$
N half-saturation	k_{sn}	15	$\mu gN\ L^{-1}$
Background light extinction	k_e'	0.2	m^{-1}
Herbivorous zooplankton:			
Grazing rate	C_{gh}	5	$L\ mgC^{-1}\ d^{-1}$
Temperature factor	θ_{gh}	1.08	
Grazing efficiency	ε_h	0.7	
Respiration rate	k_{rh}	0.1	d^{-1}
Temperature factor	θ_{rh}	1.08	
Algae half-saturation	k_{sa}	10	$\mu gChl a\ L^{-1}$
Carnivorous zooplankton:			
Grazing rate	C_{gh}	5	$L\ mgC^{-1}\ d^{-1}$
Temperature factor	θ_{gh}	1.08	
Grazing efficiency	ε_c	0.7	
Respiration rate	k_{rc}	0.04	d^{-1}
Temperature factor	θ_{rc}	1.08	
Death rate	k_{dc}	0.04	d^{-1}
Temperature factor	θ_{dc}	1.08	
Herbivore half-saturation	k_{sh}	0.4	$mgC\ L^{-1}$
Nonliving carbon:			
Particulate settling	v_p	0.2	$m\ d^{-1}$
Dissolution rate	k_p	0.1	d^{-1}
Temperature factor	θ_p	1.08	
Hydrolysis rate	k_h	0.075	d^{-1}
Temperature factor	θ_h	1.08	
Nutrients:			
Nitrification rate	k_n	0.1	d^{-1}
Temperature factor	θ_n	1.08	
Ammonia preference half-saturation	k_{am}	50	$\mu gN\ L^{-1}$

Solution: The model equations were solved numerically for 2 yr. Some of the results for the second year are summarized in Fig. 35.4. Figure 35.4*a* shows results for the food chain in the epilimnion. Note that the zooplankton are expressed in chlorophyll units to allow comparison among the variables. The results indicate that predator-prey interactions are taking place, with the peaks of the algae, herbivores, and carnivores occurring at approximately days 160, 180, and 240, respectively.

Results for epilimnetic inorganic nitrogen and phosphorus are illustrated in Fig. 35.4*b*. According to the calculation the lake is overwhelmingly phosphorus-limited

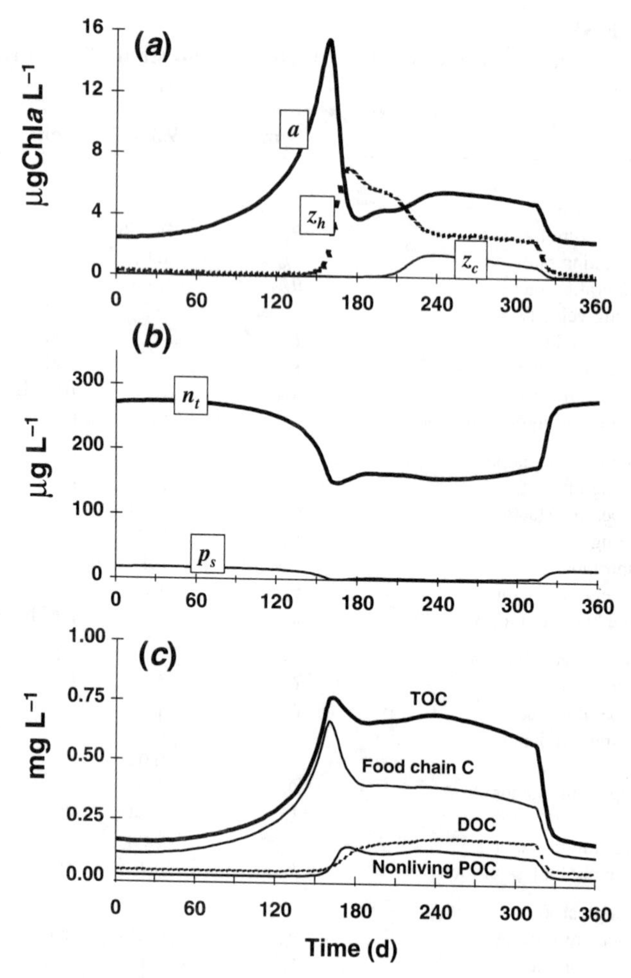

FIGURE 35.4
Simulation results for the epilimnion of Lake Ontario as computed with a nutrient/food-chain model. (*a*) Food chain with all components expressed as chlorophyll, (*b*) inorganic nutrients, and (*c*) organic carbon.

during the summer stratified period. The fact that there is excess nitrogen and phosphorus in the water reflects the low growth during the remainder of the year.

The organic carbon cycle for the epilimnion is shown in Fig. 35.4*c*. This plot indicates a clear difference between the productive summer months and the rest of the year. High organic carbon generation during the stratified period has several implications. First, its eventual decomposition can have an impact on the oxygen content of bottom waters. This would be particularly important for water bodies with smaller hypolimnion (and hence smaller oxygen reserves) than for Lake Ontario. Second, as we will see in later lectures, the transport and fate of toxic substances can be strongly associated with organic matter.

The foregoing example illustrates the three primary strengths of nutrient/food-chain models:

- *Temporal resolution of eutrophication effects.* The models provide predictions of features like peak chlorophyll levels and, hence, produce information that is extremely useful to water-quality managers. This is because the public is often most concerned with extreme events in a water body rather than in seasonal or long-term average conditions. Because of its daily time step, the nutrient/food-chain models are capable of generating both types of information.
- *Nutrient and light limitation.* Identifying the limiting nutrient or whether the system has significant light limitation is a critical step in controlling eutrophication. By mechanistically including several nutrients and light, the nutrient/food-chain models provide a means to make such identifications.
- *Organic carbon cycle.* By providing predictions of organic carbon levels, nutrient/food-chain frameworks provide a means to assess both oxygen and toxic substances in lakes.

35.4 FUTURE DIRECTIONS

The framework described in this lecture can be extended and refined in a number of ways. First, it can be extended to other water bodies such as elongated lakes, streams, and estuaries. Second, additional nutrients and food-chain compartments can be included. For example the algae might be divided into several functional groups (e.g., diatoms, blue-greens, greens, etc.), or new types of organisms such as bacteria or rooted plants might be included.

Over the coming decade one fundamental way in which the nutrient/food-chain framework could be modified is to shift the focus from chlorophyll to organic carbon. Because of concern over eutrophication, early nutrient/food-chain models focused on plant growth (Fig. 35.5*a*). For such applications, chlorophyll *a* served as a convenient measure of the effects of enhanced productivity.

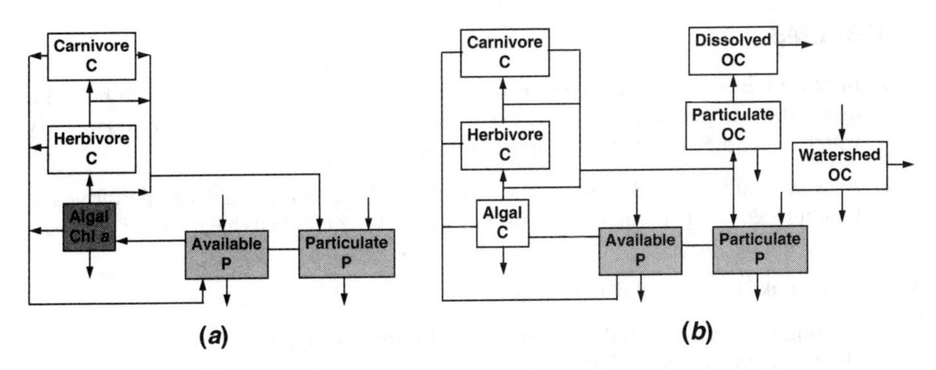

FIGURE 35.5
(*a*) Traditional nutrient/food-chain kinetics have used chlorophyll as their focus. (*b*) In the future, such production/decomposition models may shift toward a more detailed characterization of organic carbon.

TABLE 35.5
The differing reactivity of autochthonous particles in natural waters can be modeled by dividing them into "G classes" of differing reactivity (Berner 1980, Westrich and Berner 1984, Di Toro and Fitzpatrick 1993)

Particle type	Fraction of fresh particles	Decay rate (d^{-1})
G_1 (rapidly reacting)	0.65	0.035
G_2 (slowly reacting)	0.20	0.0018
G_3 (refractory)	0.15	0

Today, water-quality managers have broadened their concerns to encompass problems other than eutrophication, such as toxic substances and disinfection by-products. In such cases both dissolved and particulate organic carbon must be characterized. Further, as already discussed in Lec. 33, because of variable chlorophyll-to-carbon ratios, chlorophyll *a* may not even represent an adequate measure of particulate plant carbon. Consequently, as depicted in Fig. 35.5*b*, organic carbon may become the focus of newer modeling frameworks.

Further, distinctions will have to be made between easily decomposable and refractory forms of carbon. One case where this could be important would be short-residence-time impoundments that are dominated by allochthonous sources. Another relates to the simulation of sediment-water interactions. In such cases the distinction among varying decomposing fractions is important (Berner 1980).

Di Toro and Fitzpatrick (1993) have adopted a scheme devised for sediment diagenesis by Berner (1980) and Westrich and Berner (1984) to handle different reactivity classes of particulate matter. As listed in Table 35.5, fresh particulate organic carbon is divided into three "G fractions" corresponding to easily degradable, slowly degradable, and nondegradable fractions. By tracking each fraction separately (that is, with separate mass balances), the differential degradation can be determined.

PROBLEMS

35.1. Increase the loading of phosphorus to Lake Ontario until the spring bloom becomes limited by nitrogen or light. Express your result as an inflow concentration and identify which factor dictates the maximum algal level.

35.2. As described in Lec. 18, perform a sensitivity analysis on the simulation performed in Example 35.1. In particular determine which kinetic coefficients have the major impact on the timing and peak of the spring phytoplankton bloom.

35.3. Green Lake has the following characteristics:

Epilimnion volume $= 150 \times 10^6\ m^3$
Hypolimnion volume $= 600 \times 10^6\ m^3$
Thermocline area $= 10 \times 10^6\ m^2$
Thermocline thickness $= 3$ m
Time to establish strat. $= 30$ d
Time to end strat. $= 30$ d

Epilimnion depth $= 10$ m
Surface area $= 20 \times 10^6\ m^2$
Inflow $=$ outflow $= 150 \times 10^6\ m^3\ yr^{-1}$
Start stratification $= 150$ d
Begin. of destrat. $= 300$ d
Extinction due to water/color $= 0.2\ m^{-1}$

Data from the lake has been collected and is summarized in Table P35.3. The pond's inflow has available and unavailable phosphorus concentrations of 20 μgP L^{-1}. Assume that the lake is never limited by nitrogen. Use the same solar radiation and photoperiod as for Lake Ontario (Fig. 35.3).

(*a*) Calibrate the nutrient/food-chain model developed in this lecture to this data set.

(*b*) Determine the inflow concentration of available P needed so that the peak epilimnetic chlorophyll level does not exceed 10 μg L^{-1}.

(*c*) Determine the washout rate.

TABLE P35.3

Day	T_e (°C)	T_h (°C)	a_e (μg L^{-1})	z_{he} (mgC L^{-1})	z_{ce} (mgC L^{-1})	SRP$_e$ (μgP L^{-1})	p_{une}† (μgP L^{-1})	SRP$_h$ (μgP L^{-1})	p_{unh}† (μgP L^{-1})
15	3	4.3	4.9	0.01	0.01	13.5	1.1	15	1
45	1	2.3	6.5						
75	2	1.8	9.7						
105	2	2	12	0.05		4.0	1.8	9.0	1.6
135	4	2.8	8.7	0.2	0.005	2.8	3	7.5	2.7
165	7	5	7	0.2	0.04	1.6	4	8.5	3.4
195	15	5.5	6.8	0.07	0.16	0.4	4.3	12.7	2.7
225	20	5.7	7	0.07	0.12	0.4	3.6	15.7	1.8
255	20	5.9	6.6	0.07	0.105	0.49	3.3	17.6	1.3
285	12	6	6.3						
315	8	6.5	6.5						
345	5	6.3	4.8						

†Unavailable phosphorus.

LECTURE 36

Eutrophication in Flowing Waters

LECTURE OVERVIEW: I cover some material related to the modeling of eutrophication in flowing waters. In the first part I focus on phytoplankton models. After describing the fundamentals of simulating phytoplankton-nutrient interactions in rivers, I describe how EPA's QUAL2E model can be used to simulate steady-state eutrophication due to phytoplankton growth in vertically and laterally well-mixed streams and estuaries. I illustrate how temperature is modeled and then outline the kinetics of nutrients and plants and how they are simulated with QUAL2E. Finally I describe how fixed plants are modeled.

There are two major reasons why special modeling frameworks are necessary for flowing waters such as streams and estuaries. First, and most obviously, their physics are fundamentally different from standing waters such as lakes. In particular, horizontal transport must be considered. Second, shallower flowing waters can be dominated by fixed plants such as macrophytes and periphyton rather than the free-floating phytoplankton that we have emphasized to this point.

This lecture deals with both topics. First, I show how the phytoplankton models described in earlier lectures can be applied to deeper streams and estuaries. Then I describe some approaches for modeling fixed plants in shallower systems such as streams.

36.1 STREAM PHYTOPLANKTON/NUTRIENT INTERACTIONS

In this section I show how the lake phytoplankton models developed in earlier sections can be applied to flowing waters. I first describe a simple back-of-the-envelope calculation that is helpful in placing the subsequent numerical solutions in perspective.

36.1.1 Simple Phytoplankton/Nutrient Analysis

Thomann and Mueller (1987) have developed a simple analysis that is extremely useful for understanding the impact of point sources on stream eutrophication.

First, the analysis addresses the question of whether nutrients will be limiting in the immediate vicinity of a sewage outfall into a river. Recall from Table 28.3 that, with the exception of cases where phosphorus removal is implemented, point sources of sewage are generally nitrogen-limited. Further, nonpoint sources are usually phosphorus-limited. For cases where the receiving river above the point source is dominated by nonpoint sources, Thomann and Mueller used a simple mass balance at the outfall (recall Eq. 9.42) to draw the following general conclusions:

- If effluent flow is less than 2% of the total river flow after mixing, phosphorus would tend to be the limiting nutrient downstream.
- For effluent-to-flow ratios greater than 2%, the limiting nutrient would depend on the treatment process. In particular, if little or no phosphorus removal were used, the stream would tend to be nitrogen-limited.
- In all cases, if significant phosphorus removal is used, the stream would tend to be phosphorus-limited.

Next they investigated whether nutrient limitation would occur at the outfall. To do this they recognized that the Michaelis-Menten formulation for nutrient limitation becomes significant only when nutrient concentrations fall below a value that is 5 times the saturation constant (recall Fig. 32.2). Thus, assuming $k_{sn} = 10$ to $20\ \mu\text{gN L}^{-1}$ and $k_{sp} = 1$ to $5\ \mu\text{gP L}^{-1}$, they estimated that nitrogen and phosphorus limitation would occur only when concentrations fell below about $0.1\ \text{mgN L}^{-1}$ and $0.025\ \text{mgP L}^{-1}$, respectively. Since sewage typically has nutrient concentrations in the mg L^{-1} range (untreated sewage has about $40\ \text{mgN L}^{-1}$ and $10\ \text{mgP L}^{-1}$), they concluded that point-source-dominated streams (that is, those with effluent-to-flow ratios greater than about 1%) without significant nutrient removal would have excess nutrients at the mixing point.

Even though nutrients are in excess at the mixing point, subsequent plant growth downstream will tend to lower nutrient concentrations downstream. Thus the stream could eventually become nutrient-limited downstream. Two questions follow from this reasoning: At what distance downstream will the stream become nutrient-limited? Which nutrient will become limiting first?

To address these questions, Thomann and Mueller developed the following steady-state mass balances for a downstream stretch with constant hydrogeometric properties:

$$U\frac{da}{dx} = \frac{da}{dt*} = \left[k_g(T,\ I) - k_d - \frac{v_a}{H}\right]a = k_{\text{net}}a \tag{36.1}$$

where $t*$ = travel time (d) and k_{net} = net gain rate of phytoplankton (d^{-1}), which as shown, depends on growth, kinetic losses (respiration, excretion, death), and settling. Notice that the growth rate does not depend on nutrients as per our previous discussions. Thus the following analysis will apply only when $n \geq 0.1\ \text{mgN L}^{-1}$ and $p \geq 0.025\ \text{mgP L}^{-1}$.

If we assume a constant plant stoichiometry, balances for nitrogen and phosphorus follow,

$$\frac{dn}{dt*} = -a_{na} k_g(T, I)a \tag{36.2}$$

$$\frac{dp}{dt*} = -a_{pa} k_g(T, I)a \tag{36.3}$$

Observe that both kinetic losses and settling are assumed to be terminal losses; that is, they do not feed back to the nutrient pools.

Using boundary conditions of $a = a_0$, $n = n_0$, and $p = p_0$ at the mixing point, we can solve these equations

$$a = a_0 e^{k_{net} t*} \tag{36.4}$$

$$n = n_0 + \frac{a_{na} k_g a_0}{k_{net}} \left(1 - e^{k_{net} t*}\right) \tag{36.5}$$

$$p = p_0 + \frac{a_{pa} k_g a_0}{k_{net}} \left(1 - e^{k_{net} t*}\right) \tag{36.6}$$

The three equations are displayed in Fig. 36.1 for a stream with a significant effluent-to-flow ratio. The simulation also reflects a positive net phytoplankton growth, a significantly high algal initial condition, and an effluent with a low N:P ratio. As expected from Eqs. 36.4 through 36.6, the nutrients drop as the phytoplankton grow.

Two other features of the plot bear mention. First, nitrogen becomes limiting before phosphorus when the nitrogen level falls below 100 μgN L^{-1} at a travel time of about 7 d downstream from the source. Second, if there were more nitrogen in the effluent so that N-limitation did not occur, the plot also indicates that phosphorus

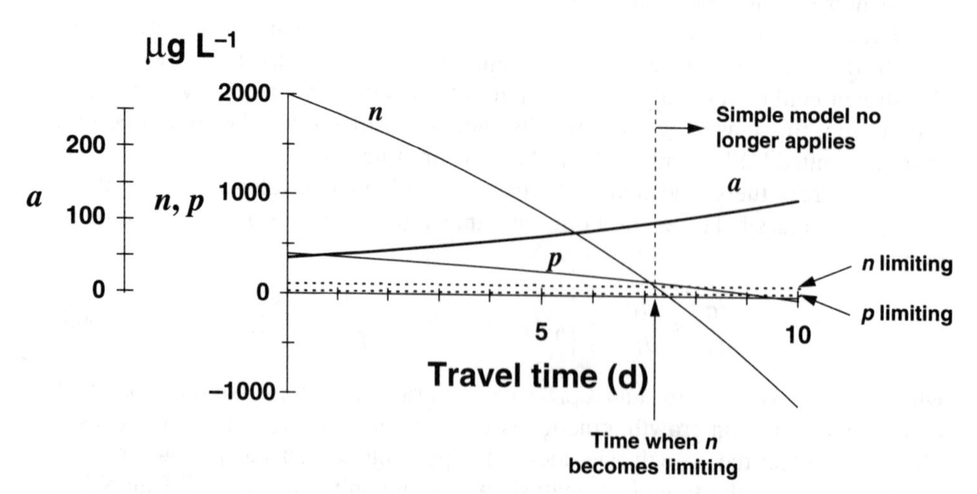

FIGURE 36.1
Plot of nutrient and phytoplankton concentrations downstream from a point source.

limitation would set in after about 8.6 d. For a stream with an average velocity of 0.15 to 0.3 mps (0.5 to 1 fps), a week's travel time corresponds to about 90 to 180 km (60 to 115 mi).

Note that Thomann and Mueller have shown that Eqs. 36.5 and 36.6 can be used to determine the critical travel times exactly,

$$t_n^* = \frac{1}{k_{net}} \ln\left(\frac{n_0' + n_0 - 25}{n_0'}\right) \qquad \text{where } n_0' = \frac{a_{an} k_g n_0}{k_{net}} \tag{36.7}$$

$$t_p^* = \frac{1}{k_{net}} \ln\left(\frac{p_0' + p_0 - 100}{p_0'}\right) \qquad \text{where } p_0' = \frac{a_{ap} k_g p_0}{k_{net}} \tag{36.8}$$

where t_n^* and t_p^* = travel times (d) for nitrogen and phosphorus limitation, respectively.

Aside from telling us where nutrient limitation begins, critical travel times also indicate whether nutrient load reduction will even have an impact on plant growth. Suppose that the stream length of interest is less than both t_n^* and t_p^*. The analysis indicates that initial nutrient load reductions would not have an impact on eutrophication. Of course, at a certain point, load reductions could bring the critical travel times within the stream length of interest. The equations could provide an estimate of the level of treatment required for this point to be reached.

36.1.2 Stream and Estuary Phytoplankton Models

Phytoplankton kinetics (Fig. 36.2) can be integrated into a general model for a one-dimensional stream or estuary, as in

$$\frac{\partial a}{\partial t} = E\frac{\partial^2 a}{\partial x^2} - U\frac{\partial a}{\partial x} + (k_g - k_{ra})a - \frac{v_a}{H}a \tag{36.9}$$

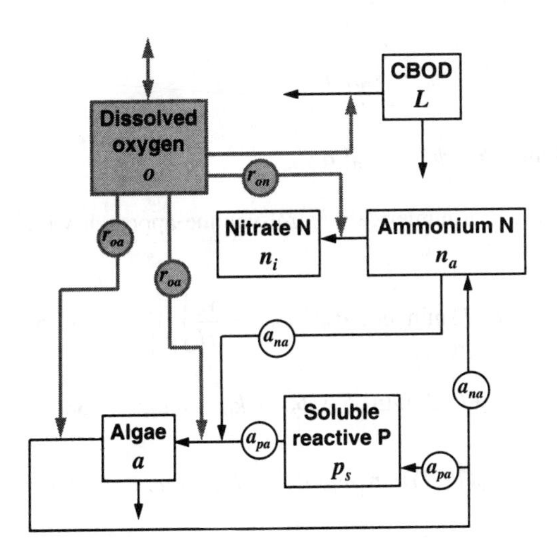

FIGURE 36.2
A simple oxygen balance for a stream including phytoplankton growth.

$$\frac{\partial L}{\partial t} = E\frac{\partial^2 L}{\partial x^2} - U\frac{\partial L}{\partial x} - k_d L - \frac{v_s}{H}L \tag{36.10}$$

$$\frac{\partial p}{\partial t} = E\frac{\partial^2 p}{\partial x^2} - U\frac{\partial p}{\partial x} - a_{pa}(k_g - k_{ra})a \tag{36.11}$$

$$\frac{\partial n_a}{\partial t} = E\frac{\partial^2 n_a}{\partial x^2} - U\frac{\partial n_a}{\partial x} - a_{na}(k_g - k_{ra})a - k_n n_a \tag{36.12}$$

$$\frac{\partial n_i}{\partial t} = E\frac{\partial^2 n_i}{\partial x^2} - U\frac{\partial n_i}{\partial x} + k_n n_a \tag{36.13}$$

$$\frac{\partial o}{\partial t} = E\frac{\partial^2 o}{\partial x^2} - U\frac{\partial o}{\partial x} - k_d L - r_{on}k_n n_a + r_{oa}(k_g - k_{ra})a + k_a(o_s - o) \tag{36.14}$$

We have developed this representation so that it is directly comparable to the back-of-the-envelope model described in the previous section. Thus it is a somewhat simplified version of a full-blown stream eutrophication model. For example it assumes that the plants can utilize only ammonium as a nitrogen source. Further it ignores the organic nitrogen and phosphorus pools that are included in more complete representations. Of course these mechanisms, as well as others such as sediment-water transfers, can be readily added. The resulting set of equations can be solved numerically, as described in the following example.

EXAMPLE 36.1. NUMERICAL STREAM PHYTOPLANKTON SIMULATION. Duplicate the calculations in Fig. 36.1, but use nutrient limitation explicitly.

Solution: The model can be made directly comparable to Thomann and Mueller's approach by omitting BOD and oxygen and combining ammonium and nitrate in a single inorganic nitrogen pool. The resulting mass balances can be written for a plug-flow stream with no light limitation as

$$\frac{da}{dt} = -U\frac{da}{dx} + \left\{k_g(T, I)\min\left[\phi_p, \phi_n\right] - k_{ra}\right\} a - \frac{v_a}{H}a$$

$$\frac{dn}{dt} = -U\frac{dn}{dx} - a_{na}\left\{k_g(T, I)\min\left[\phi_p, \phi_n\right] - k_{ra}\right\} a$$

$$\frac{dp}{dt} = -U\frac{dp}{dx} - a_{pa}\left\{k_g(T, I)\min\left[\phi_p, \phi_n\right] - k_{ra}\right\} a$$

These equations can also be expressed in a numerical control-volume approach with backward spatial differences as

$$V_i\frac{da_i}{dt} = Q_{i-1,i}a_{i-1} - Q_{i,i+1}a_i + \left\{k_g(T, I)\min\left[\phi_p, \phi_n\right] - k_{ra} - \frac{v_a}{H}\right\}_i V_i a_i \tag{36.15}$$

$$V_i\frac{dn_i}{dt} = Q_{i-1,i}n_{i-1} - Q_{i,i+1}n_i - a_{na}\left\{k_g(T, I)\min\left[\phi_p, \phi_n\right] - k_{ra}\right\}_i V_i a_i \tag{36.16}$$

$$V_i\frac{dp_i}{dt} = Q_{i-1,i}p_{i-1} - Q_{i,i+1}p_i - a_{pa}\left\{k_g(T, I)\min\left[\phi_p, \phi_n\right] - k_{ra}\right\}_i V_i a_i \tag{36.17}$$

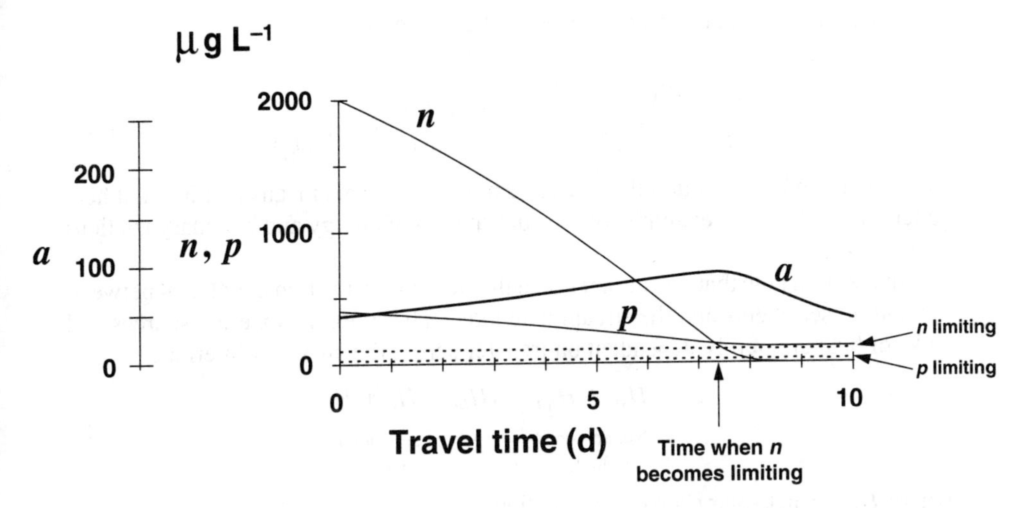

FIGURE E36.1

where i = a control volume. Along with appropriate boundary conditions, these three equations can then be integrated numerically using the solution techniques presented in Lecs. 11 through 13. Figure E36.1 shows the resulting steady-state profiles employing parameter values comparable to those used to construct Fig. 36.1.

Notice how this model behaves in an identical fashion to Fig. 36.1, up to the point that nitrogen becomes limiting at about $t* = 7.3$ d. Whereas the simple model breaks down beyond that point, the numerical model is valid because the nutrient limitation term shuts down uptake as nitrogen levels drop. Consequently both the nitrogen and phosphorus level off rather than going negative. Further the phytoplankton begin to decline since growth falls below the respiration and settling losses.

The foregoing is intended to show how nutrient/food-chain kinetics can be integrated into a one-dimensional, advective-dispersive transport framework. Now we will describe how the QUAL2E framework performs a similar calculation.

36.2 MODELING EUTROPHICATION WITH QUAL2E

This section shows how EPA's QUAL2E model can be used to simulate both temperature and nutrient/algae dynamics in flowing waters.

36.2.1 Temperature and QUAL2E

Recall that the general mass transport equation for QUAL2E (Eq. 26.28) is

$$\frac{\partial c}{\partial t} = \frac{\partial\left(A_x E \frac{\partial c}{\partial x}\right)}{A_x \partial x} dx - \frac{\partial(A_x U c)}{A_x \partial x} dx + \frac{dc}{dt} + \frac{s}{V} \tag{36.18}$$

In a similar fashion a heat balance can be written as

$$\frac{\partial T}{\partial t} = \frac{\partial\left(A_x E \frac{\partial T}{\partial x}\right)}{A_x \partial x} dx - \frac{\partial(A_x U T)}{A_x \partial x} dx + \frac{s}{\rho C V} \tag{36.19}$$

Notice that we have omitted the source term—dT/dt. This means that internal heat generation or loss (for example viscous dissipation of energy and boundary friction) is negligible.

In addition note that, because it is usually negligible, the transfer of heat between the bottom sediments and the stream is ignored. Therefore the external sources and sinks of heat are purely dependent on transfer across the air-water interface,

$$s = \underbrace{H_{sn} + H_{an}}_{\text{Net absorbed radiation}} - \underbrace{(H_{br} + H_c + H_e)}_{\text{Water-dependent terms}} \tag{36.20}$$

where H_{sn} = net solar shortwave radiation
H_{an} = net atmospheric longwave radiation
H_{br} = longwave back radiation from the water
H_c = conduction
H_e = evaporation

The individual terms in Eq. 36.20 are calculated in a fashion similar to the scheme outlined in Lec. 30. However, it should be noted that solar radiation is calculated internally based on parameters such as latitude and time of year. This is a really nice feature of the model since it obviates the need for the user to obtain such information independently.

Equation 36.20 can be substituted into Eq. 36.19 to yield the final heat balance. Within QUAL2E several additional modifications are made to (1) express the conduction exchange as a linear function of water temperature and to (2) linearize the nonlinear heat exchange terms such as back radiation and evaporation. When these modifications are made, the heat budget can be computed using QUAL2E's efficient solution algorithm.

An input file has been developed to simulate temperature for the simple CBOD/oxygen/SOD system described in detail previously in Lec. 26. This file is displayed in Fig. 36.3. The cards that must be modified to model temperature are highlighted in Fig. 36.3 and are described next.

Title data. The title information should be changed to reflect the fact that the run deals with temperature. Also, card 6 is changed to a YES so that the temperature will be simulated.

Program control (data type 1). We have chosen to modify the 6th card in this data type so that the LCD (local climatological data) and the solar data is printed. In addition a variety of new information must be added on the 14th through 18th cards of this type. These include the location (latitude, longitude, standard meridian, elevation), time (the day of the year), and some parameters related to the heat budget (evaporation and dust attenuation coefficients).

```
                                        Columns                                                    Data types
0         1         2         3         4         5         6         7         8
12345678901234567890123456789012345678901234567890123456789012345678901234567890

TITLE01             HANDS-ON 3a, QUAL-2EU WORKSHOP: TEMPERATURE
TITLE02             Steve Chapra, May 18, 1994
TITLE03    NO       CONSERVATIVE MINERAL   I
TITLE04    NO       CONSERVATIVE MINERAL  II
TITLE05    NO       CONSERVATIVE MINERAL III
TITLE06   YES       TEMPERATURE
TITLE07   YES       BIOCHEMICAL OXYGEN DEMAND
TITLE08    NO       ALGAE AS CHL-A IN UG/L
TITLE09    NO       PHOSPHORUS CYCLE AS P IN MG/L
TITLE10               (ORGANIC-P; DISSOLVED-P)
TITLE11    NO       NITROGEN CYCLE AS N IN MG/L
TITLE12               (ORGANIC-N; AMMONIA-N; NITRITE-N;' NITRATE-N)
TITLE13   YES       DISSOLVED OXYGEN IN MG/L
TITLE14    NO       FECAL COLIFORM IN NO./100 ML
TITLE15    NO       ARBITRARY NON-CONSERVATIVE
ENDTITLE
NO LIST DATA INPUT
NO WRITE OPTIONAL SUMMARY
NO FLOW AUGMENTATION
STEADY STATE
TRAPEZOIDAL CHANNELS
PRINT LCD/SOLAR DATA
NO PLOT DO AND BOD
FIXED DNSTM CONC (YES=1)=       0.      5D-ULT BOD CONV K COEF   =     0.25
INPUT METRIC            =       1.      OUTPUT METRIC            =       1.
NUMBER OF REACHES       =       6.      NUMBER OF JUNCTIONS      =       0.
NUM OF HEADWATERS       =       1.      NUMBER OF POINT LOADS    =       2.
TIME STEP (HOURS)       =       0.      LNTH. COMP. ELEMENT (KM)=        2.
MAXIMUM ROUTE TIME (HRS)=      30.      TIME INC. FOR RPT2 (HRS)=
LATITUDE OF BASIN (DEG) =      40.      LONGITUDE OF BASIN (DEG)=      105.
STANDARD MERIDIAN (DEG) =      90.      DAY OF YEAR START TIME   =     180.
EVAP. COEF.,(AE)        = 0.0000062     EVAP. COEF.,(BE)         =  .0000055
ELEV. OF BASIN (METERS) =    1670.      DUST ATTENUATION COEF.   =     0.01
ENDATA1
ENDATA1A
ENDATA1B
STREAM REACH      1. RCH= MS-HEAD             FROM       102.0      TO        100.0
STREAM REACH      2. RCH= MS100-MS080         FROM       100.0      TO         80.0
STREAM REACH      3. RCH= MS080-MS060         FROM        80.0      TO         60.0
STREAM REACH      4. RCH= MS060-MS040         FROM        60.0      TO         40.0
STREAM REACH      5. RCH= MS040-MS020         FROM        40.0      TO         20.0
STREAM REACH      6. RCH= MS020-MS000         FROM        20.0      TO          0.0
ENDATA2
ENDATA3
FLAG FIELD RCH=   1.          1.          1.
FLAG FIELD RCH=   2.         10.          6.2.2.2.2.2.2.2.2.2.
FLAG FIELD RCH=   3.         10.          2.2.2.2.2.2.2.2.2.2.
FLAG FIELD RCH=   4.         10.          6.2.2.2.2.2.2.2.2.2.
FLAG FIELD RCH=   5.         10.          2.2.2.2.2.2.2.2.2.2.
FLAG FIELD RCH=   6.         10.          2.2.2.2.2.2.2.2.2.5.
ENDATA4
HYDRAULICS RCH=   1.       0.00        2.0      2.0           10.      .0002      .035
HYDRAULICS RCH=   2.       0.00        2.0      2.0           10.      .0002      .035
HYDRAULICS RCH=   3.       0.00        2.0      2.0           10.      .0002      .035
HYDRAULICS RCH=   4.       0.00        2.0      2.0           10.      .00018     .035
HYDRAULICS RCH=   5.       0.00        2.0      2.0           10.      .00018     .035
HYDRAULICS RCH=   6.       0.00        2.0      2.0           10.      .00018     .035
ENDATA5
TEMP/LCD    RCH=  1.      1670.  0.01   0.25   25.    20.     825.  2.
ENDATA5A
REACT COEF RCH=    1.     0.00   0.000    0.000  1.   0.000  0.0000  0.0000
REACT COEF RCH=    2.     0.50   0.250    5.000  3.   0.000  0.0000  0.0000
REACT COEF RCH=    3.     0.50   0.000    0.000  3.   0.000  0.0000  0.0000
REACT COEF RCH=    4.     0.50   0.000    0.000  3.   0.000  0.0000  0.0000
REACT COEF RCH=    5.     0.50   0.000    0.000  3.   0.000  0.0000  0.0000
REACT COEF RCH=    6.     0.50   0.000    0.000  3.   0.000  0.0000  0.0000
ENDATA6
ENDATA6A
ENDATA6B
```

FIGURE 36.3
QUAL2E input file for a temperature run.

```
                                  Columns                                          Data types
0        1         2         3         4         5         6         7         8
12345678901234567890123456789012345678901234567890123456789012345678901234567890

INITIAL COND-1 RCH=   1.  22.00    8.11     0.0   0.00    0.00    0.00  0.000      0.0
INITIAL COND-1 RCH=   2.  20.59    8.11     0.0   0.00    0.00    0.00  0.000      0.0
INITIAL COND-1 RCH=   3.  20.59    8.11     0.0   0.00    0.00    0.00  0.000      0.0
INITIAL COND-1 RCH=   4.  19.72    8.11     0.0   0.00    0.00    0.00  0.000      0.0
INITIAL COND-1 RCH=   5.  19.72    8.11     0.0   0.00    0.00    0.00  0.000      0.0
INITIAL COND-1 RCH=   6.  19.72    8.11     0.0   0.00    0.00    0.00  0.000      0.0
ENDATA7
ENDATA7A
INCR INFLOW-1  RCH=   1.  0.000  00.00   0.0    0.0    0.0    0.0    0.0    0.0     0.
INCR INFLOW-1  RCH=   2.  0.000  00.00   0.0    0.0    0.0    0.0    0.0    0.0     0.
INCR INFLOW-1  RCH=   3.  0.000  00.00   0.0    0.0    0.0    0.0    0.0    0.0     0.
INCR INFLOW-1  RCH=   4.  0.000  00.00   0.0    0.0    0.0    0.0    0.0    0.0     0.
INCR INFLOW-1  RCH=   5.  0.000  00.00   0.0    0.0    0.0    0.0    0.0    0.0     0.
INCR INFLOW-1  RCH=   6.  0.000  00.00   0.0    0.0    0.0    0.0    0.0    0.0     0.
ENDATA8
ENDATA8A
ENDATA9
HEADWTR-1 HDW=    1   UPSTREAM        5.7870  20.0  7.50 2.0  00.0    0.0   0.0
ENDATA10
ENDATA10A
POINTLD-1 PTL=   1. MS0         0.00    0.463  28.0  2.00 200.0   0.0    0.0    0.0
POINTLD-1 PTL=   2. MS60        0.00    1.157  15.0  9.00   5.0   0.0    0.0    0.0
ENDATA11
ENDATA11A
ENDATA12
ENDATA13
ENDATA13A
```

FIGURE 36.3 (*continued*)

Temperature and local climatology data (data type 5a). We use a single card here to specify the key meteorological data needed to simulate a heat budget. An alternative is to specify different data for each reach.

A plot of the QUAL2E output for temperature is shown in Fig. 36.4. The temperature rise is due to the meteorology we have used for the example. Also, observe the drop in temperature induced by the tributary inflow.

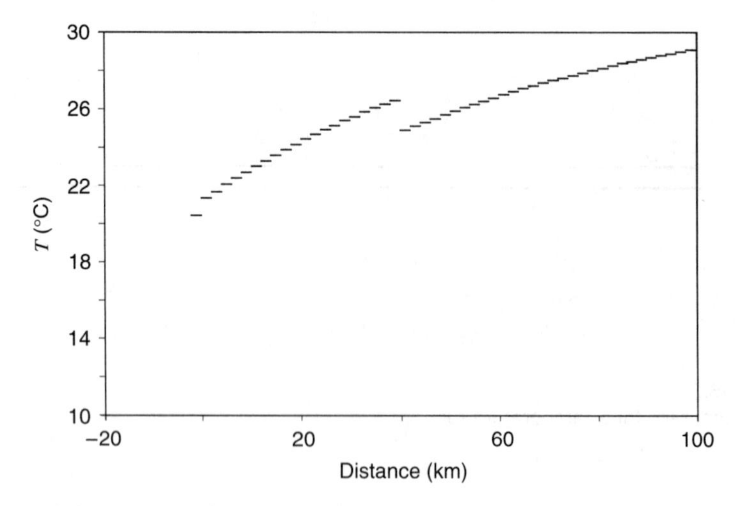

FIGURE 36.4
QUAL2E output for temperature.

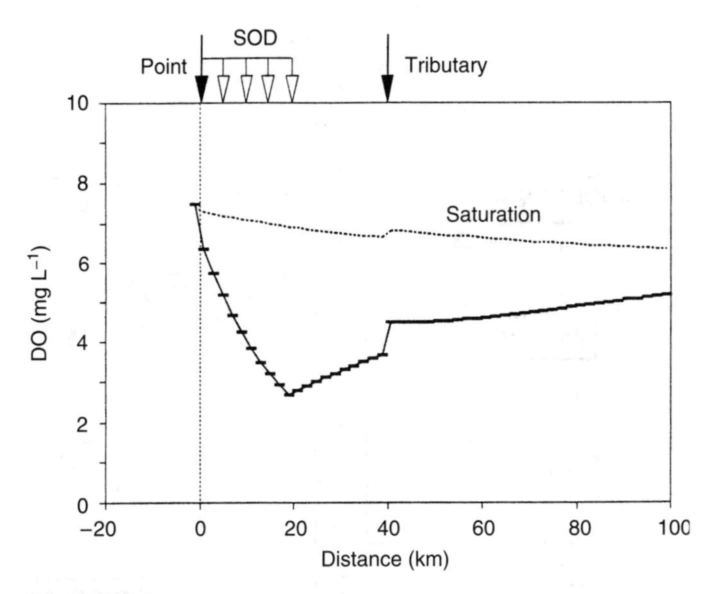

FIGURE 36.5
QUAL2E output for oxygen with variable temperature.

A plot of the QUAL2E output for oxygen is shown in Fig. 36.5. Notice how the temperature rise causes the oxygen saturation to decrease in the downstream direction.

Finally, as was mentioned in Lec. 26, a user-friendly interface for entering the input file and viewing the results is now available (Lahlou et al. 1995). The interface requires precisely the same information as in Fig. 36.3. Thus whether you employ the original version or the new interface, Figs. 36.3 and 36.7 can serve as a guide for performing the QUAL2E simulations outlined in this lecture.

36.2.2 Nutrients and Algae in QUAL2E

We will now expand our analysis to encompass nutrients and plants. As depicted in Fig. 36.6, QUAL2E simulates the kinetics of the nutrients: nitrogen and phosphorus. In addition, it calculates how these nutrients impact plant biomass.

The addition of these constituents has two effects on oxygen. First, the conversion of ammonia to nitrate in the nitrification process uses oxygen. Second, the nitrogen and phosphorus can induce plant growth. The resulting photosynthesis and respiration of the plants can add and deplete oxygen from the stream. Additional information on both these processes is provided next.

The QUAL2E kinetics for the nutrient/plant components can be written as

Algae (A):

$$\underset{\text{Accumulation}}{\frac{dA}{dt}} = \underset{\text{Growth}}{\mu A} - \underset{\text{Respiration}}{\rho A} - \underset{\text{Settling}}{\frac{\sigma_1}{H} A} \tag{36.21}$$

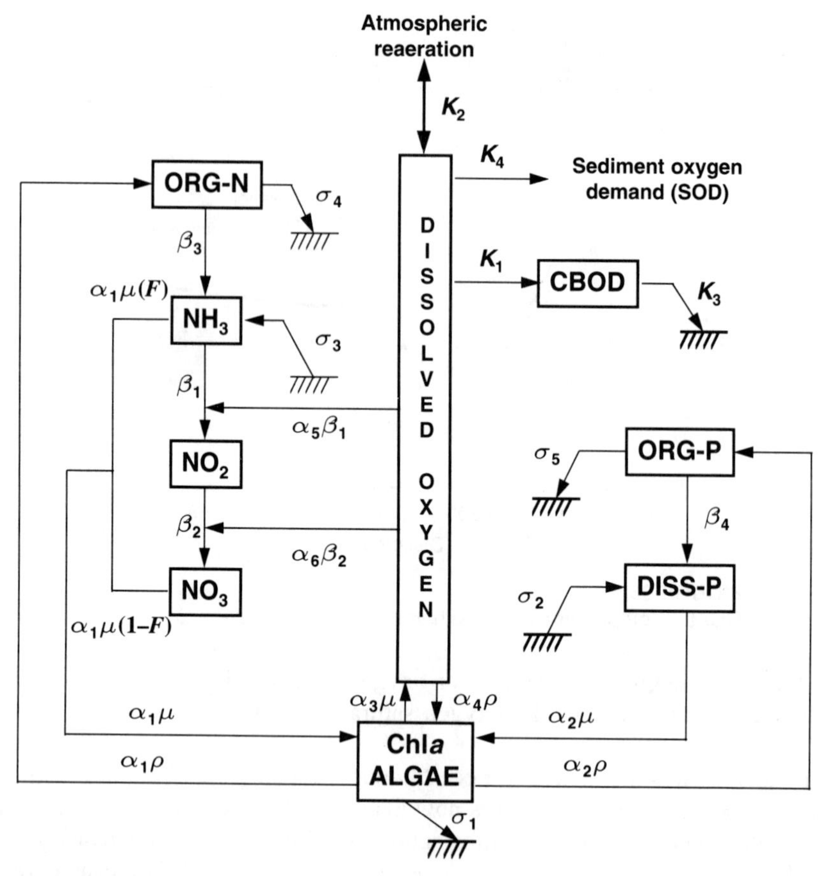

FIGURE 36.6
QUAL2E kinetics showing nutrient/plant interactions.

Organic nitrogen (N_4):

$$\underset{\text{Accumulation}}{\frac{dN_4}{dt}} = \underset{\text{Respiration}}{\alpha_1 \rho A} - \underset{\text{Hydrolysis}}{\beta_3 N_4} - \underset{\text{Settling}}{\sigma_4 N_4} \tag{36.22}$$

Ammonia nitrogen (N_1):

$$\underset{\text{Accumulation}}{\frac{dN_1}{dt}} = \underset{\text{Hydrolysis}}{\beta_3 N_4} - \underset{\text{Nitrification}}{\beta_1 N_1} + \underset{\text{Sediment}}{\frac{\sigma_3}{H}} - \underset{\text{Growth}}{F_1 \alpha_1 \mu A} \tag{36.23}$$

Nitrite nitrogen (N_2):

$$\underset{\text{Accumulation}}{\frac{dN_2}{dt}} = \underset{\text{Nitrification}}{\beta_1 N_1} - \underset{\text{Nitrification}}{\beta_2 N_2} \tag{36.24}$$

Nitrate nitrogen (N_3):

$$\underset{\text{Accumulation}}{\frac{dN_3}{dt}} = \underset{\text{Nitrification}}{\beta_2 N_2} \underset{\text{Growth}}{- (1-F)\alpha_1 \mu A} \tag{36.25}$$

Organic phosphorus (P_1):

$$\underset{\text{Accumulation}}{\frac{dP_1}{dt}} = \underset{\text{Respiration}}{\alpha_2 \rho A} - \underset{\text{Decay}}{\beta_4 P_1} - \underset{\text{Settling}}{\sigma_5 P_1} \tag{36.26}$$

Inorganic phosphorus (P_2):

$$\underset{\text{Accumulation}}{\frac{dP_2}{dt}} = \underset{\text{Decay}}{\beta_4 P_1} + \underset{\text{Sediment}}{\frac{\sigma_2}{H}} \underset{\text{Growth}}{- \alpha_2 \mu A} \tag{36.27}$$

Note that the nitrogen, phosphorus, and kinetic constituents can be simulated without computing oxygen and CBOD. However, if they are computed, the oxygen kinetics are modified to account for the effects of nitrification and plant growth/respiration,

Carbonaceous BOD (L):

$$\underset{\text{Accumulation}}{\frac{dL}{dt}} = \underset{\text{Decay}}{-K_1 L} - \underset{\text{Settling}}{K_3 L} \tag{36.28}$$

Dissolved oxygen (O):

$$\underset{\text{Accumulation}}{\frac{dO}{dt}} = \underset{\text{Reaeration}}{K_2(O_s - O)} - \underset{\text{Decomposition}}{K_1 L} - \underset{\text{SOD}}{\frac{K_4}{H}}$$

$$+ \underset{\text{Growth} - \text{respiration}}{(\alpha_3 \mu - \alpha_4 \rho)A} - \underset{\text{Nitrification}}{\alpha_5 \beta_1 N_1 - \alpha_6 \beta_2 N_2} \tag{36.29}$$

An input file has been developed to integrate plants and nutrients into the temperature program introduced earlier as displayed in Fig. 36.7. The cards that must be modified to model plants and nutrients are highlighted in Fig. 36.7 and are described next.

Title data. The title information should be changed. Also, cards 8 through 12 should be changed to YES.

Global algal, nitrogen, phosphorus, and light parameters (data type 1A). These cards include all the global parameters for the nutrient/plant interactions. Several of them bear mention:

```
                                    Columns                                                   Data types
0         1         2         3         4         5         6         7         8
12345678901234567890123456789012345678901234567890123456789012345678901234567890

TITLE01                  HANDS-ON 3b, QUAL-2EU WORKSHOP: N/P/ALGAE
TITLE02                  Steve Chapra, May 18, 1994
TITLE03    NO            CONSERVATIVE MINERAL   I
TITLE04    NO            CONSERVATIVE MINERAL  II
TITLE05    NO            CONSERVATIVE MINERAL III
TITLE06   YES            TEMPERATURE
TITLE07   YES            BIOCHEMICAL OXYGEN DEMAND
TITLE08   YES            ALGAE AS CHL-A IN UG/L
TITLE09   YES            PHOSPHORUS CYCLE AS P IN MG/L
TITLE10                    (ORGANIC-P; DISSOLVED-P)
TITLE11   YES            NITROGEN CYCLE AS N IN MG/L
TITLE12                    (ORGANIC-N; AMMONIA-N; NITRITE-N;' NITRATE-N)
TITLE13   YES            DISSOLVED OXYGEN IN MG/L
TITLE14    NO            FECAL COLIFORM IN NO./100 ML
TITLE15    NO            ARBITRARY NON-CONSERVATIVE
ENDTITLE
NO LIST DATA INPUT
NO WRITE OPTIONAL SUMMARY
NO FLOW AUGMENTATION
STEADY STATE
TRAPEZOIDAL CHANNELS
PRINT LCD/SOLAR DATA
NO PLOT DO AND BOD
FIXED DNSTM CONC (YES=1)=       0.        5D-ULT BOD CONV K COEF   =      0.25
INPUT METRIC            =       1.        OUTPUT METRIC            =        1.
NUMBER OF REACHES       =       6.        NUMBER OF JUNCTIONS      =        0.
NUM OF HEADWATERS       =       1.        NUMBER OF POINT LOADS    =        2.
TIME STEP (HOURS)       =       0.        LNTH. COMP. ELEMENT (KM)=        2.
MAXIMUM ROUTE TIME (HRS)=      30.        TIME INC. FOR RPT2 (HRS)=
LATITUDE OF BASIN (DEG) =      40.        LONGITUDE OF BASIN (DEG)=      105.
STANDARD MERIDIAN (DEG) =      90.        DAY OF YEAR START TIME   =      180.
EVAP. COEF.,(AE)        = 0.0000062       EVAP. COEF.,(BE)         =  .0000055
ELEV. OF BASIN (METERS) =    1670.        DUST ATTENUATION COEF.   =      0.01
ENDATA1
O UPTAKE BY NH3 OXID(MG O/MG N)=    3.5  O UPTAKE BY NO2 OXID(MG O/MG N)=  1.20
O PROD. BY ALGAE (MG O/MG A)   =    1.6  O UPTAKE BY ALGAE (MG O/MG A)  =   2.0
N CONTENT OF ALGAE (MG N/MG A) =   .085  P CONTENT OF ALGAE (MG P/MG A) =  .013
ALG MAX SPEC GROWTH RATE(1/DAY)=    1.5  ALGAE RESPIRATION RATE (1/DAY) =   0.1
N HALF SATURATION CONST. (MG/L)=   .155  P HALF SATURATION CONST. (MG/L)= .0255
LIN ALG SHADE CO (1/H-UGCHLA)  =  .0088  NLIN SHADE (1/H-(UGCHA/L)**2/3)= .0540
LIGHT FUNCTION OPTION (LFNOPT) =    3.0  LIGHT SATURATION COEF (INT/MIN)= .2083
DAILY AVERAGING OPTION(LAVOPT) =    3.0  LIGHT AVERAGING FACTOR (AFACT) =   .94
NUMBER OF DAYLIGHT HOURS (DLH) =     0.  TOTAL DAILY SOLAR RADTN (INT)  = 000.0
ALGY GROWTH CALC OPTION(LGROPT)=      2. ALGAL PREF FOR NH3-N (PREFN)   =   0.5
ALG/TEMP SOLAR RAD FACT(TFACT) =   0.44  NITRIFICATION INHIBITION COEF  =   0.6
ENDATA1A
ENDATA1B
STREAM REACH     1. RCH= MS-HEAD           FROM       102.0     TO        100.0
STREAM REACH     2. RCH= MS100-MS080       FROM       100.0     TO         80.0
STREAM REACH     3. RCH= MS080-MS060       FROM        80.0     TO         60.0
STREAM REACH     4. RCH= MS060-MS040       FROM        60.0     TO         40.0
STREAM REACH     5. RCH= MS040-MS020       FROM        40.0     TO         20.0
STREAM REACH     6. RCH= MS020-MS000       FROM        20.0     TO          0.0
ENDATA2
ENDATA3
FLAG FIELD RCH=  1.          1.          1.
FLAG FIELD RCH=  2.         10.          6.2.2.2.2.2.2.2.2.2.
FLAG FIELD RCH=  3.         10.          2.2.2.2.2.2.2.2.2.2.
FLAG FIELD RCH=  4.         10.          6.2.2.2.2.2.2.2.2.2.
FLAG FIELD RCH=  5.         10.          2.2.2.2.2.2.2.2.2.2.
FLAG FIELD RCH=  6.         10.          2.2.2.2.2.2.2.2.2.5.
ENDATA4
HYDRAULICS RCH=  1.       0.00       2.0   2.0         10.    .0002     .035
HYDRAULICS RCH=  2.       0.00       2.0   2.0         10.    .0002     .035
HYDRAULICS RCH=  3.       0.00       2.0   2.0         10.    .0002     .035
HYDRAULICS RCH=  4.       0.00       2.0   2.0         10.    .00018    .035
HYDRAULICS RCH=  5.       0.00       2.0   2.0         10.    .00018    .035
HYDRAULICS RCH=  6.       0.00       2.0   2.0         10.    .00018    .035
ENDATA5
TEMP/LCD    RCH= 1.     1670.  0.01   0.25   25.    20.     825.  2.
ENDATA5A
```

FIGURE 36.7
QUAL2E input file for a plant/nutrient run.

```
                                   Columns                                    Data types
0         1         2         3         4         5         6         7         8
12345678901234567890123456789012345678901234567890123456789012345678901234567890

REACT COEF RCH=   1.     0.00   0.000     0.000  1.    0.000  0.0000  0.0000
REACT COEF RCH=   2.     0.50   0.250     5.000  3.    0.000  0.0000  0.0000
REACT COEF RCH=   3.     0.50   0.000     0.000  3.    0.000  0.0000  0.0000
REACT COEF RCH=   4.     0.50   0.000     0.000  3.    0.000  0.0000  0.0000
REACT COEF RCH=   5.     0.50   0.000     0.000  3.    0.000  0.0000  0.0000
REACT COEF RCH=   6.     0.50   0.000     0.000  3.    0.000  0.0000  0.0000
ENDATA6
N AND P COEF    RCH=  1.    0.00    0.00    0.00    0.00    0.00    0.00    0.00    0.00
N AND P COEF    RCH=  2.    0.25    0.00    0.15    0.00    0.25    0.10    0.00    0.00
N AND P COEF    RCH=  3.    0.25    0.00    0.15    0.00    0.25    0.10    0.00    0.00
N AND P COEF    RCH=  4.    0.25    0.00    0.15    0.00    0.25    0.10    0.00    0.00
N AND P COEF    RCH=  5.    0.25    0.00    0.15    0.00    0.25    0.10    0.00    0.00
N AND P COEF    RCH=  6.    0.25    0.00    0.15    0.00    0.25    0.10    0.00    0.00
ENDATA6A
ALG/OTHER COEF RCH=   1.    50.0    0.00 0.1        0.00
ALG/OTHER COEF RCH=   2.    50.0    0.10 0.1        0.00
ALG/OTHER COEF RCH=   3.    50.0    0.10 0.1        0.00
ALG/OTHER COEF RCH=   4.    50.0    0.10 0.1        0.00
ALG/OTHER COEF RCH=   5.    50.0    0.10 0.1        0.00
ALG/OTHER COEF RCH=   6.    50.0    0.10 0.1        0.00
ENDATA6B
INITIAL COND-1 RCH=   1.  22.00     0.00      0.0   0.00    0.00    0.00  0.000     0.0
INITIAL COND-1 RCH=   2.  20.59     0.00      0.0   0.00    0.00    0.00  0.000     0.0
INITIAL COND-1 RCH=   3.  20.59     0.00      0.0   0.00    0.00    0.00  0.000     0.0
INITIAL COND-1 RCH=   4.  19.72     0.00      0.0   0.00    0.00    0.00  0.000     0.0
INITIAL COND-1 RCH=   5.  19.72     0.00      0.0   0.00    0.00    0.00  0.000     0.0
INITIAL COND-1 RCH=   6.  19.72     0.00      0.0   0.00    0.00    0.00  0.000     0.0
ENDATA7
INITIAL COND-2 RCH=   1.           0.000
INITIAL COND-2 RCH=   2.           0.000
INITIAL COND-2 RCH=   3.           0.000
INITIAL COND-2 RCH=   4.           0.000
INITIAL COND-2 RCH=   5.           0.000
INITIAL COND-2 RCH=   6.           0.000
ENDATA7A
INCR INFLOW-1  RCH=   1.  0.000   00.00     0.0    0.0    0.0    0.0    0.0    0.0     0.
INCR INFLOW-1  RCH=   2.  0.000   00.00     0.0    0.0    0.0    0.0    0.0    0.0     0.
INCR INFLOW-1  RCH=   3.  0.000   00.00     0.0    0.0    0.0    0.0    0.0    0.0     0.
INCR INFLOW-1  RCH=   4.  0.000   00.00     0.0    0.0    0.0    0.0    0.0    0.0     0.
INCR INFLOW-1  RCH=   5.  0.000   00.00     0.0    0.0    0.0    0.0    0.0    0.0     0.
INCR INFLOW-1  RCH=   6.  0.000   00.00     0.0    0.0    0.0    0.0    0.0    0.0     0.
ENDATA8
INCR INFLOW-2  RCH=   1.    0.00    0.00    0.00    0.00    0.00    0.00    0.00
INCR INFLOW-2  RCH=   2.    0.00    0.00    0.00    0.00    0.00    0.00    0.00
INCR INFLOW-2  RCH=   3.    0.00    0.00    0.00    0.00    0.00    0.00    0.00
INCR INFLOW-2  RCH=   4.    0.00    0.00    0.00    0.00    0.00    0.00    0.00
INCR INFLOW-2  RCH=   5.    0.00    0.00    0.00    0.00    0.00    0.00    0.00
INCR INFLOW-2  RCH=   6.    0.00    0.00    0.00    0.00    0.00    0.00    0.00
ENDATA8A
ENDATA9
HEADWTR-1 HDW=    1   UPSTREAM          5.7870  20.0  7.50 2.0  00.0    0.0    0.0
ENDATA10
HEADWTR-2 HDW=    1.   0.00 000.0   5.00   0.05   0.05   0.02   0.10   0.01   0.01
ENDATA10A
POINTLD-1 PTL=    1. MS0             0.00    0.463   28.0   2.00 200.0    0.0    0.0    0.0
POINTLD-1 PTL=    2. MS60            0.00    1.157   15.0   9.00   5.0    0.0    0.0    0.0
ENDATA11
POINTLD-2 PTL=    1.   0.00      000. 0.00   10.0   15.0   0.05   1.00   0.50   10.0
POINTLD-2 PTL=    2.   0.00      000. 1.00   0.05   0.05   0.02   0.10   0.01   0.01
ENDATA11A
ENDATA12
ENDATA13
ENDATA13A
```

FIGURE 36.7 *(continued)*

- The model allows linear, nonlinear, or no self-shading according to the values of the shading coefficients (LIN ALG SHADE CO and NLIN SHADE).
- Different functions can be employed to represent the effect of light on plant growth. This is controlled by the coefficient (LIGHT FUNCTION OPTION). In the present example we use Steele's model (option 3). Note that the interpretation of the LIGHT SATURATION COEF depends on the option.
- There are several different light averaging options (DAILY AVERAGING OPTION).
- There are several different options for trading off the light and nutrient limitations (ALGY GROWTH CALC OPTION).
- A preference for ammonia or nitrate as a nitrogen source is allowed (ALGAL PREF FOR NH3–N).

N and P coefficients (data type 6a). These cards include the reach-specific data for the nutrients.

Algae/other coefficients (data type 6b). These cards include the reach-specific data for the algae. In addition they are used to enter data for the other constituents (conservative/nonconservative) that might be modeled.

Initial conditions–2 (data type 7a). This card group, one card per reach, establishes the initial values of the system for the nutrients and algae.

Incremental inflow–2 (data type 8a). Even though we will not simulate incremental inflows, these cards must be included when plants and nutrients are simulated. As in our example, all the values can be set to zero.

Headwater sources–2 (data type 10). This card group, one card per headwater, defines the boundary conditions for plants and nutrients at the upstream ends of the system.

Point load–2 (data type 11a). This card group, one card per point source or withdrawal, defines the loading values for plants and nutrients. Note that the point load numbers are not the same as either the reach or element numbers. Rather, the point sources or withdrawals are numbered consecutively (starting at 1) from the most upstream to the most downstream.

A plot of the QUAL2E output for the algae/nutrient run is shown in Fig. 36.8. The oxygen levels are less than the previous runs because of the effect of nitrification.

36.3 FIXED PLANTS IN STREAMS

Now that we know how to simulate phytoplankton dynamics in flowing waters, we turn to the fixed plants that are particularly prevalent in shallower systems. These range from large macrophytes to small microalgae called ***periphyton***.

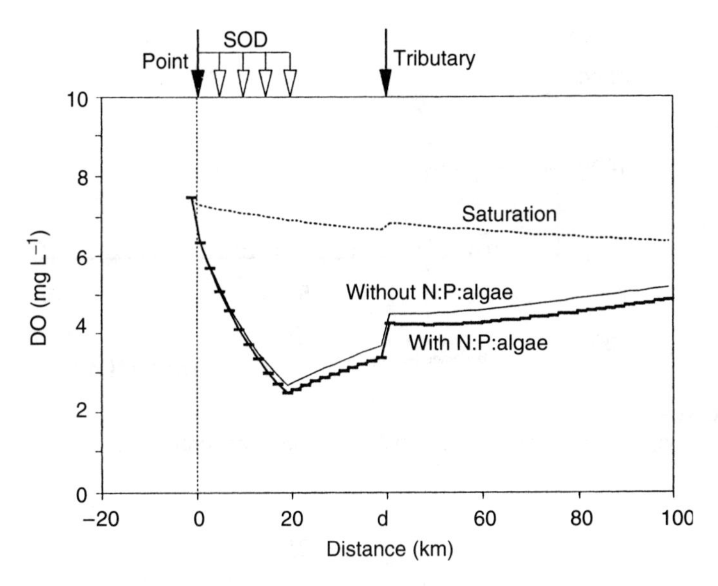

FIGURE 36.8
QUAL2E output for dissolved oxygen with and without algae and nutrients.

36.3.1 Simple Attached Plant/Nutrient Analysis

As for phytoplankton, Thomann and Mueller (1987) also have presented a simple back-of-the-envelope approach to model the effect of attached plants on stream quality. In contrast to floating plants, they assumed that the rooted plants have a constant biomass over the study stretch. Thus, rather than being modeled, the plants are assumed to act as a zero-order sink of nutrients,

$$H\frac{dn}{dt*} = -a_{na}k_g(T,\ I)a' \tag{36.30}$$

$$H\frac{dp}{dt*} = -a_{pa}k_g(T,\ I)a' \tag{36.31}$$

where $a' =$ fixed-plant biomass in areal units (μgChl m^{-2}) and $H =$ depth (m).

Using boundary conditions of $n = n_0$ and $p = p_0$ at the mixing point, these equations can be solved for

$$n = n_0 - \frac{a_{na}k_g a'}{H}t^* \tag{36.32}$$

$$p = p_0 - \frac{a_{pa}k_g a'}{H}t^* \tag{36.33}$$

Note that Eqs. 36.32 and 36.33 can be solved for the critical travel times,

$$t_n^* = \frac{n_0 - 100}{a_{na}k_g a'} \tag{36.34}$$

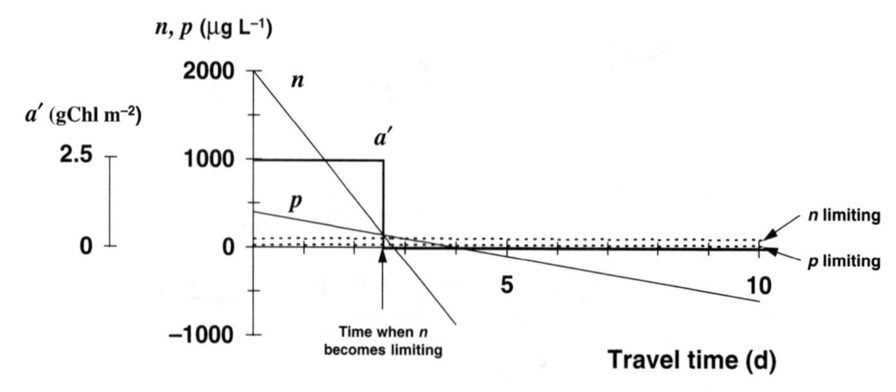

FIGURE 36.9
A simple oxygen balance for a stream including fixed-plant growth.

$$t_p^* = \frac{p_0 - 25}{a_{pa} k_g a'} \tag{36.35}$$

The model results are displayed in Fig. 36.9 for a stream with a significant effluent-to-flow ratio. The simulation also reflects an effluent with a low N:P ratio. As expected from Eqs. 36.32 through 36.33, the nutrients drop as the plants grow. However, in contrast to Fig. 36.1, the drop is linear.

Observe that the travel time to reach nutrient limitation is shorter than for Fig. 36.1. This particular result is due to the parameter values, the initial conditions chosen for the two plots, and the assumptions underlying the Thomann-Mueller models. However, it should be recognized that, all other factors being equal, attached plants and their impacts tend to be more localized than floating plants. That is, their biomass and associated effects tend to be more concentrated immediately downstream from point sources because they are not being transported downstream as they grow.

Having said this, I should stress that each situation must be assessed on a case-by-case basis. In particular the impact of light has been neglected in the foregoing analyses. Therefore, as described next, more detailed mass balance models are required for site-specific water-quality evaluations.

36.3.2 Modeling Attached Plants (Advanced Topic)

Attached plants differ from floating plants in some fundamental ways:

- As the name implies, fixed plants do not advect downstream and they don't settle. Consequently a phytoplankton model (e.g., Eqs. 36.15 through 36.17) can be adapted to attached plants by merely removing advection and settling.
- Although some attached algae can extend up through the water column, most reside on or near the bottom. Thus light attenuation must be handled differently than for phytoplankton. Recall that for the phytoplankton, the light supplied for growth

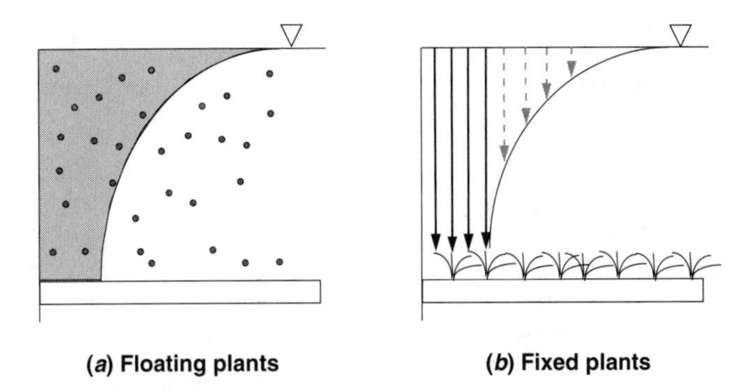

FIGURE 36.10
Contrast between how light is characterized in (*a*) floating (integrated over depth) and (*b*) fixed-plant (level determined at depth of plants) models.

was determined by integrating a growth model over time and depth (Fig. 36.10*a*). For bottom plants, growth would be dependent on the light delivered to the depth at which the plants reside (Fig. 36.10*b*). Consequently the growth attenuation factor for light could be simply determined by integrating over time for a fixed depth, as in

$$\phi_l = \frac{\int_0^{fT_p} F(I)\,dt}{T_p} \tag{36.36}$$

where $F(I)$ is one of the growth models for light. For example, if Steele's model is used, Eq. 33.19 and the Beer-Lambert law (Eq. 33.21) can be substituted into Eq. 36.36, and the result integrated to give

$$\phi_l = \frac{fI_a e^{-k_e H}}{I_s} e^{-\frac{I_a e^{-k_e H}}{I_s}+1} \tag{36.37}$$

The form of this function means that light will have a much more dramatic impact on bottom plants than on floating plants. In particular it means that water depth will have a much greater impact on whether fixed plants can even grow in the water body.

- Attached plants often require a particular bottom substrate in order to grow. Some are rooted and consequently need a fine-grained, soil-type bottom. Others attach to rocks. Thus just because there is optimal light, nutrients, and temperature, a particular plant may not grow if suitable substrate is absent.
- There is typically a maximum density for attached plants. This maximum depends on light limitation and self-shading, space constraints due to a finite area of substrate, and the fact that the plants are not mobile. Although floating plants can also reach such limits, attached plants tend to more easily approach maximum density, particularly in nutrient-rich environments with plenty of light.

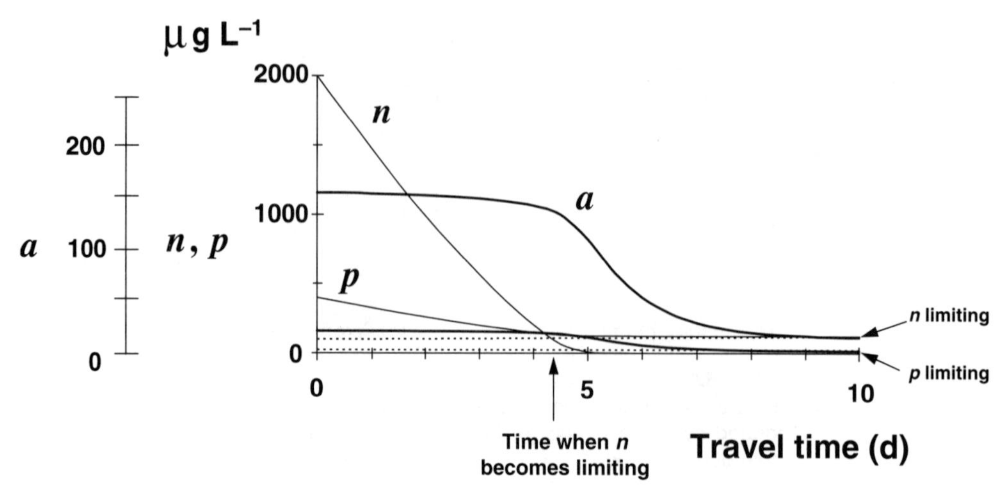

FIGURE E36.2

EXAMPLE 36.2. NUMERICAL STREAM FIXED-PLANT SIMULATION. Duplicate the calculations in Example 36.1, but fix the plants by removing the transport terms from the algal balance.

Solution: The model equations from Example 36.1 can be modified, as

$$V_i \frac{da_i}{dt} = \left\{ k_g(T, I) \min[\phi_p, \phi_n] - k_{ra} \right\}_i V_i a_i \tag{36.38}$$

$$V_i \frac{dn_i}{dt} = Q_{i-1,i} n_{i-1} - Q_{i,i+1} n_i - a_{na} \left\{ k_g(T, I) \min[\phi_p, \phi_n] - k_{ra} \right\}_i V_i a_i \tag{36.39}$$

$$V_i \frac{dp_i}{dt} = Q_{i-1,i} p_{i-1} - Q_{i,i+1} p_i - a_{pa} \left\{ k_g(T, I) \min[\phi_p, \phi_n] - k_{ra} \right\}_i V_i a_i \tag{36.40}$$

Along with appropriate boundary conditions, these three equations can then be integrated numerically using the solution techniques presented in Lecs. 11 through 13. Figure E36.2 shows the resulting steady-state profiles employing parameter values identical to those used in Example 36.1.

Notice how this model behaves in a radically different fashion from Fig. E36.1. The plants are at a higher level than in the phytoplankton model. Further, they are at a fairly constant level until the travel time when nitrogen starts to become limiting. At this point the plants die out as losses dominate growth. The resulting shape is quite similar to the assumed constant distribution of Thomann and Mueller's simplified approach (compare with Fig. 36.9).

To summarize, different model formulations are required to simulate floating and fixed plants in flowing waters. In general, given favorable light conditions, fixed plants tend to grow to higher biomass levels and to be concentrated closer to point sources than floating varieties. However, light attenuation with depth and substrate suitability are strong determinants of whether fixed varieties can thrive. Future stream eutrophication frameworks including both types should be developed. Such models could prove useful in determining whether either fixed or floating plants might dominate given changing light, nutrient, and temperature conditions.

PROBLEMS

36.1. In Sec. 36.1.1 we assume that both kinetic losses and settling are terminal; that is, they do not feed back to the nutrient pools. How could such assumptions be justified? If they are erroneous, what would be the ramifications for the subsequent conclusions of the analysis, if any?

36.2. The highest recorded phytoplankton chlorophyll levels found in temperate lakes are about 100 μg L^{-1}. The maximum macrophyte densities are about 500 g-dry weight m^{-2}. Because of their high content of structural carbon, macrophytes typically have low chlorophyll-to-carbon ratios, on the order of 10 mgChl gC^{-1}. They have typical carbon contents of about 40 gC g-dry $weight^{-1}$. If the macrophytes grow in 1 m of water, translate the maximum density into a volume-specific concentration of chlorophyll.

36.3. A phytoplankton-dominated stream has the following characteristics:

Upstream:
Flow = 0.5 cms $\quad p = 50\ \mu\text{gP L}^{-1} \quad n = 1000\ \mu\text{gN L}^{-1} \quad a = 50\ \mu\text{gChl L}^{-1}$

Point source:
Flow = 0.75 cms $\quad p = 5000\ \mu\text{gP L}^{-1} \quad n = 25{,}000\ \mu\text{gN L}^{-1} \quad a = 0\ \mu\text{gChl L}^{-1}$

Use the simplified analysis described in Sec. 36.1.1 to determine (*a*) the travel time when nutrients become limiting, (*b*) which nutrient is limiting, and (*c*) what will happen if nitrogen removal is installed and the nitrogen effluent is reduced to 5000 μgN L^{-1}. Assume the following parameter values: $k_g = 0.5\ d^{-1}$, $k_n = 0.2\ d^{-1}$, n:p:a = 7.2:1:1, $k_{sp} = 4\ \mu\text{gP L}^{-1}$, and $k_{sn} = 15\ \mu\text{gN L}^{-1}$.

36.4. Repeat Prob. 36.3, but use the control-volume approach described in Example 36.1.

36.5. Repeat Prob. 36.4, but include nitrification of ammonium to nitrate at a rate of 0.05 d^{-1}. Assume that the nitrate is unavailable for phytoplankton growth.

36.6. Using the same river as in Prob. 26.4, modify your file so that temperature is computed. Note that the simulation is to be performed for September 10, at latitude 40° N, longitude 111° W, and an elevation of 1000 m. The wind speed is 2 m s^{-1}, barometric pressure is 970 millibars (mb), and the dry- and wet-bulb air temperatures are 30 and 23°C, respectively. The dust attenuation coefficient is 0.10 and the cloudiness is 0.2. Use QUAL2E to simulate the levels of CBOD, oxygen, and temperature for this case.

36.7. Modify the file from Prob. 36.6 so that nutrients and plants are computed. The following concentrations are available for the headwater and the point load:

Constituent	Units	Headwater (*r*)	Point load (*w*)
Organic nitrogen	mgN L^{-1}	0.05	10
Ammonia nitrogen	mgN L^{-1}	0.05	15
Nitrite nitrogen	mgN L^{-1}	0.02	0.05
Nitrate nitrogen	mgN L^{-1}	0.1	1
Organic phosphorus	mgP L^{-1}	0.01	0.5
Inorganic phosphorus	mgP L^{-1}	0.01	10
Algae	μgChl L^{-1}	1.0	0.0

For all the other necessary coefficients employ the same values as were used in Fig. 36.7. Use QUAL2E to simulate the levels of CBOD, oxygen, temperature, nitrogen, phosphorus, and algae for this case.

36.8. Repeat Prob. 36.4, but fix the plants.

PART VI

Chemistry

Part VI is designed to introduce you to the rudiments of integrating aquatic chemistry into our water-quality-modeling framework.

Lecture 37 provides background on equilibrium chemistry. After an introduction to units, the subjects of chemical equilibria and pH are reviewed. Approaches for making chemical equilibria computations are also outlined.

In *Lecture 38* I describe how equilibrium chemistry can be integrated into a simple mass-balance framework. The important concept of a local equilibrium is introduced and the dissociation of ammonia is used as an example.

Finally the calculation of pH is used as a context for illustrating the basic approach in *Lecture 39.* This material focuses on simple CSTRs dominated by carbonate chemistry. CSTRs are used to show how pH is calculated in systems subject to both plant activity and gas transfer.

LECTURE 37

Equilibrium Chemistry

LECTURE OVERVIEW: I present some fundamental concepts from the area of equilibrium chemistry. After a short overview of chemical units and activity, I review the law of mass action and the notion of chemical equilibrium. This is followed by a description of pH. Finally I present some simple models of chemical equilibria that can be employed to calculate chemical levels in batch systems.

At the beginning of this book, I made the point that the transport and fate of substances in natural waters was dictated by their physics, chemistry, and biology. In the earlier lectures the chemistry and biology were represented by very simple first-order kinetics. Anyone who has ever taken a course in biology or chemistry knows that there's a lot more to these subjects. In the last several lectures we've finally begun to address the realistic modeling of biology. Now we will try to do the same for chemistry.

To do this we must first review some fundamental concepts. In particular I describe the field of equilibrium chemistry and show how calculations are made in that field.

37.1 CHEMICAL UNITS AND CONVERSIONS

To this point we have exclusively expressed the "quantity" of a substance by its mass. An alternative, which is superior from a chemical perspective, is to use moles or gram-molecular weight. A ***mole*** refers to the molecular weight in grams of a compound. For example a mole of glucose, $C_6H_{12}O_6$, can be calculated as being

$$1 \text{ mole of } C_6H_{12}O_6 = 6C\frac{12 \text{ g}}{C} + 12H\frac{1 \text{ g}}{H} + 6O\frac{16 \text{ g}}{O} = 180 \text{ g}$$

The gram-molecular weight is important because each mole includes the same number of molecules regardless of the compound. This quantity (6.02×10^{23}) is called ***Avogadro's number.*** A ***molar solution*** consists of 1 mole dissolved in enough water to make a total volume of 1 L.†

Another related way to express concentration is as equivalents per liter, or normal concentration. A one normal solution (designated as 1 *N*) contains one equivalent weight of a substance per liter of solution. Thus ***normality*** is defined as

$$\text{Normality} = \frac{\text{mass per liter}}{\text{equivalent weight}} \tag{37.1}$$

The equivalent weight is calculated as

$$\text{Equivalent weight} = \frac{\text{molecular weight}}{n} \tag{37.2}$$

where n depends on the type of substance being quantified. In natural waters it is usually based on one of three criteria:

- Charge of an ion
- Number of hydronium (protons) or hydroxyl ions transferred in an acid-base reaction
- Number of electrons transferred in an oxidation-reduction reaction

The following example illustrates the approach.

EXAMPLE 37.1. CHEMICAL CONCENTRATION. Determine the normality of 30 mg L^{-1} of carbonate (CO_3^{2-}) for the reactions:

(*a*) $Ca^{2+} + CO_3^{2-} \rightleftharpoons CaCO_3(s)$
(*b*) $HCO_3^- \rightleftharpoons CO_3^{2-} + H^+$
(*c*) $CO_3^{2-} + 2H^+ \rightleftharpoons H_2CO_3$

The molecular weight of carbonate is 60.

Solution: (*a*) For this case, the reaction involves the ionization of carbonate:

$$\text{Equivalent weight} = \frac{\text{molecular weight}}{n \text{ (ion charge)}} = \frac{60 \text{ g mole}^{-1}}{2 \text{ eq mole}^{-1}} = 30\frac{\text{g}}{\text{eq}} = 30 \text{ mg meq}^{-1}$$

$$\text{Normality} = \frac{\text{mass per liter}}{\text{equivalent weight}} = \frac{30 \text{ mg L}^{-1}}{30 \text{ mg meq}^{-1}} = 1 \text{ meq L}^{-1}$$

Therefore, the solution is 1 *N* with respect to carbonate.
(*b*) This is an acid-base reaction:

---

†This is in contrast to a molal solution, where 1 mole is dissolved in 1 kg of solvent.

$$\text{Equivalent weight} = \frac{\text{molecular weight}}{n\ (\text{number of protons})} = \frac{60 \text{ g mole}^{-1}}{1 \text{ eq mole}^{-1}} = 60\frac{\text{g}}{\text{eq}} = 60 \text{ mg meq}^{-1}$$

$$\text{Normality} = \frac{\text{mass per liter}}{\text{equivalent weight}} = \frac{30 \text{ mg L}^{-1}}{60 \text{ mg meq}^{-1}} = 0.5 \text{ meq L}^{-1}$$

Therefore the solution is 0.5 N with respect to carbonate.

(*c*) This is also an acid-base reaction:

$$\text{Equivalent weight} = \frac{\text{molecular weight}}{n\ (\text{number of protons})} = \frac{60 \text{ g mole}^{-1}}{2 \text{ eq mole}^{-1}}$$

$$= 30 \text{ g eq}^{-1} = 30 \text{ mg meq}^{-1}$$

$$\text{Normality} = \frac{\text{mass per liter}}{\text{equivalent weight}} = \frac{30 \text{ mg L}^{-1}}{30 \text{ mg meq}^{-1}} = 1 \text{ meq L}^{-1}$$

Therefore the solution is 1 N with respect to carbonate.

As in the previous example, a substance can have different equivalent weights depending on the reaction. Further, as illustrated in parts (*b*) and (*c*), an acid or a base can have more than one equivalent weight. Although this might initially seem confusing, it is offset by the advantage of expressing concentration in equivalent units. This advantage is that when substances react to form a product, the equivalents of the reactants will be the same as the equivalents of the products.

37.2 CHEMICAL EQUILIBRIA AND THE LAW OF MASS ACTION

In our discussion of reaction kinetics in Lec. 2, we presented the reaction (Eq. 2.1)

$$a\text{A} + b\text{B} \underset{r_b}{\overset{r_f}{\rightleftharpoons}} c\text{C} + d\text{D} \tag{37.3}$$

where the lowercase letters represent stoichiometric coefficients and the uppercase letters designate the reacting compounds.

The ***kinetics*** or rate of such reactions can be expressed quantitatively by the ***law of mass action,*** which states that the rate is proportional to the concentration of the reactants. A simple way to formulate the forward rate r_f of the reaction in Eq. 37.3 is

$$r_f = k_f[\text{A}]^a[\text{B}]^b \tag{37.4}$$

where the bracketed terms are molar concentrations of the reactants.[†] The rate of the backward reaction can be modeled as

$$r_b = k_b[\text{C}]^c[\text{D}]^d \tag{37.5}$$

[†] As discussed in the next section, the reaction rate for ions should be defined in terms of activity. However, concentration provides an adequate approximation when dealing with dilute aqueous solutions.

After a sufficient period of time, the backward rate will equal the forward rate,

$$k_f[\mathrm{A}]^a[\mathrm{B}]^b = k_b[\mathrm{C}]^c[\mathrm{D}]^d \tag{37.6}$$

which can be solved for

$$K = \frac{k_f}{k_b} = \frac{[\mathrm{C}]^c[\mathrm{D}]^d}{[\mathrm{A}]^a[\mathrm{B}]^b} \tag{37.7}$$

where K is the ***equilibrium constant*** of the reaction.

37.3 IONIC STRENGTH, CONDUCTIVITY, AND ACTIVITY

Although concentration provides a good approximation of the chemical activity of ions in dilute solutions, it is not adequate when dealing with concentrated solutions such as salt water. For such systems the effective concentration, or ***activity,*** is diminished by forces of attraction between positive and negative ions.

The activity of an ion, $\{c\}$, can be determined by multiplying its molar concentration $[c]$ by an activity coefficient,

$$\{c\} = \gamma[c] \tag{37.8}$$

where the activity coefficient γ is less than or equal to one.

To compute the activity of a solution, the strength of the ionic interactions must be quantified. Lewis and Randall (1921) introduced the concept of ***ionic strength*** to measure the electric field in a solution,

$$\mu = \frac{1}{2}\sum_i [c]_i Z_i \tag{37.9}$$

where Z_i = the charge of species i.

A number of correlations have also been developed that allow ionic strength to be computed without determining all the ions in solution. For example Langelier (1936) indicated that the ionic strength could be related to the total dissolved solids,

$$\mu = 2.5 \times 10^{-5}\ (\mathrm{TDS}) \tag{37.10}$$

where TDS is the concentration of total dissolved solids in mg L^{-1}. In another attempt, Russell (1976) suggested that the ionic strength could be estimated on the basis of the specific conductance,

$$\mu = 1.6 \times 10^{-5}\ (\text{specific conductance}) \tag{37.11}$$

where specific conductance must be measured in μmho cm^{-1}.

Once the ionic strength is quantified, a number of equations are available for determining the activity coefficient. One, which holds for solutions with ionic strengths less than approximately 5×10^{-3}, is called the ***Debye-Hückel limiting law,***

$$\log \gamma_i = -0.5 Z_i^2 \sqrt{u} \tag{37.12}$$

An alternative equation is also available for more concentrated solutions (ionic strength < 0.1). Called the ***extended Debye-Hückel approximation,*** it has the form,

TABLE 37.1
The size parameter *a* for various ions (Kielland 1937)

Ion	*a* (angstroms)
H^+, Al^{3+}, Fe^{3+}, La^{3+}, Ce^{3+}	9
Mg^{2+}, Be^{2+}	8
Ca^{2+}, Zn^{2+}, Cu^{2+}, Sn^{2+}, Mn^{2+}, Fe^{2+}	6
Ba^{2+}, Sr^{2+}, Pb^{2+}, $CO_3{}^{2-}$	5
Na^+, $HCO_3{}^-$, $H_2PO_4{}^-$, CH_3COO^-, $SO_4{}^{2-}$, $HPO_4{}^{2-}$, $PO_4{}^{3-}$	4
K^+, Ag^+, $NH_4{}^+$, OH^-, Cl^-, $ClO_4{}^-$, $NO_3{}^-$, I^-, HS^-	3

$$\log \gamma_i = -\frac{AZ_i^2\sqrt{u}}{1 + Ba_i\sqrt{u}} \tag{37.13}$$

where a_i = a parameter (in angstroms) corresponding to the size of the ion (Table 37.1), and

$$A = 1.82 \times 10^6(\varepsilon T)^{-3/2} \approx 0.5 \text{ (for water at 25°C)} \tag{37.14}$$

and

$$B = 50.3(\varepsilon T)^{-1/2} \approx 0.33 \text{ (for water at 25°C)} \tag{37.15}$$

where ε = the dielectric constant (= 78 for water) and T is in Kelvin.

EXAMPLE 37.2. ACTIVITY CORRECTIONS. Carbonic acid (H_2CO_3) dissociates by the following reaction,

$$H_2CO_3 \rightleftharpoons HCO_3{}^- + H^+$$

The equilibrium constant for this reaction is (25°C)

$$K_1 = 10^{-6.35} = \frac{[HCO_3{}^-][H^+]}{[H_2CO_3]}$$

If the total dissolved solids is 100 mg L^{-1} and the hydrogen ion concentration is 10^{-7}, determine the ratio of bicarbonate ($HCO_3{}^-$) to carbonic acid with and without activity corrections.

Solution: First, we can estimate the ionic strength as (Eq. 37.10)

$$\mu = 2.5 \times 10^{-5}(100) = 2.5 \times 10^{-3}$$

Because the ionic strength is less than 5×10^{-3}, the activity coefficients can be calculated as (Eq. 37.12)

$$\log \gamma_{H^+} = \log \gamma_{HCO_3{}^-} = -0.5(1)^2\sqrt{2.5 \times 10^{-3}} = -0.025$$

$$\gamma_{H^+} = \gamma_{HCO_3{}^-} = 10^{-0.025} = 0.9441$$

These values can be substituted into the equilibrium relationship,

$$10^{-6.35} = \frac{0.9441[HCO_3{}^-]0.9441\left(10^{-7}\right)}{[H_2CO_3]}$$

which can be solved for

$$\frac{[HCO_3^-]}{[H_2CO_3]} = \frac{10^{-6.35}}{0.9441^2(10^{-7})} = 5.011$$

If we had not taken activities into account, the ratio would have been

$$\frac{[HCO_3^-]}{[H_2CO_3]} = \frac{10^{-6.35}}{10^{-7}} = 4.467$$

which amounts to about an 11% difference.

37.4 pH AND THE IONIZATION OF WATER

Pure water dissociates to yield a hydrogen ion (or more properly, hydrated protons; see Stumm and Morgan 1981), H^+, and a hydroxyl ion, OH^-,

$$H_2O \rightleftharpoons H^+ + OH^- \tag{37.16}$$

If the reaction is at equilibrium, the concentrations of reactants and products are defined by

$$K = \frac{[H^+][OH^-]}{[H_2O]} \tag{37.17}$$

Because the concentration of water in dilute aqueous solutions is much greater than the ions and is decreased so little by the ionization, it may be assumed to be at a constant level. Therefore the equilibrium relationship is conventionally expressed as

$$K_w = [H^+][OH^-] \tag{37.18}$$

where K_w is known as the ***ion,*** or ***ionization, product of water,*** which at 25°C is equal to approximately 10^{-14}. The value of K_w can be computed at other temperatures by (Harned and Hamer 1933)

$$pK_w = \frac{4787.3}{T_a} + 7.1321 \log_{10} T_a + 0.010365\, T_a - 22.80 \tag{37.19}$$

This relationship predicts a mildly decreasing trend of pK_w with increasing temperature. In other words there is more dissociation at higher temperatures.

For pure water, equal amounts of $[H^+]$ and $[OH^-]$ are formed, and therefore the concentration of either is equal to approximately 10^{-7}. When an acid is added to the water, it ionizes and increases the hydrogen ion concentration. Consequently, for Equation 37.18 to hold, the hydroxyl concentration must decrease so that the product of the ions equals 10^{-14}. For example if the addition of an acid to pure water raises $[H^+]$ from 10^{-7} to 10^{-5}, then $[OH^-]$ must decrease to 10^{-9}.

Because it is somewhat cumbersome to work in terms of negative powers of 10, chemists typically employ the negative logarithm:

$$p(x) = -\log_{10} x \tag{37.20}$$

This is sometimes referred to as p(x) notation (Sawyer et al. 1994). For the hydrogen ion, the use of this notation results in the pH (short for the French *puissance d'Hydrogène,* which translates as the "power or intensity of hydrogen"),

$$pH = -\log_{10}[H^+] \tag{37.21}$$

Because of the negative sign and the logarithmic transformation of $[H^+]$, a high pH represents a low hydrogen ion activity, whereas a low pH represents a high hydrogen ion activity. Consequently the pH scale reflects the intensity of the acidic or basic (or alkaline) condition of the water. As in Figure 37.1, low pH connotes acidic and high pH connotes alkaline waters, with a neutral condition at a pH of 7.0. The pH of natural waters ranges from 1.7 for volcanic lakes to 12.0 for closed alkaline systems (Hutchinson 1957). However, the typical range is from 6.0 to 9.0.

37.5 EQUILIBRIUM CALCULATIONS

Now that we have defined our terms, we will explore how chemists use the concept of equilibrium to estimate the levels of chemical species in aquatic systems. The basic idea is to delineate the various species that exist in a solution. Once this has been done, the next step is to determine how many of these are unknowns. Then equations reflecting the interactions among the compounds must be specified. Finally, if the number of equations equals the number of unknowns, the equations can be solved for the concentrations of the species.

In these lectures we focus on acid-base systems. For such systems there are three types of equations that are used:

- Equilibrium relationships
- Concentration conditions or mass balance
- Electroneutrality conditions or charge balance

The actual approach can be elaborated on by an example.

EXAMPLE 37.3. EQUILIBRIUM CALCULATIONS. Determine the concentrations of the species in a 3×10^{-4} aqueous solution of HCl at 25°C. Note that dissociation constant for hydrochloric acid is 10^3.

Solution: First, we can list the resulting reactions,

$$H_2O \rightleftharpoons H^+ + OH^- \tag{37.22}$$

and

$$HCl \rightleftharpoons H^+ + Cl^- \tag{37.23}$$

Thus there are four species that must be determined: HCl, Cl^-, H^+, and OH^-. Consequently we require four equations.

The first two equations are provided by the equilibrium conditions for water,

$$[H^+][OH^-] = 10^{-14} \tag{37.24}$$

and for hydrochloric acid,

$$\frac{[H^+][Cl^-]}{[HCl]} = 10^3 \tag{37.25}$$

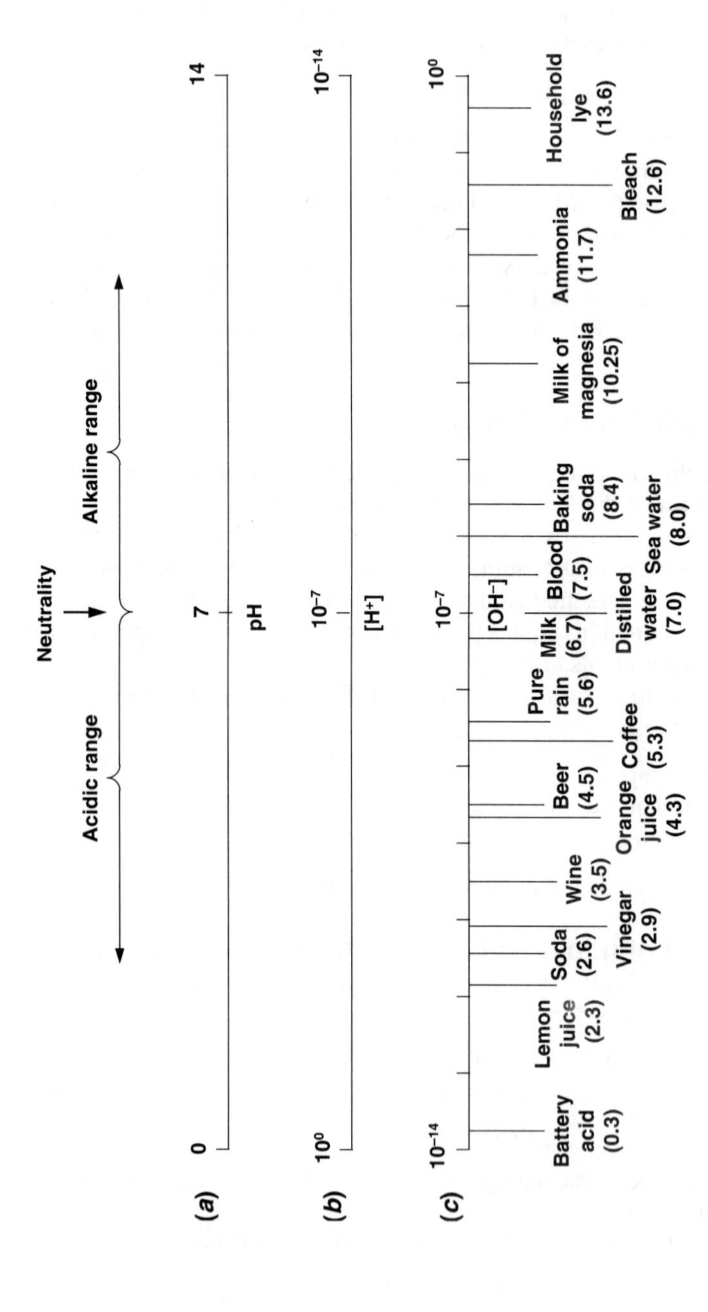

FIGURE 37.1
Three scales showing the relationship of (*a*) pH to the concentrations of (*b*) hydrogen ion and (*c*) hydroxyl ion in water. The pHs of some common substances are indicated at the bottom.

Next we recognize that this is a closed system and that the amount of chlorine in the system must be equal to the amount added originally. This is termed a ***concentration condition*** and can be written as

$$c_T = [\mathrm{HCl}] + [\mathrm{Cl^-}] = 3 \times 10^{-4} \tag{37.26}$$

where c_T = total concentration of chlorine. Note that this relationship amounts to a mass balance for chlorine.

Finally the positive and negative charges in solution must be in balance,

$$[\mathrm{H^+}] = [\mathrm{OH^-}] + [\mathrm{Cl^-}] \tag{37.27}$$

This is termed an ***electroneutrality condition***.

These four equations can now be solved for the concentrations of the species. Notice that they are a nonlinear system. Therefore some algebraic manipulation is in order. First, solve Eq. 37.25 for [HCl],

$$[\mathrm{HCl}] = \frac{[\mathrm{H^+}][\mathrm{Cl^-}]}{10^3} \tag{37.28}$$

Next, substitute this result into Eq. 37.26 and solve for

$$[\mathrm{Cl^-}] = \frac{3 \times 10^{-4}}{1 + \dfrac{[\mathrm{H^+}]}{10^3}} \tag{37.29}$$

Equation 37.24 can be solved for

$$[\mathrm{OH^-}] = \frac{10^{-14}}{[\mathrm{H^+}]} \tag{37.30}$$

Finally Eqs. 37.29 and 37.30 can be substituted into Eq. 37.27 to give

$$[\mathrm{H^+}] = \frac{10^{-14}}{[\mathrm{H^+}]} + \frac{3 \times 10^{-4}}{1 + \dfrac{[\mathrm{H^+}]}{10^3}} \tag{37.31}$$

Further algebraic manipulation yields

$$[\mathrm{H^+}]^3 + 10^3[\mathrm{H^+}]^2 - 0.3[\mathrm{H^+}] - 10^{-11} = 0 \tag{37.32}$$

Thus our manipulations have generated a cubic equation in $[\mathrm{H^+}]$ that can be solved numerically for $[\mathrm{H^+}] = 3 \times 10^{-4}$, which corresponds to a pH of 3.52. The other species can be calculated as

$$[\mathrm{OH^-}] = \frac{10^{-14}}{3 \times 10^{-4}} = 3.31 \times 10^{-11}\ M \tag{37.33}$$

$$[\mathrm{Cl^-}] = \frac{3 \times 10^{-4}}{1 + \dfrac{3 \times 10^{-4}}{10^3}} = 2.9999991 \times 10^{-4}\ M \tag{37.34}$$

$$[\mathrm{HCl}] = 3 \times 10^{4} - 2.9999991 \times 10^{-4} = 9.06 \times 10^{-11}\ M \tag{37.35}$$

PROBLEMS

37.1. Suppose that you introduce 2 mg of nitrogen in the form of the ammonium ion into a 300-mL flask. Immediately after introducing the ammonium, you fill the remainder of the volume with distilled water and stopper the flask. Determine the pH if the temperature is 25°C. Note that ammonium dissociates according to

$$NH_4^+ \rightleftharpoons NH_3 + H^+$$

where the equilibrium coefficient for the reaction is related to temperature by

$$pK = 0.09018 + \frac{2729.92}{T_a}$$

in which T_a is in Kelvin.

37.2. Determine molar concentrations of all the species in a 10^{-4} M water solution of acetic acid at a temperature of 25°C. Note that acetic acid dissociates according to

$$CH_3COOH \rightleftharpoons H^+ + CH_3COO^-$$

where the pK for the reaction is 4.70.

37.3. Repeat Example 37.2, but use the extended Debye-Hückel approximation with T = 25°C.

37.4. Determine the hydrogen and hydroxyl ion concentrations of water at a pH of 8.5 and a temperature of 30°C.

LECTURE 38

Coupling Equilibrium Chemistry and Mass Balance

LECTURE OVERVIEW: I show how the equilibrium chemistry approach can be integrated with the mass-balance models that we have used throughout this book. To do this I introduce the notion of a local equilibrium and illustrate how it provides a means to solve fast and slow reactions in tandem. Then I extend the approach to modeling chemistry in natural systems.

To this point we have addressed two different modeling perspectives that are capable of characterizing the chemistry of natural waters. (Fig. 38.1). Throughout most of the previous lectures we have used the mass-balance approach (Fig. 38.1*a*). Then in Lecture 37 we learned about equilibrium chemistry modeling (Fig. 38.1*b*). We now unify these two perspectives. Before doing this I must first present the mathematical idea that underlies this unification: the local equilibrium.

38.1 LOCAL EQUILIBRIUM

The local equilibrium idea relates to systems containing both fast and slow reactions. Suppose that you are modeling the situation depicted in Fig. 38.2. This beaker includes two reactants that are subject to two purging mechanisms that deplete mass from the system: flushing and a decay reaction. Observe that flushing acts on both reactants, whereas the decay loss affects only the second. Further, the reactants are coupled by two first-order reactions. Mass balances for the two constituents can be written as

$$V\frac{dc_1}{dt} = Qc_{1,\text{in}} - Qc_1 - k_{12}Vc_1 + k_{21}Vc_2 \qquad (38.1)$$

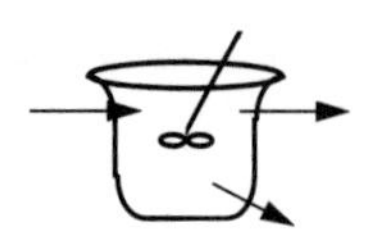

$$V\frac{dc}{dt} = Qc_{in} - Qc - kVc$$

(*a*) Mass balance

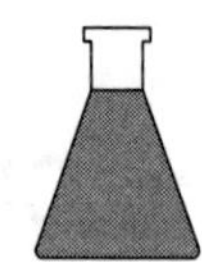

$$aA + bB \rightleftharpoons cC + dD$$

(*b*) Chemical equilibrium

FIGURE 38.1
Two modeling perspectives used to characterize the chemistry of natural waters.

$$V\frac{dc_2}{dt} = Qc_{2,\text{in}} - Qc_2 + k_{12}Vc_1 - k_{21}Vc_2 - kVc_2 \tag{38.2}$$

Given initial conditions, we can solve these equations for the concentration of the two reactants as a function of time. Now suppose that the reactions linking the reactants are much faster than the other mass transfers and transformations. For example let's say that k_{12} and k_{21} were on the order of 1 s^{-1} and all the other model rates were on the order of 1 d^{-1}. For example the residence time (V/Q) might be 1 d and the decay rate of the second reactant might be 1 d^{-1}. If we were simulating trends on a time scale of days, due to the presence of the fast coupling reactions, a very small time step would have to be used to maintain stability.

Now let's think about the situation a little more closely. If the coupling reactions are sufficiently fast, they will go to completion much quicker than the other mechanisms. In fact the backward and forward reactions should rapidly reach a balance. Thus, when viewed from the perspective of the other processes, the following equilibrium would hold:

$$k_{12}Vc_1 = k_{21}Vc_2 \tag{38.3}$$

Consequently the ratio of the two reactants should be fixed at a constant value. This of course is the idea behind an equilibrium constant. For the present example, Eq. 38.3 can be used to derive

$$K = \frac{k_{12}}{k_{21}} = \frac{c_2}{c_1} \tag{38.4}$$

where K = an equilibrium constant that specifies the ratio of the species.

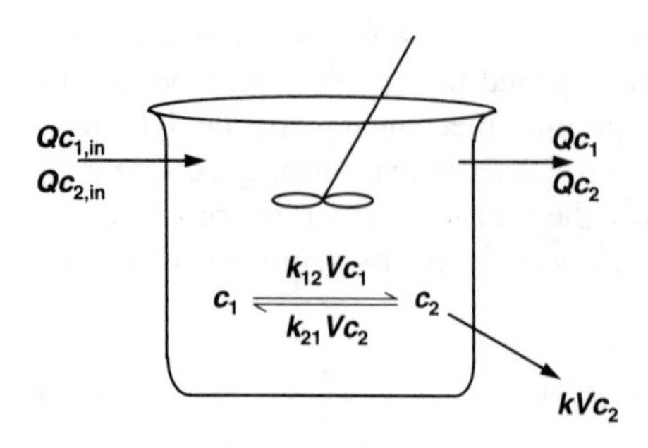

FIGURE 38.2
Two reactants in a completely mixed system.

If the above scenario is true, Di Toro (1976) has suggested that the two mass-balance equations (38.1 and 38.2) can be added to yield

$$V\frac{dc_T}{dt} = W - Qc_T - kVc_2 \tag{38.5}$$

where W = total loading of both constituents (in which $W = Qc_{1,\text{in}} + Qc_{2,\text{in}}$) and

$$c_T = c_1 + c_2 \tag{38.6}$$

Notice that the terms representing the fast reactions were canceled by the addition of Eqs. 38.1 and 38.2. This is due to the fact that if the reactions are at equilibrium and Eq. 38.3 holds, the terms $k_{12}Vc_1$ and $k_{21}Vc_2$ are equal and would cancel out.

Now at this point the mass balance, although simpler, cannot be solved, because it is a single equation with two unknowns, c_T and c_2. To rectify this difficulty, we recognize that Eqs. 38.4 and 38.6 represent a system of two equations that can be used to compute c_1 and c_2 in terms of c_T. For example Eq. 38.4 can be solved for

$$c_2 = Kc_1 \tag{38.7}$$

which can be substituted into Eq. 38.6, which in turn can be solved for

$$c_1 = F_1 c_T \tag{38.8}$$

where F_1 is the fraction of the total concentration that is represented by the first constituent,

$$F_1 = \frac{1}{1+K} \tag{38.9}$$

Equation 38.8 can be substituted back into Eq. 38.6, which can be solved for

$$c_2 = F_2 c_T \tag{38.10}$$

where F_2 is the fraction of the total concentration that is represented by the second constituent,

$$F_2 = \frac{K}{1+K} \tag{38.11}$$

Finally Eq. 38.10 can be substituted into Eq. 38.5 to give

$$V\frac{dc_T}{dt} = W - Qc_T - kVF_2c_T \tag{38.12}$$

Thus the original system of differential equations has been replaced by a single mass balance that can be solved for the total mass in the system as a function of time. Because we have eliminated the fast terms from the differential equation, a reasonable time step can be used in the solution scheme. Then at every time step, Eqs. 38.8 and 38.10 can be used to calculate the concentrations of each of the two interacting species. These species are said to be at a ***local equilibrium***.

38.2 LOCAL EQUILIBRIA AND CHEMICAL REACTIONS

In the previous section we modeled a coupled reaction that looked like

$$c_1 \underset{k_{21}Vc_2}{\overset{k_{12}Vc_1}{\rightleftharpoons}} c_2 \tag{38.13}$$

Such a reaction is directly analogous to the chemical equations developed in the previous lecture. For example the dissociation of ammonium ion can be represented as

$$NH_4^+ \underset{k_bV[NH_3][H^+]}{\overset{k_fV[NH_4^+]}{\rightleftharpoons}} NH_3 + H^+ \tag{38.14}$$

Thus the forward and backward arrows can be represented as reactions that transfer mass forward and backward between the reactants and products. If the reaction is at equilibrium, it means that

$$k_fV[NH_4^+] = k_bV[NH_3][H^+] \tag{38.15}$$

from which follows the equilibrium constant

$$K = \frac{k_f}{k_b} = \frac{[NH_3][H^+]}{[NH_4^+]} \tag{38.16}$$

Now suppose we want to model the system depicted in Fig. 38.3. Mass balances for ammonium ion and ammonia can be written as

$$V\frac{d[NH_4^+]}{dt} = Q[NH_4^+]_{in} - Q[NH_4^+] - k_fV[NH_4^+] + k_bV[NH_3][H^+] - k_nV[NH_4^+] \tag{38.17}$$

$$V\frac{d[NH_3]}{dt} = Q[NH_3]_{in} - Q[NH_3] + k_fV[NH_4^+] - k_bV[NH_3][H^+] - v_vA_s[NH_3] \tag{38.18}$$

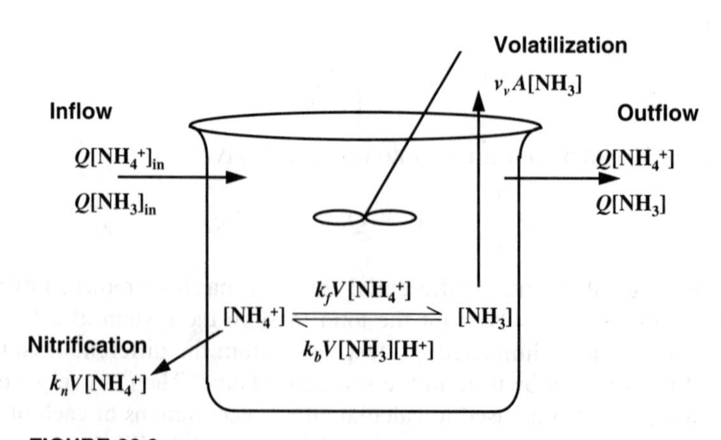

FIGURE 38.3
Ammonium ion and ammonia in a completely mixed system.

Because the chemical reaction is much faster than the other transport and transformation terms, a local equilibrium holds. Therefore the two equations can be added to yield

$$V\frac{dc_T}{dt} = Qc_{T,\text{in}} - Qc_T - k_n V[\text{NH}_4{}^+] - v_v A_s[\text{NH}_3] \tag{38.19}$$

where

$$c_T = [\text{NH}_4{}^+] + [\text{NH}_3] \tag{38.20}$$

and

$$K = \frac{[\text{NH}_3][\text{H}^+]}{[\text{NH}_4{}^+]} \tag{38.21}$$

As was done earlier in the lecture, Eqs. 38.20 and 38.21 can be solved simultaneously for

$$[\text{NH}_4{}^+] = F_i c_T \qquad \text{where } F_i = \frac{[\text{H}^+]/K}{1 + [\text{H}^+]/K} \tag{38.22}$$

and

$$[\text{NH}_3] = F_u c_T \qquad \text{where } F_u = \frac{1}{1 + [\text{H}^+]/K} \tag{38.23}$$

where F_i and F_u are the ionized and un-ionized fractions, respectively.

Finally Eq. 38.22 and 38.23 can be substituted into Eq. 38.19 to give

$$V\frac{dc_T}{dt} = Qc_{T,\text{in}} - Qc_T - k_n V F_i c_T - v_v A_s F_u c_T \tag{38.24}$$

Thus this mass balance can be solved to determine levels of both ammonium and ammonia in the system. However, observe that the pH is required to obtain solutions. This is illustrated in the following example.

EXAMPLE 38.1. AMMONIA BALANCE FOR A LAKE. A lake has the following characteristics:

Volume = 50,000 m^3
Surface area = 25,000 m^2
Mean depth = 2 m
Inflow = outflow = 7500 $m^3\ d^{-1}$
Wind speed = 3 mps

The lake's inflow has an ammonium concentration of 6 mgN L^{-1}. If the lake's pH is 9 and the temperature is 25°C, determine the steady-state concentrations of ammonium and ammonia. Note that the pK for the reaction is equal to 9.25. Also, employ a volatilization velocity of 0.1 m d^{-1} and a nitrification rate of 0.05 d^{-1}.

Solution: First, we must convert the inflow concentration to moles,

$$[\text{NH}_4{}^+]_{\text{in}} = 6\frac{\text{gN}}{\text{m}^3}\left(\frac{18\ \text{gNH}_4{}^+}{14\ \text{gN}}\right)\left(\frac{\text{moleNH}_4{}^+}{18\ \text{gNH}_4{}^+}\right)\left(\frac{\text{m}^3}{10^3\ \text{L}}\right) = 0.4286 \times 10^{-3}\ M$$

Equations 38.22 and 38.23 can be used to compute

$$F_i = \frac{[\text{H}]^+/K}{1 + [\text{H}^+]/K} = \frac{10^{-9}/10^{-9.25}}{1 + 10^{-9}/10^{-9.25}} = 0.64$$

and

$$F_u = 1 - 0.64 = 0.36$$

At steady-state, Eq. 38.24 can be solved for

$$c_T = \frac{Q}{Q + k_n V F_i + v_v A F_u} c_{T,\text{in}}$$

$$= \frac{7500}{7500 + 0.05(50{,}000)0.64 + 0.1(25{,}000)0.36} 0.4286 \times 10^{-3} = 0.3215 \times 10^{-3} M$$

The individual species can be determined as

$$[NH_4^+] = 0.64(0.3215 \times 10^{-3}) = 0.2057 \times 10^{-3}\ M$$

and

$$[NH_3] = 0.36(0.3215 \times 10^{-3}) = 0.1157 \times 10^{-3} M$$

These results can also be expressed in mass units as

$$c_i = 0.2057 \times 10^{-3} M \left(\frac{14\ \text{gN}}{\text{mole}}\right)\left(\frac{10^3\ \text{L}}{\text{m}^3}\right) = 2.88\ \text{mgN L}^{-1}$$

and

$$c_u = 0.1157 \times 10^{-3}\ M \left(\frac{14\ \text{gN}}{\text{mole}}\right)\left(\frac{10^3\ \text{L}}{\text{m}^3}\right) = 1.62\ \text{mgN L}^{-1}$$

PROBLEMS

38.1. Repeat Example 38.1, but develop a plot of total ammonia and its species versus pH.

38.2. Repeat Example 38.1, but perform a time-variable calculation over 3 d, where the pH varies sinusoidally according to

$$\text{pH} = 8 + 1.25 \cos[2\pi(t - 0.25)]$$

where t = time (d) and $t = 0$ at midnight.

38.3. A detention pond at sea level is at a temperature of 20°C. The pond has a pH of 9.5 and ammonia concentration of 5 mg L^{-1}. The gas-film exchange coefficient for the lake is 200 m d^{-1} and the Henry's constant for ammonia is 1.37×10^{-5} atm m^3 $mole^{-1}$. Compute the following (*note:* tables and figures are unacceptable; you must use formulas):
(*a*) The un-ionized ammonia concentration
(*b*) The flux of ammonia across the air-water interface

LECTURE 39

pH Modeling

LECTURE OVERVIEW: I extend the material in the preceding lectures to develop a framework for computing pH in natural waters. In this discussion I focus on systems in which the carbonate buffering system is dominant.

Now I apply the theory from the previous two lectures to simulate the pH of natural waters. Because the pH of many natural waters is dominated by the carbonate buffering system, I begin this lecture by describing the inorganic carbon system.

39.1 FAST REACTIONS: INORGANIC CARBON CHEMISTRY

The tendency of natural waters to remain within a relatively narrow band of hydrogen ion activity is due to the presence of buffers that resist pH changes. The buffers do this by scavenging H^+ and OH^- ions. In many freshwater systems, much of the buffering is related to the dissolved inorganic carbon species: carbon dioxide (CO_2), bicarbonate ion ($HCO_3{}^-$), and carbonate ion ($CO_3{}^{2-}$).

When carbon dioxide is introduced into an aqueous solution, it combines with water to form carbonic acid,

$$CO_2 + H_2O \rightleftharpoons H_2CO_3 \tag{39.1}$$

The carbonic acid, in turn, dissociates into ionic form, as in

$$H_2CO_3 \rightleftharpoons HCO_3{}^- + H^+ \tag{39.2}$$

Because the equilibrium constant for the hydration of carbon dioxide (Eq. 39.1) is so small, the proportion present as carbonic acid is negligible. Therefore the processes of hydration and dissociation conventionally are treated as a single reaction,

$$\mathrm{H_2CO_3}^* \underset{k_4}{\overset{k_3}{\rightleftharpoons}} \mathrm{HCO_3}^- + \mathrm{H}^+ \tag{39.3}$$

where

$$[\mathrm{H_2CO_3}^*] = [\mathrm{CO_2}] + [\mathrm{H_2CO_3}] \cong [\mathrm{CO_2}] \tag{39.4}$$

The equilibrium constant for the combined reaction is

$$K_1 = \frac{[\mathrm{H}^+][\mathrm{HCO_3}^-]}{[\mathrm{H_2CO_3}^*]} \tag{39.5}$$

where K_1 is called the composite acidity constant of $H_2CO_3^*$, or the first dissociation constant of carbonic acid. The value of K_1 can be computed as a function of absolute temperature by (Harned and Davis 1943)

$$\mathrm{p}K_1 = \frac{3404.71}{T_a} + 0.032786T_a - 14.8435 \tag{39.6}$$

The bicarbonate ion, in turn, dissociates to yield a carbonate and a hydrogen ion,

$$\mathrm{HCO_3}^- \underset{k_6}{\overset{k_5}{\rightleftharpoons}} \mathrm{CO_3}^{2-} + \mathrm{H}^+ \tag{39.7}$$

The equilibrium constant for this reaction is

$$K_2 = \frac{[\mathrm{H}^+][\mathrm{CO_3}^{2-}]}{[\mathrm{HCO_3}^-]} \tag{39.8}$$

where K_2 is called the second dissociation constant of carbonic acid. The value of K_2 can be computed as a function of temperature by (Harned and Scholes 1941)

$$\mathrm{p}K_2 = \frac{2902.39}{T_a} + 0.02379T_a - 6.498 \tag{39.9}$$

The inorganic carbon system delineated above consists of five unknowns: $[H_2CO_3^*]$, $[HCO_3^-]$, $[CO_2^{3-}]$, $[H^+]$, and $[OH^-]$. Therefore five simultaneous equations are needed to solve for these unknowns. Three of the five are provided by the equilibrium relationships defined by Eqs. 39.5, 39.8, and the dissociation of water (Eq. 37.18).

A fourth equation is the concentration condition,

$$c_T = [\mathrm{H_2CO_3}^*] + [\mathrm{HCO_3}^-] + [\mathrm{CO_3}^{2-}] \tag{39.10}$$

where c_T is total inorganic carbon.

Finally a fifth equation results from the fact that the solution must be electrically neutral. This can be expressed as a charge balance or electroneutrality equation,

$$c_B + [\mathrm{H}^+] = [\mathrm{HCO_3}^-] + 2[\mathrm{CO_3}^-] + [\mathrm{OH}^-] + c_A \tag{39.11}$$

where c_B and c_A are the amounts of base and acid that have been added to the system, respectively. Because it is usually impractical to measure the amounts of acid and base, a new quantity, ***alkalinity***, is defined as the acid-neutralizing capacity of the system.

Alkalinity can be formulated as[†]

$$\text{Alk} = [HCO_3^-] + 2[CO_3^{2-}] + [OH^-] - [H^+] \tag{39.12}$$

To summarize, we now have developed five equations:

$$K_1 = \frac{[H^+][HCO_3^-]}{[H_2CO_3^*]} \tag{39.13}$$

$$K_2 = \frac{[H^+][CO_3^{2-}]}{[HCO_3^-]} \tag{39.14}$$

$$K_w = [H^+][OH^-] \tag{39.15}$$

$$c_T = [H_2CO_3^*] + [HCO_3^-] + [CO_3^{2-}] \tag{39.16}$$

$$\text{Alk} = [HCO_3^-] + 2[CO_3^{2-}] + [OH^-] - [H^+] \tag{39.17}$$

with five unknowns: $[H_2CO_3^*]$, $[HCO_3^-]$, $[CO_3^{2-}]$, $[H^+]$, and $[OH^-]$.

There are a variety of ways to solve this system of equations. Stumm and Morgan (1981) have shown how Eqs. 39.13 and 39.14 can be solved for

$$[H_2CO_3^*] = \frac{[H^+][HCO_3^-]}{K_1} \tag{39.18}$$

$$[CO_3^{2-}] = \frac{[HCO_3^-]K_2}{[H^+]} \tag{39.19}$$

These results can be substituted into the mass balance (Eq. 39.16), which can be solved for

$$[H_2CO_3^*] = F_0 c_T \tag{39.20}$$

$$[HCO_3^-] = F_1 c_T \tag{39.21}$$

$$[CO_3^{2-}] = F_2 c_T \tag{39.22}$$

where F_0, F_1, and F_2 are the fractions of the total inorganic carbon in carbonic acid, bicarbonate, and carbonate, respectively:[‡]

$$F_0 = \frac{[H^+]^2}{[H^+]^2 + K_1[H^+] + K_1K_2} \tag{39.23}$$

$$F_1 = \frac{K_1[H^+]}{[H^+]^2 + K_1[H^+] + K_1K_2} \tag{39.24}$$

$$F_2 = \frac{K_1K_2}{[H^+]^2 + K_1[H^+] + K_1K_2} \tag{39.25}$$

[†] Our definition of alkalinity is applicable only to systems in which other buffers are at negligible levels. Other reactions contribute to alkalinity in fresh waters, but these are often ignored because of the overriding importance of inorganic carbon in many systems. Carbonate chemistry is not adequate for describing saltwater systems, and other buffers must be considered.

[‡] Note that Stumm and Morgan employ the parameters α_0, α_1, and α_2 to designate the fractions. We have used F's to be consistent with nomenclature used elsewhere in this text.

Now Eqs. 39.20 to 39.22 can be substituted, along with Eq. 39.15, into the charge balance to give

$$0 = F_1 c_T + 2F_2 c_T + \frac{K_w}{[H^+]} - [H^+] - \text{Alk} \tag{39.26}$$

Although it might not be apparent, Eq. 39.26 is a fourth-order polynomial in $[H^+]$. Therefore it can be solved for $[H^+]$.

EXAMPLE 39.1. pH FOR THE INORGANIC CARBON SYSTEM. Determine the pH for a system with an alkalinity of 2 meq L^{-1} and a total inorganic carbon concentration of 3 mM L^{-1}. Assume that $pK_w = 14$, $pK_1 = 6.3$, and $pK_2 = 10.3$.

Solution: First, the terms in Eq. 39.26 can be combined to give a fourth-order polynomial in $[H^+]$:

$$[H^+]^4 + (K_1 + \text{Alk})[H^+]^3 + (K_1K_2 + \text{Alk}K_1 - K_w - K_1c_T)[H^+]^2 \\ + (\text{Alk}K_1K_2 - K_1K_w - 2K_1K_2c_T)[H^+] - K_1K_2K_w = 0$$

Substituting parameter values gives

$$[H^+]^4 + 2.001 \times 10^{-3}[H^+]^3 - 5.012 \times 10^{-10}[H^+]^2 - 1.055 \times 10^{-19}[H^+] \\ -2.512 \times 10^{-31} = 0$$

This equation can be solved for $[H^+] = 2.51 \times 10^{-7}(\text{pH} = 6.6)$. Substituting this value back into Eqs. 39.20 through 39.22 gives

$$[H_2CO_3{}^*] = \frac{(2.51 \times 10^{-7})^2}{(2.51 \times 10^{-7})^2 + 10^{-6.3}(2.51 \times 10^{-7}) + 10^{-6.3}10^{-10.3}} 3 \times 10^{-3}$$
$$= 0.33304(3 \times 10^{-3}) = 0.001M$$

$$[HCO_3{}^-] = \frac{10^{-6.3}(2.51 \times 10^{-7})}{(2.51 \times 10^{-7})^2 + 10^{-6.3}(2.51 \times 10^{-7}) + 10^{-6.3}10^{-10.3}} 3 \times 10^{-3}$$
$$= 0.666562(3 \times 10^{-3}) = 0.002M$$

$$[CO_3{}^{2-}] = \frac{10^{-6.3}10^{-10.3}}{(2.51 \times 10^{-7})^2 + 10^{-6.3}(2.51 \times 10^{-7}) + 10^{-6.3}10^{-10.3}} 3 \times 10^{-3}$$
$$= 0.000133(3 \times 10^{-3}) = 1.33 \times 10^{-4}M$$

39.2 SLOW REACTIONS: GAS TRANSFER AND PLANTS

The inorganic carbon buffering system is influenced and enhanced greatly by a number of heterogeneous reactions that occur in nature. We may recall that heterogeneous reactions are those that occur between phases, as opposed to homogeneous reactions, which occur within a single phase. For the inorganic carbon system, the heterogeneous reactions include atmospheric exchange of CO_2, dissolution/precipitation of carbonate minerals such as calcium carbonate ($CaCO_3$), and photosynthesis/respiration.

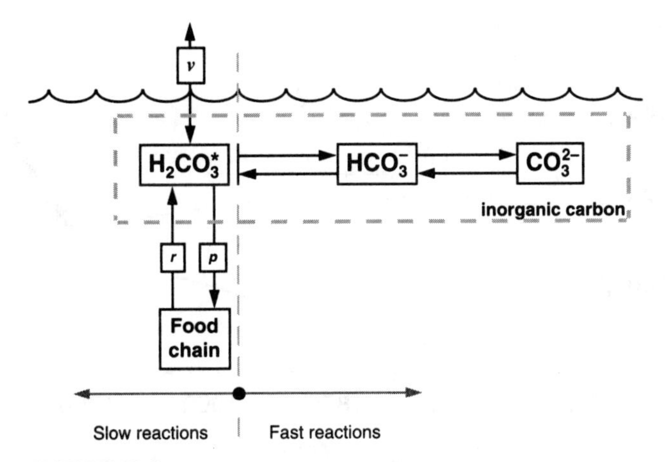

FIGURE 39.1
A simplified representation of the inorganic carbon system and its relationship with the food chain. The inorganic carbon species shown are carbonate (CO_3^{2-}), bicarbonate (HCO_3^-), and carbon dioxide/carbonic acid ($H_2CO_3^*$). The reactions between these species are relatively fast, with rates on the order of seconds to minutes. In contrast the reactions with the food chain [photosynthesis (p) and respiration (r)] and volatilization (v) are typically much slower, with rates on the order of days.

Figure 39.1 is a simple representation of how inorganic carbon interacts with the atmosphere and the food chain. Carbon dioxide enters and leaves the inorganic carbon system via two major pathways: atmospheric and biological exchange processes. Note that other heterogeneous reactions such as calcium carbonate dissolution and precipitation are neglected in this simple representation.

As can be seen in the figure, the reactions are divided into fast and slow interactions. Before developing a model, we can review these interactions.

Atmospheric exchange. The atmospheric exchange can be quantified via

$$W_{\text{atm}} = v_v A_s \left([H_2CO_3^*]_s - [H_2CO_3^*] \right) \left(\frac{10^3 \text{L}}{\text{m}^3} \right) \tag{39.27}$$

where W_{atm} = "loading" of inorganic carbon due to atmospheric exchange (mole d^{-1})

v_v = volatilization transfer coefficient (m d^{-1})

A_s = surface area of the air-water interface (m^2)

$[H_2CO_3^*]_s$ = saturation concentration of carbon dioxide (mole L^{-1})

Carbon dioxide is overwhelmingly liquid-film controlled. The transfer coefficient can be correlated with the oxygen exchange coefficient, K_l (m d^{-1}), by (Mills et al. 1982)

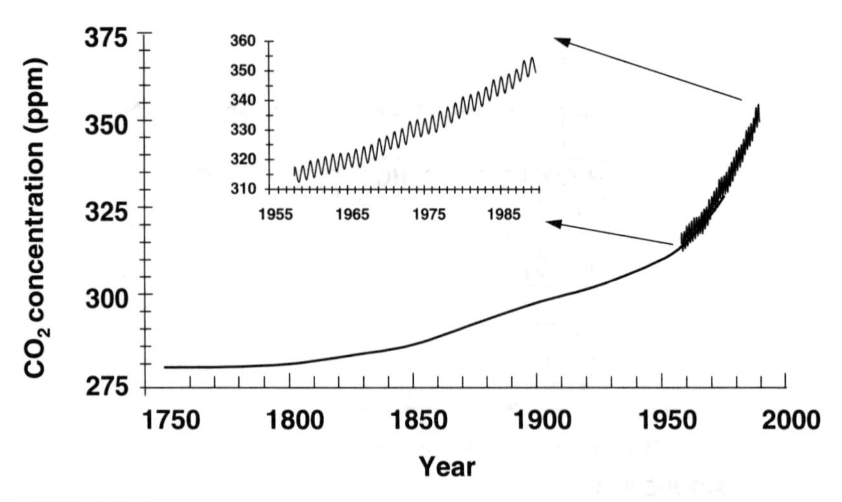

FIGURE 39.2
Recent concentrations of carbon dioxide in the atmosphere as recorded at the Mauna Loa Observatory, Hawaii. The annual cycle is superimposed on the long-term trend (from Keeling et al. 1982, 1989; Neftel et al. 1985).

$$v_v = \left(\frac{32}{M}\right)^{0.25} K_l \tag{39.28}$$

where M = molecular weight, which for carbon dioxide is 44.

The saturation concentration, $[H_2CO_3^*]_s$, is the level that the water would reach under steady-state conditions for a given temperature. This level can be estimated on the basis of Henry's law,

$$[H_2CO_3^*]_s = K_H p_{CO_2} \tag{39.29}$$

where K_H = Henry's constant [mole (L atm)$^{-1}$] and p_{CO_2} = partial pressure of carbon dioxide in the atmosphere (atm).

The value of K_H can be computed as a function of temperature by (Edmond and Gieskes 1970)

$$pK_H = -\frac{2385.73}{T_a} - 0.0152642T_a + 14.0184 \tag{39.30}$$

The partial pressure of CO_2 has been increasing, largely due to the combustion of fossil fuels (Fig. 39.2). Present values are approximately $10^{-3.45}$ atm (= 355 ppm).

Photosynthesis/Respiration. In addition to reaeration, CO_2 is introduced to and removed from a lake via biological reactions. To incorporate these mechanisms into our inorganic carbon model, we assume that photosynthesis and respiration proceed via the following simple formula:

$$6CO_2 + 6H_2O \underset{\text{respir}}{\overset{\text{photo}}{\rightleftharpoons}} C_6H_{12}O_6 + 6O_2 \tag{39.31}$$

The reaction can be formulated in a manner compatible with the carbon mass balance as

$$W_{r/p} = a_{cx}(R - P)A_s \tag{39.32}$$

where $W_{r/p}$ = net carbon load due to the combined effects of respiration and photosynthesis (mole d^{-1})

R and P = areal rates of respiration and photosynthesis (g m^{-2} d^{-1})

a_{cx} = a factor to convert the mass units to moles based on the stoichiometry of Eq. 39.31 as well as the mass units used for R and P

For example suppose that R and P are calculated in gO_2 m^{-2} d^{-1}, as is commonly done for Streeter-Phelps models. The conversion factor could be calculated as

$$a_{co} = \frac{6 \text{ mole } CO_2}{6 \times 32 \text{ g}O_2} = 0.03125 \frac{\text{mole } CO_2}{\text{g}O_2} \tag{39.33}$$

where the notation a_{co} denotes that the conversion is between carbon and oxygen units.

In summary the heterogeneous reactions with the atmosphere and the food chain proceed on time scales of hours to days and serve to introduce and remove CO_2 from the water. In turn the carbon dioxide takes part in the fast reactions among the inorganic carbon species as delineated above.

39.3 MODELING pH IN NATURAL WATERS

Now that I have delineated the major reactions affecting the inorganic carbon system, I'll show how they can be formulated to calculate the pH of natural waters. The key to this coupling is that the heterogeneous reactions governing the interrelationships between the inorganic carbon species and water are typically much faster than atmospheric transfer and photosynthesis/respiration. Therefore the computational strategy (as originally devised by Di Toro 1976) is to use a dynamic, differential equation to compute the effect of the heterogeneous reactions on the total inorganic carbon level of the system with a time step on the order of days. Because they are much more rapid, the homogeneous reactions can always be assumed to be at equilibrium on this time scale. Thus algebraic solution techniques can be employed to solve for pH and the individual inorganic carbon species at each time step.

Mass balances can be developed for the individual species involved in the inorganic carbon system,

$$V\frac{d[H_2CO_3^*]}{dt} = W_{\text{atm}} + W_{r/p} - k_3V[H_2CO_3^*] + k_4V[H^+][HCO_3^-] \tag{39.34}$$

$$V\frac{d[HCO_3^-]}{dt} = k_3V[H_2CO_3^*] - k_4V[H^+][HCO_3^-] - k_5V[HCO_3^-] + k_6V[H^+][CO_3^{2-}] \tag{39.35}$$

$$V\frac{d[CO_3^{2-}]}{dt} = k_5V[HCO_3^-] - k_6V[H^+][CO_3^{2-}] \tag{39.36}$$

Since the reactions between the inorganic carbon species are much faster than the gains and losses due to atmospheric and biotic exchange, a local equilibrium assumption can be made and Eqs. 39.34 through 39.36 can be combined to yield

$$V\frac{dc_T}{dt} = W_{\text{atm}} + W_{r/p} \tag{39.37}$$

where c_T is defined by Eq. 39.10. Equation 39.37 allows computation of the dynamics of total inorganic carbon as a function of biotic and atmospheric interactions.

EXAMPLE 39.2. pH FOR THE INORGANIC CARBON SYSTEM WITH BIOTIC AND ATMOSPHERIC INTERACTIONS. Determine the steady-state pH for the same system as in Example 39.1, except that the system is open to the atmosphere [$pK_H = 1.46$, $p_{CO_2} = 10^{-3.45}$ atm, and $v_v = 0.5$ m d^{-1}] and is subject to a level of $R-P$ of 0.5 gO m^{-2} d^{-1}. Note that this means that respiration dominates photosynthesis and that the plants are adding CO_2 to the water. As in Example 39.1 use an alkalinity of 2 meq L^{-1} and assume that $pK_w = 14$, $pK_1 = 6.3$, and $pK_2 = 10.3$.

Solution: The steady-state inorganic carbon budget can be written as

$$v_v([H_2CO_3{}^*]_s - [H_2CO_3{}^*]) + a_{co}(R - P) = 0$$

where a_{co} is 0.03125 mole CO_2/gO_2 as calculated previously in Eq. 39.33. The saturation concentration of carbon dioxide can be computed as (Eq. 39.29)

$$[H_2CO_3{}^*]_s = 10^{-1.46}10^{-3.45} = 10^{-4.91} = 1.23 \times 10^{-5}\ M$$

These values can be substituted along with other parameters into the carbon budget,

$$0.5\frac{\text{m}}{\text{d}}\left(1.23 \times 10^{-5}\frac{\text{mole}}{\text{L}} - [H_2CO_3{}^*]\right)\left(\frac{10^3\ \text{L}}{\text{m}^3}\right) + 0.03125\frac{\text{mole}}{\text{g}}0.5\frac{\text{g}}{\text{m}^2\text{d}} = 0$$

which can be solved for

$$[H_2CO_3{}^*] = 4.355 \times 10^{-5}\ \text{mole L}^{-1}$$

Thus, for this case, the carbonic acid level is fixed by the atmospheric and biotic interactions. At this point, we still have five equations, but c_T has replaced $[H_2CO_3{}^*]$ as an unknown.

Now, because we know the level of carbon dioxide, Eq. 39.20 can be solved for

$$c_T = \frac{[H_2CO_3{}^*]}{F_0} \tag{39.38}$$

which can be substituted into Eq. 39.26 to give

$$0 = ([H_2CO_3{}^*]/F_0)(F_1 + 2F_2) + \frac{K_w}{[H^+]} - [H^+] - \text{Alk}$$

Thus the problem reduces to solving this equation for $[H^+]$. This can be done numerically for $[H^+] = 1.1 \times 10^{-8}$(pH = 7.96). Substituting this value back into Eqs. 39.38, 39.21, and 39.22 gives

$$c_T = \frac{(1.1 \times 10^{-8})^2 + 10^{-6.3}(1.1 \times 10^{-8}) + 10^{-6.3}10^{-10.3}}{(1.1 \times 10^{-8})^2}4.355 \times 10^{-5}$$

$$= 0.002034\ M$$

$$[HCO_3^-] = \frac{10^{-6.3}(1.1 \times 10^{-8})}{(1.1 \times 10^{-8})^2 + 10^{-6.3}(1.1 \times 10^{-8}) + 10^{-6.3}10^{-10.3}} 2.034 \times 10^{-3}$$

$$= 0.001981\ M$$

$$[CO_3^{2-}] = \frac{10^{-6.3}10^{-10.3}}{(1.1 \times 10^{-8})^2 + 10^{-6.3}(1.1 \times 10^{-8}) + 10^{-6.3}10^{-10.3}} 2.034 \times 10^{-3}$$

$$= 9.011 \times 10^{-6}\ M$$

PROBLEMS

39.1. Develop the fourth-order polynomial in $[H^+]$ by combining the terms in Eq. 39.26.

39.2. Determine the pH for a system with an alkalinity of 3 meq L^{-1} and a total inorganic carbon concentration of 3.5 mM L^{-1}. Assume all other values are the same as in Example 39.1.

39.3. Determine the steady-state pH for the same system as in Example 39.1, except that the system is open to the atmosphere ($pK_H = 1.46$, $p_{CO_2} = 10^{-3.45}$ atm, and $v_v = 0.5$ m d^{-1}) and is not subject to significant plant activity. As in Prob. 39.2, use an alkalinity of 2 meq L^{-1}.

39.4. Repeat Example 39.2, but perform a time-variable calculation over 3 d, where the plant photosynthesis varies diurnally as a half-sinusoid with a photoperiod of 0.5 d and a peak of 3 gO m^{-2} d^{-1}. Note that respiration is 1.5 gO m^{-2} d^{-1} and the system is 3 m deep.

39.5 A wastewater treatment plant ($Q = 1 \times 10^6$ m^3 d^{-1}; $T = 25$°C; pH = 6.1; Alk = 0.003 eq L^{-1}) discharges into a river ($Q = 5 \times 10^6$ m^3 d^{-1}; $T = 10$°C; pH = 7.2; Alk = 0.0012 eq L^{-1}). Assuming instantaneous mixing, calculate the (*a*) alkalinity, (*b*) total inorganic carbon, and (*c*) pH at the mixing point.

PART VII

Toxics

The lectures in this part provide an introduction and an overview of water-quality modeling of toxic substances.

Lecture 40 is an introduction to the toxic substance problem. After an overview of the problem, I develop a mass balance for a toxic substance in a well-mixed lake. My intent is to illustrate how the modeling of conventional pollutants and of toxic substances differ. In particular the role of suspended solids and solid/liquid partitioning is introduced. Then I add a sediment layer underneath the lake, and develop a solids budget and a toxic substance model for the sediment-water system. Steady-state and time-variable solutions are generated and compared with the well-mixed lake model without sediment-water interactions.

The following lectures are devoted to the mechanisms that govern the ways in which the toxicant is transported and transformed within the environment. *Lecture 41* deals with sorption of toxicants onto particulate matter and volatilization across the air-water interface. The first part of the lecture explains the mechanism of sorption and delineates how it is quantified. The second part is devoted to volatilization. *Lecture 42* describes reactions that act to deplete toxics from water. These include photolysis, hydrolysis, and biodegradation.

Lecture 43 describes how radionuclides and heavy metals are modeled. *Lecture 44* illustrates how toxic substances are modeled in streams and estuaries. Finally *Lecture 45* deals with modeling the fate of toxicants within organisms and the food chain.

LECTURE 40

Introduction to Toxic-Substance Modeling

LECTURE OVERVIEW: After outlining major characteristics of the toxic-substance problem, I develop a simple toxic-substance budget for a well-mixed lake, emphasizing the impact of sorption and sediment-water interactions on contaminant dynamics. Steady-state and time-variable solutions are generated and compared with the well-mixed lake model without sediment-water interactions.

The following lecture is intended to provide an overview of toxic modeling. Our emphasis will be on developing simple mass balances for a well-mixed lake and its underlying sediments. In this way we can investigate the similarities and differences between conventional and toxic-pollutant models.

40.1 THE TOXICS PROBLEM

To this point the models described have almost exclusively dealt with what are called conventional pollutants. That is, pollutants that are a natural by-product of human activities and that overstimulate the natural production/decomposition cycle. Before developing toxic models it is useful to reflect on how toxics differ from conventional wastes.

40.1.1 Contrast of Conventional and Toxic Pollution

The major differences between conventional and toxic pollution relate to four areas:

1. *Natural versus alien.* The conventional pollution problem typically deals with the natural cycle of organic production and decomposition (recall Fig. 19.1). The

discharge of sewage adds both organic matter and inorganic nutrients to a water body. The decomposition of the organic matter by bacteria can result in oxygen depletion. The inorganic nutrients can stimulate excess plant growth. In both cases the problem involves an overstimulation of the natural processes governing the water body. In contrast many toxics do not occur naturally. Examples are pesticides and other synthetic organics. For such "alien" pollutants the problem is one of poisoning or interference with natural processes.

2. *Aesthetics versus health.* Although it would be an overstatement to contend that conventional pollution deals solely with aesthetics, a strong case can be made that the mitigation of "visual" pollution has been a prime motivation for conventional waste treatment. In contrast the toxic-substance issue is almost totally dominated by health concerns. Most toxic remediation focuses on the contamination of (*a*) drinking water and (*b*) aquatic food stuffs.
3. *Few versus many.* Conventional water-quality management deals with on the order of about 10 "pollutants." In contrast there are tens of thousands of organic chemicals that could potentially be introduced into our natural waters. Further, a large fraction of these are synthetic and are increasing every year (Schwarzenbach et al. 1993). If even a fraction of these proved toxic, the sheer numbers of potential toxicants will have a profound effect on the resulting control strategies. Further it is difficult to obtain detailed information on the factors governing their transport and fate in the environment. The study of problems such as dissolved-oxygen depletion and eutrophication is facilitated by the fact that they involve a few chemicals. In contrast toxicant modeling is complicated by the vast number of contaminants that are involved.
4. *Single species versus solid/liquid partitioning.* As was done in previous lectures, the conventional paradigm usually treats the pollutant as a single species. Consequently its strength in the water body is measured by a single concentration. In contrast the transport, fate, and ecosystem impact of toxicants is intimately connected with how they partition or associate with solid matter in and below the water body. Thus toxic-substance analysis must distinguish between dissolved and particulate forms. This distinction has an impact on transport and fate in the sense that certain mechanisms differentially impact the two forms. For example volatilization acts only on the dissolved component. The distinction also has importance from the standpoint of assessing ecosystem impact. For example if biotic contamination is being assessed, toxicant mass normalized to biotic mass is a better metric of contamination than toxicant mass normalized to water volume.

The four points outlined above have implications for the modeling of toxicants. Rather than an assimilative-capacity approach (recall Lec. 1), an alternative "screening" approach is sometimes adopted. In essence the strategy is designed to determine which toxicants pose the greatest threats in a particular environment. One way to focus this strategy is to divide the toxicants into classes.[†]

[†] It should be noted that compound-specific approaches are often required for adequate characterization and prediction of fate and transport. Schwarzenbach et al. (1993) is a good reference for this approach.

40.1.2 Classification of Toxics

There are many ways that toxic substances can be broken into categories. For example they can be divided by their functions (e.g., pesticides, cleaning products, fuels, etc.) or their primary industrial association (agriculture, petroleum refining, textiles, etc.). Alternatively a classification could be based on chemical criteria.

One effort at classification, that has become commonly used, was developed to categorize the so-called "priority pollutants." As the result of a court settlement involving the EPA and a number of environmental groups, the EPA published such a list. Although the list has been the subject of criticism (see Keith and Telliard 1979), it provided some structure to the toxics problem.

The list of priority pollutants consists of 129 inorganic and organic toxicants. The inorganics consisted of heavy metals and other contaminants such as asbestos and cyanide. As listed in Table 40.1, the organics were divided into nine categories. The resulting classification scheme is a mixed bag of usage (pesticides) and some broad chemical classes.

Although the categories in Table 40.1 could be significantly modified, improved, and refined, we will adopt them for use in this book. In a later section we'll attempt to use their key chemical properties in conjunction with mathematical models to assess how these chemicals could potentially impact lakes.

40.2 SOLID-LIQUID PARTITIONING

In previous lectures we have focused on conventional pollutants. In all these examples the pollutant was represented by a single concentration. Toxic substances differ from conventional pollutants in that they are divided or "partitioned" into particulate and dissolved forms.

The primary reason for solid-liquid partitioning is to produce a more mechanistically accurate characterization of the toxicant mass balance. In particular it is known that several key mechanisms act selectively on one or the other of the two forms. For example volatilization (that is, loss of the contaminant from the water to the atmosphere) acts only on the dissolved fraction. Conversely settling acts solely on the particulate-associated fraction.

Mathematically the total contaminant concentration, c (μg m^{-3}), is separated into the two components

$$c = c_d + c_p \tag{40.1}$$

where c_d = dissolved component (μg m^{-3}) and c_p = particulate component (μg m^{-3}). These components are assumed to represent fixed fractions of the total concentration, as in

$$c_d = F_d c \tag{40.2}$$

and

$$c_p = F_p c \tag{40.3}$$

where the parameters F_d and F_p are the fractions of the total contaminant that are in dissolved and particulate form, respectively. These fractions are a function of the

TABLE 40.1
The categories of the organic priority pollutants (CEQ 1978)

Pollutant	Characteristics	Sources and Remarks
Pesticides: Generally chlorinated hydrocarbons	Readily assimilated by aquatic animals, fat-soluble, concentrated through the food chain (biomagnified), persistent in soil and sediments	Direct application to farm and forest lands, runoff from lawns and gardens, urban runoff, discharge in industrial wastewater. Several chlorinated hydrocarbon pesticides already restricted by EPA: aldrin, dieldrin, DDT, DDD, endrin, heptachlor, lindane, chlordane
Polychlorinated biphenyls (PCBs): Used in electrical capacitors and transformers, paints, plastics, insecticides, other industrial products	Readily assimilated by aquatic animals, fat-soluble, subject to biomagnification, persistent, chemically similar to the chlorinated hydrocarbons	Municipal and industrial discharges disposed of in dumps and landfills. TOSCA ban on production after 6/1/79 but will persist in sediments; restrictions in many freshwater fisheries as result of PCB pollution (e.g., lower Hudson, upper Housatonic, parts of Lake Michigan)
Halogenated aliphatics (HAHs): Used in fire extinguishers, refrigerants, propellants, pesticides, solvents for oils and greases, and in dry cleaning	Largest single class of "priority pollutants"; can cause damage to central nervous system and liver, not very persistent	Produced by chlorination of water, vaporization during use. Large-volume industrial chemicals, widely dispersed, but less threat to the environment than persistent chemicals
Ethers: Used mainly as solvents for polymer plastics	Potent carcinogens; aquatic toxicity and fate not well understood	Escape during production and use. Although some are volatile, ethers have been identified in some natural waters
Phthalate esters: Used chiefly in production of polyvinyl chloride and thermoplastics as plasticizers	Common aquatic pollutant, moderately toxic but teratogenic and mutagenic properties in low concentration; aquatic invertebrates are particularly sensitive to toxic effects; persistent and can be biomagnified	Waste disposal vaporization during use (in nonplastics)

TABLE 40.1 ***(continued)***

Pollutant	Characteristics	Sources and Remarks
Monocyclic aromatics (MAHs) (excluding phenols, cresols, and phthalates): Used in the manufacture of other chemicals, explosives, dyes, and pigments, and in solvents, fungicides, and herbicides	Central nervous system depressant; can damage liver and kidneys	Enter environment during production and by-product production states by direct volatilization, wastewater
Phenols: Large-volume industrial compounds used chiefly as chemical intermediates in the production of synthetic polymers, dyestuffs, pigments, herbicides, and pesticides	Toxicity increases with degree of chlorination of the phenolic molecule; very low concentrations can taint fish flesh and impart objectionable odor and taste to drinking water; difficult to remove from water by conventional treatment; carcinogenic in mice	Occur naturally in fossil fuels; wastewater from coke ovens, oil refineries, tar distillation plants, herbicide and plastic manufacturing, can all contain phenolic compounds
Polycyclic aromatic hydrocarbons (PAHs): Used as dyestuffs, chemical intermediates, pesticides, herbicides, motor oils, and fuels	Carcinogenic in animals and indirectly linked to cancer in humans; most work done on air pollution; more is needed on the aquatic toxicity of these compounds; not persistent and are biodegradable, although bioaccumulation can occur	Fossil fuels (use, spills, and production), incomplete combustion of hydrocarbons
Nitrosamines: Used in the production of organic chemicals and rubber; patents exist on processes using these compounds	Tests on laboratory animals have shown the nitrosamines to be some of the most potent carcinogens	Production and use can occur spontaneously in food cooking operations

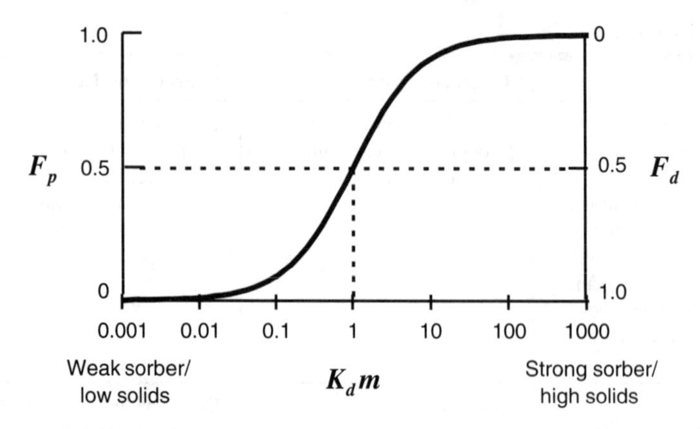

FIGURE 40.1
Relationship of the fraction of contaminant in particulate and dissolved form to the dimensionless number: $K_d m$.

contaminant's partition coefficient and the lake's suspended solids concentration, as in

$$F_d = \frac{1}{1 + K_d m} \tag{40.4}$$

and

$$F_p = \frac{K_d m}{1 + K_d m} \tag{40.5}$$

where K_d = a partition coefficient ($m^3\ g^{-1}$) and m = suspended solids concentration ($g\ m^{-3}$). Note, we will derive these formulas when we discuss sorption in the next lecture.

The partition coefficient quantifies the tendency of the contaminant to associate with solid matter,

$$F_d + F_p = 1 \tag{40.6}$$

In other words the sum of the two fractions is equal to 1.

The relationship of Eqs. 40.4 and 40.5 to suspended solids and the partition coefficient is illustrated by Fig. 40.1. Suspended solids range from approximately 1 $g\ m^{-3}$ for clear lakes to 100 $g\ m^{-3}$ for turbid rivers. Partition coefficients range from about 10^{-4} to 10 $m^3\ g^{-1}$. Therefore the dimensionless parameter group $K_d m$ would range from approximately 0.0001 to 1000. As $K_d m$ increases, an increasingly higher fraction of the toxicant will associate with solid matter. Thus for weakly sorbing contaminants (low K_d) in systems with low suspended solids (low m), the contaminant will be predominantly in dissolved form. For strong sorbers in turbid systems, the contaminant will be strongly associated with suspended solids.

40.3 TOXICS MODEL FOR A CSTR

Now that we have a way to represent solid-liquid partitioning, we can use this knowledge to develop a simple mass balance of a toxicant for a well-mixed lake or CSTR.

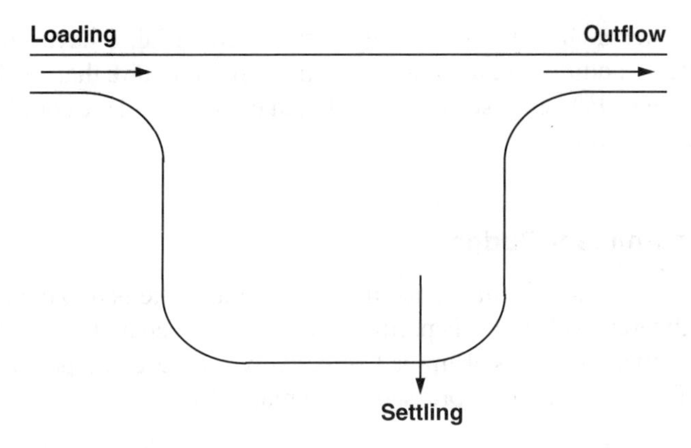

FIGURE 40.2
Schematic of a solids budget for a well-mixed lake with no sediment feedback of solids.

In this section we do this in the absence of sediment-water interactions. Then, in the following section, we add a sediment compartment and assess its effect.

40.3.1 Solids Budget

Recall that for our conventional models, a water budget was required to allow us to calculate the effect of water transport on our pollutant budget. One implication of solid-liquid partitioning is that a solids budget is now required. Just as the water-flow rate allowed us to determine how much pollutant was carried along with the water, a solids budget allows us to estimate how much contaminant will be transported along with the solids.

As in Fig. 40.2, solids are introduced to a well-mixed lake via inflowing tributaries and are purged from the system by settling and outflow. A solids balance can be written as

$$V\frac{dm}{dt} = Qm_{\text{in}} - Qm - v_s A m \tag{40.7}$$

where m_{in} = inflow suspended solids concentration (g m^{-3}) and v_s = settling velocity (m yr^{-1}).

At steady-state, Eq. 40.7 can be solved for

$$m = \frac{Qm_{\text{in}}}{Q + v_s A} \tag{40.8}$$

Alternatively the result can be expressed as the transfer coefficient

$$\beta = \frac{Q}{Q + v_s A} \tag{40.9}$$

Because of the long time horizons exhibited by many toxic substances, the steady-state assumption invoked in deriving Eq. 40.8 has conventionally been adopted. It should be noted, however, that this assumption may not be accurate for certain systems and substances. For example some lakes have highly variable solid budgets. In addition highly reactive toxics would require that a subannual time scale be adopted. For such scales a steady-state solids budget could represent an unrealistic assumption.

40.3.2 Contaminant Budget

A mass balance for a toxic substance in a well-mixed lake is depicted in Fig. 40.3. Notice that the toxic substance is partitioned into dissolved and particulate fractions and that volatilization and settling act selectively on these components. The mass balance for Fig. 40.3 can be expressed mathematically as

$$V\frac{dc}{dt} = Qc_{\text{in}} - Qc - kVc - v_v AF_d c - v_s AF_p c \tag{40.10}$$

where v_v = a volatilization mass-transfer coefficient (m yr^{-1}). Notice how we have utilized the parameters F_p and F_d to modify the mechanisms that act selectively on the particulate and dissolved fractions, respectively.

At steady-state, Eq. 40.10 can be solved for

$$c = \frac{Qc_{\text{in}}}{Q + kV + v_v AF_d + v_s AF_p} \tag{40.11}$$

The transfer coefficient can also be seen to be

$$\beta = \frac{Q}{Q + kV + v_v AF_d + v_s AF_p} \tag{40.12}$$

These results are structurally identical to the solutions for conventional pollutants. However, the toxicant model is certainly more mechanistic. For example it

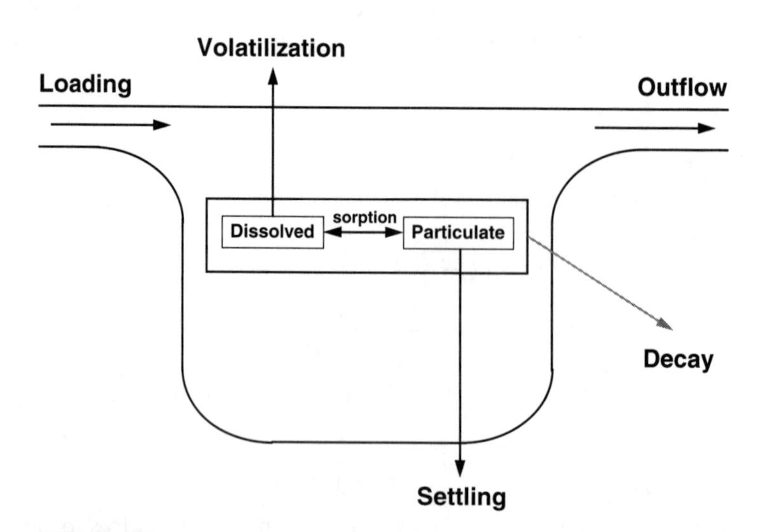

FIGURE 40.3
Schematic of a toxic substance budget for a well-mixed lake with no sediment feedback of solids.

includes a more mechanistic representation of settling and the inclusion of a volatilization loss. Together with Eqs. 40.4 and 40.5, the toxicant model provides a way to determine how the mechanisms of flushing, sorption, volatilization, and decay act to purge a toxicant from the well-mixed lake.

EXAMPLE 40.1. LAKE HURON PCB (WITHOUT SEDIMENTS). Morphometric and hydrologic information for Lake Huron (Fig. 40.4) are summarized in Table 40.2. Other relevant data needed for the computation are

Parameter	Symbol	Value	Units
Suspended solids concentration	m	0.531	g m^{-3}
Suspended solids load	W_s	6.15×10^{12}	g yr^{-1}
PCB load	W_c	5.37×10^{12}	μg yr^{-1}

If sediment feedback is neglected, determine the level of high-molecular-weight PCBs in the lake. Such toxicants have the following characteristics: $K_d = 0.0301$ m^3 g^{-1}

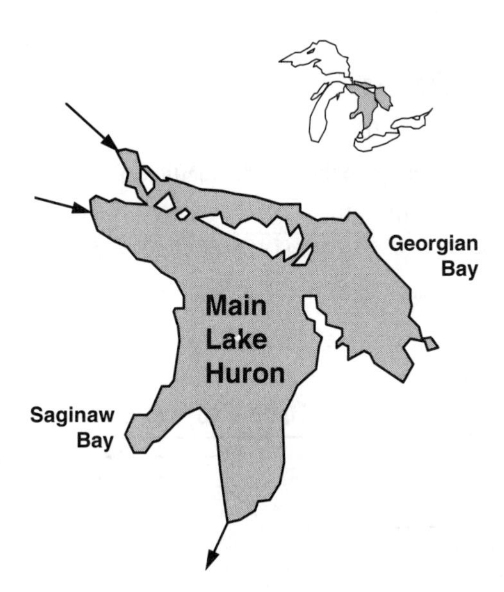

FIGURE 40.4
Lake Huron.

TABLE 40.2
Parameters for Lake Huron

Parameter	Symbol	Value	Units
Volume	V	3515×10^9	m^3
Surface area	A	$59{,}570 \times 10^6$	m^2
Depth	H	59	m
Outflow	Q	161×10^9	m^3 yr^{-1}

and $v_v = 178.45$ m yr^{-1}. In particular determine (*a*) the settling velocity of solids, (*b*) the inflow concentration of contaminant, (*c*) the transfer coefficient and the steady-state concentration, and (*d*) the 95% response time.

Solution: (*a*) At steady-state, Eq. 40.7 can be rearranged to compute a settling velocity for solids,

$$v = \frac{Qm_{\text{in}} - Qm}{Am} = \frac{6.15 \times 10^{12} - [161 \times 10^{9}(0.531)]}{59{,}570 \times 10^{6}(0.531)} = 191.82 \text{ m yr}^{-1}$$

This value is equivalent to 0.526 m d^{-1}, which is at the lower end of the range of settling velocities reported for fine-grained solids (Thomann and Mueller 1987, O'Connor 1988). One reason for the low value might be that this computation does not consider the fact that resuspension might be occurring concurrently with settling. This mechanism is included when sediment-water interactions are incorporated into our framework later in this lecture.

(*b*) The values for Lake Huron PCB loading and flow rate can be used to calculate the inflow concentration,

$$c_{\text{in}} = \frac{W}{Q} = \frac{5.37 \times 10^{12}}{161 \times 10^{9}} = 33.35\ \mu\text{g m}^{-3}$$

(*c*) Equations 40.4 and 40.5 can be used to calculate

$$F_d = \frac{1}{1 + 0.0301(0.531)} = 0.9843$$

and

$$F_p = 1 - F_d = 1 - 0.9843 = 0.0157$$

Equation 40.12 can then be employed to estimate the transfer coefficient

$$\beta = \frac{161 \times 10^{9}}{161 \times 10^{9} + [178.45(0.9843)59{,}570 \times 10^{6}] + [191.82(0.0157)59{,}570 \times 10^{6}]}$$

$$= 0.149 (= 14.9\%)$$

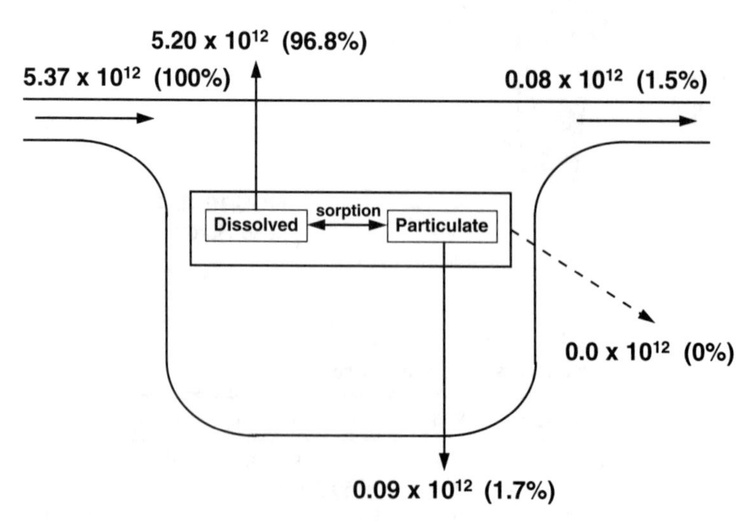

FIGURE 40.5
Budget of PCB for Lake Huron without sediment feedback.

which can be used to compute

$$c = 0.149(33.35) = 0.497\ \mu\text{g m}^{-3}$$

Thus, purging mechanisms attenuate the inflow concentration by nearly two orders of magnitude. The primary reason for this is the high level of volatilization, as is evident from inspection of the transfer coefficient. It is also clearly shown in Fig. 40.5, where volatilization accounts for approximately 97% of the PCB budget.

(*d*) The 95% response time for Lake Huron can be estimated by

$$t_{95} = \frac{3V}{Q + kV + v_v F_d A + v_s F_p A} = \frac{3(3515 \times 10^9)}{10.8 \times 10^{12}} = 0.98 \text{ yr}$$

According to our model the lake would take just under a year to show 95% recovery following a PCB load reduction. As with the steady-state prediction, the magnitude of the response time is strongly dictated by volatilization.

40.4 TOXICS MODEL FOR A CSTR WITH SEDIMENTS

Now that we have contrasted conventional pollutants and toxics for a simple CSTR, we incorporate another level of complexity into our model. This is done by simply adding a sediment layer below the lake. As in Fig. 40.6, the system is idealized as a well-mixed surface-water layer underlain by a well-mixed sediment layer. In the following discussion we designate these layers by the subscripts 1 and 2, respectively.

Before developing our model, we should recall from Lec. 17 that special parameters are required to characterize a sediment layer. These are the porosity,

$$\phi = \frac{V_{d2}}{V_2} \tag{40.13}$$

where V_d = volume of the liquid part of the sediment layer (m^3) and V_2 = total volume of the sediment layer (m^3), and the density,

$$\rho = \frac{M_2}{V_{p2}} \tag{40.14}$$

where ρ = density (g m^{-3})
M_2 = mass of the solid phase in the sediments (g)
V_{p2} = volume of the solid matter (m^{-3})

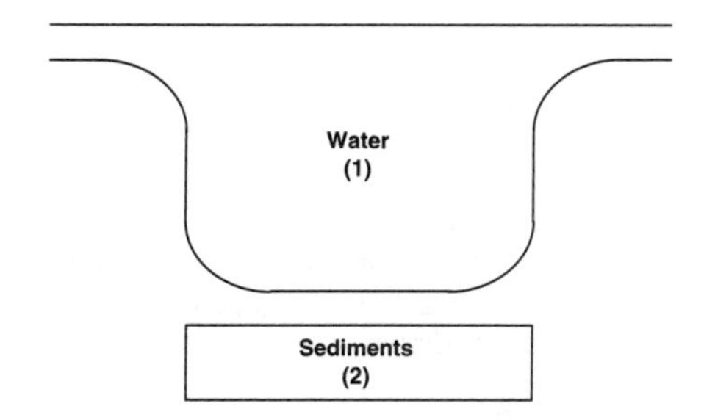

FIGURE 40.6
A well-mixed lake underlain by a well-mixed sediment layer.

As in Lec. 17, these quantities can be used to express a "suspended solids" for the sediments as

$$m_2 = (1 - \phi)\rho \tag{40.15}$$

Thus the sediment solids concentration is expressed in terms of parameters that are conventionally used to measure porous media. We now use these parameters to develop a solids budget for a sediment-water system.

40.4.1 Solids Budget

Mass balances for solids in the water and the sediment layer (Fig. 40.7) can be written as

$$V_1 \frac{dm_1}{dt} = Qm_{\text{in}} - Qm_1 - v_s A m_1 + v_r A m_2 \tag{40.16}$$

and

$$V_2 \frac{dm_2}{dt} = v_s A m_1 - v_r A m_2 - v_b A m_2 \tag{40.17}$$

where v_r = resuspension velocity (m yr^{-1}) and v_b = burial velocity (m yr^{-1}).

Equation 40.15 can be used to express sediment suspended solids, m_2, in terms of sediment porosity and density. In addition the subscript of m_1 can be dropped. The resulting solid balance equations are

$$V_1 \frac{dm}{dt} = Qm_{\text{in}} - Qm - v_s A m + v_r A(1 - \phi)\rho \tag{40.18}$$

and

$$V_2 \rho \frac{d(1 - \phi)}{dt} = v_s A m - v_r A(1 - \phi)\rho - v_b A(1 - \phi)\rho \tag{40.19}$$

At steady-state, Eq. 40.19 can be solved for

$$(1 - \phi)\rho = \frac{v_s}{v_r + v_b} m \tag{40.20}$$

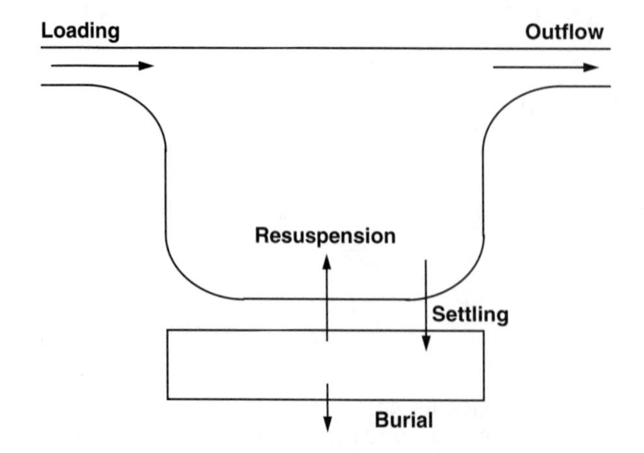

FIGURE 40.7
Schematic of a solids budget for a well-mixed lake with sediment feedback.

which can be substituted into the steady-state version of Eq. 40.18 and the result solved for

$$m = \frac{Qm_{\text{in}}}{Q + v_s A(1 - F_r)} \tag{40.21}$$

or expressing Eq. 40.21 as a transfer coefficient,

$$\beta = \frac{Q}{Q + v_s A(1 - F_r)} \tag{40.22}$$

where F_r = a resuspension factor that is defined as

$$F_r = \frac{v_r}{v_r + v_b} \tag{40.23}$$

Note that Eq. 40.21 is very similar to Eq. 40.8. The difference between the two formulations is isolated in the dimensionless parameter group F_r. This group represents the balance between the resuspension rate and the total rate at which the sediment purges itself of solids (that is, both burial and resuspension). Thus if burial dominates resuspension ($v_b \gg v_r$), $F_r \sim 0$ and Eq. 40.21 reduces to Eq. 40.8. In contrast if resuspension dominates burial ($v_b \ll v_r$), $F_r \sim 1$ and Eq. 40.21 reduces to $m = m_{\text{in}}$. In other words when resuspension is relatively dominant, the water concentration will approach the inflow concentration because everything that settles is immediately resuspended.

The above solutions are in the simulation mode, where all the parameters are known. Although the solids model can be used in this way, it is more conventional for the model to be employed to estimate some of the parameters.

Parameter estimation. The parameters in the model are ρ, ϕ, m, m_{in}, Q, A, v_s, v_r, and v_b. For the steady-state case, Eqs. 40.18 and 40.19 represent a pair of simultaneous equations. Thus given seven of the parameters, they generally provide us with a means for estimating the remaining two. Although an algorithm can be developed for this general problem, we take a different tack—we try to assess which of the parameters are least likely to be available. Then we show how the model can be employed to estimate these two.

Of the nine parameters, we assume that ρ and ϕ are known. Typical values are $\rho = 2.6 \times 10^6$ g m^{-3} and $\phi = 0.75$ to 0.95. We also assume that the flow and area, Q and A, are given.

Therefore we are left with five unknown parameters: m, m_{in}, v_s, v_r, and v_b. Among these, one is extremely difficult to measure: v_r. This will be the focus of our parameter estimation. There will be two situations that usually occur.

First, there is the case where m and m_{in} have been measured. In addition the settling velocity v_s may have been measured directly or estimated from literature values. For example a value of 2.5 m d^{-1} represents a typical value for organic and clay particles (O'Connor 1988). Ranges of settling velocities for organic and inorganic particles are also listed in Table 17.3. At steady-state, Eqs. 40.18 and 40.19 can be added to give

$$0 = Qm_{\text{in}} - Qm - v_b A(1 - \phi)\rho \tag{40.24}$$

which can be used to estimate v_b,

$$v_b = \frac{Q}{A} \frac{m_{\text{in}} - m}{(1 - \phi)\rho} \tag{40.25}$$

Second, the burial velocity is sometimes measured directly. This is best accomplished using sediment dating techniques. Once v_b has been approximated, the resuspension velocity can then be estimated by solving the steady-state version of Eq. 40.19 for

$$v_r = v_s \frac{m}{(1 - \phi)\rho} - v_b \tag{40.26}$$

40.4.2 Contaminant Budget

A mass balance for a toxic substance for the sediment-water system is depicted in Fig. 40.8. Notice that the toxic substance is partitioned into dissolved and particulate fractions in both the water and the sediments. Mechanisms such as burial and resuspension act on both components whereas diffusion acts selectively on the dissolved fraction. The mass balance for Fig. 40.8 can be expressed mathematically as

$$V_1 \frac{dc_1}{dt} = Qc_{\text{in}} - Qc_1 - k_1 V_1 c_1 - v_v A F_{d1} c_1 - v_s A F_{p1} c_1 + v_r A c_2 + v_d A (F_{d2} c_2 - F_{d1} c_1) \tag{40.27}$$

and

$$V_2 \frac{dc_2}{dt} = -k_2 V_2 c_2 + v_s A F_{p1} c_1 - v_r A c_2 - v_b A c_2 + v_d A (F_{d1} c_1 - F_{d2} c_2) \tag{40.28}$$

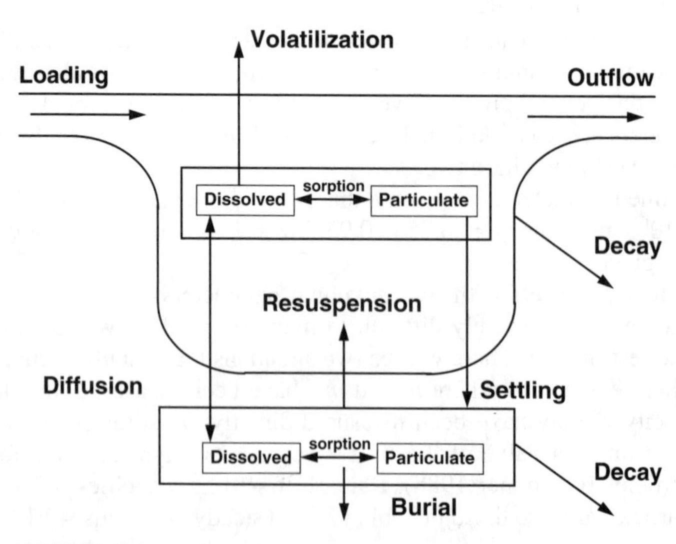

FIGURE 40.8
Schematic of a toxicant budget for a well-mixed lake with sediment feedback.

where the subscripts 1 and 2 indicate water and sediments, respectively. The fraction F_{d2} represents the ratio of the sediment pore-water concentration to the total concentration of contaminant in the sediments. This quantity is related to more fundamental sediment parameters by

$$F_{d2} = \frac{1}{\phi + K_{d2}(1-\phi)\rho} \tag{40.29}$$

where K_{d2} = the contaminant partition coefficient in the sediments ($m^3\ g^{-1}$). We will derive this formula when we discuss sorption in Lec. 41.

The diffusive mixing velocity v_d can be estimated from the empirically derived formula (Di Toro et al. 1981 as quoted in Thomann and Mueller 1987)

$$v_d = 69.35\phi M^{-2/3} \tag{40.30}$$

where M = molecular weight of the compound and v_d has units of m yr^{-1}.

Steady-state solution. At steady-state ($dc_1/dt = dc_2/dt = 0$), Eqs. 40.27 and 40.28 can be expressed as a system of two algebraic equations with two unknowns, c_1 and c_2:

$$(Q + k_1V_1 + v_vAF_{d1} + v_sAF_{p1} + v_dAF_{d1})c_1 - (v_rA + v_dAF_{d2})c_2 = Qc_{\text{in}} \tag{40.31}$$

and

$$-(v_sAF_{p1} + v_dAF_{d1})c_1 + (k_2V_2 + v_rA + v_bA + v_dAF_{d2})c_2 = 0 \tag{40.32}$$

Equation 40.32 can be solved for

$$c_2 = \frac{v_sF_{p1} + v_dF_{d1}}{k_2H_2 + v_r + v_b + v_dF_{d2}}c_1 \tag{40.33}$$

This result can be substituted into Eq. 40.31, which can be solved for

$$c_1 = \frac{Qc_{\text{in}}}{Q + k_1V_1 + v_vAF_{d1} + (1 - F_r')(v_sF_{p1} + v_dF_{d1})A} \tag{40.34}$$

where F_r' is the ratio of sediment feedback to total sediment purging,

$$F_r' = \frac{v_r + v_dF_{d2}}{v_r + v_b + v_dF_{d2} + k_2H_2} \tag{40.35}$$

Thus the transfer coefficient for this case is

$$\beta = \frac{Q}{Q + k_1V_1 + v_vAF_{d1} + (1 - F_r')(v_sF_{p1} + v_dF_{d1})A} \tag{40.36}$$

The effect of sediment-water interactions can be seen clearly by comparing Eq. 40.34 with Eq. 40.11. Notice that the settling velocity has now been supplemented by diffusion into the sediments. Also this total movement into the sediments is modified by the dimensionless number F_r'. If sediment feedback is negligible (that is, $F_r' \sim 0$), then Eq. 40.36 approaches Eq. 40.12. If sediment feedback is very high (that is, $F_r' \sim 1$), then Eq. 40.36 would approach

$$\beta = \frac{Q}{Q + k_1V_1 + v_vAF_{d1}} \tag{40.37}$$

In other words attenuation would be solely dictated by flushing, reactions, and volatilization because all settling losses would instantaneously be reintroduced into the overlying water.

Time-variable solution. For the case where all loadings are terminated, the contaminant mass balances can be divided by volume and rearranged to give

$$\frac{dc_1}{dt} = -\lambda_{11}c_1 + \lambda_{12}c_2 \tag{40.38}$$

$$\frac{dc_2}{dt} = \lambda_{21}c_1 - \lambda_{22}c_2 \tag{40.39}$$

where λ_{11} = the parameter group that reflects how the water segment purges itself,

$$\lambda_{11} = \frac{Q}{V_1} + k_1 + \frac{v_s F_{p1}}{H_1} + \frac{v_v F_{d1}}{H_1} + \frac{v_d F_{d1}}{H_1} \tag{40.40}$$

where λ_{12} = the parameter group that reflects how the water segment is affected by the sediments,

$$\lambda_{12} = \frac{v_r}{H_1} + \frac{v_d F_{d2}}{H_1} \tag{40.41}$$

where λ_{21} = the parameter group that reflects how the sediment segment is affected by the water,

$$\lambda_{21} = \frac{v_s F_{p1}}{H_2} + \frac{v_d F_{d1}}{H_2} \tag{40.42}$$

where λ_{22} = the parameter group that reflects how the sediment segment purges itself,

$$\lambda_{22} = k_2 + \frac{v_r}{H_2} + \frac{v_b}{H_2} + \frac{v_d F_{d2}}{H_2} \tag{40.43}$$

If $c_1 = c_{10}$ and $c_2 = c_{20}$, then the general solution can be developed with Laplace transforms as (recall Sec. 6.4)

$$c_1 = c_{1f}e^{-\lambda_f t} + c_{1s}e^{-\lambda_s t} \tag{40.44}$$

$$c_2 = c_{2f}e^{-\lambda_f t} + c_{2s}e^{-\lambda_s t} \tag{40.45}$$

where the λ's are eigenvalues that are defined as

$$\begin{matrix}\lambda_s \\ \lambda_f\end{matrix} = \frac{(\lambda_{11} + \lambda_{22}) \pm \sqrt{(\lambda_{11} + \lambda_{22})^2 - 4(\lambda_{11}\lambda_{22} - \lambda_{12}\lambda_{21})}}{2} \tag{40.46}$$

and the coefficients are

$$c_{1f} = \frac{(\lambda_f - \lambda_{22})c_{10} - \lambda_{12}c_{20}}{\lambda_f - \lambda_s} \tag{40.47}$$

$$c_{1s} = \frac{\lambda_{12}c_{20} - (\lambda_s - \lambda_{22})c_{10}}{\lambda_f - \lambda_s} \tag{40.48}$$

$$c_{2f} = \frac{\lambda_{21}c_{10} - (\lambda_f - \lambda_{11})c_{20}}{\lambda_s - \lambda_f} \tag{40.49}$$

$$c_{2s} = \frac{-(\lambda_s - \lambda_{11})c_{20} + \lambda_{21}c_{10}}{\lambda_s - \lambda_f} \tag{40.50}$$

It can be proven that for our model, λ_f will always be greater than λ_s. Consequently λ_f and λ_s are referred to as the "fast" and "slow" eigenvalues, respectively. This refers to the rates at which they approach zero as time progresses.

EXAMPLE 40.2. LAKE HURON PCB (WITH SEDIMENTS). Determine the level of high-molecular-weight PCBs in Lake Huron with sediment-water interactions. In particular determine (*a*) the burial and resuspension velocities of solids, (*b*) the transfer coefficient and the steady-state concentration, and (*c*) the temporal response following the termination of all loadings.

Pertinent data for the system is contained in Example 40.1. Assume that the settling velocity of solids is 2.5 m d^{-1} (912.5 m yr^{-1}). In addition the sediment porosity, density, and thickness are 0.9, 2.5×10^6 g m^{-3}, and 2×10^{-2} m, respectively. Recall from Example 40.1 that the inflow concentration of PCB is 33.35 μg m^{-3}. High-molecular-weight PCBs have the following characteristics: $M = 305.6$ gmole, $K_d = 0.0301$ g m^{-3}, and $v_v = 178.45$ m yr^{-1}.

Solution: (*a*) An inflow solids concentration can be computed as

$$m_{\text{in}} = \frac{6.15 \times 10^{12}}{161 \times 10^9} = 38.2 \text{ g m}^{-3}$$

Equation 40.25 can then be employed to estimate

$$v_b = \frac{161 \times 10^9}{59{,}570 \times 10^6} \frac{38.20 - 0.53}{(1 - 0.9)2.5 \times 10^6} = 0.000407 \text{ m yr}^{-1}$$

The resuspension velocity can be determined by Eq. 40.26,

$$v_r = 912.5\frac{0.53}{(1 - 0.9)^6 2.5 \times 10^6} - 0.000407 = 0.00153 \text{ m yr}^{-1}$$

(*b*) Equation 40.29 can be used to compute

$$F_{d2} = \frac{1}{0.9 + 0.0301(1 - 0.9)2.5 \times 10^6} = 0.000133$$

Equation 40.30 can be employed to calculate

$$v_d = 69.35(0.9)305.6^{-2/3} = 1.376 \text{ m yr}^{-1}$$

The values for Lake Huron can then be substituted into Eq. 40.35 to compute

$$F_b' = \frac{0.000407}{0.00153 + 0.000407 + 1.376(0.000133)} = 0.192$$

which can be substituted into Eq. 40.36 to calculate

$$\beta = 0.0149$$

which in turn can be employed to determine

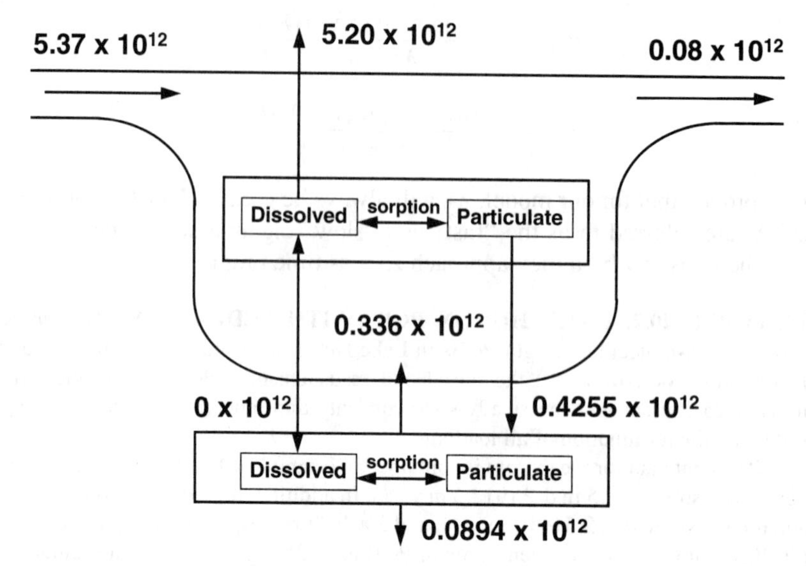

FIGURE 40.9
Budget of PCB for Lake Huron with sediment feedback.

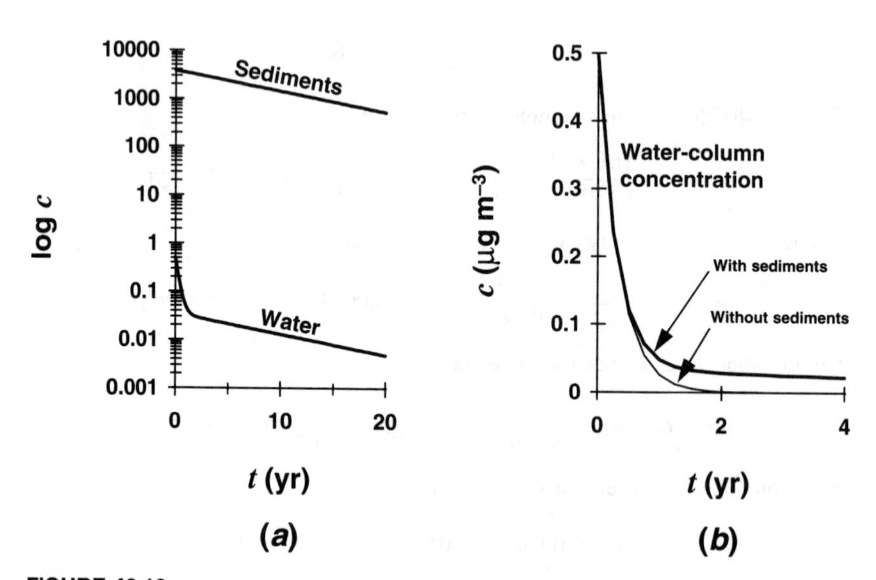

FIGURE 40.10
Temporal response of Lake Huron sediment-water system following termination of PCB loading. (*a*) Semi-log plot of the recovery. (*b*) The response for the water column with and without sediment-water interactions.

$$c_1 = 0.0149(33.35) = 0.4971\ \mu\text{g m}^{-3}$$

The sediment concentration can be determined with Eq. 40.33,

$$c_2 = \frac{912.5(0.0157) + 1.3757(0.9843)}{0.000407 + 0.00153 + 1.3757(0.000133)} 0.4971 = 3687.3\ \mu\text{g m}^{-3}$$

The results of the budget are depicted in Fig. 40.9.

(*c*) The general solution is

$$c_1 = 0.4625e^{-3.296t} + 0.0345e^{-0.0987t}$$

and

$$c_2 = -113.86e^{-3.296t} + 3801.18e^{-0.9087t}$$

These equations are used to produce Fig. 40.10*a*. Note that the water experiences a rapid initial drop due primarily to the high volatilization rate. Then the slower eigenvalue, which is primarily dictated by the sediment response, dominates the latter part of the recovery. Figure 40.10*b* shows the recovery along with the recovery for the system treated as a single CSTR (Example 40.1).

40.5 SUMMARY

In terms of analytical solutions, we have now pushed our model framework to its limits. As should be evident from Example 40.2, the inclusion of additional spatial segments (e.g., more segments) requires computers and numerical solution techniques. However, we hope that the simple models described herein have provided you with some insight regarding toxic-substance models. I urge you to implement this simple framework yourself (on a spreadsheet perhaps) so that you can experiment and build greater intuition regarding toxic-substance modeling. I believe that this is an important prerequisite for using more complicated computer packages developed by others.

PROBLEMS

40.1 Thomann and Di Toro (1983) presented the following data related to the solids budget for the Western Basin of Lake Erie:

Volume $= 23 \times 10^9$ m^2 — Area $= 3030 \times 10^6$ m^2
Solid loading $= 11.4 \times 10^{12}$ g yr^{-1} — Suspended solids $= 20$ mg L^{-1}
Flow $= 167 \times 10^9$ m^3 yr^{-1}

They assumed that the solids settle at a rate of 2.5 m/d (912.5 m/yr), and that the sediments have $\rho = 2.4$ g cm^{-3} and $\phi = 0.9$. Determine the burial and resuspension velocities.

40.2 Suppose that a toxic substance that is subject to volatilization ($v_v = 100$ m yr^{-1}) is discharged to Lake Huron with an inflow concentration of 100 μg L^{-1}. In the absence of sediment feedback, determine the concentration for three cases: (*a*) weak-sorber ($K_d = 0.002$ m^3 g^{-1}), (*b*) moderate sorber ($K_d = 0.1$), and (*c*) strong sorber ($K_d = 2$). Other necessary information should be taken from Examples 40.1 and 40.2.

40.3 A substance ($K_d = 0.02\ m^3\ g^{-1}$; $M = 300$) is discharged to a lake ($c_{in} = 100\ \mu g\ L^{-1}$) having the following characteristics:

Volume = $1 \times 10^6\ m^3$	Mean depth = 5 m
Residence time = 1 year	Suspended solids = $10\ g\ m^{-3}$
Settling velocity = $50\ m\ year^{-1}$	Sediment deposition = $100\ g\ m^{-2}\ yr^{-1}$
Sediment porosity = 0.85	Sediment density = $2.5\ g\ cm^{-3}$

(*a*) If sediment resuspension is negligible, compute the steady-state concentration for three levels of volatilization:
(i) highly soluble ($\nu_\nu = 0$)
(i) moderately soluble ($\nu_\nu = 10\ m\ yr^{-1}$)
(i) insoluble ($\nu_\nu = 100\ m\ yr^{-1}$)
(*b*) Repeat (*a*), but include the effect of sediment resuspension.

40.4. Repeat Example 40.1 for Lake Michigan. The parameters in Table P40.4 are needed for the computation.

TABLE P40.4
Parameters for Lake Michigan

Parameter	Symbol	Value	Units
Suspended solids concentration	m	0.5	$g\ m^{-3}$
Suspended solids load	W_s	4×10^{12}	$g\ yr^{-1}$
PCB load	W_c	10×10^{12}	$g\ yr^{-1}$
Volume	V	4.616×10^{12}	m^3
Surface area	A_s	57.77×10^9	m^2
Depth	H	82	m
Outflow	Q	50×10^9	$m^3\ yr^{-1}$

40.5. Repeat Example 40.2 for Lake Michigan. See Table P40.4 for relevant parameters.

LECTURE 41

Mass-Transfer Mechanisms: Sorption and Volatilization

LECTURE OVERVIEW: I explain the mechanisms of sorption and volatilization and delineate how they are quantified. I then describe the sorption isotherm, which is employed to derive the fraction of contaminant in particulate and dissolved forms for both the water and the sediments. I discuss methods for estimating partitioning and the effect of dissolved organic carbon on sorption. For volatilization, the exchange of toxicants across the air-water interface is described mathematically by the two-film theory. I also present methods for estimating Henry's constant and liquid- and gas-film transfer coefficients. Finally I outline a screening approach to assess the impact of organic contaminants on lakes in the absence of decay mechanisms.

The next two lectures deal with the mechanisms that govern the sinks of toxic substances in natural waters. In the present lecture, we deal with transport-oriented processes. First, I describe models of the sorption of toxics on solid matter. This mechanism is considered transport-related because solid-associated toxics are subject to settling losses. Second, I describe volatilization losses across the air-water interface. Then, in the following lecture, we'll turn to reaction losses that involve transformations of the toxicant.

41.1 SORPTION

Sorption is a process whereby a dissolved substance is transferred to and becomes associated with solid material. It includes both the accumulation of dissolved substances on the surface of solids (***adsorption***) and the interpenetration or intermingling of substances with solids (***absorption***). The substance that is sorbed is called the ***sorbate*** and the solid is called the ***sorbent. Desorption*** is the process whereby a sorbed substance is released from a particle.

For neutral organics, several mechanisms underlie the sorption process. These include (Schwarzenbach et al. 1993):

1. Hydrophobic effects cause the sorbate to associate with organic matter in the particulate phase because of an unfavorable free-energy cost of staying in solution.
2. Weak surface interactions via van der Waals, dipole-dipole, induced dipole, and other weak intermolecular forces.
3. Surface reactions where the sorbate actually bonds to the solid.

For charged toxins, the additional mechanism of ion exchange can occur. It should also be mentioned that, aside from organically rich materials, neutral organic chemicals are also sorbed by solids with little or no organic content. For such cases, the sorbent consists of inorganic matter such as clays. Such inorganic solid sorption is usually significant only when the organic carbon content of the solids is quite low (Schwarzenbach and Westall 1981).

In the following sections I first introduce some simple sorption models that are used to describe the process in standard water-quality models. I then demonstrate how these models can be employed to derive formulas to predict the fraction of toxicant in dissolved and particulate form. Finally I review methods for estimating sorption parameters for particular toxicants.

41.1.1 Isotherms

An experiment to measure sorption starts by mixing a mass of solids in a well-stirred vessel with a quantity of dissolved chemical (Fig. 41.1*a*). At intervals, samples are taken from the vessel, centrifuged, and the amount of chemical on the solids and in solution measured (Fig. 41.1*b*). For most chemicals, an equilibrium is rapidly reached in minutes to hours (Fig. 41.1*c*). The experiment can be repeated for different contaminant levels and the results plotted.

Figure 41.2 shows the type of plot that typically results from such an experiment. The concentration of contaminant on the solids, ν (mg g^{-1}), is graphed versus the dissolved concentration, c_d (mg m^{-3}). The plot, called an ***isotherm,*** indicates that

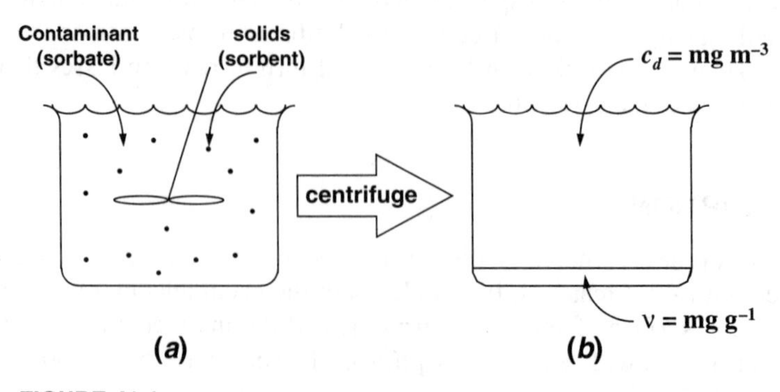

FIGURE 41.1

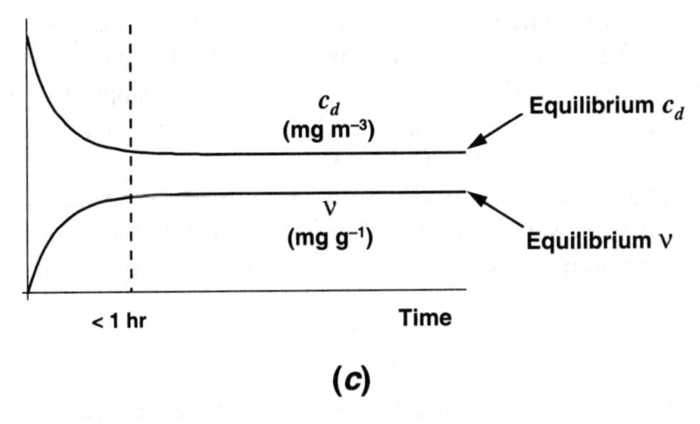

FIGURE 41.1
Experiment to determine sorption. (*a*) Solids and contaminant are mixed in a vessel. A sample is taken and centrifuged to separate the solids from the water. (*b*) The contaminant content of each phase is measured. The temporal progression of such an experiment is depicted in (*c*). An equilibrium is typically reached rapidly.

solid concentration increases with dissolved concentration until the available sites begin to become saturated. At this point the curve levels off to a maximum value ν_m (mg m^{-3}).

Several models have been proposed to mathematically represent the isotherm. Among the most popular are the

1. *Langmuir isotherm*

$$\nu = \frac{\nu_m b c_d}{1 + b c_d} \tag{41.1}$$

2. *Freundlich isotherm*

$$\nu = K_f c_d^{1/n} \tag{41.2}$$

3. *BET isotherm*

$$\nu = \frac{\nu_m B c_d}{(c_s - c_d)\left[1 + (B - 1)(c_d/c_s)\right]} \tag{41.3}$$

where b, B, c_s, K_f, and n are coefficients used to calibrate the curves to the data. Note that although the shape depicted in Fig. 41.2 occurs commonly, other patterns of sorption can occur for specific cases. For example Schwarzenbach et al. (1993) describe situations where previous sorption modifies the surface and leads to

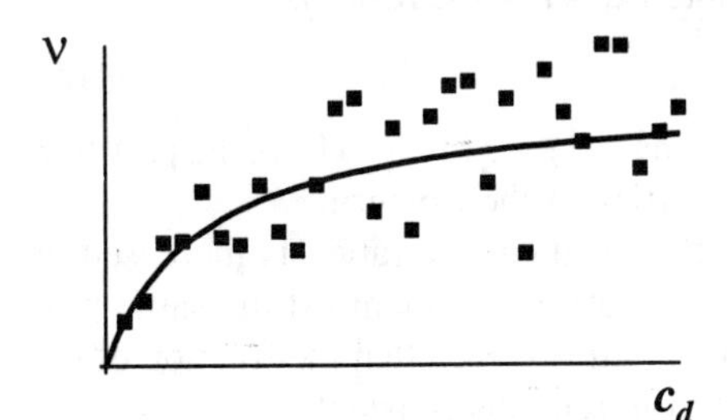

FIGURE 41.2
Sorption data and isotherm.

enhanced sorption. In such instances the isotherm could show an increasing slope with increasing dissolved concentration. The exponent in the Freundlich isotherm allows increasing (exponent > 1), decreasing (< 1), and constant slopes ($= 1$) to be modeled. The BET isotherm, as well as other models, are capable of mathematically capturing such effects (Weber and DiGiano 1996).

Because of its mathematical simplicity, the Langmuir isotherm has utility in understanding the shape of the saturating isotherm in Fig. 41.2. It can be derived by recognizing that the sorption process represents an equilibrium between the rate of adsorption and desorption,

$$R_{ad} = R_{de} \tag{41.4}$$

where the mass adsorbed per unit time, R_{ad} (mg s^{-1}), can be represented as

$$R_{ad} = k_{ad} M_s c_d (\nu_m - \nu) \tag{41.5}$$

in which k_{ad} = mass-specific volumetric rate of adsorption [m^3(mg s)$^{-1}$] and M_s = mass of solids (g).

The mass desorbed per unit time, R_{de} (mg s^{-1}), can be represented as

$$R_{de} = k_{de} M_s \nu \tag{41.6}$$

where k_{de} = a first-order desorption rate (s^{-1}).

Equations 41.5 and 41.6 can be substituted into Eq. 41.4 and the result solved for

$$\nu = \frac{\nu_m c_d}{(k_{de}/k_{ad}) + c_d} \tag{41.7}$$

which is the general form of the Langmuir model (compare with Eq. 41.1).

Equation 41.7 is plotted in Fig. 41.3*a*. Note that the curve manifests two asymptotic regions. For relatively high levels of contaminant, the curve becomes zero-order or constant. For low levels of contaminant, the curve is first-order or linear. It is this latter case that is employed in most toxic substance models because of the assumption that in most natural waters, toxicant concentrations are relatively low.

The linear region can be modeled by realizing that when $\nu \ll \nu_m$, plenty of sites are available and

$$R_{ad} = k_{ad} M_s c_d (\nu_m - \nu) \rightarrow k'_{ad} M_s c_d \tag{41.8}$$

where $k'_{ad} = k_{ad} \nu_m$ [m^3(g s)$^{-1}$]. In other words the rate of adsorption becomes independent of solid concentration and is driven only by dissolved concentration. When Eq. 41.8 is substituted along with Eq. 41.6 into Eq. 41.4, the result is

$$\nu = K_d c_d \tag{41.9}$$

where K_d = a partition coefficient = k'_{ad}/k_{de} (m^3 g^{-1}). As in Fig. 41.3*b*, the partition coefficient represents the slope of the linear portion of the isotherm.

The linear isotherm implies that if the dissolved concentration is increased or decreased a certain proportion, the solid concentration will manifest the same proportional increase or decrease. As discussed next, this means that the fraction of the contaminant in the dissolved and particulate form remains constant.

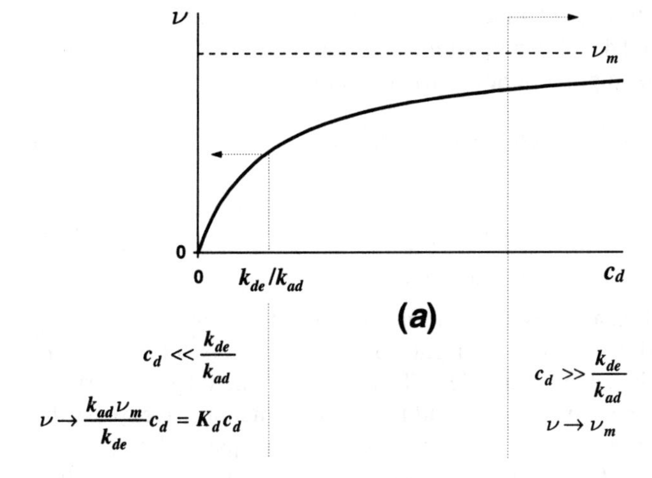

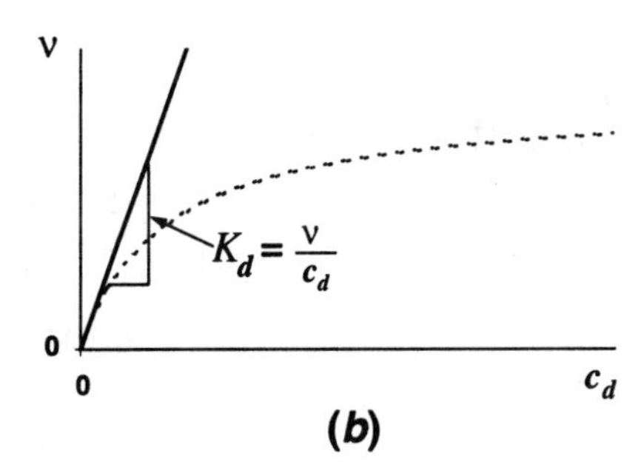

FIGURE 41.3
(*a*) A Langmuir isotherm indicating asymptotes.
(*b*) Linearization of isotherm.

41.1.2 Fraction Sorbed

Using the framework developed above, we can now derive expressions relating the contaminant fraction in dissolved (F_d) and particulate (F_p) forms to the partition coefficient. To do this, individual mass balances can be written for each fraction in a CSTR. For example for the dissolved portion of the contaminant,

$$V\frac{dc_d}{dt} = Qc_{d,\text{in}} - Qc_d + k_{de}M_s\nu - k'_{ad}M_sc_d \tag{41.10}$$

and for the particulate portion,

$$V\frac{dc_p}{dt} = Qc_{p,\text{in}} - Qc_p - v_sAc_p - k_{de}M_s\nu + k'_{ad}M_sc_d \tag{41.11}$$

where the subscripts d and p signify the dissolved and particulate contaminant, respectively. To keep this derivation simple, we have excluded decay reactions from these balances. Also, recognize that all time units are now expressed as years to be compatible with the time scale of interest for systems such as lakes.

Because sorption kinetics are so much faster than the other input-output terms in the model, Eqs. 41.10 and 41.11 can be summed (recall our general discussion of this "local equilibrium" assumption in Sec. 38.1) to yield

$$V\frac{dc}{dt} = Qc_{\text{in}} - Qc - v_s A c_p \tag{41.12}$$

where

$$c = c_d + c_p \tag{41.13}$$

and

$$c_{\text{in}} = c_{d,\text{in}} + c_{p,\text{in}} \tag{41.14}$$

Notice that the adsorption and desorption terms have been canceled by summing the two equations. In addition, note that the right side of Eq. 41.12 is not exclusively expressed in terms of the dependent variable c. That is, it also includes the variable c_p. Consequently it cannot be solved without additional information. To express the equation exclusively in terms of the unknown, recognize that particulate concentration is related to suspended solids and mass-specific contaminant concentration by

$$c_p = m\nu \tag{41.15}$$

where m = suspended solids concentration (g m^{-3}).

Equations 41.9 and 41.15 can be combined with 41.13 to yield

$$c = c_d + mK_d c_d \tag{41.16}$$

which can be solved for

$$c_d = F_d c \tag{41.17}$$

where

$$F_d = \frac{c_d}{c} = \frac{1}{1 + K_d m} \tag{41.18}$$

This result can in turn be substituted into Eq. 41.13 and the result solved for

$$c_p = F_p c \tag{41.19}$$

where

$$F_p = \frac{c_p}{c} = \frac{K_d m}{1 + K_d m} \tag{41.20}$$

Thus we have derived the fractions that were introduced previously in Eqs. 40.4 and 40.5.

Similar values can be derived for the sediments. As described in Box 41.1, these relationships differ from Eqs. 41.18 and 41.20 because of the significant fraction of the sediment volume that is solid.

BOX 41.1. Contaminant Fractions for the Sediments

The derivation of sorbed fractions for the sediments is somewhat complicated because a significant portion of the sediment volume is in solid form. This fact is accounted for by using porosity in the derivation. As with the water, the total concentration is divided into particulate and dissolved components (Eq. 41.13), where the dissolved concentration is now defined as

$$c_d = \phi c_{dp} \tag{41.21}$$

in which c_{dp} = concentration of the pore water (mg m^{-3}). Similarly the particulate concentration is defined as

$$c_p = (1 - \phi)\rho\nu \tag{41.22}$$

For the sediments, Eq. 41.9 is

$$\nu = K_{d2}c_{dp} \tag{41.23}$$

Substituting Eqs. 41.21 through 41.23 into 41.13 yields

$$c = \phi c_{dp} + (1 - \phi)\rho K_{d2}c_{dp} \tag{41.24}$$

which can be solved for

$$F_{d2} = \frac{c_{dp}}{c} = \frac{1}{\phi + \rho K_{d2}(1 - \phi)} \tag{41.25}$$

Thus we have derived the fraction that was introduced previously as Eq. 40.29.

41.1.3 Estimation of the Partition Coefficient

A variety of investigators have attempted to relate partitioning to specific properties of the chemical. Karickhoff et al. (1979) have developed the following scheme for neutral (that is, noncharged) organic chemicals.

The partition coefficient for organic contaminants, K_d (m^3 g^{-1}), is assumed to be a function of the organic-carbon content of the solids, as in

$$K_d = f_{oc}K_{oc} \tag{41.26}$$

where K_{oc} = organic-carbon partition coefficient [(mg gC^{-1})(mg m^{-3})$^{-1}$] and f_{oc} = weight fraction of the total carbon in the solid matter (gC g^{-1}).

The organic-carbon partition coefficient K_{oc}, in turn, can be estimated in terms of the contaminant's octanol-water partition coefficient K_{ow} [(mg m^{-3}_{octanol})(mg m^{-3}_{water})] by (Karickhoff et al. 1979)

$$K_{oc} = 6.17 \times 10^{-7}K_{ow} \tag{41.27}$$

Equations 41.26 and 41.27 can then be combined to yield

$$K_d = 6.17 \times 10^{-7}f_{oc}K_{ow} \tag{41.28}$$

Thomann and Mueller (1987) suggest a range of 0.001 to 0.1 for f_{oc}. However, it should be noted that the top end should be raised to account for autochthonous particles, such as living phytoplankton cells, that have an organic carbon of about 0.4.

The octanol-water partition coefficient can be obtained in a variety of ways. First, it can be measured directly. Second, it can be obtained from published tables.

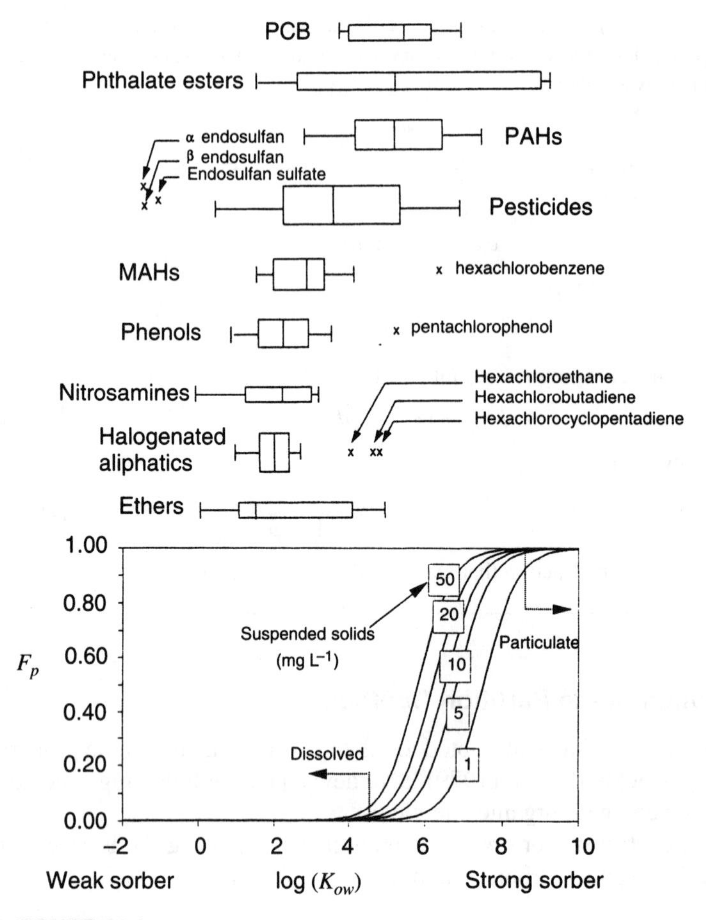

FIGURE 41.4
Plot of fraction dissolved, F_d, and fraction particulate, F_p, versus partition coefficients (Chapra 1991). Each curve represents a particular suspended solids concentration. Box-and-whisker diagrams (see Fig. 41.5) for nine classes of organic contaminants are drawn above the plot.

Third, it can be computed from formulas based on more readily available characteristics of the contaminant. For example Chiou et al. (1977) offer the following formula:

$$\log K_{ow} = 5.00 - 0.670 \log S'_w \tag{41.29}$$

where S'_w = solubility in μmole L^{-1}, which can be computed by

$$S'_w = \frac{S_w}{M} \times 10^3 \tag{41.30}$$

where S_w = solubility (mg L^{-1}) and M = molecular weight (g $mole^{-1}$). Published tables (e.g., Lyman et al. 1982, Mills et al. 1982, Schnoor et al. 1987, etc.) can be used to obtain solubility.

Other formulas are available for estimating K_d and K_{oc} from K_{ow} and S_w. Such regression approaches improve as they are applied to smaller groupings based on similar chemical structure. Schwarzenbach et al. (1993) present an excellent and comprehensive review of the various approaches.

If f_{oc} is assumed to be approximately 0.05, Eq. 41.28 can be used to estimate the partition coefficient, as in

$$K_d = 3.085 \times 10^{-8} K_{ow} \tag{41.31}$$

This estimate along with suspended solids concentration and tabulated K_{ow}'s can be substituted into Eqs. 41.18 and 41.20 to estimate the dissolved and particulate fractions. The result is displayed in Fig. 41.4. Note that we have also included box-and-whisker plots for some major classes of toxicants (recall Table 40.1). As described in Fig. 41.5, ***box-and-whisker plots*** (Tukey 1977, McGill et al. 1978) utilize robust statistics to summarize the pertinent features of a data sample. In particular they provide a one-dimensional representation of a data distribution.

Inspection of Fig. 41.4 yields some interesting conclusions. In particular, five of the groups (ethers, halogenated aliphatics, nitrosamines, phenols, and monocyclic aromatics) are overwhelmingly in dissolved form (that is, $F_d \sim 1$). In fact if a value of $K_{ow} < 10^4$ (which corresponds to $K_d < 3.2 \times 10^{-4}$ m g^{-1}) is conservatively chosen as the criterion for assuming that the contaminant is predominantly dissolved, about 65% of the 114 priority toxicants (Table 40.1) would be classified as dissolved.

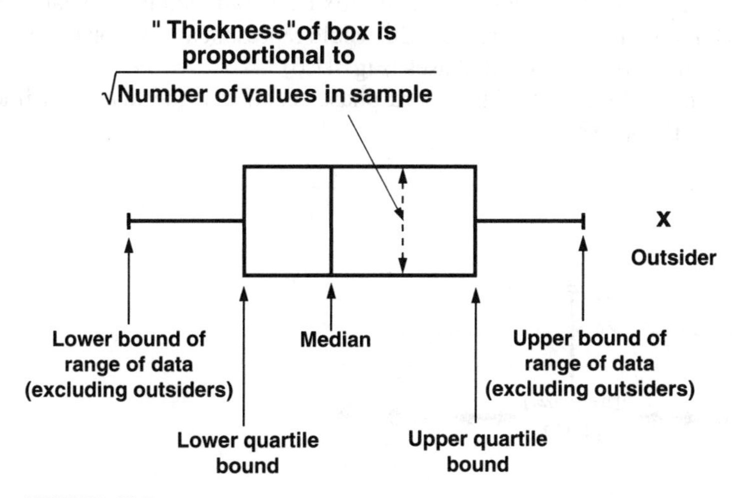

FIGURE 41.5
Tukey's (1977) box-and-whisker plot uses robust statistics such as the median and the interquartile range to summarize the distribution of a data set. In particular it provides a visual depiction of location, spread, skewness, tail length, and outlying data points (Hoaglin et al. 1983). As in the above diagram, an "outsider" refers to values that are far outside the normal range. Additional information on box-and-whisker plots and other interesting aspects of robust and exploratory data analysis can be found elsewhere (Tukey 1977, Hoaglin et al. 1983, Reckhow and Chapra 1983, etc.).

For such pollutants the analysis becomes greatly simplified because sediment-water interactions can be neglected and purging becomes solely a function of flushing, volatilization, and decomposition. For example Eq. 40.36 reduces to

$$\beta = \frac{Q}{Q + k_1 V_1 + v_v A F_{d1}} \tag{41.32}$$

Such simplifications will be explored in further detail at the end of this lecture.

41.1.4 Sorption Revisited (The "Third-Phase" Effect)

In Lec. 17, we noted that suspended solids can be categorized as allochthonous (originating in the drainage basin) or autochthonous (produced in the water via photosynthesis). To this point our sorption model would seem to be more compatible with the former perspective. That is, we characterize solids as a loading and use partition coefficients to attach some of the toxicant to the suspended particles.

At least for toxic organic compounds, this viewpoint must be broadened. In the previous section, we have seen that the partition coefficient is a function of the particle's organic carbon content. Therefore, rather than modeling suspended solids concentration, a more direct approach might be to model particulate organic carbon (POC). This would be particularly appropriate for systems with large quantities of autochthonous solids. Further, an organic carbon perspective raises the issue that productive systems also tend to have significant concentrations of dissolved (and colloidal) organic carbon (DOC). Because toxics would tend to associate with both dissolved and particulate forms, the DOC would represent a "third phase" in addition to the particulate and dissolved phases (Fig. 41.6).

To quantify the effect of this third phase recall that the dissolved fraction is modeled as (Eq. 41.18)

$$F_d = \frac{1}{1 + K_d m} \tag{41.33}$$

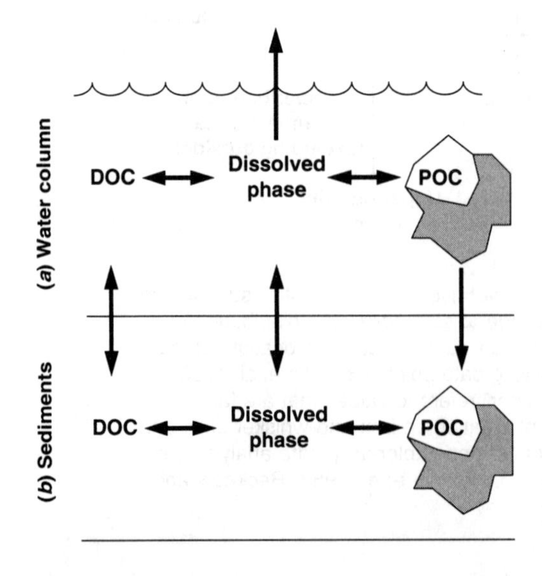

FIGURE 41.6
Depiction of how a DOC "third phase" competes with POC for dissolved toxicant in the (*a*) water and in the (*b*) sediments.

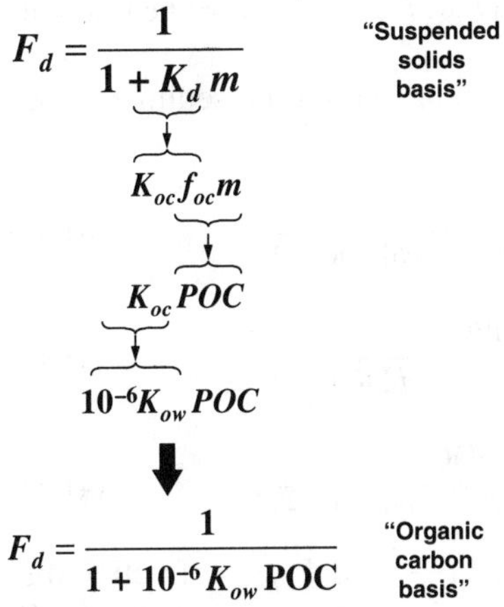

$$F_d = \frac{1}{1 + K_d m}$$ "Suspended solids basis"

$K_{oc} f_{oc} m$

$K_{oc} POC$

$10^{-6} K_{ow} POC$

$$F_d = \frac{1}{1 + 10^{-6} K_{ow} \text{POC}}$$ "Organic carbon basis"

FIGURE 41.7
Depiction of how an organic-carbon basis has supplanted the original "suspended solids" basis employed in early water-quality models of toxic organic sorption.

We have also seen that the partition coefficient can be related to the octanol-water partition coefficient by

$$K_d = 6.17 \times 10^{-7} f_{oc} K_{ow} \tag{41.34}$$

Thomann and Mueller (1987) and others have noted that this relationship can be approximated by

$$K_d \approx (1 \times 10^{-6} K_{ow}) f_{oc} \tag{41.35}$$

Substituting this approximation into Eq. 41.33 yields

$$F_d = \frac{1}{1 + (1 \times 10^{-6} K_{ow}) f_{oc} m} \tag{41.36}$$

or recognizing that POC $= f_{oc} m$,

$$F_d = \frac{1}{1 + (1 \times 10^{-6} K_{ow}) \text{POC}} \tag{41.37}$$

The process of converting from a suspended solid to an organic carbon basis is depicted in Fig. 41.7.

Now if it is assumed that the toxicant associates similarly with DOC, the third phase can be incorporated and three fractions derived,

$$F_{d1} = \frac{1}{1 + (1 \times 10^{-6} K_{ow})(\text{POC}_1 + \text{DOC}_1)} \tag{41.38}$$

$$F_{p1} = \frac{(1 \times 10^{-6} K_{ow}) \text{POC}_1}{1 + (1 \times 10^{-6} K_{ow})(\text{POC}_1 + \text{DOC}_1)} \tag{41.39}$$

$$F_{o1} = \frac{(1 \times 10^{-6} K_{ow}) \text{DOC}_1}{1 + (1 \times 10^{-6} K_{ow})(\text{POC}_1 + \text{DOC}_1)} \tag{41.40}$$

where the subscript 1 denotes the water column and F_{d1}, F_{p1}, and F_{o1} = the fractions associated with dissolved, POC, and DOC, respectively.

A similar analysis can be used to develop fractions within the sediments (see Box 41.2),

$$F_{dp2} = \frac{1}{\phi + (1 \times 10^{-6}K_{ow})(\text{POC}_2 + \phi\text{DOC}_{p2})} \tag{41.41}$$

$$F_{p2} = \frac{(1 \times 10^{-6}K_{ow})\text{POC}_2}{\phi + (1 \times 10^{-6}K_{ow})(\text{POC}_2 + \phi\text{DOC}_{p2})} \tag{41.42}$$

$$F_{op2} = \frac{(1 \times 10^{-6}K_{ow})\text{DOC}_{p2}}{\phi + (1 \times 10^{-6}K_{ow})(\text{POC}_2 + \phi\text{DOC}_{p2})} \tag{41.43}$$

where the subscript 2 denotes the sediments; F_{dp1}, F_{p2}, and F_{op2} = fraction dissolved in the pore water, associated with POC, and dissolved in pore-water DOC, respectively; DOC_{2p} = dissolved organic carbon concentration in the pore water (gC m^{-3}); and POC_2 = particulate organic carbon on a whole sediment volume basis (gC m^{-3}). Note that the last quantity is often measured as a dry-weight fraction of carbon. Therefore it is often convenient to make the following substitution in the above equations:

$$\text{POC}_2 = f_{oc,2}(1 - \phi)\rho \tag{41.44}$$

where $f_{oc,2}$ = dry-weight fraction of organic carbon in the sediments (gC g^{-1}).

BOX 41.2. Derivation of Sediment Contaminant Fractions

As for the water column, the sediment contaminant can be partitioned into three phases,

$$c = c_d + c_p + c_o \tag{41.45}$$

The dissolved fraction can be related to the dissolved contaminant in the pore water by

$$c_d = \phi c_{dp} \tag{41.46}$$

where c_{dp} = concentration of the pore water (mg m^{-3}). In a similar fashion to the derivation of Eq. 41.25, particulate concentration is defined as

$$c_p = (1 - \phi)\rho K_d c_{dp} \tag{41.47}$$

Substituting Eq. 41.35 gives

$$c_p = (1 - \phi)\rho(1 \times 10^{-6}K_{ow}) f_{oc} c_{dp} \tag{41.48}$$

Now recognizing that

$$\text{POC} = (1 - \phi)\rho f_{oc} \tag{41.49}$$

we rewrite Eq. 41.48 as

$$c_p = (1 \times 10^{-6}K_{ow})\text{POC}c_{dp} \tag{41.50}$$

For the DOC,

$$c_o = \phi c_{op} \tag{41.51}$$

where c_{op} = concentration of the toxic in the pore water DOC (mg m^{-3}). The partitioning between the dissolved and DOC forms in the pore water would be defined as

$$K_o = \frac{c_{op}/\text{DOC}_p}{c_{dp}} \tag{41.52}$$

Combining Eqs. 41.51 through 41.52,

$$c_o = \phi K_o c_{dp} \text{DOC}_p \tag{41.53}$$

or substituting Eq. 41.35,

$$c_o = \phi(1 \times 10^{-6} K_{ow}) c_{dp} \text{DOC}_p \tag{41.54}$$

Finally Eqs. 41.46, 41.50, and 41.54 can be entered into Eq. 41.45, which can be solved for

$$F_{dp2} = \frac{1}{\phi + \left(1 \times 10^{-6} K_{ow}\right)(\text{POC}_2 + \phi \text{DOC}_{p2})} \tag{41.55}$$

Further manipulation can be used to determine Eqs. 41.42 and 41.43.

What are the implications of the foregoing analysis for model calculations? They mean that the presence of dissolved organic carbon draws toxicant away from the particulate and dissolved pools. Thus mechanisms that influence these pools would be diminished. For example in the water, higher DOC would detract from the volatilization and sedimentation mechanism because less toxicant would be associated with the dissolved phase and settling particles, respectively. In the sediments higher DOC would enhance the amount of toxicant in the pore waters. Thus sediment feedback to the water column would be increased.

In conclusion, this has been a brief introduction to sorption. We have deliberately limited the discussion to the level of detail employed in commonly applied water-quality models. Further nuances and a more complete description can be found in references such as Schwarzenbach et al. (1993) and Lyman et al. (1982).

41.2 VOLATILIZATION

Many toxic substances move across a lake's surface by volatilization. From our discussion of oxygen gas transfer in Lec. 20, we are already familiar with how the process can be quantified. The flux of a gas across the air-water interface can be modeled as

$$J = v_v \left(\frac{p_g}{H_e} - c_l \right) \tag{41.56}$$

where J = mass flux (mole m^{-2} yr^{-1})
v_v = net transfer velocity across the air-water interface (m yr^{-1})
p_g = partial pressure of gas in the air over the water (atm)
H_e = Henry's constant (atm m^3 mole^{-1})
c_l = concentration of the gas in the water (mole m^{-3})

The transfer velocity can be calculated based on the two-film theory as

$$v_v = K_l \frac{H_e}{H_e + RT_a(K_l/K_g)} \tag{41.57}$$

where K_l and K_g = mass-transfer coefficients for the liquid and gaseous films ($m\ yr^{-1}$), respectively
R = universal gas constant [8.206×10^{-5} atm m^3 (K mole)$^{-1}$]
T_a = absolute temperature (K)

Thus, as depicted in Fig. 41.8, the transfer is dependent on a contaminant-specific property (Henry's constant) and two environment-specific properties (the film-transfer coefficients).

The transfer coefficients can be computed based on more fundamental parameters with two standard approaches. For quiescent systems such as lakes, a stagnant film approach is employed,

$$K_l = \frac{D_l}{z_l} \qquad K_g = \frac{D_g}{z_g} \tag{41.58}$$

where D_l and D_g = diffusion coefficients for the toxicant in the liquid and gas, respectively, and the z's = stagnant film thicknesses.

For systems such as rivers, where the liquid flow is turbulent, the surface renewal model applies,

$$K_l = \sqrt{r_l D_l} \qquad K_g = \sqrt{r_g D_g} \tag{41.59}$$

where the r's are renewal rates.

The volatilization model can be integrated into a mass-balance by multiplying the flux (Eq. 41.56) by surface area to give

$$V\frac{dc}{dt} = v_v A_s \left(\frac{p_g}{H_e} - c_d\right) \tag{41.60}$$

where V = volume of the water being modeled (m^3) and A_s = surface area. Note that we have changed the subscript for c to denote "dissolved."

Recall that when Eq. 41.60 was written for dissolved oxygen (Eq. 20.32), the result could be simplified because (*a*) oxygen is abundant in the atmosphere and (*b*) oxygen is liquid-film controlled. The result is

$$V\frac{do}{dt} = K_l A_s(o_s - o) \tag{41.61}$$

where o_s = saturation concentration corresponding to the partial pressure of oxygen

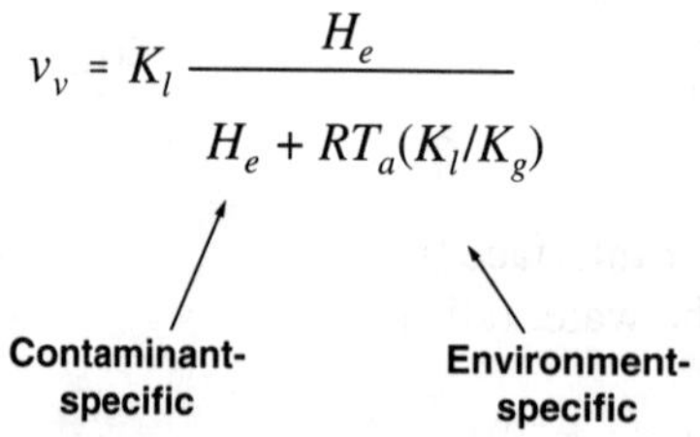

$$v_v = K_l \frac{H_e}{H_e + RT_a(K_l/K_g)}$$

Contaminant-specific

Environment-specific

FIGURE 41.8
The volatilization mass-transfer velocity is computed as a function of contaminant-specific (Henry's constant) and environment-specific (film-transfer coefficients) properties.

in the atmosphere (g m^{-3}) and o = concentration of oxygen in the water in mass units (g m^{-3}).

Toxics differ from oxygen in two ways. First, we cannot generally conclude that they are all liquid-film controlled. Second, we can assume that, in many cases, they are not abundant in the atmosphere. The latter assumption means that the gaseous concentration in Eq. 41.56 can be assumed to be zero. If this is the case, Eq. 41.56 becomes

$$J = -v_v c_d \tag{41.62}$$

If this equation is multiplied by lake surface area and the toxic substance's molecular weight, it can be expressed in the format of our mass balance as

$$V\frac{dc}{dt} = -v_v A_s c_d \tag{41.63}$$

The representation of Eq. 41.63 differs markedly from that for dissolved oxygen (Eq. 41.61). Rather than approaching an equilibrium at the saturation point, Eq. 41.63 represents a one-way loss from the system. In this sense it is just like a settling mechanism. However, rather than representing it as a velocity transport through the bottom of the lake, we represent volatilization as a velocity transport out the top.

41.2.1 Parameter Estimation

At this point the key to representing volatilization losses hinges on estimating v_v. As in Eq. 41.57, this means that we must determine values for Henry's constant and the liquid and gas–film-transfer coefficients.

Molecular weight and Henry's constant. Values of molecular weight and Henry's constant have been determined for many toxic compounds. Compendiums of this data can be found in a variety of sources (Callahan et al. 1979, Mabey et al. 1979, Mills et al. 1985, Lyman et al. 1982, Schnoor et al. 1987, Schwarzenbach et al. 1993, etc.).

Of the priority pollutants, molecular weights range from a low value of 50.6 for the halogenated aliphatic chloromethane to a high of 430 for the pesticide toxaphene. The lightest groups of compounds (< 200) are halogenated aliphatics, phenols, monocyclic aromatics, ethers, and nitrosamines. The heaviest are the pesticides followed by the phthalate esters and PCBs (> 200). The PAHs are intermediate.

Box-and-whisker diagrams for Henry's constant are depicted at the top of Fig. 41.9. In general, three classes of compounds are highly insoluble: halogenated aliphatics, monocyclic aromatics, and PCBs. The other classes have similar ranges of solubility, with some members being relatively soluble to others being moderately insoluble.

Film-transfer coefficients. The transfer coefficients are conventionally estimated by correlating them with commonly studied transfer processes. The liquid-film coefficient is typically correlated with the dissolved-oxygen-transfer coefficient. The gas-film coefficient is correlated with the evaporative transport of water vapor.

As previously noted in Lec. 20 (Eq. 20.58), the liquid-film mass-transfer coefficient (m yr^{-1}) can be related to the oxygen-transfer coefficient K_L by (Mills et al. 1982)

$$K_l = K_{l,O_2}\left(\frac{32}{M}\right)^{0.25} \tag{41.64}$$

where M = molecular weight of the compound and K_{l,O_2} = oxygen-transfer coefficient. This transfer coefficient can be estimated using the reaeration formulas from Lec. 20.

For the gas-film rate, Mills et al. (1982) have suggested

$$K_g = 168U_w\left(\frac{18}{M}\right)^{0.25} \tag{41.65}$$

where U_w = wind speed (mps). This formula is based on the gas-film transfer coefficient for water vapor. As expressed in Eq. 41.65 the units for K_g are m d^{-1}.

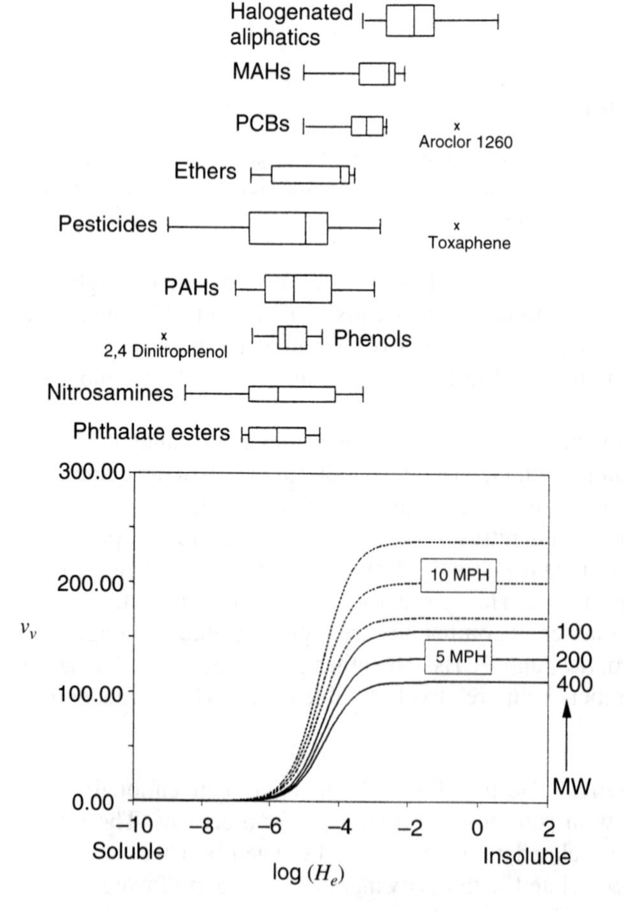

FIGURE 41.9
Plot of volatilization transfer coefficient v_v versus Henry's constant H_e (Chapra 1991). Each curve represents a particular molecular weight. Box-and-whisker diagrams (see Fig. 41.5) for nine classes of organic contaminants are drawn above the plot.

Analysis of volatilization. The impact of volatilization can be assessed by computing the net mass-transfer coefficient for a range of Henry's constants. As in Fig. 41.9, H_e ranges in value from approximately 10^{-8} to 1 atm m^3 mole^{-1}. The results indicate that at high values of the Henry's constant, the net mass-transfer coefficient approaches the liquid-film coefficient. Thus, as the terminology implies, "insoluble" gases (high H_e) are purged rapidly from the water whereas "soluble" gases (low H_e) stay in solution.

EXAMPLE 41.1. KINETIC PARAMETER ESTIMATION FOR PCB. Estimate (*a*) the sorption coefficient and (*b*) the volatilization mass-transfer velocity for high-molecular-weight PCBs (> 250 gmole). Note that parameters for these compounds can be summarized as

Toxicant	MW	log (K_{ow})	log (H_e)
Aroclor 1016	257.9	5.58	−3.48
Aroclor 1242	266.5	5.29	−2.42
Aroclor 1248	299.5	5.97	−2.45
Aroclor 1254	328.4	6.14	−2.84
Aroclor 1260	375.7	6.99	−0.60
Mean	305.6	5.99	−2.358

Assume that the wind speed is approximately 10 mph (= 4.47 m s^{-1}) and that the temperature is 10°C (= 283 K).

Solution: (*a*) Substitution of the octanol-water partition coefficient into Eq. 41.31 yields

$$K_d = 3.085 \times 10^{-8}(10^{5.99}) = 0.0301 \text{ g m}^{-3}$$

Remember that Eq. 41.31 is predicated on the assumption that $f_{oc} = 0.05$.
(*b*) The liquid-film exchange coefficient is based on the oxygen-transfer coefficient. For example, using the Banks-Herrera formula (Eq. 20.46),

$$K_{l,O_2} = 0.728(4.47)^{0.50} - 0.317(4.47) + 0.0372(4.47)^2 = 0.8655 \text{ m d}^{-1}$$

This value can be substituted into Eq. 41.64 (with appropriate conversions) to give

$$K_l = 365\left(\frac{32}{305.6}\right)^{0.25} 0.8655 = 179.7 \text{ m yr}^{-1}$$

The gas-film exchange coefficient can be computed with Eq. 41.65,

$$K_g = 61{,}320\left(\frac{18}{305.6}\right)^{0.25} 4.47 = 135{,}038 \text{ m yr}^{-1}$$

These values along with the Henry's constant can be substituted into Eq. 41.57 to give

$$v_v = 179.7\frac{10^{-2.358}}{10^{-2.358} + [8.206 \times 10^{-5}(283)(179.7/135{,}038)]} = 178.45 \text{ m yr}^{-1}$$

41.3 TOXICANT-LOADING CONCEPT

In Figs. 41.4 and 41.9 we have developed two plots that show how some classes of toxic substances sorb and volatilize. Although these plots provide insight, they suffer in that they provide an incomplete picture of the substance's transport and fate in a natural water. As in the elephant parable at the beginning of this book, the focus on the individual mechanisms does not illuminate their combined effect.

To assess the "big picture," we now integrate our knowledge regarding sorption and volatilization with the simple two-layer sediment-water model described in Sec. 40.4. We call this a "toxicant-loading model" because it is similar in spirit to the "phosphorus-loading plots" developed by Vollenweider to model eutrophication (recall Sec. 29.1). Details of the following analysis can be found in Chapra (1991).

Recall that our mass balance for the water and sediment layer of a well-mixed lake yielded the following model of the water column (Eq. 40.34):

$$c_1 = \frac{Qc_{\text{in}}}{Q + k_1V_1 + v_vAF_{d1} + (1 - F_r')(v_sF_{p1} + v_dF_{d1})A} \tag{41.66}$$

where F_r' = the ratio of recycle sediment purging to total sediment purging,

$$F_r' = \frac{v_r + v_dF_{d2}}{v_r + v_b + v_dF_{d2} + k_2H_2} \tag{41.67}$$

Now a problem with gaining insight into this model is that it depends on 14 independent parameters. Therefore we will perform some manipulations and make some simplifying assumptions to make the model more manageable. First, we reexpress the sediment feedback term F_r' in terms of the resuspension ratio F_r (Eq. 40.23). To do this, we define

$$v_r = F_rv_{sb} \tag{41.68}$$

and

$$v_b = (1 - F_r)v_{sb} \tag{41.69}$$

where v_{sb} = a scaled settling velocity,

$$v_{sb} = \frac{m}{(1 - \phi)\rho}v_s \tag{41.70}$$

By using these terms, Eq. 41.67 becomes

$$F_r' = \frac{F_rv_{sb} + v_dF_{d2}}{v_{sb} + v_dF_{d2} + k_2H_2} \tag{41.71}$$

Thus the model is no longer expressed as a function of the interdependent parameters v_r and v_b, but rather is now formulated in terms of the single parameter F_r.

Next we make two simplifying assumptions:

1. *Decay reactions are negligible (that is, $k_1 = k_2 = 0$).* Thus we ignore removal mechanisms such as photolysis, hydrolysis, and biodegradation (to be discussed in the next lecture) that act to remove the contaminant from the system through reactions. This is a conservative assumption in that any decay would tend to reduce the amount of toxicant in the system. Consequently the exclusion of such mechanisms means that our analysis will yield an upper bound on concentration.

2. *Water sorption is equal to sediment sorption (that is, $K_{d1} = K_{d2}$).* As shown by Chapra (1991) this means that diffusive sediment-water transfer will either be negligible or result in a loss from the water to the sediments. Consequently the exclusion of the diffusion mechanism also results in an upper bound prediction.

Applying these simplifying assumptions reduces Eq. 41.66 to

$$c_1 = \frac{Qc_{\text{in}}}{Q + v_T A} \tag{41.72}$$

where v_T = net toxicant loss rate (m yr^{-1}), which represents all purging mechanisms aside from flushing,

$$v_T = v_v F_{d1} + (1 - F_r) v_s F_{p1} \tag{41.73}$$

Thomann and Mueller (1987) proposed this approach as a way to separate the lake-specific mechanism of flushing from the more toxicant-specific mechanisms. If desired, Eq. 41.72 can always be used to interpret how particular values of v_T would effect water concentration for lakes with different levels of flushing.

However, for our present purposes, we focus on v_T itself. As depicted in Fig. 41.10, the net loss rate provides a means to integrate the effects of sorption, volatilization, and sediment feedback into a single number. If v_T is large, it suggests that the toxic will be purged from the water at a rapid rate. If it is small, purging would be weak.

The impact of sorption and volatilization on v_T can be assessed by mapping it on a logarithmic space defined by sorption (as represented by $K_d m$) and volatilization (as represented by H_e). To do this we assume values of T_a = 283 K, M = 200 g gmole^{-1}, U_w = 2.235 m s^{-1} (5 mph), and v_s = 91.25 m year^{-1}. Based on such an analysis, Chapra (1991) suggested that the sorption/volatilization space can be treated as three distinct regions (Fig. 41.11):

- *Air zone* (insoluble weak sorbers). Contaminants that are insoluble and sorb weakly always have high removal rates (that is, $v_T > 10$ m year^{-1}). This is because they are almost completely in dissolved form where they are subject to strong volatilization.
- *Water zone* (soluble weak sorbers). At the other extreme, compounds that are soluble and sorb weakly have extremely low removal rates (that is, $v_T < 10$ m year^{-1}). This is because they are completely in dissolved form yet are not subject to volatilization.
- *Sediment zone* (strong sorbers). These substances are independent of volatilization because they are not in dissolved form. Consequently their removal depends directly on sediment-water interactions. If resuspension is negligible, they exhibit high removal. If resuspension is high, removal is low.

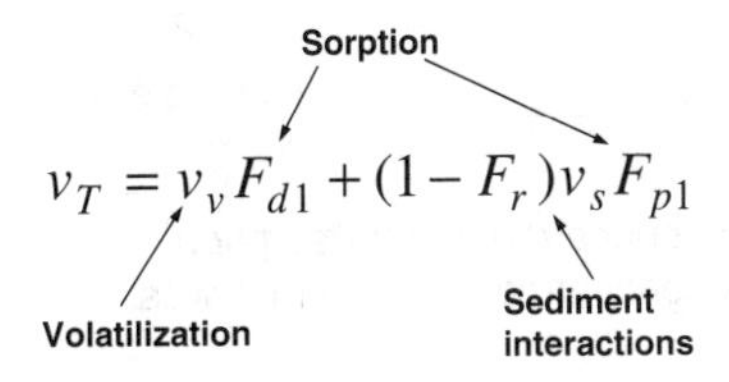

FIGURE 41.10
The net loss rate provides an integrated measure of the effects of sorption, volatilization, and sediment feedback on toxicant purging.

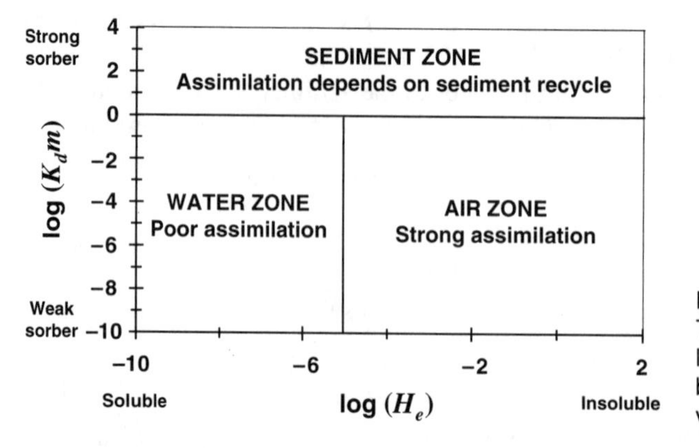

FIGURE 41.11
Three regions on a logarithmic space defined by sorption ($K_d m$) versus volatilization (H_e).

The implications of the foregoing for the organic priority pollutants (recall Table 40.1) are explored in Fig. 41.12. Here, we locate each of the pollutants on the sorption/volatilization space. For this case we need to specify the suspended solids concentration. We use a value of 5 g m^{-3} as being representative of mildly eutrophic lake conditions. The plot indicates that of the 114 organic priority pollutants, a significant portion would be subject to rapid purging due to volatilization.

I've (Chapra 1991) taken the analysis further by showing how each of the organic priority pollutant categories are arranged on the space. For the analysis a value of 5 g m^{-3} was again used for the suspended solids concentration. Although some

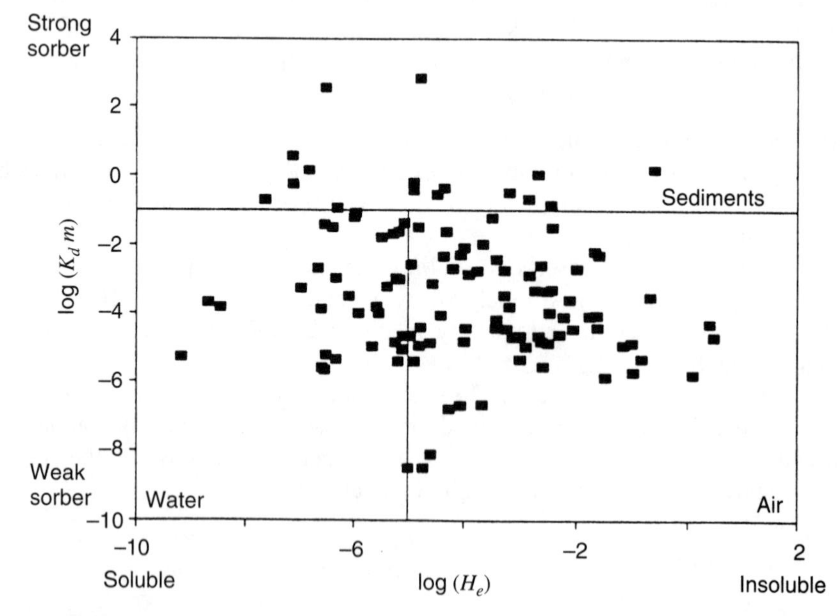

FIGURE 41.12
Mapping for organic priority pollutants on logarithmic space defined by sorption ($K_d m$) versus volatilization (H_e). A suspended solids concentration of 5 g m^{-3} was used to develop the plot.

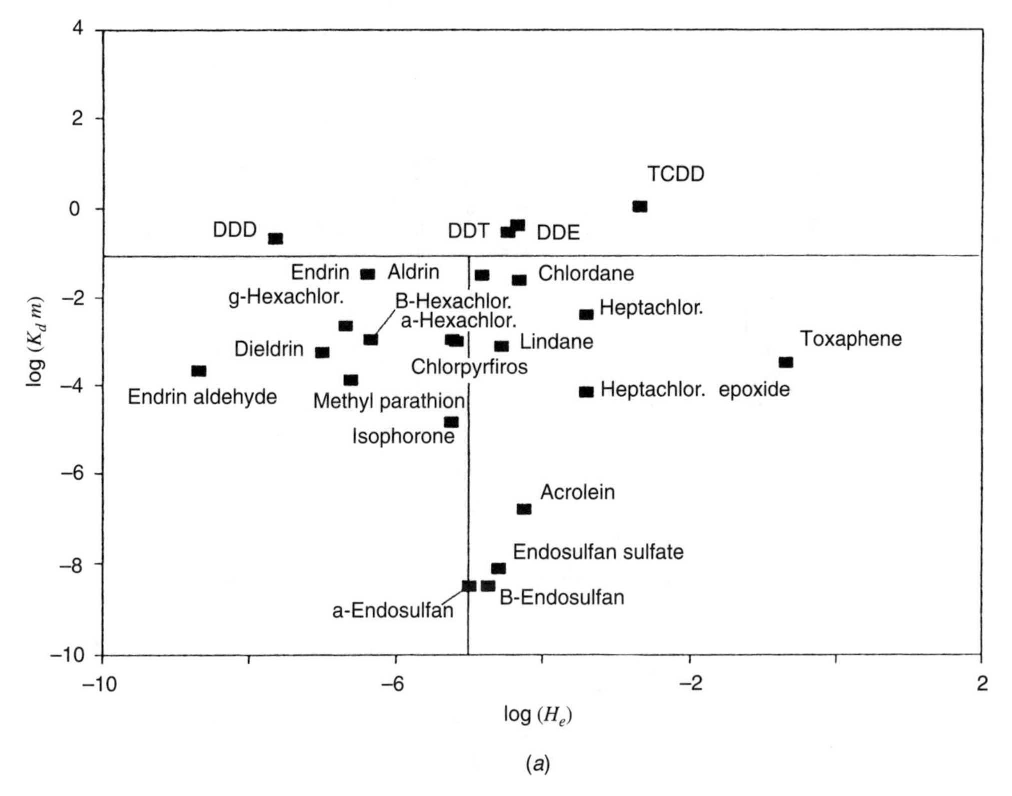

FIGURE 41.13
Mappings for (*a*) pesticides, (*b*) PCBs, and (*c*) PAHs.

of the larger groups fell across several regions (pesticides, PCBs, and PAHs), others tended to cluster within a single region. For example the halogenated aliphatics and many of the monocyclic aromatics and ethers fell within the air zone and exhibited strong volatilization losses. In contrast significant numbers of the phenols, nitrosamines, and phthalate esters fell within the water zone where purging was weak.

Figure 41.13 provides some examples of how the plots could be employed to provide further insight into the behavior within the classes. Figure 41.13*a* shows how the individual pesticides fall on the space. Such plots might prove useful in screening applications such as determining which chemicals among a large class of toxicants would have particular assimilation characteristics.

Figure 41.13*b* and *c* takes the idea farther by attaching additional information to the points. In Fig. 41.13*b* the numbers in parentheses indicate the percent chlorine for each PCB Aroclor. This plot suggests a positive correlation between sorption, volatilization, and percent chlorine. Together the two effects are manifested as a "northeast" movement as chlorine increases. Thus the benefits due to increased volatilization are somewhat negated (particularly for systems with large resuspension) by the fact that the higher chlorinated PCBs associate more strongly with particulate matter.

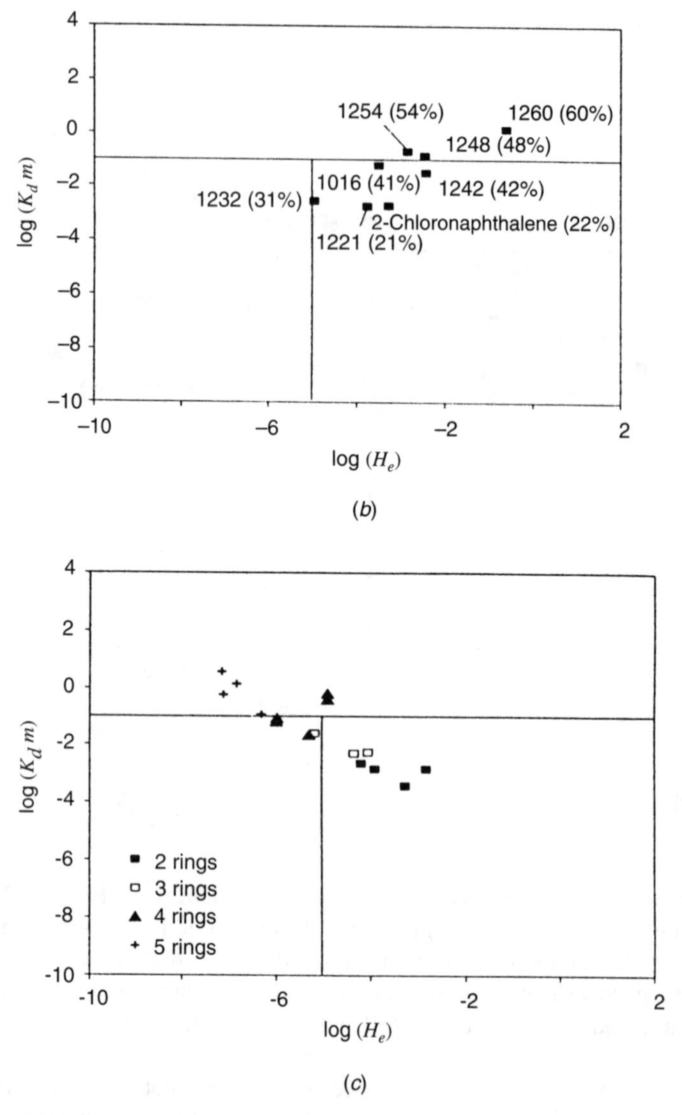

FIGURE 41.13 *(continued)*

Figure 41.13*c* classifies the PAHs according to their number of aromatic rings. For this case volatilization decreases as sorption increases. Consequently the trend is to the "northwest" as the number of rings increases. For lakes with little or no resuspension, this means that strong assimilation would still occur but that sedimentation would supplant volatilization as the primary removal mechanism. For lakes with high resuspension, assimilation would decrease as the number of rings increased.

The foregoing analysis is offered as an effort to draw general conclusions regarding the implications of sorption and volatilization on lake toxicant models for the organic priority pollutants. This analysis is based on many assumptions. It is strongly predicated on our choices for parameter values and in particular, on values for K_{ow} and H_e. The latter are a compilation of estimates from a variety of sources. As such they clearly exhibit some uncertainty. At the least they could undoubtedly be improved by additional measurements. However, because I have taken a global approach, I believe that the uncertainties in the data will not obviate the general trends that have emerged from the analysis.

In conclusion I have attempted to present a simple, visually oriented framework for assessing the assimilation of toxicants for lakes. As was done by Vollenweider in his work on lake eutrophication, my intent is to draw broad conclusions based on a global perspective.

PROBLEMS

41.1. A sorption experiment yields the following data:

c_d (μg L^{-1})	5	10	15	20	25	30	35
ν (mg g^{-1})	8	11.5	16	17	19	19	21

(*a*) Fit the data with the Langmuir isotherm. Display your resulting isotherm along with the data.

(*b*) Use the results of (*a*) to determine K_d.

(*c*) Calculate the fractions associated with particulate and dissolved form in the water based on the results from part (*b*) for a system with a suspended solids concentration of 10 g m^{-3}.

(*d*) Fit the data with a Freundlich isotherm. Display your resulting isotherm along with the data.

41.2. A lake and its underlying sediments have the following characteristics: $m = 5$ mg L^{-1}, $\phi = 0.8$, $\rho = 2.5$ g cm^{-3}. The water and sediment solids have fraction organic carbons of $f_{oc} = 0.2$ and 0.05, respectively. (*a*) Determine the fraction dissolved in the lake water (F_{d1}) and the sediment pore waters (F_{dp2}) for a toxic compound with a $K_{ow} = 10^6$. (*b*) If the total sediment contaminant concentration is 10 μg L^{-1}, compute the concentration in particulate form expressed in units of μg g^{-1}.

41.3. The particulate matter in a lake has the following characteristics: $m = 5$ mg L^{-1} and $f_{oc} = 0.4$. For a toxicant with a $K_{ow} = 10^{5.5}$: (*a*) determine the fraction in particulate form; F_{p1}. (*b*) repeat part (*a*), but include the additional effect of 5 mg L^{-1} of DOC.

41.4. Determine the volatilization velocity for DDT (MW = 350, $H_e = 10^{-4.4}$ atm m^{-3} mole^{-1}) in a lake where $U_w = 2$ mps and $T = 20$°C.

41.5. A well-mixed lake has a suspended solids concentration of 0.5 mg L^{-1} that has a carbon content of 5% and that settles at a rate of 0.2 m d^{-1}. It also has a dissolved organic carbon concentration of 1 mgC L^{-1} and its temperature is 10°C.

A contaminant in the lake has the following characteristics: $H_e = 10^{-5}$ atm m^3 mole^{-1} and $K_{oc} = 1$ m^3 gC^{-1}. Also the liquid and gas-film exchange coefficients are 179.7 m yr^{-1} and 135,038 m yr^{-1}, respectively.

Determine whether settling or volatilization will be the dominant purging mechanism. Do this by determining the ratio of settling to volatilization. *Note:* The fraction of contaminant that is associated with the DOC neither volatilizes nor associates with settling particles.

41.6. A well-mixed sediment layer has the following characteristics: $DOC_p = 10$ mg L^{-1}, $\phi = 0.9$, $v_b = 0.1$ mm yr^{-1}, and $\rho = 2.5$ g cm^{-3}. The solids in the sediments are 1% carbon. The lake has a DOC concentration of 1 mg L^{-1} and a POC concentration of 0.5 mg L^{-1} that settles at a rate of 0.1 m d^{-1}. It also has a contaminant concentration of 1 mg m^{-3}. If the log K_{ow} for a contaminant is 5, determine the concentration of contaminant in the sediment pore water and on the sediment solids. Note that the contaminant has a molecular weight of 300. Also there is no sediment resuspension, but there is diffusion between the sediment pore water and the overlying water.

41.7. A lake has the following parameters: $v_s = 100$ m yr^{-1}, $m = 5$ mg L^{-1}, $f_{oc} = 0.1$ and $T = 20$°C, and $U_w = 2$ m s^{-1}. (*a*) Determine the apparent settling velocity ($F_p v_s$) and the apparent volatilization rates ($F_d v_v$) for the contaminants listed below:

	log (K_{ow})	log (H_e)	Molecular weight
TCDD	6.84	−2.68	322
DDD	6.12	−7.66	320
Endrin aldehyde	3.15	−8.70	381
Endosulfan sulfate	−1.30	−4.59	422.9
Toxaphene	3.30	−0.68	430

(*b*) In which zones would the contaminants fall on Fig. 41.11.

41.8. Three new pesticides of equivalent toxicity are being evaluated for use in a lake watershed. The lake is shallow and has both high suspended solids ($f_{oc} = 0.05$; $m = 5$ mg L^{-1}) and significant sediment resuspension. If your only concern is protecting the lake ecosystem, determine which toxicant you would employ. Base your analysis on the toxicant loading concept.

Pesticide	log (K_{on})	log (H_e)
Sans-a-Roach	7	−4
Gnatmare	2	−8
Bugs 'r Toast	1	−1

LECTURE 42

Reaction Mechanisms: Photolysis, Hydrolysis, and Biodegradation

LECTURE OVERVIEW: I describe mechanisms that act to deplete toxics from water by transforming them into other compounds. In particular I review models of photolysis, the decomposition of organic contaminants due to the action of light. I also discuss methods to compute hydrolysis and biodegradation rates.

A number of different processes act to transform contaminants in natural waters. Some decompose due to exposure to sunlight (photolysis), through chemical reactions (e.g., hydrolysis), or by bacterial degradation. In the following lecture I provide an overview of these three processes.

42.1 PHOTOLYSIS

Photolysis refers to the breakdown of chemicals due to the radiant energy of light. Light transforms toxicants by two general modes. The first, called ***direct photolysis,*** occurs through absorption of light by the compound itself. The second, called indirect or ***sensitized photolysis,*** represents a group of processes that are initiated through light absorption by intermediary compounds. Direct decomposition predominates in systems with little extraneous dissolved or particulate matter such as clear lakes. Indirect photolysis was discovered when researchers noticed that some compounds degraded faster in natural water than in distilled water. Thus in more turbid or highly colored systems, sensitized photolysis could be a very significant decomposition mechanism for certain contaminants.

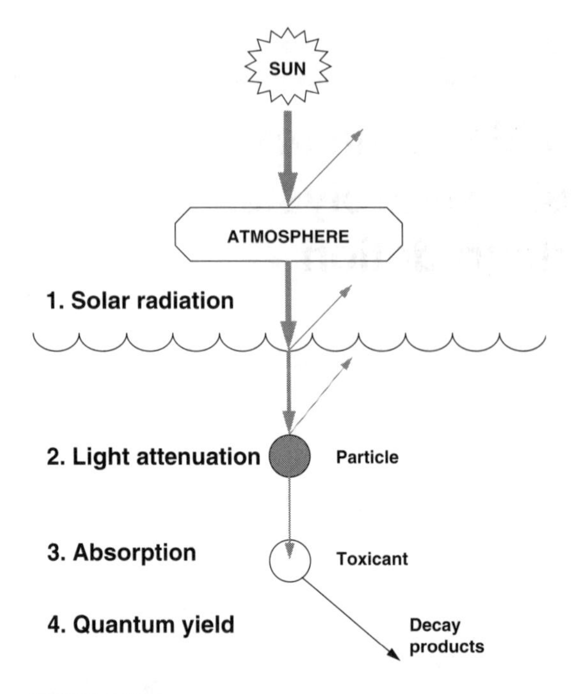

FIGURE 42.1
A simplified schematic of the four processes influencing direct photolysis in natural waters.

In this lecture I focus on direct photolysis because it is much better understood than indirect photolysis. The direct photolysis rate depends on a number of factors related to both the characteristics of the compound and the environment (Fig. 42.1):

1. *Solar radiation.* Depending on the time of year, the weather, and the geographical position of the water body, different levels of incoming solar radiation will be delivered to the surface of the water.
2. *Light attenuation in the water.* Suspended matter, color, and other factors influence the penetration and attenuation of the light. Photolysis in a quiescent, turbid lake might be limited to a thin surface layer, whereas it could extend to great depths in relatively clear water.
3. *Absorption spectrum of the chemical.* Due to its chemical structure the toxicant will absorb light energy to different degrees from various wavelengths.
4. *Quantum yield.* This refers to the fraction of absorbed photons that result in a chemical reaction.

In the following sections I will elaborate on each of these factors. Then I discuss mathematical models that predict their combined effect on compounds. However, before embarking on these subjects, I'll first present some general information concerning light.

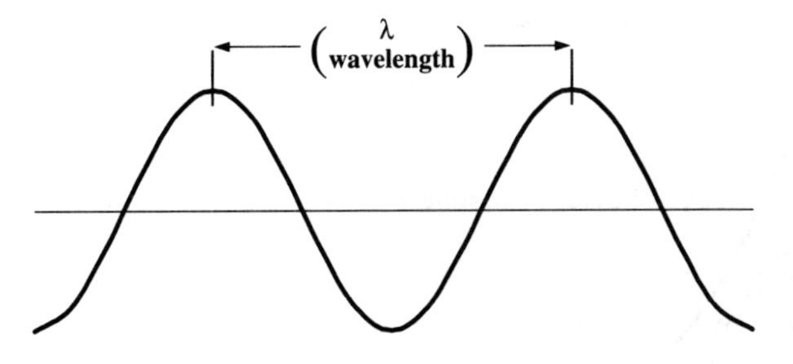

FIGURE 42.2
Wavelength is defined as the distance between peaks.

42.1.1 Light

To this point we have covered a number of environmental processes involving solar radiation. In Lec. 27 we described how light influences the die-off of bacteria. Then in Lec. 30 we learned how solar radiation contributes to the heat budget of a water body. Finally in Lec. 33 we modeled the effect of light on plant growth. In each case we treated light as a simple homogeneous entity quantified in energy units such as the langley (cal cm^{-2}). Now we take a closer look at solar radiation to understand better how it affects toxics.

Light has both wave and particle properties. From the wave perspective it can be conceptualized as oscillating electric and magnetic fields that are oriented perpendicular to each other. One way to characterize the light wave is via the distance between maxima. This distance, called the ***wavelength*** (Fig. 42.2), can be related to other fundamental properties by

$$\lambda = \frac{c}{\nu} \tag{42.1}$$

where λ = wavelength (nm)
c = speed of light in a vacuum (3.0×10^8 m s^{-1})
ν = frequency, defined here as the number of complete cycles that pass a point in a second (s^{-1})

The particle characterization of light relates to the fact that light energy consists of discrete quantities called ***photons.*** The amount of energy of a photon is related to

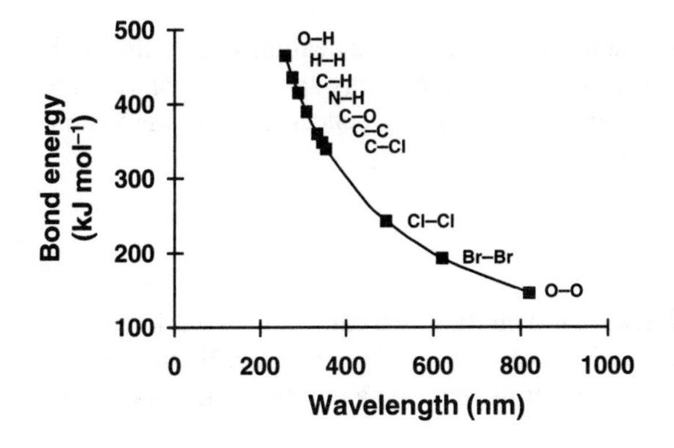

FIGURE 42.3
Relationship of bond energy to wavelength with the corresponding energy for some single bonds shown at the right (data from Schwarzenbach et al. 1993).

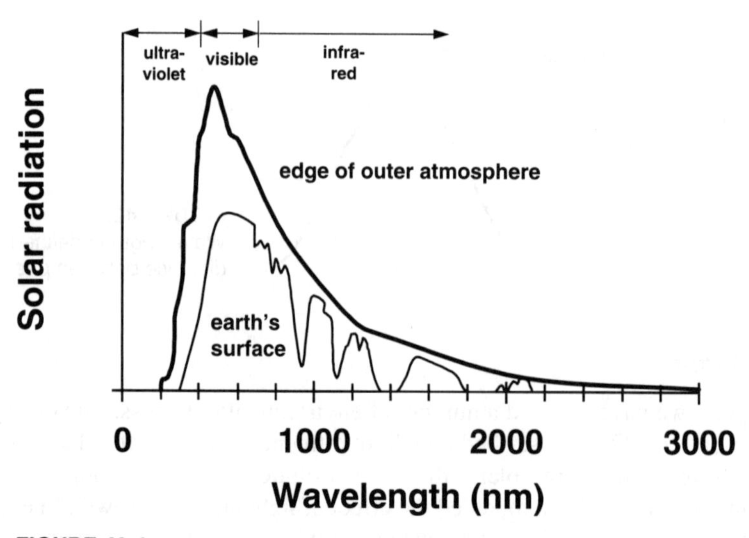

FIGURE 42.4
The effect of atmospheric absorption and scattering on delivery of solar radiation to the earth's surface (redrawn from Hutchinson 1957, Mills et al. 1985).

wavelength, as in

$$E = h\nu = h\frac{c}{\lambda} \tag{42.2}$$

where E has units of kJ and h = Planck's constant (6.63×10^{-34} J s). Because we are dealing with chemical compounds, light can be expressed on a molar basis. Thus, an ***einstein*** is defined as 6.023×10^{23} photons. Consequently the energy can be reexpressed as

$$E = 6.023 \times 10^{23} h\frac{c}{\lambda} = \frac{1.196 \times 10^5}{\lambda} \tag{42.3}$$

where E has units of kJ mole^{-1}.

According to both these formulas, shorter wavelengths have higher energy. This result is depicted in Fig. 42.3.

Finally we should mention the light source that we are dealing with in the environment: the sun. As displayed in Fig. 42.4, the solar energy spectrum spans a wide range of wavelengths. For photolysis we will be mostly concerned with the visible and ultraviolet portions of the spectrum (uv/vis).

42.1.2 Solar Radiation

Now that we have some general background on light, we can begin to work our way through the scheme outlined in Fig. 42.1. As already discussed in Lec. 30, the amount of solar radiation delivered to a natural water depends on a number of factors. These include solar altitude, atmospheric scattering and absorption, reflection, and shading. In contrast to our other models involving light, photolysis requires that these factors be considered as a function of wavelength. For example the fact that the atmosphere selectively reduces various wavelengths must be considered (Fig. 42.4).

As was discussed in the lectures on temperature modeling, reflection is important only when the solar altitude is low. Thus it is usually negligible ($< 10\%$). However, it could be significant when performing both seasonal (during certain times of the year at high latitude) and diurnal (early and late in the day) calculations.

42.1.3 Light Extinction in Natural Waters

As depicted in Fig. 42.5, natural waters selectively attenuate different wavelengths of light. The attenuation of light in natural waters can be quantified by the Beer-Lambert law,

$$I(z, \lambda) = I(0, \lambda)e^{-\alpha_D(\lambda)z} \tag{42.4}$$

where $I(z, \lambda)$ = light intensity (einstein cm^{-2} s^{-1}) at depth z (cm) and $\alpha D(\lambda)$ = apparent or diffuse attenuation coefficient (cm^{-1}). The attenuation coefficient can be determined empirically by measuring the light intensity at the surface and at a particular depth H and rearranging Eq. 42.4 to compute

$$\alpha_D(\lambda) = \frac{1}{H} \ln\left[\frac{I(0, \lambda)}{I(H, \lambda)}\right] \tag{42.5}$$

Before proceeding further we must introduce the notion that solar radiation in water does not follow a path that is strictly vertical. This is why the attenuation coefficient is called "apparent." In fact the solar irradiance in a water body actually consists of both collimated (that is, traveling as a parallel beam) and diffuse light. Perfectly diffuse light has a mean path length through water that is twice that of a beam of light. To quantify this effect the distribution function D has been developed as

$$D(\lambda) = \frac{\ell(\lambda)}{H} \tag{42.6}$$

where $\ell(\lambda)$ = mean path length of the light. For particle-free water, D has a value of approximately 1.2. Miller and Zepp (1979) have reported a mean value of 1.6 for particle-laden water. The distribution function can be used to relate the apparent coefficient to the attenuation coefficient per unit pathlength α, as in

$$\alpha_D(\lambda) = D(\lambda)\alpha(\lambda) \tag{42.7}$$

where $\alpha(\lambda)$ has units of cm^{-1}.

Now we can determine the total amount of light absorption over a layer of thickness H. This can be done by taking the difference of the incident light at the top of

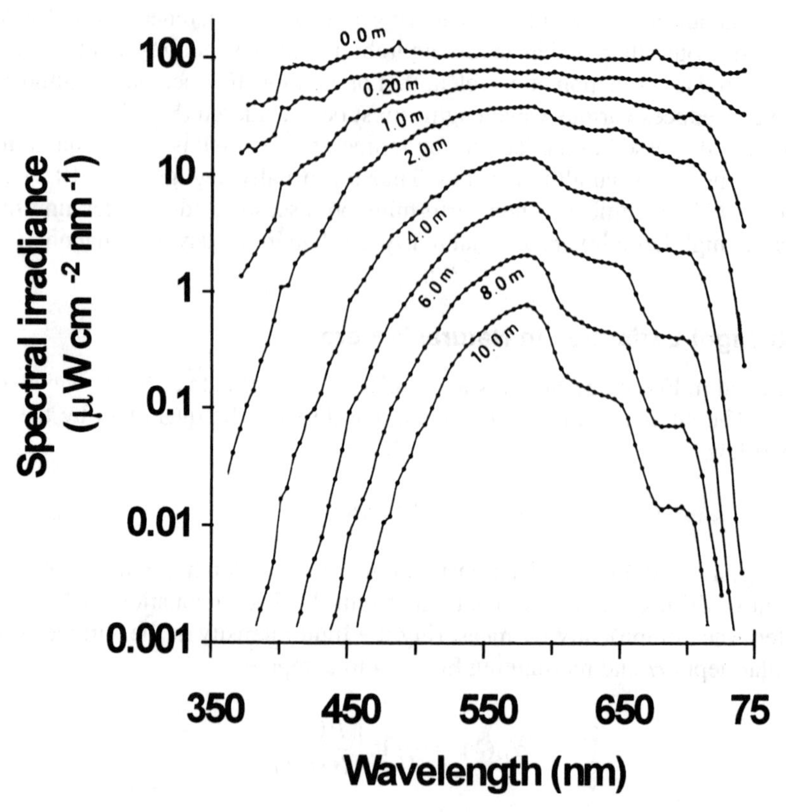

FIGURE 42.5
The attenuation of light in natural waters. Spectral irradiance at different depths in Lake San Vincente (from Tyler and Smith 1970).

the layer and the light at depth H. The result can be divided by H to yield the rate of light absorbance per unit volume,

$$I = \frac{I(0, \lambda)}{H}(1 - e^{-\alpha_D(\lambda)H}) \tag{42.8}$$

where I has units of einstein $cm^{-3}\ s^{-1}$.

42.1.4 Absorption

Now that we know how much light is absorbed in a layer, we describe how it affects compounds. The first step in this process is to characterize how compounds absorb light. To do this we employ an experiment that is used to quantify light absorption by the compound itself.

Suppose that we introduce a chemical into particle-free water and place the resulting solution in a transparent container such as a quartz cuvette. The transmission of light through such a system can be quantified by the Beer-Lambert law,

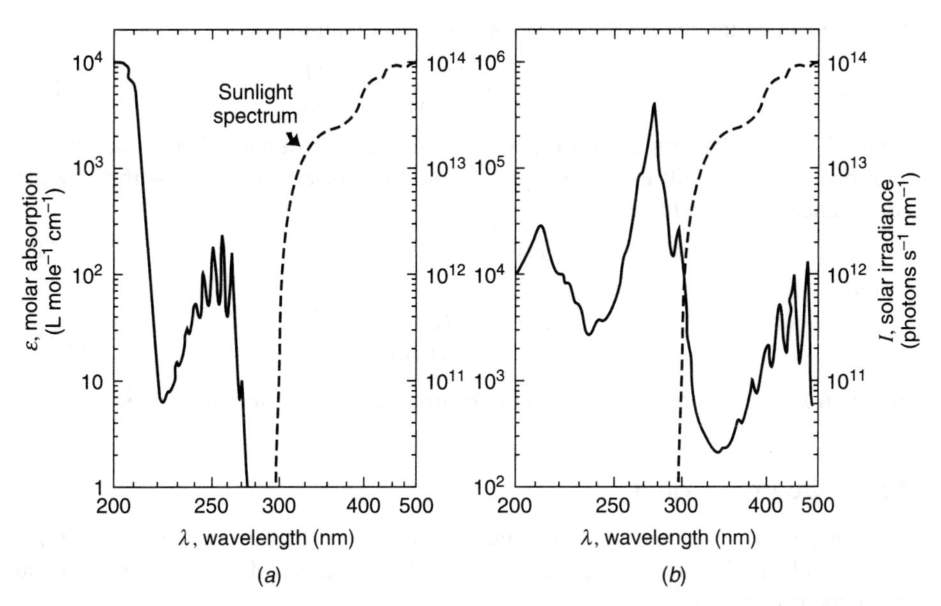

FIGURE 42.6
Electronic absorption spectra of (*a*) benzene, a compound that does not directly photolyze, and (*b*) naphthacene, a compound that does (from Mills et al. 1985). The dashed line is a sunlight spectrum taken from Burns et al. (1981).

$$I(\ell, \lambda) = I(0, \lambda)e^{-[\alpha(\lambda)+\varepsilon(\lambda)c]\ell} \tag{42.9}$$

where $I(\ell, \lambda)$ = light emerging from the system (einstein cm^{-2} s^{-1})
$I(0, \lambda)$ = incident light intensity (einstein cm^{-2} s^{-1})
$\varepsilon(\lambda)$ = molar extinction coefficient of the compound (L $mole^{-1}$ cm^{-1})
c = concentration of compound (mole L^{-1})
ℓ = pathlength of light in the solution (cm)

An experiment can be performed for each individual compound and an absorbance spectrum developed. Knowing the medium attenuation coefficient and the pathlength, we can determine the molar extinction spectrum. As in Fig. 42.6, the spectrum is usually expressed logarithmically because the attenuation coefficient often varies over several orders of magnitude.

When compared with the incident light spectrum, the absorption spectrum can be used to crudely assess whether photolysis is possible. For example benzene would not photolyze because the two spectra do not overlap (Fig. 42.6*a*). In contrast, naphthacene could photolyze because it absorbs radiation at wavelengths that are significant in natural waters (Fig. 42.6*b*).

Now that we know how much light is absorbed by the compound as a function of wavelength, we must determine how much of the light in our layer of water is absorbed by the contaminant. A simple way to do this is based on recognizing that the total light absorbance is a composite of the effects of the contaminant and other factors such as water, color, and particles. Further, in most cases, it is reasonable to assume that the absorbance due to the contaminant is dwarfed by the other factors.

Thus we can define the fraction absorbed by the contaminant as

$$F_c(\lambda) = \frac{\varepsilon(\lambda)c}{\alpha(\lambda) + \varepsilon(\lambda)c} \cong \frac{\varepsilon(\lambda)}{\alpha(\lambda)}c \tag{42.10}$$

where $F_c(\lambda)$ = fraction of absorbance due to the contaminant. This equation can then be combined with Eq. 42.8 to determine the amount of light absorbed by the contaminant in the volume,

$$I_a(\lambda) = k_a(\lambda)c \tag{42.11}$$

where

$$k_a(\lambda) = \frac{\varepsilon(\lambda)I(0, \lambda)(1 - e^{-\alpha_D(\lambda)H})}{H\alpha(\lambda)} \tag{42.12}$$

in which $k_a(\lambda)$ = specific rate of light absorbance (10^3 einstein mole^{-1} s^{-1}).

42.1.5 Quantum Yield

Just because a compound absorbs light does not mean that it will undergo significant photolysis. In fact, as depicted in Fig. 42.7, a variety of physical or chemical phenomena can occur.

The chemical changes are those that lead to an actual modification in the compound. Thus these processes are what we call ***photolysis.*** In contrast the physical processes involve energy losses. For example the molecule could return to its ground state via a loss of heat. Further, the chemical car release the energy in the form of light. This is referred to as ***luminescence.*** Finally the energy can be transferred

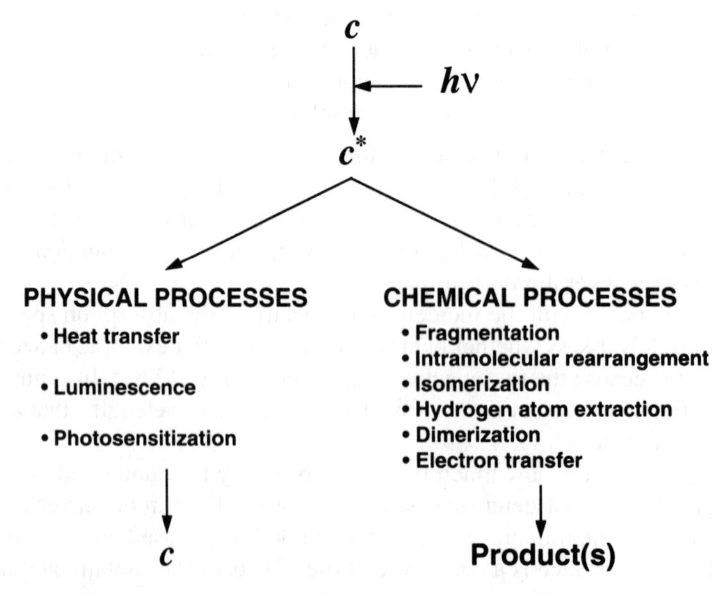

FIGURE 42.7
Once a chemical compound is excited, a variety of physical and chemical processes can occur (redrawn from Schwarzenbach et al. 1993).

to another molecule as in photosensitization. In all these phenomena the original compound is not changed chemically.

The extent of the chemical transformations is expressed by the ***quantum yield***. This is defined as

$$\Phi_r(\lambda) = \frac{\text{number of moles transformed}}{\text{total moles of photons of wavelength } \lambda \text{ absorbed}} \tag{42.13}$$

Note that $\Phi_r(\lambda)$ is referred to as the *reaction* quantum yield (mole einstein^{-1}) because it reflects the total effect of all the photolytic transformations acting on the compound. At present it must be determined experimentally.

The final step in completing the process outlined in Fig. 42.1 is to multiply Eq. 42.12 by Eq. 42.13 to give

$$k_p(\lambda) = \Phi(\lambda) k_a(\lambda) = \Phi(\lambda) \frac{\varepsilon(\lambda) I(0, \lambda)\left(1 - e^{-\alpha_D(\lambda) H}\right)}{H\alpha(\lambda)} \tag{42.14}$$

where $k_p(\lambda)$ = a first-order rate constant (s^{-1}) for photolysis as a function of wavelength. Note that the light intensity I must be expressed as 10^{-3} einstein cm^{-2} s^{-1} for the units to work out as s^{-1}.

42.1.6 Direct Photolysis Model

We can now build a direct photolysis model based on the information from the preceding sections. This is done by merely recognizing that we can simply integrate Eq. 42.14 to "sum up" the total photolysis rate across the relevant wavelengths,

$$k_p = \int_{\lambda_0}^{\lambda_1} \Phi(\lambda) \frac{\varepsilon(\lambda) I(0, \lambda)\left(1 - e^{-\alpha_D(\lambda) H}\right)}{H\alpha(\lambda)} d\lambda \tag{42.15}$$

This equation can be simplified by substituting Eq. 42.7 and assuming that the distribution function and quantum yield are wavelength-independent,

$$k_p = \Phi D \int_{\lambda_0}^{\lambda_1} \frac{\varepsilon(\lambda) I(0, \lambda)\left(1 - e^{-\alpha_D(\lambda) H}\right)}{H\alpha_D(\lambda)} d\lambda \tag{42.16}$$

or in discrete form,

$$k_p = \Phi D \sum_{\lambda=290}^{700} \frac{\varepsilon_\lambda I_{0,\lambda}\left[1 - e^{-\alpha_D(\lambda) H}\right]}{H\alpha_D(\lambda)} \tag{42.17}$$

Now, although this certainly provides a framework for calculating a single, first-order photolysis rate, it requires a great deal of information. In particular the light extinction, yield, and solar radiation must all be supplied as a function of wavelength. Further the solar radiation changes as a function of time of year and day. Although computer codes are available to facilitate the process (Zepp and Cline 1977, Zepp 1988), simplifications are also available to obtain order-of-magnitude estimates. The next section describes the most commonly used approach.

42.1.7 Near-Surface Approach

A problem with the framework developed in the previous section is that it requires the acquisition and manipulation of large quantities of information. Although this information is sometimes available, more often experimental data are reported as near-surface rate constants. It would be useful to be able to extrapolate this data to other contexts. To do this let us first look at what happens to Eq. 42.17 if $\alpha_D(\lambda)H$ is very small, as would be the case near the surface of a natural water. For such cases,

$$1 - e^{-\alpha_D(\lambda)H} \cong \alpha_D(\lambda)H \tag{42.18}$$

Therefore, Eq. 42.17 becomes

$$k_{p0} = \Phi D_0 \sum_{\lambda=290}^{700} \varepsilon_\lambda I_{0,\lambda} \tag{42.19}$$

The ratio of Eq. 42.16 to 42.17 can be formed as

$$\frac{k_p}{k_{p0}} = \frac{\Phi D \sum\limits_{\lambda=290}^{700} \dfrac{\varepsilon_\lambda I_{0,\lambda}\left(1 - e^{\alpha_D(\lambda)H}\right)}{H\alpha_D(\lambda)}}{\Phi D_0 \sum\limits_{\lambda=290}^{700} \varepsilon_\lambda I_{0,\lambda}} \tag{42.20}$$

Now to simplify further, we assume that most of the photolysis takes place at a narrow wavelength region. If this is the case, Eq. 42.20 can be simplified to give

$$k_p = k_{p0} \frac{I}{I_0} \frac{D}{D_0} \frac{1 - e^{-\alpha_D(\lambda^*)H}}{H\alpha_D(\lambda^*)} \tag{42.21}$$

TABLE 42.1
Near-surface direct photolysis parameters for a select group of toxic organics (extracted from Mills et al. 1985)

Compound	k_{p0} (d^{-1})	I_0 (ly d^{-1})	λ^* (nm)
Naphthalene	0.23	2100	310
1-Methylnaphthalene	0.76	2100	312
2-Methylnaphthalene	0.31	2100	320
Phenanthrene	2	2100	323
Anthracene	22	2100	360
9-Methylanthracene	130	2100	380
9,10-Dimethylanthracene	48	2100	400
Pyrene	24	2100	330
Chrysene	3.8	2100	320
Naphthacene	490	2100	440
Benzo(a)pyrene	31	2100	380
Benzo(a)anthracene	28	2100	340
Carbaryl	0.32	2100	313
Pentachlorophenol (anion)	0.46	600	318
3,3′-dichlorobenzidine	670	2000	280–330

where k_{p0} = direct near-surface photolysis rate (d^{-1})
I = total solar radiation at the top of the layer (ly d^{-1})
I_0 = total solar radiation at which k_{p0} was measured (ly d^{-1})
D/D_0 = ratio of the radiance distribution function to the radiance distribution function at the surface (= approximately 1.33)
$\alpha_D(\lambda^*)$ = extinction coefficient (m^{-1}) evaluated at the wavelength of maximum light adsorption λ^*

Therefore, to use this approach, three parameters are required: k_{p0}, I_0, and $\alpha_D(\lambda^*)$. Table 42.1 summarizes some values. In addition the light extinction can be calculated by

$$\alpha_D(\lambda) = D[\alpha_w(\lambda) + \alpha_a(\lambda)a + \alpha_c(\lambda)\mathrm{DOC} + \alpha_s(\lambda)m] \tag{42.22}$$

where D = 1.2 to 1.6
$\alpha_w(\lambda)$ = attenuation coefficient for water (m^{-1})
$\alpha_a(\lambda)$, $\alpha_c(\lambda)$, and $\alpha_s(\lambda)$ = concentration-specific attenuation coefficients (L $mg^{-1}m^{-1}$) for chlorophyll a, dissolved organic carbon, and inorganic suspended solids, respectively
a, DOC, and m = concentrations (mg L^{-1}) of chlorophyll a, dissolved organic carbon, and inorganic suspended solids

Table 42.2 contains values of the α's as a function of wavelength.

Figure 42.8 shows the effect of depth and water clarity on the attenuation of the surface rate. Other references (e.g., Lyman et al. 1982, Mills et al. 1985) contain parameter values for a number of constituents as well as descriptions of other techniques for estimating the photolysis rate.

TABLE 42.2
Attentuation coefficients parameterizing the dependency of light extinction on wavelength. Polynomial interpolation can be used to determine intermediate values, or a more complete version of the table can be found in Mills et al. (1985)

Wavelength center (nm)	α_w (m^{-1})	α_a (L mg^{-1} m^{-1})	α_c (L mg^{-1} m^{-1})	α_s (L mg^{-1} m^{-1})
300	0.141	69	6.25	0.35
320	0.0844	63	4.68	0.35
340	0.0561	58	3.50	0.35
360	0.0379	55	2.62	0.35
380	0.0220	46	1.96	0.35
400	0.0171	41	1.47	0.35
440	0.0145	32	0.821	0.35
400	0.0257	20	0.344	0.35
550	0.0638	10	0.167	0.35
600	0.2440	6	0.081	0.35
650	0.3490	8	—	0.35
700	0.6500	3	—	0.35
750	2.4700	2	—	0.35
800	2.0700	0	—	0.35

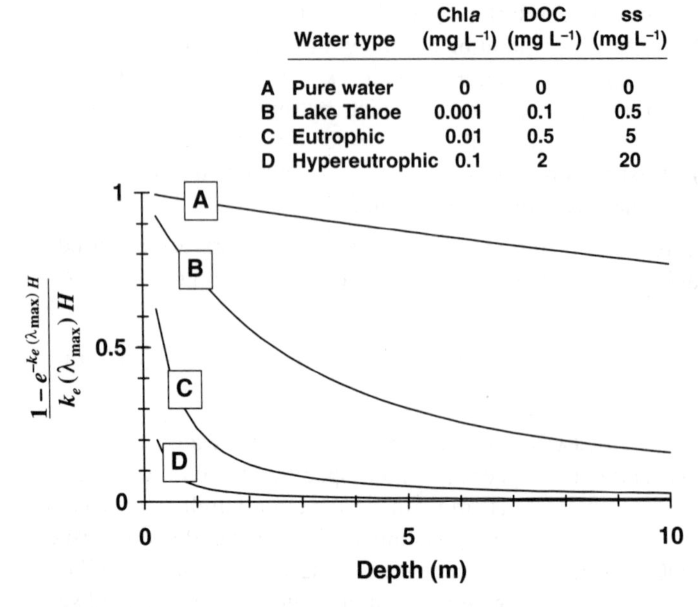

FIGURE 42.8
The effect of depth and water clarity on the attenuation of the surface photolysis rate. This plot corresponds to a wavelength of maximum light absorption of 340 nm.

EXAMPLE 42.1. NEAR-SURFACE DIRECT PHOTOLYSIS RATE. Determine the photolysis rate for naphthalene in the epilimnion (mean depth = 10 m) of (*a*) an oligotrophic and (*b*) a eutrophic lake. The average light intensity at the surface is 500 ly d^{-1}.

Solution: (*a*) From Table 42.1, the parameters for naphthalene are $k_{p0} = 0.23$ d^{-1}, $I_0 = 2100$ ly d^{-1}, and $\lambda^* = 310$ nm. The data from Table 42.2 can be evaluated with cubic interpolation to estimate the extinction coefficients for $\lambda^* = 310$ nm, with the results being $\alpha_w = 0.108$, $\alpha_a = 65.9$, $\alpha_c = 5.41$, and $\alpha_s = 0.35$. These results can be used to determine the total extinction coefficient. Because the lake has a lower solids content, we will assume that $D = 1.2$. Applying Eq. 42.22 and using the values for Lake Tahoe from Fig. 42.8 gives

$$\alpha_D(310) = 1.2[0.108 + 65.9(0.001) + 5.41(0.1) + 0.35(0.5)]$$
$$= 0.13 + 0.079 + 0.65 + 0.21 = 1.068 \text{ m}^{-1}$$

Thus, because of the low solids and DOC, the extinction is low. Now substituting this along with the other parameters into Eq. 42.21 yields

$$k_p = 0.23\frac{500}{2100}1.33\frac{1 - e^{-1.068(10)}}{10(1.068)} = 0.00682 \text{ d}^{-1}$$

This rate can be put into perspective by expressing it as a half-life of 0.693/0.00682 = 102 d.

(*b*) The calculation can be repeated for the eutrophic case from Fig. 42.8,

$$\alpha_D(310) = 1.6[0.108 + 65.9(0.01) + 5.41(0.5) + 0.35(5)]$$
$$= 0.173 + 1.05 + 4.33 + 2.80 = 8.36 \text{ m}^{-1}$$

$$k_p = 0.23\frac{500}{2100}1.33\frac{1 - e^{-8.36(10)}}{10(8.36)} = 0.00087 \text{ d}^{-1}$$

which corresponds to a half-life of 0.693/0.00087 = 795 d.

As in the previous example the productivity of the system tends to diminish the effectiveness of photolysis as a removal mechanism. This is due to the increased turbidity and color that reduces light penetration for such systems. However, it should be noted that such systems are more likely to exhibit indirect or sensitized photolysis because of the high concentrations of organic matter in eutrophic waters.

42.2 SECOND-ORDER RELATIONSHIPS

To this point we have used first-order kinetics to characterize any reactions in our models. Such reactions are formulated mathematically as

$$\text{Reaction} = kVc \tag{42.23}$$

This reaction is a special case of the more general formulation

$$\text{Reaction} = k_n V c^n \tag{42.24}$$

where n = order of the reaction. We have used the nomenclature k_n to indicate that the reaction rate will have different units depending on its order. For example the second-order reaction

$$\text{Reaction} = k_2 V c^2 \tag{42.25}$$

has the rate constant k_2, which has units ($L^3\ M^{-1}\ T^{-1}$). Thus the total expression has units of ($M\ T^{-1}$).

Although Eq. 42.25 represents one way to specify a second-order reaction, it is not the typical formulation used in toxic-substance modeling. Rather such rates are usually expressed as a function of the product of two concentrations,

$$\text{Reaction} = k_2 V b c \tag{42.26}$$

where b = a concentration ($M\,L^{-3}$) upon which the reaction depends in addition to c.

In the next two sections we discuss two mechanisms—biotransformation and hydrolysis—that are sometimes modeled with second-order reactions.

42.3 BIOTRANSFORMATION

Biotransformation refers to the microbially mediated transformation of organic contaminants. Typical degrading microorganisms are heterotrophic bacteria, actinomycetes, autotrophic bacteria, fungi, and protozoa.

The label "biotransformation" encompasses a number of distinctly different processes, including (Alexander 1979):

1. *Mineralization.* The conversion of an organic compound to inorganic products.
2. *Detoxication.* The conversion of a toxicant to innocuous by-products.
3. *Cometabolism.* The metabolism of a compound that the organisms cannot use as a nutrient (which does not result in mineralization, and organic metabolites remain).

TABLE 42.3
Second-order biodegradation rate constants for a select group of toxic organics (extracted from Mills et al. 1985)

Compound	k_{b2} (mL cell^{-1} d^{-1})
2,4-D Butoxyethyl ester	1.2×10^{-5}
Malathion	1.1×10^{-6}
Chloropropham	6.2×10^{-10}
Furadan	2.4×10^{-8}
Atrazine	2.4×10^{-8}
2-Chlorotoluene	6.5×10^{-8}
Dimethyl	1.2×10^{-4}
Di-ethyl	7.7×10^{-8}
Di-*n*-butyl	7.0×10^{-7}
Di-*n*-octyl	7.4×10^{-9}
Di-(2-ethylhexyl)	1.0×10^{-10}
Phenanthrene	3.8×10^{-6}

4. *Activation.* The conversion of a nontoxic substance to a toxic one, or an increase in a substance's toxicity, by microbial action.
5. *Defusing.* The conversion of a potential toxicant into a harmless metabolite before its potential is realized.

If the microbiological community has adapted to the contaminant, a Michaelis-Menten equation can be used to represent the rate of biotransformation as

$$k_b = \frac{\mu_{max} X}{Y(k_s + c)} \tag{42.27}$$

where μ_{max} = maximum growth rate of the culture (yr^{-1})
X = biomass concentration of the microorganisms (cells m^{-3})
Y = yield coefficient (cells produced per mass toxicant removed)
k_s = half-saturation constant (μg m^{-3})

TABLE 42.4
Typical bacterial population densities in natural waters (Mills et al. 1985)

Water-body type	Bacterial numbers
Surface waters	50–1×10^6 cells mL^{-1}
Oligotrophic lake	50–300 cells mL^{-1}
Mesotrophic lake	450–1400 cells mL^{-1}
Eutrophic lake	2000–12,000 cells mL^{-1}
Eutrophic reservior	1000–58,000 cells mL^{-1}
Dystrophic lake†	400–2300 cells mL^{-1}
Lake surficial sediments	8×10^9–5×10^{10} cells g-dry wt^{-1}
Stream sediments	10^7–10^8 cells g-dry wt^{-1}

†Dystrophic lakes are bog-like with high humic acid content.

When (as is often the case) $c \ll K_s$, Eq. 42.9 simplifies to the linear form,

$$k_b = \frac{\mu_{\max}}{Y k_s} X = k_{b2} X \tag{42.28}$$

where k_{b2} = a second-order biotransformation coefficient with units of m^3 (cells yr)$^{-1}$. For systems where the microbial population is relatively constant, Eq. 42.28 reduces to a first-order rate. Table 42.3 lists values of k_{b2} for a number of toxicants. Table 42.4 provides some estimates of bacterial concentration for a variety of aquatic environments.

Extrapolation of lab data to the environment can pose a significant problem in estimating biotransformation rates. Lyman et al. (1982), Mills et al. (1982), and Schwarzenbach et al. (1993) discuss this problem. They also contain summaries of parameter values for selected contaminants and additional information on biotransformation.

EXAMPLE 42.2. BIODEGRADATION RATE. Determine the biodegradation rate of di-*n*-butyl in (*a*) an oligotrophic and (*b*) a eutrophic lake. The second-order decay rate for this pesticide is approximately 7×10^{-7} mL cell^{-1} d^{-1} (Mills et al. 1985).

Solution: (*a*) From Table 42.3 we will assume a bacterial concentration of 100 cells mL^{-1} to be representative of an oligotrophic lake. The first-order value can be determined as

$$k_b = 7 \times 10^{-7}(100) = 0.00007 \text{ d}^{-1}$$

which corresponds to a half-life of 0.693/0.00007 = 9902 d (27 yr).

(*b*) From Table 42.3 we will assume a bacterial concentration of 5000 cells mL^{-1} to be representative of a eutrophic lake. The first-order value can be determined as

$$k_b = 7 \times 10^{-7}(5000) = 0.0035 \text{ d}^{-1}$$

which corresponds to a half-life of 0.693/0.0035 = 198 d.

As in the previous example, the water body's productivity tends to increase the effectiveness of biodegradation as a removal mechanism. This is due to the increased bacterial populations for such systems. This result is in contrast to the result for direct photolysis in Example 42.1.

42.4 HYDROLYSIS

Hydrolysis refers to reactions in which the bond of a molecule is cleaved and a new bond is formed with the hydrogen and the hydroxyl components of a water molecule. Hydrolytic reactions are catalyzed by acids or bases and, to a more limited extent, by water.[†] These catalytic effects depend on the type of reaction and

[†] Inorganic catalysts other than water (such as sulfate, nitrate, bicarbonate, etc.) can induce transformations in a similar manner to hydrolysis. Although these are often dwarfed by hydrolysis, they can be important in some systems (see Schwarzenbach et al. 1993 for details).

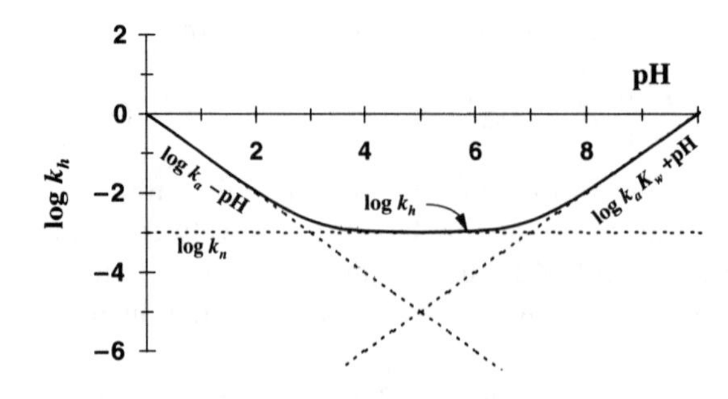

FIGURE 42.9
Effect of pH on the hydrolysis rate. The case shown is for $k_a = 1$, $k_n = 10^{-3}$, $k_b = 10^4$, and $K_w = 10^{-14}$.

the chemical structure of the compound. The pH and temperature of the solution have an influence on the rate of reaction.

The rate of hydrolysis can be expressed as

$$k_h = k_b[\mathrm{OH}^-] + k_n + k_a[\mathrm{H}^+] \tag{42.29}$$

where k_n, k_a, and k_b = coefficients parameterizing the neutral, acidic, and basic dependencies of hydrolysis. The neutral rate k_n has units of d^{-1}, whereas k_a and k_b have units of L $\mathrm{mole}^{-1}\ \mathrm{d}^{-1}$.

If the equilibrium relationship for the dissociation of water,

$$K_w = [\mathrm{H}^+][\mathrm{OH}^-] \tag{42.30}$$

is substituted into Eq. 42.29, the rate can be expressed in terms of pH as

$$k_h = k_b \frac{K_w}{10^{-\mathrm{pH}}} + k_n + k_a 10^{-\mathrm{pH}} \tag{42.31}$$

Thus, depending on the coefficient values and the pH, the rate can either be second- or first-order (Fig. 42.9).

Hydrolysis rates range from 10^{-7} to $10^{-1}\ \mathrm{d}^{-1}$. Lyman et al. (1982), Mills et al. (1985), and Schnoor et al. (1987) can be consulted to obtain parameter values for selected contaminants and additional information on hydrolysis.

EXAMPLE 42.3. HYDROLYSIS RATE. Determine the hydrolysis rate of pentachlorophenol ($k_n = 5.8 \times 10^{-3}\ \mathrm{d}^{-1}$, $k_a = 1.1 \times 10^4$ L $\mathrm{mole}^{-1}\ \mathrm{d}^{-1}$, and k_b = 3.3 L $\mathrm{mole}^{-1}\ \mathrm{d}^{-1}$) for the epilimnion (pH = 7.5 to 8.5) and hypolimnion (pH = 6.5 to 7.5) of a lake.

Solution: Using the mean pH for the epilimnion, we can estimate the hydrolysis rate as (Eq. 42.31)

$$k_h = 3.3 \frac{10^{-14}}{10^{-8}} + 5.8 \times 10^{-3} + (1.1 \times 10^4)10^{-8} = 0.0059\ \mathrm{d}^{-1}$$

The range can also be determined as 0.00615 to 0.00585 d^{-1}, which corresponds to half-lives ranging from 113 to 119 d. Thus it is fairly constant over the range of pHs encountered in the epilimnion.

For the hypolimnion, the rate at the mean pH of 7 is 0.0069 and has a variation between 0.00928 and 0.00615 d^{-1}, which corresponds to half-lives ranging from 75

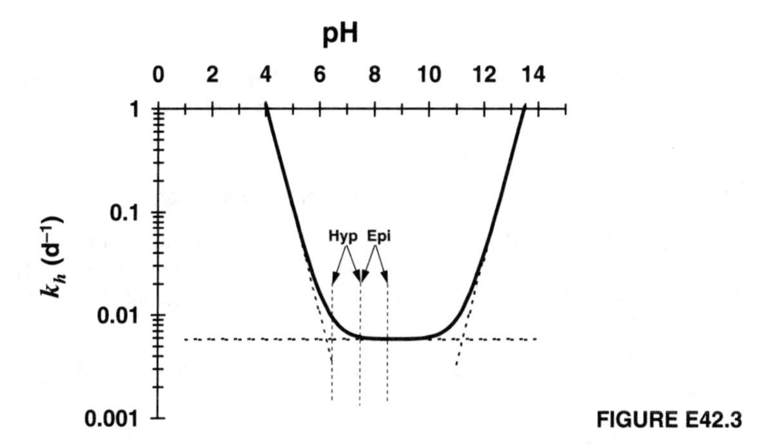

FIGURE E42.3

to 113. Consequently it seems to vary much more in the bottom waters. A plot of the decay rate versus pH provides an explanation of these results (Fig E42.3).

As shown, the epilimnion pH variations occur on the flat, neutral part of the function, whereas the hypolimnion varies across the range where the acidic constant is becoming significant. At lower pH's the effect of pH would become even more pronounced.

42.5 OTHER PROCESSES

In addition to the above processes, other mechanisms can effect the transport and fate of toxicants in aquatic environments:

Oxidation/reduction. Chemical oxidation reactions occur in natural waters when a sufficient level of oxidant is present. Common types are chlorine and ozone. The general form of the chemical oxidation rate is

$$k_O = k_O' Ox \tag{42.32}$$

where k_O' = a second-order rate constant and Ox = concentration of the oxidant. If the oxidant concentration is constant, Eq. 42.32 represents a first-order rate.

Acid-base effects. The disposition of organic acids and bases (e.g., phenol, benzidine, etc.) can be strongly effected by the hydrogen ion concentration of the water. Mills et al. (1985) can be consulted to obtain additional information on this mechanism.

Bioconcentration. The capacity of a chemical to be taken up by organisms is termed bioconcentration. This effect is treated similar to the linear sorption mechanism. It is quantified by a bioconcentration factor that is analogous to the partition coefficient (Lec. 45).

Additional information on these transformations can be found in references such as Lyman et al. (1982) and Schwarzenbach et al. (1993).

PROBLEMS

42.1. You measure values of 500 ly d^{-1} and 100 ly d^{-1} at 1 m and 9 m below the water surface of a lake, respectively. Determine the apparent attenuation coefficient.

42.2. Repeat Example 42.1, but for the compound pyrene.

42.3. Repeat Example 42.2, but for the compound 2,4-D butoxyethyl ester.

42.4. Determine the combined half-life due to hydrolysis and biodegradation for 2,4-D butoxyethyl ester in a hypereutrophic lake ($X = 50{,}000$ cells mL^{-1}) for two cases: pH = 7 and pH = 7.5. Note that for 2,4-D butoxyethyl ester, $k_a = 1.7$ L $mole^{-1}$ d^{-1}, $k_n = 0$ d^{-1}, and $k_b = 2.6 \times 10^6$ L $mole^{-1}$ d^{-1}.

42.5. Repeat Example 42.3, but for the pesticide parathion ($k_n = 3.6 \times 10^{-3}$ d^{-1}, $k_a = 1.3 \times 10^2$ L $mole^{-1}$ d^{-1}, and $k_b = 2.46 \times 10^3$ L $mole^{-1}$ d^{-1}).

42.6. Determine the 95% response time for a contaminant in a lake having the following characteristics: $Q = 50 \times 10^6$ m^3 yr^{-1}, $V = 200 \times 10^6$ m^3, $H = 10$ m. The lake has an inflow solids concentration of 20 mg L^{-1} and a suspended solids settling velocity of 0.25 m d^{-1}. The contaminant has the following characteristics: log $K_{ow} = 6$, $k_a =$ L $mole^{-1}$ d^{-1}, $k_n = 8.16 \times 10^{-4}$ d^{-1}, and $k_b = 9182.7$ L $mole^{-1}$ d^{-1}. Note that only the dissolved contaminant is subject to hydrolysis. Volatilization, sediment feedback and other decay mechanisms are negligible. The pH of the system is 7, the suspended solids are 5% organic carbon, and the dissociation constant for water is approximately 10^{-14}.

LECTURE 43

Radionuclides and Metals

LECTURE OVERVIEW: I describe the transport and fate of inorganic toxics in natural waters. After an overview of such substances, I develop and apply a model framework for radionuclides to simulate the levels of cesium-137 in Lake Michigan. I follow this with a description of two models of metals in natural waters. The first is patterned after the toxic organic frameworks described in previous lectures in that it partitions the metal into particulate and dissolved forms. The second adds metal precipitation in the sediment layer. This model illustrates how chemical equilibrium models can be integrated with mass balances in order to more effectively model metals.

To this point we have developed frameworks to model toxic organics in natural waters. We'll now cover how inorganic toxicants, namely metals and radionuclides, are simulated.

43.1 INORGANIC TOXICANTS

In this lecture I describe models for the two most common categories: heavy metals and radionuclides (Table 43.1). ***Heavy metals*** usually refer to metals between atomic numbers 21 and 84. A few nonmetals (arsenic and selenium) as well as the lighter metal aluminum are also included in this category. Most of these occur naturally. However, they can be greatly enhanced by mining and by industrial activities such as electroplating, smelting, and manufacturing.

Radionuclides (or radioactive substances) result from nuclear energy generation, nuclear weapons development, and some industrial applications. In addition certain radionuclides occur naturally (e.g., lead-210) and are employed by environmental scientists as tracers.

TABLE 43.1
Listing of some metals and radionuclides of interest in water-quality modeling

Metals				Radionuclides	
Name	**Symbol**	**Name**	**Symbol**	**Name**	**Symbol**
Aluminum	Al	Cesium	Cs	Cesium-137	^{137}Cs
Chromium	Cr	Barium	Ba	Plutonium-239, 240	$^{239,240}Pu$
Nickel	Ni	Silver	Ag	Strontium-90	^{90}Sr
Copper	Cu	Mercury	Hg	Lead-210	^{210}Pb
Zinc	Zn	Arsenic	As		
Cadmium	Cd	Selenium	Se		
Lead	Pb				

43.2 RADIONUCLIDES

Radionuclides are similar to organic pollutants in that they sorb to particulate matter. However, they differ in two fundamental ways:

- The radioactive substances of the type listed in Table 43.1 do not volatilize.
- They decompose by simple first-order radioactive decay kinetics. They are not subject to the variety of decomposition mechanisms, such as photolysis, biodegradation, etc., that act to break down organic compounds.

In addition they are measured in radioactivity units (Ci or ***curie***) rather than in mass units. One curie corresponds to the disintegration of 3.7×10^{10} atoms per second. This value was chosen because it is approximately the decay rate of 1 g of radium. A related unit is the ***becquerel*** (Bq), where 3.7×10^{10} Bq are equivalent to a Ci. Note that typical radioactivity magnitudes encountered in natural waters are quite low when expressed in Ci units. Thus, beyond the common SI prefixes used to this point (recall Table 1.1), additional prefixes corresponding to tiny magnitudes are necessary (Table 43.2).

TABLE 43.2
SI (International System of Units) prefixes commonly used in water-quality modeling of radionuclides[†]

Prefix	Symbol	Value
pico-	p	10^{-12}
femto-	f	10^{-15}
atto-	a	10^{-18}

[†] A complete list of prefixes appears in App. A.

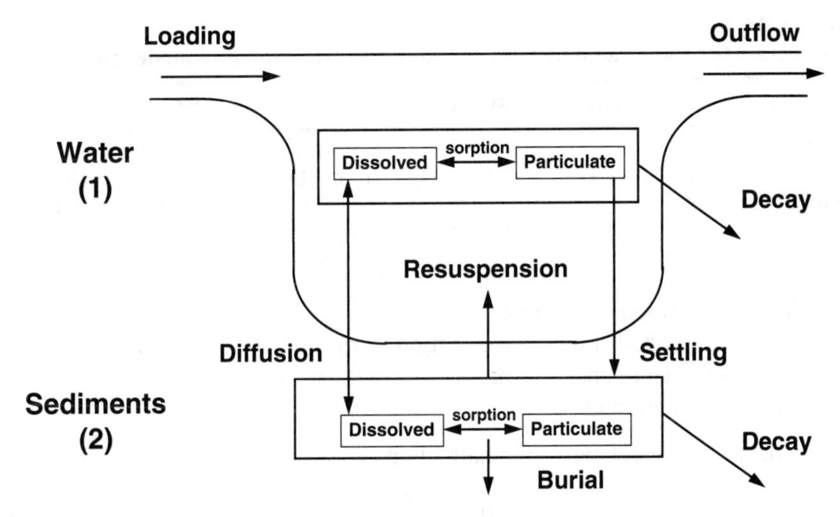

FIGURE 43.1
Schematic of a radionuclide budget for a well-mixed lake with sediment feedback.

A simple framework for a radionuclide in a well-mixed lake is depicted in Fig. 43.1. As with organic pollutants, a solids budget is required. After this is done, mass balances for each of the layers can be written as

$$V_1 \frac{dc_1}{dt} = Qc_{\text{in}} - Qc_1 - kV_1c_1 - v_sAF_{p1}c_1 + v_rAc_2 + v_dA(F_{d2}c_2 - F_{d1}c_1) \tag{43.1}$$

and
$$V_2 \frac{dc_2}{dt} = -kV_2c_2 + v_sAF_{p1}c_1 - v_rAc_2 - v_bAc_2 + v_dA(F_{d1}c_1 - F_{d2}c_2) \tag{43.2}$$

where subscripts 1 and 2 indicate water and sediments, respectively, and

t = time (yr)
c = concentration (Ci m^{-3})
V = volume (m^3)
c_{in} = inflow concentration (Ci m^{-3})
k = first-order decay coefficient
A = sediment surface area (m^2)
v_s = settling velocity of solids (m yr^{-1})
v_d = sediment-water diffusion mass-transfer coefficient (m yr^{-1})
v_r = resuspension velocity (m yr^{-1})
v_b = burial velocity (m yr^{-1})

The dissolved and particulate fractions can be calculated as previously derived in Lec. 41,

$$F_{d1} = 1 - F_{p1} = \frac{1}{1 + K_{d1}m} \tag{43.3}$$

and
$$F_{d2} = \frac{1}{\phi + K_{d2}(1 - \phi)\rho} \tag{43.4}$$

TABLE 43.3
Parameters for some radionuclides of interest in water-quality modeling

Parameter	Units	^{239,240}Pu	^{137}Cs	^{90}Sr	^{210}Pb
Half-life	yr	4.5×10^9	30	28.8	22.3
Diffusion coefficient	$cm^2\ s^{-1}$	12.1×10^{-6}	12.1×10^{-6}	3.8×10^{-6}	4.7×10^{-6}
Partition coefficient	$m^3\ g^{-1}$	0.5	0.5	2×10^{-4}	10
	$(L\ kg^{-1})$	(5×10^5)	(5×10^5)	(2×10^2)	(1×10^7)

where K_d = a partition coefficient ($m^3\ g^{-1}$)
m = suspended solids concentration ($g\ m^{-3}$)
ρ = sediment density ($g\ m^{-3}$)
ϕ = sediment porosity

The decay rate is related to the half-life by

$$k = \frac{0.693}{t_{50}} \tag{43.5}$$

Table 43.3 lists key parameters for some common radionuclides.

EXAMPLE 43.1. CESIUM-137 IN LAKE MICHIGAN. Lake Michigan (excluding Green Bay) has the following physical characteristics:

$$Q = 40 \times 10^9\ m^3 yr^{-1} \qquad A_1 = 53{,}500 \times 10^6\ m^2 \qquad V_1 = 4820 \times 10^9\ m^3$$

$$A_2 = 20{,}000 \times 10^6\ m^2 \qquad H_2 = 2\ cm$$

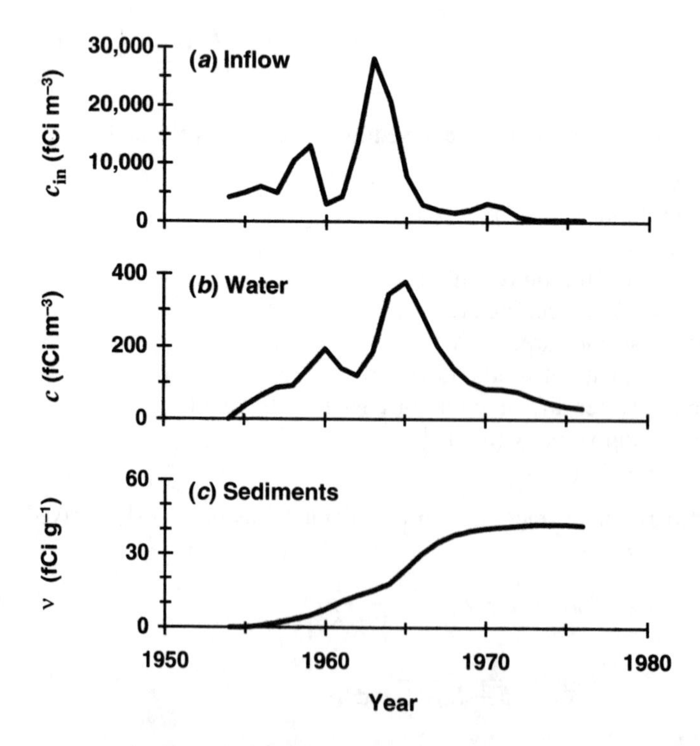

FIGURE E43.1

In addition a solids budget for the lake can be developed. The resulting parameters are

$$\phi = 0.9 \qquad \rho = 2.5 \times 10^6 \text{ g m}^{-3} \qquad m = 0.5 \text{ g m}^{-3}$$

$$W_s = 3.8 \times 10^{12} \text{ g yr}^{-1} \qquad v_s = 1.25 \text{ m d}^{-1} \qquad v_r = 0.1565 \text{ mm yr}^{-1}$$

$$v_b = 0.756 \text{ mm yr}^{-1}$$

During the late 1950s and early 1960s, nuclear weapons testing resulted in a fallout flux of a number of radionuclides to the surface of the earth (recall Example 5.3). The values for ^{239,240}Pu can be tabulated (Table 43.4). These values can be converted to fluxes of ^{90}Sr and ^{137}Cs by multiplying by approximately 57.9 and 60.9, respectively. Predict the response of Lake Michigan to the cesium flux.

Solution: Equations 43.1 and 43.2 can be integrated and the results displayed in Fig. E43.1.

TABLE 43.4
Fluxes of ^{239,240}Pu in fCi m^{-2} yr^{-1} to the Great Lakes

Year	Flux	Year	Flux	Year	Flux	Year	Flux
1954	51.5	1960	37.0	1966	36.0	1972	11.5
1955	60.7	1961	53.0	1967	253.0	1973	4.6
1956	73.6	1962	166.0	1968	20.0	1974	4.6
1957	60.7	1963	345.0	1969	25.0	1975	4.6
1958	128.8	1964	255.0	1970	39.1	1976	4.6
1959	161.0	1965	96.6	1971	32.0		

43.3 METALS

As with toxic organics and radionuclides, the fate of metals is inextricably tied together with the fate of solid matter. Thus transport mechanisms such as sorption and settling/resuspension significantly affect the fate for metals. Although there is this similarity, metals differ from organic pollutants and radionuclides in several ways:

1. *Natural levels.* Most heavy metals occur naturally. Thus we must often consider background levels when assessing anthropogenic sources.
2. *Lack of decay mechanisms.* Most metals are conservative in that the total amount of metal is not broken down by processes such as biodegradation, photolysis, or radioactive decay. The absence of these mechanisms simplifies their modeling. In addition, although most do not have gaseous phases, some of the nonmetallic inorganic toxicants (e.g., mercury) can be lost by volatilization.
3. *Inorganic sorption.* Although metals associate with particulate matter, the nature of their sorption differs from toxic organics. Recall that hydrophobic organics have an affinity for organic carbon particles. Therefore, as expressed by Schnoor et al. (1987), there is a "likes-dissolving-likes" quality to the mechanism. Metals can be complexed by organic ligands and, thus, can be sorbed to organic solids in a fashion similar to toxic organics. However, several other processes are significant including (*a*) physical adsorption to solid surfaces, (*b*) chemical sorption or binding by ligands, and (*c*) ion exchange.

4. *Chemical speciation.* Aside from sorption, metals can further speciate into different chemical forms. These species can be important in that they can exhibit differing transport and fate as well as differing toxicity.

Although some of the above features mean that metals models could be simpler than toxic organic models (e.g., lack of decay mechanisms), the other points mean that the metals frameworks could be more complex. In the following sections I'll first show how a simple metals model can be patterned after our toxic organic framework. Then I'll describe how equilibrium chemistry can be integrated with a mass balance to allow a more complex representation to be developed.

43.3.1 Simple Partitioning Models

As was the case for radionuclides, a very simple model of metal dynamics in natural waters can be developed by assuming that the metal partitions into dissolved and particulate forms. In a fashion similar to Eqs. 43.1 and 43.2, mass balances can be developed for a well-mixed system,

$$V_1 \frac{dc_1}{dt} = Qc_{\text{in}} - Qc_1 - v_s A F_{p1} c_1 + v_r A c_2 + v_d A (F_{d2} c_2 - F_{d1} c_1) \quad (43.6)$$

and

$$V_2 \frac{dc_2}{dt} = v_s A F_{p1} c_1 - v_r A c_2 - v_b A c_2 + v_d A (F_{d1} c_1 - F_{d2} c_2) \quad (43.7)$$

Notice that these equations are identical to the radionuclide model with the exception that decay is omitted.

At steady-state, these equations can be solved for

$$c_1 = \frac{Qc_{\text{in}}}{Q + v_T A} \quad (43.8)$$

and

$$v_2 = \frac{v_s F_{p1} + v_d F_{d1}}{(1 - \phi)\rho(v_r + v_b + v_d F_{d2})} c_1 \quad (43.9)$$

where v_T = net loss rate (m yr^{-1}),

$$v_T = (1 - F_r')(v_s F_{p1} + v_d F_{d1}) \quad (43.10)$$

and F_r' is the ratio of sediment feedback to total sediment purging,

$$F_r' = \frac{v_r + v_d F_{d2}}{v_r + v_b + v_d F_{d2}} \quad (43.11)$$

Thomann (1985) compiled water-column partition coefficients for copper, zinc, cadmium, chromium, lead, and nickel from 15 rivers. The values ranged from 1×10^{-4} to 0.1 m^3 g^{-1} (10^2 to 10^5 L kg^{-1}). On the basis of this data, Thomann and Mueller (1987) concluded that $F_{d1} \cong 0.8 \pm 0.2$ for metals. They also noted that the partition coefficient could be less in the sediments.

EXAMPLE 43.2. HEAVY METALS IN LAKE MICHIGAN. Using the same parameters as in Example 43.1, determine the steady-state concentrations of zinc in the water and sediments of Lake Michigan for a hypothetical inflow concentration of 1 μg L^{-1}. Assume that the partition coefficient for zinc is 2×10^5 L kg^{-1} (0.2 m^3 g^{-1}) and the sediment-water diffusion mass-transfer coefficient is 1 m yr^{-1}.

Solution: The fractions can be calculated as

$$F_{d1} = \frac{1}{1 + 0.2(0.5)} = 0.909 \qquad F_{p1} = 1 - 0.909 = 0.0909$$

$$F_{d2} = \frac{1}{0.9 + [0.2(1 - 0.9)2.5 \times 10^6]} = 2 \times 10^{-5}$$

Then the recycle fraction can be determined as (Eq. 43.11)

$$F_r' = \frac{0.000157 + 1(2 \times 10^{-5})}{0.000157 + 0.000756 + 1(2 \times 10^{-5})} = 0.1893$$

and the net loss rate as (Eq. 43.10)

$$v_T = (1 - 0.1893)[456.25(0.0909) + 1(0.909)] = 34.36 \text{ m yr}^{-1}$$

The concentrations in the water and the sediments can be determined as

$$c_1 = \frac{4 \times 10^{10}(1)}{4 \times 10^{10} + 34.36(2 \times 10^{10})} = 0.055\ \mu\text{g L}^{-1}$$

and
$$v_2 = \frac{456.25(0.0909) + 1(0.9091)}{(1 - 0.9)2.5 \times 10^6[0.000157 + 0.000756 + 1(2 \times 10^{-5})]} 0.055 \left(\frac{10^3\ \mu\text{g}}{\text{mg}}\right)$$
$$= 10 \mu\text{g g}^{-1}$$

43.3.2 Chemical Equilibria and Mass-Balance Approaches (Advanced Topic)

Although the simple sorption approaches described previously have some utility for first-order estimates, modeling frameworks must usually address key aspects of metals chemistry to obtain more refined predictions. One way to do this is to directly integrate large equilibrium codes into mass-balance frameworks. For example, Runkel et al. (1996*a, b*) combined EPA's MINTEQ model with an advection-dispersion model to study metals dynamics in a stream. Because of computing advances, such applications should become more commonplace in the future.

An alternative approach, which is closer in spirit to other models in this lecture, is to identify a few key equilibrium-chemistry mechanisms and integrate them into the mass-balance approach. Recently Dilks et al. (1995) have developed such a model for water-sediment systems. As was the case for toxic organics, this model divides the metal into particulate and dissolved forms (Fig. 43.2). However, in anoxic sediments the metals are subject to precipitation with hydrogen sulfide. This model is useful in illustrating how chemical equilibria mechanisms can be incorporated into a mass-balance approach to more effectively model metals.

Mass-balance equations can be written for the water and the sediments in a fashion identical to Eqs. 43.6 and 43.7,

$$V_1 \frac{dc_1}{dt} = Qc_{in} - Qc_1 - v_s A F_{p1} c_1 + v_r A c_2 + v_d A(F_{d2} c_2 - F_{d1} c_1) \quad (43.12)$$

and
$$V_2 \frac{dc_2}{dt} = v_s A F_{p1} c_1 - v_r A c_2 - v_b A c_2 + v_d A(F_{d1} c_1 - F_{d2} c_2) \quad (43.13)$$

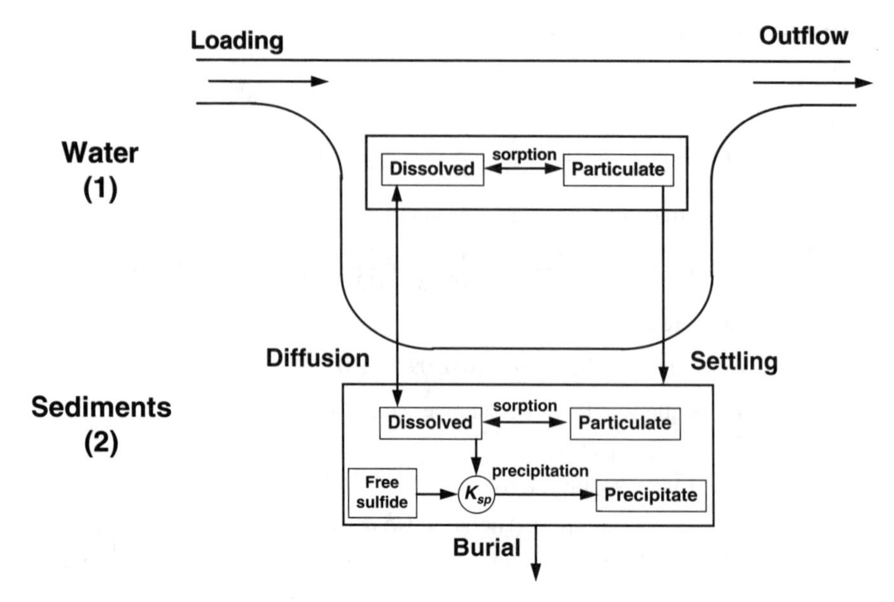

FIGURE 43.2
Schematic of a metal's budget for a well-mixed lake with sediment-water interactions.

However, now it should be noted that the total metals concentration in the sediment is written as

$$c_2 = c_{d2} + c_{p2} + c_{s2} \tag{43.14}$$

where the third form, c_{s2}, is metal that has precipitated with hydrogen sulfide. In addition, as will shortly be made evident, the factor F_{d2} is not a simple function of sorption but will also depend on precipitation reactions. All other parameters are similar to those used in the conventional metals model.

Note that to simplify the following development the sediment concentrations are all expressed on a molar basis (mole L^{-1}). In addition we drop the subscript 2 because the following derivation focuses solely on the sediments.

The chemical relationships governing the sulfur compounds are

$$H^+ + S^{2-} \rightleftharpoons HS^- \qquad \text{sulfur ion reaction} \tag{43.15}$$

$$H^+ + HS^- \rightleftharpoons H_2S \qquad \text{hydrogen sulfide reaction} \tag{43.16}$$

$$MS + H^+ \rightleftharpoons M^{2+} + HS^- \qquad \text{metal dissolution/precipitation reaction} \tag{43.17}$$

Note that the dissolved metal ion, M^{2+}, can be related to the terms in Eq. 43.14 by

$$[M^{2+}] = \frac{c_d}{\phi} \tag{43.18}$$

where division by porosity is needed to convert the concentration to a pore-water volume basis.

Now assuming that we know the pH, Eqs. 43.15 to 43.17 have five unknowns: $[S^{2-}]$, $[HS^-]$, $[H_2S]$, $[MS]$, and $[M^{2+}]$. Therefore five independent equations are required. Three are supplied by the equilibrium relationships for Eqs. 43.15 to 43.17,

$$K_1 = \frac{[HS^-]}{[H^+][S^{2-}]} \tag{43.19}$$

$$K_2 = \frac{[H_2S]}{[H^+][HS^-]} \tag{43.20}$$

$$K_s = \frac{[c_d/\phi][HS^-]}{[H^+]} \tag{43.21}$$

Two other equations are provided by mass balances. The first is the sulfur balance,

$$AVS = [S^{2-}] + [HS^-] + [H_2S] + c_s \tag{43.22}$$

where AVS = acid-volatile sulfide. This is a measure of the reactive sulfide in the system (Allen et al. 1993).

The second is the metal balance expressed by Eq. 43.14. This balance can be formulated in terms of the model variables and partition coefficients by recognizing that an equilibrium partition coefficient can be defined as (recall Eq. 41.9)

$$K_d = \frac{c_p}{c_d}\frac{\phi}{(1-\phi)\rho} \tag{43.23}$$

which can be rearranged to give

$$c_p = \frac{K_d(1-\phi)\rho}{\phi}c_d \tag{43.24}$$

which can be substituted into Eq. 43.14 to yield

$$c_2 = c_d + \frac{(1-\phi)\rho}{\phi}K_d c_d + c_s \tag{43.25}$$

Thus if pH, c_2, and AVS are given, Eqs. 43.19 to 43.22 and 43.25 represent five equations that can be solved for the five unknowns. Although computer codes are available to do this (e.g., Dilks et al. 1995), the present case is simple enough to be approached directly. First, as was done in the earlier lectures on equilibrium chemistry (recall the derivation of Eqs. 39.20 through 39.22), Eqs. 43.19 and 43.20 can be substituted into Eq. 43.22 to solve for

$$[HS^-] = F_{HS}\frac{AVS - c_s}{\phi} \quad \text{where } F_{HS} = \frac{K_1[H^+]}{1 + K_1[H^+] + K_1K_2[H^+]^2} \tag{43.26}$$

$$[S^{2-}] = F_S\frac{AVS - c_s}{\phi} \quad \text{where } F_S = \frac{1}{1 + K_1[H^+] + K_1K_2[H^+]^2} \tag{43.27}$$

$$[H_2S] = F_{H_2S}\frac{AVS - c_s}{\phi} \quad \text{where } F_{H_2S} = \frac{K_1K_2[H^+]^2}{1 + K_1[H^+] + K_1K_2[H^+]^2} \tag{43.28}$$

Equation 43.21 can be solved for

$$c_d = \frac{K_s\phi[H^+]}{[HS^-]} \tag{43.29}$$

and the result, along with Eq. 43.26, can be substituted into Eq. 43.25 to give

$$c_2 = \left[\frac{K_d(1-\phi)\rho}{\phi} + 1\right]\frac{K_s[\mathrm{H}^+]\phi^2}{F_{HS}(AVS - c_s)} + c_s \tag{43.30}$$

which can be solved as a roots-of-equations problem for c_s. This value can then be used in conjunction with Eqs. 43.26 to 43.29 to compute the other four unknowns.

Now the foregoing solution holds up to the point where c_2 exceeds the AVS. At this juncture the sulfur available for precipitation will be exhausted and c_s will become fixed at AVS. Thereafter the five simultaneous equations reduce to four equations that can be solved for the remaining unknowns.

The water-quality implications of increasing the metals beyond the AVS are great. First, it means that the excess metal, $c_2 - AVS$, will not be fixed in precipitates and, depending on the metal's sorption characteristics, a significant portion could build up in the pore waters. Beyond increasing the direct exposure to benthic organisms, higher pore water levels also means increased mobility as the dissolved metal becomes subject to diffusion.

EXAMPLE 43.3. PRECIPITATION AND METAL LEVELS IN THE SEDIMENTS. A sediment has a pH of 7, a total lead concentration of 0.0005 mol L^{-1}, and an AVS of 0.001 mol L^{-1}. Determine the dissolved lead concentration for this case. In addition calculate the levels if the total metal concentration is increased to 0.002 mol L^{-1}. Use the following parameters in your calculations: $K_s = 2.14 \times 10^{-15}$, $K_d = 0.0125\ \mathrm{m^3\ g^{-1}}$, $\phi = 0.8$, $K_1 = 6.58 \times 10^{12}$, and $K_2 = 8.73 \times 10^6$.

Solution: Parameter values can be substituted into Eq. 43.26 that can be solved for $F_{HS} = 0.5339$. This value can be substituted into Eq. 43.30 along with other parameters to give

$$c_2 = \left[\frac{0.0125(1-0.8)2.5\times 10^6}{0.8} + 1\right]\frac{2.14\times 10^{-15}(10^{-7})(0.8)^2}{0.5339(0.001 - c_s)} + c_s$$

which can be solved numerically for $c_s = 0.0005$ mol L^{-1}. Thus almost all the lead precipitates with sulfur. This result can be substituted into Eq. 43.26 to calculate

$$[\mathrm{HS}^-] = 0.5339\frac{0.001 - 0.0005}{0.8} = 3.34 \times 10^{-4}$$

which can be substituted into Eq. 43.29 to give

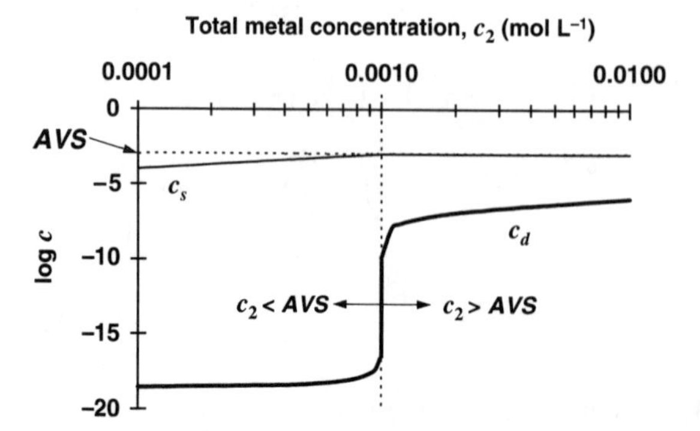

FIGURE E43.2

$$c_d = \frac{2.14 \times 10^{-15}(0.8)(10^{-7})}{3.34 \times 10^{-4}} = 5.13 \times 10^{-19}$$

which is negligible.

The calculation can then be repeated for the case where $c_2 = 0.002$. In this instance c_s is fixed at the level of the AVS (= 0.001) and Eq. 43.25 can be solved for

$$c_d = \frac{c_2 - c_s}{1 + \frac{(1-\phi)\rho}{\phi}K_d} = \frac{0.002 - 0.001}{1 + \frac{(1-0.8)2.5 \times 10^6}{0.8}0.0125} = 1.27 \times 10^{-7}$$

Thus, because the excess metal is not being precipitated, the pore-water levels increase over 11 orders of magnitude. Figure E43.2 shows how the porewater concentration changes over a range of total metals concentration.

The foregoing model can be combined with a water balance to relate environmental levels back to loadings. For example, Dilks et al. (1995) integrated it into a stream model (Box 43.1). It can also be applied to more than one metal. In such cases the metals precipitate in sequence depending on their solubility products—the ones with the lowest solubility precipitating first. As reported by Di Toro et al. (1992), the sequence is

	$-\log(K_s)$	precipitates
HgS	38.50	first
CuS	22.19	↓
PbS	14.67	↓
CdS	14.10	↓
ZnS	9.64	↓
NiS	9.23	last

This sequence continues until the AVS is exhausted. As described in Box 43.1, this can lead to some interesting results.

BOX 43.1. Multiple Metals in Stream Sediments

Dave Dilks, Joseph Helfand, and Vic Bierman of Limno-Tech have applied the metals–AVS model to streams. In addition they have included competition among five metals (Ni, Zn, Cd, Pb, and Cu) for the available sulfide.

Figure B43.1 shows two simulations they made for a stream subject to a point source of copper and cadmium. At river mile 1, a point source enters the stream. For the first simulation the metal discharge is low and, hence, the standards for both metals are not violated. Because it precipitates first, the copper levels are lower.

In the second simulation the copper discharge is increased. As would be expected the copper in the sediment increases; but it does not exceed the standard. In contrast the cadmium increases to the point that a violation occurs. Why did cadmium increase when more copper was added? By increasing the copper the AVS was exceeded. Sulfides that formerly were available to bind cadmium instead precipitate as copper sulfide. Thus the liberated cadmium causes an increase in the pore water.

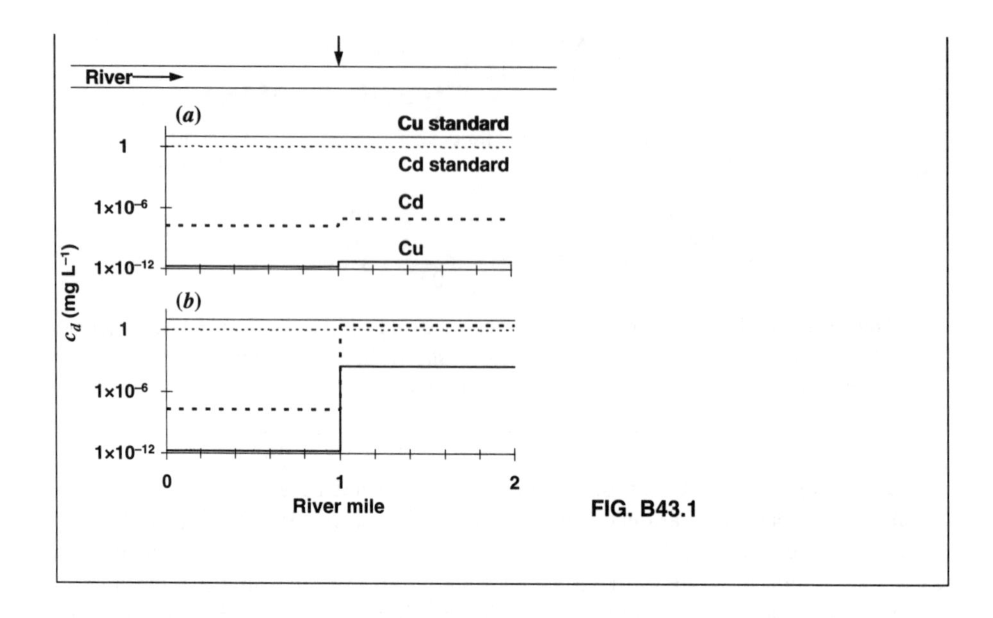

FIG. B43.1

PROBLEMS

43.1. Repeat Example 43.1, but for ^{90}Sr. Compute the inflow concentration for ^{90}Sr based on Table 43.4 and the factor (57.9) from Example 43.1.

43.2. Repeat Example 43.2 for copper. According to Mills et al. (1985), copper has a smaller partition coefficient ($K_d = 6 \times 10^4$ L kg^{-1}) than zinc.

43.3. Determine how much the zinc inflow concentration to Lake Michigan must be reduced to attain a sediment concentration of 5 μg g^{-1}. What is the resulting water concentration?

43.4. Repeat Example 43.3, but for cadmium ($\log K_s = 14.1$).

43.5. Using the parameter values from Examples 43.1 and 43.2, (*a*) determine the inflow concentration of copper to Lake Michigan to attain a water concentration of 0.1 μg L^{-1}. (*b*) What is the resulting sediment concentration in μg g^{-1}?

LECTURE 44

Toxicant Modeling in Flowing Waters

LECTURE OVERVIEW: We cover some material related to the modeling of toxicants in flowing waters. In the first part we apply the lake models from the previous lectures to rivers and streams. We first describe stream solid's budgets and introduce the notion that sediment-water interactions are simplified because the sediment compartment is stationary. We then develop a contaminant balance and describe some simple steady-state solutions for the mixing zone. Next we perform a similar analysis for estuaries. We then develop a general control-volume approach that is applicable to both streams and estuaries. Along with time-variable solutions, we illustrate how the steady-state system response matrix can be applied to toxics.

We now integrate horizontal transport into our toxics modeling scheme. Our emphasis in this lecture will be on one-dimensional streams and estuaries. However, the control-volume approach described in a later section is generally applicable to multidimensional water bodies.

44.1 ANALYTICAL SOLUTIONS

Before developing numerical approaches, I'll first describe some analytical schemes. As in other parts of this text, such solutions provide insight and can be used for quick back-of-the-envelope estimation. I'll first describe nondispersive, plug-flow systems that are applicable to many streams. Then I'll add dispersion to broaden the potential applications to encompass one-dimensional estuaries and rivers, where dispersion is significant.

44.1.1 Plug-Flow Systems

As with the lake models presented in Lec. 40, we develop both solids and contaminant balances for plug-flow rivers and streams.

Solids budget. In a fashion similar to our lake model (recall Sec. 40.4.1), a steady-state solids budget can be written for a plug-flow system with constant hydrogeometric characteristics as

$$0 = -U\frac{dm_1}{dx} - \frac{v_s}{H_1}m_1 + \frac{v_r}{H_1}m_2 \tag{44.1}$$

and for the bottom sediments as

$$0 = v_s m_1 - v_r m_2 - v_b m_2 \tag{44.2}$$

where U = stream velocity (m d^{-1})
m_1 and m_2 = suspended solids in the water (1) and sediment (2) layers (g m^{-3})
H = depth (m)
v_s, v_r, and v_b = settling, resuspension, and burial velocities (m d^{-1})

Now a key insight results from recognizing that because the sediments do not move horizontally (Eq. 44.2 has no advection term), Eq. 44.2 is identical to the sediment balance for a lake. Thus it can be solved for

$$m_2 = \frac{v_s}{v_r + v_r}m_1 \tag{44.3}$$

In other words the sediment concentration will always be a constant fraction of the concentration in the overlying water. As we will see shortly, this also applies to steady-state contaminant budgets and has beneficial ramifications for time-variable computations.

Equation 44.3 can be substituted into Eq. 44.1,

$$0 = -U\frac{dm_1}{dx} - \frac{v_s}{H_1}m_1 + \frac{v_r}{H_1}\frac{v_s}{v_r + v_b}m_1 \tag{44.4}$$

or by collecting terms,

$$0 = -U\frac{dm_1}{dx} - \frac{v_n}{H_1}m_1 \tag{44.5}$$

where v_n = the net settling velocity,

$$v_n = v_s(1 - F_r) \tag{44.6}$$

in which F_r = ratio of the resuspension velocity to the total purging velocity for the sediment layer $= v_r/(v_r + v_b)$.

Three general cases can occur (Fig. 44.1):

- $v_n = 0$. In shallow streams, there is often a negligible accumulation of sediments. Therefore water solids remain constant as resuspension balances settling.

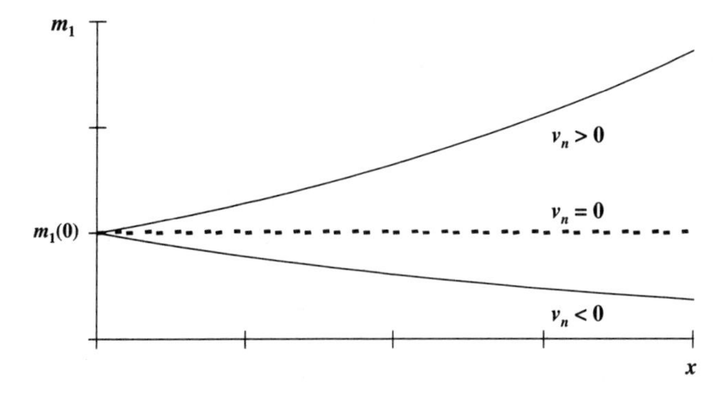

FIGURE 44.1
Suspended solids versus distance for the case where sediment solids concentration is constant.

- $v_n < 0$. In deeper streams, sediments will be deposited and there will be a net accumulation of sediments. Thus the water solids decline as the sediment solids build up.
- $v_n > 0$. During certain periods of time, scour will occur and there will be a net loss of bottom sediments. Therefore suspended solids concentration may increase in the water without external sources such as loadings, tributaries, or nonpoint runoff of solids.

Delos et al. (1984) and O'Connor (1985) have looked at the case where sediment solids concentration is a constant over the study stretch, $m_2 = (1-\phi)\rho$. For this case the solution for the water is

$$m_1 = m_1(0)e^{-\frac{v_s}{H_1 U}x} + \frac{v_r(1-\phi)\rho}{v_s}\left(1 - e^{-\frac{v_s}{H_1 U}x}\right) \tag{44.7}$$

If the initial solids concentration $m_1(0)$ is small, then the downstream solids concentration should approach a steady value of

$$m_1(\infty) = \frac{v_r(1-\phi)\rho}{v_s} \tag{44.8}$$

Based on this result, O'Connor (1985) suggested that solid profiles could be used to estimate v_r by extrapolating downstream to a stable value of $m_1(\infty)$. We explore this idea in Prob. 44.1 at the end of this lecture.

Contaminant budget. Now we extend the analysis to toxics. To do this we assume that the suspended solids are constant across the stretch of interest. If this is true a steady-state contaminant budget can be written for a plug-flow system with constant hydrogeometric characteristics as

$$0 = -U\frac{dc_1}{dx} - k_1 c_1 - \frac{v_v}{H_1}F_{d1}c_1 - \frac{v_s}{H_1}F_{p1}c_1 + \frac{v_d}{H_1}(F_{d2}c_2 - F_{d1}c_1) + \frac{v_r}{H_1}c_2 \tag{44.9}$$

and for the bottom sediments as

$$0 = v_s F_{p1}c_1 + v_d(F_{d1}c_1 - F_{d2}c_2) - k_2 H_2 c_2 - v_r c_2 - v_b c_2 \tag{44.10}$$

where k = a first-order decomposition rate (d^{-1}) and the F's are fraction of contaminant in (d)issolved and (p)articulate form in the two layers,

$$F_d = \frac{1}{1 + K_{d1}m} \qquad F_{p1} = \frac{K_{d1}m}{1 + K_{d1}m} \qquad F_{d2} = \frac{1}{\phi + K_{d2}(1 - \phi)\rho} \tag{44.11}$$

Again, because the bed does not advect downstream, Eq. 44.10 establishes a direct relationship between the sediment and the overlying water concentration,

$$c_2 = R_{21}c_1 = \frac{v_s F_{p1} + v_d F_{d1}}{v_d F_{d2} + k_2 H_2 + v_r + v_b} c_1 \tag{44.12}$$

or in terms of a mass-specific sediment concentration as

$$v_2 = \frac{R_{21}}{(1 - \phi)\rho} c_1 \tag{44.13}$$

Equation 44.12 can then be substituted into Eq. 44.9,

$$0 = -U\frac{dc_1}{dx} - \frac{v_T}{H_1} c_1 \tag{44.14}$$

where v_T = a total loss term expressed as a settling velocity (m d^{-1}),

$$v_T = k_1 H_1 + v_v F_{d1} + (v_s F_{p1} + v_d F_{d1})(1 - F_r') \tag{44.15}$$

in which F_r' = the ratio of sediment feedback to total sediment purging,

$$F_r' = \frac{v_r + v_d F_{d2}}{v_r + v_b + v_d F_{d2} + k_2 H_2} \tag{44.16}$$

Given a boundary condition of $c_1 = c_1(0)$, we can solve the water balance for

$$c_1 = c_1(0)e^{-\frac{v_T}{H_1 U}x} \tag{44.17}$$

Then Eq. 44.12 can be used to compute the sediment concentration,

$$c_2 = R_{21}c_1(0)e^{-\frac{v_T}{H_1 U}x} \tag{44.18}$$

or in terms of sediment solid concentration,

$$v_2 = \frac{R_{21}}{(1 - \phi)\rho} c_1(0)e^{-\frac{v_T}{H_1 U}x} \tag{44.19}$$

Thus the water concentration follows a simple exponential decay. The sediment traces the identical shape, but is scaled as in Eq. 44.18 and 44.19.

Critical concentration. Now an interesting calculation involves computing the initial concentrations. They are particularly germane because the maximum value occurs at the outfall. Thus this concentration represents the critical value upon which assimilative capacity calculations would be based (recall Lec 1, and particularly the discussion of Fig. 1.1).

Assuming instantaneous mixing at the injection point, we would compute the initial concentration with a simple mass balance as

$$c_1(0) = \frac{Q_r c_{1r} + Q_w c_{1w}}{Q_r + Q_w} \tag{44.20}$$

where the subscripts w and r designate the waste outfall and the receiving river, respectively.

Toxicant water-quality standards are often expressed in terms of a mass-specific sediment concentration. The initial value for the sediments can be determined as

$$\nu_2(0) = \frac{R_{21}}{(1-\phi)\rho} \frac{Q_r c_{1r} + Q_w c_{1w}}{Q_r + Q_w} \tag{44.21}$$

Now these can be applied to determine the required loading to meet the standard. One way to do this is to calculate the required waste concentration to attain a desired water concentration $c_1(0)$,

$$c_{1w} = \frac{Q_r + Q_w}{Q_w} c_1(0) - \frac{Q_r}{Q_w} c_{1,r} \tag{44.22}$$

Alternatively we could calculate the required waste concentration to attain a desired sediment concentration $\nu_2(0)$,

$$c_{1w} = \frac{Q_r + Q_w}{Q_w} \frac{(1-\phi)\rho}{R_{21}} \nu_2(0) - \frac{Q_r}{Q_w} c_{1r} \tag{44.23}$$

EXAMPLE 44.1. POINT-SOURCE ANALYSIS (PLUG-FLOW). A toxic point source discharges to a stream having the following characteristics:

$Q_r = 0.99\ \text{m}^3\ \text{s}^{-1}$	$c_{1,r} = 0\ \text{mg m}^{-3}$	$Q_w = 0.01\ \text{m}^3\ \text{s}^{-1}$	$c_{1,w} = 1000\ \text{mg m}^{-3}$
$v_s = 0.25\ \text{m d}^{-1}$	$v_v = 0.1\ \text{m d}^{-1}$	$U = 0.1$ mps	$\phi = 0.8$
$\rho = 2.5 \times 10^6\ \text{g m}^{-3}$	$H_1 = 2$ m		

In the water, the toxicant is lost by photolysis at a rate of $0.1\ \text{d}^{-1}$. All other losses are zero and the toxic associates strongly with solid matter (all F_d's = 0). Note that the suspended solids in the water is a constant $10\ \text{g m}^{-3}$. (*a*) Determine the water and sediment concentrations downstream from the source. (*b*) Calculate the needed inflow concentration so that the maximum sediment concentration is maintained at $250\ \mu\text{g g}^{-1}$.

Solution: First, we have to parameterize the solids budget. Because the suspended solids are constant, we can assume that the burial velocity is zero. Therefore settling must balance resuspension,

$$v_s m_1 = v_r(1-\theta)\rho$$

which can be solved for

$$v_r = v_s \frac{m_1}{(1-\theta)\rho} = 0.25 \frac{10}{(1-0.8)2.5 \times 10^6} = 5 \times 10^{-6}\ \text{m d}^{-1}$$

The initial water concentration can be computed as

$$c_1(0) = \frac{0.99(0) + 0.01(1000)}{0.99 + 0.01} = 10$$

The sediment-to-water concentration ratio can be computed as (note that because $F_{d2} = 0$, $k_2 = 0$, and $v_b = 0$, $F_r' = 1$)

$$R_{21} = \frac{0.25}{5 \times 10^{-6}} = 50{,}000$$

which can be used to determine

$$v_2(0) = \frac{50{,}000}{(1 - 0.8)2.5 \times 10^6} 10 \text{ mg m}^{-3} \left(\frac{10^3 \, \mu\text{g}}{\text{mg}} \right) = 1000 \, \mu\text{g g}^{-1}$$

After determining that $v_T = 0.1(2) = 0.2 \text{ m d}^{-1}$, the downstream profiles can then be computed with Eqs. 44.17 and 44.19,

$$c_1 = 10e^{-\frac{0.2}{2(8640)}x} \qquad v_2 = 1000e^{-\frac{0.2}{2(8640)}x}$$

and displayed in Fig. E44.1.

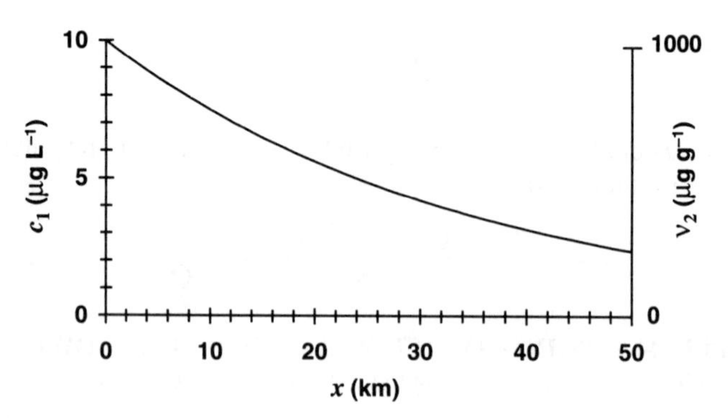

FIGURE E44.1

(*b*) Equation 44.23 can be employed to compute

$$c_{1w} = \frac{0.99 + 0.01}{0.01} \frac{(1 - 0.2)2.5 \times 10^6}{50{,}000} 0.25 - \frac{0.99}{0.01} 0 = 250 \, \mu\text{g L}^{-1}$$

Consequently the waste concentration must be reduced from 1000 to 250 μg L^{-1} to meet the standard.

Metals. For metals the model simplifies because of the omission of decay and volatilization (with the exception of mercury) reactions. In addition, if it is assumed that sorption is the same in the water and sediments, sediment-water diffusion can be omitted and the steady-state model again can be represented by

$$0 = -U\frac{dc_1}{dx} - \frac{v_T}{H_1}c_1 \tag{44.24}$$

but with the total removal rate simplified to

$$v_T = v_s F_{p1}(1 - F_r) \tag{44.25}$$

where

$$F_r = \frac{v_r}{v_r + v_b} \tag{44.26}$$

For cases where sediment resuspension is negligible, the model simplifies further to

$$0 = -U\frac{dc_1}{dx} - \frac{F_{p1}v_s}{H_1}c_1 \tag{44.27}$$

If the suspended solids are constant, the solution to this equation is merely Eq. 44.17 with $v_T = F_{p1}v_s$. However, when solids are changing, the fraction particulate will also change with distance. For this case Mills et al. (1985) provide the following solution,

$$c_1 = c_{10}e^{\left[\ln\left(K_d m_0 + e^{\frac{v_s x}{H_1 U}}\right) - \ln(K_d m_0 + 1) - \frac{v_s x}{H_1 U}\right]} \tag{44.28}$$

EXAMPLE 44.2. FLINT RIVER COPPER. Mills et al. (1985) present the following data for suspended solids and water copper concentrations in the Flint River, Michigan, in August, 1981:

x (km)	1	2.5	7	12	21	30	43	61	63
m (mg L^{-1})	11.75 (4–18)	10 (5–15)	10 (7–13.5)	8.5 (6–11)	6.75 (5.3–8)	5.5 (5.3–5.7)	11.5 (3–19.8)	13.5 (10–16.5)	11.75
c_1 (μg L^{-1})	3 (2.8–3.4)	4.2 (3.2–5.4)	5.7 (4.5–7)	5.5 (4.9–5.8)	4.2 (2.4–4.8)	4.7 (4–5.2)	8 (4–10)	6 (4.8–7.3)	5.75 (4.5–7)

where the numbers in parentheses represent observed ranges. The river has the following boundary conditions and point sources:

	Kilometers	Flow (cms)	Suspended solids (mg L^{-1})	Copper (μg L^{-1})
Upstream boundary	0	2.66	13.5	2.9
Flint WWTP	1.3	1.68	4.1	8.3
Ragone WWTP	30.9	0.69	58.7	28.5

The depth and the velocity in the first stretch (km 1.3 to 30.9) are assumed to be 0.5 m and 0.2 m s^{-1}. (*a*) Use the suspended solids to determine the settling velocity. (*b*) If copper has a partition coefficient of 0.06×10^6 L kg^{-1}, compute the copper concentration in the stream.

Solution: (*a*) First, inspection of the suspended solids data suggests that there is net sedimentation in the first stretch and zero in the second. The data for the first stretch can be used to estimate the settling velocity by determining the slope of a semi-log plot of suspended solids versus distance. The resulting settling velocity is 0.213 m d^{-1}. This value can be used to simulate the suspended solids with the results shown in the following plot:

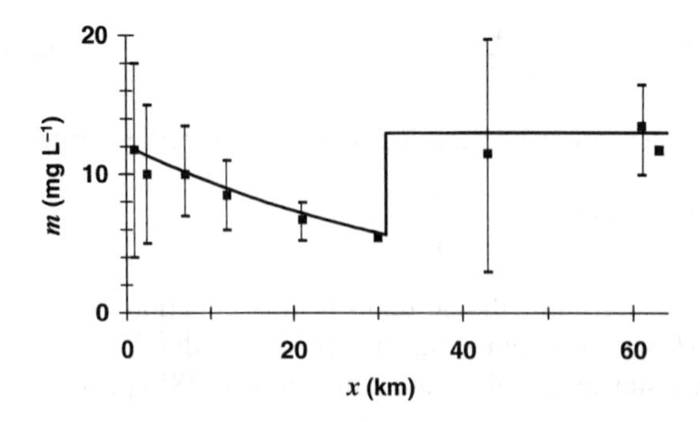

FIGURE E44.2-1

Note how the solids are reduced due to settling in the first stretch. A mass balance is then used to determine the concentration jump at the confluence with the Ragone WWTP effluent. Thereafter the concentration is constant since net settling is zero.

(*b*) For copper, Eq. 44.28 can be employed to simulate the water concentration for the first stretch. A mass balance is then used to determine the constant concentration for the second stretch. The results along with the data are displayed below:

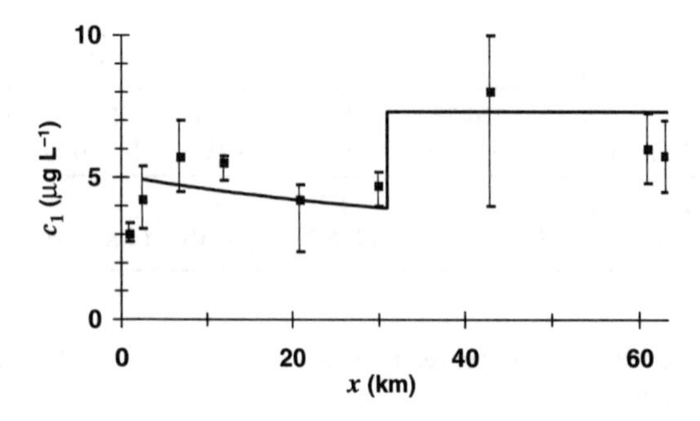

FIGURE E44.2-2

In this section we have neglected distributed flow increases along the stream. Although this can be a useful assumption, many cases involve distributed flow. O'Connor (1988*c*) has addressed this issue for both organics and metals and has developed a number of useful analytical solutions for some idealized cases. In the last section of this lecture we present a numerical approach that allows such phenomena to be incorporated into our framework.

44.1.2 Mixed-Flow Systems

Now we extend the foregoing analysis to systems where dispersion cannot be neglected. These include estuaries and certain rivers. If suspended solids are constant, a steady-state contaminant budget can be written for a mixed-flow system with constant hydrogeometric characteristics as

$$0 = E\frac{d^2c_1}{dx^2} - U\frac{dc_1}{dx} - \frac{v_T}{H_1}c_1 \qquad (44.29)$$

where the total removal velocity v_T is defined as in Eq. 44.15, and the concentration of the underlying sediments can be computed as in Eqs. 44.12 and 44.13. The solution can be computed as (recall Sec. 9.3.1)

$$c_{1-} = c_1(0)e^{\lambda_- x} \qquad x \leq 0 \tag{44.30}$$

$$c_{1+} = c_1(0)e^{\lambda_+ x} \qquad x > 0 \tag{44.31}$$

where

$$\begin{matrix}\lambda_- \\ \lambda_+\end{matrix} = \frac{U}{2E}\left(1 \pm \sqrt{1 + \frac{4v_T E}{H_1 U^2}}\right) \tag{44.32}$$

in which the subscripts "+" and "−" designate the system upstream and downstream from the source, respectively.

If it is assumed that the waste flow is much smaller than the estuary flow, a mass balance at the outfall can be determined as

$$c_1(0) = \frac{W}{Q}\frac{1}{\sqrt{1 + \dfrac{4v_T E}{HU^2}}} \tag{44.33}$$

and for the bottom sediments as

$$v_2(0) = \frac{R_{21}}{(1-\phi)\rho}\frac{W}{Q}\frac{1}{\sqrt{1 + \dfrac{4v_T E}{HU^2}}} \tag{44.34}$$

The waste concentration required to meet a water standard can be computed in a similar fashion to Eq. 44.22. If it is assumed that the receiving water upstream of the discharge has negligible toxic concentration, the result is

$$c_{1w} = \frac{Q}{Q_w}c_1(0)\sqrt{1 + 4\eta} \tag{44.35}$$

where η is the estuary number (recall Eq. 10.41), which in the present context would be defined as $v_T E/(HU^2)$. Alternatively we could calculate the required inflow concentration to attain a desired sediment concentration,

$$c_{1w} = \frac{Q}{Q_w}\frac{(1-\phi)\rho}{R_{21}}v_2(0)\sqrt{1 + 4\eta} \tag{44.36}$$

Thus we see that the term $\sqrt{1 + 4\eta}$ reflects the effect of dispersion on the assimilative capacity (compare Eq. 44.36 with 44.23).

EXAMPLE 44.3. POINT-SOURCE ANALYSIS (MIXED-FLOW). Repeat part (*b*) of Example 44.1, but assume that the system has an estuary number of 1.

Solution: Substituting the appropriate values into Eq. 44.36 gives

$$c_{1w} = \frac{1}{0.01}\frac{(1-0.8)2.5 \times 10^6}{50{,}000}0.25\sqrt{1 + 4(1)} = 250(2.236) = 559\ \mu\text{g L}^{-1}$$

Thus the dispersive mixing means that over twice as much toxicant can be discharged.

44.2 NUMERICAL SOLUTIONS

Although the analytical solutions we developed are handy for obtaining back-of-the-envelope solutions, numerical approaches provide a more general approach. In particular they provide a means to analyze multiple sources.

Mass balances can be developed for a toxicant in a control volume as (Fig. 44.2)

$$V_{1,i}\frac{dc_{1,i}}{dt} = W_i + Q_{i-1,i}\left(\alpha_{i-1,i}c_{1,i-1} + \beta_{i-1,i}c_{1,i-1}\right) - Q_{i,i+1}\left(\alpha_{i,i+1}c_{1,i} + \beta_{i,i+1}c_{1,i+1}\right) + E'_{i-1,i}\left(c_{1,i-1} - c_{1,i}\right) + E'_{i,i+1}\left(c_{1,i+1} - c_{1,i}\right) - k_{1,i}V_{1,i}c_{1,i} - v_{v,i}A_{s,i}F_{d1,i}c_{1,i} - v_{s,i}A_{s,i}F_{p1,i}c_{1,i} + v_{r,i}A_{s,i}c_{2,i} + v_{d,i}A_{s,i}\left(F_{d2,i}c_{2,i} - F_{d1,i}c_{1,i}\right) \tag{44.37}$$

and

$$V_{2,i}\frac{dc_{2,i}}{dt} = -k_{2,i}V_{2,i}c_{2,i} + v_{s,i}A_{s,i}F_{p1,i}c_{1,i} - v_{r,i}A_{s,i}c_{2,i} - v_{b,i}A_{s,i}c_{2,i} + v_{d,i}A_{s,i}\left(F_{d1,i}c_{1,i} - F_{d2,i}c_{2,i}\right) \tag{44.38}$$

These equations can be written for all the volumes in the system and then solved simultaneously to determine the distribution of contaminant throughout the system. For time-variable solutions the approaches outlined in Lecs. 12 and 13 can be used.

For the steady-state case, Eq. 44.38 can be solved for each volume in a similar fashion to Eq. 44.12,

$$c_{2,i} = R_{21,i}c_{1,i} \tag{44.39}$$

or in matrix form as

$$\{c_2\} = [R_{21}]\{c_1\} \tag{44.40}$$

where $[R_{21}]$ is a diagonal matrix. Equation 44.39 can be substituted into Eq. 44.37, and the result manipulated to yield

$$V_{1,i}\frac{dc_{1,i}}{dt} = W_i + Q_{i-1,i}\left(\alpha_{i-1,i}c_{1,i-1} + \beta_{i-1,i}c_{1,i-1}\right) - Q_{i,i+1}\left(\alpha_{i,i+1}c_{1,i} + \beta_{i,i+1}c_{1,i+1}\right) + E'_{i-1,i}\left(c_{1,i-1} - c_{1,i}\right) + E'_{i,i+1}\left(c_{1,i+1} - c_{1,i}\right) - v_{T,i}A_{s,i}c_{1,i} \tag{44.41}$$

where v_T is calculated for each volume in a similar fashion to Eq. 44.15.

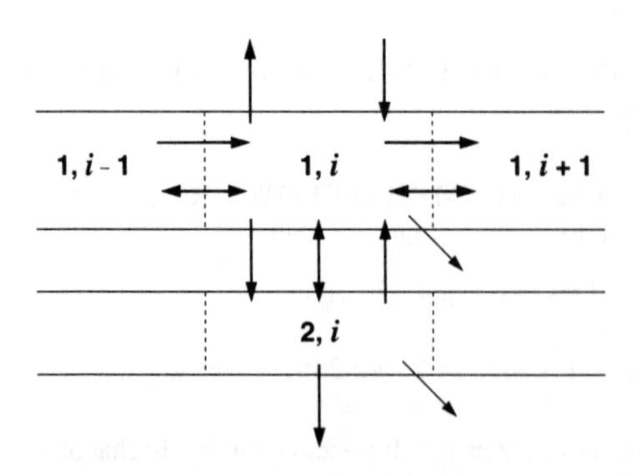

FIGURE 44.2
Segmentation scheme for control-volume approach for toxics.

Equation 44.41 can be written for all the water volumes and expressed in matrix format as (recall Eq. 11.26)

$$[A]\{c_1\} = \{W\} \tag{44.42}$$

and solved as

$$\{c_1\} = [A]^{-1}\{W\} \tag{44.43}$$

Finally it can be recognized that the sediment concentration can be determined as

$$\{v_2\} = \frac{1}{(1-\phi)\rho}[R_{21}][A]^{-1}\{W\} \tag{44.44}$$

or collecting terms,

$$\{v_2\} = [S]^{-1}\{W\} \tag{44.45}$$

where

$$[S]^{-1} = \frac{1}{(1-\phi)\rho}[R_{21}][A]^{-1} \tag{44.46}$$

Thus, the matrix inverses $[A]^{-1}$ and $[S]^{-1}$ provide the vehicles for solving inverse assimilative capacity calculations for multidimensional systems with multiple sources. In other words Eqs. 44.43 and 44.45 are the multidimensional analogs of Eqs. 44.22 and 44.23.

44.3 NONPOINT SOURCES

Although point sources are certainly important, many toxics enter streams and estuaries in a nonpoint or diffuse fashion. For example both urban and agricultural runoff can carry toxics in significant concentrations.

44.3.1 Low-Flow Nonpoint Sources

For nonpoint sources that contribute negligible flow, the water mass balance for the plug-flow, constant-parameter case becomes (compare with Eq. 44.14)

$$0 = -U\frac{dc_1}{dx} - \frac{v_T}{H_1}c_1 + S_d \tag{44.47}$$

where the total removal velocity v_T is defined as in Eq. 44.15 and S_d is the distributed loading term (mg m^{-3} d^{-1}). The closed-form solution is (recall Eq. 9.48)

$$c_1 = c_1(0)e^{-\frac{v_T}{H_1 U}x} + \frac{S_d H_1}{v_T}\left(1 - e^{-\frac{v_T}{H_1 U}x}\right) \tag{44.48}$$

The sediment concentrations follow directly from Eqs. 44.12 and 44.13.

44.3.2 Flow-Contributing Nonpoint Sources

Although nonpoint sources can contribute negligible flow, it is more likely that they are accompanied by significant volumes of water. For such cases an analysis similar to that described previously in Sec. 22.3 can be developed.

For example a flow balance would be implemented to determine segment flows and geometries in a similar fashion to that employed in Sec. 22.3. Then a steady-state, control-volume approach can be applied to the contaminant in each volume as

$$0 = W_i + Q_{i-1,i}\left(\alpha_{i-1,i}c_{1,i-1} + \beta_{i-1,i}c_{1,i-1}\right) - Q_{i,i+1}\left(\alpha_{i,i+1}c_{1,i} + \beta_{i,i+1}c_{1,i+1}\right) + E'_{i-1,i}\left(c_{1,i-1} - c_{1,i}\right) + E'_{i,i+1}\left(c_{1,i+1} - c_{1,i}\right) - v_{T,i}A_{s,i}c_{1,i} + Q_e c_{d,i} \quad (44.49)$$

where Q_e = distributed inflow for segment i ($m^3\ d^{-1}$) and $c_{d,i}$ = concentration of the contaminant in this inflow ($mg\ m^{-3}$).

This equation can be written for the n elements of a reach. Together with appropriate boundary conditions, it can be solved for the water concentrations in each element. Then Eqs. 44.12 and 44.13 can be used to determine the sediment concentration. The approach is illustrated in the following example.

EXAMPLE 44.4. FLOW-CONTRIBUTING DISTRIBUTED SOURCES OF TOXICS. A demonstration of the model can be developed for the case illustrated in Fig. 44.3. For this simulation a point source marks the beginning of the problem. For

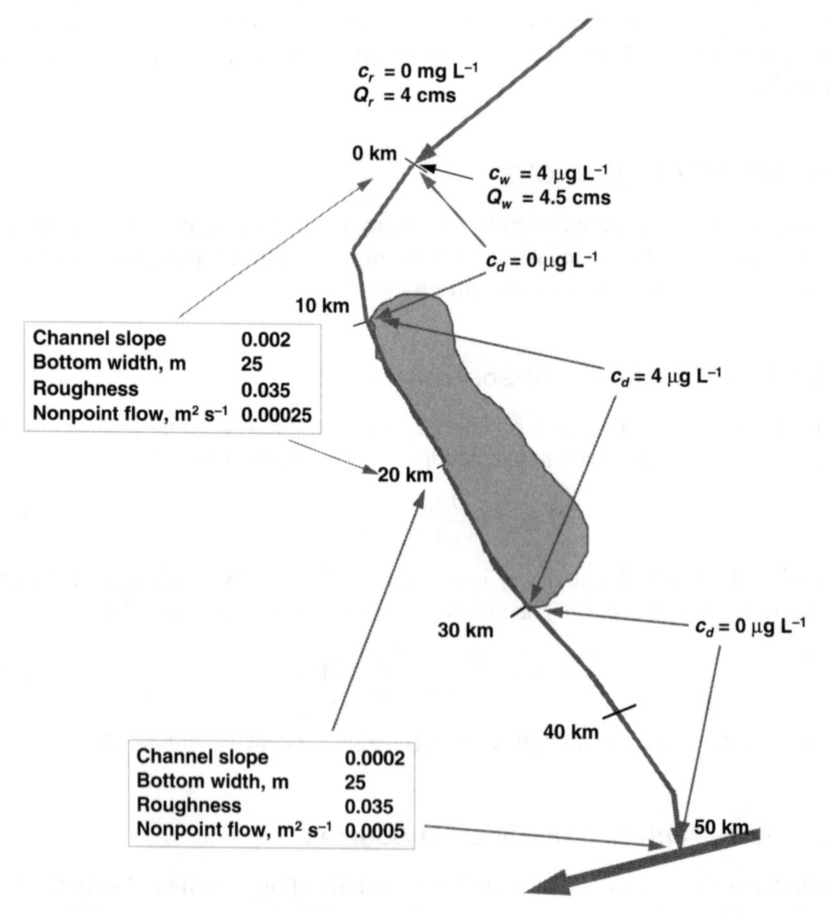

FIGURE 44.3
Example problem.

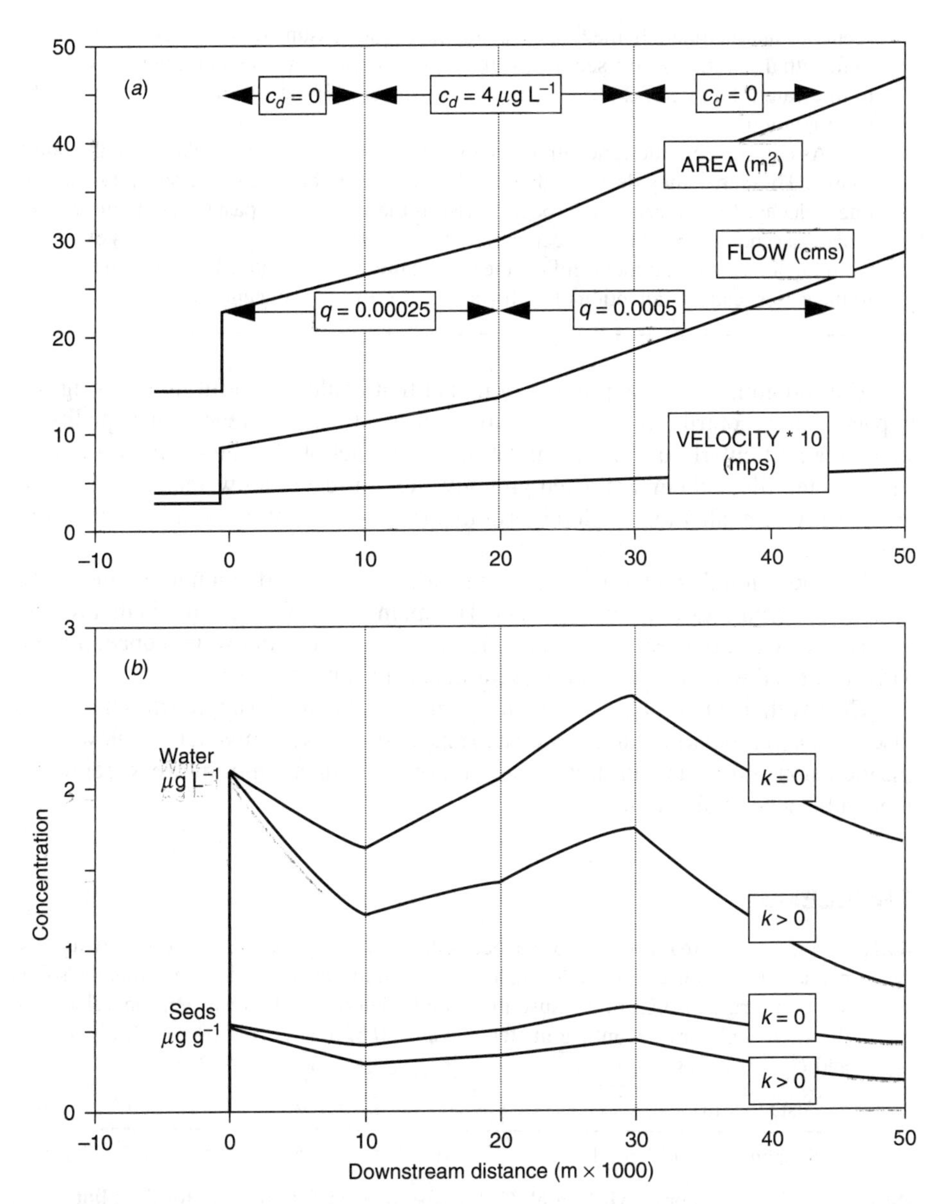

FIGURE 44.4
Simulation results. (*a*) Hydraulic variables; (*b*) concentrations.

the next 10 km, there is a constant nonpoint inflow of clean water. Then, in the stretch 10 to 30 km downstream from the point source, runoff from a landfill adds additional toxicant in a diffuse manner. Finally, for the last 20 km, the stream again receives clean water. Other model parameters are $\phi = 0.9$, $\rho = 2.5$ g cm^{-3}, $v_s = 0.05$ m d^{-1}, $v_r = 0.0006$ m d^{-1}, and $v_b = 0.0002$ m d^{-1}.

Solution: The results of the simulation are shown in Fig. 44.4. Water and sediment concentrations are expressed in μg L^{-1} and μg g^{-1}, respectively. Two contaminant decay scenarios are included. In the first the contaminants are assumed to be conservative ($k = 0$) in both the water and the sediments. In the second the contaminant is allowed to decay in the water at a rate of 2 d^{-1}. Such might be the case if the substance were subject to volatilization.

As can be seen, the concentrations abruptly increase to high levels due to the point source. Thereafter they decrease due to dilution and, in the case of the second scenario, due to decay. The concentrations begin to rise as the water flows past the nonpoint source. Once past the landfill it again decreases in response to dilution and decay. As expected from Eq. 44.13, the sediments follow the same pattern as the water. The addition of decay to the water leads to depletion of both water and sediment concentrations.

The foregoing analysis provides a model that could be useful in assessing the impact of flow-contributing diffuse sources on steady-state, stream water quality. It should be particularly useful in simulating the impact of chronic nonpoint sources, such as landfills and contaminated groundwater, on receiving waters. It also could have utility in evaluating the long-term integrated effect of urban and agricultural runoff.

In conclusion this lecture has been intended as an introduction to the modeling of toxicant dynamics in flowing waters. The main point of the lecture is that for the steady-state case, the sediments map directly on the overlying water concentration. This result is due to the fact that the sediments are stationary.

Clearly there are problems that require more sophisticated approaches than were discussed here. In particular, time-variable cases (such as stormwater overflows) and the inclusion of toxicants in stream and estuary eutrophication frameworks represent more advanced applications.

PROBLEMS

44.1. A river downstream to the confluence with a tributary manifests the following suspended solids concentration during low-flow conditions. If it is assumed that the solids settle at a rate of 0.25 m d^{-1}, use this data to determine the resuspension velocity in the stretch. Other relevant parameters are $\phi = 0.8$, $\rho = 2.5 \times 10^6$ g m^{-3}, $H_1 = 1$ m, and $U = 0.1$ mps.

Distance (m)	0	20	40	60	80	100
Suspended solids (mg L^{-1})	10	6	5	3.5	2.5	2.5

44.2. In addition to copper, Mills et al. (1985) also present data on zinc for the Flint River. The river has the following boundary conditions and point sources:

	Kilometers	**Zinc (μg L^{-1})**
Upstream boundary	0	7.7
Flint WWTP	1.3	55
Ragone WWTP	30.9	84

In a similar fashion in Example 44.2, simulate the distribution of zinc in the Flint River. Use a sorption coefficient of 0.2×10^6 L kg^{-1} for your calculation.

44.3. A point source of a toxic substance ($K_{ow} = 10^6$) is discharged to an estuary having the following characteristics:

	Value	Units
Dispersion coefficient	100×10^6	m^2 d^{-1}
Flow	15×10^3	m^3 d^{-1}
Width	75	m
Depth	10	m

The estuary has a suspended solids concentration of 5 mg L^{-1} with an $f_{oc} = 5\%$. These solids settle at a rate of 0.25 m d^{-1}. The bottom sediments have the following characteristics: thickness = 10 cm, burial velocity = 0.825 mm yr^{-1}, resuspension velocity = 1 mm yr^{-1}, porosity = 0.9, density = 2.5×10^6 g m^{-3}. There are no other loss mechanisms other than a volatilization rate of 0.2 m d^{-1}.

What mass loading of a pollutant could be input to this system under steady-state conditions if the maximum allowable sediment concentration at the outfall is 1 part per million (that is, 1 μg g^{-1})? Express your results in kg yr^{-1}. Assume complete and instantaneous lateral and vertical mixing at the outfall. Also assume that diffusive sediment-water interactions are negligible.

44.4. A stream passes by a toxic landfill for an 8-km stretch. The landfill leaches a toxicant at a rate of 50 mg per meter of stream length per day. The toxicant has an octanol-water partition coefficient of 10^5, a molecular weight of 300, and is overwhelmingly liquid-film controlled. The stream has about 1 mg L^{-1} of organic carbon that is carried in suspension. It also has a velocity of 0.1 m s^{-1}, a depth of 2 m, and a width of 30 m. If sorption is instantaneous, calculate the concentration in the water for a distance of 16 km.

LECTURE 45

Toxicant/Food-Chain Interactions

Lecture overview: I describe how toxicants interact with organisms. This interaction takes two primary forms. First, the organism can directly take up dissolved toxicant from the water. Second, the organism can consume contaminated food. I also show how both types of interaction can be integrated into a larger mass-balance framework to relate food-chain levels back to loadings.

To this point our models provide a way to predict toxicant concentration in the water and the sediments. Such information is certainly valuable because water-quality standards are often expressed in terms of concentration. For example it makes sense to express drinking water standards as concentration in the water.

However, toxics pose additional threats that extend beyond human consumption of drinking water. First, the toxic can reach harmful levels in organisms that serve as human food, such as fish and shellfish. Second, the toxic can grow to levels that interfere with the functioning of the ecosystem itself. Although both examples could be related back to a water-quality standard for concentration, a more direct approach would be to assess the concentration of the toxicant within the organisms themselves.

This lecture is devoted to introducing you to how this can be done. We first examine how individual organisms take up and purge contaminants directly from the water. Then we add food-chain transfers to show how organisms are impacted by the consumption of contaminated food.

Before proceeding we should introduce some nomenclature. ***Bioconcentration*** refers to the direct uptake of a toxicant through the gills and epithelial tissues (skin) of an aquatic organism. ***Bioaccumulation*** is a more comprehensive term referring to direct bioconcentration from the water as well as ingestion of contaminated food. ***Biomagnification*** is the process wherein bioaccumulation increases along the food chain. The following sections review modeling frameworks that have been developed to quantify these processes.

45.1 DIRECT UPTAKE (BIOCONCENTRATION)

Before writing equations we should first determine an appropriate way to quantify the toxicant within an organism. In prior models we have normalized mass either to volume or to mass. The former was appropriate for water concentration, whereas the latter was employed for the sediments. Clearly the mass-specific approach is preferable for organisms.

However, now the question arises: What mass do we normalize to? In early toxicant-organism models, the mass of organisms was expressed on a wet-weight basis. This is in contrast to particles that, as we've seen in previous lectures, are expressed as dry weight. This perspective has been further refined for organic chemicals. For these toxicants it has been recognized that the association is governed by the quantity of lipids in the organism (Mackay 1982, Connolly and Pedersen 1988, and Thomann 1989). Thus, one way to quantify the concentration of a contaminant in an organism is

$$v = \frac{v_{wt}}{f_L} \tag{45.1}$$

where v = organism toxicant concentration on a lipid basis (μg-chemical kg-lipid^{-1})
v_{wt} = organism toxicant concentration on a wet-weight basis (μg-chemical kg-wet^{-1})
f_L = fraction lipid (kg-lipid kg-wet^{-1})

Bioconcentration is often measured by placing an organism into a vessel where the concentration of a contaminant is kept at a constant level. The process can be modeled by writing a mass balance around an individual organism. As in Fig. 45.1, the primary processes are (1) direct uptake by diffusive exchange across organism membranes and (2) losses,

$$\underset{\text{Accumulation}}{\frac{dv_m}{dt}} = \underset{\text{Uptake}}{k_u w_L (c_d - c_B)} - \underset{\text{Losses}}{K v_m} \tag{45.2}$$

where v_m = toxicant whole-body burden (μg-toxic per organism)
k_u = chemical uptake rate (L kg-lipid^{-1} d^{-1})
w_L = organism lipid weight (kg-lipid)
c_d = "freely available" dissolved chemical in the surface or pore waters (μg L^{-1})
c_B = "free" concentration of chemical in the organism's blood (μg L^{-1})
K = losses (that is, losses from skin surface, excretion, and chemical metabolism) (d^{-1})

Next, if it is assumed that the partitioning between the tissue concentration and the blood concentration is proportional to the octanol-water partition coefficient, we can define

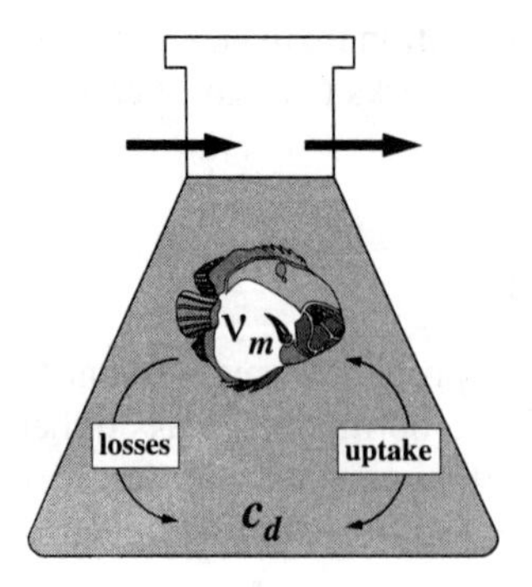

FIGURE 45.1
A simplified schematic of an experiment to determine bioconcentration. Inputs and outputs are shown to indicate that the dissolved toxicant level in the CSTR is kept at a constant level.

$$\frac{v}{c_B} = N_{wl} \approx K_{ow} \tag{45.3}$$

where N_{wl} = a lipid concentration to blood concentration partition coefficient (L kg-lipid^{-1}). This result can be substituted into Eq. 45.2 to give

$$\frac{dv_m}{dt} = k_u w_L \left(c_d - \frac{v}{K_{ow}} \right) - K v_m \tag{45.4}$$

The whole body burden is related to the chemical concentration by

$$v_m = v w_L \tag{45.5}$$

which can be substituted into Eq. 45.4 to give

$$\frac{d(v w_L)}{dt} + k_u w_L \left(c_d - \frac{v}{K_{ow}} \right) - K v_m \tag{45.6}$$

If the organism's lipid mass is constant, we can take it outside of the differential. However, for organisms such as fish, this is not a good assumption. Therefore, we must expand the differential,

$$v \frac{dw_L}{dt} + w_L \frac{dv}{dt} = k_u w_L \left(c_d - \frac{v}{K_{ow}} \right) - K v_m \tag{45.7}$$

or rearranging terms,

$$\underset{\text{Accumulation}}{w_L \frac{dv}{dt}} = \underset{\text{Uptake}}{k_u w_L \left(c_d - \frac{v}{K_{ow}} \right)} - \underset{\text{Losses}}{K w_L v} - \underset{\text{Growth dilution}}{G w_L v} \tag{45.8}$$

where G = growth rate of the organism (d^{-1}),

$$G = \frac{dw_L/dt}{w_L} \tag{45.9}$$

The differential equation is now written to determine the temporal changes of the mass-specific concentration (rather than the contaminant mass as in Eq. 45.2). Consequently, along with uptake and depuration, the change is also dictated by the rate at which the organism gains biomass. This latter effect can be thought of as a "dilution" effect because the increase means that the mass of contaminant will be normalized to a large biomass (and lipid content).

Equation 45.8 can be simplified further as

$$\frac{dv}{dt} + K'v = k_u c_d \tag{45.10}$$

where K' = the sum of the membrane losses, the other losses, and the growth dilution rates,

$$K' = \frac{k_u}{K_{ow}} + K + G \tag{45.11}$$

For zero K and G, this result suggests that the total losses would be inversely proportional to the octanol-water partition coefficient. That is, higher K_{ow} compounds would be retained better than lower ones.

Assuming that the organism is uncontaminated at the beginning of the experiment ($v = 0$ at $t = 0$), we can solve Eq. 45.10 for

$$v = \frac{k_u}{K'} c_d (1 - e^{-K't}) \tag{45.12}$$

Thus the mass-specific concentration starts at zero and asymptotically approaches a steady-state concentration of

$$v = \frac{k_u}{K'} c_d \tag{45.13}$$

Note that this steady-state level can be expressed as

$$N_w = \frac{v}{c_d} = \frac{k_u}{K'} = \frac{k_u}{k_u/K_{ow} + K + G} \tag{45.14}$$

$$\left[\text{units: } \frac{\mu\text{g-contaminant kg-lipid}^{-1}}{\mu\text{g-contaminant L-water}^{-1}} = \frac{\text{L-water}}{\text{kg-lipid}}\right]$$

where N_w = a bioconcentration factor (BCF). This representation is directly analogous to a partition coefficient. However, rather than representing the association of a toxicant with suspended solids, it parameterizes the association of a contaminant with an organism.

Note that an interesting case results when losses and growth are zero. For this case Eq. 45.14 reduces to

$$N_w = K_{ow} \tag{45.15}$$

EXAMPLE 45.1. BIOCONCENTRATION. A zooplankton is placed in an aquarium with a contaminant concentration of 1 μg L^{-1}. Compute its steady-state concentration given the following parameters: $K_{ow} = 10^5$, k_u = 31,000 L kg-lipid^{-1} d^{-1}, G = 0.15 d^{-1}, and K = 0.01 d^{-1}.

Solution: The organism's total loss rate can be computed as (Eq. 45.11)

$$K' = \frac{31{,}000}{10^5} + 0.01 + 0.15 = 0.47\ \text{d}^{-1}$$

and the BCF can be calculated with Eq. 45.14,

$$N_w = \frac{31{,}000}{0.47} = 65{,}957 \text{ L kg-lipid}^{-1}$$

Finally the organism's concentration can be determined as

$$v = N_w c_d = 65{,}957 \text{ L kg-lipid}^{-1}\left(1\frac{\mu\text{g}}{\text{L}}\right) = 65{,}957\ \mu\text{g kg-lipid}^{-1}$$

45.2 FOOD-CHAIN MODEL (BIOACCUMULATION)

Although bioconcentration factors are certainly useful, they ignore a second way in which organisms can take up toxic substances, namely by ingesting contaminated prey.

A number of possible mathematical formulations characterize interactions among the elements of the food chain. As described in Lec. 35 fairly elaborate nonlinear models are typically used to simulate seasonal dynamics of biomass. However, for long-term computations, these seasonal models are costly to implement and available field data for toxics rarely support the use of such fine-scaled frameworks. One alternative is to integrate a seasonal model over the annual cycle to estimate transfer rates for a long-term food-chain model. Hydroscience (1973) used such an approach to calibrate a linear, donor-dependent model for cadmium in western Lake Erie. A second alternative is to directly develop a long-term model based on constant average annual rates. The following description follows this approach and is derived primarily from the research of Thomann and Connolly (Thomann 1981, 1989; Connolly 1991; Thomann et al. 1992; Thomann and Connolly 1992).

As a first approximation, the pelagic food chain can be divided into four compartments (Fig. 45.2). The phytoplankton are treated in a fashion similar to other small particles. That is, membrane exchange is omitted and the following balance results:

$$\frac{dv_1}{dt} = k_{u1} c_d - K_1 v_1 \tag{45.16}$$

where the subscript 1 denotes the phytoplankton.

For the higher organisms, a balance accounting for membrane exchange as well as grazing gains from the phytoplankton can be written as

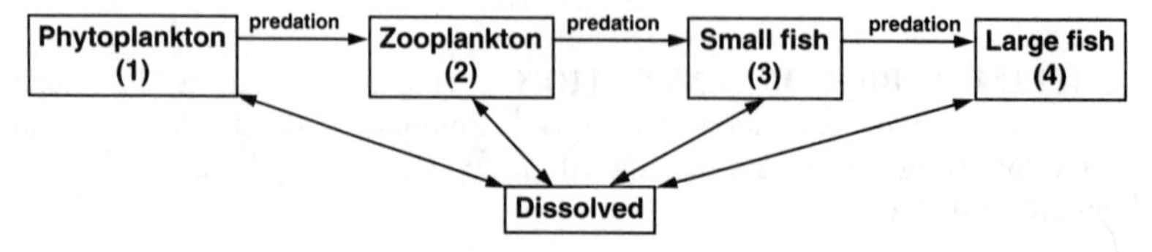

FIGURE 45.2
A simplified representation of a pelagic food chain originally proposed by Thomann (1981). Note that the term ***pelagic*** refers to the open-water region beyond the influence of the shore and bottom.

$$\frac{dv_2}{dt} = k_{u2}c_d + \alpha_{21}I_{L2}v_1 - K_2'v_2 \tag{45.17}$$

where $\alpha_{j,k}$ = chemical assimilation efficiency (μg-chemical absorbed per μg-chemical ingested) for jth predator feeding on kth prey and I_{L2} = the lipid-specific consumption rate (kg-lipid prey per kg-lipid predator per d).

At steady-state, Eq. 45.16 can be solved for

$$v_1 = \frac{k_{u,1}}{K_1}c_d \tag{45.18}$$

or, as a water bioconcentration factor,

$$N_{1w} = \frac{v_1}{c_d} = \frac{k_{u1}}{K_1} \tag{45.19}$$

For Eq. 45.17, the steady-state solution is

$$v_2 = N_{2w}c_d + g_{21}v_1 \tag{45.20}$$

where

$$N_{2w} = \frac{v_2}{c_d} = \frac{k_{u2}}{K_2'} \tag{45.21}$$

and g_{21} is a dimensionless food-chain multiplier,

$$g_{21} = \frac{\alpha_{21}I_{L2}}{K_2'} \tag{45.22}$$

Note that Eq. 45.18 can be substituted into Eq. 45.20 so that the level in the zooplankton is directly related to the water concentration, as in

$$v_2 = N_2c_d \tag{45.23}$$

where N_2 = overall bioaccumulation factor for the zooplankton,

$$N_2 = N_{2w} + g_{21}N_{1w} \tag{45.24}$$

Notice that for phytoplankton,

$$N_1 = N_{1w} \tag{45.25}$$

since these organisms do not receive contaminant by grazing, whereas for zooplankton,

$$N_2 \neq N_{2w} \tag{45.26}$$

Similar equations can be developed for the other compartments of the food chain. For example the small fish can be represented as

$$N_3 = N_{3w} + g_{32}N_{2w} + g_{32}g_{21}N_1 \tag{45.27}$$

and the large fish as

$$N_4 = N_{4w} + g_{43}N_{3w} + g_{43}g_{32}N_{2w} + g_{43}g_{32}g_{21}N_1 \tag{45.28}$$

A generalized relationship can be expressed as

$$N_n = N_{nw} + \sum_{j=1}^{n-1}\left[\left(\prod_{i=j+1}^{n} g_{i,i-1}\right)N_{iw}\right] \tag{45.29}$$

where N_n is the bioaccumulation factor (BAF) for the nth organism.

45.3 PARAMETER ESTIMATION

In the previous framework there are chemical and organism parameters. We now describe how each of these is estimated. We then apply them to compute the concentration of a toxicant in a natural water.

45.3.1 Chemical Parameters

Uptake. The chemical uptake rate depends on the organism's respiration rate and the efficiency of transfer across the organism membrane. Connolly (1991) has suggested

$$k_u = 10^6 \frac{r_{oc} a_{cd} \rho}{a_{wd} f_L o_w} \frac{E_c}{E_o} \tag{45.30}$$

where r_{oc} = oxygen-to-carbon ratio ($\cong$ 2.67 kgO kgC^{-1})
a_{cd} = carbon-to-dry-weight ratio ($\cong$ 0.4 kgC kg-dry weight^{-1})
ρ = organism oxygen respiration rate (d^{-1})
a_{wd} = ratio of wet-to-dry-weight (kg-wet kg-dry^{-1})
o_w = oxygen concentration in the water (mg L^{-1})
E_c = chemical transfer efficiency
E_o = oxygen transfer efficiency for the organism

Note that the 10^6 is included so that the units of k_u work out to L kg^{-1}.

It has been suggested that the chemical transfer efficiency depends on a number of factors, including the lipid partition coefficient, molecular weight, etc. (see McKim et al. 1985 for a review). Thomann (1989) has developed some piecewise functions to compute the chemical transfer efficiency based on the octanol-water partition coefficient and organism size. Thomann and Connolly (1992) have developed a simplified version for the ratio E_c/E_o that is independent of organism size. It can be represented mathematically as

$$\begin{aligned}
\log \frac{E_c}{E_o} &= -2.6 + 0.5 \log K_{ow} && 2 \le \log K_{ow} \le 4 \\
\frac{E_c}{E_o} &= -3.339 + 0.8976 \log K_{ow} && 4 \le \log K_{ow} \le 4.5 \\
\frac{E_c}{E_o} &= 0.7 && 4.5 \le \log K_{ow} \le 6.5 \\
\frac{E_c}{E_o} &= 3.3 - 0.4 \log K_{ow} && 6.5 \le \log K_{ow} \le 8 \\
\log \frac{E_c}{E_o} &= 7 - \log K_{ow} && 8 \le \log K_{ow} \le 9
\end{aligned} \tag{45.31}$$

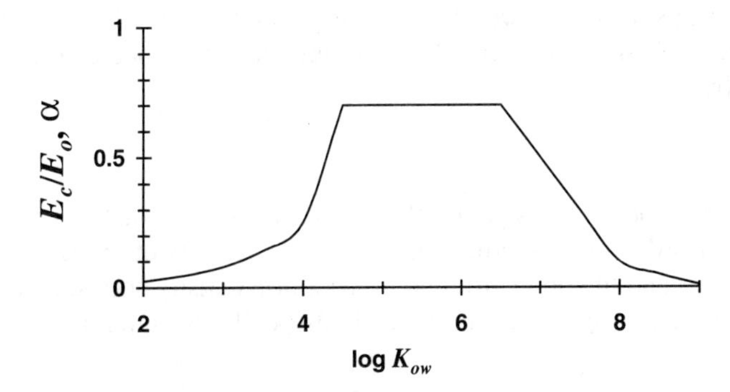

FIGURE 45.3
Ratio of chemical to oxygen transfer efficiency (E_c/E_o) versus octanol-water partition coefficient. The chemical transfer efficiency (α) is assumed to follow the same function.

As displayed in Fig. 45.3, the efficiency increases, levels off, and then decreases with K_{ow}.

Membrane and excretion losses. Using Eq. 45.11, the membrane and excretion losses are defined as

$$\frac{k_u[E_c(K_{ow}), \rho]}{K_{ow}} + K \tag{45.32}$$

where the nomenclature $k_u[E_c(K_{ow}), \rho]$ denotes that the uptake is computed with Eq. 45.30.

Chemical assimilation efficiency. The assimilation of chemical through food ingestion should be similar to the factors governing gill membrane transport. Consequently Thomann and Connolly (1992) have assumed that α is a function of K_{ow}. As a first approximation they employed the same function as in Fig. 45.3.

45.3.2 Organism Parameters

Growth and respiration. Both organism growth and respiration rates can be developed to a first approximation by correlation with organism weight. Thomann (1989) has proposed the following relationships for growth,

$$G = 0.002512w^{-0.2} \tag{45.33}$$

and for respiration,

$$\rho = 0.009043w^{-0.2} \tag{45.34}$$

where w = organism weight (kg-wet weight).

Consumption rate. According to Connolly (1991), the organism's energy usage rate can be computed, as in

$$P_i = \lambda_i(\rho_i + G_i) \tag{45.35}$$

where P_i = energy usage rate of organism i (cal kg-wet^{-1} d^{-1}) and λ_i = caloric density of organism i (cal kg-wet^{-1}). The energy intake rate is this usage rate divided by the fraction of ingested energy that is assimilated. The food consumption rate is then given by

$$I_i = \frac{\lambda_i}{\lambda_{i-1}} \frac{\rho_i + G_i}{a_{i,i-1}} \tag{45.36}$$

where the subscript $i - 1$ designates the prey and $a_{i,i-1}$ = fraction of ingested prey that is assimilated. If it is assumed that the caloric density of dry tissue is the same for the predator and the prey, differences in caloric density should be related to the wet-weight-to-dry-weight ratio. Thus the lipid-specific consumption rate can be calculated as

$$I_{Li} = \frac{a_{wd,i-1}}{a_{wd,i}} \frac{f_{L,i-1}}{f_{L,i}} \frac{\rho_i + G_i}{a_{i,i-1}} \tag{45.37}$$

where a_{wd} = wet-to-dry-weight ratio (kg-wet kg-dry^{-1}).

EXAMPLE 45.2. TOXICANT/FOOD CHAIN. Thomann et al. (1992) presented the following data for the pelagic food chain in Lake Ontario:

Organism	No.	Weight (kg-wet)	f_L (% lipid)	a	a_{wd}	G (d^{-1})	ρ (d^{-1})	I_L (d^{-1})
Phyto/detritus	1	—	1	—	10	—	—	—
Zooplankton	2	1×10^{-5}	5	0.3	5	0.025	0.09	0.154
Forage fish	3	0.1	8	0.8	4	0.004	0.014	0.016
Piscivorous fish	4	1	20	0.8	4	0.0025	0.009	0.0058

Note that the last three columns of the table have been calculated with Eqs. 45.33, 45.34, and 45.37. (*a*) Determine the levels of contaminant in the four organisms for a toxicant with a K_{ow} = 10^5 and a total water concentration of 1 μg L^{-1}. Assume that the phytoplankton behave as simple particles, and

$$N_{1w} \approx K_{ow}$$

The lake has a phytoplankton/detritus concentration of about 1 mg L^{-1} with a carbon content of 20%. Also assume that excretion losses, K, are negligible. (*b*) Plot the concentration in the four compartments versus log K_{ow} from 2 to 9.

Solution: Using relationships for partitioning developed in Lec. 41 (e.g., Eq. 41.37), we can first calculate the concentration in dissolved form as

$$c_d = \frac{1}{1 + 10^{-6}(10^5)0.2(1)} 1 = 0.9804 \ \mu\text{g L}^{-1}$$

Recall that the 10^{-6} in the denominator is necessary because the suspended solids are expressed as mg L^{-1}. The solids concentration of the small organic particles (that is, phytoplankton, detritus) can be computed as

$$\nu_1 = N_{1w} c_d = 10^5(0.9804) = 98{,}039 \ \mu\text{g kg-lipid}^{-1}$$

Next we can determine the zooplankton concentration. First, using a value of 0.7 for E_c/E_o (from Fig. 45.3), we can calculate the uptake rate as

$$k_u = 10^6 \frac{2.67(0.4)0.09}{5(0.05)8.5} 0.7 = 31{,}814 \text{ L kg-lipid}^{-1} \text{ d}^{-1}$$

and the total loss rate as

$$K_2' = \frac{31{,}814}{10^5} + 0.025 = 0.3433 \text{ d}^{-1}$$

which can be used to compute the bioconcentration factor as

$$N_{2w} = \frac{31{,}814}{0.3433} = 92{,}682 \text{ L kg-lipid}^{-1}$$

The food-chain multiplier can be calculated as

$$g_{21} = \frac{0.7(0.154)}{0.3433} = 0.31418$$

This value, along with the BCF, can be used to determine the zooplankton concentration as

$$\nu_2 = 92{,}682(0.9804) + 0.31418(98{,}039) = 90{,}865 + 30{,}802 = 121{,}667 \ \mu\text{g kg-lipid}^{-1}$$

Results for the rest of the food chain can be computed in a similar fashion, and the results are summarized in the following table and plot:

				ν_{fc} (components)					
Organism	N_w (BCF)	ν_w	g	Phyto	Zoo	Forage	ν_{fc} (total)	ν	N (BAF)
Phyto/detritus	98,039	98,039	–	–	–	–	–	98,039	100,000
Zooplankton	92,682	90,865	0.314	30,802	–	–	30,802	121,667	124,100
Forage fish	90,821	89,040	0.289	8890	35,117	–	44,006	133,047	135,708
Piscivorous fish	79,830	78,265	0.325	2887	11,403	43,205	57,496	135,751	138,476

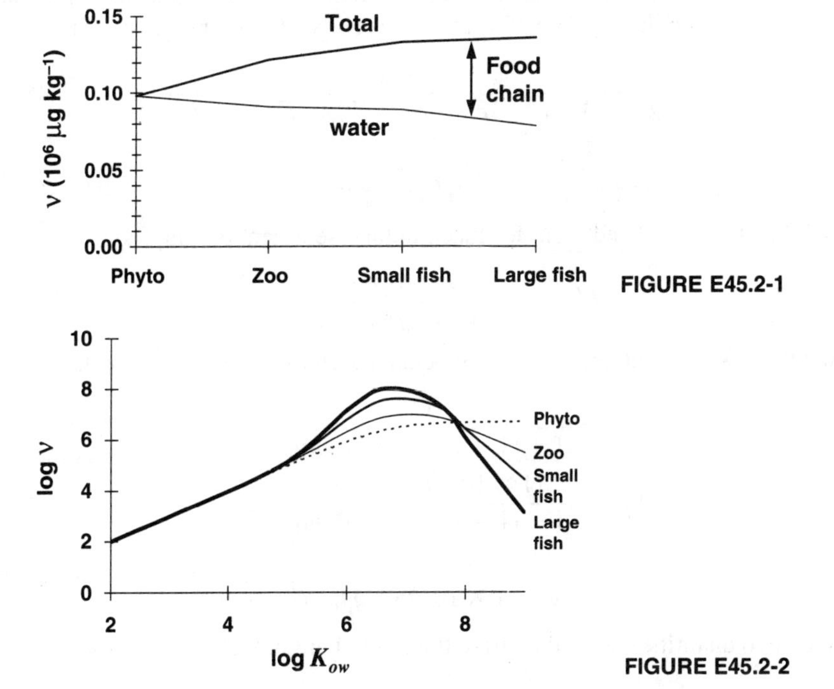

FIGURE E45.2-1

FIGURE E45.2-2

(*b*) Results for a range of K_{ow}'s are displayed in Fig. E45.2.2. Notice how the peak occurs at a value of about $K_{ow} = 6$ to 7. This result is determined by the relationship of both α and E_c to K_{ow} (Fig. 45.3) along with the fact that more toxicant is associated with the food as K_{ow} increases.

It should be noted that it is believed that phytoplankton do not actually behave as simple particles, and that N_{1w} is not equal to K_{ow} for all organic chemicals. For example Thomann (1989) suggests that the proportionality holds up to about $\log K_{ow} = 6$. Above that point the N_{1w} appears to plateau at a level of about $\log N_{1w} \cong 6.5$. We will explore the implications of this leveling off in one of the problems at the end of this lecture.

45.4 INTEGRATION WITH MASS BALANCE

In the previous section we have illustrated how food-chain-toxicant concentrations can be determined as a function of a given water concentration. Integration of the food-chain model into a mass-balance framework would have the added utility of tying the biotic concentrations back to toxicant loadings.

If it can be assumed that the toxicant mass retained in the nonphytoplankton parts of the food chain are negligible, we can merely apply the toxicant mass-balance models developed to this point to calculate the toxicant concentrations in the dissolved and particulate phases. If the latter is assumed to represent food (that is, phytoplankton and organic detritus), these levels can then serve as the forcing functions for the food-chain model.

For example recall that the steady-state solution for a well-mixed lake underlain by a single sediment layer can be formulated as (Eqs. 40.33 through 40.35)

$$c_1 = \frac{Qc_{\text{in}}}{Q + k_1 V_1 + v_v A F_{d1} + (1 - F_r')(v_s F_{p1} + v_d F_{d1})A} \tag{45.38}$$

and

$$c_2 = \frac{v_s F_{p1} + v_d F_{d1}}{k_2 H_2 + v_r + v_b + v_d F_{d2}} c_1 \tag{40.39}$$

where F_r' is the ratio of sediment feedback to total sediment purging,

$$F_r' = \frac{v_r + v_d F_{d2}}{v_r + v_b + v_d F_{d2} + k_2 H_2} \tag{45.40}$$

and the fractions are defined in terms of octanol-water partition coefficients as

$$F_{d1} = \frac{1}{1 + (1 \times 10^{-6} K_{ow}) f_{oc} m_1} \tag{41.41}$$

$$F_{p1} = \frac{(1 \times 10^{-6} K_{ow}) f_{oc} m_1}{1 + (1 \times 10^{-6} K_{ow}) f_{oc} m_1} \tag{41.42}$$

$$F_{dp2} = \frac{1}{\phi + (1 \times 10^{-6} K_{ow}) f_{oc2} (1 - \phi) \rho} \tag{41.43}$$

Thus the two quantities needed to drive the food-chain model can be computed as

$$c_{d1} = F_{d1}c_1 \tag{45.44}$$

and

$$v_1 = \frac{F_{p1}c_1}{f_{oc,1}m_1} \tag{45.45}$$

EXAMPLE 45.3. TOXICANT/FOOD-CHAIN MODEL AND MASS BALANCE. Lake Ontario has a flow of $Q = 212 \times 10^9$ m^3 yr^{-1} and a deposition zone of approximately $10{,}000 \times 10^6$ m^2. Assume that the suspended solids in the lake settle at a rate of 100 m yr^{-1}. Also assume that the contaminant is subject to no losses other than settling and flushing and that resuspension and sediment-water diffusion are negligible. All other parameters and assumptions are the same as Example 45.2. If the contaminant has a K_{ow} of 10^5, compute the inflow concentration required so that large fish concentration is maintained at a level of 10^5 μg kg-lipid^{-1}.

Solution: For no resuspension, sediment-water diffusion, and decay reactions, Eq. 45.38 simplifies to

$$c_1 = \frac{Qc_{\text{in}}}{Q + v_s F_{p1} A}$$

From Example 45.2 we have determined a BAF of 138,476 for the top predators. Thus this result can be multiplied by the dissolved concentration in the water to determine the large fish concentration,

$$v_4 = N_4 F_{d1} \frac{Qc_{\text{in}}}{Q + v_s F_{p1} A}$$

which can be rearranged to determine the target inflow concentration

$$c_{\text{in}} = \frac{v_4(Q + v_s F_{p1} A)}{N_4 F_{d1} Q} = \frac{v_4}{N_4 F_{d1}} \left[1 + \frac{v_s(1 - F_{d1})}{q_s}\right]$$

where $q_s = Q/A$, which for Lake Ontario is $212 \times 10^9/10{,}000 \times 10^6 = 21.2$ m yr^{-1}. Substituting this along with the other parameter values yields

$$c_{\text{in}} = \frac{10^5}{138{,}476(0.9804)} \left[1 + \frac{100(0.0196)}{21.2}\right] = 0.668\ \mu\text{g L}^{-1}$$

which represents about a $\frac{1}{3}$ reduction.

45.5 SEDIMENTS AND FOOD WEBS (ADVANCED TOPIC)

The framework presented above can be extended to incorporate sediment-water interactions and food webs. Figure 45.4 illustrates how both can be implemented by including an organism that feeds on bottom sediments.

Balances for the five food-chain compartments can be written as

$$\begin{aligned}
0 &= k_{u1}c_d - K_1' v_1 \\
0 &= k_{u2}c_d + \alpha_{21} I_{L2} v_1 - K_2' v_2 \\
0 &= k_{u3}c_d + p_{32}\alpha_{32} I_{L3} v_2 + p_{35}\alpha_{35} I_{L3} v_5 - K_3' v_3 \\
0 &= k_{u4}c_d + \alpha_{43} I_{L4} v_3 - K_4' v_4 \\
0 &= [k_{u5}(b_{5s}c_s + b_{5d}c_d)] + [p_{5s}\alpha_{5s} I_{Loc} v_s + p_{51}\alpha_{51} I_{L5} v_l] - K_5' v_5
\end{aligned} \tag{45.46}$$

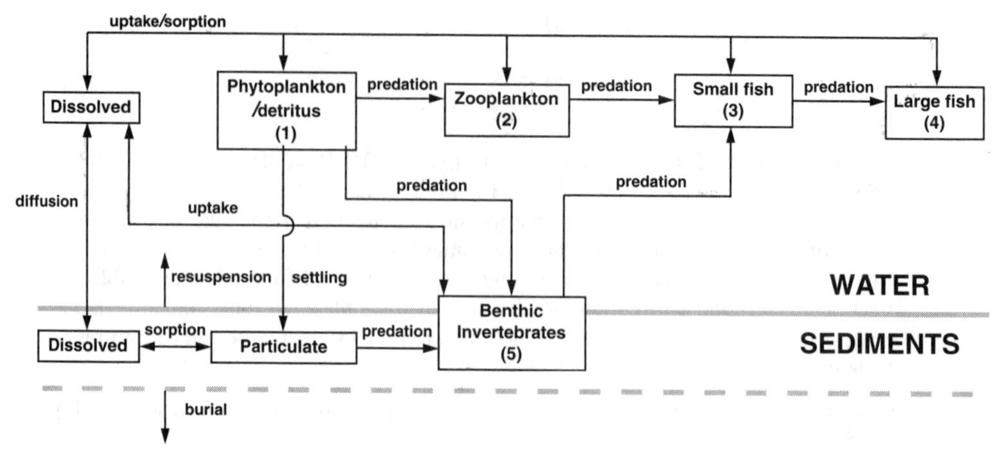

FIGURE 45.4
A five-compartment food-web model including sediment-water interactions (modified from Thomann et al. 1992).

where b_{5s} and b_{5d} = the fraction of the uptake from sediment and overlying water ($b_{5s} + b_{5d} = 1$); p_{32} and p_{35} = preference factors of forage fish for zooplankton and benthic invertebrates, respectively ($p_{32} + p_{35} = 1$); p_{5s} and p_{51} = preference factors of benthic invertebrates for sediments and phytoplankton, respectively ($p_{5s} + p_{51} = 1$); and I_{Loc} = rate of kilograms of organic carbon ingested per kilogram lipid of the predator (kg-orgC kg-lipid^{-1} d^{-1}) that can be computed with (compare with Eq. 45.37)

$$I_{Loci} = \frac{f_{oc,i}}{f_{L,i}} \frac{\rho_i + G_i}{a_{i,i-1} a_{wd,i}} \tag{45.47}$$

We have highlighted the terms in Eq. 45.46 that make this model different from the purely pelagic model depicted in Fig. 45.2. These differences are

- The use of preference factors to split the grazing of both the benthic invertebrates and the forage fish between different food sources. It is this splitting that makes the system a food web rather than a food chain.
- The connection of the benthic macroinvertebrates with the sediment pore water (uptake) and the sediment solids (ingestion). In the former case the b terms are used to divide the uptake between pore and overlying waters.
- The reformulation of the benthic organism consumption rate (Eq. 45.47) to reflect the fact that the solid contaminant in the sediments is expressed on a carbon rather than a lipid basis.

Thomann et al. (1992) applied this framework to Lake Ontario and illustrated how the benthic pathway can contribute significantly to the contaminant levels in the pelagic organisms.

In summary we have attempted to introduce you to how contaminant/food-chain interactions can be modeled and integrated with the mass-balance approach. Although our focus has been on organic toxicants, models of other contaminants such as metals and radionuclides follow a similar approach. That is, direct uptake and predation provide the pathways by which chemicals move and concentrate in the biota.

PROBLEMS

45.1. Repeat Example 45.1, but compute the BCF and lipid-specific concentration for the forage fish (see Example 45.2 for parameters).

45.2. Repeat part (*a*) of Example 45.2 for a contaminant with a log K_{ow} of 6.

45.3. Repeat part (*b*) of Example 45.2, but use the following relationship for N_{1w}:

$$N_{1w} = K_{ow} \qquad 2 \le \log K_{ow} \le 6.5$$

$$N_{1w} = 10^{6.5} \qquad 6.5 \le \log K_{ow} \le 9$$

45.4. Repeat Example 45.3, but for a compound with a log $K_{ow} = 6$.

APPENDIX A

Conversion Factors

Metric and SI Prefixes

Factor	Prefix	Symbol
10^{12}	tera-	T
10^{9}	giga-	G
10^{6}	mega-	M
10^{3}	kilo-	k
10^{2}	hecto-	h
10^{1}	deka-	da
10^{-1}	deci-	d
10^{-2}	centi-	c
10^{-3}	milli-	m
10^{-6}	micro-	μ
10^{-9}	nano-	n
10^{-12}	pico-	p
10^{-15}	femto-	f
10^{-18}	atto-	a

Time

1 year (yr) = 31.536×10^6 s
1 average month (mo) = 30.42 d
1 day (d) = 1440 min
= 86,400 seconds (s)

Length

1 foot (ft) = 0.3048 m
1 yard (yd) = 0.9144 m
1 inch (in) = 2.54 cm
1 mile (mi) = 1.609344 km
1 nautical mile = 1.852 km
1 meter (m) = 3.2808 ft
= 39.37 in
1 angstrom (Å) = 10^{-10} m
= 0.1 nm

Area

1 square inch (in^2) = 6.452 cm^2
1 hectare (ha) = 10,000 m^2
= 2.4710 acres
1 square kilometer (km^2) = 0.3861 mi^2
1 acre (ac) = 43,560 ft^2
= 0.404685 ha

Volume

1 cubic meter (m^3) = 35.3147 ft^3
= 264.172 gal (U.S.)
= 1000 liters (L)
1 cubic foot (ft^3) = 7.480 gal
1 acre-feet (ac-ft) = 43,560 ft^3
= 1,233.49 m^3
= 325,851 gal (U.S.)
1 gallon (U.S.) = 3.785 L
1 MG = 10^6 gallons
1 barrel oil (bbl) = 0.15899 m^3
= 42 gal (U.S.)
1 fluid ounce = $\frac{1}{128}$ gal (U.S.)

Velocity

1 meter/second ($m\ s^{-1}$)	= 3.2808 $ft\ s^{-1}$
	= 86.4 $km\ d^{-1}$
	= 2.237 mph
1 foot/second ($ft\ s^{-1}$)	= 0.6818 mph
	= 0.3048 $m\ s^{-1}$
	= 16.364 $mi\ d^{-1}$
1 mile per hour (mph)	= 1.467 $ft\ s^{-1}$
	= 0.4470 $m\ s^{-1}$
	= 1.609 $km\ hr^{-1}$

Dispersion/diffusion

1 square centimeter/second ($cm^2\ s^{-1}$)	= 3.336×10^{-6} $mi^2\ d^{-1}$
	= 93 $ft^2\ d^{-1}$
1 square mile/day ($mi^2\ d^{-1}$)	= 0.2998×10^6 $cm^2\ s^{-1}$

Flow

1 cubic meter/second ($m^3\ s^{-1}$)	= 35.3147 cfs
	= 2118.9 $ft^3\ min^{-1}$ (cfm)
	= 22.8245 MGD
	= 70.07 ac-ft d^{-1}
1 cubic feet/second (cfs)	= 0.028316 $m^3\ s^{-1}$
	= 7.4806 gal min^{-1}
	= 0.0646 MGD
1 million gallons/day (MGD)	= 0.0438 $m^3\ s^{-1}$
	= 1.5472 cfs
1 gallon/minute (gpm)	= 6.30902×10^{-5} $m^3\ s^{-1}$

Mass

1 kg	= 2.2046 lb_m
	= 35.274 ounce
1 metric ton (tonne)	= 1000 kg
	= 2204.6 lb_m
1 pound (lb)	= 453.6 g

Density

1 $g\ cm^{-3}$	= 1000 $kg\ m^{-3}$
	= 62.428 $lb_m\ ft^{-3}$

Concentration

1 $g\ L^{-1}$	= 1 ppt
	= 1 ‰
1 $mg\ L^{-1}$	= 8.34 lb_m $(10^6\ gal)^{-1}$
	= 1 $g\ m^{-3}$
	= 1 ppm
1 $\mu g\ L^{-1}$	= 1 $mg\ m^{-3}$
	= 1 ppb
1 $ng\ L^{-1}$	= 1 $\mu g\ m^{-3}$
	= 1 pptr

Energy

1 calorie (cal)	= 4.1868 J
	= 0.003968 Btu
1 joule (J)	= 1 W-s
	= 1 N-m
1 kilowatt-hour	= 3600 kJ
	= 860 kcal
	= 3412 Btu
1 British thermal unit (Btu)	= 1.055056 kJ
	= 252 cal
	= 0.2930 W-hr
	= 778 ft-lb
1 erg	= 10^{-10} kJ

Temperature

Degrees Celsius (°C)	$= \frac{5}{9}[T(°F) - 32]$
	$= \frac{[T(°F) + 40]}{1.8} - 40$
Degrees Fahrenheit (°F)	$= \frac{9}{5}T(°C) + 32$
	$= 1.8[T(°C) + 40] - 40$
Kelvins (K)	$= T(°C) + 273.15$
Degrees Rankine (°R)	$= T(°F) + 459.67$

Degrees Celsius (°C)	**Degrees Fahrenheit (°F)**
0	32
5	41
10	50
15	59
20	68
25	77
30	86
35	95
40	104
45	113
50	122

Heat flux

1 langley/d (ly d^{-1}) = 1 cal cm^{-2} d^{-1}
1 Btu ft^{-2} hr^{-1} = 3.154591 W m^{-1}

Pressure

1 kPa = 0.0098692 atmosphere
= 0.296134 inHg
= 7.5006151 torr (mmHg)
= 0.1450377 psi
= 0.01 bar
= 0.0101972 kg_f cm^{-2}

Force

1 newton (N) = 1 kg m s^{-2}
= 0.2248 lb_f
= 7.233 lb_m ft s^{-2}
= 1×10^5 dyne
1 dyne (dyn) = 1 g cm s^{-2}

Power

1 Watt (W) = 1 joule s^{-1} (J s^{-1})
= 1 kg m^2 s^{-3}
= 3.412 Btu hr^{-1}
= 1.34×10^{-3} hp

Viscosity (dynamic)

1 Pa s = 1 kg m^{-1} s^{-1}
= 1000 centipose (cp)
= 0.672 lb_m ft^{-1} s^{-1}

Heat capacity

1 kJ (kg ° C)$^{-1}$ = 0.2388 cal (g °C)$^{-1}$
= 0.2388 Btu (lb_m °F)$^{-1}$

Thermal conductivity

1 W (m K)$^{-1}$ = 0.5778 Btu (hr ft °F)$^{-1}$
= 2.388×10^{-3} cal (s cm °C)$^{-1}$

Heat transfer coefficient

1 W (m^2 K)$^{-1}$ = 0.17611 Btu (ft^2 hr °F)$^{-1}$

Gas law constant

R = 8.206×10^{-5} m^3 atm (mol K)$^{-1}$
= 8.3143 J (mol K)$^{-1}$
= 1.9872 cal (mol K)$^{-1}$
= 10.73 ft^3 psi (lb_{mol} R)
= 0.7302 ft^3 atm (lb_{mol} R)$^{-1}$

APPENDIX B

Oxygen Solubility

Temperature (°C)	Chloride concentration ($g\ L^{-1}$) 0	5	10	15	20	25
0	14.621	13.726	12.885	12.096	11.356	10.660
1	14.216	13.354	12.544	11.782	11.068	10.396
2	13.830	12.999	12.217	11.482	10.792	10.143
3	13.461	12.659	11.904	11.195	10.528	9.900
4	13.108	12.334	11.605	10.920	10.275	9.668
5	12.771	12.023	11.319	10.656	10.032	9.445
6	12.448	11.726	11.045	10.404	9.800	9.231
7	12.139	11.441	10.782	10.161	9.577	9.025
8	11.843	11.167	10.530	9.929	9.362	8.828
9	11.560	10.906	10.288	9.706	9.157	8.639
10	11.288	10.654	10.056	9.492	8.959	8.456
11	11.027	10.413	9.834	9.286	8.769	8.281
12	10.777	10.182	9.620	9.089	8.587	8.113
13	10.537	9.960	9.414	8.898	8.411	7.950
14	10.306	9.746	9.216	8.715	8.242	7.794
15	10.084	9.540	9.026	8.539	8.079	7.643
16	9.870	9.342	8.843	8.370	7.922	7.498
17	9.665	9.152	8.666	8.206	7.770	7.358
18	9.467	8.968	8.496	8.048	7.624	7.223
19	9.276	8.791	8.332	7.896	7.483	7.092
20	9.092	8.621	8.173	7.749	7.347	6.966
21	8.915	8.456	8.020	7.607	7.215	6.843
22	8.744	8.297	7.872	7.470	7.088	6.725
23	8.578	8.143	7.729	7.337	6.964	6.611
24	8.418	7.994	7.591	7.208	6.845	6.499
25	8.263	7.850	7.457	7.083	6.729	6.392
26	8.114	7.710	7.327	6.962	6.616	6.287
27	7.968	7.575	7.201	6.845	6.507	6.186
28	7.828	7.444	7.079	6.731	6.401	6.087
29	7.691	7.316	6.960	6.621	6.298	5.991
30	7.559	7.193	6.845	6.513	6.198	5.898
31	7.430	7.073	6.733	6.409	6.100	5.807
32	7.305	6.956	6.623	6.307	6.005	5.718
33	7.183	6.842	6.517	6.208	5.913	5.632
34	7.065	6.731	6.414	6.111	5.822	5.548
35	6.949	6.623	6.313	6.016	5.734	5.465
36	6.837	6.518	6.214	5.924	5.648	5.385
37	6.727	6.415	6.118	5.834	5.564	5.306
38	6.620	6.315	6.024	5.746	5.481	5.229
39	6.515	6.217	5.932	5.660	5.401	5.153
40	6.413	6.121	5.842	5.576	5.322	5.079

APPENDIX C

Water Properties

Temperature (°C)	Density ($g\ cm^{-3}$)	Absolute viscosity [10^{-2} gm $(cm\ s)^{-1}$]	Kinematic viscosity $10^{-2}\ cm^2\ s^{-1}$	Vapor pressure (mmHg)
0	0.99987	1.7921	1.7923	4.58
2	0.99997	1.6740	1.6741	5.29
4	1.00000	1.5676	1.5676	6.10
6	0.99997	1.4726	1.4726	7.01
8	0.99988	1.3872	1.3874	8.04
10	0.99973	1.3097	1.3101	9.21
12	0.99952	1.2390	1.2396	10.52
14	0.99927	1.1748	1.1756	11.99
16	0.99897	1.1156	1.1168	13.63
18	0.99862	1.0603	1.0618	15.48
20	0.99823	1.0087	1.0105	17.54
22	0.99780	0.9608	0.9629	19.83
24	0.99733	0.9161	0.9186	22.38
26	0.99681	0.8746	0.8774	25.21
28	0.99626	0.8363	0.8394	28.34
30	0.99568	0.8004	0.8039	31.81

APPENDIX D

Chemical Elements

	Symbol	Atomic number	Atomic weight†		Symbol	Atomic number	Atomic weight†
Actinium	Ac	89	[227]	Europium	Eu	63	151.96
Aluminum	Al	13	26.9815	Fermium	Fm	100	[257]
Americium	Am	95	[243]	Fluorine	F	9	18.9984
Antimony	Sb	51	121.75	Francium	Fr	87	[223]
Argon	Ar	18	39.948	Gadolinium	Gd	64	157.25
Arsenic	As	33	74.9216	Gallium	Ga	31	69.72
Astatine	At	85	[210]	Germanium	Ge	32	72.61
Barium	Ba	56	137.33	Gold	Au	79	196.967
Berkelium	Bk	97	[247]	Hafnium	Hf	72	178.49
Beryllium	Be	4	9.0122	Helium	He	2	4.0026
Bismuth	Bi	83	208.980	Holmium	Ho	67	164.930
Boron	B	5	10.811	Hydrogen	H	1	1.0079
Bromine	Br	35	79.904	Indium	In	49	114.82
Cadmium	Cd	48	112.41	Iodine	1	53	126.9045
Calcium	Ca	20	40.078	Iridium	Ir	77	192.22
Californium	Cf	98	[251]	Iron	Fe	26	55.847
Carbon	C	6	12.011	Krypton	Kr	36	83.80
Cerium	Ce	58	140.12	Lanthanum	La	57	138.91
Cesium	Cs	55	132.905	Lawrencium	Lr	103	[260]
Chlorine	Cl	17	35.453	Lead	Ph	82	207.2
Chromium	Cr	24	51.996	Lithium	Li	3	6.941
Cobalt	Co	27	58.9332	Lutetium	Lu	71	174.97
Copper	Cu	29	63.546	Magnesium	Mg	12	24.305
Curium	Cm	96	[247]	Manganese	Mn	25	54.9380
Dysprosium	Dy	66	162.50	Mendelevium	Md	101	[258]
Einsteinium	Es	99	[252]	Mercury	Hg	80	200.59
Erbium	Er	68	167.26	Molybdenum	Mo	42	95.94

† Values in brackets denote the mass number of the longest lived or best-known isotope.

	Symbol	Atomic number	Atomic weight†
Neodymium	Nd	60	144.24
Neon	Ne	10	20.180
Neptunium	Np	93	[237]
Nickel	Ni	28	58.69
Niobium	Nb	41	92.906
Nitrogen	N	7	14.0067
Nobelium	No	102	[259]
Osmium	Os	76	190.2
Oxygen	O	8	15.9994
Palladium	Pd	46	106.4
Phosphorus	P	15	30.9738
Platinum	Pt	78	195.08
Plutonium	Pu	94	[244]
Polonium	Po	84	[209]
Potassium	K	i9	39.098
Praseodymium	Pr	59	140.9077
Promethium	Pm	61	[145]
Protactinium	Pa	91	[231]
Radium	Ra	88	[226]
Radon	Rn	86	[222]
Rhenium	Re	75	186.2
Rhodium	Rh	45	102.905
Rubidium	Rb	37	85.4678
Ruthenium	Ru	44	101.07
Samarium	Sm	62	150.4
Scandium	Sc	21	44.956
Selenium	Se	34	78.96
Silicon	Si	14	28.086
Silver	Ag	47	107.868
Sodium	Na	11	22.9898
Strontium	Sr	38	87.62
Sulfur	S	16	32.066
Tantalum	Ta	73	180.948
Technetium	Tc	43	[98]
Tellurium	Te	52	127.60
Terbium	Tb	65	158.925
Thallium	Tl	81	204.38
Thorium	Th	90	232.038
Thulium	Tm	69	168.934
Tin	Sn	50	118.71
Titanium	Ti	22	47.88
Tungsten	W	74	183.85
Unnilhexium	Unh	106	[263]
Unnilpentium	Unp	105	[262]
Unnilquadium	Unq	104	[261]
Unnilseptium	Uns	107	[262]
Uranium	U	92	238.03
Vanadium	V	23	50.94
Xenon	Xe	54	131.29
Ytterbium	Yb	70	173.04
Yttrium	Y	39	88.9059
Zinc	Zn	30	65.39
Zirconium	Zr	40	91.22

† Values in brackets denote the mass number of the longest lived or best-known isotope.

APPENDIX E

Numerical Methods Primer

Numerical methods involve solving mathematical problems with the computer. The following material summarizes some simple numerical methods that are useful in environmental modeling. Additional details on the following methods, plus additional techniques, can be found in Chapra and Canale (1988). The reader is advised to consult this and other references to understand nuances of the following techniques that are beyond the scope of this primer.

ROOTS OF EQUATIONS

The root of an algebraic or transcendental equation is the value of a variable that makes the equation equal to zero. That is, it is the value of x that yields

$$f(x) = 0 \tag{E.1}$$

Numerical methods for finding roots are based on making a guess (or guesses) at the root and having the computer systematically refine the estimate. There are two fundamental approaches for accomplishing this objective: bracketing and open methods.

Bracketing Method: Bisection

As the name implies, bracketing methods begin with two guesses that bracket the root. That is, they define an interval within which there is a single root. This can be done by determining whether a sign change occurs in the interval (a sign change means that the function passes through zero within the interval),

$$f(x_l)f(x_u) < 0 \tag{E.2}$$

where x_l and x_u = lower and upper guesses. If two such guesses can be developed, a root estimate can be simply computed as occurring at the middle of the interval,

$$x_r = \frac{x_l + x_u}{2} \tag{E.3}$$

and x_r is the root estimate.

Once the root estimate is obtained, a new bracket must be defined that encompasses the root. To do this, it is determined whether a sign change occurs between the lower guess and the root estimate. If this is true the new upper bound is defined as the root estimate. If it is not, the new lower bound is defined as the root estimate. Once the bracket is redefined, a new midpoint can be computed and the process repeated to obtain finer estimates of the root.

The process is iterated until an error estimate falls below some prescribed stopping criterion. One such way to estimate the error is

$$\epsilon_a = \left| \frac{x_r^{\text{new}} - x_r^{\text{old}}}{x_r^{\text{new}}} \right| (100\%) \tag{E.4}$$

where ϵ_a = approximate error and x_r^{new} and x_r^{old} = present and previous root estimates. Some pseudocode to implement this algorithm is displayed in Fig. E.1.

```
SUB Bisect(xl, xu, xr, iter, ea)

  This subroutine must have access to
  the function being evaluated, f(x)

  ASSIGN imax, es
  iter = 0
  DO
    xrold = xr
    xr = (xl+xu)/2
    iter = iter + 1
    IF xr≠0 THEN
      ea = abs((xr-xrold)/xr)*100
    END IF
    test = f(xl)*f(xr)
    IF test = 0 THEN
      ea = 0
    IF test<0 THEN
      xu = xr
    ELSE
      xl = xr
    END IF
  LOOP UNTIL ea<es OR iter>imax

END Bisect
```

FIGURE E.1
Bisection algorithm.

Open Methods: Newton-Raphson and Modified Secant

Open methods do not require guesses that bracket the root. As a consequence they often converge faster than bracketing methods. However, they can also diverge from the root. The speed and possibility of convergence usually depends on the quality of the initial guess. To a lesser extent it depends on the type of function being evaluated.

The most commonly employed and (when it works) the most efficient open method is the Newton-Raphson approach. It is based on using the function's derivative to define a path to the next root estimate. This process can be expressed by the following iterative formula,

$$x_{i+1} = x_i - \frac{f(x_i)}{f'(x_i)} \tag{E.5}$$

This formula is repeated until the percent relative error (computed with a formula like Eq. E.4) falls below a specified level. Some pseudocode to implement this algorithm is displayed in Fig. E.2.

One problem with the Newton-Raphson approach is that it requires that the derivative of the function be determined. An alternative that does not require a derivative evaluation is the modified secant method. This approach uses a finite difference to estimate the derivative,

$$f'(x_i) \cong \frac{f(x_i + \epsilon x_i) - f(x_i)}{\epsilon x_i} \tag{E.6}$$

where ϵ = a small perturbation fraction ($\cong 0.01$). Equation E.6 can be substituted into Eq. E.5 to give the modified secant formula,

$$x_{i+1} = x_i - \frac{\epsilon x_i f(x_i)}{f(x_i + \epsilon x_i) - f(x_i)} \tag{E.7}$$

The Newton-Raphson formula (Eq. E.5) can be replaced by Eq. E.7 to convert the algorithm in Fig. E.2 to the modified secant method.

```
SUB Newton(xr, iter, ea)

  This subroutine must have access to the
  function being evaluated, f(x), and its
  derivative, f'(x).

  ASSIGN imax, es
  iter = 0
  DO
    xrold = xr
    xr = xr - f(xr)/f'(xr)
    iter = iter+1
    IF xr≠0 THEN
      ea = abs((xr-xrold)/xr)*100
    END IF
  LOOP UNTIL ea<es OR iter>imax

END Newton
```

FIGURE E.2
Newton-Raphson method.

LINEAR ALGEBRAIC EQUATIONS

This class of problems involves solving n linear algebraic equations with n unknowns. An example of such a system is

$$\begin{aligned} a_{11}x_1 + a_{12}x_2 + a_{13}x_3 &= b_1 \\ a_{21}x_1 + a_{22}x_2 + a_{23}x_3 &= b_2 \\ a_{31}x_1 + a_{32}x_2 + a_{33}x_3 &= b_3 \end{aligned} \tag{E.8}$$

where the a_{ij}'s are constant coefficients in which i denotes the row (or equation) and j denotes the column (or unknown), the b's are constants, and the x's are the unknowns. There are two fundamental approaches for solving for the unknowns: elimination (or decomposition) methods and iterative methods.

Elimination Methods: Gauss Elimination and LU Decomposition

Gauss elimination is based on multiplying the equations by constant factors and then subtracting them. As the name implies, the object is to eliminate unknowns from the equations. After elimination, the last equation will have one unknown, the next-to-the-last two unknowns, etc. Back-substitution can then be used to compute values for the unknowns. The process is depicted in Fig. E.3. To avoid division by zero, a partial pivoting strategy can be included in the algorithm. This involves switching equations to avoid division by zero.

Although Gauss elimination certainly represents a sound way to solve simultaneous equations, it becomes inefficient when solving equations with the same coefficients (the a's), but with different right-hand-side constants (the b's). Because this is

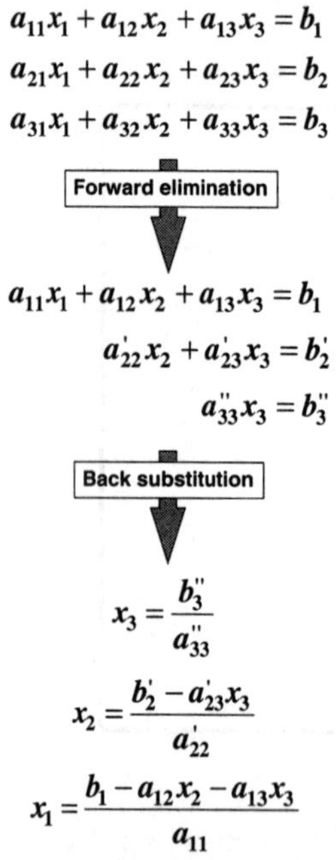

$$\begin{aligned} a_{11}x_1 + a_{12}x_2 + a_{13}x_3 &= b_1 \\ a_{21}x_1 + a_{22}x_2 + a_{23}x_3 &= b_2 \\ a_{31}x_1 + a_{32}x_2 + a_{33}x_3 &= b_3 \end{aligned}$$

Forward elimination

$$\begin{aligned} a_{11}x_1 + a_{12}x_2 + a_{13}x_3 &= b_1 \\ a'_{22}x_2 + a'_{23}x_3 &= b'_2 \\ a''_{33}x_3 &= b''_3 \end{aligned}$$

Back substitution

$$x_3 = \frac{b''_3}{a''_{33}}$$

$$x_2 = \frac{b'_2 - a'_{23}x_3}{a'_{22}}$$

$$x_1 = \frac{b_1 - a_{12}x_2 - a_{13}x_3}{a_{11}}$$

FIGURE E.3
Schematic representation of the solution strategy underlying Gauss elimination.

often the case in water-quality modeling, the Gauss-elimination algorithm can be alternatively expressed as what is called an ***LU decomposition*** format. In this expression, the constant factors used to eliminate unknowns are first multiplied by the coefficients on the left-side of the equations. They are then saved so that they can be used later to multiply the individual right-hand-side constants. This can be done by a forward substitution. Thereafter a back substitution can be employed to obtain solutions.

An algorithm to implement an LU decomposition expression of Gauss elimination is listed in Fig. E.4.

Three features of this algorithm bear mention:

- The factors used for the forward pass are stored in the lower part of the matrix. This can be done because these are converted to zeros anyway and are unnecessary for the final solution. This storage saves space.
- This algorithm keeps track of pivoting by using an order vector o. This greatly speeds up the algorithm because only the order vector (as opposed to the whole row) is pivoted.
- The algorithm checks for singularity and near-singularity. If it passes back a value of er $= -1$, a singular matrix has been detected and the computation should be terminated.

One important application of the algorithm in Fig. E.4 is to calculate the matrix inverse. This can be done by attaining solutions with unit vectors as the right-hand-side constants. For example if the right-hand-side constant has a 1 in the first position and zeros elsewhere,

$$\{b\} = \begin{Bmatrix} 1 \\ 0 \\ 0 \end{Bmatrix} \tag{E.9}$$

the resulting solution will be the first column of the matrix inverse. Similarly if a unit vector with a 1 at the second row is used, $\{b\}^T = \{0 \quad 1 \quad 0\}$, the result will be the second column of the matrix inverse. Pseudocode to generate the matrix inverse in this way is shown in Fig. E.5.

The final elimination method involves the solution of tridiagonal systems,

$$\begin{aligned}
f_1x_1 + g_1x_2 \qquad\qquad\qquad\qquad\qquad\qquad &= r_1 \\
e_2x_1 + f_2x_2 + g_2x_3 \qquad\qquad\qquad\qquad &= r_2 \\
e_3x_2 + f_3x_3 + g_3x_4 \qquad\qquad\qquad &= r_3 \\
\cdot \quad \cdot \quad \cdot \qquad\qquad\qquad &\quad \cdot \\
\cdot \quad \cdot \quad \cdot \qquad\qquad &\quad \cdot \\
\cdot \quad \cdot \quad \cdot \qquad &\quad \cdot \\
e_{n-1}x_{n-2} + f_{n-1}x_{n-1} + g_{n-1}x_n &= r_{n-1} \\
e_nx_{n-1} + f_nx_n &= r_n
\end{aligned} \tag{E.10}$$

Such equations apply to several areas of water-quality modeling, notably the solution

```
SUB Decomp (a(), o(), n, eps, er)
  sets up order vector
  DOFOR i = 1 TO n
    o(i) = i
  END DO
  performs decomposition
  DOFOR i = 1 TO n - 1
    CALL Pivot(a(), o(), n, i)
    DOFOR k = i + 1 TO n
      factor = a(o(k), i) / a(o(i), i)
      a(o(k), i) = factor
      DOFOR j = i + 1 TO n
        a(o(k), j) = a(o(k), j) - factor * a(o(i), j)
      END DO
    END DO
  END DO
  checks for singularity
  DOFOR i = 1 TO n
    IF ABS(a(o(i), i)) < eps THEN
      er = -1
      EXIT DOFOR
    END IF
  END DO
END Decomp

SUB Pivot (a(), o(), n, i)
  p = i
  big = ABS(a(o(i), i))
  DOFOR k = i + 1 TO n
    dummy = ABS(a(o(k), i))
    IF dummy > big THEN
      big = dummy
      p = k
    END IF
  END DO
  dummy = o(p)
  o(p) = o(i)
  o(i) = dummy
END Pivot

SUB Subst (a(), o(), n, b(), x(), eps)
  forward substitution
  DOFOR i = 1 TO n - 1
    DOFOR k = i + 1 TO n
      factor = a(o(k), i)
      b(o(k)) = b(o(k)) - factor * b(o(i))
    END DO
  END DO
  back substitution
  DOFOR i = n TO 1 STEP -1
    sum = 0
    DOFOR j = i + 1 TO n
      sum = sum + a(o(i), j) * x(j)
    END DO
    x(i) = (b(o(i)) - sum) / a(o(i), i)
  END DO
END Subst
```

FIGURE E.4
Gauss elimination/LU decomposition.

```
Driver program to generate matrix inverse, ai()
  ENTER a(), n
  eps = .000001
  CALL Ludcmp(a(), o(), n, eps, er)
  IF er = 0 THEN
    FOR i = 1 TO n
      FOR j = 1 TO n
        IF i = j THEN
          b(j) = 1
        ELSE
          b(j) = 0
        END IF
      NEXT j
      CALL Substitute(a(), o(), n, b(), x(), eps)
      FOR j = 1 TO n
        ai(j, i) = x(j)
      NEXT j
    NEXT i
    Output ai(), if desired
  ELSE
    PRINT "ill-conditioned system"
  END IF
END
```

FIGURE E.5
Matrix inverse.

of partial differential equations. Because so many elements are zero, the algorithms for solving such systems are extremely efficient. In addition such systems often do not require pivoting. An algorithm for solving tridiagonal systems, which is in the LU decomposition format, is shown in Fig. E.6.

```
SUB Decomp(e(), f(), g(), n)
  DO FOR i = 2 TO n
    e(i) = e(i)/f(i)
    f(i) = f(i) - e(i)*g(i-1)
  ENDDO
END Decomp

SUB Subst(e(), f(), g(), r(), x(), n)
  forward substitution
  DO FOR i = 2 TO n
    r(i) = r(i) - e(i)*r(i-1)
  ENDDO
  back substitution
  x(n) = r(n)/f(n)
  DO FOR i = n-1 TO 1 STEP -1
    x(i) = (r(i) - g(i)*x(i+1))/f(i)
  ENDDO
END Subst
```

FIGURE E.6
Tridiagonal solver.

Iterative Methods: Gauss-Seidel

The Gauss-Seidel method provides an approximate approach to solve linear simultaneous equations that is very well suited for many environmental modeling problems. It can be illustrated for the three simultaneous equations from Eq. E.8. To implement the technique, we merely solve the first equation for x_1, the second for x_2, and the third for x_3,

$$\begin{aligned} x_1 &= \frac{b_1 - a_{12}x_2 - a_{13}x_3}{a_{11}} \\ x_2 &= \frac{b_2 - a_{21}x_1 - a_{23}x_3}{a_{22}} \\ x_3 &= \frac{b_3 - a_{31}x_1 - a_{32}x_2}{a_{33}} \end{aligned} \tag{E.11}$$

These equations can be solved iteratively until the estimates converge on the solution. Convergence can be tested by a formula like Eq. E.4. A code for making these computations is listed in Fig. E.7.

The Gauss-Seidel method will be assured of converging only when the system being solved is diagonally dominant. This means that the absolute value of the diagonal element in each equation must be greater than the summation of the absolute values of the off-diagonal elements. Although this might appear to be a severe limitation, it just happens that many linear systems confronted in environmental modeling (particularly those related to partial differential equations) have this attribute. Thus the Gauss-Seidel approach has broad utility.

```
SUB Gseid(a(), b(), n, x())

  ASSIGN imax, es
  iter = 0
  DO
    flag = 0
    iter = iter + 1
    DOFOR i = 1 TO n
      old = x(i)
      sum = b(i)
      DOFOR j = 1 TO n
        IF i ≠ j THEN sum = sum - a(i,j)*x(j)
      END DO
      x(i) = sum/a(i,i)
      IF flag = 0 AND x(i) ≠ 0 THEN
        ea = ABS((x(i) - old)/x(i))*100
        IF (ea > es) THEN flag = 1
      END IF
    END DO
  LOOP UNTIL flag = 0 OR iter > imax

END Gseid
```

FIGURE E.7
Gauss-Seidel algorithm.

Further, it is very well suited for what are called sparse systems. These are sets of simultaneous equations that have mostly zero elements. For such cases the Gauss-Seidel approach requires only that the nonzero elements be stored and manipulated. Again, because many environmental modeling problems exhibit sparseness, this makes the technique useful.

CURVE FITTING

Fitting a curve to data points is widely employed in environmental engineering. Two different approaches are among the most commonly employed. For cases where the data are uncertain (that is, they exhibit variability), regression is used to generate a "best fit" line through the points. When the data are known precisely (e.g., as in a table of physical properties of water), interpolation is used to generate curves connecting the points. In this case the object is to estimate values between the points.

Linear Regression

Given a set of points, $(x_1, y_1), (x_2, y_2), \ldots, (x_n, y_n)$, estimates of the slope and intercept of a straight line that minimizes the sum of the squares of the residuals can be computed. The equation for the straight line is

$$y = a_0 + a_1 x \tag{E.12}$$

where the slope is computed by

$$a_1 = \frac{n \sum x_i y_i - \sum x_i \sum y_i}{n \sum x_i^2 - (\sum x_i)^2} \tag{E.13}$$

where all the summations are from $i = 1$ to n. The intercept is calculated with

$$a_0 = \overline{y} - a_1 \overline{x} \tag{E.14}$$

where $\overline{x}$ and $\overline{y}$ = means of the x's and y's, respectively.

Note that two statistics can be used to assess the goodness of fit. The first is the ***coefficient of determination*** r^2, which represents the fraction of the error that is explained by the regression ($r^2 = 1$ for perfect fit and $r^2 = 0$ for no fit),

$$r^2 = \frac{S_t - S_r}{S_t} \tag{E.15}$$

where S_t = total sum of the squares,

$$S_t = \sum_{i=1}^{n} (y_i - \overline{y})^2 \tag{E.16}$$

and S_r = residual sum of squares around the regression line,

$$S_r = \sum_{i=1}^{n} (y_i - a_0 - a_1 x_i)^2 \tag{E.17}$$

```
SUB Regres(x(), y(), n, a1, a0, syx, r2)

  sumx = 0: sumxy = 0: st = 0
  sumy = 0: sumx2 = 0: sr = 0
  DO FOR i = 1 TO n
    sumx = sumx + x(i)
    sumy = sumy + y(i)
    sumxy = sumy + x(i)*y(i)
    sumx2 = sumy + x(i)*x(i)
  ENDDO
  xm = sumx/n
  ym = sumy/n
  a1 = (n*sumxy*sumx*sumy)/(n*sumx2-sumx*sumx)
  a0 = ym - a1*xm
  DO FOR i = 1 TO n
    st = st +(y(i)-ym)^2
    sr = sr + (y(i) - a1*x(i) - a0)^2
  END DO
  syx = (sr/(n - 2))^0.5
  r2 = (st - sr)/st

END Regres
```

FIGURE E.8
Linear regression.

The second statistic is the ***standard error of the estimate*** $s_{y/x}$,

$$s_{y/x} = \sqrt{\frac{S_r}{n-2}} \tag{E.18}$$

A code to implement linear regression is listed in Fig. E.8.

Polynomial Interpolation

An nth-order polynomial can be fit to a given set of $n + 1$ points: (x_0, y_0), (x_1, y_1), ..., (x_n, y_n). This polynomial can then be used to calculate intermediate values within the range. The Lagrange interpolating polynomial provides a convenient computational formula for doing this,

$$f_n(x) = \sum_{i=0}^{n} L_i(x) f(x_i) \tag{E.19}$$

where

$$L_i(x) = \prod_{\substack{j=0 \\ j \neq i}}^{n} \frac{x - x_j}{x_i - x_j} \tag{E.20}$$

in which $\prod$ designates "the product of." A code to implement the algorithm and evaluate the function at a specific value of x is displayed in Fig. E.9.

COMPUTER CALCULUS

The simplest forms of computer calculus involve simple integration and differentiation of tabular data.

```
FUNCTION Lagrng(x(), y(), n, x)

  sum = 0
  DOFOR i = 0 TO n
    product = y(i)
    DOFOR j = 0 TO n
      IF i ≠ j THEN
        product = product*(x - x(j))/(x(i)-x(j))
      ENDIF
    ENDDO
    sum = sum + product
  ENDDO
  Lagrng = sum

END Lagrng
```

FIGURE E.9
Lagrange interpolation.

Integration

Integration involves solving the general problem

$$I = \int_a^b f(x)\,dx \tag{E.21}$$

As depicted graphically in Fig. E.10*a,* the problem can be visualized as determining the area under the curve defined by the function.

A simple numerical approach consists of approximating the area as a series of trapezoids (Fig. E.10*b*). Thus the sum of the trapezoids represents an estimate of the area,

$$I \cong (x_1 - x_0)\frac{f(x_0) + f(x_1)}{2} + (x_2 - x_1)\frac{f(x_1) + f(x_2)}{2} + \cdots + (x_n - x_{n-1})\frac{f(x_{n-1}) + f(x_n)}{2} \tag{E.22}$$

A code for making this evaluation is listed in Fig. E.11.

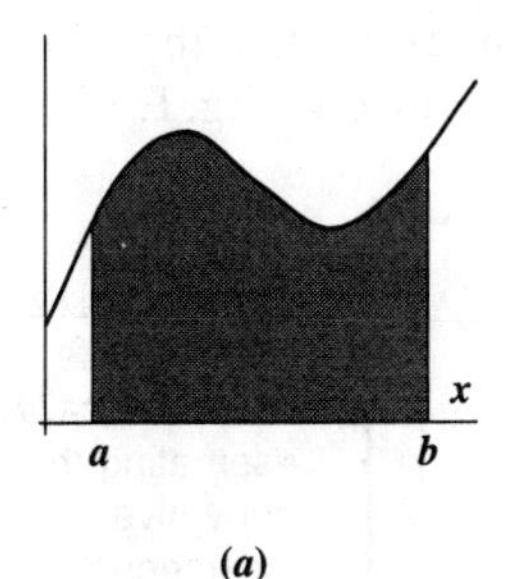

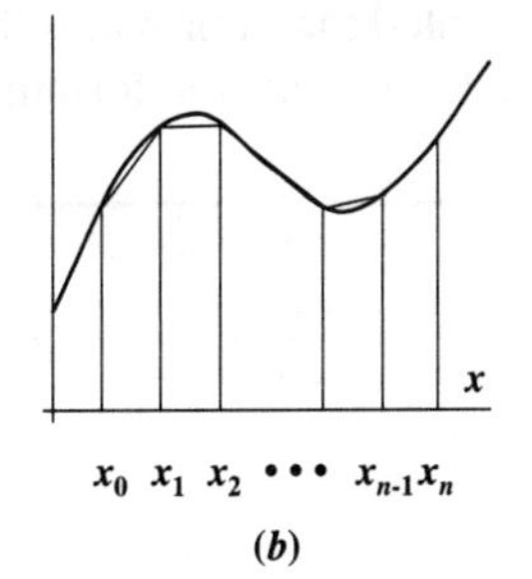

FIGURE E.10
(*a*) The integral of a function can be represented as the area under the function.
(*b*) The trapezoidal rule approximates the integral by summing trapezoids that approximate the area.

```
FUNCTION Area (x(), y(), n)

  area = 0
  DOFOR i = 1 TO n
    area = area + (x(i)-x(i-1))*(y(i-1)+y(i))/2
  ENDDO

END Area
```

FIGURE E.11 Pseudocode for the multiple-segment trapezoidal rule.

Two other widely used for formulas (for equally spaced points) are ***Simpson's 1/3 rule,***

$$I \cong (x_2 - x_0)\frac{f(x_0) + 4f(x_1) + f(x_2)}{6} \tag{E.23}$$

and ***Simpson's 3/8 rule,***

$$I \cong (x_3 - x_0)\frac{f(x_0) + 3f(x_1) + 3f(x_2) + f(x_3)}{8} \tag{E.24}$$

Differentiation

Numerical differentiation involves determining a derivative (or slope) estimate from discrete data points. For example for equally spaced data, a simple approach is a centered difference,

$$f'(x_1) \cong \frac{f(x_2) - f(x_0)}{x_2 - x_0} \tag{E.25}$$

Although this approach has utility, it is limited because (*a*) it is limited to equally spaced data and (*b*) it yields the derivative estimate only at a discrete point x_1. A more powerful method involves fitting an interpolating polynomial to the points and differentiating the resulting formula. The result for the case of a quadratic applied to three points is

$$f'(x) \cong f(x_{i-1})\frac{2x - x_i - x_{i+1}}{(x_{i-1} - x_i)(x_{i-1} - x_{i+1})} + f(x_i)\frac{2x - x_{i-1} - x_{i+1}}{(x_i - x_{i-1})(x_i - x_{i+1})}$$
$$+ f(x_{i+1})\frac{2x - x_{i-1} - x_i}{(x_{i+1} - x_{i-1})(x_{i+1} - x_i)} \tag{E.26}$$

Note that in contrast to Eq. E.25, this formula can be used for unequally spaced points and provides a derivative estimate anywhere within the interval containing the points. However, it should be noted that it provides the best results near the middle of the interval. A simple code to evaluate the formula is provided in Fig. E.12.

```
FUNCTION Deriv(x0, y0, x1, y1, x2, y2, x)

  t1 = y0 * (2*x-x1-x2)/(x0-x1)/(x0-x2)
  t2 = y1 * (2*x-x0-x2)/(x1-x0)/(x1-x2)
  t3 = y2 * (2*x-x0-x1)/(x2-x0)/(x2-x1)
  Deriv = t1 + t2 + t3

END Deriv
```

FIGURE E.12 Pseudocode for numerically estimating the derivative for unequally spaced data.

APPENDIX F

Bessel Functions

The following pseudocodes (abstracted from programs developed by Press et al. 1992) define program functions to evaluate the modified Bessel functions with a computer program. In addition I have included tables of values that are useful for solving simple problems.

```
FUNCTION bessi0 (x)

DOUBLE p1, p2, p3, p4, p5, p6, p7
DOUBLE q1, q2, q3, q4, q5, q6, q7
DOUBLE q8, q9
DOUBLE y, d

p1 = 1.; p2 = 3.5156229
p3 = 3.0899424; p4 = 1.2067492
p5 = .2659732; p6 = .0360768
p7 = .0045813
q1 = .39894228; q2 = .01328592
q3 = .00225319; q4 = -.00157565
q5 = .00916281; q6 = -.02057706
q7 = .02635537; q8 = -.01647633
q9 = .00392377

IF ABS(x) < 3.75 THEN
  y = (x / 3.75) ^ 2
  d = p6 + y * p7
  d = p5 + y * d
  d = p4 + y * d
  d = p3 + y * d
  d = p2 + y * d
  d = p1 + y * d
  bessi0 = d
ELSE
  ax = ABS(x)
  y = 3.75 / ax
  d = q8 + y * q9
  d = q7 + y * d
  d = q6 + y * d
  d = q5 + y * d
  d = q4 + y * d
  d = q3 + y * d
  d = q2 + y * d
  d = q1 + y * d
  bessi0 = EXP(ax) / SQRT(ax) * d
END IF

END FUNCTION
```

```
FUNCTION bessk0 (x)

DOUBLE p1, p2, p3, p4, p5, p6, p7
DOUBLE q1, q2, q3, q4, q5, q6, q7
DOUBLE y, d

 p1 = -.57721566; p2 = .4227842
 p3 = .23069756; p4 = .0348859
 p5 = .00262698; p6 = .0001075
 p7 = .0000074
 q1 = 1.25331414; q2 = -.07832358
 q3 = .02189568; q4 = -.01062446;
 q5 = .00587872; q6 = -.0025154
 q7 = .00053208

  IF x ≤  2 THEN
    y = x * x / 4
    d = p6 + y * p7
    d = p5 + y * d
    d = p4 + y * d
    d = p3 + y * d
    d = p2 + y * d
    d = p1 + y * d
    bessk0 = (-LN(x/2)*bessi0(x)) + d

  ELSE
    y = 2 / x
    d = q6 + y * q7
    d = q5 + y * d
    d = q4 + y * d
    d = q3 + y * d
    d = q2 + y * d
    d = q1 + y * d
    bessk0 = EXP(-x) / SQRT(x) * d
  END IF

  END FUNCTION
```

FIGURE F.1
Pseudocodes to evaluate modified Bessel functions (adapted from Press et al. 1992).

TABLE F.1
Modified Bessel functions of the first kind, $I_0(x)$

x	0	0.1	0.2	0.3	0.4	0.5	0.6	0.7	0.8	0.9
0	1.000000	1.002502	1.010025	1.022627	1.040402	1.063483	1.092045	1.126303	1.166515	1.212985
1	1.266066	1.326160	1.393726	1.469278	1.553395	1.646723	1.749981	1.863965	1.989560	2.127741
2	2.279586	2.446283	2.629143	2.829606	3.049256	3.289839	3.553268	3.841650	4.157296	4.502747
3	4.880790	5.294488	5.747203	6.242625	6.784807	7.378196	8.027676	8.738607	9.516876	10.36894
4	11.30191	12.32355	13.44244	14.66795	16.01041	17.48114	19.09259	20.85842	22.79363	24.91473
5	27.23981	29.78879	32.58352	35.64802	39.00869	42.69453	46.73743	51.17239	56.03794	61.37637
6	67.23420	73.66256	80.71765	88.46126	96.96131	106.2925	116.5369	127.7849	140.1356	153.6984
7	168.5932	184.9522	202.9205	222.6579	244.3400	268.1601	294.3308	323.0860	354.6829	389.4044
8	427.5620	469.4983	515.5903	566.2526	621.9417	683.1595	750.4587	824.4476	905.7950	995.2379
9	1093.586	1201.733	1320.659	1451.446	1595.284	1753.481	1927.479	2118.866	2329.388	2560.967

TABLE F.2
Modified Bessel functions of the second kind, $K_0(x)$

x	0	0.1	0.2	0.3	0.4	0.5	0.6	0.7	0.8	0.9
0	∞	2.427069	1.752704	1.372460	1.114529	0.924419	0.777522	0.660520	0.565347	0.486730
1	0.421024	0.365602	0.318508	0.278248	0.243655	0.213806	0.187955	0.165496	0.145931	0.128846
2	0.113894	0.100784	0.089269	0.079140	0.070217	0.062348	0.055398	0.049255	0.043820	0.039006
3	0.034740	0.030955	0.027595	0.024611	0.021958	0.019599	0.017500	0.015631	0.013966	0.012482
4	0.011160	0.009980	0.008927	0.007988	0.007149	0.006400	0.005730	0.005132	0.004597	0.004119
5	0.003691	0.003308	0.002966	0.002659	0.002385	0.002139	0.001919	0.001721	0.001544	0.001386
6	0.001244	0.001117	0.001003	0.000900	0.000808	0.000726	0.000652	0.000586	0.000526	0.000473
7	0.000425	0.000382	0.000343	0.000308	0.000277	0.000249	0.000224	0.000201	0.000181	0.000163
8	0.000146	0.000132	0.000118	0.000107	0.000096	0.000086	0.000078	0.000070	0.000063	0.000057
9	0.000051	0.000046	0.000041	0.000037	0.000033	0.000030	0.000027	0.000024	0.000022	0.000020

APPENDIX G

Error Function and Complement

b	erf(b)	erfc(b)	b	erf(b)	erfc(b)
0.00	0.000000	1.000000	1.00	0.842701	0.157299
0.05	0.056372	0.943628	1.05	0.862436	0.137564
0.10	0.112463	0.887537	1.10	0.880205	0.119795
0.15	0.167996	0.832004	1.15	0.896124	0.103876
0.20	0.222703	0.777297	1.20	0.910314	0.089686
0.25	0.276326	0.723674	1.25	0.922900	0.077100
0.30	0.328627	0.671373	1.30	0.934008	0.065992
0.35	0.379382	0.620618	1.35	0.943762	0.056238
0.40	0.428392	0.571608	1.40	0.952285	0.047715
0.45	0.475482	0.524518	1.45	0.959695	0.040305
0.50	0.520500	0.479500	1.50	0.966105	0.033895
0.55	0.563323	0.436677	1.55	0.971623	0.028377
0.60	0.603856	0.396144	1.60	0.976348	0.023652
0.65	0.642029	0.357971	1.65	0.980376	0.019624
0.70	0.677801	0.322199	1.70	0.983790	0.016210
0.75	0.711155	0.288845	1.75	0.986672	0.013328
0.80	0.742101	0.257899	1.80	0.989091	0.010909
0.85	0.770668	0.229332	1.85	0.991111	0.008889
0.90	0.796908	0.203092	1.90	0.992790	0.007210
0.95	0.820891	0.179109	1.95	0.994179	0.005821

References

Ahrens, C. D. 1988. *Meteorology Today.* West Pub. Co., St. Paul, MN.

Alberts, J. J. and Wahlgren, M. A. 1981. "Concentrations of $^{239,240}Pu$, ^{137}Cs, and ^{90}Sr in the Waters of the Laurentian Great Lakes. Comparison of 1973 and 1976 Values." *Environ. Sci. Tech.* 15:94–98.

Alexander, M. 1979. "Biodegradation of Toxic Chemicals in Water and Soil." In *Dynamics, Exposure and Hazard Assessment of Toxic Chemicals,* R. Haque, ed., Ann Arbor Science, Ann Arbor, MI.

Allen, D. L. 1973. *Wolves of Minong: Their Vital Role in a Wild Community.* Houghton-Mifflin, Boston.

Allen, H. E., Fu, G., and Deng, B. 1993. "Analysis of Acid Volatile Sulfide (AVS) and Simultaneously Extracted Metals (SEN) for the Estimation of Potential Toxicity in Aquatic Sediments." *Environ. Toxicol. and Chem.* 12(8):1441–1453.

American Heritage Dictionary. 1987. Houghton-Mifflin, Boston.

APHA (American Public Health Association). 1992. *Standard Methods for the Examination of Water and Wastewater,* 18 ed., Washington, DC.

Auer, M. T. 1979. "The Dynamics of Fixed Inorganic Nitrogen Nutrition in Two Species of Chlorophycean Algae." Ph.D. Dissertation, University of Michigan, Ann Arbor, MI.

Auer, M. T. 1982. Ecology of Filamentous Algae, special issue. *J. Great Lakes Res.* 8(1).

Baity, H. G. 1938. "Some Factors Affecting the Aerobic Decomposition of Sewage Sludge Deposits." *Sewage Works J.* 10(2):539–568.

Banks, R. B. 1975. "Some Features of Wind Action on Shallow Lakes." *J. Environ. Engr. Div. ASCE* 101(EE5): 813–827.

Banks, R. B. and Herrera, F. F. 1977. "Effect of Wind and Rain on Surface Reaeration." *J. Environ. Engr. Div. ASCE* 103(EE3): 489–504.

Bannister, T. T. 1974. "Prediction Equations in Terms of Chlorophyll Concentration, Quantum Yield, and Upper Limit to Production." *Limnol. Oceanogr.* 19(1):1–12.

Barnes, R. O. and Goldberg, E. D. 1976. "Methane Production and Consumption in Anoxic Marine Sediments." *Geology* 4:297–300.

Bartsch, A. F. and Gakstatter, J. H. 1978. Management Decisions for Lake Systems on a Survey of Trophic Status, Limiting Nutrients, and Nutrient Loadings in American-Soviet Symposium on Use of Mathematical Models to Optimize Water Quality Management, 1975, U.S. Environmental Protection Agency Office of Research and Development, Environmental Research Laboratory, Gulf Breeze, FL, pp. 372–394. EPA-600/9-78-024.

Bartsch, A. F. and Ingram, W. F. 1967. Biology of Water Pollution, U.S. Dept. of Interior, Water Pollution Control Administration.

Bates, S. S. 1976. "Effects of Light and Ammonium on Nitrate Uptake by Two Species of Estuarine Phytoplankton." *Limnol. Oceanog.* 21:212–218.

Beeton, A. M. 1958. "Relationship Between Secchi Disk Readings and Light Penetration in Lake Huron." *American Fisheries Society Trans.* 87:73–79.

Berner, R. A. 1980. *Early Diagenesis.* Princeton University Press, Princeton, NJ.

Bierman, V. J., Jr. 1976. "Mathematical Model of the Selective Enrichment of Blue-Green Algae by Nutrient Enrichment." In *Modeling of Biochemical Processes in Aquatic Ecosystems,* R. P. Canale, ed., Ann Arbor Science, Ann Arbor, MI, p. 1.

Bowie, G. L., Mills, W. B., Porcella, D. B., Campbell, C. L., Pagenkopf, J. R., Rupp, G. L., Johnson, K. M., Chan, P. W. H., Gherini, S. A. and Chamberlin, C. E. 1985. Rates, Constants, and Kinetic Formulations in Surface Water Quality Modeling. U.S. Environmental Protection Agency, ORD, Athens, GA, ERL, EPA/600/3-85/040.

Boyce, F. M. 1974. "Some Aspects of Great Lakes Physics of Importance to Biological and Chemical Processes." *J. Fish. Res. Bd. Can.* 31:689–730.

Boyce, F. M. and Hamblin, P. F. 1975. "A Simple Diffusion Model of the Mean Field Distribution of Soluble Materials in the Great Lakes." *Limnol. Oceanogr.* 20(4):511–517.

Bradbury, J. and Waddington, J. C. B. 1973. "Stratigraphic Record of Pollution in Shagawa Lake, Northeastern Minnesota." In *Symposium on Quartenary Plant Ecology,* H. J. B. Birks and R. G. West, eds., Blackwell, London, pp. 289–307.

Brady, D. K., Graves, W. L., and Geyer, J. C. 1969. Surface Heat Exchange at Power Plant Cooling Lakes, Cooling Water Discharge Project Report, No. 5, Edison Electric Inst. Publication No. 69-901, New York.

Bras, R. L. 1990. *Hydrology.* Addison-Wesley, Reading, MA.

Broecker, H. C., Petermann, J., and Siems, W. 1978. The Influence of Wind on CO_2 Exchange in a Wind-Wave Tunnel, *J. Marine. Res.* 36(4):595–610.

Brown, L. C. and Barnwell, T. O., Jr. 1987. The Enhanced Stream Water Quality Models QUAL2E and QUAL2E-UNCAS: Documentation and Users Manual, U.S. Environmental Protection Agency, Athens, GA, Report EPA/600/3-87/007.

Burns, L. A., Cline, D. M., and Lassiter, R. R. 1981. Exposure Analysis Modeling System (EXAMS): User Manual and System Documentation. Environmental Research Laboratory, Environmental Protection Agency, Athens, GA.

Burns, N. M. and Rosa, F. 1980. "In Situ Measurement of the Settling Velocity of Organic Carbon Particles and 10 Species of Phytoplankton." *Limnol. Oceanogr.* 25:855–864.

Butts, T. A. and Evans, R. L. 1983. Effects of Channel Dams on Dissolved Oxygen Concentrations in Northeastern Illinois Streams, Circular 132, State of Illinois, Dept. of Reg. and Educ., Illinois Water Survey, Urbana, IL.

Callahan, M. A., et al. 1979. *Water-Related Fate of 129 Priority Pollutants,* U.S. Environmental Protection Agency, Washington, DC, Report EPA/440/4-79-029b, 2 vols.

Canale, R. P. and Auer, M. T. 1982. "Ecological Studies and Mathematical Modeling of *Cladophora* in Lake Huron: 7. Model Verification and System Response." *J. Great Lakes Res.* 8(1):134–143.

Canale, R. P. and Effler, S. W. 1996. A Model for the Impact of Zebra Mussels on River Water Quality (draft manuscript).

Canale, R. P. and Vogel, A. H. 1974. "Effects of Temperature on Phytoplankton Growth." *J. Environ. Engr. Div. ASCE* 100(EE1):231–241.

Canale, R. P., DePalma, L. M., and Vogel, A. H. 1976. "A Plankton-Based Food Web Model for Lake Michigan." In *Modeling Biochemical Processes in Aquatic Ecosystems,* R. P. Canale, ed., Ann Arbor Science, Ann Arbor, MI, p. 33.

Canale, R. P., Hinemann, D. F., and Nachippan, S. 1974. "A Biological Production Model for Grand Traverse Bay." University of Michigan Sea Grant Program, Technical Report No. 37 (Ann Arbor, MI: University of Michigan Sea Grant Program).

Canale, R. P., Owens, E. M., Auer, M. T., and Effler, S. W. 1995. "The Validation of a Water Quality Model for the Seneca River, New York." *J. Water Resour. Plan. Management* 121(3):241–250.

Carnahan, B., Luther, H. A., and Wilkes, J. O. 1969. *Applied Numerical Methods.* Wiley, New York.

Carslaw, H. S. and Jaeger, J. C. 1959. *Conduction of Heat in Solids.* Oxford University Press, London.

CEQ. 1978. *Environmental Quality, The Ninth Annual Report of the Council on Environmental Quality.* U.S. Government Printing Office, Washington, DC.

Cerco, C. F. and Cole, T. 1993. "Three-Dimensional Eutrophication Model of Chesapeake Bay." *J. Environ. Engr. ASCE* 119(6):1006–1025.

Cerco, C. F. and Cole, T. 1994. Three-Dimensional Eutrophication Model of Chesapeake Bay. Vol. 1: Main Report. U.S. Army Corps of Engineers, Waterways Experiment Station, Tech. Report EL-94-4.

Chapra, S. C. 1975. "Comment on 'An Empirical Method of Estimating the Retention of Phosphorus in Lakes,' by W. B. Kirchner and P. J. Dillon." *Water Resour. Res.* 11: 1033–1034.

Chapra, S. C. 1977. "Total Phosphorus Model for the Great Lakes." *J. Environ. Engr. Div. ASCE* 103(EE2):147–161.

Chapra, S. C. 1979. "Applying Phosphorus Loading Models to Embayments." *Limnol. Oceanogr.* 24(1):163–168.

Chapra, S. C. 1980. "Application of the Phosphorus Loading Concept to the Great Lakes." In *Phosphorus Management Strategies for Lakes,* R. C. Loehr et al., eds., Ann Arbor Science, Ann Arbor, MI.

Chapra, S. C. 1982. "Long-Term Models of Interactions Between Solids and Contaminants in Lakes." Ph.D. Dissertation, The University of Michigan, Ann Arbor, MI.

Chapra, S. C. 1991. "A Toxicant Loading Concept for Organic Contaminants in Lakes." *J. Environ. Engr.* 117(5):656–677.

Chapra, S. C. and Canale, R. P. 1988. *Numerical Methods for Engineers.* 2d ed., McGraw-Hill, New York.

Chapra, S. C. and Canale, R. P. 1991. "Long-Term Phenomenological Model of Phosphorus and Oxygen in Stratified Lakes." *Water Research* 25(6):707–715.

Chapra, S. C. and Di Toro, D. M. 1991. "The Delta Method for Estimating Community Production, Respiration and Reaeration in Streams." *J. Environ. Engr.* 117(5):640–655.

Chapra, S. C. and Reckhow, K. H. 1983. *Engineering Approaches for Lake Management, Vol. 2: Mechanistic Modeling.* Butterworth, Woburn, MA.

Chapra, S. C. and Sonzogni, W. C. 1979. "Great Lakes Total Phosphorus Budget for the Mid 1970s." *J. Water Poll. Control Fed.* 51:2524–2533.

Chapra, S. C. and Tarapchak, S. J., 1976. "A Chlorophyll *a* Model and Its Relationship to Phosphorus Loading Plots for Lakes." *Water Resour. Res.* 12:1260–1264.

Chaudry, M. H. 1993. *Open-Channel Flow.* Prentice-Hall, Englewood Cliffs, NJ.

Chen, C. W. 1970. "Concepts and Utilities of Ecological Models." *J. San. Engr. Div. ASCE* 96(SA5): 1085–1086.

Chen, C. W. and Orlob, G. T. 1975. "Ecological Simulation for Aquatic Environments." In *Systems Analysis and Simulation in Ecology, Vol. III,* B. C. Patton. ed., Academic, New York, p. 475.

Chiou, C. T., et al. 1977. "Partition Coefficient and Bioaccumulation of Selected Organic Chemicals." *Env. Sci. Tech.* 11:475–478.

Chow, V. T. 1959. *Open-Channel Hydraulics.* McGraw-Hill, New York.

Chow, V. T., Maidment, D. R., and Mays, L. W. 1988. *Applied Hydrology.* McGraw-Hill, New York.

Churchill, M. A., Elmore, H. L., and Buckingham, R. A. 1962. "Prediction of Stream Reaeration Rates." *J. San. Engr. Div. ASCE* SA4:1, Proc. Paper 3199.

Cole, T. M. and Buchak, E. M. 1995. CE-QUAL-W2: A Two-Dimensional, Laterally Averaged, Hydrodynamic and Water Quality Model, Version 2. U.S. Army Corps of Engineers, Waterways Experiment Station, Tech. Report EL-95-x.

Connolly, J. P. 1991. "Application of a Food Chain Model to Polychlorinated Biphenyl Contamination of the Lobster and Winter Flounder Food Chains in New Bedford Harbor." *Environ. Sci. Technol.* 25(4):760–770.

Connolly, J. P. and Pedersen, C. J. 1988. "A Thermodynamic Based Evaluation of Organic Chemical Accumulation in Aquatic Organisms." *Environ. Sci. Technol.* 22(1):99–103.

Covar, A. P. 1976. "Selecting the Proper Reaeration Coefficient for Use in Water Quality Models." Presented at the U.S. EPA Conference on Environmental Simulation and Modeling, April 19–22, Cincinnati, OH.

Dailey, J. E. and Harleman, D. R. F. 1972. Numerical Model for the Prediction of Transient Water Quality in Estuary Networks. R.M. Parsons Laboratory, Tech. Report No. 158. MIT, Cambridge, MA.

Daily, J. W. and Harleman, D. R. F. 1966. *Fluid Dynamics.* Addison-Wesley, Reading, MA.

Dalton, J. 1802. "Experimental Essays on the Constitution of Mixed Gases; on the Force of Steam or Vapor from Water and Other Liquids, Both in a Torricellium Vacuum and in Air; on Evaporation; and on the Expansion of Gases by Heat." *Manchester Literary and Philosophical Society Proceedings* 5:536–602.

Danckwerts, P. V. 1951. "Significance of Liquid-Film Coefficients in Gas Absorption." *Ind. Eng. Chem.* 43(6):1460–1467.

Danckwerts, P. V. 1953. "Continuous Flow Systems. Distribution of Residence Times." *Chem. Engr. Sci.* 2(1):1–13.

Deininger, R. A. 1965. "Water Quality Management—The Planning of Economically Optimal Control Systems." Proc. of the First Annual Meeting of the American Water Resources Assoc.

Delos, C. G., et al. 1984. Technical Guidance Manual for Performing Wasteload Allocations, Book II. Streams and Rivers, chapter 3, Toxic Substances. USEPA, Washington, D.C, EPA-440/4-84-022.

deRegnier, D. P., et. al. 1989. "Viability of *Giardia* Cysts in Lake, River, and Tap Water." *Appl. Environ. Microbiol.* 55(5):1223–1229.

Dickinson, W. T. 1967. Accuracy of Discharge Determinations. Hydrology Paper 20, Colorado State University, Fort Collins, CO.

Dilks, D. W., Bierman, V. J., Jr., and Helfand, J. S. 1995. "Development and Application of Models to Determine Sediment Quality Criteria-Driven Permit Levels for Metals." In *Toxic Substances in Water Environments: Assessment and Control. WEF Specialty Conf. Series Proceedings,* Cincinnati, OH, Water Environ. Fed., Alexandria, VA.

Dillon, P. J. and Rigler, F. H. 1974. "The Phosphorus-Chlorophyll Relationship for Lakes." *Limnol. Oceanogr.* 19:767–773.

Dillon, P. J. and Rigler, F. H. 1975. "A Simple Method for Predicting the Capacity of a Lake for Development Based on Lake Trophic Status." *J. Fish. Res. Bd. Can.* 31(9):1519–1531.

Di Toro, D. M. 1972*a*. "Recurrence Relations for First-Order Sequential Reactions in Natural Waters." *Water Resour. Res.* 8(1):50–57.

Di Toro, D. M. 1972*b*. "Line Source Distribution in Two Dimensions: Applications to Water Quality." *Water Resour. Res.* 8(6):1541–1546.

Di Toro, D. M. 1976. "Combining Chemical Equilibrium and Phytoplankton Models." In *Modeling Biochemical Processes in Aquatic Ecosystems,* R. P. Canale, ed., Ann Arbor Science, Ann Arbor, MI, pp. 233–255.

Di Toro, D. M. 1978. "Optics of Turbid Estuarine Waters: Approximations and Applications." *Water Res.* 12:1059–1068.

Di Toro, D. M. and Fitzpatrick, J. J. 1993. Chesapeake Bay Sediment Flux Model. U.S. Army Corps of Engineers, Waterways Experiment Station, Tech. Report EL-93-2.

Di Toro, D. M., Mahony, J. D., Hansen, D. J., Scott, K. J., Carlson, A. R., and Ankley, G. T. 1992. "Acid Volatile Sulfide Predicts the Acute Toxicity of Cadmium and Nickel in Sediments." *Environ. Sci. Technol.* 26(1):96–101.

Di Toro, D. M., O'Connor, D. J., Thomann, R. V., and St. John, J. P. 1981. Analysis of Fate of Chemicals in Receiving Waters. Phase 1. Chemical Manufact. Assoc., Washington, D.C. Prepared by HydroQual Inc., Mahwah, NJ.

Di Toro, D. M., Paquin, P. R., Subburamu, K., and Gruber, D. A. 1990. "Sediment Oxygen Demand Model: Methane and Ammonia Oxidation." *J. Environ. Engr. Div. ASCE* 116(5):945–986.

Di Toro, D. M., Thomann, R. V., and O'Connor, D. J. 1971. "A Dynamic Model of Phytoplankton Population in the Sacramento–San Joaquin Delta." In *Advances in Chemistry Series 106: Nonequilibrium Systems in Natural Water Chemistry,* R. F. Gould, ed., American Chemical Society, Washington, DC, p.131.

Droop, M. R. 1974. "The Nutrient Status of Algal Cells in Continuous Culture." *J. Mar. Biol. Assoc.* (U.K.) 54:825–855.

Eagleson, P. S. 1970. *Dynamic Hydrology.* McGraw-Hill, New York.

Edinger, J. E., Brady, D. K., and Geyer, J. C. 1974. Heat Exchange and Transport in the Environment. Report No. 14, EPRI Pub. No. EA-74-049-00-3, Electric Power Research Institute, Palo Alto, CA.

Edmond, J. M. and Gieskes, J. A. T. M. 1970. "On the Calculation of the Degree of Saturation of Sea Water with Respect to Calcium Carbonate Under *In Situ* Conditions." *Geochim. Cosmochim. Acta* 34:1261–1291.

Effler, S. W. and Siegfried, C. 1994. "Zebra Mussel Populations in the Seneca River: Impact on Oxygen Resources." *Environ. Sci. Technol.* 28(12):2216–2221.

Emerson, K., et al. 1975. "Aqueous Ammonia Equilibrium Calculations: Effect of pH and Temperature." *J. Fish. Res. Bd. Can.* 32:2379.

Emerson, S. 1975. "Gas Exchange in Small Canadian Shield Lakes." *Limnol. Oceonogr.* 20:754.

Engelhardt, W. V. 1977. *The Origin of Sediments and Sedimentary Rocks.* Halstead, New York.

EPA. 1989. National Primary Drinking Water Regulations; Filtration and Disinfection; Turbidity; *Giardia lamblia,* Viruses, *Legionella,* and Heterotrophic Bacteria. *Fed. Regist.* 54:27486–27541.

Eppley, R. W. 1972. "Temperature and Phytoplankton Growth in the Sea." *Fishery Bulletin* 70(4):1063–1085.

Erdmann, J. B. 1979*a*. "Systematic Diurnal Curve Analysis." *J. Water Pollut. Control Fed.* 15(1), 78–86.

Erdmann, J. B. 1979*b*. "Simplified Diurnal Curve Analysis." *J. Environ. Engr. Div. ASCE* 105(6), 1063–1074.

Fair, G. M., Moore, E. W., and Thomas, H. A. 1941. "The Natural Purification of River Muds and Pollutional Sediments." *Sewage Works J.* 13(2):270–307; 13(4):756–779; 13(6):1209–1228.

Fillos, J. and Molof, A. H. 1972. "Effect of Benthal Deposits on Oxygen and Nutrient Economy of Flowing Waters." *J. Water Poll. Control Fed.* 44(4):644–662.

Fischer, H. B. 1968. "Dispersion Predictions in Natural Streams." *J. San. Engr. Div. ASCE* 94(SA5): 927–944.

Fischer, H. B., List, E. J., Koh, R. C. Y., Imberger, J., and Brooks, N. H. 1979. *Mixing in Inland and Coastal Waters.* Academic, New York.

Fogler, H. S. 1986. *Elements of Chemical Reaction Engineering.* Prentice-Hall, Englewood Cliffs, NJ.

Ford, D. E. and Johnson, L. S. 1986. "An Assessment of Reservoir Mixing Processes." U.S. Army Corps of Engineers, Waterways Experiment Station, Tech. Report. EL-86-7.

FWPCA. 1966. Delaware Estuary Comprehensive Study, Preliminary Report and Findings. Federal Water Pollution Control Federation, Dept. of the Interior, Philiadelphia, PA.

Gameson, A. L. H. and Gould, D. J. 1974. Effects of Solar Radiation on the Mortality of Some Terrestial Bacteria in Sea Water. *Proc. Int. Symp. on Discharge of Sewage from Sea Outfalls, London,* Pergamon, Great Britain, Paper No. 22.

Gardiner, R. D., Auer, M. T., and Canale, R. P. 1984. "Sediment Oxygen Demand in Green Bay (Lake Michigan)." Proc. 1984 Specialty Conf., ASCE, 514–519.

Gassmann, L. and Schwartzbrod, J. 1991. "Wastewater and *Giardia* Cysts." *Wat. Sci. Tech.* 24(2): 183–186.

Gaudy, A. F. and Gaudy, E. T. 1980. *Microbiology for Environmental Scientists and Engineers.* McGraw-Hill, New York.

Gelda, R. K., Auer, M. T., Effler, S. W., Chapra, S. C., and Storey, M. L. 1996. "Determination of Reaeration Coefficients: A Whole Lake Approach." *J. Environ. Engr.* (in press)

Goldman, J. C., Oswald, W. J., and Jenkins, D. 1974. "The Kinetics of Inorganic Carbon Limited Algal Growth." *J. Water Poll. Control Fed.* 46:554–573.

Goldman, J. C., Tenore, D. R., and Stanley, H. I. 1973. "Inorganic Nitrogen Removal from Wastewater: Effect on Phytoplankton Growth in Coastal Marine Waters." *Science* 180:955–956.

Gotham, I. J. and G. Y. Rhee. 1981. "Comparative Kinetic Studies of Phosphate Limited Growth and Phosphate Uptake in Phytoplankton in Continuous Culture." *J. Phycol.* 17:257–265.

Grady, C. P. L., Jr. and Lim, H. C. 1980. *Biological Wastewater Treatment.* Marcel Dekker, New York.

Graham, J. M., Auer, M. T., Canale, R. P., and Hoffmann, J. P. 1982. "Ecological Studies and Mathematical Modeling of *Cladophora* in Lake Huron: 4. Photosynthesis and Respiration as Functions of Light and Temperature." *J. Great Lakes Res.* 8(1):100–111.

Grant, R. S. and Skavroneck, S. 1980. "Comparison of Tracer Methods and Predictive Equations for Determination of Stream Reaeration Coefficients on Three Small Streams in Wisconsin." USGS. Water Resources Investigation 80-19. Madison, WI.

Grenney, W. J., Bella, D. A., and Curl, H. C., Jr. 1973. "A Mathematical Model of the Nutrient Dynamics of Phytoplankton in a Nitrate-Limited Environment." *Biotech. BioEngin.* 15:331–358.

Gundelach, J. M. and Castillo, J. E. 1970. "Natural Stream Purification Under Anaerobic Conditions." *J. Water Poll. Contr. Fed.* 48(7):1753–1758.

Gupta, R. S. 1989. *Hydrology and Hydraulic Systems.* Prentice-Hall, Englewood Cliffs, NJ.

Hall, C. A. and Porshing, T. A. 1990. *Numerical Analysis of Partial Differential Equations.* Prentice-Hall, Englewood Cliffs, NJ.

Hansen, J. S. and Ongerth, J. E. 1991. "Effects of Time and Watershed Characteristics on the Concentration of *Cryptosporidium* Oocysts in River Water." *Appl. Environ. Microbiol.* 57(10): 2790-2795.

Harned, H. S. and Davis, R., Jr. 1943. "The Ionization Constant of Carbonic Acid in Water and the Solubility of Carbon Dioxide in Water and Aqueous Salt Solutions from 0 to 50°C." *J. Am. Chem. Soc.* 65:2030–2037.

Harned, H. S. and Hamer, W. J. 1933. "The Ionization Constant of Water." *J. Am. Chem. Soc.* 51:2194.

Harned, H. S. and Scholes, S. R. 1941. "The Ionization Constant of HCO_3." *J. Am. Chem. Soc.* 63:1706–1709.

Harvey, H. S. 1955. *The Chemistry and Fertility of Seawater.* Cambridge University Press, Cambridge, England.

Henderson-Sellers, B. 1984. *Engineering Limnology.* Pitman, Boston.

Higbie, R. 1935. "The Rate of Adsorption of a Pure Gas Into a Still Liquid During Short Periods of Exposure." *Trans. Amer. Inst. Chem. Engin.* 31:365–389.

Hoffman, J. D. 1992. *Numerical Methods for Engineers and Scientists.* McGraw-Hill, New York.

Hornberger, G. M. and Kelly, M. G. 1972. "The Determination of Primary Production in a Stream Using an Exact Solution to the Oxygen Balance Equation." *Water Resour. Bull.* 8(4), 795–801.

Hornberger, G. M. and Kelly, M. G. 1975. "Atmospheric Reaeration in a River Using Productivity Analysis." *J. Environ. Engr. Div. ASCE* 101(5), 729–739.

Hutchinson, G. E. 1957. *A Treatise on Limnology, Vol. I, Geography, Physics and Chemistry.* Wiley, New York.

Hydroscience, Inc. 1971. Simplified Mathematical Modeling of Water Quality, prepared for the Mitre Corporation and the USEPA, Water Programs, Washington, D.C., Mar. 1971.

Hydroscience, Inc. 1972. Addendum to Simplified Mathematical Modeling of Water Quality, USEPA, Washington, D.C.

Hydroscience, Inc. 1973. Limnological Systems Analysis of the Great Lakes: Phase I—Preliminary Model Design. Hydroscience, Inc., Westwood, NJ (now HydroQual, Inc., Mahwah, NJ).

Ijima, T. and Tang, F. L. W. 1966. "Numerical Calculations of Wind Waves in Shallow Water." In *Proc. 10th Coastal Engr. Conf.,* Tokyo, pp. 38–45.

Imboden, D. M. 1974. "Phosphorus Models of Lake Eutrophication." *Limnol. Oceanogr.* 19:297–304.

Imboden, D. M. and Gachter, R. 1978. "A Dynamic Model for Trophic State Prediction." *Ecol. Model.* 4:77–98.

International Joint Commission. 1979. "Great Lakes Water Quality 1978, Appendix D, Radioactivity Subcommittee Report." International Joint Commission, Windsor, Ontario.

Jamil, A. 1971. "Raw Sewage Characteristics in Greater Beirut." M.S. Thesis, American University of Beirut, Lebanon.

Kang, S. W., Sheng, Y. P., and Lick, W. 1982. "Wave Action and Bottom Shear Stresses in Lake Erie." *J. Great Lakes Res.* 8(3):482–494.

Karickhoff, S. W., Brown, D. S., and Scott, T. A. 1979. "Sorption of Hydrophobic Pollutants on Natural Sediments." *Water Research* 13:241–248.

Kavanaugh, M. C. and Trussell, R. R. 1980. "Design of Aeration Towers to Remove Volatile Contaminants from Drinking Water." *J. Am. Water Works Assoc.* 72(12):684–692.

Keeling, C. D., et al. 1982. As presented in M. C. MacCracken and H. Moses, The First Detection of Carbon Dioxide Effects: Workshop Summary 8–10 June 1981, Harpers Ferry, WV. *Bull. Am. Meteorol. Soc.* 63:1165.

Keeling, C. D., et al. 1989. "A Three Dimensional Model of Atmospheric CO_2 Transport Based on Observed Winds: Observational Data and Preliminary Analysis." *Aspects of Climate Variability in the Pacific and the Western Americas.* Geophysical Monograph, American Geophysical Union, Vol. 55.

Keith, L. H. and Telliard, W. A. 1979. "Priority Pollutants, I. A Perspective View." *Environ. Sci. Technol.* 13(4):416–423.

Kelly, M. G., Hornberger, G. M., and Cosby, B. J. 1974. "Continuous Automated Measurement of

Photosynthesis and Respiration in an Undisturbed River Community." *Limnol. Oceanog.* 19(2), 305–312.

Kenner, B. A. 1978. "Fecal Streptococcal Indicators." In *Indicators of Viruses in Water and Food,* G. Berg, ed., Ann Arbor Science, Ann Arbor, MI.

Kielland, J. 1937. "Individual Activity Coefficients of Ions in Aqueous Solutions." *J. Am. Chem. Soc.* 59:1675–1678.

King, D. L. 1972. "Carbon Limitation in Sewage Lagoons." In *Nutrients and Eutrophication Special Symposia, Vol. I* (Am. Soc. Limnol. and Oceanog.), pp. 98–112.

King, D. L. and J. T. Novak. 1974. "The Kinetics of Inorganic Carbon-Limited Algal Growth." *J. Water Poll. Control Fed.* 46:1812–1816.

Kreider, J. F. 1982. *The Solar Heating Design Process.* McGraw-Hill, New York.

Kreith, F. and Bohn, M. S. 1986. *Principles of Heat Transfer,* 4th ed. Harper & Row, New York.

Lahlou, M., Choudhury, S., Wu, Yin, and Baldwin, K. 1995. QUAL2E Windows Interface Users Manual. EPA-823-B-95-003. U.S. Environmental Protection Agency, Washington, DC.

Lam, D. C. L., Schertzer, W. M., and Fraser, A. S. 1984. Modeling the Effects of Sediment Oxygen Demand in Lake Erie Water Quality Conditions Under the Influence of Pollution Control and Weather Variations. As quoted in Bowie et al. (1985).

Langbien, W. B. and Durum, W. H. 1967. The Aeration Capacity of Streams. USGS, Washington, DC, Circ. 542.

Langelier, W. F. 1936. "The Analytical Control of Anti-Corrosion Water Treatment." *J. Am. Water Works Assoc.* 28:1500.

Larsen, D. P. and Malueg, K. W. 1976. "Limnology of Shagawa Lake, Minnesota, Prior to Reduction of Phosphorus Loading." *Hydrobiol.* 50:177–189.

Larsen, D. P. and Malueg, K. W. 1981. "Whatever Became of Shagawa Lake?" In *Restoration of Lakes and Inland Waters. International Symposium on Inland Waters and Lake Restoration.* U.S. Environmental Protection Agency, EPA-440/4-81-010, pp. 67–72.

Larsen, D. P. and Mercier, K. W. 1976. "Limnology of Shagawa Lake, Minnesota, Prior to Reduction of Phosphorus Loading." *Hydrobiologia* 50(2):177–189.

Larsen, D. P., Malueg, K. W., Schults, D. W., and Brice, R. M. 1975. "Response of Eutrophic Shagawa Lake, Minnesota, U.S.A., to Point Source Phosphorus Reduction." *Verh. Int. Verein. Limnol.* 19:884–892.

Larsen, D. P., Mercier, H. T., and Malueg, K. W. 1973. "Modeling Algal Growth Dynamics in Shagawa Lake, Minnesota, with Comments Concerning Projected Restoration of the Lake." In *Modeling the Eutrophication Process,* E. J. Middlebrooks, D. H. Falkenborg, and T. E. Maloney, eds., Ann Arbor Science, Ann Arbor, MI, p. 15.

Larsen, D. P., Schults, D. W., and Malueg, K. W. 1981. "Summer Internal Phosphorus Supplies in Shagawa Lake, Minnesota." *Limnol. Oceanogr.* 26:740–753.

Larsen, D. P., Van Sickle, J., Malueg, K. W., and Smith, P. D. 1979. "The Effect of Wastewater Phosphorus Removal on Shagawa Lake, Minnesota: Phosphorus Supplies, Lake Phosphorus, and Chlorophyll *a*." *Water Res.* 13:1259–1272.

Laws, E. A. and Chalup, M. S. 1990. "A Microalgal Growth Model." *Limnol. Oceanogr.* 35(3):597–608.

LeChevallier, M. W. and Norton, W. D. 1995. "*Giardia* and *Cryptosporidium* in Raw and Finished Water." *J. Am. Water Works Assoc.* 87(9):54–68.

LeChevallier, M. W., et al. 1991. "Occurrence of *Giardia* and *Cryptosporidium* spp. in Surface Water Supplies." *Appl. Environ. Microbiol.* 57(9):2610–2616.

Lehman, T. D., Botkin, D. B., and Likens, G. E. 1975. "The Assumptions and Rationales of a Computer Model of Phytoplankton Population Dynamics." *Limnol. Oceanogr.* 20: 343–364.

Leopold, L. B. and Maddock, T. 1953. *The Hydraulic Geometry Channels and Some Physiographic Implications.* Geological Survey Professional Paper 252, Washington, D.C.

Lerman, A. 1972. "Strontium 90 in the Great Lakes: Concentration-Time Model." *J. Geophys. Res.* 77:3256–3264.

Lewis, G. N. and Randall, M. 1921. "The Activity Coefficient of Strong Electrolytes." *J. Am. Chem. Soc.* 43:1112–1154.

Lewis, W. K. and Whitman, W. G. 1924. Principles of Gas Absorption. *Ind. Eng. Chem.* 16(12):1215–1220.

Liss, P. S. 1975. "Chemistry of the Sea Surface Microlayer." In *Chem. Oceanogr.,* J. P. Riley and G. Skirrow, eds., Academic, London.

Loucks, D. P., Revelle, C. S., and Lynn, W. R. 1967. "Linear Programming Models for Water Pollution Control." *Management Science* 14(4):B166–B181.

Ludyanskiy, M. L., McDonald, D., and MacNeil, D. 1993. "Impact of the Zebra Mussel, A Bivalve Invader." *Bioscience* 43:533–544.

Lung, W. S. 1994. *Water Quality Modeling, Vol. 3 Application to Estuaries.* CRC, Boca Raton, FL.

Lyman, W. J., Reehl, W. F., and Rosenblatt, D. H. 1982. *Handbook of Chemical Property Estimation Methods, Environmental Behavior of Organic Compounds.* McGraw-Hill, New York.

Mabey, W. R., et al. 1979. *Aquatic Fate Process Data for Organic Priority Pollutants,* U.S. Environmental Protection Agency, Washington, DC, Report EPA/440/4-81-014.

MacCormack, R. W. 1969. "The Effect of Viscosity in Hypervelocity Impact Cratering." *Am. Inst. Aeronaut. Astronaut.* Paper 69–354.

Mackay, D. 1977. "Volatilization of Pollutants from Water." In *Aquatic Pollutants: Transformations and Biological Effects,* O. Hutzinger et al., eds., Pergamon, Amsterdam, p. 175.

Mackay, D. 1982. "Fugacity Revisited." *Environ. Sci. Technol.* 16(12): 654A–660A.

Mackay, D. and Yeun, A. T. K. 1983. "Mass Transfer Coefficients Correlations for Volatilization of Organic Solutes from Water." *Environ. Sci. Technol.* 17:211–233.

Mackie, G. L. 1991. "Biology of the Exotic Zebra Mussel, *Dreissena polymorpha,* in Relation to Native Bivalves and Its Potential Impact on Lake St. Clair." *Hydrobiologia* 219: 251–268.

Madore, M. S., et al. 1987. "Occurrence of *Cryptosporidium* Oocysts in Sewage Effluents and Select Surface Waters." *J. Parasitol.* 73:702.

Malueg, K. W., Larsen, D. P., Schults, D. W., and Mercier, H. T. 1975. "A Six Year Water, Phosphorus and Nitrogen Budget of Shagawa Lake." *J. Environ. Qual.* 4:236–242.

Mancini, J. L. 1978. Numerical Estimates of Coliform Mortality Rates Under Various Conditions. *J. Water Poll. Control Fed.* 50(11).

McGill, R., Tukey, J. W., and Larsen, W. A. 1978. "Variations of Box Plots." *Am. Stat.* 32: 12–16.

McKim, J. P., Schmeider, P., and Veith, G. 1985. "Absorption Dynamics of Organic Chemical Transport Across Trout Gills as Related to Octanol-Water Partition Coefficients." *Toxicol. Appl. Pharacol.* 77:1–10.

McQuivey, R. S. and T. N. Keefer, 1974. Simple Method for Predicting Dispersion in Streams, *J. Environ. Engr. Div. ASCE* 100(EE4):997–1011.

Metcalf and Eddy, Inc. 1979. *Wastewater Engineering.* McGraw-Hill, New York.

Metcalf and Eddy, Inc. 1991. *Wastewater Engineering.* McGraw-Hill, New York.

Miller, G. C. and Zepp, R. G. 1979. "Effects of Suspended Sediments on Photolysis Rates of Dissolved Pollutants." *Water Res.* 13:453–459.

Millero, F. J. and Poisson, A. 1981. "International One-Atmosphere Equation of State for Sea Water." *Deep-Sea Res.* (Part A). 28(6A):625–629.

Mills, W. B., et al. 1985. *Water Quality Assessment: A Screening Procedure for Toxic and Conventional Pollutants, Part I.* Tetra Tech, Inc., Env. Res. Lab., Office of Research and Devel., USEPA, Athens, GA, EPA-600/6-82-004a.

Moran, J. M., Morgan, M. D., and Wiersma, J. H. 1986. *Introduction to Environmental Science,* 2d ed. Freeman, New York.

Mortimer, C. H. 1941. "The Exchange of Dissolved Substances Between Mud and Water. I and II." *J. Ecol.* 29:280–329.

Mortimer, C. H. 1942. "The Exchange of Dissolved Substances Between Mud and Water. III." *J. Ecol.* 30:147–201.

Mortimer, C. H. 1974. Lake Hydrodynamics. *Mitt. Internat. Verein. Limnol.* 20:124–197.

Mortimer, C. H. 1975*a*. "Environmental Status of Lake Michigan Region, Vol. 2. Physical Limnology of Lake Michigan, Part I. Physical Characteristics of Lake Michigan and Its Response to Applied Forces." Argonne National Laboratory, Argonne, IL ANL/Es-40, vol. 2.

Mortimer, C. H. 1975*b*. "Modeling of Lakes as Physico-Biochemical Systems—Present Limitations and Needs." In *Modeling of Marine Systems,* J. C. J. Nihoul, ed., Elsevier, New York, p. 217.

Mueller, J. A. 1976. Accuracy of Steady-State Finite Difference Solutions. Technical Memorandum, Hydroscience, Inc. (now HydroQual, Inc., Mahwah, NJ).

Munk, W. H. and Anderson, E. R. 1948. "Notes on a Theory of the Thermocline." *J. Marine Res.* 7:276–295.

Murthy, C. R. 1976. "Horizontal Diffusion Characteristics in Lake Ontario." *J. Physic. Oceanogr.* 6:76–84.

Nakamura, Y. and Stefan, H. G. 1994. "Effect of Flow Velocity on Sediment Oxygen Demand: Theory." *J. Environ. Engr.* 120(5):996–1016.

Neftel, A., Moor, E., Oeschger, H., and Stauffer, B. 1985. "Evidence from Polar Ice Cores for the Increase in Atmospheric CO_2 in the Past Two Centuries." *Nature.* 315: May 2.

O'Connor, D. J. and Dobbins, W. E. "Mechanism of Reaeration in Natural Streams." *ASCE Trans.* 86(SA3):35–55.

O'Connor, D. J. 1960. "Oxygen Balance of an Estuary." *J. San. Engr. Div. ASCE* 86(SA3): 35–55.

O'Connor, D. J. 1962. "The Bacterial Distribution in a Lake in the Vicinity of a Sewage Discharge." In *Proceedings of the 2nd Purdue Industrial Waste Conference,* West Lafayette, IN.

O'Connor, D. J. 1967. "The Temporal and Spatial Distribution of Dissolved Oxygen in Streams." *Water Resour. Res.* 3(1):65–79.

O'Connor, D. J. 1976. "The Concentration of Dissolved Solids and River Flow." *Water Resour. Res.* 12(2):279–294.

O'Connor, D. J. 1985. Modeling Frameworks, Toxic Substances Notes, Manhattan College Summer Institute in Water Pollution Control, Manhattan College, Bronx, NY.

O'Connor, D. J. 1988*a*. "Models of Sorptive Toxic Substances in Freshwater Systems. I: Basic Equations." *J. Envir. Engr.* 114(3):507–532.

O'Connor, D. J. 1988*b*. "Models of Sorptive Toxic Substances in Freshwater Systems. II: Lakes and Reservoirs." *J. Envir. Engr.* 114(3):533–551.

O'Connor, D. J. 1988*c*. "Models of Sorptive Toxic Substances in Freshwater Systems. III: Streams and Rivers." *J. Envir. Engr.* 114(3):552–574.

O'Connor, D. J. 1989. "Seasonal and Long-Term Variations of Dissolved Solids in Lakes and Reservoirs." *J. Environ. Engr. Div. ASCE* 115(6):1213–1234.

O'Connor, D. J. and Di Toro. D. M. 1970. "Photosynthesis and Oxygen Balance in Streams." *J. San. Engr. Div. ASCE* 96(2), 547–571.

O'Connor, D. J. and Dobbins, 1958. "Mechanism of Reaeration in Natural Streams." *Trans. Am. Soc. Civil Engin.* 123:641–666.

O'Connor, D. J. and Mueller, J. A. 1970. "A Water Quality Model of Chloride in the Great Lakes." *J. San. Engr. Div. ASCE* 96(SA4):955–975.

O'Loughlin, E. M. and K. H. Bowmer, 1975. "Dilution and Decay of Aquatic Herbicides in Flowing Channels." *J. Hydrol.* 26:217–235

O'Melia, C. R. 1972. "An Approach to the Modeling of Lakes." *Schweiz. Z. Hydrol.* 34:1–34.

Odum, H. T. 1956. "Primary Production in Flowing Waters." *Limnol. Oceanog.* 1(2), 102–117.

Officer, C. B., 1976. *Physical Oceanography of Estuaries (and Associated Coastal Waters).* Wiley, New York.

Officer, C. B. 1983. "Physics of Estuarine Circulation." In *Ecosystems of the World, Estuaries and Enclosed Seas,* B. H. Ketchum, ed., Elsevier, Amsterdam, pp. 15–41.

Okubo, A. 1971. "Oceanic Diffusion Diagrams." *Deep-Sea Res.* 18:789–802.

Omernik, J. M., 1977. Non-Point Source-Stream Nutrient Level Relationships: A Nationwide Study, Corvallis ERL. ORD. USEPA, Corvallis, OR. EPA-600/3-77-105.

Oreskes, N., Shrader-Frechette, K., and Belitz, K. 1994. "Verification, Validation and Confirmation of Numerical Models in the Earth Sciences." *Science* 263(5147):641–646.

Orlob, G. T. 1977. "Mathematical Modeling of Surface Water Impoundments, Vols. I and II." NTIS, PB-293-204.

Oswald, T. and Roth, C. 1988. "Make Us a River." *Rod and Reel* May/June:28.

Owens, M., Edwards, R., and Gibbs, J. 1964. Some Reaeration Studies in Streams. *Int. J. Air Water Poll.* 8:469–486.

Peterson, R. O., Page, R. E., and Dodge, K. M. 1984. "Wolves, Moose, and the Allometry of Population Cycles." *Science* 224:1350–1352.

Pielou, E. C. 1969. *An Introduction to Mathematical Ecology.* Wiley-Interscience, New York.

Ponce, V. M. 1989. *Engineering Hydrology: Principles and Practices.* Prentice-Hall, Englewood Cliffs, NJ.

Press, W. H., Teukolsky, S. A., Vetterling, W. T., and Flannery, B. P. 1992. *Numerical Recipes in FORTRAN,* 2d ed. Cambridge University Press, New York.

Rast, W. and Lee, G. F. 1978. Summary Analysis of the North American Project (US portion) OECD Eutrophication Project: Nutrient Loading-Lake Response Relationships and Trophic State Indices, USEPA Corvallis Environmental Research Laboratory, Corvallis, OR, EPA-600/3-78-008.

Rathbun, R. E., Stephens, D. W., Shultz, D. J., and Tai, D. Y. 1978. "Laboratory Studies of Gas Tracers for Reaeration." *J. Environ. Engr. Div. ASCE* 104(EE1):215–229.

Raudkivi, A. J. 1979. *Hydrology.* Pergamon, Oxford, England.

Ravelle, C., Loucks, D. P., and Lynn, W. R. (1967) "A Management Model for Water Quality Control." *J. Water Poll. Control Fed.* 39(7):1164–1183.

Rawson, D. S. 1955. "Morphometry as a Dominant Factor in the Productivity of Large Lakes." *Verh. Int. Ver. Limnol.* 12:164–175.

Reckhow, K. H. 1977. Phosphorus Models for Lake Management. Ph.D. Dissertation, Harvard University, Cambridge, MA.

Reckhow, K. H. 1979. "Empirical Lake Models for Phosphorus Development: Applications, Limitations, and Uncertainty." In *Perspectives on Lake Ecosystem Modeling,* D. Scavia and A. Robertson, eds., Ann Arbor Science, Ann Arbor, MI, pp. 183–222.

Reckhow, K. H. and Chapra, S. C. 1979. "Error Analysis for a Phosphorus Retention Model." *Water Resour. Res.* 15:1643–1646.

Reckhow, K. H. and Chapra, S. C. 1983. *Engineering Approaches for Lake Management, Vol. 1: Data Analysis and Empirical Modeling.* Butterworth, Woburn, MA.

Redfield, A. C., Ketchum, B. H., and Richards, F. A. 1963. "The Influence of Organisms on the Composition of Seawater." In *The Sea,* M. N. Hill, ed., Wiley-Interscience, New York, pp. 26–77.

Regli, S., et al. 1991. "Modeling the Risk from *Giardia* and Viruses in Drinking Water."*J. Am. Water Works Assoc.* 80:74–77.

Reynolds, T. D. 1982. *Unit Operations and Processes in Environmental Engineering,* Brooks/ Cole, Monterey, CA.

Richardson, L. F. 1926. "Atmospheric Diffusion Shown on a Distance-Neighborhood Graph." *Proc. Royal Soc.* (*A*) 110:709–727.

Riley, G. A. 1946. "Factors Controlling Phytoplankton Population on Georges Bank." *J. Mar. Res.* 6:-104–113.

Riley, G. A., 1956. Oceanography of Long Island Sound 1952-1954. II. Physical Oceanography, *Bull. Bingham. Oceanog. Collection 15,* pp. 15–16.

Riley, G. A. 1963. "Theory of Food-Chain Relations in the Ocean." In *The Sea,* M. N. Hill, ed., Wiley-Interscience, New York, pp. 438–463.

Robbins, J. A. and Edgington, D. N. 1975. "Determination of Recent Sedimentation Rates in Lake Michigan Using Pb-210 and Cs-137." *Geochim. Cosmochim. Acta* 39:285–301.

Roesner, L. A., Giguerre, P. R., and Evenson, D. E. 1981*a*. Computer Program Documentation for Stream Quality Modeling (QUAL-II), U.S. Environmental Protection Agency, Athens, GA, Report EPA/600/9-81-014.

Roesner, L. A., Giguerre, P. R., and Evenson, D. E. 1981*b*. Users Manual for Stream Quality Modeling (QUAL-II), U.S. Environmental Protection Agency, Athens, GA, Report EPA/600/9-81-015.

Rose, J. B. 1988. "Occurrence and Significance of *Cryptosporidium* in Water." *J. Am. Water Works Assoc.* 80:53–58.

Rose, J. B., et al. 1991. "Risk Assessment and the Control of Waterborne Pathogens." *Am. J. Public Health,* 81:709.

Runkel, R. L. 1996. Personal communication.

Runkel, R. L., Bencala, K. E., Broshears, R. E., and Chapra, S. C. 1996*a*. "Reactive Solute Transport in

Small Streams: I. Development of an Equilibrium-Based Simulation Model." *Water Resour. Res.* (in press).
Runkel, R. L., McKnight, D. M., Bencala, K. E., and Chapra, S. C. 1996*b*. "Reactive Solute Transport in Small Streams: II. Simulation of a pH-Modification Experiment." *Water Resour. Res.* (in press).
Russell, L. L. 1976. "Chemical Aspects of Groundwater Recharge with Wastewaters." Ph.D. Dissertation, Univ. Calif., Berkeley, CA.
Salvato, J. A. 1982. *Environmental Engineering and Sanitation.* Wiley, New York.
Sawyer, C. N., McCarty, P. L., and Parkin, G. F. 1994. *Chemistry for Environmental Engineering.* McGraw-Hill, New York.
Scavia, D. 1979. "The Use of Ecological Models of Lakes in Synthesizing Available Information and Identifying Research Needs." In *Perspectives on Lake Ecosystem Modeling,* D. Scavia and A. Robertson, eds., Ann Arbor Science, Ann Arbor, MI, p. 109.
Scavia, D. 1980. "An Ecological Model for Lake Ontario." *Ecol. Model.* 8:49–78.
Scavia, D. and Bennett, J. R. 1980. "Spring Transition Period in Lake Ontario—A Numerical Study of the Causes of the Large Biological and Chemical Gradients." *Can. J. Fish. Aquat. Sci.* 37(5):-823–833.
Schmidt, G. D. and Roberts, L. S. 1977. *Foundations of Parasitology.* C. V. Mosby, St. Louis, MO.
Schnoor, J. L., Sato, C., McKechnie, D., and Sahoo, D. 1987. *Processes, Coefficients, and Models for Simulating Toxic Organics and Heavy Metals in Surface Waters,* U.S. Environmental Protection Agency, Athens, GA, Report EPA/600/3-87/015, 1987.
Schurr, J. M. and Ruchti, J. 1975. "Kinetics of O_2 Exchange, Photosynthesis, and Respiration in Rivers Determined from Time-Delayed Correlations Between Sunlight and Dissolved Oxygen." *Schweiz. Z. Hydrol.* 37(1), 144–174.
Schurr, J. M. and Ruchti, J. 1977. "Dynamics of O_2 and CO_2 Exchange, Photosynthesis, and Respiration in Rivers From Time-Delayed Correlations with Ideal Sunlight." *Limnol. Oceanog.* 22(2): 208–225.
Schuster, R. J. 1987. Colorado River Simulation System Documentation: System Overview, U.S. Bureau of Reclamation, Denver, CO.
Schwarzenbach, R. P. and Westall, J. 1981. "Transport of Nonpolar Organic Compounds from Surface Water to Groundwater: Laboratory Sorption Studies." *Environ. Sci. Technol.* 15:1360–1367.
Schwarzenbach, R. P., Gschwend, P. M., and Imboden, D. M. 1993. *Environmental Organic Chemistry.* Wiley-Interscience, New York.
Shah, I., 1970. *Tales of the Dervishes.* Dutton, New York.
Shapiro, J. 1973. "Blue-Green Algae: Why They Became Dominant." *Science* 179:382–384.
Shugart, H. H., Goldstein, R. A., O'Neill, R. V., and Mankin, J. B. 1974. "TEEM: A Terrestrial Ecosystem Energy Model for Forests." *Oecol. Plant.* 9(3):231–264.
Simons, T. J. and Lam, D. C. L. 1980. "Some Limitations of Water Quality Models for Large Lakes: A Case Study of Lake Ontario." *Water Resour. Res.* 16:105–116.
Smith, J. W. and Wolfe, M. S. 1980. "Giardiasis." *Amer. Rev. Med.* 31:373–383.
Smith, V. H. and Shapiro, J. 1981. A Retrospective Look at the Effects of Phosphorus Removal in Lakes, in Restoration of Lakes and Inland Waters. USEPA, Office of Water Regulations and Standards, Washington, DC. EPA-440/5-81-010.
Snodgrass, W. J. 1974. A Predictive Phosphorus Model for Lakes: Development and Testing. Ph.D. Dissertation, University of North Carolina, Chapel Hill, NC.
Snodgrass, W. J. and O'Melia, C. R. 1975. Predictive Model for Phosphorus in Lakes. *Environ. Sci. Technol.* 9:937–944.
Steele, J. H. 1962. "Environmental Control of Photosynthesis in the Sea." *Limnol. Oceanogr.* 7:137–150.
Steele, J. H. 1965. "Notes on Some Theoretical Problems in Production Ecology." In *Primary Production in Aquatic Environments,* C. R. Goldman, ed., University of California Press, Berkeley, CA.
Stefan, H. G., Cardoni, J. J., Schiebe, F. R., and Cooper, C. M. 1983. "Model of Light Penetration in a Turbid Lake." *Water Resour. Res.* 19(1):109–120.
Streeter, H. W. and Phelps, E. B. 1925. A Study of the Pollution and Natural Purification of the Ohio River, III. Factors Concerning the Phenomena of Oxidation and Reaeration. U.S. Public Health Service, Pub. Health Bulletin No. 146, February, 1925. Reprinted by U.S., DHEW, PHA, 1958.
Stumm, W. and Morgan, J. J. 1981. *Aquatic Chemistry.* Wiley-Interscience, New York.

Sverdrup, H. U., Johnson, M. W., and Fleming, R. H. 1942. *The Oceans.* Prentice-Hall, Englewood Cliffs, NJ.

Syskora, J. L., et al. 1991. "Distribution of *Giardia* Cysts in Wastewater." *Water Res.* 24(2):187–192.

Taylor, G. I. 1953. "Dispersion of Soluble Matter in Solvent Flowing Slowly Through a Tube." *Proc. Royal Soc. London Ser. A* 219:186–203.

Texas Water Development Board 1970. Simulation of Water Quality in Streams and Canals, Program Documentation and User's Manual, Austin, TX.

Thackston, E. L. and Krenkel P. A. 1966. "Reaeration Predictions in Natural Streams." *J. San. Engr. Div. ASCE* 95(SA1):65–94.

Thomann, R. V. 1963. "Mathematical Model for Dissolved Oxygen." *J. San. Engr. Div. ASCE* 89(SA5): 1–30.

Thomann, R. V. 1972. *Systems Analysis and Water Quality Management.* McGraw-Hill, New York.

Thomann, R. V. 1981. "Equilibrium Model of Fate of Microcontaminants in Diverse Aquatic Food Chains." *Can. J. Fish. Aquat. Sci.* 38:280–296.

Thomann, R. V. 1985. "A Simplified Heavy Metals Model for Streams." (in preparation). Manhattan College, Bronx, NY.

Thomann, R. V. 1989. "Bioaccumulation Model of Organic Chemical Distribution in Aquatic Food Chains." *Environ. Sci. Technol.* 23:699–707.

Thomann, R. V. and Connolly, J. P. 1992. "Modeling Accumulation of Organic Chemicals in Aquatic Food Webs." In *Chemical Dynamics in Freshwater Ecosystems,* F. A. P. C. Gobas and J. A. McCorquodale, eds., Lewis, Boca Raton, FL, pp. 153–186.

Thomann, R. V. and Di Toro, D. M. 1983. "Physico Chemical Model of Toxic Substances in the Great Lakes." *J. Great Lakes Res.* 9(4), 474–496.

Thomann, R. V. and Fitzpatrick, J. F. 1982. Calibration and Verification of a Mathematical Model of the Eutrophication of the Potomac Estuary; report by Hydroqual, Inc., Mahwah, NJ, to DES, Dist. Col.

Thomann, R. V. and Mueller, J. A. 1987. *Principles of Surface Water Quality Modeling and Control.* Harper & Row, New York.

Thomann, R. V. and Sobel, M. J. 1964. "Estuarine Water Quality Management and Forecasting." *J. San. Engr. Div. ASCE* 90(SA5):9–36.

Thomann, R. V., Connolly, J. P., and Parkerton, T. F. 1992. "An Equilibrium Model of Organic Chemical Accumulation in Aquatic Food Webs with Sediment Interactions." *Environ. Toxicol. Chem.* 11:615–629.

Thornton, K. W. and Lessem, A. S. 1978. "A Temperature Algorithm for Modifying Rates." *Trans. Am. Fish. Soc.* 107(2):284–287.

Tsivoglou, E. C. and Neal, L. A. 1976. "Tracer Measurements of Reaeration: III. Predicting the Reaeration Capacity of Inland Streams." *J. Water Poll. Control Fed.* 48(12): 2669–2689.

Tsivoglou, E. C. and Wallace, S. R. 1972. Characterization of Stream Reaeration Capacity, USEPA, Report No. EPA-R3-72-012.

Tukey, J. W. 1977 *Exploratory Data Analysis.* Addison-Wesley, Reading, MA.

Tyler, J. E. and Smith, R. C. 1970. *Measurement of Spectral Irradiance Under Water.* Gordon and Breach, New York.

Velz, C. J., 1938. "Deoxygenation and Reoxygenation." *Proc. Am. Soc. Civ. Engr.* 65(4): 677–680.

Velz, C. J., 1947. "Factors Influencing Self-Purification and Their Relation to Pollution Abatement." *Sewage Works J.* 19(4):629–644.

Vollenweider, R. A. 1968. "The Scientific Basis of Lake and Stream Eutrophication with Particular Reference to Phosphorus and Nitrogen as Eutrophication Factors." Technical Report DAS/DSI/68.27, Organization for Economic Cooperation and Development, Paris.

Vollenweider, R. A. 1975. "Input-Output Models with Special Reference to the Phosphorus Loading Concept in Limnology." *Schweiz. Z. Hydrol.* 37:53–84.

Vollenweider, R. A. 1976. "Advances in Defining Critical Loading Levels for Phosphorus in Lake Eutrophication." *Mem. Ist. Ital. Idrobiol.* 33:53–83.

Walker, W. W., Jr. 1977. "Some Analytical Methods Applied to Lake Water Quality Problems." Ph.D. Dissertation, Harvard University, Cambridge, MA.

Walker, W. W., Jr. 1980. "Variability of Trophic State Indicators in Reservoirs." In *Restoration of Lakes and Inland Waters,* U.S. Environmental Protection Agency, EPA-440/5-81-010, pp. 344–348.

Walker, W. W., Jr. 1983. Personal communication quoted in Brown and Barnwell (1987).

Walsh, J. J. and Dugdale, R. C. 1972. "Nutrient Submodels and Simulation Models of Phytoplankton Production in the Sea." In *Nutrients in Natural Waters,* H. E. Allen and J. R. Kramer, eds., Wiley, New York, p. 171.

Wanninkhof, R., Ledwell, J. R., and Crusius, J. 1991. "Gas Transfer Velocities on Lakes Measured with Sulfur Hexafluoride." In *Symposium Volume of the Second International Conference on Gas Transfer at Water Surfaces,* S. C. Wilhelms and J. S. Gulliver, eds., Minneapolis, MN.

Weber, W. J., Jr. 1972. *Physicochemical Processes for Water Quality Control.* Wiley, New York.

Westrich, J. T. and Berner, R. A. 1984. "The Role of Sedimentary Organic Matter in Bacterial Sulfate Reduction: The G Model Tested." *Limnol. Oceanogr.* 29(2):236–249.

Wetzel, R. G. 1975. *Limnology.* Saunders. Philadelphia.

Wetzel, R. G. 1983. *Limnology.* 2d ed. Saunders, Philadelphia.

Whitman, W. G. 1923. "The Two-Film Theory of Gas Absorption." *Chem. Metallurg. Eng.* 29(4):146–148.

Wickramanayake, G. B., Rubin, A. J., and Sproul, O. J. 1985. "Effect of Ozone and Storage Temperature on *Giardia* Cysts." *J. Am. Water Works Assoc.* 77:74–77.

Wilcox, R. J. 1984*a*. "Methyl Chloride as a Gas-Tracer for Measuring Stream Reaeration Coefficients — I. Laboratory Studies." *Water Res.* 18(1):47–52.

Wilcox, R. J. 1984*b*. "Methyl Chloride as a Gas-Tracer for Measuring Stream Reaeration Coefficients — II. Stream Studies." *Water Res.* 18(1):53–57.

Wright, R. M. and McDonnell, A. J. 1979. "In-stream Deoxygenation Rate Prediction." *ASCE J. Env. Eng. Div.* 105:323–335.

Yotsukura, N. 1968. As referenced in preliminary report *Techniques of Water Resources Investigations of the U.S. Geological Survey,* Measurement of Time of Travel and Dispersion by Dye Tracing, Book 3, Chapter A9, by F. A. Kilpatrick, L. A. Martens, and J. F. Wilson, 1970.

Zepp, R. G., "Environmental photoprocesses involving natural organic matter." In *Humic Substances and Their Role in the Environment,* F. H. Frimmel and R. F. Christman, eds., Wiley, New York, 1988 pp. 193–214.

Zepp, R. G. and D. M. Cline, "Rates of direct photolysis in aquatic environment." *Environ. Sci. Technol.* 11:359–366 (1977).

Zison, S. W., Mills, W. B., Diemer, D., and Chen, C. W. 1978. Rates, Constants, and Kinetic Formulations in Surface Water Quality Modeling. U.S. Environmental Protection Agency, ORD, Athens, GA, ERL, EPA/600/3-78-105.

Acknowledgments

For complete citations, see References, pages 821-834.

Fig. 2.6 Reprinted from Chapra and Canale (1988; Fig. 17.7, p. 534) by permission of McGraw-Hill, New York, NY.

Fig. 5.7 From Fogler (1986; Fig. 9-5, p. 464); Prentice-Hall, Englewood Cliffs, NJ.

Fig. B18.1 From Canale and Auer (1982; Figs. 3 and 8, p. 137); reprinted with permission of *Journal of Great Lakes Research;* International Association for Great Lakes Research, Toronto, Ontario, Canada.

Figs. 24.3-24.7 From Chapra and Di Toro (1991; Figs. 1, 2, 4-6, p. 642, 644, 647, 649); *Journal of Environmental Engineering,* American Society of Civil Engineers, New York, NY.

Figs. 25.4, 25.5, 25.9 From Di Toro et al. (1990; Figs. 2-4, p. 950, 951, 960). *Journal of Environmental Engineering,* American Society of Civil Engineers, New York, NY.

Figs. 29.10-29.13 From Chapra and Canale (1991; Figs. 4-7, p. 712, 713;); "Long-Term Phenomenological Model of Phosphorus and Oxygen in Stratified Lakes." Water Research. 25 (6): 707-715, with kind permission from Elsevier Science Ltd., The Boulevard, Langford Lane, Kidlington OX5 1BG, UK.

Fig. 33.3 From Canale and Vogel (1974; Fig. 1, p. 239); *Journal of Environmental Engineering,* American Society of Civil Engineers, New York, NY.

Figs. 41.4, 41.9, 41.11-41.13 From Chapra (1991: Figs. 2, 3, 5, 6, 9, p. 664, 666, 668, 668, 669); *Journal of Environmental Engineering,* American Society of Civil Engineers, New York, NY.

Fig 42.7 From Schwarzenbach et al. (1993; Fig. 13.6, p. 447); Wiley-Interscience, New York, NY.

Fig 45.4 From Thomann et al. (1992; Fig. 1, p. 616); "An Equilibrium Model of Organic Chemical Accumulation in Aquatic Food Webs with Sediment Interactions." *Environmental Toxicology and Chemistry,* 11:615-629, from Pergamon Press, Elsevier Science Ltd., The Boulevard, Langford Lane, Kidlington OX5 1GB, UK.

Fig F.1 From Press et al. (1992; p. 230, 231); Cambridge University Press, New York, NY.

Index

Absorption, 715
Accumulation, 48
Accuracy, 212
Acid-base effects, 755
Acid-volatile sulfide, 765–767
Activation, 752
Activity, 670–672
Adsorption, 715
Advection, 138, 180
  impulse spills, 180
  QUAL2E, 488–489
Advection number, 216
AHOD, 452, 543–544
Algae, 635–636
  growth of, on a limiting nutrient, 600–601
  in QUAL2E, 653–658
Algebraic method, 57
Alkalinity, 684
Allochthonous, 298
Ammonia toxicity, 420–421, 430–431, 680–682
Anaerobic, 399–401
Analytical solutions:
  for a diffuse source that contributes both flow and mass, 410–415
  and organic decomposition, 428–429
  and QUAL2E, 649–653
  and total SOD, 468–470
  and water quality, 419–421
Annual hydrograph, 236
Areal hypolimnetic oxygen demand, 452, 543–544
Arrhenius equation, 40
Assimilation factor, 12
Atmospheric moisture, 566–568
Attached plants (*see* Fixed plants)
Auer, Marty, 326
Autochthonous, 298
Autotrophic, 347
Avogadro's number, 668
AVS (*see* Acid-volatile sulfide)

"Back-of-the-envelope" calculations, 79
Backward difference, 193
Bacteria, 289, 504–511
Bacterial growth, 590–591
Bacterial loss rate, 506–510
  light, 507–508
  natural mortality and salinity, 506
  settling, 508–509
  total loss rate, 509–510
Baseflow, 236
Batch reactor, 24, 596–597
  reactions with feedback in, 114–116
Bathymetry, 278
Becquerel (Bq), 758
Bed effects, and BOD removal, 359–360
Beer-Lambert law, 507, 611, 661, 744
Bernoulli's distribution, 179
Bessel functions, 288–292, 817–819
Binomial distribution, 179
Bioaccumulation, 784, 788–790
Biochemical oxygen demand (*see* BOD)
Bioconcentration, 755, 784, 785–788
Biomagnification, 785
Biotransformation, 751–753
"Blue babies," 430
BOD, 353–355
  carbonaceous, 355
  5-day, 357
  loadings, concentrations, and rates, 357–360
  model for a stream, 355–357
  nitrogenous, 355
  no-flow sources, 407–409
BOD_5, 357
Bode diagrams, 76
Bottom scour, 313
Bottom sediments, 302–304
  as a distributed system, 307–311
Boulder Creek, CO, 23, 246, 257–259, 449
Boundary conditions, 194–195
  Dirichlet, 194
  Neumann, 194
Box-and-whisker plot, 723
Brackish water, 260
Budget:
  contaminant, 702–705, 708–713, 771–772
  heat, 560–576
  solids, 701–702, 706–707, 770–771
Budget models, 536–538

Calcite precipitation, 298
Calibration, and Streeter-Phelps model, 398–399
  (*See also* Model calibration)
Canale, Ray, 326, 453
Carbon, 526–527
  nonliving organic, 636
Carbonaceous SOD, 466–468
Carbon dioxide, 686–689
"Cartoon" modeling, 54

Cascade, 89–91
Cascade Reservoir, ID, 293
CBOD, 355
Celerity, 254
Centered-difference approach, 198–201
CE-QUAL-W2, 321
Channelization, effect of, on hydrographs, 238
Characteristic, 414
Characteristic curve, 175
Chemistry, 665–691
  equilibrium, 25, 667–682
  inorganic carbon, 683–686
  and mass balance, 677–682
  units and conversions, 667–669
Chlorophyll, 540–543
  Secchi-disk depth correlations, 541–543
  variable models, 615–621
Churchill reaeration formula, 380
Civil engineers, 4
Cladophora, in Great Lakes, 325–327
Closed systems, reactions with feedback in, 114–116
Closed-form solutions, 83
Coliforms, 504–511
Colorado River, 118–119
Column, 104
Column vectors, 104
Cometabolism, 751
Completely mixed systems, 1–133
Computer methods, 482–500
  QUAL2E model, 486–487
  steady-state system response matrix, 482–486
  well-mixed reactors, 120–131
Concentration, 6–8
Concentration condition, 675
Condensation, 571
Conduction, 151, 570
Conductivity, 670
Conservation of mass, 13
Constant-parameter analysis, estuary dispersion, 263–264
Continuity equation, 9
Continuous spills, 183–186
Continuously stirred tank reactor (*See* CSTR)
Control-volume approach:
  extension to two- and three-dimensional systems, 208–209
  steady-state solutions, 192–209
  time-variable solutions, 215–216
Convection, 570–571
  forced, 152
  free, 152
Courant condition, 216
Courant number, 216
Cramer's rule, 102
Crank-Nicolson method, 227
Cryptosporidiosis, 512
Cryptosporidium, 512–516
  parvum, 512
CSTR, 47, 598–600
  comparison with plug-flow reactor, 159–160
  reactions with feedback in, 116
  toxics model for, 700–713
Curie (Ci), 758
Cysts, 512

Dalton's law of evaporation, 284, 567, 571
Dam, effect of, on stream reaeration, 382
Damkohler number, 163
Danckwerts, 162
Danckwerts boundary conditions, 162
Debye-Hückel limiting law, 670
Decay, of impulse spills, 183
Decomposition, 347
  nitrification and, 428–429
Defusing, 752
Delaware estuary, 110
Delta method, 442–448
Denitrification, 420, 478
Density
  in bottom sediments, 303–304
  particulate matter, 299
  water, 271, 299
Dependent variable, 51
Depth, flow, and velocity, 247–250
Design mode, 11
Desorption, 716
Detoxification, 751
Dew-point temperature, 567
Diagenesis, 457–459
Differential method, 31
Diffuse sources:
  with flow, 410–417
  with no flow, 407–410
Diffusion, 137–152, 149–150, 177, 180
  defined, 138, 139
  experiment for, 138–141
  impulse spills, 180
  molecular, 138
  turbulent, 138, 149–150
  vertical, 580–585
Diffusion number, 214
Dimension, 104
Dirac delta function, 66
Dirichlet boundary conditions, 194
Discharge coefficients, 248
Dispersion, 150–151, 181
  estuary, 263–267, 269
  impulse spills, 181

lateral, 246–247
longitudinal, 245–246
numerical, 201–203
QUAL2E, 489
stream, 245–247, 489
Dispersion coefficient, 263–270
Dissolved oxygen, 345–500
experiment, 351–353
nitrogen and, 421–423
no-flow sources, 409
saturation, 361–364
Dissolved oxygen deficit, 390
balance at the discharge point, 391–392
Dissolved oxygen sag, 348–351
Distributed-parameter systems:
time-variable solutions, 173–190
steady-state solutions, 156
Distributed source
in estuaries, 170–171
parameterization of, 405–407
in streams, 166–167
Streeter-Phelps model, 405–417
Diurnal oxygen swings, 387
Dreissena polymorpha, 453
"Drunkard's" walk, 177–180

Eigenvalue, 58, 111, 710
Einstein (mole of photons), 742
Electroneutrality condition, 675
Element matrix, 104
Embayment model, 143–149
estimation of diffusion, 143–145
steady-state solution, 145–146
time-variable solution, 146–149
Engineers, and water quality, 4–6
Entrainment rate, 313–315
Environments, 233–344
Ephemeral flow regime, 236
Equilibrium:
calculations, 673–675
chemistry, 25, 667–682
local, 677–682
Equilibrium approach, 115
Error function, 185, 820
Estuaries, 260–272
application of the MFR model to, 168–171
comparison with streams, lakes, and the ocean, 268–270
dispersion coefficient, 263–270
net flow, 262
reaeration, 383–385
slack-tide sampling, 268
SOD in, 479
Streeter-Phelps model, 401–403
transport in, 260–262
vertical stratification, 270–272

Estuary number, 189–190, 309
Euler's method, 121–123
Eutrophication, 420, 519–663
cultural, 521
in flowing waters, 644–663
modeling with QUAL2E, 649–658
nutrients, 521–533
temperature, 560–589
Evaporation, 567, 571
in lakes, 283–285
Explicit method, 212–214
stability, 214–215
Exponential loading, 71–73
Extended Debye-Hückel approximation, 670
Extensive property, 6, 561

Fall overturn, 579
Fecal coliform (FC), 504–505
Fecal streptococci (FS), 505
Feedback, 113–116
(*See also* Reversible reactions)
Feedback systems of reactors, 101–116
Feedforward reactions in a single reactor, 95–99
Feedforward systems of reactors, 86–99
mass balance and steady-state, 86–91
time variable, 91–95
Fick, Adolf, 141
Fick's first law, 141–143
Fick's second law, 181
Finite-difference analysis, estuary dispersion, 264–268
Finite-difference equation
Crank-Nicolson, 227
forward-time/centered-space (FTCS), 214
First-order error analysis, 327–332
First-order reactions, 27
Fixed plants, 326–328, 434, 658–662
Flint River, 775–776
Flow, depth and velocity, 247–250
Flow regime:
ephemeral, 236
perennial, 236
Flux, 8–9
Focusing, 297
Food chain, 635–636
Food webs, 795–797
Food-chain/toxicant interactions (*see* Toxicant/food-chain interactions)
Forcing function, 51
Forward-time/centered-space (FTCS) difference equation, 214
Fourier series, 80–83
Fundamental frequency, 80
Fundamental qualities, 6–10

Gas transfer, 369–388
  experiment, 389–391
  and plants, 686–689
  surface renewal model, 375–378
  Whitman's two-film theory, 371–375
Gauss quadrature, 240
General solution, to mass balance equation, 58–60
Giardia, 512–516
Giardiasis, 512
Gradient, 140
Grand River, MI, 447
Great Lakes, 10, 93–95, 99–100, 117, 325
Green Bay, 154

Half-life, 34, 61
Half-saturation constants, 592
  algae, 627
  light, 610, 616
  nutrients, 607–609
  toxics, 752
Hanes equation, 593
Harmonics, 80
Heat and temperature, 561–562
Heat balance, 563–565
  multilayer, 585–588
  time variable, 573–575
Heat budget, 560–576
Heat exchange, 565–571
Henry's law, 360–361
Heterogeneous reactions, 25
Heterotrophic, 347
Hofstee equation, 593
Homogeneous reaction, 25
Huen's method, 124–126
Human water use, effect of, on hydrographs, 238
Hydrogeometry, 238
Hydrolysis, 753–755

Ideal gas law, 360–361
Ideal reactors, 156–164
  plug-flow reactor, 157–159
Identity matrix, 104
Implicit approaches, 223–228
Impounding, 238
Impoundments, 276–293
Impulse function, 66
  loading, 66–68
  spills, 180–183
Incompletely mixed systems, 135–294
Increment function, 126
Independent variables, 51
Indicator organisms, 504–506
  concentrations, 505–506
Input-output modeling, 110
Instantaneous or "impulse" spills, 180–183
Integral method, 30
Intensive property, 6, 561
Inverse, 106
Ionic strength, 670
Ionization product of water, 672
Irreversible reaction, 25
Isle Royale, 631–632
Isotherms, 716–719

Kinematic wave, 251–257
Kinetic approach, 115
Kinetic segmentation, 557–559
Kinetics, 26, 669
  QUAL2E, 489–491

Lake, 276–293
  evaporation, 283–285
  morphometry, 278–282
  near-shore models, 287–293
  reaeration, 383–385
  water balance, 282–287
Lake Erie, 292, 589, 713
Lake Huron, 144–147, 703–705, 711–713
Lake Michigan, 20, 78, 154, 317, 714, 760–761, 762–763
Lake Ontario, 12, 63, 278, 308, 555–556, 580, 582–583, 637–640
Lake Sammamish, WA, 559
Lake St. Clair, 100
Lake Washington, 132
Lake Youngs, WA, 293
Large systems of reactors, 103–107
Lateral mixing, 245
Law of mass action, 26, 669–670
Lead-210, sediment dating, 312
Leontiff, W. W., 110
Liebig's law of the minimum, 608
Light, 435–437, 741–742
  absorption, 744–746
  effect of, on bacteria, 507–508
  extinction in natural waters, 441, 507, 586, 611, 743–744
  and phytoplankton growth, 439–442, 609–612, 615–621
Light-bottle/dark-bottle technique, 438–439
Limits, 603–605
  and nonpredatory losses, 613–615
  and nutrients, 607–609
  to phytoplankton growth, 603–621
  rate model, 612–613
  and temperature, 605–607
Limnologist, 276
Linear algebraic equations, 103

Linear loading, 70–71
Linear programming, 110
Lineweaver-Burk plot, 593
Loading, 8, 13, 49
  exponential, 71–73
  impulse, 66–68
  linear ("ramp"), 70–71
  sinusoidal, 73–76
  step, 68–69
Local equilibrium, 116, 677–682
  and chemical reactions, 680–682
Longitudinal mixing, 245
Lotka-Volterra equations, 622–626
Low-flow analysis, 243–244
Luminescence, 746
Lumped-parameter systems, 156

MacCormack method, 229–230
Macronutrients, 522
  carbon, 526–527
  nitrogen, 524–526, 530–533
  phosphorus, 522–524, 530–533
  silicon, 527
Macrophytes (*see* Fixed plants)
Manning equation, 248–250, 252, 413, 488–489
Marks, Dave, 110
Mass, 6–8
Mass balance, 47–51
  for a diffuse source that contributes both flow and mass, 410–415
  for a well-mixed lake, 47
Mass balance equation, 52
  feedback systems of reactors, 101–116
  particular solutions, 65–83
  solutions for feedforward systems of reactors, 86–99
  steady-state, 86–91, 101–103
  steady-state solutions, 52–57
  total solution, 76–79
Mass flux rate, 9
Mass loading, 8
  rate, 8
  volumetric flow, 9
Mass-transfer mechanisms, 715–738
Mathematical models, 10–14
Matrix, 104
  algebra, 104–107
  inversion, 106
  multiplication, 105
Measurement methods, photosynthesis/respiration, 437–448
Metalimnion, 579
Metals, 757, 761–768, 774–776
Methane balance, 460–464
Methane bubble formation, 474–475
Methemoglobinemia, 430
Method of excess, 35
Method of half-lives, 34
Method of initial rates, 34
MFR (*see* Mixed flow reactor)
Michaelis-Menten model, 592, 607–610, 616, 627, 752
  (*See also* Half saturation constants)
Microbe/substrate modeling, 590–602
Micronutrients, 522
Mineralization, 751
MINTEQ, 763
Mixed flow reactor, 160–164
  application of, to estuaries, 168–171
Mixed-flow systems, 776–777
Mixed-layer approach, for vertical temperature distributions, 586
Mixing, 245–247
  lateral, 246–247
Mixing length, 147
Model:
  assessing performance, 335–338
  defined, 10
  empirical, 12
  implementations, 11
  mechanistic, 12
Model calibration, 336–338, 398–399
Model resolution, 339–341
Model selection, 318–322
  numerical specification and validation, 321–322
  theoretical development, 319–321
Model sensitivity, 327–335
  Monte Carlo analysis, 332–335
"Modeling" environment, 318–342
Molar solution, 668
Monod model, 592
Monte Carlo analysis, 333–336
Morphometry, lake, 278–282
Multiple point sources, Streeter-Phelps model, 393–396

Natural mortality, 506
NBOD (*see* Nitrogenous BOD)
Near-shore models, 287–293
  steady-state case in a bounded fluid, 291–293
  steady-state case in an infinite fluid: advection along the shoreline, 290
  no advection, 288–290
Net estuarine flow, 262
Neumann boundary conditions, 194
Nitrate, 430, 477–478
Nitrate pollution, 420, 430

Nitrification, 39, 355, 420, 421–423
  inhibition, 427–428
  modeling, 426–428
  sediments, 458–459, 468–474, 477
Nitrobacter, 422
Nitrogen, 419–432, 458–459, 524–526, 530–533
  ammonium, 637
  nitrate, 637
Nitrogen:phosphorus ratio, 530–533
Nitrogenous BOD, 355, 424–425
Nitrosomonas, 421
No-flow sources:
  BOD, 407–409
  dissolved oxygen, 409
Normal distribution, 333
Normality, 668
Null zone, 297
Numerical differentiation, 31
Numerical dispersion, 201–203, 231
  in FTCS, 216–221
Numerical methods, 120
  for diffuse sources with flow, 415–417
Nutrient/food-chain interactions, 629–632
  modeling, 633–643
Nutrients, 522–527, 607–609, 637
  in QUAL2E, 653–658

Ockham, 322
O'Connor, Donald, 83
O'Connor-Dobbins formula, 379–380
Okubo diffusion diagram, 150, 269, 343
Onondaga Lake, 22, 44
Oocysts, 512
Open systems, reactions with feedback in, 116
Orbital velocity, 313
Origin, of lake, defined, 277
Outflow, 49
Owens-Gibbs formula, 380
Oxidation/reduction, 755
Oxygen
    (*see also* Sediment oxygen demand; BOD)
  areal hypolimnetic demand, 452, 543–544
Oxygen demand:
  areal hypolimnetic, 452
  biochemical, 353–360, 407–409
  sediment, 450–479
Oxygen equivalents, in SOD process, 458
Oxygen reaeration, 369–388
  formulas, 379–386

Parameter, 51
  perturbation, 328
Particular solutions for mass balance equation, 65–83
Particulate organic matter (POM), 457
Partition coefficient, 508, 697, 700, 721–724
Pathogens, 503–516
  bacterial loss rate, 506–510
  indicator organisms, 504–506
  protozoans, 512–516
  sediment-water interactions, 510–511
Peclet number, 163
Penetration theory, derivation, 376–377
Perennial flow regime, 236
Periphyton, 658
    (*See also* Fixed plants)
Perturbation, 327
PFR (*see* Plug-flow reactor)
pH:
  and ionization of water, 672–673
  modeling, 683–691
Phosphate, 477, 478–479
Phosphorus, 522–524, 530–533
  budget models, 536–538
  loading concept, 534–559
  sediments, 478–479
  soluble reactive, 637
  trophic-state correlations, 539–544
Photolysis, 739–751
  direct model, 747
  near-surface approach, 748–751
Photons, 741
Photoperiod, 436
Photosynthesis, 347, 433–448
  delta method, 442–448
  fundamentals, 433–434
  light, 435–437
  measurement methods, 437–448
Photosynthesis/respiration, 442–448
  measurement methods, 437–448
  units and parameterization, 434–435
Phytoplankton, 603
  stream and estuary models, 647–649
Phytoplankton growth, 603–621
  and light, 609–612
  limits to, 603–605
  and nonpredatory losses, 613–615
  and nutrients, 607–609
  and temperature, 605–607
Phytoplankton/nutrient interactions, stream, 644–649
Planimeter, 278
Plant:
  fixed, in streams, 658–662
  and gas transfer, 686–689
Plant activity, 434
Plant growth, 603–621
  and nonpredatory losses, 613–615
Plant stoichiometry, 527–530

Plug-flow reactor (PFR), 157–160
comparison with CSTR, 159–160
model, 164–167
time-variable solution, 173–177
Plug-flow systems, 770–776
Point estimate, 238
Point source, 8, 207–208
in estuaries, 168–170
segmentation around, 207–208
in streams, 164–166
Streeter-Phelps model, 389–403
Point-slope method, 122
Pollutant, routing, 256–257
Pollutant reduction
temporal aspects, 57–62
Porosity:
bottom sediments, 302–303
defined, 303
Porous media, 303
as sources of BOD, 407
Positivity, 203–205
Post-audit, 324–327
Predator-prey interactions, 622–632
Predictor-corrector method, 125
Pressure, effect of, on oxygen saturation, 363–364
Priority pollutants, 698–699
Problem specification, 318
Product of two matrices, 105
Protozoans, 512–516
concentrations in natural waters, 513
drinking-water treatment and acceptable risk levels, 514
Giardia/Cryptosporidium model, 514–516
loadings, 512–513

QUAL2E, 486–487, 649–658
application of, 493–500
kinetics, 489–491
numerical algorithm, 491–493
reaeration, 491
spatial discretization and model overview, 487–488
Quantum yield, 746–748
Q_{10}, 41

Radiation, 565–566
atmospheric longwave, 570
net absorbed, 568–570
solar shortwave, 568–569
water longwave, 570
Radionuclides, 93–95, 312, 757–761
"Ramp" loading, 70–71
Random walk, 177–180
mathematics, 179–180
Rate, 8–10
mass flux, 9
Rate data, analysis of, 29–38
differential method, 31
integral method, 30
method of excess, 35
method of half-lives, 34
method of initial rates, 34
Rate law, 26
Reach estimates, 242–243
Reaction, 13, 50
with feedback, 113–116
heterogeneous, 25
homogeneous, 25
irreversible, 25
reversible, 25
Reaction kinetics, 24–42
defined, 26
Reaction mechanisms, 734–756
Reaction order, 26
Reaction types, 25–26
Reaeration, 349, 376–386
extrapolating other gases, 385–386
measurement of, with tracers, 386–388
oxygen, 369–388
photosynthesis/respiration, 445
QUAL2E, 491
standing waters and estuaries, 383–385
Recovery from pollutant, 57–62
Residence time, 55
Respiration, 347, 433–448
delta method, 442–448
fundamentals, 433–434
measurement methods, 437–448
units and parameterization, 434–435
Response time, 60–62
Resuspension, 312–315
bottom scour, 313
orbital velocity, 313
shear stress, 313
suspended solids concentration and entrainment rate, 313–315
wave height, period, and length, 312–313
Reversible reaction, 25
in batch reaction (closed system), 114–116
in CSTR, 116
Richardson number, 578, 588
Rivers, 235–257
reaeration, 379–382
types, 235–238
Robust, 324
Roughness coefficients, 249

Routing:
  pollutants, 256–257
  water, 251–256
  and water quality, 250–257
Rule of 72, 62
Runge-Kutta methods, 126–128

Saginaw Bay, 144–147
Salinity:
  effect of, on bacteria, 506
  effect on oxygen saturation, 363
Saturation concentration of oxygen, 361
Seasonal cycle, simulation, 637–641
Second-order reactions, 29
Second-order, 751
  (*See also* Reversible reactions)
Sediment, 296–316, 795–797
Sediment dating, 311
Sediment diagenesis, 457–459
Sediment oxygen demand (SOD), 450–479
  observations, 451–455
  (*See also* SOD model)
Sediment solids concentration, in bottom sediments, 303–304
Sediment-water interactions, 545–550
  bacteria, 510–511
  hypolimnetic oxygen model, 546
  total phosphorus model, 545
Segmentation, 14, 340–342
  kinetic, 634–637
  spatial, 633–634
Sensitivity analysis, 322, 327
  first-order, 327–332
Settling, 50, 300
  of bacteria, 508–509
  of particles, 301
  effects on BOD removal, 359
7Q10, 243
"Sewage fungus," 453
Shagawa Lake, 43, 293, 546–551
Shape, of lake, defined, 277
Shear stress, 313
Silicon, 527
Simulation mode, 11
Sinusoidal loading, 73–76
Size, of lake, defined, 277
Sobal, Matt, 110
SOD model:
  analytical modeling, 459–470
  carbonaceous, 466–468
  methane balance, 460–464
  methane bubble formation, 474–475
  "naive" Streeter-Phelps, 455–457
  nitrate, 477–478
  nitrogen and, 468–470
  numerical model, 470–474
  oxidation in the aerobic zone, 464–466
  phosphate, 477, 478–479
  time-variable calculations, 475–476
  water boundary layer, 476–477
Sodium sulfite, 387
Solar radiation, 568–570, 742–743
Solid-liquid partitioning, 697–700
Solids budget:
  simple, 304–307
  parameter estimation, 305–307
Solids, suspended, 297–302
  (*See also* Suspended solids)
Sorbate, 716
Sorbent, 716
Sorption, 697, 700, 715–727
  "third-phase" effect, 724–727
Sphaerolitus, 452, 453
Spill models, 180–186
  instantaneous, 180–183
  impulse spills, 180–183
Spills, 66–68, 180–183
  continuous, 183–186
Square matrices, 104
St. Venant equations, 251
Stability, 212, 214–215
  explicit method, 214–215
Standing waters, reaeration, 383–385
State-space representation, 625
Steady-state, 14
  bottom sediments, 309–310
  control-volume approach, 192–209
  distributed-parameter systems, 156–171
  embayment model, 145–146
  near-shore models, 288–290
  plug-flow reactor (PFR), 157–159
  solutions, 52, 145–146
  system response matrix, 107–110, 482–486
  for two CSTRs with feedback, 101–103
Stefan-Boltzmann law, 566
Step function, 68
Step input, 68
Step loading, 68–69
Stoichiometry, 38–40
  plant, 527–530
  variable, 609
Stokes' law, 301, 316, 515
"Stream of consciousness" modeling, 54
Stream, 235–257
  application of the plug-flow reactor model to, 164–167
  BOD model for, 355–357
  hydrogeometry, 238–243
  reaeration, 379–382

Streeter-Phelps equation:
point-source, 391
steady-state system response matrix, 482–486
Streeter-Phelps model, 389–403
anaerobic condition, 399–401
analysis of, 396–397
distributed sources, 405–417
estuary, 401–403
point sources, 389–403
SOD, 455–457
total, 409–410
Streptococcus:
bovis, 505
equinus, 505
faecalis, 505
Strontium 90, 93–95
Surface heat exchange, 565–571
Surface renewal model, 375–378
Surface renewal theory, 377
Suspended solids, 297–302, 313–315
in bottom sediments, 303–304
properties, 297–300
settling and Stokes' law, 300–302
System response matrix, 107–110, 197–198, 482–486
Systems of equations, numerical methods, 128–130

Temperate lakes, 577–580
Temperature:
effect of, on oxygen saturation, 362
modeling, 571–575
Temperature effects on reaction kinetics, 40–42
Temporal aspects of pollutant reduction, 57–62
Thermal stratification, 577–589
in temperate lakes, 577–580
Thermocline, 579
Thomann, Bob, 110
Time-domain representation, 624
Time shifts, 76–79
Time-variable calculations, 475–476
Time-variable solutions, 146–149
advanced, 223–232
bottom sediments, 309–311
control-volume approach, 215–216
distributed-parameter systems, 173–190
embayment model, 146–149
plug-flow reactor, 173–177
simple, 212–221
two reactors, 111–113
water balance for lakes, 285–287
Tolt Reservoir, WA, 280–281
Total balance, 50
Total coliform (TC), 504
Total Kjeldahl nitrogen, 424
Total solution, to mass balance equation, 76–79
Toxicant/food-chain interactions, 784–797
chemical parameters, 790–791
integration with mass balance, 794–795
Toxics, 693–797
classification of, 697–699
contrasted with conventional pollution, 695–697
inorganic, 757
loading concept, 732–737
model for a CSTR, 700–713
modeling, 695–714
modeling in flowing waters, 769–783
nonpoint sources, 779–782
Tracer studies, 186–189, 384–386
Transfer function, 55
Transport, 13
in estuaries, 260–262
Triangular distribution, 333
Trophic-state correlations, 539–544
Turbulent diffusion, 149–150
approach to modeling temperature distribution, 586
Two-film theory, 371–375

Uniform distribution, 333
Upstream difference, 193
Uptake, 609
Urbanization, effect of, on hydrographs, 238

Vapor pressure, 284, 567
Velocity, flow, depth, and, 247–250
Vertical transport, 580–585
Volatilization, 369–376, 727–731
Vollenweider loading plots, 534–536
Volumetric flow rate, 9

Washout, 598
Water balance, steady-state solution, for lake, 282–283
Water quality:
engineers and, 4–6
introduction, 3–20
nitrogen and, 419–421
routing and, 250–257
Water-quality models, historical development of, 14–18
Water routing, 251–256
density, 271, 579
Waterfalls, effect of, on stream reaeration, 382

Water-quality-modeling process, 317–327
  calibration, 322–324
  confirmation and robustness, 324
  management applications, 324
  model selection, 318–322
  post-audit, 324–326
  preliminary application, 322
  problem specification, 318
Wave, height, period, and length, 312–313
Wavelength, 741
Well-mixed reactors, computer methods, 120–131
"Whitings," 300
Whitman's two-film theory, 371–375

Zebra mussel, 453
Zebra mussel oxygen demand (ZOD), 453–454
Zero-order reactions, 27
Zooplankton:
  carnivorous, 636
  herbivorous, 636
  interactions with phytoplankton, 626–629